Hyperbolic Problems: Theory, Numerics, Applications

Sylvie Benzoni-Gavage • Denis Serre
Editors

Hyperbolic Problems: Theory, Numerics, Applications

Proceedings of the Eleventh International Conference on Hyperbolic Problems held in Ecole Normale Supérieure, Lyon, July 17–21, 2006

Springer

Sylvie Benzoni-Gavage
Institut Camille Jordan
UMR CNRS 5208
Université Claude Bernard Lyon 1
43, Bd du 11 novembre 1918
69622 Villeurbanne cedex, France
benzoni@math.univ-lyon1.fr

Denis Serre
UMPA, UMR CNRS 5669
Ecole Normale Supérieure de Lyon
46, allée d'Italie
69364 Lyon cedex 07, France
denis.serre@umpa.ens-lyon.fr

ISBN 978-3-540-75711-5 e-ISBN 978-3-540-75712-2

Library of Congress Control Number: 2007937290

Mathematics Subject Classification (2000): 35L40, 35L60, 35L65, 35L67, 35Q35, 35Q72, 35Q75, 65M06, 65M50, 83C55

Cover Design: WMXDesign GmbH, Heidelberg

Printed on acid-free paper

9 8 7 6 5 4 3 2 1

springer.com

Preface

This volume contains papers that were presented at HYP2006, the International Conference on "Hyperbolic Problems: Theory, Numerics and Applications" held at École Normale Supérieure de Lyon, France, July, 17–21, 2006.

HYP2006 is the 11th meeting in a biennial series that was initiated in 1986 in Saint-Étienne and that has become one of the most important international events in Applied Mathematics. The very first conference in Saint-Étienne had been the occasion to gather western and eastern researchers working on hyperbolic PDEs. At that time, all participants had of course in mind various applications of those PDEs, but the meeting was mainly focused on analytical tools. Since then, as computers were becoming more and more powerful, the interplay between theory, modelling, and numerical algorithms, gained considerable impact, and the scope of HYP conferences expanded accordingly. The field is nowadays in interaction with a lot of scientific domains, including physics, chemistry, biology and engineering. For example, many effective numerical methods, originally developed in the context of Computational Fluid Dynamics, have found new applications in the recent years. Applications of hyperbolic PDEs now range from fluid dynamics to biology, from road or network traffic to electro-magnetism, from meso/nano-scale material properties of semi-conductors to multi-phase flows, including phase transitions, geophysical flows, and from chemical or industrial processes to astro-physical problems where general relativity is relevant - this list is of course not limitative. Of course, other kinds of PDEs are also concerned, especially parabolic and dispersive equations, as far as, either they tell something about the hyperbolic theory, or the hyperbolic theory helps in studying them. Probabilistic tools (particle systems) also bring another valuable point of view.

A total of about 240 participants attended HYP2006. We had the great pleasure to welcome leading researchers from many different areas, who addressed theoretical, modelling and computational issues involving hyperbolic PDEs, with applications to a variety of domains, including everyday life problems (e.g. traffic flow). More than one hundred selected papers are collected in this volume. They reflect the high quality of contributions, invited or not, to HYP2006, and cover a wide range of topics. We hope that HYP2006 and its present Proceedings will contribute to encourage younger generations to work in the field, to generate future cooperations, and to promote cross-disciplinary interactions.

We would like to acknowledge financial support from the following institutions:

- École Normale Supérieure de Lyon,
- Office of Naval Research,
- Centre National de la Recherche Scientifique (CNRS),
- Ministère de l'Education Nationale et de la Recherche,
- CNRS laboratories: UMPA (UMR # 5569) and ICJ (UMR # 5208),
- Commissariat à l'Energie Atomique,
- CNRS groupement de recherche CHANT (GDR # 2900),
- Conseil de la Région Rhône-Alpes,
- Conseil du Département du Rhône,
- ACI "Etudes mathématiques de paramétrisations en océanographie",
- Ministère des Affaires etrangères.

Lyon,
July 2007

Sylvie Benzoni-Gavage
Denis Serre

Contents

Part III Mini-Symposium on Hydraulics

Part IV Contributed Talks

Contributors

David Vanden Abeele
Von Karman Institute for Fluid Dynamics (VKI)
72 Chaussée de Waterloo
1640 Rhode-Saint-Genèse, Belgium
david.vanden.abeele@vki.ac.be

Rémi Abgrall
University of Bordeaux
Mathématiques Appliquées
351 Cours de la Libération
33 405 Talence Cedex, France
remi.abgrall@math.u-bordeaux1.fr

Alessandro Alaia
Dipartimento di Matematica
corso Duca degli Abruzzi 24
10129 Torino, Italy

A. Alshabu
Shock Wave Laboratory
RWTH Aachen University
52062 Aachen, Germany

Debora Amadori
Dipartimento di Matematica Pura e Applicata
Universitá degli Studi dell'Aquila
Via Vetoio, loc. Coppito
67010 L'Aquila, Italy
amadori@univaq.it

David Ambrose
Department of Mathematical Sciences
Clemson University, Martin Hall
Clemson, SC 29634, USA
dambros@clemson.edu

Annalisa Ambroso
DEN/DM2S/SFME CEA Saclay
Gif-sur-Yvette, France

Boris Andreianov
Département de mathématiques
UFR des sciences et techniques
16 route de Gray
25030 Besançon Cedex, France
borisa@math.univ-fcomte.fr

Denise Aregba-Driollet
MAB-IMB, Université Bordeaux 1
351 cours de la Libération
33405 Talence Cedex, France
aregba@math.u-bordeaux1.fr

Paul Arminjon
Centre de Recherches Mathématiques
Université de Montréal,
C.P. 6128 Succ. Centre Ville,
Montréal, Québec, Canada H3C 3J7
arminjon@crm.umontreal.ca

B. Audebert
Office National d'Etudes et
Recherches Aérospatiales
29 Avenue de la Division Leclerc
92322 Châtillon Cedex, France
bruno.audebert@ensta.org

Josef Ballmann
Mechanics Department
Lehr- und Forschungsgebiet für
Mechanik
RWTH Aachen University
Templergraben 64
52056 Aachen, Germany
ballmann@lufmech.
rwth-aachen.de

Alfredo Bermúdez de Castro
Departamento de Matemática
Aplicada
Facultad de Matemáticas
Universidade de Santiago de
Compostela
15782 Santiago de Compostela
A Coruña, Spain
mabermud@usc.es

Devaki Bhaya
Department of Plant Biology
Carnegie Institution of Washington
Stanford University
Stanford, CA 94305, USA
dbhaya@stanford.edu

Stefano Bianchini
SISSA-ISAS,
Via Beirut 2–4
34014 Trieste, Italy
bianchin@sissa.it

Philipp Birken
Fachbereich für Mathematik und
Informatik
Universität Kassel, Heinrich Plett
Str. 40 (AVZ)
34132 Kassel, Germany
birken@mathematik.
uni-kassel.de

Alexander M. Blokhin
Sobolev Institute of Mathematics
Koptyug avenue 4
Novosibirsk 630090, Russia
blokhin@math.nsc.ru

François Bouchut
Département de Mathématiques
et Applications
École Normale Supérieure
45 rue d'Ulm
75230 Paris Cedex 05, France
francois.bouchut@ens.fr

Andrei Bourchtein
Institute of Physics
and Mathematics
Pelotas State University
Rua Anchieta 4715, bloco K,
ap. 304, Pelotas 96015-420
Brazil
burstein@terra.com.br

Ludmila Bourchtein
Institute of Physics
and Mathematics
Pelotas State University
Brazil
burstein@terra.com.br

Benjamin Boutin
Université Pierre et Marie
Curie-Paris 6
UMR 7598 LJLL
Paris 75005, France
CNRS
and
CEA-Saclay
91191 Gif-sur-Yvette Cedex, France

Didier Bresch
LMC-IMAG UMR5223
51, rue des mathématiques, BP 53
38041 Grenoble Cedex, France
didier.bresch@imag.fr

Miroslav Čada
Seminar for Applied Mathematics
ETH-Zürich HG, Rämistrasse 101
8092 Zürich, Switzerland
cada@math.ethz.ch

Donna Calhoun
Commissariat à l'Énergie Atomique
DM2S/SFME/LTMF Bt. 454
Centre de Saclay
91191 Gif-sur-Yvette, France
donna@semt2.smts.cea.fr

Cristóbal E. Castro
Laboratory of Applied Mathematics
Faculty of Engineering
University of Trento
38050 Mesiano di Povo
Trento, Italy
cristobal.castro@ing.unitn.it

Fausto Cavalli
Dipartimento di Matematica
Università di Milano
via Saldini 50
20133 Milano, Italy
cavalli@mat.unimi.it

Christophe Chalons
Laboratoire Jacques Louis Lions,
U.M.R. 7598, Université Pierre et
Marie Curie Paris 6
Boîte courrier 187
75252 Paris Cedex 05, France
chalons@math.jussieu.fr

Alina Chertock
Department of Mathematics
North Carolina State University
Campus Box 8205/HA 243
Raleigh, NC 27695, USA
chertock@math.ncsu.edu

Guillaume Chiavassa
École Centrale de Marseille
Technopole de Chateau-Gombert
13451 Marseille Cedex 20, France
guillaume.chiavassa
@ec-marseille.fr

Cleopatra Christoforou
Department of Mathematics
Northwestern University
2033 Sheridan Rd.,
Evanston, IL 60208, USA
cleo@math.northwestern.edu

Giuseppe Maria Coclite
Dipartimento di Matematica
Universitadi Bari
Via E. Orabona 4
70125 Bari, Italy
coclitegm@dm.uniba.it

Rinaldo M. Colombo
Dipartimento di Matematica
Università degli Studi di Brescia
Via Branze 38
25123 Brescia, Italy
rinaldo.colombo@ing.unibs.it

Frédéric Coquel
CNRS et Laboratoire Jacques Louis
Lions
Université Pierre et Marie Curie
175 rue du Chevaleret
75013 Paris, France
coquel@ann.jussieu.fr

Andrea Corli
Dipartimento di Matematica
Università di Ferrara
Via Machiavelli 35
44100 Ferrara, Italy
andrea.corli@unife.it

Jean-François Coulombel
Laboratoire Paul Painlevé
Batiment M2, Cité scientifique
59655 Villeneuve d'Asq Cedex
France
jfcoulom@math.univ-lille1.fr

Gianluca Crippa
Scuola Normale Superiore
Piazza dei Cavalieri 7
56126 Pisa, Italy
g.crippa@sns.it

Carlota Maria Cuesta
Department of Theoretical Mechanics
School of Mathematical Sciences
University of Nottingham, University Park
Nottingham NG7 2RD, UK
pmxcc@maths.nottingham.ac.uk

Herman Deconinck
Von Karman Institute for Fluid Dynamics
Chaussée de Waterloo 72
Rhode-Saint-Genèse, Belgium
deconinck@vki.ac.be

Andreas Dedner
Mathematisches Institut
Universität Freiburg
Herrmann-Herder-Str. 10
79104 Freiburg, Germany
mario@mathematik.uni-freiburg.de

Camillo De Lellis
Institut für Mathematik
Universität Zürich
Winterthurerstrasse 190
8057 Zürich, Switzerland
delellis@math.unizh.ch

Bruno Després
Commissariat à l'Energie Atomique
CEA/DIF/DSSI/SNEC,
94195 Bruyères le Chatel BP 12
France
bruno.despres@cea.fr

Manuel Jezús Castro Díaz
Dpto. Análisis Matemático
Universidad de Málaga
Campus de Teatinos s/n
Málaga, Spain
castro@anamat.cie.uma.es

Christian Dickopp
Lehr- und Forschungsgebiet f.
Mechanik der RWTH Aachen
Templergraben 64
52056 Aachen, Germany
dickopp@lufmech.rwth-aachen.de

Giacomo Dimarco
Dipartimento di Matematica
University of Ferrara
Via Machiavelli 35
44100 Ferrara, Italy
giacomo.dimarco@unife.it

Jiří Dobeš
Department of Technical Mathematics
Faculty of Mechanical Engineering
Czech Technical University
Karlovo Namesti 13
121 35 Praha 2, Czech Republic
jiri.dobes@fs.cvut.cz

Vít Dolejši
Faculty of Mathematics and Physics
Charles University
Sokolovska 83
186 75 Prague, Czech Republic
dolejsi@karlin.mff.cuni.cz

Rosa Donat
Dept. Matematica Aplicada
C/Dr. Moliner 50
46100 Burjassot (Valencia), Spain
donat@uv.es

Donatella Donatelli
Dipartimento di Matematica Pura ed Applicata
Università di L'Aquila
67100 L'Aquila, Italy
donatell@univaq.it

Alexander Dressel
Department of Mathematics
University of Stuttgart
Pfaffenwaldring 57
70569 Stuttgart, Germany
adressel@mathematik.
uni-stuttgart.de

Kamal El Kadi
Cemagref
Unité de Recherches
Hydrologie–Hydraulique
3 bis quai Chauveau CP 220
69336 Lyon Cedex 09, France
elkadi@lyon.cemagref.fr

Volker Elling
Division of Applied Mathematics
Brown University
182 George Street
Providence, RI 02912, USA
volker@dam.brown.edu

Erik E. Endres
Department of Mathematics
Pennsylvania State University
McAllister Building
University Park, PA 16802-6401
USA
eee121@psu.edu

Chevallier Estelle
Laboratoire d'Ingénierie Circulation
Transport (ENTPE/INRETS)
Rue Maurice Audin
Vaulx-en-Velin, France
leclercq@entpe.fr

Steinar Evje
Centre of Mathematics for Applications
University of Oslo
P.O. Box 1053, Blindern
0316 Oslo, Norway
sev@rf.no
and
International Research Institute of Stavanger
P.O. Box 8046
4068 Stavanger, Norway
steinar.evje@irisresearch.no

Kirsten Fagnan
Department of Applied Mathematics
University of Washington
Box 352420
Seattle, WA 98195, USA
kfagnan@amath.washington.edu

Eduard Feireisl
Mathematical Institute ASCR
Zitna 25
115 67 Prague, Czech Republic
feireisl@math.cas.cz

Miloslav Feistauer
Faculty of Mathematics and Physics
Charles University Prague
Sokolovska 83
186 75 Praha 8, Czech Republic
feist@karlin.mff.cuni.cz

Enrique Fernández-Nieto
Dpto. Matemática Aplicada I, ETS Arquitectura
Universidad de Sevilla
Avda. Reina Mercedes 2
41012 Sevilla, Spain
edofer@us.es

Ana M. Ferreiro
Dpto. Análisis Ecuaciones Diferenciales y Análisis Numérico
Universidad de Sevilla
C/Tarfia S/N
Sevilla, Spain
anafefe@us.es

Jose A. Font
Departamento de Astronomia y Astrofisica
Universidad de Valencia
Dr. Moliner 50
46100 Burjassot (Valencia), Spain
j.antonio.font@uv.es

Heinrich Freistühler
Mathematisches Institut
Universität Leipzig
Augustusplatz 10-11
04109 Leipzig, Germany

Naoki Fujino
Graduate School of Pure and Applied Sciences
University of Tsukuba
305-8571 Ibaraki, Japan
nfujino@math.tsukuba.ac.jp

François Gallaire
CNRS
Laboratoire Dieudonné
UMR 6621, Université de Nice
Parc Valrose
06108 Nice Cedex 02, France
gallaire@math.unice.fr

Jose Gallardo
Dept. Analisis Matematico
Facultad de Ciencias
Universidad de Malaga
Campus de Teatinos s/n
29071 Malaga, Spain
gallardo@anamat.cie.uma.es

Thierry Gallouët
LATP, CMI, Université de Marseille
39 rue Joliot Curie
13453 Marseille Cedex 13, France
gallouet@cmi.univ-mrs.fr

Mauro Garavello
Dipartimento di Matematica e Applicazioni
Università di Milano-Bicocca
Via Roberto Cozzi 53
20125 Milano, Italy
mauro.garavello@unimib.it

David George
Department of Mathematics
University of Utah
155 S 1400 E RM
233 Salt Lake City, UT 84112-0090
USA
dave.jorge@gmail.com

David Gérard-Varet
CNRS
DMA, UMR 8853
École Normale Supérieure
45 rue d'Ulm
750005 Paris, France
david.gerard-varet@ens.fr

Hans-Peter Gittel
Department of Mathematics
University of Leipzig
Augustusplatz 10/11
04109 Leipzig, Germany
hans-peter.gittel@math.uni-leipzig.de

Paola Goatin
Laboratoire d'Analyse Non linéaire Appliquée et Modélisation
I.S.I.T.V., Université du Sud Toulon-Var
Avenue Georges Pompidou, B.P. 56
83162 La Valette du Var Cedex
France
goatin@univ-tln.fr

Edwige Godlewski
UPMC, Laboratoire Jacques Louis Lions
175 rue du Chevaleret
75013 Paris, France
godlewski@ann.jussieu.fr

Sergei K. Godunov
Sobolev Institute of Mathematics
Siberian Branch of the Russian Academy of Sciences
Acad. Koptyug av., 4
Novosibisk 630090, Russia
godunov@math.nsc.ru

González-Vida
Dept. Matemática Aplicada
Universidad de Málaga
Campus de Teatinos s/n
29071 Málaga, Spain

Graziano Guerra
Dipartimento di Matematica e Applicazioni
Università di Milano
Bicocca Via Roberto Cozzi 53
20125 Milano, Italy
graziano.guerra@unimib.it

Yan Guo
Division of Applied Mathematics
Box F
Brown University
Providence, RI 02912, USA
guoy@dam.brown.edu

Seung-Yeal Ha
Department of Mathematical Sciences
Seoul National University
Seoul 151-747, Republic of Korea
syha@snu.ac.kr

Bernard Hanouzet
Mathématiques Appliquées de Bordeaux
Université Bordeaux 1
351 cours de la Libération
33405 Talence, France
bernard.hanouzet@math.u-bordeaux1.fr

Petr Havlík
Faculty of Nuclear Science and Physical Engineering
Czech Technical University in Prague Trojanova 13
120 00 Praha 2, Czech Republic
xhavlik@fermi.fjfi.cvut.cz

Christiane Helzel
Institut für Angewandte Mathematik
University of Bonn
Bonn, Germany
helzel@iam.uni-bonn.de

V. Hermes
Shock Wave Laboratory
RWTH Aachen University
52062 Aachen, Germany

Michael Herty
Fachbereich Mathematik
TU Kaiserslautern
67653 Kaiserslautern, Germany
herty@rhrk.uni-kl.de

Helge Holden
Department of Mathematical Sciences
Norwegian University of Science and Technology
Alfred Getz vei 1
7491 Trondheim, Norway

and

Department of Mathematics
Centre of Mathematics for Applications
University of Oslo
P.O. Box 1053, Blindern
0316 Oslo, Norway
holden@math.ntnu.no

Ling Hsiao
Academy of Mathematics and Systems Science
Chinese Academy of Sciences
Beijing 100080, P.R. China
hsiaol@amss.ac.cn

Rolf Jeltsch
Seminar for Applied Mathematics
ETH-Zurich, Switzerland
jeltsch@math.ethz.ch

Helge Kristian Jenssen
Department of Mathematics
Pennsylvania State University
McAllister Building, University Park
State College, PA 16802, USA
hkj1@psu.edu
jenssen@math.psu.edu

Laval Jorge
Laboratoire d'Ingénierie Circulation
Transport (ENTPE/INRETS)
Rue Maurice Audin
Vaulx-en-Velin, France

Kenneth H. Karlsen
Centre of Mathematics for Applications
University of Oslo
P.O. Box 1053, Blindern
0316 Oslo, Norway
kennethk@math.uio.no

Smadar Karni
Department of Mathematics
University of Michigan
Ann Arbor, MI 48109-1043, USA
karni@umich.edu

Eugene Kashdan
Division of Applied Mathematics
Brown University
Providence, RI 02912, USA
eugene_kashdan@brown.edu

Shuichi Kawashima
Faculty of Mathematics
Kyushu University
Fukuoka 812-8581, Japan
kawashim@math.kyushu-u.ac.jp

Friedemann Kemm
Inst. f. Ang. Math. u. Wiss. Rechnen
Brandenburgische Technische
Universität Cottbus (BTU)
Konrad-Zuse-Straße 1
03044 Cottbus, Germany
kemm@math.tu-cottbus.de

David Ketcheson
University of Washington
9408 Lake City Way NE # 301
Seattle, WA 98115, USA
dketch@gmail.com

Igor Klioutchnikov
Shock Wave Laboratory
RWTH Aachen University
52062 Aachen, Germany
klioutchnikov@swl.rwth-aachen.de

Victor Kovenya
Institute of computational technologies
Novosibirsk, Russia
kovenya@ict.nsc.ru

Tatyana Kozlinskaya
Novosibirsk State University
Pirogov Street 4, 218
630090 Novosibirsk, Russia
ktv-sun@mail.ru

Stanislav Kračmar
Department of Technical Mathematics
Faculty of Mechanical Engineering
Czech Technical University
Karlovo nám. 13
121 35 Prague 2, Czech Republic
kracmar@marian.fsik.cvut.cz

Marcus Kraft
Hamburg University of Technology
Institute of Numerical Simulation
Schwarzenbergstrasse 95
21073 Hamburg, Germany
kraft@tu-harburg.de

Dietmar Kröner
Abteilung für angewandte Mathematik
Universität Freiburg
79104 Freiburg, Germany

Václav Kučera
Faculty of Mathematics and Physics
Charles University Prague
Sokolovská 83
186 75 Praha 8, Czech Republic
vaclav.kucera@email.cz

Milan Kucharik
Los Alamos National Laboratory
Los Alamos, NM 87545, USA
kucharik@lanl.gov

Alexander Kurganov
Mathematics Department
Tulane University
6823 St. Charles Ave.
New Orleans, LA 70118, USA
kurganov@math.tulane.edu

Marc Laforest
Dépt. de math et génie industriel
École Polytechnique de Montréal
C.P. 6079 succ. Centre-ville
Montréal, Québec, Canada H3C 3A7
marc.laforest@polymtl.ca

Wanderson Lambert
IMPA – Instituto de Matemática Pura e Aplicada
Estrada Dona Castorina 110
CEP:22460-320
Rio de Janeiro, RJ, Brazil
lambert@fluid.impa.br

Andrea Lani
Von Karman Institute for Fluid Dynamics (VKI)
72 Chaussée de Waterloo
1640 Rhode-Saint-Genèse, Belgium
lani@vki.ac.be

Corrado Lattanzio
Dipartimento di Matematica Pura ed Applicata
Sezione di Matematica per l'Ingegneria
Universitá degli Studi dell'Aquila
Piazzale E. Pontieri 2
67040 Monteluco di Roio (L'Aquila), Italy
corrado@univaq.it

Ludovic Leclercq
Laboratoire d'Ingénierie Circulation Transport
ENTPE/LICIT, Rue Maurice Audin
69518 Vaulx-en-Velin Cedex, France
leclercq@entpe.fr

Philippe G. LeFloch
Laboratoire Jacques Louis Lions and CNRS UMR 7598
University of Paris VI
75252 Paris, France
lefloch@ann.jussieu.fr

Randall J. LeVeque
Department of Applied Mathematics
University of Washington
Box 352420
Seattle, WA 98195-2420, USA
rjl@amath.washington.edu

Doron Levy
Department of Mathematics
Stanford University
Stanford, CA 94305-2125, USA
dlevy@math.stanford.edu

Fucai Li
Department of Mathematics
Nanjing University
Nanjing 210093, P.R. China
fli@nju.edu.cn

Hai-Liang Li
Department of Mathematics
Capital Normal University
Beijing 100037, P.R. China
hailiang.li.math@gmail.com

Jing Li
Institute of Applied Mathematics
AMSS, Academia Sinica, P.R. China
ajingli@gmail.com

Yong Li
Department of Mathematical Sciences
Tsinghua University
Beijing 100084, P.R. China
ygli@math.tsinghua.edu.cn
and
College of Applied Sciences
Beijing University of Technology
Beijing 100022, P.R. China
yli@bjut.edu.cn

Richard Liska
Faculty of Nuclear Sciences and Physical Engineering
Czech Technical University in Prague
Brehova 7
115 19 Prague 1, Czech Republic
liska@siduri.fjfi.cvut.cz

Tai-Ping Liu
Department of Mathematics
Stanford University
450 Serra Mall, Building 380
Stanford, CA 94305, USA
liu@math.stanford.edu

Raphael Loubere
Universite Paul-Sabatier Toulouse 3
Maths pour l'Industrie et la Physique
MIP
31062 Toulouse Cedex 9, France
loubere@mip.ups-tlse.fr

Maria Lukacova
Institut für Numerische Simulation
TU Hamburg-Harburg
Schwarzenbergstr. 95
21073 Hamburg, Germany
lukacova@tu-harburg.de

Brian MacConaghy
Applied Physics Lab
University of Washington
Box 355640
Seattle, WA 98195, USA
brianm@apl.washington.edu

Aziz Madrane
Airbus/Institute for Aerospace-Technology
Mosel Strasse 93
28199 Bremen, Germany
aziz.madrane@airbus.com

Anne Mangeney
Département de modélisation physique et numérique
IPGP, 4, pl. Jussieu
75232 Paris Cedex 05, France
and
Équipe de Sismologie
Institut de Physique du Globe de Paris
4 place Jussieu
75252 Paris Cedex 05, France
mangeney@ipgp.jussieu.fr

Pierangelo Marcati
Dipartimento di Matematica Pura ed Applicata
Università degli Studi dell' Aquila
67100 L'Aquila, Italy
marcati@univaq.it

Fabien Marche
6 rue Joseph le Brix, Apt 30
33000 Bordeaux, France
and
Mathématiques Appliquées de Bordeaux
Université Bordeaux 1
351 cours de la libération
33405 Talence, France
fabien.marche@math.u-bordeaux1.fr

Dan Marchesin
Instituto Nacional de Matemática Pura e Aplicada (IMPA)
Estrada Dona Castorina 110
22460-320 Rio de Janeiro, RJ, Brazil
marchesi@fluid.impa.br

Sébastien Martin
Laboratoire de Mathématiques
Université Paris-Sud-Bât. 425
91405 Orsay Cedex, France
sebastien.martin@math.u-psud.fr

Corrado Mascia
Dipartimento di Matematica "G. Castelnuovo"
Università di Roma "La Sapienza"
Piazzale Aldo Moro 2
00185 Roma, Italy
mascia@mat.uniroma1.it

Vitor Matos
Faculdade de Economia da Universidade do Porto (FEP)
Rua Dr. Roberto Frias
4200-464 Porto, Portugal
vmatos@fep.up.pt

Thomas J. Matula
Applied Physics Lab
University of Washington
Box 355640
Seattle, WA 98195, USA
matula@apl.washington.edu

Andrea Mentrelli
Centro Interdipartimentale di Ricerca per le Applicazioni della Matematica (CIRAM)
University of Bologna
via Saragozza 8
40123 Bologna, Italy
mentrelli@ciram.unibo.it

Siddhartha Mishra
Center of Mathematics for Applications
University of Oslo
P.O. Box 1053, Blindern
0316 Oslo, Norway
siddharm@cma.uio.no

Tomas Morales
École Normale Supérieure – DMA
45 rue d'Ulm
75230 Paris Cedex 05, France
morales@dma.ens.fr

Alessandro Morando
Dipartimento di Matematica
Facoltà di Ingegneria
Università di Brescia
Via Valotti 9
25133 Brescia, Italy
morando@ing.unibs.it

Salissou Moutari
Laboratoire J.-A. Dieudonné, UMR CNRS 6621
Université de Nice-Sophia Antipolis
Parc Valrose
06108 Nice Cedex 2, France
salissou@math.unice.fr

Rafael Muñoz-Sola
Departamento de Matemática Aplicada
Facultad de Matemáticas
Universidade de Santiago de Compostela
15782 Santiago de Compostela
Spain
rafams@usc.es

Giovanni Naldi
Dipartimento di Matematica
Università di Milano
via Saldini 50
20133 Milano, Italy
naldi@mat.unimi.it

Roberto Natalini
Istituto per le Applicazioni del Calcolo "Mauro Picone"
CNR, Viale del Policlinico 137
00161 Roma, Italy
r.natalini@iac.cnr.it

Šárka Nečasová
Mathematical Institute ASCR
Zitna 25
11567 Prague 1, Czech Republic
matus@math.cas.cz

Q.-Long Nguyen
Département Mathématiques Appliquées
Institut Français du Pétrole
1 et 4 avenue de Bois-Préau
92852 Rueil-Malmaison Cedex
France
q-long.nguyen@ifp.fr

Se Eun Noh
Department of Mathematical Sciences
Seoul National University
Seoul 151-747, Republic of Korea
senoh@math.snu.ac.kr

Martin Nolte
Abteilung für angewandte Mathematik
Universität Freiburg
Hermann-Herder-Straße 10
79104 Freiburg, Germany
nolte@mathematik.uni-freiburg.de

Myunghyun Oh
Department of Mathematics
University of Kansas
Lawrence, KS 66045, USA
moh@math.ku.edu

Mario Ohlberger
Mathematisches Institut
Universität Freiburg
Hermann-Herder-Str. 10
79104 Freiburg i.Br., Germany
mario@mathematik.uni-freiburg.de

H. Oliver
Shock Wave Laboratory
RWTH Aachen University
52062 Aachen, Germany

Evgeniy Yu. Panov
Mathematical Analysis Department
Novgorod State University
B.St-Peterburgskaya 41
173003 Velikiy Novgorod, Russia
pey@novsu.ac.ru

André Paquier
Cemagref, Unité de Recherches Hydrologie–Hydraulique
3 bis, quai Chauveau, CP 220
69336 Lyon Cedex 09, France
paquier@lyon.cemagref.fr

Carlos Parés
Departamento de Análisis Matemático
Facultad de Ciencias
Universidad de Málaga
Campus de Teatinos s/n
29080 Málaga, Spain
pares@anamat.cie.uma.es

Lorenzo Pareschi
Department of Mathematics
University of Ferrara
Ferrara, Italy
lorenzo.pareschi@unife.it

Yu.Yu. Pashinin
Novosibirsk State University
Pirogova street 2
Novosibirsk 630090, Russia

Marica Pelanti
Département de Mathématiques
et Applications
École Normale Supérieure
45 rue d'Ulm
75230 Paris Cedex 05, France
marica.pelanti@ens.fr

Patrick Penel
Department of Mathematics
University of Sud Toulon-Var
BP 20132
83957 La Garde Cedex, France
penel@univ-tln.fr

Yue-Jun Peng
Laboratoire de Mathématiques,
CNRS UMR 6620
Université Blaise Pascal
(Clermont-Ferrand 2)
63177 Aubière Cedex, France
peng@math.univ-bpclermont.fr

Guergana Petrova
Department of Mathematics
Texas A&M University
College Station, TX 77843, USA
gpetrova@math.tamu.edu

Benedetto Piccoli
Istituto per le Applicazioni del
Calcolo "M. Picone" C. N. R.
Viale del Policlinico 137
00161 Roma, Italy
b.piccoli@iac.cnr.it

Sandra Pieraccini
Dipartimento di Matematica
corso Duca degli Abruzzi 24
10129 Torino, Italy
sandra.pieraccini@polito.it

Ramón G. Plaza
Max-Planck-Institute for
Mathematics in the Sciences
Inselstr. 22–26
04103 Leipzig, Germany
and
Mathematisches Institut
Universität Leipzig
Augustusplatz 10–11
04109 Leipzig, Germany
plaza@math.uni-leipzig.de

Marie Postel
Laboratoire Jacques Louis Lions
Université Pierre et Marie Curie
Boîte courrier 187
75252 Paris Cedex 05, France
postel@ann.jussieu.fr

Marilena N. Poulou
National Technical University
Zografou Campus
157 80 Athens, Hellas
mpoulou@math.ntua.gr

Gabriella Puppo
Dipartimento di Matematica
Politecnico di Torino
Corso Duca degli Abruzzi 24
10129 Torino, Italy
gabriella.puppo@polito.it

Shamsul Qamar
Institut für Analysis und Numerik
Otto-von-Guericke University
Magdeburg
Universitätsplatz 2
39106 Magdeburg, Germany
shamsul.qamar@mathematik.
uni-magdeburg.de

Michel Rascle
Laboratoire J.-A. Dieudonné
UMR CNRS 6621
University of Nice-Sophia
Antipolis, France
rascle@math.unice.fr

Xavier Raynaud
Department of Mathematical Sciences
Norwegian University of Science and Technology
NTNU Alfred Getz vei 1
7491 Trondheim, Norway
raynaud@math.ntnu.no

Tiago Requeijo
Department of Mathematics
Stanford University
Stanford, CA 94305-2125, USA
requeijo@math.stanford.edu

Felix Rieper
BTU Cottbus, Lehrstuhl NMWR
Karl-Marx-Str. 17
03044 Cottbus, Germany
rieper@math.tu-cottbus.de

Carmen Rodríguez
Departamento de Matemática Aplicada
Facultad de Matemáticas
Universidade de Santiago de Compostela
15782 Santiago de Compostela
A Coruña, Spain
carmenri@usc.es

Christian Rohde
Institut für Angewandte Analysis und Numerische Simulation
Universität Stuttgart
Pfaffenwaldring 57
70569 Stuttgart, Germany
crohde@mathematik.uni-stuttgart.de

Jens Rottmann-Matthes
Fakultaet fuer Mathematik
Universitaet Bielefeld
Postfach 100131
33501 Bielefeld, Germany
jrottman@math.uni-bielefeld.de

Olivier Rouch
Université de Montréal
C.P. 6128, Succ. Centre ville
Montréal, Québec, Canada H3C 3J7
rouch@dms.umontreal.ca

Frédéric Rousset
CNRS, Laboratoire Dieudonné, UMR 6621
Université de Nice
Parc-Valrose
06108 Nice Cedex 02, France
frousset@math.unice.fr

Olga Rozanova
Department of Differential Equations
Mechanics and Mathematics Faculty
Moscow State University, GZ
Moscow 119992, Russia
rozanova@mech.math.msu.su

Tommaso Ruggeri
Research Centre of Applied Mathematics (CIRAM)
University of Bologna
Via Saragozza 8
40123 Bologna, Italy
ruggeri@ciram.unibo.it

Giovanni Russo
Department of Mathematics and Computer Science
Università di Catania
Viale Andrea Doria 6
95125 Catania, Italy
russo@dmi.unict.it

Karima Sbihi
UFR Sciences et techniques
département de mathématiques
16 route de Gray
25030 Besançon Cedex, France
ksbihi@math.univ-fcomte.fr

Paolo Secchi
Dipartimento di Matematica
Facoltà di Ingegneria
Via Valotti 9
25133 Brescia, Italy
paolo.secchi@ing.unibs.it

Nicolas Seguin
UMR 7598, Laboratoire JL Lions
Université Paris et Marie
Curie-Paris 6
Boîte courrier 187
75252 Paris Cedex 05, France
seguin@ann.jussieu.fr

Matteo Semplice
Dipartimento di Matematica
Università di Milano
Via Saldini 50
20133 Milano, Italy
semplice@mat.unimi.it

Susana Serna
Math Science
Institute of Geophysics
and Planetary Physics
University of California Los Angeles
520 Portola Plaza
Los Angeles, CA 90095-1555, USA
serna@math.ucla.edu

Denis Serre
Unité de Mathématiques Pures et
Appliquées
UMR CNRS 5669
École Normale Supérieure de Lyon
46 Allée d'Italie
69364 Lyon Cedex 07, France
serre@umpa.ens-lyon.fr

Mikhail Shashkov
Los Alamos National Laboratory
Los Alamos, NM 87545, USA
shashkov@lanl.gov

Vladimir M. Shelkovich
Department of Mathematics
St. Petersburg State Architecture
and Civil Engineering
University
2 Krasnoarmeiskaya 4
190005 St. Petersburg, Russia
shelkv@vs1567.spb.edu

Wen Shen
Department of Mathematics
Pennsylvania State University
223A McAllister
University Park, PA 16803, USA
shen_w@math.psu.edu

Keh-Ming Shyue
Department of Mathematics
National Taiwan University
Taipei 106, Taiwan
shyue@math.ntu.edu.tw

Ivan Sofronov
Keldysh Institute of Applied
Mathematics RAS
Miusskaya sq. 4
125047 Moscow, Russia
sofronov@spp.keldysh.ru

Aparecido de Souza
Departamento de Matematica e
Estatistica
Universidade Federal de Campina
Grande
Av. Aprigio Veloso 882
Caixa Postal 10044
58109-970 Campina Grande, PB
Brazil
cido@dme.ufcg.edu.br

Nikolaos Stavrakakis
Department of Mathematics
School of Applied Mathematics
and Physical Sciences
National Technical University
Zografos Campus
157-80 Athens, Greece
nikolas@central.ntua.gr

Marie-Odette St-Hilaire
Université de Montréal
Montréal, Québec, Canada H3C 3J7
sthilair@dms.umontreal.ca

Franck Sueur
Laboratoire Jacques Louis Lions
175 rue du Chevaleret
75005 Paris, France
fsueur@ann.jussieu.fr

Eitan Tadmor
Department of Mathematics
Center for Scientific Computation
and Mathematical Modeling
(CSCAMM)
Institute of Physical Science
and Technology
University of Maryland
College Park, MD 20742, USA
tadmor@cscamm.umd.edu

Andrea Terracina
Dipartimento di Matematica
"G. Castelnuovo"
Università di Roma "La Sapienza"
Piazzale Aldo Moro 2
00185 Roma, Italy
terracin@mat.uniroma1.it

Nils Tiemann
Fakultät für Mathematik
Universität Bielefeld
Postfach 100 131, Germany
nils.tiemann@uni-bielefeld.de

Vladimir A. Titarev
University of Trento
38050 Mesiano di Povo
Trento, Italy
titarev@ing.unitn.it

Dmitriy L. Tkachev
Sobolev Institute of Mathematics
Koptyug avenue 4
630090 Novosibirsk, Russia
tkachev@math.nsc.ru

Eleuterio F. Toro
Laboratory of Applied Mathematics
Faculty of Engineering
University of Trento
Via Mesiano 77, Mesiano di Povo
38050 Trento, Italy
toro@ing.unitn.it

Manuel Torrilhon
Applied and Computational
Mathematics
Princeton University
Princeton, NJ 08544-1000, USA
mtorrilh@princeton.edu

Yuri Trakhinin
Sobolev Institute of Mathematics
Koptyg avenue 4
630090 Novosibirsk, Russia
trakhin@math.nsc.ru

Quang-Huy Tran
Département Mathématiques
Appliquées
Institut Français du Pétrole, 1 et 4
avenue de Bois-Préau
92852 Rueil-Malmaison Cedex
France
q-huy.tran@ifp.fr

Paola Trebeschi
Dipartimento di Matematica
Facoltà di Ingegneria
Università di Brescia
Via Valotti 9
25133 Brescia, Italy
paola.trebeschi@ing.unibs.it

Konstantina Trivisa
Department of Mathematics
University of Maryland
College Park, MD 20742, USA
trivisa@math.umd.edu

Sergey D. Ustyugov
Keldysh Institute of Applied Mathematics
Russian Academy of Sciences
4 Miusskaya sq.
Moscow 125047, Russia
ustyugs@keldysh.ru

Miguel Ángel Vilar
Departamento de Matemática Aplicada
Universidade de Santiago de Compostela
Escuela Politécnica Superior
27002 Lugo, Spain
miguel@lugo.usc.es

Jean-Pierre Vilotte
Équipe de Sismologie
Institut de Physique du Globe de Paris
4 place Jussieu
75252 Paris Cedex 05, France
vilotte@ipgp.jussieu.fr

Julien Vovelle
ENS Cachan IRMAR/ENS Cachan
Antenne de Bretagne, Campus de Ker Lann
35170 Bruz, France
vovelle@bretagne.ens-cachan.fr

Shu Wang
College of Applied Sciences
Beijing University of Technology
Pingleyuan 100
Beijing 100022, P.R. China
wangshu@bjut.edu.cn

Gerald Warnecke
Institute for Analysis and Numerics
Otto-von-Guericke University PSF 4120
39106 Magdeburg, Germany
shamsul.qamar@mathematik.uni-magdeburg.de

Jeroen A.S. Witteveen
Faculty of Aerospace Engineering
Delft University of Technology
Kluyverweg 1
2629 HS Delft, The Netherlands
j.a.s.witteveen@tudelft.nl

Zhouping Xin
Institute of Mathematical Science
The Chinese University of Hong Kong
Shatin, Hong Kong
zpxin@ims.cuhk.edu.hk

Mehmet Sarp Yalim
Von Karman Institute for Fluid Dynamics (VKI)
72 Chaussee de Waterloo
1640 Rhode-Saint-Genese, Belgium
yalim@vki.ac.be

Mitsuru Yamazaki
Graduate School of Pure and Applied Sciences
University of Tsukuba
305-8571 Ibaraki, Japan
yamazaki@math.tsukuba.ac.jp

Wen-An Yong
Zhou Pei-Yuan Center for Applied Mathematics
Tsinghua University
Beijing 100084, P.R. China
wayong@mail.tsinghua.edu.cn

Seok-Bae Yun
Department of Mathematical Sciences
Seoul National University
Seoul, Republic of Korea
sbyun@math.snu.ac.kr

Nickolai A. Zaitsev
Keldysh Institute of Applied Mathematics
RAS
Moscow, Russia
zaitsev@keldysh.ru

Xu Zhang
Departamento de Matemáticas
Facultad de Ciencias
Universidad Autónoma de Madrid
28049 Madrid, Spain

and

Key Laboratory of Systems
and Control
Academy of Mathematics
and Systems Sciences
Chinese Academy of Sciences
Beijing 100080, P.R. China
xuzhang@amss.ac.cn

Weigang Zhong
Department of Mathematics
Center for Scientific Computation
and Mathematical Modeling
(CSCAMM)
University of Maryland
College Park, MD 20742, USA
wzhong@math.umd.edu

Enrique Zuazua
Departamento de Matemáticas
Facultad de Ciencias
Universidad Autónoma de Madrid
28049 Madrid, Spain
enrique.zuazua@uam.es

Kevin Zumbrun
Indiana University
Rawles Hall
Bloomington, IN 47405, USA
kzumbrun@indiana.edu

Part I

Plenary Lectures

General Relativistic Hydrodynamics and Magnetohydrodynamics: Hyperbolic Systems in Relativistic Astrophysics

J.A. Font

1 Introduction

Einstein's theory of general relativity plays a major role in astrophysics, particularly in scenarios involving compact objects such as neutron stars and black holes. Those include gravitational collapse, γ-ray burts, accretion, relativistic jets in active galactic nuclei, or the coalescence of compact neutron star (or black hole) binaries. Astronomers have long been scrutinizing these systems using the complete frequency range of the electromagnetic spectrum. Nowadays, they are the main targets for ground-based laser interferometers of gravitational radiation. The direct detection of these elusive ripples in the curvature of space–time, and the wealth of new information that could be extracted thereof, is one of the driving motivations of present-day research in relativistic astrophysics.

Theoretical astrophysics has long relied on numerical simulations as a formidable way to improve our understanding of the dynamics of astrophysical systems. For the case we are concerned with in this paper, the mathematical framework upon which such simulations are based is nowadays developed to high levels of sophistication. The equations governing the dynamics of relativistic astrophysical systems are an intricate set of coupled, time-dependent partial differential equations (PDEs), comprising the general relativistic hydrodynamics and magnetohydrodynamics equations (GRHD/GRMHD hereafter) and Einstein's gravitational field equations. Simplifications can be made when the "test-fluid" approximation holds, in which the fluid's self-gravity is neglected against the background gravitational field. Additionally, descriptions employing ideal hydrodynamics (inviscid fluids) and ideal MHD (infinite conductivity), are also fairly standard choices in numerical astrophysics. On the other hand, there are also situations where the number of equations must be augmented instead, as to account for radiative processes or microphysics (finite-temperature equations of state (EOS) and nuclear physics).

This article aims at presenting a brief overview of the equations of GRHD/GRMHD within the $3 + 1$ formalism of general relativity in ways suitable for numerical work. Space constraints limit the level of detail of the presentation, which the interested reader can complement with the help of the available literature on the subject (see [An89, MM03, Fo03, An06] and references therein). The article also discusses the different numerical approaches designed to solve hyperbolic systems of conservation laws such as the GRHD/GRMHD equations. Some examples in the numerical solution in scenarios of relativistic astrophysics are mentioned in Sect. 5.

2 General Relativistic Hydrodynamics

The GRHD equations are the local conservation laws of momentum and energy, encoded in the stress-energy tensor $T^{\mu\nu}$, and of the matter density, J^{μ} (the continuity equation)

$$\nabla_{\mu} T^{\mu\nu} = 0, \qquad \nabla_{\mu} J^{\mu} = 0, \tag{1}$$

where ∇_{μ} stands for the four-dimensional covariant derivative. (Throughout Greek indices run from 0 to 3 and Latin indices from 1 to 3; geometrized units $G = c = 1$ are used; G is Newton's gravitational constant and c is the speed of light.) The density current reads $J^{\mu} = \rho u^{\mu}$, where u^{μ} is the 4-velocity of the fluid and ρ its rest-mass density. We assume a perfect fluid stress-energy tensor

$$T^{\mu\nu} = \rho h u^{\mu} u^{\nu} + p g^{\mu\nu}, \tag{2}$$

where p is the pressure, h is the specific enthalpy, $h = 1 + \varepsilon + p/\rho$, ε being the specific internal energy, and $g_{\mu\nu}$ is the space–time metric tensor.

The previous system of equations is closed once a EOS is chosen, i.e. a constitutive relation of the form $p = p(\rho, \varepsilon)$. In the so-called test-fluid approximation the dynamics of the matter fields is completely described by the previous conservation laws and the EOS. If such approximation does not hold, these equations must be solved in conjunction with Einstein's equations for the gravitational field which describe the evolution of a dynamical space–time.

The approach most commonly employed to solve Einstein's equations in numerical relativity is the so-called Cauchy or $3+1$ formulation (IVP) (see [Al04] and references therein for details). In this formulation space–time is foliated into a set of nonintersecting space-like hypersurfaces, for which the lapse function α measures the rate of advance of time along a time-like unit vector n^{μ} normal to a surface, and the space-like shift vector β^{i} describes how coordinates move between different hypersurfaces. Introducing a coordinate chart (x^0, x^i) the $3+1$ line element reads

$$ds^2 = -(\alpha^2 - \beta_i \beta^i) dx^0 dx^0 + 2\beta_i dx^i dx^0 + \gamma_{ij} dx^i dx^j, \tag{3}$$

where γ_{ij} is the spatial 3-metric induced on each space-like slice. Hence, the above conservation equations, (1), read

$$\frac{\partial}{\partial x^\mu}\sqrt{-g}J^\mu = 0, \qquad \frac{\partial}{\partial x^\mu}\sqrt{-g}T^{\mu\nu} = -\sqrt{-g}\Gamma^\nu_{\mu\lambda}T^{\mu\lambda}, \tag{4}$$

where $g = \det(g_{\mu\nu}) = \alpha\sqrt{\gamma}$, $\gamma = \det(\gamma_{ij})$ and $\Gamma^\nu_{\mu\lambda}$ are the so-called Christoffel symbols.

Chronologically, the first attempt to formulate and solve the equations of relativistic Eulerian hydrodynamics in multidimensions was due to Wilson [Wil72], who wrote the system as a coupled set of advection equations within the $3+1$ formalism. This approach sidestepped an important guideline for the formulation of nonlinear hyperbolic equations, namely the preservation of their conservation form. This is a necessary feature to guarantee correct evolution in regions of entropy generation (shocks). As a result, some amount of numerical dissipation (artificial viscosity terms) had to be used to stabilize the numerical solution at discontinuities.

The main practical limitation of such nonconservative system was the numerical inability to handle ultrarelativistic flows [NW86]. This handicap posed a tremendous challenge to the numerical modeling of relativistic astrophysical sources where flow velocities as large as 99% of the speed of light or higher are known to exist. Paradigmatic examples of such sources are the jets associated with active galactic nuclei as well as with γ-ray bursts, the most luminous events in the Universe after the Big Bang. On the other hand, the presence of the Lorentz factor ($W \equiv \alpha u^0$) in the convective (transport) terms of the GRHD equations (and of the pressure in the specific enthalpy) make the relativistic equations much more coupled than their Newtonian counterparts. In an attempt to capture more accurately such coupling [NW86] proposed the use of implicit schemes. While some progress was achieved limitations on the fluid speeds attainable persisted, with a maximum value for W of about 10. Ultrarelativistic flows could only be handled (and with explicit schemes) once conservative formulations were adopted.

Since the early 1990s conservative formulations of the GRHD equations, well adapted to numerical methodology, were developed: [MIM91] ($1+1$, general EOS), [EM95] (covariant, perfect fluid EOS), [Ban97] ($3+1$, general EOS), and [PF99] (covariant, general EOS). Numerically, the hyperbolic and conservative nature of the GRHD equations allows to design a solution procedure based on the characteristic speeds and fields of the system (i.e., Riemann solvers), translating to relativistic hydrodynamics existing tools of computational fluid dynamics. This procedure departs from earlier approaches [Wil72], most notably in avoiding the need for artificial dissipation terms to handle discontinuous solutions as well as implicit schemes as proposed by [NW86].

The extension of modern *high-resolution shock-capturing* schemes (HRSC hereafter) from classical fluid dynamics to relativistic hydrodynamics was accomplished in three steps: (a) casting the GRHD equations as a system of conservation laws; (b) identifying the suitable vector of unknowns; and (c) building up an approximate Riemann solver. For brevity we focus next on the approach taken by [Ban97]. The interested reader is addressed to the previous references for specific details on additional formulations.

In [Ban97] the GRHD equations were written as a first-order, flux-conservative hyperbolic system, amenable to numerical work:

$$\frac{1}{\sqrt{-g}}\left(\frac{\partial\sqrt{\gamma}\mathbf{U}(\mathbf{w})}{\partial x^0}+\frac{\partial\sqrt{-g}\mathbf{F}^i(\mathbf{w})}{\partial x^i}\right)=\mathbf{S}(\mathbf{w}). \quad (5)$$

With respect to an Eulerian observer and in terms of the *primitive variables*, $\mathbf{w}=(\rho,v^i,\varepsilon)$, where v^i is the 3-velocity of the fluid, the state vector $\mathbf{U}$ (conserved variables) and the vectors of fluxes $\mathbf{F}$ and source terms $\mathbf{S}$ are given by

$$\mathbf{U}(\mathbf{w})=(D,S_j,\tau), \quad (6)$$

$$\mathbf{F}^i(\mathbf{w})=\left(D\tilde{v}^i,S_j\tilde{v}^i+p\delta^i_j,\tau\tilde{v}^i+pv^i\right), \quad (7)$$

$$\mathbf{S}(\mathbf{w})=\left(0,T^{\mu\nu}\left(\frac{\partial g_{\nu j}}{\partial x^\mu}-\Gamma^\delta_{\nu\mu}g_{\delta j}\right),\alpha\left(T^{\mu 0}\frac{\partial \ln\alpha}{\partial x^\mu}-T^{\mu\nu}\Gamma^0_{\nu\mu}\right)\right), \quad (8)$$

with $\tilde{v}^i=v^i-\beta^i/\alpha$ and δ^i_j being the Kronecker delta. The conserved quantities are the relativistic densities of mass, momenta, and energy, defined as $D=\rho W$, $S_j=\rho hW^2v_j$, and $\tau=\rho hW^2-p-D$.

HRSC schemes based on Riemann solvers use the local characteristic structure of the hyperbolic system of equations. The wave structure of the GRHD equations (5) analyzed in [Ban97] (see also [Fo00]) showed that the eigenvalues (characteristic speeds) of the corresponding Jacobian matrices are real and there exists a complete set of right-eigenvectors. System (5) satisfies, thus, the definition of hyperbolicity.

It is worth mentioning a key difference of the wave structure of the GRHD equations with respect to the Newtonian case. This is also apparent in the absence of gravity, that is, in special relativity, $g_{\mu\nu}=\mathrm{diag}(-1,1,1,1)$, which we will adopt next for simplicity. In this case the eigenvalues (along the x-direction) read [Fo94]

$$\lambda_0=v^x \qquad \text{(triple)}, \quad (9)$$

$$\lambda_\pm=\frac{1}{1-v^2c_s^2}\left(v^x(1-c_s^2)\pm c_s\sqrt{(1-v^2)[1-v^xv^x-(v^2-v^xv^x)c_s^2]}\right), \quad (10)$$

where c_s is the speed of sound and $v^2=v^iv_i$. Thus, the eigenvalues along the x-direction involve the coupling of the components of the velocity transverse to the chosen direction (and similarly for the other two directions). Even in the purely one-dimensional case, $\mathbf{v}=(v^x,0,0)$, the eigenvalues read $\lambda_0=v^x$ and $\lambda_\pm=\frac{v^x\pm c_s}{1\pm v^xc_s}$, the latter involving a Lorentz addition of the fluid velocity and the sound speed, as opposed to the Galilean addition of the velocities appearing in the Newtonian case ($\lambda_\pm=v^x\pm c_s$).

A distinctive feature of the numerical solution of the RHD equations is that while the numerical algorithm updates the vector of conserved quantities $\mathbf{U}$ (see below), the numerical code makes extensive use of the primitive variables $\mathbf{w}$. Those would appear repeatedly in the solution procedure, e.g., in the characteristic fields, in the solution of the Riemann problem, and in the computation of the numerical fluxes. For space-like foliations of the space–time (i.e., 3 + 1) the relation between the two sets of variables turns out to be implicit. Therefore, iterative (root finding) algorithms are required to recover the primitive variables. Those have been developed for all existing formulations [EM95, Ban97, PF99]. This feature, absent in Newtonian hydrodynamics, may lead to accuracy losses in regions of low density and small velocities, apart from being computationally inefficient. Only for *null* foliations of the space–time, the procedure of connecting primitive and conserved variables is explicit for a perfect fluid EOS, a direct consequence of the particular form of the Bondi–Sachs metric [PF99].

3 General Relativistic Magnetohydrodynamics

General relativistic MHD is concerned with the dynamics of relativistic, electrically conducting fluids (plasma) in the presence of magnetic fields. Here, we concentrate on purely *ideal* GRMHD, neglecting the presence of viscosity and heat conduction in the limit of infinite conductivity (perfect conductor fluid). As the GRHD equations discussed before, the GRMHD equations can also be cast in first-order, flux-conservative hyperbolic form. The discussion reported here follows the derivation of these equations as presented in [An06] to which the reader is addressed for details (see also references therein).

In terms of the (Faraday) electromagnetic tensor $F^{\mu\nu}$, Maxwell's equations read

$$\nabla_\nu {}^*F^{\mu\nu} = 0, \qquad \nabla_\nu F^{\mu\nu} = \mathcal{J}^\mu, \tag{11}$$

where $F^{\mu\nu} = U^\mu E^\nu - U^\nu E^\mu - \eta^{\mu\nu\lambda\delta} U_\lambda B_\delta$, its dual ${}^*F^{\mu\nu} = \frac{1}{2}\eta^{\mu\nu\lambda\delta} F_{\lambda\delta}$, and $\eta^{\mu\nu\lambda\delta} = \frac{1}{\sqrt{-g}}[\mu\nu\lambda\delta]$, where $[\mu\nu\lambda\delta]$ is the completely antisymmetric Levi-Civita symbol. E^μ and B^μ stand for the electric and magnetic fields measured by an observer with 4-velocity U^μ, and $\mathcal{J}^\mu$ is the electric 4-current, $\mathcal{J}^\mu = \rho_q u^\mu + \sigma F^{\mu\nu} u_\nu$ where ρ_q is the proper charge density and σ is the electric conductivity.

Maxwell's equations can be simplified if the fluid is a perfect conductor. In this case σ is infinite and, to keep the current finite, the term $F^{\mu\nu}u_\nu$ must vanish, which results in $E^\mu = 0$ for a comoving observer. This case corresponds to the so-called ideal MHD condition. Under this assumption the electric field measured by the Eulerian observer has components

$$E^0 = 0, \qquad E^i = -\alpha\eta^{0ijk} v_j B_k, \tag{12}$$

and Maxwell's equations $\nabla_\nu {}^*F^{\mu\nu} = 0$ reduce to the divergence-free condition plus the induction equation for the evolution of the magnetic field

$$\frac{\partial(\sqrt{\gamma}B^i)}{\partial x^i} = 0, \qquad \frac{1}{\sqrt{\gamma}}\frac{\partial}{\partial x^0}(\sqrt{\gamma}B^i) = \frac{1}{\sqrt{\gamma}}\frac{\partial}{\partial x^j}\{\sqrt{\gamma}[\alpha\tilde{v}^i B^j - \alpha\tilde{v}^j B^i]\}. \tag{13}$$

For a fluid endowed with a magnetic field the stress-energy tensor is the sum of that of the fluid and that of the electromagnetic field, $T^{\mu\nu} = T^{\mu\nu}_{\rm Fluid} + T^{\mu\nu}_{\rm EM}$, where $T^{\mu\nu}_{\rm Fluid}$ is given by (2) for a perfect fluid. On the other hand $T^{\mu\nu}_{\rm EM}$ can be obtained from the Faraday tensor as follows:

$$T^{\mu\nu}_{\rm EM} = F^{\mu\lambda}F^{\nu}_{\lambda} - \frac{1}{4}g^{\mu\nu}F^{\lambda\delta}F_{\lambda\delta}, \tag{14}$$

which, in ideal MHD, can be rewritten as

$$T^{\mu\nu}_{\rm EM} = \left(u^{\mu}u^{\nu} + \frac{1}{2}g^{\mu\nu}\right)b^2 - b^{\mu}b^{\nu}, \tag{15}$$

where b^{μ} is the magnetic field measured by the observer comoving with the fluid and $b^2 = b^{\nu}b_{\nu}$. The total stress-energy tensor is thus given by

$$T^{\mu\nu} = \rho h^* u^{\mu}u^{\nu} + p^* g^{\mu\nu} - b^{\mu}b^{\nu}, \tag{16}$$

with the definitions $p^* = p + b^2/2$ and $h^* = h + b^2/\rho$.

Following [An06] the conservation equations for the energy–momentum tensor, together with the continuity equation and the equation for the evolution of the magnetic field, can be written as a first-order, flux-conservative, hyperbolic system equivalent to (5). The state vector and the vector of fluxes of the GRMHD system of equations read:

$$\mathbf{U}(\mathbf{w}) = (D, S_j, \tau, B^k), \tag{17}$$

$$\begin{aligned}\mathbf{F}^i(\mathbf{w}) = (&D\tilde{v}^i, S_j\tilde{v}^i + p^*\delta^i_j - b_jB^i/W,\\ &\tau\tilde{v}^i + p^*v^i - \alpha b^0 B^i/W, \tilde{v}^iB^k - \tilde{v}^kB^i),\end{aligned} \tag{18}$$

where the conserved quantities are now defined as $D = \rho W$, $S_j = \rho h^* W^2 v_j - \alpha b^0 b_j$, and $\tau = \rho h^* W^2 - p^* - \alpha^2(b^0)^2 - D$. The corresponding vector of sources coincides with the one given by (8) save for the use of the complete (fluid plus electromagnetic field) stress-energy tensor (the magnetic field evolution equation is source-free).

The hyperbolic structure of those equations is discussed in [An06]. In the classical MHD case the wave structure was analyzed by [BW88]. There are seven physical waves: two Alfvén waves (with eigenvalues $\lambda_{a\pm} = v_x \pm v_a$, v_x and v_a being the fluid and Alfvén speeds, respectively), two fast and two slow magnetosonic waves ($\lambda_{f\pm} = v_x \pm v_f$, $\lambda_{s\pm} = v_x \pm v_s$), and one entropy wave ($\lambda_e = v_x$), ordered such that $\lambda_{f_-} < \lambda_{a_-} < \lambda_{s_-} < \lambda_e < \lambda_{s+} < \lambda_{a+} < \lambda_{f_+}$. The expressions for the Alfvén and magnetosonic speeds read

$$v_a = \sqrt{\frac{B_x^2}{\rho}}, \tag{19}$$

$$v_{f,s}^2 = \frac{1}{2}\left\{c_s^2 + \frac{B_x^2 + B_y^2 + B_z^2}{\rho} \pm \sqrt{\left(c_s^2 + \frac{B_x^2 + B_y^2 + B_z^2}{\rho}\right)^2 - 4v_a^2c_s^2}\right\}. \tag{20}$$

The corresponding wave structure for relativistic MHD was thoroughly analyzed by [An89]. The investigation of the roots of the characteristic equation showed that only the entropic waves and the Alfvén waves can be (explicitly) obtained in closed form, while the magnetosonic waves are given by the numerical solution of a quadratic equation.

For the GRMHD formulation of [An06] the characteristic speed of the entropic waves propagating in the x-direction reads

$$\lambda_e = \alpha v^x - \beta^x. \tag{21}$$

For Alfvén waves, there are two solutions corresponding to different speeds of the waves,

$$\lambda_{a\pm} = \frac{b^x \pm \sqrt{\rho h + b^2} u^x}{b^0 \pm \sqrt{\rho h + b^2} u^0}. \tag{22}$$

Just as in the classical case, the relativistic MHD equations have degenerate states in which two or more wave speeds coincide, which breaks the strict hyperbolicity of the system. [Ko99] has reviewed the properties of these degeneracies. In the fluid rest frame, the degeneracies in both classical and relativistic MHD are the same: either the slow and Alfvén waves have the same speed as the entropy wave when propagating perpendicularly to the magnetic field (Degeneracy I), or the slow or the fast wave (or both) have the same speed as the Alfvén wave when propagating in a direction aligned with the magnetic field (Degeneracy II). These degeneracies have been characterized by [An06] in terms of the components of the magnetic field 4-vector normal and tangential to the Alfvén wavefront, $\mathbf{b}_n$, $\mathbf{b}_t$. When $\mathbf{b}_n = 0$, the system falls within Degeneracy I, while Degeneracy II is reached when $\mathbf{b}_t = 0$. In addition, [An06] have also worked out a single set of right and left eigenvectors which are regular and span a complete basis in any physical state, including degenerate states. Such renormalization procedure is a relativistic generalization of the work performed by [BW88] in classical MHD.

On the other hand, as for the case of the GRHD equations discussed before, iterative (root finding) algorithms are also required for the GRMHD equations to recover the primitive variables from the state vector. The recovery procedure is in this case more involved than for unmagnetized flows. For the GRMHD formulation discussed in this section [An06] find the roots of an eighth-order polynomial using a two-dimensional Newton–Raphson scheme. The interested reader is addressed to [No06] for a comparison of different methods.

We end this section by pointing out that major advances on the physical understanding of the wave structure of the relativistic hydrodynamics equations have been possible in recent years, remarkably thanks to the derivation of exact solutions of the Riemann problem both in special relativistic hydrodynamics and MHD [MM94, Po00, RZ02, Ro05, GR06].

4 Solution Procedure for the GRHD/GRMHD Equations

Just as their Newtonian counterparts, the GRHD/GRMHD equations are non-linear hyperbolic systems of conservation laws. A distinctive feature of such systems is that smooth initial data can develop discontinuities during the time evolution. It is well known that standard finite difference schemes show deficiencies when dealing with discontinuous solutions. Typically, first-order accurate schemes are too dissipative across discontinuities while second-order (or higher) schemes produce spurious oscillations near discontinuities.

Finite difference schemes provide numerical solutions of the discretized version of the PDEs. Therefore, convergence properties under grid refinement must be enforced on such schemes to guarantee the validity of the numerical result. The Lax–Wendroff theorem states that for hyperbolic systems of conservation laws, schemes written in *conservation form* converge to one of the so-called *weak solutions* of the PDEs. However, the class of all weak solutions is too wide as there is no uniqueness for the IVP. Thus, among all weak solutions, the numerical scheme must guarantee convergence to the *physically admissible solution*, a property whose mathematical characterization was given by Lax for hyperbolic systems of conservation laws.

A conservative scheme for system (5) can be straightforwardly devised by using the corresponding integral form:

$$\int_{\Omega} \frac{1}{\sqrt{-g}} \frac{\partial \sqrt{\gamma}\mathbf{U}}{\partial x^0} d\Omega + \int_{\Omega} \frac{1}{\sqrt{-g}} \frac{\partial \sqrt{-g}\mathbf{F}^i}{\partial x^i} d\Omega = \int_{\Omega} \mathbf{S} d\Omega, \tag{23}$$

where Ω is a region of the four-dimensional manifold enclosed within a three-dimensional surface $\partial\Omega$ which is bounded by two space-like surfaces $\Sigma_{x^0}, \Sigma_{x^0+\Delta x^0}$ and two time-like surfaces $\Sigma_{x^i}, \Sigma_{x^i+\Delta x^i}$. For numerical purposes the above relation can be written as:

$$\begin{aligned}
\bar{\mathbf{U}}_{t+\Delta t} - \bar{\mathbf{U}}_t = & -\left(\int_{\Sigma_{x^1+\Delta x^1}} \sqrt{-g}\hat{\mathbf{F}}^1 dx^0 dx^2 dx^3 - \int_{\Sigma_{x^1}} \sqrt{-g}\hat{\mathbf{F}}^1 dx^0 dx^2 dx^3 \right) \\
& - \left(\int_{\Sigma_{x^2+\Delta x^2}} \sqrt{-g}\hat{\mathbf{F}}^2 dx^0 dx^1 dx^3 - \int_{\Sigma_{x^2}} \sqrt{-g}\hat{\mathbf{F}}^2 dx^0 dx^1 dx^3 \right) \\
& - \left(\int_{\Sigma_{x^3+\Delta x^3}} \sqrt{-g}\hat{\mathbf{F}}^3 dx^0 dx^1 dx^2 - \int_{\Sigma_{x^3}} \sqrt{-g}\hat{\mathbf{F}}^3 dx^0 dx^1 dx^2 \right) \\
& + \int_{\Omega} \mathbf{S} d\Omega,
\end{aligned} \tag{24}$$

where

$$\bar{\mathbf{U}} = \frac{1}{\Delta V} \int_{x^1}^{x^1+\Delta x^1} \int_{x^2}^{x^2+\Delta x^2} \int_{x^3}^{x^3+\Delta x^3} \sqrt{\gamma}\mathbf{U} dx^1 dx^2 dx^3 \tag{25}$$

and

$$\Delta V = \int_{x^1}^{x^1+\Delta x^1} \int_{x^2}^{x^2+\Delta x^2} \int_{x^3}^{x^3+\Delta x^3} \sqrt{\gamma} dx^1 dx^2 dx^3. \tag{26}$$

The main advantage of this procedure is that those variables which obey a conservation law are conserved during the evolution, as long as the balance between the fluxes at the boundaries of the computational domain and the source terms are zero. The *numerical fluxes* appearing in (24) are calculated at cell interfaces where the flow conditions can be discontinuous. Those numerical fluxes are approximations to the time-averaged fluxes across an interface, i.e.

$$\hat{\mathbf{F}}_{i+\frac{1}{2}} = \frac{1}{\Delta x^0} \int_{x^{0\,n}}^{x^{0\,n+1}} \mathbf{F}(\mathbf{U}(x_{i+\frac{1}{2}}, x^0)) dx^0, \tag{27}$$

where the flux integral depends on the solution at the numerical interfaces, $\mathbf{U}(x_{i+1/2}, x^0)$, during a time step. Godunov first proposed to calculate $\mathbf{U}(x_{i+1/2}, x^0)$ by exactly solving Riemann problems at every cell interface to obtain $\mathbf{U}(x_{i+1/2}, x^0) = \mathbf{U}(0; \mathbf{U}_i^n, \mathbf{U}_{i+1}^n)$, which denotes the Riemann solution for the (left and right) states $\mathbf{U}_i^n$, $\mathbf{U}_{i+1}^n$ along the ray $x^i/x^0 = 0$. This was a procedure of far-reaching consequences as it was incorporated in the design of numerical schemes for solving the Euler equations of classical gas dynamics in the presence of shock waves, which led to major advances in the field.

The derivation of the exact Riemann solution involves the computation of the full wave speeds to find where they lie in state space. This is a computationally expensive procedure, particularly for complex EOS and in multidimensions. Furthermore, for relativistic multidimensional flows, the coupling of all velocity components through the Lorentz factor results in the increase in the number of algebraic Rankine–Hugoniot conditions to consider in the case of shock waves and in solving a system of ODEs for the rarefaction waves. In spite of this the exact solution of the Riemann problem in *special* relativistic hydrodynamics has been derived [MM94, Po00]. Nevertheless, the computational inefficiency involved in the use of the exact solver in long-term numerical simulations motivated the gradual development of *approximate Riemann solvers*. These, being much cheaper than the exact solver yield equally accurate results.

The spatial accuracy of the numerical solution can be increased by reconstructing the primitive variables at the cell interfaces before the actual computation of the numerical fluxes. Diverse cell-reconstruction procedures are available in the literature (see references in [To97, MM03]) and have been straightforwardly applied in relativistic hydrodynamics. Correspondingly, the temporal accuracy of the scheme can be improved by advancing in time the equations in integral form using the method of lines in tandem with a high order, conservative Runge–Kutta method.

The main approaches extended from computational fluid dynamics to build HRSC schemes in relativistic hydrodynamics can be divided in the following broad categories:

1. HRSC schemes based on *Riemann solvers* (upwind methods): Developments include both solvers relying on the *exact solution* of the Riemann problem: [MM94, Po00], relativistic PPM [MM96], Glimm's random choice method [WPL97], and two-shock approximation [Ba94, DW97], as well as *linearized solvers* based on local linearizations of the Jacobian matrices of the flux-vector Jacobians, e.g., Roe-type Riemann solvers (Roe-Eulderink: [EM95]; local characteristic approach: [MIM91, Fo94, Ban97]), primitive-variable formulation: [FK96], and Marquina flux formula: [Do98].
2. HRSC schemes sidestepping the use of characteristic information (*symmetric schemes* with nonlinear numerical dissipation): Various approaches have been undertaken recently, including those by [Ko98] (Lax–Wendroff scheme with conservative TVD dissipation terms), [DZB02] (Lax–Friedrichs or HLL schemes with third-order ENO reconstruction algorithms), [AF02] (nonoscillatory central differencing), and [LS04, SF05] (semidiscrete central scheme of Kurganov–Tadmor [KT]).

Other approaches worth mentioning include: (1) artificial viscosity [NW86, AF02], (2) flux-corrected transport scheme [Sch93], and (3) smoothed particle hydrodynamics [CM97, SR99]. The interested reader is addressed to the review article by [MM03] for a complete list of references on this topic as well as for an in-depth comparison of the performance of these various approaches.

The numerical advantage of using (24) for the hydrodynamical variables is not apparent for the magnetic field components, as there is no guarantee that such procedure conserves the divergence of the magnetic field during an evolution. The main physical implication of the divergence constraint is that the magnetic flux through a closed surface is zero. This property is essential to the design of the so-called constrained transport method [EH88, To00], a common choice among the methods designed to solve the induction equation while preserving the divergence of the magnetic field [An06].

The current approaches to solve the RMHD equations within HRSC schemes also fall within the categories mentioned above, yet the development is somewhat more limited here than in the purely hydrodynamical case. Methods based on Riemann solvers have been initiated in special relativity by [Ko99] (includes eigenvector sets for degenerate states), [Ba01] (reconstruction not done on primitive variables), [Ko02] (right and left eigenvectors in covariant variables, but one-dimensional), and [An06] (right and left eigenvectors in conserved variables, complete set even for degenerate states), as well as in general relativity by [Ko05, An06]. On the other hand symmetric schemes (namely HLL and Kurganov–Tadmor) are being currently employed by a growing number of groups in GRMHD [Ko98, GMT03, Du05, SS05, An06]. All references listed here use conservative formulations of the RMHD equations. Artificial viscosity approaches are advocated by [Yo93, DVH03].

5 Applications in Relativistic Astrophysics

Numerical HD/MHD simulations are an essential tool in theoretical astrophysics, both to model classical and relativistic sources. In the latter case the progress achieved during the last few decades as a result of ever-increasing computational improvements as well as greater understanding of the mathematical aspects of the equations and of the numerical schemes to solve them, has been outstanding. Furthermore, its scope involves a large number of scenarios at the forefront of research in astrophysics which has only been possible to start approaching in recent times.

Examples include heavy ion collisions (in the special relativistic limit), formation and propagation of jets associated with both active galactic nuclei and γ-ray burst progenitors, gravitational stellar collapse to neutron stars and black holes, pulsations and instabilities of rotating relativistic stars, accretion on to black holes, and binary neutron star mergers. Such a list of applications is too large to allow for an adequate coverage within the space constraints of this article. Hence, only a paradigmatic example will be briefly discussed here, namely gravitational stellar core collapse to a neutron star, addressing the interested reader to [Fo03, Fo05] and references there in for more extended discussions.

The gravitational collapse of massive stars is a distinctive example in relativistic astrophysics involving self-gravitating fluids whose dynamics is governed by the GRHD/GRMHD equations coupled to Einstein's gravitational field equations. Stars with initial masses larger than $\sim 9M_\odot$ (where $M_\odot$ is the mass of the Sun) end their thermonuclear evolution developing a core composed of iron group nuclei, which is dynamically unstable against gravitational collapse. The core collapses to a neutron star releasing gravitational binding energy of the order $\sim 3\times 10^{53}$ erg $(M/M_\odot)^2(R/10\ \mathrm{km})^{-1}$, sufficient to power a supernova explosion. Numerical simulations show how sensible the explosion mechanism is to the details of the postbounce evolution: gravity, the nuclear EOS and the properties of the nascent neutron star, the treatment of the neutrino transport, and the neutrino-matter interaction. Only recently simulations including state-of-the-art neutrino transport, in which the Boltzmann equation is solved in connection with the hydrodynamics, are becoming possible (see [Bu03] and references therein). Relativistic simulations of microphysically detailed core collapse beyond spherical symmetry are not yet available.

Steps toward that final goal are however being taken. Numerical simulations of (axisymmetric) relativistic rotational core collapse, approximating Einstein's equations for a conformally flat 3-metric, were first reported in [Di02]. These did not include the necessary microphysics involved in supernova modeling as they aimed at computing the gravitational radiation from core collapse, to highlight the differences in the dynamics and waveforms between Newtonian and relativistic gravity. The gravitational wave signal is characterized by a burst associated with the hydrodynamical bounce followed by the proto-neutron star ringdown phase. While the central density reaches higher values in relativistic gravity the gravitational wave signals are of comparable amplitudes, which constraints the chances for detection to galactic events. Further simulations have improved the approximation in the metric equations [Ce05] or use the full Einstein equations [SS04]. First attempts toward simulating GRMHD core collapse are currently being taken.

6 Summary

Formulations of the equations of (inviscid) general relativistic hydrodynamics and (ideal) magnetohydrodynamics have been discussed, along with methods for their numerical solution. Upon the explicit choice of an Eulerian observer and suitable fluid and magnetic field variables, it is possible to cast both systems of equations as first-order, hyperbolic systems of conservation laws. During the last 15 years, the so-called (upwind) high-resolution shock-capturing schemes based on Riemann solvers have been extended from classical to relativistic fluid dynamics (both special and general), to the point that GRHD simulations in relativistic astrophysics are routinely performed nowadays. While such advances also hold true in the case of the MHD equations, the development still awaits here for a thorough numerical exploration. The article has also presented a brief overview of numerical techniques, providing examples of their applicability to general relativistic fluids and magneto-fluids in scenarios of relativistic astrophysics.

It is worth spending a last comment to mention the long-term, numerically stable formulations of Einstein's equations (or accurate enough approximations) that have been proposed by several numerical relativity groups worldwide in recent years. The paradigm which the numerical relativist is currently confronted with has suddenly changed for the best. Accurate and long-term stable, *coupled* evolutions of the GRHD/GRMHD equations and Einstein's equations are just becoming possible in three dimensions (benefited from the steady increase in computing power), allowing for the study of interesting relativistic astrophysics scenarios for the first time, such as gravitational collapse, accretion onto black holes, and binary neutron star mergers.

Acknowledgment

Research was supported by the Spanish *Ministerio de Educación y Ciencia* (grant AYA2004-08067-C03-01).

References

[An89] Anile, A.M.: Relativistic fluids and magneto-fluids. Cambridge University Press (1989)

[Fo03] Font, J.A.: Numerical hydrodynamics in general relativity. Liv. Rev. Relativ., **6**, 4 (2003)

[MM03] Martí, J.M. and Müller, E.: Numerical hydrodynamics in special relativity. Liv. Rev. Relativ., **6**, 7 (2003)

[An06] Antón, L., Zanotti, O., Miralles, J.A., Martí, J.M., Ibáñez, J.M., Font, J.A., and Pons, J.A.: Numerical $3+1$ general relativistic magnetohydrodynamics: A local characteristic approach. Astrophys. J., **637**, 296–312 (2006)

[Al04] Alcubierre, M.: Brief introduction to numerical relativity. AIP Conf. Proc., **758**, 193–207 (2005)

[Wil72] Wilson, J.: Numerical study of fluid flow in a Kerr space. Astrophys. J., **173**, 431–438 (1972)

[NW86] Norman, M.L. and Winkler, K-H.: Why ultrarelativistic numerical hydrodynamics is difficult? In Astrophysical Radiation Hydrodynamics, Reidel Publishing Company, 449–475 (1986)

[MIM91] Martí, J.M., Ibáñez, J.M., and Miralles, J.A.: Numerical relativistic hydrodynamics: Local characteristic approach. Phys. Rev. D, **43**, 3794–3801 (1991)

[EM95] Eulderink, F. and Mellema, G.: General relativistic hydrodynamics with a Roe solver. Astron. Astrophys. Suppl. Ser., **110**, 587–623 (1995)

[Ban97] Banyuls, F., Font, J.A., Ibáñez, J.M., Martí, J.M., and Miralles, J.A.: Numerical $3+1$ general relativistic hydrodynamics: A local characteristic approach. Astrophys. J., **476**, 221–231 (1997)

[PF99] Papadopoulos, P. and Font, J.A.: Relativistic hydrodynamics on spacelike and null surfaces: Formalism and computations of spherically symmetric spacetimes. Phys. Rev. D, **61**, 024015 (1999)

[Fo00] Font, J.A., Miller, M., Suen, W-M., and Tobias, M.: Three-dimensional numerical general relativistic hydrodynamics: Formulations, methods and code tests. Phys. Rev. D, **61**, 044011 (2000)

[BW88] Brio, M. and Wu, C.C.: An upwind differencing scheme for the equations of ideal magnetohydrodynamics. J. Comput. Phys., **75**, 400–422 (1988)

[Ko99] Komissarov, S.S.: A Godunov-type scheme for relativistic magnetohydrodynamics, MNRAS, **303**, 343–366 (1999)

[No06] Noble, S.C., Gammie, C.F., McKinney, J.C., Del Zanna, L.: Primitive variable solvers for conservative general relativistic magnetohydrodynamics. Astrophys. J., **641**, 626–637(2006)

[MM94] Martí, J.M. and Müller, E.: The analytical solution of the Riemann problem in relativistic hydrodynamics. J. Fluid Mech., **258**, 317–333 (1994)

[Po00] Pons, J.A., Martí, J.M., and Müller, E.: The exact solution of the Riemann problem with non-zero tangential velocities in relativistic hydrodynamics. J. Fluid Mech., **422**, 125–139 (2000)

[RZ02] Rezzolla, L. and Zanotti, O.: An improved exact Riemann solver for relativistic hydrodynamics. J. Fluid Mech., **449**, 395–411 (2001)

[Ro05] Romero, R., Martí, J.M., Pons, J.A., Miralles, J.A., and Ibáñez, J.M.: The exact solution of the Riemann problem in relativistic magnetohydrodynamics with tangential magnetic fields. J. Fluid Mech., **544**, 323–338 (2005)

[GR06] Giacomazzo, B. and Rezzolla, L.: The exact solution of the riemann problem in relativistic MHD. J. Fluid Mech., **544**, 323–338 (2005)

[MM96] Martí, J.M. and Müller, E.: Extension of the piecewise parabolic method to one-dimensional relativistic hydrodynamics. J. Comput. Phys., **123**, 1–14 (1996)

[WPL97] Wen, L., Panaitescu, A., and Laguna, P.: A shock-patching code for ultrarelativistic fluid flows. Astrophys. J., **486**, 919–927 (1997)

[Ba94] Balsara, D.: Riemann solver for relativistic hydrodynamics. J. Comput. Phys., **114**, 284–297 (1994)

[DW97] Dai, W. and Woodward, P.: A High-order Godunov-type scheme for shock interactions in ideal magnetohydrodynamics. SIAM J. Sci. Comput., **18**, 957–981 (1997)

[Fo94] Font, J.A., Ibáñez, J.M., Martí, J.M., and Marquina, A.: Multidimensional relativistic hydrodynamics: Characteristic fields and modern high-resolution shock-capturing schemes. Astron. Astrophys., **282**, 304–314 (1994)

[FK96] Falle, S.A.E.G. and Komissarov, S.S.: An upwind numerical scheme for relativistic hydrodynamics with a general equation of state. Mon. Not. R. Astron. Soc., **278**, 586–602 (1996)

[Do98] Donat, R., Font, J.A., Ibáñez, J.M., and Marquina, A.: A Flux-split algorithm applied to relativistic flows. J. Comput. Phys., **146**, 58–81 (1998)

[Ko98] Koide, S., Shibata, K., and Kudoh, T.: General relativistic magnetohydrodynamic simulations of jets from black hole accretion disks: Two-component jets driven by nonsteady accretion of magnetized disks. Astrophys. J., **495**, L63–L66 (1998)

[DZB02] Del Zanna, L. and Bucciantini, N.: An efficient shock-capturing central-type scheme for multidimensional relativistic flows. I. Hydrodynamics. Astron. Astrophys., **390**, 1177–1186 (2002)

[AF02] Anninos, P. and Fragile, P.C.: Non-oscillatory central difference and artificial viscosity schemes for relativistic hydrodynamics. Astrophys. J. Suppl. Ser., **144**, 243–257 (2002)

[LS04] Lucas-Serrano, A., Font, J.A., Ibáñez, J.M., and Martí, J.M.: Assessment of a high-resolution central scheme for the solution of the relativistic hydrodynamics equations. Astron. Astrophys., **428**, 703–715 (2004)

[SF05] Shibata, M. and Font, J.A.: Robustness of a high-resolution central scheme for hydrodynamic simulations in full general relativity. Phys. Rev. D, **72**, 047501 (2005)

[KT] Kurganov, A. and Tadmor, E.: Solution of two-dimensional Riemann problems for gas dynamics without Riemann problem solvers. J. Comput. Phys., **160**, 214 (2000)

[Sch93] Schneider, V., Katscher, V., Rischke, D.H., Waldhauser, B., Marhun, J.A., and Munz, C.-D.: New algorithms for ultra-relativistic numerical hydrodynamics. J. Comput. Phys., **105**, 92–107 (1993)

[CM97] Chow, E. and Monaghan, J.J.: Ultrarelativistic SPH. J. Comput. Phys., **134**, 296–305 (1997)

[SR99] Siegler, S. and Riffert, H.: Smoothed particle hydrodynamics simulations of ultra-relativistic shocks with artificial viscosity. Astrophys. J., **531**, 1053-1066 (2000)

[To97] Toro, E.F.: Riemann solvers and numerical methods for fluid dynamics. Springer (1997)

[EH88] Evans, C. and Hawley, J.F.: Simulation of magnetohydrodynamic flows: a constrained transport method. Astrophys. J., **332**, 659–677 (1988)

[To00] Tóth, G.: The $\nabla \cdot B = 0$ constraint in shock-capturing magnetohydrodynamics codes. J. Comput. Phys., **161**, 605–652 (2000)

[Ba01] Balsara, D.: Total variation diminishing scheme for relativistic magnetohydrodynamics. Astrophys. J. Suppl. Ser., **132**, 83–101 (2001)

[Ko02] Koldoba, A.V., Kuznetsov, O.A., and Ustyugova, G.V.: An approximate Riemann solver for relativistic magnetohydrodynamics. MNRAS, **333**, 932–942 (2002)

[Ko05] Komissarov, S.S.: Observations of the Blandford-Znajek process and the magnetohydrodynamic Penrose process in computer simulations of black hole magnetospheres. MNRAS, **359**, 801–808 (2005)

[GMT03] Gammie, C.F., McKinney, J.C., and Tóth, G.: HARM: A Numerical scheme for general relativistic magnetohydrodynamics. Astrophys. J., **589**, 444–457 (2003)

[Du05] Duez, M.D., Liu, Y.T., Shapiro, S.L., and Stephens, B.C.: Relativistic magnetohydrodynamics in dynamical spacetimes: Numerical methods and tests. Phys. Rev. D, **72**, 024028 (2005)

[SS05] Shibata, M. and Sekiguchi, Y.: Magnetohydrodynamics in full general relativity: Formulation and tests. Phys. Rev. D, **72**, 044014 (2005)

[Yo93] Yokosawa, M.: Energy and angular momentum transport in magnetohydrodynamical accretion onto a rotating black hole. Publ. Astron. Soc. Jpn., **45**, 207–218 (1993)

[DVH03] De Villiers, J. and Hawley, J.F.: A numerical method for general relativistic magnetohydrodynamics. Astrophys. J., **589**, 458–480 (2003)

[Fo05] Font, J.A.: General relativistic hydrodynamics and magnetohydrodynamics and their applications. Plasma Phys. Control. Fusion, **47**, B679–B690 (2005)

[Bu03] Buras, R., Rampp, M., Janka, H-T., and Kifonidis, K.: Improved Models of Stellar Core Collapse and Still No Explosions: What Is Missing? Phys. Rev. Lett., **90**, 241101 (2003)

[Di02] Dimmelmeier, H., Font, J.A., and Müller, E.: Relativistic simulations of rotational core collapse II. Collapse dynamics and gravitational radiation. Astron. Astrophys., **393**, 523–542 (2002)

[Ce05] Cerdá-Durán, P., Faye, G., Dimmelmeier, H., Font, J.A., Ibáñez, J.M., Müller, E., and Schäfer, G.: CFC+: improved dynamics and gravitational waveforms from relativistic core collapse simulations. Astron. Astrophys., **439**, 1033–1055 (2005)

[SS04] Shibata, M. and Sekiguchi, Y.: Gravitational waves from axisymmetric rotating stellar core collapse to a neutron star in full general relativity. Phys. Rev. D, **69**, 084024 (2004)

On Approximations for Overdetermined Hyperbolic Equations

S.K. Godunov*

Summary. Some aspects of approximations for overdetermined systems of hyperbolic equations are considered. The formulations of extended overdetermined systems of the thermodynamically consistent equations are presented for fluid dynamics equations, magnetohydrodynamics equations, the Maxwell equations and the elasticity equations. An approach for constructing discrete models of such systems is discussed.

1 Introduction

We focus upon difficulties that have to be overcome in the creation of algorithms for solving overdetermined systems of equations called **hyperbolic**. In general, a system is said to be hyperbolic if the number of its equations is equal to the number of unknowns and its characteristic conoid has a specific structure (definition of I.G. Petrovskii) or the system can be written using coefficients that form symmetric (or Hermitian) matrices (definition of Friedrichs). One of these matrices must be positive definite. It is useful to emphasize that a hyperbolic system in the sense of Petrovskii not necessarily can be written as a hyperbolic system in the sense of Friedrichs. An example demonstrating this fact was shown in 1984 by V.V. Ivanov. This example, however, is not well known because it was only published in Russian (see [1]).

We use Friedrichs' definition: a system of m equations with m unknowns functions, which form the m-dimensional vector $u(t, x) = (u_1, \ldots, u_m)$, $t > 0$, $x = (x_1, \ldots, x_d) \in R^d$, is hyperbolic if it can be written in the form

$$A \frac{\partial u}{\partial t} + B_k \frac{\partial u}{\partial x_k} = f,$$

here A and B_k are $m \times m$ Hermitian matrices such that $A = A(x) = A^* > 0$ and $B_k = B_k^*$ $(k = 1, \ldots, d)$ are either constant or smooth functions of u.

* Joint work with Dmitrii P. Babii, Olga B. Feodoritova, Victor T. Zhukov

Consider the simplest system of hydrodynamics equations in the Lagrange coordinates

$$\begin{gathered} \frac{\partial u}{\partial t} + \frac{\partial p}{\partial m} = 0, \quad E = E(v,s), \\ -\frac{\partial v}{\partial t} + \frac{\partial u}{\partial m} = 0, \quad p = -E_v(v,s), \\ \frac{\partial s}{\partial t} = 0. \end{gathered} \tag{1}$$

It is just the system for which the scheme [2] (known now as the Godunov scheme) was elaborated in winter of 1953–1954; it was published only in 1959. Introducing the Hamiltonian

$$H(u,p,s) = E + pv + \frac{u^2}{2}, \qquad dH = udu - vdp + Tds$$

the system (1) can be written in the form suggested by Friedrichs

$$\begin{pmatrix} H_{uu} & H_{up} & H_{us} \\ H_{pu} & H_{pp} & H_{ps} \\ H_{su} & H_{sp} & H_{ss} \end{pmatrix} \frac{\partial}{\partial t} \begin{pmatrix} u \\ p \\ s \end{pmatrix} + \begin{pmatrix} 0 & 1 & 0 \\ 1 & 0 & 0 \\ 0 & 0 & 0 \end{pmatrix} \frac{\partial}{\partial m} \begin{pmatrix} u \\ p \\ s \end{pmatrix} = 0.$$

Using the notation $\widehat{H} = pu$ and taking into account the equality $\widehat{H}_s = 0$, one can easily verify that this form is equivalent to the system

$$\begin{array}{ll} u & \dfrac{\partial H_u}{\partial t} + \dfrac{\partial \widehat{H}_u}{\partial m} = 0, \\ p & \dfrac{\partial H_p}{\partial t} + \dfrac{\partial \widehat{H}_p}{\partial m} = 0, \\ T & \dfrac{\partial H_s}{\partial t} + \dfrac{\partial \widehat{H}_s}{\partial m} = 0. \end{array} \tag{2}$$

The linear combination of these equations with the multipliers u, p, T (standing to the left of the equations) results in the additional equality consistent with (2), namely,

$$\frac{\partial (E + \frac{u^2}{2})}{\partial t} + \frac{\partial pu}{\partial m} = 0, \tag{3}$$

$$E + \frac{u^2}{2} = uH_u + pH_p + sH_s - H.$$

Thus, adding (3) to the system (2) we obtain a consistent overdetermined system of four equations. It is interesting to remark that symmetric hyperbolic subsystem can be extracted from this overdetermined system in other way. Let this subsystem be composed from the equations

$$
\begin{array}{ll}
-\dfrac{u}{T} & \dfrac{\partial u}{\partial t} + \dfrac{\partial p}{\partial m} = 0, \\[2ex]
-\dfrac{p}{T} & \dfrac{\partial v}{\partial t} + \dfrac{\partial u}{\partial m} = 0, \\[2ex]
\dfrac{1}{T} & \dfrac{\partial (E + \frac{u^2}{2})}{\partial t} + \dfrac{\partial pu}{\partial m} = 0.
\end{array}
\tag{4}
$$

The linear combination of these equations with the coefficients, standing to the left of the equations, results in "entropy conservation law" $\partial s/\partial t = 0$. Using the thermodynamic identity

$$
ds = -\frac{u}{T}du - \frac{p}{T}d(-v) + \frac{1}{T}d\left(E + \frac{u^2}{2}\right),
$$

$$
d\left(\frac{E + pv - Ts - \frac{u^2}{2}}{T}\right) = ud\left(\frac{u}{T}\right) - vd\left(-\frac{p}{T}\right) + \left(E + \frac{u^2}{2}\right)d\frac{1}{T}
$$

and the notation $\quad q_1 = -\dfrac{u}{T}, \quad q_2 = -\dfrac{p}{T}, \quad q_0 = \dfrac{1}{T},$

$$
L = \frac{E + pv - Ts - \frac{u^2}{2}}{T} = L(q_0, q_1, q_2), \qquad L^1 = \frac{q_1 q_2}{q_0},
$$

we can rewrite (4) in the form

$$
\begin{array}{ll}
q_0 & \dfrac{\partial L_{q_0}}{\partial t} + \dfrac{\partial L^1_{q_0}}{\partial m} = 0, \\[2ex]
q_1 & \dfrac{\partial L_{q_1}}{\partial t} + \dfrac{\partial L^1_{q_1}}{\partial m} = 0, \\[2ex]
q_2 & \dfrac{\partial L_{q_2}}{\partial t} + \dfrac{\partial L^1_{q_2}}{\partial m} = 0.
\end{array}
\tag{5}
$$

Since

$$
q_0 L_{q_0} + q_1 L_{q_1} + q_2 L_{q_2} - L = s, \qquad q_0 L^1_{q_0} + q_1 L^1_{q_1} + q_2 L^1_{q_2} - L = 0,
$$

the entropy conservation law $\partial s/\partial t = 0$ can be obtained by taking the linear combination of (5) with the coefficients q_0, q_1, q_2. For the convex generating potential $L(q_0, q_1, q_2)$, the quasilinear form of (5), namely,

$$
L_{q_i q_j}\frac{\partial q_i}{\partial t} + L^1_{q_i q_j}\frac{\partial q_j}{\partial m} = 0,
$$

indicates their hyperbolicity.

The system (4), which is composed of the conservation laws for momentum, volume, and energy, underlies the Godunov scheme to unconditionally satisfy the requirement that the entropy be nondecreasing. This property allows us to use this scheme for calculating discontinuous solutions, that is, shock waves. The entropy grows in shock waves and the overdetermined system consisting of (1) and (3) can not be assumed to be consistent.

This fact forces to define the class of thermodynamically consistent equations (for instance, see [3]–[5]) which present symmetric hyperbolic systems composed of divergences (conservation laws)

$$\frac{\partial L_{q_i}}{\partial t} + \frac{\partial M^j_{q_i}}{\partial x_j} = 0$$

and consistent (on smooth solutions) with the additional conservation laws

$$\frac{\partial (q_i L_q - L)}{\partial t} + \frac{\partial [q_i M^j_{q_i} - M^j]}{\partial x_j} = 0.$$

However, we have later found out that not all the equations of classical mathematical physics can be reduced to this form. For instance, consider the analogous (but nondivergent) formalization for magnetohydrodynamics equations

$$\begin{aligned} \frac{\partial L_{q_o}}{\partial t} + \frac{\partial (u_k L) q_0}{\partial x_k} &= 0, \\ \frac{\partial L_{h_i}}{\partial t} + \frac{\partial (u_k L)_{h_i}}{\partial x_k} - L_{h_k} \frac{\partial u_i}{\partial x_k} &= 0, \\ \frac{\partial L_{u_i}}{\partial t} + \frac{\partial (u_k L)_{u_i}}{\partial x_k} - L_{h_k} \frac{\partial h_i}{\partial x_k} &= 0. \end{aligned} \tag{6}$$

Although system (6) can be rewritten in the symmetric hyperbolic form [5], it does not include the conservation laws of momentum and energy. These are the laws (divergent equalities)

$$\frac{\partial L_{u_i}}{\partial t} + \frac{\partial [(u_k L)_{u_i} - h_i L_{h_k}]}{\partial x_k} = 0,$$

$$\frac{\partial (q_0 L_{q_0} + h_i L_{h_i} + u_i L_{u_i} - L)}{\partial t} + \frac{\partial [u_k (q_0 L_{q_0} + h_i L_{h_i} + u_i L_{u_i}) - u_i h_i L_{h_k}]}{\partial x_k} = 0$$

and they are not valid for all the solutions of the system (6); they hold only for the solutions satisfying the additional equation

$$\partial L_{h_i} / \partial x_i \equiv div\ L_{\mathbf{h}} = 0,$$

which is consistent with the system (6) according to its consequence

$$\frac{\partial}{\partial t} \left[\frac{1}{L_{q_0}} \; \frac{\partial L_{h_i}}{\partial x_i} \right] + u_k \frac{\partial}{\partial x_k} \left[\frac{1}{L_{q_0}} \; \frac{\partial L_{h_i}}{\partial x_i} \right] = 0.$$

A similar situation occurs in nonlinear elasticity theory if the equations are written in the Euler coordinates. The emphasized circumstance is a reason of difficulties in numerical solution of magnetohydrodynamics problems. An overview of these difficulties and the description of various techniques for overcoming them one can find in monograph [6] which we have read after professor van Leer had focused our attention about the difficulties caused by the accumulation of discretization errors. Such an accumulation, it seems, mainly occurs in long-time calculations of large smooth fields, whereas the papers reviewed in [6] paid main attention of the construction of schemes adapted to the calculation of zones containing shock waves. We propose to put off temporarily considerations of discontinuous solutions and speculate, firstly, how the smooth solutions of consistent overdetermined systems must be calculated. Investigations of discontinuous solutions can be started just after the situation with smooth solutions will be clear.

Below, we report on our study. Although it is at its initial stage, we have already come across new understanding. For constructing numerical algorithms we propose to do the next steps. Firstly, based on analysis of a mathematical model we define three objects: a symmetric hyperbolic subsystem ("background" system), physically substantial conservation laws consistent with this system, and a set of additional relations (for instance, divergent restrictions). Remark the theory of symmetric hyperbolic systems and discretization techniques of such systems are deeply developed. Therefore for discretization it is better to use an appropriate symmetric hyperbolic system. Such background system must be determined (a number of unknown functions coincides with a number of equations). A choice of a background subsystem is not unique but important. Estimation of quality of a background system must be obtained in numerical experiments by investigation of discretization errors of the additional relations. If these errors do not grow significantly in time we might believe that conservation laws hold.

So let a background hyperbolic system be chosen and has a form

$$A\frac{\partial u}{\partial t}+B\frac{\partial u}{\partial x}+C\frac{\partial u}{\partial y}+D\frac{\partial u}{\partial z}=F, \tag{7}$$

where the square matrices $A = A^* > 0$, $B = B^*$, $C = C^*$ and $D = D^*$ may depend on the variables x, y, z and t; in the quasi-linear case, they also may depend on the solution u. The proposed three-dimensional scheme consists of utilization of a one-dimensional scheme along each of the space directions. For simplicity, we consider the one-dimensional system with a zero right-hand side and constant matrices A and B:

$$A\frac{\partial u}{\partial t}+B\frac{\partial u}{\partial x}=0. \tag{8}$$

It is known that there is a linear transformation $u = \Lambda v$ such that $\Lambda^* A\Lambda = I$ is the identity matrix, and $\Lambda^* B\Lambda = \widehat{B}$ is the diagonal matrix. The transformed system has the canonical form

$$\frac{\partial v}{\partial t} + \widehat{B}\frac{\partial v}{\partial x} = 0, \tag{9}$$

where v is the so-called Riemann invariant vector. It is easily to construct a few discrete approximations for this system. Using the one-dimensional scheme we can write a few three-dimensional schemes of the first and higher order of accuracy. One of such schemes is presented in Sect. 3.

2 The Acoustic Equations

Consider the simplest example of the acoustic equations in R^2

$$\begin{aligned} \frac{\partial u}{\partial t} + \frac{\partial p}{\partial x} &= 0, \\ \frac{\partial v}{\partial t} + \frac{\partial p}{\partial y} &= 0, \\ \frac{\partial p}{\partial t} + \frac{\partial u}{\partial x} + \frac{\partial v}{\partial y} &= 0, \end{aligned} \tag{10}$$

$$\text{in } \Omega = \{x, y) : 0 \le x, y \le \pi\}, \quad t \ge 0.$$

For simplicity we consider the problem with the condition $p = 0$ imposed on the boundary $\partial\Omega$ of the domain and the initial conditions given at $t = 0$. The additional equations consistent with this system have the form

$$\begin{aligned} \frac{\partial}{\partial t}\frac{u^2 + v^2 + p^2}{2} + \frac{\partial pu}{\partial x} + \frac{\partial pv}{\partial y} &= 0, \\ \frac{\partial}{\partial t}(u_y - v_x) &= 0. \end{aligned} \tag{11}$$

It is well known that the main difficulty here is to provide a reasonable accuracy of (11) on a discrete level. To extend our system we add the new unknown potential ω and the new equation

$$\frac{\partial \omega}{\partial t} = p(x, y, t), \qquad \omega(x, y, 0) = 0. \tag{12}$$

One can easily verify that ω satisfies the wave equation

$$\frac{\partial^2 \omega}{\partial t^2} = \Delta\omega.$$

Here, Δ is the Laplacian, and $u(x, y, t)$, $v(x, y, t)$ can be presented in the form

$$\begin{aligned} u(x, y, t) &= u(x, y, 0) - \omega_x(x, y, t), \\ v(x, y, t) &= v(x, y, 0) - \omega_y(x, y, t). \end{aligned} \tag{13}$$

Two stages are utilized to perform one time step $t \to t+\tau$. At the first stage, for the symmetric hyperbolic system (chosen as the background one)

$$\begin{aligned}
&\frac{\partial u}{\partial t} + \frac{\partial p}{\partial x} = 0,\\
&\frac{\partial v}{\partial t} + \frac{\partial p}{\partial y} = 0,\\
&\frac{\partial p}{\partial t} + \frac{\partial u}{\partial x} + \frac{\partial v}{\partial y} = 0,\\
&\frac{\partial \omega}{\partial t} = p
\end{aligned} \tag{14}$$

is used. We compute grid approximations of $u(x,y,t+\tau)$, $v(x,y,t+\tau)$, $p(x,y,t+\tau)$ by a discrete scheme. On the second stage, the new values $u(x,y,t+\tau)$, $v(x,y,t+\tau)$ are computed by the discrete approximations of (13). After this replacement the next time step is to be implemented. The functions u, v computed on the first stage are viewed as the intermediate details.

In fact, the described procedure can be considered as the solution of the wave equation

$$\frac{\partial^2 p}{\partial t^2} = \Delta p$$

and (12) for the potential and then calculation of u, v using (13).

Numerical examples for the acoustic equations in a more general form to illustrate the efficiency of this approach are shown in [7].

3 The Maxwell Equations

Here we describe a choice of a symmetric hyperbolic system for the Maxwell equations and the construction of the second-order difference approximation. We introduce a uniform grid with a mesh size h. For the one-dimensional system (8) we write the conventional Godunov scheme (see [6], [8]) as a predictor:

$$\frac{v_i^{n+1} - v_i^n}{\tau} + \widehat{B}\frac{V_{i+1/2} - V_{i-1/2}}{h} = 0, \tag{15}$$

where $\widehat{B} = A^{-\frac{1}{2}}BA^{-\frac{1}{2}}$, v_i^n, $V_{i\pm1/2}$ are grid vector functions. The superscript $n = 0,1,\ldots$ is an index of time layer, the integer subscript $i = 0,1,\ldots I$ denotes the value of the grid function corresponding to the center of the cell, the half-integer subscripts $i-1/2$ and $i+1/2$ denote the respective values on the left and right boundary of the ith cell. Each value $V_{i\pm1/2}$ is the intercell numerical flux defined by solution of the corresponding Riemann problem for the boundaries $i\pm1/2$. The step size τ with respect to time is determined by the known stability condition. For simplicity we consider the periodic boundary conditions. The scheme (15) is a first-order scheme with respect to space and time and serves as a predictor for the second-order scheme described below.

On the correction stage the intercell fluxes are computed:

$$\widetilde{V}_{i+1/2} = \frac{v_{i-1}^n + v_{i+1}^n + v_{i-1}^{n+1} + v_{i+1}^{n+1}}{4}. \tag{16}$$

Next the values v_i^{n+1} are recomputed by using the scheme (15)

$$\frac{v_i^{n+1} - v_i^n}{\tau} + \widehat{B}\frac{\widetilde{V}_{i+1/2} - \widetilde{V}_{i-1/2}}{h} = 0. \tag{17}$$

The scheme for the three-dimensional system (7) is obtained by applying of the algorithm (15)–(17) to each of three directions. We write this scheme in the operator form

$$\frac{v^{n+1} - v^n}{\tau} + R_1 v^n + R_2 v^n + R_3 v^n = F^{n+1/2}, \tag{18}$$

where the operators R_1, R_2, R_3 correspond to the schemes along the directions x, y, z, respectively.

Now we describe the choice of the background hyperbolic system for the classical Maxwell equation

$$\frac{\partial \mathbf{E}}{\partial t} = rot\mathbf{H} - \mathbf{j}, \quad \frac{\partial \mathbf{H}}{\partial t} = -rot\mathbf{E}, \quad div\ \mathbf{H} = 0. \tag{19}$$

A consequence of the system is a relation

$$\frac{\partial}{\partial t}(div\ \mathbf{E}) = -div\mathbf{j}. \tag{20}$$

Define the charge density ρ by formula $\rho = div\mathbf{E}$. Then, the charge must obey the equality (the conservation law)

$$\frac{\partial \rho}{\partial t} + div\mathbf{j} = 0. \tag{21}$$

When approximating the background system for (19) we ignore this additional relation and take it into account later. We focus on the equation $div\mathbf{H} = 0$ involved in the overdetermined (but consistent) system (19). For the Maxwell equations the role of potentials play the vector and scalar functions $\mathbf{A}$ and Φ, which satisfy the vector and the scalar wave equations, respectively:

$$\frac{\partial^2 \mathbf{A}}{\partial t^2} - \Delta\mathbf{A} = \mathbf{j}, \qquad \frac{\partial^2 \Phi}{\partial t^2} - \Delta\Phi = \rho.$$

These potentials can be calculated using a symmetric hyperbolic system. The scalar potential Φ has a certain degree of freedom (gauge invariance). It is related to $\mathbf{A}$ through the equalities

$$\frac{\partial \mathbf{A}}{\partial t} + grad\ \Phi = -\mathbf{E}, \qquad \frac{\partial \Phi}{\partial t} + div\mathbf{A} = 0, \tag{22}$$

In fact the first equation is the definition of $\mathbf{A}$. The magnetic field vector $\mathbf{H}$ and the vector potential $\mathbf{A}$ are connected via the equality $\mathbf{H} = rot\mathbf{A}$, which automatically provides the relation $div\mathbf{H} = 0$.

Thus, we suggest to use the hyperbolic system for $\mathbf{E}$, $\mathbf{H}$ and $\mathbf{A}$, Φ

$$\frac{\partial \mathbf{E}}{\partial t} - rot\mathbf{H} = -\mathbf{j}, \qquad \frac{\partial \mathbf{H}}{\partial t} + rot\mathbf{E} = 0, \tag{23}$$

$$\frac{\partial \mathbf{A}}{\partial t} + grad\ \Phi = -\mathbf{E}, \qquad \frac{\partial \Phi}{\partial t} + div\ \mathbf{A} = 0$$

as a background system. From the second equation in (22) it follows that the right-hand sides of the wave equations for Φ, $\mathbf{A}$ satisfy the conservation law (21). This circumstance explains the absence of the charge density ρ in the right-hand sides of (23). Let us consider some details of our time-integration procedure. At the first stage we begin from $\mathbf{E}, \mathbf{H}, \mathbf{A}, \Phi$ at the moment t and calculate them at the next time step $t + \tau$. We apply the correction to $\mathbf{H}$ and compute new values $\mathbf{H}$ by the relation $\mathbf{H} = rot\mathbf{A}$ at $t + \tau$. All derivatives of the components A_k of the grid vector function $\mathbf{A}$ involved in the expression $rot\mathbf{A}$ are computed by the central differences

$$\frac{A_k(x_i + h) - A_k(x_i - h)}{2h},$$

here x_i $(i = 1, 2, 3)$ is one of the space variables. The corrected vector $\mathbf{H}$ may not exactly equal to the vector calculated from (23). It is fruitful to correct the charge density ρ by difference approximation of the equality $\rho = div\mathbf{E}$.

We suggest to check smoothness of the discrete solution by computation of finite differences up to some higher order and to check the residuals of the system (23) and the additional equations

$$\mathbf{H} - rot\mathbf{A} = 0, \qquad div\mathbf{H} = 0, \qquad div\mathbf{E} = \rho, \qquad \frac{\partial \rho}{\partial t} + div\mathbf{j} = 0. \tag{24}$$

Similar approach can be used for the Maxwell equations for the material medium. Let $K = K(x_1, x_2, x_3) = K^* > 0,\ \ M = M(x_1, x_2, x_3) = M^* > 0$ be the matrices that describe the dielectric polarization and the magnetic permeability, respectively. They relate $\mathbf{E}$ and $\mathbf{H}$ to $\mathbf{D}$ and $\mathbf{B}$: $\mathbf{D} = K\mathbf{E}, \quad \mathbf{B} = M\mathbf{H}$. Consider the Maxwell equations and the proper equations for the potentials:

$$\frac{\partial K\mathbf{E}}{\partial t} - rot\mathbf{H} = -\mathbf{j}, \qquad \frac{\partial M\mathbf{H}}{\partial t} + rot\mathbf{E} = 0,$$

$$\frac{\partial \mathbf{A}}{\partial t} + grad\ \Phi = -\mathbf{E}, \qquad \frac{\partial \Phi}{\partial t} + div\mathbf{A} = 0, \qquad \rho = div(K\mathbf{E}).$$

They imply the relations

$$\frac{\partial}{\partial t}(rot\mathbf{A}) + rot\mathbf{E} = 0, \qquad \frac{\partial}{\partial t}div(K\mathbf{E}) = -div\mathbf{j},$$

$$\frac{\partial}{\partial t}(M\mathbf{H} - rot\mathbf{A}) = 0, \qquad \frac{\partial \rho}{\partial t} + div\mathbf{j} = 0,$$

which provide consistency of the system with the additional equations

$$H = M^{-1}rot\mathbf{A}, \qquad div\mathbf{B} = 0, \qquad \frac{\partial \rho}{\partial t} + div\mathbf{j} = 0.$$

The first of these equations allows us to extend the approach described above for the electrodynamics equations for the material medium.

To demonstrate the approach we solve the extended three-dimensional Maxwell equations (23) in the three-dimensional cube $[-2; 2]\times[-2; 2]\times[-2; 2]$ with the periodic boundary conditions and the zero initial conditions

$$\mathbf{E}|_{t=0} = \mathbf{H}|_{t=0} = \mathbf{A}|_{t=0} = 0, \qquad \Phi|_{t=0} = 0.$$

We choose the source $\mathbf{j}$ in the form

$$\mathbf{j} = [j_x \;\; j_y \;\; j_z]^T = t^2\left(cos\frac{\pi x}{4}cos\frac{\pi y}{4}cos\frac{\pi z}{4}\right)^2 [1 \quad 1/2 \quad 2]^T,$$

and take diagonal matrices M, K: $M = diag\{\mu, \mu, \mu\}$, $K = diag\{\epsilon, \epsilon, \epsilon\}$, where $\mu = 1.1,\ \epsilon = 1.2$

All calculations are performed on the cubic grids of sizes $30 \times 30 \times 30$, $60 \times 60 \times 60$ with the number of time steps equal to 25 and 50, respectively. Figures 1–3 present the profiles of the grid functions along the diagonal $x = y = z$ at the final time. The results are given for $\mu = 1,\ \epsilon = 1$ (Test 1) and $\mu = 1.1,\ \epsilon = 1.2$ (Test 2). Figure 1 shows the profiles of the components of

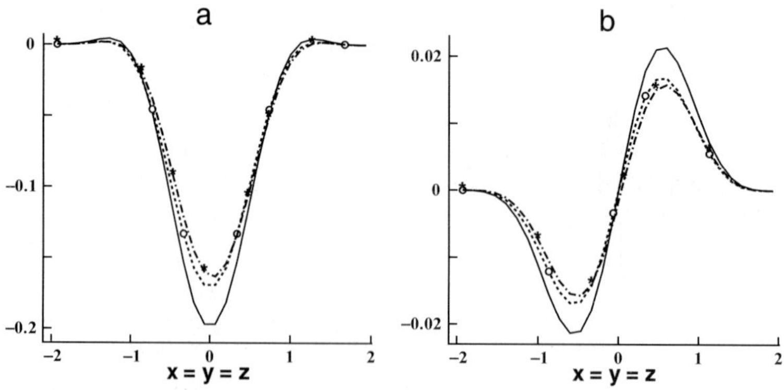

Fig. 1. Profiles E_y (**a**) and H_z (**b**)

the electric and the magnetic fields (parts (a) and (b), respectively). The solid line corresponds to the solution for Test 1 on the $30 \times 30 \times 30$ grid, the symbol $\circ$ and $*$ indicate the solution for Test 2 on the $30 \times 30 \times 30$ and $60 \times 60 \times 60$ grids, respectively. To evaluate the accuracy we calculate the residuals of the equations

$$\frac{\partial \mathbf{D}}{\partial t} - rot\mathbf{H} = -\mathbf{j}, \qquad \frac{\partial \mathbf{B}}{\partial t} + rot\mathbf{E} = 0,$$

using three time layers for time approximation and central differences for space approximation. Figure 2a,b plots the profiles of the residual of the first and second equations, respectively.

The residual of the equation $div\mathbf{H} = 0$ and the conservation law (21) is shown in Fig. 3a,b, respectively. The equality $div\mathbf{H} = 0$ is true practically exactly (the scale is 10^{-16}). In addition Fig. 3b confirms that the proposed scheme achieves the second order of accuracy.

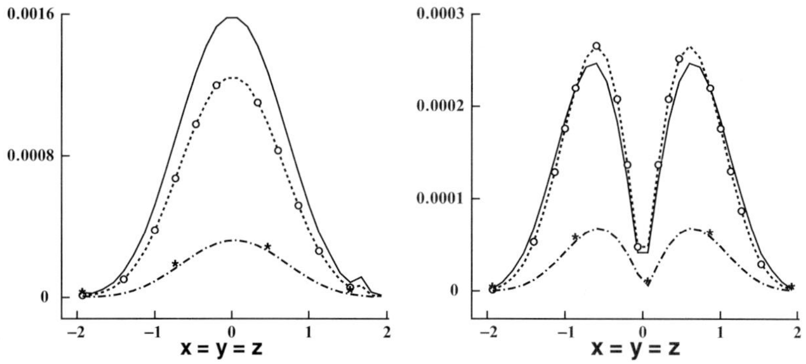

Fig. 2. (**a**) Residual of $\frac{\partial \mathbf{D}}{\partial t} - rot\mathbf{H} = -\mathbf{j}$; (**b**) Residual of $\frac{\partial \mathbf{B}}{\partial t} + rot\mathbf{E} = 0$

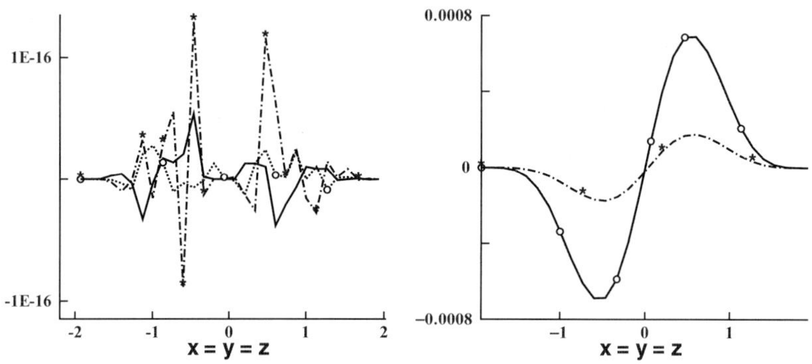

Fig. 3. (**a**) Residual of $div\ \mathbf{H} = 0$; (**b**) Residual of $\frac{\partial \rho}{\partial t} + div\mathbf{j} = 0$

4 On Solution of the Magnetohydrodynamics Equations

Apparently, similar approach can be applied to the magnetohydrodynamics equations (6):

$$\frac{\partial \mathbf{A}}{\partial t} + grad\Phi = \mathbf{u} \times L_{\mathbf{h}}, \quad \frac{\partial \Phi}{\partial t} + div\mathbf{A} = 0,$$
$$\frac{\partial L_{q_o}}{\partial t} + \frac{\partial (u_k L) q_0}{\partial x_k} = 0,$$
$$\frac{\partial L_{\mathbf{h}}}{\partial t} + \frac{\partial (u_k L)_{\mathbf{h}}}{\partial x_k} - L_{\mathbf{h}} \frac{\partial \mathbf{u}}{\partial x_k} = 0,$$
$$\frac{\partial L_{\mathbf{u}}}{\partial t} + \frac{\partial (u_k L)_{u_i}}{\partial x_k} - L_{\mathbf{h}} \frac{\partial \mathbf{h}}{\partial x_k} = 0,$$
$$L_{\mathbf{h}} = rot\mathbf{A}.$$

In this case we refuse from simplification, namely, elimination of the potential of the electromagnetic field Φ. However, in the considered model it provides simply the additional equality $div\ L_{\mathbf{h}} = 0$. We can still select the gauge factor, which determines the characteristic cone of the equations added to system (6).

5 On Solution of the Elasticity Equations

Now, we outline the way how the equations of the linear elasticity theory can be adapted to the suggested approach. Consider the symmetric hyperbolic system

$$\frac{\partial H_{u_i}}{\partial t} + \frac{\partial \sigma_{ik}}{\partial x_k} = 0,$$
$$\frac{\partial H_{\sigma_{ik}}}{\partial t} + \frac{1}{2}\left(\frac{\partial u_i}{\partial x_k} + \frac{\partial u_k}{\partial x_i}\right) = 0, \tag{25}$$

where

$$H = \rho_0 \frac{u_1^2 + u_2^2 + u_3^3}{2} + \frac{1}{\sigma(3\lambda + 2\mu)}(\sigma_{11} + \sigma_{22} + \sigma_{33})^2 +$$
$$\frac{1}{\mu}\left[\left(\sigma_{11} - \frac{\sigma_{11} + \sigma_{22} + \sigma_{33}}{2}\right)^2 + \left(\sigma_{22} - \frac{\sigma_{11} + \sigma_{22} + \sigma_{33}}{2}\right)^2 +\right.$$
$$\left.\left(\sigma_{33} - \frac{\sigma_{11} + \sigma_{22} + \sigma_{33}}{2}\right)^2 + 2(\sigma_{12}^2 + \sigma_{23}^2 + \sigma_{31}^2)\right].$$

It is usually supplemented with the six Saint-Venant equations, which are consistent with this system. Two of these equations have the form

$$\frac{\partial^2 \varepsilon_{11}}{\partial x_2^2} - \frac{\partial^2 \varepsilon_{12}}{\partial x_1 \partial x_2} + \frac{\partial^2 \varepsilon_{22}}{\partial x_1^2} = 0, \tag{26}$$
$$\frac{\partial^2 \varepsilon_{12}}{\partial x_2^2} - \frac{\partial^2 \varepsilon_{13}}{\partial x_3 \partial x_2} - \frac{\partial^2 \varepsilon_{32}}{\partial x_1 \partial x_3} + 2\frac{\partial^2 \varepsilon_{33}}{\partial x_1 \partial x_2} = 0.$$

The remaining four equations are obtained from (26) by cyclically changing the indices, that is, (123) $\to$ (231) $\to$ (312). The coordinates σ_{ij} of the stress tensor and the coordinates ε_{ij} of the strain tensor are connected by the linear relations

$$\varepsilon_{ik} = H_{\sigma_{ik}}. \tag{27}$$

Supplementing system (25) with (26), we obtain an extended overdetermined system. If Saint-Venant relations are fulfilled for initial data, then they are valid at any time. This fact follows from equality (27) and the second equation in system (25). A convenient way to ensure that the Saint-Venant conditions are always fulfilled (to a reasonable accuracy) in the numerical process is to extend the elasticity equations by the equations for displacement ζ_i:

$$\frac{\partial \zeta_i}{\partial t} = u_i. \tag{28}$$

These displacements serve as the potentials for representation of the strain tensor as

$$\varepsilon_{ik} = \frac{1}{2}\left(\frac{\partial \zeta_i}{\partial \xi_k} + \frac{\partial \zeta_k}{\partial \xi_i}\right). \tag{29}$$

The structure of the extended overdetermined system for linear elasticity theory allows us to apply the correction technique to the numerical calculations. At the first stage of the time step $t \to t+\tau$ we solve the system (25) and calculate the displacements ζ_i from the system (28). All values σ_{ik} calculated at time $t+\tau$ are removed from values u_i, σ_{ik}, ζ_i and replaced on σ_{ik} calculated with relations (27), (29) and values ζ_i at the moment $t+\tau$.

6 Conclusion

The suggestions how to construct discrete schemes for the overdetermined system of classical mathematical physics are not final yet. We only try to outline certain unconventional research directions. One of them arisen in our numerous experiments performed over the last 2 years.

We solved the problem of eigenspace calculation for the two-dimensional acoustic equations. In these equations for each nonzero frequency the corresponding eigenvector has zero vorticist. And the zero frequency has an indefinite invariant subspace with zero or constant pressure. Vorticist does not depend on time and can be completely arbitrary. For calculation of the leading non-zero frequencies and the corresponding eigenfunctions we used the time-integration scheme for the non-stationary acoustic equations with a special resonance periodic source to enhance smooth eigensolutions and to damp the parasitic vortices. The discretization was based on the first-order Godunov scheme, which gives rise to parasitic vortices. The correction procedure based on the Ritz method allows us to erase non-physical eigensolutions. The simplest central differences were used to approximate the derivatives in the corresponding variational functional for eigenfunctions. As admissible functions we used difference solutions and linear combinations of the residual functions. These functions are computed by substitution of difference solutions (obtained by the conventional Godunov scheme) into the second-order accuracy scheme based on central differences. The whole procedure computes a subspace contained a set of smooth eigenfunctions in spite of that the second-order scheme has a lot of parasitic eigenfunctions. This fact is described in [9]–[11] for the acoustic and elasticity equations. Because of such parasitic oscillations the second-order scheme has never used in practice. This scheme worked in our case due to we used only smooth solutions obtained with the first-order scheme and it was the reason of positive final results.

All situations we observed in our experiments give us opportunity to elaborate a large group of new experiments in which we try to utilize different discrete models on different stages of calculation for similar variables. In initial stage the main goal is to obtain "smooth" discrete solutions of the difference scheme. On the correction stage we "differentiate" these smooth solutions by the central difference relations. In such a way the obtained difference approximations of derivatives are used for correction and for computing the residual errors both of the basic equations and the additional ones included in the enlarged system.

In this paper we reported on an attempt (which we regard as a successful one) to apply this approach as a part of the numerical technique for solving overdetermined hyperbolic systems.

Acknowledgments

I would like to thank Prof. Denis Serre to give me opportunity to present this work on the XIth International Conference on Hyperbolic Problems (HYP2006). This research is supported by the Russian Foundation for Basic Research and the Netherlands Organization for Scientific Research (project no. 047.016.003-NWO), Russian Federation under grant "Leading Scientific Schools" 9019.2006.1.

References

1. Ivanov, V.V. Strictly hyperbolic polynomials without any hyperbolic symmetrization. Transactions of The International Conference on Partial Differential Equations. Novosibirsk: Nauka, Siberian Branch, 84–93, 1986 (in Russian).
2. Godunov, S.K. Difference method for numerical calculations of discontinuous solutions of gas-dynamic equations. Matem. Sbornik, **47**(89), 271–306, 1959 (in Russian)
3. Godunov, S.K., Gordienko, V.M. The simplest Galilean-invariant and thermodynamically consistent conservation laws. Journal of Applied Mechanics and Technical Physics, **43**(1), 1–12, 2002.

4. Godunov, S.K., Gordienko, V.M. Complicated structures of Galilean-invariant conservation laws. Journal of Applied Mechanics and Technical Physics, **43**(2), 175–189, 2002.
5. Godunov, S.K. New version of the thermodynamically consistent model of maxwell viscosity. Journal of Applied Mechanics and Technical Physics, **45**(6), 775–783, 2004.
6. Kulikovskii, A.G., Pogorelov, N.V., Semenov, A. Yu. Mathematical Aspects of Numerical Solution of Hyperbolic Systems equations. Physmatlit, Moscow, 2001 (in Russian).
7. Babii, D.P., Godunov, S.K., Feodoritova, O.B., Zhukov, V.T. On difference approximations for overdetermined hyperbolic equations of classical mathematical physics. Journal of Computational Mathematics and Mathematical Physics, Pleiades Publishing, Inc., 2007. To be published.
8. Godunov, S.K., Nauka, M. (ed.). Numerical solution of multidimensional gas-dynamics problems. (1976). (French transl.: Resolution numerique des problemes multidimensionnels de la dynamique des gaz, Moscou: Mir, 1979).
9. Godunov, S.K., Selivanova, S.V. Experiments involving resonance for spectral analysis of skew-symmetric operators. Sib. Zh. Vychisl. Mat., **9**(2), 2006 (in Russian).
10. Godunov, S.K., Feodoritova, O.B., Zhukov, V.T. A method for computing invariant subspaces of symmetric hyperbolic systems. Journal of Computational Mathematics and Mathematical Physics, **46**(6), 971–982, 2006.
11. Godunov, S.K., Feodoritova, O.B., Zhukov, V.T. Computation of eigenspaces of hyperbolic system. The IV International Conference on Computational Fluid Dynamics. July 10–14. Ghent, Belguim, 2006.

Stable Galaxy Configurations

Y. Guo

Summary. Dynamics of galaxies can be described by the Vlasov–Poisson system, where stars interact only with the gravitational field they create collectively. In this note, recent developments in mathematical study of stability of galaxy configurations are discussed.

1 Introduction

There are about 10^{10}–10^{12} stars in a typical galaxy. To study large scale properties of a galaxy, keeping track of the dynamics of every star inside the galaxy is neither feasible nor even desired. Instead, the evolution of the averaged mass distribution of the galaxy is the issue. Such a statistical description of a large ensemble of gravitationally interacting mass points leads to a mathematical problem which is far more tractable. A galaxy or a globular cluster can then be modeled as an ensemble of particles, i.e., stars, which interact only by the gravitational field which they create collectively, collisions among the stars being sufficiently rare to be neglected. The time evolution of a galaxy is given by the Vlasov–Poisson system for the distribution function $f(t, x, v)$:

$$\partial_t f + v \cdot \nabla_x f - \nabla_x U \cdot \nabla_v f = 0, \tag{1}$$

$$\Delta U = 4\pi \int_{\mathbf{R}^3} f(t, x, v) dv. \tag{2}$$

ere $f(t, x, v) \geq 0$, and its integral over any region of the phase space $(x, v) \in \mathbf{R}^3 \times \mathbf{R}^3$ gives the number of stars which have the phase space coordinates in the region at time t.

There are two important conservation laws for the Vlasov–Poisson system:

$$\frac{1}{2} \int\int |v|^2 f - \frac{1}{8\pi} \int |\nabla_x U_f|^2 \equiv \text{constant} \tag{3}$$

$$\int\int Q(f) \equiv \text{constant} \tag{4}$$

The first identity (3) is the conservation for the total energy; it should be noted that, in contrast to the Vlasov–Poisson system for plasmas, the gravitational energy $-\frac{1}{8\pi}\int|\nabla_x U_f|^2$ is not positive. This property creates both mathematical difficulties and interesting features of such a Vlasov–Poisson system. The second generalized mass identity (4) is valid for *all* smooth $Q(\cdot)$ such that $Q(0) = 0$. This incredible property leads to preservation of all L^p norms for f, and it makes such a kinetic model more regular [LP], [P], and also more stable than its fluid counterparts.

A *symmetric* solution satisfies $f(t, x, v) \equiv f(t, Ox, Ov)$ for all orthogonal matrix O. Let

$$r = |x|, \qquad v_r = \frac{v \cdot x}{|x|}, \qquad L = |x \times v|^2 = |x|^2|v|^2 - (x \cdot v)^2.$$

It is more convenient to denote a symmetric solution by $f(t, r, v_r, L)$, which satisfies the following symmetric Vlasov–Poisson system:

$$\partial_t f + v_r \partial_r f + \left(\frac{L^2}{r^3} - \partial_r U(t,r)\right)\partial_{v_r} f = 0, \tag{5}$$

$$\partial_{rr} U + \frac{2}{r}\partial_r U = 4\pi \int_{\mathbf{R}^3} f(t, r, v_r, L)dv. \tag{6}$$

In the presence of the symmetry, there is an additional conservation law

$$\int_{\mathbf{R}^3\times\mathbf{R}^3} Q(f(t), L)dxdv = \text{constant}, \tag{7}$$

for *any* smooth function Q such that $Q(0, L) = 0$.

2 Steady Galaxy Configurations

There are many steady galaxy models of the simplest form

$$f_0(x, v) \equiv f_0(E),$$

where the particle energy

$$E \equiv \frac{1}{2}|v|^2 + U_0(x),$$

and $U_0(x)$ satisfies the self-consistent Poisson system

$$\Delta U_0 = 4\pi \int_{\mathbf{R}^3} f_0(E)dv.$$

Among them, there are two important types of steady states. The first type is so-called (normalized) polytrope:

$$f_0(E) = (E_0 - E)_+^k, \tag{8}$$

where $(\cdot)_+$ denotes the positive part, E_0 is a nonpositive constant, and $0 < k \leq \frac{7}{2}$. For $0 < k < \frac{7}{2}$, $E_0 < 0$, then f_0 has compact support in both x and v. On the other hand, in the limiting case of $k = \frac{7}{2}$ (the Plummer sphere), f_0 has an unbounded support but with finite total mass [BFH]. By some elementary computations, the pressure of such a polytrope (8) satisfies $p(\rho) = c\rho^{1+\frac{1}{n}}$, where $n = k + \frac{3}{2}$. This is exactly the famous polytropic gas law in the theory of compressible gas dynamics, from which the polytrope is named after.

The second important type of steady states is the (normalized) King model, which takes the form

$$f_0(E) = \left(e^{E_0 - E} - 1\right)_+, \tag{9}$$

where $E_0 < 0$ is a given negative constant. The existence of such a model is discussed in [RR]. The King model describes isothermal galaxies (formally corresponding to $k = \infty$ in the polytropes $(E_0 - E)_+^k$), which provides a canonical form for many galaxy models widely used in astronomy.

One of the central questions, which has attracted considerable attention in the astrophysics literature, cf [BT] [FP] and the references there, is the *dynamically stability* for galaxy models $f_0(E)$. Assume $f_0' < 0$, then it has been well known that $f_0(E)$ is a formal critical point of the following Casimir-Energy functional (as a Liapunov functional)

$$\mathcal{H}(f) \equiv \int\int Q_0(f) + \frac{1}{2}\int\int |v|^2 f - \frac{1}{8\pi}\int |\nabla_x U_f|^2, \tag{10}$$

which is constant along the time evolution. If $\int\int f = \int\int f_0$,

$$\begin{aligned}\mathcal{H}(f) - \mathcal{H}(f_0) &= \int_{\mathbf{R}^3\times\mathbf{R}^3} [Q_0(f) - Q_0(f_0) + (E - E_0)(f - f_0)]dxdv \\ &\quad - \frac{1}{8\pi}\int_{\mathbf{R}^3} |\nabla U_f - \nabla U_{f_0}|^2 dx.\end{aligned}$$

Here the particular Casimir function Q_0 satisfies

$$Q_0'(f_0(E)) \equiv -E.$$

It follows that the first variation at f_0 is zero $\mathcal{H}^{(1)}(f_0(E)) = 0$ (on the support of $f_0(E)$), and the (formal) second order variation of $\mathcal{H}$ at f_0 is

$$\mathcal{H}_{f_0}^{(2)}[g] \equiv \frac{1}{2}\int\int_{f_0>0} \frac{g^2}{-f_0'(E)} - \frac{1}{8\pi}\int |\nabla_x U_g|^2, \tag{11}$$

where $Q''(f_0) = \frac{1}{-f_0'(E)}$. It is natural to expect that the positivity of such a quadratic form should imply stability for $f_0(E)$. Ever since the seminal work of Antonov [An] around 1960, there have been enormous efforts in the astrophysics community over the last four decades, to study the positivity of the $\mathcal{H}^{(2)}_{f_0}[g]$. Define the Lie bracket as

$$\{f_1, f_2\} \equiv \nabla_x f_1 \cdot \nabla_v f_2 - \nabla_v f_1 \cdot \nabla_x f_2, \tag{12}$$

the best result along this direction is given by the Kandrup and Sygnet Lemma:

Lemma 1. *Let $f_0' < 0$. Let $h \in C_c^\infty$ with support inside the set of $f_0 > 0$ such that $h(x,v) = -h(x,-v)$. Then*

$$\mathcal{H}^{(2)}_{f_0}[-\{f_0(E), h\}] \geq -\frac{1}{2}\int_{f_0>0} f_0'(E)\left[|rv_r|^2\left|\left\{E, \frac{h}{rv_r}\right\}\right|^2 + \frac{|h|^2}{r}U_0'\right] dxdv \geq 0.$$

Lemma 1 leads to linearized stability for the Vlasov–Poisson system (1) and (2) around the steady state $f_0(E)$. Despite the significance of such a result, the Kandrup and Sygnet Lemma [K] is still quite a distance away from the true dynamical stability of $f_0(E)$. There are at least two serious mathematical difficulties. First of all, it is very challenging to use the positivity of $\mathcal{H}^{(2)}_{f_0}[g]$ to control higher order remainder in $\mathcal{H}(f) - \mathcal{H}(f_0)$ to conclude stability [Wa1]. This is due to the nonsmooth nature of $f_0(E)$ in all important examples. Second of all, even if one can succeed in controlling the nonlinearity, the positivity of $H^{(2)}_{f_0}[g]$ in Lemma 1 is *only* valid for certain perturbation of the form $g = \{f_0, h\}$. It is not clear at all if $\{f_0, h\}$ can be *any* arbitrary, general perturbation.

3 Characteristics for $\{E, h\}$

Given a steady state $f_0(E)$. In the presence of symmetry,

$$\{E, h\} = \left[v_r\partial_r + \left(\frac{L}{r^3} - \partial_r U_0(t,r)\right)\partial_{v_r}\right]h.$$

For fixed constants $E \leq E_0 < 0$ and $L > 0$, consider the corresponding characteristic curves $(r, v_r(r))$ given by

$$\frac{1}{2}v_r^2 + U_0(r) + \frac{L}{2r^2} \overset{\text{define}}{\equiv} \frac{1}{2}v_r^2 + \mathcal{C}(r,L) = E. \tag{13}$$

Along the characteristic (13) parametrized by r, $v_r(r) = \pm\sqrt{2E - 2\mathcal{C}(r,L)}$,

$$\{E, h\} = v_r \frac{d}{dr}h. \tag{14}$$

In order to analyze the characteristic (13), the function

$$\mathcal{C}(r,L) \equiv U_0(r) + \frac{L}{2r^2} \tag{15}$$

is studied. It can be shown that $\mathcal{C}(r,L)<0$ for large r and $\lim_{r\to\infty}\mathcal{C}(r,L)=0$ [BFH]. Moreover $\lim_{r\to 0}\mathcal{C}(r,L) = +\infty$ for $L>0$. Let r_0 be the unique critical point of $\mathcal{C}(r,L)$ which is both the local and global minimizer of $\mathcal{C}(r,L)$. It follows that for any constant $E\leq E_0$, there exist exactly two points $r_-(L,E)<r_0(L)<r_+(L,E)$ such that the characteristic curve can be parametrized as:

$$[r:\mathcal{C}(r,L)=E]=[r_-(E,L)<r<r_+(E,L)].$$

4 Stability of the Polytropes

To prove nonlinear stability, a direct variational approach was initiated by Wolansky [Wo1], then further developed systematically by Guo and Rein in [G1], [G2], [GR1]–[GR4], [RG]. Their method avoids entirely the delicate analysis of the second-order variation $\mathcal{H}^{(2)}_{f_0}$ in (11), which has led to first rigorous nonlinear stability proof for a large class of $f_0(E)$. The high point of such a program is the nonlinear stability proof for every polytrope [GR3] $f_0(E) = (E_0-E)^k_+$ for $0<k\leq\frac{7}{2}$. Consider the variational problem for $0<k<\frac{7}{2}$:

$$\min\left(\frac{1}{2}\int\int|v|^2 f-\frac{1}{8\pi}\int|\nabla_x U_f|^2\right)$$

subject to the constraint

$$\int\int Q_0(f)+f=\text{constant},$$

where $Q_0(f)=\frac{k}{k+1}f^{1+\frac{1}{k}}$. It was shown that the polytrope $f_0=(E_0-E)^k_+$ is a minimizer of such a variational problem, and for any minimizing sequence f_n, the corresponding gravitational field $\nabla_x U_{f_n}$ is compact in L^2, up to translations in x. This compactness leads to the stability.

In light of (10), a natural measurement of stability between f and f_0 is

$$d(f,f_0)=\int_{\mathbf{R}^3\times\mathbf{R}^3}[Q_0(f)-Q_0(f_0)+(E-E_0)(f-f_0)]dxdv. \tag{16}$$

The integrand above is nonnegative if $f_0'<0$. In fact, on the set $f_0\equiv 0$ (i.e., $E-E_0>0$),

$$Q_0(f)-Q_0(f_0)+(E-E_0)(f-f_0)=Q_0(f)+(E-E_0)f\geq 0$$

since $f\geq 0$. On the other hand, on the set $f_0>0$ (i.e., $E-E_0\leq 0$), the integrand is also nonnegative since Q_0 is convex.

Theorem 1 (GR3). *For $0<k<\frac{7}{2}$, for any $\varepsilon>0$ there exists $\delta>0$ such that for any classical solution $f(t,x,v)$ to the Vlasov–Poisson system with*

$f(0,x,v) \in C_c^1$ and $\int\int[Q_0(f(0,x,v)) + f(0,x,v)]dxdv = \int\int[Q_0(f_0) + f_0]$ $dxdv$, the initial estimate

$$d(f(0), f_0) + \frac{1}{8\pi}\int |\nabla U_{f(0)} - \nabla U_0|^2 dx < \delta$$

implies that for any $t \geq 0$ there is a shift vector a such that

$$d(T^a f(t), f_0) + \frac{1}{8\pi}\int |T^a \nabla U_{f(t)} - \nabla U_0|^2 dx < \varepsilon,$$

where $T^a f(t) = f(t, x + a, v)$.

For the case of the Plummer sphere $k = \frac{7}{2}$, additional scaling transformations must be included to deduce stability, see [GR4]. It is also possible to prove stability for generalized polytropes, thanks to [Sch].

As a by-product of such a variational approach, many new stable galaxy models are discovered. In particular, f_0 can depend on E and L [GR2,3], or even on one component of $x \times v$ [RG]. The latter is an example of axisymmetric galaxy which are more common in the nature.

The proof of such a compactness result encounters the usual analytical difficulty: translations can make any minimizing sequence noncompact. In the case with symmetry (5), one can use simple scaling $f(t,x,v) \to af(t,bx,cv)$, and splitting argument to deduce such compactness in a straightforward fashion [G1], [G2]. In the original proof in [GR3], a careful estimate for the potential ∇U_{f_n} takes care of the translations for general $f(t,x,v)$ without symmetry.

A general result was proven by Burchard and Guo [BG], to conclude compactness, up to possible translations in the general case by studying the symmetric rearrangements. Recently, Hadzic [H] gave an elegant stability proof for polytropes, based on such a new approach. There are also recent new proofs [SS], [LMR] based on Lions' concentration compactness principle. See also [DSS], [Wa1], [Wa2], [Wo2] for related references, and an excellent monograph [R].

5 Stability of the King Model

Unfortunately, despite its huge success, the King model is out of the reach of such a variational approach. The Casimir function for a *normalized* King model is

$$Q_0(f) = (1+f)\ln(1+f) - 1 - f, \tag{17}$$

which has very slow growth for $f \to \infty$. As a result, the direct variational method fails. A new method to study its stability has been developed.

Consider a class of measuring preserving perturbations:

$$\mathcal{S}_{f_0} \equiv \left[f(t,r,v_r,L) \geq 0 : \int Q(f,L) = \int Q(f_0,L), \text{ for } Q \in C_c^\infty \right.$$
$$\left. \text{and } Q(0,L) \equiv 0. \right] \tag{18}$$

The main stability theorem is proven by Guo and Rein [GR5]:

Theorem 2. *For any $\varepsilon > 0$, there is a $\delta > 0$, such that if $f(0,x,v) \in \mathcal{S}_{f_0} \cap C_c^1(\mathbf{R}^3 \times \mathbf{R}^3)$, and $d(f(0), f_0) < \delta$, then* $\sup_{0 \leq t \leq \infty} d(f(t), f_0) < \varepsilon$.

The main ingredient of the proof is based on the following:

Theorem 3. *There exists $\delta_0 > 0$, and $C_{\delta_0} > 0$ such that if $f(x,v) \in \mathcal{S}_{f_0}$, and $d(f,f_0) \leq \delta_0$, then*

$$\mathcal{H}(f) - \mathcal{H}(f_0) \geq C_{\delta_0} ||\nabla U_f - \nabla U_{f_0}||_{L^2}^2. \tag{19}$$

The dynamical stability (Theorem 2) then follows immediately from Theorem 3 as in [GR3]. The proof of Theorem 3 is based on an indirect, contradiction argument. There are two main ingredients: The first part is a general argument to establish that, if (19) fails, then one can find a nonzero function g such that

$$\mathcal{H}_{f_0}^{(2)}[g] = 0,$$

and for all smooth Q,

$$\int\int Q'(E,L)g \equiv 0.$$

The second part is to use a detailed analysis along the characteristic curves, to establish for every E and L,

$$\int_{r_-(E,L)}^{r_+(E,L)} g(r,E,L) \frac{dr}{\sqrt{2E - 2\mathcal{C}(r,L)}} = 0.$$

It is then possible to construct a function h such that $g = \{f_0, h\}$. It should be pointed out that the arbitrary dependence of L in Q plays a crucial role in such a construction of h. This leads to a contradiction to Lemma 1 for the positivity of $\mathcal{H}_{f_0}^{(2)}$.

Very recently, Guo and Lin [GL] were able to get rid of the restriction of $\mathcal{S}_{f_0}$, to show that the King model (9) is dynamically stable among all symmetric perturbations. Moreover, an abstract linear instability criterion was also established in [GL].

References

[An] Antonov, V.A. *Remarks on the problem of stability in stellar dynamics.* Soviet Astr, AJ., 4, 859–867 (1961)

[Al] Aly, J.J. *On the lowest energy state of a collisionless selfgravitating system under phase space volume constraint.* Mon. Not. R. Astron. Soc. 241 (1989), 15–27.

[BFH] Batt, J., Faltenbacher and Horst, E. *Stationary spherically symmetric models in stellar dynamics.* Arch. Ration. Mech. Anal. 93 (1986), 159–183.

[BT] Binney, J., Tremaine, S., *Galactic Dynamics.* Princeton University Press, 1987.

[BG] Burchard, A.; Guo, Y. *Compactness via symmetrization.* J. Funct. Anal. 214 (2004), no. 1, 40–73.

[DSS] Dolbeault, J.; Sánchez, Ó.; Soler, J. *Asymptotic behaviour for the Vlasov-Poisson system in the stellar-dynamics case.* Arch. Ration. Mech. Anal. 171 (2004), no. 3, 301–327.

[FP] Fridman, A., Polyachenko, V., *Physics of Gravitating System I., Equilibrium and Stability*, Springer-Verlag, 1984.

[G1] Guo, Y., *Variational method for stable polytropic galaxies*, Arch. Rational Mech. Anal., 147, 225–243, 1999.

[G2] Guo, Y., *On generalized Antonov stablility criterion for polytropic steady states*, Contem. Math., 263, 85–107, 1999.

[GL] Guo, Y., Lin, Z. *in preparations.*

[GR1] Guo, Y., Rein, G., *Stable steady states in stellar dynamics*, Arch. Rational Mech. Anal., 147, no. 3, 225–243, (1999).

[GR2] Guo, Y., Rein, G., *Existence and stability of Camm type steady states in galactic dynamics*, Indiana U. Math. J., 48, 1237–1255, 1999.

[GR3] Guo, Y., Rein, G., *Isotropic steady states in stellar dynamics*, Commun. Math. Phys., 219, 2001.

[GR4] Guo, Y., Rein, G., *Isotropic steady states in stellar dynamics revisited.*, Los Alamos Preprint, 2002.

[GR5] Guo, Y., Rein, G. *A non-variational approach to nonlinear stability in stellar dynamics applied to the King model.* Commun. Math. Phys. to appear.

[H] Hadzic, M. *Compactness and stability of some systems of nonlinear PDE-s in galatic dynamics.* Diplomarbeit. 2005. See also Quart. Appl. Math., to appear.

[K] Kandrup, H.; Signet, J.F.; *A simple proof of dynamical stability for a class of spherical clusters.* The Astrophys. J. 298 (1985) 1, 27–33.

[LMR] Lemou, M.; Mehats, F.; Raphael, P. *On the orbital stability of the ground states and the singularity formation for the gravitational Vlasov Poisson system.* Preprint 2005.

[LP] Lions, P.-L., Perthame, B., *Propagation of moments and the regularity for the 3-dimensional Vlasov-Poisson system.* Invent. Math. 105 (1991), 415–430.

[P] Pfaffelmoser, K., Global classical solutions of the Vlasov-Poisson system in three dimensions for general initial data. J. Diff. Equ. 95 (1992), 281–303.

[R] Rein, G.: *Collisionless Kinetic Equations from Astrophysics - The Vlasov-Poisson system.* Handbook of Differential Eqautions. Evolutionary Equations, vol 3. Edited by C.M. Dafermos and E. Feireisl. Elsevier B.V. 2006.

[R1] Rein, G., *Flat steady states in stellar dynamics - existence and stability.* Commun. Math. Phys. 205 (1999) 229–247.

[R2] Rein, G. *Stability of spherically symmetric steady states in galactic dynamics against general perturbations.* Arch. Ration. Mech. Anal. 161 (2002), no. 1, 27–42.

[RG] Rein, G., Guo, Y.: *Stable elliptical galaxies*, Mon. Notice. Royal Astron. Soc. 344, (2003) 1296–1306.

[RR] Rein, G., Rendall, A. *Compact support of spherically symmetric equilibria in non-relativistic and relativistic galatic dynamics.* Math. Proc. Camb. Phil. Soc. 128 (2000) 363–380.

[SS] Sanchez, O., Soler, J.: *Orbital stability for polytropic galaxies.* Preprint 2003.

[Sch] Schaeffer, J. *Steady states in galactic dynamics.* Arch. Ration. Mech. Anal. 172 (2004) 1–19.

[Wa1] Wan, Y-H., *On onlinear stability of isotropic models in stellar dynamics.* Arch. Rational. Mech. Anal., 147, (1999) 245–268.

[Wa2] Wan, Y-H., *Nonlinear stability of spherical systems in galactic dynamics,* preprint, 2000.

[Wo1] Wolansky, G., *On nonlinear stability of polytropic galaxies.* Ann. Inst. Henri Poincare. (1999), 16, 15–48.

[Wo2] Wolansky, G., *Static solutions of the Vlasov-Einstein system.* Arch. Rational. Mech. Anal., 156 (2001) 3, 205–230.

Dissipative Structure of Regularity-Loss Type and Applications

S. Kawashima

Summary. We discuss dissipative structure for partial differential equations. A useful characterization of the dissipative structure was well known for symmetric systems such as symmetric hyperbolic systems, symmetric hyperbolic–parabolic coupled systems, and so on. Recently, a new dissipative structure was found in the study of the dissipative Timoshenko system. This new dissipativity is of regularity-loss type and causes difficulties in solving the nonlinear problems. We present a time-weighted energy method to resolve the difficulties.

1 Introduction

In the study of asymptotic stability problems for nonlinear partial differential equations, it is crucial to derive quantitative decay estimates for the corresponding linearized equations. Such decay estimates could be obtained when the linearized equations have suitable dissipative structure.

To give a characterization of the dissipative structure, we consider linear partial differential equations of the form $P(\partial_t, \partial_x)u = 0$, where $t \geq 0$ and $x \in \mathbf{R}^n$. The corresponding characteristic equation is $P(\lambda, i\xi) = 0$, where $\lambda \in \mathbf{C}$ and $\xi \in \mathbf{R}^n$. The characteristic roots $\lambda = \lambda(i\xi)$ are called eigenvalues and the relation $\lambda = \lambda(i\xi)$ is sometimes called the dispersion relation. We say that the equation is dissipative if $\operatorname{Re}\lambda(i\xi) \leq 0$ for any $\xi \in \mathbf{R}^n$. This dissipativity is, however, so weak to get quantitative decay estimates. The useful notion of dissipativity is the strict dissipativity, which is characterized by the relation that $\operatorname{Re}\lambda(i\xi) < 0$ for any $\xi \in \mathbf{R}^n$ with $\xi \neq 0$. Here we consider the following two cases:

$$\text{Type (I):}\quad \operatorname{Re}\lambda(i\xi) \leq -c|\xi|^2/(1+|\xi|^2) \quad \text{for any } \xi \in \mathbf{R}^n,$$

$$\text{Type (II):}\quad \operatorname{Re}\lambda(i\xi) \leq -c|\xi|^2/(1+|\xi|^2)^2 \quad \text{for any } \xi \in \mathbf{R}^n,$$

where c is a positive constant. More generally, the Type (II) should cover the case where $\operatorname{Re}\lambda(i\xi) \leq -c|\xi|^2/(1+|\xi|^2)^m$ $(m > 1)$ for any $\xi \in \mathbf{R}^n$.

For the strict dissipativity of Type (I), we have $\operatorname{Re}\lambda(i\xi) \sim -c|\xi|^2$ for $|\xi| \to 0$ and $\operatorname{Re}\lambda(i\xi) \sim -c$ for $|\xi| \to \infty$. There is a complete characterization of this type of strict dissipativity for symmetric hyperbolic systems and for symmetric hyperbolic–parabolic coupled systems, which was done in [6] for more than 20 years ago. Also, this characterization was generalized to a class of symmetric hyperbolic–elliptic coupled systems (see [4]). In these cases we obtain useful quantitative decay estimates similar to the one for the heat equation (see [7]).

On the other hand, for the strict dissipativity of Type (II), we have $\operatorname{Re}\lambda(i\xi) \sim -c|\xi|^2$ for $|\xi| \to 0$ and $\operatorname{Re}\lambda(i\xi) \sim -c|\xi|^{-2}$ for $|\xi| \to \infty$. Therefore, $\lambda(i\xi)$ may approach the imaginary axis $\operatorname{Re}\lambda = 0$ for $|\xi| \to \infty$. Consequently, this type of strict dissipativity is so weak in high frequency region and we only obtain decay estimates of regularity-loss type. This was recently observed in the study of the dissipative Timoshenko system ([5] and [2]) and the modified radiating gas model system ([1]).

In this article we first give a survey of general theory of the strict dissipativity of Type (I) in a class of symmetric hyperbolic systems. Then, as an example of the strict dissipativity of Type (II), we study the dissipative Timoshenko system

$$\begin{aligned} &w_{tt} - (w_x - \psi)_x = 0, \\ &\psi_{tt} - a^2\,\psi_{xx} - (w_x - \psi) + \gamma\psi_t = 0, \end{aligned}$$

where a and γ are positive constants, and observe that this dissipativity is of regularity-loss type. Also, we study the following simple nonlinear model system as an example of type (II):

$$\begin{aligned} &u_t + \left(u^2/2\right)_x + q_x = 0, \\ &(\partial_x^4 - \partial_x^2 + 1)q + u_x = 0. \end{aligned}$$

This is a modified radiating gas model. For this model system we present a time-weighted energy method to overcome the difficulties caused by the dissipativity of regularity-loss type. Finally, we give some open questions concerning the strict dissipativity of Type (II).

2 General Theory for Strict Dissipativity of Type (I)

The strict dissipativity of Type (I) was completely characterized in [6] and [4] for symmetric hyperbolic systems, symmetric hyperbolic–parabolic coupled systems and symmetric hyperbolic–elliptic coupled systems. This general theory covers the following typical examples:

Example 1. Dissipative wave equation: $\phi_{tt} - \phi_{xx} + \phi_t = 0$, which is equivalent to the hyperbolic system

$$v_t - u_x = 0, \qquad u_t - v_x + u = 0,$$

where $u = \phi_t$ and $v = \phi_x$.

Example 2. The compressible Navier–Stokes equation:

$$v_t - u_x = 0, \qquad u_t - v_x = u_{xx}.$$

This is a hyperbolic–parabolic system and is equivalent to $\phi_{tt} - \phi_{xx} - \phi_{xxt} = 0$, where $u = \phi_t$ and $v = \phi_x$.

Example 3. Radiating gas model:

$$u_t + q_x = 0, \qquad -q_{xx} + q + u_x = 0.$$

This is a hyperbolic–elliptic system and is equivalent to $\phi_t - \phi_{xx} - \phi_{xxt} = 0$, where $q = \phi_t$ and $u = -\phi_x$.

Here we survey the general theory of the strict dissipativity of Type (I) in a class of symmetric hyperbolic systems:

$$A^0 u_t + \sum_{j=1}^{n} A^j u_{x_j} + Lu = 0, \tag{1}$$

where $u = u(x,t)$ is an m-vector function of $x = (x_1, \cdots, x_n) \in \mathbf{R}^n$ and $t \geq 0$, and the coefficient matrices verify the following conditions:

(a) A^0 is symmetric and positive definite

(b) A^j is symmetric for any j

(c) L is symmetric and nonnegative definite

Take the Fourier transform of (1) to get

$$A^0 \widehat{u}_t + i|\xi| A(\omega)\widehat{u} + L\widehat{u} = 0, \tag{2}$$

where

$$A(\omega) = \sum_{j=1}^{n} A^j \omega_j, \quad \omega = \xi/|\xi| \in S^{n-1}.$$

Let $\lambda = \lambda(i\xi)$ be the eigenvalues, which are the solutions to the characteristic equation

$$\det(\lambda A^0 + A(i\xi) + L) = 0.$$

To characterize the dissipative structure of the symmetric hyperbolic systems (1), we formulate the following two conditions:

Stability condition: Let $\mu \in \mathbf{R}$, $\varphi \in \mathbf{R}^m$, $\omega \in S^{n-1}$, and let $\mu A^0 \varphi + A(\omega)\varphi = 0$ and $L\varphi = 0$. Then $\varphi = 0$.

Condition (K): There exists $K(\omega)$ with the following properties:

(i) $K(\omega)A^0$ is skew-symmetric for any $\omega \in S^{n-1}$
(ii) $\big[K(\omega)A(\omega)\big]' + L$ is (symmetric and) positive definite for any $\omega \in S^{n-1}$, where $[X]'$ denotes the symmetric part of X.

For the symmetric hyperbolic systems (1), the strict dissipativity is completely characterized by the following theorem.

Theorem 1. *([6]) The following four conditions are equivalent.*

(a) Stability condition
(b) Condition (K)
(c) $\operatorname{Re}\lambda(i\xi) \le -c|\xi|^2/(1+|\xi|^2)$ *for any* $\xi \in \mathbf{R}^n$
(d) $\operatorname{Re}\lambda(i\xi) < 0$ *for any* $\xi \in \mathbf{R}^n$ *with* $\xi \neq 0$

Moreover, under the stability condition, it is shown that the equation (2) has a Lyapunov function of the form

$$E[\widehat{u}] = (A^0\widehat{u},\, \widehat{u}) - \frac{\alpha|\xi|}{1+|\xi|^2}(iK(\omega)A^0\widehat{u},\, \widehat{u}), \tag{3}$$

where α is a small positive constant, and $(\cdot,\cdot)$ is the inner product of $\mathbf{C}^m$. In fact, we have the following inequality:

$$\frac{\partial}{\partial t}E[\widehat{u}] + \frac{c|\xi|^2}{1+|\xi|^2}\,|\widehat{u}|^2 + c|(I-P)\widehat{u}|^2 \le 0, \tag{4}$$

where P is the orthogonal projection onto the null space $\mathcal{N}(L)$ of L. This ordinary differential inequality easily gives an energy estimate for (1). In fact, multiplying (4) by $(1+|\xi|^2)^s$ and integrating with respect to t and ξ, we obtain

$$\|u(t)\|_{H^s}^2 + \int_0^t \|\partial_x u(\tau)\|_{H^{s-1}}^2 + \|(I-P)u(\tau)\|_{H^s}^2\, d\tau \le C\|u_0\|_{H^s}^2, \tag{5}$$

where $s \ge 1$, and u_0 denotes the initial data. Note that this energy inequality contains the dissipative estimate without loss of regularity.

The ordinary differential inequality (4) also gives a decay estimate for (2). In fact, we have from (4) that

$$\frac{\partial}{\partial t}E[\widehat{u}] + c\eta(\xi)E[\widehat{u}] \le 0, \tag{6}$$

where $\eta(\xi) = |\xi|^2/(1+|\xi|^2)$. Solving this ordinary differential inequality, we have the following pointwise estimate of solutions to (2).

Theorem 2. *([7]) Under the stability condition we have*

$$|\widehat{u}(\xi,t)| \le Ce^{-c\eta(\xi)t}|\widehat{u}_0(\xi)|, \tag{7}$$

where $\eta(\xi) = |\xi|^2/(1+|\xi|^2)$.

This pointwise estimate yields the following energy decay estimate for solutions to (1).

Corollary 1. *([7]) Under the stability condition we have*

$$\|\partial_x^k u(t)\|_{L^2} \le Ce^{-ct}\|\partial_x^k u_0\|_{L^2} + C(1+t)^{-n/4-k/2}\|u_0\|_{L^1}, \tag{8}$$

where $k \ge 0$.

Note that this decay estimate holds without loss of regularity.

3 Dissipative Timoshenko System

As an example of the strict dissipativity of Type (II), we consider the dissipative Timoshenko system:

$$\begin{cases} w_{tt} - (w_x - \psi)_x = 0, \\ \psi_{tt} - a^2\,\psi_{xx} - (w_x - \psi) + \gamma\psi_t = 0, \end{cases} \tag{9}$$

where a and γ are positive constants. Putting $u = w_t$, $v = w_x - \psi$, $y = \psi_t$, and $z = a\psi_x$, we transform (9) into the equivalent first order system:

$$\begin{cases} v_t - u_x + y = 0, \\ u_t - v_x = 0, \\ z_t - ay_x = 0, \\ y_t - az_x - v + \gamma y = 0. \end{cases} \tag{10}$$

The system is written in the vector form as

$$U_t + AU_x + LU = 0, \tag{11}$$

where

$$U = \begin{pmatrix} v \\ u \\ z \\ y \end{pmatrix}, \quad A = -\begin{pmatrix} 0\,1\,0\,0 \\ 1\,0\,0\,0 \\ 0\,0\,0\,a \\ 0\,0\,a\,0 \end{pmatrix}, \quad L = \begin{pmatrix} 0\,0\,0\,1 \\ 0\,0\,0\,0 \\ 0\,0\,0\,0 \\ -1\,0\,0\,\gamma \end{pmatrix}.$$

Since A is symmetric, this is a symmetric hyperbolic system with L being not symmetric but nonnegative definite. Moreover, this system verifies the stability condition formulated in the previous section. In fact, we have:

Claim. Though L is not symmetric, the dissipative Timoshenko system (11) satisfies the stability condition: If $\mu\varphi + A\varphi = 0$ and $L\varphi = 0$ for $\mu \in \mathbf{R}$ and $\varphi \in \mathbf{R}^4$, then $\varphi = 0$.

To study the dissipative structure of the above Timoshenko system (11), we compute the eigenvalues $\lambda(i\xi)$, which are the solutions to the characteristic equation $\det(\lambda + i\xi A + L) = 0$. A direct computation shows that

- If $a = 1$, then $\mathrm{Re}\,\lambda(i\xi) \le -c\xi^2/(1+\xi^2)$ for any $\xi \in \mathbf{R}$.
- If $a \neq 1$, then $\mathrm{Re}\,\lambda(i\xi) \le -c\xi^2/(1+\xi^2)^2$ for any $\xi \in \mathbf{R}$.

This implies that the dissipative Timoshenko system (11) is strict dissipative in the sense of Type (I) for $a = 1$ and of Type (II) for $a \neq 1$. More precisely, when $a \neq 1$, one can verify that the eigenvalues $\lambda(i\xi)$ behave as

$$
\begin{aligned}
\lambda(i\xi) &= \pm\, i\xi \pm \frac{\sigma}{2}(i\xi)^{-1} + \sigma^2\gamma(i\xi)^{-2} + O(|\xi|^{-3}),\\
\lambda(i\xi) &= \pm\, ai\xi - \frac{\gamma}{2} + O(|\xi|^{-1})
\end{aligned}
$$

for $|\xi| \to \infty$, where $\sigma = (a^2-1)^{-1}$. The above computations show that the general theory of the strict dissipativity of Type (I) reviewed in the previous section can not be applied to the case where L is not symmetric.

Although the dissipative Timoshenko system (11) is classified into Type (II) for $a \neq 1$, it admits a Lyapunov function of the form

$$
E[\hat{U}] = |\hat{U}|^2 + \frac{\alpha_1}{1+\xi^2}\Big\{-\mathrm{Re}\big(\hat{v}\bar{\hat{y}} + a\hat{u}\bar{\hat{z}}\big) + \frac{\alpha_2\xi}{1+\xi^2}\,\mathrm{Re}\big(i\hat{v}\bar{\hat{u}} + i\hat{y}\bar{\hat{z}}\big)\Big\}, \tag{12}
$$

where $a \neq 1$, and α_1 and α_2 are small positive constants. In fact, by straightforward computations, we can show that

$$
\frac{\partial}{\partial t}E[\hat{U}] + cF[\hat{U}] \le 0 \tag{13}
$$

for $a \neq 1$, where

$$
F[\hat{U}] = \frac{\xi^2}{(1+\xi^2)^2}\big(|\hat{u}|^2 + |\hat{z}|^2\big) + \frac{1}{1+\xi^2}|\hat{v}|^2 + |\hat{y}|^2.
$$

This ordinary differential inequality gives an energy estimate for $a \neq 1$. In fact, multiplying (13) by $(1+\xi^2)^s$ and integrating with respect to t and ξ, we have

$$
\|U(t)\|_{H^s}^2 + \int_0^t \|\partial_x(u,z)(\tau)\|_{H^{s-2}}^2 + \|v(\tau)\|_{H^{s-1}}^2 + \|y(\tau)\|_{H^s}^2\, d\tau \le C\|U_0\|_{H^s}^2, \tag{14}
$$

where $s \ge 2$, and U_0 denotes the initial data. In this energy inequality for $a \neq 1$, the dissipative estimate for the three components v, u, and z is of regularity-loss type and this regularity-loss property would be a typical phenomena in the strict dissipativity of type (II).

To derive a decay estimate for dissipative Timoshenko system (11) with $a \neq 1$, we rewrite (13) as

$$\frac{\partial}{\partial t} E[\,\widehat{U}\,] + c\rho(\xi) E[\,\widehat{U}\,] \leq 0, \tag{15}$$

where $\rho(\xi) = \xi^2/(1+\xi^2)^2$. As in the derivation of (7), solving this ordinary differential inequality, we can obtain the following pointwise estimate for $\widehat{U}$.

Theorem 3. *When $a \neq 1$, we have*

$$|\widehat{U}(\xi,t)| \leq Ce^{-c\rho(\xi)t}|\widehat{U}_0(\xi)|, \tag{16}$$

where $\rho(\xi) = \xi^2/(1+\xi^2)^2$.

The above pointwise estimate yields the corresponding energy decay estimate of solutions to (11).

Corollary 2. *When $a \neq 1$, we have*

$$\|\partial_x^k U(t)\|_{L^2} \leq C(1+t)^{-l/2}\|\partial_x^{k+l}U_0\|_{L^2} + C(1+t)^{-1/4-k/2}\|U_0\|_{L^1}, \tag{17}$$

where k, $l \geq 0$.

This result shows that we have the decay rate $t^{-l/2}$ for $t \to \infty$ only with the lth order regularity-loss on the initial data. This regularity-loss property in the energy decay estimate would be the typical phenomena in the strict dissipativity of Type (II).

Proof. By applying the Plancherel theorem and the pointwise estimate (16), we have

$$\begin{aligned}\|\partial_x^k U(t)\|_{L^2}^2 &= \int \xi^{2k}|\widehat{U}(\xi,t)|^2 d\xi \leq C\int \xi^{2k}e^{-c\rho(\xi)t}|\widehat{U}_0(\xi)|^2 d\xi \\ &= \int_{|\xi|\leq 1} + \int_{|\xi|\geq 1} =: I_1 + I_2.\end{aligned}$$

In the low frequency region $|\xi| \leq 1$, we have $\rho(\xi) \geq c\xi^2$ and therefore we can estimate the term I_1 as

$$\begin{aligned}I_1 &\leq C\int_{|\xi|\leq 1} \xi^{2k}e^{-c\xi^2 t}|\widehat{U}_0(\xi)|^2 d\xi \\ &\leq C\sup_{|\xi|\leq 1}|\widehat{U}_0(\xi)|^2 \int_{|\xi|\leq 1} \xi^{2k}e^{-c\xi^2 t} d\xi \leq C(1+t)^{-1/2-k}\|U_0\|_{L^1}^2.\end{aligned}$$

On the other hand, in the high frequency region $|\xi| \geq 1$, we have $\rho(\xi) \geq c\xi^{-2}$. Therefore, the term I_2 can be estimated as

$$\begin{aligned}I_2 &\leq C\int_{|\xi|\geq 1} \xi^{2k}e^{-ct/\xi^2}|\widehat{U}_0(\xi)|^2 d\xi \\ &\leq C\sup_{|\xi|\geq 1}\frac{e^{-ct/\xi^2}}{\xi^{2l}} \int_{|\xi|\geq 1} \xi^{2(k+l)}|\widehat{U}_0(\xi)|^2 d\xi \leq C(1+t)^{-l}\|\partial_x^{k+l}U_0\|_{L^2}^2.\end{aligned}$$

These observations prove the desired estimate (17). □

4 Modified Radiating Gas Model

In this section, we consider the following modified radiating gas model system:

$$\begin{cases} u_t + \left(u^2/2\right)_x + q_x = 0, \\ (\partial_x^4 - \partial_x^2 + 1)q + u_x = 0. \end{cases} \tag{18}$$

Here we modified the original radiation gas model by introducing the fourth order elliptic term $\partial^4 q$ in the second equation. First we verify that this model system has the dissipative structure of Type (II). To see this, we consider the corresponding linearized system

$$\begin{cases} u_t + q_x = 0, \\ (\partial_x^4 - \partial_x^2 + 1)q + u_x = 0. \end{cases} \tag{19}$$

Taking the Fourier transform and eliminating q, we get

$$\widehat{u}_t + \rho(\xi)\widehat{u} = 0,$$

where $\rho(\xi) = \xi^2/(1+\xi^2+\xi^4)$. This shows that the eigenvalue of the system is $\lambda(i\xi) = -\rho(\xi)$, which is of Type (II).

Let e^{tA} be the semigroup associated with the linearized system (19). We have the expression

$$e^{tA}\varphi = \mathcal{F}^{-1}e^{-\rho(\xi)t}\mathcal{F}\varphi. \tag{20}$$

This semigroup satisfies the following decay estimate of regularity-loss type that is just the same as (17).

Claim. We have

$$\|\partial_x^k e^{tA}\varphi\|_{L^2} \le C(1+t)^{-l/2}\|\partial_x^{k+l}\varphi\|_{L^2} + C(1+t)^{-1/4-k/2}\|\varphi\|_{L^1}, \tag{21}$$

where $k,\, l \ge 0$.

The main purpose of this section is to explain the difficulty in showing the global solvability of the nonlinear equation (18) satisfying the strict dissipativity of Type (II). First, we mention that the standard energy method is not useful enough to derive the desired a priori estimates of solutions to (18). To see this, we apply to (18) the standard energy method, obtaining

$$\frac{d}{dt}\|\partial_x^k u\|_{L^2}^2 + 2\|\partial_x^k q\|_{H^2}^2 \le C\|u_x\|_{L^\infty}\|\partial_x^k u\|_{L^2}^2, \tag{22}$$

where $k \ge 0$ (when $k = 0$, we have the equality with zero right hand side). Note that the term on the right hand side of (22) comes from the nonlinearity of (18). This yields the energy estimate

$$\|u(t)\|_{H^s}^2 + 2\int_0^t \|q\|_{H^{s+2}}^2 d\tau \le \|u_0\|_{H^s}^2 + C\int_0^t \|u_x\|_{L^\infty} \|\partial_x u\|_{H^{s-1}}^2 d\tau, \qquad (23)$$

where $s \ge 1$, and u_0 denotes the initial data. On the other hand, we have from the second equation of (18) that $\|\partial_x u\|_{H^{s-2}} \le C\|q\|_{H^{s+2}}$, which together with (23) gives

$$\|u(t)\|_{H^s}^2 + \int_0^t \|\partial_x u\|_{H^{s-2}}^2 d\tau \le C\|u_0\|_{H^s}^2 + C\int_0^t \|u_x\|_{L^\infty} \|\partial_x u\|_{H^{s-1}}^2 d\tau, \quad (24)$$

where $s \ge 2$. We note that the dissipative estimate here is of the regularity-loss type and is essentially the same as (14). One can estimate the nonlinearity in (24) as

$$C\int_0^t \|u_x\|_{L^\infty} \|\partial_x u\|_{H^{s-1}}^2 d\tau \le C \sup_{0\le\tau\le t} \|u_x(\tau)\|_{L^\infty} \int_0^t \|\partial_x u\|_{H^{s-1}}^2 d\tau,$$

but it can not be controlled by the dissipativity because of the loss of regularity. Also, one can estimate as

$$C\int_0^t \|u_x\|_{L^\infty} \|\partial_x u\|_{H^{s-1}}^2 d\tau \le C \sup_{0\le\tau\le t} \|\partial_x u(\tau)\|_{H^{s-1}}^2 \int_0^t \|u_x\|_{L^\infty} d\tau.$$

However, we can not control this nonlinearity because the expected optimal decay estimate $\|u_x(t)\|_{L^\infty} \le C(1+t)^{-1}$ is not integrable in t.

To remove the difficulty caused by the loss of regularity in the dissipative estimate, we introduce a time-weighted energy method with the weight function $(1+t)^\alpha$, where $\alpha = -1/2$. Such a weight with negative exponent produces an artificial dissipativity in the energy estimates and this idea combined with the optimal decay for $\|u_x\|_{L^\infty}$ can prove the global existence of solutions. In fact, we have:

Theorem 4. *([1]) If u_0 is small in $H^s \cap L^1$, where $s \ge 7$, then there exists a unique global solution (u, q) to (18) such that*

$$\begin{aligned} \|\partial_x^k u(t)\|_{L^2} &\le CE_0(1+t)^{-1/4-k/2}, \\ \|\partial_x^l q(t)\|_{H^2} &\le CE_0(1+t)^{-3/4-l/2}, \end{aligned} \qquad (25)$$

where $0 \le k \le [(s-1)/2]-1$, $0 \le l \le [(s-1)/2]-2$, and $E_0 = \|u_0\|_{H^s} + \|u_0\|_{L^1}$.

The key to the proof of the above theorem is to derive the desired a priori estimates of solutions. We use the following quantities

$$E(t)^2 = \sum_{j=0}^{[s/2]} \sup_{0\le\tau\le t} (1+\tau)^{j-1/2} \|\partial_x^j u(\tau)\|_{H^{s-2j}}^2,$$

$$D(t)^2 = \sum_{j=0}^{[s/2]} \int_0^t (1+\tau)^{j-3/2} \|\partial_x^j u(\tau)\|_{H^{s-2j}}^2 \, d\tau,$$

$$M(t) = \sum_{j=0}^{s_1} \sup_{0 \le \tau \le t} (1+\tau)^{1/4+j/2} \|\partial_x^j u(\tau)\|_{L^2},$$

where $s_1 = [(s-1)/2]-1$. $E(t)$, $D(t)$, and $M(t)$ are corresponding to the weighted energy, the weighted dissipativity, and the optimal decay, respectively.

Proposition 1. *Let $s \ge 7$. Then we have*

$$E(t)^2 + D(t)^2 \le C\|u_0\|_{H^s}^2 + CM(t)D(t)^2. \tag{26}$$

Proposition 2. *Let $s \ge 5$. Then we have*

$$M(t) \le C(\|u_0\|_{H^{s-2}} + \|u_0\|_{L^1}) + CM(t)^2 + CM(t)E(t). \tag{27}$$

We combine these two propositions and deduce that $E(t)+D(t)+M(t) \le CE_0$, provided that E_0 is suitably small. This gives the desired a priori estimate of solutions, which combined with a local existence theorem proves Theorem 4

Proof. We prove Proposition 1 by applying the time-weighted energy method. We multiply (22) by $(1+t)^\alpha$ and integrate with respect to t. This yields

$$\begin{aligned} &(1+t)^\alpha \|\partial_x^k u(t)\|_{L^2}^2 + 2\int_0^t (1+\tau)^\alpha \|\partial_x^k q\|_{H^2}^2 \, d\tau \\ &\le \|\partial_x^k u_0\|_{L^2}^2 + \alpha \int_0^t (1+\tau)^{\alpha-1} \|\partial_x^k u\|_{L^2}^2 \, d\tau \\ &\quad + C\int_0^t (1+\tau)^\alpha \|u_x\|_{L^\infty} \|\partial_x^k u\|_{L^2}^2 \, d\tau, \end{aligned} \tag{28}$$

where $k \ge 0$ and $\alpha \in \mathbf{R}$.

First, we put $\alpha = -1/2$ in (28) and add for $0 \le k \le s$. This gives the following energy estimate with an artificial dissipativity:

$$\begin{aligned} &(1+t)^{-1/2}\|u(t)\|_{H^s}^2 + \int_0^t (1+\tau)^{-1/2}\|q\|_{H^{s+2}}^2 d\tau + \int_0^t (1+\tau)^{-3/2}\|u\|_{H^s}^2 d\tau \\ &\le C\|u_0\|_{H^s}^2 + C\int_0^t (1+\tau)^{-1/2}\|u_x\|_{L^\infty}\|u\|_{H^s}^2 d\tau \\ &\le C\|u_0\|_{H^s}^2 + CM(t)D(t)^2, \end{aligned}$$

where we have estimated the nonlinearity by using $M(t)$ and $D(t)$ as

$$C\int_0^t (1+\tau)^{-1/2}\|u_x\|_{L^\infty}\|u\|_{H^s}^2\,d\tau$$
$$\le CM(t)\int_0^t (1+\tau)^{-3/2}\|u\|_{H^s}^2\,d\tau \le CM(t)D(t)^2.$$

On the other hand, it follows from the second equation of (18) that $\|\partial_x u\|_{H^{s-2}} \le C\|q\|_{H^{s+2}}$. Therefore, we obtain for $s \ge 2$,

$$\int_0^t (1+\tau)^{-1/2}\|\partial_x u\|_{H^{s-2}}^2\,d\tau$$
$$\le C\int_0^t (1+\tau)^{-1/2}\|q\|_{H^{s+2}}^2\,d\tau \le C\|u_0\|_{H^s}^2 + CM(t)D(t)^2.$$

Next, we put $\alpha = 1/2$ in (28) and add for $1 \le k \le s-1$, obtaining

$$(1+t)^{1/2}\|\partial_x u(t)\|_{H^{s-2}}^2 + \int_0^t (1+\tau)^{1/2}\|\partial_x q\|_{H^s}^2\,d\tau$$
$$\le C\|\partial_x u_0\|_{H^{s-2}}^2 + C\int_0^t (1+\tau)^{-1/2}\|\partial_x u\|_{H^{s-2}}^2\,d\tau$$
$$+ C\int_0^t (1+\tau)^{1/2}\|u_x\|_{L^\infty}\|\partial_x u\|_{H^{s-2}}^2\,d\tau$$
$$\le C\|u_0\|_{H^s}^2 + CM(t)D(t)^2.$$

Also, using the second equation of (18), we have

$$\int_0^t (1+\tau)^{1/2}\|\partial_x^2 u\|_{H^{s-4}}^2\,d\tau$$
$$\le C\int_0^t (1+\tau)^{1/2}\|\partial_x q\|_{H^s}^2\,d\tau \le C\|u_0\|_{H^s}^2 + CM(t)D(t)^2,$$

where $s \ge 4$. Repeating this argument, we arrive at the desired inequality (26). $\square$

Proof. We prove Proposition 2 by applying the decay estimate (21). We eliminate q from (18) and rewrite the resulting equation for u by using the simigroup e^{tA} as

$$u(t) = e^{tA}u_0 + \int_0^t e^{(t-\tau)A}\partial_x g(\tau)\,d\tau, \tag{29}$$

where $g = -u^2/2$. We apply ∂_x^k to (29). Then, applying the decay estimate (21) with various choice of l, we obtain

$$
\begin{aligned}
&\|\partial_x^k u(t)\|_{L^2}^2 \le C(1+t)^{-1/4-k/2}\|u_0\|_{L^1} + C(1+t)^{-(k+1)/2}\|\partial_x^{2k+1}u_0\|_{L^2} \\
&+ C\int_0^{t/2}(1+t-\tau)^{-3/4-k/2}\|g\|_{L^1}d\tau + C\int_0^{t/2}(1+t-\tau)^{-(k+1)/2}\|\partial_x^{2k+2}g\|_{L^2}d\tau \\
&+ C\int_{t/2}^{t}(1+t-\tau)^{-3/4}\|\partial_x^k g\|_{L^1}d\tau + C\int_{t/2}^{t}(1+t-\tau)^{-1/2}\|\partial_x^{k+2}g\|_{L^2}d\tau \\
&=: I_1 + \cdots + I_6,
\end{aligned}
$$

where $0 \le k \le [(s-1)/2] - 1$. We can estimate each term on the right hand side as follows. For the first two terms, we have

$$
I_1 + I_2 \le C(\|u_0\|_{H^{s-2}} + \|u_0\|_{L^1})(1+t)^{-1/4-k/2}.
$$

The terms I_3 and I_5 are estimated in terms of $M(t)$ as

$$
\begin{aligned}
I_3 &\le CM(t)^2\int_0^{t/2}(1+t-\tau)^{-3/4-k/2}(1+\tau)^{-1/2}d\tau \\
&\le CM(t)^2(1+t)^{-1/4-k/2}, \\
I_5 &\le CM(t)^2\int_{t/2}^{t}(1+t-\tau)^{-3/4}(1+\tau)^{-1/2-k/2}d\tau \\
&\le CM(t)^2(1+t)^{-1/4-k/2},
\end{aligned}
$$

where we used the fact that $\|\partial_x^j g\|_{L^1} \le C\|u\|_{L^2}\|\partial_x^j u\|_{L^2}$. On the other hand, noting that $\|\partial_x^j g\|_{L^2} \le C\|u\|_{L^\infty}\|\partial_x^j u\|_{L^2}$ and using the weighted energy $E(t)$, we can estimate I_4 and I_6 as

$$
\begin{aligned}
I_4 &\le CM(t)E(t)\int_0^{t/2}(1+t-\tau)^{-1/2-k/2}(1+\tau)^{-3/4}d\tau \\
&\le CM(t)E(t)(1+t)^{-1/4-k/2}, \\
I_6 &\le CM(t)E(t)\int_{t/2}^{t}(1+t-\tau)^{-1/2}(1+\tau)^{-3/4-k/2}d\tau \\
&\le CM(t)E(t)(1+t)^{-1/4-k/2}.
\end{aligned}
$$

Combining all these estimates gives the desired inequality (27). □

5 Generalization and Open Questions

Here we present several open problems concerning the strict dissipativity of Type (II). The first problem is on the global solvability of the nonlinear dissipative Timoshenko system

$$
\begin{cases}
w_{tt} - (w_x - \psi)_x = 0, \\
\psi_{tt} - \sigma(\psi_x)_x - (w_x - \psi) + \gamma\psi_t = 0,
\end{cases}
\tag{30}
$$

where $\sigma(z)$ is a smooth and strictly increasing function of z and γ is a positive constant. Let $a := \sqrt{\sigma'(0)}$ and consider the case $a \neq 1$, which is the case of the strict dissipativity of Type (II). This problem is expected to be solved as in Theorem 4 and the theory is now in progress.

The second problem is on the global solvability of the following modified radiating gas model system:

$$\begin{cases} u_t + f(u)_x + q_x = 0, \\ (-1)^m \partial_x^{2m} q + q + u_x = 0, \end{cases} \tag{31}$$

where $f(u)$ is a smooth and strictly convex function of u and $m \geq 2$ is an integer. The eigenvalue of this system is given by $\lambda(i\xi) = -\xi^2/(1+\xi^{2m})$ so that it is of Type (II). The theory for this problem is now in progress and a generalization of Theorem 4 will be established.

The last problem is on the characterization of the strict dissipativity of Type (II) in a general situation. For example, let us consider the symmetric hyperbolic systems (1) in the following situation:

(a) A^0 is symmetric and positive definite,
(b) A^j is symmetric for any j,
(c) L is *not symmetric* but is nonnegative definite.

An important problem is to formulate a condition that guarantees the strict dissipativity of type (II) in this situation. The theory should cover the dissipative structure of the dissipative Timoshenko system (11).

References

1. T. Hosono and S. Kawashima, *Decay property of regularity-loss type and application to some hyperbolic-elliptic systems*, to appear in Math. Models Meth. Appl. Sci.
2. K. Ide, K. Haramoto and S. Kawashima, *Decay property of the dissipative Timoshenko system*, (in preparation).
3. T. Iguchi and S. Kawashima, *On space-time decay properties of solutions to hyperbolic-elliptic coupled systems*, Hiroshima Math. J., **32** (2002), 229–308.
4. S. Kawashima, Y. Nikkuni and S. Nishibata, *The initial value problem for hyperbolic-elliptic coupled systems and applications to radiation hydrodynamics*, Analysis of Systems of Conservation Laws (H. Freistühler, ed.), Chapman & Hall/CRC, 1998, 87–127.
5. J.E.M. Revera and R. Racke, *Global stability for damped Timoshenko systems*, Discrete and Continuous Dynamical Systems, **9** (2003), 1625–1639.
6. Y. Shizuta and S. Kawashima, *Systems of equations of hyperbolic-parabolic type with applications to the discrete Boltzmann equation*, Hokkaido Math. J., **14** (1985), 249–275.
7. T. Umeda, S. Kawashima and Y. Shizuta, *On the decay of solutions to the linearized equations of electro-magneto-fluid dynamics*, Japan J. Appl. Math., **1** (1984), 435–457.

Dissipative Hyperbolic Systems: the Asymptotic Behavior of Solutions

S. Bianchini, B. Hanouzet, and R. Natalini

1 Introduction

In this talk we shall review some recent results about qualitative behavior of smooth solutions to the Cauchy problem for general hyperbolic symmetrizable m-dimensional systems of balance laws of the form

$$u_t + \sum_{\alpha=1}^{m} (f_\alpha(u))_{x_\alpha} = g(u), \tag{1}$$

with the initial condition

$$u(x,0) = u_0(x), \tag{2}$$

where $u = (u_1, u_2) \in \Omega \subseteq \mathbb{R}^{n_1} \times \mathbb{R}^{n_2}$, with $n_1 + n_2 = n$. We assume that there are n_1 conservation laws in the system, namely that we can take

$$g(u) = \begin{pmatrix} 0 \\ q(u) \end{pmatrix}, \quad \text{with } q(u) \in \mathbb{R}^{n_2}. \tag{3}$$

As well known, even for nice initial data, smooth solutions may break down in finite time, due to the appearance of singularities. On the other hand, sometimes dissipative mechanisms induced by the source term can prevent the formation of singularities, at least for small initial data, as for instance for the compressible Euler equations with damping [Ni78, STW03].

Recently, it was proposed in [HN03] a quite general framework of sufficient conditions to have the global existence in time of smooth solutions. Actually, for systems which are endowed with a strictly convex entropy function $\mathcal{E} = \mathcal{E}(u)$, a first natural assumption is the *entropy dissipation condition*, see [CLL94]. Unfortunately, it is easy to see that this condition is too weak to prevent the formation of singularities. A quite natural supplementary condition can be imposed to entropy dissipative systems, following the classical approach by Shizuta and Kawashima [Kw87, SK85]. It is possible to prove that this condition, which is satisfied in many interesting examples, is also sufficient to establish a general result of global existence for small perturbations of equilibrium constant states, see [HN03, Yo04]. These results will be shortly reviewed in Sect. 2.

Here, we intend to present a short survey of the new results of [BRS06], about the asymptotic behavior in time of the global solutions, then always assuming the existence of a strictly convex entropy and the Shizuta and Kawashima condition. However, let us recall that, starting from the seminal paper by Liu [Li87], there were some previous studies on 2×2 systems, see [Na99] for some references. For more general models, we recall [Ze99], for a model of gas dynamics in thermal nonequilibrium, and [RS02], about stability of constant equilibrium states for general hyperbolic systems in one space dimension, under zero-mass perturbations.

The paper is organized as follows: Section 2 is devoted to recall some basic results about hyperbolic systems with entropy dissipation and the Shizuta–Kawashima condition. In this section we also introduce the decomposition of the linearized system, which will be called the onservative–dissipative form, which will be necessary to cast the decay results in the optimal way. Section 3 contains some precise results about the asymptotic behavior of the Green kernel for the linearized problem. Finally, Sect. 4 is devoted to the study of the decay properties of the nonlinear system. Not only we shall present the decay results for both the conservative and the dissipative part of the solution, but we shall show also that the conservative variable approaches the conservative part of the solution of the corresponding linearized problem, faster that the decay of the heat kernel for $m \geq 2$. Also, it is possible to see that the solution of the parabolic problem given by the Chapman–Enskog expansion, is a good approximation of the conservative part of the solution of the nonlinear hyperbolic system. For $m \geq 2$ the Chapman–Enskog operator is linear, while, in one space dimension, the decay of the nonlinear part has a stronger influence, and so we can only show the faster convergence toward the solution of a parabolic equation with quadratic nonlinearity. Let us point out that the present results can be useful to design more accurate numerical approximations of these problems, which are increasingly accurate for large times, see [ABN06] for some preliminary results in this direction.

Finally, we notice again that these results were obtained by assuming all the time the condition (SK). Unfortunately, this condition is not satisfied by many models, as for instance for the model studied in [Ze99], where however global existence of solutions has been established. For the Kerr–Debye system, studied in [HH00, CH106, CH206], condition (SK) is not always verified for $m = 1$, see [HN03], and never verified for $m \geq 2$.

2 A Short Review of Some Basic Structures for the Entropy Dissipative Hyperbolic Systems

In the following we shall consider a general m-dimensional system of balance laws given by (1), with the source term $g = g(u)$ given by (3). According to the general theory of hyperbolic systems of conservation laws [Da00], we shall assume that the system satisfies an entropy principle: There exists a strictly convex function $\mathcal{E} = \mathcal{E}(u)$, the entropy density, and some related entropy-flux functions $\mathcal{F}_\alpha = \mathcal{F}_\alpha(u)$, such that for every smooth solution $u \in \Omega$ to system (1), there holds

$$\mathcal{E}(u)_t + \sum_{\alpha=1}^{m} \left(\mathcal{F}_\alpha(u)\right)_{x_\alpha} = \mathcal{G}(u) \ , \tag{4}$$

where ${\mathcal{F}_\alpha}' = (f'_\alpha)^T \mathcal{E}'$ and $\mathcal{G} = \mathcal{E}' \cdot g$. Let us introduce the set γ of the equilibrium points:

$$\gamma = \{u \in \Omega; g(u) = 0\}.$$

We shall assume that system (1) is *nondegenerate* in the sense that around a given $\bar{u} \in \gamma$, it holds

$$q_{u_2}(\bar{u}) \text{ is nonsingular.} \tag{5}$$

We shall also assume that system (1) is *entropy dissipative* around the given equilibrium point $\bar{u} \in \gamma$ and $u \in \Omega$, in the sense that we have

$$(\mathcal{E}'(u) - \mathcal{E}'(\bar{u})) \cdot g(u) \leq 0. \tag{6}$$

Following [Go61, LF71], it is now possible to symmetrize our system by introducing a new variable, the *entropy variable*, which is just given by $W = \mathcal{E}'(u)$. This change of variable is very useful in the statement of the global existence results. Actually, since $\mathcal{E}$ is a strictly convex function, we can inverse $\mathcal{E}'$ to recover the original variable u by the inverse map $\Phi \doteq (\mathcal{E}')^{-1}$. Let us set now $A_0(W) \doteq \Phi'(W)$, $C_\alpha(W) \doteq Df_\alpha(\Phi(W))A_0(W)$, and $G(W) = g(\Phi(W)) = \begin{pmatrix} 0 \\ Q(W) \end{pmatrix}$. It is easy to see that the matrix $A_0(W)$ is symmetric positive definite and, for every $\alpha = 1.\dots, m$, $C_\alpha(W)$ is symmetric. Then, selecting W as the new variable, our system reads

$$A_0(W)W_t + \sum_{\alpha=1}^{m} C_\alpha(W)W_{x_\alpha} = G(W). \tag{7}$$

In the following, without loss of generality, we can always suppose $\bar{u} = 0 \in \gamma$ and consider system (1) with $g(0) = 0$, $f_\alpha(0) = 0$, and endowed with a quadratic entropy function $\mathcal{E}$. Next, we focus our investigation on a slightly restricted class of entropy dissipative nondegenerate systems, namely the systems such that

$$Q(W) = D(W)W_2, \text{ with } D(0) \text{ negative definite.} \tag{8}$$

To continue our analysis of smooth solutions for dissipative hyperbolic systems, we need some supplementary coupling conditions to avoid shock formation. A very natural condition was first introduced by Shizuta and Kawashima in [SK85], for hyperbolic–parabolic systems. Let us introduce this condition for the original unknown, i.e., for system (1).

Definition 1. *The system (1) verifies condition (SK), if every eigenvector of* $\sum_{\alpha=1}^{m} Df_\alpha(0)\xi_\alpha$ *is not in the null space of* $Dg(0)$*, for every* $\xi \in \mathbb{R}^m \setminus \{0\}$.

About the existence of a solution, we recall the following result [HN03, Yo04].

Theorem 1. *Assume that system (1) is strictly entropy dissipative and condition (SK) is satisfied. Then there exists* $\delta > 0$ *such that, if* $\|u_0\|_s \le \delta$*, with* $s \ge [m/2] + 2$*, there is a unique global solution* u *of (1)–(2), which verifies*

$$u \in C^0([0,\infty); H^s(\mathbb{R}^m)) \cap C^1([0,\infty); H^{s-1}(\mathbb{R}^m)),$$

and such that, in terms of the entropy variable $W = (W_1, W_2)$*,*

$$\sup_{0 \le t < +\infty} \|W(t)\|_s^2 + \int_0^{+\infty} \left(\|\nabla W_1(\tau)\|_{s-1}^2 + \|W_2(\tau)\|_s^2 \right) d\tau \le C(\delta) \|W_0\|_s^2, \quad (9)$$

where $C(\delta)$ *is a positive constant.*

Let us consider now the linearized version of system (1), namely, setting $A_\alpha = Df_\alpha(0)$ and $B = Dg(0)$,

$$u_t + \sum_{\alpha=1}^{m} A_\alpha u_{x_\alpha} = Bu, \qquad u \in \mathbb{R}^n, x \in \mathbb{R}^m, t \in \mathbb{R}^+, \quad (10)$$

with B of the form

$$B = \begin{bmatrix} 0 & 0 \\ D_1 & D_2 \end{bmatrix}, \qquad D_1 \in \mathbb{R}^{n_1 \times n_2}, D_2 \in \mathbb{R}^{n_2 \times n_2}, \quad (11)$$

with $n = n_1 + n_2$. Set $A(\xi) = \sum_{\alpha=1}^{m} A_\alpha \xi_\alpha$. According to the previous discussion, we can assume that:

(H1) There is a symmetric positive definite matrix A_0 such that $A_\alpha A_0$ is symmetric, for every $\alpha = 1, \dots, m$, and

$$BA_0 = \begin{bmatrix} 0 & 0 \\ 0 & D \end{bmatrix},$$

where $D \in \mathbb{R}^{n_2 \times n_2}$ is negative definite.

(H2) Any eigenvector of $A(\xi)$ is not in the null space of B, for every $\xi \in \mathbb{R}^m \setminus \{0\}$.

To give a precise picture of the asymptotic behavior of the solutions for large times, we have to transform the original system in a suitable form, using a special projection of the original variables.

Definition 2. *The partially dissipative system* (10) *is in conservative–dissipative form (C–D form) if the matrices A_α are symmetric and there exists a negative definite matrix $D \in \mathbb{R}^{n_2 \times n_2}$, such that*

$$B = \begin{pmatrix} 0 & 0 \\ 0 & D \end{pmatrix}. \tag{12}$$

In the following $w_1 \doteq w_c$ is called the conservative variable, while $w_2 \doteq w_d$ is the dissipative one.

We shall prove in the following that there exists a linear change of variable such that (10) takes the C–D form under assumptions (H1). We shall construct the C–D variables using the projection Q_0 on the null space and the complementing projection $Q_- = I - Q_0$. We can compute Q_0 by using the classical representation formula, which yields

$$Q_0 = A_0^{1/2} \begin{bmatrix} (A_{0,11})^{-1} & 0 \\ 0 & 0 \end{bmatrix} A_0^{1/2}.$$

Note that, due to the assumptions on A_0, this projector is symmetric. In particular we can choose left and right projectors $L_0 \in \mathbb{R}^{n \times n_1}$, $R_0 \in \mathbb{R}^{n_1 \times n}$, so that $Q_0 = R_0 L_0$, $L_0 R_0 = I \in \mathbb{R}^{n_1 \times n_1}$, $L_0 = R_0^T$, and they are uniquely given by

$$R_0 = A_0^{1/2} \begin{bmatrix} (A_{0,11})^{-1/2} \\ 0 \end{bmatrix}, \qquad L_0 = \begin{bmatrix} (A_{0,11})^{-1/2} & 0 \end{bmatrix} A_0^{1/2}. \tag{13}$$

We define the complementary projection Q_- such that $Q_- \doteq I - Q_0 = R_- L_-$, $L_- R_- = I \in \mathbb{R}^{n_2 \times n_2}$, $L_- = R_-^T$. The matrices R_-, L_- are uniquely given by

$$R_- = A_0^{-1/2} \begin{bmatrix} 0 \\ ((A_0^{-1})_{22})^{-1/2} \end{bmatrix}, \qquad L_- = \begin{bmatrix} 0 & ((A_0^{-1})_{22})^{-1/2} \end{bmatrix} A_0^{-1/2}. \tag{14}$$

If we set $w_1 = L_0 A_0^{-1/2} u$ and $w_2 = L_- A_0^{-1/2} u$, we obtain the following result.

Proposition 1. *If u is a solution to system* (10), *then, under assumption (H1),*

$$w = Mu = \begin{bmatrix} (A_{0,11})^{-1/2} & 0 \\ ((A_0^{-1})_{22})^{-1/2} (A_0^{-1})_{21} & ((A_0^{-1})_{22})^{1/2} \end{bmatrix} u \tag{15}$$

is a solution of a system in C–D form.

In the following, we are always going to assume that the unknown u is chosen in such a way that (10) is in conservative–dissipative form. In this case, we say that also system (1) is in conservative–dissipative form and we shall set $u = (u_c, u_d) \in \mathbb{R}^{n_1} \times \mathbb{R}^{n_2}$.

Example 1. The p-system with relaxation. Let us consider system

$$\begin{cases} \partial_t u + \partial_x v = 0 \ , \\ \partial_t v + \partial_x \sigma(u) = h(u) - v, \end{cases} \tag{16}$$

with $\sigma'(u) > 0$. One symmetrizer for the linearized version of (16) is given by $A_0 = \begin{pmatrix} 1 & a \\ a & \lambda^2 \end{pmatrix}$, with $\lambda = \sqrt{\sigma'(0)}$ and $a = h'(0)$. The matrix A_0 is positive definite if it holds the subcharacteristic condition $\lambda > |a|$. It is easy to verify that assumption (H1) is verified and the proper C–D form of the system is found for the new unknowns

$$\begin{pmatrix} u_c \\ u_d \end{pmatrix} = M \begin{pmatrix} u \\ v \end{pmatrix} = \begin{pmatrix} 1 & 0 \\ -a(\lambda^2 - a^2)^{-\frac{1}{2}} & (\lambda^2 - a^2)^{-\frac{1}{2}} \end{pmatrix} \begin{pmatrix} u \\ v \end{pmatrix},$$

see formula (2.29) in [BRS06].

3 The Green Kernel for Linear Dissipative Systems

3.1 The One-Dimensional Case

We want to characterize the Green kernel $\Gamma(t)$ for a linear dissipative hyperbolic system. The fact that we are in dimension one will help us in inverting the Fourier transforms, hence giving explicit form to the principal parts of $\Gamma(t)$. We can consider directly a system in C–D form, according to the results of Sect. 2. So we write our system as

$$w_t + Aw_x = Bw, \tag{17}$$

where $w = (w_c, w_d) \in \mathbb{R}^{n_1} \times \mathbb{R}^{n_2}$. We assume that the matrix A is symmetric and the matrix B verifies (12). We assume also that (17) verifies condition (SK). We want to present a precise description of the Green kernel $\Gamma(t,x)$ of (17), which is the solution to problem

$$\begin{cases} \Gamma_t + A\Gamma_x = B\Gamma \\ \Gamma(0,x) = \delta(x)I, \end{cases} . \tag{18}$$

by extending the approach used in [Ze99].

Consider the entire function $E(z) = B - zA$. The Fourier transform of the solution to (18) is given by $\hat{\Gamma}(t,\xi) = e^{E(i\xi)t}$. Notice that also $\hat{\Gamma}(t,z) = e^{E(z)t}$ is an entire function of z. In general, if z is not an exceptional point, the function $E(z)$, is represented by

$$E(z) = \sum_j \lambda_j(z)P_j(z) + \sum_j D_j(z), \tag{19}$$

where λ_j are the eigenvalues of $E(z)$, $P_j(z)$ the corresponding eigenprojections and D_j are the nilpotent matrices, due to the fact that in general E is not diagonalizable, since we are not assuming that the matrix B is symmetric. We now study which consequences have the assumptions (H1), (H2) on $E(z)$ and $\hat{\Gamma}(t,z)$ near the point $z = 0$ and $z = \infty$. Both points are in general exceptional. For $z \to 0$, n_1 eigenvalues different from 0 converges to 0. When $|z| \to \infty$, the matrix A is diagonalizable, but it can have common eigenvalues: then the perturbation B/z will in general remove part of this degeneracy.

First we consider the case near $z = 0$. It has been shown in [BRS06], by standard representation formula, that we can expand the projector in the following way:

$$P(z) = \begin{bmatrix} I + \mathcal{O}(z^2) & zA_{12}D^{-1} + \mathcal{O}(z^2) \\ zD^{-1}A_{21} + \mathcal{O}(z^2) & z^2D^{-1}A_{21}A_{12}D^{-1} + \mathcal{O}(z^3) \end{bmatrix}. \tag{20}$$

We introduce the right and left eigenprojectors of $P(z)$, $R(z) \in \mathbb{R}^{n\times n_1}$, $L(z) \in \mathbb{R}^{n_1\times n}$, which verify $P(z) = R(z)L(z)$, $L(z)R(z) = I$. We can find the power series of $L(z)$ and $R(z)$: $L(z) = L_0 + zL_1 + z^2L_1 + \mathcal{O}(z^3)$, and $R(z) = L_0 + zR_1 + z^2R_1 + \mathcal{O}(z^3)$. We compute L_0 and R_0 by

$$L_0 = R_0^T = \begin{bmatrix} I_{n_1} & 0 \end{bmatrix}. \tag{21}$$

The following terms can be also computed explicitely. So, we can decompose $E(z)$ according to the right and left operators:

$$E(z) = R(z)F(z)L(z) + E_1(z), \tag{22}$$

where $F(z) \doteq L(z)E(z)R(z) \in \mathbb{R}^{n_1\times n_1}$ and E_1 is a rest with a faster decay. We have

$$\begin{aligned} F(z) &= \big(L_0 + zL_1 + \mathcal{O}(z^2)\big)(B - zA)\big(R_0 + zR_1 + \mathcal{O}(z^2)\big) \\ &= -zA_{11} - z^2A_{12}D^{-1}A_{21} + \mathcal{O}(z^3). \end{aligned} \tag{23}$$

The matrix A_{11} is symmetric, from assumption (H1), so that we can write for some eigenvalues λ_j^1, with multiplicity m_j', and left and right eigenprojections $l_j \in \mathbb{R}^{m_j'\times n_1}$, $r_j \in \mathbb{R}^{n_1\times m_j'}$, with $l_j = r_j^T$,

$$A_{11} = \sum_{j=1}^{m'} \lambda_j^1 r_j l_j. \tag{24}$$

By a further decomposition, we obtain, for some projectors P_{jk} and D_{jk}, the projection of $E(z)$ on the null eigenvalue as

$$R(z)F(z)L(z) = \sum_{jk} \Big(-z\lambda_j^1 I - z^2c_{jk}I + \mathcal{O}(z^3)\Big)P_{jk}(z) + D_{jk}(z). \tag{25}$$

Proposition 2. *We have the following decomposition near $z = 0$*

$$E(z) = \sum_{jk} \big(\Lambda_{jk}(z) P_{jk}(z) + D_{jk}(z) \big) + E_1(z), \tag{26}$$

where the Λ_{jk} are diagonal $n \times n$ matrices composed by the n_1 eigenvalues λ_{jk}, the coefficients c_{jk} having strictly negative real part, thanks to assumption (H2).

We can perform a similar, but simpler, analysis for $|z| \to \infty$, since we can write

$$E(z) = B - zA = z\left(-A + \frac{1}{z}B \right) = z\tilde{E}(1/z),$$

where $\tilde{E}(\zeta) = -A + \zeta B$.

Proposition 3. *We have the following decomposition near $z = \infty$*

$$E(z) = \sum_{jk} \big(\Upsilon_{jk}(z) \mathcal{P}_{jk}(z) + \mathcal{D}_{jk}(z) \big), \tag{27}$$

where Υ_{jk} is the diagonal matrix whose entries are the eigenvalues of the jk family, the coefficients b_{jk} having strictly negative real part, thanks to assumption (H2).

Now, we are ready to estimate the global behavior for large t of the Green kernel $\Gamma(t, x)$ using the local expansions contained in Propositions 2 and 3. We associate a diffusive operator with Green function $K(t, x)$ to the expansion (26), and a dissipative transport operator with Green function $\mathcal{K}(t, x)$ to (27), and we estimate the remainder term $R(t, x) = \Gamma(t, x) - K(t, x) - \mathcal{K}(t, x)$. In the following we shall consider the Green kernel as composed of four parts, acting on w_c, w_d:

$$\Gamma(t, x) = \begin{bmatrix} L_0 \Gamma(t, x) R_0 & L_0 \Gamma(t, x) R_- \\ L_- \Gamma(t, x) R_0 & L_- \Gamma(t, x) R_- \end{bmatrix} = \begin{bmatrix} \Gamma_{00}(t, x) & \Gamma_{0-}(t, x) \\ \Gamma_{-0}(t, x) & \Gamma_{--}(t, x) \end{bmatrix}. \tag{28}$$

We can associate to each term of P_{jk} in (25), the solution g_{jk} to the parabolic equation

$$w_t + \lambda_j^1 w_x = -\big(c_{jk} I + d_{jk} \big) w_{xx}, \qquad w \in \mathbb{R}^{m_j'}, \tag{29}$$

where $c_{jk} = -\mu_{jk} - i\nu_{jk}$, with $\mu_{jk} > 0$. The solution g_{jk} can be computed explicitly, and we have in any case that for some $c > 0$

$$|g_{jk}(t, x)| \le \frac{\mathcal{O}(1)}{\sqrt{t}} e^{-(x - \lambda_j^1 t)^2/(ct)} \quad \forall k, \ (t, x) \in \mathbb{R}^+ \times \mathbb{R}. \tag{30}$$

The function

$$K(t, x) \doteq \sum_{jk} \begin{bmatrix} r_j \big(g_{jk}(t, x) p_{jk} \big) r_j^T & -r_j \left(\dfrac{d g_{jk}}{dx} p_{jk} \right) r_j^T A_{12} D^{-1} \\ -D^{-1} A_{21} r_j \left(\dfrac{d g_{jk}}{dx} p_{jk} \right) p_{jk} r_j^T & D^{-1} A_{21} r_j \left(\dfrac{d^2 g_{jk}}{dx^2} p_{jk} \right) r_j^T A_{12} D^{-1} \end{bmatrix} \tag{31}$$

collects the principal parts of each component (28) of the Green kernel $\Gamma(t,x)$. Near $z=\infty$, we can associate to each projectors, the Fourier transform of the Green kernel of the transport equation

$$w_t + \lambda_j w_x = (b_{jk} I + \tilde{d}_{jk}) w, \qquad w \in \mathbb{R}^{m_{jk}}. \tag{32}$$

Since $\Re(b_{jk}) \le -c < 0$, we can estimate its solution $\tilde{g}_{jk}$ by

$$|\tilde{g}_{jk}(t,x)| \le C\delta(x-\lambda_j t)e^{-ct} \quad \forall k,\ (t,x) \in \mathbb{R}^+ \times \mathbb{R}. \tag{33}$$

We associate to the kernels $\tilde{g}_{jk}$, the hyperbolic Green function

$$\mathcal{K}(t,x) = \sum_{jk} \delta(x-\lambda_j t) e^{b_{jk}t} e^{t\mathcal{D}_{jk}(\infty)} \mathcal{P}_{jk}(\infty). \tag{34}$$

Theorem 2. *Let $\Gamma(t,x)$ be the Green function of system* (17)*, under the assumptions (H1) and (H2). Let $K(t,x)$ be the Green function of the diffusive operator given by* (31) *and $\mathcal{K}(t,x)$ the Green function of the dissipative transport operator given by* (34)*. Then, we have the decomposition*

$$\Gamma(t,x) = K(t,x)\chi\{\underline{\lambda} t \le x \le \bar{\lambda} t, t \ge 1\} + \mathcal{K}(t,x) + R(t,x)\chi\{\underline{\lambda} t \le x \le \bar{\lambda} t\}, \tag{35}$$

where χ is the characteristic function, and $\bar{\lambda} := \max_j \lambda_j$, $\underline{\lambda} := \min_j \lambda_j$. The matrix $R(t,x)$ can be written as

$$R(t,x) = \sum_j \frac{e^{-(x-\lambda_j^1 t)^2/Ct}}{1+t} \begin{bmatrix} \mathcal{O}(1) & \mathcal{O}(1)(1+t)^{-1/2} \\ \mathcal{O}(1)(1+t)^{-1/2} & \mathcal{O}(1)(1+t)^{-1} \end{bmatrix}. \tag{36}$$

where $R(t,x)$ can for some constant c. Here λ_j^1 are the eigenvalues of the symmetric block A_{11} of the matrix A.

3.2 The Multidimensional Green Function

We want to state an analogous result for multidimensional systems. We consider the Cauchy problem for the linear relaxation system in the conservative–dissipative form

$$w_t + \sum_{\alpha=1}^{m} A_\alpha w_{x_\alpha} = Bw, \qquad w \in \mathbb{R}^{n_1+n_2}, \tag{37}$$

$$w(0,\cdot) = w^0. \tag{38}$$

We assume that A_α, $\alpha = 1,\ldots,m$, are symmetric matrices and we have assumptions (H1) and (H2). Set, for $\xi \in \mathbb{R}^m$,

$$A(\xi) := \sum_{\alpha=1}^{m} \xi_\alpha A_\alpha, \quad E(i\xi) = B - iA(\xi). \tag{39}$$

Let us introduce the polar coordinates in $\mathbb{R}^m$, to say $\xi = \rho\zeta$, with $\rho = |\xi|$ and $\zeta \in S^{m-1}$, and set $E(i\rho, \zeta) = E(i\rho\zeta)$. As before, we want to study the Green kernel $\Gamma(t,x)$ of (37). The solution of the Cauchy problem (37)–(38) is given by $w(t,\cdot) = \Gamma(t,\cdot) * w^0$. Let us study the expansion of $E(z,\zeta) = B - zA(\zeta)$ near $z = 0$. We can use the result of Sect. 3, noting that the matrix A is simply replaced by $A(\zeta)$. We introduce the total projector $P(z,\zeta)$ corresponding to all the eigenvalues near 0, and $P_-(z,\zeta) = I - P(z,\zeta)$ is the projector corresponding to the whole family of the eigenvalues with strictly negative real part. As in (23), we have

$$F(z,\zeta) \doteq L(z,\zeta)E(z,\zeta)R(z,\zeta) = -zA_{11}(\zeta) - z^2 A_{12}(\zeta)D^{-1}A_{21}(\zeta) + \mathcal{O}(z^3) \tag{40}$$

$$F_-(z) \doteq L_-(z,\zeta)E(z,\zeta)R_-(z,\zeta) = D - zA_{22}(\zeta) + \mathcal{O}(z^2). \tag{41}$$

So we make a decomposition of the Green operator:

$$\Gamma(t) = K(t) + \mathcal{K}(t), \tag{42}$$

where, for a given constant a small enough,

$$\widehat{K}(t) \doteq \chi(|\xi| \leq a)R(i\xi)e^{F(i\xi)t}L(i\xi),$$

$$\widehat{\mathcal{K}}(t) \doteq \chi(|\xi| \leq a)R_-(i\xi)e^{F_-(i\xi)t}L_-(i\xi) + \chi(|\xi| > a)\, e^{E(i\xi)t}.$$

Theorem 3. *Consider the linear PDE in the conservative–dissipative form*

$$w_t + \sum_{\alpha=1}^{m} A_\alpha w_{x_\alpha} = Bw, \tag{43}$$

where A_α, B satisfy the assumption (SK) of Definition 1, and let $Q_0 = R_0 L_0$, $Q_- = I - Q_0 = R_- L_-$ be the eigenprojectors on the null space and the negative definite part of B. Then, for any function $w^0 \in L^1 \cap L^2(\mathbb{R}^m, \mathbb{R}^n)$, the solution $w(t) = \Gamma(t)w^0$ of (37), (38) can be decomposed as

$$w(t) = \Gamma(t)w^0 = K(t)w^0 + \mathcal{K}(t)w^0, \tag{44}$$

where the following estimates hold: for any multiindex β and for every $p \in [2, +\infty]$,

$$\begin{aligned} \|L_0 D^\beta K(t)w^0\|_{L^p} \leq{}& C(|\beta|)\min\{1, t^{-\frac{m}{2}(1-\frac{1}{p})-|\beta|/2}\}\|L_0 w^0\|_{L^1} \\ &+ C(|\beta|)\min\{1, t^{-\frac{m}{2}(1-\frac{1}{p})-1/2-|\beta|/2}\}\|L_- w^0\|_{L^1}, \end{aligned} \tag{45}$$

$$\begin{aligned} \|L_- D^\beta K(t)w^0\|_{L^p} \leq{}& C(|\beta|)\min\{1, t^{-\frac{m}{2}(1-\frac{1}{p})-1/2-|\beta|/2}\}\|L_0 w^0\|_{L^1} \\ &+ C(|\beta|)\min\{1, t^{-\frac{m}{2}(1-\frac{1}{p})-1-|\beta|/2}\}\|L_- w^0\|_{L^1}, \end{aligned} \tag{46}$$

$$\|D^\beta \mathcal{K}(t)w^0\|_{L^2} \leq Ce^{-ct}\|D^\beta w^0\|_{L^2}. \tag{47}$$

Example 2. Consider for example the linearized isentropic Euler equations with damping

$$\begin{cases} \rho_t + \mathrm{div} v = 0 \\ \\ v_t + \nabla \rho = -v \end{cases} \tag{48}$$

To fix the ideas, take $m = 3$, $n = 4 = n_1 + n_2 = 1 + 3$. Clearly the system is already in the conservative–dissipative form and condition (SK) is satisfied. We decompose $K(t, x)$ as

$$K(t,x) = \begin{bmatrix} G(t,x) & (\nabla G(t,x))^T \\ \nabla G(t,x) & \nabla^2 G(t,x) \end{bmatrix} + R_1(t,x), \tag{49}$$

where $G(t, x)$ is the heat kernel for $u_t = \Delta u$, and $R_1(t, x)$ satisfies the bound

$$R_1(t,x) = \frac{e^{-c|x|^2/t}}{(1+t)^2} \begin{bmatrix} \mathcal{O}(1) & \mathcal{O}(1)(1+t)^{-1/2} \\ \mathcal{O}(1)(1+t)^{-1/2} & \mathcal{O}(1)(1+t)^{-1} \end{bmatrix}. \tag{50}$$

4 Decay Estimates in the Nonlinear Case and More Accurate Asymptotic Behavior

In this section we study the time decay properties of the global smooth solutions to a nonlinear entropy strictly dissipative relaxation system in conservative–dissipative form.

4.1 Nonlinear Decay Estimates

We now prove the decay estimates in $L^p(\mathbb{R}^m; \mathbb{R}^n)$, $p \in [2, +\infty]$, for the solution u with initial data in $L^1 \cap H^s$, with s sufficiently large, for the nonlinear equation

$$u_t + \sum_{\alpha=1}^{m} (f_\alpha(u))_{x_\alpha} = g(u), \tag{51}$$

with $f_\alpha(0) = g(0) = 0$ and initial condition

$$u(x, 0) = u_0(x). \tag{52}$$

We shall assume that the system (51) is strictly entropy dissipative and condition (SK) is satisfied. Under the assumptions of Theorem 1, we consider the global solution u of (51)–(52).

Theorem 4. *Let $u(t)$ be a smooth global solution to problem* (51), (52). *Let $E_s = \max\Big\{ \|u(0)\|_{L^1}, \|u(0)\|_{H^s} \Big\}$, and assume $E_{[m/2]+2}$ small enough. Let $p \in [2, +\infty]$. The following decay estimate holds*

$$\|D^\beta u(t)\|_{L^p} \le C\min\Big\{1, t^{-\frac{m}{2}(1-\frac{1}{p})-|\beta|/2}\Big\}E_{|\beta|+[m/2]+1}, \tag{53}$$

with $C = C(E_{|\beta|+\sigma})$, *for* σ *large enough. For* $m = 1$, *this estimate holds for* $p \in [1,+\infty]$.

We now study the faster decay of the dissipative variables $u_d(t)$.

Theorem 5. *Under the assumptions of Theorem 4, we have the following decay estimates for the dissipative part of* u*:*

$$\|D^\beta u_d(t)\|_{L^p} \le C\min\Big\{1, t^{-\frac{m}{2}(1-\frac{1}{p})-1/2-|\beta|/2}\Big\}E_{|\beta|+[m/2]+1}, \tag{54}$$

with $C = C(E_{|\beta|+\sigma})$, *for* σ *large enough, and* $p \in [2,+\infty]$. *For* $m = 1$, *this estimate holds for* $p \in [1,+\infty]$.

4.2 Decay to the Linearized Profile

We consider here the difference among the solution of the nonlinear equation (51) and the linearized one

$$u_t + \sum_{\alpha=1}^{m} Df_\alpha(0)u_{x_\alpha} = \begin{pmatrix} 0 \\ D_{u_d}q(0)u_d \end{pmatrix}. \tag{55}$$

We can show that, if the dimension $m \ge 2$, then the solution to the nonlinear equation converges to the linearized solution. Using the linear estimates, and with a special argument for the case $m = 2$, we have the following result. The following result does not hold for $m = 1$.

Theorem 6. *Let* u_l *be the solution of problem* (55), (52), *under the assumptions of Theorem 4, for* $m \ge 2$ *and* $p \in [2,\infty]$, *we have the following decay estimate*

$$\|D^\beta(u(t) - u_l(t))\|_{L^p} \le C\min\Big\{1, t^{-\frac{m}{2}(1-\frac{1}{p})-|\beta|/2-1/2}\Big\}E_{|\beta|+[m/2]+1}, \tag{56}$$

with $C = C(E_{|\beta|+\sigma})$, *for* σ *large enough.*

4.3 The Chapman–Enskog Expansion

Next we show how the solutions to the parabolic Chapman–Enskog expansion approximate the conservative part of the solutions to the nonlinear hyperbolic problem (1). We use the conservative–dissipative decomposition of u to yields

$$u_{c,t} + \sum_{\alpha=1}^{m} A_{\alpha,11}(0)u_{c,x_\alpha} + \sum_{\alpha=1}^{m} A_{\alpha,12}(0)u_{d,x_\alpha} = L_0 \sum_{\alpha=1}^{m} \big(A_\alpha(0)u - f_\alpha(u)\big)_{x_\alpha}; \tag{57}$$

$$u_{d,t} + \sum_{\alpha=1}^{m} A_{\alpha,21}(0)u_{c,x_\alpha} + \textstyle\sum_{\alpha=1}^{m} A_{\alpha,22}(0)u_{d,x_\alpha} = D_{u_d}q(0)u_d$$
$$+ L_- \sum_{\alpha=1}^{m} (A_\alpha(0)u - f_\alpha(u))_{x_\alpha} + (q(u) - D_{u_d}q(0)u_d)\,. \tag{58}$$

Now, consider the linear parabolic equation

$$w_t + \sum_{\alpha=1}^{m} A_{\alpha,11}(0)w_{x_\alpha} + \sum_{\beta=1}^{m}\sum_{\alpha=1}^{m} A_{\alpha,12}(0)(D_{u_d}q(0))^{-1}A_{\beta,21}(0)w_{x_\alpha x_\beta} = 0, \tag{59}$$

and denote by $u_p(t)$ the solution of the weakly parabolic equation (59) with

$$u_p(0) = L_0 u(0). \tag{60}$$

Compute u_d using (58) and decay estimates of Theorems 4 and 5. This yields the following sharper decay result.

Theorem 7. *Let u_p be the solution of problem* (59), (60), *under the assumptions of Theorem 4, for $m \geq 2$ and $p \in [2, \infty]$, we have the following decay estimate*

$$\|D^\beta(u_c(t) - u_p(t))\|_{L^p} \leq C \min\Big\{1, t^{-\frac{m}{2}(1-\frac{1}{p})-|\beta|/2-1/2}\Big\} E_{|\beta|+[m/2]+1}, \tag{61}$$

with $C = C(E_{|\beta|+\sigma})$, for σ large enough.

Example 3. Consider the isentropic dissipative Euler equations

$$\begin{cases} \rho_t + \mathrm{div}(\rho v) = 0, \\ (\rho v)_t + \mathrm{div}(\rho v \otimes v) + \frac{1}{\gamma}\nabla \rho^\gamma = -v. \end{cases} \tag{62}$$

We can linearize the system around the constant state $(\bar{\rho}, \bar{v}) = (1, 0)$, so obtaining system (48) of Example 2. In that case we can immediately apply Theorems 4–7. In particular, by eliminating v in (48), we obtain the estimate

$$\|D^\beta(\rho(t) - \rho_w(t))\|_{L^p} + \|D^\beta(\rho(t) - \rho_p(t))\|_{L^p} \leq C \min\Big\{1, t^{-\frac{m}{2}(1-\frac{1}{p})-|\beta|/2-1/2}\Big\},$$

where ρ_w and ρ_p are, respectively, the solutions of the m-dimensional dissipative wave equation equation and the m-dimensional heat equation

$$\rho_{w,t} + \rho_{w,tt} - \Delta\rho_w = 0, \;\; \rho_{p,t} - \Delta\rho_p = 0.$$

These estimates improve on previous results in [STW03] and [CG04].

The Chapman–Enskog expansion in the case $m=1$. For $m=1$, the decay of u^2 convoluted with the linear kernel and integrated in time gives the same decay estimate of u, so we have to keep a nonlinear term in the diffusive expansion. Set $\tilde{A}=\frac{1}{2}\left(L_0 D^2_{u_c} f(0) - A_{12}(0)(D_{u_d} q(0))^{-1} D^2_{u_c} q(0)\right)$ and $\tilde{B} = A_{12}(0)(D_{u_d} q(0))^{-1} A_{21}(0)$. We rewrite (57) as

$$u_{c,t} + \Big(A_{11}(0)u_c + \tilde{A}(u_c,u_c)\Big)_x + \tilde{B}u_{c,xx} = S_x, \tag{63}$$

where S is some given faster decaying term. We replace (59) by the nonlinear diffusive equation

$$w_t + \Big(A_{11}(0)w + \tilde{A}(w,w)\Big)_x + \tilde{B}w_{xx} = 0. \tag{64}$$

Theorem 8. *For $m=1$, let u_p be the solution of problem (64), (60), under the assumptions of Theorem 4. For $p\in[1,\infty]$, and a fixed $\mu\in[0,1/2)$, if E_1 sufficiently small with respect to $(1/2-\mu)$, then we have*

$$\|D^\beta(u_c(t)-u_p(t))\|_{L^p} \le C\min\Big\{1, t^{-\frac{1}{2}(1-\frac{1}{p})-\mu-\beta/2}\Big\}F_{\beta+4}, \tag{65}$$

where $C=C(\mu,F_{\beta+\sigma})$, for σ large enough, where $F_1=E_1$ and, if $\beta\ge 1$,

$$F_{\beta+1} = \begin{cases} E_{\beta+1}, & \text{if } p\in[2,\infty], \\ E_{\beta+1} + \|D^\beta u(0)\|_{L^1}, & \text{otherwise.} \end{cases}$$

Example 4. The p-system with relaxation. We can apply Theorem 8 to the Example 1. In this case, as already shown in [Ch95], the Chapman–Enskog expansion is given by the semilinear parabolic equation

$$u_{p,t} + h'(0)u_{p,x} + \frac{1}{2}h''(0)(u_p^2)_x - (\lambda^2 - a^2)u_{p,xx} = 0. \tag{66}$$

References

[ABN06] D. Aregba-Driollet, M. Briani, R. Natalini, Asymptotic high-order schemes for dissipative hyperbolic systems, in preparation.

[BRS06] S. Bianchini, B. Hanouzet, R. Natalini, Asymptotic behavior of smooth solutions for partially dissipative hyperbolic systems with a convex entropy, IAC Report 79 (11/2005); to appear in Comm. Pure and Appl. Math.

[CH106] G. Carbou, B. Hanouzet, Comportement semi-linéaire d'un système hyperbolique quasi-linéaire: le modèle de Kerr-Debye. (French) [Semilinear behavior for a quasilinear hyperbolic system: the Kerr-Debye model] C. R. Math. Acad. Sci. Paris **343** (2006), no. 4, 243–247.

[CH206] G. Carbou, B. Hanouzet, Relaxation approximation of some nonlinear Maxwell initial-boundary value problem. Commun. Math. Sci. **4** (2006), no. 2, 331–344.

[CLL94] G.-Q. Chen, C.D. Levermore, and T.-P. Liu, Hyperbolic conservation laws with stiff relaxation terms and entropy, Comm. Pure and Appl. Math. **47** (1994), 787–830.

[Ch95] I.L. Chern, *Long-time effect of relaxation for hyperbolic conservation laws*, Commun. Math. Phys. **172** (1995), 39–55.

[CG04] J.-F. Coulombel and T. Goudon, The strong relaxation limit of the multidimensional isothermal Euler equations, preprint 2004, to appear in Trans. AMS.

[Da00] C.M. Dafermos, Hyperbolic conservation laws in continuum physics, (Springer-Verlag, Berlin, 2000) xvi+443 pp.

[LF71] K.O. Friedrichs and P.D. Lax, Systems of conservation equations with a convex extension, Proc. Nat. Acad. Sci. U.S.A. **68** (1971), 1686–1688.

[Go61] S.K. Godunov, An interesting class of quasi-linear systems, Dokl. Akad. Nauk SSSR **139** (1961), 521–523.

[HH00] B. Hanouzet and Ph. Huynh, Approximation par relaxation d'un système de Maxwell non linéaire, C. R. Acad. Sci. Paris Sér. I Math. **330** (2000), no. 3, 193–198.

[HN03] B. Hanouzet and R. Natalini, Global existence of smooth solutions for partially dissipative hyperbolic systems with a convex entropy, Arch. Ration. Mech. Anal. **169** (2003), no. 2, 89–117.

[Kw87] S. Kawashima, Large-time behaviour of solutions to hyperbolic-parabolic systems of conservation laws and applications, Proc. Roy. Soc. Edinburgh Sect. A **106** (1987), no. 1-2, 169–194.

[Li87] T.-P. Liu, Hyperbolic conservation laws with relaxation, Comm. Math. Phys. **108** (1987), 153–175.

[Na99] R. Natalini, *Recent results on hyperbolic relaxation problems*, Analysis of systems of conservation laws (Aachen, 1997), Chapman & Hall/CRC, Boca Raton, FL, 1999, pp. 128–198.

[Ni78] T. Nishida, Nonlinear hyperbolic equations and related topics in fluid dynamics, Département de Mathématique, Université de Paris-Sud, Orsay, 1978, Publications Mathématiques d'Orsay, No. 78-02.

[RS02] T. Ruggeri and D. Serre, Stability of constant equilibrium state for dissipative balance laws system with a convex entropy. Quart. Appl. Math. **62** (2004), no. 1, 163–179.

[SK85] Y. Shizuta and S. Kawashima, Systems of equations of hyperbolic-parabolic type with applications to the discrete Boltzmann equation, Hokkaido Math. J. **14** (1985), no. 2, 249–275.

[STW03] T. Sideris, B. Thomases, and D. Wang, Decay and singularities of solutions of the three-dimensional Euler equations with damping, Comm. Partial Differential Equations 28 (2003), no. 3-4, 795–816.

[Yo04] W.-A. Yong, Entropy and global existence for hyperbolic balance laws, Arch. Ration. Mech. Anal. **172** (2004), no. 2, 247–266.

[Ze99] Y. Zeng, Gas dynamics in thermal nonequilibrium and general hyperbolic systems with relaxation, Arch. Ration. Mech. Anal. **150** (1999), no. 3, 225–279.

Part II

Invited Lectures

Higher Order Numerical Schemes for Hyperbolic Systems with an Application in Fluid Dynamics

V. Dolejší

Summary. We deal with a numerical solution of the (nonlinear hyperbolic) system of the Euler equations, which describe a motion of inviscid compressible flow. We present a higher order numerical scheme with respect to the space as well as time coordinates. This scheme is based on the discontinuous Galerkin method for the space semi-discretization and the backward difference formula for the time discretization. We employ a suitable linearization of inviscid fluxes and an explicit extrapolation in nonlinear terms, which preserve a high order of accuracy and lead to a linear algebraic problem at each time step. Moreover, we discuss a use of non-reflecting boundary conditions at inflow/outflow parts of boundary and present a stabilization technique that avoid spurious oscillations of numerical solution in vicinity of shock waves. Finally, two numerical examples of unsteady compressible flow demonstrating an efficiency of the scheme are presented.

1 Introduction

Our aim is to developed an efficient, robust, and accurate numerical scheme for a simulation of unsteady compressible flow. It seems to be promising to carry out the space semi-discretization with the aid of the *discontinuous Galerkin method* (DGM), which is based on a piecewise polynomial but discontinuous approximation, for a DGM survey see [CKS00]. It is possible to use a discontinuous approximation also for the time discretization (see [VV02]) but the most usual approach is an application of the method of lines. In this case, the Runge–Kutta methods are very popular for their simplicity and a high order of accuracy, see [BR97], [CKS00], [HH02]. Their disadvantage is a strong restriction to the length of the time step. To avoid this drawback it is possible to use an implicit time discretization, but a fully implicit scheme leads to a necessity to solve a nonlinear algebraic system at each time step, which should be rather expensive. Therefore, we proposed in [DF04b] a semi-implicit method, which is based on a suitable linearization of inviscid fluxes. The linear terms were treated implicitly whereas the nonlinear ones explicitly, which leads to a linear algebraic problem at each time step. In this paper we generalize this approach to a higher order approximation with respect to the time using the *backward difference formulae* (BDF). Furthermore, we present a new type of nonreflecting boundary conditions based on the solution of the local Riemann problem and the new stabilization technique based on the jump indicator.

The contents of the rest of the paper is the following. In Sect. 2 we introduce the considered problem and in Sect. 3 we carried out its discretization with the aid of the higher order semi-implicit DGM and BDF. Sections 4 and 5 are devoted to the presentation of the boundary conditions and the stabilization technique, respectively. Section 6 contains examples of unsteady subsonic as well as transonic flow simulations. A short summary is given in Sect. 7.

2 Problem Formulation

The system of the *Euler equations* describing a 2D motion of an inviscid compressible fluid can be written in the form

$$\frac{\partial \boldsymbol{w}}{\partial t} + \sum_{s=1}^{2} \frac{\partial \boldsymbol{f}_s(\boldsymbol{w})}{\partial x_s} = 0 \quad \text{in } Q_T = \Omega \times (0,T), \tag{1}$$

where $\Omega \subset I\!\!R^2$ is a bounded polygonal domain occupied by gas, $T > 0$ is the length of a time interval,

$$\boldsymbol{w} = (w_1, \dots, w_4)^{\mathrm{T}} = (\rho,\ \rho v_1,\ \rho v_2,\ e)^{\mathrm{T}} \tag{2}$$

is the *state vector* and

$$\boldsymbol{f}_s(\boldsymbol{w}) = (\rho v_s,\ \rho v_s v_1 + \delta_{s1} p,\ \rho v_s v_2 + \delta_{s2} p,\ (e+p)\, v_s)^{\mathrm{T}}, \ \ s = 1,2 \tag{3}$$

are the *inviscid (Euler) fluxes*. We use the following notation: ρ, density; p, pressure; e, total energy; $\boldsymbol{v} = (v_1, v_2)$, velocity; δ_{sk}, Kronecker symbol (if $s = k$, then $\delta_{sk} = 1$, else $\delta_{sk} = 0$). The equation of state implies that

$$p = (\gamma - 1)\,(e - \rho|\boldsymbol{v}|^2/2). \tag{4}$$

Here $\gamma > 1$ is the Poisson adiabatic constant.

The system (1)–(4) is *hyperbolic*. It is equipped with the initial condition

$$\boldsymbol{w}(\boldsymbol{x}, 0) = \boldsymbol{w}^0(x), \quad x \in \Omega, \tag{5}$$

and the boundary conditions

$$B(\boldsymbol{w}) = 0 \quad \text{on } \partial\Omega \times (0,T), \tag{6}$$

chosen in such a way that problem (1)–(6) is linearly well-posed. (See, e.g. [FFS03], Sect. 3.3.6.) To this end, the boundary $\partial\Omega$ is formed by disjoint parts Γ_{IO} and Γ_W representing the inflow/outflow and impermeable walls, respectively. On Γ_W we prescribe the impermeability condition

$$\boldsymbol{v} \cdot \boldsymbol{n} = 0 \quad \text{on } \Gamma_W, \tag{7}$$

where $\boldsymbol{n}$ denotes the unit outer normal to $\partial\Omega$. To determine boundary conditions on Γ_{IO}, we define the matrix $\boldsymbol{P}(\boldsymbol{w}, \boldsymbol{n}) \equiv \sum_{s=1}^{2} \boldsymbol{A}_s(\boldsymbol{w}) n_s$, where $\boldsymbol{n} = (n_1, n_2) \in I\!R^2$, $n_1^2 + n_2^2 = 1$, and

$$\boldsymbol{A}_s(\boldsymbol{w}) \equiv \frac{D\boldsymbol{f}_s(\boldsymbol{w})}{D\boldsymbol{w}}, \quad s = 1, 2 \tag{8}$$

are the Jacobi matrices of the mappings $\boldsymbol{f}_s$. Then we prescribe m_n quantities characterizing the state vector $\boldsymbol{w}$, where m_n is the number of negative eigenvalues of the matrix $\boldsymbol{P}(\boldsymbol{w}, \boldsymbol{n})$ and extrapolate m_p quantities of $\boldsymbol{w}$ from interior of Ω, where $m_p = 4 - m_n$ is the number of nonnegative eigenvalues of $\boldsymbol{P}(\boldsymbol{w}, \boldsymbol{n})$. For details see, e.g., [FFS03].

3 Discretization

3.1 Broken Sobolev Space

Let $\mathcal{T}_h$ $(h > 0)$ denote a triangulation of the closure $\overline{\Omega}$ of the domain Ω into a finite number of closed elements (triangles or quadrilaterals) K with mutually disjoint interiors.

We set $h = \max_{K \in \mathcal{T}_h} \operatorname{diam}(K)$. Let I be a suitable index set such that $\mathcal{T}_h = \{K_i\}_{i \in I}$. If two elements K_i, $K_j \in \mathcal{T}_h$ contain a nonempty open part of their faces, we put $\Gamma_{ij} = \Gamma_{ji} = \partial K_i \cap \partial K_j$. For $i \in I$ we set $s(i) = \{j \in I; \Gamma_{ij} \text{ exists}\}$. The boundary $\partial\Omega$ is formed by a finite number of faces of elements K_i adjacent to $\partial\Omega$. We denote all these boundary faces by S_j, where $j \in I_b$ is a suitable index set and put $\gamma(i) = \{j \in I_b; S_j \text{ is a face of } K_i\}$, $\Gamma_{ij} = S_j$ for $K_i \in \mathcal{T}_h$ such that $S_j \subset \partial K_i$, $j \in I_b$. Further we define two disjoint subsets $\gamma_{IO}(i)$ and $\gamma_W(i)$ corresponding to the boundary parts Γ_{IO} and Γ_W, respectively. Obviously, $\gamma(i) = \gamma_{IO}(i) \cup \gamma_W(i)$. Moreover we put $S(i) = s(i) \cup \gamma(i)$ and $\boldsymbol{n}_{ij} = ((n_{ij})_1, (n_{ij})_2)$ is the unit outer normal to ∂K_i on the face Γ_{ij}.

Over the triangulation $\mathcal{T}_h$ we define the *broken Sobolev space*

$$H^k(\Omega, \mathcal{T}_h) = \{v; v|_K \in H^k(K) \; \forall K \in \mathcal{T}_h\}, \tag{9}$$

where $H^k(K) = W^{k,2}(K)$ denotes the (classical) Sobolev space on element K. For $v \in H^1(\Omega, \mathcal{T}_h)$ we set

$$\begin{aligned} v|_{\Gamma_{ij}} &= \text{trace of } v|_{K_i} \text{ on } \Gamma_{ij}, \\ v|_{\Gamma_{ji}} &= \text{trace of } v|_{K_j} \text{ on } \Gamma_{ji}, \end{aligned} \tag{10}$$

denoting the *traces* of v on $\Gamma_{ij} = \Gamma_{ji}$, which are different in general. Moreover,

$$[v]_{\Gamma_{ij}} = v|_{\Gamma_{ij}} - v|_{\Gamma_{ji}} \tag{11}$$

denotes the jump of function v over the edge Γ_{ij}.

The approximate solution of problem (1)–(6) is sought in the space of vector-valued discontinuous piecewise polynomial functions $\boldsymbol{S}_h$ defined by

$$\boldsymbol{S}_h \equiv [S_h]^4, \quad S_h \equiv \{v; v|_K \in P^p(K)\ \forall K \in \mathcal{T}_h\},$$

where p is a positive integer and $P^p(K)$ denotes the space of all polynomials on K of degree at most p. Obviously, $S_h \subset H^1(\Omega, \mathcal{T})$.

3.2 Space Semi-Discretization

To derive the discrete problem, we multiply (1) by a test function $\boldsymbol{\varphi} \in [H^1(\Omega, \mathcal{T}_h)]^4$, integrate over any element K_i, $i \in I$, apply Green's theorem and sum over all $i \in I$. In this way we obtain the integral identity

$$\frac{\partial}{\partial t} \sum_{K_i \in \mathcal{T}_h} \int_{K_i} \boldsymbol{w} \cdot \boldsymbol{\varphi} \, \mathrm{d}x \tag{12}$$

$$= \sum_{K_i \in \mathcal{T}_h} \int_{K_i} \sum_{s=1}^{2} \boldsymbol{f}_s(\boldsymbol{w}) \cdot \frac{\partial \boldsymbol{\varphi}}{\partial x_s} \, \mathrm{d}x - \sum_{K_i \in \mathcal{T}_h} \sum_{j \in S(i)} \int_{\Gamma_{ij}} \sum_{s=1}^{2} \boldsymbol{f}_s(\boldsymbol{w}) \cdot \boldsymbol{\varphi} \, (n_{ij})_s \, \mathrm{d}S,$$

which represents a weak form of the Euler equations in the sense of the broken Sobolev space $H^1(\Omega, \mathcal{T}_h)$ defined by (9).

Now we shall introduce the discrete problem approximating identity (12) with the aid of DGM. To evaluate the boundary integrals in (12) we use the approximation

$$\int_{\Gamma_{ij}} \sum_{s=1}^{2} \boldsymbol{f}_s(\boldsymbol{w}(t)) \, (n_{ij})_s \cdot \boldsymbol{\varphi} \, \mathrm{d}S \approx \int_{\Gamma_{ij}} \boldsymbol{H}(\boldsymbol{w}(t)|_{\Gamma_{ij}}, \boldsymbol{w}(t)|_{\Gamma_{ji}}, \boldsymbol{n}_{ij}) \cdot \boldsymbol{\varphi} \, \mathrm{d}S, \tag{13}$$

where $\boldsymbol{H}$ is a *numerical flux*, $\boldsymbol{w}(t)|_{\Gamma_{ij}}$ and $\boldsymbol{w}(t)|_{\Gamma_{ji}}$ are the values of $\boldsymbol{w}$ on Γ_{ij} considered from the interior and the exterior of K_i, respectively, and at time t. If $j \in \gamma_W(i)$, then we use the standard approach when the pressure is extrapolated on Γ_{ij} from K_i, for details see, e.g., [FFS03]. The case $j \in \gamma_{IO}$ is discussed in the Sect. 4.

For $\boldsymbol{w}_h, \boldsymbol{\varphi}_h \in \boldsymbol{S}_h$ we introduce the forms

$$(\boldsymbol{w}_h, \boldsymbol{\varphi}_h) = \int_{\Omega} \boldsymbol{w}_h(\boldsymbol{x}) \cdot \boldsymbol{\varphi}_h(\boldsymbol{x}) \, \mathrm{d}\boldsymbol{x}, \tag{14}$$

$$\tilde{\boldsymbol{b}}_h(\boldsymbol{w}_h, \boldsymbol{\varphi}_h) = \sum_{K_i \in \mathcal{T}_h} \Bigg(\sum_{j \in S(i)} \int_{\Gamma_{ij}} \boldsymbol{H}(\boldsymbol{w}_h|_{\Gamma_{ij}}, \boldsymbol{w}_h|_{\Gamma_{ji}}, \boldsymbol{n}_{ij}) \cdot \boldsymbol{\varphi}_h \mathrm{d}S$$

$$- \int_{K_i} \sum_{s=1}^{2} \boldsymbol{f}_s(\boldsymbol{w}_h) \cdot \frac{\partial \boldsymbol{\varphi}_h}{\partial x_s} \, \mathrm{d}\boldsymbol{x} \Bigg).$$

Now we can introduce the *semi-discrete problem*.

Definition 1. *Function* $\boldsymbol{w}_h$ *is a* **semi-discrete solution** *of the problem* (1)–(6), *if*

$$\begin{aligned} &(a)\ \ \boldsymbol{w}_h \in C^1([0,T]; \boldsymbol{S}_h) \\ &(b)\ \ \left(\frac{\partial \boldsymbol{w}_h(t)}{\partial t}, \boldsymbol{\varphi}_h\right) + \tilde{\boldsymbol{b}}_h(\boldsymbol{w}_h(t), \boldsymbol{\varphi}_h) = 0 \quad \forall \boldsymbol{\varphi}_h \in \boldsymbol{S}_h \ \forall t \in (0,T) \\ &(c)\ \ \boldsymbol{w}_h(0) = \boldsymbol{w}_h^0, \end{aligned} \tag{15}$$

where $\boldsymbol{w}_h^0 \in \boldsymbol{S}_h$ *denotes an* S_h*-approximation of the initial condition* $\boldsymbol{w}^0$ *from (5).*

Here $C^1([0,T]; \boldsymbol{S}_h)$ is the space of continuously differentiable mappings of the interval $[0,T]$ into $\boldsymbol{S}_h$.

3.3 Full Space–Time Discretization

The problem (15), (a)–(c) exhibits a system of ordinary differential equations (ODE) for $\boldsymbol{w}_h(t)$, which has to be discretized by a suitable method. We follow the approach presented in [DF04b] where a semi-implicit discretization of (15), (a)–(c) was presented. We define a suitable linearization of $\tilde{\boldsymbol{b}}_h(\cdot,\cdot)$ by the form

$$\boldsymbol{b}_h(\boldsymbol{u}_h^1, \boldsymbol{u}_h^2, \boldsymbol{u}_h^3), \quad \boldsymbol{u}_h^1, \boldsymbol{u}_h^2, \boldsymbol{u}_h^3 \in \boldsymbol{S}_h, \tag{16}$$

which is linear with respect to its second and third argument and it is consistent with the form $\tilde{\boldsymbol{b}}_h(\cdot,\cdot)$ in the following way

$$\boldsymbol{b}_h(\boldsymbol{u}_h, \boldsymbol{u}_h, \boldsymbol{\varphi}_h) = \tilde{\boldsymbol{b}}_h(\boldsymbol{u}_h, \boldsymbol{\varphi}_h) \quad \forall \boldsymbol{u}_h, \boldsymbol{\varphi}_h \in \boldsymbol{S}_h. \tag{17}$$

This linearization is based on the homogeneity property of the Euler fluxes $\boldsymbol{f}_s$, $s = 1, 2$, and a suitable choice of the numerical flux $\boldsymbol{H}(\cdot,\cdot,\cdot)$, for detail we refer to the original paper [DF04b].

The main idea of the semi-implicit discretization is to treat the linear part of $\boldsymbol{b}_h$ (represented by its second argument) implicitly and the nonlinear part of $\boldsymbol{b}_h$ (represented by its first argument) explicitly. To obtain a sufficiently accurate approximation with respect to the time coordinate we use the so-called *backward difference formula* (BDF) for the solution of the ODE problem (15), (a)–(c). Moreover, for the nonlinear part of $\boldsymbol{b}_h(\cdot,\cdot,\cdot)$ we employ a suitable explicit higher order extrapolation that preserve a given order of accuracy and does not destroy the linearity of the problem at each time level.

Let $0 = t_0 < t_1 < \cdots < t_r = T$ be a partition of the interval $(0,T)$ and $\tau_k \equiv t_{k+1} - t_k$, $k = 0, 1, \ldots, r-1$.

Definition 2. *We define the* **approximate solution** *of problem (1)–(6) as functions* $\boldsymbol{w}_h^k$, $k = 1, \ldots, r$, *satisfying the conditions*

Table 1. Values of coefficients α_l and β_l for constant time step

n	$\alpha_l,\ l=0,\dots,n$	$\beta_l,\ l=1,\dots,n$
1	$1,\ -1$	1
2	$\frac{3}{2},\ -2,\ \frac{1}{2}$	$2,\ -1$
3	$\frac{11}{6},\ -3,\ \frac{3}{2},\ -\frac{1}{3}$	$3,\ -3,\ \ 1$

$$\text{(a)}\ \ \boldsymbol{w}_h^{k+1} \in \boldsymbol{S}_h \tag{18}$$

$$\text{(b)}\ \ \frac{1}{\tau_k}\left(\sum_{l=0}^{n} \alpha_l \boldsymbol{w}_h^{k+1-l}, \boldsymbol{\varphi}_h\right) + \boldsymbol{b}_h\left(\sum_{l=1}^{n} \beta_l \boldsymbol{w}_h^{k+1-l}, \boldsymbol{w}_h^{k+1}, \boldsymbol{\varphi}_h\right) = 0$$

$$\forall\, \boldsymbol{\varphi}_h \in \boldsymbol{S}_h,\ k = n-1,\dots,r-1$$

(c) $\boldsymbol{w}_h^0$ *is* S_h *approximation of* $\boldsymbol{w}^0$

(d) $\boldsymbol{w}_h^l \in \boldsymbol{S}_h,\ l = 1,\dots,n-1$ *are given by a suitable one-step method*

where $n \geq 1$ *is the degree of the BDF scheme, the coefficients* $\alpha_l,\ l = 0,\dots,n$ *and* $\beta_l,\ l = 1,\dots,n$ *depend on time steps* $\tau_{k-l},\ l = 0,\dots,n$.

The relations for the coefficients $\alpha_l,\ l = 0,\dots,n$ and $\beta_l,\ l = 1,\dots,n$ were derived in [Dol05] for $n = 1, 2, 3$ and their values for constant time step $\tau_k = \tau$, $k = 1,\dots,r$ are given in Table 1. The problem (18), (a)–(d) represents a system of linear algebraic equations for each $k = n-1,\dots,r-1$, which is solved by a suitable iterative solver (e.g., GMRES method).

Based on numerical experiments carried out for a scalar nonlinear convection–diffusion equation in [Dol05] we suppose that the orders of convergence of scheme (18), (a)–(d) are

$$O(h^{\bar{p}} + \tau^n), \quad \bar{p} = \begin{cases} p & \text{for } p \text{ even} \\ p+1 & \text{for } p \text{ odd} \end{cases} \tag{19}$$

in $L^\infty((0,T); L^2(\Omega))$-norm and

$$O(h^p + \tau^n) \tag{20}$$

in $L^\infty((0,T); H^1(\Omega))$ for a sufficiently regular exact (weak) solution.

4 Inflow/Outflow Boundary Conditions

If $\Gamma_{ij} \subset \Gamma_{IO}$ then it is necessary to specify the boundary state vector $\boldsymbol{w}|_{\Gamma_{ji}}$, which is an argument of the numerical flux $\boldsymbol{H}$ in (14). In virtue of the end of Sect. 2 it is necessary to prescribe m_n quantities and the others ($m_p = 4 - m_n$) have to be extrapolated. There is $m_n = 3$ for a subsonic inflow and $m_n = 1$

for subsonic outflow. The usual approach is to extrapolate the pressure and prescribe the other quantities at the subsonic inflow and to prescribe the pressure and extrapolated other quantities at the subsonic outflow. Although this approach gives satisfactory results for transonic flow regimes (see e.g., [FFS03]), its application to subsonic and low speed flows does not give reasonable results. In this case some unphysical reflections of waves from an artificial inlet/outlet part of the boundary are observed. Therefore, it is necessary to use *nonreflecting boundary conditions* transparent for acoustic effects coming from inside of Ω.

In [DFK04] we presented the nonreflecting boundary conditions based on a solution of the *local linearized Riemann problem* considered on each boundary edge. Although this approach gives a very satisfactory result, we observed some small instabilities near far-field boundary for some flow regimes and some types of numerical approximations. Here we propose a new type of boundary condition (based on the solution of the *local nonlinearized Riemann problem*) which do not suffer from the mentioned drawback.

Let $\Gamma_{ij} \subset \Gamma_{IO}$, $\boldsymbol{w}_{ij} = \boldsymbol{w}|_{\Gamma_{ij}}$, and $\boldsymbol{w}_{BC}$ be a prescribed boundary state at the inflow or outflow. Using the rotational invariance, we transform the Euler equations (1) to the coordinates $\tilde{x}_1$, parallel with the normal direction $\boldsymbol{n}$ to Γ_{ij}, and $\tilde{x}_2$, tangential and neglect the derivative with respect to $\tilde{x}_2$. Then we obtain the following problem

$$\frac{\partial \boldsymbol{q}}{\partial t} + \boldsymbol{A}_1(\boldsymbol{q})\frac{\partial \boldsymbol{q}}{\partial \tilde{x}_1} = 0, \tag{21}$$

for the transformed vector-valued function $\boldsymbol{q} = \boldsymbol{Q}(\boldsymbol{n}_{ij})\boldsymbol{w}$, considered in the set $(-\infty, \infty) \times (0, \infty)$ and equipped with the initial condition

$$\begin{aligned} \boldsymbol{q}(\tilde{x}_1, 0) &= \boldsymbol{q}_{ij}, \quad \tilde{x}_1 < 0, \\ \boldsymbol{q}(\tilde{x}_1, 0) &= \boldsymbol{q}_{BC}, \quad \tilde{x}_1 > 0, \end{aligned} \tag{22}$$

where $\boldsymbol{q}_{BC} = \boldsymbol{Q}(\boldsymbol{n}_{ij})\boldsymbol{w}_{BC}$. Here $\boldsymbol{Q}(\boldsymbol{n}_{ij})$ denotes the 4×4 transformation matrix. The solution of the problem (21)–(22) can be found numerically by a suitable iterative process, for more details see, e.g., [Tor97]. The sought boundary state $\boldsymbol{w}|_{\Gamma_{ji}}$ is finally defined by

$$\boldsymbol{w}|_{\Gamma_{ji}} = \boldsymbol{Q}^{-1}(\boldsymbol{n}_{ij})\boldsymbol{q}(\tilde{x}_1 = 0), \tag{23}$$

where $\boldsymbol{q}(\tilde{x}_1 = 0)$ is the solution of (21)–(22) at $\tilde{x}_1 = 0$ and $t > 0$.

5 Stabilization

Application of the numerical scheme (18), (a)–(d) to transonic flow leads to spurious overshoots and undershoots in computed quantities near shock waves. We present a stabilization technique that combines ideas of the artificial viscosity approach first introduced in [JJS95] and the jump indicator presented in [DFS03].

Similarly as in [DFS03], we define the quantity

$$g_{K_i}(\boldsymbol{w}_h) \equiv \frac{\sum_{j\in s(i)} \int_{\Gamma_{ij}} [w_{h,1}]^2 \,\mathrm{d}S}{|K_i|^{3/4} \sum_{j\in s(i)} |\Gamma_{ij}|}, \quad K_i \in \mathcal{T}_h, \tag{24}$$

where $w_{h,1}$ denotes the first component (density) of $\boldsymbol{w}_h \in \boldsymbol{S}_h$, $[\cdot]$ is the interelement jump defined by (11), $|\Gamma_{ij}|$ is the length of Γ_{ij}, and $|K_i|$ is the area of element K_i. This parameter measures the interelement jump of the piecewise polynomial function $w_{h,1}$. Furthermore, we put

$$g_{\Gamma_{ij}}(\boldsymbol{w}_h) \equiv \frac{1}{2}\left(g_{K_i}(\boldsymbol{w}_h) + g_{K_j}(\boldsymbol{w}_h)\right), \quad j \in s(i),\ i \in I. \tag{25}$$

Moreover, we define the forms

$$d_h(\boldsymbol{u}_h, \bar{\boldsymbol{u}}_h, \boldsymbol{\varphi}_h) \equiv \sum_{i\in I} h_{K_i} g_{K_i}(\boldsymbol{u}_h) \int_{K_i} \nabla \bar{\boldsymbol{u}}_h^2 \cdot \nabla \boldsymbol{\varphi}_h \,\mathrm{d}x \tag{26}$$

and

$$J_h(\boldsymbol{u}_h, \bar{\boldsymbol{u}}_h, \boldsymbol{\varphi}_h) \equiv \sum_{i\in I} \sum_{j\in s(i)} \frac{g_{\Gamma_{ij}}(\boldsymbol{u}_h)}{|\Gamma_{ij}|} \int_{\Gamma_{ij}} [\bar{\boldsymbol{u}}_h] \cdot [\boldsymbol{\varphi}_h] \,\mathrm{d}S, \tag{27}$$

where $\boldsymbol{u}_h, \bar{\boldsymbol{u}}_h, \boldsymbol{\varphi}_h \in \boldsymbol{S}_h$.

Our proposed stabilization technique is performed by adding the term

$$d_h\left(\sum_{l=1}^{n} \beta_l \boldsymbol{w}_h^{k+1-l}, \boldsymbol{w}_k^{k+1}, \boldsymbol{\varphi}_h\right) + J_h\left(\sum_{l=1}^{n} \beta_l \boldsymbol{w}_h^{k+1-l}, \boldsymbol{w}_k^{k+1}, \boldsymbol{\varphi}_h\right) \tag{28}$$

to the left-hand-side of (18), (c). The form d_h exhibits the so-called artificial viscosity (see [JJS95]) and the form J_h the interior penalty. Both forms vanish in region where $\boldsymbol{w}_h$ is smooth since $g(\boldsymbol{w}_h)$ is also vanishing.

6 Numerical Example

6.1 Subsonic Unsteady Flow

We present an example of an *unsteady subsonic flow* through the GAMM channel, see Fig. 1, left. As an initial condition we take a steady state solution for the inflow Mach number $M = 0.5$ (see Fig. 1, right showing the isolines of pressure), which was obtained using the boundary condition described in Sect. 4 with $\boldsymbol{w}_{BC} = (1, 1, 0, 7.642857)$ at the inflow and outflow parts of the channel, which using (4) gives the inflow/output pressure $p_0 = 2.857142857$.

Fig. 1. GAMM channel, used triangulation (*left*) and steady-state solution – isolines of pressure (*right*)

At $t = 0$ we start to periodically modify the pressure at outflow of the channel according the formulae

$$p = p_0(1 + 0.1\sin(\pi\, t)) \tag{29}$$

and consequently (using (4)) the value of $\boldsymbol{w}_{BC}$ at the outflow. Figure 2 shows the isolines of pressure at several time instants. These results were achieved on relatively coarse triangular grid having 598 triangles (Fig. 1, left) with piecewise cubic polynomial approximation (i.e., $p = 3$ in (12)) and the third order time discretization (i.e., $n = 3$ in (18)). We observe a periodical propagation of pressure waves from the right to the left of the channel and also any reflection of these waves from the inflow, which verifies the correctness of the boundary conditions presented in Sect. 4.

6.2 Transonic Unsteady Flow

We deal with a numerical simulation of an *unsteady transonic flow* through this channel. We take a steady state solution for the inflow Mach number $M = 0.67$ as the initial condition, (see Fig. 3, right showing the isolines of pressure). The steady-state solution was obtained using the boundary conditions described in Sect. 4 with $\boldsymbol{w}_{BC} = (1, 1, 0, 4.377978)$ at the inflow and outflow parts of the channel, which gives using (4) the inflow/output pressure $p_0 = 1.5911912$. At $t = 0$ we start again to periodically modify the pressure at outflow of the channel according (29) with $p_0 = 1.5911912$.

Figure 4 shows the isolines of pressure at several time instants. These results were achieved on relatively coarse triangular grid having 1,683 triangles (Fig. 3, left), which was refined with the aid of the anisotropic adaptation technique from [DF04a]. We use piecewise cubic polynomial approximation (i.e., $p = 3$ in (12)) and the third order time discretization (i.e., $n = 3$ in (18)). We observe a periodical propagation of pressure waves from the right to the left of the channel without any instabilities of the numerical solution, thanks to the stabilization technique from Sect. 5.

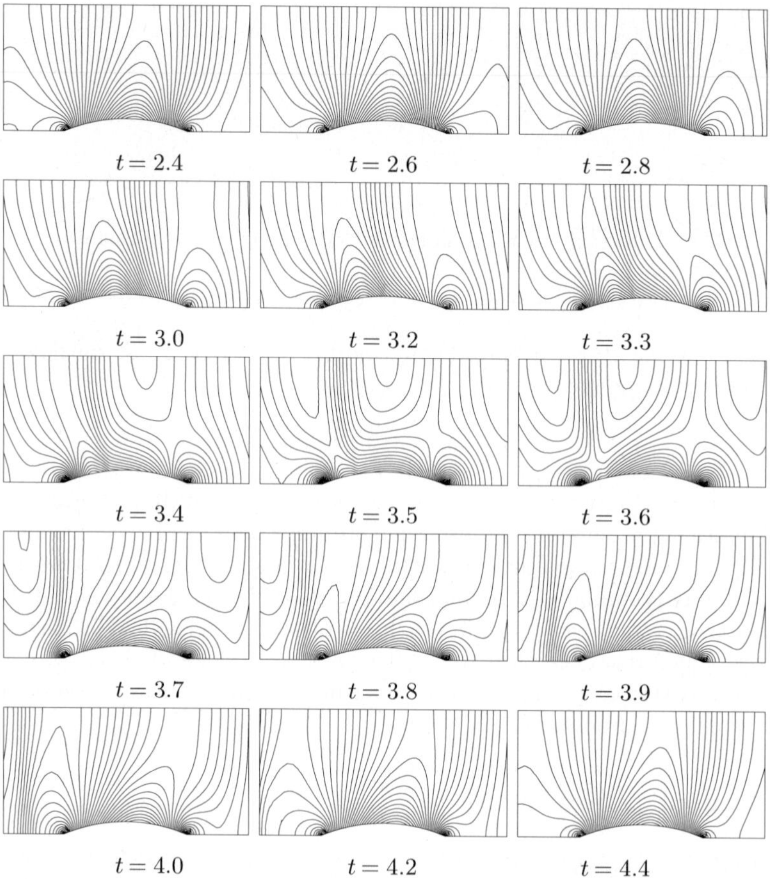

Fig. 2. GAMM channel with oscillating outflow pressure, isolines of pressure at several time instants

Fig. 3. GAMM channel, used triangulation (*left*) and steady-state solution – isolines of pressure (*right*)

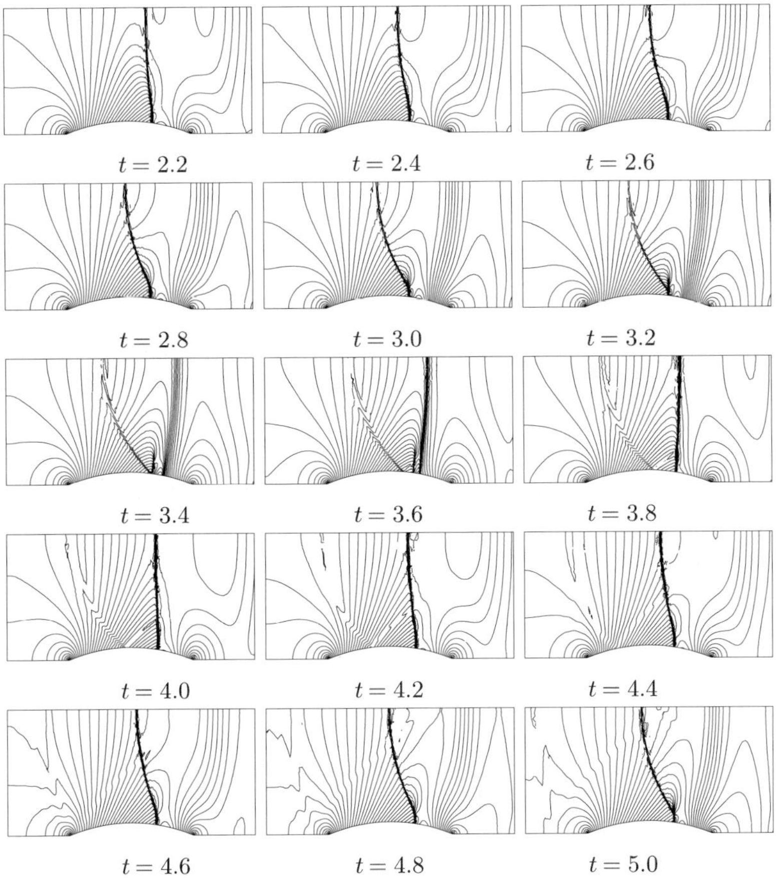

Fig. 4. Transonic flow through the GAMM channel with oscillating outflow pressure, isolines of pressure at several time instants

7 Conclusion

We numerically solved the hyperbolic system of the Euler equations by the higher order semi-implicit scheme, which is based on the combination of the discontinuous Galerkin method for the space semi-discretization and the backward difference formula for the time discretization. A suitable choice of the boundary conditions and the stabilization technique allow efficient simulations of subsonic as well as transonic flow.

Acknowledgments

This work is a part of the research project MSM 0021620839 financed by the Ministry of Education of the Czech Republic and it was partly supported by the Grant No. 316/2006/B-MAT/MFF of the Grant Agency of the Charles University Prague.

References

[BR97] Bassi, F., Rebay, S.: High-order accurate discontinuous finite element solution of the 2D Euler equations. J. Comput. Phys., **138**, 251–285 (1997)

[CKS00] Cockburn, B., Karniadakis, G.E., (eds.), C.-W.S.: Discontinuous Galerkin methods. In Lecture Notes in Computational Science and Engineering 11. Springer, Berlin (2000)

[Dol05] Dolejší, V.: Higher order semi-implicit discontinuous Galerkin finite element schemes for compressible flow simulation. In Software and Algorithms of Numerical Mathematics ((submitted) 2005).

[DF04a] Dolejší, V., Felcman, J.: Anisotropic mesh adaptation and its application for scalar diffusion equations. Numer. Methods Partial Differ. Equations, **20**, 576–608 (2004)

[DF04b] Dolejší, V., Feistauer, M.: Semi-implicit discontinuous Galerkin finite element method for the numerical solution of inviscid compressible flow. J. Comput. Phys., **198**, 727–746 (2004)

[DFS03] Dolejší, V., Feistauer, M., Schwab, C.: On some aspects of the discontinuous Galerkin finite element method for conservation laws. Math. Comput. Simul., **61**, 333–346 (2003)

[DFK04] Feistauer, M., Dolejší, V., Kučera, V.: On a semi-implicit discontinuous Galerkin FEM for the nonstationary compressible Euler equations. In: Asukara, F., Aiso, H., Kawashima, S., Matsumura, A., Nishibata, S., Nishihara, K. (eds.) Hyperbolic Probelms: Theory Numerics and Applications (Tenth international conference in Osaka 2004), Yokohama Publishers, Inc., 391–398 (2006)

[FFS03] Feistauer, M., Felcman, J., Straškraba, I.: Mathematical and Computational Methods for Compressible Flow. Oxford University Press, Oxford (2003)

[HH02] Hartmann, R., Houston, P.: Adaptive discontinuous Galerkin finite element methods for the compressible Euler equations. J. Comput. Phys., **183**, 508–532 (2002)

[JJS95] Jaffre, J., Johnson, C., Szepessy, A.: Convergence of the discontinuous Galerkin finite element method for hyperbolic conservation laws. Math. Models Methods Appl. Sci, **5**, 367–286 (1995)

[Tor97] Toro, E.F.: Riemann Solvers and Numerical Methods for Fluid Dynamics. Springer-Verlag, Berlin (1997)

[VV02] van der Vegt, J.J.W., van der Ven, H.: Space-time discontinuous Galerkin finite element method with dynamic grid motion for inviscid compressible flows. I: General formulation. J. Comput. Phys., **182**, 546–585 (2002)

A Penalization Technique for the Efficient Computation of Compressible Fluid Flow with Obstacles

G. Chiavassa and R. Donat

1 Introduction

The simulation of moderate to high Reynolds number flows around solid obstacles of arbitrary shape is one of the most important problems in aerodynamics, from a practical point of view. Flows over the wings and fuselage of airplanes, off-shore drilling or wind engineering of buildings are relevant examples of situations where this simulations could be of interest.

The presence of obstacles in the flow is a complex issue that has been handled by a variety of techniques, from coordinate transformations to body fitted grids and fictitious domain approaches [GPWZ96, KPAC00].

In recent years a volume penalization technique introduced by Arquis and Caltagirone [AC84] has been considered by various authors in numerical simulations involving solid obstacles in incompressible flows. The physical idea of the penalization technique in [AC84] is to model the obstacle as a porous medium with porosity tending to zero. For incompressible flows, it leads to a Brinkman-type model with variable permeability, where the fluid domain has a large permeability in front of one of the porous medium.

Penalization methods were developed in order to be able to avoid the use of boundary fitted grids, and other nonstandard adjustments. The penalized system is solved in an obstacle-free domain, hence fast and effective methods for Cartesian grids can be used. Different numerical simulations of incompressible viscous flows using adaptive wavelet methods, pseudospectral methods, or finite difference/volume methods (see, e.g., [ABF99, KPAC00, PCLS, SF02, KV05] and references therein), have put in evidence the efficiency of this technique for incompressible flow simulations.

Compressible flow simulations need numerical schemes that can provide accurate numerical solutions in the presence of shocks, which should be represented by sharp numerical profiles. For compressible viscous flows at high Reynolds numbers, an adequate numerical treatment of shock waves can easily be achieved by using the high-resolution shock capturing (HRSC henceforth) technology in the discretization of the convective fluxes.

We propose to apply the HRSC schemes developed by Shu and Osher in [SO89], which have been extensively used in the literature. It is well known, however, that these schemes need to be implemented on uniform Cartesian grids (or smooth deviations of them). In order to avoid the computational difficulties associated with the implementation of no-slip boundary conditions on obstacles in Cartesian meshes, we propose a Brinkman-type penalization method for the numerical simulation of *compressible* Navier–Stokes flows. Following the guidelines of the incompressible case, a penalization term is added to the momentum and energy equations with the aim of enforcing the boundary conditions on the obstacle. Dirichlet boundary conditions on the temperature can be handled similarly.

We show, via a formal expansion on the penalization parameter, that the solution of the penalized system should satisfy the prescribed boundary conditions up to first-order terms. There is no theoretical proof of the convergence of this formal expansion, however our computational experience indicates that the penalization technique provides numerical results which are consistent with the physical characteristics of the flow, and the computed solutions comply with the prescribed boundary conditions. Thus, as in the incompressible case, this technique could be potentially very useful in a wide variety of applications involving complex industrial flows.

The paper is organized as follows. In Sect. 2 we recall the physical setting and the governing equations. The penalization method is described in Sect. 3 and in Sect. 4 we give specific details on the numerical technique we use in the simulations. In Sect. 5 we display several numerical experiments that show that the computed solution does indeed satisfy the no-slip boundary conditions on the velocity and the Dirichlet condition on the temperature. Finally, we draw some conclusions in Sect. 6.

2 Physical Setting

Let Ω be a regular open set in $\mathbb{R}^2$ (the computational domain) containing N fixed regular obstacles Ω_s^n, $1 \le n \le N$. We set

$$\Omega_s = \bigcup_{n=1}^{N} \Omega_s^n, \qquad \Omega_f = \Omega \backslash \overline{\Omega}_s,$$

such that $\overline{\Omega}_s$ is the region occupied by the solid bodies and Ω_f is the compressible-fluid domain.

Our initial boundary value problem is the following:

$$\begin{cases} \partial_t \vec{U} + \nabla \vec{\mathcal{F}}(\vec{U}) = \dfrac{1}{Re} \nabla \vec{\mathcal{F}}_V(\vec{U}), & (X,t) \in \Omega_f \times \mathbb{R}^+, \\ \vec{U}(X, t=0) = \vec{U_0}(X) & \text{in } \Omega_f, \\ \text{boundary conditions} & \text{on } \Gamma_s = \partial \Omega_s. \end{cases} \tag{1}$$

The compressible Navier–Stokes equations[1] hold on the fluid domain. As usual, $\overrightarrow{U} = (\rho, \rho u, \rho v, E)^T$ is the vector of dimensionless conservative variables, ρ being the density of the fluid, u and v the components of fluid velocity, and E the total fluid energy. The Reynolds number is defined as usual through a reference velocity u_0, density ρ_0, length l_0, and viscosity μ_0 by $Re = \frac{\rho_0 l_0 u_0}{\mu_0}$. The convective flux-vector $\overrightarrow{\mathcal{F}} = (f(\overrightarrow{U}), g(\overrightarrow{U}))$, is defined as

$$
\begin{aligned}
f(\overrightarrow{U}) &= (\rho u, \rho u^2 + p, \rho uv, (E+p)u)^T \\
g(\overrightarrow{U}) &= (\rho v, \rho uv, \rho v^2 + p, (E+p)v)^T,
\end{aligned} \tag{2}
$$

while the viscous flux-vector $\overrightarrow{\mathcal{F}}_V = (f_V(\overrightarrow{U}), g_V(\overrightarrow{U}))$ is defined by

$$
f_V(\overrightarrow{U}) = \left(0, \tau_{xx}, \tau_{xy} u\tau_{xx} + v\tau_{xy} + \frac{\gamma\mu}{Pr} e_x\right)^T \tag{3}
$$

$$
g_V(\overrightarrow{U}) = \left(0, \tau_{yx}, \tau_{yy}, u\tau_{yx} + v\tau_{yy} + \frac{\gamma\mu}{Pr} e_y\right)^T, \tag{4}
$$

where τ is the stress tensor, with the usual definitions corresponding to a Newtonian fluid (see, e.g., [A82] for specific details), $\gamma = \frac{C_p}{C_v}$ is the constant specific heat ratio and Pr is the Prandtl number.

System (1) is closed by the equation of state for an ideal polytropic gas:

$$
p = (\gamma - 1)\rho e, \tag{5}
$$

where e stands for the internal fluid energy, linked to the temperature T and to the total energy E by the relations

$$
e = C_v T, \qquad E = \rho e + \frac{1}{2}\rho(u^2 + v^2). \tag{6}
$$

The prescribed boundary conditions are as follows: On the exterior boundary of the computational domain Ω, outflow or inflow boundary conditions, to be described in Sect. 5, are imposed. On the boundary of each solid body Ω_s^n, denoted by Γ_s^n, a no-slip condition is imposed on the velocity field:

$$
u_{|\Gamma_s^n} = v_{|\Gamma_s^n} = 0. \tag{7}
$$

and a Dirichlet boundary condition for the temperature: The temperature of the solid body is fixed, $T_{\Omega_s^n}$, hence the wall temperature is also fixed

$$
T_{|\Gamma_s^n} = T_{\Omega_s^n}. \tag{8}
$$

[1] ∇ is the spatial divergence operator.

3 The Penalization Method

The penalization technique proposed in [AC84] for incompressible flow considers the entire domain Ω as a fluid domain. An additional term is introduced in the governing equations that is activated only at the solid body in order to enforce the natural no-slip boundary conditions. Assuming χ_{Ω_s} to be the characteristic (or mask) function of the solid domain Ω_s, in this so-called Brinkman-penalization technique, the term

$$\frac{1}{\eta}\,\chi_{\Omega_s}\begin{pmatrix} u \\ v \end{pmatrix} \tag{9}$$

is added to the incompressible Navier–Stokes equations. The resulting system is then solved in the entire computational domain.

The analysis carried out in [ABF99] shows that the solution of the penalized equations converges to the exact solution of the incompressible Navier–Stokes system in the limit as the penalization parameter tends to zero. The main advantage of this penalization technique, compared with other penalization methods, is that the error can be estimated rigorously in terms of the penalization parameter.

In the case of incompressible flows, this so-called Brinkman-penalization technique has been applied successfully in many configurations (fixed and moving obstacles) (see, e.g., [ABF99, KPAC00, SF02, KV05] and references therein).

For compressible Navier–Stokes flow the boundary condition includes also the temperature (8). Following the same ideas we propose a natural penalization technique for system (1): We consider the whole computational domain Ω as an *extended fluid* domain and define a new *penalized problem* as

$$\begin{cases} \partial_t \vec{U}_\eta + \nabla\vec{\mathcal{F}}(\vec{U}_\eta) + \dfrac{1}{\eta}\,\chi_{\Omega_s}\begin{pmatrix} 0 \\ \rho_\eta u_\eta \\ \rho_\eta v_\eta \\ E_\eta - E_{\Omega_s} \end{pmatrix} = \dfrac{1}{Re}\nabla\vec{\mathcal{F}}_V(\vec{U}_\eta), \quad (X,t) \in \Omega \times I\!R^+ \\ \vec{U}_\eta(X, t=0) = \vec{U_0}(X) \ \text{ in } \Omega, \\ + \text{ boundary conditions on } \partial\Omega, \end{cases} \tag{10}$$

where $\vec{U}_\eta$ represents the vector of unknowns, depending now on the parameter η. E_{Ω_s} is the energy of the solid body Ω_s and it is computed as

$$E_{\Omega_s} = \rho C_v T_{\Omega_s},$$

where T_{Ω_s} is the fixed temperature of the solid body.

At $t = 0$, we need to prescribe also the conserved quantities on the solid body. Here, and in our numerical experiments, we shall assume constant density and pressure around the solid body at $t = 0$, and use these values in the extension to the interior of the solid.

As in [ABF99], we can carry out a formal analysis based on an expansion of the new variables in terms of the penalization parameter. We can choose the density and the internal energy (for example) as the thermodynamic variables and write

$$\begin{aligned} \rho_\eta &= \rho + \eta\tilde{\rho}, & u_\eta &= u + \eta\tilde{u} \\ e_\eta &= e + \eta\tilde{E} & v_\eta &= v + \eta\tilde{v} \end{aligned} \tag{11}$$

(η is small, $0 < \eta << 1$). Then, we easily get that $\overrightarrow{U}_\eta = \overrightarrow{U} + \eta\overrightarrow{\tilde{U}}$ and $p_\eta = p + \eta\tilde{p}$. After introducing these quantities in system (10) and identifying the terms of order η^{-1} we get

$$\chi_{\Omega_s}\left(0, \rho u, \rho v, E - E_{\Omega_s}\right)^T = 0. \tag{12}$$

Since $\chi_{\Omega_s}(x,y) \equiv 1$ for $(x,y) \in \overline{\Omega}_s$, (12) implies

$$u = v = 0 \quad \text{and} \quad E = E_{\Omega_s}, \qquad (x,y) \in \overline{\Omega}_s. \tag{13}$$

From the $O(1)$ terms, we get

$$\partial_t\overrightarrow{U} + \nabla\overrightarrow{\mathcal{F}}(\overrightarrow{U}) + \chi_{\Omega_s}\begin{pmatrix} 0 \\ \rho\tilde{u} + \tilde{\rho}u \\ \rho\tilde{v} + \tilde{\rho}v \\ \tilde{E} \end{pmatrix} = \frac{1}{Re}\nabla\overrightarrow{\mathcal{F}}_V(\overrightarrow{U}). \tag{14}$$

Relation (12) implies that the first three equations of this system become

$$\partial_t\rho = 0, \qquad \nabla p + \rho(\tilde{u}, \tilde{v})^T = 0 \quad \text{in } \Omega_s. \tag{15}$$

Thus, the density does not vary in time inside the solid. Then, inside the solid

$$E = \rho C_v T = E_{\Omega_s} = \rho C_{\mathrm{v}} T_{\Omega_s} \tag{16}$$

and we recover the temperature boundary condition.

On the other hand, $\chi_{\Omega_s}(x,y) = 0$ for $(x,y) \in \Omega_s)$, hence outside of the solid equation (14) reduces to the original Navier–Stokes system (1).

Notice that the second equation in (15) implies that velocity vector $(\tilde{u}, \tilde{v})$ inside the solid satisfies a Darcy-type law in Ω_s, as in the incompressible case. Thus, obstacles would still be associated to porous media.

4 Numerical Algorithm

The entire computational domain Ω is discretized using a uniform Cartesian grid $\mathcal{G} = \{(x_i, y_j),\ 0 \le i \le N_x,\ 0 \le j \le N_y\}$ covering Ω, with step sizes $\delta x, \delta y$.

We discretize the penalized system by a *method of lines*, hence we consider the following system of ODE

$$\frac{d\overrightarrow{U_{ij}}}{dt} + D(\overrightarrow{U_{ij}}) + K(\overrightarrow{U_{ij}}) + H(\overrightarrow{U_{ij}}) = 0, \tag{17}$$

where $D(\overrightarrow{U_{ij}})$ is the numerical divergence associated to the inviscid fluxes (see Sect. 4.1), $K(\overrightarrow{U_{ij}})$ the finite difference approximation of the viscous fluxes (see Sect. 4.2) and $H(\overrightarrow{U_{ij}})$ the approximation of the penalization term (see Sect. 4.3). We have omitted the subscript η in the variables for simplicity.

The system (17) will be solved by a Runge–Kutta time integrator described in Sect. 4.4.

4.1 Discretization of the Inviscid Fluxes

Shocks are correctly and accurately computed by using a high-resolution shock-capturing scheme (HRSC) technique to compute the numerical flux functions involved in the convective term $D(\overrightarrow{U}_{ij})$. Hence, our discretization of this term is of the form

$$D(\overrightarrow{U_{ij}}) = \frac{\overrightarrow{F}_{i+1/2,j} - \overrightarrow{F}_{i-1/2,j}}{\delta x} + \frac{\overrightarrow{G}_{i,j+1/2} - \overrightarrow{G}_{i,j-1/2}}{\delta y}. \tag{18}$$

with $\overrightarrow{F}_{i+1/2,j} = \overrightarrow{F}(\overrightarrow{U}_{i-k,j}, \cdots, \overrightarrow{U}_{i+m,j})$, $\overrightarrow{G}_{i,j+1/2} = \overrightarrow{G}(\overrightarrow{U}_{i,j-k}, \cdots, \overrightarrow{U}_{i,j+m})$, where $\overrightarrow{F}$ and $\overrightarrow{G}$ are HRSC numerical flux functions, consistent with the physical inviscid flux functions (2). Here we use a third-order HRSC scheme described in [DM96].

In order to reduce the computational cost associated to a HRSC scheme such as the one used in this paper, we use a cost-reduction technique based on a multilevel evaluation of the numerical divergence (see [CD01] for details) Our multilevel scheme is based on an interpolating wavelet transform and is capable to obtain the high quality expected of a HRSC with a significant reduction in the global computational effort and, hence, in the CPU time of the simulation (for details and results see [CD01, RCD00]).

Our multilevel technique reduces the effective CPU time associated to the computation of the numerical divergence (which is the main time-consuming part of the full algorithm) in our simulation by an average factor of 5.5 for uniform grids with $1,536^2$ points, of 4 for $1,024^2$, and of 3 for 512^2 grid points.

4.2 Discretization of the Viscous Fluxes

We use centered discretizations for the viscous fluxes. For consistency with the convective part, fourth order accurate discretizations are used. For the first space derivatives we take

$$\begin{aligned} (u_x)_{ij} &= \frac{u_{i-2,j} - 8u_{i-1,j} + 8u_{i+1,j} - u_{i+2,j}}{12\delta x} + \mathcal{O}(\delta x^4), \\ (u_y)_{ij} &= \frac{u_{i,j-2} - 8u_{i,j-1} + 8u_{i,j+1} - u_{i,j+2}}{12\delta y} + \mathcal{O}(\delta y^4), \end{aligned} \tag{19}$$

and lengthy, but straightforward, algebra gives $K(\overrightarrow{U_{ij}})$, the space approximation of the viscous part of (10).

4.3 Discretization of the Penalization Term

Once the parameter η is fixed, the penalization term is treated as a source term and computed at each grid point $(x_i, y_j) \in \mathcal{G}$ by

$$\begin{cases} H(\overrightarrow{U_{ij}}) = 0 & \text{if } (x_i, y_j) \in \Omega_f, \\ H(\overrightarrow{U_{ij}}) = \frac{1}{\eta} \begin{pmatrix} 0 \\ (\rho u)_{ij} \\ (\rho v)_{ij} \\ (E - E_{\Omega_s})_{ij} \end{pmatrix} & \text{if } (x_i, y_j) \in \overline{\Omega}_s. \end{cases} \tag{20}$$

This term requires only the knowledge of the geometry of the solid domain and adds no extra computation.

4.4 The Time Discretization

The time integration of the semidiscrete system of equations (17) is based on a third-order TVD Runge–Kutta scheme defined in [SO89]. Since we are interested in simulations of supersonic flows at large Reynolds numbers, there is no need to treat the viscous terms implicitly. The time step restriction due to stability for this scheme is $\delta t = \min(\delta t_e, \delta t_v)$, with δt_e the inviscid time step restriction

$$\delta t_e = C_1 \frac{\min(\delta x, \delta y)}{S^n_{max}}, \tag{21}$$

where S^n_{max} is the maximum wave speed throughout the domain at the time level n, and δt_v the viscous time step restriction

$$\delta t_v = C_2 \big(\min(\delta x, \delta y)\big)^2 Re \frac{Pr}{\gamma \mu}. \tag{22}$$

For the simulations we use $C_1 = 0.8$ and $C_2 = 0.5$ and the time step is adapted automatically during the computation. For $Re >> 1$, we easily get $\delta t_e < \delta t_v$, so that an explicit discretization of the viscous fluxes does not affect stability.

On the other hand, an explicit discretization of the penalization term $H(\overrightarrow{U_{ij}})$ will lead to a stability restriction of the form $\delta t \sim \eta$, which is intractable for the value of η considered in practical simulations (see Sect. 5). Hence, we treat implicitly the penalization term. Let

$$L(\overrightarrow{U_{ij}}) = D(\overrightarrow{U_{ij}}) + K(\overrightarrow{U_{ij}}), \tag{23}$$

we consider the following three steps time integrator for (17):

$$\begin{aligned}
\vec{U}^{*} &= \vec{U}^{n} - \delta t\ L(\vec{U}^{n}) - \delta t\ H(\vec{U}^{*}),\\
\vec{U}^{**} &= \tfrac{1}{4}\big(3\vec{U}^{n} + \vec{U}^{*} - \delta t\ L(\vec{U}^{*}) - \delta t\ H(\vec{U}^{**})\big),\\
\vec{U}^{n+1} &= \tfrac{1}{3}\big(\vec{U}^{n} + 2\vec{U}^{**} - 2\delta t\ L(\vec{U}^{**}) - 2\delta t\ H(\vec{U}^{n+1})\big).
\end{aligned} \tag{24}$$

The implicit treatment leads to a diagonal system, hence *no extra work* is involved in practice in comparison with a fully explicit scheme.

The penalized system has to be solved also in the solid domain. However, except for a thin layer around the solid body boundary, the flow is constant inside of the solid (see Sect. 5) therefore the multilevel technique allows for a considerable reduction of the associated computational expense.

Notice that when the point $(x_i, y_j) \in \Omega_f$ the discrete function $H(\cdot)$ is zero, and we recover the classical third-order TVD Runge–Kutta scheme.

5 Numerical Results

Here we display a sample computation of a shock–cylinder interaction, as described in Fig. 1, in order to show that the penalization technique effectively imposes the no-slip conditions on the velocity at the boundary of the solid body, as well as the Dirichlet condition on the temperature.

The computational domain Ω is prescribed to be a square ($0 \leq x \leq 2$, $0 \leq y \leq 2$). The solid body is a circle of radius $r = 0.2$, centered at $(0.5; 1)$. A Mach-3 shock wave ($M_s = 3$), initially located at $x = 0.1$, travels from left to right to interact with the solid. Supersonic inlet flow quantities are fixed at $x = 0$ while nonreflecting boundary conditions (outflow) are prescribed for the other boundaries.

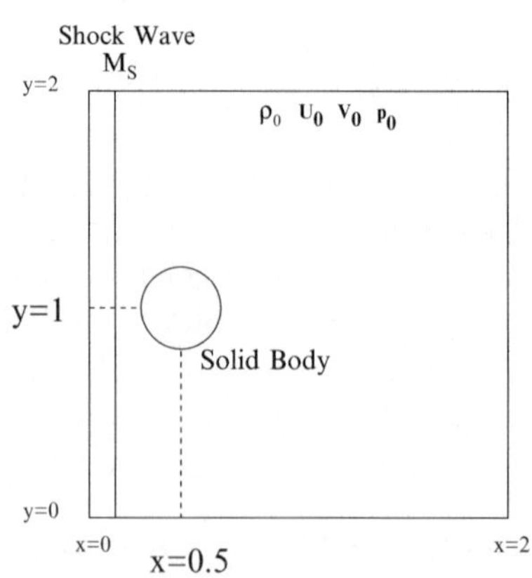

Fig. 1. Initial layout for the shock–cylinder interaction

In the flow domain Ω_f, variables at the right of the shock at time $t = 0$ are $(\rho_0, p_0, u_0, v_0) = (1, 1, 0, 0)$ and quantities on the left of the shock are computed from the ones at the right side and the Mach number M_s using classical shock considerations.

Inside the obstacle, i.e., in $\overline{\Omega}_s$, velocity is set to zero, density to ρ_0, and the temperature is $T_{\Omega_s} = 3$. Energy and pressure are computed from relations (5) and (16).

All the quantities are dimensionless, the Reynolds number is fixed to $Re = 5 \cdot 10^4$ and the Prandtl number to 0.72 for all the simulations.

A Schlieren plot of the density obtained with our numerical scheme after the interaction, has taken place is displayed on Fig. 2. Upstream of the obstacle, we clearly observe a well resolved reflected (detached) shock wave, while a more complex structure develops downstream.

In Fig. 3 we display the profiles of the different flow variables inside of the cylinder. The velocity norm is computed as $\|(u, v)_{ij}\| = (u_{ij}^2 + v_{ij}^2)^{1/2}$. We clearly observe that the velocity becomes zero at the boundaries of the solid body (vertical lines) and the temperature is equal to the prescribed value of $T_{\Omega_s}=3$. Pressure (and energy) are nearly constants while the density increases rapidly in front of the solid. This effect is due to the rapid decrease in temperature, and can be explained by the perfect gas law where the density is proportional to p/T.

Notice also that the values inside of the solid domain remain nearly constant, even though they are not relevant from a physical point of view. This feature implies in turn that, due to the multi-level computation of the numerical divergence, no costly HRSC computations are being made inside the obstacle.

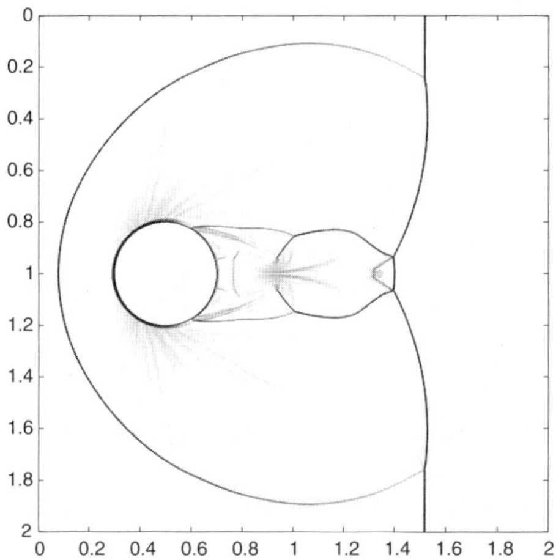

Fig. 2. Shock–cylinder interaction for $M_s = 3$. Density at time $t = 0.4$ for 1024^2 grid points and penalization parameter $\eta = 10^{-12}$ (Numerical Schlieren image)

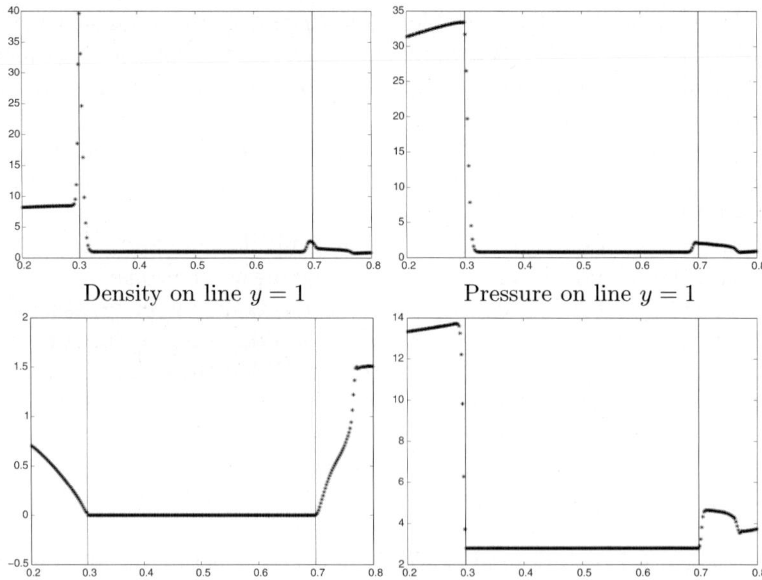

Density on line $y = 1$ — Pressure on line $y = 1$

Velocity norm on line $y = 1$ — Temperature on line $y = 1$

Fig. 3. Profiles of the density, pressure, velocity norm, temperature, and energy for the shock–cylinder interaction ($M_s = 3$ and $\eta = 10^{-12}$). The *vertical lines* define the limits of the solid body

We address now the issue of the influence of the value of η on the computed solution and the boundary conditions. For this, we perform numerical computations of the shock–cylinder test for different values of η and compute the relative velocity and energy errors inside the obstacle as:

$$\|(u,v)\|_{L_2(\Omega_s)} = \frac{1}{M_{\mathrm{v}}}\left(\delta x \delta y \sum_{(x_i,y_j)\in\overline{\Omega}_s} \|(u,v)_{ij}\|^2\right)^{1/2}, \tag{25}$$

$$\|E - E_{\Omega_s}\|_{L_2(\Omega_s)} = \frac{1}{M_E}\left(\delta x \delta y \sum_{(x_i,y_j)\in\overline{\Omega}_s} |E_{ij} - E_{\Omega_s}|^2\right)^{1/2} \tag{26}$$

with $M_{\mathrm{v}} = \max\{\|(u,v)_{ij}\|, (x_i,y_j) \in \overline{\Omega}\}$ and $M_E = \max\{|E_{ij}|, (x_i,y_j) \in \overline{\Omega}\}$.

The results shown in the left plot of Fig. 4 indicate that $\|(u,v)\|_{L_2(\Omega_s)}$ and $\|E - E_{\Omega_s}\|_{L_2(\Omega_s)}$ tend to zero as $O(\eta)$. Notice also the scale on Fig. 4 right. We clearly observe that $\|(u,v)\|_{L_2(\Omega_s)}$ is of the order of η at the boundary. In the interior of the solid body it is essentially zero.

We recall that the theoretical estimates for incompressible flows obtained in [ABF99], provide a convergence rate of order $O(\eta^{1/4})$, but in subsequent papers (see, e.g., [KV05, PCLS, SF02] and references therein), it was found

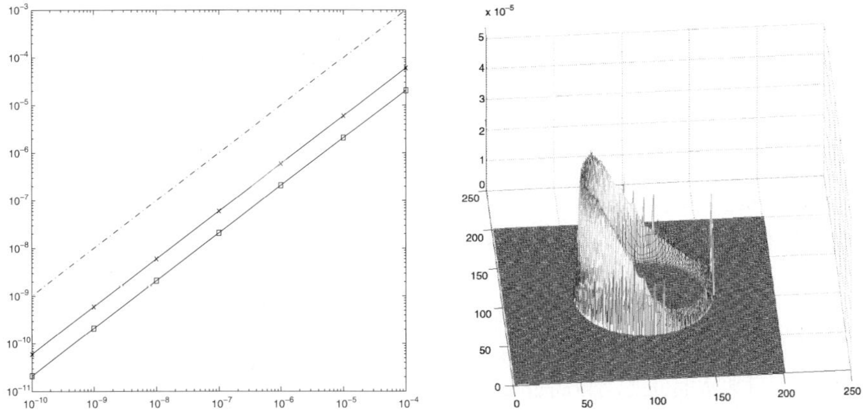

Fig. 4. *Left*: Relative velocity (*crosses*) and energy (*squares*) norms (25, 26) inside the cylinder with respect to η (*dot lines*: $f(\eta) = 10 \cdot \eta$, logarithmic scales). *Right*: $||(u, v)_\eta||_{L^2(\Omega_s)}$ for a sample computation with $\eta = 10^{-5}$

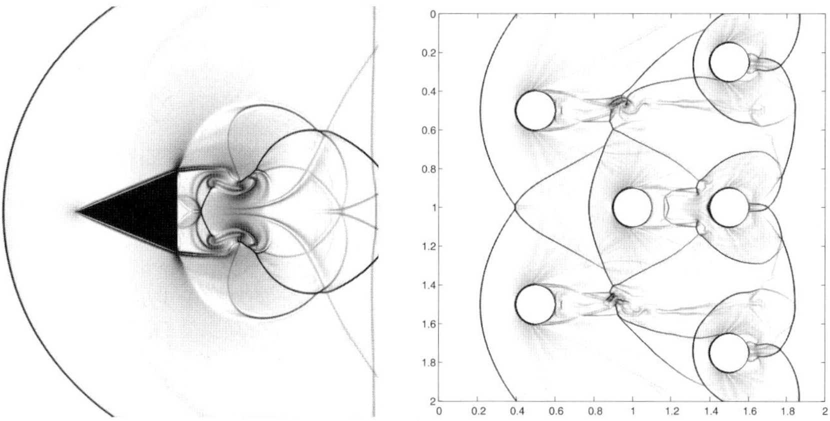

Fig. 5. Schlieren density plots of density. *Left*: Mach 1.7. *Right* Mach 3

that the effective rate inside the solid body in numerical simulations was of $O(\eta)$. Figure 4 indicates that for compressible flows the same rate is reached.

6 Conclusions

Numerical simulations of shocked flows with obstacles can be handled in an efficient manner by a penalization technique, similar to the penalization technique for compressible flows introduced in [AC84] and analyzed in [ABF99].

Together with an HRSC multilevel evaluation of the convective terms, highly accurate numerical simulations can be easily obtained, completely avoiding body fitted mesh structures. The penalization algorithm only needs an explicit description of the characteristic function of the solid domain, hence the inclusion of many bodies does not represent an increase in the computational cost. We display in Fig. 5 a very fine numerical simulation of a the interaction of a shock with a triangular shape (left) and the interaction of a Mach 3 flow with six cylinders (right). In the latter case, absolutely no extra cost is added with respect to the one-obstacle computation. More numerical simulations using our penalization technique can be found in [CD06].

References

[ABF99] P. Angot, C.-H. Bruneau, P. Fabrie: A penalization method to take into account obstacles in incompressible flows. Numer. Math., **81**, 497–520 (1999)

[A82] J.D. Anderson Jr.: Modern Compressible Flow. Series in Mechanical Engineering. McGraw-Hill, (1982)

[AC84] E. Arquis, J.P. Caltagirone: Sur les conditions hydrodynamiques au voisinage d'une interface milieu fluide-milieu poreux: application à la convection naturelle. C.R. Acad. Sci. Paris II, **299**, 1–4 (1984)

[CD01] G. Chiavassa, R. Donat: Point Value Multiscale Algorithms for 2d Compressible Flows. SIAM J. Sci. Comput., **23-3**, 805–823 (2001)

[CD06] G. Chiavassa, R. Donat: A penalization Method for Compressible Fluid Flow. Procedings of Fifth International Conf. on Engineering Computational Technology ECT2006, B.H.V Topping, G. Montero and R. Montenegro, Civil-Comp Press, 2006

[DM96] R. Donat, A. Marquina: Capturing Shock Reflections: An improved Flux Formula. J. Comput. Phys., **125**, 42–58 (1996)

[GPWZ96] R. Glowinski, T.W. Pan, R.O. Wells Jr., X. Zhou: Wavelet and Finite Element Solutions for the Neumann Problem using Fictitious Domains. J. Comput. Phys., **126(1)**, 40–51 (1996)

[KV05] N.K.R. Kevlahan, O.V. Vasilyev: An Adaptive Wavelet collocation Method for Fluid-Structure Interaction at High Reynolds Number. SIAM J. Sci. Compt., **26-6**, (2005).

[KPAC00] K. Khadra, S. Parneix, P. Angot, J.P. Caltagirone: Fictitious domain approach for numerical modelling of Navier-Stokes equations. Int. J. Num. Mth. Fluids, **34**, 651–684 (2000)

[PCLS] A. Paccou, G. Chiavassa, J. Liandrat, K. Schneider: A penalization method applied to the wave equation. C.R. Acad. Sci. Paris, **329**, (2003).

[RCD00] A. Rault, G. Chiavassa, R. Donat: Shock-Vortex Interaction at High Mach Numbers. J. Sci. Comput., **19**, 347–372 (2003)

[SO89] C.W. Shu, S.J. Osher: Efficient Implementation of Essentially Non-Oscillatory Shock Capturing Schemes II. J. Comput. Phys., **83**, 32–78 (1989)

[SF02] K. Schneider, M. Farge: Adaptive wavelt simulation of a flow around an impulsively started cylinder using penalization, Appl. Comput. Harm., **12**, 374–380 (2002)

Exact Solutions to Supersonic Flow onto a Solid Wedge

V. Elling and T.-P. Liu

1 Introduction

Consider compressible fluid flow onto a solid wedge, symmetric in the flow direction (see Fig. 1). It is known from experiments and numerics that for supersonic flow, the time-asymptotic (steady) state consists of one straight shock on each side, emanating downstream. Each shock separates two constant-state regions, the upstream, and downstream area. The downstream area has higher density, but lower velocity, with direction tangential to the wedge surface. An alternative point of view is to consider flow parallel to a wall, up to a concave corner.

Straight shocks must satisfy the *Rankine–Hugoniot conditions*, a system of several algebraic equations. For full (nonisentropic) Euler flow they have the form

$$[\rho \mathbf{v} \cdot \mathbf{n}] = 0, \tag{1}$$

$$[\rho(\mathbf{v} \cdot \mathbf{n})^2 + p] = 0, \tag{2}$$

$$[\mathbf{v} \cdot \mathbf{t}] = 0, \tag{3}$$

$$[(\mathbf{v} \cdot \mathbf{n})(\rho e + p)] = 0, \tag{4}$$

where p is pressure, $e = q + \frac{1}{2}|\mathbf{v}|^2$ energy-per-mass, and q internal energy-per-mass. $[Q]$ indicates the difference of the upstream and downstream value of a quantity Q.

The starting point of our research is a rather surprising classical observation: for a sharp wedge and large upstream Mach number there are two choices for each shock. One is called *strong shock*; it approaches a normal shock (perpendicular to the flow velocity) as the wedge angle vanishes. The other solution is the (comparatively) *weak shock*; for vanishing wedge it becomes tangential to the wedge surface.

This is obvious from the *shock polar* (see Fig. 2), the curve of possible downstream velocities when holding the upstream data (velocity, density) fixed

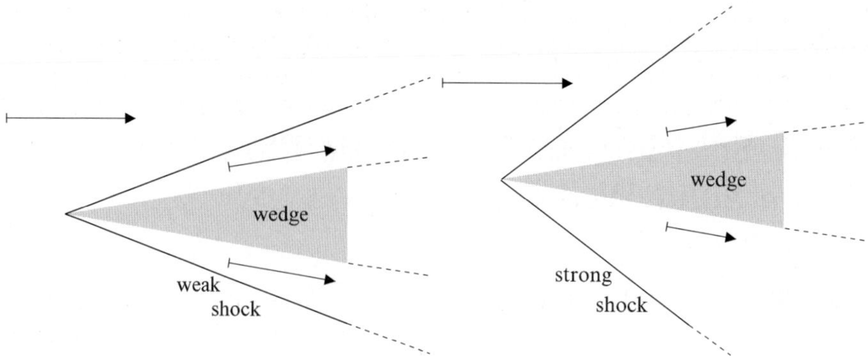

Fig. 1. Two steady solutions of supersonic flow onto a solid wedge: weak (*left*) and strong shocks (*right*)

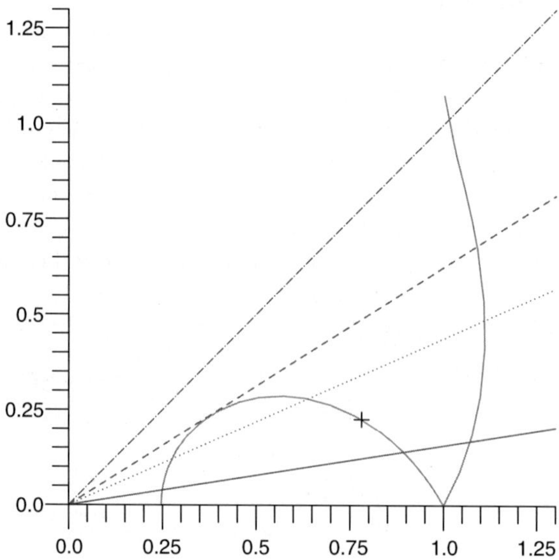

Fig. 2. Red curve: possible downstream velocities for varying shock normal. The blue lines represent various choices of wedge surface. The upstream velocity is in $(1, 0)$. The respective shock normal is the vector from downstream to upstream velocity. Example for $\gamma = 7/5$ and Mach 1.5

and the shock steady, while varying the shock *normal*. We seek a downstream velocity that forms a fixed angle θ (half the wedge angle) to the upstream velocity. As is apparent from Fig. 2, for small θ there are three points on the shock polar, corresponding to three different shock normals. The rightmost point is part of the *unphysical* branch of the shock polar: $\mathbf{v} \cdot \mathbf{n}$ increases across the shock that violates the Lax entropy condition. The other two points, however, are physically admissible.

To our knowledge, prior to our work no mathematical argument has been found that prefers the strong or the weak shock. There have been several hypotheses, however. The strong shock has been believed to be unstable under perturbations, such as a small change to the wedge surface or other physical parameters, or small changes in the initial conditions.

2 Experiments

To settle the question, we have devised several numerical experiments. In the first experiment (Fig. 3), the strong shock was perturbed by adding a large amount of mass in a square area near the corner; surprisingly it remains stable. In the second experiment (Fig. 4) the strong shock was truncated at some distance, corresponding to changing the boundary data at infinity. It turns out that the strong shock is unstable under *this* perturbation: it is swept away and the weak shock appears at the corner. The same happens, less drastically, if the downstream density is decreased no matter how little. Increasing it causes the shock to detach from the corner and to wander upstream (not shown). Hence, to have a strong shock it is necessary to control the downstream flow perfectly. In a numerical experiment this means enforcing an appropriate downstream boundary condition. In a physical situation or experiment one would need

Fig. 3. Strong shock stable under perturbation

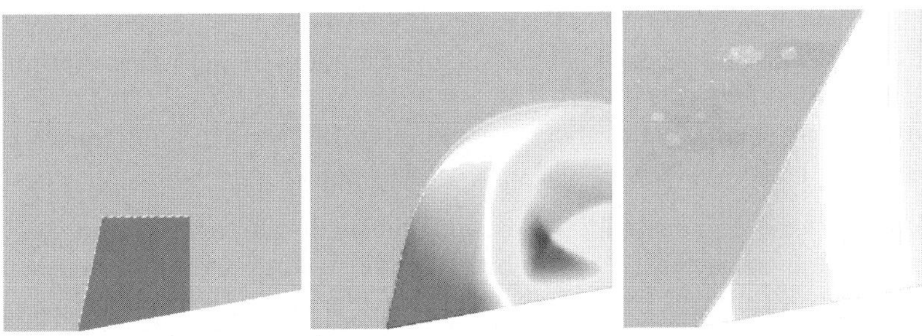

Fig. 4. Strong shock truncated away from corner; it is swept away and the weak shock appears

Fig. 5. The weak shock appears at a solid wall corner (*bottom left*)

precise control over some parameter, such as the cross-section or shape of a downstream nozzle, to achieve a strong shock. In either case it is likely that the observed shock would not exactly be a strong shock, but a small perturbation, such as a slightly detached shock or a small weak-shock segment followed by a flow pattern with a strong shock.

The final experiment is an "unbiased" test: at $t = 0$ the entire domain is filled with the constant upstream state. Whichever shock appears at positive times should be considered the physical choice. It turns out that the weak shock appears spontaneously and eventually fills out the entire numerical domain (Fig. 5).

3 Exact Solutions

It is desirable to settle such and other questions mathematically rather than by experiments. Although numerics and experiments work well in the absence of deep mathematical insight and effort,

- They are often restricted in the range of parameters that can be considered,
- They include extraneous effects such as physical or numerical dissipation and perhaps kinetic effects,
- And most importantly they are not suitable for studying unstable or nearly-unstable solutions and other pathologies.

In our work we construct an exact solution to the initial-value problem corresponding to the last experiment. We chose compressible potential flow as model. It is obtained from the isentropic Euler equations by assuming that the vorticity $\nabla \times \mathbf{v}$ is zero. *Smooth* solutions of potential flow are actually *exact* solutions, since vorticity satisfies a linear transport equation along flow lines; weak shocks with small curvature are also fine.

Potential flow is derive from the isentropic Euler equations of compressible gas dynamics in two space dimensions:

$$\rho_t + \nabla \cdot (\rho \mathbf{v}) = 0 \tag{5}$$
$$(\rho \mathbf{v})_t + \nabla \cdot (\rho \mathbf{v} \otimes \mathbf{v}) + \nabla(p(\rho)) = 0, \tag{6}$$

here ρ is the density, $\mathbf{v} = (u, v)$ is the velocity, $\mathbf{x} = (x^1, x^2) = (x, y)$ is the space variables, and $\nabla = \partial_{\mathbf{x}}$ is the spatial gradients. We consider the polytropic gases so that, after some rescaling, the pressure function $p(\rho)$ is given by

$$p(\rho) = \rho^\gamma, \ \gamma \in (1, \infty).$$

If $\nabla \times \mathbf{v} = 0$ (and the domain is topologically trivial), then there exists a scalar potential function ϕ such that

$$\mathbf{v} = \nabla\phi.$$

From this and the Euler equations, (5) and (6), the following Bernoulli law holds for smooth flows:

$$\begin{aligned} 0 &= \phi_{x_i t} + \nabla\phi_{x_i} \cdot \nabla\phi + \pi(\rho)_{x_i} \\ &= (\phi_t + \frac{|\nabla\phi|^2}{2} + \pi(\rho))_{x_i}, \end{aligned}$$

where $\pi(\rho)$ is related to the sound speed $c(\rho)$ by

$$\pi_\rho = c^2(\rho)/\rho, \ c^2(\rho) = p_\rho(\rho).$$

Thus, for some constant A,

$$\rho = \pi^{-1}(A - \phi_t - \frac{|\nabla\phi|^2}{2}).$$

Substituting this into (5) yields the potential flow equation:

$$\phi_{tt} + 2\phi_x\phi_{tx} + 2\phi_y\phi_{ty} + [(\phi_x)^2 - c^2]\phi_{xx} + 2\phi_x\phi_y\phi_{xy} + [(\phi_x)^2 - c^2]\phi_{yy} = 0. \tag{7}$$

The stationary potential flow equation

$$[c^2 - (\phi_x)^2]\phi_{xx} - 2\phi_x\phi_y\phi_{xy} + [c^2 - (\phi_x)^2]\phi_{yy} = 0 \tag{8}$$

is elliptic when the flow is subsonic, $M := |\nabla\phi|/c < 1$, and hyperbolic when it is supersonic: $M > 1$.

The numerical solution to our last experiment is apparently *self-similar*: it grows at a constant rate; density and velocities are functions of $\boldsymbol{\xi} = t^{-1}\mathbf{x}$ alone. Such solutions are very characteristic for initial-value problems for hyperbolic conservation laws: shocks and rarefaction waves are 1D self-similar solutions. For potential flow, the selfsimilarity assumption corresponds to the ansatz

$$\phi(x, y, t) = t\psi(\xi, \eta), \ \xi = x/t, \ \eta = y/t.$$

The self-similar potential flow equation is

$$[c^2 - (\psi_\xi - \xi)^2]\psi_{\xi\xi} - 2(\psi_\xi - \xi)(\psi_\eta - \eta)\psi_{\xi\eta} + [c^2 - (\psi_\eta - \eta)^2]\psi_{\eta\eta} = 0. \quad (9)$$

Using a different variable

$$\chi = \psi - \frac{1}{2}|\boldsymbol{\xi}|^2$$

we obtain the convenient form

$$(c^2 - \chi_\xi)^2\chi_{\xi\xi} - 2\chi_\xi\chi_\eta\psi_{\xi\eta} + (c^2 - \chi_\eta^2)\chi_{\eta\eta} = |\nabla\chi|^2 - 2c^2. \quad (10)$$

The velocity is

$$\nabla\psi = \mathbf{v}.$$

Like steady potential flow, the self-similar case is of mixed type. It is hyperbolic if the flow is pseudo-supersonic: $L > 1$ with the *pseudo-Mach number* L defined as $L := |\mathbf{v} - \boldsymbol{\xi}|/c$. It is elliptic if $L < 1$, parabolic if $L = 1$.

Constructing Fig. 6 as an exact solution faces a number of technical problems. Most difficult are the parabolic circular arcs that are characteristic too: standard elliptic regularity estimates break down. The linearization has infinite gradients or at least second derivatives, so that small perturbation arguments do not apply. This is compounded by the shock being a free boundary, the nonlinearity of equation and boundary conditions, as well as the shock-parabolic corners where all problems combine.

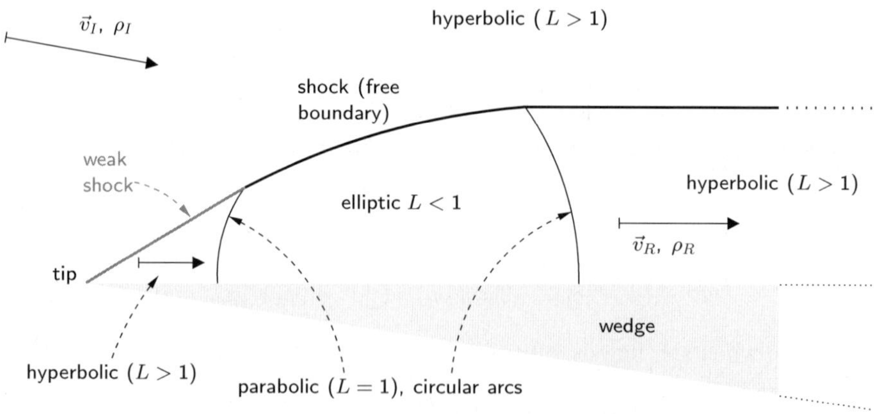

Fig. 6. Mathematical structure of the flow pattern in Fig. 5

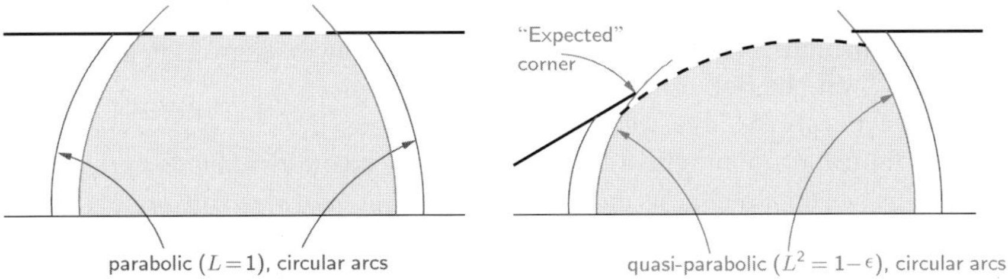

Fig. 7. *Left*: known unperturbed solution; *right*: after perturbation

4 Leray–Schauder Theory

To cope with the degeneracy at the arcs, we *regularize* the problem: we impose boundary conditions $L = \sqrt{1-\epsilon}$, for $\epsilon > 0$ small, on *quasi-parabolic arcs*, which are chosen so that the extension of the hyperbolic downstream regions would satisfy the new boundary condition.

There is an *unperturbed* solution, a straight shock reflected from an infinite wall, where the entire downstream solution including elliptic region is known explicitly. We solve the regularized problems by perturbation (see Fig. 7). Our elliptic region solution candidates are in a set $\mathcal{F}$, which is defined by many constraints: no vacuum $\rho > 0$, ellipticity $L < 1$, $C^{2,\alpha}_\beta$ regularity (with a weight in the shock-arc corners), limits on the shock location, corner angle bounds, etc.

On $\mathcal{F}$ a two-step iteration $\mathcal{K}$ is defined: in the first step we solve a fixed-boundary elliptic problem, using one of the shock conditions as boundary conditions; in the second step we adjust the shock according to the second boundary condition. The elliptic problem is chosen so that any fixed-point of $\mathcal{K}$ is a solution of our regularized problem. $\mathcal{K}$ should be considered a family of iterations, parameterized continuously by γ, M_I, ϵ, θ, etc.

Most of our work is concerned with showing that $\mathcal{K}$ cannot have fixed points on $\partial\mathcal{F}$. Thus the Leray–Schauder degree of $\mathcal{K}$ on $\mathcal{F}$ is constant under continuous parameter changes. To show existence of a fixed point for each choice of parameters, it is sufficient to prove nonzero degree for *one* choice. The unperturbed problem, with $\gamma = 1$, turns out to be simple enough to compute the degree explicitly.

5 Ellipticity Principle

Since the equation is nonlinear and mixed-type, it is necessary to ensure that the elliptic region in Fig. 6 is indeed elliptic. It is easy to find artificial equations where even small perturbations cause an arbitrary variety of hyperbolic or parabolic "bubbles" to appear, the explicit construction of which would be utterly hopeless. Fortunately (10) satisfies a strict maximum principle: the pseudo-Mach number L cannot have a maximum in the interior of a sufficiently smooth elliptic region. To bound L above away from 1, it is sufficient to control L on the boundary of the elliptic region.

On the parabolic arcs we have already imposed $L = \sqrt{1-\epsilon}$. The wall becomes interior upon reflection across the horizontal axis. For the shock we can derive an analogous maximum principle: it rules out that L has a maximum greater than $1 - \delta$, where $\delta > 0$ depends continuously on the shock strength. However, a lower bound for the shock strength must be found separately.

We also observe another crucial fact: L attains its global maximum – with respect to the elliptic region – in each point on the quasi-parabolic arcs.

Finally, the ellipticity principle [EL05] actually holds in a somewhat stronger version: not only is

$$L(\boldsymbol{\xi}) \leq \sup_{\text{arcs}} L,$$

but

$$L(\boldsymbol{\xi}) \leq \sup_{\text{arcs}} L - b(\boldsymbol{\xi}) = \sqrt{1-\epsilon} - b(\boldsymbol{\xi})$$

for a smooth positive function b. Thus, for every $K \subset \Omega$ with a positive distance from the arcs, L is bounded above by a constant <1, which does *not* depend on ϵ. This allows us to obtain *uniform* in ϵ regularity in any $C^{k,\alpha}$ norm, away from the arcs.

6 Parabolic Arcs

We aim to construct a weak solution to the entire problem. To this end it is necessary to have C^1 continuity across the parabolic arcs. This corresponds to imposing *two* boundary conditions

$$\lim_{\text{elliptic}} \chi = \lim_{\text{hyperbolic}} \chi \qquad \lim_{\text{elliptic}} \chi_n = \lim_{\text{hyperbolic}} \chi_n$$

(a third condition, χ_t continuity, is obtained by taking the tangential derivative of the first one.) Here n, t are outer unit normal and counterclockwise unit tangent for the arcs. The *expected* behavior is that χ is constant and $\chi_n = -c$ (for n the outer normal to the elliptic region).

However, in elliptic problems we cannot impose more than one condition on each boundary segment. Therefore, we need to impose one condition and *prove* the rest. In our approach it is sufficient if the differences of $\chi, \nabla\chi$ on the hyperbolic and elliptic sides converge to 0 as $\epsilon \downarrow 0$. Fortunately this can be shown:

On the quasi-parabolic arcs we impose the boundary condition $L = \sqrt{1-\epsilon}$. In each point on the quasi-parabolic arcs, the second derivatives of χ satisfy two linear equations: the interior equation (10), which can be written

$$(c^2 - \chi_t^2)\chi_{tt} - 2\chi_t\chi_n\chi_{tn} + (c^2 - \chi_n^2)\chi_{nn} = \text{lower-order terms},$$

and the tangential derivative of $L = \sqrt{1-\epsilon}$, which can be written

$$\chi_t \chi_{tt} + \chi_n \chi_{tn} = \text{lower-order terms.}$$

The crucial observation is that, by virtue of the ellipticity principle, L has a global maximum in each point on the quasi-parabolic arcs, and so we obtain an additional *inequality*: the normal derivative of L is nonnegative, which can be written

$$\chi_t \chi_{nt} + \chi_n \chi_{nn} \geq \text{lower-order terms.}$$

Combining the three relations, we obtain one inequality for each second derivative of χ.

Particularly important is the one for χ_{tt}. Focusing on the isothermal case with $c = 1$ constant, a lengthy calculation yields

$$\chi_{tt} \geq \sqrt{x} - (1+\epsilon)x, \qquad x = 1 - \frac{\chi_t^2}{1-\epsilon}. \tag{11}$$

The right-hand side (see Fig. 8) is an even function of χ_t. We expect and want to show $\chi_t \approx 0$. For $\chi_t = 0$, the right-hand side of the inequality is negative, but there are two zeros $\chi_t = \pm\chi_{t*} = O(\sqrt{\epsilon})$.

When starting in $\chi_t \in (-\chi_{t*}, \chi_{t*})$ and moving counterclockwise (positive t direction) along the arcs, the ODE inequality (11) does not allow us to reach $\chi_t < -\chi_{t*}$; on the other hand it is possible to each $\chi_t \geq \chi_{t*}$, but impossible to return to $\chi_t < \chi_{t*}$. Therefore, if $\chi_t \in (-\chi_{t*}, \chi_{t*})$ in the *endpoints* of each arc, then *in between* as well.

Since $\chi_{t*} = O(\sqrt{\epsilon})$, this control is tight enough. Another ODE argument then proves that χ is also within $O(\sqrt{\epsilon})$ of its desired value (some constant). The boundary condition $1-\epsilon = L^2 = \chi_t^2 + \chi_n^2$ yields $\chi_n = -\sqrt{1-\epsilon} + O(\sqrt{\epsilon})$, which matches the hyperbolic side as well. These bounds are sufficient for taking $\epsilon \downarrow 0$ and obtaining a weak solution to the original problem.

It remains to check whether $\chi_t \in (-\chi_{t*}, \chi_{t*})$ in the endpoints. One of the endpoints is the wall-arc corner where C^1 regularity and the no-slip condition $\chi_n = 0$ on the wall yield $\chi_t = 0$ on the arc (right angle). However, the other

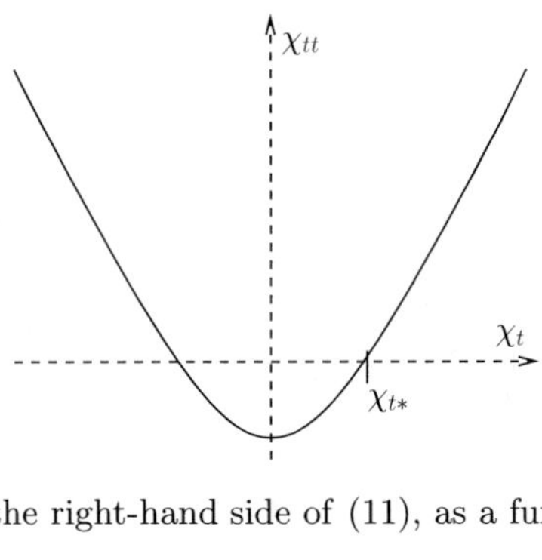

Fig. 8. Graph of the right-hand side of (11), as a function of (small) χ_t

endpoint is the corner between arc and shock, which is free to move along the circular extension of the arc. If that corner is at its "expected" location, which is the meeting point with the straight hyperbolic-side shock, then $\chi_t = 0$ in the corner.

It remains to show that the shock-parabolic corner is indeed at or "near" its expected location. A lower bound is obtained as follows: we have C^1 regularity in the corner, by adapting a result of Lieberman [Lie88]. Therefore, χ and $\nabla\chi$ in the corner are determined by the two boundary conditions on the shock and the one condition on the arc, as functions of the corner location. If the right (e.g.) corner is more than $C\chi_{t*}$ below its expected location (for a certain constant C), then a detailed calculation shows $\chi_t < -\chi_{t*}$. t being the counterclockwise direction, we cannot pass from $\chi_t = 0$ in the right wall-arc corner to $\chi_t < -\chi_{t*}$ in the right arc-shock corner without violating (11).

For an upper bound, an entirely different argument is needed: we argue that ψ cannot attain a minimum anywhere in the domain if the corners are above their expected locations, which yields a contradiction. This provides the final piece of the arc treatment.

This technique works because the right-hand side of (11) is $O(\epsilon)$ for $\chi_t = 0$. If it was instead something large, then χ_{t*} would be large and the inequality would not imply tight bounds on $\chi, \nabla\chi$ on the arcs.

7 Shock Control

Another problem is to control the location of the shock, a free boundary. We use another set of maximum principles: the density cannot have minima in the interior of an elliptic region, or at a straight wall, and it cannot have minima at the shock, except in *pseudo-normal* points where $\nabla\chi \cdot \mathbf{t} = 0$. In these points it can be shown that the shock has positive curvature (as part of the elliptic region boundary). If the *global* (with respect to the elliptic region) density minimum is attained in a shock point, then the rest of the shock must be below the tangent through that point. If not, then there would have to be a second point with parallel but *lower* tangent, meaning lower density – contradiction! Since the shock ends in the shock-arc corners, this means the minimum point must be *above* the line through the two corners (see Fig. 9). If the circle with radius c_I and center $\mathbf{v}_I$ is completely below that line, then it can be shown that $\rho \geq \rho_I + \delta$ in the minimum point.

The alternative, a global density minimum on either of the arcs (including corners) is also fine, as our arc analysis shows it must be within $O(\sqrt{\epsilon})$ of the hyperbolic side values $\rho_L, \rho_R \gg \rho_I$.

This means the shock has uniform strength everywhere; in particular a density vacuum is not possible. $\rho \geq \rho_I + \delta$ is also an ingredient of a maximum principle for L at the shock.

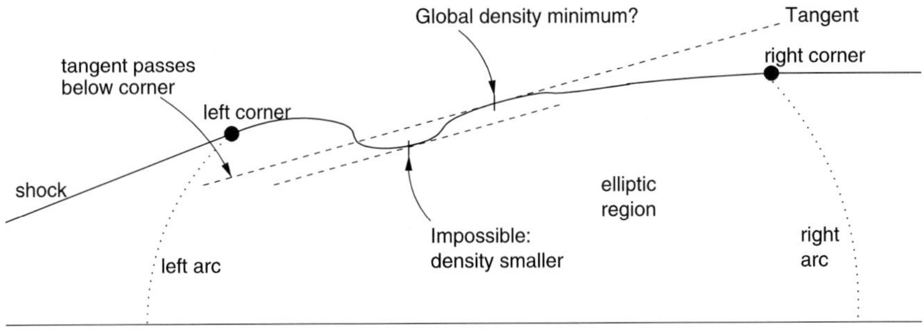

Fig. 9. Argument for uniform shock strength

An upper bound for the shock location, as well as upper bounds for the shock-arc corner angles, is obtained from separate maximum principles for components of $\mathbf{v}$, the velocity.

8 Conclusion

Using many techniques in addition to the most important ones presented here, we have constructed an exact solution of the problem, with a weak shock at the wedge tip, for all $\gamma \geq 1$, $M_I > 1$, $M_I^y \in (-\infty, 0)$, $\rho_I, c_I \in (0, \infty)$ that satisfy two conditions:

1. The weak shock is supersonic–supersonic.
2. The line through the expected corners of the elliptic region does not meet the circle with center $\mathbf{v}_I$ and radius c_I.

The first condition is physical: if the weak shock is supersonic–*sub*sonic, we observe in experiments that the solution of the initial value problem has a shock detached from the wedge tip. The second condition is technical and we expect that additional research can remove it. For sufficiently large upstream Mach number, both conditions are satisfied by at least one wedge angle.

The techniques developed here are not restricted to the wedge flow problem; many other applications with self-similar initial data (for example regular reflection [CFar]) can be solved now.

Acknowledgments

This material is based upon work supported by an SAP/Stanford Graduate Fellowship and by the National Science Foundation under Grant no. DMS 0104019. Any opinions, findings, and conclusions or recommendations expressed in this material are those of the author and do not necessarily reflect the views of the National Science Foundation.

References

[CFar] Gui-Qiang Chen and M. Feldman, *Global solutions to shock reflection by large-angle wedges for potential flow*, Annals of Math. (to appear).

[EL05] V. Elling and Tai-Ping Liu, *The ellipticity principle for selfsimilar potential flow*, J. Hyper. Diff. Eqns. **2** (2005), no. 4, 909–917, See also arxiv:math.AP-0509332.

[Lie88] G. Lieberman, *Hölder continuity of the gradient at a corner for the capillary problem and related results*, Pac. J. Math. **133** (1988), no. 1, 115–135.

Resonance and Nonlinearities

T. Gallouët

1 Introduction

Let $p \in \mathbb{N}^\star$ and A be a real $p \times p$ matrix. One considers the following Cauchy problem, where the unknown is the function $\mathbf{W}$ from $\mathbb{R} \times \mathbb{R}_+$ to $\mathbb{R}^p$:

$$\begin{aligned} &\partial_t \mathbf{W}(x,t) + A\partial_x \mathbf{W}(x,t) = 0, (x,t) \in \mathbb{R} \times \mathbb{R}_+, \\ &\mathbf{W}(x,0) = \mathbf{W}_0(x),\ x \in \mathbb{R}. \end{aligned} \tag{1}$$

A weak solution of Problem (1), for $\mathbf{W}_0 \in L^1_{loc}(\mathbb{R}, \mathbb{R}^p)$, is a function $u \in L^1_{loc}(\mathbb{R} \times \mathbb{R}_+, \mathbb{R}^p)$ such that, for all $\varphi \in C_c^\infty(\mathbb{R} \times \mathbb{R}_+, \mathbb{R}^p)$,

$$\int_0^\infty \int_{\mathbb{R}} (\mathbf{W} \cdot \partial_t \varphi + A\mathbf{W}\partial_x \varphi)(x,t)dxdt + \int_{\mathbb{R}} \mathbf{W}_0(x)\varphi(x,0)dx = 0. \tag{2}$$

If A is diagonalizable in $\mathbb{R}$, we will say that the first equation of Problem (1) is a linear genuinely hyperbolic system. In this case, if $q \in [1,\infty]$ and $\mathbf{W}_0 \in L^q(\mathbb{R}, \mathbb{R}^p)$ (the Lebesgue space on $\mathbb{R}$ with value in $\mathbb{R}^p$)), Problem (1) has a unique weak solution $\mathbf{W}$, and $\mathbf{W}(\cdot,t) \in L^q(\mathbb{R}, \mathbb{R}^p)$ for a.e. $t \in \mathbb{R}^+$.

If the matrix A has only real eigenvalues but is not diagonalizable, the first equation of Problem (1) is said to be a linear hyperbolic resonant system. In this case, Problem (1) is ill posed in the sense that if $\mathbf{W}_0 \in L^q(\mathbb{R}, \mathbb{R}^p)$ (for some $q \in [1,\infty]$), it has, in general, no weak solution $\mathbf{W}$ (and, in particular, no weak solution with $\mathbf{W}(\cdot,t) \in L^q(\mathbb{R}, \mathbb{R}^p)$ for a.e. $t \in \mathbb{R}^+$). (However, Problem (1) is well posed in C^∞, it has a unique solution in $C^\infty(\mathbb{R} \times \mathbb{R}_+, \mathbb{R}^p)$ if the initial datum $\mathbf{W}_0$ belongs to $C^\infty(\mathbb{R}, \mathbb{R}^p)$.) This ill posedness is due to the fact that there is a lack of regularity between $\mathbf{W}(\cdot,t)$ (for $t > 0$) and $\mathbf{W}_0$. For instance, the Riemann problem, that is Problem (1) with $\mathbf{W}_0(x) = \mathbf{w}_l$ for $x < 0$ and $\mathbf{W}_0(x) = \mathbf{w}_r$ for $x > 0$ (and $\mathbf{w}_l$, $\mathbf{w}_r \in \mathbb{R}^p$), does not have a weak solution (in the sense given before) except for very particular choices of $\mathbf{W}_0$, but it has a solution in a greater space. In the case $p = 2$, it has a (unique) solution in a space allowing $\mathbf{W}(\cdot,t)$ to be, for $t > 0$, a measure on the bounded sets of $\mathbb{R}$, see Sect. 2 below (in the case $p \geq 3$, the solution $\mathbf{W}(\cdot, t)$ may even be less regular).

One considers now that the matrix A in (1) is depending on $\mathbf{w}$, leading to the following nonlinear system:

$$\begin{aligned} &\partial_t \mathbf{W}(x,t) + A(\mathbf{W}(x,t))\partial_x \mathbf{W}(x,t) = 0, (x,t) \in \mathbb{R} \times \mathbb{R}_+, \\ &\mathbf{W}(x,0) = \mathbf{W}_0(x), \ x \in \mathbb{R}. \end{aligned} \tag{3}$$

The unknown $\mathbf{W}$ is supposed to take values in an admissible set $D \subset \mathbb{R}^p$.

If the matrix $A(w)$ is diagonalizable in $\mathbb{R}$ for all $w \in D$, we say that the first equation of Problem (3) is a nonlinear genuinely hyperbolic system. Problem (3) is expected to have a unique solution, with $\mathbf{W}(\cdot, t)$ belonging to a Lebesgue space, in a convenient sense (including, for instance, an entropy condition). This result could be suggested by the fact that the linear problem (1) with $A = A(\mathbf{w})$ is well posed in Lebesgue spaces for any $\mathbf{w} \in D$.

Assume now that there exists $R \subset D$, $R \neq \emptyset$, such that the matrix $A(\mathbf{w})$ is diagonalizable in $\mathbb{R}$ for all $\mathbf{w} \in D \setminus R$ and has only real eigenvalues but is not diagonalizable if $\mathbf{w} \in R$. Then, the first equation of Problem (3) is said to be a nonlinear resonant hyperbolic system. The linear problem (1) has, in general, no weak solution (in the sense given before) if $A = A(\mathbf{w})$, for any $\mathbf{w} \in R$ (since it corresponds to a linear resonant hyperbolic system). In this case, two questions seem of interest:

1. Is it possible to have an existence and uniqueness result (in Lebesgue spaces) for this nonlinear resonant hyperbolic system?
2. What is the behavior of numerical schemes using a linearization of the system (and then, possibly, using some linear resonant systems)?

There are many recent works on nonlinear resonant hyperbolic systems, in particular, for proving an existence and uniqueness result for the Riemann problem. See, for instance, [GL04], [GS06] (for an example in phase transition), and [CLS04] (for the case of shallow water with topography). There are also papers devoted to the study of numerical schemes for nonlinear resonant hyperbolic systems. See [AGG04] for a quite general study and, for the case of shallow water with topography, [CLS04], [KL02], [ABB04], and [GHS03]. In this latter case, it seems possible to use linearized Riemann problems for the design of numerical schemes, even if the linearized system is resonant for the computation of some fluxes (see [GHS03]).

In this paper, we focus on a simple example, coming from the modelization of a two phase flow in an heterogeneous porous medium. It leads to a scalar equation with a flux function discontinuously depending on the spatial variable. Then, it can be seen as a nonlinear hyperbolic system and this system is resonant for some values of the unknown. For this problem, it is possible to prove, for a large class of initial data, an existence and uniqueness result of an entropy weak solution (including cases where the initial datum belongs to R, the set corresponding to resonant systems, for all $x \in \mathbb{R}$), along with the convergence of numerical schemes, following [SV03], [BV06], [Bac04], [Bac05] (for an existence and uniqueness result) and [Bac05], [Bac06] (for the convergence of numerical schemes). Many other papers are devoted to this case of a scalar conservation law with discontinuous coefficients, see, for instance, [Tow00] and [KRT03]. Some contributions on this problem are in the present proceedings.

One considers in this paper that the space variable x belongs to $\mathbb{R}$, but some extensions to $x \in \mathbb{R}^d$, $d = 2$ or 3, are possible.

2 Linear Resonant Systems

Let $p = 2$ and A be a real 2×2 matrix, which has only real eigenvalues but is not diagonalizable. Then, using a change of unknown, the Riemann problem for the linear problem (1) can be put under the following form, with some $\lambda \in \mathbb{R}$ (which is the unique eigenvalue of A):

$$\begin{bmatrix} u \\ v \end{bmatrix}_t + \begin{bmatrix} \lambda & 1 \\ 0 & \lambda \end{bmatrix} \begin{bmatrix} u \\ v \end{bmatrix}_x = 0,$$

$$\begin{bmatrix} u(x,0) \\ v(x,0) \end{bmatrix} = \begin{bmatrix} u_l \\ v_l \end{bmatrix}, \text{ if } x < 0, \text{ and } \begin{bmatrix} u_r \\ v_r \end{bmatrix}, \text{ if } x > 0,$$

with u_l, u_r, v_l, $v_r \in \mathbb{R}$. The second equation of the system and the second initial condition are decoupled from the first ones. Then, the unique solution for v (uniqueness holds even in the larger possible space of distributions) is $v(\cdot, t) = v(\cdot - \lambda t, 0)$ for all $t > 0$. It is now possible to give the solution for u (which is also unique in in the larger possible space of distributions), it is, for all $t > 0$:

$$u(\cdot, t) = u_l 1_{\{x \in \mathbb{R},\, x < \lambda t\}} + u_r 1_{\{x \in \mathbb{R},\, x > \lambda t\}} + t(v_l - v_r)\delta_{\lambda t},$$

where 1_B is the characteristic function of B, for $B \subset \mathbb{R}$, and δ_a is the Dirac mass at point a, for $a \in \mathbb{R}$. In this example, the problem has no weak solution with $u(\cdot, t)$ in a Lebesgue space but it has a unique solution in a space allowing $u(\cdot, t)$ to be a measure on the bounded sets of $\mathbb{R}$.

If $p > 2$, the (unique) solution of the Riemann problem for a linear hyperbolic resonant system may be even less regular. Indeed, the regularity of the solution depends on the difference between the algebraic and the geometric multiplicity of the eigenvalues.

To conclude this section, one also presents the Riemann problem for the simplest example of nonlinear resonnant hyperbolic system:

$$\begin{aligned} u_t + (au)_x &= 0, \\ a_t &= 0, \end{aligned} \tag{4}$$

$$\begin{bmatrix} u(x,0) \\ a(x,0) \end{bmatrix} = \begin{bmatrix} u_l \\ a_l \end{bmatrix}, \text{ if } x < 0, \text{ and } \begin{bmatrix} u_r \\ a_r \end{bmatrix}, \text{ if } x > 0, \tag{5}$$

Problems (4) and (5) have no weak solution (in the natural sense, similar to (2), and even in a weaker sense allowing the solution to takes values, for $t > 0$, in a distribution space) if $a_l > 0$, $a_r < 0$ and $a_l u_l \neq a_r u_r$ and has infinitely many weak solution with $u(\cdot, t) \in L^\infty(\mathbb{R})$ for a.e. t, if $a_l < 0$ and $a_r > 0$. See [BJ98] for the study of such problems.

Problems (4) and (5) correspond to a nonlinear hyperbolic resonant system since the system is equivalent (for regular solution) to:

$$\begin{bmatrix} u \\ a \end{bmatrix}_t + \begin{bmatrix} a & u \\ 0 & 0 \end{bmatrix} \begin{bmatrix} u \\ a \end{bmatrix}_x = 0.$$

Then, resonance occurs, for this system, when $a = 0$ and $u \neq 0$ and, as it is said before, an existence and uniqueness result of a weak solution for the Riemann problem does not hold for this nonlinear system provided that 0 is between a_r and a_l (except for some particular data).

3 Hyperbolic Equation With a Discontinuous Coefficient

The example of a nonlinear resonant hyperbolic system studied in this paper is given by a two phase flow in an heterogeneous porous medium, considering only gravity effect (without capillarity and with a total flux equal to zero). The unknown is the saturation, which is a function $u : \mathbb{R} \times \mathbb{R}_+ \to [0,1] \subset \mathbb{R}$. The equation is (forgetting the variable (x, t)):

$$\partial_t u + \partial_x (k g(u)) = 0, \text{ in } \mathbb{R} \times \mathbb{R}_+, \tag{6}$$

where $k(x) = k_l$, for $x < 0$, and $k(x) = k_r$, for $x > 0$, $k_l, k_r > 0$, $k_l \neq k_r$, the function $g : [0,1] \to \mathbb{R}$ is Lipschitz continuous, nonnegative and such that $g(0) = g(1) = 0$. A typical example, studied in [SV03], is $g(u) = u(1-u)$.

This hyperbolic equation with a discontinuous coefficient can be viewed has a conservative 2×2 system, adding k has an unknown and the equation $k_t = 0$:

$$\begin{aligned} u_t + (k g(u))_x &= 0, \\ k_t &= 0. \end{aligned}$$

Then, with $\mathbf{W} = \begin{bmatrix} u \\ k \end{bmatrix}$ and $F(\mathbf{W}) = \begin{bmatrix} k g(u) \\ 0 \end{bmatrix}$, this system is:

$$\mathbf{W}_t + (F(\mathbf{W}))_x = 0,$$

or equivalently (for regular solutions), with $A(\mathbf{w}) = DF(\mathbf{w})$ for $\mathbf{w} \in \mathbb{R}^2$:

$$\mathbf{W}_t + A(\mathbf{W})\mathbf{W}_x = 0.$$

This leads to problem (3) with $p = 2$, $\mathbf{W} = \begin{bmatrix} u \\ k \end{bmatrix}$, $A(\mathbf{W}) = \begin{bmatrix} k g'(u) & g(u) \\ 0 & 0 \end{bmatrix}$.

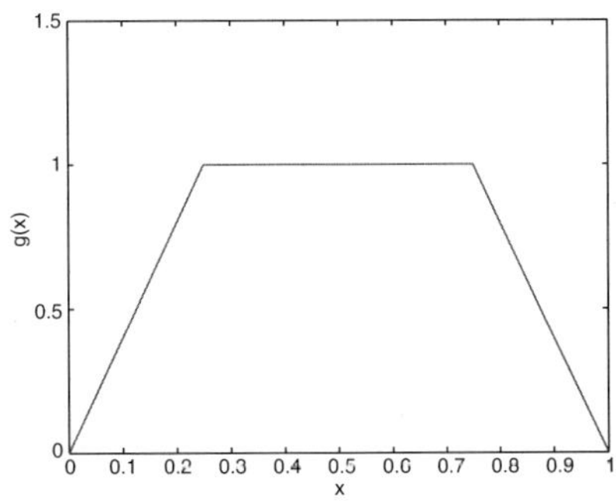

Fig. 1. Resonance occurs for all (k, u) with $u \in (\frac{1}{4}, \frac{3}{4})$

The admissibility domain is $D = \{(u,k)^t, u \in [0,1], k > 0\}$. Assuming that $g \in C^1$ (which is not a necessary hypothesis for this problem), let $R = \{(u,k)^t \in D,\ g'(u) = 0,\ g(u) \neq 0\}$. The matrix $A(\mathbf{w})$ is diagonalizable in $\mathbb{R}$ for $\mathbf{w} = (u,k)^t \in D \setminus R$ and has only 0 as eigenvalue but is not diagonalizable if $\mathbf{w} \in R$. In the case $g(u) = u(1-u)$, $R = \{1/2\} \times \mathbb{R}_+^\star$. But the domain R corresponding to resonance may be larger. In the case corresponding to Fig. 1, $R = \{(u,k)^t \in D,\ g'(u) = 0,\ g(u) \neq 0\}$ contains $(1/4, 3/4) \times \mathbb{R}_+^\star$.

Despite this resonance phenomenon, it is possible to prove existence and uniqueness of an "entropy weak solution" of (6) with an initial condition u_0, provided that $u_0 \in L^\infty(\mathbb{R})$ takes its values in $[0,1]$. Actually, it is proven in [BV06] (previous partial results were, for instance, in [SV03], [KRT03], and [Bac04]) that there exists a unique solution of the following weak entropic formulation of (6) with the initial condition u_0:

$$
\begin{aligned}
&u \in L^\infty(\mathbb{R}_+ \times \mathbb{R}),\ 0 \le u \le 1 \text{ a.e.,}\\
&\int_{\mathbb{R}_+}\int_{\mathbb{R}} [|u(x,t) - \kappa|\partial_t\varphi(x,t) + k(x)\phi(u(x,t),\kappa)\partial_x\varphi(x,t)]\,dxdt\\
&\quad + \int_{\mathbb{R}} |u_0(x) - \kappa|\varphi(x,0)dx + |k_r - k_l| \int_{\mathbb{R}_+} g(\kappa)\varphi(0,t)dt \ge 0,\\
&\qquad \forall \kappa \in [0,1],\ \forall \varphi \in C_c^\infty(\mathbb{R} \times \mathbb{R}_+, \mathbb{R}_+),
\end{aligned} \tag{7}
$$

where $\phi(s,\kappa) = \operatorname{sign}(s-k)(g(s)-g(\kappa))$ for $s \in [0,1]$. This definition of entropy weak solution was previously given in [Tow00]. Of course, as usual, an entropy weak solution of (6) with the initial condition u_0 (that is a solution of (7)) is a weak solution, that is satisfies:

$$
\begin{aligned}
&u \in L^\infty(\mathbb{R}_+ \times \mathbb{R}),\ 0 \le u \le 1 \text{ a.e.,}\\
&\int_{\mathbb{R}_+}\int_{\mathbb{R}} (u(x,t)\partial_t\varphi(x,t) + k(x)g(u(x,t)))\partial_x\varphi(x,t))dxdt\\
&\quad + \int_{\mathbb{R}} (u_0(x)\varphi(x,0)dx \ge 0,\ \ \forall \varphi \in C_c^\infty(\mathbb{R} \times \mathbb{R}_+, \mathbb{R}_+).
\end{aligned} \tag{8}
$$

A crucial property (see Sect. 4 for a sketch of proof) is that the constant functions 0 and 1 are solutions of (6). Without this hypothesis, it is sometimes possible to obtain an existence (and, possibly, with uniqueness) result for (6) with the initial condition u_0, but the solution does not takes (in general) its values in $[0, 1]$ and is not a weak solution of (6) (actually, the jump condition at point $x = 0$ is not the jump condition implicitely given in (8)). Some contributions in this direction are in the present proceedings.

The fact that the flux function has, with respect to u, the same form for $x < 0$ and $x > 0$ (namely $k_l g(u)$ and $k_r g(u)$) is not necessary. A similar result of existence and uniqueness was proven recently in [Bac05] when $k(x)g(u)$ is replaced by $g(x, u)$ with $g(x, u) = g_l(u)$ pour $x < 0$ and $g(x, u) = g_r(u)$, where g_l and g_r are some Lipschitz continuous functions from $[[0, 1]$ to $\mathbb{R}$. But, for this generalization, a main hypothesis is still that $g(\cdot, 0)$ and $g(\cdot, 1)$ are some constants functions, that is to say $g_l(0) = g_l(0)$ and $g_l(1) = g_l(1)$ (and these values are not necessarily equal to zero).

In the following section, one gives a sketch of proof of this existence and uniqueness result.

4 An Existence and Uniqueness Result

In Sect. 3, we saw the following theorem (which is proven in [BV06] and generalized in [Bac05] to a larger class of flux function):

Theorem 1. *Let $k_l, k_r > 0$, k defined (from $\mathbb{R}$ to $\mathbb{R}$) by $k(x) = k_l$, for $x < 0$, and $k(x) = k_r$, for $x > 0$, and $g : [0, 1] \to \mathbb{R}$ be a Lipschitz continuous function, nonnegative and such that $g(0) = g(1) = 0$. Let $u_0 \in L^\infty(\mathbb{R})$ be such that $u_0(x) \in [0, 1]$ for a.e. $x \in \mathbb{R}$. Then, there exists a unique solution to (7).*

One gives now a sketch of proof of this theorem.

There are similar methods for proving the existence part of Theorem 1. For instance:

1. Replace, in (6), k by a regular function k_ϵ (then, existence and uniqueness of the solution u_ϵ is classical, following Krushkov theory) converging pointwise to k (as $\epsilon \to 0$) and pass to limit as $\epsilon \to 0$.
2. Add a viscous term $-\epsilon \partial_x^2 u$ to (6) (with $\epsilon > 0$, then existence and uniqueness of the solution u_ϵ is also classical) and pass to limit as $\epsilon \to 0$.
3. Pass to the limit (as the discretization parameters go to 0) on the approximate solution given by "monotone" numerical schemes (such as the Godunov sheme). We will call also u_ϵ the approximate solution.

The first method was used in [SV03], [Bac04], and [BV06]. The second and third methods are very closed since the monotonicity of a numerical scheme leads to some viscosity term in the approximate equation. The third method is used in [Bac05] and [Bac06].

For these three methods, an L^∞ estimate on u_ϵ is quite easy. Actually, it holds $0 \le u_\epsilon \le 1$ a.e.. Then, it is possible to assume, at least for a subsequence of a sequence of approximate solutions, that $u_\epsilon \to u$ for the weak-$\star$ topology of $L^\infty(\mathbb{R} \times \mathbb{R}_+)$ and $0 \le u \le 1$ a.e.

The main difficulty (even if u_0 is regular) in order to pass to the limit (as ϵ goes to 0) and to obtain the existence part of Theorem 1 is to prove the a.e. convergence of u_ϵ towards u, at least also for a subsequence of a sequence of approximate solutions. This a.e. convergence is useful for proving that $h(u_\epsilon)$ converges towards $h(u)$ for all bounded continuous function h from $\mathbb{R}$ to $\mathbb{R}$.

The main difficulty for the uniqueness part of Theorem 1 is to prove the existence of traces for u on the line $\{(0,t), t > 0\}$.

To get rid of this two difficulties, some authors ([Tow00], [KRT03], [SV03], [Bac04]...) use the following hypothesis of genuine nonlinearity for the flux function (where "meas" stands for the Lebesgue measure on $\mathbb{R}$):

$$g \in C^2 \text{ and } \operatorname{meas}(\{x \in [0,1];\, g''(s) = 0\}) = 0. \tag{9}$$

With this hypothesis, the existence part of Theorem 1 follows by proving the a.e. convergence of u_ϵ towards u, using some tools as "Temple function" or "compensated compacness." The uniqueness part of Theorem 1 is obtain using the existence of traces for an entropy weak solution on the line $\{(0,t), t > 0\}$.

Without Assumption (9) on g, the proof of Theorem 1 is much more tricky (see [BV06] and [Bac05]). Passing to the limit as ϵ goes to 0 leads to the existence of a solution in a very weak sense, namely it gives a "kinetic process solution" (see the definition below). Then, an uniqueness result proves the fact that this kinetic process solution is indeed an entropy weak solution (and that this entropy weak solution is unique). This gives the existence part and the uniqueness part of Theorem 1. A by product of the proof is that u_ϵ converges towards u (as ϵ goes to 0) in $L^p_{loc}(\mathbb{R} \times \mathbb{R}_+)$, for all $1 \le p < \infty$ (and then a.e., at least for subsequences of sequences of approximate solutions). One gives now some details on this proof.

Since a sequence of approximate solutions is bounded in $L^\infty(\mathbb{R} \times \mathbb{R}_+)$, one can assume that, up to a subsequence, it converges towards a Young measure (see, for instance, [DiP85]) or equivalently towards some $u \in L^\infty(\mathbb{R} \times \mathbb{R}_+ \times (0,1))$ in the "nonllnear weak-$\star$ sense," using the following result (see [EGH00] and [EGH95]):

Theorem 2. *Let $N \ge 1$, Ω be an open set of $\mathbb{R}^N$ and $(u_n)_{n\in\mathbb{N}}$ be a bounded sequence of $L^\infty(\Omega)$. Then, there exists a subsequence, still denoted by $(u_n)_{n\in\mathbb{N}}$, and there exists $u \in L^\infty(\Omega \times (0,1))$ such that, for all $\psi \in L^1(\Omega)$ and all $\theta \in C(\mathbb{R},\mathbb{R})$:*

$$\int_\Omega \theta(u_n(y))\psi(y)dy \to \int_0^1 \int_\Omega \theta(u(y,\alpha))\psi(y)dyd\alpha, \ \text{as } n \to \infty.$$

Then, it is quite easy to prove that this function u is an entropy process solution, that is a solution of

$$
\begin{aligned}
& u \in L^\infty(\mathbb{R}_+ \times \mathbb{R} \times (0,1)),\ 0 \le u \le 1 \text{ a.e.},\\
& \int_0^1 \int_{\mathbb{R}_+} \int_{\mathbb{R}} [|u(x,t,\alpha) - \kappa| \partial_t \varphi(x,t) + k(x)\phi(u(x,t,\alpha),\kappa)\partial_x \varphi(x,t)]\, dx dt d\alpha \\
& \qquad + \int_{\mathbb{R}} |u_0(x) - \kappa| \varphi(x,0) dx + |k_r - k_l| \int_{\mathbb{R}_+} g(\kappa)\varphi(0,t) dt \ge 0,\\
& \qquad\qquad \forall \kappa \in [0,1],\ \forall \varphi \in C_c^\infty(\mathbb{R} \times \mathbb{R}_+, \mathbb{R}_+).
\end{aligned}
$$

If k is a regular function (then, with the hypotheseses of Theorem 1, k is a constant function, but similar results are also true if k is not a constant function), it is possible to prove an uniqueness result for this entropy process solution and the fact that u does not depends on α, by using the doubling variable technique of Krushkov (see [EGH00], for instance). This gives that u is an entropy weak solution and concludes the proof of Theorem 1 for k regular (i.e., constant). Unfortunately, the doubling variable technique does not seem easily generalizable to the case of a discontinuous function k.

To overcome this difficulty, one introduces once again a new variable, denoted ξ, and one remarks that u is also a "kinetic process solution." It means that, for all $\xi \in \mathbb{R}$, there exist two positive Radon measures on $\mathbb{R} \times \mathbb{R}_+$, denoted $m_{\xi,\pm}$, continuously depending on ξ in the sense that $\xi \mapsto \int \varphi dm_{\xi,\pm}$ is continuous for all $\varphi \in C_c(\mathbb{R} \times \mathbb{R}_+)$, such that, for all $\varphi \in C_c^\infty(\mathbb{R} \times \mathbb{R}_+ \times \mathbb{R})$:

$$
\begin{aligned}
& \int_0^1 \int_{\mathbb{R}} \int_{\mathbb{R}_+} \int_{\mathbb{R}} h_\pm(x,t,\alpha,\xi)(\partial_t \varphi(x,t,\xi) + k(x)a(\xi)\partial_x \varphi(x,t,\xi)) dx dt d\xi d\alpha \\
& \quad + \int_{\mathbb{R}} \int_{\mathbb{R}} h_\pm^{(0)}(x,\xi)\varphi(x,0,\xi) dx d\xi + \int_{\mathbb{R}} \int_{\mathbb{R}_+} (k_r - k_l)^\pm a(\xi)\varphi(0,t,\xi) dt d\xi \\
& \qquad = \textstyle\int_{\mathbb{R}} \int_{\mathbb{R}\times\mathbb{R}_+} \partial_\xi \varphi(x,t,\xi) dm_{\xi,\pm}(x,t) d\xi,
\end{aligned}
$$

where $a(\xi) = g'(\xi)$, $h_\pm(x,t,\alpha,\xi) = \text{sign}_\pm(u(t,x,\alpha) - \xi)$ and $h_\pm^{(0)}(x,\xi) = \text{sign}_\pm(u_0(x) - \xi)$ (with $\text{sign}_+(s) = 1$ if $s > 0$ and 0 if $s < 0$, $\text{sign}_-(s) = -1$ if $s < 0$ and 0 if $s > 0$). The functions $h_\pm$ and $h_\pm^{(0)}$ are the equilibrium functions associated to u and u_0. This definition of a kinetic process solution is a natural generalization of the definition of a kinetic solution for a nonlinear hyperbolic equation, see, for instance, [Per98] (the generalization appears in the variable α and in the discontinuity of k). One can now adapt the proof of uniqueness of the kinetic solution given in [Per98]. It gives here that u does not depends on α and is the unique kinetic solution of our problem and then the unique entropy weak solution (i.e., the unique solution of (7)). This concludes the proof of Theorem 1. Note also that the preceding proof gives that u_ϵ converges towards u (as ϵ goes to 0) in $L^p_{loc}(\mathbb{R} \times \mathbb{R}_+)$, for all $1 \le p < \infty$ (and then a.e., at least for subsequences of sequences of approximate solutions).

5 Numerical Schemes and Numerical Results

The presentation of the numerical schemes is restricted here to the case of a system under a conservative form, which is the case of the simple system presented in Sect. 3. An additional work has to be done for a system with a nonconservative term (this is the case for Shallow Water with topography, see [GHS03] for instance). Then, the system reads, with the same notations are before:

$$\begin{aligned} &\partial_t \mathbf{W} + \partial_x F(\mathbf{W}) = 0, \\ &\mathbf{W}(\cdot, 0) = \mathbf{W}_0, \end{aligned} \tag{10}$$

where F is a Lipschitz continuous function from $D \subset \mathbb{R}^p$ to $\mathbb{R}^p$. Recall that the unknown $\mathbf{W}$ is a function from $\mathbb{R} \times \mathbb{R}_+$ to $D \subset \mathbb{R}^p$, where D is the so-called admissible domain.

The time and space steps are denoted by δt and δx. For simplicity, they are assumed to be constant. Let $t_n = n\delta t$ and $x_{i+1/2} = ih$ for $n \in \mathbb{N}$ and $i \in \mathbb{Z}$. The approximate solution is defined by the family $\{\mathbf{w}_i^n,\ i \in \mathbb{Z},\ n \in \mathbb{N}\} \subset \mathbb{R}$, where $\mathbf{w}_i^n$ is the value of the approximate solution for $t \in (t_n, t_{n+1})$ and in the control volume $M_i = (x_{i-1/2}, x_{i+1/2})$.

The initial condition is used to compute $\{\mathbf{w}_i^0,\ i \in \mathbb{Z}\}$:

$$\mathbf{w}_i^0 = \frac{1}{\delta x} \int_{x_{i-\frac{1}{2}}}^{x_{i+\frac{1}{2}}} \mathbf{W}_0(x) dx, \ \text{ for } i \in \mathbb{Z}. \tag{11}$$

One describes now two possibilities for the computation of $\{\mathbf{w}_i^{n+1},\ i \in \mathbb{Z}\}$ using $\{\mathbf{w}_i^n,\ i \in \mathbb{Z}\}$. The first one uses the resolution of the Riemann problem associated to (10), it is the Godunov scheme. The second one uses a linearized Riemann problem.

Godunov Scheme

Let $\mathbf{w}_l,\ \mathbf{w}_r \in D$. The Riemann problem associated to $\mathbf{w}_l$ and $\mathbf{w}_r$ is (10) with $\mathbf{W}(x, 0) = \mathbf{w}_l$ if $x < 0$ and $\mathbf{W}(x, 0) = \mathbf{w}_r$ if $x > 0$. One assumes that this Riemann problem has a self similar function, which one denotes by $\mathbf{W}(x, t) = \mathbf{R}(x/t, \mathbf{w}_l, \mathbf{w}_r)$ and that it is possible to compute this solution. This is the case, in particular, for the three nonlinear resonant hyperbolic systems given in Sect. 1 (phase transition [GS06], Shallow Water with topography [CLS04], and two phase flow in an heterogeneous porous medium [Bac05]), although this solution is sometimes not unique (in the case of shallow water, see [CLS04]). One sets $\mathbf{w}^{\star,\pm}(\mathbf{w}_l, \mathbf{w}_r) = \mathbf{R}(0^\pm, \mathbf{w}_l, \mathbf{w}_r)$. The values $\mathbf{w}^{\star,\pm}(\mathbf{w}_l, \mathbf{w}_r)$ are always well defined, even if $\{(0, t),\ t > 0\}$ is a line of discontinuity for $\mathbf{W}$. Then, the Godunov scheme is defined by

$$\frac{\mathbf{w}_i^{n+1} - \mathbf{w}_i^n}{k} + \mathbf{F}_{i+\frac{1}{2}}^n - \mathbf{F}_{i-\frac{1}{2}}^n = 0,\ i \in \mathbb{Z},\ n \in \mathbb{N}, \tag{12}$$

with $\mathbf{F}^n_{i+\frac{1}{2}} = F(\mathbf{w}^{n,\pm}_{i+\frac{1}{2}})$ and $\mathbf{w}^{n,\pm}_{i+\frac{1}{2}} = \mathbf{w}^{\star,\pm}(\mathbf{w}^n_i, \mathbf{w}^n_{i+1})$. The definition of $\mathbf{F}^n_{i+\frac{1}{2}}$ is correct, since $F(\mathbf{w}^{n,+}_{i+\frac{1}{2}}) = F(\mathbf{w}^{n,-}_{i+\frac{1}{2}})$ even if $\mathbf{w}^{n,+}_{i+1/2} \neq \mathbf{w}^{n,-}_{i+1/2}$, thanks to the Rankine-Hugoniot condition for the solution of the Riemann problem. This scheme is very efficient. It uses, as usual for an explicit scheme, a CFL condition, which reads $\delta t \leq c\delta x$ where c is computed with the eigenvalues of $A(\mathbf{w}) = DF(\mathbf{w})$ ($DF(\mathbf{w})$ is the jacobian matrix of F at point $\mathbf{w} \in D$, assuming F continuously differentiable). It is sometimes too expansive and it is the reason of the introduction of a modified scheme, using a linearized Riemann problem.

The VFRoe-ncv Scheme

Assuming, for simplicity, that F is continuously differentiable, one sets $A(\mathbf{w}) = DF(\mathbf{w})$ for $\mathbf{w} \in D$, where $DF(\mathbf{w})$ is the jacobian matrix of F at point $\mathbf{w} \in D$. Let ϕ be a regular function of $D \subset \mathbb{R}^p$ to $\mathbb{R}^p$. It is not necessary to assume that ϕ is one-to-one from D to $Ra(\phi) = \{\phi(\mathbf{w}),\ \mathbf{w} \in D\}$, but one assumes that there exists a continuous function C, from D to the set of $p \times p$ matrix with real entries, and a continuous function $\tilde{F}$, from $Ra(\phi)$ to $\mathbb{R}^p$ such that $D\phi(\mathbf{w})A(\mathbf{w}) = C(\mathbf{w})D\phi(\mathbf{w})$ and $F(\mathbf{w}) = \tilde{F}(\phi(\mathbf{w}))$ for all $\mathbf{w} \in D$.

Let $\mathbf{W} : \mathbb{R} \times \mathbb{R}_+ \to D$ be a regular solution of $\partial_t \mathbf{W} + A(\mathbf{W})\partial_x \mathbf{W} = 0$. Then, $\mathbf{Y} = \phi(\mathbf{W})$ satisfy $\partial_t \mathbf{Y} + D\phi(\mathbf{W})A(\mathbf{W})\partial_x \mathbf{W} = 0$ and, thanks to the hypothesis on ϕ, the function $\mathbf{Y}$ satisfies

$$\partial_t \mathbf{Y} + C(\mathbf{W})\partial_x \mathbf{Y} = 0. \tag{13}$$

It is now possible to describes the VFRoe-ncv scheme associated to ϕ. For $\mathbf{w}_l,\ \mathbf{w}_r \in \mathbb{R}^p$, one sets $\mathbf{w}_{l,r} = (\mathbf{w}_r + \mathbf{w}_l)/2$ (it is possible to take another mean value between $\mathbf{w}_l$ and $\mathbf{w}_r$) and considers the following linear Riemann problem:

$$\begin{aligned} &\partial_t \mathbf{Y} + C(\mathbf{w}_{l,r})\partial_x \mathbf{Y} = 0, \\ &\mathbf{Y}(x,0) = \begin{cases} \mathbf{y}_l = \phi(\mathbf{w}_l) \text{ if } x < 0, \\ \mathbf{y}_r = \phi(\mathbf{w}_r) \text{ if } x > 0. \end{cases} \end{aligned} \tag{14}$$

If $C(\mathbf{w}_{l,r})$ is diagonalizable in $\mathbb{R}$, Problem (14) has a unique solution. It is a self similar function: $\mathbf{Y}(x,t) = \mathbf{R}_\phi(\frac{x}{t}, \mathbf{y}_l, \mathbf{y}_r)$. Then one sets

$$\mathbf{y}^{\star,\pm}(\mathbf{w}_l, \mathbf{w}_r) = \mathbf{R}_\phi(0^\pm, \mathbf{y}_l, \mathbf{y}_r).$$

If $C(\mathbf{w}_{l,r})$ has only real eigenvalues but is not diagonalizable in $\mathbb{R}$, the first equation of (14) is a linear resonant hyperbolic system. In this case, Problem (14) has also a unique solution but it is not, in general, a function (in the example of Sect. 2, if $\lambda = 0$, there is a Dirac mass at $x = 0$ for any $t > 0$). However, $\mathbf{R}_\phi(0^\pm, \mathbf{y}_l, \mathbf{y}_r)$ is always well defined (forgetting the Dirac mass in the example of Sect. 2) and it is also possible to set $\mathbf{y}^{\star,\pm}(\mathbf{w}_l, \mathbf{w}_r) = \mathbf{R}_\phi(0^\pm, \mathbf{y}_l, \mathbf{y}_r)$. The VFRoe-ncv scheme associated to ϕ is (11)–(12) but with $\mathbf{F}^n_{i+\frac{1}{2}} = (1/2)(\tilde{F}(\mathbf{y}^{n,+}_{i+\frac{1}{2}}) + \tilde{F}(\mathbf{y}^{n,-}_{i+\frac{1}{2}}))$ (assuming that $\mathbf{y}^{n,\pm}_{i+\frac{1}{2}} \in$

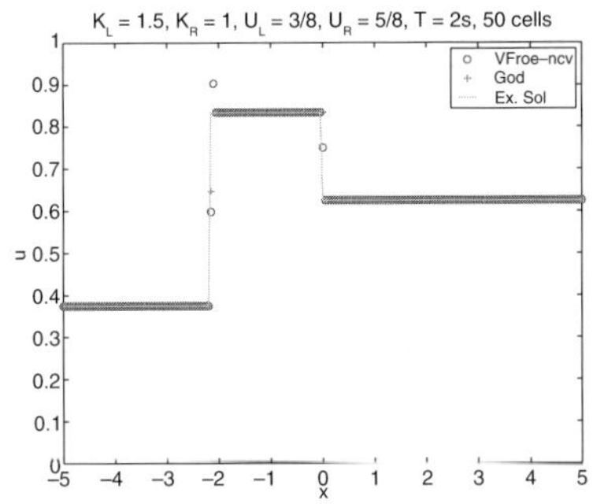

Fig. 2. Numerical results for the Godunov scheme and the VFRoe-ncv scheme

$Ra(\phi))$, $\mathbf{y}^{n,\pm}_{i+\frac{1}{2}} = \mathbf{y}^{\star,\pm}(\mathbf{w}^n_i, \mathbf{w}^n_{i+1})$. A possible drawback of the method seems to be the fact that the numerical flux of the scheme is not a continuous function of its arguments when an eigenvalue changes sign (namely, $\mathbf{F}^n_{i+1/2}$ does not depend continuously of $\mathbf{w}^n_i$ and $\mathbf{w}^n_{i+1}$). In practice, this drawback does not seem to be so important. As for the Godunov scheme, the scheme uses a CFL condition which reads $\delta t \leq c\delta x$.

In the case studied in Sect. 3, for $\mathbf{w} = (u,k)^t \in D = [0,1] \times \mathbb{R}^\star_+$, one has $F(\mathbf{w}) = (kg(u), 0)^t$. A simple choice of ϕ is $\phi(\mathbf{w}) = (kg(u), k)^t$ for $\mathbf{w} = (u,k)^t$. With this choice of ϕ, the matrix $C(\mathbf{w})$ is, for any $\mathbf{w} \in D$, diagonal and System (13) is not a resonant system.

One presents in Fig. 2 a numerical result (given in [Bac05]) with this two schemes, the function g given in Fig. 1, and $u_0(x) = 3/8$, for $x < 0$, $u_0(x) = 5/8$ for $x > 0$. This result shows the good behavior of the two schemes (with only one "wrong point" with the VFRoe-ncv scheme).

References

[AGG04] Amadori, D., Gosse, L., Guerra, G.: Godunov-type approximation for a general resonant balance law with large data. J. Differential Equations **198**, n° 2, 233–274 (2004)

[ABB04] Audusse, E., Bouchut, F., Bristeau, M.-O., Klein, R., Perthame, B.: A fast and stable well-balanced scheme with hydrostatic reconstruction for shallow water flows. SIAM J. Sci. Comp., **25**, 2050–2065 (2004)

[Bac04] Bachmann, F.: Analysis of a scalar conservation law with a flux function with discontinuous coefficients. Advances in Differential Equations, n° 11–12, 1317–1338 (2004)

[Bac05] Bachmann, F.: PhD thesis, univ. Marseille (2005) and submitted paper

[Bac06] Bachmann, F.: Finite volume schemes for a nonlinear hyperbolic conservation law with a flux function involving discontinuous coefficients. Int. J. on Finite Volume (electronic), **3**, n° 1 (2006)

[BV06] Bachmann, F., Vovelle, J.: Existence and uniqueness of entropy solution of scalar conservation laws with a flux function involving discontinuous coefficients. Comunications in PDE, **31**, 371–395 (2006)

[BJ98] Bouchut, F., James, F.: One-dimensional transport equations with discontinuous coefficients. Nonlinear Analysis, TMA, **32**, 891–933 (1998)

[CLS04] Chinnayya, A., LeRoux, A.Y., Seguin N.: A well-balanced numerical scheme for shallow-water equations with topography: resonance phenomenon. Int. J. on Finite Volume (electronic), **1**, n° 1 (2004)

[DiP85] DiPerna, R.J.: Measure-valued solutions to conservation laws. Arch. Rational Mech. Anal., **88**, n° 3, 223–270 (1985)

[EGH95] Eymard, R., Gallouët, T., Herbin, R.: Existence and uniqueness of the entropy solution to a nonlinear hyperbolic equation. Chin. Ann. of Math., 16B: 1, 1–14 (1995)

[EGH00] Eymard, R., Gallouët, T., Herbin, R.: Finite Volume Methods. In: Ciarlet, P.G., Lions, J.L. (ed) Handbook of Numerical Analysis, Vol. VII, 713–1020, North-Holland (2000)

[GHS03] Gallouët, T., Hérard, J.M., Seguin, N.: Some approximate Godunov schemes to compute shallow-water equations with topography. Comput. Fluids, **32**, n° 4, 479–513 (2003)

[GL04] Goatin, P., LeFloch, P.G.: The Riemann problem for a class of resonant nonlinear systems of balance laws. Ann. Inst. H. Poincaré - Analyse Nonlinéaire **21**, 881–902 (2004)

[GS06] Godlewski, E., Seguin, N.: The Riemann problem for a simple model of phase transition. Commun. Math. Sci., **4**, n° 1, 227–247 (2006)

[KRT03] Karlsen, K.H., Risebro, N.H., Towers, J.D.: L^1 stability for entropy solutions of nonlinear degenerate parabolic convection-diffusion equations with discontinuous coefficients. Sbr. K. Nor. Vid. Sel., n° 3, 1–49 (2003)

[KL02] Kurganov, A., Levy, D.: Central-Upwind Schemes for the Saint-Venant System. Math. Mod. and Num. An., **36**, 397–425 (2002)

[Per98] Perthame, B.: Uniqueness and error estimates in first order quasilinear conservation laws via the kinetic entropy defect measure. J. Math. Pures Appl., **77**, n° 10, 1055–1064 (1998)

[SV03] Seguin, N., Vovelle, J.: Analysis and approximation of a scalar conservation law with a flux function with discontinuous coefficients. Math. Models Methods Appl. Sci., **13**, n° 2, 221–257 (2003)

[Tow00] Towers, J.D.: Convergence of a difference scheme for conservation laws with a discontinuous flux. SIAM J. Numer. Anal., **38**, n° 2, 681–698 (2000)

L^p-Stability Theory of the Boltzmann Equation Near Vacuum

S.-Y. Ha, M. Yamazaki, and S.-B. Yun

1 Introduction

The spatially inhomogeneous Boltzmann equation describes the phase space evolution of a one-particle distribution function $f = f(x, \xi, t)$ of moderately dilute gas particles at physical location $x \in \mathbf{R}^3$ and with velocity $\xi \in \mathbf{R}^3$ at time $t \in \mathbf{R}_+$. In the absence of external forces, it reads

$$\begin{aligned} &\partial_t f + \xi \cdot \nabla_x f = Q(f, f), \qquad (x, \xi, t) \in \mathbf{R}^3 \times \mathbf{R}^3 \times \mathbf{R}_+, \\ &f(x, \xi, 0) = f_0(x, \xi), \end{aligned} \tag{1}$$

where $Q(f, f)$ is the collision operator, which only acts on the velocity variable ξ:

$$Q(f, f) \equiv \frac{1}{\kappa} \int_{\mathbf{R}^3 \times \mathbf{S}^2_+} B(\xi - \xi_*, \omega)(f' f'_* - f f_*) d\omega d\xi_* \tag{2}$$

with the usual notations

$$f' = f(x, \xi', t), \quad f'_* = f(x, \xi'_*, t), \quad f = f(x, \xi, t) \quad \text{and} \quad f_* = f(x, \xi_*, t),$$

where

$$\xi' = \xi - [(\xi - \xi_*) \cdot \omega]\omega \quad \text{and} \quad \xi'_* = \xi_* + [(\xi - \xi_*) \cdot \omega]\omega \tag{3}$$

are the postcollisional velocities of two molecules colliding with velocities ξ and ξ_* respectively. Here κ denotes the Knudsen number which is the ratio between mean free path and the characteristic length of a flow and $\mathbf{S}^2_+ \equiv \{\omega \in \mathbf{S}^2 : (\xi - \xi_*) \cdot \omega > 0\}$.

Following the notations of Kaniel-Shinbrot [16], we use auxiliary functions $f^\sharp$ and $Q^\sharp(f, f)$ evaluated along the particle trajectory

$$f^\sharp(x, \xi, t) \equiv f(x + t\xi, \xi, t) \quad \text{and} \quad Q^\sharp(f, f)(x, \xi, t) \equiv Q(f, f)(x + t\xi, \xi, t).$$

We integrate (1) along the particle path to get a mild form

$$f^\sharp(x,\xi,t) = f_0(x,\xi) + \int_0^t Q^\sharp(f,f)(x,\xi,s)ds. \tag{4}$$

The definitions of mild solutions and classical solutions can be stated as follows.

Definition 1. *1. Let T be a given positive number. A nonnegative function $f \in C([0,T); L^1_+(\mathbf{R}^3 \times \mathbf{R}^3))$ is a mild solution of (1) with a nonnegative initial datum f_0 if and only if for all $t \in [0,T)$ and a.e. $(x,\xi) \in \mathbf{R}^3 \times \mathbf{R}^3$, f satisfies the integral equation (4) in pointwise sense.*

2. A function $f \in C(\mathbf{R}^3 \times \mathbf{R}^3 \times [0,T))$ is a classical solution of (1) with a nonnegative initial datum f_0 if and only if f is continuously differentiable with respect to (x,t) and f satisfies the equation (1) in pointwise sense.

For L^p-estimates and stability theory to the Boltmzann equation, there are two types of results available in previous literatures. The first type of L^p results is concerned with the space homogeneous Boltzmann equation in [1, 2, 5, 7, 8, 19, 26, 27]. In contrast, the second type of result is for the space-inhomogeneous case near global Maxwellian, where the energy method [9, 17, 24, 25] is effectively used for the linearized Boltzmann equation around a background global Maxwellian. However, for the space-inhomogeneous Boltzmann equation near vacuum, L^p-type estimates are not available, which is a stark difference compared to a near Maxwellian regime. The standard L^p-estimates based on the energy method do not seem to work in this regime, due to the triviality of the linearized collision operator.

In the sequel, we denote by C the generic constant independent of time t. In this short article, we briefly present main result and idea in authors' recent paper [13] for a robust L^p-stability theory for the space inhomogeneous Boltzmann equation. The L^p-type stability of classical solutions are subject to the existence of the following nonlinear functional $\mathcal{H}_M(t) \equiv \mathcal{H}_M(f(t),\bar{f}(t)),\ M \in \mathbf{Z}_+$ satisfying

1. $\mathcal{H}_M(t)$ is equivalent to $\displaystyle\sum_{p=1}^{M} ||f(t)-\bar{f}(t)||^p_{L^p}$ in the sense that

$$\frac{1}{C}\sum_{p=1}^{M} ||f(t)-\bar{f}(t)||^p_{L^p} \le \mathcal{H}_M(t) \le C\sum_{p=1}^{M} ||f(t)-\bar{f}(t)||^p_{L^p}.$$

2. $\mathcal{H}_M(t)$ satisfies the uniform stability estimate:

$$\mathcal{H}_M(t) + C\int_0^t \Lambda_{dM}(s)ds \le C\mathcal{H}_M(0), \qquad t \ge 0.$$

This yields the uniform summational L^p-type stability estimate

$$\sup_{0\leq t<\infty} \sum_{p=1}^{M} ||f(t)-\bar{f}(t)||_{L^p}^p \leq C\sum_{p=1}^{M} ||f_0-\bar{f}_0||_{L^p}^p. \tag{5}$$

Here we used a simplified notation for L^p-distance: For $1 \leq p \leq \infty$,

$$||f(t)||_{L^p} \equiv ||f(\cdot,\cdot,t)||_{L^p(\mathbf{R}^6)}.$$

We now explain why such an L^p-type stability estimate makes sense for the Boltzmann equation near vacuum. As is well-known in [3, 10, 11, 14, 15, 18, 20, 21, 22, 23], the dynamics of the Boltzmann equation near vacuum is equivalent to that of the free molecular motion time-asymptotically:

$$\begin{aligned} &\partial_t g + \xi\cdot\nabla_x g = 0, \qquad (x,\xi,t) \in \mathbf{R}^3\times\mathbf{R}^3\times\mathbf{R}_+,\\ &g(x,\xi,0) = g_0(x,\xi). \end{aligned} \tag{6}$$

Hence, we can regard the Boltzmann equation (1) in a close-to-a vacuum regime as the small perturbation of the free transport equation (6), and reasonably expect that the solutions of (1) behave like solutions of (6) qualitatively as long as the perturbation is small enough. Let g and $\bar{g}$ be the $L^p_{x,\xi}$-weak solutions to (6) corresponding to $L^p_{x,\xi}$-initial data g_0 and $\bar{g}_0$. Then the method of characteristics yields explicit representations for g and $\bar{g}$:

$$g(x,\xi,t) = g_0(x-t\xi,\xi) \qquad \text{and} \qquad \bar{g}(x,\xi,t) = \bar{g}_0(x-t\xi,\xi).$$

These imply that the L^p-distance between g and $\bar{g}$ is time-invariant, i.e.,

$$||g(t)-\bar{g}(t)||_{L^p} = ||g_0-\bar{g}_0||_{L^p}, \qquad t\geq 0. \tag{7}$$

Therefore we might expect that the solutions f and $\bar{f}$ to the Boltmzann equation (1) satisfy

$$||f(t)-\bar{f}(t)||_{L^p} \leq C||f_0-\bar{f}_0||_{L^p}, \qquad t\geq 0, \tag{8}$$

or some variants of (8). So far it is unknown whether $L^p_{x,\xi}$-mild solutions to the Boltzmann equation satisfy uniform $L^p_{x,\xi}$-stability (8) or not.

Below we discuss the framework for L^p-stability theory. We set the standard bounding functions decaying algebraically: For $\mu_1,\mu_2,\delta > 0$,

$$h_{\mu_1}(x) \equiv \frac{1}{(1+|x|^2)^{\frac{\mu_1}{2}}} \qquad \text{and} \qquad m_{\mu_2}(\xi) \equiv \frac{1}{(1+|\xi|^2)^{\frac{\mu_2}{2}}},$$

and define a set $\mathcal{S}(\mu_1,\mu_2,\delta)$ and a norm $|||\cdot|||$:

$$\begin{aligned} &|||f||| \equiv \sup_{x,\xi,t} f^\sharp(x,\xi,t)h_{\mu_1}^{-1}(x)m_{\mu_2}^{-1}(\xi),\\ &\mathcal{S}(\mu_1,\mu_2,\delta) \equiv \{f\in C(\mathbf{R}^3\times\mathbf{R}^3\times\mathbf{R}_+):\ |||f||| < \delta\}. \end{aligned}$$

Note that for $\mu_1 > 3, \mu_2 > 3$, $\mathcal{S}(\mu_1, \mu_2, \delta)$ is a subset of $L^\infty(\mathbf{R}_+; (L^1 \cap L^\infty)(\mathbf{R}^6))$. The main structural assumptions $\mathcal{A}$ employed in this paper are as follows.

Main assumptions $\mathcal{A}$

- The collision kernel satisfies an angular cut-off assumption, and the intermolecular force satisfies an inverse power law:

$$B(\xi - \xi_*, \omega) = |\xi - \xi_*|^\gamma b_\gamma(\theta), \quad -2 < \gamma \leq 1 \quad \text{and} \quad \frac{b_\gamma(\theta)}{\cos\theta} \leq B_* < \infty,$$

where θ denotes a scattering angle between $\xi - \xi_*$ and ω, i.e.,

$$\theta \equiv \cos^{-1}\left(\frac{(\xi - \xi_*)\cdot\omega}{|\xi - \xi_*|}\right).$$

- The parameters in $\mathcal{S}(\mu_1, \mu_2, \delta)$ satisfy

$$0 < \delta \ll \min\{\kappa, 1\}, \qquad \mu_1 > 4 \qquad \text{and} \qquad \mu_2 > 7.$$

The rest of the paper is organized as follows. In Sect. 2, we present explicit generalized collision potentials and their time-decay estimates along smooth Boltmzann flow. Section 3 is devoted to the uniform L^p-type stability estimates for classical solutions.

2 Generalized Collision Potentials

In this section, we introduce a generalized $(p, 1)$-type collision potential $\mathcal{D}^{p,1}(f(t))$ measuring all possible *future collisions after time* t between particles with densities f^p and f respectively. Since estimates in this section can be generalized to continuous mild solutions using the standard mollification procedure as in [4, 12], we only consider classical solutions.

Note that f^p and f satisfy

$$\partial_t f^p + \xi\cdot\nabla_x f^p = p f^{p-1} Q(f, f), \qquad p \in \mathbf{Z}_+, \tag{9}$$

$$\partial_t f + \xi\cdot\nabla_x f = Q(f, f). \tag{10}$$

For notational simplicity, we suppress the t-dependence in f and $Q(f, f)$:

$$f(x, \xi) \equiv f(x, \xi, t), \qquad Q(f, f)(x, \xi) \equiv Q(f, f)(x, \xi, t).$$

We define a generalized $(p, 1)$-type collision potential and its corresponding production rate: For $M \in \mathbf{Z}_+$,

$$
\begin{aligned}
\mathcal{D}_M(f(t)) &\equiv \mathcal{D}^{1,1}(f(t)), \quad M = 1, \\
&\equiv \sum_{p=2}^{M} \mathcal{D}^{p,1}(f(t)) + K_1 \mathcal{D}^{1,1}(f(t)), \quad M \geq 2,
\end{aligned}
$$

$$
\begin{aligned}
\mathcal{D}^{p,1}(f(t)) &\equiv \int_{\mathbf{R}^6} f^{p\sharp}(x,\xi) \\
&\quad \times \Big[\int_{\mathbf{R}^3 \times \mathbf{R}_+} |\xi - \xi_*|^{\gamma-1} f^{\sharp}(x + t(\xi - \xi_*) + \tau\omega(\xi,\xi_*), \xi_*) d\tau d\xi_* \Big] d\xi dx,
\end{aligned} \tag{11}
$$

$$
\Lambda^{p,1}(f(t)) \equiv \int_{\mathbf{R}^9} |\xi - \xi_*|^{\gamma} f^{p\sharp}(x,\xi) f^{\sharp}(x + t(\xi - \xi_*), \xi_*) d\xi_* d\xi dx, \tag{12}
$$

where K_1 is a large positive constant to be determined later, and $\omega(\xi, \xi_*)$ is the unit vector:

$$
\omega(\xi,\xi_*) \equiv \frac{\xi - \xi_*}{|\xi - \xi_*|}, \qquad \xi \neq \xi_*.
$$

The weight K_1 in front of $\mathcal{D}^{1,1}$ was introduced to modulate $\mathcal{D}^{p,1}$ in time-evolution estimate.

Notice that for any function $f \in \mathcal{S}(\mu_1, \mu_2, \delta)$, the above collision potentials are well-defined, i.e.,

$$
\mathcal{D}^{p,1}(f(t)) \leq \delta^{p-1} \Big[\frac{8\mu_1 \delta \pi (\gamma + 5)\mu_2 - 9}{3(\mu_1 - 1)(\gamma + 2)(\mu_2 - 3)} \Big] ||f_0||_{L^1} < \infty.
$$

Since $\mathcal{D}^{1,1}$ is the collision potential introduced in [4, 10, 11, 12], it is non-increasing along mild and classical solutions to (1) under the smallness assumption $0 < \delta \ll \min\{\kappa, 1\}$.

Proposition 1. [10, 11] *Suppose that main assumptions $\mathcal{A}$ in Sect. 1 hold, and let f be a classical solution in $\mathcal{S}(\mu_1, \mu_2, \delta)$ of (1) corresponding to initial datum f_0. Then $\mathcal{D}^{1,1}(f(t))$ satisfies a Lyapunov estimate*

$$
\frac{d}{dt} \mathcal{D}^{1,1}(f(t)) \leq -C_0 \Lambda^{1,1}(f(t)), \qquad t \geq 0,
$$

where C_0 is a positive constant independent of time t.

Next proposition shows that although $\mathcal{D}^{p,1}, (p \geq 2)$ may not be non-increasing along classical solutions of (1), however positive terms in their time-evolution estimates can be controlled by definite negative terms in the time-evolution of $\mathcal{D}^{1,1}$, which results in the nonincreasing of $\mathcal{D}_M$.

Proposition 2. [13] *Suppose that main assumptions $\mathcal{A}$ in Sect. 1 hold, and let f be a classical solution of (1) in $\mathcal{S}(\mu_1, \mu_2, \delta)$ corresponding to initial datum f_0. Then we have*

$$
(i) \;\; \frac{d}{dt} \mathcal{D}^{p,1}(f(t)) \leq -\Lambda^{p,1}(f(t)) + p\delta^p \frac{C_1}{\kappa} \Lambda^{1,1}(f(t)), \qquad t \geq 0, \quad p \geq 2.
$$

$$
(ii) \;\; \frac{d}{dt} \mathcal{D}_M(f(t)) \leq -\sum_{p=2}^{M} \Lambda^{p,1}(f(t)) - C_2 \Lambda^{1,1}(f(t)),
$$

where $C_1 = C_1(\mu_1, \mu_2, \gamma, B_*)$ and C_2 are positive constants independent of time t.

3 Uniform L^p-Stability Estimates

In this section, we present a new L^p-type stability estimate for classical solutions to (1) in a close-to-a vacuum regime in [13]. Let f and $\bar{f}$ be two classical solutions in $\mathcal{S}(\mu_1, \mu_2, \delta)$ to (1). For L^p-stability analysis, we devise a new nonlinear functional $\mathcal{H}^p(t) \equiv \mathcal{H}^p(f(t), \bar{f}(t))$ which is equivalent to p-th power of L^p-distance between f and $\bar{f}$. Note that the difference $|f - \bar{f}|$ satisfies a differential inequality:

$$\partial_t |f - \bar{f}| + \xi \cdot \nabla_x |f - \bar{f}| \le R(f, \bar{f}). \tag{13}$$

Here $R(f, \bar{f})$ is given by

$$\begin{aligned} R(f, \bar{f})(x, \xi) \equiv \frac{1}{2\kappa} \int_{\mathbf{R}^3 \times \mathbf{S}^2_+} & |\xi - \xi_*|^\gamma b_\gamma(\theta) \Big[|f' - \bar{f}'|(f'_* + \bar{f}'_*) + |f'_* - \bar{f}'_*|^\sharp \\ & \times (f'^\sharp + \bar{f}'^\sharp) + |f - \bar{f}|^\sharp (f^\sharp_* + \bar{f}^\sharp_*) + |f_* - \bar{f}_*|^\sharp (f^\sharp + \bar{f}^\sharp) \Big] d\omega d\xi_*. \end{aligned}$$

For a given positive integer $M \ge 2$, we define

$$\begin{aligned} \mathcal{D}_{dM}(t) &\equiv \sum_{p=1}^M \mathcal{D}_d^{p,1}(t), \qquad \mathcal{H}_M(t) \equiv \sum_{p=1}^M \mathcal{H}^p(t), \\ \mathcal{D}_d^{p,1}(t) &\equiv \int_{\mathbf{R}^6} |f - \bar{f}|^{p\sharp}(x, \xi) \Big[\int_{\mathbf{R}^3 \times \mathbf{R}_+} |\xi - \xi_*|^{\gamma - 1} \\ & \quad \times (f^\sharp + \bar{f}^\sharp)(x + t(\xi - \xi_*) + \tau\omega(\xi, \xi_*), \xi_*) d\tau d\xi_* \Big] d\xi dx, \\ \mathcal{H}^p(t) &\equiv \begin{cases} ||f(t) - \bar{f}(t)||^p_{L^p} + \mathcal{D}_d^{p,1}(t), & p \ge 2, \\ ||f(t) - \bar{f}(t)||_{L^1} + K_2 \mathcal{D}_d^{1,1}(t), & p = 1, \end{cases} \end{aligned}$$

where $K_2(>1)$ is a large positive constant to be determined later.

Notice that $\mathcal{D}_d^{p,1}$ is the $(p,1)$-type collision potential between $|f - \bar{f}|$, f and $\bar{f}$ introduced in Sect. 2. On the other hand, the functional $\mathcal{H}^p$ can be rewritten as

$$\begin{aligned} \mathcal{H}^p(t) \le \int_{\mathbf{R}^6} & |f - \bar{f}|^{p\sharp}(x, \xi) \Big[1 + K_2 \int_{\mathbf{R}^3 \times \mathbf{R}_+} |\xi - \xi_*|^{\gamma - 1} \\ & \times (f^\sharp + \bar{f}^\sharp)(x + t(\xi - \xi_*) + \tau\omega(\xi, \xi_*), \xi_*) d\tau d\xi_* \Big] d\xi dx. \end{aligned}$$

It is easy to see that the nonlinear functional $\mathcal{H}^p(t)$ is equivalent to $||f(t) - \bar{f}(t)||^p_{L^p}$ in the sense that

$$||f(t) - \bar{f}(t)||^p_{L^p} \le \mathcal{H}^p(t) \le C_3 ||f(t) - \bar{f}(t)||^p_{L^p}, \qquad t \ge 0.$$

We next define a collision production functional resulting from the time-decay estimate of $\mathcal{H}^p$:

$$\Lambda_{dM}(t) \equiv \sum_{p=1}^{M} \Lambda^{p,1}_d(t),$$

$$\Lambda^{p,1}_d(t) \equiv \int_{\mathbf{R}^9} |\xi - \xi_*|^\gamma |f - \bar{f}|^{p\sharp}(x,\xi)(f^\sharp + \bar{f}^\sharp)(x + t(\xi - \xi_*), \xi_*) d\xi_* d\xi dx.$$

We next study the time-variation of the aforementioned functionals. Unlike the time-decay of the functional $\mathcal{D}_M$, the accumulative collision potential $\mathcal{D}_{dM}$ might be increasing in t due to the asymmetry ($|f - \bar{f}|$ and $f, \bar{f}$), however the possible increment in $\frac{d}{dt}\mathcal{D}_{dM}(t)$ can be controlled by the time-integrability of the positive collision operator $Q_+(f,f)$.

Proposition 3. [13] *Suppose the main assumptions $\mathcal{A}$ in Sect. 1 hold, and let f and $\bar{f}$ be two classical solutions of (1) in $\mathcal{S}(\mu_1, \mu_2, \delta)$. Then $\mathcal{D}^{p,1}_d$ satisfies*

$$\text{(i)} \quad \frac{d}{dt}\sum_{p=1}^{M} ||f(t) - \bar{f}(t)||^p_{L^p} \le \frac{4\pi B_*}{\kappa(1-2\delta)^2} \Lambda^{1,1}_d(t).$$

$$\text{(ii)} \quad \frac{d}{dt}\mathcal{D}^{p,1}_d(t) \le -\Lambda^{p,1}_d(t) + \frac{p(2\delta)^{p-1}\delta C_4}{\kappa}\Lambda^{1,1}_d(t) + \Big[\mathcal{E}(f(t)) + \mathcal{E}(\bar{f}(t))\Big] ||f(t) - \bar{f}(t)||^p_{L^p}, \quad p \ge 1,$$

where $C_4 = C_4(\mu_1, \mu_2, B_*)$ is a positive constant independent of t.

Theorem 1. [13] *Suppose the main assumptions $\mathcal{A}$ hold, and let f and $\bar{f}$ be two classical solutions in $\mathcal{S}(\mu_1, \mu_2, \delta)$ of (1) corresponding to initial data f_0 and $\bar{f}_0$ respectively. Then we have*

$$\sum_{p=1}^{M} ||f(t) - \bar{f}(t)||^p_{L^p} \le G_1 \sum_{p=1}^{M} ||f_0 - \bar{f}_0||^p_{L^p}, \quad \textit{for all } M \in \mathbf{Z}_+,$$

where G_1 is a positive constant independent of t and M.

Remark 1. 1. For $M = 1$, the above stability estimate is reduced to the well-known L^1-stability estimate given in [4, 6, 10, 11, 12]:

$$\sup_{0 \le t < \infty} ||f(t) - \bar{f}(t)||_{L^1} \le G_1 ||f_0 - \bar{f}_0||_{L^1}.$$

2. The above stability estimate implies the following L^p-stability type estimate

$$\sup_{0 \le t < \infty} ||f(t) - \bar{f}(t)||_{L^p} \le \left(G_1 \sum_{k=1}^{p} ||f_0 - \bar{f}_0||^k_{L^k} \right)^{\frac{1}{p}}, \quad 1 \le p < \infty.$$

Acknowledgments

Research of S.-Y Ha is supported by the grant of KRF-2006-312-C00042, research of M. Yamazaki is supported in part by Grant-in-Aid for Scientific Research (C), Japan Society for the Promotion of Science and the research of S.-B Yun is supported by the BK21 and Seoul Science Fellowship.

References

1. Arkeryd, L.: Stability in L^1 for the spatially homogeneous Boltzmann equation. Arch. Rational Mech. Anal. **103**, 151–168 (1988)
2. Arkeryd, L.: L^∞-estimates for the space homogeneous Boltzmann equation. J. Stat. Phys. **31**, 347–361 (1982)
3. Bellomo, N., Toscani, G.: On the Cauchy problem for the nonlinear Boltzmann equation: global existence, uniqueness and asymptotic behavior. J. Math. Phys. **26**, 334–338 (1985)
4. Chae, M., Ha, S.-Y.: Stability estimates of the Boltzmann equation with quantum effects. Contin. Mech. Thermodyn. **17**, 511–524 (2006)
5. Desvillettes, L., Mouhot, C.: About L_p estimates for the spatially homogeneous Boltzmann equation. Ann. Inst. H. Poincare Anal. Non Lineaire **22**, 127–142 (2005)
6. Duan, R., Yang, T., Zhu, C.: L^1 and BV-type stability of the Boltzmann equation with external forces. J. Differential Equations **227**, 1–28 (2006)
7. Gustafsson, T.: L^p-estimates for the nonlinear spatially homogeneous Boltzmann equation. Arch. Rational Mech. Anal. **92**, 23–57 (1997)
8. Gustafsson, T.: Global L^p-properties for the spatially homogeneous Boltzmann equation. Arch. Rational. Mech. Anal. **103**, 1–38 (1988)
9. Guo, Y.: The Boltzmann equation in the whole space. Indiana Univ. Math. J. **53**, 1081–1094 (2004)
10. Ha, S.-Y.: Nonlinear functionals of the Boltzmann equation and uniform stability estimates. J. Differential Equations **215**, 178–205 (2005)
11. Ha, S.-Y.: L^1 stability of the Boltzmann equation for the hard sphere model. Arch. Rational Mech. Anal. **173**, 279–296 (2004)
12. Ha, S.-Y., Yun, S.-B.: Uniform L^1-stability estimates of the Boltzmann equation near a local Maxwellian. Physica D **120**, 79–97 (2006)
13. Ha, S.-Y., Yamazaki, M., Yun, S.-B.: Uniform L^p-stability estimate for the spatially inhomogeneous Boltzmann equation near vacuum. Submitted.
14. Hamdache, K.: Existence in the large and asymptotic behavior for the Boltzmann equation. Japan. J. Appl. Math. **2**, 1–15 (1984)
15. Illner, R., Shinbrot, M.: Global existence for a rare gas in an infinite vacuum. Commun. Math. Phys. **95**, 217–226 (1984)
16. Kaniel, S., Shinbrot, M.: The Boltzmann equation 1: Uniqueness and local existence. Commun. Math. Phys. **58**, 65–84 (1978)
17. Liu, T.-P., Yang, T., Yu, S.-H.: Energy method for Boltzmann equation. Physica D **188**, 178–192 (2004)
18. Lu, X.: Spatial decay solutions of the Boltzmann equation: converse properties of long time limiting behavior. SIAM J. Math. Anal. **30**, 1151–1174 (1999)

19. Mouhot, C., Villani, C.: Regularity theory for the spatially homogeneous Boltzmann equation with cut-off. Arch. Rational Mech. Anal. **173**, 169–212 (2004)
20. Polewczak, J.: Classical solution of the nonlinear Boltzmann equation in all R^3: asymptotic behavior of solutions. J. Stat. Phys. **50**, 611–632 (1988)
21. Toscani, G.: Global solution of the initial value problem for the Boltzmann equation near a local Maxwellian. Arch. Rational Mech. Anal. **102**, 231–241 (1988)
22. Toscani, G.: H-theorem and asymptotic trend of the solution for a rarefied gas in a vacuum. Arch. Rational Mech. Anal. **100**, 1–12 (1987)
23. Toscani, G.: On the nonlinear Boltzmann equation in unbounded domains. Arch. Rational Mech. Anal. **95**, 37–49 (1986)
24. Ukai, S., Yang, T.: The Boltzmann equation in the space $L^2 \cap L^\infty_\beta$: global and time-periodic solutions. Anal. Appl. (Singap.) **4**(3), 263–310 (2006)
25. Ukai, S., Yang, T.: Mathematical theory of Boltzmann equation. Lecture notes series-no. 8, Liu Bie Ju Center for Math. Sci. City University of Hong Kong, 2006
26. Wennberg, B.: Stability and exponential convergence for the Boltzmann equation. Arch. Rational Mech. Anal. **130**, 103–144 (1995)
27. Wennberg, B.: Stability and exponential convergence in L^p for the spatially homogeneous Boltzmann equation. Nonlinear Anal. TMA, **20**, 935–964 (1993)

A Relaxation Scheme for the Two-Layer Shallow Water System

R. Abgrall and S. Karni

1 Introduction

The shallow water equations are commonly used to describe flows that are nearly horizontal. In stratified flows, one considers layers that are distinguished by different densities, ρ_1 and ρ_2, for example due to different water salinity, with the lighter fluid placed on top of the heavy one, $\rho_2/\rho_1 = r \le 1$, and $r \approx 1$. The model is given by

$$
\begin{aligned}
&\frac{\partial}{\partial t}(\rho_1 h_1) + \frac{\partial}{\partial x}(\rho_1 h_1 u_1) = 0\\
&\frac{\partial}{\partial t}(\rho_1 h_1 u_1) + \frac{\partial}{\partial x}\left(\rho_1 h_1 u_1^2 + g\rho_1 \frac{h_1^2}{2} + g\rho_2 h_1 h_2\right) = \rho_2 g h_2 \frac{\partial h_1}{\partial x} - \rho_1 g h_1 \frac{\partial B}{\partial x}\\
&\frac{\partial}{\partial t}(\rho_2 h_2) + \frac{\partial}{\partial x}(\rho_2 h_2 u_2) = 0\\
&\frac{\partial}{\partial t}(\rho_2 h_2 u_2) + \frac{\partial}{\partial x}\left(\rho_2 h_2 u_2^2 + g\rho_2 \frac{h_2^2}{2}\right) = -\rho_2 g h_2 \frac{\partial h_1}{\partial x} - \rho_2 g h_2 \frac{\partial B}{\partial x}
\end{aligned}
\tag{1}
$$

with h_i and u_i denoting the respective layers depth and (averaged) velocity, and $B(x)$ denoting the bottom elevation. The notation and setup for the two-layer system is illustrated in Fig. 1.

The system contains source terms due to bottom topography, and nonconservative products describing momentum exchange between the layers. The presence of nonconservative products has major consequences for the analysis of solutions and for their numerical computations, both of which rely heavily on the conservation form. Using

$$
\begin{aligned}
W &= (\rho_1 h_1, \rho_1 h_1 u_1, \rho_2 h_2, \rho_2 h_2 u_2) \quad \text{and}\\
S &= (0, -\rho_1 g h_1 B'(x), 0, -\rho_2 g h_2 B'(x))
\end{aligned}
$$

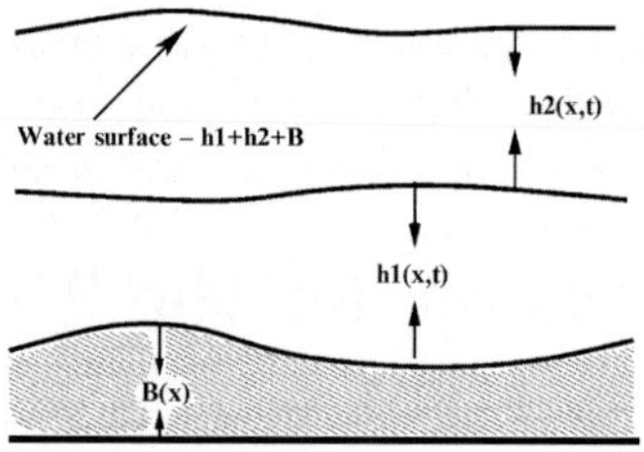

Fig. 1. Schematic for the two layer shallow water model

to denote the solution vector and the vector containing the geometric source due to bottom topography, the system may be written in quasilinear form

$$W_t + A(W)W_x = S(x, W)$$

with

$$A(W) = \begin{pmatrix} 0 & 1 & 0 & 0 \\ gh_1 - u_1^2 & 2u_1 & gh_1 & 0 \\ 0 & 0 & 0 & 1 \\ rgh_2 & 0 & gh_2 - u_2^2 & 2u_2 \end{pmatrix}, \tag{2}$$

where the coefficient matrix $A(W)$ in (2) contains flux gradient terms as well as the nonconservative products.

The characteristic polynomial for the two layer system

$$\left(\lambda^2 - 2u_1\lambda + u_1^2 - gh_1\right)\left(\lambda^2 - 2u_2\lambda + u_2^2 - gh_2\right) = rg^2h_1h_2 \tag{3}$$

does not provide analytic expression for the eigenvalues. To first order approximation in $u_2 - u_1$, the eigenvalues are given by [SCH53, LNG56]

$$\begin{aligned} \lambda_{ext}^{\pm} &= U_m \pm \sqrt{g(h_1 + h_2)} \\ \lambda_{int}^{\pm} &= U_c \pm \sqrt{g' \frac{h_1h_2}{h_1 + h_2}\left[1 - \frac{(u_1 - u_2)^2}{g'(h_1 + h_2)}\right]} \end{aligned} \tag{4}$$

where

$$U_m = \frac{h_1u_1 + h_2u_2}{h_1 + h_2}, \qquad U_c = \frac{h_1u_2 + h_2u_1}{h_1 + h_2},$$

and $g' = (1 - r)g$ is the *reduced gravity*. Equation (4) reveals that the system (1) is only conditionally hyperbolic, provided

$$\left(u_1 - u_2\right)^2 < g'\left(h_1 + h_2\right). \tag{5}$$

Violation of (5) is associated with shear layer interfacial instability, for which the vertically averaged shallow water model (1) may not be suitable.

Smooth steady state solutions of (1) are characterized by four constants

$$\begin{aligned} Q_1 &= h_1 u_1, & H_1 &= \tfrac{1}{2}u_1^2 + g\big(h_1 + B\big) + rgh_2, \\ Q_2 &= h_2 u_2, & H_2 &= \tfrac{1}{2}u_2^2 + g\big(h_1 + h_2 + B\big), \end{aligned} \tag{6}$$

representing the (specific) mass flow rate and energy in the respective layers. For a given choice of parameters, the (smooth) steady flow may be solved by nonlinear root finding. The structure of steady shallow water flow bears great resemblance to steady flow in a converging-diverging channel, with the role of the *Mach number* in channel flow assumed by the *Froude number* in shallow water flow. The *internal* Froude numbers, F_i, and *composite* Froude number, G, are given respectively by

$$F_i^2 = \frac{u_i^2}{g' h_i}, \qquad G^2 = F_1^2 + F_2^2 - (1-r)F_1^2 F_2^2.$$

2 A Relaxation Model

Even within the hyperbolic regime, the eigenstructure of (1) is given only implicitly. As a result, numerical methods based on characteristic field decomposition, need to compute the eigenvalues by numerical rootfinding, and the corresponding eigenvectors by solving numerically a linear system (see [CMP01]). Availability of analytic expressions for the e-vectors may facilitate building desirable properties into the scheme (e.g., accuracy near steady-state). In the present section, we propose a relaxation approach to address these issues.

We define two auxiliary variables, h_1' and h_2', and propose the relaxation model

$$\begin{aligned}
&\frac{\partial}{\partial t}(\rho_1 h_1) + \frac{\partial}{\partial x}(\rho_1 h_1 u_1) = 0 \\
&\frac{\partial}{\partial t}(\rho_1 h_1 u_1) + \frac{\partial}{\partial x}\left(\rho_1 h_1 u_1^2 + \rho_1 g \frac{h_1^2}{2} + \boxed{\rho_2 g h_1 h_2'}\right) = \boxed{\rho_2 g h_2 \frac{\partial h_1'}{\partial x}} - \rho_1 g h_1 B'(x) \\
&\frac{\partial}{\partial t}(\rho_2 h_2) + \frac{\partial}{\partial x}(\rho_2 h_2 u_2) = 0 \\
&\frac{\partial}{\partial t}(\rho_2 h_2 u_2) + \frac{\partial}{\partial x}\left(\rho_2 h_2 u_2^2 + \rho_2 g \frac{h_2^2}{2}\right) = -\boxed{\rho_2 g h_2 \frac{\partial h_1'}{\partial x}} - \rho_2 g h_2 B'(x) \\
&\frac{\partial}{\partial t} h_1' + U_1^* \frac{\partial}{\partial x} h_1' = \frac{h_1' - h_1}{\epsilon} \\
&\frac{\partial}{\partial t} h_2' + U_2^* \frac{\partial}{\partial x} h_2' = \frac{h_2' - h_2}{\epsilon}
\end{aligned} \tag{7}$$

where ϵ is a relaxation parameter of $h_i' \to h_i$, and U_i^* are propagation speeds associated with the auxiliary variables h_i', which are at our disposal (we have used $U_i^* = u_i$). For convenience, we have identified the terms in the original system (1) that are affected by the introduction of the auxiliary variables.

The extended system (7) can be written in quasi-linear form

$$W_t + A(W)W_x = S(x, W) + \epsilon^{-1}R(W)$$

with

$$W = \left(\rho_1 h_1,\ \rho_1 h_1 u_1,\ \rho_2 h_2,\ \rho_2 h_2 u_2,\ h_1',\ h_2'\right)^T$$

$$S = \left(0,\ -\rho_1 g h_1 B'(x),\ 0,\ -\rho_2 g h_2 B'(x),\ 0,\ 0\right)^T$$

$$R = \left(0,\ 0,\ 0,\ 0,\ h_1' - h_1,\ h_2' - h_2\right)^T$$

$$A = \begin{pmatrix} 0 & 1 & 0 & 0 & 0 & 0 \\ c_1^2 - u_1^2 & 2u_1 & 0 & 0 & -\rho_2 g h_2 & \rho_2 g h_1 \\ 0 & 0 & 0 & 1 & 0 & 0 \\ 0 & 0 & c_2^2 - u_2^2 & 2u_2 & \rho_2 g h_2 & 0 \\ 0 & 0 & 0 & 0 & U_1^* & 0 \\ 0 & 0 & 0 & 0 & 0 & U_2^* \end{pmatrix}, \qquad (8)$$

where we have introduced the *sound speeds*

$$c_1^2 = gh_1 + rgh_2', \qquad c_2^2 = gh_2.$$

The extended model "hyperbolizes" the system and has easily accessible eigenstructure, given by

$$\mu_{1,2} = u_1 \mp c_1, \qquad \mu_{3,4} = u_2 \mp c_2, \qquad \mu_5 = U_1^*, \qquad \mu_6 = U_2^*. \qquad (9)$$

and the matrix of right eigenvectors is

$$R = \begin{pmatrix} 1 & 1 & 0 & 0 & a & c \\ u_1 - c_1 & u_1 + c_1 & 0 & 0 & U_1^* a & U_2^* c \\ 0 & 0 & 1 & 1 & b & 0 \\ 0 & 0 & u_2 - c_2 & u_2 + c_2 & U_1^* b & 0 \\ 0 & 0 & 0 & 0 & 1 & 0 \\ 0 & 0 & 0 & 0 & 0 & 1 \end{pmatrix} \qquad (10)$$

with

$$a = \frac{\rho_2 g h_2'}{c_1^2 - (u_1 - U_1^*)^2}, \quad b = \frac{-\rho_2 g h_2}{c_2^2 - (u_2 - U_1^*)^2}, \quad c = \frac{-\rho_2 g h_1}{c_1^2 - (u_1 - U_2^*)^2}.$$

2.1 Stability

Stability of relaxation models in the limit $\epsilon \to 0$ is studied through the Chapman-Enskog expansion [CHP70]. In the classical setup for stability analysis, the conservation law

$$\frac{\partial w}{\partial t} + \frac{\partial f(w)}{\partial x} = 0 \tag{11}$$

is approximated by the relaxation system

$$\frac{\partial w^\epsilon}{\partial t} + \frac{\partial v}{\partial x} = 0, \qquad \frac{\partial v}{\partial t} + a^2 \frac{\partial w^\epsilon}{\partial x} = \frac{1}{\epsilon}\Big(v - f(w^\epsilon)\Big).$$

for some parameter a. As $\epsilon \to 0$, $v = f(w)$ to leading order in ϵ, and the equilibrium conservation law is recovered. Next order corrections yield the advection-diffusion equation

$$\frac{\partial w^\epsilon}{\partial t} + \frac{\partial f(w^\epsilon)}{\partial x} = \epsilon \frac{\partial}{\partial x}\left\{\left(a^2 - f'(w^\epsilon)^2\right)\frac{\partial w^\epsilon}{\partial x}\right\}, \tag{12}$$

which is well-posed provided it is dissipative, that is if

$$-a \le f'(w^\epsilon) \le a. \tag{13}$$

This is known as the *subcharacteristic condition*, which under suitable assumptions, implies convergence of the relaxation system (see, for example [CLL94, JX95]. The relaxation model (7), however, does not fall into the above framework of relaxation models, and the relationship between the subcharacteristic condition and dissipativity is less clear.

The subcharacteristic condition may be loosely interpreted as a *CFL-like* condition, requiring that the domain of dependence of solutions for the relaxation system (7) contains the domain of dependence of the underlying equilibrium system (1). Indeed, it is not likely that the relaxation system will be stable if it cannot support the stable propagation of waves of the underlying equilibrium system (1), given approximately in (4). We adopt this loose notion of stability, and address it at the numerical level.

3 Numerical Scheme

We solve (7) using a split-step method:

- *Propagation Step:* Solve $W_t + A(W)W_x = S(x, W)$.
- *Relaxation Step:* Solve $W_t = \epsilon^{-1}R(W)$ in the limit of instantaneous relaxation $\epsilon \to 0$. This step amounts to resetting the auxiliary variables to their relaxed values $h_1' := h_1, \;\; h_2' := h_2$.

Conservative Linearization

For the propagation step, we use a Roe-type method. The eigenvalues and eigenvectors of the relaxation system (7) are given in (9) and (10). In the absence of momentum exchange terms, the system is in conservation form and has a Roe linearization [PLR81] satisfying $\Delta F = \overline{A}\Delta W$. The averages are given by

$$\overline{h_i} = \frac{h_i^L + h_i^R}{2}, \quad \overline{u_i} = \frac{\sqrt{h_i^L} u_i^L + \sqrt{h_i^R} u_i^R}{\sqrt{h_i^L} + \sqrt{h_i^R}}, \qquad i = 1, 2.$$

The geometrical source terms are upwinded by projecting them onto the eigenvectors of the coefficient matrix [PLR86] $S = \sum_{k=1}^{6} \beta_k r_k$, with strengths

$$\beta_1 = \frac{\rho_1 g h_1 B'(x)}{2c_1} = -\beta_2, \quad \beta_3 = \frac{\rho_2 g h_2 B'(x)}{2c_2} = -\beta_4, \quad \beta_5 = \beta_6 = 0. \qquad (14)$$

Steady States

Time changes in the solution arise if the flux gradient and the source terms are out of balance. Schemes that are able to recognize such a balance often give superior results when computing near steady state solutions. Typically, one aims at respecting steady states either exactly or to the order of the numerical approximation. One trivial steady state is the steady state of rest

$$u_1 = u_2 = 0, \qquad h_1 + B = Const, \qquad h_2 = Const.$$

A straight forward calculation shows that if the source term S is approximated using

$$\overline{B'(x)} = \frac{\Delta B}{\Delta x}, \qquad \overline{h_1} = \frac{(h_1)_L + (h_1)_R}{2}$$

then the total fluctuation

$$\lambda_k \alpha_k - \Delta x \beta_k, \qquad k = 1, \ldots, 4$$

vanishes identically for steady-states of rest, resulting in zero change to the solution, thus ensuring that the steady state of rest is perfectly respected.

Stability

If the subcharacteristic condition is violated, waves supported by the equilibrium system (1) travel faster than those of the relaxation system (7). In this case, the numerical viscosity associated with the scheme will not generate enough dissipation to stabilize the faster waves of the underlying equilibrium system (1).

For the upwind scheme, given here in the form of wave signal

$$\Delta F^{\pm} = A^{\pm}\Delta W = \sum_k \frac{\lambda_k \pm |\lambda_k|}{2}\alpha_k r_k, \tag{15}$$

the amount of numerical viscosity is controlled by $|\lambda_k|$, and the method is stable for wave speeds up to λ_k. To stablize the method for possibly faster waves that may be present in the underlying equilibrium system, we replace $|\lambda_k|$ in the numerical viscosity by

$$|\tilde{\lambda}_k| = \max\Big\{|\lambda_k|,\ |\mu_k|\Big\}, \qquad k = 1,2,3,4. \tag{16}$$

Here μ_k are the approximate eigenvalues of the original system (4), and λ_k are the eigenvalues of the relaxation system (9).

4 Numerical Results

We consider an example, taken from [CMP01], featuring a flow that is close to the boundary of the hyperbolic regime. For this flow, even small numerical errors may "push" the solution outside the hyperbolic limit and trigger instabilities. In this example, the topography is flat, $B(x) = 0$, $g = 10$ and $r = 0.98$. The number of grid points is 200 and the CFL number is 0.7. Initial data is given by

$$\begin{array}{llll} (h_1)_L = 0.5, & (q_1)_L = 1.25, & (h_1)_R = 0.45, & (q_1)_R = 1.125, \\ (h_2)_L = 0.5, & (q_2)_L = 1.25, & (h_2)_R = 0.55, & (q_2)_R = 1.375, \end{array}$$

generating predominantly an internal (interfacial) wave moving from its initial location at $x = 0.3$ to the right. Figure 2 shows the results by the propagation-relaxation method described above, with the propagation step (15) disregarding the subcharacteristic condition. As the wave moves along, an instability develops, which rather quickly contaminates the solution and causes the solution to break down. Figure 3 shows the computation of the same problem, with the modified viscosity (16). The computation is clearly stablized by the modified viscosity. Furthermore, a fine grid calculation, using 10,000 grid points, reveals that what appeared to be a single wave is, in fact, a four-wave structure, all clearly visible on the velocity plot. Figure 4 shows the respective sets of eigenvalues for this calculation, $\{\lambda_k\}_k$ and $\{\mu_k\}_k$, where it may be observed that the eigenvalues of the relaxation system μ_k indeed slightly underestimated the external eigenvalues of the underlying equilibrium system, λ_k. The viscosity used in Fig. 3 is the larger of the two (16).

The relaxation scheme respects steady state of rest, $u_1 = u_2 = 0$, $h_1 + B(x) = 1$ and $h_2 = 1$, as can be seen in Fig. 5.

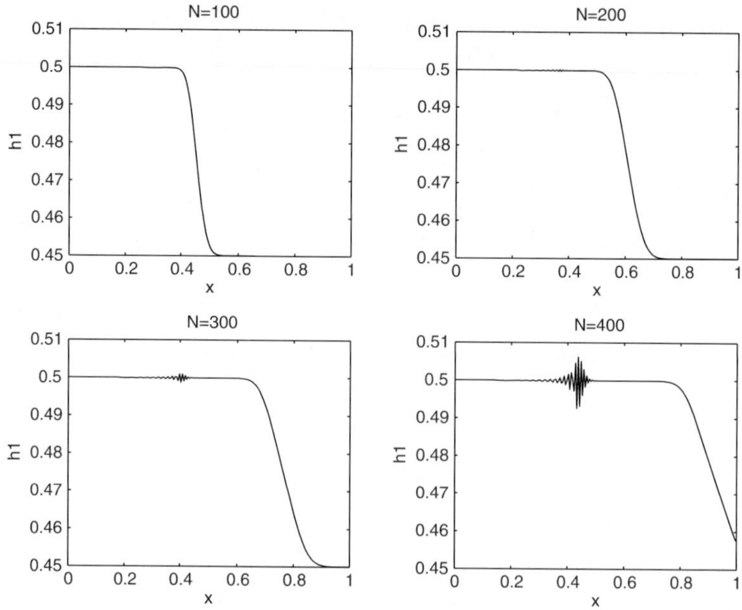

Fig. 2. Interfacial instability – Relaxation model

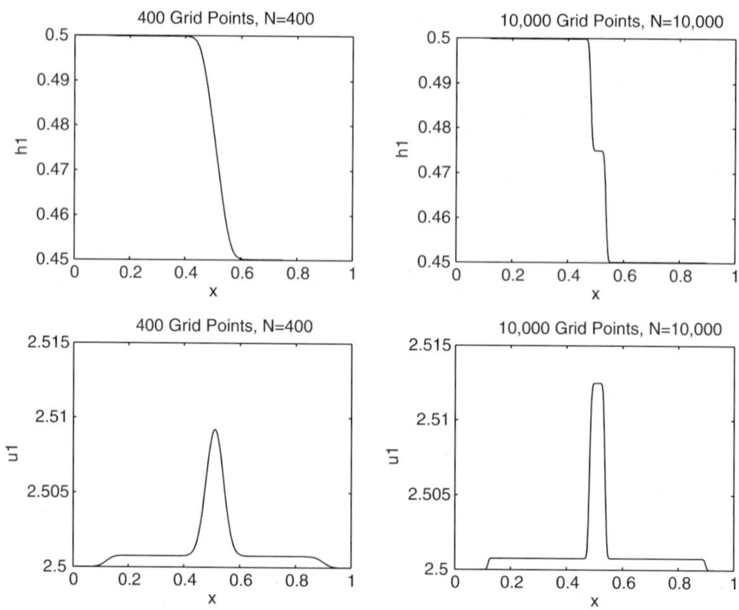

Fig. 3. Interfacial Instability – Relaxation Model with "subcharacteristic" viscosity

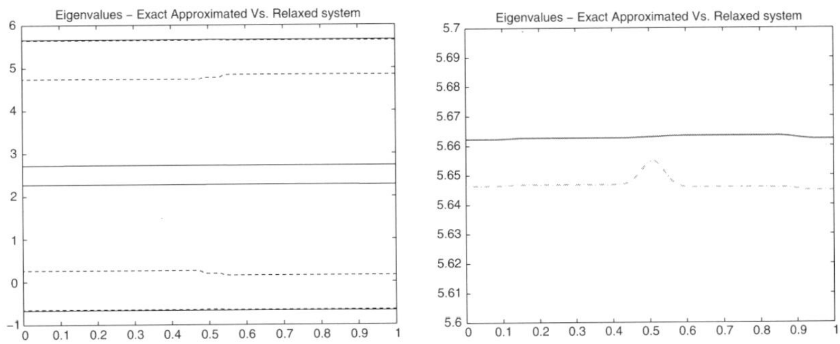

Fig. 4. Left: Eigenvalues $\{\mu_k\}$ in (9) (*dashed line*) and $\{\lambda_k\}$ in (4) (*solid line*); Right: Zoom

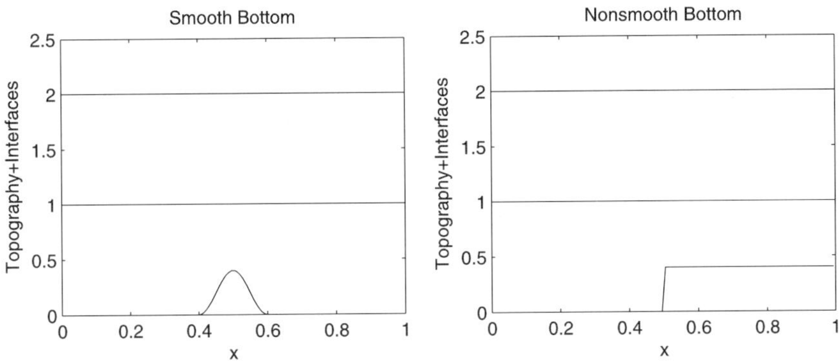

Fig. 5. Steady state of rest over a smooth bump (*left*) and a step (*right*) is respected

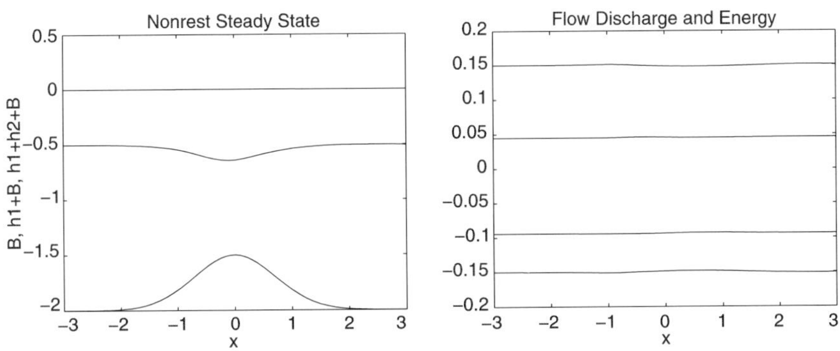

Fig. 6. Smooth subcritical steady state: Interface and water surface (*left*); Flow steady state constants $Q_{1,2}$ and $H_{1,2}$ (*right*)

Finally, we computed a nontrivial subcritical steady state corresponding to flow rates $Q_1 = 0.15$ and $Q_2 = -0.15$ at inflow, and the layers depths $h_1 = 1.5$ and $h_2 = 0.5$ at outflow. The interface and water level are shown in Fig. 6. The flow rates $Q_{1,2}$ and energies $H_{1,2}$ should be constant at the steady state. They are also shown in Fig. 6, and seem reasonably constant.

References

[CMP01] M. Castro, M., Macias, J., Pares, C.: A Q-scheme for a class of systems of coupled conservation laws with source terms. Application to a two-layer 1-D shallow water system, Mathematical Modelling and Numerical Analysis (M2AN), **35**, 107–127 (2001)

[CHP70] Chapman, S., Cowling, T.G.: The Mathematical Theory of Nonuniform Gases, Cambridge Univ. Press, 3rd Edition, (1970)

[CLL94] Chen, G.-Q., Levermore, D., Liu, T.-P.: Hyperbolic conservation laws with stiff relaxation terms and entropy, Comm. Pure Appl. Math., **47**, 787–830 (1994)

[JX95] Jin, S., Xin, Z.: The relaxation scheme for systems of conservation laws in arbitrary space dimensions, Comm. Pure Appl. Math., **48**, 235–276 (1995)

[LNG56] Long, R.R.: Long waves in a two-fluid system, J. Met., **13**, 70–74 (1956)

[PLR81] Roe, P.L.: Approximate Riemann solvers, parameter vectors and difference schemes, J. Comp. Phys., **43**, 357–372 (1981)

[PLR86] Roe, P.L.: Upwind differencing schemes for hyperbolic conservation laws with source terms, Lecture Notes in Math., **1270**, 41–51, Springer, Berlin (1987)

[SCH53] Schijf, J.B., Schonfeld J.C.: Theoretical considerations on the motion of salt and fresh water, Proc. of the Minn. Int. Hydraulics Conv., Joint meeting IAHR and Hyd. Div. ASCE. September 1953, 321–333 (1953)

Group Dynamics of Phototaxis: Interacting Stochastic Many-Particle Systems and Their Continuum Limit

D. Bhaya, D. Levy, and T. Requeijo

Summary. In this chapter, we introduce new models for describing the motion of phototaxis, i.e., bacteria that move toward light. Following experimental observations, the first model describes the locations of bacteria, an internal property of the bacteria that is related to the group dynamics, and the interaction between the bacteria and the medium in which it resides. The second model is a new multiparticle system following the same quantities as in the first model. The main theorem shows how to obtain a new system of PDEs, which we refer to as *the phototaxis system* as the limit dynamics of the multiparticle system. Numerical simulations are provided.

1 Introduction

Microbes live in environments that are fluctuating and often limiting for growth. Thus they have evolved several sophisticated mechanisms to sense minute changes in important environmental parameters such as light and nutrients. Most bacteria also have complex appendages that allow them to move, so they can swim or crawl into optimal conditions. This combination of sensing changes in the immediate environment and transducing these changes to a flagellar motor allows for movement in a particular direction: a phenomenon known as chemotaxis or phototaxis [1]. Many of the molecular players in this pathway of signal transduction are now known and the flagella which is used for motility is also extremely well characterized [7, 12]. Cyanobacteria are a lineage of ancient, ubiquitous photosynthetic microbes that use a different surface appendage for motility. They use thin, long flexible pili which are referred to as Type IV pili [2]. The pili adhere to various surfaces and are alternately extended and retracted to slowly pull the cell body in a particular direction. This movement requires energy in the form of ATP which is provided to a motor protein encoded by *pilT*. Interestingly, pili are multifunctional organelles involved in a host of important biological processes such as DNA uptake, pathogenicity, and biofilm formation [6]. Yet we know much less about these organelles that we do about flagella synthesis or regulation.

Cyanobacteria track light direction and quality to optimize conditions for photosynthesis. The motility toward a light source is called “phototaxis” and requires (1) a photoreceptor, (2) a signal transduction event, and (3) the motility apparatus or pili. Many of the molecular components for pilus biosynthesis and signal transduction have been identified in the recent years [3, 4]. We have also taken a genetic approach to identify mutants in the process of phototaxis. Time-lapse video microscopy can also be used to monitor the movement of individual cells and groups of cells [5]. These movies suggest that there is both a group dynamic between cells, as well as the ability of single cells to move directionally. The various patterns of movement that we observe appear to be a complex function of cell density, surface properties, and genotype. Very little is known about the interactions between these parameters. Some models have been proposed to explain pilus-dependent social motility in *Myxococcus xanthus* which have led to important insights into group behavior [10, 9]. Considering the widespread use of pilus-dependent surface motility, an attempt to model this motility in Cyanobacteria may provide novel insights into an important biological phenomenon.

In the past several decades, starting from the Keller–Segel model [11], there has been a lot of activity in the mathematical community in studying chemotaxis (i.e., bacteria that move in the direction of a chemical attractant), see, e.g., [13, 16] and the references therein. At the same time, no mathematical models have been developed for describing the somewhat more mysterious motion of phototaxis. This chapter is the first attempt in that direction. It is structured in the following way. In Sect. 2.3, we introduce a mathematical model that is based on experimental observations. Following these observations we assume that the bacteria, while being able to move individually, do require an internal property, which we refer to as an *excitation* to reach a certain threshold in order for a motion to develop. This excitation accumulates as a result of a group-like influence. In addition, we assume that the bacteria are more likely to move on surfaces that were already traveled on by other bacteria, i.e., we take into account also an interaction between the bacteria and the medium on which they move. We proceed in Sect. 3 by deriving a multiparticle system approach to the first model considered in Sect. 2.3. The main result in this chapter is the limit dynamics to the multiparticle system, given by the system of PDEs (20), which is discussed in the final part of Sect. 3. We conclude with several numerical simulations that demonstrate the properties of our models.

2 The Framework

2.1 Observations

The time-lapse video microscopy we used to track the movement of cells [5] has led us to the following observations regarding the characteristics of the motion:

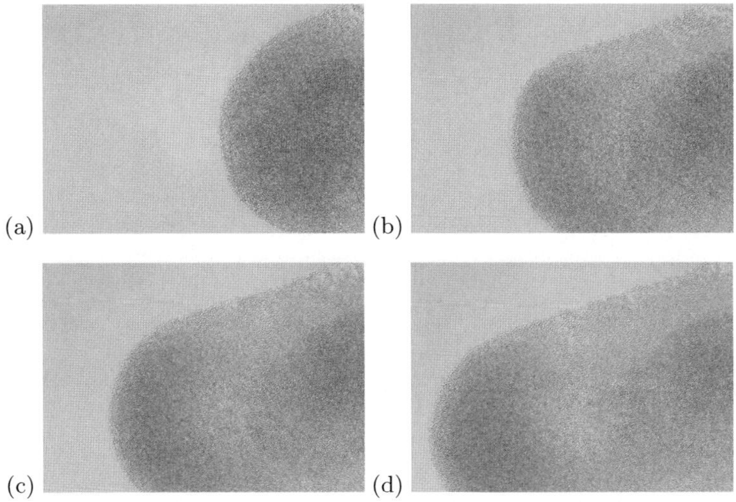

Fig. 1. Bacteria motion when the density is high. The snapshots were taken at increasing times, starting from (**a**) and ending at (**d**). The light comes from the *left* of the domain

(1) *Motion when the density is high.* When the density of the cells is homogeneously high, all bacteria move as if they were forming a solid block (see Fig. 1). Worth noting is how bacteria do not scatter around. Instead, we observe a somewhat well-defined interface between the group of bacteria and the medium.
(2) *Fingering.* When the density of cells is not homogeneous, we observe a faster movement in areas of higher bacteria density. This faster movement of bacteria toward light creates *fingers* as bacteria in lower density areas tend to remain still or move slower. This behavior is shown in Fig. 2. In this case, the light source is from the upper-right corner of the domain.
(3) *Pinching.* Pinching is related to the fingering effect described above. Pinching develops when the density of cells is high enough to form a finger but as the finger is formed and bacteria move toward the light source, the density behind the leading tip decreases (if there are not enough bacteria present), and the tip eventually detaches.
(4) *Existence of an external substance.* The movies suggest that, when the cells move, they leave behind a trail. It is unclear what is the nature of the trail, e.g., they might segregate a substance that adheres to the medium. Our observations indicate that this substance either has an extremely slow decay or does not decay at all (with the time-frame of the movies, typically several hours). In either case, it is clear that this is an important factor in the dynamics as cells that revisit locations that were already traveled by other bacteria are likely to follow a similar pattern of motion. This phenomenon is demonstrated in Fig. 3.

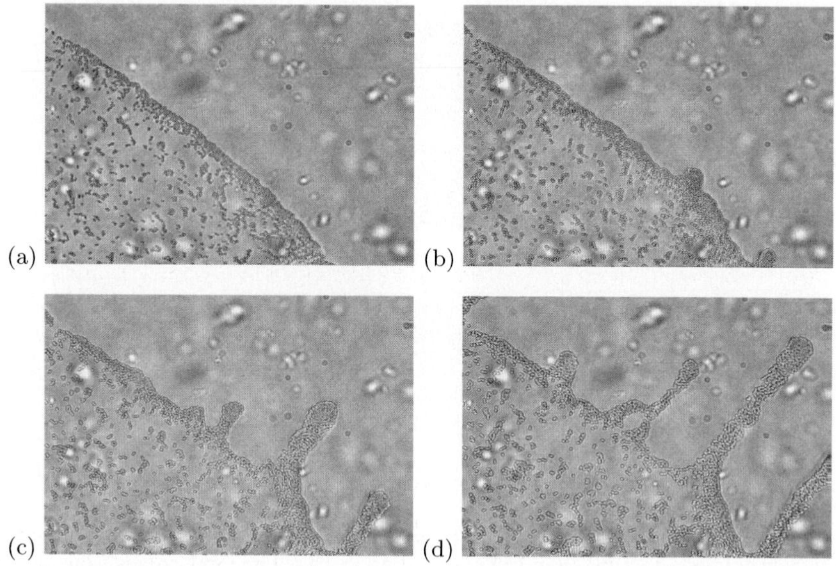

Fig. 2. Creation of fingers with a light source at the *upper-right* corner of the domain. Part (**a**) shows the edge of the colony with single cells showing as *dark dots*. Part (**b**) shows the initial phase of creating the fingers. Parts (**c**) and (**d**) taken several hours later show the bacteria moving toward the light source

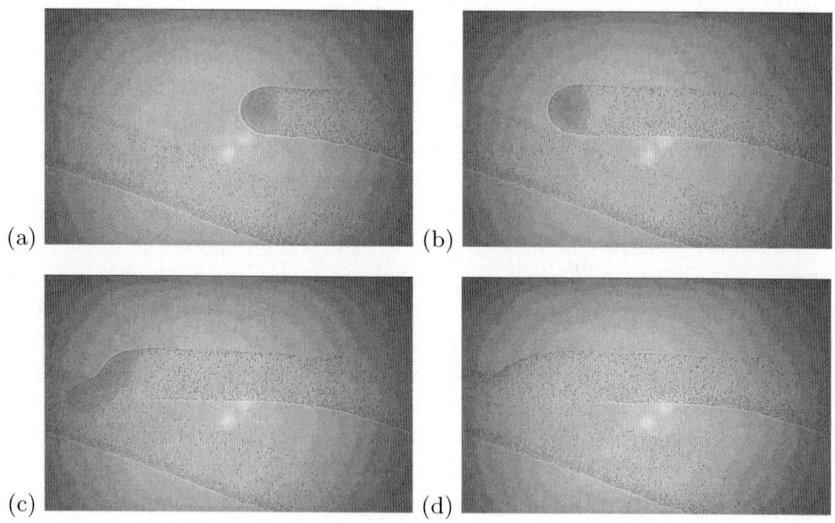

Fig. 3. Bacteria follow a similar pattern of motion on locations that were traveled by other bacteria. Shown are snapshots taken at consecutive times starting from (**a**) and ending at (**d**)

2.2 Interpretation and Assumptions

As mentioned in item (4) above, we assume bacteria leave some substance behind. The evolution from the second to third pictures in Fig. 3 suggests that, when in contact with a surface that was previously occupied, bacteria seem more predisposed to moving. Also, from the same sequence, it seems the substance left behind diffuses slowly or does not diffuse at all (if there was a considerably amount of diffusion, the interfaces would not be so well defined).

Observing Fig. 2 leads us to believe that there exists some sort of communication between bacteria since those tend to aggregate and move faster or slower depending on the number of bacteria around themselves. Assuming this mechanism is present, then the pinching phenomenon is nothing more than an effect of such mechanism; as mentioned before, if the density is high enough to make a group of bacteria move faster, but the density drops considerably when that group starts moving, then we expect that the initial group will detach from its neighbors.

The mechanism proposed to describe such communication among bacteria consists on the existence of some excitation level associated to each bacterium (this will be called *excitation*). In chemotaxis, bacteria segregate a substance that diffuses on the medium and serves as a communication mechanism; bacteria flow toward higher concentrations of this substance which, in turn, decays over time. In our case this is not a desirable mechanism; it would mean that moving particles would leave yet another substance behind, so all the *excitation* would be passed to bacteria immediately behind and we would always observe aggregation instead of a detachment in low densities (which is what we observe in experiments). We also assume that bacteria can sense how excited to moving the neighboring bacteria are. This leads to propagation in high densities, if the excitation of a particular bacterium is high, it will cause an immediate increase in the excitation of its neighbors.

Another difference with chemotaxis resides on the behavior of the external substance. With chemotaxis, the external substance does not only decays and diffuses over time, but is also consumed by bacteria. In our phototaxis system, the external substance is assumed to be persistent, i.e., there is no consumption, decay, or diffusion. Thus, this substance functions as a memory effect; its existence or not at a specific point in space and time t signals if any bacterium was present at that point, at previous times $s \leq t$.

2.3 The Mathematical Framework

To derive a model that is based on the observations and assertions above, we need to specify three different stochastic processes. The first two concern the position of each bacterium and its excitation at a specific moment in time, while the third process is related to the extra substance at a specific point (on the medium) and time. To that end, let N denote the number of bacteria present in a free boundary medium (assume the medium is $\mathbb{R}^2$) and denote by $X_i(t) \in \mathbb{R}^2$ the position of bacterium i at time $t \geq 0$.

From the assumptions above, it is natural to assume that the process L, describing the external substance, is a pure jump process given by

$$L(t;x,y) = \max_{\substack{0\leq s\leq t \\ i=1,\ldots,N}} \delta_{(x,y)}(X_i(t)), \tag{1}$$

where $\delta_{(x,y)}$ is the Dirac delta function in $\mathbb{R}^2$.

Remark 1. It would be more realistic to assume that the external substance is produced in a continuous manner, rather than by a pure jump process. However, in the setting of the multiparticle system discussed in Sect. 3, this difference does not play an important role. Also, according to our assumptions, the quantity of such a substance should quickly increase when it is in contact with bacteria (up to a certain level). Hence, even from practical considerations, when simulating the model, the best way to discretize this process is according to (1).

Denote the excitation process for particle i by S_i and let $\mu_i(t)$ be some weighted average of the total excitation around particle i at time t. This means μ_i is a quantity that describes how excited bacteria around bacterium i are. We will assume that $S_i(t)$ is given by a mean-reverting process

$$\frac{dS_i(t)}{S_i(t)} = (\mu_i(t) - S_i(t))dt + \sigma dW_i(t), \tag{2}$$

where σ is a quantity inherent to the this kind of bacteria (hence constant) and W_i, $i = 1, \ldots, N$ are independent Brownian motions.

With $\mu_i(t)$ and $S_i(t)$ defined this way, we know that $S_i(t) > 0$ for all $t \geq 0$ and also that $S_i(t)$ tends to move toward the mean-reverting level $\mu_i(t)$. Hence, controlling $\mu_i(t)$ will implicitly control $S_i(t)$ (in particular, since μ_i is bounded then the same will hold for S_i almost surely).

To define the position processes $X_i(t)$, we assume that the light source is always present, that it is of uniform intensity, and that it is located to the left of the bacteria. Together with bacteria sensitivity to the extra substance, this can be encoded into a C^∞ function $g\colon \mathbb{R}_0^+ \times \mathbb{R}^2 \to [0,1]$ satisfying:

(1) g is strictly increasing (in both variables).
(2) $\lim_{s\to\infty} g(s,w) = 1$.
(3) $g(0,\cdot) = 0$.

We can thus define

$$\begin{aligned} dX_i(t) = &-v_s \left\{ g\left((S_i(t) - K)^+, \nabla L^N(t;x,y)\right), 0\right\} dt \\ &+ v_g \nabla \rho^N(t;x,y)dt + v_r d\tilde{W}_i(t), \end{aligned} \tag{3}$$

where $\tilde{W}_i(t)$ are independent two-dimensional Brownian motions, and v_s, v_g, and v_r are the maximum velocity components for (1) excitation and sensitivity to external substance, (2) the density gradient, and (3) the random phenomena, respectively. For each $N > 0$, L^N and ρ^N are the stochastic process obtained from L and $\frac{1}{N}\sum_i \delta_{X_i}$ by a convolution with a mollifier. This model thus accounts for sensitivity to the extra substance, sensitivity to light, natural group dynamics (not coming from reaction to light), and random phenomena.

Remark 2. As mentioned above, we assume that the light source is always present, and bacteria sense it coming from the left. These assumptions can be easily dropped by:

- Substituting the quantity $\left\{g\left((S_i(t)-K)^+, \nabla L^N(t;x,y)\right), 0\right\}$, appearing in $dX_i(t)$, by $g\left((S_i(t)-K)^+, \nabla L^N(t;x,y)\right)\xi_t$, where ξ_t is the unit vector representing the direction from which bacteria sense the light at time t
- Changing the processes $S_i(t)$ either by making them decay fast when the light source is not present or even by making them jump to values below the threshold K at the moment the light source vanishes

Remark 3. A major difference of this work compared with the work of Stevens [16] resides in the existence of a the *excitation* quantity, which is an internal property of each bacteria. Equation (2) has the desired effect that an individual's excitation will evolve toward a surrounding neighborhood trend (given by $\mu_i(t)$).

3 Many-Particle System

Based on the model introduced in Sect. 2.3, we now introduce a new particle system resembling the models in [16] and [14]. In particular, we consider an initial population of approximately N particles that can move in $\mathbb{R}^2$, die, or give birth to new particles. As the initial population size N tends to infinity, we rescale the interaction between individuals in a moderate way. This means that the instantaneous change on a particular particle depends on the configuration of the remaining particles in a neighborhood, which is macroscopically small and microscopically large. That is, the volume tends to 0 as $N \to \infty$ and it contains an arbitrarily large number of individuals as $N \to \infty$.

For $N \in \mathbb{N}$, we consider a population in $\mathbb{R}^2$ consisting of approximately N particles, which is divided into three subpopulations: bacteria, excitation, and external substance. From now on, the indexes u, v, and l will refer, respectively, to each of these subpopulations and we will refer to those subpopulations as type u, v, or l. Denote by $M(N,r,t)$, where $r = u,v,l$, the set of all individuals belonging at time t to population type r. Also, let $M(N,t) = \bigcup_{r=u,v,l} M(N,r,t)$ be the set of all individuals alive at time t.

For $k \in M(N,t)$, let $P_N^k(t) \in \mathbb{R}^2$ denote the position of particle k at time t, and consider the measure-valued processes

$$t \to S_{N,r}(t) = \frac{1}{N}\sum_{k\in M(N,r,t)} \delta_{P_N^k}(t), \tag{4}$$

where $r = u,v,l$ and δ_x denotes the Dirac measure at $x \in \mathbb{R}^2$.

Following Sect. 2.3, excitation particles should be associated with a particular bacterium, hence we have to be careful in numbering the set $M(N, v, t)$. To that end, define $M_w(N, v, t) \subset M(N, v, t)$ as the set of excitation particles associated with bacterium $w \in M(N, u, t)$. For $w \in M(N, u, t)$, define the measure-valued process

$$t \to S_{N,v,w}(t) = \sum_{k \in M_w(N,v,t)} \delta_{P_N^k}(t). \tag{5}$$

Note that since $\{M_w(N, v, t)\}_{w \in M(N,u,t)}$ is a partition of $M(N, v, t)$, then $S_{N,v}(t) = \frac{1}{N} \sum_{w \in M(N,u,t)} S_{N,v,w}(t)$.

3.1 Densities

We introduce now smoothed versions of the empirical processes above. For a fixed symmetric and sufficiently smooth function W_1, let

$$W_N(x) = \alpha_N^d W_1(\alpha_N x), \qquad \hat{W}_N(x) = \hat{\alpha}_N^d W_1(\hat{\alpha}_N x),$$

where $\alpha_N = N^{\alpha/d}$ and $\hat{\alpha}_N = N^{\hat{\alpha}/d}$ for fixed scaling exponents α and $\hat{\alpha}$. Conditions for α and $\hat{\alpha}$ are given in [16].

Define for $r = u, v, l$

$$\begin{aligned} s_{N,r}(t, x) &= (S_{N,r}(t) * W_N * W_N)(x) \\ \hat{s}_{N,r}(t, x) &= (S_{N,r}(t) * W_N * \hat{W}_N)(x) \end{aligned} \tag{6}$$

and, for $v \in M_w(N, v, t)$,

$$\begin{aligned} s_{N,v,w}(t, x) &= (S_{N,v,w}(t) * W_N * W_N)(x) \\ \hat{s}_{N,v,w}(t, x) &= (S_{N,v,w}(t) * W_N * \hat{W}_N)(x). \end{aligned} \tag{7}$$

These functions formally represent the density or concentration of each subpopulation type near x at time t. We introduce two density versions of each type (s and $\hat{s}$) for technical reasons. A more detailed discussion and technical details can be found in [14].

3.2 Dynamics

In our model, we assume that particles not only move in $\mathbb{R}^2$, but also cause discontinuous changes to the population; they can die or give birth to new particles. We first describe how existing particles move. For $t \geq 0$ and $k \in M(N, u, t)$, let

$$\begin{aligned} dP_N^k(t) &= g\left(\hat{s}_{N,v,k}(t, P_N^k(t)), \nabla s_{N,l}(t, P_N^k(t))\right) dt \\ &\quad + \tilde{g}\left(\nabla s_{N,u}(t, P_N^k(t))\right) dt + \mu dW^k(t) \\ &= g_N^k\left(t, P_N^k(t)\right) dt + \tilde{g}_N\left(t, P_N^k(t)\right) dt + \mu dW^k(t), \end{aligned} \tag{8}$$

where $g_N^k(t,x) = g(\hat{s}_{N,v,k}(t,x), \nabla s_{N,l}(t,x))$ and $\tilde{g}_N(t,x) = \tilde{g}(\nabla s_{N,u}(t,x))$. For $k \in M_w(N,v,t) \subset M(N,v,t)$, we impose

$$dP_N^k(t) = dP_N^w(t). \tag{9}$$

Finally, for $k \in M(N,l,t)$,

$$dP_N^k(t) = 0. \tag{10}$$

Note that this is consistent with Sect. 2.3; any excitation particle moves together with a specific bacterium and extra substance particles do not move at all.

We assume that any individual $k \in M(N,u,t)$ at position $dP_N^k(t) = y$ may induce discontinuous changes in the v and l subpopulations; namely:

- Give birth to type l particles, with intensity $\lambda_N(t,y)$
- Give birth to type v particles, with intensity $\beta_{N,k}(t,y)$

and that any individual $k \in M(N,v,t)$ at position $dP_N^k(t) = y$ may cause the death of type v particles, with intensity $\gamma_{N,k}(t,y)$.

We also assume that these intensities depend on the densities of the N particle system,

$$\begin{aligned} \beta_{N,k}(t,x) &= \beta(\hat{s}_{N,v}(t,x), \hat{s}_{N,v,k}(t,x)) \\ \gamma_{N,k}(t,x) &= \gamma(\hat{s}_{N,v}(t,x), \hat{s}_{N,v,w}(t,x)) \\ \lambda_N(t,x) &= \lambda(\hat{s}_{N,u}(t,x), \hat{s}_{N,l}(t,x)) \end{aligned} \tag{11}$$

To describe this behavior, we introduce the processes

$$\begin{aligned} \beta_N^k(t) &= Q_N^{\beta,k}\left(\int_0^t \mathbf{1}_{M(N,u,\tau)}(k)\beta_{N,k}(\tau, P_N^k(\tau))d\tau\right) \\ \gamma_N^k(t) &= Q_N^{\gamma,k}\left(\int_0^t \mathbf{1}_{M(N,v,\tau)}(k)\gamma_{N,k}(\tau, P_N^k(\tau))d\tau\right) \\ \lambda_N^k(t) &= Q_N^{\lambda,k}\left(\int_0^t \mathbf{1}_{M(N,l,\tau)}(k)\lambda_N(\tau, P_N^k(\tau))d\tau\right) \end{aligned} \tag{12}$$

where $Q_N^{\cdot}$ are independent standard Poisson processes. Thus the point processes $\beta_N^k(t)$, $\gamma_N^k(t)$, and $\lambda_N^k(t)$ have intensities $\mathbf{1}_{M(N,u,\tau)}(k)\beta_{N,k}(t, P_N^k(t))$, $\mathbf{1}_{M(N,u,t)}(k)\gamma_{N,k}(t, P_N^k(t))$, and $\mathbf{1}_{M(N,u,t)}(k)\lambda_N(t, P_N^k(t))$, respectively, for a jump of size 1 at time t.

3.3 The Limit Dynamics

Let $\langle \mu, f \rangle = \int_{\mathbb{R}^2} f(x)\mu(dx)$ for any measure μ and real-valued function f in $\mathbb{R}^2$. Using Itô's formula, we obtain, for $f \in C_b^{1,2}(\mathbb{R}^+ \times \mathbb{R}^2)$ and $k \in M(N,u,t)$,

$$f\left(t, P_N^k(t)\right) = f\left(0, P_N^k(0)\right) + \int_0^t \mu \nabla f\left(\tau, P_N^k(\tau)\right) \cdot dW^k(t) + \int_0^t \Big[\nabla f\left(\tau, P_N^k(\tau)\right) \cdot \left(g_N^k + \tilde{g}_N\right)\left(\tau, P_N^k(\tau)\right) + \partial_\tau f\left(\tau, P_N^k(\tau)\right) + \mu \Delta f\left(\tau, P_N^k(\tau)\right)\Big] d\tau \tag{13}$$

and, since there is no birth or death of type u particles,

$$\begin{aligned} \langle S_{N,u}(t), f(t,\cdot)\rangle =& \langle S_{N,u}(0), f(0,\cdot)\rangle \\ &+ \frac{1}{N}\int_0^t \sum_{k\in M(N,u,\tau)} \mu \nabla f\left(\tau, P_N^k(\tau)\right) \cdot dW^k(t) \\ &+ \frac{1}{N}\int_0^t \sum_{k\in M(N,u,\tau)} \Big[\nabla f\left(\tau, P_N^k(\tau)\right) \cdot \left(g_N^k + \tilde{g}_N\right)\left(\tau, P_N^k(\tau)\right) \\ &\qquad + \partial_\tau f\left(\tau, P_N^k(\tau)\right) + \mu \Delta f\left(\tau, P_N^k(\tau)\right)\Big] d\tau \end{aligned} \tag{14}$$

For $k \in M(N, v, t)$, we have to take into account birth and decay of particles. We also have $dP_N^k(t) = dP_N^w(t)$ for $k \in M_w(N, v, t)$, thus

$$\begin{aligned} \langle S_{N,v}(t), f(t,\cdot)\rangle &= \frac{1}{N} \sum_{k\in M(N,v,t)} f\left(t, P_N^k(t)\right) \\ &= \frac{1}{N} \sum_{k\in M(N,u,t)} N_k(t) f\left(t, P_N^k(t)\right) + \text{birth/decay terms} \end{aligned}$$

where $N_k(t) = |M_w(N, v, t)| = S_{N,v,w}(t)(P_N^k(t))$ for $k \in M_w(N, v, t)$. Hence,

$$\begin{aligned} \langle S_{N,v}(t), f(t,\cdot)\rangle =& \langle S_{N,v}(0), f(0,\cdot)\rangle \\ & \frac{1}{N}\int_0^t + \sum_{k\in M(N,u,\tau)} N_k(\tau) \mu \nabla f\left(\tau, P_N^k(\tau)\right) \cdot dW^k(t) \\ &+ \frac{1}{N}\int_0^t \sum_{k\in M(N,u,\tau)} N_k(\tau) \Big[\nabla f\left(\tau, P_N^k(\tau)\right) \cdot \left(g_N^k + \tilde{g}_N\right)\left(\tau, P_N^k(\tau)\right) \\ &\qquad + \partial_\tau f\left(\tau, P_N^k(\tau)\right) + \mu \Delta f\left(\tau, P_N^k(\tau)\right)\Big] d\tau \\ &+ \frac{1}{N}\int_0^t \sum_{k\in M(N,u,\tau)} f\left(\tau, P_N^k(\tau)\right) \beta_N^k(d\tau) \\ &- \frac{1}{N}\int_0^t \sum_{k\in M(N,v,\tau)} f\left(\tau, P_N^k(\tau)\right) \gamma_N^k(d\tau) \end{aligned} \tag{15}$$

Finally, $dP_k^N(t) = 0$ for type l individuals leads to

$$\begin{aligned}\langle S_{N,l}(t), f(t,\cdot)\rangle = \frac{1}{N}\int_0^t \sum_{k\in M(N,l,\tau)} \partial_\tau f\left(\tau, P_N^k(\tau)\right) d\tau \\ + \frac{1}{N}\int_0^t \sum_{k\in M(N,u,\tau)} f\left(\tau, P_N^k(\tau)\right) \lambda_N^k(d\tau)\end{aligned} \tag{16}$$

We introduce the processes

$$\begin{aligned}
M_u^N(t,f) &= \frac{1}{N}\int_0^t \sum_{k\in M(N,u,t)} \mu\nabla f\left(\tau, P_N^k(\tau)\right)\cdot dW^k(t) \\
M_v^N(t,f) &= \frac{1}{N}\int_0^t \sum_{k\in M(N,v,t)} \mu\nabla f\left(\tau, P_N^k(\tau)\right)\cdot dW^k(t) \\
M_{v,\gamma}^N(t,f) &= \frac{1}{N}\int_0^t \sum_{k\in M(N,u,\tau)} f\left(\tau, P_N^k(\tau)\right)\left(\gamma_N^k(d\tau) - \gamma_{N,k}\left(\tau, P_N^k(\tau)\right) d\tau\right) \\
M_{v,\beta}^N(t,f) &= \frac{1}{N}\int_0^t \sum_{k\in M(N,v,\tau)} f\left(\tau, P_N^k(\tau)\right)\left(\beta_N^k(d\tau) - \beta_{N,k}\left(\tau, P_N^k(\tau)\right) d\tau\right) \\
M_l^N(t,f) &= \frac{1}{N}\int_0^t \sum_{k\in M(N,u,\tau)} f\left(\tau, P_N^k(\tau)\right)\left(\lambda_N^k(d\tau) - \lambda_N\left(\tau, P_N^k(\tau)\right) d\tau\right)
\end{aligned} \tag{17}$$

These processes are martingales with respect to the natural filtration generated by the processes $t \to \left(P_N^k(t), \mathbf{1}_{M(N,r,t)}(k)\right)\mathbf{1}_N^k(t)$, where $\mathbf{1}_N^k(t)$ is the indicator function of the lifetime of individual k. We assume that the quadratic variation of each of the martingales above tends to 0 as $N \to \infty$, hence we can neglect them when passing to the limit dynamics.

From Sect. 3.1, we see that in the sense of distributions

$$\lim_{N\to\infty} W_N = \lim_{N\to\infty} \hat{W}_N = \delta_0.$$

We assume that, for $r = u, v, l$ and $t \geq 0$ (see [16]),

$$\lim_{N\to\infty} S_{N,r}(t) = S_r(t),$$

where the measure $S_r(t)$ has smooth density $r(t,\cdot)$. It follows that

$$\begin{aligned}
\lim_{N\to\infty} s_{N,r}(t,\cdot) &= \lim_{N\to\infty} \hat{s}_{N,r}(t,\cdot) = r(t,\cdot), \\
\lim_{N\to\infty} \nabla s_{N,r}(t,\cdot) &= \lim_{N\to\infty} \nabla\hat{s}_{N,r}(t,\cdot) = \nabla r(t,\cdot).
\end{aligned}$$

Let $u_0(\cdot)$, $v_0(\cdot)$, and $l_0(\cdot)$ be the densities of $S_u(0)$, $S_v(0)$, and $S_l(0)$, respectively. Define $\tilde{g}_\infty(\tau,x) = \tilde{g}(\nabla u(\tau,x))$ and $\lambda_\infty(\tau,x) = \lambda(u(\tau,x),l(\tau,x))$. To define g_∞, β_∞, and γ_∞, first note that as $N \to \infty$, we assume $|M_w(N,v,t)| \simeq |M_{\tilde{w}}(N,v,t)|$ when $P_N^w(t) = P_N^{\tilde{w}}(t)$. Thus, as $N \to \infty$, $S_{N,v,w}(t)$ should be approximated by the number of excitation particles at $P_N^w(t)$ divided by the number of bacteria particles at $P_N^w(t)$, i.e.,

$$S_{N,v,w}(t)(P_N^w) \simeq \frac{\sum_{k\in S_v(N,t)} \delta_{P_N^k}(P_N^w)}{\sum_{k\in S_u(N,t)} \delta_{P_N^k}(P_N^w)} = \frac{NS_{N,v}(t)(P_N^w)}{NS_{N,u}(t)(P_N^w)} = \frac{S_{N,v}(t)(P_N^w)}{S_{N,u}(t)(P_N^w)}.$$

Since $S_{N,r}$ are sums of a finite number of delta functions, $\lim \hat{s}_{N,v,x}(t) = \frac{v(t,x)}{u(t,x)}$. Hence $g_\infty(\tau,\cdot) = g\left(\frac{v(\tau,\cdot)}{u(\tau,:)}, \nabla l(\tau,\cdot)\right)$, $\gamma_\infty(\tau,\cdot) = \gamma\left(v(\tau,\cdot), \frac{v(\tau,\cdot)}{u(\tau,\cdot)}\right)$, and $\beta_\infty(\tau,\cdot) = \beta\left(v(\tau,\cdot), \frac{v(\tau,\cdot)}{u(\tau,\cdot)}\right)$. We can now take the limit and obtain

$$\begin{aligned}
\langle u(t,\cdot), f(t,\cdot)\rangle &= \langle u_0(\cdot), f(0,\cdot)\rangle \\
&\quad + \int_0^t \langle u(t,\cdot), \nabla f(\tau,\cdot)\cdot(g_\infty + \tilde{g}_\infty)(\tau,\cdot) + \partial_\tau f(\tau,\cdot) + \mu\Delta f(\tau,\cdot)\rangle d\tau \\
\langle v(t,\cdot), f(t,\cdot)\rangle &= \langle v_0(\cdot), f(0,\cdot)\rangle \\
&\quad + \int_0^t \langle v(t,\cdot), \nabla f(\tau,\cdot)\cdot(g_\infty + \tilde{g}_\infty)(\tau,\cdot) + \partial_\tau f(\tau,\cdot) + \mu\Delta f(\tau,\cdot)\rangle d\tau \\
&\quad + \int_0^t \langle u(t,\cdot), \beta_\infty(\tau,\cdot) f(\tau,\cdot)\rangle\, d\tau - \int_0^t \langle v(t,\cdot), \gamma_\infty(\tau,\cdot) f(\tau,\cdot)\rangle d\tau \\
\langle l(t,\cdot), f(t,\cdot)\rangle &= \langle l_0(\cdot), f(0,\cdot)\rangle \\
&\quad + \int_0^t \langle u(t,\cdot), \lambda_\infty(\tau,\cdot) f(\tau,\cdot)\rangle\, d\tau + \int_0^t \langle l(t,\cdot), \partial_\tau f(\tau,\cdot)\rangle\, d\tau
\end{aligned} \tag{18}$$

Integrating (18) by parts, we obtain the system corresponding to

$$\begin{aligned}
\partial_t u &= \mu\Delta u - \nabla\cdot\Big(\big(g(v/u,\nabla l) + \tilde{g}(\nabla u)\big)u\Big) \\
\partial_t v &= \mu\Delta v - \nabla\cdot\Big(\big(g(v/u,\nabla l) + \tilde{g}(\nabla u)\big)v\Big) + \beta(v,v/u)u - \gamma(v,v/u)v \\
\partial_t l &= \lambda(u,l)u
\end{aligned} \tag{19}$$

The system (19) can be simplified by rewriting it as

$$\begin{aligned}
\partial_t u &= \mu\Delta u - \nabla\cdot\big(g(u,v,\nabla u,\nabla l)u\big) \\
\partial_t v &= \mu\Delta v - \nabla\cdot\big(g(u,v,\nabla u,\nabla l)v\big) + \beta(u,v)u - \gamma(u,v)v \\
\partial_t l &= \lambda(u,l)u
\end{aligned} \tag{20}$$

where the new function g captures the effect of both g and $\tilde{g}$ in (19).

Remark 4. The structure of the system (20) is not surprising, when we take into account the discussion in Sect. 2.3. In fact, from (3) we expect the rate of change on bacteria density u to be given by a diffusion part (coming from the Brownian motion term) plus a term representing sensitivity to light and external substance. Similarly, the *excitation* density v can be expected to follow the same pattern as u with the addition of birth and decay terms.

Remark 5. The phototaxis system (20) is similar to the chemotaxis system

$$\begin{aligned} \partial_t u &= \mu \Delta u - \nabla \cdot \big(\chi(u,v) u \nabla v\big) \\ \partial_t v &= \eta \Delta v + \beta(u,v) u - \gamma(u,v) v \end{aligned} \tag{21}$$

where u is the bacteria density and v the density of chemoattractant (which is an external substance). Using the analogy to our problem, we can easily deduce how the phototaxis system would look, if we were to assume that the external substance can also diffuse and decay. In that case the last equation in (20) should be replaced by $\partial_t l = \eta_\lambda \Delta l + \lambda(u,l)u - \gamma_\lambda(u,l)l$.

Remark 6. Our work is based on the assumption that, as $N \to \infty$, $|M_w(N,v,t)| \simeq |M_{\tilde{w}}(N,v,t)|$ if $P_N^w(t) = P_N^{\tilde{w}}(t)$. That is, as $N \to \infty$, our model looks like a reaction–diffusion system, which allows us to use the methods discussed in [16] and [14]. As mentioned above, when rescaling the interaction in a moderate way, we are looking at neighborhoods that are microscopically large. Hence, as $N \to \infty$, we expect to have an arbitrarily large number of individuals in such neighborhoods. From (2) and the definition of $\mu_i(t)$ in Sect. 2.3, this means that as $N \to \infty$ the *excitation* of particle i should quickly converge to μ_i, a weighted average of all the excitations of particles in the interacting neighborhood. This behavior is assumed in our derivation of the limit dynamics.

4 Simulation

We used the model described in Sect. 2.3 to simulate the behavior of this bacteria population. The simulation consists of building a cellular automaton by discretizing (2) and (3) for small time increments. The functions and parameters that were used in our simulations are $\Delta t = 0.1$, $\sigma = 0.3$, $K = 0.1$, $v_s = 3.0$, $v_g = 0.1$, and $v_r = 0.05$, $g(s,w) = \big(1 - exp(s)\big)w \cdot \{-1, 0\}$, and $\mu_i(t) = \dfrac{1}{N} \sum_{j=1}^{N} \left[\left(1 - \dfrac{d(X(j),X(i))}{r}\right)^+ S(j) \right]$, where $r = 2$.

In Fig. 4, we show snapshots at different times of a bacteria motion in the direction of a light source that is located to the left of the domain. Bacteria surrounded by a higher number of individuals tend to move faster. Darker colors correspond to higher bacteria density.

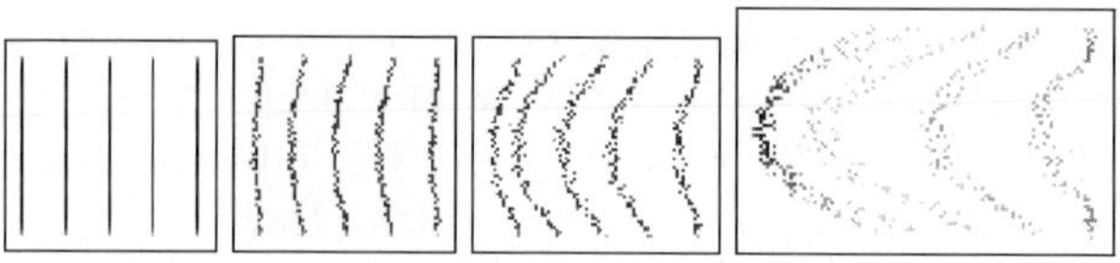

Fig. 4. A simulation of a bacteria motion toward the light source, on the *left*. Bacteria surrounded by a higher number of individuals tend to move faster. *Darker colors* correspond to higher bacteria density

Acknowledgments

The work of D. Bhaya was supported in part by the NSF under Grant 0110544. The work of D. Levy was supported in part by the NSF under Career Grant DMS-0133511. We would like to thank Matthew Burriesci for generating the images and movies.

References

1. J.P. Armitage. Bacterial tactic responses. *Adv Microb Physiol.*, 41:229–89, 1999.
2. D. Bhaya. Light matters: Phototaxis and signal transduction in unicellular cyanobacteria. *Mol. Microbiol.*, 53:745–754, 2004.
3. D. Bhaya, N.R. Bianco, D. Bryant, and A.R. Grossman. Type IV pilus biogenesis and motility in the cyanobacterium *Synechocystis* sp. PCC6803. *Mol. Microbiol.*, 37:941–951, 2000.
4. D. Bhaya, A. Takahashi, and A.R. Grossman. Light regulation of Type IV pilus-dependent motility by chemosensor-like elements in *Synechocystis* PCC 6803. *Proc. Natl. Acad. Sci. U.S.A.*, 98:7540–7545, 2001.
5. M. Burriesci and D. Bhaya. Tracking the motility of single cells and groups of Synechocystis sp. strain pcc6803 during phototaxis. *in preparation*, 2006.
6. L.L. Burrows. Weapons of mass retraction. *Mol Microbiol.*, 57(4):878–88, Aug 2005.
7. W. Chiu, M.L. Baker, and S.C. Almo. Structural biology of cellular machines. *Trends Cell Biol.*, 16(3):144–50, Mar 2006.
8. B. Davis. Reinforced random walks. *Probability Theory Related Fields*, 84(1): 203–229, 1990.
9. A. Gallegos, B. Mazzag, and A. Mogilner. Two continuum models for the spreading of myxobacteria swarms. *Bull Math Biol.*, 68(4):837–61, May 2006.
10. O.A. Igoshin, A. Goldbeter, D. Kaiser, and G. Oster. A biochemical oscillator explains several aspects of *Myxococcus xanthus* behavior during development. *Proc Natl Acad Sci U. S. A.*, 101(44):15760–5, Nov 2 2004.
11. E.F. Keller and L.A. Segel. Traveling band of chemotactic bacteria: a theoretical analysis. *J. Theor Biol.*, 30:235–48, 1971.
12. L.L. McCarter. Regulation of flagella. *Curr Opin Microbiol.*, 9(2):180–6, Apr 2006.
13. J.D. Murray. *Mathematical Biology.* Springer, Berlin Heidelberg New York, third edition, 2003.

14. K. Oelschläger. On the derivation of reaction-diffusion equations as limit dynamics of systems of moderately interacting stochastic many-particle processes. *Probability Theory Related Fields*, 82(1):565–586, 1989.
15. B. Øksendal. *Stochastic Differential Equations.* Springer, fifth edition, 2002.
16. A. Stevens. The derivation of chemotaxis equations as limit dynamics of moderately interacting stochastic many-particle systems. *SIAM Journal of Applied Mathematics*, 61(1):183–212, 2000.
17. A. Stevens. A stochastic cellular automaton modeling gliding and aggregation of myxobacteria. *SIAM Journal of Applied Mathematics*, 61(1):172–182, 2000.

Vacuum Problem of One-Dimensional Compressible Navier–Stokes Equations

H.-L. Li, J. Li, and Z. Xin

1 Introduction

In the description of motion of compressible flow in the presence of vacuum, the compressible isentropic Navier–Stokes equations with density-dependent viscosity cause more attention

$$\begin{cases} \rho_t + \mathrm{div}(\rho u) = 0, \\ (\rho u)_t + \mathrm{div}(\rho u \otimes u) - 2\mathrm{div}(\mu D(u)) - \nabla(\xi \mathrm{div} u) + \nabla p(\rho) = 0, \end{cases} \tag{1}$$

where $x \in \Omega \subset \mathbf{R}^N, t \in (0,T)$, $D(u) = (\nabla u + (\nabla u)^{\mathrm{T}})/2$, and $p(\rho) = a\rho^\gamma, a > 0, \gamma \geq 1$, the density-dependent viscosity coefficients $\mu = \mu(\rho), \xi = \xi(\rho)$ are assumed to satisfy $\mu \geq 0$ and $\xi + 2\mu/N \geq 0$.

For constant viscosity coefficients μ and ξ, there is huge literature on the studies of the global existence and/or behavior of classical or weak solutions to (1), refer to [24], [17], [7], [3], [14], [5], and references therein. In spite of the important progress, the regularity, uniqueness, and behavior of the weak solutions remain largely open. As emphasized in [8], [9], [12], [23], and [26], the possible appearance of vacuum is one of the major difficulties when trying to prove global existence and strong regularity results. It was proven [9] that weak solutions of the one-dimensional compressible Navier–Stokes equations do not exhibit vacuum states in a finite time so long as no vacuum is present initially. Such a phenomenon was observed for the spherically symmetric case in [28]. However, it was shown [26] that there is no global smooth solution to Cauchy problem for (1) with a nontrivial compactly supported initial density, which gives results for finite time blowup in the presence of vacuum; moreover, even one point initial vacuum shall cause the blowup of global strong solutions (1) for large time [12], [13].

The independence of viscosity on density makes it possible to trace either the particle path or the trajectory of vacuum state. This, however, leads to the failure of continuous dependence of weak solutions containing vacuum state on initial data [8]. For the case that the density changes continuously across the interfaces separating the gas and vacuum, the global existence and uniqueness of weak solutions were obtained in [16], where the authors obtained that velocity is smooth enough up to the interfaces which are particle paths separating the gas from the vacuum and that the support of gas density expands outside as well as interface connecting gas and vacuum moves at an algebraic rate.

Thus, viscous compressible fluids near vacuum should be better modeled by the compressible Navier–Stokes equations with density-dependent viscosities, as was derived in the fluid-dynamical approximation of Boltzmann equation for dilute gases. Further, as was first pointed out and investigated by Liu–Xin–Yang in [15], in the derivation of the compressible Navier–Stokes equations from the Boltzmann equation by the Chapman–Enskog expansions, the viscosity depends on the temperature, which is translated into the dependence of the viscosity on the density for isentropic flow. Moreover, an one-dimensional compressible flow model, called the viscous Saint–Venant system for laminar shallow water, derived rigorously from incompressible Navier–Stokes system with a moving free surface [6]

$$\begin{cases} \rho_t + (\rho u)_x = 0, \\ (\rho u)_t + (\rho u^2)_x - a(\rho u_x)_x + (\rho^2)_x = 0, \end{cases} \tag{2}$$

with the viscosity coefficients given by $\mu(\rho) = \xi(\rho) = a\rho/3$ for a given constant a. Indeed, such models appear naturally and often in geophysical flows [1], [2].

In the case of one-dimensional problem with $\mu = \xi = a\rho^\alpha$ for some positive constants a and α, the well-posedness of the Cauchy problem has been studied by many authors for initially compact-supported density. Indeed, the local (in time) well-posedness of weak solutions to this problem was first established by Liu–Xin–Yang in [15], where the initial density was assumed to be connected to vacuum with discontinuities. This property, as shown in [15], can be maintained for some finite time. And the global existence of weak solutions, together with $\alpha \in (0, 1)$ and the density function connecting to vacuum with discontinuities, was considered by many authors, see [21], [29], [10], and the references therein. It is noted that the above analysis is based on the uniform positive lower bound of the density with respect to the construction of the approximate solutions.

On the other hand, if the density function connects to vacuum continuously, there is no positive lower bound for the density and the viscosity coefficient vanishes at vacuum. This degeneracy in the viscosity coefficient gives rise to new analysis difficulties because of the less regularizing effect on the solutions. Yang et al. [30] first obtained a local existence result for this case under the free boundary condition with $\alpha > 1/2$. The authors in [31], [25], and [4] obtained the global existence of weak solutions with $\alpha \in (0, 1/2)$.

However, almost all above results concern mainly with free boundary problems, and for the global existence results, the choices of viscosity do not fit the important physical model, the shallow water equation (2) with $\mu(\rho) = \rho$

(namely $\alpha = 1$). For the constant viscosity case, one has known not only that the vacuum state will not develop later on time if there is no vacuum state initially [9], but also that the separate two initial vacuum states shall not meet together in a finite time [28], and that one point initial vacuum causes strong solutions to blowup [12] at infinity as well. However, little is known on the dynamics of the vacuum states of weak solutions to the compressible Navier–Stokes equations (1) with density-dependent viscosity on bounded domain. And in particular, it is not clear yet how the vacuum states evolve with respect to time and whether the initial vacuum states shall exist all the time or not for weak solutions. The study of these important dynamical problems about vacuum states is rather difficult because the nonlinear diffusion is degenerate as vacuum appears, which is quite different from the case of constant viscosity. This causes the loss of information about the velocity and makes it difficult to trace the evolution of vacuum states in general. Across the interface (or vacuum boundary), it is usually difficult to obtain enough information about velocity even if considering specific cases such as point vacuum or continuous vacuum of one piece. It is important to get enough information about the velocity since the flow particles transport usually along particle path, and all interfaces of vacuum, such as free boundaries [16], [10] which can be observed and dealt with, also move along the trajectories determined by velocity field.

We consider the dynamics of vacuum state of global weak solutions of the compressible Navier–Stokes equations for general large initial data with finite entropy and vacuum in the present chapter. We prove that the entropy weak solutions for general large initial data satisfying finite initial entropy exist globally in time, and for more regular initial data, there is a global entropy weak solution which is unique and regular with well-defined velocity field for short time, and the initial structure of vacuum state is maintained for this short time and the interface of vacuum state propagates along particle path. Then, we show that, for any global entropy weak solution, any (possibly existing) vacuum state must vanish within finite time. The velocity (even if regular enough and well defined) blows up in finite time as the vacuum states vanish. After the vanishing of vacuum states, the global entropy weak solution becomes a strong solution and tends to the nonvacuum equilibrium state exponentially in time.

2 Main Results

We consider the IBVP for the one-dimensional compressible Navier–Stokes equations with density-dependent viscosity

$$\rho_t + (\rho u)_x = 0, \tag{3}$$

$$(\rho u)_t + (\rho u^2 + p(\rho))_x - (\mu(\rho) u_x)_x = 0, \tag{4}$$

with $\rho \geq 0$ the density and ρu the momentum. For simplicity, we only take the compressible Navier–Stokes equation for the motion of shallow water, i.e., both pressure $p = p(\rho)$ and viscosity $\mu = \mu(\rho)$ of the following forms

$$p(\rho) = \rho^2, \ \ \mu(\rho) = \rho.$$

The Dirichlet boundary condition is imposed for (3) and (4) as

$$\rho u(0,t) = \rho u(1,t) = 0, \quad t \geq 0. \tag{5}$$

Remark 1. For the case of Dirichlet boundary (5), the boundary is given by the physical observable momentum instead of velocity. It is natural to employ such boundary condition as considering the dynamics of (global in time) weak solutions to the IBVP for the compressible Navier–Stokes equations with possible vacuum states included since it is usually the case that, at vacuum states, the momentum is zero and is observed and controllable, but almost nothing is known yet for the velocity for weak solutions. Note here that for any (weak) solution away from vacuum at the boundary, the boundary condition (5) reduces to $u(0,t) = u(1,t) = 0$, $t \geq 0$.

The initial data are given for density ρ and momentum ρu

$$\rho(x,0) = \rho_0(x) \geq 0, \quad \rho u(x,0) = m_0(x), \quad x \in \Omega, \tag{6}$$

where the domain Ω is chosen as unit interval denoting the spatial domain $(0,1)$ or periodic domain with period length one, and throughout the present chapter the initial data are assumed to satisfy

$$\begin{cases} \rho_0 \geq 0 \text{ a.e. in } \Omega, \quad \rho_0 \in L^1(\Omega), \quad (\sqrt{\rho_0}\,)_x \in L^2(\Omega), \\ m_0 = 0, \quad \text{a.e. on } \{x \in \Omega \,|\, \rho_0(x) = 0\}, \ \frac{|m_0|^2}{\rho_0} \in L^1(\Omega). \end{cases} \tag{7}$$

To define the weak solutions to the IBVP for the compressible Navier–Stokes equations (3) and (4) with initial data (6) and boundary condition (5), we introduce the set of test functions as

$$\Psi \triangleq C_0^\infty(\overline{\Omega} \times [0,T)) \quad \text{and} \quad \Phi \triangleq C_0^\infty(\Omega \times [0,T)).$$

We define the weak solutions to the IBVP for the compressible Navier–Stokes equations (3) and (4) as follows.

Definition 1 (Global entropy weak solutions). *For any $T > 0$, (ρ, u) is said to be a weak solution to the IBVP problem for the compressible Navier–Stokes equations* (3)–(6) *in $\Omega \times (0,T)$, if*

$$\begin{cases} 0 \leq \rho \in L^\infty(0,T; L^1(\Omega) \cap L^2(\Omega)), \quad (\sqrt{\rho})_x \in L^\infty(0,T; L^2(\Omega)), \\ \sqrt{\rho} u \in L^\infty(0,T; L^2(\Omega)), \quad \rho u_x \in L^2(0,T; W^{-1,1}_{\text{loc}}(\Omega)), \end{cases} \tag{8}$$

and (ρ, u) *satisfies*

$$\int_\Omega \rho_0 \psi(x,0)dx + \int_0^T \int_\Omega \rho \psi_t dxdt + \int_0^T \int_\Omega \sqrt{\rho}\sqrt{\rho} u \psi_x dxdt = 0 \tag{9}$$

for any $\psi \in \Psi$, *and*

$$\begin{aligned} &\int_\Omega m_0 \varphi(x,0)dx + \int_0^T \int_\Omega \sqrt{\rho}(\sqrt{\rho}u)\varphi_t dxdt \\ &+ \int_0^T \int_\Omega \left((\sqrt{\rho}u)^2 + \rho^2\right) \varphi_x dxdt - \langle \rho u_x, \varphi_x \rangle = 0 \end{aligned} \tag{10}$$

for all $\varphi \in \Phi$, *and the following uniform entropy inequality holds*

$$\begin{aligned} &\sup_{0\le t\le T} \int_\Omega \left(|\sqrt{\rho}u|^2 + |(\sqrt{\rho})_x|^2 + \rho^2\right)(x,t)\, dx \\ &+ \int_0^T \int_\Omega (|\rho_x|^2 + |\sqrt{\rho}u_x|^2)(x,t)\, dxdt \\ \le & C_0 \int_\Omega (\tfrac{|m_0|^2}{\rho_0} + |(\sqrt{\rho_0}\,)_x|^2 + \rho_0^2)(x)\, dx =: C_0 E_0 \end{aligned} \tag{11}$$

with a positive constant C_0 *independent of* T. *The nonlinear diffusion term* $\rho u_x = \sqrt{\rho}(\sqrt{\rho}u_x)$ *makes sense as expressed as*

$$\langle \rho u_x, \varphi \rangle = -\int_0^T \int \sqrt{\rho}\sqrt{\rho}u\varphi_x dxdt - 2\int_0^T \int (\sqrt{\rho})_x \sqrt{\rho}u\varphi dxdt \tag{12}$$

for any $\varphi \in \Phi$, *where* $\rho \in L^\infty(\Omega \times (0,T))$ *due to* (8).

Remark 2. According to (9) and (8), the weak solution (ρ, u) satisfies (3) in the sense of distribution and justifies the boundary condition (5) in the sense that $\int_I \sqrt{\rho}\sqrt{\rho}u(x,s)ds \to 0$ for any time interval $I \subset [0,T]$ as $x \to \partial\Omega$. If, in addition, it holds $\sqrt{\rho}\sqrt{\rho}u \in L^p(0,T;W^{1,q}(\Omega))$ for some $p \ge 1, q > 1$, then the Dirichlet boundary condition (5) is satisfied in the sense of trace because it follows from (9) that $\sqrt{\rho}\sqrt{\rho}u \in L^p(0,T;W_0^{1,q}(\Omega))$.

We have the following result on the existence of global entropy weak solutions.

Theorem 1 (Global existence). *Assume that in addition the initial data* (ρ_0, m_0) *satisfy* (7) *and* $\frac{|m_0|^{2+\nu}}{\rho_0^{1+\nu}} \in L^1(\Omega)$ *for some positive constant* $\nu \in (0,3)$. *Then for any* $T > 0$, *there exists a global entropy weak solution* (ρ, u) *to the IBVP problem* (3)–(6) *in* $\Omega \times (0,T)$ *in the sense of Definition 1. Moreover, the weak solution* (ρ, u) *also satisfies*

$$\sqrt{\rho}(\sqrt{\rho}u) \in L^2(0,T;W_0^{1,(4+2\nu)/(4+\nu)}(\Omega)), \tag{13}$$

i.e., $\sqrt{\rho}(\sqrt{\rho}u)$ *satisfies the Dirichlet boundary condition* (5) *in the sense of trace.*

Proof (Outline of the proof). The theorem can be done by the construction of approximate solution and compactness argument. The approximate classical solution is constructed by solving the approximate compressible Navier–Stokes equations

$$\rho_{\varepsilon t} + (\rho_\varepsilon u_\varepsilon)_x = 0, \tag{14}$$

$$(\rho_\varepsilon u_\varepsilon)_t + (\rho_\varepsilon u_\varepsilon^2 + (\rho_\varepsilon^2)_x - (\mu_\varepsilon(\rho_\varepsilon) u_{\varepsilon x})_x = 0, \tag{15}$$

$$(\rho_\varepsilon, \rho_\varepsilon u_\varepsilon)(x,0) = (\rho_{0\varepsilon}, m_{0\varepsilon})(x), \tag{16}$$

where the viscosity is modified as $\mu_\varepsilon(\rho) = \rho + \varepsilon\rho^\theta,\ \varepsilon > 0,\ \theta \in (0,1/2)$, the boundary condition is replaced by

$$u_\varepsilon(0,t) = u_\varepsilon(1,t) = 0, \tag{17}$$

and the initial data $\rho_{0\varepsilon}, m_{0\varepsilon} \in C^\infty(\overline{\Omega})$ satisfy

$$\begin{cases} \rho_{0\varepsilon} \to \rho_0 \text{ in } L^1(\Omega),\ \rho_{0\varepsilon}^{\alpha-1/2} \to \rho_0^{\alpha-1/2} \text{ in } H^1(\Omega), \\ \frac{m_{0\varepsilon}^2}{\rho_{0\varepsilon}} \to \frac{m_0^2}{\rho_0},\quad \frac{|m_{0\varepsilon}|^{2+\nu}}{\rho_{0\varepsilon}^{1+\nu}} \to \frac{|m_0|^{2+\nu}}{\rho_0^{1+\nu}} \text{ in } L^1(\Omega) \end{cases} \tag{18}$$

as $\varepsilon \to 0^+$, and

$$\rho_{0\varepsilon} \geq c_0 \varepsilon^{1/(2\alpha-2\theta)} \tag{19}$$

for some c_0 independent of ε.

It can be proven that, for any classical solution $(\rho_\varepsilon, u_\varepsilon)$ of the IBVP problem (14)–(17), the density ρ_ε is bounded from above and below uniformly with respect to time. Using these uniform bounds, any local in time classical solution can be extended the local classical solution globally in time by standard arguments. Moreover, the classical solution satisfies [1], [18] for all $t > 0$,

$$\begin{aligned} &\int_\Omega (\rho_\varepsilon |u_\varepsilon|^{2+\nu} + |\sqrt{\rho_\varepsilon} u_\varepsilon|^2 + |\tfrac{\mu(\rho_\varepsilon)_x}{\sqrt{\rho_\varepsilon}}|^2 + |\tfrac{\varepsilon((\rho_\varepsilon)^\theta)_x}{\sqrt{\rho_\varepsilon}}|^2 + \pi(\rho_\varepsilon))(x,t)dx \\ &\quad + \textstyle\int_0^t \int_\Omega (\mu(\rho_\varepsilon)|u_{\varepsilon x}|^2 + 2\mu'(\rho_\varepsilon)|\rho_{\varepsilon x}|^2)(x,s)dxds \leq CE_0, \end{aligned} \tag{20}$$

with C independent of t and ε.

By employing above the a priori estimates and compactness argument, one can prove that there is (ρ, u) so that

$$\rho_\varepsilon \to \rho \text{ in } C([0,T] \times \overline{\Omega}), \quad \text{as} \quad \varepsilon \to 0, \tag{21}$$

$$(\rho_\varepsilon^{1/2})_x \rightharpoonup (\rho^{1/2})_x \text{ weakly in } L^2(\Omega \times (0,T)), \tag{22}$$

$$\sqrt{\rho_\varepsilon} u_\varepsilon \to \sqrt{\rho} u, \quad \rho_\varepsilon u_\varepsilon \to \rho u \text{ in } L^{2+\nu/2}(\Omega \times (0,T)). \tag{23}$$

Furthermore, one can prove that (ρ, u) is indeed a global entropy weak solution of the IBVP problem (3)–(6) in $\Omega \times (0,T)$ in the sense of Definition 1. □

We next show that there is a global entropy weak solution (ρ, u) in the sense of Definition 1 for which the vacuum states and the structure of interface, if existing initially, can be preserved for a short time, so long as the initial data have additional regularity besides (7) and "smooth" connection with vacuum. For simplicity, we consider the case of one point vacuum state contained at $x = x_0 \in (0,1)$ in the initial data $(\rho_0, m_0) = (\rho_0, \rho_0 u_0)$ with additional regularity

$$A_0|x - x_0|^\sigma \le \rho_0(x) \le A_1|x - x_0|^\sigma, \text{ for any } x \in \Omega, \tag{24}$$

$$u_0 \in C^1(\bar{\Omega}), \quad (\rho_0^{1+1/2j})_x \in L^{2j}(\Omega), \ \rho_0^{-1+1/2j}(\rho_0 u_{0x})_x \in L^{2j}(\Omega) \tag{25}$$

for $j = 1, n$ with $n \ge 2$ an integer. Here, σ, A_0, A_1, and B_0, B_1 are positive constants, and the power $\sigma \in (\sigma_-, \sigma_+)$ with positive constants $\sigma_\pm$ to be determined by (30) later. We also require that the initial data $(\rho_0, m_0) = (\rho_0, \rho_0 u_0)$ given by (6) are consistent with boundary value for Dirichlet boundary condition.

We have the result on short time structure of global entropy weak solutions.

Theorem 2 (Short time structure of vacuum states). *Besides the assumptions of Theorem 1, assume in addition that there is one point vacuum state in initial data* (ρ_0, u_0) *with* (24)–(25) *satisfied. Then, there exists a global entropy weak solution* (ρ, u) *to the IBVP problem* (3)–(6) *in the sense of Definition 1.*

Moreover, there is a short time $T_* > 0$, *so that the global entropy weak solution* (ρ, u) *is unique*[1] *and regular on the domain* $\Omega \times [0, T_*]$, *and the initial structure of vacuum states is maintained for* $t \in [0, T_*]$ *in the following sense. That is, the solution* (ρ, u) *is regular and unique on the domain* $\bar{\Omega} \times [0, T_*]$,

$$(\rho, u) \in C^0(\bar{\Omega} \times [0, T_*]), \quad u_x \in L^\infty(0, T_*; C^0(\bar{\Omega})), \tag{26}$$

$$\|u\|_{L^\infty(\bar{\Omega} \times [0,T_*])} + \|u_x\|_{L^\infty([0,T_*];C^0(\bar{\Omega}))} \le C(T_*). \tag{27}$$

The one point vacuum state propagates along particle path, namely, there is one particle path $x = X_0(t) : [0, T_*] \to \overline{\Omega}$ *with* $X_0(t) \in C([0, T_*])$ *defined by*

$$\dot{X}_0(t) = u(X_0(t), t), \quad X_0(0) = x_0 \in (0,1), \tag{28}$$

so that

$$a_-|x - X_0(t)|^\sigma \le \rho(x,t) \le a_+|x - X_0(t)|^\sigma \tag{29}$$

for $(x,t) \in \Omega \times [0, T_*]$, *where the two positive constants* $a_\pm$ *are independent of time* T_*.

Proof (Outline of the proof). The proof of short time structure of the global entropy weak solution consists of two parts, i.e., the construction of short time unique solutions maintaining the initial structure of vacuum states and the extension of this solution to the global one. The proof is rather complicated, the reader can refer to [11], we omit the details. □

[1] Here the uniqueness is specified for density ρ and momentum $\rho u = \sqrt{\rho}\sqrt{\rho}u$ for continuous vacuum states of one piece.

Remark 3. The constants $\sigma_\pm = \beta_\pm/(1-\beta_\pm) > 0$ with $\beta_\pm$ determined by

$$\beta_- = \tfrac{1}{2}(1 - \tfrac{1}{2n}), \qquad \beta_+ = (1 - \tfrac{1}{2n}). \tag{30}$$

The regularity assumptions (24) and (25) are satisfied for the following initial data

$$\rho_0(x) = \tfrac{1}{2}(A_0 + A_1)(|x - x_0|^2)^{(\sigma_- + \sigma_+)/4}, \quad u_0(x) = 0, \quad x \in \Omega.$$

Next, we prove that for any global entropy weak solution (ρ, u) to the IBVP problem (3)–(6) together with boundary condition (5) in the sense of Definition 1, even though for some cases that the vacuum states may exist for some finite time, any possible vacuum state has to vanish within finite time after which the density is always away from vacuum.

Theorem 3 (Vanishing of vacuum states). *Let (ρ, u) be any global entropy weak solution to the IBVP problem (3)–(6) in the sense of Definition 1. Then, there exists some time $T_0 > 0$ (depending on initial data) and a constant ρ_- so that*

$$\inf_{x \in \Omega} \rho(x,t) \geq \rho_- > 0, \quad t \geq T_0, \tag{31}$$

and the global entropy weak solution (ρ, u) becomes a unique strong solution (ρ, u) for $t \geq T_0$ and satisfies

$$\begin{cases} \rho \in L^\infty(T_0, t; H^1(\Omega)), & \rho_t \in L^\infty(T_0, t; L^2(\Omega)), \\ u \in H^1(T_0, t; L^2(\Omega)) \cap L^2(T_0, t; H^2(\Omega)), \end{cases} \tag{32}$$

with velocity u and nonlinear diffusion term given by

$$u \triangleq \frac{\sqrt{\rho} u}{\sqrt{\rho}}, \quad (\rho u_x)_x = (\sqrt{\rho}(\sqrt{\rho} u_x))_x, \tag{33}$$

respectively. In addition, there exist two positive constants μ_0 and c_0 both depending on initial data (ρ_0, m_0) and ρ_-, such that

$$\|(\rho - \bar{\rho_0}, u - u_s)(\cdot, t)\|_{L^2(\Omega)} \leq c_0 e^{-\mu_0 (t - T_0)}, \quad t > T_0, \tag{34}$$

where $\bar{\rho_0} = \frac{1}{|\Omega|}\int_\Omega \rho_0(x) dx$.

Proof (Outline of the proof). Let $T \in (0, \infty)$ be fixed. Choosing a constant $b \geq 3$, one can derive from (11) and conservation of mass that

$$\sup_{0 \leq t \leq T} \left(\|\rho\|_{L^\infty} + \left\| (\rho^b)_x \right\|_{L^2} \right) + \int_0^T \left\| (\rho^b)_x \right\|_{L^2}^2 dt \leq C. \tag{35}$$

which implies

$$\left\| \left(\rho^b - \overline{\rho^b} \right)(\cdot, t) \right\|_{C(\bar\Omega)} \leq C \left\| \left(\rho^b - \overline{\rho^b} \right)(\cdot, t) \right\|_{L^4(\Omega)}^{2/3} \|(\rho^b)_x(\cdot, t)\|_{L^2}^{1/3}$$

$$\leq C\left\|\left(\rho^b - \overline{\rho^b}\right)(\cdot, t)\right\|_{L^4(\Omega)}^{2/3} \to 0, \text{ as } t \to \infty, \tag{36}$$

where $\overline{\rho^b}(t) = \frac{1}{|\Omega|}\int_\Omega \rho^b(x,t)dx$. This is enough to prove (31) due to the following simple fact $\overline{\rho^b}(t) \geq \bar{\rho}^b(t) \equiv \overline{\rho_0}^b > 0$, for any $t \geq 0$. Then, one can define the velocity after the vanishing of vacuum as $u \triangleq \frac{\sqrt{\rho}u}{\sqrt{\rho}}$ and gain its regularity, and then recover the nonlinear diffusion. It is then straightforward to verify that the weak solution becomes a strong solutions after the vanishing of vacuum and tends to equilibrium state exponentially. The reader can refer to [11], we omit the details. □

Finally, for any global entropy weak solution (ρ, u) to the IBVP problem (3)–(6) in the sense of Definition 1, the density is continuous, i.e., $\rho \in C(\overline{\Omega} \times [0,T])$ for any $T > 0$, due to (8) and (9). Thus, the continuity of ρ and Theorem 3 imply that if the density contains vacuum states at least at one point, then there exists some critical time $T_1 \in [0, T_0)$ with $T_0 > 0$ given by (31) and a nonempty subset $\Omega^0 \subset \bar{\Omega}$ such that

$$\begin{cases} \rho(x,T_1) = 0, & \forall\, x \in \Omega^0 \\ \rho(x,T_1) > 0, & \forall\, x \in \bar{\Omega} \setminus \Omega^0, \\ \rho(x,t) > 0, & \forall\, (x,t) \in \bar{\Omega} \times (T_1, T_0]. \end{cases} \tag{37}$$

We deduce from (32) easily that, for any $\delta > 0$, it holds

$$\int_{T_1+\delta}^{T_0} \|u_x\|_{L^\infty} ds < \infty. \tag{38}$$

Under the condition that vacuum states appear, we shall prove that the spatial derivative of velocity (if regular enough and definable) blows up in finite time as the vacuum states vanish, even if the solution is regular enough for short time so that the velocity field and its derivatives are bounded as shown by Theorem 2.

Theorem 4 (Finite time blowup). *Let (ρ, u) be any global entropy weak solution, which contains vacuum states at least at one point for some finite time, to the IBVP for the compressible Navier–Stokes equations* (3)–(6) *with boundary condition* (5) *in the sense of Definition 1. Let $T_0 > 0$ and $T_1 \in [0, T_0)$ be the time such that* (31) *and* (37) *hold, respectively.*

Then, the solution (ρ, u) blows up as vacuum states vanish. Namely, for T_1 satisfying (37) *and for given any fixed $\eta > 0$, it holds*

$$\lim_{t \to T_1^+} \int_t^{T_1+\eta} \|u_x\|_{L^\infty} ds = \infty. \tag{39}$$

On the other hand, if there exists some $T_2 \in (0, T_0)$ such that the weak solution (ρ, u) satisfies

$$\|u\|_{L^1(0,T_2;W^{1,\infty}(\Omega))} < \infty,$$

then, there is a time $T_3 \in [T_2, T_0)$ *so that the blowup phenomena happen for* (ρ, u), *i.e.*,

$$\lim_{t \to T_3^-} \int_0^t \|u_x\|_{L^\infty} ds = \infty. \tag{40}$$

Proof (Outline of the proof). One can easily understand this blowup phenomena due to the fact that away from vacuum the particle transports along particle path. If no blowup (39) (or (40)) happens, then it is able to connect one vacuum state (or vacuum interface) with one nonvacuum point through particle path. It is a contradiction since the density function is always zero at any vacuum point, but strictly positive away from vacuum state. One can refer to [11]. □

Remark 4. Theorems 1–4 provide a complete dynamical description on the vanishing of vacuum states and blowup phenomena for the global entropy weak solutions to the compressible Navier–Stokes equations with density-dependent viscosity. That is, a global entropy weak solution exists for general large initial data with finite entropy. For short time, such weak solution is unique and regular with well-defined velocity field subject to additional initial regularity, and any existing vacuum state is maintained with the same interface structure as initial. Then, within finite time the vacuum states vanish definitely and the velocity blows up (even if it is regular enough and definable along the interfaces). After the vanishing of vacuum states, the global entropy weak solution becomes a strong one and tends to the nonvacuum equilibrium state exponentially in time. However, before the time of vacuum-vanishing, the uniqueness of the global entropy weak solution to the compressible Navier–Stokes equations with density-dependent viscosity subject to the initial data is not known yet.

Remark 5. All theories established in Theorems 1–4 can be generalized to general compressible Navier–Stokes equation with both γ-law pressure and density-dependent viscosity coefficient (refer to [11]). For additional related problems to the compressible Navier–Stokes equations, the reader can refer to [11].

References

1. Bresch, D.; Desjardins, B. Existence of global weak solutions for a 2D viscous shallow water equations and convergence to the quasi-geostrophic model, *Comm. Math. Phys.* **238** (2003), 211–223.
2. Bresch, D.; Desjardins, B. *On the existence of global weak solutions to the Navier-Stokes equations for viscous compressible and heat conducting fluids.* preprint 2005
3. Danchin, R. Global existence in critical spaces for compressible Navier-Stokes equations. *Invent. Math.* **141** (2000), 579–614.

4. Fang, D.; Zhang, T. Compressible Navier-Stokes equations with vacuum state in one dimension. *Comm. Pure Appl. Anal.* **3** (2004), 675–694.
5. Feireisl, E.; Novotny, A.; Petzeltová, H. On the existence of globally defined weak solutions to the Navier-Stokes equations. *J. Math. Fluid Mech.* **3** (2001), no. 4, 358–392.
6. Gerbeau, J.-F.; Perthame, B. Derivation of viscous Saint-Venant system for laminar shallow water; numerical validation. *Discrete Contin. Dyn. Syst. Ser. B1*, (2001), 89–102.
7. Hoff, D. Strong convergence to global solutions for multidimensional flows of compressible, viscous fluids with polytropic equations of state and discontinuous initial data. *Arch. Rational Mech. Anal.* **132** (1995), 1–14.
8. Hoff, D.; Serre, D. The failure of continuous dependence on initial data for the Navier-Stokes equations of compressible flow. *SIAM J. Appl. Math.* **51** (1991), no. 4, 887–898.
9. Hoff, D.; Smoller, J., Non-formation of vacuum states for compressible Navier-Stokes equations. *Comm. Math. Phys.* **216** (2001), no. 2, 255–276.
10. Jiang, S.; Xin, Z.; Zhang, P. *Golobal weak solutions to 1D compressible isentropic Navier-Stokes equations with density-dependent viscosity*, Methods and Applications of Analysis, to appear 2005.
11. Li H.-L.; Li J.; Xin Z. *Vanishing of Vacuum States and Blow-up Phenomena of the Compressible Navier-Stokes Equations*, preprint 2006.
12. Li, J. *Qualitative behavior of solutions to the compressible Navier-Stokes equations and its variants.* PhD Thesis, Chinese University of Hong Kong, 2004.
13. Li, J.; Xin Z. Some uniform estimates and blowup behavior of global strong solutions to the Stokes approximation equations for two-dimensional compressible flows. *J. Differential Equations* **221** (2006), 275–308.
14. Lions, P.-L., *Mathematical topics in fluid mechanics. Vol. 2. Compressible models.* Oxford University Press, New York, 1998.
15. Liu, T.-P.; Xin, Z.; Yang, T. Vacuum states for compressible flow. *Discrete Contin. Dynam. Systems* **4** (1998), 1–32.
16. Luo, T.; Xin, Z.; Yang, T. Interface behavior of compressible Navier-Stokes equations with vacuum. *SIAM J. Math. Anal.* **31** (2000), 1175–1191.
17. Matsumura, A.; Nishida, T. The initial boundary value problems for the equations of motion of compressible and heat-conductive fluids. *Comm. Math. Phys.* **89** (1983), 445–464.
18. Mellet, A.; Vasseur, A. *On the isentropic compressible Navier-Stokes equations*, preprint 2005.
19. Okada, M. Free boundary problem for the equation of one-dimensional motion of viscous gas. *Japan J. Appl. Math.* **6** (1989), 161–177.
20. Okada, M.; Makino, T. Free boundary problem for the equation of spherically symmetric motion of viscous gas *Japan J. Appl. Math.* **10** (1993), 219–235.
21. Okada, M.; Matsusu-Necasova, S.; Makino, T. Free boundary problem for the equation of one-dimensional motion of compressible gas with density-dependent viscosity. *Ann. Univ. Ferrara Sez. VII (N.S.)* **48** (2002), pp. 1–20.
22. Straškraba, I; Zlotnik, A. Global properties of solutions to 1D-viscous compressible barotropic fluid equations with density dependent viscosity. *Z. Angew. Math. Phys.* **54** (2003), no. 4, 593–607.
23. Salvi, R.; Straškraba I. Global existence for viscous compressible fluids and their behavior as $t \to \infty$. *J. Fac. Sci. Univ. Tokyo Sect. IA, Math.,* **40** (1993), 17–51.

24. Serre, D. On the one-dimensional equation of a viscous, compressible, heat-conducting fluid. *C. R. Acad. Sci. Paris Sér. I Math.* **303** (1986), no. 14, 703–706.
25. Vong, S.-W.; Yang, T.; Zhu, C. Compressible Navier-Stokes equations with degenerate viscosity coefficient and vacuum. II. *J. Differential Equations* **192** (2003), no. 2, 475–501.
26. Xin, Z. Blowup of smooth solutions to the compressible Navier-Stokes equation with compact density. *Comm. Pure Appl. Math.* **51** (1998), pp. 229–240.
27. Xin, Z. On the behavior of solutions to the compressible Navier-Stokes equations. 159–170, *AMS/IP Stud. Adv. Math.* 20, AMS, Providence, RI, 2001.
28. Xin, Z.; Yuan, H. *Vacuum state for spherically symmetric solutions of the compressible Navier-Stokes equations*, submitted 2005.
29. Yang T.; Yao Z.; Zhu, C. Compressible Navier-Stokes equations with density-dependent viscosity and vacuum. *Comm. Partial Differential Equations* **26** (2001), no. 5–6, 965–981.
30. Yang, T.; Zhao, H. A vacuum problem for the one-dimensional compressible Navier-Stokes equations with density-dependent viscosity. *J. Differential Equations* **184** (2002), pp. 163–184.
31. Yang, T.; Zhu, C. Compressible Navier-Stokes equations with degenerate viscosity coefficient and vacuum. *Comm. Math. Phys.* **230** (2002), pp. 329–363.

Stability and Instability Issues for Relaxation Shock Profiles

C. Mascia

A hyperbolic system with relaxation is a system of partial differential equations of hyperbolic type with a zero-order term, describing the relaxation mechanism toward a given equilibrium. After hyperbolic rescaling, the system can be thought as a dynamical system with two time scales: the fast one is governed mainly by the kinetic part of the system itself and it drives the solution toward the equilibrium manifold, the slow one is described by a reduced hyperbolic system of conservation laws.

Many physical systems, in particular in continuum mechanics, possess such structure, therefore systems with relaxation have been studied for a long time (see [Whi74]). The attempt of building up a general theory on this class of evolution equations is more recent. A fundamental contribution in this construction, following the analysis of hyperbolic systems with damping, is due to Liu and is the content of the pioneering paper [TLiu87]. In that paper, the problem of stability of nonlinear waves (shocks and rarefactions) for systems with relaxation is addressed for the 2×2 case, in analogy with the case of system of conservation laws with viscosity, whose analysis started some years before. After that, the attention of the conservation laws' community on this field has grown up, giving raise to a large amount of articles on hyperbolic systems with relaxation in general.

The aim of this chapter is to briefly present the state of the art on the analysis of stability of relaxation profiles and to draw attention on possible directions for future research. Both the aspects follow the personal point of view of the author and, therefore, are biased by taste and personal experience.

1 Basic Framework

Given smooth functions $F, G : \mathbb{R}^N \to \mathbb{R}^N$, we consider the hyperbolic system of balance laws for the unknown variable $w \in \mathbb{R}^N$

$$w_t + F(w)_x = G(w), \tag{1}$$

with $x \in \mathbb{R}$ and $t > 0$. The function F is assumed to satisfy the usual (strictly) hyperbolicity assumption: $\sigma(\mathrm{d}F(w))$ has N real, distinct eigenvalues for any w (here $\sigma(A)$ denotes the spectrum of the operator A). In the semilinear case, i.e., if the flux function F is linear, system (1) is called *discrete kinetic model.*

System (1) has a *relaxation structure* if:

- The unknown variable w can be represented in the couple $w = (u, v) \in \mathbb{R}^n \times \mathbb{R}^r$ so that $F = (f, g)$ and $G = (0_n, q)$, for some $f \in C^1(\mathbb{R}^N; \mathbb{R}^n)$ and $g, q \in C^1(\mathbb{R}^N; \mathbb{R}^r)$. The variable u (respectively, v) is a *conservative* (respectively, *non conservative*) variable.
- The equilibrium manifold $\mathcal{E} := \{w : G(w) = 0\}$ can be represented as a graph of the function u, i.e., $\mathcal{E} = \{(u, v) : v = v^*(u)\}$ for some $v^* \in C^1(\mathbb{R}^n; \mathbb{R}^r)$.
- The equilibrium manifold $\mathcal{E}$ is exponentially attractive with respect to the kinetic $w' = G(w)$, i.e., $\sigma(\mathrm{d}_v q)\big|_{\mathcal{E}} \subset \{\mathrm{Re}\lambda < 0\}$.

After hyperbolic rescaling $(x, t) \mapsto (x/\varepsilon, t/\varepsilon)$, system (1) takes the form

$$\varepsilon\,(w_t + F(w)_x) = G(w). \tag{2}$$

Thus, the large-time/far-distance behavior of solutions to (1) is formally described by the singular limit of (2), obtained by setting $\varepsilon = 0$, i.e., by the *relaxed system* (usually assumed to be hyperbolic)

$$u_t + f^*(u)_x = 0, \qquad \text{where} \quad f^*(u) := f(u, v^*(u)). \tag{3}$$

Proving rigorously the validity of the singular limit $\varepsilon \to 0^+$ is a somewhat complementary to the problem of large-time behavior and it is subject of active research still (in this direction, we suggest the fundamental reference [CLL94] and the more recent articles [Bia06, Ser00]).

The *Chapman–Enskog expansion* gives an ε–order approximation of rescaled version (2). This expansion is based on the idea that the variable v is indeed a fast variable with respect to u when ε is small. Therefore, it makes sense to consider the dynamics of v as determined by the conservative variable u. Precisely, this is done by considering the ansatz

$$v = v^*(u) + \varepsilon\, v^1 + o(\varepsilon) \qquad \text{as} \quad \varepsilon \to 0^+.$$

Inserting in (2) and equating the ε–order coefficients, we get

$$v^1 = -B^*(u)\, u_x, \qquad \text{where} \quad B^*(u) = (\mathrm{d}_v q^*)^{-1}(\mathrm{d}v^*\, \mathrm{d}f^* - \mathrm{d}g^*),$$

where $\mathrm{d}_v q^*, f^*, g^*$ denote the functions $\mathrm{d}_v q, f, g$ calculated at the equilibrium manifold $(u, v^*(u))$. Plugging the formal expansion of v in the system, we get

$$u_t + f^*(u)_x = \varepsilon(B^*(u)\, u_x)_x + o(\varepsilon),$$

thus suggesting that hyperbolic systems with relaxation can be seen as high-order perturbations of a system of conservation laws. In this sense, they belong to a large class of similar problems, arising when considering other high-order perturbations: viscosity, dispersion, capillarity, thin film, radiating gas dynamics, etc. Almost all the questions that can be asked for any one of this perturbed systems of conservation laws can be asked for all the others, mutatis mutandis.

2 Relaxation Shock Profiles

A *shock wave* for the reduced hyperbolic system (3) is a solution of the form

$$u(x,t) := \begin{cases} u_- & x < st \\ u_+ & x > st \end{cases} \tag{4}$$

for some $u_\pm \in \mathbb{R}^n$, $s \in \mathbb{R}$. To fix ideas, let us assume that (4) defines a $p-$Lax shock, i.e.,

$$\lambda^*_{p-1}(u_-) < s < \lambda^*_p(u_-), \qquad \lambda^*_p(u_+) < s < \lambda^*_{p+1}(u_+),$$

where $\lambda^*_1(u) < \cdots < \lambda^*_n(u)$ are the (real, distinct) eigenvalues of $df^*(u)$ and the pth characteristic field of (3) is genuinely nonlinear.

A *relaxation shock profile* for the system (1) is a solution of the form

$$W = W(x - st) \qquad \text{with} \quad W(\pm\infty) = W_\pm := (u_\pm, v^*(u_\pm)) \tag{5}$$

for some $s \in \mathbb{R}$. Given the asymptotic values $W_\pm$, by integrating the ODE system satisfied by the profile W, it is immediate to check that the speed s satisfies the Rankine–Hugoniot condition for the relaxed system (3). In the following $W = (U, V)$ where U (respectively, V) is the conserved (respectively, nonconserved) component of W.

The existence of a relaxation profile corresponding to a given shock wave for the relaxed system can be interpreted as a sort of consistency of the original model (1) with respect to the limiting reduced one (3). Following the terminology used in the context of conservation laws with viscosity, the problem of showing the existence of relaxation shock profiles corresponding to a shock wave is sometimes referred to as the *Gel'fand problem.*

Apart for specific models, for which relaxation shock profiles can be determined explicitly, existence of such special solutions has been proved for the general case in [YZ00] under the assumption of weak shock, i.e., for $|u_+ - u_-|$ small. Small relaxation shocks are *smooth*, thanks to a regularization effect due to the kinetic/relaxation term G, bringing into play viscosity effects. In general, large relaxation shocks may present subshocks inside their structure.

Applying the change of coordinates $x \mapsto x - st$, the relaxation profile W becomes a stationary solution of (1). Hence, without loss of generality, we assume from now on that the propagation speed s is zero.

Once the existence of a relaxation profile has been established, it is natural to ask for conditions guaranteeing asymptotic stability, that is to answer to the following question consider the solution $\tilde{w}$ corresponding to the Cauchy problem determined by the initial datum

$$\tilde{w}_0 = W + w_0 = (U + u_0, V + v_0), \tag{6}$$

does the perturbation $w := \tilde{w} - W$ decay as $t \to +\infty$, if w_0 is sufficiently small? Due to translation invariance of the equation (1), $W_\delta(x) := W(x + \delta)$ is a relaxation shock profile for any $\delta \in \mathbb{R}$. In other words, the stationary solution W belongs to a one-dimensional manifold of stationary solutions, parametrized by the shift δ. Thus asymptotic stability of a single profile cannot be expected for general perturbations. Hence, either it is possible to choose appropriate initial perturbation so that asymptotic shift is not allowed, or it is necessary to study the stability of the whole family of translations of the given shock profile W. The latter statement means to prove that

$$\lim_{t \to +\infty} \operatorname{dist}(w, \{W_\delta \,:\, \delta \in \mathbb{R}\}) = 0.$$

Once stability is proved, more precise results on the asymptotic behavior can be asked for: to determine the temporal rate of decay, to find the large-time shape of the decaying perturbations. Actually, the literature on the subject – with no difference with respect to all of the mathematics in general – does not always pass from the easiest statement to the more detailed one: many results mix the different levels, proposing strategies that are inherently much more finer than proving a crude asymptotic stability result.

3 Two (Classical) Examples

3.1 The Broadwell System (1964)

This system was proposed by Broadwell in [Br64] as a reduced Boltzmann equation to model a gas whose molecules move with a finite set of velocities. In the one-dimensional case, the system reads

$$\begin{cases} f_t^+ + f_x^+ = (f^0)^2 - f^+ f^-, \\ f_t^0 = -\frac{1}{2}\left[(f^0)^2 - f^+ f^-\right], \\ f_t^- - f_x^- = (f^0)^2 - f^+ f^-. \end{cases} \tag{7}$$

in which f^+, f^0, f^- represent the density of particles moving with speeds $+1, 0, -1$ in the x-direction. In the variables $\rho := f_+ + 4f_0 + f_-$, $\rho u := f_+ - f_-$ e $v := f_+ + f_-$, the system takes the form of a hyperbolic system with relaxation

$$\begin{cases} \rho_t + (\rho u)_x = 0, \\ (\rho u)_t + v_x = 0, \\ v_t + (\rho u)_x = \frac{1}{8} \left\{ (\rho - v)^2 - 4(v^2 - (\rho u)^2) \right\}, \end{cases}$$

whose relaxed system is

$$\begin{cases} \rho_t + (\rho u)_x = 0, \\ (\rho u)_t + (\rho F(u))_x = 0, \end{cases} \qquad \text{where} \quad F(u) := \frac{1}{3}(2\sqrt{1 + 3u^2} - 1).$$

System (7) exhibits (explicit) relaxation shock profiles. The stability of these fronts has been considered in [KM85, CL88].

The first article, [KM85], one of the milestones in the analysis of stability of shock profiles, deals with the case of zero-integral hydrodynamical moments.

Theorem 1. *[KM85] Suppose that the initial condition (6) is such that* $v_0 \in H^1$ *and* $z_0 \in H^2$ *where*

$$z_0(x) := \int_{-\infty}^{x} u_0(y)\, dy. \tag{8}$$

Then if $|u_+ - u_-|$, $|z_0|_{H^2}$, $|w_0|_{H^1}$, *are sufficiently small, then there is a unique global solution* w *and this solution satisfies*

$$\lim_{t \to +\infty} |w(\cdot, t)|_{L^\infty}(t) = 0.$$

The assumption $z_0 \in H^2$, in particular, implies that the initial perturbation u_0 of the conserved quantity u has zero mean value, leading to the convergence of the perturbed solution to the original profile with no shift.

In [CL88], the zero-mass assumption is removed by introducing an appropriate wave decomposition of the disturbances obtained by constructing diffusion waves, as in the case of viscous conservation laws.

Theorem 2. *[CL88] If* $w_0 \in L^\infty(\mathbb{R})$ *and* $|u_+ - u_-|$, $|w_0|_{L^1 \cap H^2}$ *are sufficiently small then the solution to (7) with initial condition (6) is global in time and there exists* x_0 *such that*

$$\lim_{t \to \infty} |\tilde{w}(\cdot, t) - W(\cdot + x_0)|_{L^\infty} = 0.$$

3.2 The Jin–Xin System (1995)

In 1995, Jin and Xin proposed a new class of numerical schemes for systems of conservation law, based on the construction of an augmented semilinear hyperbolic system with relaxation. In the subsequent 10 years, more than one hundred of papers has been published, quoting [JX95] in the references, witnessing the great impact of this class of models in the field. In the case of one-dimensional space variable, the Jin–Xin system is

$$\begin{cases} u_t + v_x = 0, \\ v_t + a^2\, u_x = f^*(u) - v. \end{cases} \tag{9}$$

where u, v are vector-valued functions, $a > 0$. Differentiating the first equation with respect to t and the second with respect to x, we obtain the *modified equation* for u:

$$u_t + f^*(u)_x = \Box\, u := a^2\, u_{xx} - u_{tt},$$

i.e., a system of conservation laws, perturbed by the wave operator.

The study of stability of relaxation profiles for the Xin–Jin model has been addressed in many papers. These concern either the *scalar case*, i.e., u, v are scalar [MN96, LWY97, LWY98, Ser04], or the case of systems [God01, HLiu03, Hum03]. The first group of articles uses special features of the scalar case. In [MN96, Ser04], the L^1-contraction property is the keypoint of the analysis in the same spirit of analogous results concerning viscous scalar conservation laws (details are given in [Ser04]; see also [FS98]). In [LWY97, LWY98], following the strategy of [TLiu87], the auxiliary quantity z with $z_x = u$ is considered and energy estimates are applied for z and v. The integration step is allowed in full generality because of the scalar case and thus the disturbances can be assumed to have zero-mass. The same procedure for systems needs a restriction on the class of initial data.

For systems, articles [God01] and [Hum03] give somewhat complementary results: the former determines necessary conditions for stability using the Evans function framework, the latter shows that weak shock profiles are spectrally stable, i.e., the linearized operator at the relaxation profile has no point spectrum in the unstable halfplane. The result by Liu [HLiu03] concerns nonlinear stability under the basic assumption of zero-mass perturbation.

Theorem 3. *[HLiu03] Let the subcharacteristic condition $|\lambda_k(u)| < a$ for $k = 1, \dots, n$ be satisfied for any u under consideration.*

Then, if $|u_+ - u_-|$, $|z_0|_{H^3}$ and $|v_0|_{H^2}$ are sufficiently small (z_0 defined as in (8)), system (9) with initial condition (6) has a unique global solution and this solution satisfies

$$\lim_{t\to\infty} |w(\cdot,t)|_{L^\infty} = 0.$$

The key point is to introduce the integrated quantity z and to deal with the better behaved integrated system. Fine energy estimates, based also on the introduction of appropriated weighted matrices (in order to separate principal modes from the remaining ones), are then applied. At the present time, this result seems to be the best achievement of the arguments used in [TLiu87] for the 2×2 case. We are not aware of any analogous result applying the same strategy to general hyperbolic systems with relaxation.

Finally, let us quote also [LS98], where the linear stability of weak profiles is proved for a 3×3 viscoelastic system by energy estimates.

4 Stability Through Linearization

A different approach to stability follows a very classical and natural strategy: to study the linearized operator $\mathcal{L}$ at the relaxation profile W and to use Duhamel formula to deduce the nonlinear stability result, taking advantage from precise estimates on the semigroup $e^{\mathcal{L}t}$.

This strategy was used, in the case of 2×2 relaxation systems, in [Che95] in order to prove asymptotic stability of constant states. The powerfulness of the method was evident after [LZ97], where detailed analysis of the Green function of the linearized operator was used to prove asymptotic stability of constant states for a class of hyperbolic–parabolic systems. Still for constant states, same kind of analysis was settled down for relaxation systems in [Zen99] and [BHN06]. Studying stability of constant states, the linearized operator has constant coefficients and it is natural to apply Fourier transform to change space derivatives into multiplication by the factor $i\,\xi$. The linear equation is transformed in a family of evolutionary ODE parametrized by ξ, whose solution has to be appropriately decomposed to obtain a significant asymptotic description in the frequency variable. The final step is to apply the inverse Fourier transform and determine estimates on the Green function of the original linearized problem by shifting contour of integrations in an optimal way.

In the case of stability of traveling waves, the linearized operator is a differential operator with variable coefficients and the use of Fourier transform does not suit anymore to the problem. This obstruction motivated Zumbrun and coauthors to deal with the linearized problem by applying of Laplace transform to replace time derivative with multiplication by a factor λ, hence considering the resolvent equation determined by the operator $\lambda I - \mathcal{L}$. The first paper in this direction is [ZH98] and it deals with uniformly parabolic systems of conservation laws. After that, the approach has been extended to different situations, including equations with real viscosity or dispersion. Very recently, also the case of combustion models has been approached by this technique, see [LTRZ06]. Here, we present the results on the relaxation case, contained in [MZ02, MZ05].

4.1 Linear Stability

Let W be a stationary shock profile of (1). Looking for solution of the form $W + w$, we obtain

$$w_t = \mathcal{L}w + \mathcal{R}(w), \tag{10}$$

where $\mathcal{L}w := -(\mathrm{d}F(W)\,w)_x + \mathrm{d}G(W)\,w$ and $\mathcal{R}$ is the rest.

Differentiating the equation solved by the profile W, we get $\mathcal{L}W' = 0$. Since W' is exponentially decaying at infinity, 0 is an eigenvalue for the operator

$\mathcal{L}$ in any reasonable space, e.g., in L^p, $p \geq 1$. Moreover, $\mathcal{L}$ is asymptotically constant, i.e.,

$$\mathcal{L} \to \mathcal{L}_\pm \qquad \text{as} \quad x \to \pm\infty,$$

with coefficients converging exponentially fast to their asymptotic limits under the assumption of noncharacteristic shocks, i.e., if $0 \notin \sigma(\mathrm{d}f_*(u_\pm))$, with f_* in (3). The operator $\mathcal{L}$ can be thought as a (compact) perturbation of the piecewise constant operator defined by $\mathcal{L}_-$, respectively, $\mathcal{L}_+$, for $x < 0$, respectively, $x > 0$. As a consequence, the location of the essential spectrum of $\mathcal{L}$ is controlled by the spectra of $\mathcal{L}_\pm$, that is by the stability property of the asymptotic states $W_\pm = (u_\pm, v^*(u_\pm))$. This is encoded in the *Kawashima–Shizuta condition*:

$$\mathrm{Re}\,\sigma(-i\xi\,\mathrm{d}F(W_\pm) + \mathrm{d}G(W_\pm)) \leq -\frac{\theta|\xi|^2}{1+|\xi|^2} \qquad \forall\xi \in \mathbb{R}$$

for some $\theta > 0$. Under this assumption, the essential spectrum is confined in the stable halfplane $\mathrm{Re}\,\lambda \leq 0$ and intersect the imaginary axis $i\mathbb{R}$ solely at $\lambda = 0$. In particular, there is no spectral gap between the eigenvalue $\lambda = 0$ and the rest of the spectrum of $\mathcal{L}$. This property is strictly related to the conservative structure of the systems, since the asymptotic states are never isolated constant solutions of the system. Moreover, the absence of spectral gap is translated into the fact the rate of decay of the perturbations turns out to be of algebraic nature and not of exponential type.

In order to have stability, it is necessary that the multiplicity of the eigenvalue $\lambda = 0$ is one and to control the location of the remaining point spectrum σ_p. The former is controlled by *dynamical* and *structural stability* conditions corresponding to the so-called *Liu–Majda determinant condition*, related only to the limiting system (3), and to transversality of the connection in the ODE for the relaxation profile W, respectively (for details, see [MZ02, MZ05]). The latter control is encoded in the *spectral stability condition*

$$\sigma_\mathrm{p}(\mathcal{L}) \cap \{\mathrm{Re}\lambda \geq 0\} = \{0\}. \tag{11}$$

For completeness, let us stress that the eventual eigenvectors of the linear operator $\mathcal{L}$ are always exponentially decaying, hence the point spectrum relative to L^p does not depend on the summability exponent p.

Under the above assumptions (dynamical, structural and spectral stability), in [MZ02], it is proved that the Green distribution $\mathcal{G}$ for the linearized operator $\mathcal{L}$ can be decomposed into the sum of three terms

$$\mathcal{G} = \mathcal{E} + \tilde{\mathcal{G}} + \mathcal{H}, \tag{12}$$

where the term $\mathcal{E}$ (*excited term*) concerns the contribution in the direction of the zero eigenfunction W', hence it contains the information relative to the shift of the wave at the linearized level, the term $\tilde{\mathcal{G}}$ is approximately a sum of gaussian signals scattered by the shock layer and, finally, the term $\mathcal{H}$, containing delta distributions exponentially decaying in time, describes propagation of signals along hyperbolic characteristics. The main difference with respect to the uniformly parabolic case, considered in [ZH98], is the presence of these delta terms in $\mathcal{H}$, arising in the large frequencies analysis.

A representation for the Green function $\mathcal{G}$, analogous to (12), except for the absence of the excited term $\mathcal{E}$, is given in [Zen99] and [BHN06] in the case of the linearized operator at a fixed constant state. As explained previously, the methods in the two situations are different, being based on Laplace transform/resolvent kernel estimates in the case of shock profiles, and on Fourier transform/evolutionary ODE in the case of constant states.

An immediate consequence of the decomposition (12) is the linear stability of the shock profiles together with decay rates: at the linear level, perturbations of the traveling wave decay in time, except for the one-dimensional component relative to the eigenspace of 0. Since the large time behavior is mainly determined by the term $\widetilde{G}$, rates of decay are the same as for the viscous case, which are those of the heat kernel.

4.2 Nonlinear Stability

To deal with nonlinear stability, it is natural to use Duhamel formula, to give an implicit representation of the solution to (10). The main problem, here, resides in the presence of the nondecaying term $\mathcal{E}$. To get rid of that, a shift variable $\delta = \delta(t)$ is introduced and the Duhamel formula is written for the shifted perturbation quantity $w(x + \delta(t), t)$. Choosing δ so that nondecaying terms are eliminated in the equation for the shifted pertubation permits to obtain estimates on both the unknowns w and δ. A nonlinear iteration argument, combining the linearized decay rates with energy estimates, permits to prove the following result.

Theorem 4. *[MZ05] Under the previous assumptions, if w_0 is sufficiently small in $L^1 \cap H^2$, then the solution $\tilde{w}$ to (1) with initial condition (6) is global in time and satisfies, for some function δ,*

$$|\tilde{w}(\cdot, t) - W(\cdot - \delta(t))|_{L^p} \leq C|w_0|_{L^1 \cap H^2}(1+t)^{-\frac{1}{2}(1-\frac{1}{p})},$$

$$|\delta(t)| \leq C|w_0|_{L^1 \cap H^2}, \qquad |\dot{\delta}(t)| \leq C|w_0|_{L^1 \cap H^2}(1+t)^{-\frac{1}{2}}.$$

for any $p \in [1, +\infty]$.

Differently from the analysis in [CL88], diffusion waves do not appear in the Green function approach, used for proving Theorem 4. This can be explained thinking back to the heat equation: in that case, to find decay rates of perturbations to constant states is possible either to estimate directly the (explicit) Green function or to determine the asymptotic profile of the solutions. In term of decay rates, the two approaches are equivalent. Nevertheless, the second one gives more precise information on the large-time behavior, describing the shape of the perturbation itself. In general, the additional information that is gained through the determination of the asymptotic profile has to be paid either in terms of technical difficulties in the proof or in terms of stronger assumptions.

In [MZ02], a nonlinear stability result was proved under the stronger assumptions either of small amplitude of the profile for general relaxation systems or of large-amplitude shocks for discrete kinetic models. Theorem 4 extends this previous result to the same assumptions considered in the linearized analysis, hence with no direct restriction on the size of the shock. Thus, the main achievement of Theorem 4 is that it asserts that *the eigenvalue problem of the linearized operator at the profile is the main issue in the stability analysis of relaxation shock profiles.*

Dynamical and structural stability depend, respectively, on the properties of the relaxed system (3) and of the associated ODE for the profiles. The failure of one of these conditions can be clearly interpreted. Spectral stability is a more delicate condition. The analysis of this condition may reveal interesting new features of the relaxation dynamics, especially in case of instability.

4.3 Spectral Stability

The spectral stability condition is weaker than stability with respect to zero-mass perturbations, i.e., stability with respect to initial perturbations $w_0 = (u_0, v_0)$ where u_0 has zero integral over all $\mathbb{R}$. In particular, the combination of Theorems 1, 3, and 4 guarantees asymptotic stability of weak relaxation shock profiles for the Broadwell and the Jin–Xin systems.

Direct proof of spectral stability of relaxation shock profiles for the Jin–Xin system with small amplitude has been obtained in [Hum03] by using energy estimates. The case of small shocks for general relaxation system has been considered in [PZ04], where the spectral stability is proved by using Evans function tools under the additional assumption of $\mathrm{d}F \neq 0$. The aim of [MZ06] is to prove spectral stability for weak relaxation profiles by energy estimates, removing the technical assumption of [PZ04].

So far, the situation can be summarized by stating that: *small-amplitude relaxation shock profiles are (spectrally/linearly/nonlinearly) stable.*

What can be said for large shocks? In the scalar Jin–Xin case (i.e., $u, v \in \mathbb{R}$), thanks to L^1-contraction, it is possible to prove directly nonlinear stability. Moreover, following [Hum03], it is possible to give an easy direct proof of spectral stability. Assume by contradiction that λ is an eigenvalue of the linearized operator $\mathcal{L}$ with $\mathrm{Re}\,\lambda \geq 0$. Let z be such that $z' = u$, then

$$\lambda z - sz' + v = 0, \qquad \lambda v - sv' + a^2 z'' = \mathrm{d}f^*(U)z' - v.$$

Multiplying the second equation by $\bar{z}$ and by $\bar{v}$, and taking real parts, we obtain

$$a^2 |z'|^2_{L^2} < |v|^2_{L^2}, \qquad\qquad |v|^2_{L^2} \leq \int_{\mathbb{R}} |\mathrm{d}f^*(U) z' \bar{v}|.$$

Applying Young inequality and, recalling the subcharacteristic condition $|\mathrm{d}f^*(u)| < a$, we get a contradiction. This argument permits to prove spectral stability for any size of the profile only in the scalar case. In the case of systems, smallness of the profile is essential in controlling interaction terms. As a consequence, the situation for large profiles is far from being well understood and it is one of the most interesting topics for future research in the field.

5 Onset of Instability

In the 1990s, many authors started using the complex-valued Evans function $D = D(\lambda)$ as a useful tool for stability analysis of traveling waves in the context of reaction–diffusion equations and conservation laws. This is an analytic function encoding information on the singularities of the resolvent kernel $(\lambda I - \mathcal{L})^{-1}$ as $D(\lambda) = 0$ if and only if $\lambda \in \sigma_{\mathrm{p}}(\mathcal{L})$. Since $D(0) = 0$ and D has real values for real arguments, a sufficient condition for instability is $D'(0)\,D(+\infty) < 0$. In [God01], it is shown that this instability condition can be written in a more concrete form for the Jin–Xin model: for $u, v \in \mathbb{R}^2$, the instability condition for a 2-shock reads as

$$\det(r_1^-, u^+ - u^-) \cdot \det(r_1^-, r_2^-) < 0,$$

where r_1^-, r_2^- are eigenvectors of (3) and r_2^- is suitably oriented (see [God01] for the general case). Later, in [GL03] an explicit unstable relaxation shock profile has been exhibited, confirming that the spectral stability condition does not hold in general for large shocks.

At this point, all of the analysis and the results reported here suggest many open problems in the study of relaxation profiles. Here is a list of possible questions (following the personal taste of the author):

- What is the simplest example of unstable relaxation shock (i.e., the minimal dimension for the conservative/non conservative variables u and v)?
- Is instability possible if (3) is a scalar equation (i.e., $u \in \mathbb{R}$)?
- Does monotonicity of the profiles play a rôle in stability/instability properties also in the system case?
- Is there any relation between stability/instability of relaxation profiles and the viscous profiles of the corresponding Chapman–Enskog expansion?

Wider directions that would be interesting to investigate in next years are classification of instabilities and determination of patterns emerging in case of instability. This research direction is well summarized in a sentence taken from [PW02]: *a long-term goal of the study of nonlinear waves is to sort out what kinds of structures and interactions are significant and robust.*

References

[Bia06] Bianchini, S.: Hyperbolic limit of the Jin-Xin relaxation model, Comm. Pure Appl. Math. **59** (2006), no. 5, 688–753.

[BHN06] Bianchini, S.; Hanouzet, B.; Natalini, R.: Asymptotic behavior of smooth solutions for partially dissipative hyperbolic systems with a convex entropy, Comm. Pure Appl. Math., to appear.

[Br64] Broadwell, J.E.: Shock structure in a simple discrete velocity gas, Physics Fluids **7** (1964), no. 8, 1243–1247.

[CL88] Caflisch, R.E.; Liu, T.-P.: Stability of shock waves for the Broadwell equations. Comm. Math. Phys. **114** (1988), no. 1, 103–130.

[CLL94] Chen, G.Q.; Levermore, C.D.; Liu, T.-P.: Hyperbolic conservation laws with stiff relaxation terms and entropy, Comm. Pure Appl. Math. **47** (1994), no. 6, 787–830.

[Che95] Chern, I.-L.: Long-time effect of relaxation for hyperbolic conservation laws, Comm. Math. Phys. **172** (1995), no. 1, 39–55.

[FS98] Freistühler, H.; Serre, D.: L^1 stability of shock waves in scalar viscous conservation laws. Comm. Pure Appl. Math. **51** (1998), no. 3, 291–301.

[God01] Godillon, P.: Linear stability of shock profiles for systems of conservation laws with semi-linear relaxation, Phys. D **148** (2001), no. 3–4, 289–316.

[GL03] Godillon, P.; Lorin, E.: A Lax shock profile satisfying a sufficient condition of spectral instability, J. Math. Anal. Appl. **283** (2003), no. 1, 12–24.

[Hum03] Humpherys, J.: Stability of Jin-Xin relaxation shocks, Quart. Appl. Math. **61** (2003), no. 2, 251–263.

[JX95] Jin, S.; Xin, Z.P.: The relaxation schemes for systems of conservation laws in arbitrary space dimensions, Comm. Pure Appl. Math. 48 (1995), no. 3, 235–276.

[KM85] Kawashima, S.; Matsumura, A.: Asymptotic stability of traveling wave solutions of systems for one-dimensional gas motion, Comm. Math. Phys. **101** (1985), no. 1, 97–127.

[HLiu03] Liu, H.: Asymptotic stability of relaxation shock profiles for hyperbolic conservation laws, J. Differential Equations **192** (2003), no. 2, 285–307.

[LWY97] Liu, H.; Woo, C.-W.; Yang, T.: Decay rate for travelling waves of a relaxation model, J. Differential Equations **134** (1997), no. 2, 343–367.

[LWY98] Liu, H.; Wang, J.; Yang, T.: Stability of a relaxation model with a nonconvex flux, SIAM J. Math. Anal. **29** (1998), no. 1, 18–29

[TLiu87] Liu, T.-P.: Hyperbolic conservation laws with relaxation. Comm. Math. Phys. **108** (1987), no. 1, 153–175.

[LZ97] Liu, T.-P.; Zeng, Y.: Large time behavior of solutions for general quasilinear hyperbolic-parabolic systems of conservation laws, Mem. Amer. Math. Soc. **125** (1997), no. 599.

[LS98] Luo, T.; Serre, D.: Linear stability of shock profiles for a rate-type viscoelastic system with relaxation, Quart. Appl. Math. **56** (1998), no. 3, 569–586.

[LTRZ06] Lyng, G.; Texier, B.; Raoofi M.; Zumbrun K.: Pointwise Green function bounds and stability of traveling waves for Majda's model of reacting flow, in preparation.

[MN96] Mascia, C.; Natalini, R.: L^1 nonlinear stability of traveling waves for a hyperbolic system with relaxation, J. Differential Equations **132** (1996), no. 2, 275–292.

[MZ02] Mascia, C.; Zumbrun, K.: Pointwise Green's function bounds and stability of relaxation shocks, Indiana Univ. Math. J. **51** (2002), no. 4, 773–904.

[MZ05] Mascia, C.; Zumbrun, K.: Stability of large-amplitude shock profiles of general relaxation systems, SIAM J. Math. Anal. **37** (2005), no. 3, 889–913

[MZ06] Mascia, C.; Zumbrun, K.: Spectral stability of weak relaxation shock profiles, in preparation.

[PW02] Pego, R.L.; Warchall, H.A.: Spectrally stable encapsulated vortices for nonlinear Schrödinger equations, J. Nonlinear Sci. **12** (2002), no. 4, 347–394.

[PZ04] Plaza, R.; Zumbrun, K.: An Evans function approach to spectral stability of small-amplitude shock profiles, Discrete Contin. Dyn. Syst. **10** (2004), no. 4, 885–924.

[Ser04] Serre, D.: L^1-stability of nonlinear waves in scalar conservation laws. In: Evolutionary equations. Vol. I, 473–553, Handb. Differ. Equ., North-Holland, Amsterdam, 2004.

[Ser00] Serre, D.: Relaxations semi-linéaire et cinétique des systemes de lois de conservation, Ann. Inst. H. Poincaré Anal. Non Linéaire **17** (2000), no. 2, 169–192.

[Whi74] Whitham, J.: Linear and nonlinear waves. New York: Wiley, 1974.

[YZ00] Yong, W.-A.; Zumbrun, K.: Existence of relaxation shock profiles for hyperbolic conservation laws, SIAM J. Appl. Math. **60** (2000), no. 5, 1565–1575.

[Zen99] Zeng, Y.: Gas dynamics in thermal nonequilibrium and general hyperbolic systems with relaxation, Arch. Ration. Mech. Anal. **150** (1999), no. 3, 225–279.

[ZH98] Zumbrun, K.; Howard, P.: Pointwise semigroup methods and stability of viscous shock waves, Indiana Univ. Math. J. **47** (1998), no. 3, 741–871.

A New *hp*-Adaptive DG Scheme for Conservation Laws Based on Error Control

A. Dedner and M. Ohlberger

Summary. In this contribution we present a new *hp*-adaptive Discontinuous Galerkin approximation for scalar conservation laws. The *hp*-adaptivity of the scheme is based on a rigorous a posteriori error estimate for a generalized class of Discontinuous Galerkin schemes presented in [DMO06]. Numerical experiments demonstrate the efficiency and stability of the scheme in multiple space dimensions.

1 Introduction

In this article we present a new *hp*-adaptive version of the Runge–Kutta Discontinuous Galerkin approximation of Cockburn and Shu (see [CS91, CHS90, CS01]) for nonlinear scalar conservation laws in several space dimensions. As a prototype conservation law, consider the Cauchy initial value problem

$$\partial_t u + \nabla \cdot f(u) = 0 \qquad \text{in } \mathbb{R}^d \times \mathbb{R}^+, \tag{1}$$

$$u(x,0) = u_0(x) \quad \text{in } \mathbb{R}^d. \tag{2}$$

Here $u : \mathbb{R}^d \times \mathbb{R}^+ \to \mathbb{R}$ denotes the dependent solution variable, $f \in C^1(\mathbb{R})$ denotes the flux function, and $u_0 \in \mathrm{BV}(\mathbb{R}^\mathrm{d}) \cap \mathrm{L}^\infty(\mathbb{R}^\mathrm{d})$ the initial data satisfying $u_0 \in [A, B]$ a. e.

If we denote by $\widetilde{u}_h$ a postprocessed solution of the Discontinuous Galerkin approximation to (1) and (2), the *hp*-adaptivity is based on a posteriori error control of the form

$$||u - \widetilde{u}_h|| \leq \sum_{n,j} \rho_j^n(h_j, \overline{u}_h - \widetilde{u}_h)\; R_j^n(\widetilde{u}_h), \tag{3}$$

where $\rho_j^n(h_j, \overline{u}_h - \widetilde{u}_h)$ is a weighting factor depending on the deviation of $\widetilde{u}_h$ from its piecewise average $\overline{u}_h$, and $R_j^n(\widetilde{u}_h)$ reflects the element, jump, and projection residuals of the approximate solution. The general idea in the new *hp*-adaptive algorithm is to choose the local polynomial degree in order to optimize the weighting factor ρ_j^n, and to choose the local mesh size h_j in order to get an equal distribution of the local indicator values $\rho_j^n(h_j, \overline{u}_h - \widetilde{u}_h)\, R_j^n(\widetilde{u}_h)$. The underlying a posteriori error estimate, together with an one-dimensional realization of the *hp*-adaptive algorithm were presented in [DMO06].

For further results on a posteriori error control and adaptive solution algorithms for Runge–Kutta Discontinuous Galerkin schemes we refer to Hartmann and Houston [HH02], Larson and Barth [LB00], Süli, Houston et al. [HS01, HSS02, SH03], and Adjerid et al. [ADFK02].

The article is organized as follows. In Sect. 2 we introduce a general class of DG methods on adaptive meshes. The a posteriori error estimate from [DMO06] is stated in Sect. 3. In Sect. 4 we describe the full hp-adaptive algorithm derived from the a posteriori error estimate. Some implementation issues for the multidimensional method are discussed in Sect. 5. Finally, in Sect. 6 we give new numerical experiments for the hp-adaptive algorithm in two-space dimensions.

2 A General Class of DG Methods

In order to define the RK-DG methods let us introduce some notation. Let $\mathcal{T}$ define a partition of $\mathbb{R}^d$ of control volumes $T \in \mathcal{T}$ satisfying $\cup_{T\in\mathcal{T}}\overline{T} = \mathbb{R}^d$. For two distinct control volumes T_i and T_j in $\mathcal{T}$, the intersection is either an oriented edge (two-dimensional) or face (three-dimensional) S_{ij} with oriented normal ν_{ij} or else a set of measure at most $d-2$.

On $\mathcal{T}$ we define the space of discontinuous piecewise polynomials of degree p by

$$V_h^p := \{v_h \in BV(\mathbb{R}^d) |\, v_j := v_h|_{T_j} \in \mathbb{P}_p \text{ for all } T_j \in \mathcal{T}\}. \qquad (4)$$

Let us denote with $\Pi_{V_h^p}$ the L^2-projection into V_h^p.

The semidiscrete DG-finite element scheme is the basis of the definition of RK-DG methods.

Definition 1 (Space-discrete DG approximation). $u_h : C^1(0, T_{\max}; V_h^p)$ *is called a semidiscrete DG approximation of (1-2), if*

$$u_h(0) = \Pi_{V_h^p}(u_0), \qquad (5)$$

$$\frac{d}{dt}(u_j(t), v_j)_{T_j} - (f(u_j(t)), \nabla v_j)_{T_j} + \sum_{l\in N(j)} (f_{jl}(u_j(t), u_l(t)), v_j)_{S_{jl}} = 0, \qquad (6)$$

$$\textit{for all } v_j \in \mathbb{P}_p, T_j \in \mathcal{T}.$$

Here $(\cdot,\cdot)$ denotes the L^2-inner product and f_{jl} is a monotone numerical flux.

In the literature of DG methods the stabilization due to the "upwinding" of the discrete fluxes is usually accompanied by extra artificial "shock capturing" terms as in [JJS95, JS95, CG96] or limiting projections as in [CS01]. The RK-DG methods introduced by Cockburn and Shu are based on a combination of limiting projections and an Runge–Kutta discretization of (6). Therefore, in the next step we are going to introduce limiting projections in the discretization. Thus, let $\{0 = t^0, \dots, t^N = T_{\max}\}$ denote a partition of $(0, T_{\max})$ and let us define the time increment $\Delta t^n := t^{n+1} - t^n$. To this end we associate with each time interval $(t^n, t^{n+1}]$ a (possibly different) finite element space V_h^p denoted by

$$V_{h,n}^p := \{v_h \in BV(\mathbb{R}^d) |\, v_h|_T \in \mathbb{P}_p \text{ for all } T \in \mathcal{T}_n\}. \tag{7}$$

The associated index set of $\mathcal{T}_n$ is denoted by I^n.

To define a local projection operator we proceed as follows: We define $\overline{v_h}$ through $\overline{v_j} := \Pi_{V_h^0}(v)|_{T_j}$ for any $v \in L^2(\Omega)$, i.e., $\overline{v_h}$ is the element wise average of v_h. Furthermore, with each n we associate projections $\Lambda_h^{n,t}$ with the following properties: The projection $\Lambda_h^{n,t}$ is supposed to be a continuous function with respect to t on the interval $[t^n, t^{n+1}]$ which maps $V_{h,n}^p \to V_{h,n}^p$ for $t \in (t^n, t^{n+1}]$ and $V_{h,n-1}^p \to V_{h,n}^p$ for $t = t^n$. The projection is element wise conservative, i.e., $\overline{\Lambda_h^{n,t}(v_h(\cdot, t))} = \overline{v_h(t)}$ and lead to no increase in the gradient between time steps, i.e, $\|\Lambda_h^{n,t^n}(u_h) - \overline{u_h}\|_\infty \le \|\Lambda_h^{n-1,t^n}(u_h) - \overline{u_h}\|_\infty$.

Note that $\Lambda_h^{n,t}$ accounts for both limiting projections and projections to the new spaces. We define the restriction of $\Lambda_h^{n,t}$ on the element T_j by $\Lambda_j^{n,t}$:

$$\Lambda_j^{n,t} \equiv \Lambda_h^{n,t} \quad \text{in } T_j \times [t^n, t^{n+1}],\ j \in I^n.$$

We now define the generalized semidiscrete DG approximation.

Definition 2 (Generalized semidiscrete DG approximation). *Let us suppose that a projection $\Lambda_h^{n,t}$ with the above properties is given. In addition assume that the discrete fluxes f_{ij} are monotone. The function u_h is called a generalized semidiscrete DG approximation of (1)–(2), if for $u_h^{-1} := \Pi_{V_{h,0}^p}(u_0)$ u_h satisfies:*

For $n = 0, \dots, N-1$, $u_h^n|_{[t^n, t^{n+1}]} \in C^1(t^n, t^{n+1}; V_{h,n}^p)$ is defined through

$$u_h^n(t^n) := \Lambda_h^{n,t^n}(u_h^{n-1}(t^n)), \tag{8}$$

$$\begin{aligned}\frac{d}{dt}(u_j^n(t), v_j)_{T_j} &= -\sum_{l \in N(j)} (f_{jl}(\Lambda_j^{n,t}(u_h^n(t)), \Lambda_l^{n,t}(u_h^n(t))), v_j)_{S_{jl}} \\ &\quad + (f(\Lambda_j^{n,t}(u_h^n(t))), \nabla v_j)_{T_j}, \text{ for all } v_j \in \mathbb{P}_p, j \in I^n, t \in (t^n, t^{n+1}).\end{aligned} \tag{9}$$

The global approximation $u_h \in L^\infty(0, T_{\max}; V_{h,n}^p)$ is defined through $u_h(0) := u_h^{-1}$, and $u_h|_{(t^n, t^{n+1}]} := u_h^n|_{(t^n, t^{n+1}]}$.

For time integration explicit strongly stability preserving Runge–Kutta methods are used to solve the ode (9) [GST01, Shu01]. In order to get a continuous evaluation of the approximate solution in time, the so called natural continuous extension (NCE) of Runge–Kutta methods, introduced by Zennaro [Zen86], is used; this leads to an approximation $u_h \in C^0(0, T_{\max}; V_{h,n}^p)$

3 A Posteriori Error Estimate for the General DG Method

Theorem 1. *Let u_h be given by the semidiscrete generalized DG method (8)–(9). For $\widetilde{u}_h(t) := \Lambda_h^t(u_h(\cdot,t))$ we have the following a posteriori error estimate*

$$||(u-\widetilde{u}_h)(T_{\max})||_{L^1(B_R(x_0))} \leq \eta_h := \eta_0 + R_h \tag{10}$$

with

$$\eta_0 = \sum_{j\in J^0}\int_{T_j}|u_0 - \widetilde{u}_j^0(0)|, \quad R_h^2 = 2\sum_n\sum_{j\in J^n}\rho_j^n\Big(R_{T,j}^n + R_{S,j}^n + R_{\Lambda,j}^n\Big), \tag{11}$$

$$\text{and} \quad \rho_j^n := K_1 h_j + K_2 \max_{k\in\{j,\, l\in N(j)\}} ||\overline{u}_k^n - \widetilde{u}_k^n||_{L^\infty((t^n,t^{n+1})\times T_k)}. \tag{12}$$

The element, projection and jump residuals are given as

$$R_{T,j}^n := \int_{t^n}^{t^{n+1}}\int_{T_j}\Big|\partial_t\widetilde{u}_j + \nabla\cdot f(\widetilde{u}_j)\Big|, \quad R_{\Lambda,j}^n := \int_{T_j}|\widetilde{u}^n(t^{n+1}) - \widetilde{u}^{n+1}(t^{n+1})|, \tag{13}$$

$$R_{S,j}^n := \int_{t^n}^{t^{n+1}}\sum_{l\in N(j)}\int_{S_{jl}}|2f_{jl}(\widetilde{u}_j,\widetilde{u}_l) - f_{jl}(\widetilde{u}_j,\widetilde{u}_j) - f_{jl}(\widetilde{u}_l,\widetilde{u}_l)|. \tag{14}$$

Proof. For a detailed proof and assumptions on the data we refer to [DMO06]. □

4 A New *hp*-Adaptive Scheme

In this section we are going to describe a new *hp*-adaptive DG scheme in multispace dimensions that is constructed with the use of the a posteriori error estimate given in Theorem 1. For the description of the main idea in our approach, let us recall that the local contribution on $T_j \times [t^n, t^{n+1}]$ of the error estimator from Theorem 1 is

$$\eta_j^n := \rho_j^n\Big(R_{T,j}^n + R_{S,j}^n + R_{\Lambda,j}^n\Big),$$

where $\rho_j^n := \rho_j^n(h_j, \overline{u}_h^n x - \widetilde{u}_h^n)$ is a weighting factor and $R_{T,j}^n, R_{S,j}^n, R_{\Lambda,j}$ denote the element, jump, and projection residuals, respectively. The principal idea for the p-adaptive step is to choose the projection operator locally in such a way, that $\rho_j^n = \mathcal{O}(h_j)$. In a second step the local mesh size is chosen in order to equilibrate the sizes of η_j^n among all grid cells $T_j \in \mathcal{T}_n$. Finally, the time step size is determined via the CFL-condition for SSP-Runge–Kutta schemes. In the following sections we describe the method in more detail.

4.1 Choice of the Projection Operator/p-Adaptivity

Let $p_{\max} \in \mathbb{N}$ be given. For some $v \in \mathbb{P}_{p_{\max}}(T_j^n), j \in I^n$ let us denote the L^2–projection onto $\mathbb{P}_p(T_j^n), 0 \leq p \leq p_{\max}$ by $\Pi_{j,n}^p(v)$.

In order to stabilize the scheme and in order to get a convergent error estimator we suggest to choose the projection operator such that the following bound holds

$$||\overline{u}_j^n - \widetilde{u}_j^n||_{L^\infty(T_j)}(t) \leq \lambda_j^n \tag{15}$$

with λ_j^n defined as

$$\lambda_j^n := \frac{h_j}{\left(h_j + \frac{h_j}{|T|\Delta t^n}\left(R_{T,j}^n + R_{S,j}^n\right)\right)^2}. \tag{16}$$

In order to achieve the bound (15) we define the projection operator in the following way:

Let $0 \leq p^* \leq p_{\max}$ denote the maximal index such that

$$|\Pi_{j,n}^0(u_j^n) - \Pi_{j,n}^{p^*}(u_j^n)| \leq \lambda_j^n, \qquad \text{for all } x \in T_j.$$

Then, the p-adaptive projection on the cell T_j is defined through

$$\tilde{\Lambda}_j^{n,t}(u_h(\cdot,t)) := \Pi_{j,n}^{p^*}(u_j^n). \tag{17}$$

The resulting projections fall into the class of moment limiters as introduced in [BDF94]. In Fig. 1 the limiting procedure is sketched for a simple one-dimensional situation.

4.2 Choice of the Local Mesh Size/h-Adaptivity

The mesh adaptation is done with a *solve – estimate – mark – adapt* algorithm that is based on an equidistribution strategy of the error indicator η_h

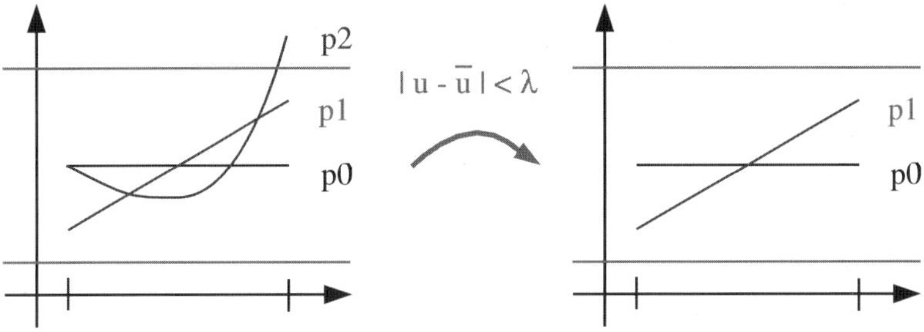

Fig. 1. Sketch of the general approach for the polynomial adaption procedure within the hp-adaptive scheme. A local polynomial of degree two (p2) is shown together with its projection to linears (p1) and constants (p0). Depending on the bound from our p-adaptive strategy, the polynomial degree is locally reduced (*right-hand side*)

of Theorem 1. The goal of the mesh adaptation is to guarantee that for a given tolerance TOL we achieve $\eta_h \leq \mathrm{TOL}$. One possibility to obtain such an algorithm for a given tolerance TOL and some $\theta \in (0,1)$ is the so called equal distribution strategy, defined as follows:

Let a computational grid $\mathcal{T}_{n-1}$ and an approximation u_h^{n-1} on $\mathcal{T}_{n-1}$ at time t^{n-1} be given such that

$$\eta_h^{n-1}(u_h) \leq \frac{\Delta t^{n-1}}{T_{\max}} \mathrm{TOL}.$$

Compute $\mathcal{T}_n$ and u_h^n according to the following algorithm:

1. Set $\mathcal{T}_n = \mathcal{T}_{n-1}$.
2. Compute u_h^n, $\eta_h^n(u_h)$ on $\mathcal{T}_n$.
3. (a) Refine or coarsen the grid locally such that for all $T_j \in \mathcal{T}_n$ it holds

$$\eta_j^n(u_h) \leq \frac{\Delta t^n}{T_{\max}|\mathcal{T}_n|} \mathrm{TOL} \quad \text{and, if possible,} \quad \theta \, \frac{\Delta t^n}{T_{\max}|\mathcal{T}_n|} \mathrm{TOL} \leq \eta_j^n(u_h).$$

   This results in an updated grid $\mathcal{T}_n$.

   (b) If $\eta_h^n(u_h) > \frac{\Delta t^n}{T_{\max}} \mathrm{TOL}$ proceed with step 2.
4. Proceed with time step t^{n+1}.

For more details on the algorithm with respect to the projection errors and for an efficient modification of the equidistribution strategy we refer to [DMO06].

5 Implementation Within a Generic Software Framework

The implementation of the Discontinuous Galerkin method is done within the Distributed and Unified Numerics Environment (DUNE) [BDE+03]. After semidiscretization in space as given through (8)–(9) the discrete evolution equation on a time slab is written in the form

$$\frac{d}{dt} \mathbf{U}_h^n(t) = \mathcal{L}_h^n[\mathbf{U}_h^n(t)],$$

where $\mathcal{L}_h^n[\mathbf{U}_h^n(t)]$ may be decomposed into two operators corresponding to limiting projection and evolution in time. In detail we have

$$\mathcal{L}_h^n[\mathbf{U}_h^n(t)] = \Pi_S \circ \mathcal{L}_{h,2} \circ \mathcal{L}_{h,1}[\mathbf{U}_h^n(t)]$$

with

$$\begin{aligned} \mathcal{L}_{h,1}[\mathbf{U}_h^n(t)] &= (\tilde{\mathbf{U}}_h^n(t), \mathbf{U}_h^n(t)), \\ \mathcal{L}_{h,2}[\tilde{\mathbf{U}}_h^n(t), \mathbf{U}_h^n(t)] &= (\mathbf{W}_h^n(t), \tilde{\mathbf{U}}_h^n(t), \mathbf{U}_h^n(t)), \\ \Pi_S[\mathbf{W}_h^n(t), \tilde{\mathbf{U}}_h^n(t), \mathbf{U}_h^n(t)] &= \mathbf{W}_h^n(t), \end{aligned}$$

where $\mathbf{W}_h^n(t)$ is defined through the right-hand side of (9). The implementation of the decomposed operators is done within the general framework for evolution equation in DUNE which allows a flexible application of the stabilization into more complex applications (see [BDD+05]).

6 Numerical Experiments

In this section we examine numerically the RK-DG method defined through Definition 2 together with the projection from Sect. 4.1 and the local adaptive grid refinement from Sect. 4.2. We study the convergence behavior of the resulting p-, and hp-adaptive methods. As test problems, we look at two-dimensional linear transport with smooth and nonsmooth initial data, and at a combination of linear and nonlinear advection where the nonlinearity comes from the Buckley–Leverett equation. All tests described here are performed on nonconform simplex meshes with highly irregular macrotriangulation using the ALUGrid Library [ALU].

6.1 Linear Advection

As a first numerical example we look at the linear transport equation

$$\partial_t u + a\nabla u = 0 \ , \qquad u(\cdot,0) = u_0(\cdot)$$

with the constant transport velocity $a = (2.25, 0.22)$. For fixed initial data the solution u is then given by $u(x,t) = u_0(x - at)$. We study the setting on $\Omega = [0,2] \times [0,0.5]$ with smooth initial data given as

$$u_0(x) := \begin{cases} 16(0.25\cos(\pi|x-p|/r) + 0.25)^4, & \text{for } |x-p| < r, \\ 0, & \text{else,} \end{cases}$$

and nonsmooth initial data given as

$$u_0(x) := \begin{cases} 1 - |x-p|/r, & \text{for } |x-p| < r, \\ 0, & \text{else.} \end{cases}$$

In Fig. 2 we plotted the error $||u - u_h||_{L^1(\Omega)}$ at $t = 0.5$ vs. the overall cpu time for the hp-adaptive algorithm with $p_{\max} = 0, 1, 2$. For smooth and nonsmooth initial data, the benefit of the higher order methods are clearly visible. In the nonsmooth case, the convergence rates of the hp-adaptive algorithm are a little bit smaller for $p_{\max} = 1, 2$, if they are compared with the convergence rates for smooth initial data.

Buckley–Leverett Type Problem in Two-Dimensional

As a second example we look at a combination of linear transport in y-direction and the Buckley–Leverett transport in x-direction. In detail, we consider a solution u of the form $u(x,y,t) = v(x, y - at, t)$ of the nonlinear conservation law

$$\partial_t u(x,y,t) + \partial_x f(u(x,y,t)) + a\partial_y u(x,y,t) = 0$$

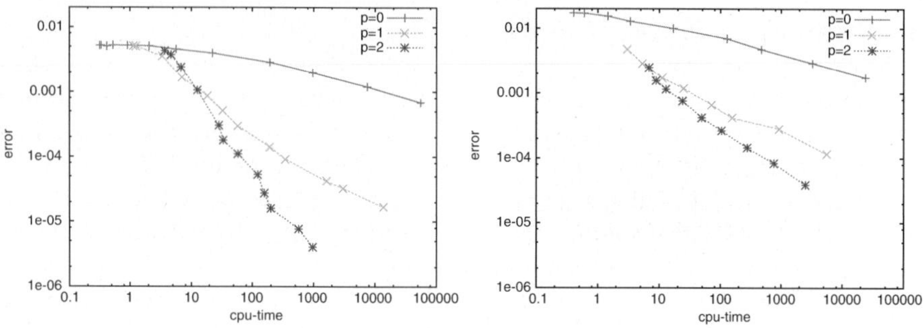

Fig. 2. Convergence study of the hp-adaptive algorithm for the two-dimensional linear transport problems. On the *left-hand* side the error vs. runtime is plotted for the problem with smooth initial data for $p_{\max} = 0, 1, 2$, while the convergence for the nonsmooth initial data is given on the *right-hand side*

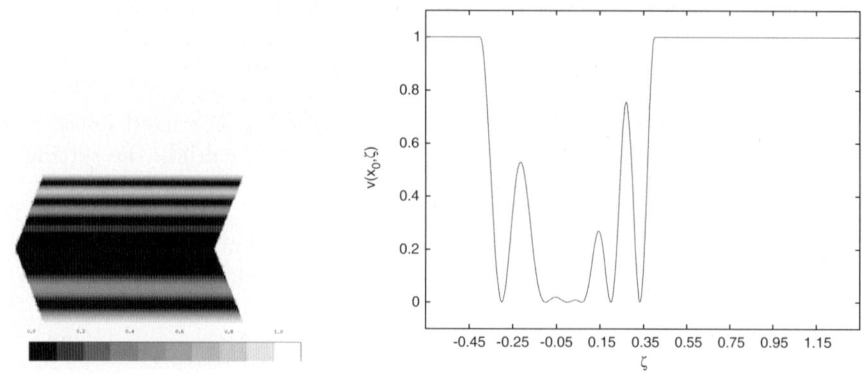

Fig. 3. Initial data for two-dimensional Buckley–Leverett type problem (*left*). The right plot shows a cut parallel to the ζ axis at $x = -0.1$

with $a = 1.3$, $f(u) = \frac{u^2}{u^2+\frac{1}{2}(1-u)^2}$, and $(x, y) \in [-1, 1] \times [-0.65, 1.35]$ and $t \in [0, 0.35]$. For u to be a solution of (18) the function $v(x, \zeta, t)$ for ζ fixed has to satisfy

$$\partial_t v(x, \zeta, t) + \partial_x f(u(x, \zeta, t)) = 0,$$

i.e., has to be a solution of the one-dimensional conservation law with initial data $v(x, \zeta, 0) = v_0(x, \zeta)$. For $\zeta \in [-0.4, 0.4]$ we choose v_0 to be a combination of two Riemann problems:

$$v_0(x, \zeta) = \begin{cases} v_-(\zeta) \text{ for } \quad x - 0.4|\zeta| \in [-0.8, 0.3] \quad \text{and} \quad \zeta \in [-0.4, 0.4] \\ \qquad 1 \text{ otherwise.} \end{cases}$$

We thus have one Riemann problem from 1 to $v_-(\zeta)$ and one from $v_-(\zeta)$ to 1. We take

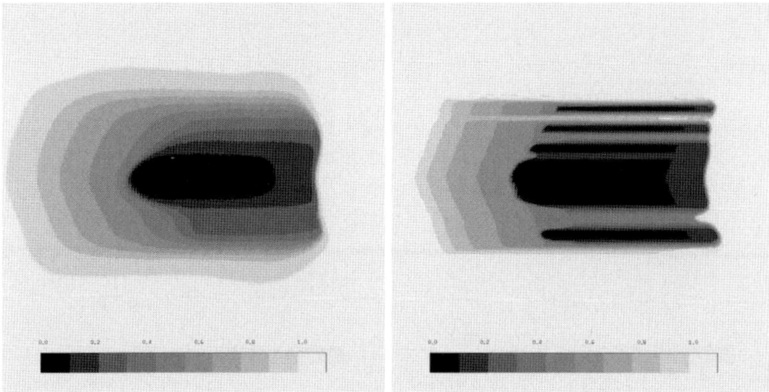

Fig. 4. P-adaptive solution on a uniform grid for the two-dimensional Buckley–Leverett type problem. The solution for $p_{\max} = 0$ is shown on the *left-hand side* and the solution for $p_{\max} = 2$ on the *right-hand side*

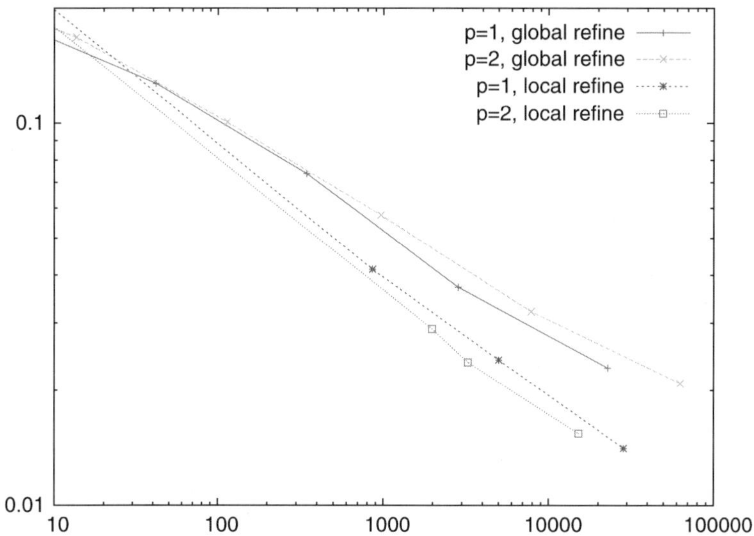

Fig. 5. Convergence study of the hp-adaptive algorithm for the Buckley–Leverett type problem in two-space dimensions

$$v_-(\zeta) = \frac{1}{2}(1 - \cos(\pi\zeta/0.4))\cos(2\pi\zeta/0.4)^2 \quad \text{for} \quad \zeta < 0,$$
$$v_-(\zeta) = \frac{1}{2}(1 - \cos(\pi\zeta/0.4))\cos(4\pi\zeta/0.4)^2 \quad \text{for} \quad \zeta > 0.$$

The function v_0 is shown in Fig. 3 (left), while v_- is plotted in Fig. 3 (right).

In Fig. 4 the p-adaptive solutions are given on an uniform grid for $p_{\max} = 0$ (left) and $p_{\max} = 2$ (right). As expected, the lower order scheme has too much numerical viscosity such that for instance the structure in y-direction is

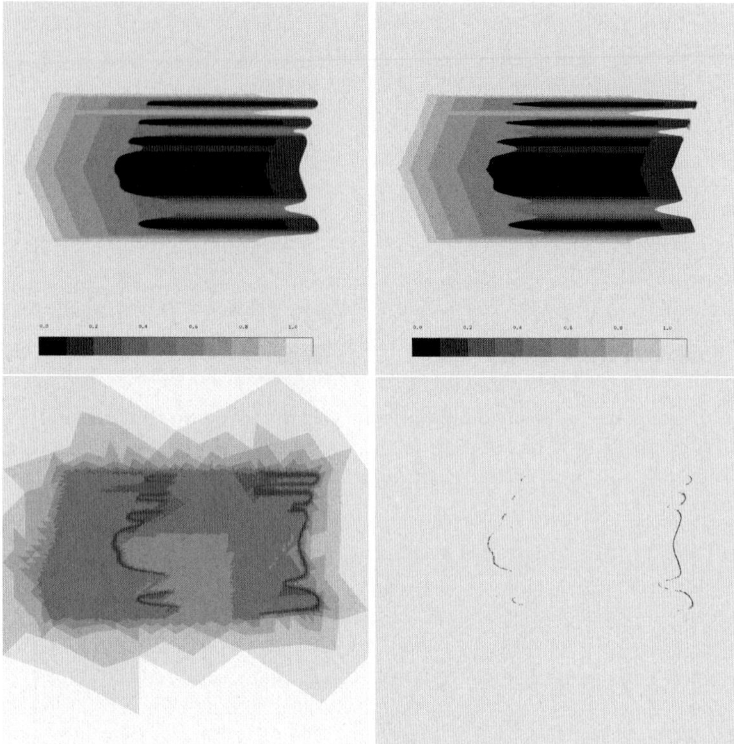

Fig. 6. hp-adaptive solution of the two-dimensional Buckley–Leverett problem. In the *top* row, the numerical solution (*left*) and the exact solution projected to the adaptive grid (*right*) are shown. In the *bottom* row, a shading of the levels of the adaptive grid (*left*) and the distribution of the polynomial degree (*right*) are plotted

completely smeared out. On the other hand the shocks within the x-direction are kept very well. In contrast to this, the higher order method with $p_{\max} = 2$ is able to capture both, the profiles in x and y-direction (a reference solution is shown on the top right of Fig. 6. Please notice that due to our p-adaptive scheme the higher order scheme is stable and converges toward the right entropy solution for the Buckley–Leverett type problem. The convergence history for the purely p-adaptive algorithm (uniform grid refinement) in comparison with the hp-adaptive case is plotted in Fig. 5 for $p_{\max} = 1, 2$. We see that the hp-adaptive solutions converge with a better rate than the solutions on uniform grids. In addition, on uniform grids we do not benefit from the higher polynomial order, as the results for $p = 1$ are better than those for $p = 2$. This is not the case for the hp-adaptive algorithm.

Finally in Fig. 6 the numerical solution with $p_{\max} = 2$ is given for the hp-adaptive algorithm. In the top row the resulting numerical solution (left) is compared with the exact solution projected to the adaptive numerical grid (right). In the bottom row a color shading of the levels of the adaptive grid is plotted on the right-hand side. Each jump in the grey value corresponds to a transition of one grid level. The grid levels correspond to a nonconforming red refinement of the triangles. Thus, the local mesh size h is coupled to the level information through $h \approx (0.5)^{\text{level}}$. Finally, on the right-hand side of the bottom row, the local polynomial degree is plotted. It can be seen that $p = p_{\max}$ on large parts of the domain and that the polynomial degree is only reduced in the strong shocks, i.e., where the normal to the shock is parallel to the x-axis. In contrast, the polynomial degree is not reduced in shocks with normals almost parallel to the y-axis, which are "almost" contact discontinuities.

7 Conclusion

In this contribution we have shown the good numerical performance of our new hp-adaptive RK-DG scheme for scalar conservation laws in two-space dimensions. The hp-adaptivity is based on the a posteriori error estimate from [DMO06]. The numerical results show that we are able to benefit from the higher order methods, even in the case where shocks or contact discontinuities are involved.

References

[ADFK02] S. Adjerid, K.D. Devine, J.E. Flaherty, and L. Krivodonova. A posteriori error estimation for discontinuous Galerkin solutions of hyperbolic problems. *Comput. Methods Appl. Mech. Engrg.*, 191:1097–1112, 2002.

[ALU] ALUGrid Library homepage. http://www.mathematik.uni-freiburg.de/IAM/Research/alugrid.

[BDD+05] A. Burri, A. Dedner, D. Diehl, R. Klöfkorn, and M. Ohlberger. A general object oriented framework for discretizing nonlinear evolution equations. In *Proceedings of the 1st Kazakh-German Advanced Research Workshop on Computational Science and High Performance Computing, Almaty, Kazakhstan, September 25 - October 1*, 2005.

[BDE+03] P. Bastian, M. Droske, C. Engwer, R. Klöfkorn, T. Neubauer, M. Ohlberger, and M. Rumpf. Towards a unified framework for scientific computing. In *Proceedings of the 15th International Conference on Domain Decomposition Methods, Berlin, July 21-25*, 2003.

[BDF94] R. Biswas, K.D. Devine, and J.E. Flaherty. Parallel, adaptive finite element methods for conservation laws. *Appl. Numer. Math.*, 14:255–283, 1994.

[CG96] B. Cockburn and P.A. Gremaud. Error estimates for finite element methods for scalar conservation laws. *SIAM J. Num. Anal.*, 33:522–554, 1996.

[CHS90] B. Cockburn, S. Hou, and C.-W. Shu. The Runge-Kutta local projection discontinuous Galerkin finite element method for conservation laws. IV. The multidimensional case. *Math. Comp.*, 54(190):545–581, 1990.

[CS91] B. Cockburn and C.-W. Shu. The Runge-Kutta local projection P^1-discontinuous-Galerkin finite element method for scalar conservation laws. *RAIRO Modél. Math. Anal. Numér.*, 25(3):337–361, 1991.

[CS01] B. Cockburn and C.-W. Shu. Runge–Kutta discontinuous Galerkin methods for convection-dominated problems. *J. Sci. Comput.*, 16(3): 173–261, 2001.

[DMO06] A. Dedner, C. Makridakis, and M. Ohlberger. Error control for a class of Runge Kutta Discontinuous Galerkin methods for nonlinear conservation laws. Accepted for publication in *SIAM J. Numer. Anal.*, 2006.

[GST01] S. Gottlieb, C.-W. Shu, and E. Tadmor. Strong stability-preserving high-order time discretization methods. *SIAM Rev.*, 43(1):89–112, 2001.

[HH02] R. Hartmann and P. Houston. Adaptive discontinuous Galerkin finite element methods for nonlinear hyperbolic conservation laws. *SIAM J. Sci. Comput.*, 24(3):979–1004, 2002.

[HS01] P. Houston and E. Süli. *hp*-adaptive discontinuous Galerkin finite element methods for first-order hyperbolic problems. *SIAM J. Sci. Comput.*, 23(4):1226–1252, 2001.

[HSS02] P. Houston, B. Senior, and E. Süli. *hp*-discontinuous Galerkin finite element methods for hyperbolic problems: error analysis and adaptivity. *Internat. J. Numer. Methods Fluids*, 40(1-2):153–169, 2002. ICFD Conference on Numerical Methods for Fluid Dynamics (Oxford, 2001).

[JJS95] J. Jaffré, C. Johnson, and A. Szepessy. Convergence of the discontinuous Galerkin finite element method for hyperbolic conservation laws. *Math. Models Methods Appl. Sci.*, 5(3):367–386, 1995.

[JS95] C. Johnson and A. Szepessy. Adaptive finite element methods for conservation laws based on a posteriori error estimates. *Commun. Pure Appl. Math.*, 48:199–234, 1995.

[LB00] M.G. Larson and T.J. Barth. A posteriori error estimation for adaptive discontinuous Galerkin approximations of hyperbolic systems. In *Discontinuous Galerkin methods (Newport, RI, 1999)*, volume 11 of *Lect. Notes Comput. Sci. Eng.*, pages 363–368. Springer, Berlin, 2000.

[SH03] E. Süli and P. Houston. Adaptive finite element approximation of hyperbolic problems. In *Error estimation and adaptive discretization methods in computational fluid dynamics*, volume 25 of *Lect. Notes Comput. Sci. Eng.*, pages 269–344. Springer, Berlin, 2003.

[Shu01] C.-W. Shu. A survey of strong stability preserving high order time discretizations. Scientific Computing Report Series (2001-18), Brown University, 2001.

[Zen86] M. Zennaro. Natural continuous extensions of Runge-Kutta methods. *Math. Comput.*, 46:119–133, 1986.

Elliptic and Centrifugal Instabilities in Incompressible Fluids

F. Gallaire, D. Gérard-Varet, and F. Rousset

1 Introduction

We consider the incompressible Navier–Stokes equation

$$\partial_t u + u \cdot \nabla u + \nabla p = \frac{1}{\mathrm{Re}} \Delta u + F, \quad \nabla \cdot u = 0 \tag{1}$$

for large Reynolds number Re, in the domain $\mathcal{D} = \Sigma \times \mathbb{T}$, where Σ is a two-dimensional domain without boundary ($\mathbb{R}^2$ or $\mathbb{T}^2$ for example). The source term F is an exterior forcing. Since we are interested in the dynamics at large Reynolds numbers, we set for convenience

$$\frac{1}{\mathrm{Re}} = \nu \varepsilon^2,$$

where $\nu > 0$ is fixed and $\varepsilon > 0$ will be a small parameter. We study the instability of stationary solutions $\mathbf{u^s}$ of (1), which are two-dimensional, i.e., $\mathbf{u}^s(x) = (u^s(x_1, x_2), 0)$ when they are submitted to three-dimensional perturbations. This is a classical problem, with a long history in fluid mechanics. A particularly interesting class of stationary solutions is made of the two-dimensional stationary solutions of the Euler equation. In general they are not solutions of the homogeneous Navier–Stokes equation, so in this case, the exterior forcing F is used to prevent the slow motion induced by the small viscosity. A perturbation v of a stationary solution $\mathbf{u}^s$ evolves according to the nonlinear equation

$$\partial_t v + \mathbf{u}^s \cdot \nabla v + v \cdot \nabla \mathbf{u}^s + \nabla p + v \cdot \nabla v = \nu \varepsilon^2 \Delta v, \quad \nabla \cdot v = 0. \tag{2}$$

At first step, we consider that v has small amplitude and we neglect the nonlinear term to study only the linear equation

$$\partial_t v + \mathbf{u}^s \cdot \nabla v + v \cdot \nabla \mathbf{u}^s + \nabla p = \nu \varepsilon^2 \Delta v, \quad \nabla \cdot v = 0. \tag{3}$$

The linear instability of $\mathbf{u}^s$ is linked to the existence of unstable spectrum for the linear operator

$$L^\varepsilon v = \mathbb{P}\Big(-\mathbf{u}^s \cdot \nabla v - v \cdot \nabla \mathbf{u}^s + \nu\varepsilon^2 \Delta v\Big),$$

where $\mathbb{P}$ is the Leray projection on divergence free vector fields. For the linearized Euler operator (i.e., when $\nu = 0$), a classical way to find linear instabilities that were introduced by Lifschitz–Hameiri [11] and Friedlander–Vishik [6] is to study high frequency oscillations: we look for solutions of (3) under the form

$$v = a(t,x)e^{i\frac{\varphi(t,x)}{\varepsilon}}. \tag{4}$$

By using Lagrangian coordinates, we get a system of ordinary differential equations (the bicharacteristic amplitude system):

$$\frac{dx}{dt} = \mathbf{u}^s(x), \tag{5}$$

$$\frac{d\xi}{dt} = -\nabla \mathbf{u}^s \cdot \xi, \tag{6}$$

$$\frac{da}{dt} = \Big(2\frac{\xi \otimes \xi}{|\xi|^2} - I\Big) D\mathbf{u}^s \cdot a, \tag{7}$$

with the constraint $a \cdot \nabla \xi = 0$ where $\xi(t) = \nabla\varphi(t, x(t, x_0))$. To avoid confusion, we recall that Du stands for the matrix $(\partial_j u_i)_{i,j}$ whereas $\nabla u = {}^tDu$. Then, one can define a Lyapounov exponent for this system:

$$\mu = \lim_{t\to+\infty} \frac{1}{t} \log \sup\Big\{a(t, x_0, \xi_0, a_0), \quad |\xi_0| = |a_0| = 1,\ a_0 \cdot \xi_0 = 0\Big\}.$$

It was proved in [17] that if $\mu > 0$ then there is linear instability in L^2 for (3). More recently, it was shown that in two dimension the essential spectrum of the linearized operator is a vertical band and that the width of the band is determined by the Lyapounov exponent of the bicharacteristic-amplitude system [16], [14]. An important consequence of these results is that even in 2D every flow with an hyperbolic stagnation point (see [6]) is unstable in L^2 since the Lyapounov exponent μ is positive.

The meaning of this kind of instability for the nonlinear equation is not completely clear. Indeed, there has been a lot of recent activity towards the justification of the fact that linear instability implies nonlinear instability for the Euler or Navier–Stokes equation, see [5], [9], [10], [12]. In particular, it was established in [12] that for the two-dimensional Euler equation the existence of an unstable eigenvalue implies nonlinear instability. Moreover, the importance of the norm in which the instability is stated has been emphasized in [12]: there exists flows that are linearly unstable in the L^2 norm of the velocity (because there is an hyperbolic stagnation point so that the above criterion applies) but are nonlinearly stable in the L^2 norm of the vorticity (which is equivalent to the H^1 norm of the velocity). The consequence of this example is that it seems difficult to deduce a general nonlinear instability result from the existence of unstable L^2 essential spectrum. The aim of this note is to describe the results of [7] where it was proven that in some special cases the presence of L^2 unstable essential spectrum for the linearized Euler operator implies some nonlinear instability for the Navier–Stokes equation with small viscosity.

2 Main Results

Besides flows with hyperbolic stagnation points, there are other classical examples in fluid mechanics where the bicharacteristic-amplitude system can be analyzed. We consider the case where u^s is a stationary solution of the 2D Euler equation, then we can assume that the vorticity $\omega^s = \partial_1 u_2^s - \partial_2 u_1^s$ and the streamfunction ψ^s such that $u^s = \nabla^\perp \psi^s$ are linked through the relation $\omega^s = F(\psi^s)$. If u^s has closed streamlines, both the geometry of the streamlines and the distribution of the vorticity through the properties of F may contribute to some instability in the bicharacteristic-amplitude system. These are the well-known elliptic and centrifugal instabilities.

Let us first describe the elliptic instability. We assume that there is a stagnation point x_0 such that $A = D\mathbf{u}^s(x_0)$ has purely imaginary eigenvalues. In this case, in the vicinity of the stagnation point, the flow $\mathbf{u}^s$ is well-approximated by the linear flow Ax. By using this approximation we get a simpler bicharacteristic-amplitude system:

$$\begin{cases} \frac{dx}{dt} = Ax, \\ \frac{d\xi}{dt} = -A^*\xi, \\ \frac{da}{dt} = \left(2\frac{\xi\otimes\xi}{|\xi|^2} - I\right) A\, a. \end{cases} \tag{8}$$

In this approximation, the vorticity is constant in the vicinity of the stagnation point so that the only ingredient that may contribute to the instability is the geometry of the streamlines. The equation for the amplitude (the third equation of (8)) is a linear ordinary differential equation with periodic coefficients and hence can be analyzed with Floquet theory. We easily get that the Floquet exponents are given by $\zeta_0(\xi_0)$, $-\zeta_0(\xi_0)$, and 0. Our assumption of linear elliptic instability is

- **(H1)** There exists ξ_0 such that the real part $\sigma_0(\xi_0)$ of $\zeta_0(\xi_0)$ is positive.

This assumption has been checked numerically in fluid mechanics papers, [2], [13] and was later checked analytically by Waleffe [18]: (H1) is actually verified for all elliptical noncircular flows.

The other classical example that we want to analyze is the centrifugal instability. We assume that u^s has locally closed streamlines, but we study the dynamics in the vicinity of a given closed streamline. We use a local curved coordinate system $(\rho, \theta) \to y(\rho, \theta) = (x_1(\rho, \theta), x_2(\rho, \theta))$ for $\rho \in (\rho_0 - \delta, \rho_0 + \delta)$

such that the stream function is constant on the curves $\theta \mapsto y(\rho, \theta)$, i.e., $\psi^s(y(\rho,\theta)) = \rho$. Moreover, since we assume that the streamlines are locally closed, we have that $\theta \to y(\rho,\theta)$ is $T(\rho)$ periodic. In a local orthonormal basis (e_ρ, e_θ), u^s is given by

$$u^s(y) = U^s(\rho,\theta)e_\theta, \quad \forall \rho \in (\rho_0 - \delta, \rho_0 + \delta). \tag{9}$$

Thanks to this structural assumption on our basic flow, we can reduce the study of the Lyapounov exponent of the amplitude equation (7). Indeed, we first notice that (5) reduces to

$$\theta'(t) = \frac{U^s}{h}(\theta(t),\rho), \quad \rho'(t) = 0.$$

Since U^s is periodic, θ will be periodic. Next, we restrict our study to the simple phase

$$\xi(\rho,\theta(t)) = e_z. \tag{10}$$

With this special choice of the phase, we can rewrite the equation for the amplitude as

$$\frac{da}{dt} = G(t,\theta_0,\rho)a. \tag{11}$$

Again, the Floquet exponents for this system are $\zeta_0(\rho)$, $-\zeta_0(\rho)$, 0 (we can easily check that they do not depend on θ_0 because of the periodicity of θ). Our assumption for linear centrifugal instability reads:

- **(H1′)** There exists ρ_0 such that $\sigma_0(\rho_0) > 0$.

This assumption can be checked numerically, see [15] for example. Moreover, there are some criteria that allow to predict the presence of centrifugal instability ([1], [8]). In some specific situations, explicit computations are possible. In the case of circular vortices $u^s = U^s(r)e_\theta$, where (r,θ) are the standard polar coordinates in the plane, the matrix G does not depend on time and hence we just have to compute its eigenvalues. This yields a famous Rayleigh criterion for centrifugal instability, the flow is unstable if the sign of the vorticity changes. This inherently 3D instability differs from the 2D azimuthal shear instability. A necessary condition for this purely 2D shear instability, also due to Rayleigh, is the extension to circular geometry of the inflection point theorem and requires a change in the sign of $dW(r)/dr$ where W is the vorticity. Consequently, we see on this simple example that there are flow that are stable in 2D but unstable in 3D.

Our aim is to prove that (H1) or (H1′) implies a nonlinear instability for (1), which is localized in the vicinity of the stagnation point or the unstable close. The main difficulty in this justification is the competition between the elliptical or centrifugal instability and the possible 2D instabilities that may occur. We shall assume that the elliptical or centrifugal instability dominates all the possible 2D instabilities in a certain class. More precisely, let us set

$$\sigma = \sigma_0 - \alpha\nu, \quad \alpha = \frac{1}{T}\int_0^T |\xi|^2,$$

where σ_0 is given by (H1) or (H1′), we consider a space X made of smooth functions of at most two variables, which contains the unstable solution of the amplitude equation given by (H1) or (H1′) and is stable by derivation and multiplication. We shall denote by $||\cdot||$ the L^2 norm. Our second assumption reads

- **(H2)** There exists $C > 0$ such that for every $\varepsilon \in (0,1)$ and for every $f(t, x_1, x_2) \in X$ such that

$$||f|| + ||\text{ curl } f|| \leq C_f e^{\gamma t} \tag{12}$$

with $\gamma \geq 2\sigma$, then the two-dimensional solution v of

$$\partial_t v + u^s \cdot \nabla v + v \cdot \nabla u^s + \nabla p - \nu\varepsilon^2 \Delta v = f, \ \nabla \cdot v = 0, \tag{13}$$

which vanishes at $t = 0$, is in X and satisfies the estimate

$$||v|| + ||\text{ curl } v|| \leq CC_f e^{\gamma t}. \tag{14}$$

This assumption means that the instability that we have detected dominates the two-dimensional instability in a certain class X. The meaning of the space X is that we can use the symmetries of the problem to check (H2). For example, in the case of the centrifugal instability of the circular vortices $u^s(r) = U(r)e_\theta$ in polar coordinates, we can choose X as the space of smooth functions in the vicinity of r_0, which are independent of θ. Note that in this space we cannot see the 2D instabilities: in this case, the solutions of (13) also depend only on r, so that (13) reduces to the heat equation

$$\partial_t v - \nu\varepsilon^2 \Delta v = f$$

and hence the assumption (H2) is clearly verified. In the general case (H2) can be difficult to check analytically, there is an example in $\mathbb{T}^2$ of a genuine 2D flow for which this assumption can be checked analytically, we refer to [7]. The competition between the centrifugal instability and the shear instability has been studied numerically in details in [3] and it was seen that the centrifugal instability dominates the 2D shear waves, so that our assumption (H2) seems to be generically verified.

We are now able to state our main results. We begin with the nonlinear elliptic instability:

Theorem 1 (Nonlinear elliptic instability). *Consider $\mathbf{u}^s(x) = (u_s(x_1, x_2), 0)$ a stationary solution of (1) and assume that there exists a stagnation point x_0 such that $A = D\mathbf{u}^s(x_0)$ has only purely imaginary eigenvalues. Then if $\Sigma = \mathbb{R}^2$ or $\mathbb{T}^2$ and under the assumptions (H1) and (H2), $\mathbf{u}^s$ is nonlinearly unstable in the following sense:*

there exists ν_0 *such that for every* $\nu \in (0, \nu_0]$ *and for every* $N \in \mathbb{N}$, $s \in \mathbb{N}$, *and* $\beta \in (1/3, 1)$, *there exists* ε_0 *and* $\eta > 0$ *such that, for every* $\varepsilon \in (0, \varepsilon_0)$, *there exists an initial data* $v^{0,\varepsilon}$ *and a time* T^ε *such that*

$$||v^{0,\varepsilon}||_{H^s} \leq \varepsilon^N$$

and the solution of the equation (2) verifies

$$||v^\varepsilon(T^\varepsilon)||_{L^2(D(x_0,\varepsilon^\beta)\times\mathbb{T})} \geq \eta\varepsilon^{\beta+1}.$$

Moreover, we also have

$$\sup_{t\in[0,T^\varepsilon]} ||v^\varepsilon(t)||_{L^\infty(D(x_0,\varepsilon^\beta)\times\mathbb{T})} \geq \eta\varepsilon.$$

This theorem is a nonlinear instability result: it states that a small perturbation u^ε can be amplified from the amplitude ε^N, for arbitrary N, until it reaches the amplitude $\varepsilon^{\beta+1}$ in the L^2 norm and ε in the L^∞ norm locally in the small disk $D(x_0, \varepsilon^\beta)$ around the stagnation point. Note that in the same disk $D(x_0, \varepsilon^\beta)$, we have that

$$||\mathbf{u}^s||_{L^2(D(x_0,\varepsilon^\beta)\times\mathbb{T})} \leq C\varepsilon^{2\beta}, \quad ||\mathbf{u}^s||_{L^\infty(D(x_0,\varepsilon^\beta)\times\mathbb{T})} \leq C\varepsilon^\beta.$$

Since β can be chosen arbitrarily close to 1, this yields that locally the perturbation u^ε can almost reach the amplitude of the reference flow $\mathbf{u}^s$. In this way we almost recover a classical instability result. There was a lot of recent activity towards the justification of the fact that linear instability implies nonlinear instability for the Euler or Navier–Stokes equation, see [5], [9], [10], [12]. Note that here the instability does not come from the existence of an unstable eigenmode. We have already explained the importance of the norm in which the instability is stated, here our result has a simple and clear physical interpretation: there is a violent amplification of the amplitude of the perturbation of the velocity in the L^2 and Sup norm in the vicinity of the stagnation point.

In a similar way, we can prove nonlinear centrifugal instability

Theorem 2 (Nonlinear centrifugal instability). *Consider* $\mathbf{u}^s(x) = (u_s(x_1, x_2), 0)$ *a two-dimensional stationary solution of (1) and assume that there exists closed streamlines in the vicinity of* ρ_0. *Then if* $\Sigma = \mathbb{R}^2$ *or* $\mathbb{T}^2$ *and under the assumption (H1′) and (H2),* $\mathbf{u}^s$ *is nonlinearly unstable in the following sense: there exists* ν_0 *such that for every* $\nu \in (0, \nu_0]$ *and for every* $N \in \mathbb{N}$, $s \in \mathbb{N}$, *there exists* ε_0 *and* $\eta > 0$ *such for every* $\varepsilon \in (0, \varepsilon_0)$ *there exists an initial data* $v^{0,\varepsilon}$ *and a time* T^ε *such that*

$$||v^{0,\varepsilon}||_{H^s} \leq \varepsilon^N$$

and the solution of the equation (2) verifies

$$||v^\varepsilon(T^\varepsilon) \cdot e_\rho||_{L^2(\rho_0-\sqrt{\varepsilon},\rho_0+\sqrt{\varepsilon})} \geq \eta\varepsilon^{\frac{5}{4}}.$$

Moreover, we also have

$$\sup_{t\in[0,T^\varepsilon]} ||v^\varepsilon(t)\cdot e_\rho||_{L^\infty(\rho_0-\sqrt{\varepsilon},\rho_0+\sqrt{\varepsilon})} \geq \eta\varepsilon.$$

This theorem states that the perturbation v^ε normal to the streamlines of u^s can be arbitrarily amplified.

3 Sketch of the Proof

The detailed proofs can be found in [7]. Here we just give the main steps, we shall focus on the proof of Theorem 1. As in [9], [4], we look for a high order unstable solution. We set $u = u^s + \varepsilon v$ so that v solves

$$\partial_t v + u^s\cdot\nabla v + v\cdot\nabla u^s + \nabla p = \nu\varepsilon^2 \Delta v - \varepsilon v\cdot\nabla v, \quad \nabla\cdot v = 0. \tag{15}$$

Without loss of generality, we assume that the stagnation point $x_0 = 0$, so that we can write

$$u^s = Ay + v^s(y), \quad v^s = \mathcal{O}(|y|^2)$$

in a vicinity of the stagnation point. We set $U = (v, p)$, and we rewrite (15) as

$$L^\varepsilon U + \varepsilon v\cdot\nabla v = 0, \quad \nabla\cdot v = 0 \tag{16}$$

where

$$L^\varepsilon U = \partial_t v + u^s\cdot\nabla v + v\cdot\nabla u^s + \nabla p - \nu\varepsilon^2\Delta v.$$

The first step is to look for an expansion under the form

$$U^{app} = \varepsilon^N \sum_{k=0}^{M} \varepsilon^{Nk} U^k, \quad U^k = (V^k, P^k). \tag{17}$$

Plugging this ansatz in (16), we find

$$\sum_{k=0}^{M} \varepsilon^{Nk} L^\varepsilon U^k + \varepsilon^{N+1} \sum_{0\leq k,l\leq M} \varepsilon^{N(k+l)} V^k\cdot\nabla V^l = 0,$$

$$\sum_{k=0}^{M} \varepsilon^{Nk} \operatorname{div} V^k = 0.$$

We solve approximately this system by setting

$$\begin{cases} L^\varepsilon U^0 = \varepsilon^N R^0, \\ \operatorname{div} V^0 = \varepsilon^N d^0 \end{cases} \tag{18}$$

and for $k \geq 1$

$$\begin{cases} L^\varepsilon U^k = -R^{k-1} + \varepsilon^N R^k + \varepsilon \sum_{j+l=k-1} V^j \cdot \nabla V^l, \\ \text{div } V^k = -d^{k-1} + \varepsilon^N d^k. \end{cases} \tag{19}$$

The terms R^k, $k \geq 0$ will be uniformly bounded with respect to ε on a sufficiently large interval of time $[0, T_\varepsilon]$ for ε sufficiently small. If we had a complete understanding of the linear operator L^ε we could solve the system with $R^k = 0$. Since this is not the case here, we use a WKB expansion of U^k to solve (18), (19). We set $\beta = 1 - \frac{1}{\mu}$ with $\mu \geq 3$. We look for a WKB expansion U^0 under the form

$$U^0(t,x) = \mathcal{U}^0\Big(t, \frac{y}{\varepsilon^\beta}, \frac{\varphi(t, x/\varepsilon^\beta)}{\varepsilon^{\frac{1}{\mu}}}\Big)$$

with the expansion

$$\mathcal{U}^0 = \sum_{l=0}^{\mu(N+1)} \varepsilon^{\frac{l}{\mu}} \mathcal{U}^{0,l}\Big(t, \frac{y}{\varepsilon^\beta}, \frac{\varphi(t, x/\varepsilon^\beta)}{\varepsilon^{\frac{1}{\mu}}}\Big), \tag{20}$$

where $\mathcal{U}^{0,l}(t, Y, \lambda)$ are compactly supported in $D(0,1) \subset \mathbb{R}^2$ in Y and periodic in λ. Note that U^0 depends on the x_3 variable only through the phase. Moreover, we also require that

$$\int_\lambda \mathcal{U}^{0,l}(t, Y, \lambda)\, d\lambda = 0. \tag{21}$$

We easily get the equations for the profiles, the first term solves the bicharacteristic-amplitude system so that we find a growing solution by our assumption (H1). To get a profile compactly supported, we can use an arbitrary smooth compactly supported function.

The main difficulty arises when we want to solve (19) for $k \geq 1$. Indeed, because of the nonlinear terms, we can split the right hand side into an oscillating part $\tilde{Q}^k(t, Y, \lambda)$ with $\int \tilde{Q}^k d\lambda = 0$ and a nonoscillating part $\overline{Q}^k(t, Y)$. The oscillating part can be easily absorbed by using again an expansion like (20). The difficulty comes from the nonoscillating part, for example, for $k = 1$, we have to solve the two-dimensional linearized Navier–Stokes equation

$$\partial_t v + u^s \cdot \nabla v + v \cdot \nabla u^s + \nabla p = \nu \varepsilon^2 \Delta v + \overline{Q}^1(t, y/\varepsilon^\beta), \quad \nabla \cdot v = 0, \tag{22}$$

with a source term that verifies $\varepsilon^{\beta|\alpha|} |\partial^\alpha \overline{Q}| \lesssim e^{2\sigma t}$. Our expansion (17) will be meaningful to prove a nonlinear instability like in [9] only if the terms U^k are bounded by $e^{k\sigma t}$ so that in the expansion the dominant term is really given by the first one. In particular, we want a solution of (22) which is bounded by $e^{2\sigma t}$. Because of the global constraint $\nabla \cdot v = 0$, it does not seem possible to find an approximate solution based on a WKB expansion with solutions localized in a vicinity of the stagnation point and hence to reduce the problem to a simpler one as we do for the oscillating part. Consequently, at this point, we work directly with the exact solution of (22) and hence we use our assumption (H2). Note that (H2) gives that the H^1 norm of v satisfies the required estimates. Nevertheless, in order to construct U^{app}, we also need the same control for high order derivatives of v. At this point of the analysis, the presence of the small but nonzero viscosity $\nu\varepsilon^2$ is crucial. Indeed for the linearized Euler equation, in general, the derivatives have a faster growth. Indeed, if u^s has a hyperbolic stagnation point the derivative ∂^α curl v has a growth $e^{|\alpha|\mu t}$. When we keep some viscosity, we can prove that the penalized derivatives $\varepsilon^{|\alpha|}\partial^\alpha$ are still bounded by $e^{2\sigma t}$. Moreover, locally in the vicinity $D(x_0, \varepsilon^\beta)$ of the stagnation point, it is possible to control $\varepsilon^{\beta|\alpha|}\partial^\alpha$, which is the good weight compatible with the expansion (20). Once these estimates are established, the nonlinear instability follows by using the same method as in [9]

References

1. Bayly, B. Three-dimensional centrifugal type instability in an inviscid two-dimensional flow. *Phys. Fluids 31* (1988), 56–64.
2. Bayly, B. J. Three-dimensional instability of elliptical flow. *Phys. Rev. Lett. 57*, 17 (1986), 2160–2163.
3. Billant, P., and Gallaire, F. Generalized rayleigh criterion for non axisymmetric centrifugal instabilities. *J. Fluid Mech. 542* (2005), 365–379.
4. Desjardins, B., and Grenier, E. Linear instability implies nonlinear instability for various types of viscous boundary layers. *Ann. Inst. H. Poincaré Anal. Non Linéaire 20*, 1 (2003), 87–106.
5. Friedlander, S., Strauss, W., and Vishik, M. Nonlinear instability in an ideal fluid. *Ann. Inst. H. Poincaré Anal. Non Linéaire 14*, 2 (1997), 187–209.
6. Friedlander, S., and Vishik, M. M. Instability criteria for the flow of an inviscid incompressible fluid. *Phys. Rev. Lett. 66*, 17 (1991), 2204–2206.
7. Gallaire, F., Gérard-Varet, D., and Rousset, F. Three-dimensional instability of planar flows. *Preprint* (2006).
8. Gallaire, F., and Rousset, F. Shortwave centrifugal instability in the vicinity of vanishing total vorticity streamlines. *Phys. Fluids 18* (2006), 058102.
9. Grenier, E. On the nonlinear instability of Euler and Prandtl equations. *Comm. Pure Appl. Math. 53*, 9 (2000), 1067–1091.
10. Hwang, H. J., and Guo, Y. On the dynamical Rayleigh-Taylor instability. *Arch. Ration. Mech. Anal. 167*, 3 (2003), 235–253.
11. Lifschitz, A., and Hameiri, E. Local stability conditions in fluid dynamics. *Phys. Fluids A 3*, 11 (1991), 2644–2651.
12. Lin, Z. Some stability and instability criteria for ideal plane flows. *Comm. Math. Phys. 246*, 1 (2004), 87–112.
13. Pierrehumbert, R. T. Universal short-wave instability of two-dimensional eddies in an inviscid fluid. *Phys. Rev. Lett. 57*, 17 (1986), 2157–2159.
14. Shvydkoy, R., and Latushkin, Y. Essential spectrum of the linearized 2D Euler equation and Lyapunov-Oseledets exponents. *J. Math. Fluid Mech. 7*, 2 (2005), 164–178.

15. Sipp, D., and Jacquin, L. Three-dimensional centrifugal-type instabilities of two-dimensional flows in rotating systems. *Phys. Fluids 12*, 7 (2000), 1740–1748.
16. Vishik, M. Spectrum of small oscillations of an ideal fluid and Lyapunov exponents. *J. Math. Pures Appl. (9) 75*, 6 (1996), 531–557.
17. Vishik, M. M., and Friedlander, S. Dynamo theory methods for hydrodynamic stability. *J. Math. Pures Appl. (9) 72*, 2 (1993), 145–180.
18. Waleffe, F. On the three-dimensional instability of strained vortices. *Phys. Fluids A 2*, 1 (1990), 76–80.

On Compressible Current–Vortex Sheets

Y. Trakhinin

Summary. Recent results for ideal compressible current–vortex sheets (tangential MHD discontinuities) are surveyed. A sufficient condition for the weak stability of planar current–vortex sheets is first found for a general case of the unperturbed flow. In astrophysics, this condition can be interpreted as the sufficient condition for the macroscopic stability of the heliopause. The crucial role in finding this stability condition is played by a new symmetric form of the MHD equations. The linear variable coefficients problem for nonplanar current–vortex sheets is studied as well. The fact that the Kreiss–Lopatinski condition is satisfied only in a weak sense yields losses of derivatives in a priori estimates. We prove an a priori tame estimate that is a necessary step to show the local-in-time existence of "stable" current–vortex sheet solutions to the nonlinear equations of ideal compressible MHD by a suitable Nash–Moser-type iteration scheme. Since the tangential discontinuity is characteristic, the functional setting is provided by the anisotropic weighted Sobolev spaces H^m_*.

1 Introduction

We consider the system of ideal compressible magnetohydrodynamics (MHD):

$$\begin{gathered} \partial_t\rho + \operatorname{div}(\rho\mathbf{v}) = 0, \\ \partial_t(\rho\mathbf{v}) + \operatorname{div}(\rho\mathbf{v}\otimes\mathbf{v} - \mathbf{H}\otimes\mathbf{H}) + \nabla q = 0, \\ \partial_t\mathbf{H} - \nabla\times(\mathbf{v}\times\mathbf{H}) = 0, \\ \partial_t\big(\rho e + (|\mathbf{H}|^2/2)\big) + \operatorname{div}\big((\rho e + p)\mathbf{v} + \mathbf{H}\times(\mathbf{v}\times\mathbf{H})\big) = 0, \end{gathered} \tag{1}$$

where ρ, $\mathbf{v}$, $\mathbf{H}$, and p are the density, the fluid velocity, the magnetic field, and the pressure, respectively, $q = p + (|\mathbf{H}|^2/2)$ is the total pressure, $e = E + (|\mathbf{v}|^2/2)$. With a state equation of medium, $E = E(\rho, S)$, (1) is a closed system, where S is the entropy obeying the Gibbs relation. As the unknown we can fix, for example, the vector $\mathbf{U} = (p, \mathbf{v}, \mathbf{H}, S)$. System (1) is supplemented by the divergent constraint

$$\operatorname{div}\mathbf{H} = 0 \tag{2}$$

on the initial data $\mathbf{U}(0,\mathbf{x}) = \mathbf{U}_0(\mathbf{x})$, $\mathbf{x} = (x_1, x_2, x_3)$.

Let $\Gamma(t) = \{x_1 - f(t,\mathbf{y}) = 0\}$ be a smooth hypersurface in $\mathbb{R}\times\mathbb{R}^3$ ($\mathbf{y} = (x_2, x_3)$ are tangential coordinates). We assume that $\Gamma(t)$ is a surface of tangential discontinuity [3, 4] (*current–vortex sheet*) for solutions of the MHD system (1). This is the type of contact discontinuities for which the MHD Rankine–Hugoniot conditions are satisfied in the following way:

$$\partial_t f = v_{\mathrm{N}}^{\pm}, \quad H_{\mathrm{N}}^{\pm} = 0, \quad [q] = 0, \tag{3}$$

where $H_{\mathrm{N}} = (\mathbf{H}, \mathbf{N})$, $v_{\mathrm{N}} = (\mathbf{v}, \mathbf{N})$, $[g] = g^+ - g^-$, $g^\pm := g(t, f(t,\mathbf{y}) \pm 0, \mathbf{y})$, and $\mathbf{N} = (1, -\partial_2 f, -\partial_3 f)$ is the space normal vector to $\Gamma(t)$. The tangential components of the velocity and the magnetic field may undergo any jump on $\Gamma(t)$. It is worth noting that (2) as well as the boundary conditions $H_{\mathrm{N}}^{\pm} = 0$ can be regarded as the restrictions only on the initial data

$$f(0,\mathbf{y}) = f_0(\mathbf{y}), \quad \mathbf{y}\in\mathbb{R}^2; \quad \mathbf{U}(0,\mathbf{x}) = \mathbf{U}_0(\mathbf{x}), \quad \mathbf{x}\in\Omega^\pm(0), \tag{4}$$

where $\Omega^\pm(t) = \{x_1 \gtrless f(t,\mathbf{y})\}$.

Our final goal is to find conditions on the initial data (4) providing the existence of current–vortex sheet solutions to the MHD system, i.e., the existence of a solution $(\mathbf{U}, f)$ of the free boundary value problem (1), (3), (4), where $\mathbf{U}$ is smooth in the domains $\Omega^\pm(t)$. Because of the general properties of hyperbolic conservation laws it is natural to expect only the local-in-time existence of current–vortex sheet solutions. Therefore, the question on the nonlinear Lyapunov's stability of an ideal current–vortex sheet has no sense. At the same time, the study of the linearized stability of current–vortex sheets is not only a necessary step to prove local-in-time existence but also is of independent interest in connection with various astrophysical applications. In particular, the ideal compressible current–vortex sheet is used for modeling the *heliopause*, which is caused by the interaction of the supersonic solar wind plasma with the local interstellar medium (in some sense, the heliopause is the *boundary of the solar system*). The model of heliopause was suggested in [2], and the heliopause is in fact a current–vortex sheet separating the interstellar plasma compressed at the bow shock from the solar wind plasma compressed at the termination shock wave. Note that in December 2004 the spacecraft *Voyager 1* has crossed the termination shock at the distance of 93 AU from the Sun.

Piecewise constant solutions of (1) satisfying (3) on a planar discontinuity are a simplest case of current–vortex sheet solutions. In astrophysics the linear stability of a planar compressible current–vortex sheet is usually interpreted as the macroscopic stability of the heliopause. One can show that for the constant coefficients linearized problem for planar current–vortex sheets the uniform Kreiss–Lopatinski condition is never satisfied [13]. That is, planar current–vortex sheets can be only neutrally (*weakly*) stable or violently unstable. In the 1960–90s, in a number of works motivated by astrophysical applications (see [10] and references therein) the linear stability of planar compressible current–vortex sheets was examined by the normal modes analysis.

But, because of insuperable technical difficulties neither stability nor instability conditions were found for a general case of the unperturbed flow. We propose an alternative energy method that has first enabled one to find a sufficient condition [13] for the weak stability of compressible current–vortex sheets. For the most general case when for the unperturbed flow $[\mathbf{v}] \neq 0$ and $\mathbf{H}^+ \times \mathbf{H}^- \neq 0$ this condition reads

$$|[\mathbf{v}]| < |\sin(\varphi^+ - \varphi^-)| \min\left\{ \frac{c^+ c_{\mathrm{A}}^+}{|\sin\varphi^-|\sqrt{(c^+)^2 + (c_{\mathrm{A}}^+)^2}}, \frac{c^- c_{\mathrm{A}}^-}{|\sin\varphi^+|\sqrt{(c^-)^2 + (c_{\mathrm{A}}^-)^2}} \right\}, \tag{5}$$

where c and $c_{\mathrm{A}} = |\mathbf{H}|/\sqrt{\rho}$ are the sound velocity and the Alfvén velocity, respectively (for the unperturbed flow), φ^+ (φ^-) is the angle between the vector $\mathbf{H}^+$ ($\mathbf{H}^-$) and the jump $[\mathbf{v}]$.

Thus, our goal is to prove the local-in-time existence of current–vortex sheet solutions to the MHD system (1), provided that the initial data (4) satisfy the stability condition (5) together with all the other necessary conditions (hyperbolicity condition, compatibility conditions, etc.). The basic a priori estimate for the linearized variable coefficients problem for nonplanar current–vortex sheets was obtained in [13]. Since the Kreiss–Lopatinski condition is satisfied only in a weak sense there appears a loss of derivatives phenomena. In this paper we present an a priori *tame* estimate that can be used to achieve nonlinear local-in-time existence by Nash–Moser iterations.[1] Note that recently the Nash–Moser method was successfully used in [5] for 2D compressible vortex sheets. In our case there is an additional principal difficulty in comparison with compressible vortex sheets. The point is that in MHD the loss of control on normal derivatives cannot be compensated as it was done in [5] for vortex sheets by estimating missing normal derivatives through a vorticity-type linearized equation. Therefore, the natural functional setting is provided by the anisotropic weighted Sobolev spaces H^m_* (see, e.g., [11, 12]).

2 A Secondary Generalized Friedrichs Symmetrizer for the Compressible MHD Equations

The crucial role in obtaining the stability condition (5) and proving the a priori estimate for the linearized problem associated to (1), (3), (4) (see Sect. 3 and 4) is played by a new symmetric form [13] of the MHD equations. Using the linear analog of this symmetrization makes the linearized constant coefficients

[1] The work towards the proof of nonlinear local-in-time existence by Nash–Moser iterations has been recently finished [15].

boundary conditions conservative. The hyperbolicity condition for the corresponding linearized interior equations gives us the sufficient stability condition (5).

Let us first consider a general situation. Consider a system of N conservation laws ($t > 0$, $\mathbf{x} \in \mathbb{R}^n$)

$$\partial_t \boldsymbol{\mathcal{P}}^0(\mathbf{U}) + \sum_{j=1}^{n} \partial_j \boldsymbol{\mathcal{P}}^j(\mathbf{U}) = 0,$$

which is rewritten as the quasilinear system

$$B_0(\mathbf{U})\partial_t \mathbf{U} + \sum_{j=1}^{n} B_j(\mathbf{U})\partial_j \mathbf{U} = 0, \tag{6}$$

where $\boldsymbol{\mathcal{P}}^\alpha = (\mathcal{P}_1^\alpha, \ldots, \mathcal{P}_N^\alpha)$, $\mathbf{U}(t, \mathbf{x}) = (u_1, \ldots, u_N)$, $B_\alpha = (\partial \boldsymbol{\mathcal{P}}^\alpha / \partial \mathbf{U})$. Using an *additional* (and usually a priori known) conservation law

$$\partial_t \Phi^0(\mathbf{U}) + \operatorname{div} \boldsymbol{\Phi}(\mathbf{U}) = 0 \qquad (\boldsymbol{\Phi} = (\Phi^1, \ldots, \Phi^n)),$$

we can perform Godunov's symmetrization [6]: $\mathbf{U} \to \mathbf{Q} = \partial \Phi^0 / \partial \boldsymbol{\mathcal{P}}^0$. At the same time, it gives us the Friedrichs symmetrizer $S = (\partial \mathbf{Q} / \partial \mathbf{U})^{\mathsf{T}}$. That is, multiplying (6) from the left by the matrix S we get the symmetric system

$$A_0(\mathbf{U})\partial_t \mathbf{U} + \sum_{j=1}^{n} A_j(\mathbf{U})\partial_j \mathbf{U} = 0, \tag{7}$$

with $A_\alpha = SB_\alpha = A_\alpha^{\mathsf{T}}$.

The situation is a little bit different if system (6) is supplemented by a set of K *divergent constrains*

$$\operatorname{div} \boldsymbol{\Psi}_j(\mathbf{U}) = 0, \quad j = \overline{1, K} \qquad (\boldsymbol{\Psi}_j = (\Psi_j^1, \ldots, \Psi_j^n))$$

(e.g., for the MHD system (1) we have the sole divergent constraint (2)). In this case we have to introduce a *generalized Friedrichs symmetrizer* that can be found from the modified Godunov's symmetrization [7]. Such a symmetrizer is now the set $\mathbb{S} = (S, \mathbf{R}_1, \ldots, \mathbf{R}_K)$ containing the same matrix S as above and the vectors $\mathbf{R}_j$ determined from the relations

$$\mathbf{R}_j = S \frac{\partial r_j}{\partial \mathbf{Q}}, \qquad d\Phi^k = (\mathbf{Q}, d\boldsymbol{\mathcal{P}}^j) + \sum_{j=1}^{K} r_j \, d\Psi_j^k, \quad k = \overline{1, n}.$$

Multiplying (6) from the left by the matrix S and adding to the result the sum $\sum_{j=1}^{K} \mathbf{R}_j \operatorname{div} \boldsymbol{\Psi}_j(\mathbf{U})$ we come to the symmetric system (7).

Note that for the MHD system (1) we do not need to proceed in this way because it can be trivially symmetrized by rewriting it in the nonconservative form

$$\frac{1}{\rho c^2}\frac{dp}{dt} + \operatorname{div}\mathbf{v} = 0\,, \qquad \rho\frac{d\mathbf{v}}{dt} - (\mathbf{H},\nabla\mathbf{H}) + \nabla q = 0\,,$$
$$\frac{d\mathbf{H}}{dt} - (\mathbf{H},\nabla)\mathbf{v} + \mathbf{H}\operatorname{div}\mathbf{v} = 0\,, \qquad \frac{dS}{dt} = 0, \tag{8}$$

where constraint (1) was taken into account, $d/dt = \partial_t + (\mathbf{v},\nabla)$, and c is the sound speed. System (8) is written in the symmetric form (7) for the vector $\mathbf{U} = (p, \mathbf{v}, \mathbf{H}, S)$. The hyperbolicity condition is $A_0 > 0$, i.e.,

$$\rho > 0\,, \quad c^2 > 0\,. \tag{9}$$

Suppose now that for system (6) there exists a generalized Friedrichs symmetrizer $\mathbb{S}_1$ different from the *basic* symmetrizer

$$\mathbb{S}_0 = \left(\left(\frac{\partial \mathbf{Q}}{\partial \mathbf{U}}\right)^{\mathsf{T}}, \left(\frac{\partial \mathbf{Q}}{\partial \mathbf{U}}\right)^{\mathsf{T}} \frac{\partial r_1}{\partial \mathbf{Q}}, \ldots, \left(\frac{\partial \mathbf{Q}}{\partial \mathbf{U}}\right)^{\mathsf{T}} \frac{\partial r_K}{\partial \mathbf{Q}} \right)$$

(we exclude the uninteresting case when $\mathbb{S}_1 = \mathrm{const}\,\mathbb{S}_0$). Then, the symmetrizer $\mathbb{S}_2$ such that $\mathbb{S}_1 = \mathbb{S}_2 \circ \mathbb{S}_0$ is a *secondary* generalized Friedrichs symmetrizer for the symmetric system (7). Indeed, system (7) was already symmetric, but the result of the application of the secondary symmetrizer $\mathbb{S}_2$ is again a symmetric system. It is quite natural that the hyperbolicity condition for the resulting symmetric system can be more restrictive than that for system (7).

For the symmetric MHD system (8) a secondary generalized Friedrichs symmetrizer was proposed in [13] and has the form $\mathbb{S}_2 = (\widetilde{S}, \mathbf{R})$, with

$$\widetilde{S} = \begin{pmatrix} 1 & \frac{\lambda H_1}{\rho c^2} & \frac{\lambda H_2}{\rho c^2} & \frac{\lambda H_3}{\rho c^2} & 0 & 0 & 0 & 0 \\ \lambda H_1 \rho & 1 & 0 & 0 & -\rho\lambda & 0 & 0 & 0 \\ \lambda H_2 \rho & 0 & 1 & 0 & 0 & -\rho\lambda & 0 & 0 \\ \lambda H_3 \rho & 0 & 0 & 1 & 0 & 0 & -\rho\lambda & 0 \\ 0 & -\lambda & 0 & 0 & 1 & 0 & 0 & 0 \\ 0 & 0 & -\lambda & 0 & 0 & 1 & 0 & 0 \\ 0 & 0 & 0 & -\lambda & 0 & 0 & 1 & 0 \\ 0 & 0 & 0 & 0 & 0 & 0 & 0 & 1 \end{pmatrix}, \quad \mathbf{R} = -\lambda \begin{pmatrix} 1 \\ 0 \\ 0 \\ 0 \\ H_1 \\ H_2 \\ H_3 \\ 0 \end{pmatrix},$$

and the function $\lambda = \lambda(\mathbf{U})$ is arbitrary. Applying $\mathbb{S}_2$ to system (8), we get the symmetric system

$$A_0(\mathbf{U})\partial_t \mathbf{U} + \sum_{j=1}^{3} A_j(\mathbf{U})\partial_j \mathbf{U} = 0 \tag{10}$$
$$\left(= \widetilde{S}A_0\partial_t\mathbf{U} + \sum_{j=1}^{3} \widetilde{S}A_j\partial_j\mathbf{U} + \mathbf{R}\operatorname{div}\mathbf{H}\right),$$

where

$$\mathcal{A}_0 = \widetilde{S}A_0 = \begin{pmatrix} \frac{1}{\rho c^2} & \frac{\lambda H_1}{c^2} & \frac{\lambda H_2}{c^2} & \frac{\lambda H_3}{c^2} & 0 & 0 & 0 & 0 \\ \frac{\lambda H_1}{c^2} & \rho & 0 & 0 & -\rho\lambda & 0 & 0 & 0 \\ \frac{\lambda H_2}{c^2} & 0 & \rho & 0 & 0 & -\rho\lambda & 0 & 0 \\ \frac{\lambda H_3}{c^2} & 0 & 0 & \rho & 0 & 0 & -\rho\lambda & 0 \\ 0 & -\rho\lambda & 0 & 0 & 1 & 0 & 0 & 0 \\ 0 & 0 & -\rho\lambda & 0 & 0 & 1 & 0 & 0 \\ 0 & 0 & 0 & -\rho\lambda & 0 & 0 & 1 & 0 \\ 0 & 0 & 0 & 0 & 0 & 0 & 0 & 1 \end{pmatrix};$$

for the concrete form of the matrices $\mathcal{A}_k$ we refer to [13]. Note that system (10) coincides with (8) if $\lambda = 0$. The symmetric system (10) is hyperbolic if $\mathcal{A}_0 > 0$ (this also guarantees that $\det \widetilde{S} \neq 0$). Direct calculations show that the last condition is satisfied if inequalities (9) hold together with the additional requirement

$$\rho\lambda^2 < \frac{1}{1 + (c_{\mathrm{A}}^2/c^2)}. \tag{11}$$

Of course, the hyperbolicity condition for system (10) is much more restrictive than the usual natural assumptions (9). It should be also noted that condition (11) guarantees the equivalence of systems (8) and (10) on smooth solutions, provided that $\lambda(\mathbf{U})$ is a smooth function of $\mathbf{U}$, and the analogous assertion can also be proved for current–vortex sheet solutions.

The new symmetric form (10) for the MHD equations was guessed by considering the magnetoacoustics system and using for it a "compressible" counterpart of the so-called cross-helicity integral $d/dt\,\left(\int_{\mathbb{R}^3}(\mathbf{v},\mathbf{H})\,d\mathbf{x}\right) = 0$ taking place for incompressible MHD. This gives a new conserved integral for the linearized constant coefficient MHD equations and, respectively, a concrete form of the matrix $\mathcal{A}_0$, etc.

3 Linear Stability of Planar Current–Vortex Sheets

To reduce the free boundary value problem (1), (3), (4) to that in fixed domains we make, as usual (see, e.g., [8]), the change of variables in $\mathbb{R}\times\mathbb{R}^3$: $\tilde{t} = t\,,\ \tilde{x}_1 = x_1 - f(t,\mathbf{y})\,,\ \tilde{\mathbf{y}} = \mathbf{y}$. Then, $\widetilde{\mathbf{U}}(\tilde{t},\tilde{\mathbf{x}}) := \mathbf{U}(t,\mathbf{x})$ is a smooth vector-function for $\tilde{\mathbf{x}} \in \mathbb{R}^3_\pm$, and problem (1), (3), (4) is reduced to the following problem (we omit tildes to simplify the notation):

$$L(\mathbf{U},\nabla_{t,\mathbf{y}}f)\mathbf{U} = 0 \quad \text{in } [0,T]\times(\mathbb{R}^3_+\cup\mathbb{R}^3_-), \tag{12}$$

$$v_{\mathrm{N}}^\pm = f_t\,, \quad H_{\mathrm{N}}^\pm = 0\,, \quad [q] = 0 \quad \text{on } [0,T]\times\{x_1=0\}\times\mathbb{R}^2, \tag{13}$$

$$\mathbf{U}|_{t=0} = \mathbf{U}_0 \quad \text{in } \mathbb{R}^3_+ \cup \mathbb{R}^3_-, \qquad f|_{t=0} = f_0 \quad \text{in } \mathbb{R}^2. \tag{14}$$

Here

$$L = A_0(\mathbf{U})\partial_t + A_\nu(\mathbf{U},\mathbf{F})\partial_1 + A_2(\mathbf{U})\partial_2 + A_3(\mathbf{U})\partial_3;$$

$A_\nu = A_1(\mathbf{U}) - f_t A_0(\mathbf{U}) - f_{x_2} A_2(\mathbf{U}) - f_{x_3} A_3(\mathbf{U})$ is the *boundary matrix.*

Since the boundary matrix A_ν is singular at $x_1 = 0$ (see [13]), compressible current–vortex sheets are *characteristic* discontinuities.

Let $(\widehat{\mathbf{U}}(t,\mathbf{x}),\ \hat{f}(t,\mathbf{y}))$ be a given vector-function, where $\widehat{\mathbf{U}} = (\hat{p}, \hat{\mathbf{v}}, \widehat{\mathbf{H}}, \widehat{S})$ is supposed to be smooth for $\mathbf{x} \in \mathbb{R}^3_\pm$. Then the linearization of (12)–(14) results in a variable coefficients problem for determining small perturbations $(\delta\mathbf{U}, \delta f)$ (below we drop δ). The interior equations for this problem contain first-order terms for f. To avoid this difficulty we make the change of unknowns (see [1])

$$\bar{\mathbf{U}} = \mathbf{U} - f\widehat{\mathbf{U}}_{x_1}. \tag{15}$$

In terms of the "good unknown" (15) (below we omit bars) and after dropping the zero-order term $f\partial_1\{L(\widehat{\mathbf{U}}, \nabla_{t,\mathrm{y}} f)\widehat{\mathbf{U}}\}$ in the interior equations (as was recommended in [1]) we get the following linear problem

$$L(\widehat{\mathbf{U}}, \nabla_{t,\mathrm{y}}\hat{f})\mathbf{U} + \widehat{C}\mathbf{U} = \mathbf{f} \quad \text{in } [0,T] \times (\mathbb{R}^3_+ \cup \mathbb{R}^3_-), \tag{16}$$

$$\begin{pmatrix} f_t + \hat{v}_2^+ f_{x_2} + \hat{v}_3^+ f_{x_3} - (\hat{v}_\mathrm{N})^+_{x_1} f - v_\mathrm{N}^+ \\ f_t + \hat{v}_2^- f_{x_2} + \hat{v}_3^- f_{x_3} - (\hat{v}_\mathrm{N})^-_{x_1} f - v_\mathrm{N}^- \\ \widehat{H}_2^+ f_{x_2} + \widehat{H}_3^+ f_{x_3} - (\widehat{H}_\mathrm{N})^+_{x_1} f - H_\mathrm{N}^+ \\ \widehat{H}_2^- f_{x_2} + \widehat{H}_3^- f_{x_3} - (\widehat{H}_\mathrm{N})^-_{x_1} f - H_\mathrm{N}^- \\ [q] + f[\hat{q}_{x_1}] \end{pmatrix} = \mathbf{g} \qquad \text{if} \quad x_1 = 0. \tag{17}$$

Here $v_\mathrm{N} = (\mathbf{v}, \widehat{\mathbf{N}})$, $(\hat{v}_\mathrm{N})^\pm_{x_1} = (\hat{v}_\mathrm{N})_{x_1}|_{x_1=\pm 0}$, $q = p + (\widehat{\mathbf{H}}, \mathbf{H})$, etc.; $\widehat{\mathbf{N}} = (1, -\hat{f}_{x_2}, -\hat{f}_{x_2})$; the matrix $\widehat{C}$ depends on $\widehat{\mathbf{U}}$, $\widehat{\mathbf{U}}_t$, $\nabla\widehat{\mathbf{U}}$, $\nabla_{t,\mathrm{y}}\hat{f}$ and can be explicitly written out (see [13]). We introduced the source terms $\mathbf{f}(t,\mathbf{x}) = \mathbf{f}^\pm(t,\mathbf{x})$ for $\mathbf{x} \in \mathbb{R}^n_\pm$ and $\mathbf{g}(t,\mathbf{y})$ to make the interior equations and the boundary conditions inhomogeneous because this is needed to attack the nonlinear problem. Note that if we make suitable assumptions for $\mathbf{f}$, then we can prove that the third and the fourth boundary conditions in (17) with zero source terms as well as the divergent constraint $\partial_1 H_\mathrm{N} + \partial_2 H_2 + \partial_3 H_3 = 0$ are just the restrictions on the initial data for problem (16), (17). This is necessary to have a correct number of boundary conditions (the property of maximality) and to use the divergent constraint for getting a priori estimates for problem (16), (17).[2]

[2] It follows from the divergent constraint that H_N is one of the "noncharacteristic" unknowns. The other ones are q and v_N. If we, however, omit zero-order terms for f in (17) (that is actually inadmissible if we want to use a priori estimates for Nash–Moser iterations), then the divergent constraint just gives us an additional regularity for H_N but we do not really need it for getting a priori estimates.

Let us first freeze the coefficients in (16), (17), omit zero-order terms, and consider the case of homogeneous interior equations and boundary conditions. Then we have the problem that is the result of the linearization of (12), (13) with respect to the piecewise constant solution $\widehat{\mathbf{U}} = \widehat{\mathbf{U}}^{\pm}$ for $x_1 \gtrless 0$ for the planar current–vortex sheet $x_1 = 0$ (without loss of generality we take $\hat{f} = 0$). Since the constant coefficients linearized problem for compressible current–vortex sheets is of independent interest in connection with astrophysical applications mentioned in Sect. 1, we do not introduce in it artificial source terms. At the same time, the a priori estimates proved in [13] for this problem (see just below) can be easily generalized to the case of inhomogeneous problem.

We can show that current–vortex sheets cannot be *uniformly* stable [13].

Lemma 1. *For the constant coefficients linearized problem for planar current–vortex sheets (see (16), (17) with frozen coefficients), the uniform Kreiss–Lopatinski condition is never satisfied.*

Since the linearized problem for current–vortex sheets is a hyperbolic problem with characteristic boundary, there appears a loss of control on derivatives in the normal (x_1-)direction. Therefore, in the theorem below we use the following "nonsymmetric" Sobolev norm:

$$\|\mathbf{U}(t)\|^2_{\widetilde{H}^1(\mathbb{R}^3_+\cup\mathbb{R}^3_-)} = \sum_{\pm}\sum_{|\alpha|\le 1,\ \alpha_1=0}\left(\|\partial^\alpha \mathbf{U}(t)\|^2_{L_2(\mathbb{R}^3_\pm)} + \|\partial_1 \mathbf{U}_{\rm n}(t)\|^2_{L_2(\mathbb{R}^3_\pm)}\right),$$

where $\partial^\alpha = \partial_1^{\alpha_1}\partial_2^{\alpha_2}\partial_3^{\alpha_3}$, $\mathbf{U}_{\rm n} = (q, v_1, H_1)$.

Theorem 1. *Let* $\widehat{\mathbf{H}}^+ \times \widehat{\mathbf{H}}^- \neq 0$ *and either* $[\hat{\mathbf{v}}] = 0$ *(current-sheet) or (5) is satisfied for* $\widehat{\mathbf{U}}$*. Then, for the constant coefficients linearized problem for planar current–vortex sheets the Lopatinski condition is satisfied and the a priori estimates*

$$\|\mathbf{U}(t)\|_{\widetilde{H}^1(\mathbb{R}^3_+\cup\mathbb{R}^3_-)} \le C\|\mathbf{U}_0\|_{\widetilde{H}^1(\mathbb{R}^3_+\cup\mathbb{R}^3_-)}, \tag{18}$$

$$\|f(t)\|_{H^1(\mathbb{R}^2)} \le \|f_0\|_{L_2(\mathbb{R}^2)} + C\|\mathbf{U}_0\|_{\widetilde{H}^1(\mathbb{R}^3_+\cup\mathbb{R}^3_-)} \tag{19}$$

hold for any $t \in [0,T]$, *where* $T > 0$, $C = C(T) = \text{const} > 0$.

The case when $\widehat{\mathbf{H}}^+ \times \widehat{\mathbf{H}}^- = 0$, $\widehat{\mathbf{H}}^{\pm} \times [\hat{\mathbf{v}}] = 0$ *and either*

$$|[\hat{\mathbf{v}}]| < \max\left\{\max\{\gamma^+,\gamma^-\},\, 2\min\{\gamma^+,\gamma^-\}\right\}, \qquad \gamma^{\pm} = \frac{\hat{c}^{\pm}\hat{c}^{\pm}_{\rm A}}{\sqrt{(\hat{c}^{\pm})^2 + (\hat{c}^{\pm}_{\rm A})^2}},$$

or $[\hat{\mathbf{v}}] = 0$ *corresponds to the transition to violent instability, and for the function* $f(t,\mathbf{x}')$ *we have the weaker estimate*

$$\|f(t)\|_{L_2(\mathbb{R}^2)} \le \|f_0\|_{L_2(\mathbb{R}^2)} + C\|\mathbf{U}_0\|_{\widetilde{H}^1(\mathbb{R}^3_+\cup\mathbb{R}^3_-)}.$$

For the detailed proof of Theorem 1 we refer to [13]. The proof is based on the use of system (10). For a certain choice of the constants $\lambda^{\pm} = \lambda(\widehat{\mathbf{U}}^{\pm})$ the boundary conditions are *dissipative* (even conservative). The hyperbolicity condition (11) for the linearized system (10) for $x_1 \gtrless 0$ for chosen $\lambda^{\pm}$ gives the sufficient stability condition (5). Note also that the process of getting the a priori estimates (18), (19) can be formalized by introducing the notations of *dissipative p-symmetrizers* [14]. In fact, for the constant coefficients problem the dissipative (but *not* strictly dissipative [14]) 0-symmetrizer $\mathbb{S} = \left\{\tilde{S}(\widehat{\mathbf{U}}^{+}), \tilde{S}(\widehat{\mathbf{U}}^{-}), \mathbf{R}(\widehat{\mathbf{U}}^{+}), \mathbf{R}(\widehat{\mathbf{U}}^{-})\right\}$ (cf. (10)) has been constructed.

4 The Variable Coefficients Analysis

The main difficulties in the variable coefficients analysis are connected with zero-order terms for f in the boundary conditions (17). Moreover, for constant coefficients we could work in usual Sobolev spaces. In the variable coefficients analysis we have to require a little bit more regularity for solutions. In fact, the natural functional setting is provided by the anisotropic weighted Sobolev spaces H^m_* (see [11, 12] and references therein).

The function space $H^m_*(\Omega)$ (in our case $\Omega = \mathbb{R}^3_+ \cup \mathbb{R}^3_-$) is defined as follows:

$$H^m_*(\Omega) := \left\{u \in L_2(\Omega) \,:\, | \quad \partial^{\alpha}_* \partial^k_1 u \in L_2(\Omega) \quad \text{if} \quad |\alpha| + 2k \leq m \right\},$$

where $\partial^{\alpha}_* = (\sigma(x_1)\partial_1)^{\alpha_1}\partial_2^{\alpha_2}\partial_3^{\alpha_3}$, and $\sigma(x_1) \in C^{\infty}(\mathbb{R}_+) \cap C^{\infty}(\mathbb{R}_-)$ is a monotone increasing function for $x_1 > 0$ and monotone decreasing for $x_1 < 0$ such that $\sigma(x_1) = |x_1|$ in a neighborhood of the origin and $\sigma(x_1) = 1$ for $|x_1|$ large enough. The space $H^m_*(\Omega)$ is normed by

$$\|u\|^2_{m,*} = \sum_{\pm} \sum_{|\alpha|+2k \leq m} \|\partial^{\alpha}_* \partial^k_1 u\|^2_{L_2(\mathbb{R}^3_{\pm})} \,.$$

For solutions of problem (16), (17) we use also the norm

$$|||\mathbf{U}(t)|||^2_m = |||\mathbf{U}(t)|||^2_{m,*} + |||\partial_1 \mathbf{U}_{\mathrm{n}}(t)|||^2_{m-1,*} \,,$$

where $|||(\cdot)(t)|||^2_{m,*} = \sum_{j=0}^k \|\partial^j_t(\cdot)(t)\|^2_{k-j,*}$; $\mathbf{U}_{\mathrm{n}} = (q, v_{\mathrm{N}}, H_{\mathrm{N}})$ is the "noncharacteristic" unknown. Here we use the notations from [11].

For the basic state $(\widehat{\mathbf{U}}, \hat{f})$, we assume that there exists a constant $K > 0$ such that

$$\|\widehat{\mathbf{U}}\|_{W^2_{\infty}(\Omega_T)} + \|\hat{f}\|_{W^3_{\infty}(\partial\Omega_T)} + \|\partial_1 \widehat{\mathbf{U}}^{\pm}\|_{W^2_{\infty}(\partial\Omega_T)} \leq K \,,$$

where $\Omega_T = [0,T] \times \Omega$, $\partial\Omega_T = [0,T] \times \mathbb{R}^2$. We also suppose that the basic state satisfies the Rankine–Hugoniot conditions, the divergent constraint, and the hyperbolicity condition (9). We are now in a position to formulate the main result from [13], but unlike [13] we formulate it for the inhomogeneous problem ($\mathbf{f} \neq 0$, $\mathbf{g} \neq 0$).

Theorem 2. *Let the basic state $(\widehat{\mathbf{U}}, \hat{f})$ satisfy all the assumptions above. Let also there exists a positive constant δ such that*

$$\inf_{\partial\Omega_T} |\hat{\mathbf{h}}^+ \times \hat{\mathbf{h}}^-| \geq \delta > 0 \tag{20}$$

and the condition

$$r^\pm(t, \mathbf{y}) < b^\pm(t, \mathbf{y}) \tag{21}$$

holds for all $t \in [0, T]$ at each point $\mathbf{y} \in \mathbb{R}^2$ such that $\hat{\mathbf{u}}^+(t, \mathbf{y}) \neq \hat{\mathbf{u}}^-(t, \mathbf{y})$, where

$$\hat{\mathbf{h}} = (\widehat{H}_{\mathrm{N}}, \widehat{H}_2, \widehat{H}_3), \quad \hat{\mathbf{u}} = (\hat{v}_{\mathrm{N}} - \hat{f}_t, \hat{v}_2, \hat{v}_3), \quad r^\pm(t, \mathbf{x}) = \sqrt{\hat{\rho}^\pm \left(1 + \frac{(\hat{c}_{\mathrm{A}}^\pm)^2}{(\hat{c}^\pm)^2}\right)},$$

$$b^\pm(t, \mathbf{x}') = \frac{|\hat{\mathbf{h}}^\pm|\,|\sin(\varphi^+ - \varphi^-)|}{|[\hat{\mathbf{u}}]|\,|\sin \varphi^\mp|}, \quad \cos \varphi^\pm(t, \mathbf{x}') = \frac{([\hat{\mathbf{u}}], \hat{\mathbf{h}}^\pm)}{|[\hat{\mathbf{u}}]|\,|\hat{\mathbf{h}}^\pm|}.$$

Then, for problem (16), (17) the a priori estimate

$$\begin{aligned} &|||\mathbf{U}(t)|||_1 + \|f\|_{H^1(\partial\Omega_T)} \\ &\leq C\left\{\|\mathbf{f}\|_{H^1(\Omega_T)} + \|\mathbf{g}\|_{H^2(\partial\Omega_T)} + |||\mathbf{U}_0|||_1 + \|f_0\|_{H^1(\mathbb{R}^2)}\right\} \end{aligned} \tag{22}$$

holds for any $t \in [0, T]$. Here $C = C(T, K)$ is a positive constant independent of the data $(\mathbf{U}_0, f_0, \mathbf{f}, \mathbf{g})$.

For the detailed proof of Theorem 2 we refer to [13]. Inequality (21) appearing in this theorem is the analogue of the stability condition (5) for variable coefficients.[3]

5 The A Priori Tame Estimate

We now just formulate the recent result concerning an a priori *tame* estimate for problem (16), (17). We consider the case of zero initial data for this problem that is usual assumption, and postpone the case of nonzero initial data to the nonlinear analysis (construction of a so-called approximate solution, etc.).

Following [11], we define the space

$$\mathfrak{L}_T^2(H_*^m) = \bigcap_{k=0}^{m} H^k([0, T], H_*^{m-k})$$

equipped with the norm $[u]_{m,*,T}^2 = \int_0^T |||u(t)|||_{m,*}^2\, dt.$

[3] Unlike [13] we assume that the stability condition for variable coefficients is satisfied only on the boundary at each point of the nonplanar current–vortex sheet. This is possible thanks to the use of a kind of cut-off function for $\lambda(\mathbf{U})$, i.e., roughly speaking, the hyperbolicity condition (11) can be imposed only on the boundary.

Theorem 3. *Let $T > 0$ and m is an even number, $m \geq 6$. Assume that the basic state $(\widehat{\mathbf{U}}, \hat{f}) \in \mathfrak{L}^2_T(H^{m+4}_*(\Omega_T)) \times H^{m+4}(\partial\Omega_T)$ satisfies the hyperbolicity condition (9), the Rankine–Hugoniot conditions (3), the divergent constraint* $\operatorname{div} \hat{\mathbf{h}} = 0$, *the assumption (20), the stability condition (21), and*

$$[\widehat{\mathbf{U}}]_{10,*,T} + \|\hat{f}\|_{H^{10}(\partial\Omega_T)} \leq K,$$

where $K > 0$ is a constant. Assume also that the data $(\mathbf{f}, \mathbf{g}) \in \mathfrak{L}^2_T(H^m_*(\Omega_T)) \times H^{m+1}(\partial\Omega_T)$ vanish in the past. Then there exists a positive constant K_0, which does not depend on m and T, and there exists a constant $C = C(K_0) > 0$ such that, if $K \leq K_0$, then there exists a unique solution $(\mathbf{U}, f) \in \mathfrak{L}^2_T(H^m_*(\Omega_T)) \times H^m(\partial\Omega_T)$ to problem (16), (17) that vanishes in the past and obeys the following a priori tame estimate for T small enough:

$$\begin{aligned}[\mathbf{U}]_{m,*,T} + \|f\|_{H^m(\partial\Omega_T)} \leq Ce^{CT}\Big\{&[\mathbf{f}]_{m,*,T} + \|\mathbf{g}\|_{H^{m+1}(\partial\Omega_T)} \\ &+ \left([\mathbf{f}]_{6,*,T} + \|\mathbf{g}\|_{H^7(\partial\Omega_T)}\right)\left([\widehat{\mathbf{U}}]_{m+4,*,T} + \|\hat{f}\|_{H^{m+4}(\partial\Omega_T)}\right)\Big\}\,. \end{aligned} \tag{23}$$

The proof of Theorem 3 is based on the use of Moser-type inequalities following from the Gagliardo–Nirenberg inequality for H^m_* presented in [1].[4] The tame estimate (23) is the basic tool to prove the local-in-time existence of "stable" current–vortex sheet solutions to the nonlinear MHD equations by the Nash–Moser method. The proof has been recently finished [15].

Acknowledgments

This work was supported by the EPSRC research grants No. GR/R79753/01 and GR/S96609/01. The last part of this work (see Sect. 5) was done during the fellowship of the author at the Landau Network–Centro Volta–Cariplo Foundation spent at the Department of Mathematics of the University of Brescia, Italy. The author gratefully thanks Paolo Secchi for many helpful discussions. Also, the author thanks him, Alessandro Morando, Paola Trebeschi, and other people from the Department of Mathematics of the University of Brescia for their kind hospitality.

References

1. Alinhac, S.: Existence d'ondes de raréfaction pour des systèmes quasi-linéaires hyperboliques multidimensionnels. Comm. Partial Differential Equations, **14**, 173–230 (1989)

[4] The Gagliardo–Nirenberg inequality for H^m_* is valid if m is even. This is why we have to assume that m is even. Note also that to work globally in x_1 and to avoid assumptions about compact support we use in the proof the reduction to fixed domains proposed in [9] (see (4.1.2)), which differs from that in the beginning of Sect. 3.

2. Baranov, V.B., Krasnobaev, K.V., Kulikovsky, A.G.: A model of interaction of the solar wind with the interstellar medium. Sov. Phys. Dokl., **15**, 791–793 (1970)
3. Blokhin, A., Trakhinin, Yu.: Stability of strong discontinuities in fluids and MHD. In: Friedlander, S., Serre, D. (eds) Handbook of mathematical fluid dynamics, vol. 1, North-Holland, Amsterdam (2002)
4. Blokhin, A., Trakhinin, Yu.: Stability of strong discontinuities in magnetohydrodynamics and electrohydrodynamics. Nova Science Publishers, New York (2003)
5. Coulombel, J.-F., Secchi, P.: Nonlinear compressible vortex sheets in two space dimensions. Preprint, Seminario Matematico, Brescia (2005)
6. Godunov, S.K.: An interesting class of quasi-linear systems. Soviet Math. Dokl., **2**, 947–948 (1961)
7. Godunov, S.K.: Symmetric form of the equations of magnetohydrodynamics. In: Numerical methods for continuum mechanics, vol. 3, Computer Center of the Siberian Branch of the USSR Academy of Sciences, Novosibirsk (1972)
8. Majda, A.: The stability of multi-dimensional shock fronts. Mem. Amer. Math. Soc., **41**(275) (1983)
9. Métivier, G.: Stability of multidimensional shocks. In: Advances in the Theory of Shock Waves, Progr. Nonlinear Differential Equations Appl., 47, Birkhäuser Boston, Boston (2001)
10. Ruderman, M.S., Fahr H.J.: The effect of magnetic fields on the macroscopic instability of the heliopause. II. Inclusion of solar wind magnetic fields. Astron. Astrophys., **299**, 258–266 (1995)
11. Secchi, P.: Linear symmetric hyperbolic systems with characteristic boundary. Math. Methods Appl. Sci., **18**, 855–870 (1995)
12. Secchi, P.: Some properties of anisotropic Sobolev spaces. Arch. Math., **75**, 207–216 (2000)
13. Trakhinin, Yu.: On existence of compressible current-vortex sheets: variable coefficients linear analysis. Arch. Rational Mech. Anal., **177**, 331–366 (2005)
14. Trakhinin, Yu.: Dissipative symmetrizers of hyperbolic problems and their applications to shock waves and characteristic discontinuities. SIAM J. Math. Anal., **37**, 1988–2024 (2006)
15. Trakhinin, Yu.: The existence of current-vortex sheets in ideal compressible magnetohydrodynamics. In preparation

On the Motion of Binary Fluid Mixtures

K. Trivisa

1 Introduction

A multidimensional model is introduced for the investigation of the dynamic behavior of binary (or multicomponent) mixtures of viscous, compressible reacting fluids. Gas mixtures have a variety of applications in science and engineering, often used to describe phase transition, combustion, the evolution of stars in astrophysics, or the dynamic behavior of semiconductors. In this setting the mixture can be thought as a *continuum* occupying in a certain time t and certain domain $\Omega \in \mathbf{R}^3$.

The state of the mixture is characterized by the density $\varrho = \varrho(t,x)$, the velocity $\mathbf{u} = \mathbf{u}(t,x)$, the temperature $\vartheta = \vartheta(t,x)$, and the density of the individual component $\varrho_s = \varrho_s(t,x), s, s' \in \mathcal{S}$, the collection of species in the mixture. The motion of the mixture is governed by the Navier–Stokes equations, which represent the following:

- *Conservation of mass:*

$$\partial_t \varrho + div(\varrho \mathbf{u}) = 0, \tag{1}$$

- *Balance of momentum:*

$$\partial_t(\varrho \mathbf{u}) + div(\varrho \mathbf{u} \otimes \mathbf{u}) + \nabla p = div\mathbf{S} + \varrho \nabla_x \Psi, \tag{2}$$

- *Poisson equation:*

$$-\Delta \Psi = G, \varrho \quad G > 0, \tag{3}$$

- *Balance of entropy*

$$\partial_t(\varrho_s) + div(\varrho_s \mathbf{u}) + div\left[-\frac{k(\vartheta)}{\vartheta}\nabla\vartheta - (\varrho_s s_z) M d \nabla \log(\varrho_s \vartheta)\right] = \mathbf{r}, \tag{4}$$

- *Species conservation equation*

$$\partial_t(\varrho_s) + div(\varrho_s \mathbf{u}) = w + div\mathcal{F}. \tag{5}$$

Here $s = s(\varrho, \vartheta, \varrho_s)$ denotes the *specific entropy*

$$s = s_F + s_z, \tag{6}$$

s_F the *standard* specific entropy of the fluid and s_z the part that accounts for the presence of individual components; both parts will be specified in the sequel. In addition, $\mathbf{r}$ denotes the *entropy production rate*, $\mathcal{F}$ the *diffusion flux*, and w the *reaction rate*. The quantity

$$\mathbf{q} = \left\{ -\frac{k(\vartheta)}{\vartheta} \nabla\vartheta - (\varrho_s s_z) M d \nabla \log(\varrho_s \vartheta) \right\} \tag{7}$$

is the *heat flux*, which in the present context consists of two parts, the first given by *Fourier's law*

$$\mathbf{q_F} = -\frac{k(\vartheta)}{\vartheta} \nabla\vartheta,$$

while the second part

$$\mathbf{q_z} = -(\varrho_s s_z) M d \nabla \log(\varrho_s \vartheta)$$

takes into consideration the presence of the individual components in the mixture and accounts for the effects of *enthalpy*. Here, and in what follows, M denotes a constant to be specified in the sequel. In addition,

- $\varrho_s(x,t)$ is the density of the individual component

$$\varrho_s = \varrho z, \quad \varrho = \varrho_s + \varrho_{s'}, \tag{8}$$

with ϱ_s, $\varrho_{s'}$ the densities of the two distinct individual components and z the species mass fraction.
- w_s is the reaction rate function denoting the mass of the species produced per unit volume per time unit. In accordance with the *conservation of mass* we have

$$w_s + w_{s'} = 0 \tag{9}$$

where ϱ is the *total mass density*, $\mathbf{u}\otimes\mathbf{u}$ is the tensor product of *velocity* vectors, K the *reaction rate*, and d the *species diffusion* coefficient, which is a function that depends only on the *absolute temperature.* In this article we consider the phase transition of a mixture of fluids (or gases) consisting of two (or more) components. The reaction function w determines the chemical reaction that yields to *phase transition* as soon as the temperature rises enough, and is chosen in such a way so that to guarantee that the entropy production rate that will be specified in the sequel is positive.

We assume that the mixture occupies a bounded domain $\Omega \subset \mathbf{R}^N$, $N = 3$ and the whole system is both mechanically and thermally isolated as described by the following boundary conditions

$$\mathbf{u}|_{\partial\Omega} = 0, \quad \mathbf{q}\cdot\mathbf{n}|_{\partial\Omega} = 0, \quad \mathcal{F}\cdot\mathbf{n}|_{\partial\Omega} = 0. \tag{10}$$

The *total energy* now is given by

$$E = \frac{1}{2}\varrho|\mathbf{u}|^2 + \varrho e + h\varrho_s, \tag{11}$$

where $e = e(\varrho, \vartheta, z)$ is the *internal energy* of the mixture, while h represents the difference of the stoichiometric coefficients for components appearing as reactant and product in the reaction [11].

The total energy is constant of motion, in the sense

$$\frac{d}{dt}\int_\Omega E(t)\,dx = 0. \tag{12}$$

The multicomponent character of the mixture is expressed through rather complex constitutive relations and thermodynamic relations.

The *internal energy* e is related to the *specific entropy* s through the following thermodynamic relation

$$\vartheta\mathbf{D}s = \mathbf{D}e + p\mathbf{D}\left(\frac{1}{\varrho}\right) + B(\vartheta, \varrho_s)\mathbf{D}z, \tag{13}$$

with B an appropriate function that typically depends on the constitutive relations and the specific physical context.

1.1 Constitutive Relations and Structural Conditions

- The viscous stress tensor $\mathbf{S}$ is given by the *Newton's viscosity formula*

$$\mathbf{S} = \mu(\vartheta)\left(\nabla\mathbf{u} + \nabla\mathbf{u}^T - \frac{2}{3}div\mathbf{u}\,\mathbf{I}\right) + \zeta(\vartheta)\,div\mathbf{u}\,\mathbf{I}, \tag{14}$$

where the *shear viscosity* μ and the *bulk viscosity* ζ are supposed to be nonnegative and continuously differentiable functions of the absolute temperature satisfying

$$\begin{cases} 0 < \underline{\mu}(1+\vartheta^\alpha) \le \mu(\vartheta) \le \bar{\mu}(1+\vartheta^\alpha), \\ 0 < \underline{\zeta}\vartheta^\alpha \le \zeta(\vartheta) \le \bar{\zeta}(1+\vartheta^\alpha) \end{cases} \tag{15}$$

for $\alpha \ge \frac{1}{2}$.

- The *diffusion flux* $\mathcal{F}_s$ is determined by the law

$$\mathcal{F}_s = -D\varrho_s\mathcal{V}_s, \tag{16}$$

with $\mathcal{V}_s$ denoting the so–called *diffusion velocities* and are given by

$$\mathcal{V}_s = \nabla\log(\varrho_s\vartheta), \tag{17}$$

and D the species diffusion coefficient, which in accordance with the kinetic theory of gases is a function of the form

$$D(\varrho_s) \approx \frac{d}{\varrho_s}.$$

Taking the above into consideration, here and in what follows, the diffusion flux $\mathcal{F}_s$ is of the form

$$\mathcal{F}_s = -d\nabla \log(\varrho_s \vartheta). \tag{18}$$

In accordance with the *conservation of mass* the diffusion fluxes $\mathcal{F}_s$ typically compensate with each other, in the sense that

$$\mathcal{F}_s + \mathcal{F}_s' = 0. \tag{19}$$

In the present framework this is not the case, however, because of the effect of chemical reaction in the mixture, which often results to the elimination of certain components as soon as the mixing begins. Remarks on the properties of the diffusion velocities (17) and their compatibility with the conservation of mass and in particular relation (19) can be found in [11] (see also the discussion in [14], [15], [16]). In fact (19) is a consequence of a constraint typically imposed on some coefficients in the *diffusion fluxes* $\mathcal{F}_s$ known in the literature as *species thermal diffusion ratios*, which for simplicity of the presentation are omitted in the above relations (we refer the reader to Ern and Giovangili [7], Weldmann and Trübenbacher [17] for further details).
The *species diffusion coefficient* d is a continuously differentiable function depending only on the absolute temperature in the following way

$$0 < \underline{d} < d(\vartheta) \leq \bar{d}(1 + \vartheta^3) \tag{20}$$

for all $\vartheta > 0$.

- The pressure p satisfies a rather general equation of state

$$p = p_R + p_B + p_e, \tag{21}$$

where the term p_B satisfies the *Boyle's law* for the individual species

$$p_B = \frac{\varrho_s R\vartheta}{m_s} + \frac{\varrho_{s'} R\vartheta}{m_{s'}}, \tag{22}$$

with m_i denotes the molecular weight of the ith component, $p_e = p_e(\varrho)$ is the so-called *elastic pressure* whose properties will be discussed in the sequel, while p_R accounts for *radiation effects*. We remark that the *elastic pressure* may include higher order terms as in the *Beattie–Bridgman model,* where the state equation for the pressure includes an *elastic* component of the form

$$p_e(\varrho) = \beta_1 \varrho^2 + \beta_2 \varrho^3 + \beta_3 \varrho^4,$$

for appropriate constants β_i, [1], [8].
The *radiation pressure* on the other hand help us accommodate the possible presence of *photons* in our mixture accounting for the *quantum effects* associated with the binary gas mixture and is given through the *Stefan–Boltzmann law*

$$p_R = \frac{a}{3}\vartheta^4, \quad \text{with} \quad a > 0 \quad \textit{a constant.}$$

The underlying assumption here (cf. [6], [10], [13]) is that the *high temperature radiation* is at thermal equilibrium with the fluid.
Analogously, standard thermodynamic relations require that the specific internal energy of the fluid be also augmented, by the term

$$e_R = e_R(\varrho, \vartheta) = \frac{a}{\varrho}\vartheta^4.$$

We remark that *radiation effects* are of particular interest in *astrophysical plasma models.*
The pressure therefore satisfies, in the present context, the general law

$$p = \frac{a}{3}\vartheta^4 + \frac{\varrho_s R\vartheta}{m_s} + \frac{\varrho_{s'} R\vartheta}{m_{s'}} + p_e(\varrho).$$

Taking into consideration (8), the pressure law now takes the form

$$p = \frac{a}{3} + \frac{\varrho}{m_{s'}} R\vartheta + R\left(1 - \frac{m_s}{m_{s'}}\right)\frac{\varrho_s}{m_s}\vartheta + p_e(\varrho).$$

Without loss of generality, we assume that the molecular weight of the first component is significantly less than the molecular weight of the other, namely $m_s << m_{s'}$, which implies that the constant

$$M = \left(1 - \frac{m_s}{m_{s'}}\right)\frac{1}{m_s} > 0 \tag{23}$$

and so the pressure law becomes

$$p = \frac{a}{3}\vartheta^4 + \frac{\varrho}{m_{s'}} R\vartheta + MR\varrho_s\vartheta + p_e(\varrho). \tag{24}$$

Motivated by the above discussion, here and in what follows, we consider the following rather general equation of state for the pressure

$$p(\varrho, \vartheta, \varrho_s) = p_{radiation} + p_{thermal} + p_{elastic}$$

$$\begin{cases} p(\varrho, \vartheta, \varrho_s) = p_R(\vartheta) + \vartheta\left[\frac{R}{m_2} p_\theta(\varrho) + MRp_\theta(\varrho_s)\right] + \left[p_e(\varrho) + p_e(\varrho_s)\right] \\ \qquad p_R(\vartheta) = \frac{a}{3}\vartheta^4, \end{cases} \tag{25}$$

with p_e the *elastic pressure* and p_θ the *thermal pressure*, which are functions satisfying the following properties.
The *elastic* pressure p_e and the *thermal* pressure p_θ are continuously differentiable functions of the density and in addition satisfy certain coercivity properties as in [6], [5], namely

$$\begin{cases} p_e(0) = 0, \ p'_e(\varrho) \geq a_1\varrho^{\gamma-1} - c_1, \\ p_e(\varrho) \leq a_2\varrho^{\gamma} + c_2, \quad a_1 > 0. \end{cases} \begin{cases} p_\theta(0) = 0, \ p'_\theta(\varrho) \geq 0 \\ p_\theta(\varrho) \leq a_3\varrho^{\Gamma} + c_3, \end{cases} \tag{26}$$

with

$$\gamma \geq 2 \quad \gamma > \frac{4\Gamma}{3}.$$

• The *internal energy* e satisfies a general constitutive law $e = e(\varrho, \varrho_i, \vartheta)$, namely

$$e = e(\varrho, \vartheta, \varrho_s) = [P_e(\varrho) + P_e(\varrho_s)] + \vartheta \left[\frac{c_v}{m_2} + Mc_v\frac{\varrho_s}{\varrho}\right] + a\frac{\vartheta^4}{\varrho}, \tag{27}$$

with

$$P_e(\varrho) = \int_1^{\varrho} \frac{p_e(s)}{s^2}\, ds \tag{28}$$

denoting the so–called *elastic potential* and c_v the *specific heat*, which can be either or a Lipschitz function of the mass fractions z as in [5].
In the present physical context the *balance of energy* reads

$$\partial_t(\varrho e) + div(\varrho e\mathbf{u}) + div(\tilde{\mathbf{q}}) = \mathbf{S} : \mathbf{u} - p div\mathbf{u}, \tag{29}$$

with $\tilde{\mathbf{q}}$ given by

$$\tilde{\mathbf{q}} = -k(\vartheta)\nabla\vartheta - \frac{9\gamma - 5}{4} M d\varrho_s\vartheta\nabla\log(\varrho_s\vartheta). \tag{30}$$

Multiplying the *continuity equation* by $(\varrho P_e(\varrho))'$ we obtain

$$\partial_t(\varrho P_e(\varrho)) + div(\varrho P_e(\varrho)\mathbf{u}) + p_e(\varrho) div\mathbf{u} = 0 \tag{31}$$

and so the *balance of energy* yields the *thermal equation*

$$\begin{aligned} \partial_t \left[a\vartheta^4 + \frac{c_v}{m_{s'}}\varrho\vartheta + c_v M\varrho_s\vartheta\right] &+ div\left[\left(a\vartheta^4 + \frac{c_v}{m_{s'}}\varrho\vartheta + c_v M\varrho_s\vartheta\right)\mathbf{u}\right] \\ &+ div(\tilde{\mathbf{q}}) = \mathbf{S} : \nabla\mathbf{u} - \left[\frac{a}{3}\vartheta^4 + \frac{R}{m_{s'}}\vartheta p_\theta(\varrho) + M\vartheta p_\theta(\varrho_s)\right] div\mathbf{u} \\ &+ (\varrho_s P_e(\varrho_s))' div(\nabla\log(\varrho_s\vartheta)). \end{aligned} \tag{32}$$

• In the specific physical context described here and in accordance with (7) the *specific entropy* reads

$$s(\varrho, \vartheta, z) = s_F(\varrho, \vartheta) + s_z(\varrho, \vartheta, z), \tag{33}$$

with

$$\left\{ \begin{array}{l} s_F = \frac{4a}{3}\frac{\vartheta^3}{\varrho} + \frac{c_v}{m_{s'}} \log(\vartheta) - P_\theta(\varrho), \\ \varrho s_z = M c_v \varrho_s \log(\vartheta) - M R \varrho_s p_\theta(\varrho_s) - M R \varrho_s \log(\varrho_s), \end{array} \right\} \tag{34}$$

where

$$P_\theta(\varrho) = \int_1^\varrho \frac{p_\theta(s)}{s^2}\, ds.$$

Note that in the case of one component fluid, the part s_z of the *entropy* is typically not present. The reader should contrast the form of the entropy in the present context to the one presented in [6], [4], [5], [8].

• For *smooth* solutions the *entropy production rate* $\mathbf{r}$ is now expressed by

$$\mathbf{r} = \frac{1}{\vartheta}\left(\mathbf{S} : \nabla \mathbf{u} + \frac{k(\vartheta)|\nabla\vartheta|^2}{\vartheta} + M d \vartheta |\nabla \log(\varrho_s \vartheta)|^2 \right), \tag{35}$$

and the entropy equation reads

$$\partial_t(\varrho_s) + div(\varrho s \mathbf{u}) + div\left[-\frac{k(\vartheta)}{\vartheta}\nabla\vartheta - (\varrho_s s_z) M d \nabla \log(\varrho_s \vartheta) \right] = \mathbf{r}.$$

We refer the reader to [14], [15] for further discussion on the derivation of the *balance of entropy equation.*

In the case, of a general nonsmooth motion now, and in the spirit of the second law of thermodynamics, we can only assert that

$$\partial_t[\varrho\,(s_F + s_z)] + div\,[\varrho\,(s_F + s_z)\mathbf{u}] + div\left(-\frac{k(\vartheta)}{\vartheta}\nabla\vartheta - (\varrho_s s_z) d M \nabla \log(\varrho_s \vartheta) \right)$$
$$\geq \frac{1}{\vartheta}\left(\mathbf{S} : \nabla \mathbf{u} + \frac{k(\vartheta)|\nabla\vartheta|^2}{\vartheta} + d M \vartheta |\nabla \log(\varrho_s \vartheta)|^2 \right). \tag{36}$$

Here, the *heat conductivity* $k(\vartheta)$ obeys the rule

$$\left\{ \begin{array}{l} k(\vartheta) = k_C(\vartheta) + \sigma\vartheta^3, \\ 0 < \underline{k}_C \leq k_C(\vartheta) \leq \overline{k}_C(1 + \vartheta^3), \end{array} \right. \tag{37}$$

where the term $\{\sigma\vartheta^3\}$ with $\sigma > 0$ accounts for the *radiative* effects.

The goal of this work is to establish the global existence of weak solutions to an initial boundary value problem with the boundary conditions (12) and *large* initial data given by

$$\left\{ \begin{array}{ll} \varrho(0, \cdot) = \varrho_0, & (\varrho\,\mathbf{u})(0, \cdot) = \mathbf{m}_0, \\ (\varrho\,\vartheta)(0, \cdot) = \chi_0, & (\varrho_s)(0, \cdot) = \psi_0, \end{array} \right. \tag{38}$$

together with the compatibility condition

$$\mathbf{m}_0 = 0,\ \chi_0 = 0,\ \psi_0 = 0 \quad \text{on the set} \quad \{x \in \Omega |\ \varrho_0(x) = 0\}. \tag{39}$$

1.2 Variational Formulation and Main Results

For the solvability of the initial boundary value problem described above we rely on the concept of *variational solutions*, in the sense of Feireisl [8]. We remark that this concept is in the spirit of the notion of a *Leray solution* in the context of incompressible fluids (we refer the reader also to P.-L. Lions [12] for relevant discussion).

Definition 1. *Let $\Omega \subset \mathbf{R}^3$ be a bounded domain. We say that $\{\varrho, \mathbf{u}, \vartheta, \varrho_s\}$ is a variational solution to on $(0,T) \times \Omega$ satisfying the initial conditions*

$$\varrho(0,\cdot) = \varrho_0,\ \ \varrho_s(0,\cdot) = \varrho_{s,0},\ \ \mathbf{u}(0,\cdot) = \mathbf{u}_0,\ \ \vartheta(0,\cdot) = \vartheta_0, \tag{40}$$

provided the following holds:
• *The continuity equation (1) is satisfied in the sense of renormalized solutions, specifically,*

$$\rho \in C([0,T]; L^1(\Omega)) \cap L^\infty(0,T; L^\gamma(\Omega)),\ \rho(0,\cdot) = \rho_0$$

satisfying the continuity equation (1) in the sense of $\mathcal{D}'((0,T) \times \mathbf{R}^3)$ provided that $\rho, \mathbf{u}$ were extended to be zero outside Ω. Now,

$$\mathbf{u} \in L^2(0,T; W_0^{1,2}(\Omega; R^3)),\ \ \varrho\mathbf{u} \in L^\infty(0,T; L^1(\Omega; R^3)),$$

and the integral identity

$$\int_0^T \int_\Omega \left(\varrho B(\varrho)\partial_t \varphi + \varrho B(\varrho)\mathbf{u} \cdot \nabla_x \varphi - b(\varrho) div_x \mathbf{u}\right)\, dxdt = \\ -\int_\Omega \varrho_0 B(\varrho_0)\varphi(0,\cdot)\, dx \tag{41}$$

holds for any test function $\varphi \in \mathcal{D}([0,T) \times \bar{\Omega})$ and any

$$b \in BC[0,\infty),\ \ B(\varrho) = B(1) + \int_1^\varrho \frac{b(z)}{z^2}\, \mathrm{d}z. \tag{42}$$

• *The density of the individual component ϱ_s is a nonnegative measurable function belonging to the space*

$$\varrho_s \in L^\infty(0,T; L^\gamma(\Omega)),\ \log(\varrho_s) \in L^2(0,T; W^{1,2}(\Omega)),$$

while the temperature ϑ is a nonnegative function such that

$$\vartheta, \log(\vartheta) \in L^2(0,T; W^{1,2}(\Omega)),$$

and the integral identity

$$\int_0^T \int_\Omega \varrho_s \partial_t \varphi + \varrho_s \mathbf{u} \cdot \nabla_x \varphi + \mathcal{F} \cdot \nabla_x \varphi \, dxdt = \int_0^T \int_\Omega \omega \varphi \, dxdt$$
$$- \int_\Omega \varrho_{s,0} \varphi(0, \cdot) dx, \tag{43}$$

holds for any test function $\varphi \in \mathcal{D}'([0,T) \times \bar{\Omega})$.

• *The momentum balance equation (2) is satisfied in the sense of distributions. Moreover, the pressure* $p \in L^1((0,T) \times \Omega)$ *is related to the state variables* ϱ, ϱ_s, *and* ϑ *through the constitutive equation (24), the viscous stress tensor* $\mathbf{S} \in \mathbf{L^1(0,T;L^1(\Omega;R^{3\times 3}))}$ *is given by Newton's law of viscosity (16),*

$$\varrho \mathbf{u} \otimes \mathbf{u} \in L^1(0,T;L^1(\Omega;R^{3\times 3})),$$

while the gravitational potential Ψ *determined by (3) considered on the whole space* $\mathbf{R}^3$.

• *The entropy* ϱs *is determined by the formula (34), the density of the individual component* ϱ_s *as well as the absolute temperature are positive a.a. on* $(0,T) \times \Omega$ *and the integral inequality*

$$\int_0^T \int_\Omega \{(\varrho s)\, \partial_t \varphi + (\varrho s) \mathbf{u} \cdot \nabla \varphi + \mathbf{q} \cdot \nabla \varphi\} \, dxdt$$
$$\leq \int_0^T \int_\Omega \left\{ -\frac{\mathbf{S} : \nabla \mathbf{u}}{\vartheta} - \frac{k(\vartheta)|\nabla \vartheta|^2}{\vartheta^2} - Md(\vartheta)|\nabla \log(\varrho_s \vartheta)|^2 \right\} \varphi \, dx \, dt, \tag{44}$$

holds for any nonnegative function $\varphi \in \mathcal{D}((0,T) \times \mathbf{R^N})$.

Moreover,

$$\operatorname{ess}\lim_{t\to 0+} \int_\Omega \varrho s(t) \varphi \, dx \geq \int_\Omega \varrho_0 s_0 \varphi \, dx, \quad \text{for any nonnegative } \varphi \in \mathcal{D}(\Omega),$$

where

$$\varrho_0 s_0 = \frac{4a}{3} \vartheta_0^3 + \frac{c_v}{m_{s'}} \varrho_0 \log(\vartheta_0) - \varrho_0 P_\theta(\varrho_0) + Mc_v \varrho_{s,0} \log(\vartheta_0)$$
$$-MR\varrho_{s,0} \log(\vartheta_0) - MR\varrho_{s,0} \log(\varrho_{s,0}).$$

• *The total energy of the system is conserved, specifically,*

$$E(t) = \int_\Omega \left(\frac{1}{2} \varrho |\mathbf{u}|^2 + \varrho e(\varrho, \varrho_s, \vartheta) + h \varrho_s \right)(t) dx =$$
$$\int_\Omega \left(\frac{1}{2} \varrho_0 |\mathbf{u}_0|^2 + \varrho_0 e(\varrho_0, \varrho_{s,0}, \vartheta_0) + h \varrho_{s,0} \right) dx = E_0, \tag{45}$$

where the internal energy density ϱe *is given by (27).*

2 Main Result

We are now ready to state the existence result for the initial boundary value problem introduced in Sect. 1.

Theorem 1. *Let $\Omega \subset \mathbf{R}^3$ be a bounded domain with a boundary $\partial\Omega \in C^{2+\nu}, \nu > 0$. Suppose that the pressure p is determined by the equation of state (25), with $a > 0$, p_e and p_θ satisfying (26). In addition, let the viscous stress tensor $\mathbf{S}$ be given by (14), where μ and ζ are continuous differentiable globally Lipschitz functions of ϑ satisfying (17) for $\frac{1}{2} \leq \alpha \leq 1$. Similarly, let the heat flux $\mathbf{q}$ be given by (7) with $k(\vartheta)$ satisfying (37). Finally, assume that the initial data $\varrho_0, \mathbf{m}_0, \vartheta_0, \varrho_{s,0}$ satisfy*

$$\begin{cases} \varrho_0 \geq 0, \ \varrho_0 \in L^\gamma(\Omega), \\ \mathbf{m}_0 \in [L^1(\Omega)]^3, \ \frac{|\mathbf{m}_0|^2}{\varrho_0} \in L^1(\Omega), \\ \vartheta_0 \in L^\infty(\Omega), \ 0 < \underline{\vartheta} \leq \vartheta_0(x) \leq \bar{\vartheta} \ \text{for a.e. } x \in \Omega, \\ \varrho_{s,0} \geq 0, \ \varrho_{s,0} \in L^\gamma(\Omega). \end{cases} \tag{46}$$

Then, for any given $T > 0$ the initial boundary value problem (1)–(5) together with (10)–(46) has a variational solution on $(0, T) \times \Omega$.

Proof. For the proof we follow similar line of argument to the one presented in [15]. We mention here the main ingredients of this proof, which involve the construction of a three level approximating scheme obtained by the addition of special ε-terms (*artificial viscosity*) and δ-terms (*artificial pressure*) in the original system of equations. The former terms guarantee that the *a priori* estimates hold true, while the latter ensure that the pressure estimates and the estimates on the temperature are not lost as the *artificial viscosity* ε vanishes. At the first level the solution sequence $(\varrho_n, \mathbf{u}_n, \vartheta_n, \varrho_{s_n})$ is smooth. The resulting sequence of approximate problems can be resolved in the spirit of the discussion presented in [14], [15], where a relevant system is considered (see also [6], [8], [4], [5]) and therefore the details are here omitted. At this level, energy estimates can be obtained as in [14]. These estimates are independent of n and therefore are valid for the sequence $(\varrho_\varepsilon, \mathbf{u}_\varepsilon, \vartheta_\varepsilon, \varrho_{s_\varepsilon})$, which is obtained as $n \to \infty$. These bounds on our solution sequence are crucial in order to obtain the necessary compactness. Refined estimates are then needed in order to improve the integrability properties of the pressure, which at this point belongs in the nonreflexive space

$$p \in L^\infty(0, T, L^1(\Omega)).$$

Next we let the artificial viscosity ε go to zero and we recover the original system by letting δ go to zero . Both processes are very delicate due to the *oscillation* effects on ϱ and *concentration* effects on the temperature ϑ and pressure p. To deal with these difficulties we employ a variety of techniques in the spirit of Feireisl [8] and P.-L. Lions [12] by accommodating them appropriately to the *new* context. This work extends the earlier work [4], [5] on phase transition models since it now accommodates mixtures of compressible fluids (*gases*) that consist of two (or more) individual species taking into consideration the distinct feature of the individual components as given by their *density fraction* as well as their *molecular weights.* We refer the reader also to the article [14] where a model for the dynamics of *liquid–vapor* phase transition has been analyzed. A relevant one-dimensional model was introduced by Chen, Hoff and Trivisa [3] for the investigation of viscous, compressible, polytropic gases. For related articles in the literature we refer the reader to Ducomet and Feireisl [6], Feireisl [8], [9], and Feireisl and Novotný [10]. □

2.1 Other Equations of State

Other equations of state for multicomponent gas mixtures are presented in [16], where the global existence of solutions with large initial data is also obtained. In this context the state equation for the pressure is given by

$$p = \sum_{s=1}^{n} \frac{2}{3} \frac{\varrho_s e}{m_s} + \frac{a}{3}\vartheta^4, \quad s \in S. \tag{47}$$

If we denote by m_j the molecular weight of the heaviest component in the mixture, that is $m_j >> m_s$ far all s, then

$$p = \frac{2}{3}\frac{\varrho e}{m_j} + \sum_{j \neq s=1}^{n} \frac{2}{3} L_s \varrho_s e + \frac{a}{3}\vartheta^4, \tag{48}$$

where

$$L_s = \frac{1}{m_s}\left(1 - \frac{m_s}{m_j}\right).$$

Taking the above discussion into consideration the pressure $p = p(\varrho, \vartheta, \varrho_s)$ of the mixture is given as the sum of three parts

$$p = p_F + p_s + p_R, \tag{49}$$

where

$$\begin{cases} p_F(\varrho,\vartheta) = \frac{2}{3}\varrho e_F, \quad \text{or} \quad p_F(\varrho,\vartheta) = \vartheta^{5/2} P\left[\frac{\varrho}{\vartheta^{3/2}}\right], \\ p_s(\varrho_s,\vartheta) = \sum_k L_k \vartheta^{5/2} P_s\left[\frac{\varrho_s}{\vartheta^{5/2}}\right], \\ p_R(\vartheta) = \frac{a}{3}\vartheta^4, \end{cases} \tag{50}$$

where the term p_R accounts for *radiation effects* and p_k accounts for the presence of the individual components. In the above relations, P and P_s are given functions that satisfy the following assumptions

$$\begin{cases} P \in C^1[0,\infty),\, P(0) = 0, P'(Y) \geq 0, \quad \text{for all} \quad Y > 0, \\ 0 < \frac{5}{3}P(Y) - P'(Y)Y \leq cY^{\frac{1}{3}} \quad \text{for all} \quad Y > 0, \\ \lim_{Y\to\infty} \frac{P(Y)}{Y^{\frac{5}{3}}} \equiv P_\infty. \end{cases} \tag{51}$$

In accordance with the above discussion and with standard thermodynamic principles the *specific internal energy* is now given as

$$e = e_F + e_s + e_R \tag{52}$$

with

$$e_F(\varrho, \vartheta) = \frac{3}{2}\frac{p_F(\varrho, \vartheta)}{\varrho}, \quad e_s(\varrho_s, \vartheta) = \frac{3}{2}\frac{p_s(\varrho_s, \vartheta)}{\varrho}, \quad e_R(\varrho, \vartheta) = \frac{a}{\varrho}\vartheta^4. \tag{53}$$

The detailed analysis is presented in [16].

References

1. E. Becker. *Gasdynamik.* Teubner–Verlag, Stuttgart, 1966.
2. G.-Q. Chen, D. Hoff, and K. Trivisa, On the Navier–Stokes equations for exothermically reacting compressible fluids, *Acta Math. Appl. Sinica*, **18** (2002), 15–36.
3. G.-Q.Chen, D. Hoff, and K. Trivisa, Global solutions to a model for exothermically reacting, compressible flows with large discontinuous initial data, *Arch. Ration. Mech. Anal.* **166** (2003), 321–358.
4. D. Donatelli and K. Trivisa, A multidimensional model for the combustion of compressible fluids. Arch. Ration. Mech. Anal. **185, no. 3** (2007), 379–408.
5. D. Donatelli and K. Trivisa, On the motion of a viscous, compressible, radiative–reacting gas. To appear *Comm. Math. Phys.* (2006).
6. B. Ducomet and E. Feireisl, On the Dynamics of Gaseous Stars. *Arch. Ration. Mech. Anal.*, **174**, (2004), 221–266.
7. A. Ern and V. Giovangigli, *Multicomponent Transport Algorithms,* Lecture Notes in Physics, New Series "Monographs", **m 24**, Springer–Verlag, Berlin, (1994).
8. E. Feireisl, *Dynamics of viscous compressible fluids.* Oxford University Press, Oxford, 2004.
9. E. Feireisl, *On the motion of a viscous, compressible and heat conducting fluid.* Indiana Univ. Math. J. **53**, no. 6 (2004) 1705–1738.
10. E. Feireisl, A. Novotný. On a simple model of reacting flows arising in astrophysics. *Proc. Roy. Soc. Edinburgh, Sect. A* **135** (2005) 1169–1194.
11. V. Giovangigli, *Multicomponent Flow Modeling*, Modeling and Simulation in Science, Engineering and Technology, Birkhäuser (1999).
12. P.L. Lions, *Mathematical Topics in Fluid Mechanics*, Vol. **2**, Oxford University Press: New York, 1998.
13. J. Oxenius. *Kinetic Theory of Particles and Photons*, Springer Series in Electrophysics **20**, Springer–Verlag (1986).
14. K. Trivisa. On the Dynamics of Liquid-Vapor Phase Transition. To appear in SIAM J. Math. Anal. (2007).
15. K. Trivisa. On binary fluid mixtures. To appear in Contemp. Math. Amer. Math. Soc. (2007).
16. K. Trivisa. On multicomponent fluid mixtures. In preparation.
17. L. Waldmann und E. Trübenbacher, *Formale Kinetische Theorie von Gasgemischen aus Anregbaren Molekülen,* Zeitschr. Naturforschg., **17a**, (1962), pp. 363–376.

On the Optimality of the Observability Inequalities for Kirchhoff Plate Systems with Potentials in Unbounded Domains

X. Zhang and E. Zuazua

Summary. In this paper, we derive a sharp observability inequality for Kirchhoff plate equations with lower order terms in an unbounded domain Ω of $\mathbb{R}^n$. More precisely, when the observation is assumed to be located in a subdomain ω such that $\Omega \setminus \overline{\omega}$ is bounded and the observation time $T > 0$ is sufficiently large, we establish an observability estimate with an explicit observability constant for Kirchhoff plate systems with an arbitrary finite number of components and in any space dimension with lower order bounded potentials. Also, when $\Omega = \mathbb{R}^n$, by means of the Meshkov construction for the bi-Laplacian equation, we prove the optimality of this estimate for systems of more than two components and in even space dimensions $n \geq 2$.

1 Introduction

Let $n \geq 1$ and $N \geq 1$ be two integers. Let D be a bounded convex domain in $\mathbb{R}^n$ with C^4 boundary Γ, $\Omega = \mathbb{R}^n \setminus \overline{D}$, and ω be an open subset in Ω so that $\Omega \setminus \overline{\omega}$ is bounded. Obviously, when $D = \emptyset$, one has $\Omega = \mathbb{R}^n$. Let $T > 0$ be given and sufficiently large. Put $Q \stackrel{\triangle}{=} (0, T) \times \Omega$ and $\Sigma \stackrel{\triangle}{=} (0, T) \times \Gamma$. For simplicity, we will use the notation $y_i = \frac{\partial y}{\partial x_i}$, where x_i is the ith coordinate of a generic point $x = (x_1, \cdots, x_n)$ in $\mathbb{R}^n$. Throughout this paper, we will use $C = C(T, \Omega, \omega)$ to denote generic positive constants depending on their arguments, which may vary from line to line.

Denote by $\mathcal{D}'(\Omega)$ the usual space of distributions in Ω. Set

$$\mathcal{H} \stackrel{\triangle}{=} \left\{ (\varphi, \psi) \in \mathcal{D}'(\Omega) \times \mathcal{D}'(\Omega) \middle| \; \varphi, \Delta\varphi, \psi \in H_0^1(\Omega), \; \Delta\psi \in L^2(\Omega) \right\}^N.$$

Clearly, $\mathcal{H}$ is a Hilbert space with the following norm:

$$||(\varphi, \psi)||_{\mathcal{H}} \stackrel{\triangle}{=} \sqrt{||\varphi||^2_{(H^1(\Omega))^N} + ||\Delta\varphi||^2_{(H^1(\Omega))^N} + ||\psi||^2_{(H^1(\Omega))^N} + ||\Delta\psi||^2_{(L^2(\Omega))^N}}.$$

We consider the following $\mathbb{R}^N$-valued plate system with a potential

$$a \in L^\infty(0,T; L^p(\Omega; \mathbb{R}^{N\times N})) \quad \text{for some} \quad p \in [n/3, \infty]:$$

$$\begin{cases} y_{tt} + \Delta^2 y - \Delta y_{tt} + ay = 0 & \text{in } Q, \\ y = \Delta y = 0 & \text{on } \Sigma, \\ y(0) = y^0, \quad y_t(0) = y^1 & \text{in } \Omega, \end{cases} \tag{1}$$

where $y = (y_1, \cdots, y_N)^\top$, and the initial datum (y^0, y^1) is supposed to belong to $\mathcal{H}$, the state space of system (1). It is easy to show that system (1) admits one and only one weak solution $y \in C([0,T]; \mathcal{H})$.

In what follows, we shall denote by $|\cdot|$ and $||\cdot||_p$ the (canonical) norms on $\mathbb{R}^N$ and $L^\infty(0,T; L^p(\Omega; \mathbb{R}^{N\times N}))$, respectively.

We shall study the observability constant $K(a)$ of system (1), defined as the smallest constant such that the following observability estimate for system (1) holds:

$$\begin{aligned} &||(y^0, y^1)||^2_{\mathcal{H}} \\ &\leq K(a) \int_0^T \int_\omega \left(|y|^2 + |\nabla y|^2 + |\Delta y|^2 + |\nabla \Delta y|^2\right) dxdt, \quad \forall\, (y^0, y^1) \in \mathcal{H}. \end{aligned} \tag{2}$$

This inequality, the so-called *observability inequality*, allows estimating the total energy of solutions in terms of the energy localized in the observation subdomain ω. It is relevant for control problems. In particular, in this linear setting, this (observability) inequality is equivalent to the so-called exact controllability property, i.e., that of driving solutions to rest by means of control forces localized in ω (see [4, 10]). This type of inequality, with explicit estimates on the observability constant, is also relevant for the control of semilinear problems ([9]). Similar inequalities are also useful for solving a variety of Inverse Problems ([7]).

Obviously the observability constant $K(a)$ in (2) not only depends on the potential a, but also on the domains Ω and ω and on time T. The main purpose of this paper is to analyze only its explicit and sharp dependence on the potential a.

The main tools to derive the explicit observability estimates are the so-called *Carleman inequalities*. Here we have chosen to work in the space $\mathcal{H}$ in which Carleman inequalities can be applied more naturally. But some other choices of the state space are possible. For example, one may consider similar problems in state spaces of the form $(H^1_0(\Omega))^N \times (L^2(\Omega))^N$ or $\left\{\varphi \in \mathcal{D}'(\Omega) \middle|\ \varphi \in H^1_0(\Omega),\ \Delta\varphi \in L^2(\Omega)\right\}^N \times (H^1_0(\Omega))^N$, where the Kirchhoff plate system is also well posed. But the corresponding analysis on the observability constants, in turn, is technically more involved.

The same problem was considered in [8] in the case of bounded domains. In that paper, under some assumptions on the observation domain and time, an observability inequality of the form (2) was established with the following estimate on the observability constant:

$$K(a) \leq C \exp\left(C||a||_p^{\frac{1}{3-5n/2p}}\right). \tag{3}$$

In this paper we show that the same estimate holds in unbounded domains. Similar (boundary and/or internal) observability estimates (in suitable spaces) have been established for the heat and wave equations in [2], and for the Euler–Bernoulli plate equations in [3]. Furthermore, in [2, 3], the optimality of these estimates has been proved based on the Meshkov construction (see [6, 2]) of highly concentrated solutions of elliptic equations with potentials. Unfortunately, in [8], we failed to show the optimality of (3) for bounded domains, because of the difficulty one encounters to correct the rescaled Meshkov-type solutions of the bi-harmonic equation so that they fulfill the homogeneous boundary conditions. In our present case, we shall see that this difficulty disappears when we consider the problem in the whole space, i.e., $\Omega = \mathbb{R}^n$. However, for general unbounded domains optimality of the observability constants is an open problem.

The rest of this paper is organized as follows. In Sect. 2 we give some preliminary energy estimates for Kirchhoff plate systems, and recall some fundamental weighted pointwise estimates for the wave and elliptic operators. In Sect. 3 we present the sharp observability estimate for the Kirchhoff plate system. Finally, Sect. 4 is devoted to the analysis of the optimality of the observability estimate.

2 Preliminaries

In this section, we collect some preliminary energy estimates for Kirchhoff plate systems, and weighted pointwise estimates for the wave and elliptic operators.

2.1 Energy Estimates for Kirchhoff Plate Systems

Denote the energy of system (1) by

$$\begin{aligned} E(t) &= \tfrac{1}{2}(||(y, y_t)||^2_{\mathcal{H}} + ||\nabla y_t||^2_{(L^2(\Omega))^{nN}}) \\ &\equiv \tfrac{1}{2}\int_\Omega \left(|y|^2 + |\nabla y|^2 + |\Delta y|^2 + |\nabla \Delta y|^2 + |y_t|^2 + 2|\nabla y_t|^2 + |\Delta y_t|^2\right) dx. \end{aligned}$$

By system (1), it is easy to verify the following energy law:

$$\frac{dE(t)}{dt} = \int_\Omega (y - ay) \cdot (y_t - \Delta y_t) dx.$$

For $s_0 = \frac{n}{3p}$, consider also the modified energy function:

$$\mathcal{E}(t) = E(t) + \frac{1}{2}||a||_p^{\frac{2}{2-s_0}} |y(t, \cdot)|^2_{(L^2(\Omega))^N}.$$

It is clear that both energies are equivalent. Indeed,

$$E(t) \leq \mathcal{E}(t) \leq C\left(1+||a||_p^{\frac{2}{2-s_0}}\right)E(t).$$

Similar to [8], the following estimate holds for the modified energy:

Lemma 1. *Let $a \in L^\infty(0,T;L^p(\Omega;\mathbb{R}^{N\times N}))$ for some $p \in [n/3,\infty]$. Then there is a constant $C_0 = C_0(\Omega,p,n) > 0$, independent of T, such that*

$$\mathcal{E}(t) \leq C_0 e^{C_0||a||_p^{\frac{1}{2-s_0}}|t-s|}\mathcal{E}(s), \qquad \forall\, t,s \in [0,T]. \tag{4}$$

2.2 Pointwise Weighted Estimates for the Wave and Elliptic Operators

In this subsection, we present some pointwise weighted estimates for the wave and elliptic equations that will play a key role when deriving the sharp observability estimates for the Kirchhoff plate system.

As in [8], one of the key points to derive inequality (2) for system (1) is the possibility of decomposing the Kirchhoff plate operator $\partial_t^2+\Delta^2-\partial_t^2\Delta$ as follows:

$$\partial_t^2+\Delta^2-\partial_t^2\Delta = (\partial_{tt}-\Delta)(I-\Delta)+\Delta, \tag{5}$$

where I is the identity operator. Actually, we set

$$z = y-\Delta y, \tag{6}$$

where y is the solution of (1). By the first equation of (1) and noting (5), it follows that

$$-ay = y_{tt}+\Delta^2 y-\Delta y_{tt} = (\partial_{tt}-\Delta)(y-\Delta y)+\Delta y = z_{tt}-\Delta z+y-z.$$

Therefore, the Kirchhoff plate system (1) can be written equivalently as the following coupled elliptic-wave system

$$\begin{cases} \Delta y+z-y=0 & \text{in } Q,\\ z_{tt}-\Delta z+y-z+ay=0 & \text{in } Q,\\ y=z=0 & \text{on } \Sigma,\\ z(0)=y^0-\Delta y^0, \quad z_t(0)=y^1-\Delta y^1 & \text{in } \Omega. \end{cases} \tag{7}$$

Consequently, to derive the desired observability inequality (2) for system (1), it is natural to proceed in cascade by applying the global Carleman estimates to the second order operators in the two equations in system (7).

In what follows, to simplify the argument to derive the desired observability inequality (2), we assume $D \neq \emptyset$ (When $D=\emptyset$, we can modify an argument in [5, Case 2 in the proof of Theorem 5.1] to derive the same result).

For any (large) $\lambda > 0$, any $x_0 \in D$ and $c \in \mathbb{R}$, set

$$\ell \equiv \ell(t,x) = \lambda\Big(|x-x_0|^2 - c\Big(t-\frac{T}{2}\Big)^2\Big), \qquad \theta = e^{\ell}. \tag{8}$$

Without loss of generality, we may assume $\overline{D} \cap \overline{\omega} = \emptyset$ (Otherwise one uses a smaller observation domain, say $\omega \setminus \Big\{x \in \mathbb{R}^n \,\Big|\, \mathrm{dist}\,(x, \overline{D}) \le 1\Big\}$, to replace the original ω). Put

$$\Omega_0 = \Omega \setminus \omega, \qquad \Omega_1 = \Big\{x \in \mathbb{R}^n \,\Big|\, \mathrm{dist}\,(x, \overline{\Omega_0}) \le 1\Big\} \setminus \overline{D}. \tag{9}$$

Clearly, $\Omega_0 \subset \Omega_1$ and both of them are bounded domains in $\mathbb{R}^n$.

Since $x_0 \in D \subset \mathbb{R}^n \setminus \overline{\Omega_1}$, it holds

$$0 < R_0 \stackrel{\triangle}{=} \min_{x \in \Omega_1} |x - x_0| < R_1 \stackrel{\triangle}{=} \max_{x \in \Omega_1} |x - x_0|. \tag{10}$$

Also, for any $\beta > 0$, we set

$$\Theta = \Theta(t) \stackrel{\triangle}{=} \exp\Big\{-\frac{\beta R_1}{t} - \frac{\beta R_1}{T-t}\Big\}, \qquad 0 < t < T. \tag{11}$$

It is easy to see that $\Theta(t)$ decays rapidly to 0 as $t \to 0$ or $t \to T$. We recall the following two known pointwise Carleman-type estimates (with singular weight Θ) for the wave and elliptic operators, respectively:

Lemma 2. ([8]) *Let $u \in C^2([0,T] \times \overline{\Omega_1})$ and $v = \theta u$. Then there exist four constants $T_0 > 0$, $\lambda_0 > 0$, $\beta_0 > 0$, and $c_0 > 0$, independent of u, such that for all $T \ge T_0$, $\beta \in (0, \beta_0)$, and $\| \ge \|_0$ it holds*

$$\begin{aligned}
&\theta^2\Theta|u_{tt} - \Delta u|^2 + 2\Big\{\Theta\Big(\ell_t(v_t^2 + |\nabla v|^2) - 2(\nabla\ell)\cdot(\nabla v)v_t - \Psi v v_t \\
&\qquad\qquad + (A+\Psi)\ell_t v^2\Big)\Big\}_t \\
&+2\Theta\sum_{i=1}^n \Big\{2v_i(\nabla\ell)\cdot(\nabla v) - \ell_i|\nabla v|^2 + \Psi v v_i - 2\ell_t v_t v_i + \ell_i v_t^2 \\
&\qquad\qquad - (A+\Psi)\ell_i v^2\Big\}_i \\
&\ge c_0\lambda\theta^2\Theta(u_t^2 + |\nabla u|^2 + \lambda^2 u^2),
\end{aligned} \tag{12}$$

where

$$\begin{cases} \Psi \stackrel{\triangle}{=} \lambda(2n - 2c - 1 + k), \\ A = 4\lambda^2[c^2(t-T/2)^2 - |x-x_0|^2] + \lambda(4c + 1 - k). \end{cases} \tag{13}$$

Lemma 3. ([8]) *Let $p = p(t,x) \in C^2([0,T] \times \overline{\Omega_1})$, and set $q = \theta p$. Then there exist two constants $\lambda_0 > 0$ and $c_0 > 0$, independent of p, such that for all $T > 0$, $\beta > 0$, and $\lambda \ge \lambda_0$ it holds*

$$\begin{aligned}&\theta^2\Theta|\Delta p|^2+2\Theta\sum_{i=1}^{n}\Big\{2q_i(\nabla\ell)\cdot(\nabla q)-\ell_i|\nabla q|^2+\widetilde{\Psi}qq_i-(\widetilde{A}+\widetilde{\Psi})\ell_i q^2\Big\}_i\\&\geq c_0\lambda\theta^2\Theta(|\nabla p|^2+\lambda^2p^2),\end{aligned}\tag{14}$$

where

$$\begin{cases}\widetilde{\Psi}\stackrel{\triangle}{=}\lambda(2n-1), \quad \widetilde{A}=-4\lambda^2|x-x_0|^2+\lambda,\\ \widetilde{B}=40\lambda^3|x-x_0|^2+O(\lambda^2), \quad \textit{uniformly w.r.t. } t\in[0,T].\end{cases}\tag{15}$$

3 Sharp Observability Estimate

In this section we establish a sharp observability estimate for system (1).

One of the main results in this paper is the following observability inequality with explicit dependence of the observability constant on the potential a for system (1):

Theorem 1. *Let D be a bounded convex domain in $\mathbb{R}^n$, $\Omega=\mathbb{R}^n\setminus\overline{D}$ and ω an open subset in Ω so that $\Omega\setminus\overline{\omega}$ is bounded. Let $p\in[5n/2,\infty]$. Then there is a constant $C>0$ such that for any $T>T_0$, with T_0 as in Lemma 2, and any $a\in L^\infty(0,T;L^p(\Omega;\mathbb{R}^{N\times N}))$, the weak solution y of system (1) satisfies estimate (2) with the observability constant $K(a)>0$ verifying (3).*

Remark 1. If D is not assumed to be convex, the result in Theorem 1 still holds provided that ω is further assumed to contain a set of the form

$$\Big\{x\in\Gamma\;\Big|\;(x-x_0)\cdot\nu(x)>0\Big\}$$

for some $x_0\in\mathbb{R}^n$, where $\nu(x)$ is the unit outward normal vector of Ω at $x\in\Gamma$.

We now sketch the main points in the proof of Theorem 1. As in [8], we decompose the Kirchhoff plate equation into a coupled system of wave and elliptic equations as in (7) and apply the pointwise estimates of the previous section in cascade. However, in the present setting in which the system holds in an unbounded domain, we need to use a cut-off argument. Indeed, roughly speaking, the observed quantity, i.e., the right hand side terms of inequality (2), provides full information on the solution in the observed subdomain ω. One then needs to recover the missing information in $\Omega\setminus\omega$. It is therefore natural to split the solution into two parts, i.e., to distinguish the restriction of the solution to ω, which is known, and the unknown restriction to $\Omega\setminus\omega$. This argument, needed for deriving the observability on unbounded domains, was used for instance in [1] for parabolic equations in unbounded domains.

To be more precise, we fix any smooth cut-off function ϕ so that

$$\begin{cases}\phi\equiv 1 \quad \text{in } \Omega_0,\\ \operatorname{supp}\phi\subset\Omega_1\cup\overline{D}.\end{cases}\tag{16}$$

Put

$$Y = \phi y, \qquad Z = \phi z. \tag{17}$$

First, we apply Lemma 2 to Z. Integrating (12) in $Q_1 = (0,T)\times\Omega_1$, noting that $\Theta(t)$ decays rapidly to 0 as $t \to 0+$ or $t \to T-$, by (16)–(17) and recalling that $Z|_\Sigma = 0$ (and hence $\nabla Z = \frac{\partial Z}{\partial \nu}\nu$ and $Z_i = \frac{\partial Z}{\partial \nu}\nu_i$ on Σ), using the fact that $(x-x_0)\cdot\nu(x) \le 0$ on Σ (because we assume D is convex), that

$$Z_{tt} - \Delta Z = Z - Y - aY - 2\nabla\phi\cdot\nabla z - z\Delta\phi \qquad \text{in } Q_1,$$

and that

$$\int_0^T\int_{\partial\Omega_1\setminus\Gamma}\theta^2\Theta\left|\frac{\partial Z}{\partial\nu}\right|^2(x-x_0)\cdot\nu(x)dxdt = 0$$

(because, by the second condition in (16), one has $Z\equiv 0$ on $(0,T)\times(\partial\Omega_1\setminus\Gamma)$), one may deduce that

$$\begin{aligned}
&\lambda\int_{Q_1}\theta^2\Theta(|Z_t|^2+|\nabla Z|^2)dxdt + \lambda^3\int_{Q_1}\theta^2\Theta|Z|^2dxdt\\
&\le C\Big\{\int_{Q_1}\theta^2\Theta|Z_{tt}-\Delta Z|^2dxdt + 4\lambda\Big(\int_\Sigma\theta^2\Theta\|\tfrac{\partial Z}{\partial\nu}\|^2(x-x_0)\cdot\nu(x)dxdt\\
&\qquad\qquad + \int_0^T\int_{\partial\Omega_1\setminus\Gamma}\theta^2\Theta\left|\tfrac{\partial Z}{\partial\nu}\right|^2(x-x_0)\cdot\nu(x)dxdt\Big)\Big\}\\
&\le C\int_{Q_1}\theta^2\Theta|Z_{tt}-\Delta Z|^2dxdt\\
&\le C\int_{Q_1}\theta^2\Theta[|aY|^2+|Y|^2+|Z|^2+|\nabla\phi\cdot\nabla z|^2+|z\Delta\phi|^2]dxdt\\
&\le C\Big\{\int_{Q_1}\theta^2\Theta[|aY|^2+|Y|^2+|Z|^2]dxdt\\
&\qquad\qquad + e^{C\lambda}\int_0^T\int_\omega(|z|^2+|\nabla z|^2)dxdt\Big\}.
\end{aligned} \tag{18}$$

Similarly, applying Lemma 3 respectively to Y and Y_t, we deduce that

$$\begin{aligned}
&\lambda\int_{Q_1}\theta^2\Theta|\nabla Y|^2dxdt + \lambda^3\int_{Q_1}\theta^2\Theta|Y|^2dxdt\\
&\le C\Big\{\int_{Q_1}\theta^2\Theta(|Y|^2+|Z|^2)dxdt + e^{C\lambda}\int_0^T\int_\omega(|y|^2+|\nabla y|^2)dxdt\Big\},
\end{aligned} \tag{19}$$

and

$$\begin{aligned}
&\lambda\int_{Q_1}\theta^2\Theta|\nabla Y_t|^2dxdt + \lambda^3\int_{Q_1}\theta^2\Theta|Y_t|^2dxdt\\
&\le C\Big\{\int_{Q_1}\theta^2\Theta[|Y_t|^2+|Z_t|^2+|\nabla\phi\cdot\nabla y_t|^2+|y_t\Delta\phi|^2]dxdt\Big\}\\
&\le C\Big\{\int_{Q_1}\theta^2\Theta[|ay|^2+|Y_t|^2+|Z_t|^2]dxdt\\
&\qquad + e^{C\lambda}\int_0^T\int_\omega\left(|y|^2+|\nabla y|^2+|\Delta y|^2+|\nabla\Delta y|^2\right)dxdt\Big\}.
\end{aligned} \tag{20}$$

Similar to [8], from (18)–(20), it follows

$$
\begin{aligned}
&\lambda \int_{Q_1} \theta^2 \Theta[|\Delta Y_t|^2 + |\nabla \Delta Y|^2] dxdt + \lambda^2 \int_{Q_1} \theta^2 \Theta |\nabla Y_t|^2 dxdt \\
&\quad + \lambda^3 \int_{Q_1} \theta^2 \Theta[|\Delta Y|^2 + |Y_t|^2] dxdt + \lambda^4 \int_{Q_1} \theta^2 \Theta |\nabla Y|^2 dxdt \\
&\quad + \lambda^6 \int_{Q_1} \theta^2 \Theta |Y|^2 dxdt \\
&\leq C\Big\{ ||\theta\sqrt{\Theta} ay||^2_{L^2(Q)} \\
&\qquad + e^{C\lambda} \int_0^T \int_\omega \left(|y|^2 + |\nabla y|^2 + |\Delta y|^2 + |\nabla \Delta y|^2\right) dxdt \Big\}.
\end{aligned} \tag{21}
$$

Adding both sides of (21) by

$$
\begin{aligned}
\int_0^T \int_{\mathbb{R}^n \setminus \overline{\Omega_0 \cup D}} \theta^2 \Theta \Big\{ &\lambda[|\Delta Y_t|^2 + |\nabla \Delta Y|^2] + \lambda^2 |\nabla Y_t|^2 \\
&+ \lambda^3 [|\Delta Y|^2 + |Y_t|^2] + \lambda^4 |\nabla Y|^2 + \lambda^6 |Y|^2 \Big\} dxdt,
\end{aligned}
$$

one may deduce that

$$
\begin{aligned}
&\lambda \int_Q \theta^2 \Theta[|\Delta y_t|^2 + |\nabla \Delta y|^2] dxdt + \lambda^2 \int_Q \theta^2 \Theta |\nabla y_t|^2 dxdt \\
&\quad + \lambda^3 \int_Q \theta^2 \Theta[|\Delta y|^2 + |y_t|^2] dxdt + \lambda^4 \int_Q \theta^2 \Theta |\nabla y|^2 dxdt \\
&\quad + \lambda^6 \int_Q \theta^2 \Theta |y|^2 dxdt \\
&\leq C\Big\{ ||\theta\sqrt{\Theta} ay||^2_{L^2(Q)} \\
&\qquad + e^{C\lambda} \int_0^T \int_\omega \left(|y|^2 + |\nabla y|^2 + |\Delta y|^2 + |\nabla \Delta y|^2\right) dxdt \Big\}.
\end{aligned} \tag{22}
$$

Finally, using the same argument as in [8], the desired estimates (2) and (3) follow from (22) and Lemma 1.

4 Optimality of the Observability Constant for Kirchhoff Plate Systems in the Whole Space

This section is devoted to analyze the optimality of the observability inequality for system (1).

4.1 Optimality in the Whole Space

First, we shall show that when $p = \infty$, the term $||a||_p^{\frac{1}{3-5n/2p}}$ (i.e., $||a||_\infty^{1/3}$) in the estimate (3) is sharp in what concerns the exponential dependence on the potential a for systems with at least two equations in $\mathbb{R}^n$ for even $n \geq 2$. More precisely, the following holds:

Theorem 2. *Assume that* $\Omega = \mathbb{R}^n$ *(i.e.,* $D = \emptyset$*),* $n \geq 2$ *is even and that* $N \geq 2$*. Let* ω *be an open nonempty subset of* $\mathbb{R}^n$ *such that* $\mathbb{R}^n \setminus \overline{\omega} \neq \emptyset$*. Then, there exist a constant* $c > 0$*, a family of time-independent potentials* $\{a_R\}_{R>0} \subset L^\infty(\Omega; \mathbb{R}^{N\times N})$ *satisfying*

$$||a_R||_\infty \to \infty, \qquad as\ R \to \infty$$

and a family of initial data $\{(y_R^0, y_R^1)\}_{R>0} \in \mathcal{H}$ *such that for any* $T > 0$, *the corresponding weak solutions* $\{y_R\}_{R>0}$ *of (1) satisfy*

$$\lim_{R\to\infty} \left\{ \frac{||(y_R^0, y_R^1)||_{\mathcal{H}}^2}{\exp\left(c||a_R||_\infty^{1/3}\right) \int_0^T \int_\omega \left(|y|^2 + |\nabla y|^2 + |\Delta y|^2 + |\nabla \Delta y|^2\right) dtdx} \right\} = \infty. \tag{23}$$

The main idea to prove this optimality result is the same as that in [2], which is based on a suitable construction of u and q satisfying the following bi-Laplacian equation:

$$\Delta^2 u = qu, \qquad \text{in } \mathbb{R}^n, \tag{24}$$

which decays at infinity sufficiently fast. More precisely, following Meshkov's construction [6, 2, 3], we have the following result on u and q for (24):

Lemma 4. *Let* $n \geq 2$ *be even. Then there exist two nontrivial complex-valued functions:*

$$u \in C^\infty(\mathbb{R}^n; \mathbb{C}), \quad q \in C^\infty(\mathbb{R}^n; \mathbb{C}) \bigcap L^\infty(\mathbb{R}^n; \mathbb{C})$$

such that (24) is satisfied, and for some constant C:

$$|u(x)| + |\nabla u(x)| + |\Delta u(x)| + |\nabla \Delta u(x)| \leq Ce^{-|x|^{4/3}}, \qquad \forall\, x \in \mathbb{R}^n. \tag{25}$$

Since we assume that $\Omega = \mathbb{R}^n$, the proof of Theorem 2 via Lemma 4 is quite easy. Indeed, consider the solution u and potential q on $\mathbb{R}^n$ given by Lemma 4. Recalling that both u and q are complex-valued, by setting

$$u_R(x) = \begin{pmatrix} \operatorname{Re} u(Rx) \\ \operatorname{Im} u(Rx) \end{pmatrix}, \qquad a_R(x) = -R^4 \begin{pmatrix} \operatorname{Re} u(Rx) & -\operatorname{Im} q(Rx) \\ \operatorname{Im} q(Rx) & \operatorname{Re} q(Rx) \end{pmatrix}, \tag{26}$$

we obtain a family of potentials $\{a_R\}_{R>0}$ and solutions $\{u_R\}_{R>0}$ satisfying

$$\Delta^2 u_R + a_R(x) u_R = 0 \qquad \text{in } \mathbb{R}^n, \tag{27}$$

and

$$\begin{aligned} &|u_R(x)| + |\nabla u_R(x)| + |\Delta u_R(x)| + |\nabla \Delta u_R(x)| \\ &\leq C \exp\Big(- R^{4/3}|x|^{4/3}\Big) \qquad \text{in } \mathbb{R}^n. \end{aligned} \tag{28}$$

Furthermore, for some constant $C > 0$, the potential a_R is such that

$$C^{-1} R^4 \leq ||a_R||_\infty \leq CR^4, \quad \text{if } n \text{ is even.} \tag{29}$$

Set

$$\psi_R(t,x) = u_R(x), \qquad (t,x) \in [0,T] \times \mathbb{R}^n. \tag{30}$$

The functions $\{\psi_R\}_{R>0}$ may also be viewed as a family of stationary solutions of the Cauchy problem

$$\begin{cases} \psi_{R,\, tt} + \Delta^2 \psi_R - \Delta \psi_{R,\, tt} + a_R \psi_R = 0 & \text{in } (0,T)\times \mathbb{R}^n, \\ \psi_R(0) = u_R, \quad \psi_{R,\, t}(0) = 0 & \text{in } \mathbb{R}^n, \end{cases} \tag{31}$$

with potentials $a_R = a_R(x)$ as in (26).

By (28) and (30), we have

$$\begin{array}{l} |\psi_R(x,t)| + |\nabla \psi_R(x,t)| + |\Delta \psi_R(x,t)| + |\nabla \Delta \psi_R(x,t)| \\ \le C \exp\Big(- R^{4/3}|x|^{4/3}\Big) \quad \text{in } \mathbb{R}^n. \end{array} \tag{32}$$

Without loss of generality, assume that $\omega \subset \mathbb{R}^n \setminus B$, where B is the unit ball in $\mathbb{R}^n$. Then

$$\begin{array}{l} \int_0^T \int_\omega \Big(|\psi_R|^2 + |\nabla \psi_R|^2 + |\nabla \Delta \psi_R|^2 + |\Delta \psi_R|^2 \Big) dt dx \\ \le C \int_{\mathbb{R}^n \setminus B} \exp(-2R^{4/3}|x|^{4/3}) dx \\ \le C \exp(-R^{4/3}) \int_{\mathbb{R}^n \setminus B} \exp(-R^{4/3}|x|^{4/3}) dx \le C \exp(-R^{4/3}). \end{array} \tag{33}$$

Now, combining (26), (29), and (33), one establishes (23) immediately.

4.2 An Open Problem: Optimality for General Domains

One could expect to be able to use Lemma 4 to establish similar optimality results in general domains (i.e., $\Omega \neq \mathbb{R}^n$). However, this is an open problem. Indeed, as in the above subsection, based on the construction of u and q in Lemma 4, one can find a family of rescaled potentials $a_R(x) = R^4 q(Rx)$ with an L^∞-norm of the order of R^4 and a family of solutions $u_R(x) = u(Rx)$ of the corresponding bi-harmonic problem in Ω. These solutions can be regarded also as solutions of the Kirchhoff plate system for suitable initial data. However, they do not fulfill homogeneous boundary conditions. Therefore, one needs to compensate them by subtracting the solution taking their boundary data and zero initial ones. In turn, one has to show that these solutions are as small as $\exp\Big(- ||a_R||_\infty^{1/3}\Big)$ in the energy space $\mathcal{H}$. However, by inequality (4) in Lemma 1, the energy estimate yields an exponential growth $\exp\Big(T||a_R||_\infty^{1/2}\Big)$ for the energy evolution, and it has to be used in the whole time duration $[0,T]$. Since one needs to take the time T to be large enough, this breaks down the concentration effect that Meshkov's construction guarantees.

Acknowledgments

The work is supported by the Grant MTM2005-00714 of the Spanish MEC, the DOMINO Project CIT-370200-2005-10 in the PROFIT program of the MEC (Spain), the SIMUMAT projet of the CAM (Spain), the EU TMR Project "Smart Systems," and the NSF of China under grants 10371084 and 10525105.

References

1. V. R. Cabanillas, S. B. de Menezes, E. Zuazua, *Null controllability in unbounded domains for the semilinear heat equation with nonlinearities involving gradient terms, J. Optim. Theory Appl.*, **110** (2001), 245–264.
2. T. Duyckaerts, X. Zhang, E. Zuazua, *On the optimality of the observability inequality for parabolic and hyperbolic systems with potentials,* Ann. Inst. H. *Poincaré Anal. Non Linéaire,* to appear.
3. X. Fu, X. Zhang, E. Zuazua, *On the optimality of the observability inequalities for plate systems with potentials,* in *Phase Space Analysis of PDEs*, A. Bove, F. Colombini, and D. Del Santo, eds., Birkhäuser, 2006, 117–132.
4. J. L. Lions, *Contrôlabilité exacte, stabilisation et perturbations de systèmes distribués, Tome 1*, RMA **8**, Paris, 1988.
5. A. López, X. Zhang and E. Zuazua, *Null controllability of the heat equation as singular limit of the exact controllability of dissipative wave equations, J. Math. Pures Appl.*, **79** (2000), 741–808.
6. V. Z. Meshkov, *On the possible rate of decrease at infinity of the solutions of second-order partial differential equations, (Russian), Mat. Sb.*, **182** (1991), 364–383; *translation in Math. USSR-Sb.*, **72** (1992), 343–361.
7. M. Yamamoto, X. Zhang, *Global uniqueness and stability for an inverse wave source problem with less regular data, J. Math. Anal. Appl.*, **263** (2001), 479–500.
8. X. Zhang and E. Zuazua, *A sharp observability inequality for Kirchoff plate systems with potentials, Comput. Appl. Math.*, **25**, **no. 2–3** (2006), 353–373.
9. E. Zuazua, *Exact controllability for semilinear wave equations in one space dimension, Ann. Inst. H. Poincaré Anal. Non Linéaire,* **10** (1993), 109–129.
10. E. Zuazua, *Controllability and observability of partial differential equations: some results and open problems,* in *Handbook of Differential Equations: Evolutionary Differential Equations, vol. 3, C. Dafermos and E. Feireisl, eds.*, Elsevier Science, to appear.

Part III

Mini-Symposium on Hydraulics

High Order Finite Volume Methods Applied to Sediment Transport and Submarine Avalanches

D. Bresch, M.J.C. Díaz, E.D. Fernández-Nieto, and A.M. Ferreiro, and A. Mangeney

1 Introduction

In this work different numerical models related with the sediment transport process are presented. Two kinds of models are presented: the first one uses a continuity equation for the sediment layer in order to study bed-load sediment transport phenomena. The second one is a model for submarine avalanches.

All models presented in this work are particular cases of a general class of hyperbolic systems of conservations laws with coupled and source terms and they could be rewritten under the form of hyperbolic systems in nonconservative form. In the 1D case we have the system

$$\frac{\partial W}{\partial t} + \mathcal{A}(W)\frac{\partial W}{\partial x} = 0, \quad x \in \mathbf{R},\ t > 0. \tag{1}$$

The nonconservative product $\mathcal{A}(W)W_x$ makes difficult the definition of weak solutions for this kind of systems. After the theory developed by Dal Maso, LeFloch, and Murat, a definition of nonconservative products as Borel measures is introduced, which is based on the selection of a family of paths in the phases space.

It is well-known that the presence of source or coupling terms can affect the quality of the results obtained when steady or nearly steady state solutions are approximated. To handle such problems, the concept of *well-balanced* schemes, that is, schemes that preserves all equilibria of the system or a family of them, has been considered by several authors (see [Chacon03]). Recently, Parés and Castro (see [Pares04]) have extended the concept of well-balancing to nonconservative systems of the form (1), for which generalized Roe schemes (first order) have been developed.

A high-resolution well-balanced extension of the generalized Roe schemes for solving the 1D coupled system shallow-water/bedload transport model using polynomial reconstructions of the conserved variabled is considered (see [Castro06]). Also a high order 2D extension of the numerical scheme using the line-method together with polynomial reconstructions of the conserved variables will be presented. To test the performances of the high-resolution schemes, some numerical experiments are presented.

Finally, we consider a well-balanced numerical scheme using an extension of the MUSTA scheme (see [Toro04]) for nonconservative hyperbolic systems applied to the submarine avalanches. Some numerical experiments will be presented.

2 Bedload Sediment Transport

To solve the problem of sediment transport by a fluid, a coupled mathematical model can be used with a hydrodynamical component and a morphodynamical component.

For the hydrodynamical component we consider Shallow Water equations, which simulates the movement of a fluid in rivers, channels, coastal areas, etc. We will denote by $q(x,t)$ and $h(x,t)$ the discharge and the height of the water column, respectively.

The morphodynamical component consists on a continuity equation. The expression of the conservation sediment volume equation is given by

$$\frac{\partial z_b}{\partial t} + \xi \frac{\partial q_b}{\partial x} = 0, \tag{2}$$

where $\xi = 1/(1-\rho_0)$ and ρ_0 is the porosity of the sediment layer. By $q_b = q_b(h,q)$ we denote the solid transport discharge.

Different estimations of the solid transport discharge q_b have been obtained by empirical methods (see [CFF06]). For example, Grass ([Grass81]) proposed the following formula for the solid transport discharge,

$$q_b = A_g \frac{q}{h} \left|\frac{q}{h}\right|^{m_g - 1}, \quad 1 \leqslant m_g \leqslant 4, \tag{3}$$

where the constant A_g (s^2/m) is usually obtained by experimental data and takes into account the grain diameter and the kinematic viscosity. The value of the exponent m_g is set to $m_g = 3$, as usually.

Meyer-Peter & Müller [MP&M] developed one of the most known formulae for the solid transport discharge, based on median grain diameter d_{50}, given by

$$\frac{q_b}{\sqrt{(G-1)gd_i^3}} = \operatorname{sgn}(u) 8 \left(\tau_* - \tau_{*c}\right)^{3/2}, \tag{4}$$

where τ_{*c} usually is set to 0.047.

If the variable $S = H - z_b$ is defined, as $\dfrac{\partial S}{\partial t} = -\dfrac{\partial z_b}{\partial t}$, we can rewrite the coupled system as

$$\begin{cases} \dfrac{\partial h}{\partial t} + \dfrac{\partial q}{\partial x} = 0, \\ \dfrac{\partial q}{\partial t} + \dfrac{\partial}{\partial x}\left(\dfrac{q^2}{h} + \dfrac{1}{2}gh^2\right) = gh\dfrac{\partial S}{\partial x} - ghS_f, \\ \dfrac{\partial S}{\partial t} - \xi\dfrac{\partial q_b}{\partial x} = 0 \end{cases} \tag{5}$$

where S_f is the friction term.

2.1 High Order Well-Balanced Finite Volume Method

There exist different techniques to obtain high order methods. A possibility is to use state reconstructions. First a state reconstruction operator P is considered, that is, an operator that associates, to a given sequence $\{W_i(t)\}$, two new sequences $\left\{W^-_{i+1/2}(t)\right\}$, $\left\{W^+_{i+1/2}(t)\right\}$ in such a way that, whenever

$$W_i(t) = \frac{1}{\Delta x}\int_{I_i} W(x,t)dx,\ \forall i \in \mathrm{Z};$$

for some regular function W, then

$$W^{\pm}_{i+1/2}(t) = W(x_{i+1/2},t) + O(\Delta x^p),\ \forall i \in \mathrm{Z}.$$

In [Castro06] the following numerical scheme is proposed,

$$W_i^{'} = -\frac{1}{\Delta x}\left(\mathcal{A}^+_{i-1/2}\left(W^+_{i-1/2}(t) - W^-_{i-1/2}(t)\right)\right.$$

$$\left. + \mathcal{A}^-_{i+1/2}\left(W^+_{i+1/2}(t) - W^-_{i+1/2}(t)\right) + \int_{x_{i-1/2}}^{x_{i+1/2}} \mathcal{A}\left[P^t(x)\right]\frac{d}{dx}P^t(x)dx\right), \tag{6}$$

where $\mathcal{A}^{\pm} = (\mathcal{A} \pm D)/2$. D is the viscosity matrix and $\mathcal{A}$ is the Roe matrix. The definition of the roe matrix for a nonconservative system is associated to a choice of paths (see [Pares04]). Matrix D is a defined (or semi-defined) positive matrix with the same eigenvectors like matrix $\mathcal{A}$. Moreover, if by λ and d we denote the eigenvalues of $\mathcal{A}$ and D then the following relation must be verified

$$\left(\frac{\Delta t}{\Delta x}\lambda\right)^2 \leq \frac{\Delta t}{\Delta x}d \leq 1.$$

For $d = |\lambda|$ we obtain Roe method. For $d = \varphi|\lambda| + (1-\varphi)\frac{\Delta t}{\Delta x}\lambda^2$ a flux-limiter version that is second order in regular areas, at least for linear problems, being φ a flux-limiter function (see [CFF06]).

2.2 Numerical Test

This test consists on comparing a numerical solution with an asymptotic analytical solution obtained by Hudson and Sweby in [1], for Grass model when the interaction constant A_g is smaller than 10^{-2}. In this case, the layer sediment $\tilde{z}_b$ is over all computational domain and fluid is moving slowly with a constant discharge $q = q_0 \leq 10$. Under these hypothesis it is possible to obtain the following analytical solution,

$$h = A_r - z_b(x,t), \quad q = q_0$$

A_r being a fixed reference level, $q_0 = 10$ a constant value, and

$$z_b(x,t) = \begin{cases} \sin^2\left(\dfrac{\pi\,(x_o - 300)}{200}\right) + z_0 & \text{if } 300 \leq x_o \leq 500, \\ z_0 & \text{otherwise,} \end{cases} \tag{7}$$

where x_o is the solution of the equation

$$\begin{cases} x = x_o + A_g\,\xi\, m_g\, q_0^{m_g}\, t\left(A_r - \sin^2\left(\dfrac{\pi\,(x_o - 300)}{200}\right)\right)^{-(m_g+1)} & \text{if } 300 \leq x_o \leq 500 \\ x = x_o + A_g\,\xi\, m_g\, q_0^{m_g} t A_r^{-(m_g+1)} & \text{otherwise} \end{cases} \tag{8}$$

and $z_= = 0.01$.

The usual value of $m_g = 3$ is considered.

We consider a computational domain whose length is $L = 1{,}000$ meters, discretized with 250 nodes. CFL condition is set to 0.8. The sediment porosity is set to $\rho_0 = 0.4$ and the constant A_g of Grass formula (3) is set to 0.001, which corresponds to a weak interaction.

The initial conditions are

$$h(x,0) = 10 - z_b(x,0), \quad q(x,0) = 10.$$

As boundary condition, the flux and the depth of the sediment is imposed upstream, while free boundary conditions are imposed downstream.

In Fig. 1 we compare the analytical solution (continuous line) and numerical solution obtained with the different schemes corresponding to sediment layer evolution at instant $t = 238{,}080\,\text{s}$.

In Fig. 1a we compare Roe method with the second order method using flux limiters. In Fig. 1b we use the high order by state reconstruction method (6). We compare the numerical approximation when we consider Weno2 and Weno3 state reconstruction operators. For the time discretization we use TVD Runge–Kutta of second and third order, respectively.

We can observe that all numerical schemes show a good sediment layer localization, Roe scheme being the most diffusive (Fig. 1a). Moreover, the scheme that gives the best approximation is the high order generalized Roe scheme with Weno state reconstructions of order 3 (Fig. 1b).

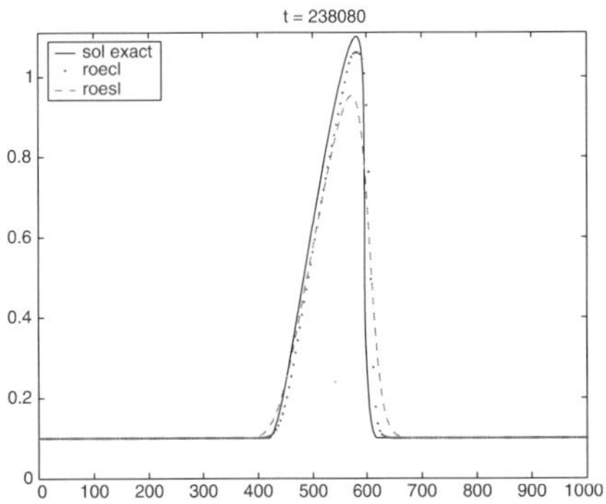

(a) Roe–Flux limiters (*dotted line*). Euler–Roe (*dash line*). Sediment layer thickness

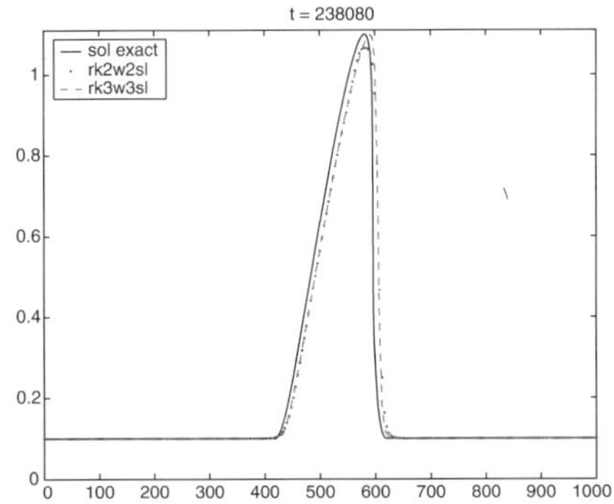

(b) Weno2-Rk2 (*dotted line*). Weno3-Rk3 (*dash line*)

Fig. 1. Sediment layer thickness. *Continuous line*: exact solution

3 2D High Order Finite Volume Methods by State Reconstructions

We consider the 2D nonconservative system

$$\frac{\partial W}{\partial t} + \mathcal{A}_1(W)\frac{\partial W}{\partial x_1} + \mathcal{A}_2(W)\frac{\partial W}{\partial x_2} = 0. \tag{9}$$

Our objetive is to design a high order finite volume method based on state reconstructions for (9).

First we introduce the notation that we use. We suppose that the computational domain is subdivided into cells or control volumes, $V_i \subset \mathbb{R}^2$, that we suppose to be defined by a closed poligone. We use the following notation: For a given control volume V_i, $N_i \in \mathbb{R}^2$ is its associated center point, $\mathcal{N}_i$ is the set of index j such that V_j is a neighbor of V_i, E_{ij} is the common edge between the two control volumes V_i and V_j, and $|E_{ij}|$ is its length, $\eta_{ij} = (\eta_{ij,1}, \eta_{ij,2})$ is the unitary normal vector to the edge E_{ij} outward to V_i (see Fig. 2). Moreover each control volume is divided into the subcells $\{V_{ij}\}_{j\in\mathcal{N}_i}$ (see Fig. 2). V_{ij} is the triangle defined by the vertex N_i with the edge E_{ij}. By $|V_i|$ and $|V_{ij}|$ we denote the volume of the cell V_i and the subcell V_{ij}, respectivelly. Finally W_i represents the approximation

$$W_i^n \cong \frac{1}{|V_i|}\int_{V_i} W(x, t^n).$$

Moreover, if $W(\mathbf{x}, t)$ is the exact solution then by $\overline{W}_i^{\,n}$ we denote its mean value over the control volume V_i at time $t = t^n$: $\overline{W}_i^{\,n} = \dfrac{1}{|V_i|}\displaystyle\int_{V_i} W_i(\mathbf{x}, t^n)d\mathbf{x}$, that is $W_i^n \simeq \overline{W}_i^{\,n}$.

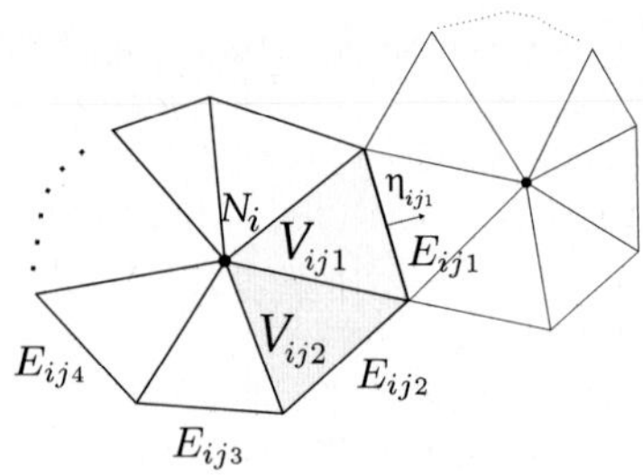

Fig. 2. General finite control volume

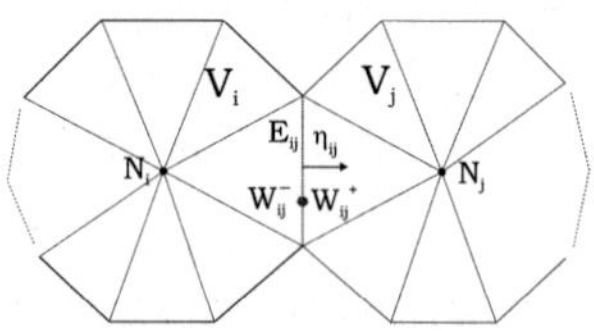

Fig. 3. Sketch of state reconstruction of the edge E_{ij}

With respect to the state reconstruction operator we use the following notation: For a given control volume V_i we denote by P_i the state reconstruction operator over it.

For a given vector η_{ij} from V_i to V_j and normal to the edge E_{ij} (Fig. 3) we denote by $W_{ij}^-(t, \mathbf{x}_{ij})$ and $W_{ij}^+(t, \mathbf{x}_{ij})$ $\forall \mathbf{x}_{ij} \in E_{ij}$, to the limit of P_i (P_j respectively) when $\mathbf{x}$ tends to $\mathbf{x}_{ij}$ by the interior of V_i (V_j respectively):

$$\lim_{\substack{\mathbf{x} \to \mathbf{x}_{ij} \\ \mathbf{x} \cdot \eta_{ij} < k_{ij}}} P_i^t(\mathbf{x}) = W_{ij}^-(t), \qquad \lim_{\substack{\mathbf{x} \to \mathbf{x}_{ij} \\ \mathbf{x} \cdot \eta_{ij} > k_{ij}}} P_j^t(\mathbf{x}) = W_{ij}^+(t), \tag{10}$$

where the edge $E_{ij} = V_i \cap V_j$ is contained on the line,$\mathbf{x} \cdot \eta_{ij} = x_1 \eta_{ij,1} + x_2 \eta_{ij,2} = k_{ij}$.

We suppose that P_i is order p over the boundary of V_i, it is order q at the interior of the control volume V_i, and ∇P_i is an approximation of order m of the gradient of the solution (see [Ferreiro06]).

3.1 Proposed Scheme

We proposee the following numerical scheme:

$$\begin{aligned} W_i'(t) = -\frac{1}{|V_i|} \Bigg[& \sum_{j \in \mathcal{N}_i} |E_{ij}| \sum_{l=1}^{n(\bar{r})} w_l \mathcal{A}_{ij,l}^-(W_{ij,l}^+, W_{ij,l}^-, \eta_{ij})(W_{ij,l}^+ - W_{ij,l}^-) \\ & + \int_{V_i} \left(\mathcal{A}_1(P_i(\mathbf{x})) \frac{\partial P_i}{\partial x_1}(\mathbf{x}) + \mathcal{A}_2(P_i(\mathbf{x})) \frac{\partial P_i}{\partial x_2}(\mathbf{x}) \right) d\mathbf{x} \Bigg], \end{aligned} \tag{11}$$

where w_l, $l = 1, \ldots, n(\bar{r})$ are the weights of a quadrature formula associated to the 1D integral over the edge E_{ij}. If by x_l we denote the points over the edge E_{ij} of the quadrature formula then $W^{\pm}_{ij,l} = W^{\pm}_{ij}(\mathbf{x}_l)$. In practice this formula is choosen in function of the order of the state reconstruction: If by $\bar{r}$ we denote the order of the quadrature formula, then $\bar{r} > p$.

By $\mathcal{A}_{ij}$ we denote Roe matrix associated to the 1D nonconservative projected problem over η_{ij}. Also a family of paths in this case must be chosen.

The following result hold (See [Ferreiro06]).

Theorem 1. *We suppose that $\mathcal{A}_1$ and $\mathcal{A}_2$ are of classes $\mathcal{C}^2$ with bounded derivatives and $\mathcal{A}_{ij}$ is bounded for all i, j. Then,*

$$\begin{aligned}
&\sum_{j\in\mathcal{N}_i}\left(|E_{ij}|\sum_{l=1}^{n(\bar{r})} w_l \mathcal{A}^-_{\Psi,ij,l}(W^+_{ij,l}, W^-_{ij,l}, \eta_{ij})(W^+_{ij,l}(\mathbf{x},t) - W^-_{ij,l}(\mathbf{x},t))\right)\\
&+\int_{V_i}\left(\mathcal{A}_1(P_i(\mathbf{x}))\frac{\partial P_i}{\partial x_1}(\mathbf{x}) + \mathcal{A}_2(P_i(\mathbf{x}))\frac{\partial P_i}{\partial x_2}(\mathbf{x})\right) d\mathbf{x} \qquad (12)\\
&=\int_{V_i}\left(\mathcal{A}_1(W(\mathbf{x}))\frac{\partial W(\mathbf{x})}{\partial x_1} + \mathcal{A}_2(W(\mathbf{x}))\frac{\partial W(\mathbf{x})}{\partial x_2}\right) d\mathbf{x} + \mathcal{O}(\Delta^\delta);
\end{aligned}$$

with $\delta = \min\{p+1, q+2, m+2\}$.

3.2 2D Shallow Water Equations with Sediment Transport

In this case we consider the coupled system formed by the 2D Shallow Water equations and a 2D continuity equation:

$$\begin{cases}
\dfrac{\partial h}{\partial t} + \dfrac{\partial q_1}{\partial x} + \dfrac{\partial q_2}{\partial x} = 0,\\[2mm]
\dfrac{\partial q_1}{\partial t} + \dfrac{\partial}{\partial x_1}\left(\dfrac{q_1^2}{h} + \dfrac{1}{2}gh^2\right) + \dfrac{\partial}{\partial x_2}\left(\dfrac{q_1 q_2}{h}\right) = gh\dfrac{\partial S}{\partial x_1} - ghS_{f,x_1},\\[2mm]
\dfrac{\partial q_2}{\partial t} + \dfrac{\partial}{\partial x_1}\left(\dfrac{q_1 q_2}{h}\right) + \dfrac{\partial}{\partial x_2}\left(\dfrac{q_2^2}{h} + \dfrac{1}{2}gh^2\right) = gh\dfrac{\partial S}{\partial x_2} - ghS_{f,x_2},\\[2mm]
\dfrac{\partial S}{\partial t} - \xi\dfrac{\partial q_{b,x_1}}{\partial x_1} - \xi\dfrac{\partial q_{b,x_2}}{\partial x_2} = 0,
\end{cases} \qquad (13)$$

where the unknowns are the height of the water column $h(\mathbf{x}, t)$, the discharge $q(\mathbf{x}, t) = (q_1(\mathbf{x}, t), q_2(\mathbf{x}, t))$ $S(\mathbf{x}, t) = H(\mathbf{x}) - z_b(\mathbf{x}, t)$, where H is the bathimetry of the fixed bottom and z_b is the height of the sediment layer column. By q_{b,x_1} and q_{b,x_2} we denote the solid transport discharge for the two components x_1 and x_2, respectively. The definition of q_b depends on the considered model. For example, we can consider Grass' model and Meyer-Peter & Müller's model (See [Ferreiro06]). By S_{f,x_1} and S_{f,x_2} we denote the friction terms.

The system of equations (13) can be rewritten under the structure of a 2D nonconservative hyperbolic system (9).

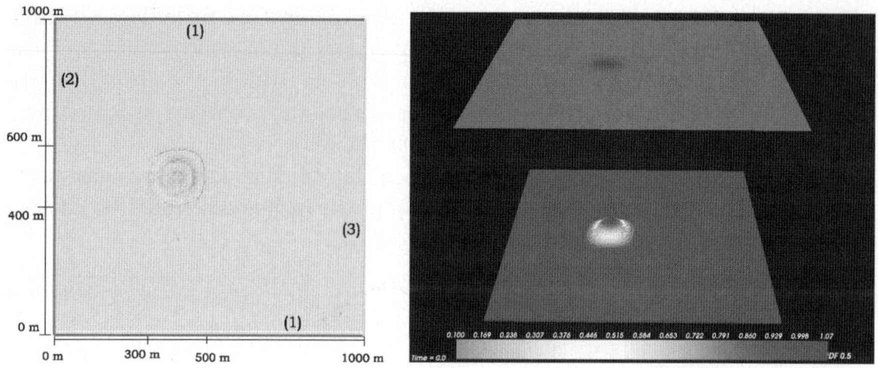

(a) Up view (b) Lateral view

Fig. 4. Initial condition

3.3 Numerical Test: Two-Dimensional Simulation of a Conical Dune of Sand Evolution

In this test we study the evolution of a sand conical dune in a channel of dimensions $1,000 \times 1,000\,\mathrm{m}^2$ (see [1]). Initial conditions are

$$h(x,y,0) = 10.1 - z_b(x,y,0),\ q_x(x,y,0) = 10,\ q_y(x,y,0) = 0;$$

and initial sediment layer is a sand dune with a conical form (See Fig. 4),

$$z_b(x,y,0) = \begin{cases} 0.1 + \sin^2\left(\dfrac{\pi(x-300)}{200}\right)\sin^2\left(\dfrac{\pi(y-400)}{200}\right) & \text{if } \begin{array}{l} 300 \le x \le 500, \\ 400 \le y \le 600, \end{array} \\ 0.1 & \text{otherwise.} \end{cases}$$

We study the sediment layer evolution by bedload transport using Grass's formulae. We use the numerical scheme (11) with MUSCL state reconstructions (see [Ferreiro06]). We consider a weak interaction between fluid and sediment layer, imposing $A_g = 0.001$. Sediment porosity considered is 0.4. CFL condition is equal to 0.8. We use a mesh of 7,600 edge finite volumes. We impose a discharge $\mathbf{q} = (10,0)$ and sediment layer thickness $z_b = 0.1$ in the boundary corresponding to $x = 0$ and free boundary condition corresponding to $x = 1,000$. At lateral walls we impose sliding condition $\mathbf{q} \cdot \eta = 0$. With this data the experiment evolutes until 100 h. In Fig. 6 are shown different instants of sediment layer evolution. We can appreciate that bedload sediment layer provoques a deformation of the dune in a star form. Sediment layer is deformed gradually towards a start form, expanding along time with a certain angle. It is possible to get an analytical approximation angle, supposing a weak interaction between sediment layer and fluid, that is, $A_g < 0.01$ (see [1]). An analytical approximation of this expansion angle is

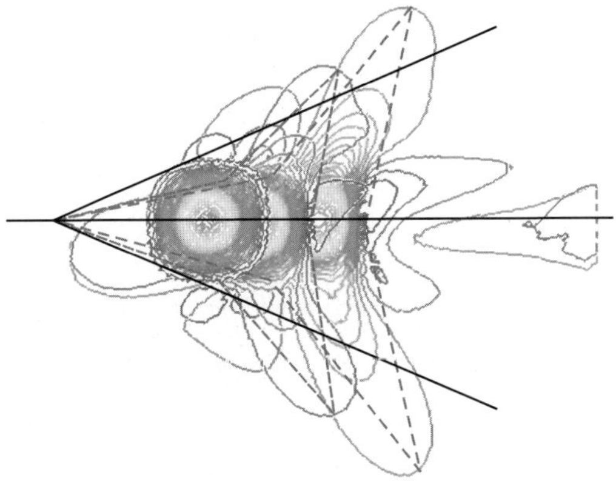

Fig. 5. Spread angle stimulation (using 20 level curves)

$$\alpha = \tan^{-1}\left(\frac{3\sqrt{3}}{13}\right) = 21.786789^o.$$

To measure the expansion angle we have superposed level curves obtained in different instants (0 s, 120,000 s y 295,000 s), and expansion angle obtained is 23° (Fig. 5).

4 Submarine Avalanches

In this section, we present a model to study submarine avalanches, by a two-layer Savage–Huter type model (See [SH91]).

By h_i and q_i we denote the height and discharge of layer i. By index 1 we denote the fluid layer and by index 2 the sediment layer. If ρ_i is the density of layer i, then we denote $r = \rho_1/\rho_2$.

If we consider only friction terms of Coulomb type, we obtain system

$$\begin{cases} \partial_t h_1 + \partial_x(q_1) = 0, \\ \partial_t(q1) + \partial_x(\frac{q_1^2}{h_1} + g\frac{h_1^2}{2}) = -gh_1\partial_x z_b - gh_1\partial_x h_2, \\ \partial_t h_2 + \partial_x(q_2) = 0, \\ \partial_t(q2) + \partial_x(\frac{q_2^2}{h_2} + g\frac{h_2^2}{2}) = -gh_2\partial_x z_b - rgh_2\partial_x h_1 + \mathcal{T}, \end{cases} \tag{14}$$

where by $\mathcal{T}$ we denote the Coulomb friction term. We observe that this term must be understood as

$$\text{If } |\mathcal{T}| \geq \sigma_c \quad \Rightarrow \quad \mathcal{T} = -g(1-r)h_2\frac{U_2}{|U_2|}\tan(\delta_0). \tag{15}$$

$$\text{If } |\mathcal{T}| < \sigma_c \quad \Rightarrow \quad q_2 = 0, \tag{16}$$

where $\sigma_c = g(1-r)h_2\tan(\delta_0)$.

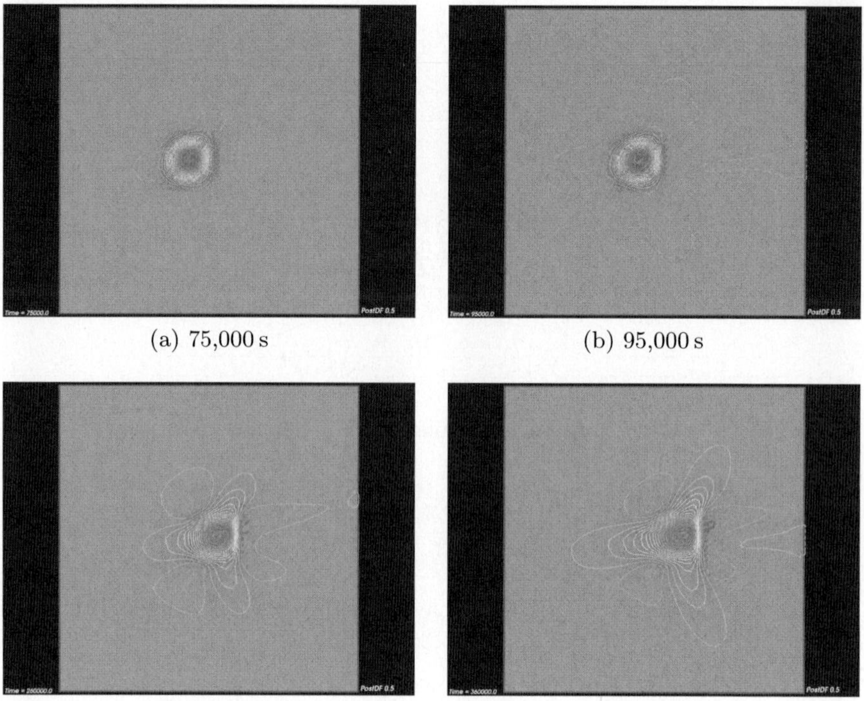

(a) 75,000 s (b) 95,000 s

(c) 26,000 s (d) 360,000 s

Fig. 6. Sediment layer evolution. View of level curves behavior (20 level curves)

One of the problems that present this model in order to be discretized is the numerical treatment of the Coulomb term. We consider a well-balanced numerical scheme using an extension of the MUSTA scheme (see [Toro04]) for nonconservative hyperbolic systems applied to the submarine avalanches. The multi-step character of the numerical scheme is relevant to dealing with the difficulty of the imposition of the Coulomb source term, without loosing the well-balanced character of the numerical scheme.

And it is by this term that we consider a MUSTA method (see [Toro04]), which is a two step by time method.

4.1 Numerical Test

We consider a domain of 20 m length with a flat bottom. We simulate an avalanche when the initial condition is water at rest and a rectangular profile of the sediment layer. Concretely, as initial condition we impose

$$q_1 = 0, \quad q_2 = 0, \quad h_1 = 2 - h_2, \quad h_1 = \begin{cases} 1 & \text{if } -2 \leq x \leq 2 \\ 0 & \text{otherwise} \end{cases}$$

In following pictures we present the evolution from the initial condition to the stationary solution.

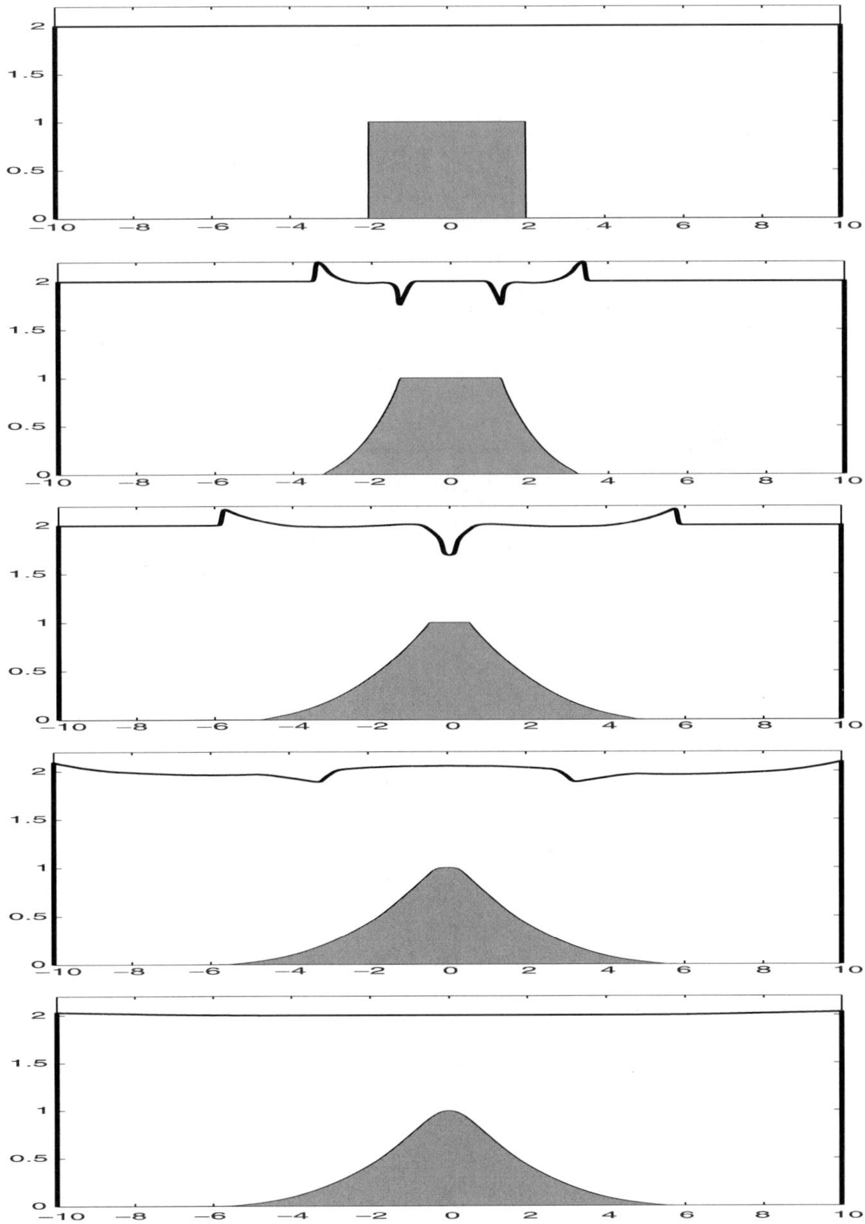

References

[CFF06] M.J. Castro, E.D. Fernández-Nieto, A.M. Ferreiro. Sediment transport models in Shallow Water equations and numerical approach by high order finite volume methods. Submitted (2006).

[Castro06] M.J. Castro, J.M. Gallardo and Carlos Parés. High order finite volume schemes based on reconstruction of states for solving hyperbolic systems with nonconservative products. Applications to shallow-water systems. Math. Comp. **75**(255): 1103–1134, (2006).

[Chacon03] T. Chacón Rebollo, A. Domínguez Delgado, E.D. Fernández Nieto, *A family of stable numerical solvers for Shallow Water equations with source terms.* Comput. Methods Appl. Mech. Eng. 192, 203–225, (2003).

[Ferreiro06] A.M. Ferreiro. Desarrollo de técnicas de post-proceso de flujos hidrodinámicos, modelización de problemas de transporte de sedimentos y simulación numérica mediante técnicas de volúmenes finitos. Thesis, University of Sevilla, Spain (2006).

[Grass81] A.J. Grass. Sediments transport by waves and currents. SERC London Cent. Mar. Technol., Report No. FL29, (1981).

1. J. Hudson. Numerical technics for morphodynamic modelling. Thesis, University of Whiteknights, (2001).

[MP&M] E. Meyer-Peter, R. Müller. Formulas for bed-load transport. Rep. 2nd Meet. Int. Assoc. Hydraul. Struct. Res., Stockholm: 39–64, (1948).

[Pares04] C. Parés, M.J. Castro. On the well-balanced property of Roe's method for nonconservative hyperbolic systems. Applications to shallow water systems. ESAIM: M2AN **38**(5): 821–852, (2004).

[SH91] S.B. Savage, K. Hutter. The dynamics of avalanches of granular materials frominitiation to run-out. Acta Mech. 86, 201–223 (1991).

[Toro04] E.F. Toro, V.A. Titarev. MUSTA schemes for systems of conservation laws. Technical Report NI04033, Isaac Newton Institute for Mathematical Sciences, Univ. Cambridge, (2004).

On a Well-Balanced High-Order Finite Volume Scheme for the Shallow Water Equations with Bottom Topography and Dry Areas

J.M. Gallardo, M. Castro, C. Parés, and J.M. González-Vida

1 Introduction

The shallow water equations are widely used in ocean and hydraulic engineering to model flows in rivers, reservoirs, or coastal areas, among others. These equations constitute a hyperbolic system of conservation laws with a source term due to the bottom topography.

In the recent years, there has been an increasing interest concerning the design of high-order numerical schemes for the shallow water equations (see, e.g., [CGP06], [GHS03], [NPGN06], [VS02], and the references therein). These kind of schemes compute solutions with high order of accuracy (both in space and time) in the regions where they are smooth, while at the same time shock discontinuities are properly captured. Usually, these schemes are based on high-order reconstructions of numerical fluxes or states. However, when a source term is present, the schemes must also satisfy a balance between the flux and the source terms, to properly compute stationary or almost stationary solutions. This property is known as *well balancing*, and it is currently an active subject of research (see, e.g., [BV94], [Bou04], [CGP06], [GHS03], [Gos00], [GL96], [LV98], [NPGN06], [PC04], [VS02]).

In [CGP06], a high-order well-balanced finite volume scheme was developed in a general nonconservative framework. The scheme uses high-order reconstruction of states and it is based on the generalized Roe schemes introduced in [Tou92], whose well-balance properties have been studied in [PC04]. In particular, it was successfully applied to the shallow water equations with bottom topography, using fifth-order WENO reconstructions in space and a third-order TVD Runge–Kutta scheme to advance in time.

An important difficulty that arises in the simulation of free surface flows is the appearance of dry areas, due to the initial conditions or as a result of the fluid motion. Examples are numerous: flood waves, dambreaks, breaking of waves on beaches, etc. If no modifications are made, standard numerical schemes may fail in the presence of wet/dry situations, producing unphysical results. Several methods can be found in the literature that overcome this problem: see [Bou04] and [Toro01] for a review.

When applied to the shallow water equations, the Roe schemes introduced in [PC04] lose their well-balance properties in the presence of wet/dry transitions. Moreover, they may produce negative values of the thickness of the water layer in the proximities of the wet/dry front. Recently, a new technique for treating wet/dry fronts in the context of Roe schemes has been presented in [CGVP06]. It consists of replacing, at the intercells where a wet/dry transition has been detected, the corresponding linear Riemann problem by an adequate nonlinear one.

The main idea in the present work is to properly combine the scheme developed in [CGP06] with the treatment of wet/dry fronts introduced in [CGVP06]. This is by no means a trivial task, as many difficulties appear. In particular, the numerical fluxes must be modified accordingly to the kind of wet/dry transition found. Moreover, the variables to be reconstructed have to be properly chosen to maintain the well-balance property and, at the same time, to preserve the positivity of the water height. In particular, the hyperbolic reconstruction method introduced in [Mar94] has been considered here, as it provides a monotone reconstruction at every computational cell. On the other hand, polynomial reconstructions may introduce oscillations that lead to the appearance of negative values of the water height.

2 Shallow Water Equations, Nonconservative Hyperbolic Systems, and Generalized Roe Schemes

We are interested in the one-dimensional shallow water system given by

$$\begin{cases} \dfrac{\partial h}{\partial t} + \dfrac{\partial q}{\partial x} = 0, \\ \dfrac{\partial q}{\partial t} + \dfrac{\partial}{\partial x}\left(\dfrac{q^2}{h} + \dfrac{g}{2}h^2\right) = gh\dfrac{\mathrm{d}H}{\mathrm{d}x}, \end{cases} \tag{1}$$

which are the equations governing the flow of a shallow layer of fluid through a straight channel with constant rectangular cross section. The variable x refers to the axis of the channel and t is time; $h(x,t)$ and $q(x,t)$ represent the thickness and the discharge, respectively; g is the gravity constant; finally, $H(x)$ is the depth function measured from a fixed level of reference. The fluid is supposed to be homogeneous and inviscid. The terms modeling bottom friction or wind effects are not considered here for simplicity.

Addition of the trivial equation ([Gos00], [GL96])

$$\frac{\partial H}{\partial t} = 0$$

to (1) allows us to write the system in *nonconservative* form

$$W_t + \mathcal{A}(W)W_x = 0, \quad x \in \mathbb{R},\ t > 0. \tag{2}$$

The state variable is given by

$$W = \begin{bmatrix} h \\ q \\ H \end{bmatrix}$$

and the matrix $\mathcal{A}(W)$ is defined as

$$\mathcal{A}(W) = \begin{bmatrix} 1 & 0 & 0 \\ -u^2 + c^2 & 2u & -c^2 \\ 0 & 0 & 0 \end{bmatrix},$$

where $u = q/h$ is the averaged velocity and $c = \sqrt{gh}$.

As long as dry zones do not occur, the variable W takes its values in the open convex set $\Omega = \{[h, q, H]^T : h > 0, q \in \mathbb{R}, H \in \mathbb{R}\}$. Although we will be eventually interested in the case where W belongs to $\Omega^* = \Omega \cup \{[0, 0, H]^T : H \in \mathbb{R}\}$, thus allowing the presence of dry zones, we will assume along this section that the condition $h > 0$ holds. The treatment of wet/dry zones will be considered in Sect. 4.

The nonconservative product $\mathcal{A}(W)W_x$ does not make sense, in general, within the framework of the theory of distributions. More precisely, the term ghH_x does not make sense if both h and H have discontinuities at the same points. However, following the theory developed by Dal Maso, LeFloch, and Murat ([DLM95]), it is possible to give a sense to the nonconservative product as a Borel measure. We refer to [DLM95] for the technical details.

Many of the usual numerical schemes designed for systems of conservation laws can be adapted to the discretization of the nonconservative system (2). This is the case of Roe schemes, whose general definition is based on the concept of *Roe linearization* introduced in [Tou92]: the interested reader is referred to this reference and to [PC04] for a complete description of the generalized Roe schemes. Here, we restrict ourselves to the particular case in which the Roe matrices are given, for arbitrary states $W_0, W_1 \in \Omega$, by

$$\mathcal{A}(W_0, W_1) = \begin{bmatrix} 0 & 1 & 0 \\ -\tilde{u}^2 + \tilde{c}^2 & 2\tilde{u} & -\tilde{c}^2 \\ 0 & 0 & 0 \end{bmatrix},$$

where

$$\tilde{u} = \frac{\sqrt{h_0}u_0 + \sqrt{h_1}u_1}{\sqrt{h_0} + \sqrt{h_1}}, \quad \tilde{c} = \sqrt{g\frac{h_0 + h_1}{2}} \tag{3}$$

are the usual Roe states.

The spatial domain is decomposed in computing cells $I_i = [x_{i-1/2}, x_{i+1/2}]$. For the sake of simplicity, it is assumed that the cells have constant size Δx and that $x_{i+1/2} = i\Delta x$; thus, $x_i = (i - 1/2)\Delta x$ is the center of the cell I_i.

Let Δt be the time step and $t^n = n\Delta t$. Denote by W_i^n the approximation to the cell averages on I_i of the exact solution, at time t^n.

Given the states W_i^n and W_{i+1}^n, the intermediate Roe matrix is defined as

$$\mathcal{A}_{i+1/2} = \mathcal{A}(W_i^n, W_{i+1}^n).$$

Then, the numerical scheme can be written as follows:

$$W_i^{n+1} = W_i^n - \frac{\Delta t}{\Delta x}\left(\mathcal{A}_{i-1/2}^{+}(W_i^n - W_{i-1}^n) + \mathcal{A}_{i+1/2}^{-}(W_{i+1}^n - W_i^n)\right). \tag{4}$$

Concerning the stability requirements, a CFL-type condition has to be considered. Finally, the entropy-fix technique of Harten–Hyman has also been applied.

As it is well known, the presence of source or coupling terms in general systems of conservation laws can affect the quality of the numerical solution when steady or nearly steady state solutions are approximated. To handle such problems, the concept of *well-balanced* schemes has been considered by many authors (see, e.g., [BV94], [Bou04], [Gos00], [GL96], [LV98], etc.). In [PC04], a general definition of well balancing in the nonconservative framework is provided.

In the context of the shallow water equations (1), a scheme is *well balanced with order* γ if it approximates the steady state solutions with order γ, and it is *exactly well balanced* if the steady state solutions are exactly computed. In [PC04] it is proved that the scheme (4) is well balanced with order 2 for general stationary solutions. Moreover, it exactly solves stationary solutions corresponding to water at rest, i.e., it has the so-called C-property introduced and proved in [BV94].

3 High-Order Roe Schemes

The generalized Roe schemes introduced in Sect. 2 are only first-order accurate. To properly capture shock discontinuities, it is convenient to develop methods with a higher order of accuracy. Recently, in [CGP06] a high-order extension of the scheme (4) based on high-order reconstructions of the state variable W was presented. We give here a brief review about the construction of such extension. It will be assumed throughout this section that no dry zones are present.

For a given time t, we consider reconstructing functions of order p, $\mathcal{R}_{i+1/2}^{-,t}(x)$ and $\mathcal{R}_{i+1/2}^{+,t}(x)$, defined on $[x_i, x_{i+1/2}]$ and $[x_{i+1/2}, x_{i+1}]$, respectively. We denote by $W_{i+1/2}^{\pm}(t)$ the corresponding reconstructions at the intercell $x_{i+1/2}$. Then, the following semidiscrete formulation of the high-order extension of the numerical scheme (4) is considered ([CGP06])

$$
\begin{aligned}
W_i'(t) = -\frac{1}{\Delta x}\Bigg(& \mathcal{A}^+_{i-1/2}(W^+_{i-1/2}(t) - W^-_{i-1/2}(t)) \\
& + \mathcal{A}^-_{i+1/2}(W^+_{i+1/2}(t) - W^-_{i+1/2}(t)) \\
& + \int_{x_{i-1/2}}^{x_i} \mathcal{A}(\mathcal{R}^{+,t}_{i-1/2}(x)) \frac{\mathrm{d}}{\mathrm{d}x} \mathcal{R}^{+,t}_{i-1/2}(x)\, dx \\
& + \int_{x_i}^{x_{i+1/2}} \mathcal{A}(\mathcal{R}^{-,t}_{i+1/2}(x)) \frac{\mathrm{d}}{\mathrm{d}x} \mathcal{R}^{-,t}_{i+1/2}(x)\, dx \Bigg),
\end{aligned} \tag{5}
$$

where $\mathcal{A}_{i+1/2}$ is the intermediate Roe matrix corresponding to the states $W^-_{i+1/2}(t)$ and $W^+_{i+1/2}(t)$.

Remark 1. As we stated in Sect. 2, in our framework the nonconservative product $\mathcal{A}(W)W_x$ is interpreted as a Borel measure. Roughly speaking, the two first terms in the right-hand side of (5) are related to the singular part of the measure, while the integral terms are associated to its regular part. The technical details can be found in [CGP06].

The order of the scheme (5) is given by Theorem 3.2 in [CGP06]. In that article, it is also proved that the scheme is exactly well balanced for still water solutions and well balanced with order p for general stationary solutions.

Finally, to have a fully discretized scheme, an adequate high-order scheme has to be applied to (5) for time stepping.

4 The MRoe Scheme for the Shallow Water Equations

Recently, in [CGVP06] a modification of the Roe scheme for dealing with the appearance of wet/dry areas when solving the shallow water system (1) has been developed. This technique consists in replacing, at the intercells where a wet/dry transition is detected, the approximate linear Riemann problem by a nonlinear one, which is exactly solved. The numerical fluxes must be accordingly modified: see [CGVP06] for details. The resulting scheme will be called the *Modified Roe* (MRoe) scheme.

Two important properties should be remarked:

- The MRoe scheme is well balanced in the sense that it solves exactly the steady state solutions corresponding to water at rest, including or not wet/dry situations.
- The positivity of the values of h calculated at the intercells where a wet/dry front has been detected is assured except in some very special cases. In those cases, a restriction on the CFL condition can be considered to warranty the positivity of h.

5 High-Order Extension of the MRoe Scheme

We develop in this section a well-balanced finite volume scheme for solving (1) that is able to handle wet/dry situations, being at the same time high-order accurate in smooth wet areas.

The ingredients to construct the scheme have been developed in Sects. 3 and 4. The idea is to combine the high-order scheme (5) with the MRoe technique for treating wet/dry situations. This high-order extension of the MRoe scheme will be called the *HMRoe scheme.*

First of all, we have to select adequate variables to be reconstructed. As it was done in [CGP06], to have an exactly well-balanced scheme for water at rest solutions, the surface elevation $\eta = h - H$ has to be reconstructed. On the other hand, it is also important to choose a reconstruction of the water height h that preserves positivity. Thus, the variables considered are (h, q, η); if $(\tilde{h}, \tilde{q}, \tilde{\eta})$ represent the reconstructed values, then the reconstructed depth function is defined as $\tilde{H} = \tilde{h} - \tilde{\eta}$.

We can give now a complete description of the semidiscrete HMRoe scheme. Assume that the cell averages $W_i = [h_i, q_i, H_i]^T$ at a given time t are known (for the sake of clarity, we will drop the dependence on time). Then:

- Define the cell averages $\eta_i = h_i - H_i$. If $h_i < h_\varepsilon$, where h_ε is a given tolerance, we set $h_i = 0$ and $q_i = 0$.
- Use the cell averages on an adequate stencil to build reconstructing functions for each variable $\alpha \in \{h, q, \eta\}$
$$r^-_{\alpha,i+1/2}(x), \quad x \in [x_i, x_{i+1/2}]; \quad r^+_{\alpha,i+1/2}(x), \quad x \in [x_{i+1/2}, x_{i+1}],$$
and compute the reconstructed values at the intercell
$$\alpha^\pm_{i+1/2} = r^\pm_{\alpha,i+1/2}(x_{i+1/2}), \quad \alpha \in \{h, q, \eta\}.$$
Define also $r^\pm_{H,i+1/2}(x) := r^\pm_{h,i+1/2} - r^\pm_{\eta,i+1/2}(x)$ and consider the reconstructed states at the intercell $W^\pm_{i+1/2} = [h^\pm_{i+1/2}, q^\pm_{i+1/2}, H^\pm_{i+1/2}]^T$, where $H^\pm_{i+1/2} := h^\pm_{i+1/2} - \eta^\pm_{i+1/2}$. If a stencil contains a dry cell, we simply take $W^-_{i+1/2} = W^n_i$ or $W^+_{i+1/2} = W^n_{i+1}$, depending on the case.
- In the wet-bed case, the numerical fluxes are defined as in (5).
- If a wet/dry transition is detected, the technique introduced in Sect. 4 is applied. Some modifications on the numerical fluxes must be made depending on the kind of wet/dry transition found (see [CGVP06] for details).

Once the semidiscrete scheme has been defined, the system (5) is discretized in time by using a standard solver.

Following the results in [CGP06] and [CGVP06], the HMRoe scheme results to be exactly well balanced for water at rest solutions (including or not dry areas), and well balanced with the same order of the reconstruction operator for general stationary solutions.

6 Choice of the Reconstruction Operators: A Hyperbolic HMRoe Scheme

In [CGP06], parabolic WENO reconstruction operators (see [Shu97] for a report) were considered for the implementation of the high-order scheme (5) in wet domains, providing very good results. However, when dry or almost dry areas are present, the nonmonotone character of parabolas may lead to negative (nonphysical) values of the water height h (even though the WENO technique provides a way for damping the oscillations, this is not sufficient in general for avoiding the appearance of negative values of h). Although in the experiments performed the negative values of h are relatively small; they may lead the program to crash.

Instead of reconstructions of parabolic type, we will focus here on third-order *hyperbolic* reconstructions. Specifically, the piecewise hyperbolic method (PHM) introduced in [Mar94] will be applied. This method prescribes at each cell I_i a hyperbola that preserves the cell average, interpolates the lateral derivative of the solution with smaller absolute value, and assigns as the central derivative the harmonic mean of the lateral derivatives. This method is LTVB (local total variation bounded), i.e., the total variation of each hyperbola is bounded by $M\Delta x$, for some constant M. About the details on the method, see [Mar94].

The main reasons for using hyperbolas are monotonicity (it avoids the appearance of negative h), third order of accuracy on the whole cell (this leads to a third-order scheme), compact stencil, and lower total variation than parabolas. The main drawback of the method is the loss of total variation at local extrema.

With respect to the time discretization, a third-order TVD Runge–Kutta method (see [Shu97]) has been considered.

7 Numerical Results

We test the performances of the hyperbolic HMRoe scheme defined in Sect. 6. The CFL number is set to 0.9 unless otherwise stated. The wet/dry tolerance h_ε has been taken as 10^{-6} and the gravity constant is $g = 9.81$.

7.1 Exact Well-Balance Property for Still Water Solutions

We consider a simple test to verify that water at rest solutions is exactly computed with the HMRoe scheme. The bottom consists in a rectangular obstacle defined by the depth function

$$H(x) = \begin{cases} 1 & \text{if } x \in [6, 10], \\ 3 & \text{otherwise,} \end{cases}$$

Table 1. Steady subcritical flow

No. cells	L^1 error h	L^1 order h	L^1 error q	L^1 order q
80	2.36×10^{-3}	–	7.67×10^{-3}	–
160	3.07×10^{-4}	2.94	1.32×10^{-3}	2.54
320	1.40×10^{-5}	4.45	5.65×10^{-5}	4.55
640	9.76×10^{-7}	3.84	3.76×10^{-6}	3.91
1,280	6.45×10^{-8}	3.92	2.42×10^{-7}	3.96
2,560	4.28×10^{-9}	3.91	1.50×10^{-8}	4.01

in the domain $[0, 20]$. The initial water height has been taken as 1.5 to the left of the obstacle and 1 to its right. A mesh with 400 nodes has been considered.

As expected, the HMRoe scheme exactly preserves this stationary solution up to machine accuracy. The same can be said if the initial water height is 2 on the whole domain (wet-bed case) or if it is zero (no water).

7.2 Well-Balance Property for General Stationary Solutions

The approximate well-balance property of the HMRoe scheme for general stationary solutions is tested in this section. We consider the depth function

$$H(x) = 2 - 0.2\exp(-0.16(x-10)^2), \quad x \in [0, 20],$$

and the initial condition corresponding to the steady subcritical flow with discharge $q(x, 0) = 4.42$. This test has been studied in several works, e.g., [GHS03] or [VS02]. As expected, this solution is preserved by the HMRoe scheme with third order of accuracy (see Table 1); indeed, the order of approximation is close to 4 in this case.

7.3 Dambreak over a Plane

The ability of the HMRoe scheme for computing the advance of wet/dry fronts is tested in the experiments considered in this section that were previously analyzed in [CGVP06]. The computational domain considered is the interval $[-15, 15]$ and the bottom is given by the depth function

$$H(x) = 1 - x\tan(\alpha),$$

for some angle α. The initial conditions considered are

$$q(x, 0) = 0, \qquad h(x, 0) = \begin{cases} H(x) & \text{if } x < 0, \\ 0 & \text{otherwise.} \end{cases}$$

With regard to the boundary conditions, the discharge $q(0, t) = 0$ is imposed at $x = -15$ while a free boundary condition is considered at $x = 15$.

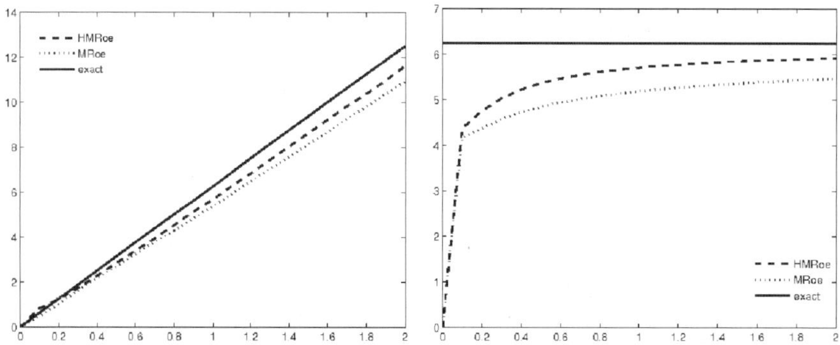

(a) Wet/dry front position (b) Wet/dry front velocity

Fig. 1. Dambreak problem over a plane. Case $\alpha = 0$

For this problem, the exact position and velocity of the advancing wet/dry front can be exactly computed (see [CGVP06]).

Depending on the sign of $\tan(\alpha)$, three different kinds of wet/dry fronts may appear. As in [CGVP06], we have considered the values $\alpha = 0$ (flat bottom), $\alpha = \pi/60$ (emerging topography), and $\alpha = -\pi/60$ (bottom with increasing depth).

The time evolution of the position and velocity of the wet/dry front is shown in Fig. 1a, b, respectively. Similar results are obtained for $\alpha = \pm\pi/60$.

7.4 Dry Bed Generation

We first consider an experiment over flat bottom proposed by Toro in [Toro01]. The initial conditions are

$$h(x,0) = 0.1, \qquad q(x,0) = \begin{cases} -0.3 & \text{if } x \geq 5, \\ 0.3 & \text{if } x < 5. \end{cases}$$

In this case, a dry bed is formed in the middle of two rarefaction waves traveling in opposite directions. The generation of the dry bed makes this problem numerically difficult.

The results obtained at time $t = 1$ with both the HMRoe and the MRoe schemes are compared with the exact solution in Fig. 2. The computational domain is the interval $[0, 10]$ and the space step is $\Delta x = 0.05$. In this case, the CFL number has been reduced to 0.8 to avoid the appearance of negative values of the water height. As it can be observed, the MRoe scheme leaves a small wet zone between the two rarefaction waves.

Let us consider now a modification of the previous test by including a nontrivial topography. Specifically, we consider the test proposed in [GHS03], where the depth function is given by

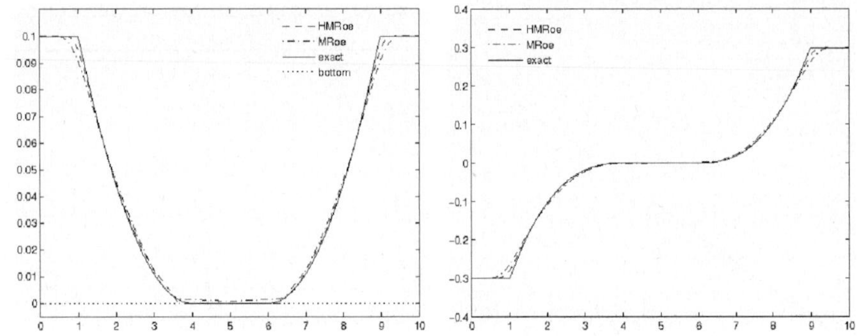

(a) Free surface elevation (b) Discharge q

Fig. 2. Dry bed generation on a flat bottom

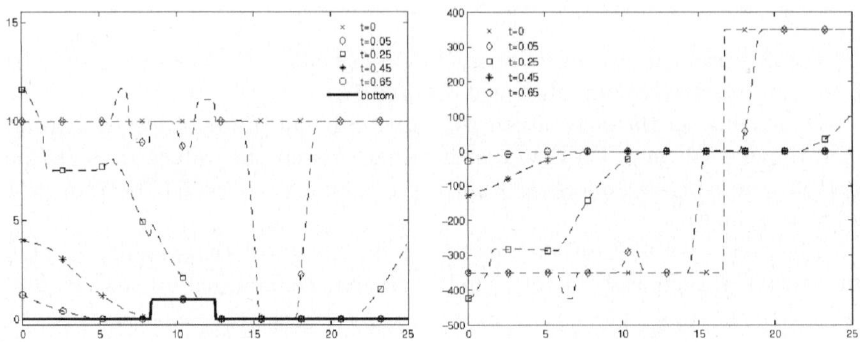

(a) Free surface elevation (b) Discharge q

Fig. 3. Dry bed generation over a nonflat bottom

$$H(x) = \begin{cases} 13 & \text{if } 25/3 < x < 12.5, \\ 14 & \text{otherwise,} \end{cases}$$

in the domain $[0, 25]$. The initial water height has been initialized to 10 and the initial discharge is

$$q(x, 0) = \begin{cases} -350 & \text{if } x < 50/3, \\ 350 & \text{otherwise.} \end{cases}$$

In Fig. 3, the results obtained at times 0, 0.05, 0.25, 0.45, and 0.65, with 300 nodes and CFL number 0.8 are shown. These results are in good agreement with those obtained in [GHS03].

7.5 Drain on a Nonflat Bottom

This experiment was proposed in [GHS03]. The bottom topography is defined by

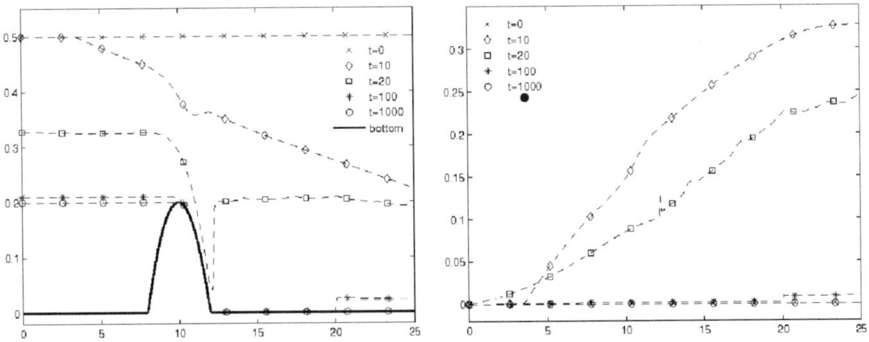

(a) Free surface elevation (b) Discharge q

Fig. 4. Drain on a nonflat bottom

$$H(x) = \begin{cases} 0.05(x-10)^2 & \text{if } 8 \leq x \leq 12, \\ 0.2 & \text{otherwise,} \end{cases}$$

in the domain $[0, 25]$, with initial conditions $h(x, 0) = H(x) + 0.3$ and $q(x, 0) = 0$. We impose a free boundary condition at the left boundary, and an outlet condition on a dry bed at the right one. The flow reaches a stationary state with $h = 0.2$ to the left of the bump and $h = 0$ to its right.

The HMRoe scheme has been applied with 300 nodes and CFL number 0.8, at times $t = 0$, 10, 20, 100, and 1,000. The results obtained are shown in Fig. 4, and can be directly compared with those presented in [GHS03].

Acknowledgment

This research has been partially supported by the Spanish Government Research project BFM2003-07530-C02-02.

References

[BV94] Bermúdez, A., Vázquez, M.E.: Upwind methods for hyperbolic conservation laws with source terms. Comp. & Fluids, **23**, 1049–1971 (1994)

[Bou04] Bouchut, F.: Nonlinear Stability of Finite Volume Methods for Hyperbolic Conservation Laws and Well-Balanced Schemes for Sources. Birkäuser (2004)

[CGP06] Castro, M., Gallardo, J.M., Parés, C.: High order finite volume schemes based on reconstruction of states for solving hyperbolic systems with nonconservative products. Applications to shallow water systems. Math. Comp., **75**, 1103–1134 (2006)

[CGVP06] Castro, M.J., González-Vida, J.M., Parés, C.: Numerical treatment of wet/dry fronts in shallow flows with a modified Roe scheme. Math. Mod. Meth. App. Sci., **16**, 897–931 (2006)

[DLM95] Dal Maso, G., LeFloch, Ph., Murat, F.: Definition and weak stability of nonconservative products. J. Math. Pures Appl., **74**, 483–548 (1995)

[GHS03] Gallouët T., Hérard, J.-M., Seguin, N.: Some approximate Godunov schemes to compute shallow-water equations with topography. Comp. & Fluids, **32**, 479–513 (2003)

[Gos00] Gosse, L.: A well-balanced flux-vector splitting scheme designed for hyperbolic systems of conservation laws with source terms. Comput. Math. Appl., **39**, 135–159 (2000)

[GL96] Greenberg, J.M., LeRoux, A.Y.: A well-balanced scheme for the numerical processing of source terms in hyperbolic equations. SIAM J. Numer. Anal., **33**, 1–16 (1996)

[LV98] LeVeque, R.J.: Balancing source terms and flux gradients in high-resolution Godunov methods: the quasi-steady wave-propagation algorithm. J. Comput. Phys., **146**, 346–365 (1998)

[Mar94] Marquina, A.: Local piecewise hyperbolic reconstructions for nonlinear scalar conservation laws. SIAM J. Sci. Comput., **15**, 892–915 (1995)

[NPGN06] Noelle, S., Pankratz, N., Puppo, G., Natvig, J.R.: Well-balanced finite volume schemes of arbitrary order of accuracy for shallow water flows. J. Comput. Phys., **213**, 474–499 (2006)

[PC04] Parés, C., Castro, M.: On the well-balanced property of Roe's method for nonconservative hyperbolic systems. Applications to shallow water systems. ESAIM: M2AN, **38**,, 821–852 (2004)

[Shu97] Shu, C.-W.: Essentially non-oscillatory and weighted essentially non-oscillatory schemes for hyperbolic conservation laws. ICASE Report 97-65 (1997)

[Toro01] Toro, E.F.: Shock-Capturing Methods for Free-Surface Shallow Flows. J. Wiley & Sons (2001)

[Tou92] Toumi, I.: A weak formulation of Roe's approximate Riemann solver. J. Comput. Phys., **102**, 360–373 (1992)

[VS02] Vukovic, S., Sopta, L.: ENO and WENO schemes with the exact conservation property for one-dimensional shallow water equations. J. Comput. Phys., **179**, 593–621 (2002)

A Simple Well-Balanced Model for Two-Dimensional Coastal Engineering Applications

F. Marche

1 Introduction

The description of nearshore and coastal hydrodynamic is a complex topic. The main aim is to determine the water level and the discharge at a specific given place and time. Reliable mathematical models as well as fast, accurate and robust numerical simulations are important for suitable prediction of the morpho-dynamical evolution of the coasts or the transport of pollutants for instance.

Using the shallow water equations, the wave breaking phenomenon is not numerically modeled in details, as it can be done with the Boussinesq-like equations, but can be taken into account through the natural steepening of the solution and the development of discontinuities. However, most of the numerical models used in coastal engineering rely on finite-difference methods, classically unable to compute rapidly evolving flows and the propagation of bores. In the better case, the numerical schemes are modified, introducing some artificial dissipation through the propagating fronts [11], which can be empirically calibrated using the shock-wave theory. Some improvements have been made since a decade to rationally apply the shock-wave theory through finite-volume framework [3], considering the shallow water equations as an hyperbolic system of conservation laws, eventually with a source term, and using shock-capturing finite-volume methods.

Nonetheless, recent studies by Brocchini et al. [5] have shown that computation of two-dimensional bore propagation over a complex topography is a difficult task with specific and acute numerical requirements. Actually, dissipation in two-dimensional bores crossing breakwaters leads to the generation of large scale vorticity with vertical axes. Numerical investigations by Brocchini et al. [5] revealed that this kind of study demands a high-order accuracy. Moreover, the treatment of the bed slope source term requires an adapted method to accurately compute the interactions between incident propagating waves and the topography variations, leading to longshore variations in waves breaking and generating particular coastal circulations.

Besides, one of the most important problems faced when modeling nearshore flows is to obtain a suitable description of the shoreline boundary of the domain, crucial for the implementation of a robust model for nearshore flows. Modeling the flow near the shoreline is far from trivial but is fundamental, since the region of the swash and run-up of waves is where a noticeable part of the sediment transport occurs. Actually, flow properties change rapidly as the water depth vanishes and the shoreline location is always evolving, as a part of the solution itself. Various methods have been proposed in the concerned literature to cope with these difficulties, using for instance coordinate-transformations [17] or extrapolation methods combined with smoothing and filtering procedures [13]. But these methods often lead to complex algorithms, are computationally expensive and not sufficiently flexible for realistic two-dimensional simulations. The simple idea of using a conservative scheme on a fixed Cartesian grid have of course been introduced [4] but the authors faced several limitations in practice, due to the use of fractional step methods for the treatment of the bed-slope source term and the lake of positivity preserving properties concerning the water depth.

From these observations, the idea of applying a high-order accuracy finite-volume well-balanced scheme is attractive. On one hand we have some classical and robust well-balanced approaches, initially introduced for hydraulic applications [10, 8]. These schemes have been proved to accurately deal with wet/dry propagating fronts, even on a nonflat topography, and can easily be extended to second-order accuracy. On the other hand, some recent improvements have been introduced concerning very high order well-balanced schemes [16, 19] which are unable to deal with wet/dry fronts due to the WENO reconstruction processes. At this level, it is important to highlight that in the study of propagating bores in the swash zone, we must deal with random motions of the shoreline, involving run-up and back-wash processes. This situation is far more complex to simulate than the classical hydraulic test cases usually used to assess the ability of schemes to deal with wet/dry fronts, since even a small instability generated at the beginning of the process will be dramatically amplified during the evolution. Consequently, several schemes acknowledged for their ability in handling flooding and drying processes fail in providing accurate enough results when studying run-up of bores [14] and a strict preservation of the positivity of the water depth must be achieved.

Following the quest of deriving a very high order well-balanced finite-volume scheme which can accurately handle time varying wet/dry fronts, we introduce here a new finite-volume model for the solution of two-dimensional shallow water equations [14]. The equations are solved using a linearized Riemann solver which relies on a symmetrizing change of variables and enables to accurately deal with dry areas [9]. This solver is combined with the recent “hydrostatic reconstruction” [1] for the treatment of the topography variations. In addition, an efficient and simple algorithm is proposed to accurately simulate wetting and drying phenomena. The whole model is robust enough to deal with periodic wet/dry fronts, even in the two-dimensional framework.

A second-order reconstruction based on the MUSCL formalism is also performed, preserving the well-balanced properties and the ability to handle moving shoreline. This raises an accurate well-balanced model which can easily handle bore propagation over strong varying topography.

A new model '*SURF_WB*' based on this method has been applied to a number of wave propagation situations. First, the ability to accurately handle periodic moving shoreline is highlighted through comparisons with analytical solutions. Then, we investigate some simulations related to coastal engineering. We first study briefly the propagation of one-dimensional regular broken waves in the surf zone. Then we focus on the two-dimensional run-up of a solitary wave on a sloping beach with longshore variations.

2 Numerical Model

2.1 Basic Equations

We consider first the following one-dimensional hyperbolic system with source term:

$$\mathbf{U}_{,t} + F(\mathbf{U})_{,x} = S(\mathbf{U}), \tag{1}$$

$$\mathbf{U} = \begin{pmatrix} h \\ hu \end{pmatrix}, \; F(\mathbf{U}) = \begin{pmatrix} hu \\ hu^2 + \dfrac{g}{2}h^2 \end{pmatrix}, \; S(\mathbf{U}) = \begin{pmatrix} 0 \\ -g\,h\,d_x \end{pmatrix},$$

where $()_{,x}$ stands for the derivate along the x-direction, u is the depth-averaged velocity, and h is the local water depth. $\mathbf{U}$ is the vector for the conservative variables, $F(\mathbf{U})$ stands for the flux functions and $S(\mathbf{U})$ the bed slope source term. We recall briefly the well-balanced approach used by Marche et al. [14]. A FV discrimination of this system is written as follows:

$$\frac{\mathbf{U}_i^*(t) - \mathbf{U}_i^n(t)}{\Delta t} + \frac{1}{\Delta x}\left(\mathbf{F}^{n,-}_{i+\frac{1}{2}} - \mathbf{F}^{n,+}_{i-\frac{1}{2}}\right) = \mathbf{S}^n_{c,i}, \tag{2}$$

with left and right numerical fluxes through the mesh interfaces:

$$\mathbf{F}^{n,-}_{i+\frac{1}{2}} = \mathbf{F}^n_{i+\frac{1}{2}} + \mathbf{S}^n_{i+\frac{1}{2}-}, \qquad \mathbf{F}^{n,+}_{i+\frac{1}{2}} = \mathbf{F}^n_{i+\frac{1}{2}} + \mathbf{S}^n_{i+\frac{1}{2}+}, \tag{3}$$

$$\mathbf{F}^n_{i+\frac{1}{2}} = F(\mathbf{U}_{i+\frac{1}{2}}(0, \mathbf{U}^n_{i+\frac{1}{2}-}, \mathbf{U}^n_{i+\frac{1}{2}+})). \tag{4}$$

2.2 Interface Values

The interface value $\mathbf{U}_{i+\frac{1}{2}}(0, \mathbf{U}_{i+\frac{1}{2}-}, \mathbf{U}_{i+\frac{1}{2}+})$ between cell i and $i+1$ is obtained from a VFRoe-ncv linearized Riemann solver. This solver relies on the symmetrizing change of variable $\mathbf{W}(\mathbf{U}) = {}^t(2c, u)$ which preserves the water depth positivity at least for interface values [8]. The formalism of this solver is not recalled in details. It relies on the exact resolution of a linearized problem around an averaged state $\tilde{\mathbf{W}}$. The eigenvalues of the linearized convection matrix are

$$\tilde{\lambda}_1 = \tilde{u} - \tilde{c}, \quad \tilde{\lambda}_2 = \tilde{u} + \tilde{c},$$

where $c = \sqrt{gh}$. Then, the exact solution of the linearized Riemann problem is given by the sign of the eigenvalues $\tilde{\lambda}_k$:

- If $\tilde{\lambda}_1 > 0$ or $\tilde{\lambda}_2 < 0$, then the flow is super critical and we recover a classical upwinding. The interface value is defined as follows

$$\mathbf{W}^*_{i+\frac{1}{2},j} = \begin{cases} \mathbf{W}_i \text{ if } \tilde{\lambda}_k > 0 \,\forall k \\ \mathbf{W}_{i+1} \text{ if } \tilde{\lambda}_k < 0 \,\forall k. \end{cases} \tag{5}$$

- If $\tilde{\lambda}_1 < 0$ and $\tilde{\lambda}_2 > 0$ then the flow is sub-critical and we obtain

$$c^*_{i+\frac{1}{2},j} = \tilde{c} - \frac{1}{4}(u_{i+1,j} - u_{i,j}) \text{ and } u^*_{i+\frac{1}{2},j} = \tilde{u} - (c_{i+1,j} - c_{i,j}). \tag{6}$$

We can finally recover conservative variables, using the inverse change of variable. The key point is that this interface value is computed from values $\mathbf{U}_{i+\frac{1}{2}-}$ and $\mathbf{U}_{i+\frac{1}{2}+}$ issued from a combined limited linear/hydrostatic reconstruction.

2.3 Variable Reconstruction

More precisely, considering the cell i, we compute first linear reconstructions $\mathbf{U}_{i,r}$ and $\mathbf{U}_{i,l}$, respectively, at $i+\frac{1}{2}-$ and $i-\frac{1}{2}+$, using a *minmod* limiter. Values of $H_{i,l}$ and $H_{i,r}$, where $H = h + d$, are also reconstructed, and we deduce reconstructions of $d_{i,l}$ and $d_{i,r}$. Then, interface topography values $d_{i+\frac{1}{2},j}$ are defined with $d_{i+\frac{1}{2}} = max(d_{i,r}, d_{i+1,l})$.

The positivity preserving hydrostatic reconstruction of the water height is thus defined as follows:

$$h_{i+\frac{1}{2}-} = max(0, h_{i,r} + d_{i,r} - d_{i+\frac{1}{2}}), \;\; h_{i+\frac{1}{2}+} = max(0, h_{i+1,l} + d_{i+1,l} - d_{i+\frac{1}{2}}),$$

and we deduce reconstructed values on each side of the interface:

$$\mathbf{U}_{i+\frac{1}{2}-} = \begin{pmatrix} h_{i+\frac{1}{2}-} \\ h_{i+\frac{1}{2}-}\, u_{i,r} \end{pmatrix}, \quad \mathbf{U}_{i+\frac{1}{2}+} = \begin{pmatrix} h_{i+\frac{1}{2}+} \\ h_{i+\frac{1}{2}+}\, u_{i+1,l} \end{pmatrix}. \tag{7}$$

Finally we introduce:

$$\mathbf{S}^n_{i+\frac{1}{2}-} = \begin{pmatrix} 0 \\ \frac{g}{2} h^2_{i+\frac{1}{2}-} - \frac{g}{2} h^2_{i,r} \end{pmatrix}, \quad \mathbf{S}_{i-\frac{1}{2}+} = \begin{pmatrix} 0 \\ \frac{g}{2} h^2_{i,l} - \frac{g}{2} h^2_{i-\frac{1}{2}+} \end{pmatrix}, \tag{8}$$

and a centered source term discrimination is added to achieve second order well balancing:

$$\mathbf{S}_{c,i} = \begin{pmatrix} 0 \\ g\frac{h_{i,l}+h_{i,r}}{2}(d_{i,l}-d_{i,r}) \end{pmatrix}. \tag{9}$$

The extension to the two-dimensional framework on Cartesian meshes is straightforward.

2.4 Wetting and Drying Procedure

No special tracking procedure has been used in this model. A distinction is made between wet and dry cells with the introduction of an artificial threshold value $h_{tol} = 10^{-10}$ m for the definition of a dry cell. The detailed wetting/drying procedure is simple. At the beginning of each time step, we search for dry cells, i.e., cells for which the water depth is less than h_{tol}, and set them to $h = h_{tol}$. The key point is to apply this "wetting procedure" to the quantity which are issued from the hydrostatic reconstruction instead of the natural quantities, leading to a better accuracy at the shoreline and a better robustness in practice.

Thereafter, the shoreline is constructed from the mesh interface between wet and dry cells, without any interpolation. We have three possible types of internal edge, which are *wet/wet*, *wet/dry*, and *dry/dry*. In the case *dry/dry*, nothing is done. The cases *wet/dry* and *wet/wet* are automatically handle by the Riemann solver and it is worth mentioning that in practice, the case *wet/dry* reduces to the treatment of a reflection at a solid wall. So, thanks to the robustness of the VFRoe-ncv solver, the computation is directly performed, without any distinction between these possible configurations and the shoreline tracking is naturally included into the computation.

3 Numerical Assessments

We propose here some validations relying on analytical solutions.

3.1 The Periodic Carrier and Greenspan Solution

We compare here numerical results (solid lines) with the periodic Carrier and Greenspan one-dimensional analytical solutions [6] (in dots). This solution represents the motion of a periodic wave traveling shoreward and being reflected out to sea, generating a standing wave which is let run-up and run-down on a plane beach. Focusing on the numerical way to handle wet/dry fronts, experience shows that test cases involving oscillating wet/dry front are far more difficult to compute than the classical cases found in [8] for instance. Numerical investigations have shown that even the smallest spurious oscillation is immediately amplified during the next period, leading to an instability of the computation. Furthermore, this case is only an idealized view of the run-up and back-wash processes, since the oscillations are periodic with no wave

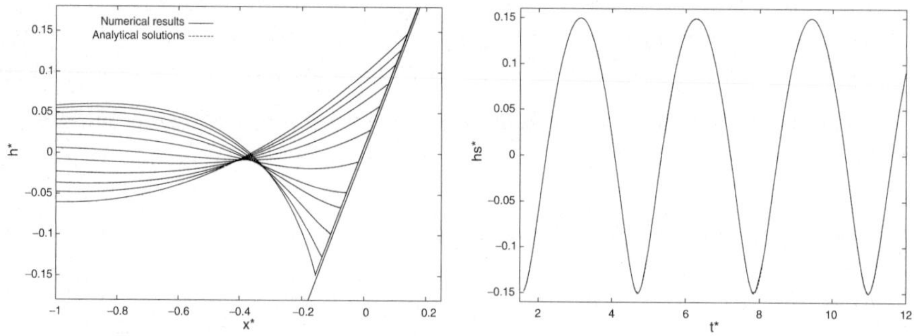

Fig. 1. The carrier and greenspan solution. Comparison between numerical and analytical results. *Left*: surface elevation versus the onshore coordinate for different values of time. *Right*: time series of the shoreline motion

breaking. In reality, incident bores may generate random run-up with possible resonance phenomena. However, it has been observed that even though waves may break just before they reach the shoreline, the tip of the resulting swash motion has many similarities with the nonbreaking run-up of waves and this test case is generally acknowledged to be a suitable assessment of shoreline motions computation. Fig. 1 highlights the ability of this new model. The analytical solution is used as a left inlet boundary condition and we use $\Delta x = 0.01$ and $CFL = 0.7$. The shoreline oscillations are accurately computed even after a large number of period. Evaluation of the L^2-error for this test can be found in [14]. We stress out that many other well-balanced methods, highly validated in hydraulic situations, are unable to provide such results [14].

3.2 The Two-Dimensional Thacker Solution

We perform a comparison with the analytical solution of Thacker [18], describing free-boundary free-surface oscillations in a two-dimensional parabolic basin. This case is perhaps one of the more difficult case to handle for a numerical model, since it involves a two-dimensional periodic moving shoreline with a no radially symmetrical configuration. The shoreline is a circle in the (x, y) plane and the motion is such that the center of the circle orbits the center of the basin, while the surface remains planar with constant gradient at any given instant. We use here $\Delta x = \Delta y = 0.01\,\mathrm{m}$ and $CFL = 0.7$.

Tiny distortions can be observed on Fig. 2 but the planar aspect of the surface is preserved even after several periods.

It is more difficult to model accurately the velocity. As highlighted in [14] the L^2-error is always larger for the velocity and we can see on Fig. 3 that the largest discrepancies are located close to the moving wet/dry interface and where the water depth is vanishing. It is in agreement with the fact that the velocity is obtained here from the discharge. Therefore, even a small error on

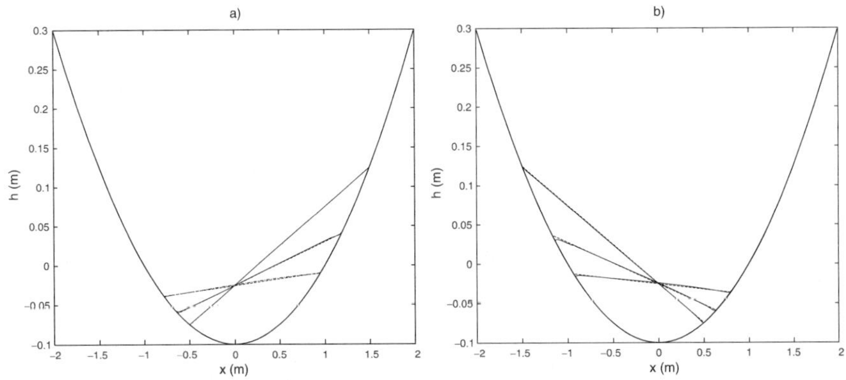

Fig. 2. Thacker's planar solution. Comparison between numerical results (in *solid lines*) and analytical solution (in *dotted lines*). Centerline free surface profiles for the surface elevation h are plotted versus the x coordinate, for $y = 0$, at (**a**) $\mathrm{t} = 2\mathrm{T}$, $\mathrm{t} = 2\mathrm{T} + \mathrm{T}/6\mathrm{T}$, $\mathrm{t} = 2\mathrm{T} + \mathrm{T}/5\mathrm{T}$ and (**b**) $\mathrm{t} = 2\mathrm{T} + \mathrm{T}/4$, $\mathrm{t} = 2\mathrm{T} + \mathrm{T}/3$, $\mathrm{t} = 2\mathrm{T} + \mathrm{T}/2$

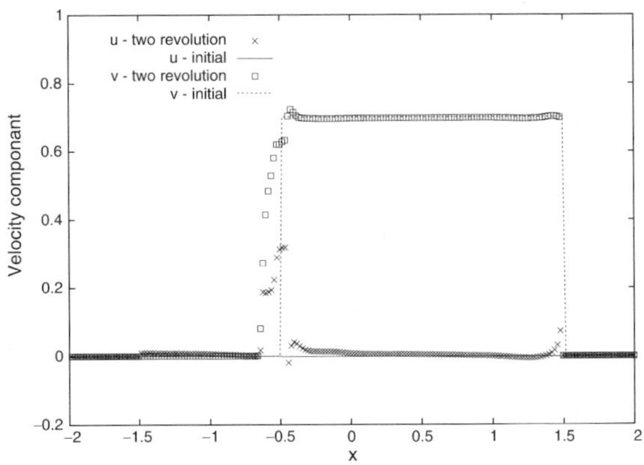

Fig. 3. Thacker's planar solution. Comparison between initial centerline profiles for the two components (in *solid* and *dashed lines*) of the velocity and the same centerline profile after two complete revolutions (in *dots*)

the vanishing water depth leads to a larger error on the velocity. However, the computation of the discharge during the oscillations is as accurate as for the water depth. The numerical results for u and v are considered satisfactory and these discrepancies are not amplified for long time simulations and do not disturb the accuracy of the moving shoreline predictions. Note that these results are almost as accurate as those presented in the literature, relying on far more complex or expensive algorithms, like extrapolation and filtering [13] or even adaptive refinement method [12].

4 Coastal Engineering Applications

4.1 Breaking Waves Propagation in the Surf Zone

We show here the ability of the model to compute the transformation of regular broken waves in the surf zone. We show a comparison, inspired from [2], between numerical results and a set of data based on the experiment of Cox [7]. Measurements of the surface elevation and velocity were taken at four locations inside the ISZ. The seaward boundary condition is given by time series of water depth measured at the first location, using an inlet generating/absorbing boundary condition and assuming the flow to be subcritical. The computational domain is discretized by 200 nodes and the simulation was run with a constant time step $\Delta t = 0.01$. The solution settles to a quasisteady state shortly after several bores have climbed up the beach. Once this steady state has been reached, it yields relatively small shoreline motions. We can observe in Fig. 5 the comparison between computed and measured time series of the surface elevation at four locations inside the surf zone. It appears that our model provides accurate results with small discrepancies. The nonlinear wave distortion is well computed and the wave height is accurately predicted. The minimum wave elevation and the periodicity of the waves are accurately predicted. In addition, there is no phase shift between numerical results and measurements. We can observe in Fig. 4 numerical results for the

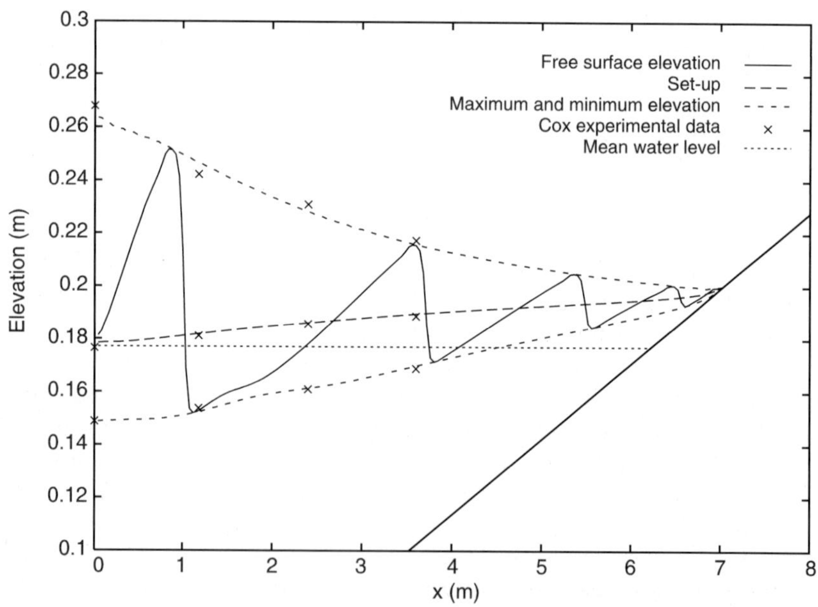

Fig. 4. Propagation of breaking waves. Spatial evolution of the wave train elevation. Comparison between numerical results and the experimental data of Cox for the set-up and the maximum and minimum elevation

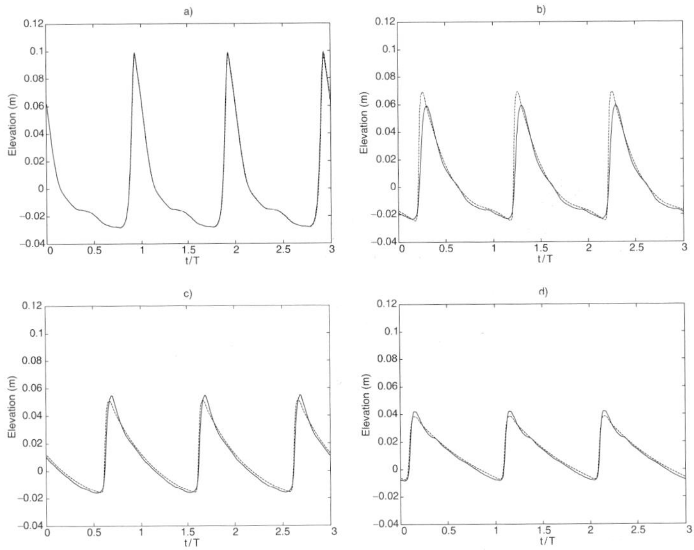

Fig. 5. Propagation of breaking waves. Time series of wave elevation at different locations in the inner surf zone. Comparison between the numerical model (in *dashed line*) and experiments by Cox (*solid line*) at (**a**) $x = 0\,\mathrm{m}$, (**b**) $x = 1.2\,\mathrm{m}$, (**c**) $x = 2.4\,\mathrm{m}$, and (**d**) $x = 3.6\,\mathrm{m}$

wave elevation profile, the maximum and minimum wave elevation and the set-up.

4.2 Run-Up of a Solitary Wave

To further illustrate the ability of the model in the computation of moving shoreline, a strongly two-dimensional test is performed, which combines several aspects of importance in the processes of run-up. This test was performed by Zelt [20] in the course of studying the response of harbors to long wave excitation. It involves the run-up of a large solitary wave, regarded as a tsunami, in a bay with a sloping bottom. The test geometry combines a curved (sinusoidal) still water shoreline with a sloping nearshore bathymetry that merges with a constant depth region offshore (see Fig. 6). We generate the solitary wave at the offshore boundary and study the corresponding run-up and back-wash. The shape of the wave depends on the nonlinear parameter α, which is the wave height to the offshore water height ratio. The lateral boundary conditions are solid walls. During the run-up process and when the wave amplitude is large enough, the wave is expected to bend toward the lateral boundaries because of refraction. Run-up first occurs at these boundaries because of the steep slope. While the wave still propagates shoreward at the center of the bay, where the slope is more gentle, a back-wash can be observed from the lateral boundaries and reflection may tend to direct wave

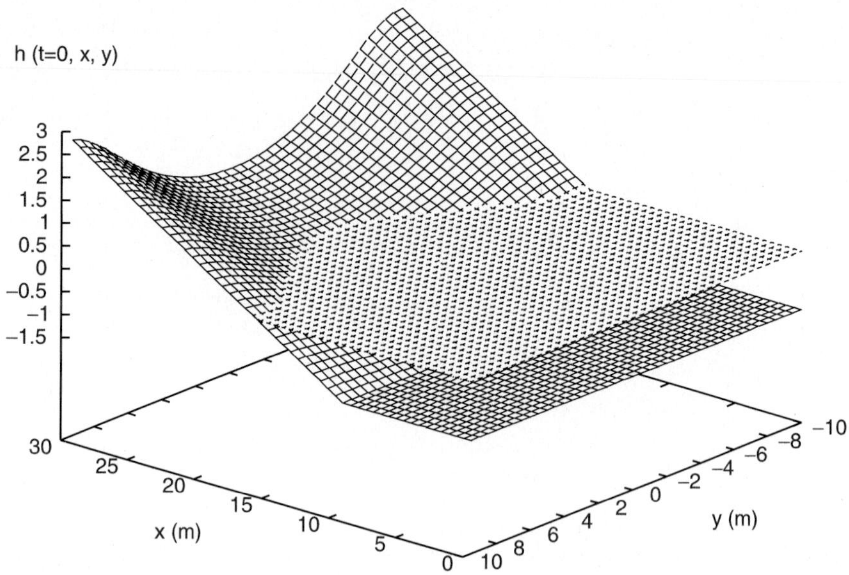

Fig. 6. Zelt's test for the run-up of large tsunami wave: bottom topography of the basin and initial free surface

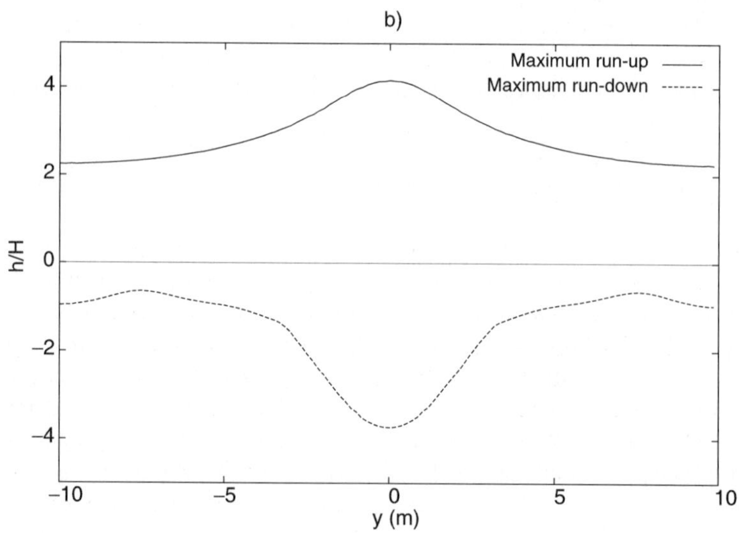

Fig. 7. Run-up of a large tsunami wave. The maximum run-up and the minimum run-down as a function of the longshore position. The run-up is normalized by the solitary wave amplitude H

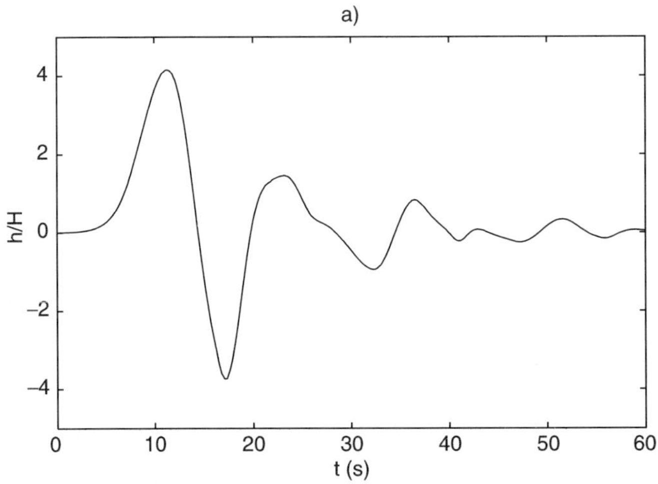

Fig. 8. Run-up of a large tsunami wave. Time series of the run-up perpendicular to the shoreline at a centered longshore location

energy toward the shoreline at the center of the bay. The combination of these various processes may enhanced the run-up processes at the center of the bay.

The results of the simulations are presented, for a value $\alpha = 0.02$, as time series of the surface elevation of the shoreline in the cross shore direction at the centered locations in the longshore direction (Fig. 8), and as the normalized maximum run-up and back-wash as a function of the longshore location (Fig. 7). The normalization by the amplitude of the solitary wave gives a good representation of the absolute magnitude of the shoreline motion within the bay.

The numerical results provided by the *SURF_WB* model are qualitatively similar to those obtained by Zelt, with a finite-element and computationally expensive Lagrangian description of the flow. Even if it is not shown here, we observe a good behavior with respect to the nonlinear parameter α. The well-balanced method enables us to compute the convergence towards the steady state at rest, after the evacuation of all the refracted and reflected waves from the computational domain (approximatively 150 s). It may provide useful estimations for the characteristic time scales of the phenomenon.

5 Conclusion

From these validations and applications, this new model appears as a promising and efficient tool for the study of many coastal hydrodynamic processes, including wave breaking and two-dimensional tsunami propagation over complex topography. The two-dimensional physical processes involved in the last application can be accurately reproduced and the finite-volume framework, with accurate shock-capturing abilities, clearly enables to handle the propagation and run-up of breaking waves, which was not possible with finite-element and finite-difference approaches. Moreover, the shoreline tracking method proposed here is far more simpler to implement and less computationally expensive than many methods found in the literature. However, we need to validate the model with laboratory and observations-based data to assess the real improvements brought up by the well-balanced approach. Several other applications, more related to coastal engineering, have also been investigated, like the study of wave-induced mean currents or even the generation of macrovorticity due to the longshore variations of energy dissipation through the breaking waves [15].

References

1. Audusse, E., Bouchut, F., Bristeau, M.O., Klein, R., Perthame, B. A fast and stable well-balanced scheme with hydrostatic reconstruction for shallow water flows. *SIAM J. Sci. Comp.*, **25**(6), 2050–2065 (2004).
2. Bonneton, P., Marieu, V., Dupuis, H., Sénéchal, N., Castelle, B. Wave transformation and energy dissipation in the surf zone: Comparison between a non-linear model and field data. *J. Coast. Res.*, **39**, 329–3333 (2004).
3. Brocchini, M., Bernetti, R., Mancinelli, A., Albertini, G. An efficient solver for nearshore flows based on the WAF method. *Coast. Eng.*, **43**, 105–129 (2001).
4. Brocchini, M., Svendsen, I.A., Prasad, R.S., Belloti, G. A comparison of two different types of shoreline boundary conditions. *Comput. Methods Appl. Mech. Eng.*, **191**, 4475–4496 (2002).
5. Brocchini, M., Mancinelli, A., Soldini, L., Bernetti, R. Structure-generated macrovortices and their evolution in very shallow depths. *Proc. Coastal Eng.*, 772–783 (2002).
6. Carrier, G.F., Greenspan, H.P. Water waves of finite amplitude on a sloping beach. J. Fluid Mech., **4**, 97–109 (1958).
7. Cox, D.T. Experimental and numerical modelling of surf zone hydrodynamics. PhD Dissertation, University of Delaware, Newark (1995).
8. Gallouet, T., Herard, J.M., Seguin, N. Some approximate Godunov schemes to compute shallow-water equations with topography. *Comput. Fluids*, **32**, 479–513 (2003).
9. Gallouet, T., Herard, J.M., Seguin, N. On the use of some symetrizing variables to deal with vacuum. *Calcolo*, **40**, 163–194 (2003).
10. Greenberg, J.M., Leroux, A.Y. A well balanced scheme for the numerical processing of source terms in hyperbolic equations. *SIAM J. Numer. Anal.*, **33**(1), 1–16 (1996).
11. Hibberd, S., Peregrine, D.H. Surf and run-up on a beach: A uniform bore. *J. Fluid. Mech.*, **95**(2), 323–345 (1979).
12. Hubbard, M.E., Dodd, N. A 2D numerical model of wave run-up and overtopping. *Coast. Eng.*, **47**, 1–26 (2002).
13. Lynett, P.J., Wu, T.S., Liu, P.L.F. Modeling wave runup with depth-integrated equations. *Coast. Eng.*, **46**, 89–107 (2002).

14. Marche, F., Bonneton, P., Fabrie, P., Seguin, N. Evaluation of well-balanced bore-capturing schemes for 2D wetting and drying procedure. *Int. J. Numer. Methods Fluids*, **53**(5), 867–894 (2006); available on the Web site.
15. Marche, F., Bonneton P., Fabrie, P., Seguin, N. Evaluation of well-balanced bore-capturing schemes for 2D wetting and drying processes. *Internat. J. Numer. Methods Fluids*, **53**(5), 867–894 (2007).
16. Noelle, S., Pankratz, N., Puppo, G., Natvig, J.R. Well-balanced finite volume schemes of arbitrary order of accuracy for shallow water flows, *J. Comp. Phys.* **213**, 474–499 (2006).
17. Prasad, R.S., Svendsen, I.A. Moving shoreline boundary condition for nearshore models, *Coast. Eng.*, **49**, 239–261 (2003)
18. Thacker, W.C. Some exact solutions to the nonlinear shallow-water wave equations. *J. Fluid Mech.*, **107**, 499–508 (1981).
19. Xing, Y., Shu, C.W. A New Approach of high order well-balanced finite volume WENO schemes and discontinuous Galerkin methods for a class of hyperbolic systems with source terms, *Com. Comput. Phys.* 2006, **1**(1), 101–135.
20. Zelt, J.A. Tsunamis: The response of harbors with sloping boundaries to long wave exitation. *Tech. Rep. KH-R-47* 1986; California Institute of Technology.

A Model for Bed-Load Transport and Morphological Evolution in Rivers: Description and Pertinence

A. Paquier and K. El Kadi

1 Introduction

Sediment transport is one of the main process that should be considered in studying "natural" rivers. Even if sediment load is low, exchanges are always occurring between banks, bottom and the water flow. This process is at the origin of the morphology of the rivers and may generate huge consequences on human activities and environment. Most part of the sediment transport occurs during high flows. Flood events and thus unsteady flows should be modelled in a detailed way to assess the main morphological changes.

The model called RubarBE was developed for simulating flow and sediment transport in mountain streams. It is based on a one-dimensional description of the river bed topography using a set of cross sections along the main axis of the river.

Here after, the various parts of the model are described. Then, the numerical scheme is detailed and its stability is discussed. Finally, one example is provided to demonstrate the coherence of the model with physical processes and its applicability to simulate real conditions.

2 Description of the Governing Equations

The model relies on the de Saint-Venant equations (1) and (2) to describe the water flow propagation:

$$\partial_t A + \partial_x Q = q, \tag{1}$$

$$\partial_t Q + \partial_x (\beta \frac{Q^2}{A}) + gA\partial_x z = -g\frac{Q^2}{K^2 A R^{4/3}} + kq\frac{Q}{A}, \tag{2}$$

in which t is time (s), x stream-wise coordinate (m), A cross-sectional flow area (m^2), Q water discharge (m^3/s), q lateral water flow per unit of length (m^2/s), R hydraulic radius (m), z water surface elevation (m), g acceleration due to gravity (m/s^2), K Manning–Strickler coefficient ($m^{1/3}/s$), β coefficient of quantity of movement and k ratio between the velocity of the main flow and the axis velocity of the lateral flow.

The mean evolution of the bed geometry comes from the sediment continuity equation (3):

$$(1-p)\partial_t A_s + \partial_x Q_s = q_s \tag{3}$$

in which A_s is bed-material area (m^2), Q_s sediment discharge (m^3/s), q_s lateral sediment flow per unit of length (m^2/s) and p porosity.

2.1 Space Lag Equation

The sediment discharge Q_s is calculated from the maximum sediment transport capacity C_s through a space lag equation (4) (non-equilibrium sediment transport approach with exponential trend).

$$\partial_x Q_s = \frac{(C_s - Q_s)}{D_{char}} \tag{4}$$

in which D_{char} is a distance that characterizes the ability of sediment transport Q_s to reach the value of the sediment transport capacity C_s.

For trapping suspended load, Fig. 1 shows the results of calculation using laboratory experiment data from [van Rijn, 1987]: rectangular channel, width: 0.5 m, $d_{50} = 0.16$ mm, D_{char} about 0.8 m, $Q_s = 0.02$ kg/s, $Q = 0.009945\, m^3/s$, $\Delta x = 0.5$ m, $\Delta t = 10$ s.

For bed load transport in rivers, the space lag D_{char} is generally very short (a few meters), which means that it can be often neglected when compared to space step.

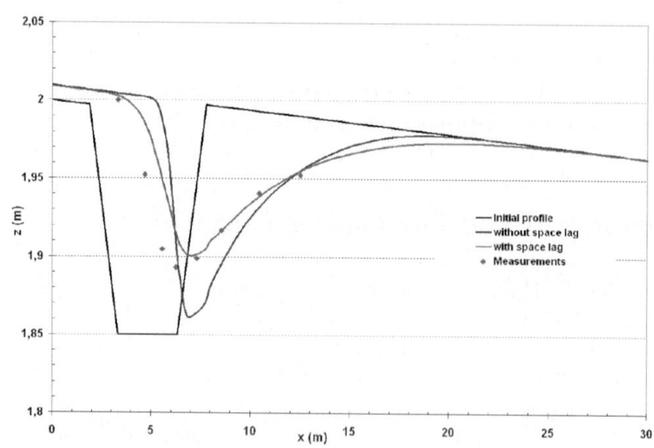

Fig. 1. [van Rijn, 1987] Experiment: calculated and measured bed level profiles at 7.5 h along with initial channel bed profile (from [El Kadi, 2006])

2.2 Sediment Transport Capacity

First, RubarBE software used the classical relation (5) that [Meyer-Peter and Müller, 1948] proposed for bed load transport:

$$C_s = \frac{8L_a\sqrt{g}}{(\rho_s - \rho)\sqrt{\rho}}(\rho JR - 0.047d_{50}(\rho_s - \rho))^{3/2} \quad (5)$$

in which C_s is sediment transport capacity (m^3/s), d_{50} median diameter of sediment (m), J friction slope, L_a active width (m), ρ_s density of sediment (kg/m^3), ρ density of water (kg/m^3).

For practical use, this relation can be modified considering a change in the value of the parameters, particularly the non-dimensional ones: critical shear stress that is set at 0.047 in (5) and the multiplying factor of capacity that is set at 8 in (5). Other sediment transport capacity equations have been also implemented in RubarBE to take into account the cases in which suspended load should also be considered: [Bagnold, 1966] and [Engelund and Hansen, 1967].

3 First Complementary Model: Sediment Description

3.1 Representative Parameters of Sediment Mixture

Mountain stream beds are usually composed of non-uniform sediment mixture of clay, sand, gravel, cobbles, and boulders up to 2 m in the case of steep slopes consolidated by stone blocks. A consequence is that the commonly representation of sediment mixture by one single diameter d_r, such as d_{50} (d_N is the grain size for which N percent of sediment is finer by weight), is not appropriate (for instance, because of armouring) even during extreme floods when all the sediment particles move together. Then, in addition to a representative diameter d_r, a second parameter σ (defined by the ratio d_{84}/d_{50} or d_{50}/d_{16}) is used by the model RubarBE to account for the effect of grain size distribution as [Shih and Komar, 1990] validate this assumption in case of relatively homogeneous sediments. This parameter is equivalent to the standard deviation of the lognormal particle size distribution if this distribution is assumed to be Gaussian and centred at d_{50}. This is a relevant approximation if the sediment mixture is quite homogeneous, which implies that the succession of floods would have created a quite regular grain size distribution. One of the questions that arise then stands in: with such a model, is it possible to provide the downstream variation of d_r? It is necessary to add some empirical relations to link the diameter d_r of a sediment mixture to the diameters of these particles that are mixed and oppositely the one of the deposited (or eroded) sediments when the process involves only part of the sediments. Clearly, there is a contradiction between the hypothesis of homogeneous sediment mixture and this attempt, which means that no exact mathematical solution exists.

Thus, the used approximation consists in averaging both d_r (assimilated to d_{50}) and σ when a mixture occurs. The opposite operation is set when a division occurs but an additional problem is the choice of the characteristics of one of the two sediments of the division.

3.2 The Complementary Model Used in RubarBE

The model assumes that the d_{84} (=σ d_{50}) of the initial sediment is kept (for both sediments) and that the d_{50} of the coarser sediment increases to this d_{84} following an exponential law.

During the various steps of calculation, sediments are mixed or shared into two fractions of different characteristics. The relations used for mixing are specific averages ((6)–(8) for respectively the mass, the diameter and the standard deviation).

$$M = M_1 + M_2 \tag{6}$$

$$d = d_1^{\frac{M_1}{(M_1+M_2)}} d_2^{\frac{M_2}{(M_1+M_2)}} \tag{7}$$

$$\sigma = \sigma_1^{\frac{M_1}{(M_1+M_2)}} \sigma_2^{\frac{M_2}{(M_1+M_2)}} \tag{8}$$

in which indexes 1 and 2 refer to the quantities of sediments that are added and no index to the mixture, M the mass, d the mean diameter and σ the standard deviation. These relations are the only ones for which addition of once a double mass and addition of twice a unit mass are strictly equivalent.

For sharing, the diameter d_1 of the coarser sediment can be calculated from a relation as (9). The standard deviation σ_1 can be calculated from (10) in order to keep d_{84}.

$$d_1 = d \times e^{\frac{M-M_1}{M} \frac{\sigma-1}{\sigma} \frac{\Delta x}{D_{chard}}} \tag{9}$$

$$\sigma_1 = \sigma \times e^{-\frac{M-M_1}{M} \frac{\sigma-1}{\sigma} \frac{\Delta x}{D_{char\sigma}}} \tag{10}$$

Symmetrically, the parameters of the finer sediments are calculated from (11) and (12) to obtain that the mixing of coarser and finer sediments will go back to the starting sediment.

$$d_2 = d \times e^{-\frac{M-M_2}{M} \frac{\sigma-1}{\sigma} \frac{\Delta x}{D_{chard}}} \tag{11}$$

$$\sigma_2 = \sigma \times e^{\frac{M-M_2}{M} \frac{\sigma-1}{\sigma} \frac{\Delta x}{D_{char\sigma}}} \tag{12}$$

3.3 Example of Sharing When Deposition

This idealized case consists in a rectangular channel initially in dynamic equilibrium at constant slope (1 m to 1 km) with bed sediments (d = 1 mm, $\sigma = 1.6$). The increase of the upstream sediment discharge (17.5–25 kg/s) with coarser sediments (d = 10 mm, $\sigma = 5$) is supposed to produce deposits in horizontal layers. After obtaining the new equilibrium slope of the channel, the diameter of the deposits are 10 mm for infinite D_{chard} (equalled to $D_{char\sigma}$), 10.9 mm for D_{chard} = 400 m and 14.6 mm for D_{chard} = 100 m. The orientation of this result is what could be expected but clearly the additional parameters D_{chard} and $D_{char\sigma}$ should be calibrated from comparison between measurements and calculation results.

4 Second Complementary Model: Evolution of the Shape of One Cross Section

4.1 The Submodels Provided in RubarBE

The basic hypothesis is that the one-dimensional structure is not changing including the cross sections location. It means that every cross section can be considered independently from any other. To simplify the algorithm, it is supposed that any node defining the cross section is moving only vertically and no new node is created. Taking into account this assumption, the sediment layers constituting the bed are described only on the vertical direction. Locally, one sediment layer is described using its thickness and the characteristics of the sediments (d_{50} and σ). In one cross section, one part of the sediments that are moving is exchanged with the bed (this part constitutes the active layer) and the remaining part is transported by flow. After computing the mass balance in the cross section, if there are too much sediments in the active layer, deposition occurs. Conversely, if there are too few sediments in the active layer, erosion occurs.

Erosion or deposition is distributed in the cross section according to the local shear stress. For the change of shape of the cross section, various alternative methods were tested. In the case of erosion, the movable bed under water lowers in relation with local shear stress (according to $(\tau_j - \tau_{cj})^m$). In the case of deposition, either the volume of deposited sediment is spread across the wetted perimeter, starting from the bottom, or the thickness of the deposit is related to the local shear stress (either according to ${\tau_j}^m$ or to $(\tau_{cj} - \xi\tau_j)^m$) in which τ_j is the local shear stress, τ_{cj} is the local critical shear stress taking into account the slope from [Ikeda, 1982], ξ a coefficient and m the exponent of the sediment transport capacity relation (for instance 1.5 for (5)). The various submodels for computing changes in the river morphology make possible to adapt the model to the real characteristics of the case.

4.2 Calculation of Shear Stress and Example of Evolution of a Cross Section

The local shear stress is either kept constant or calculated using the merged perpendicular method (MPC) [Khodashenas and Paquier, 1999]. Figure 2 shows a comparison between observed and calculated (non-dimensionalized) shear stress in a trapezoidal channel (experiments from [Ghosh and Roy,

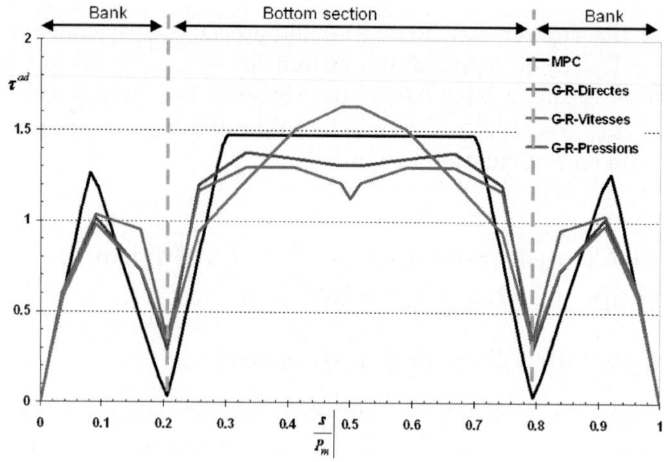

Fig. 2. Observed and calculated shear stresses in trapezoidal cross section (experiment from [Ghosh and Roy, 1970]) (from [El Kadi, 2006])

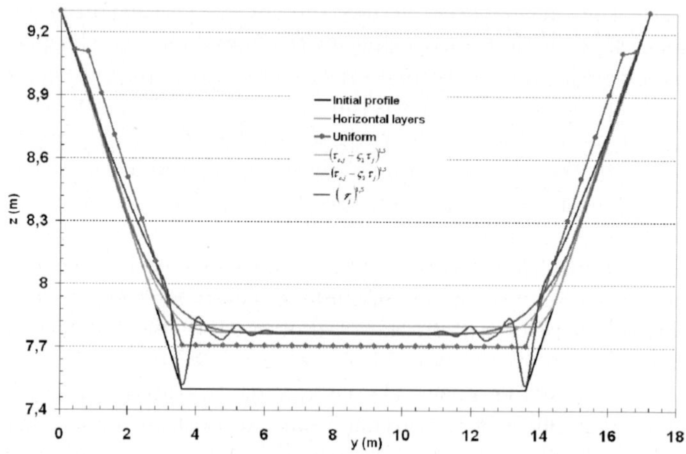

Fig. 3. Stable cross section obtained from deposition in trapezoidal cross section (from [El Kadi, 2006])

1970]). The use of three methods (direct measurement of stress, using velocities, using pressures) to estimate the observed shear stress shows clearly that the predicted values are relatively accurate.

Figures 3 and 4 show the evolution of two cross sections to a stabilized profile using various methods for calculation of deposits. The differences between the results demonstrate the adaptability of the methods to be calibrated on real rivers. However, applying the method to real rivers shows the necessity to integrate also the curvature of the river that explains a part of the variability inside the cross section [Paquier and Khodashenas, 2002].

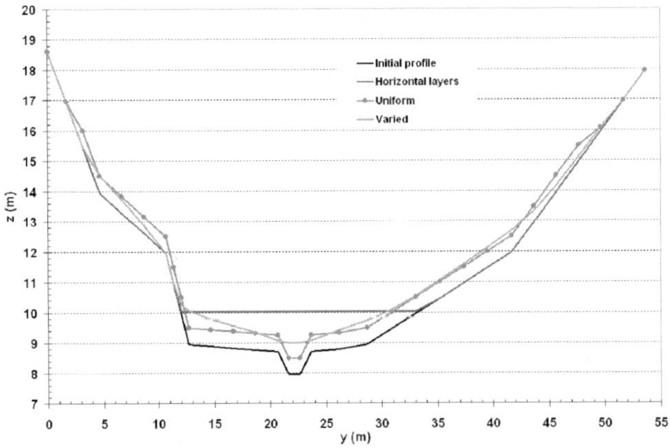

Fig. 4. Stable cross section obtained from deposition in irregular cross section (from [El Kadi, 2006])

5 Numerical Scheme

5.1 Description of Numerical Scheme

To deal with fast unsteady flows, the de Saint Venant equations are solved by a second-order Godunov type scheme. The scheme used includes four steps [Paquier, 1995]:

1. The slopes of the variables are computed from the values at the middle of the cell by some kind of minmod relation. On each of the scalar variable Q discharge or h water depth, independently, the slope is computed. For h, a supplementary limitation through water level is imposed.
2. From these slopes and a time discretization of Euler type on one cell, the values on the limits of the cell are computed at time $t_n + 0.5\Delta t$. Two values are thus obtained: one from the left cell and one for the right cell.
3. As values at the same limit are generally different when computed from the left cell and from the right cell, a Riemann problem is solved in an approximate way by a Roe-type linearization [Roe, 1981].
4. The value at the middle of the cell is computed from the difference of the fluxes for the conservative part of the equations and from an estimate of second member at time $t_n + 0.5\Delta t$. Second member of the equations that is taken into account as a correction is mainly constituted of two terms: a topographical term that is computed as a difference of pressures at constant water level and a friction term that is computed in an implicit way.

Sediment equations are solved by a first-order Euler type scheme, coupled with de Saint-Venant equations at every time step through the calculation of the sediment discharge and the update of the river bed geometry. The sediment transport capacity is calculated solving the spatial lag equation inside a cell. The sediment discharge downstream the cell is calculated from the upstream sediment discharge while distinguishing the sediments that are only transferred to the ones that are interfering with the sediments previously present in the cell (active layer). Then the sediment continuity (3) is applied to every cell. This leads to a change of A_s that should be transformed in a change of the shape of the cross section. This change will then modify the water elevation if the hypothesis of no change in the water depth and velocity is applied. The active layer has its thickness fixed from the sediment transport capacity, the velocity of the flow and the space step. The deposits and erosions occur when this active layer is, respectively, too thick or too shallow. Then the upper substrate layer is increased or decreased; moreover, one layer can be created or can disappear.

5.2 Stability of Numerical Scheme

First, it should be noted that, following [Balayn, 2001], to improve stability, the sediment cell (on which the sediment continuity equation is calculated) is moved by 0.5 Δx from the water cell (on which the water governing equations are calculated).

For de Saint-Venant equations in uniform fixed geometry, the stability condition is a Courant number below 1. For the complete set of equations (flow and sediment), it is assumed that this condition is not too much reduced except in case of strong geometrical irregularities and high sediment transport.

For a dam break problem with mobile bed in laboratory [Muramoto, 1987], Figs. 5 and 6 show the bed and water profiles at t = 0.2 s with the influence of time step. Calculations were performed with either constant Courant number (up to 5 but Courant numbers higher than 1 could not provide results) or constant time steps (Courant number was also below 1). This test confirms

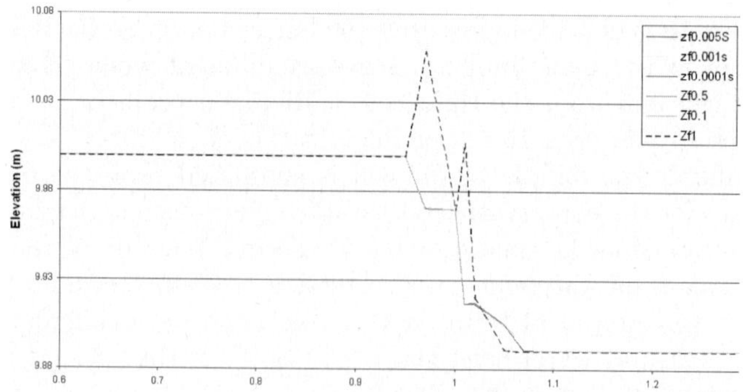

Fig. 5. Influence of time step on bed elevation in experiments from [Muramoto, 1987]

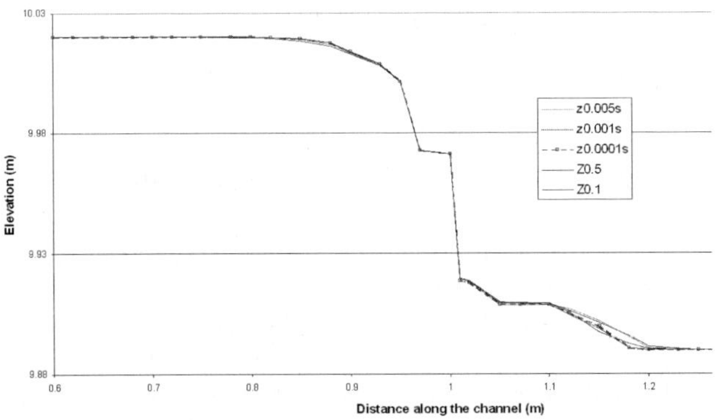

Fig. 6. Influence of time step on water surface elevation in experiments from [Muramoto, 1987]

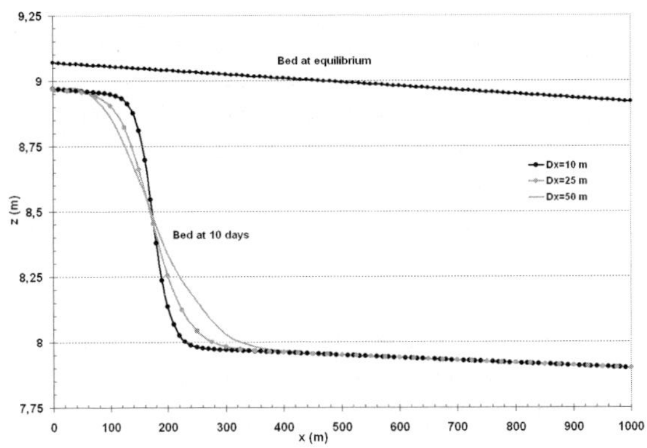

Fig. 7. Influence of space step in the deposition in irregular cross section (from [El Kadi, 2006])

the necessity to limit Courant number to less than 1 to obtain convenient results and then the stability of results (only a slight advance of water front for higher time steps).

Figure 7 shows the influence of space step. During the propagation of the front in the example of the channel with the cross section of Fig. 4, smoothening increases with Δx but final equilibrium solution does not depend on the space step.

6 Pertinence of the Model

A set of idealized cases was used to check that the model reproduces the main processes of sediment transport in channels with various geometries. Both the correctness of the final state and the stability of the rate of erosion or deposition were checked (as shown in the previous section). For checking the numerical stability in more real and severe conditions, the case of an historical dam break event was selected.

On July 19–21, 1996, a major storm system stalled over St Lawrence area and dropped record amounts of rain causing widespread flooding in southern Quebec. The Ha!Ha! River is located in the mountainous and forested area of the Saguenay basin, from Pk35.70 (the Ha!Ha! Lake) to Pk0.00 (the Ha!Ha! Bay). The Ha!Ha! river bed slope changes downstream from 0.2 to 0.0016 m/m and reflects the irregular profile of the river as the channel crosses bedrock and non-alluvial areas. On Lake Ha!Ha!, the flood of July 19–21, 1996 led to overtopping and failure of an earth dyke (discharge up to 900 m^3/s) [Lapointe et al., 1998]. From the lake to the river mouth, the Ha!Ha! riverbed was dramatically modified. The event along the whole reach was modelled using several geometry update methods of RubarBE (see [El Kadi and Paquier, 2004] for more details). Figures 8 and 9 focus in the central part in which the flow bypassed the Chute-à-Perron rapids.

Results confirm that the model can capture the main features for a major part of the river but numerical results are not fully in agreement with observations.

7 Conclusions

A one-dimensional hydro-morphodynamic model to study non-equilibrium sediment transport is presented. The model solves a complete set of water and sediment equations by using a finite difference explicit scheme. The model has the capability of computing combination of subcritical and supercritical

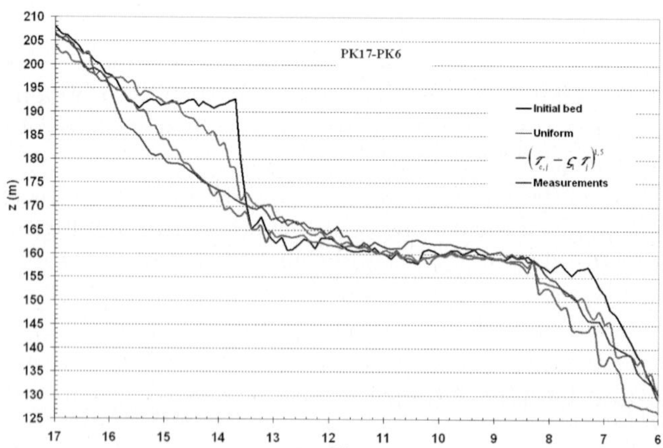

Fig. 8. Longitudinal profile from Ha!Ha! dam break wave propagation (from [El Kadi, 2006])

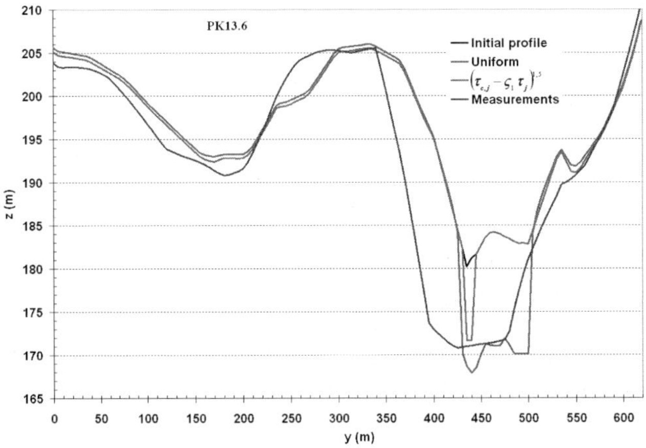

Fig. 9. Cross section from Ha!Ha! dam break wave propagation (from [El Kadi, 2006])

flows. Furthermore, it can handle irregular cross sections and the availability of multiple cross section updating methods enables the user to select the more relevant channel morphological evolution. On some idealized cases, calculations using RubarBE are not sensitive to time and space steps within expected limits. However, the lag distance is an important parameter for the evaluation of sediment transport processes. The model was tested on the Ha!Ha! dam break wave propagation showing that the model could be used to simulate historic high flows that are at the origin of the main morphological changes.

References

[Bagnold, 1966] Bagnold, R.A. (1966). *An approach to the sediment transport problem from general physics*, Vol. 422-I. United States Geological Survey, Washington DC.

[Balayn, 2001] Balayn, P. (2001). *Contribution à la modélisation numérique de l'évolution morphologique des cours d'eau aménagés lors de crues.* Phd, Université Claude Bernard Lyon 1.

[El Kadi, 2006] El Kadi, K. (2006). *Evolution d'un lit de rivière en fonction des apports.* Phd, Université Claude Bernard.

[El Kadi and Paquier, 2004] El Kadi, K. and Paquier, A. (2004). Sediment transport and morphological changes in the ha! ha! river induced by the flood event of July 1996. In *EC Contract EVG1-CT-2001-00037 IMPACT Investigation of Extreme Flood Processes and Uncertainty, 4th Project Workshop*, volume CDRom, p. 13, Zaragoza, Spain.

[Engelund and Hansen, 1967] Engelund, F. and Hansen, E. (1967). *A monograph on sediment transport in alluvial streams.* Teknisk Forlag, Copenhagen.

[Ghosh and Roy, 1970] Ghosh, S.N. and Roy, N. (1970). Boundary shear distribution in open channel flow. *Journal of the Hydraulics Division*, 96(HY4):967–994.

[Ikeda, 1982] Ikeda, S. (1982). Incipient motion of sand particles on side slopes. *Journal of the Hydraulics Division*, 108(HY1):95–114.

[Khodashenas and Paquier, 1999] Khodashenas, S.R. and Paquier, A. (1999). A geometrical method for computing the distribution of boundary shear stress across irregular straight open channels. *Journal of Hydraulic Research*, 37(3):381–388.

[Lapointe et al., 1998] Lapointe, M., Secretan, Y., Driscoll, S., Bergeron, N., and Leclerc, M. (1998). Response of the ha! ha! river to the flood of july 1996 in the saguenay region of quebec: Large-scale avulsion in a glaciated valley. *Water Resources Research*, 34(9):2383–2392.

[Meyer-Peter and Müller, 1948] Meyer-Peter, E. and Müller, R. (1948). Formulas for bed-load transport. In *Report on second meeting of IARH*, pp. 39–64, Stockholm, Sweden. IAHR.

[Muramoto, 1987] Muramoto, Y. (1987). Water and sediment outflow from a reservoir by dam collapse. In *XXII IAHR Congress*, Lausanne, Switzerland. IAHR.

[Paquier, 1995] Paquier, A. (1995). *Modélisation et simulation de la propagation de l'onde de rupture de barrage (Modelling and simulating the propagation of dam-break wave)*. Phd, Université Jean Monnet de Saint Etienne.

[Paquier and Khodashenas, 2002] Paquier, A. and Khodashenas, S.R. (2002). River bed deformation calculated from boundary shear stress. *Journal of Hydraulic Research*, 40(5):603–609.

[Roe, 1981] Roe, P.L. (1981). Approximate riemann solvers, parameter vectors and difference schemes. *Journal of Computational Physics*, 43:357–372.

[Shih and Komar, 1990] Shih, S.-M. and Komar, P.D. (1990). Hydraulic controls of grain-size distributions of bedload gravels in oak creek, Oregon, USA. *Sedimentology*, 37:367–376.

[van Rijn, 1987] van Rijn, L.C. (1987). Mathematical modeling of morphological processes in the case of suspended sediment transport. Technical report, Delft University of Technology.

Part IV

Contributed Talks

Homogenization of Conservation Laws with Oscillatory Source and Nonoscillatory Data

D. Amadori

1 Introduction

We consider the equation

$$u_t^\varepsilon + f(u^\varepsilon)_x = \frac{1}{\varepsilon} V'\left(\frac{x}{\varepsilon}\right), \qquad x \in \mathbb{R},\ t > 0,\ \varepsilon > 0. \tag{1}$$

We assume that

(H1) $f : \mathbb{R} \to \mathbb{R}$ is $\mathbf{C}^2$, $f(u) \to +\infty$ as $|u| \to \infty$;
(H2) $f'' > 0$;
(H3) $V \in \mathbf{C}^2(\mathbb{R})$, periodic with period 1;
(H4) V attains its minimum value at a single point in $\mathbb{R}/\mathbb{Z}$.

Without loss of generality, we can assume that $\min V(x) = 0$ and $f(0) = \min_{\mathbb{R}} f = 0$. As a consequence of **(H2)**, one has

(H2)′ $uf'(u) > 0$ if $u \neq 0$.

Given the initial data $u^\varepsilon(x, 0) = u_o(x)$, we will analyze the limiting behavior of the sequence u^ε. To describe the problem, we first recall the results obtained in [AS06, Ama06]. Let us write the equation (1) for $\varepsilon = 1$:

$$u_t + f(u)_x = V'(x). \tag{2}$$

We introduce the family $\mathcal{S}$ of 1-periodic, steady solutions to (2):

$$\mathcal{S} = \{\psi : \mathbb{R}/\mathbb{Z} \to \mathbb{R}; \qquad \psi = \psi(x) \text{ is a weak entropy solution to (2)}\};$$

the properties of $\mathcal{S}$, resulting by the assumptions above, will be described in Sect. 2. We denote by $\langle\cdot\rangle$ the average over the period of a periodic function. Theorem 1 is concerned with the large time behavior of the periodic solutions to (2).

Theorem 1. *([AS06, E92]) Assume* **(H1)**, **(H1)**, **(H3)**. *Let $u_o \in \mathbf{L}^\infty(\mathbb{R})$ be 1-periodic. Let u be the entropic solution to the Cauchy problem $u_t + f(u)_x = V'(x)$, $u(x,0) = u_o(x)$. Then there exists a steady state solution $\psi \in \mathcal{S}$ such that*

$$\|u(t,\cdot) - \psi\|_{\mathbf{L}^1(\mathbb{R}/\mathbb{Z})} \to 0 \qquad \textit{as} \;\; t \to +\infty.$$

The limit state ψ has the property $\langle\psi\rangle = \langle u_o\rangle$. If moreover **(H4)** *holds, then ψ is uniquely identified by its mean value.*

Now let us consider the Cauchy problem for (1) with initial data

$$u^\varepsilon(x,0) = u_o\left(x, \frac{x}{\varepsilon}\right), \tag{3}$$

where $u_o \in \mathbf{L}^\infty(\mathbb{R} \times (\mathbb{R}/\mathbb{Z}))$. After [LPV87], it is known that the sequence $\{u^\varepsilon\}_{\varepsilon>0}$ converges to a function $\bar{u}$ weak * in $\mathbf{L}^\infty{}_{loc}(\mathbb{R} \times \mathbb{R}_+)$, where $\bar{u}$ is the unique entropy solution of

$$\bar{u}_t + \bar{f}(\bar{u})_x = 0\,, \qquad \bar{u}(x,0) = \int_0^1 u_o(x,y)\,dy \tag{4}$$

and $\bar{f}$, the so-called *effective flux*, is a well-defined function on $\mathbb{R}$, see Sect. 2.

Theorem 2. *([Ama06]) Assume* **(H1)**, **(H2)′**, **(H3)**, **(H4)** *and that f is convex in an arbitrarily small neighborhood of* 0. *Let u^ε be the unique solution to (1), (3) and assume that the initial data $u_o \in \mathbf{L}^\infty(\mathbb{R} \times (\mathbb{R}/\mathbb{Z}))$ satisfy*

$$y \mapsto u_o(x,y) \in \mathcal{S}, \qquad \text{for a.e. } x. \tag{5}$$

Let $\bar{u}$ be the entropy solution to (4) and let $U : \mathbb{R} \times \mathbb{R}_+ \times (\mathbb{R}/\mathbb{Z}) \to \mathbb{R}$ be uniquely defined, for a.e. (x,t), as follows:

$$y \mapsto U(x,t,y) \in \mathcal{S}, \qquad \langle U(x,t,\cdot)\rangle = \bar{u}(x,t)\,. \tag{6}$$

Then, as $\varepsilon \to 0$,

$$u^\varepsilon(x,t) - U\left(x, t, \frac{x}{\varepsilon}\right) \to 0 \qquad \textit{in} \;\; \mathbf{L}^2_{loc}(\mathbb{R} \times (0,+\infty)). \tag{7}$$

In this note, we are going to extend the result of Theorem 2, by relaxing the assumption (5) on the initial data. We focus on the special case of $u^\varepsilon(x,0) = u_o(x)$: the initial data do not depend on ε. In particular, they are not prepared, in the sense that do not satisfy the strong assumption (5).

Without assuming the strict convexity of f, we may expect that the asymptotic profile U depends on both x/ε, t/ε, as can be seen by simple counterexamples with linear flux (see [E92]). Observe that the (global) convexity is not required in Theorem 2, but the assumption on the initial data prevents the formation of initial layers.

Here we will assume that the flux is uniformly convex, **(H2)**. The main result is the following.

Theorem 3. *Assume* **(H1)**, **(H2)**, **(H3)**, **(H4)** *and that* $u_o \in \mathbf{BV}_{loc}(\mathbb{R}) \cap \mathbf{L}^\infty(\mathbb{R})$. *Let* u^ε *be the entropy solution to the Cauchy problem (1),* $u^\varepsilon(x,0) = u_o(x)$ *and* $\bar{u}$ *be the entropy solution to*

$$\bar{u}_t + \bar{f}(\bar{u})_x = 0\,, \qquad \bar{u}(x,0) = u_o(x). \tag{8}$$

Let $U : \mathbb{R} \times \mathbb{R}_+ \times (\mathbb{R}/\mathbb{Z}) \to \mathbb{R}$ *be defined by (6). Then, as* $\varepsilon \to 0$ *one has (7).*

We remark that, since u^ε, U are uniformly bounded in $\mathbf{L}^\infty$, the convergence (7) actually holds in $\mathbf{L}^p_{loc}$ for every $p \geq 1$.

After some preliminary arguments, object of Sect. 2, the proof of Theorem 3 will be given in Sect. 3. It makes use of both Theorems 1 and 2. The key point is the introduction of a sequence of solutions to (1) with suitably prepared initial data, with the same asymptotic representation as u^ε. Then we proceed by using a localization argument, as done in [EE93] for conservation laws with smooth oscillatory data.

See [TT97] for related problems and [Tar86], plus references therein, as a general reference on the study of oscillations in nonlinear partial differential equations, with an application to one-dimensional scalar conservation laws.

2 Preliminaries

We start by reporting some properties of $\mathcal{S}$. Clearly, if $\psi \in \mathcal{S}$, then ψ satisfies $f(\psi(y)) - V(y) = C$ for some constant C; $\psi^\varepsilon(x) = \psi(x/\varepsilon)$ is a steady solution to (1); ψ is bounded because of **(H1)**.

2.1 The Effective Flux

We recall the definition of $\bar{f}$ ([E92, LPV87]). If there exists a 1-periodic function w such that

$$f(p + w'(y)) - V(y) = const.$$

independently of y, then $\bar{f}(p)$ is defined as the constant value on the right-hand side. This procedure leads to a well-defined function $\bar{f}$; an explicit formula in terms of f, V is given by (2.8) in [E92]. See Fig. 1 for an example.

Proposition 1. *(a) Assume* **(H1)**, **(H2)$'$**, **(H3)**. *Then, for every* $p \in \mathbb{R}$, *there exists a map* $\psi_p(y) \in \mathcal{S}$ *such that*

$$\langle \psi_p \rangle = p, \qquad f(\psi_p(y)) - V(y) = \bar{f}(p) \qquad \forall\, y \in \mathbb{R}/\mathbb{Z}. \tag{9}$$

(b) In addition, if **(H4)** *holds, there exists a unique* ψ_p *satisfying (9), hence the map* $\mathcal{S} \ni \psi \mapsto \langle \psi \rangle \in \mathbb{R}$ *is bijective. Moreover, the following monotonicity property holds:*

$$p_1 \leq p_2 \quad \Rightarrow \quad \psi_{p_1}(y) \leq \psi_{p_2}(y), \qquad \forall y.$$

See [Ama06] for details.

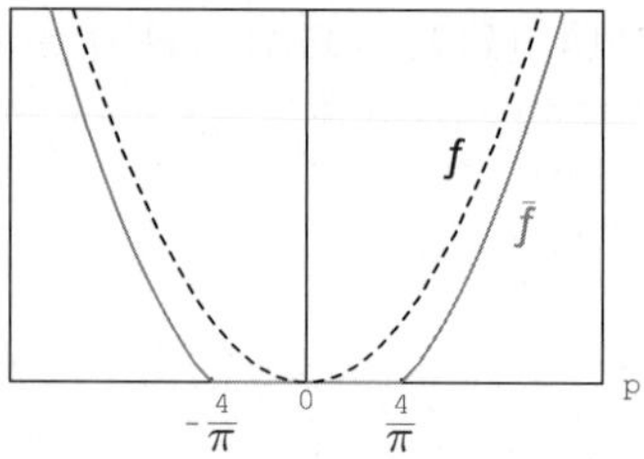

Fig. 1. Graph of $\bar{f}$ in the case $f(u) = u^2/2$, $V(y) = 1 + \sin(2\pi y)$

2.2 Modified Initial Data

From now on, assume **(H1)**, **(H2)**$'$, **(H3)**, **(H4)**. Given $u_o \in \mathbf{L}^\infty(\mathbb{R})$ we can define

$$w_o(x,y) \doteq \psi_{u_o(x)}(y) \,. \tag{10}$$

That is, w_o is defined by the two following properties: $y \to w_o(x,y) \in \mathcal{S}$, for a.e. x, and

$$\langle w_o(x,\cdot)\rangle = \int_0^1 \psi_{u_o(x)}(y)\,dy = u_o(x). \tag{11}$$

Thanks to assumption **(H4)**, w_o is well defined in $\mathbf{L}^\infty(\mathbb{R} \times (\mathbb{R}/\mathbb{Z}))$. Denote by $w^\varepsilon(x,t)$ the solution to (1) with initial data $w_o\left(x, \frac{x}{\varepsilon}\right)$:

$$w_t^\varepsilon + f(w^\varepsilon)_x = \frac{1}{\varepsilon} V'\left(\frac{x}{\varepsilon}\right) \qquad w^\varepsilon(x,0) = w_o\left(x, \frac{x}{\varepsilon}\right). \tag{12}$$

Observe that $w_o(x, x/\varepsilon)$ is a prepared initial data, with pointwise average u_o.

2.3 Uniform $\mathbf{L}^\infty$ Bounds

As in [ES92, Ama06], we can prove that u^ε, w^ε are bounded independently of ε in $\mathbf{L}^\infty(\mathbb{R} \times \mathbb{R}_+)$, by bounding the initial data from above and below with suitable steady solutions (which are bounded because of **(H1)**) and then by using a comparison argument.

Also, the map $U(x,t,y)$ defined by (6) is bounded in $\mathbf{L}^\infty(\mathbb{R} \times \mathbb{R}_+ \times (\mathbb{R}/\mathbb{Z}))$: indeed, $y \mapsto U(x,t,y) \in \mathcal{S}$ and $\langle U(x,t,\cdot)\rangle = \bar{u}(x,t)$, solution to (8), whose values belong to a bounded set.

2.4 An Approximation Lemma

Let $I = (0,1)$, $N \in \mathbb{N}$, $h = 1/N$ and define

$$\begin{aligned} A_j &= (jh,(j+1)h)\,, & j &= 0,\ldots,N-1 \\ u_o^h(x) &= \frac{1}{h}\int_{A_j} u_o(y)\,dy\,, & x &\in A_j \\ w_o^h(x,y) &= \psi_{u_o^h(x)}(y), & x &\in A_j,\ y\in\mathbb{R}. \end{aligned}$$

Lemma 1. *Let $u_o \in \mathbf{BV}_{loc}(\mathbb{R}) \cap \mathbf{L}^\infty(\mathbb{R})$ and w_o defined by (10). Then, for every $\delta > 0$ the following holds. There exist $N_o \in \mathbb{N}$ and $\varepsilon_o > 0$ such that, for all $N = 1/h \geq N_o$ and $0 < \varepsilon \leq \varepsilon_o$ one has*

$$\int_{\cup A_j} \left| w_o\left(x, \frac{x}{\varepsilon}\right) - w_o^h\left(x, \frac{x}{\varepsilon}\right)\right| dx < \delta.$$

Proof. Set $M_j = \sup_{A_j} u_o$, $m_j = \inf_{A_j} u_o$ (we choose the right continuous representative of u_o to avoid ambiguity). Thanks to the monotonicity property of ψ_p w.r.t. p, we have for $x \in A_j$

$$\begin{aligned} \left| w_o\left(x, \frac{x}{\varepsilon}\right) - w_o^h\left(x, \frac{x}{\varepsilon}\right)\right| &= \psi_{\max\{u_o(x),u_o^h(x)\}}\left(\frac{x}{\varepsilon}\right) - \psi_{\min\{u_o(x),u_o^h(x)\}}\left(\frac{x}{\varepsilon}\right) \\ &\leq \psi_{M_j}\left(\frac{x}{\varepsilon}\right) - \psi_{m_j}\left(\frac{x}{\varepsilon}\right) \end{aligned}$$

and hence

$$\begin{aligned} \int_{A_j} \left| w_o\left(x, \frac{x}{\varepsilon}\right) - w_o^h\left(x, \frac{x}{\varepsilon}\right)\right| dx &\leq \int_{A_j} \psi_{M_j}\left(\frac{x}{\varepsilon}\right) - \psi_{m_j}\left(\frac{x}{\varepsilon}\right) dx \\ &\leq \varepsilon\left(\left[\frac{h}{\varepsilon}\right]+1\right)\int_I \left[\psi_{M_j}(y) - \psi_{m_j}(y)\right] dy \\ &\leq (h+\varepsilon)(M_j - m_j) \leq (h+\varepsilon)\,\text{Tot.Var.}\{u_o, A_j\} \end{aligned}$$

so that

$$\int_{\cup A_j} \left| w_o\left(x, \frac{x}{\varepsilon}\right) - w_o^h\left(x, \frac{x}{\varepsilon}\right)\right| dx \leq (h+\varepsilon)\,\text{Tot.Var.}\{u_o, I\} < \delta$$

for h, ε sufficiently small. This concludes the proof of the Lemma. □

3 Proof of the Main Theorem

In this section we prove Theorem 3. Since u^ε, U are uniformly bounded, it is enough to prove that the convergence (7) holds in $\mathbf{L}^1_{loc}(\mathbb{R}\times\mathbb{R}_+)$.

Step 1: Constant initial data. Assume that $u_o(x) = \bar{u}_o \in \mathbb{R}$. Then one has $u^\varepsilon(x,t) = v\left(\frac{x}{\varepsilon}, \frac{t}{\varepsilon}\right)$ where v is the solution of $u_t + f(u)_x = V'(x)$, $u(x,0) = \bar{u}_o$. Applying Theorem 1, there exists a limit state ψ such that

$$v(y,\tau) - \psi(y) \to 0 \qquad \text{in } \mathbf{L}^1(\mathbb{R}/\mathbb{Z}) \qquad \text{as } \tau\to\infty. \tag{13}$$

Observe that $\langle\psi\rangle = \bar{u}_o$ and that the solution to (8) is constant: $\bar{u}(x,t) \equiv \bar{u}_o$. Hence the asymptotic profile U, introduced at (6), is given here by $U = U(y) \doteq \psi(y)$.

Then, let $K > 0$, $t_2 > t_1 > 0$ and define $M(\varepsilon) = \left[\frac{K}{\varepsilon}\right] + 1$, so that $K < \varepsilon M(\varepsilon) < K + 1$ as $\varepsilon \to 0$. We easily get

$$\begin{aligned}
&\iint_{[t_1,t_2]\times[-K,K]} |u^\varepsilon(x,t) - \psi(x/\varepsilon)| dxdt \\
&\quad \leq 2\varepsilon M(\varepsilon) \int_{[t_1,t_2]} \int_{[0,1]} |v(y,t/\varepsilon) - \psi(y)| dydt \\
&\quad \leq 2\varepsilon M(\varepsilon)\,(t_2 - t_1) \int_{[0,1]} |v(y,t_1/\varepsilon) - \psi(y)| dy \to 0 \qquad \text{as } \varepsilon \to 0.
\end{aligned}$$

Remark that, for this part of the proof, **(H4)** is not required: the profile ψ is that one, among the ones having mean value $\bar{u}_o$, that satisfies (13) and whose existence is guaranteed by Theorem 1. In other words, it is the limit profile *chosen* by the constant initial data $\bar{u}_o$. In a similar way, one proves also that

$$\int_I |u^\varepsilon(x,t) - \psi(x/\varepsilon)|\, dx \to 0 \qquad \text{as } \varepsilon \to 0, \tag{14}$$

for every closed interval $I = [a,b]$ and every $t > 0$.

Step 2: Modified initial data. In general, with $u_o \in \mathbf{BV}_{loc}(\mathbb{R}) \cap \mathbf{L}^\infty(\mathbb{R})$, we consider the solution w^ε of the Cauchy problem (12).

By [LPV87], and thanks to (11), the sequences u^ε, w^ε have the same weak limit $\bar{u}$, solution to (8). Also, the corresponding asymptotic profile U is the same, see (6), since it depends only on the limit $\bar{u}$. Then, by applying Theorem 2, we get that $w^\varepsilon(x,t) - U(x,t,x/\varepsilon) \to 0$ in $\mathbf{L}^1_{loc}$.

Now we evaluate $u^\varepsilon - w^\varepsilon$. We proceed similarly to [EE93], proof of Theorem 4.1, and show that for any fixed $t > 0$

$$u^\varepsilon(\cdot,t) - w^\varepsilon(\cdot,t) \to 0 \qquad \text{in } \mathbf{L}^1_{loc}(\mathbb{R}). \tag{15}$$

By using (15) and the $\mathbf{L}^1$-contraction, we can easily prove that $u^\varepsilon - w^\varepsilon \to 0$ in $\mathbf{L}^1_{loc}(\mathbb{R}\times\mathbb{R}_+)$.

Step 3: Proof of (15). Let I be a closed, bounded interval. Without loss of generality in the following arguments, we can assume $I = [0,1]$.

Fix a $\delta > 0$. Let $t_o > 0$, $N \in \mathbb{N}$ to be chosen later, $h = 1/N$. Define, for $j = 0,\dots,N-1$

$$\begin{aligned}
A_j &= (jh, (j+1)h)\,, & B_j &= (jh + Lt_o, (j+1)h - Lt_o) \\
A &= \cup_{j=0}^{N-1} A_j, & B &= \cup_{j=0}^{N-1} B_j \\
u_o^h(x) &= \frac{1}{h}\int_{A_j} u_o(\xi)\, d\xi & w_o^h(x,y) &= \psi_{u_o^h(x)}(y) \quad \text{if } x \in A_j,
\end{aligned}$$

being $L \doteq \sup_{|z|\le M} |f'(z)|$, M a bound on $\|u^\varepsilon\|_\infty$, $\|w^\varepsilon\|_\infty$. Similarly, define

$$\begin{aligned} j &= 0,\dots,N, & x_j &= \left(j+\frac{1}{2}\right)h \\ C_j &= (x_{j-1}, x_j)\,, & E_j &= (x_{j-1}+Lt_o, x_j - Lt_o) \\ C &= \cup_{j=0}^N C_j, & E &= \cup_{j=0}^N E_j \\ v_o^h(x) &= \frac{1}{h}\int_{C_j} u_o(\xi)\,d\xi & W_o^h(x,y) &= \psi_{v_o^h(x)}(y) \quad \text{if } x\in C_j. \end{aligned}$$

By Lemma 1, we can choose $N = 1/h$ so large that the following holds: there exists a $\varepsilon_o > 0$ such that, for all $0 < \varepsilon \le \varepsilon_o$, one has

$$\int_A |u_o(x) - u_o^h(x)|\,dx + \int_A \left|w_o\left(x,\frac{x}{\varepsilon}\right) - w_o^h\left(x,\frac{x}{\varepsilon}\right)\right|\,dx < \delta, \tag{16}$$

$$\int_C |u_o(x) - v_o^h(x)|\,dx + \int_C \left|w_o\left(x,\frac{x}{\varepsilon}\right) - W_o^h\left(x,\frac{x}{\varepsilon}\right)\right|\,dx < \delta.$$

Then, we choose t_o such that: $Lt_o < h/4$. With this choice of t_o, one has $I \subset B \cup E$.

Now, let $u^{\varepsilon,h}(x,t)$ and $w^{\varepsilon,h}(x,t)$ be the solutions to (1) with initial data, respectively, $u_o^h(x)$ and $w_o^h(x, x/\varepsilon)$. Using (16) and the $\mathbf{L}^1$-contraction property, one has

$$\begin{aligned} &\int_B \left|u^\varepsilon(x,t_o) - u^{\varepsilon,h}(x,t_o)\right|\,dx + \int_B \left|w^\varepsilon(x,t_o) - w^{\varepsilon,h}(x,t_o)\right|\,dx \\ &\quad \le \int_A \left|u_o(x) - u_o^h(x)\right|\,dx + \int_A \left|w_o\left(x,\frac{x}{\varepsilon}\right) - w_o^h\left(x,\frac{x}{\varepsilon}\right)\right|\,dx < \delta\,. \end{aligned} \tag{17}$$

Now, observe that

$$w^{\varepsilon,h}(x,t_o) = w_o^h\left(x,\frac{x}{\varepsilon}\right) = \psi_{u_o(x_j)}\left(\frac{x}{\varepsilon}\right) \qquad \text{if } x \in B_j,$$

because the initial data w_o^h coincides, on A_j, with a steady state, hence $w^{\varepsilon,h}$ coincides on B_j with the same steady state. Nevertheless, $u^{\varepsilon,h}(x,t_o)$ coincides, on B_j, with the solution to (1) corresponding to the constant initial data $u_o(x_j)$. Then, we can use (14) and obtain

$$\begin{aligned} &\int_{B_j} \left|u^{\varepsilon,h}(x,t_o) - w^{\varepsilon,h}(x,t_o)\right|\,dx \\ &\qquad = \int_{B_j} \left|u^{\varepsilon,h}(x,t_o) - \psi_{u_o(x_j)}\left(\frac{x}{\varepsilon}\right)\right|\,dx \to 0 \quad \text{as } \varepsilon \to 0. \end{aligned}$$

Summing up over j, one can conclude that there exists $\varepsilon_o > 0$ such that, for all $0 < \varepsilon < \varepsilon_o$,

$$\int_B \left| u^{\varepsilon,h}(x,t_o) - w^{\varepsilon,h}(x,t_o) \right| \, dx < \delta. \tag{18}$$

In conclusion, using (17), (18), one gets that $\int_B |u^\varepsilon(x,t_o) - w^\varepsilon(x,t_o)| \, dx \le 2\delta$ for any ε sufficiently small.

We repeat the same argument to estimate $\int_E |u^\varepsilon(x,t_o) - w^\varepsilon(x,t_o)| dx$ and finally get that, for any ε sufficiently small

$$\begin{aligned} &\int_I |u^\varepsilon(x,t_o) - w^\varepsilon(x,t_o)| \, dx \\ &\le \int_B |u^\varepsilon(x,t_o) - w^\varepsilon(x,t_o)| \, dx + \int_E |u^\varepsilon(x,t_o) - w^\varepsilon(x,t_o)| \, dx \le 4\delta. \end{aligned}$$

To conclude the proof of (15), let $I = [a,b]$ be a given interval and $t > 0$ a given time. Define $J = [a - Lt, b + Lt]$. By the previous argument, for any positive δ there exist $\varepsilon_1 > 0$, $t_o > 0$ (that can be chosen less than t) such that

$$\int_J |u^\varepsilon(x,t_o) - w^\varepsilon(x,t_o)| \, dx \le \delta \qquad \forall 0 < \varepsilon < \varepsilon_1 .$$

Since $\int_I |u^\varepsilon(x,t) - w^\varepsilon(x,t)| \, dx \le \int_J |u^\varepsilon(x,t_o) - w^\varepsilon(x,t_o)| \, dx$, the conclusion follows. □

References

[Ama06] Amadori, D.: On the homogenization of conservation laws with resonant oscillatory source. Asymptotic Anal., **46**, 53–79 (2006)

[AS06] Amadori, D., Serre, D.: Asymptotic behavior of solutions to conservation laws with periodic forcing. J. Hyperbolic Differ. Equ., **3**, 387–401 (2006)

[E92] E, W.: Homogenization of scalar conservation laws with oscillatory forcing terms. SIAM J. Appl. Math., **52**, 959–972 (1992)

[ES92] E, W., Serre, D.: Correctors for the homogenization of conservation laws with oscillatory forcing terms. Asymptotic Anal., **5**, 311–316 (1992)

[EE93] Engquist, B., E, W.: Large time behavior and homogenization of solutions of two-dimensional conservation laws. Comm. Pure Appl. Math., **46**, 1–26 (1993)

[LPV87] Lions, P.L., Papanicolaou, G., Varadhan, S.R.S.: Homogenization of Hamilton–Jacobi equations. Preprint (1987)

[TT97] Tadmor, E., Tassa, T.: On the homogenization of oscillatory solutions to nonlinear convection-diffusion equations. Adv. Math. Sci. Appl., **7**, 93–117 (1997)

[Tar86] Tartar, L.: Oscillations in nonlinear partial differential equations: compensated compactness and homogenization. In: Nonlinear systems of partial differential equations in applied mathematics, Part 1 (Santa Fe, N.M., 1984), Lectures in Appl. Math. **23** (1986)

Short-Time Well-Posedness of Free-Surface Problems in Irrotational 3D Fluids

D.M. Ambrose

Summary. We discuss the proof of short-time well-posedness for free-surface problems in irrotational three-dimensional fluids. We consider three situations: the vortex sheet with surface tension, the water wave, and Darcy flow. A common framework is described for treating each of these problems. In this framework, we choose convenient parameterizations and variables. In each case, we arrive at a system of evolution equations which is amenable to the use of the energy method. The work on the vortex sheet and the water wave is joint with Nader Masmoudi.

1 Introduction

Much progress has been made in the past decade in the study of initial value problems in fluid dynamics when free surfaces are present. In particular, free surface problems such as the water wave, the vortex sheet with surface tension, and Hele–Shaw and Darcy flows have been proved to be well posed in two- and three-space dimensions.

While many papers had addressed the well-posedness of water waves in restricted settings (such as [Cra85, Yos82], and many others), the first proof of well-posedness for the water wave for a very wide class of initial data (i.e., arbitrarily large data in Sobolev spaces, with the interface height being possibly multivalued) was by Wu in two dimensions [Wu97]; this was followed by her proof of the corresponding theorem in three dimensions [Wu99]. Similarly, for the vortex sheet with surface tension in two-dimensional fluids, there was a proof of well-posedness in the case of small data with single-valued height [ITT97], but the first proof for large data and multivalued height was by the author in [Amb03]. Subsequently, in [AM05], the author and N. Masmoudi have provided a new proof of well-posedness of the water wave in two-dimensions, arriving at the water wave without surface tension by taking the zero surface tension limit of water waves with surface tension. In [Amb04], the author has proved well-posedness of two-phase Hele–Shaw flow without surface tension if a condition on the data is satisfied. This condition essentially states that the more viscous fluid must displace the less viscous fluid. Other authors had proved well-posedness for Hele–Shaw problems in other cases, but the case of two fluids without surface tension had previously been an open problem. It is interesting to note that while [Amb04] proves short-time well-posedness with arbitrarily large initial data (subject to a few conditions), it had previously been proved that the problem had global solutions with small data [SCH04]. We remark that Hele–Shaw flow is inherently two-dimensional, but the equations of motion (primarily Darcy's law) can be understood for three-dimensional fluids as well. Thus, for the three-dimensional version of the Hele–Shaw results, we use the name "Darcy Flow" rather than "Hele–Shaw."

In the present chapter, we describe the author's recent work proving well-posedness of the vortex sheet with surface tension, the water wave, and Darcy flows without surface tension in three-dimensional fluids. The first two of these are joint work with Masmoudi. As in the author's works for two-dimensional fluids, the method is to first reformulate the problem using a favorable choice of parameterizations and a convenient choice of dependent variables. Also, as in the two-dimensional case, one of the most important steps is to find a useful formula for the Birkhoff–Rott integral. That is, the Birkhoff–Rott integral must be understood as being equal to a collection of important terms plus a smooth remainder; the important terms should be expressed using well-understood singular integral operators, such as the Riesz transforms. After the choice of variables and parameterizations has been made, and after the Birkhoff–Rott integral has been written in this way, it is possible to perform energy estimates on the resulting system. As is usually the case, these energy estimates then lead to well-posedness.

In Sect. 2 we describe the choice of parameterization, variables, and the evolution equation for the mean curvature of the free surface. This information is common to all three of the physical settings. In Sects. 3–5 we discuss the vortex sheet with surface tension, the water wave, and Darcy flow without surface tension, respectively; these are the problems treated in [AM06a, AM06b, Amb06]. In Sect. 6 we offer some concluding remarks.

The author was supported by National Science Foundation grant DMS-0610898.

2 General Formulation

We consider a free surface $\mathbf{X}(\alpha, \beta, t)$. This surface is the boundary between two fluids in the case of the vortex sheet or Darcy flow, and it is the boundary between a single fluid and a vacuum in the case of a water wave. We consider the case in which $(\alpha, \beta) \in \mathbf{R}^2$, and the surface decays to the horizontal plane at infinity. We have the unit tangent and normal vectors

$$\hat{\mathbf{t}}^{\mathbf{1}} = \frac{\mathbf{X}_\alpha}{|\mathbf{X}_\alpha|}, \qquad \hat{\mathbf{t}}^{\mathbf{2}} = \frac{\mathbf{X}_\beta}{|\mathbf{X}_\beta|}, \qquad \hat{\mathbf{n}} = \hat{\mathbf{t}}^{\mathbf{1}} \times \hat{\mathbf{t}}^{\mathbf{2}}.$$

The first fundamental form of a surface $\mathbf{X}(\alpha, \beta)$ is defined as (E, F, G), where

$$E = \mathbf{X}_\alpha \cdot \mathbf{X}_\alpha, \qquad F = \mathbf{X}_\alpha \cdot \mathbf{X}_\beta, \qquad G = \mathbf{X}_\beta \cdot \mathbf{X}_\beta. \tag{1}$$

The surfaces we are considering are evolving, and we say $\mathbf{X}$ evolves with velocity

$$\mathbf{X}_t = U\hat{\mathbf{n}} + V_1\hat{\mathbf{t}}^{\mathbf{1}} + V_2\hat{\mathbf{t}}^{\mathbf{2}}. \tag{2}$$

The normal velocity, U, must be determined by the fluid dynamics. The tangential velocities, V_1 and V_2, can be chosen arbitrarily, as this choice just serves to reparameterize the surface. We choose the tangential velocities so as to enforce the parameterization by isothermal coordinates,

$$E = G, \qquad F = 0. \tag{3}$$

This choice of parameterizations yields a pair of elliptic equations for the V_i. That is, if we differentiate the equations in (3) with respect to time, and if then use (1) and (2), we will find a pair of elliptic equations for the V_i. We use these equations to define the tangential velocities V_i; thus, if our surface is initially parameterized according to (3), then such a parameterization will be maintained at positive times by this choice of tangential velocity.

An important feature of this parameterization is that the function E is smoother than it might first appear. In particular, if $\mathbf{X}$ is in the Sobolev space H^ℓ, then one might guess $E \in H^{\ell-1}$. In fact, instead we have $E \in H^\ell$ also. This can be seen by the simple calculation that $\Delta E = (\mathbf{X}_\beta \cdot \mathbf{X}_\beta)_{\alpha\alpha} + (\mathbf{X}_\alpha \cdot \mathbf{X}_\alpha)_{\beta\beta}$, and to leading order this is the same as $(\mathbf{X}_\alpha \cdot \mathbf{X}_\beta)_{\alpha\beta}$. Of course, the latter expression vanishes to leading order since $\mathbf{X}_\alpha \cdot \mathbf{X}_\beta = 0$. This leads to a gain of one derivative in E.

The first step in finding a system of evolution equations for the problems under consideration is to make a choice of dependent variables. In particular, we will need a position-like variable and a velocity-like variable. For the position-like variable, we will use the mean curvature of the surface. This choice is made because in the problems with surface tension, the jump in pressure across the free surface is the constant surface tension coefficient times the mean curvature of the surface (this is the Laplace–Young boundary condition). It turns out, however, that this choice is well-suited even to problems without surface tension. The choice of the other dependent variable (i.e., the velocity-like variable) depends on the particular problem being treated. In the remainder of this section, we will discuss the evolution of the mean curvature.

The second fundamental form of $\mathbf{X}$ is (L, M, N), where

$$L = \mathbf{X}_{\alpha\alpha} \cdot \hat{\mathbf{n}}, \qquad M = \mathbf{X}_{\alpha\beta} \cdot \hat{\mathbf{n}}, \qquad N = \mathbf{X}_{\beta\beta} \cdot \hat{\mathbf{n}}.$$

Using the isothermal coordinates, the mean curvature is then $\kappa = (L+N)/2E$. The evolution equation for κ in terms of U and the V_i can be written

$$\kappa_t = \frac{1}{2E}\Delta U + \frac{V_1}{\sqrt{E}}\kappa_\alpha + \frac{V_2}{\sqrt{E}}\kappa_\beta + f, \tag{4}$$

where f is a collection of lower-order terms. The regularity of κ and μ is different in each different physical setting, but in all of the cases currently being considered, the terms which constitute f are indeed lower order and can be treated routinely in the energy estimates. (In the sequel, we will continue to use f in this fashion. In particular, the exact terms represented by f will change from line to line, but these terms will always be a collection of lower-order terms which can be treated routinely in the energy estimates.)

We now discuss U; we see from (4) that we must have a good understanding of U in order to understand κ_t. Whereas the tangential velocities of the interface were taken to be artificial (in order to preserve a favorable parameterization), the normal velocity must be determined by the fluid dynamics. The normal velocity, U, must be the normal component of the Birkhoff–Rott integral, $U = \mathbf{W} \cdot \hat{\mathbf{n}}$. We will now define $\mathbf{W}$.

We use the notation $\boldsymbol{\alpha} = (\alpha, \beta)$ and $\boldsymbol{\alpha}' = (\alpha', \beta')$. The Birkhoff–Rott integral, $\mathbf{W}$, is

$$\mathbf{W}(\boldsymbol{\alpha}) = \frac{1}{4\pi}\mathrm{PV}\iint (\mu_\beta(\boldsymbol{\alpha}')\mathbf{X}_\alpha(\boldsymbol{\alpha}') - \mu_\alpha(\boldsymbol{\alpha}')\mathbf{X}_\beta(\boldsymbol{\alpha}')) \times \frac{\mathbf{X}(\boldsymbol{\alpha}) - \mathbf{X}(\boldsymbol{\alpha}')}{|\mathbf{X}(\boldsymbol{\alpha}) - \mathbf{X}(\boldsymbol{\alpha}')|^3}\, d\boldsymbol{\alpha}'.$$

The Birkhoff–Rott integral can be found by the usual process of recovering velocity from vorticity. Since the fluids we are considering are irrotational in the bulk of the fluids, but there is a jump in velocity across the free surface, the vorticity is measure-valued, with support on the free surface. Using this fact with the Biot–Savart law, and taking the limit at the free surface, we get the Birkhoff–Rott integral. The interested reader might consult [Saf95] for further details of the Birkhoff–Rott integral.

To better understand $\mathbf{W}$, we use Taylor expansions for $\mathbf{X}(\boldsymbol{\alpha}) - \mathbf{X}(\boldsymbol{\alpha}')$. This must be done carefully so that we are sure that all of the integrals we deal with are convergent; the full details are given in [AM06a]. The result is the following formula:

$$(\mathbf{W}_\alpha \cdot \hat{\mathbf{n}})_\alpha + (\mathbf{W}_\beta \cdot \hat{\mathbf{n}})_\beta = \frac{1}{2}\Lambda\left(\sqrt{E}\left(\frac{\mu_\alpha}{E}\right)_\alpha + \sqrt{E}\left(\frac{\mu_\beta}{E}\right)_\beta\right) + f. \tag{5}$$

The operator Λ has symbol $|\xi|$; that is, $\Lambda = H_1 D_\alpha + H_2 D_\beta$, and Λ could be expressed as $\Lambda = \sqrt{-\Delta}$. Formulas similar to (5) can be found for the tangential components of $\nabla\mathbf{W}$.

With formula (5) for the leading-order behavior of ΔU, we can now state the evolution equation for κ:

$$\kappa_t = \frac{1}{4E}\Lambda\left(\sqrt{E}\left(\frac{\mu_\alpha}{E}\right)_\alpha + \sqrt{E}\left(\frac{\mu_\beta}{E}\right)_\beta\right) + \frac{V_1 - \mathbf{W}\cdot\hat{\mathbf{t}}^1}{\sqrt{E}}\kappa_\alpha + \frac{V_2 - \mathbf{W}\cdot\hat{\mathbf{t}}^2}{\sqrt{E}}\kappa_\beta + f. \tag{6}$$

Having established this evolution equation for κ, we now turn to the discussion each particular problem, starting in Sect. 3 with the vortex sheet.

3 The Vortex Sheet

For the vortex sheet with surface tension, we consider the case in which the two fluids have equal density. The proof of [AM06a] applies equally well in the case of different densities, but the exposition is simpler in the density-matched case.

Given (6) above, we now need to discuss μ. We can infer an evolution equation for μ from the Euler equations. In particular, μ is the jump in the velocity potential across the free surface, and the potential on each side satisfies a Bernoulli equation. We can take the limit at the interface and take the jump, as in [BMO82].

When stating the evolution equation for μ, it is more illuminating to first apply the operator Λ; doing so, we have the equation

$$\begin{aligned}\Lambda(\mu)_t &= \tau\Lambda(\kappa) + \frac{1}{\sqrt{E}}(\mu_\alpha H_1 + \mu_\beta H_2)^2(\kappa)\\ &\quad + \frac{V_1 - \mathbf{W}\cdot\hat{\mathbf{t}}^1}{\sqrt{E}}\Lambda(\mu)_\alpha + \frac{V_2 - \mathbf{W}\cdot\hat{\mathbf{t}}^2}{\sqrt{E}}\Lambda(\mu)_\beta + f. \qquad (7)\end{aligned}$$

This equation and the evolution equation for the mean curvature form the main system of equations for the vortex sheet with surface tension. Now that we have arrived at the system, we see that what we have is (in a generalized sense) a quasilinear hyperbolic system. Well-posedness can be proved, then, by symmetrizing the equations and performing energy estimates. The details of this procedure can be found in [AM06a]. After symmetrizing, the variables we have are κ and A, where $A = (A_1, A_2)$, with

$$A_1 = \Lambda^{1/2}\left(\frac{\mu_\alpha}{\sqrt{E}}\right), \qquad A_2 = \Lambda^{1/2}\left(\frac{\mu_\beta}{\sqrt{E}}\right).$$

We are able to prove energy estimates with $\kappa \in H^{s-1}$ and $A \in H^{s-1}$. This corresponds to $\mathbf{X} \in H^{s+1}$ and $\mu \in H^{s+1/2}$. We must take s larger than a certain positive value, i.e., we are assuming that s is large enough. We are then able to prove that if the initial data is in this same space, satisfying a natural non-self-intersection condition, then there is a time interval on which a solution exists in this space and which satisfies the same non-self-intersection condition. This solution is unique, and the solution depends continuously on the initial data.

We make a comment about the important terms on the right-hand side of (7). The leading-order term on the right-hand side is $\tau\Lambda\kappa$, and this term has a good sign as long as τ is positive, which it always is as long as surface tension is present. The next most important term has a zero-order operator, $(\mu_\alpha H_1 + \mu_\beta H_2)^2$, applied to κ, and this term is destabilizing. That is, this term can be seen to be responsible for the Kelvin–Helmholtz instability when $\tau = 0$. If we were to hope for a hyperbolic system in the case without surface tension, we would find that this term has the wrong sign. Thus, in the absence of surface tension, this leading-order term leads us to a quasilinear elliptic system; as elliptic equations have ill-posed initial value problems, we see that the vortex sheet without surface tension is ill-posed in Sobolev spaces.

4 The Irrotational Water Wave

For the water wave, in [AM06b] we find that in 3D, as in 2D (see [AM05]), μ is not the most satisfactory choice of dependent variable. Instead, we introduce δ_1 and δ_2:

$$\delta_1 = \frac{\mu_\alpha}{2\sqrt{E}} + \mathbf{W}\cdot\hat{\mathbf{t}}^{\mathbf{1}} - V_1, \qquad \delta_2 = \frac{\mu_\beta}{2\sqrt{E}} + \mathbf{W}\cdot\hat{\mathbf{t}}^{\mathbf{2}} - V_2.$$

There are two useful ways of viewing these terms: they can either be seen as essentially being μ_α and μ_β plus some corrections, or they can be viewed as being the differences between the Lagrangian tangential velocities on the free surface and the artificial tangential velocities. The dependent variables for the water wave problem are taken to be κ and a variable which is related to the divergence of (δ_1, δ_2); we call this variable B. With these variables we are able to find (in both the case with and the case without surface tension) a quasilinear hyperbolic system of evolution equations. In particular, we find the following system of evolution equations:

$$\kappa_t = \frac{1}{2\sqrt{E}}B - \frac{\delta_1}{\sqrt{E}}\kappa_\alpha - \frac{\delta_2}{\sqrt{E}}\kappa_\beta + G + f,$$

$$B_t = -\tau\mathcal{L}\mathcal{L}^*(\kappa) - c\Lambda(\kappa) - \kappa\Lambda(c) - \frac{\delta_1}{\sqrt{E}}B_\alpha - \frac{\delta_2}{\sqrt{E}}B_\beta + f.$$

The operator $\mathcal{L}$ and its adjoint $\mathcal{L}^*$ each act like 3/2 of a derivative, c is discussed below, and G is a collection of terms which is lower order than the leading-order terms, but cannot be treated routinely in the estimates.

While only the density-matched case was treated explicitly in [AM06a], the proof of well-posedness of vortex sheets with surface tension extends to the case of fluids with different densities. Well-posedness of water waves with surface tension is a special case; with this proof, the interval of existence vanishes with surface tension. In [AM06b], we establish new estimates for the problem which are uniform in surface tension. (It is worth mentioning that the estimates for the water wave with surface tension are interesting in their own right.) We are then able to take the limit as surface tension vanishes, giving a new proof of well-posedness of water waves without surface tension.

The water wave is well-posed only when the generalized Taylor condition is satisfied. This condition is on the normal derivative of the pressure,

$$-\nabla p\cdot\hat{\mathbf{n}} = c(\boldsymbol{\alpha}, t) \geq \bar{c} > 0,$$

for some constant $\bar{c}$ and for all $\boldsymbol{\alpha}$. Wu has proved in [Wu97] and [Wu99] that the condition is satisfied as long as the surface is non-self-intersecting. In the two-dimensional case, this condition is discussed in [BHL93].

5 Darcy Flow

In the case of Darcy flow, we do not have an evolution equation for μ. In this case, as before, μ is the jump in velocity potential across the free surface. For Darcy flow, the velocity potential is proportional to the pressure, with the constant of proportionality involving the viscosity and density. Since the viscosity and density may jump across the free surface, there is a jump in the potential, even without surface tension. If we write $\mathbf{X} = (x, y, z)$, then we have the following integral equations:

$$\mu_\alpha = -2A_\nu\sqrt{E}\mathbf{W}\cdot\hat{\mathbf{t}}^{\mathbf{1}} - Rz_\alpha,$$
$$\mu_\beta = -2A_\nu\sqrt{E}\mathbf{W}\cdot\hat{\mathbf{t}}^{\mathbf{2}} - Rz_\beta,$$

where A_ν is proportional to the viscosity difference and R is proportional to the density difference.

Plugging these formulas into the equation for κ, and collecting lower-order terms together, we get

$$\kappa_t = -k\Lambda\kappa + \left(\frac{V_1 - \mathbf{W}\cdot\hat{\mathbf{t}}^{\mathbf{1}}}{\sqrt{E}} - \frac{A_\nu\mu_\alpha}{2E}\right)\kappa_\alpha + \left(\frac{V_2 - \mathbf{W}\cdot\hat{\mathbf{t}}^{\mathbf{2}}}{\sqrt{E}} - \frac{A_\nu\mu_\beta}{2E}\right)\kappa_\beta + f. \tag{8}$$

The quantity $k = k(\boldsymbol{\alpha}, t)$ is very important to the well-posedness. It is

$$k(\boldsymbol{\alpha}, t) = \frac{2A_\nu U + Rh}{\sqrt{E}}, \qquad h = \hat{\mathbf{n}}\cdot(0, 0, 1).$$

The equation (8) is clearly parabolic; the sign of k determines whether it is a well-posed or ill-posed parabolic problem. In order to guarantee short-time well-posedness, we take the condition

$$k(\boldsymbol{\alpha}) > \bar{k} > 0, \tag{9}$$

for some constant $\bar{k}$, uniformly at the initial time. (The case in which such a condition is satisfied is known as the "stable case" of the problem.) That (9) is satisfied essentially says that the more viscous fluid must displace the less viscous fluid, although this may not need to be strictly true, depending on gravity.

The proof of well-posedness is similar to the proof of well-posedness for the vortex sheet with surface tension in that it relies on energy estimates. It should be noted that since we are proving existence of smooth solutions, and since condition (9) is satisfied at the initial time, it will continue to be satisfied at least for a positive amount of time.

6 Conclusion

The method described in this chapter allows for proof of short-time well-posedness of irrotational free-surface flows in three-dimensional fluids. It is a generalization of the method used by the author (and Masmoudi) to prove short-time well-posedness of free-surface flows in two-dimensional fluids; this method in two dimensions grew out of prior numerical work [HLS94]. It is expected that with the formulation described in this paper for three-dimensional problems, further progress in numerical methods for these problems may be possible. Furthermore, it is possible that this kind of formulation may help shed light on analytical studies of the behavior of these systems at long times.

References

[AM05] D. Ambrose and N. Masmoudi. The zero surface tension limit of two-dimensional water waves. *Comm. Pure Appl. Math.*, 58:1287–1315, 2005.

[AM06a] D. Ambrose and N. Masmoudi. Well-posedness of 3D vortex sheets with surface tension. 2006. Submitted.

[AM06b] D. Ambrose and N. Masmoudi. The zero surface tension limit of three-dimensional water waves. 2006. In preparation.

[Amb03] D. Ambrose. Well-posedness of vortex sheets with surface tension. *SIAM J. Math. Anal.*, 35:211–244, 2003.

[Amb04] D. Ambrose. Well-posedness of two-phase Hele-Shaw flow without surface tension. *European J. Appl. Math.*, 15:597–607, 2004.

[Amb06] D. Ambrose. Well-posedness of two-phase Darcy flow in 3D. 2006. In preparation.

[BHL93] J.T. Beale, T. Hou, and H. Lowengrub. Growth rates for the linearized motion of fluid interfaces away from equilibrium. *Comm. Pure Appl. Math.*, 46:1269–1301, 1993.

[BMO82] G. Baker, D. Meiron, and S. Orszag. Generalized vortex methods for free-surface flow problems. *J. Fluid Mech.*, 123:477–501, 1982.

[Cra85] W. Craig. An existence theory for water waves and the Boussinesq and Korteweg-de Vries scaling limits. *Comm. Partial Differential Equations*, 10:787–1003, 1985.

[HLS94] T. Hou, J. Lowengrub, and M. Shelley. Removing the stiffness from interfacial flows with surface tension. *J. Comput. Phys.*, 114:312–338, 1994.

[ITT97] T. Iguchi, N. Tanaka, and A. Tani. On the two-phase free-boundary problem for two-dimensional water waves. *Math. Ann.*, 309:199–223, 1997.

[Saf95] P.G. Saffman. *Vortex Dynamics*. Cambridge University Press, 1995.

[SCH04] M. Siegel, R. Caflisch, and S. Howison. Global existence, singular solutions, and ill-posedness for the Muskat problem. *Comm. Pure Appl. Math.*, 57:1374–1411, 2004.

[Wu97] S. Wu. Well-posedness in Sobolev spaces of the full water-wave problem in 2-D. *Invent. Math.*, 130:39–72, 1997.

[Wu99] S. Wu. Well-posedness in Sobolev spaces of the full water wave problem in 3-D. *J. Amer. Math. Soc.*, 12:445–495, 1999.

[Yos82] H. Yosihara. Gravity waves on the free surface of an incompressible perfect fluid of finite depth. *Publ. Res. Inst. Math. Sci.*, 18:49–96, 1982.

Mathematical Study of Static Grain Deep-Bed Drying Models

D. Aregba-Driollet

1 Introduction

Considerable amount of agricultural crops, and more particularly grains, are dried artificially using near ambient or high-temperature air in various grain drying systems. The speed and efficiency of drying depend on the drying air characteristics: relative humidity, temperature, and velocity. In order to preserve the quality of the product it is important to control its temperature and moisture content during the drying process.

Here we study nonequilibrium models for static grain deep-bed drying: motionless Piles of grain are exposed to drying air and there is no heat and mass equilibrium between the drying air and the grain through the bed.

Let us consider the drier as a vertical column of length L in the x-direction, $x > 0$. The drying air comes vertically from the bottom $x = 0$ with a constant temperature T_{ab} and a positive constant speed V_a. We denote respectively X_a and X_p the moisture contents of the air and of the product to dry, T_a and T_p the temperatures of the air and of the product. They are functions of (x,t) and have to take nonnegative values. All the functions here below are defined for nonnegative values of the unknowns.

The grain is supposed to be hygroscopic: when dried, its moisture content tends to a (nonzero) equilibrium state $X_{eq}(X_a)$, which is specific of each product. For cereals X_{eq} is a smooth strictly increasing function defined on a bounded interval $[0, X_{max}[$ and such that

$$X_{eq}(0) = 0, \qquad \lim_{X \to X_{max}} X_{eq}(X) = +\infty.$$

The drying kinetic is deduced from experiments on thin layers of the considered products. Here we have:

$$\partial_t X_p = -K(T_p)\,[X_p - X_{eq}(X_a)]$$

with $K(T_p) = d \exp\left(-\, c/T_p\right)$, d and c being positive constants.

Then, the models are derived by mass and energy balance. We focus our attention on Spencer's model [Sp]:

$$\begin{cases} \partial_t X_a + \dfrac{V_a}{\epsilon}\partial_x X_a = \dfrac{\alpha}{\epsilon} K(T_p)\left[X_p - X_{eq}(X_a)\right], \\[2ex] \partial_t T_a + \dfrac{V_a}{\epsilon}\partial_x T_a = -\dfrac{\beta(X_a,T_a)}{\epsilon}(T_a - T_p), \\[2ex] \partial_t X_p = -K(T_p)\left[X_p - X_{eq}(X_a)\right], \\[2ex] \partial_t T_p = \Psi_1(X_a,T_a,X_p)(T_a-T_p) - \Psi_2(X_p,T_p)\left[X_p - X_{eq}(X_a)\right]. \end{cases} \tag{1}$$

Here, α is a positive constant, β, Ψ_1, Ψ_2 are positive smooth functions which do not detail here. The positive constant ϵ is the void fraction of the bed. The system is supplemented with initial data for $x \in]0, L[$

$$X_a(x,0) = X_{a0}, \quad T_a(x,0) = T_{a0}, \quad X_p(x,0) = X_{p0}, \quad T_p(x,0) = T_{p0}, \tag{2}$$

and boundary data for $t > 0$

$$X_a(0,t) = X_{ab}, \quad T_a(0,t) = T_{ab}. \tag{3}$$

In most cases, authors neglect the accumulation terms in (1), see [BBH] for example:

$$\begin{cases} \partial_x X_a = \dfrac{\alpha}{V_a} K(T_p)\left[X_p - X_{eq}(X_a)\right], \\[2ex] \partial_x T_a = -\dfrac{\beta(X_a,T_a)}{V_a}(T_a - T_p), \\[2ex] \partial_t X_p = -K(T_p)\left[X_p - X_{eq}(X_a)\right], \\[2ex] \partial_t T_p = \Psi_1(X_a,T_a,X_p)(T_a-T_p) - \Psi_2(X_p,T_p)\left[X_p - X_{eq}(X_a)\right]. \end{cases} \tag{4}$$

The boundary conditions (3) remain the same while only the initial data (X_{p0}, T_{p0}) is requested.

The frequent use of model (4) in the literature can be explained by the fact that system (4) can be discretized by methods for which the use of time step control procedures is easy. However, it has been shown that in many realistic situations, models (1) and (4) produce different results, see [AN] for comparisons.

As a first approximation, we consider the diffusion coefficient K as a constant and study the following 2×2 reduced system:

$$\begin{cases} \partial_t X_a + \dfrac{V_a}{\epsilon}\partial_x X_a = \dfrac{\alpha}{\epsilon} K\left[X_p - X_{eq}(X_a)\right], \\ \partial_t X_p = -K\left[X_p - X_{eq}(X_a)\right]. \end{cases} \tag{5}$$

The numerical experiments show that the asymptotic behavior of the full model is qualitatively the same as the one of the reduced system, see Sect. 3.

The same arguments lead us to consider the reduced version of system (4):

$$\begin{cases} \partial_x X_a = \dfrac{\alpha}{V_a} K \left[X_p - X_{eq}(X_a) \right], \\ \partial_t X_p = -K \left[X_p - X_{eq}(X_a) \right]. \end{cases} \tag{6}$$

The plan of the paper is as follows: in Sect. 2 we prove global existence and convergence to the dry state equilibrium $X_{eq}(X_a) = X_p$ for the IBVP problem associated to (5). For the Goursat problem related to system (6) the global existence has been proved in [NT] for a similar problem arising in a different context. The asymptotic behavior can be proved with the same ideas as in Sect. 2, we do not give the details here. In Sect. 3 we present some numerical experiments illustrating the drying process on the full model (1).

2 Global Existence and Asymptotic Behavior for (5)

We denote $u = X_a$, $v = X_p$, $g = X_{eq}$, $c = \frac{V_a}{\epsilon}$ and without loss of generality we set the other coefficients equal to 1. The system can then be written as

$$\begin{cases} \partial_t u + c \partial_x u = v - g(u) \\ \partial_t v = -v + g(u) \end{cases} \quad x \in]0, L[, \quad t > 0. \tag{7}$$

The constant c is positive and g is a smooth increasing function defined on $[0, 1[$ such that:

$$g(0) = 0, \quad \lim_{u \to 1} g(u) = +\infty.$$

The initial and boundary data are

$$\begin{cases} u(x, 0) = u_0(x), \\ v(x, 0) = v_0(x), \quad x \in]0, L[, \quad t > 0, \\ u(0, t) = u_b(t) \end{cases} \tag{8}$$

where (u_0, v_0) and u_b are bounded measurable functions satisfying

$$0 < u_0, \quad 0 < u_b, \quad \max(\|u_0\|_\infty, \|u_b\|_\infty) < 1. \tag{9}$$

By the general theory of semilinear hyperbolic systems, uniqueness, and local existence of solutions hold in $C^0([0, T], L^1(]0, L[)) \cap L^\infty(]0, T[\times]0, L[) \cap C^0([0, L], L^1(]0, T[))$. Moreover, if the maximal time of existence T^* is finite, then

$$\lim_{t \to T^*} \|(u(t), v(t))\|_{L^\infty(]0,L[)} = +\infty$$

see for example [Su]. To obtain uniform bounds and hence global existence, we remark that system (7) is quasimonotone as defined in [HN]. This allows us to compare solutions by using the ideas of [Te].

Proposition 1. *Let* (u_1, v_1) *and* (u_2, v_2) *two solutions of problem (7)(8) defined on* $[0, L] \times [0, T]$*, respectively associated to data* (u_{10}, v_{10}, u_{1b}) *and* (u_{20}, v_{20}, u_{2b})*. For all* $(x, t) \in [0, L] \times [0, T]$

$$\int_0^x [u_1 - u_2]_+(y,t) + [v_1 - v_2]_+(y,t)dy + c\int_0^t [u_1 - u_2]_+(x,s)ds \tag{10}$$
$$\leq \int_0^x [u_{10} - u_{20}]_+(y) + [v_{10} - v_{20}]_+(y)dy + c\int_0^t [u_{1b} - u_{2b}]_+(s)ds.$$

Proof. We give the idea of the proof. We write the system satisfied by $(u_1 - u_2, v_1 - v_2)$ and multiply by $(Y(u_1 - u_2), Y(v_1 - v_2))$, where Y is the Heavyside function:

$$\begin{cases} \partial_t[u_1 - u_2]_+ + c\partial_x[u_1 - u_2]_+ = Y(u_1 - u_2)(v_1 - v_2 - g(u_1) + g(u_2)) \\ \partial_t[v_1 - v_2]_+ = -Y(v_1 - v_2)(v_1 - v_2 - g(u_1) + g(u_2)) \end{cases} \tag{11}$$

Due to quasimonotonicity, the sum of the right-hand sides is always nonpositive, so that

$$\partial_t([u_1 - u_2]_+ + [v_1 - v_2]_+) + c\partial_x[u_1 - u_2]_+ \leq 0.$$

Integrating this inequality on $[0, x] \times [0, t]$ gives (10). □

Corollary 1. *Under the assumptions of Proposition 1, the following inequality holds for* $t \in [0, T]$*:*

$$\|u_1(t) - u_2(t)\|_{L^1(0,L)} + \|v_1(t) - v_2(t)\|_{L^1(0,L)} \tag{12}$$
$$\leq \|u_{10} - u_{20}\|_{L^1(0,L)} + \|v_{10} - v_{20}\|_{L^1(0,L)} + c\|u_{1b} - u_{2b}\|_{L^1(0,T)}.$$

We can now prove global existence:

Theorem 1. *Let* u_0*,* u_b *be bounded measurable functions satisfying (9). Let* v_0 *be a measurable function. We denote*

$$v_m = \min(\inf_x g(u_0(x)), \inf_t g(u_b(t))), \quad v_M = \max(g(\|u_0\|_\infty), g(\|u_b\|_\infty)).$$

We assume that

$$v_m \leq v_0(x) \leq v_M \ \ ae. \tag{13}$$

The problem (7)(8) owns a global solution and

$$g^{-1}(v_m) \leq u \leq g^{-1}(v_M) \text{ and } v_m \leq v \leq v_M. \tag{14}$$

Proof. By (9) $v_m, v_M \in [0, +\infty[$. We have to find a uniform bound for any local solution. Let U be a real number. For the following initial boundary data

$$u(x, 0) = u(0, t) = U, \quad v(x, 0) = g(U)$$

the solution of (7) is global and given by

$$u(x,t) = U, \quad v(x,t) = g(U).$$

If $U = g^{-1}(v_m)$ then $U \leq u_0$, $U \leq u_b$ et $g(U) \leq v_0$. By Proposition 1:

$$g^{-1}(v_m) \leq u(x,t) \ \text{ and } \ v_m \leq v(x,t).$$

If $U = g^{-1}(v_M)$ then $u_0 \leq U$, $u_b \leq U$ and $v_0 \leq g(U)$. So

$$u \leq g^{-1}(v_M) \ \text{ and } \ v \leq v_M.$$

Hence the solution is global. □

To study the asymptotic behavior of the solutions, we restrict ourselves to the realistic physical situation where the moisture content u_b of the air at the inlet of the drier and the initial moisture contents (u_0, v_0) of the air and of the product are constant functions. Moreover we suppose that

$$u_0, u_b \in]0,1[, \quad g(u_b) \leq v_0 \leq g(u_0). \tag{15}$$

As a consequence we have global existence.

Theorem 2. *Let u_0, v_0, u_b be constant functions satisfying (15). Then the solution of problem (7)(8) tends to the equilibrium in $L^1(]0,L[)$:*

$$\lim_{t \to +\infty} \int_0^L |v(x,t) - g(u(x,t))| dx = 0. \tag{16}$$

Proof. We first construct a sequence $(u_0^\epsilon, v_0^\epsilon)_{\epsilon > 0}$ of C^∞ initial data converging to (u_0, v_0) in $L^1(]0,L[)$, such that

$$u_0^\epsilon(0) = u_b, \ v_0^\epsilon(0) = g(u_b), \ \frac{d^k u_0^\epsilon}{dx^k}(0) = 0 \text{ for } k \geq 1, \ \frac{du_0^\epsilon}{dx} \geq 0, \ \frac{dv_0^\epsilon}{dx} \geq 0,$$

and

$$\|u_0^\epsilon\|_\infty = u_0, \quad \|v_0^\epsilon\|_\infty = v_0.$$

The solutions are global and

$$u_b \leq u^\epsilon(x,t) \leq u_0, \quad g(u_b) \leq v^\epsilon(x,t) \leq g(u_0).$$

By Corollary 1 and the fact that g is Lipschitz continuous on $[u_b, u_0]$, it is sufficient to prove the theorem for the data $(u_0^\epsilon, v_0^\epsilon)$, u_b. As far as it is not confusing, in the sequel we omit the superscript ϵ for the data and for the solution.

The above conditions ensure that the solution of (7)(8) is C^1. We first prove that $U = \partial_x u \geq 0$ and $V = \partial_x v \geq 0$. These functions satisfy the system

$$\begin{cases} \partial_t U + c\partial_x U = V - g'(u)U \\ \partial_t V = -V + g'(u)U \end{cases} \quad x \in]0,L[, \quad t > 0.$$

The initial values $U(x,0)$, $V(x,0)$ are nonnegative by assumption and at the boundary

$$U(0,t) = \frac{1}{c}(v(0,t) - g(u_b)). \tag{17}$$

Writing v as

$$v(x,t) = v_0(x)e^{-t} + \int_0^t e^{-t+s} g(u(x,s))ds,$$

we see that $U(0,t) = 0$. We then use a comparison method:

$$\partial_t([-U]_+ + [-V]_+) + c\partial_x[-U]_+ = (V - g'(u)U)[Y(-V) - Y(-U)]$$

and one can prove that the right-hand side is nonpositive. Integration as in the proof of Proposition 1 gives

$$U \geq 0, \ \ V \geq 0.$$

As a consequence:

$$\partial_t(v - g(u)) + c\partial_x(v - g(u)) \geq \partial_t(v - g(u)) - c\partial_x g(u) = (-1 - g'(u))(v - g(u)).$$

Using (17) with the fact that $U(0,t) = 0$ we integrate along the characteristics coming from the boundary $x = 0$ and we find

$$v(x,t) \geq g(u(x,t)), \ \ \forall t > x/c.$$

Let us denote

$$\int_0^L |v(x,t) - g(u(x,t))| dx = y(t).$$

As $\partial_t \int_0^L v dx = \int_0^L (-v + g(u))dx$, when $t \to +\infty$ the limit v_∞ of $\int_0^L v(x,t)dx$ exists and

$$v_\infty = \int_0^L v(x,t)dx - \int_t^{+\infty} y(s)ds, \ \ \forall t > L/c.$$

Therefore $y \in L^1(]L/c, +\infty[)$. To prove that $\lim_{t\to+\infty} y(t) = 0$, it is sufficient to prove that

$$y' \in L^1(]L/c, +\infty[). \tag{18}$$

For $t > L/c$:

$$y'(t) = \int_0^L (-1 - g'(u))(v - g(u))dx + c[g(u(L,t)) - g(u_b)].$$

Moreover

$$g(u(L,t)) - g(u_b) = \int_{t-L/c}^t g'(u)(v - g(u))(L - c(t-s), s)ds$$

which leads to (18) and ends the proof. □

3 Numerical Results

We present here some numerical results for the full model (1), obtained with a second-order semi-implicit finite volume method. The right-hand side is discretized by a second-order MUSCL extension of upwind scheme, the non-linear source-term is treated in a semi-implicit way which allows a fully explicit formulation. We refer the reader to [AA] for details.

Here $L = 0.6\,\mathrm{m}$, $V_a = 2\,\mathrm{ms}^{-1}$, T_{ab} is constant: $T_{ab} = 45°\mathrm{C}$, the relative humidity at the boundary is also constant: $H_{ra,b} = 30\%$. X_{ab} is a known function of T_{ab} and $H_{ra,b}$. Initially, all the data are uniform: the temperature of the air and of the product is 25°C, the relative humidity of the air is $H_{ra,0} = 60\%$ and the moisture content of the product is $X_{p0} = 0.6$.

Figure 1 shows the time evolution of the air moisture content (left) and of the air temperature (right) for three points: near the bottom ($x = 0.05$), in the middle ($x = 0.25$), and near the top ($x = 0.59$) of the drier. Figure 2 shows the time evolution of the grain moisture content (left) and of the grain temperature (right) at the same position. One can observe the strong variations at the beginning of the drying process, which makes the computations rather long and delicate. Figure 3 shows the evolution of $X_p - X_{eq}(X_a)$ for the same three

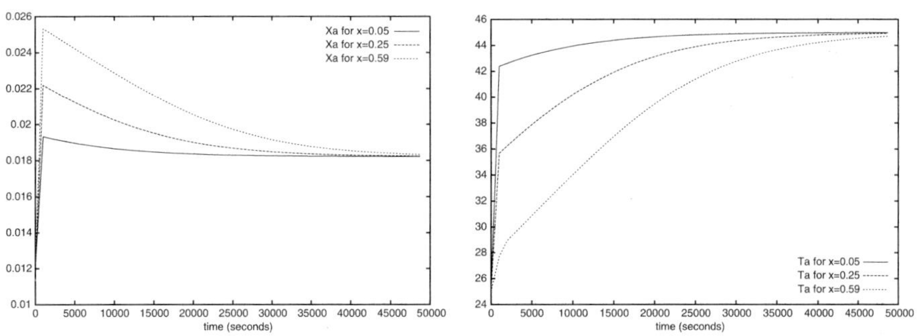

Fig. 1. Air moisture content (*left*) and temperature (*right*) evolution for three points

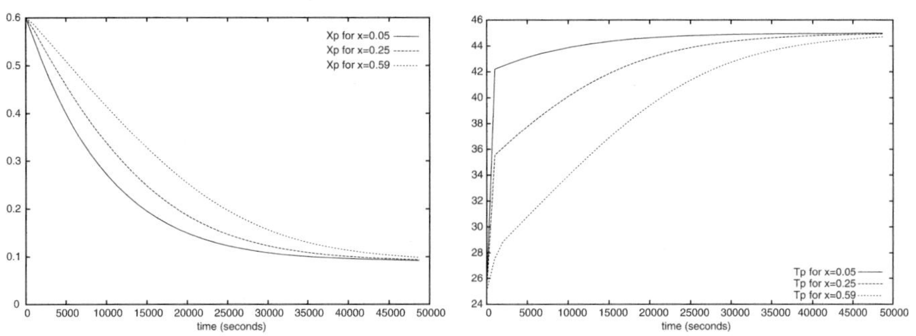

Fig. 2. Grain moisture content (*left*) and temperature (*right*) evolution for three points

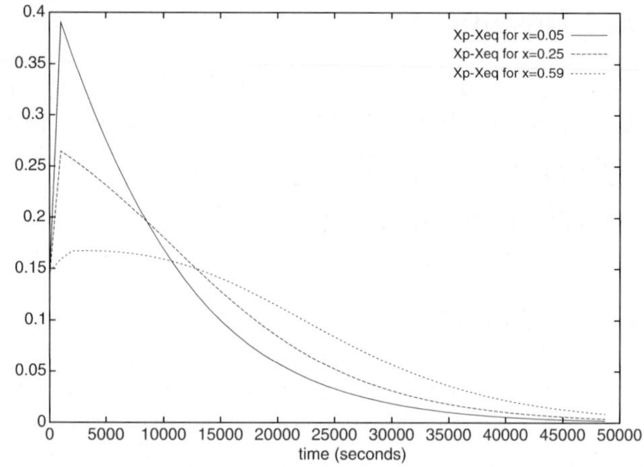

Fig. 3. Convergence to equilibrium for three points

points and the convergence to the dry state equilibrium. One can remark that this quantity remains positive, as in the proof of Theorem 2 for the reduced 2×2 system.

References

[AA] Aregba, A.W., Aregba-Driollet, D.: Modelisation and simulation of static grain deep-bed drying. In: Bermùdez de Castro, A. et al (ed) Numerical Mathematics and advanced Applications. Springer-Verlag, Berlin Heidelberg (2006)

[AN] Aregba, A.W., Nadeau, J.P.: Comparison of two non-equilibrium models for static grain deep-bed drying by numerical simulations. Journal of Food Engineering **78**, no. 4, 1174–1187 (2007)

[BBH] Brooker, D.B., Bakker-Arkema, F.W., Hall, C.W.: Drying cereal grains. Westport, Connecticut, AVI (1974)

[HN] Hanouzet, B., Natalini, R.: Weakly coupled systems of quasilinear hyperbolic equations. Differential Integral Equations **9**, no. 6, 1279–1292 (1996)

[NT] Natalini, R., Tesei, A.: On the Barenblatt model for non-equilibrium two phase flow in porous media. Arch. Ration. Mech. Anal. **150**, no. 4, 349–367 (1999)

[Sp] Spencer, H.B.: A mathematical simulation of grain drying. J. Agric. Eng. Res. **14** (3), 226–235 (1969)

[Su] Sueur, F.: Couches limites semilinéaires. Ann. Fac. Sci. Toulouse Math. **6** 15, no. 2, 323–380 (2006)

[Te] Terracina, A.: Comparison properties for scalar conservation laws with boundary conditions. Nonlinear Anal. **28**, no. 4, 633–653 (1997)

Finite Volume Central Schemes for Three-Dimensional Ideal MHD

P. Arminjon and R. Touma

Summary. We present second-order accurate central finite volume methods adapted here to three-dimensional problems in ideal magnetohydrodynamics. These methods alternate between two staggered grids, thus leading to Riemann solver-free algorithms with relatively favorable computing times.

The original grid considered in this paper is Cartesian, while the dual grid features either Cartesian or diamond-shaped oblique dual cells.

The $\text{div} \cdot \mathbf{B} = 0$ constraint on the magnetic field is enforced with a suitable adaptation of the constrained transport method to our central schemes.

Numerical experiments show the feasibility of these methods and our results are in good agreement with existing results in the literature.

1 Introduction

The equations of ideal MHD, which consist of the conservation laws for the mass density ρ, momentum $\rho\mathbf{v}$, total energy ρe as well as Faraday's induction law, can be written as [JT64]

$$\frac{\partial}{\partial t}\begin{bmatrix}\rho\\ \rho\mathbf{v}\\ \rho e\\ \mathbf{B}\end{bmatrix} + \nabla\cdot\begin{bmatrix}\rho\mathbf{v}\\ \rho\mathbf{v}\mathbf{v} + I(p+\frac{\mathbf{B}\cdot\mathbf{B}}{2}) - \mathbf{B}\mathbf{B}\\ (\rho e + p + \frac{\mathbf{B}\cdot\mathbf{B}}{2})\mathbf{v} - (\mathbf{v}\cdot\mathbf{B})\mathbf{B}\\ \mathbf{v}\mathbf{B} - \mathbf{B}\mathbf{v}\end{bmatrix} = 0, \tag{1}$$

where $\mathbf{B}$ is the magnetic field and I is the (3×3) identity matrix; the thermal pressure is computed from an ideal gas equation of state,

$$P = (\gamma - 1)(\rho e - \frac{1}{2}\rho|\mathbf{v}|^2 - \frac{1}{2}|\mathbf{v}|^2), \tag{2}$$

where γ denotes the ratio of specific heats.

2 3D Central Schemes with Cartesian or Diamond-Shaped Dual Cells

The 3D-numerical schemes considered here are based on our two-dimensional finite volume schemes [AV95][ASV95], inspired from the Nessyahu–Tadmor one-dimensional central scheme [NT90].

We consider the three-dimensional hyperbolic system:

$$\vec{U}_t + \nabla \cdot \mathbf{F} = \vec{U}_t + \vec{f}_x + \vec{g}_y + \vec{h}_z = 0, \quad t > 0, \tag{3}$$

with initial condition $\vec{U}(x, y, z, t = 0) = \vec{U}_0(x, y, z)$.

The computational domain is a uniform parallelepiped-shaped grid. We consider here the case of Cartesian dual cells. Starting from cell-average values $\vec{U}^n_{ijk}$ defined, at time t^n on the Cartesian cells $C_{i,j,k} \equiv [x_{i-1/2}, x_{i+1/2}] \times [y_{j-1/2}, y_{j+1/2}] \times [z_{k-1/2}, z_{k+1/2}]$ of the original grid, we compute new values $\vec{U}^{n+1}_{i+1/2,j+1/2,k+1/2}$ defined, at time t^{n+1}, on the staggered Cartesian dual cells $D_{i+1/2,j+1/2,k+1/2} = [x_i, x_{i+1}] \times [y_j, y_{j+1}] \times [z_k, z_{k+1}] \equiv \mathcal{D}_{i,j,k}$ for simplicity (Fig. 1, dashed line cube) by integrating (3) on $\mathcal{D}_{i,j,k} \times [t^n, t^{n+1}]$ and applying Green's theorem, obtaining

$$Vol(\mathcal{D}_{i,j,k})\vec{U}^{n+1}_{i+1/2,j+1/2,k+1/2}) \equiv \int_{\mathcal{D}_{i,j,k}} \vec{U}(x, y, z, t^{n+1})dV =$$
$$\int_{\mathcal{D}_{i,j,k}} \vec{U}(x, y, z, t^n)dV - \int_{t_n}^{t_n+1} \int_{\partial\mathcal{D}_{i,j,k}} (\vec{f} n_x + \vec{g} n_y + \vec{h} n_z)dAdt \tag{4}$$

where $\vec{n} = (n_x, n_y, n_z)$ = unit outer normal to $\partial\mathcal{D}_{i,j,k}$ (boundary of $\mathcal{D}_{i,j,k}$).

Applying piecewise linear interpolants in x, y, z with slopes controlled by van Leer's MC-θ limiter, and a midpoint quadrature for the time integration leads to second-order accuracy; the intermediate time value $\vec{U}^{n+1/2}$ is

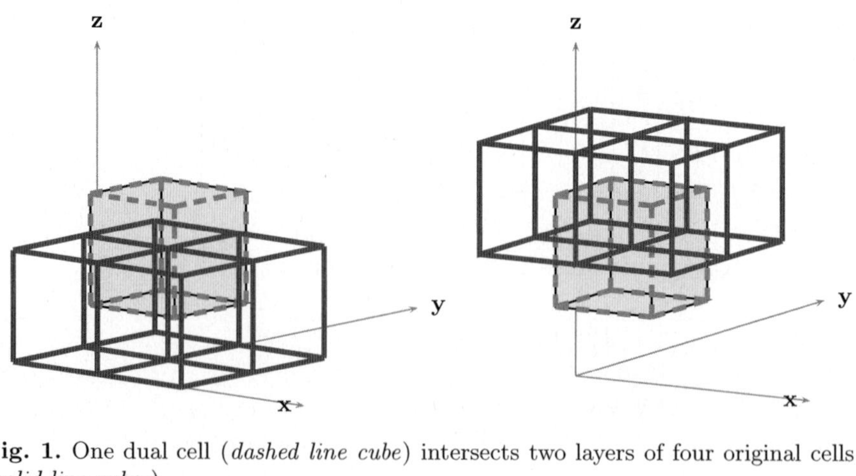

Fig. 1. One dual cell (*dashed line cube*) intersects two layers of four original cells (*solid line cubes*)

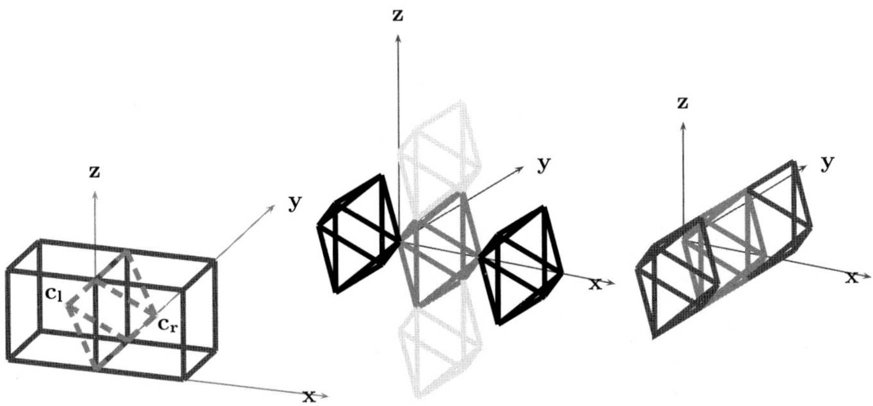

Fig. 2. (*Left*) Two Cartesian cells $C_{i,j,k}$, $C_{i+1,j,k}$ and dual diamond cell $D_{i+1/2,j,k}$. (*Middle and right*) Six diamond dual cells along the x and z directions (*middle*) and the y direction (*right*) are required to compute $\nabla \cdot B^{n+1}_{i+1/2,j,k}$

obtained using an explicit Euler time discretization of (3) ([T05]). A similar second time step then uses the values $\overrightarrow{U}^{n+1}_{i+1/2,j+1/2,k+1/2}$ to construct values U^{n+2}_{ijk} on the original cells C_{ijk}. The scheme using diamond-shaped dual cells $\mathcal{D}_{i+1/2,j,k}$ (or $\mathcal{D}_{i,j+1/2,k}$ and $\mathcal{D}_{i,j,k+1/2}$, respectively) (see Fig. 2, left) proceeds in a similar way, see [AT05, TA06].

3 A Constrained Transport Divergence Treatment for 3D Central Schemes (CTCS)

Based on experimental observations, the expression of the magnetic field $\mathbf{B}$ given by Biot and Savart's law [PP55] leads to the existence of a magnetic vector potential $\mathbf{A}$ such that curl$\mathbf{A} = \mathbf{B}$ and therefore to Maxwell's equation $\nabla \cdot \mathbf{B} = 0$, which must be satisfied. Faraday's law guarantees that if the initial magnetic field is solenoidal, it remains divergence-free at upcoming time. But the accumulation of numerical errors most often leads to a numerical solution that does not satisfy the $\nabla \cdot \mathbf{B} = 0$ constraint. Among many useful methods to restore a solenoidal $\mathbf{B}$, Evans and Hawley's Constrained Transport (CT) approach [EH88] has proven to be very efficient. Since our schemes use two staggered grids, none of the existing versions of the CT method can be applied directly. Our own approach to satisfy the divergence constraint, inspired from the CT method, is based on a specific discretization of the induction equation on both the original and staggered dual grids. In this paper we describe our Constrained Transport method for Central Schemes (CTCS) in the case of diamond-shaped dual cells $\mathcal{D}_{i+1/2,j,k}$.

Let $\vec{U}^n_{ijk}$ denote the solution at time t^n on the Cartesian cell C_{ijk}, and $\vec{U}^{n+1}_{i+1/2,j,k}$ the solution at time t^{n+1} on the staggered dual cell $\mathcal{D}_{i+1/2,j,k}$. We suppose that the constraint $\nabla \cdot \mathbf{B}^n_{ijk} = 0$ is satisfied, i.e., the central difference discretization of the divergence operator satisfies

$$\begin{aligned}\nabla \cdot \mathbf{B}^n_{i,j,k} \approx \frac{B^{n,x}_{i+1,j,k} - B^{n,x}_{i-1,j,k}}{2\Delta x} + \frac{B^{n,y}_{i,j+1,k} - B^{n,y}_{i,j-1,k}}{2\Delta y} \\ + \frac{B^{n,z}_{i,j,k+1} - B^{n,z}_{i,j,k-1}}{2\Delta z} = 0.\end{aligned} \tag{5}$$

Performing the step $t^n \to t^{n+1}$, we obtain a solution $\vec{U}^{n+1}_{i+1/2,j,k}$ for the dual cells $\mathcal{D}_{i+1/2,j,k}$ whose magnetic field part, denoted by $\mathbf{B}^*$, must be treated to obtain a divergence-free magnetic field $\mathbf{B}^{n+1}$. We first compute the electric field $\mathbf{E}^{n+1/2}_{i+1/2,j,k} = (\Omega^x, \Omega^y \Omega^z)^{n+1/2}_{i+1/2,j,k}$ at time $t^{n+1/2}$ using the data at time t^n and t^{n+1}, on the original and the dual staggered grids, respectively, as follows:

$$\begin{aligned}\mathbf{E}^{n+1/2}_{i+1/2,j,k} &= -(\mathbf{v} \times \mathbf{B})^{n+1/2}_{i+1/2,j,k} \\ &\cong -\frac{1}{2}\left[(\mathbf{v}^{n+1} \times \mathbf{B}^*)_{i+1/2,j,k} + \frac{1}{2}\{(\mathbf{v} \times \mathbf{B})^n_{i,j,k} + (\mathbf{v} \times \mathbf{B})^n_{i+1,j,k}\}\right].\end{aligned} \tag{6}$$

This discretization will ensure second-order accuracy with respect to time. We then discretize the induction equation $\partial_t \mathbf{B} + \nabla \times \mathbf{E} = 0$ on the $\mathcal{D}_{i+1/2,j,k}$-type dual cells using the following centered differences:

$$\begin{aligned}\mathbf{B}^{n+1,x}_{i+1/2,j,k} = \frac{1}{2}(B^{n,x}_{i,j,k} + B^{n,x}_{i+1,j,k}) - \Delta t \frac{\Omega^{n+1/2,z}_{i+1/2,j+1,k} - \Omega^{n+1/2,z}_{i+1/2,j-1,k}}{2\Delta y} \\ + \Delta t \frac{\Omega^{n+1/2,y}_{i+1/2,j,k+1} - \Omega^{n+1/2,y}_{i+1/2,j,k-1}}{2\Delta z},\end{aligned} \tag{7}$$

with similar expressions for the y and z components [TA06]. This special discretization of the induction equation and the above interpolation for the electric field at the intermediate time $t^{n+1/2}$ preserve the second-order accuracy of the base scheme.

It can be shown that these choices lead to

$$\nabla \cdot \mathbf{B}^{n+1}_{i+1/2,j,k} = \frac{1}{2}\{\nabla \cdot \mathbf{B}^n_{i,j,k} + \nabla \cdot \mathbf{B}^n_{i+1,j,k}\} = 0, \tag{8}$$

since the magnetic field $\mathbf{B}$ was assumed to be solenoidal at time t^n. A similar treatment of the magnetic field on the cells of type $\mathcal{D}_{i,j+1/2,k}$ and $\mathcal{D}_{i,j,k+1/2}$ then guarantees a divergence-free magnetic field $\mathbf{B}^{n+1}$ at time t^{n+1}.

The CTCS divergence treatment in the case of Cartesian dual cells proceeds in a similar way [TA06].

4 Numerical Experiments

Because of the use of two staggered dual grids, which results in a restriction on the time step, our numerical experiments have been performed with a CFL number of 0.475.

1. Our first test is an Orszag–Tang-type problem [DW98, T00, Z04], which is itself a modification of a problem proposed by Nodes et al. [NGL04]. The initial data are as follows: $\rho(x,y,z) = \rho_0$, $p(x,y,z) = p_0$, $\mathbf{u}(x,y,z) = -\sin y \sin z\ \mathbf{i} + \sin x \sin z\ \mathbf{j}$, $\mathbf{B}(x,y,z) = -\sin y \sin z\ \mathbf{i} + \sin(2x) \sin z\ \mathbf{j} + \sin(2x) \sin y\ \mathbf{k}$, with $0 \leq x, y, z \leq 2\pi$, $\rho_0 = 25/36$ and $p_0 = 5/3$. $\mathbf{i}$, $\mathbf{j}$, and $\mathbf{k}$ are the unit vectors in the x, y, and z directions, respectively. We have computed the solution at time $t = 0.5$ on 100^3 gridpoints using the Cartesian dual cell scheme; thanks to our CTCS divergence treatment, the maximum absolute value of the divergence observed for this Orszag–Tang vortex problem is of the order of 10^{-14}. Figure 3 (left) shows the contours of the mass density in the plane $x = \pi/2$ at time $t = 0.5$. If we do not apply the CTCS divergence treatment the base scheme can still reach the final time without showing instabilities: we have also solved this problem using the diamond dual cell scheme on 50^3 gridpoints, without CTCS divergence treatment. Figure 3 (right) shows the plots along the line $y = z = \pi/2$ of the energy obtained with the Cartesian dual cell with CTCS treatment and on both 100^3 grid (solid line) and 50^3 grid (dashed line), and with the diamond dual scheme without CTCS divergence treatment (dotted line). We find that the numerical results obtained without CTCS treatment are still reasonable, and the maximum of the absolute value of $\nabla \cdot \mathbf{B}$ observed in this case is 3.124×10^{-1}. Comparing the dotted line with the dashed line, the necessity of a divergence treatment is more apparent in the neighborhood of local extremas.

2. Our second test is a three-dimensional adaptation of a classical 2D MHD shock-cloud interaction problem [10,24,27,20]. The computational domain $(x,y,z) \in [0,1]^3$ is uniformly discretized using 100^3 gridpoints. Two constant states U_l = [3.86859, 11.2536, 0, 0, 167.345, 0, 2.1826182, −2.1826182] and U_r = [1, 0, 0, 0, 1, 0, 0.56418958, 0.56418958] are separated by the plane

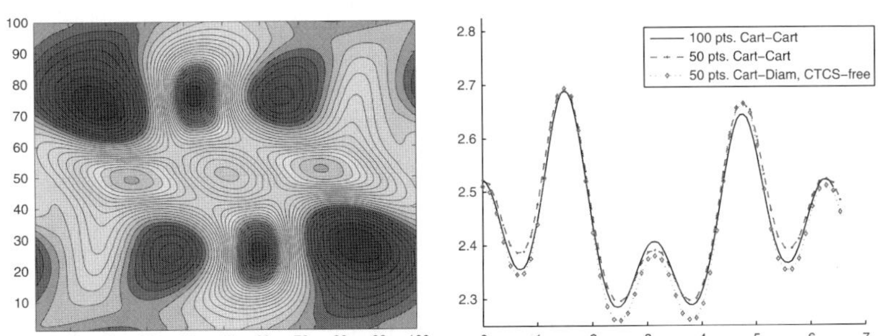

Fig. 3. (*Left*) Mass density contours in the plane $z = \pi/2$. (*Right*) Energy along the line $y = z = \pi/2$ obtained with (*solid line* and *dashed line*) or without (*dotted line*) the aid of the CTCS treatment

$x = 0.05$; here $U = (\rho, u_x, u_y, u_z, p, B_x, B_y, B_z)$. A 10 times denser spherical cloud centered at (0.25,0.5,0.5) with a radius $r = 0.15$ is in hydrostatic equilibrium with the surrounding state. The profile of the initial mass density is shown in Fig. 4 (left). The numerical solution is computed at time $t = 0.06$ using the Cartesian dual cell scheme along with its corresponding CTCS divergence treatment; the maximum absolute value of the divergence observed remains within the 10^{-12} values (Fig. 5d). An equivalent variant of

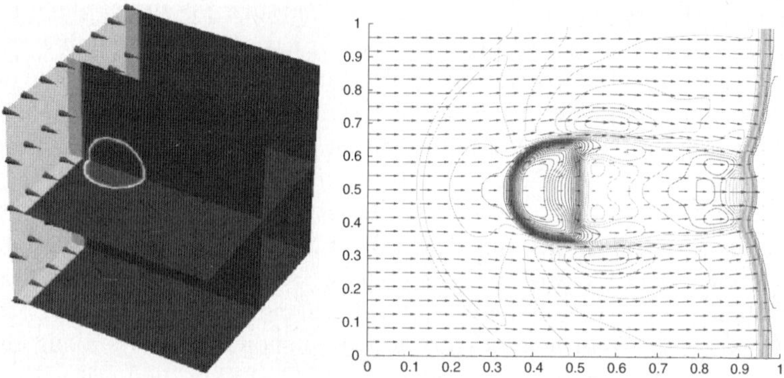

Fig. 4. (*Left*) Initial mass density profile for the 3D shock-cloud interaction problem; we also see the velocity field magnitude as a cone plot. (*Right*) Several contour lines of the mass density logarithm at time $t = 0.06$

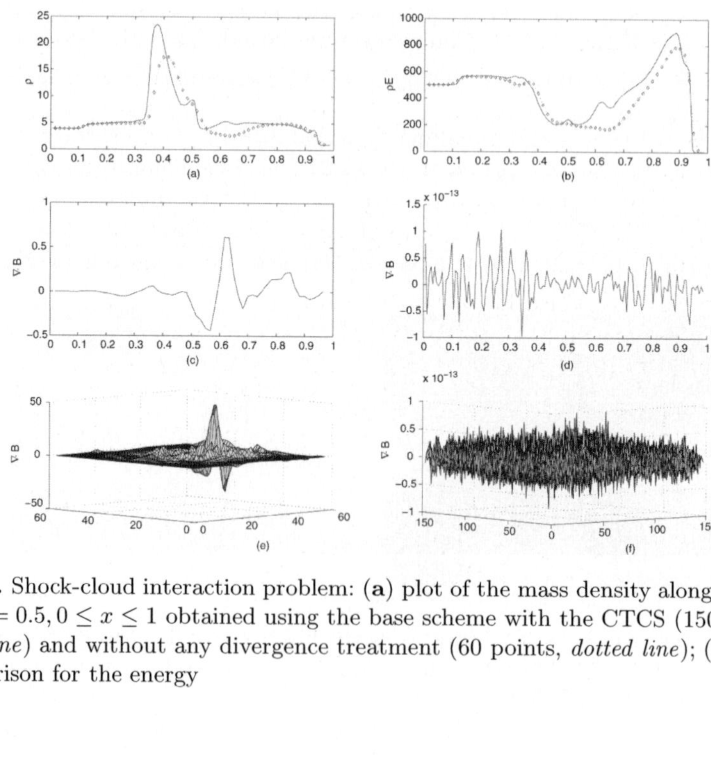

Fig. 5. Shock-cloud interaction problem: (**a**) plot of the mass density along the line $y = z = 0.5, 0 \leq x \leq 1$ obtained using the base scheme with the CTCS (150 points, *solid line*) and without any divergence treatment (60 points, *dotted line*); (**b**) same comparison for the energy

this three-dimensional problem is considered in [Z04]. Figure 4 (right) shows several contour lines of the mass density. Figure 5a shows a plot of the mass density using the Cartesian dual cell scheme with the CTCS divergence treatment and 150^3 gridpoints; the results are very similar to the reference solution presented in [27]. We have also included in this figure our results using the diamond dual cell scheme with 60^3 gridpoints and no CTCS divergence treatment. Figure 5b shows the same comparison for the energy. We observe that the base scheme does not become unstable; although $\nabla \cdot \mathbf{B}$ is not negligible (Fig. 5c, showing $\nabla \cdot \mathbf{B}$ along the line $y = z = 0.5$), the numerical solution still gives a somewhat reasonable overall approximation, with more significant deviations at gridpoints where the divergence is non-negligible (Fig. 5a,b).

5 Conclusion

We have presented three-dimensional, second-order accurate finite volume central schemes for solving systems of hyperbolic equations in the context of MHD problems. The resolution of the Riemann problems at the cell interfaces is avoided, thanks to the use of two staggered dual grids at alternate time steps. The cells of the original grid are Cartesian while those of the dual grid are either Cartesian or diamond-shaped.

To maintain a divergence-free magnetic field, we have constructed a method inspired from Evans and Hawley's Constrained Transport approach, which treats, at the end of each time step, the magnetic field obtained using our numerical base scheme. This Constrained Transport method for Central Schemes ("CTCS") applies to both Cartesian and diamond-shaped dual cell schemes and preserves the second-order accuracy of the base scheme. The divergence of the magnetic field is thus maintained under 10^{-11}.

For the ideal MHD problems we have considered here, both numerical schemes can still reach the final time and even yield reasonable results without generating instabilities, when we do not apply the CTCS procedure; but in this case, the magnetic field does not remain solenoidal and the CTCS method should in fact be applied for optimal results.

Our numerical results obtained with both base schemes are very similar, and in very good agreement with other results in the literature.

References

[AV95] Arminjon, P., Viallon, M.C.: Généralisation du schéma de Nessayahu-Tadmor pour une équation hyperbolique à deux dimensions d'espace. In: C.R. Acad. Sci. Paris, **320** (I), 85–88 (1995)

[ASV95] Arminjon, P., Stanescu, D., Viallon, M.C.: A two-dimensional finite volume extension of the Lax-Friedrichs and Nessyahu-Tadmor schemes for compressible flows. In: Hafez, M., Oshima, K. (Eds.) Proc. 6th. Int. Symp. on Comp. Fluid Dynamics, Vol. 4, 7–14 (1995)

[AT05] Arminjon, P., Touma, R.: Central Finite volume methods with constrained transport divergence treatment for ideal MHD. J. Comp. Phys., **204**, 737–759 (2005)

[DW98] Dai, W., Woodward, P.R.: A simple finite difference scheme for multidimensional magnetohydro-dynamical equations. J. Comp. Phys., **142**, 331–369 (1998)

[EH88] Evans, C.R., Hawley, J.F.: Simulation of magnetohydrodynamic flows: A constrained transport method. Astrophys. J., **332**, 659–677 (1988)

[JT64] Jeffrey, A., Taniuti, T.: Non-Linear Wave Propagation. Academic Press, New York (1964)

[NT90] Nessyahu, H., Tadmor, E.: Non-oscillatory central differencing for hyperbolic conservation laws. J. Comp. Phys., **87**, 408–463 (1990)

[NGL04] Nodes, C., Gritschneder, G.T., Lesch, H.: Radio emission and particle acceleration in plerionic supernova remnants. Astronomy and Astrophysics **423**, 13–19 (2004).

[PP55] Panofsky, W.K.H, Phillips, M.: Classical Electricity and Magnetism. Addisson-Wesley, Reading (Mass.), (1955)

[T05] Touma, R.: Méthodes de volumes finis pour les systèmes d'équations hyperboliques: applications en aérodynamique et en magnétohydrodynamique. PhD Thesis, Université de Montréal, Montréal (2005)

[TA06] Touma, R., Arminjon, P.: Central finite volume schemes with constrained transport divergence treatment for three-dimensional ideal MHD. J. Comp. Phys., **212**, 617–636 (2006)

[T00] Tóth, G.: The $\nabla \cdot \mathbf{B} = 0$ Constraint in Shock-Capturing Magnetohydrodynamics Codes. J. Comp. Phys., **161**, 605–652 (2000)

[Z04] Ziegler, U.: A central-constrained transport scheme for ideal magnetohydrodynamics. J. Comp. Phys., **192**, 393–416 (2004)

Finite Volume Methods for Low Mach Number Flows under buoyancy

P. Birken

1 Introduction

The efficient simulation of low Mach number flows still poses difficulties. While the flow is often incompressible, in a lot of applications, the Mach number or the compressibility properties vary strongly in time or space, for example in nozzle flow or laminar combustion. Here, we will focus on fires in road tunnels, where the strong heat sources make parts of the flow compressible. Standard compressible flow solvers have accuracy and convergence problems at low Mach numbers. On the other hand, standard incompressible flow solvers cannot deal with strong temperature or strong density gradients.

In recent years there were a number of deadly incidents in car tunnels, for example, in the Mont-Blanc tunnel 1999 or the Kaprun tunnel 2000. This demonstrates the need for more efficient safety measures. There exists a variety of codes dealing with these problems, see for example the survey [11]. The models applied in praxis are often quite primitive, but lead to fast methods. The increase in computer power makes the use of the much more accurate tools provided by modern CFD feasible.

The first characterizing property here is the low Mach number. One important idea in this context was the preconditioner of Turkel [12] for compressible equations. This scheme incorporates a preconditioning of the time derivative, thus allowing faster convergence to steady state. Along these lines, other preconditioners also were proposed [13, 3]. In this article, we will concentrate on the flux preconditioning approach of Guillard and Viozat [6], where only the dissipation within the numerical flux is changed. This has the advantage that the implementation is very simple. However, it turns out that the stability region of an explicit preconditioned method deteriorates as the Mach number tends to zero [2]. Nevertheless, this can be overcome by implicit methods.

The other defining property for the tunnel fire problem is buoyancy. A correct computation of this, in particular of near hydrostatic flows, is mandatory for a successful simulation. The addition of the relevant source terms for gravitation and heat is possible using operator splittings. It turns out that the interplay between the numerical schemes for low Mach number and gravitation is nontrivial and some care has to be taken. This will be discussed, together with numerical results, toward the end of the article.

A more comprehensive description of the techniques used and the results described in this article can be found in [1].

2 The Governing Equations

We consider the two-dimensional Euler equations on an open domain $D \subset \mathbb{R}^2$:

$$\partial_t \mathbf{u} + \sum_{j=1}^{2} \partial_{x_j} \mathbf{f}_j(\mathbf{u}) = \mathbf{g} \quad \text{in } D \times \mathbb{R}^+, \tag{1}$$

where $\mathbf{u} = (\rho, m_1, m_2, \rho E)^T$ represents the vector of conserved variables. The flux functions $\mathbf{f}_j$ and the source term $\mathbf{g}$ are given by

$$\mathbf{f}_j(\mathbf{u}) = \begin{pmatrix} m_j \\ m_j v_1 + \delta_{1j} p \\ m_j v_2 + \delta_{dj} p \\ H m_j \end{pmatrix}, \; j = 1, 2, \; \mathbf{g}(\mathbf{u}) = \begin{pmatrix} 0 \\ \rho g_1 / Fr^2 \\ \rho g_2 / Fr^2 \\ Qq/M^2 \end{pmatrix}.$$

The quantities ρ, $\mathbf{v} = (v_1, v_2)^T$, $\mathbf{m} = (m_1, m_2)^T$, E and $H = E + \frac{p}{\rho}$ describe the density, velocity, momentum per unit volume, total energy per unit mass, and total enthalpy per unit mass, respectively. The equations are closed by the equation of state for a perfect gas $p = (\gamma - 1)\rho(E - \frac{1}{2}|\mathbf{v}|^2)$, where γ denotes the ratio of specific heats, taken as 1.4 for air.

This form of the Euler equations derives from the usual reference values $\hat{x}_{ref}$, $\hat{\rho}_{ref}$, and $\hat{v}_{ref}$. Thus, the nondimensional characteristic number M appearing in the energy equation is defined as $M = \frac{\hat{v}_{ref}}{\hat{c}_{ref}}$ with $\hat{c}_{ref} = \sqrt{\hat{p}_{ref}/\hat{\rho}_{ref}}$. The other nondimensional characteristic numbers are the Froude number $Fr = \frac{\hat{v}_{ref}}{\sqrt{\hat{x}_{ref}\hat{g}_{ref}}}$ and the strength of the heatsource $Q = \frac{\hat{q}_{ref}\hat{x}_{ref}}{\hat{p}_{ref}\hat{v}_{ref}}$.

3 The Numerical Method for Low Mach Numbers

Introducing the cell average $\mathbf{u}_i(t) = \frac{1}{|\Omega_i|}\int_{\Omega_i} \mathbf{u}(\mathbf{x}, t)\, d\mathbf{x}$, a finite volume method for the Euler equations with respect to a control volume Ω_i can be written as

$$\frac{d}{dt}\mathbf{u}_i(t) = -\frac{1}{|\Omega_i|} \sum_{e_{ij} \subset \partial\Omega_i} \int_{e_{ij}} \sum_{\ell=1}^{2} \mathbf{f}_\ell(\mathbf{u}(\mathbf{x}, t)) n_\ell \, ds + \frac{1}{|\Omega_i|} \int_{\Omega_i} \mathbf{g}(\mathbf{u}) d\mathbf{x}, \tag{2}$$

where e_{ij} denotes the edge between the adjacent control volumes Ω_i and Ω_j and $\mathbf{n} = (n_1, n_2)^T$ represents the outer unit normal vector on the cell.

The convective parts $\mathbf{f}$ and the source terms $\mathbf{g}$, like gravity and heat, will be treated separately via *fractional step* or *operator splitting* methods.

To approximate the boundary integral in (2), we employ a numerical flux function of Lax–Friedrichs type

$$H(\mathbf{u}_i, \mathbf{u}_j, \mathbf{n}) = \frac{1}{2}\left(\sum_{\ell=1}^{2} (\mathbf{f}_\ell(\mathbf{u}_i) + \mathbf{f}_\ell(\mathbf{u}_j))\, n_\ell - \mathbf{D}(\mathbf{u}_i, \mathbf{u}_j, \mathbf{n})(\mathbf{u}_j - \mathbf{u}_i)\right). \quad (3)$$

Numerical methods of this type differ only in the dissipation term $\mathbf{D}$. We prefer a matrix-valued term $\mathbf{D}$, which was proposed by Friedrich [4]. It was shown in [8] that the use of this Lax–Friedrichs-type flux function can lead to an unphysical pressure distribution. In particular, variations of the first order pressure field are generated on the space scale $\mathbf{x}$, which is in violation of the results of an asymptotic analysis of the Euler equations by Klainermann and Majda [7]. To extend the validity of the numerical method we utilize a preconditioning technique originally proposed by Guillard and Viozat [6] and later on derived in the context of the Lax–Friedrichs method in [9]. Therefore, the dissipation matrix is defined in the sense of $\mathbf{D} = \mathbf{P}^{-1}\,|\mathbf{PF}|$:

$$\mathbf{D}(\mathbf{u}_i, \mathbf{u}_j, \mathbf{n}) = \mathbf{P}^{-1}\left(\frac{\mathbf{u}_j + \mathbf{u}_i}{2}\right)\mathbf{R}\left(\frac{\mathbf{u}_j + \mathbf{u}_i}{2}, \mathbf{n}\right)|\mathbf{\Lambda}|\mathbf{R}^{-1}\left(\frac{\mathbf{u}_j + \mathbf{u}_i}{2}, \mathbf{n}\right).$$

Herein, $\mathbf{R}(\mathbf{u}, \mathbf{n})$ represents the matrix of the right eigenvectors of the corresponding preconditioned Jacobian

$$\mathbf{F}^P(\mathbf{u}, \mathbf{n}) = \mathbf{P}(\mathbf{u})\mathbf{F}(\mathbf{u}, \mathbf{n})$$

and $|\mathbf{\Lambda}|(\mathbf{u}_i, \mathbf{u}_j, \mathbf{n})$ denotes the diagonal matrix defined by

$$|\mathbf{\Lambda}| = \operatorname{diag}\left\{\max_{\mathbf{u}\in\left\{\mathbf{u}_i, \mathbf{u}_j, \frac{\mathbf{u}_i+\mathbf{u}_j}{2}\right\}} |\lambda_1(\mathbf{u}, \mathbf{n})|, \ldots, \max_{\mathbf{u}\in\left\{\mathbf{u}_i, \mathbf{u}_j, \frac{\mathbf{u}_i+\mathbf{u}_j}{2}\right\}} |\lambda_4(\mathbf{u}, \mathbf{n})|\right\},$$

where $\lambda_i(\mathbf{u}, \mathbf{n})$, $i = 1, \ldots, 4$ are chosen to be the eigenvalues of the matrix $\mathbf{F}^P(\mathbf{u}, \mathbf{n})$. The properties of the derived method strongly depend on the preconditioning matrix used. Arranging $\mathbf{P}(\mathbf{u}, \mathbf{n})$ to be the identity yields the scheme proposed by Friedrich [4]. Using the so-called entropy variables $\mathbf{w} = (p, v_1, v_2, s)^T$, whereby s denotes the entropy determined as $s = \ln \frac{p}{\rho^\gamma}$, and following Turkel [12] we introduce

$$\mathbf{P}(\mathbf{u}) = (\mathbf{UQW})(\mathbf{u}), \quad (4)$$

where $\mathbf{U} = \frac{\partial \mathbf{u}}{\partial \mathbf{w}}$, $\mathbf{W} = \frac{\partial \mathbf{w}}{\partial \mathbf{u}}$, and

$$\mathbf{Q} = \operatorname{diag}(\beta^2, 1, 1, 1) \text{ with } \beta \neq 0,\ \beta = \mathcal{O}_S(M),\ M \to 0. \quad (5)$$

It was recently proven in [10] that utilizing the preconditioned Lax–Friedrichs flux (3) for the choice (5), within the finite volume method associated with

(2) without source terms, yields a pressure distribution satisfying the asymptotic properties of the continuous equations in a discrete sense. Furthermore, a discrete divergence constraint corresponding to the results known for the continuous equations is shown for this scheme in the absence of compression and expansion over the boundary of the computational domain.

4 Stability of the Preconditioned Method

As can be seen in the following theorem, the preconditioned method combined with an explicit time integration has unfavorable stability properties due to the modified dissipation term.

Theorem 1. *A necessary condition to ensure stability of the linearized preconditioned Lax–Friedrichs scheme with* $\beta = \mathcal{O}_S(M)$, $M \to 0$, $\beta \neq 0$, *and arbitrary* α *is*

$$\Delta t = \mathcal{O}(M^2), \quad M \to 0.$$

Proof. The core method of the proof is a von Neumann stability analysis, where the difficult part is the computation of the eigenvalues of the dissipation matrix. This is made possible by choosing an appropriate similarity transform. We refer to [2] for the details. □

We illustrate the theorem on a test case, namely a NACA0012 profile at zero angle of attack with varying inflow Mach numbers. A C-Type grid with 7,487 cells was employed. We used a first order spatial discretization and the explicit Euler scheme in time. The results can be seen in the first tabular. They show exactly the behavior predicted by the theorem.

It should be pointed out that, for the implicit Euler scheme, no bound on the CFL number is observed and none is obtained by an L_2-stability analysis.

5 A First Test Case

To analyze the behavior of our methods for flows subject to buoyancy, we consider a two-dimensional section of a tunnel of five meter height. For the initial velocity we use $v_1 \equiv 1$ and $v_2 \equiv 0$. We choose an initial pressure and

Table 1. Preconditioned and unpreconditioned timestep at different Mach numbers

M	CFL prec.	Δt prec.	CFL unprec.	Δt unprec.	CFL implicit
0.1	0.1	$7 \times$ 10E-7	0.9	$6 \times$ 10E-5	5,000
0.01	0.01	10E-8	0.9	$7 \times$ 10E-6	5,000
0.001	0.001	10E-10	0.9	$7 \times$ 10E-7	5,000
0.0001	0.0001	10E-12	0.9	$7 \times$ 10E-8	5,000

density distribution that is constant in vertical direction, and in horizontal direction we choose $p(x_2)$ and $\rho(x_2)$ according to the hydrostatic pressure:

$$p(x_2) = p_r \left(1 - \frac{\Gamma x_2}{T_r}\right)^{c_p/R}. \tag{6}$$

For a reference pressure of 101,325 Pa, the exact pressure difference from top to bottom is 63.4 Pa and the Mach number $Ma = 0.0036$.

5.1 Numerical Experiments

To compute the steady state, we use a first order discretization in space and the explicit Euler method in time, to eliminate possible influences from inner iterations. At the tunnel ends, Neumann boundary conditions are used. We employ 20 cells in x_2-direction.

At first, we use the unpreconditioned Lax–Friedrichs flux and a CFL number of 0.9. Already after 5,000 time steps, a correct nearly linear distribution of the density is obtained. The pressure distribution is slightly changed, while preserving the pressure difference of 63.4 Pa. That change is probably due to the coarse grid and the fact that the discretization is only of first order.

Then we use the low Mach preconditioned Lax–Friedrichs flux. Here, also a nearly linear distribution of the density is obtained, but the pressure difference reduces to 3 Pa after 5,000 time steps with a CFL number of 0.0036. Then it increases slowly up to 54 Pa after 100.000 time steps.

The unpreconditioned method thus produces a physically reasonable result, while the preconditioned method does not. To illustrate this, we perform only one time step without source terms for both cases. For a cell in the middle of the tunnel the physical flux update in the first step is $\mathbf{f} = (0, 0, \Delta p, 0)^T$. The numerical mass and momentum flux approximate this vector very well. However, the preconditioned energy flux is -1.83711 and the unpreconditioned is -0.0034101. The energy flux in both cases is thus too big, but much worse so in the preconditioned case.

5.2 Analysis of the Energy Flux

To understand this phenomenon, we look at the flux updates in an interior cell in the first step. The fluxes in x_1-direction cancel out, but along horizontal borders they differ. We have for the preconditioned flux function with $\mathbf{n} = (0, 1)^T$

$$\mathbf{f}^{LFP}(\mathbf{u}_L, \mathbf{u}_R; \mathbf{n}) = \frac{1}{2}(\mathbf{f}_L + \mathbf{f}_R - \mathbf{D}\Delta\mathbf{u}) = \frac{1}{2}\left(\begin{pmatrix} 0 \\ 0 \\ 2p \\ 0 \end{pmatrix} - \mathbf{D}\begin{pmatrix} \Delta\rho \\ \Delta\rho \cdot v_1 \\ 0 \\ \Delta E \end{pmatrix}\right).$$

The unpreconditioned flux is the same though with a different dissipation matrix. The energy component we are interested in depends only on the dissipation term. In this testcase, we have $v_n = v_2 = 0$ and thus, lengthy but straightforward computations lead to

$$f_4^{LFP} = \frac{H(\gamma - 1)}{\beta c} \left(\frac{|\mathbf{v}|^2}{2} \Delta\rho + v_1 \Delta(\rho v_1) + \Delta(\rho E) \right). \tag{7}$$

The formula in the unpreconditioned case is obtained by setting $\beta = 1$. Thus, the vertical fluxes differ by a factor of $1/\beta$. The update of a nonboundary cell is formed by the difference between the flux from the top of the cell and the flux from the bottom of the cell, as the numerical solution does not vary in x_1-direction. Although this is even in the simple testcase a complicated nonlinear function of the initial data, we can deduct an asymptotic behavior. The energy E and the enthalpy H are $\mathcal{O}(M^{-2})$ as the Mach number tends to zero. Thus, the update in the preconditioned case increases by fourth order with the Mach number going to zero and in the unpreconditioned by third order. This can indeed be confirmed by numerical results.

This effect can lead to steady states where the balance between the preconditioned flux and the gravitational source term is achieved at a very small and unphysical pressure difference. There are two ways to remedy this, as can be seen from (7). We can decrease the last factor and thereby the flux by decreasing the differences in the data either by refining the mesh or by using a method of higher order in space. In the same test case as above but using MUSCLE interpolation, the energy flux is decreased by a factor of 10^6, which allows to compute the correct pressure distribution up to an error of a few Pascal for moderately small Mach numbers. For $M = 10^{-5}$ a refinement of the grid is necessary, despite the higher order. The dependence on the Mach number is otherwise again as predicted.

5.3 Gravitation and Heat

In a more complicated testcase, we place a circular package of heat near the entrance of a short tunnel with two sharp bends. For the initial data, we use hydrostatic pressure and a Mach number of 0.01. The initial temperature in the hot zone is up to 450 K. After 3.7 s, the package has reached the second bend, which can be seen in Fig. 1. It is now more boomerang shaped, because the hottest air with a peak temperature of 285 K is in the center and flows toward the ceiling faster than the cooler parts. This is what was expected to happen and thus we move on to fire events.

6 Simulation of a Fire Event

The final test case is similar to a fire event with a burning vehicle in a tunnel. A rectangular heat source of the size $3 \times 5\,\text{m}^2$ with a total power of 1 MW is placed in the middle of a tunnel, which is one kilometer long, 5 meters high,

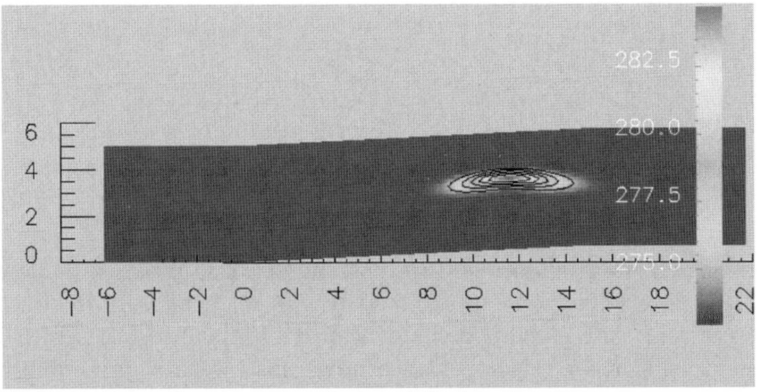

Fig. 1. Temperature distribution in bended tunnel after 3.7 s

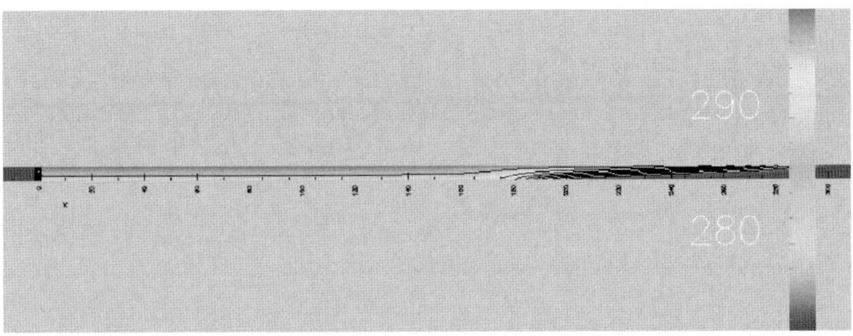

Fig. 2. Temperature for 1 MW after 60 s

and has no slope. At the boundary, we prescribe hydrostatic pressure (6). The initial conditions are again obtained by computing the steady state if the heat source is not active.

If we assume that the tunnel is 10 m wide, the heat source is distributed over a volume of 150 m^3. With an inflow Mach number of $Ma = 0.01$ this leads to a nondimensional parameter Q of 32,350 and the Froud number for a reference length of $\hat{x}_{ref} = 5$ m is $Fr = 0.488$. This setting is similar to that described in [5].

For the Euler equations, we use the implicit midpoint rule. We start with a CFL number of 0.01, which is increased up to 1.5. Figure 2 shows the middle part of the tunnel after 1 min. It can be clearly seen that the heat concentrates on the ceiling, due to the buoyancy. Furthermore, it slowly drifts downstream, at about the rate of 3.6 m s^{-1}. The air circulates between bottom and ceiling, but in a steady way, which was expected.

7 Conclusions

A class of preconditioning techniques was used to extend the validity of a density based flow solver into the low Mach number regime. A von Neumann stability analysis showed that the stability region of the corresponding explicit scheme gets smaller with M^2, as M tends to zero. Thus, implicit time integration methods have to be used in this context.

The method was then applied to problems with gravitation. There, it was shown that a first order discretization needs an unacceptably fine grid in this context and that higher order discretizations have to be used. Finally, in addition to gravitation, a heat source term was included to simulate tunnel fires. Several test cases demonstrate the feasibility of the method.

References

1. P. Birken. *Numerical Simulation of Flows at Low Mach Numbers with Heat Sources.* Dissertation, Fachbereich Mathematik/Informatik, Universität Kassel, 2005.
2. P. Birken and A. Meister. Stability of Preconditioned Finite Volume Schemes at Low Mach Numbers. *BIT*, 45(3), 2005.
3. D. Choi and C.L. Merkle. The Application of Preconditioning in Viscous Flows. *J. Comput. Phys.*, 105:207–223, 1993.
4. O. Friedrich. Weighted Essential Non-Oscillatory Schemes for the Interpolation of Mean Values on Unstructured Grids. *J. Comput. Phys.*, 144:194–212, 1998.
5. I. Gasser, J. Struckmeier, and I. Teleaga. Modelling and Simulation of Fires in Vehicle Tunnels. *Int. J. for Numerical Methods in Fluids*, 44(3):277–296, 2004.
6. H. Guillard and C. Viozat. On the Behaviour of Upwind Schemes in the Low Mach Number Limit. *Computers and Fluids*, Vol. 28:63–86, 1999.
7. S. Klainerman and A. Majda. Singular limits of quasilinear hyperbolic system with large parameters and the incompressible limit of compressible fluids. *Comm. Pure Appl. Math.*, 34:481–524, 1981.
8. A. Meister. *Analyse und Anwendung Asymptotik-basierter numerischer Verfahren zur Simulation reibungsbehafteter Strömungen in allen Mach-Zahlbereichen.* Habilitation, Universität Hamburg, 2001.
9. A. Meister. Asymptotic based preconditioning technique for low Mach number flows. *Z. Angew. Math. Mech.*, 83(1):3–25, 2003.
10. A. Meister. Viscous Flow Fields at all Speeds: Analysis and Numerical Simulation. *Journal of Applied Mathematics and Physics (ZAMP)*, 54:1010–1049, 2003.
11. S.M. Olenick and D.J. Carpenter. An Updated International Survey of Computer Models for Fire and Smoke. *J. of Fire Protection Engineering*, 13:87–110, 2003.
12. E. Turkel. Preconditioned Methods for Solving the Incompressible and Low Speed Compressible Equations. *J. Comput. Phys.*, 72:277–298, 1987.
13. B. van Leer, W.-T. Lee, and P. Roe. Characteristic Time-Stepping or Local Preconditioning of the Euler Equations. *AIAA Paper*, AIAA-91-1552, 1991.

Time Splitting with Improved Accuracy for the Shallow Water Equations

A. Bourchtein and L. Bourchtein

1 Introduction

The shallow water equations (SWE) are frequently used for simulation of different specific phenomena (dynamics of coastal oceans, lakes, rivers, etc.) as well for study of some regimes of atmosphere and ocean. Besides, the SWE keep essential properties of more complex 3D systems, which makes these equations an important testing system for computational fluid dynamics, especially for atmosphere and ocean modeling.

In the last two decades the semi-Lagrangian semi-implicit method became one of the most efficient and popular technique for numerical solution of the differential equations of atmosphere dynamics [Dur99, LLS00, SC91]. It allows to overcome the Courant–Friedrichs–Lewy (CFL) condition with respect to both gravity and inertial waves at relatively low cost of solving the trajectory equations and linear elliptic problems at each time step.

To circumvent the remaining computationally expensive problem of solution of elliptic equations, different time splitting techniques have been applied in the context of the atmospheric models [Bat84, CDI85, KTM94, YR86]. However, additional splitting error decreases the accuracy of such models when the time step exceeds the CFL advection criterion [Bat84, CDI85, KTM94, YR86]. To reduce the splitting errors, Douglas et al. have recently proposed small modifications to the splitting method in the case of parabolic equations [DK01, DKL03]. In this study we apply the last technique in the context of semi-Lagrangian semi-implicit approach for the SWE. Performed linear analysis of numerical stability shows that the constructed scheme is stable for the time steps up to 2 h and its accuracy is very close to that of semi-Lagrangian Crank–Nicholson scheme. The numerical experiments with actual atmospheric fields of pressure and wind showed that forecast skill of the scheme is comparative to performance of unsplit methods, while computational cost is reduced due to substitution of solution of 2D elliptic problems by a set of 1D ones.

2 Time Discretization of the Linearized SWE

Linearized SWE in rotating reference system can be written as follows:

$$d_t u = fv - \partial_x \Phi, \quad d_t v = -fu - \partial_y \Phi, \quad d_t \Phi = -\bar{\Phi}(\partial_x u + \partial_y v), \tag{1}$$

where

$$d_t \phi = \partial_t \phi + a\partial_x \phi + b\partial_y \phi, \quad \phi = u, v, \Phi.$$

The equations are written using Cartesian spatial coordinates x, y, time coordinate t, and common notations for unknown functions and parameters: u and v are the velocity components, $\Phi = gz$ is the geopotential, g is the gravitational acceleration, z is the height of pressure surface, $\bar{\Phi}$ is the mean value of the geopotential, $f = constant$ is the Coriolis parameter, a and b are the constant advection velocities.

Standard two-time-level semi-Lagrangian semi-implicit time discretization to problem (1) can be written in the form [GSG01, McD86, TS87]

$$\frac{u^{n+1}-u^n}{\tau} = f\frac{v^{n+1}+v^n}{2} - \partial_x \frac{\Phi^{n+1}+\Phi^n}{2}, \quad \frac{v^{n+1}-v^n}{\tau} = -f\frac{u^{n+1}+u^n}{2} - \partial_y \frac{\Phi^{n+1}+\Phi^n}{2},$$

$$\frac{\Phi^{n+1}-\Phi^n}{\tau} = -\bar{\Phi}\left(\partial_x \frac{u^{n+1}+u^n}{2} + \partial_y \frac{v^{n+1}+v^n}{2}\right). \tag{2}$$

Here

$$\phi^n = \phi(t_n, x, y), \quad \phi^{n+1} = \phi(t_{n+1}, x + a\tau, y + b\tau),$$

τ is the time step, n is the time step number, $t_n = n\tau$, $t_{n+1} = (n+1)\tau$. Evidently, it is Crank–Nicholson second-order discretization performed on the points of the rectilinear trajectories $\partial_t x = a, \partial_t y = b$. This method is sufficiently accurate and absolutely stable, allowing the use of time step chosen on a base of accuracy consideration rather then stability CFL restriction as in explicit and semi-implicit Eulerian schemes.

At each time step the system (2) is usually reduced to solution of a 2D elliptic problem for geopotential function, which turns to be computationally expensive task for the fine spatial grids (generating the sparse but not compactly sparse matrices of the order 10^4–10^8). To circumvent this problem different splitting methods have been employed, but, applied to atmospheric dynamics, such numerical schemes lose accuracy for large time steps due to additional splitting error [Bat84, CDI85, YR86]. Recently, Douglas with coauthors have proposed an approach to decrease the splitting error, which has shown promising results for parabolic problems [DK01, DKL03]. Applying their approach to the system (1), we obtain the following time splitting scheme

$$\frac{u^{n+1/2}-u^n}{\tau/2} = fv^{n+1/2} - \partial_x \Phi^{n+1/2} + A, \quad \frac{v^{n+1/2}-v^n}{\tau/2} = -fu^n - \partial_y \Phi^n,$$

$$\frac{\Phi^{n+1/2}-\Phi^{n}}{\tau/2}=-\bar{\Phi}\left(\partial_x u^{n+1/2}+\partial_y v^{n}\right); \quad (3)$$

$$\frac{u^{n+1}-u^{n+1/2}}{\tau/2}=fv^{n+1/2}-\partial_x\Phi^{n+1/2}+A,\ \frac{v^{n+1}-v^{n+1/2}}{\tau/2}=-fu^{n+1}-\partial_y\Phi^{n+1},$$

$$\frac{\Phi^{n+1}-\Phi^{n+1/2}}{\tau/2}=-\bar{\Phi}\left(\partial_x u^{n+1/2}+\partial_y v^{n+1}\right). \quad (4)$$

Here

$$A=-\frac{\tau f^2}{4}\left(u^n-u^{n-1}\right)-\frac{\tau f}{4}\partial_y\left(\Phi^n-\Phi^{n-1}\right)-\frac{\tau\bar{\Phi}}{4}\partial_{xy}\left(v^n-v^{n-1}\right)$$

is an additional modification term as compared with classic ADI (alternating direction implicit) scheme, ϕ^{n-1} denotes the values at the past time level $t_{n-1}=(n-1)\tau$, and $\phi^{n+1/2}=\phi(t_{n+1/2},x+a\tau/2,y+b\tau/2)$ are intermediate values, assigned to the halfway time points $t_{n+1/2}=(n+1/2)\tau$. Equations (3) and (4) can be reduced to the form without intermediate values

$$\frac{u^{n+1}-u^{n}}{\tau}=f\frac{v^{n+1}+v^{n}}{2}-\partial_x\frac{\Phi^{n+1}+\Phi^{n}}{2}$$

$$+\left[\frac{\tau f^2}{4}\left(u^{n+1}-2u^{n}+u^{n-1}\right)+\frac{\tau f}{4}\partial_y\left(\Phi^{n+1}-2\Phi^{n}+\Phi^{n-1}\right)+\frac{\tau\bar{\Phi}}{4}\partial_{xy}\left(v^{n+1}-2v^{n}+v^{n-1}\right)\right],$$

$$\frac{v^{n+1}-v^{n}}{\tau}=-f\frac{u^{n+1}+u^{n}}{2}-\partial_y\frac{\Phi^{n+1}+\Phi^{n}}{2},$$

$$\frac{\Phi^{n+1}-\Phi^{n}}{\tau}=-\bar{\Phi}\left(\partial_x\frac{u^{n+1}+u^{n}}{2}+\partial_y\frac{v^{n+1}+v^{n}}{2}\right). \quad (5)$$

Evidently, the scheme (5) is of the second order of accuracy, but the additional splitting errors in the brackets are of the third order in time, while the classic ADI approach results in the additional terms of the second order. The last two equations in (5) have no splitting terms neither in the modified nor in classic ADI splitting. Each of splitting systems (3) and (4) can be reduced to a set of 1D elliptic problems solved efficiently by Thomas algorithm.

Let us analyze accuracy and stability properties of the scheme (3) and (4). First, recall that general analytical solution to primitive equations (1) with periodic boundary conditions can be found in the form of Fourier series consisting of the individual waves

$$(u,v,\Phi)=(U,V,H)e^{i\sigma t}e^{i(mx+ly)},$$

where σ and U,V,H are the frequency and amplitudes of separate wave with wave numbers m,l. For each wave vector (m,l) there exist three different frequencies corresponding to slow advection and two fast gravity modes:

$$\sigma_1=-am-bl,\ \sigma_{2,3}=-am-bl\pm K,\ K=\sqrt{\bar{\Phi}(m^2+l^2)+f^2}.$$

Since all frequencies are real numbers, the amplitudes of the individual waves are constant in time.

The wave solutions to semi-Lagrangian Crank–Nicholson scheme (2) can be found in similar form:

$$(u, v, \Phi)^n = (U, V, H)\mu^n e^{i(mx+ly)},$$

where μ is the amplification factor of the scheme. Substitution of this solution in (2) readily gives the following amplification factors:

$$\mu_1 = e^{-i(am+bl)\tau}, \ \mu_{2,3} = e^{-i(am+bl)\tau}\frac{1 \pm i\tau K/2}{1 \mp i\tau K/2}.$$

Therefore, all amplitudes are constant in time and the frequency relations are as follows:

$$\sigma_1 = -am - bl, \ \sigma_{2,3} = -am - bl \pm \frac{2}{\tau}\tan^{-1}(\tau K/2).$$

Finally, substitution of a separate wave in the modified splitting scheme (5) results in the following characteristic equation for parameter $\eta = \mu e^{i(am+bl)\tau}$:

$$(\eta - 1)\, P(\eta) = 0, \tag{6}$$

$$P(\eta) = \eta(\eta-1)^2 + \left(\frac{\tau^2\bar{\Phi}}{4}\left(m^2 + l^2\right) + \frac{\tau^2 f^2}{4}\right)\eta(\eta+1)^2$$
$$-\frac{\tau^2 f^2}{4}(\eta-1)^3 + \frac{\tau^4\bar{\Phi}^2}{16}m^2 l^2(\eta-1)(\eta+1)^2.$$

The first root $\eta_1 = 1$ in (6) is stable and corresponds to slow advective solution. Applying Routh–Hurwitz theory it can be shown that the polynomial $P(\eta)$ has the roots within the unit disk if and only if

$$\tau^2 f^2 \leq 2. \tag{7}$$

Therefore, this is the stability condition of the modified splitting scheme (3) and (4). It is quite weak restriction corresponding to accuracy requirements. Greater time steps are generally not used in atmospheric models.

It can also be shown that there is exactly one real root representing unphysical mode of computational solution and other two roots approximate the amplitude and frequency of gravity waves. The form of solutions imply that all physical modes of computational solution represent exactly the advective part of differential solution due to applied semi-Lagrangian approximation.

The inertial solutions of two schemes coincide with the respective differential mode, which is important property because slow modes contain the main part of the atmosphere and ocean energy, while the amplitudes of gravity waves are sufficiently small. Some characteristics of gravity modes are shown

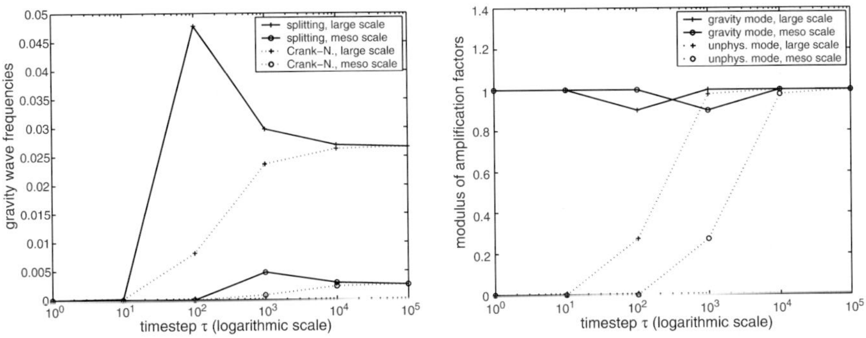

Fig. 1. (**a**) Frequencies of gravity modes; (**b**) Moduli of computational modes

in Fig. 1: the differences between computational and differential frequencies in Fig. 1a and the ratio between amplitudes in Fig. 1b.

All graphs are plotted as functions of time step (given in seconds) for two different values of the wave numbers corresponding to wavelength of 100 and 1,000 km (meso and synoptic scales, respectively). Figure 1a shows the frequency differences for splitting scheme (solid lines) and Crank–Nicholson scheme as well (dotted lines). Figure 1b shows the moduli of gravity mode (solid lines) and unphysical mode (dotted lines). It can be seen that accuracy properties of the modified splitting and Crank–Nicholson schemes are very close and both schemes approximate well gravity waves of large spatial scale.

3 Numerical Algorithm and Tests for the Nonlinear SWE

The full system of SWE can be written as follows:

$$d_t u = fv - \Phi_x,\; d_t v = -fu - \Phi_y,\; d_t \Phi = -\Phi(u_x + v_y)\,, \tag{8}$$

where

$$d_t \phi = \phi_t + u\phi_x + v\phi_y,\;\; \phi = u, v, \Phi.$$

Hereafter we use notations introduced in Sect. 2.

The algorithm of numerical solution of (8) consists of two steps. First, the trajectory equations

$$d_t x = u, d_t y = v \tag{9}$$

with given positions of arrival points (the nodes of uniform spatial grid) are solved with the second order of accuracy for departure points by applying the standard iterative algorithm [Dur99, SC91, TS87].

On the second step the modified splitting approach is applied to gravity and Coriolis terms

$$\frac{u^{n+1/2}-u^n}{\tau/2} = fv^{n+1/2} - \partial_x \Phi^{n+1/2} + A, \quad \frac{v^{n+1/2}-v^n}{\tau/2} = -fu^n - \partial_y \Phi^n,$$

$$\frac{\Phi^{n+1/2}-\Phi^n}{\tau/2} = -\bar{\Phi}\partial_x u^{n+1/2} - (\Phi^n - \bar{\Phi})\partial_x u^n - \Phi^n \partial_y v^n; \tag{10}$$

$$\frac{u^{n+1}-u^{n+1/2}}{\tau/2} = fv^{n+1/2} - \partial_x \Phi^{n+1/2} + A, \quad \frac{v^{n+1}-v^{n+1/2}}{\tau/2} = -fu^{n+1} - \partial_y \Phi^{n+1},$$

$$\frac{\Phi^{n+1}-\Phi^{n+1/2}}{\tau/2} = -\Phi^{n+1/2}\partial_x u^{n+1/2} - \bar{\Phi}\partial_y v^{n+1} - (\Phi^{n+1/2} - \bar{\Phi})\partial_y v^{n+1/2}. \tag{11}$$

The linear analysis of this scheme is given in Sect. 2. The performed numerical tests with atmospheric fields show that small deviations from linear terms treated explicitly in the Φ-equations do not affect practically neither stability nor accuracy of the scheme.

To evaluate the performance of the modified splitting scheme (9)–(11), the numerical experiments with actual atmospheric data were carried out on the spatial domain of $5,000 \times 5,000\,\text{km}^2$ centered at the point $(35^0 S, 55^0 W)$. The initial and boundary conditions for geopotential and wind components of the 500 hPa pressure surface were obtained from the objective analysis and global forecasts of the National Centers for Environmental Prediction (NCEP). The standard central difference approximation on staggered spatial grid C with mesh size $h = 50\,\text{km}$ was applied [Dur99]. The 24-h forecasts with time step 1 h were computed and the scheme accuracy was evaluated in inner window of size about $2,000 \times 2,000\,\text{km}^2$ in order to eliminate an influence of the boundary conditions. Similar integrations were made with semi-Lagrangian Crank–Nicholson scheme using the same time step of 1 h. The obtained statistical measures of the forecast skill (the root-mean-square error ϵ given in meters and correlation coefficient between observed and forecast tendencies ρ) are presented in Table 1 along with computational time required for one forecast (T_{CPU} given in seconds). The used measures are standard for evaluation of short-range forecasts and obtained results are quite typical for SWE model [AKH89]. One can see that for chosen rather large time step the accuracy of both schemes is quite similar, but splitting method is of less computational cost.

Finally, the charts of an example of 24-h forecast are presented in Figs. 2 and 3. The former shows the initial field of geopotential height at 500 hPa surface and the latter the 24-h forecast of cyclonic formation within the evaluation window according to the modified splitting scheme. This is the case of rather successful forecast with $\epsilon = 40.9$ and $\rho = 0.94$.

Table 1. Accuracy of 24-h forecasts of geopotential height at 500 hPa

Scheme	ϵ	ρ	T_{CPU}
Modified splitting	46.2	0.83	1.8
Crank–Nicholson	46.0	0.83	2.9

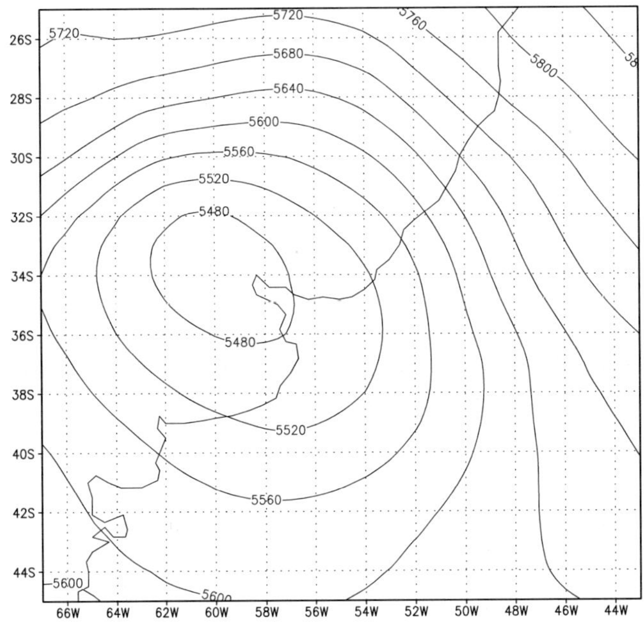

Fig. 2. 500 hPa height field from NCEP analysis for 0000 UTC 06 August 2003

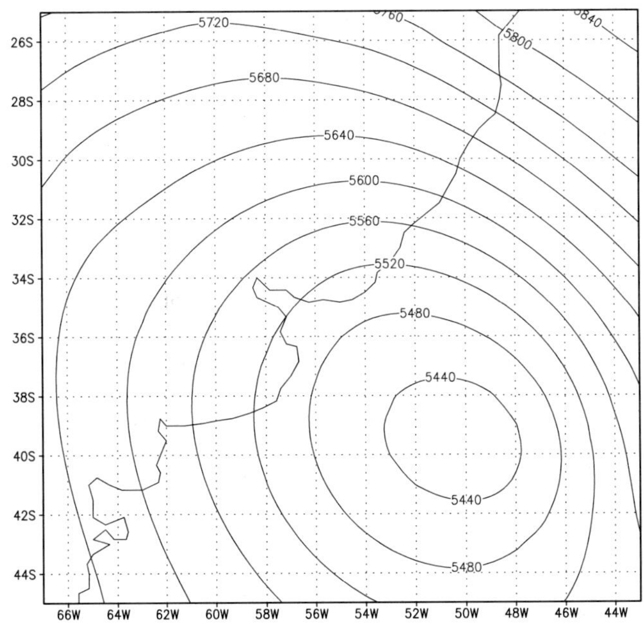

Fig. 3. 24-h forecast of 500 hPa height field for 0000 UTC 07 August 2003

Acknowledgements

This research was supported by Brazilian science foundation CNPq.

References

[AKH89] Anthes, R.A., Kuo, Y.H., Hsie, E.Y., Low-Nam, S., Bettge, T.W.: Estimation of skill and uncertainty in regional numerical models. Q. J. R. Meteor. Soc., **115**, 763–806 (1989)

[Bat84] Bates, J.R.: An efficient semi-Lagrangian and alternating direction implicit method for integrating the shallow water equations. Mon. Wea. Rev., **112**, 2033–2047 (1984)

[CDI85] Cohn, S.E., Dee, D., Isaacson, E., Marchesin, D., Zwas, G.: A fully implicit scheme for the barotropic primitive equations. Mon. Wea. Rev., **113**, 436–448 (1985)

[DK01] Douglas, J., Kim, S.: Improved accuracy for locally one-dimensional methods for parabolic equations. Math. Mod. Meth. Appl. Sci., **11**, 1563–1579 (2001)

[DKL03] Douglas, J., Kim, S., Lim, H.: An improved alternating-direction method for a viscous wave equation. Current Trends in Scientific Computing, Contemporary Mathematics, **329**, 99–104 (2003)

[Dur99] Durran, D.: Numerical Methods for Wave Equations in Geophysical Fluid Dynamics. Springer-Verlag, New York (1999)

[GSG01] Gospodinov, I.G., Spiridonov, V.G., Geleyn, J.F.: Second-order accuracy of two-time-level semi-Lagrangian schemes. Q. J. R. Meteor. Soc., **127**, 1017–1033 (2001)

[KTM94] Kar, S.K., Turco, R.P., Mechoso, C.R., Arakawa, A.: A locally one-dimensional semi-implicit scheme for global gridpoint shallow-water models. Mon. Wea. Rev., **122**, 205–222 (1994)

[LLS00] LeRoux, D.Y., Lin, C.A., Staniforth, A.: A semi-implicit semi-Lagrangian finite-element shallow-water ocean model. Mon. Wea. Rev., **128**, 1384–1401 (2000)

[McD86] McDonald, A.: A semi-Lagrangian and semi-implicit two time level integration scheme. Mon. Wea. Rev., **114**, 824–830 (1986)

[SC91] Staniforth, A., Cote, J.: Semi-Lagrangian integration schemes for atmospheric models - A review. Mon. Wea. Rev., **119**, 2206–2223 (1991)

[TS87] Temperton, C., Staniforth, A.: An efficient two-time-level semi-Lagrangian semi-implicit integration scheme. Q. J. R. Meteor. Soc., **113**, 1025–1039 (1987)

[YR86] Yakimiw, E., Robert, A.: Accuracy and stability analysis of a fully implicit scheme for shallow water equations. Mon. Wea. Rev., **114**, 240–244 (1986)

Compact Third-Order Logarithmic Limiting for Nonlinear Hyperbolic Conservation Laws

M. Čada, M. Torrilhon, and R. Jeltsch

Summary. To achieve high order accurate numerical approximation to nonlinear smooth functions, we employ and generalize the idea of double-logarithmic reconstruction for the numerical solution of hyperbolic equations. The result is a class of efficient third-order schemes with a compact stencil. These methods handle discontinuities as well as local extrema within the standard semi-discrete MUSCL algorithm using only a single limiter function.

1 Introduction

For simplicity we consider the numerical approximations to the scalar initial value problem

$$u_t + f(u)_x = 0, \qquad u(x,0) = u_0(x), \tag{1}$$

where u_0 is a piecewise smooth function with compact support. We cover the uniform computational region with control cells $C_i^n = [x_{i-\Delta x/2}, x_{i+\Delta x/2}] \times [t^n, t^{n+1}]$, with $t^{n+1} = t^n + \Delta t$ and $x_{i\pm 1} = x_i \pm \Delta x$. Integrating the conservation law (1) over the control volume C_i^n, we obtain the standard finite volume update

$$\frac{d}{dt}\bar{u}_i = L_i(\hat{u}_i) = \frac{1}{\Delta x}\left[\mathcal{F}(\hat{u}^{(-)}_{i-\frac{1}{2}}, \hat{u}^{(+)}_{i-\frac{1}{2}}) - \mathcal{F}(\hat{u}^{(-)}_{i+\frac{1}{2}}, \hat{u}^{(+)}_{i+\frac{1}{2}})\right], \tag{2}$$

for the cell average $\bar{u}$ based on the numerical flux function $\mathcal{F}$. The evolution of $\bar{u}$ is governed by the left and right limits $\hat{u}^{(\pm)}_{i+\frac{1}{2}}$ – the interface values – of the reconstructed function $\hat{u}(x)$. The calculation of the interface values from the known cell mean values is the essential reconstruction task.
Since we consider a local reconstruction, i.e., a three point stencil we set the interface values

$$\hat{u}^{(-)}_{i+\frac{1}{2}} = L(\bar{u}_{i-1}, \bar{u}_i, \bar{u}_{i+1}), \qquad \hat{u}^{(+)}_{i+\frac{1}{2}} = R(\bar{u}_i, \bar{u}_{i+1}, \bar{u}_{i+2}). \tag{3}$$

To obtain high order nonoscillatory reconstructions, the interpolation function (3) is a priori nonlinear.

2 Standard TVD-MUSCL Methods

We adopt the classical TVD-MUSCL scheme, where the time and space discretization is decoupled. This approach assumes that a piecewise linear interpolation is reconstructed from the average values $\bar{u}_i(t)$ based on central differences. In the classical setting slope limiters are used to avoid spurious oscillation:

$$\hat{u}^{(-)}_{i+\frac{1}{2}} = \bar{u}_i + \frac{\Delta x}{2}\sigma_i = \bar{u}_i + \frac{1}{2}\phi(\theta_i)\delta_{i+\frac{1}{2}}, \qquad (4)$$
$$\hat{u}^{(+)}_{i-\frac{1}{2}} = \bar{u}_i - \frac{\Delta x}{2}\sigma_i = \bar{u}_i - \frac{1}{2}\phi(\theta_i)\delta_{i+\frac{1}{2}}.$$

The reconstructed functions $\hat{u}^{(\mp)}_{i\pm\frac{1}{2}}$ can be expressed in terms of nonlinear limiter functions using the downwind slope

$$\sigma_i = (\frac{\delta_{i+\frac{1}{2}}}{\Delta x})\phi(\theta_i), \qquad (5)$$

where

$$\delta_{i+\frac{1}{2}} = \bar{u}_{i+1} - \bar{u}_i$$

is the difference across a cell interface, and

$$\theta_i = \frac{\delta_{i-\frac{1}{2}}}{\delta_{i+\frac{1}{2}}}, \quad \delta_{i+\frac{1}{2}} \neq 0$$

is a local smoothness measurement. If the limiter function satisfies $\phi(1+\Delta x) = 1+\frac{\Delta x}{2}+\mathcal{O}(\Delta x^2)$ for $\bar{u} \in C^\infty$, the reconstructed values (5) converge to central differences and hence are of second order accuracy in space. Classical limiter functions, such as minmod, super-bee, monotonized central difference limiter (MC), etc. (see e.g. [Sw84], [Le03]) are only defined for $\theta \geq 0$.
For $\theta < 0$ the reconstruction reduces to the constant cell average itself, and therefore extrema for which $\theta < 0$ holds can not be reconstructed and are limited to first order. Hence this method is of second order away from local extrema and discontinuities.

3 Compact Third-Order Reconstruction

The classical TVD-MUSCL scheme uses a linear interpolation to obtain a second order spatial reconstruction with a local three point stencil (3). However, a quadratic interpolation can also be recovered from a three point stencil, resulting in a local third-order reconstruction

$$\hat{u}^{(-)}_{i+\frac{1}{2}} = \frac{5}{6}\bar{u}_i - \frac{1}{6}\bar{u}_{i-1} + \frac{1}{3}\bar{u}_{i+1}, \qquad (6)$$
$$\hat{u}^{(+)}_{i-\frac{1}{2}} = \frac{5}{6}\bar{u}_i - \frac{1}{6}\bar{u}_{i+1} + \frac{1}{3}\bar{u}_{i-1}.$$

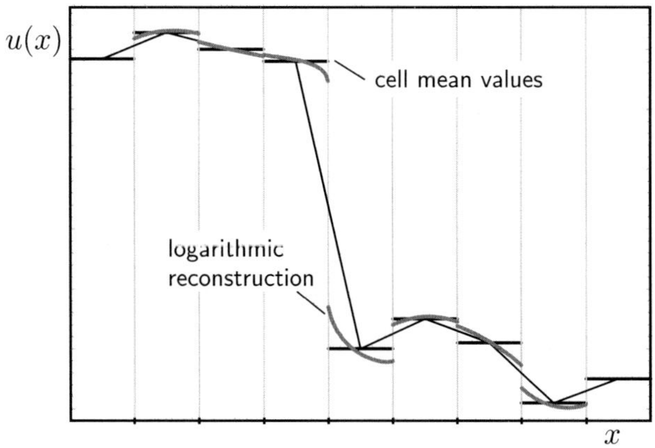

Fig. 1. Example of logarithmic reconstruction.

To avoid oscillation we have to reformulate (6) similar to (4) with a suitable limiter function, taking local extrema into account.

3.1 Logarithmic Limiter

Based on the idea of nonlinear and non-polynomial interpolation, Artebrant and Schroll [AS06] developed a local double logarithmic reconstruction (LDLR). It is essentially of third order away from discontinuities. Because of the logarithmic nature it is able to reconstruct local extrema without loss of accuracy and can handle very large slopes followed by flat regions (see Fig. 1). Because of its local, symmetric nature, it turns out that the whole reconstruction procedure can be written in a limiter formulation that resembles (4). After some algebraic reformulation we find

$$\begin{aligned} \hat{u}^{(-)}_{i+\frac{1}{2}} &= \bar{u}_i + \frac{1}{2}\phi(\theta_i)\delta_{i+\frac{1}{2}}, \\ \hat{u}^{(+)}_{i-\frac{1}{2}} &= \bar{u}_i - \frac{1}{2}\phi(\theta_i^{-1})\delta_{i-\frac{1}{2}}. \end{aligned} \tag{7}$$

This formulation uses only a single limiter function

$$\phi(\theta) = \frac{2p((p^2 - 2p\theta + 1)\ln p - (1-\theta)(p^2-1))}{(p^2-1)(p-1)^2}, \tag{8}$$

with

$$p = p(\theta) = 2\frac{|\theta|^q}{1+|\theta|^{2q}}. \tag{9}$$

However, in this formulation we have to use the limiter function also with inverse input and hence also with reverse slopes $\delta_{i\pm\frac{1}{2}}$. The parameter $q > 1$ controls the total variation of the reconstruction as derived in [AS06]. The

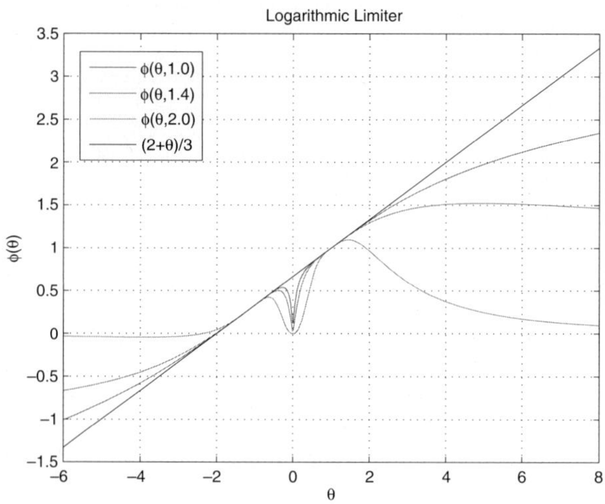

Fig. 2. Shape of the limiter function for different values of q. The larger q the less total variation appears. $q = 1.4$ is used in [AS06]

function (8) has a removable singularity at $p = \pm 1$ by setting $\phi(1) = 1$ and $\phi(-1) = \frac{1}{3}$. This ensures that we do not limit local extrema for which $\theta = \pm 1$ holds.

The shape of the log-limiter function $\phi(\theta)$ is shown in Fig. 2 for different values of q. The limiter function asymptotically approaches the linear function $\phi(\theta) = \frac{2+\theta}{3}$ for $q \to 1$. Inserting this linear function into (7) results in the known third-order interpolation (6). Indeed, the smaller q the more variation is produced in LDLR. The nonlinear limiter function smoothly extends into the range of $\theta < 0$ and vanishes for $\theta \to \pm\infty$ and $\theta \to 0$, which is essential for limiting discontinuities.

3.2 Simplified Third-Order Limiter

The logarithmic limiter is derived from double logarithmic ansatz functions, which are conservative, third-order accurate, and essentially local variation bounded.

However, both the LDLR and the logarithmic limiter function are complicated and computationally expensive. Further on the LDLR is very sensitive to q and Artebrant and Schroll [AS06] suggested to set $q = 1.4$ to ensure stability. To look for a simplification of (8) we consider the formulation (7) and demand the limiter to satisfy the following:

1. $\phi(1+\Delta x) = 1 + \frac{\Delta x}{3} + \mathcal{O}(\Delta x^m)$ and $\phi(-1+\Delta x) = \frac{1}{3} + \frac{\Delta x}{3} + \mathcal{O}(\Delta x^n)$ with $n, m \geq 2$
2. $\lim_{\theta \to 0} \phi(\theta) \to 0$ and $\lim_{\theta \to \pm\infty} \phi(\theta) \to 0$

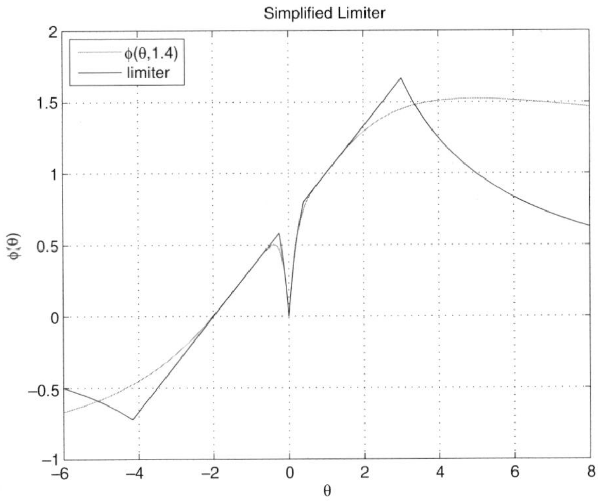

Fig. 3. Simplified limiter function matching for a special q

The first condition guaranties a better resolution of local extrema, especially for $\theta < 0$ and the second ensures (8) to be local variation bounded.

A simple function matching these conditions is, e.g., given by

$$\hat{\phi}(\theta) = \max\Big(\frac{3}{-|\theta|}, \min\big(\frac{2+\theta}{3}, \frac{2|\theta|}{0.6+|\theta|}, \frac{5}{|\theta|}\big)\Big). \tag{10}$$

Figure 3 compares the simplified limiter $\hat{\phi}$ with the original logarithmic limiter function $\phi(q = 1.4)$. The new limiter stays in the context of known TVD-limiter (see [Sw84]), since it is a discontinuous (max, min)-function that acts as a switch depending on the smoothness measure θ. Its rigorous cutoffs for $\theta \to \pm 0$ and $\theta \to \pm\infty$ ensure sharp resolution of discontinuous functions. Further discussion with different limiter functions will be given in [CT07].

4 Time Integration

The finite volume update (2) can be based on any numerical flux function. To reach third-order resolution the time integration has to be adjusted. We use an explicit three stage TVD Runge–Kutta method of [GS98]:

$$\begin{aligned} \bar{u}_i^{(1)} &= \bar{u}_i^n + \Delta t L_i(\bar{u}^n), \\ \bar{u}_i^{(2)} &= \frac{3}{4}\bar{u}_i^n + \frac{1}{4}(\bar{u}_i^{(1)} + \Delta t L_i(\bar{u}^{(1)})), \\ \bar{u}_i^{n+1} &= \frac{1}{3}\bar{u}_i^n + \frac{2}{3}(\bar{u}_i^{(2)} + \Delta t L_i(\bar{u}^{(2)})). \end{aligned} \tag{11}$$

Although it consists of three update calls, this method can be realized with only one temporary variable and hence only requires two storage units per ODE $\bar{u}_i^{n+1}$ and $\bar{u}^n$.

This method generates TVD updates under the usual CFL conditions. A thorough stability analysis in [CT07] assuming unlimited third-order reconstruction (6) shows that, neglecting the TVD condition, the stability region can be enlarged to CFL ≤ 1.62.

5 Numerical Experiments

We have already conducted several accuracy test. Here we only shortly present a test for the linear advection equation with periodic boundary condition and initial data as proposed in [JS96]. The initial signal consists of a smooth, but narrow combination of Gaussians, a square wave, a triangle wave, and a half ellipse.

Figure 4 illustrates the very weak performance of classical second order TVD-limiters as described in Sect. 2. A second order TVD Runge–Kutta method [Le03] is used for the time integration. All limiters, especially minmod, show a very weak performance. The approximation illustrates rather no symmetry and neither the smooth maxima nor the sharp corners are resolved sufficiently. The strong nonsymmetry is due to the one-sided reconstruction of (4).

In Fig. 5 we compare our new limiter function $\hat{\phi}$ (LIM3) with Marquinas [Ma94] third-order local hyperbolic reconstruction (LHHR), third-order ENO method using a five point stencil and with LDLR. All schemes use the same third-order Runge–Kutta method for time integration.

Especially in contrast to the classical TVD-MUSCL scheme, the new method gives a good approximation of the advection profile. LIM3 resolves corners sharper, is symmetric, and does not produce any spurious oscillation such as LDLR [AS06]. Especially compared to ENO3 and to LHHR it shows significant improvement of resolution for the narrow Gaussian peak and the

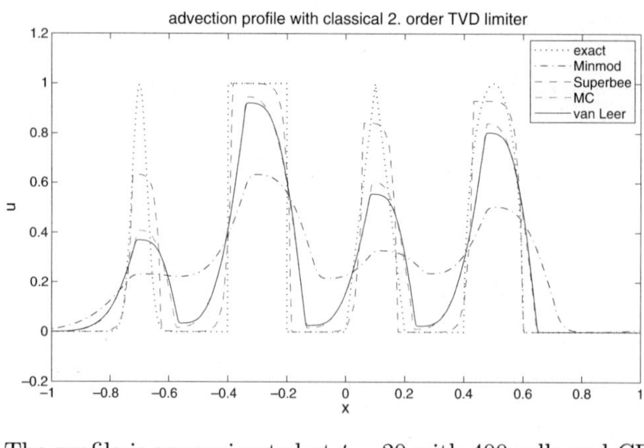

Fig. 4. The profile is approximated at $t = 20$ with 400 cells and CFL $= 0.8$

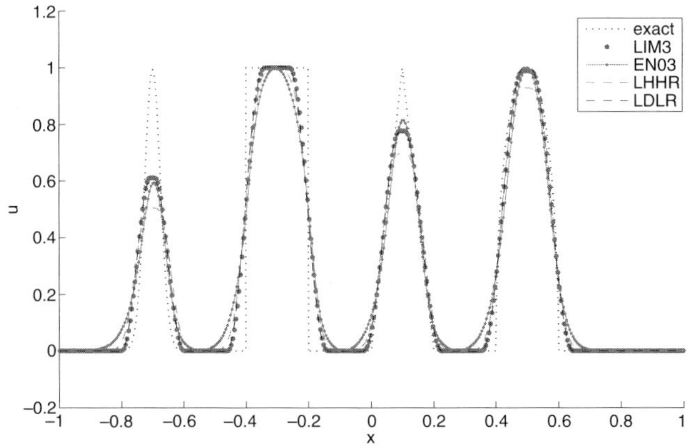

Fig. 5. The profile is approximated at $t = 20$ with 400 cells and CFL = 0.8. ENO3 used CFL = 0.6. For details we refer to [AS06]

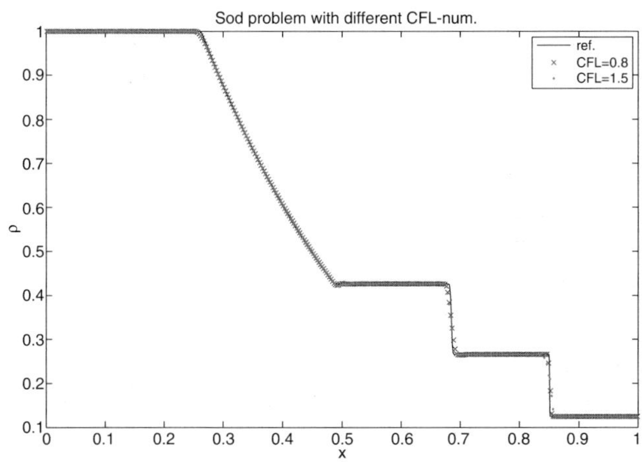

Fig. 6. Density plot of Sod problem with 200 cells at $t = 0.2$

triangle wave. It produces the sharpest result for the corners of the square wave.

Finally, the Sod problem (see e.g. [JS96]) of the nonlinear Euler equations is solved to show the stability for high CFL numbers (see Fig. 6.). We use Roe's approximative Riemann solver with Godunov's numerical flux function. The limiter is applied to the conservative variables.

The scheme remains stable and accurate even for CFL = 1.5. The simplified limiter $\hat{\phi}$ introduces slight undershoots in the rarefaction, which can be eliminated by limiting the characteristic variables instead of the conservative. Shock and contact waves are resolved very accurately. The slight oscillations in the rarefaction wave increase with larger CFL numbers. We also observe very small shock wave dispersion when reaching the limit of the stability region.

Besides the promising results of this scheme, the main advantage are the simplicity and the computational costs. Since we use a single rather simple limiter function, we do not have to preprocess the slopes such as LHHR requires or evaluate ln-functions such as LDLR. Our simple smoothness measure θ is very simple and very fast to be computed. Existing second order TVD-MUSCL schemes can easily be extended. Unlike the ENO3 scheme, the method is compact and has a less strict CFL condition.

References

[Ma94] Marquina, A.: Local Hyperbolic Reconstruction of Numerical Fluxes for Nonlinear Scalar Conservation Laws. SIAM J. Sci. Comput., **15**/4, pp. 892–915 (1994)

[GS98] Gottlieb, S., Shu, C.-W.: Total Variation Diminishing Runge-Kutta Schemes. Math. Comput., **67**/221, pp. 73–85 (1998)

[AS06] Artebrant, R., Schroll, H.J.: Limiter-free third order logarithmic reconstruction. SIAM J. Sci. Comput., **28**/1, pp. 359–381 (2006)

[JS96] Jiang, G.-S., Shu, C.-W.: Efficient implementation of weighted ENO schemes. J. Comput. Phys., 126, pp. 202–228 (1996)

[Sw84] Sweby, J.C.: High resolution schemes using flux limiters for hyperbolic conservation laws. SIAM J. Num. Anal., **21**, pp. 995–1011 (1984)

[CT07] Cada, M., Torrilhon, M.: New compact third order logarithmic limiter for FV-Methods. (in preparation)

[Le03] LeVeque, R.J.: Finite Volume Methods for Hyperbolic Problems. Cambridge Texts in Applied Mathematics (2003)

A Finite Volume Grid for Solving Hyperbolic Problems on the Sphere

D. Calhoun, C. Helzel, and R.J. LeVeque

1 Introduction

Uniform Cartesian grids are well suited for solving problems in rectangular domains. Mapped grids and domain embedding techniques are often used to apply rectangular grids to more general domains. In particular, the use of the logically rectangular polar grid is widely used for problems in circular or spherical domains. Other grids of this type include the standard latitude–longitude grid used for the sphere, and the spherical grid used for the ball.

However, these three standard grids – the polar grid, the spherical grid, and the latitude–longitude grid – all suffer from the problem that the ratio of the largest cells to the smallest grows as the grid is refined. As a result, when using explicit time stepping schemes with these grids, one is forced to take a time step that is much smaller than desirable, to respect the CFL stability limits imposed by the smallest cells. A second but related problem with these grids is that the grid resolution is poorly distributed. The polar grid, for example, has cells with very large aspect ratios near the outer edge and tiny cells near the center. For many problems of practical interest, this distribution of large and small cells does not match the physical requirements of the problem.

Because of the limitations imposed by these standard grids, several researchers have proposed grid mappings which fix the problem of cell size distribution. One popular approach is to map multiple grids in patchwork fashion to the disk, the sphere or the ball. One example of such a grid is the *gnomic projection* grid (sometimes called the "cubed-sphere" grid), first described by [6]. Using this approach, one patches together six logically Cartesian grids to cover the sphere. In a related approach, one maps five grids to the disk, or seven cubes to the ball. This approach can lead to good results when used with appropriate solvers (e.g. see [5]). However, the chief disadvantage of these mappings is that they require that one manage a multiblock structure with nontrivial communication between bordering cells of the patches.

Furthermore, modifying a standard adaptive mesh refinement code for use in a multiblock setting presents many technical challenges.

In the work presented here, we focus our attention on describing a grid mapping for the sphere that attempts to fix the cell size distribution problem. Unlike the multiblock grid for the sphere described above, our mapping covers the entire sphere with a single logically Cartesian grid. Furthermore, the resulting mesh has the desirable property that the area of the largest to smallest cell is about 2, and it is easily adapted for use with an existing adaptive mesh refinement code. We show that this works well in practice when used with explicit finite-volume schemes for hyperbolic problems.

2 The Mappings

In this section, we describe a grid mapping for the unit disk and the unit sphere. Our mapping for the sphere (e.g., the surface of the solid ball) is based on the mapping for the unit disk, and so we start with a description of that mapping. In Sec. 2.2, we extend this mapping to the sphere.

2.1 Mapping the Square to the Unit Disk

In this section, we describe a mapping which maps the computational domain $[-1,1]\times[-1,1]$ to the unit disk. To describe the mapping, we focus our attention on the region of the square in which computational coordinates (ξ,η) satisfy $|\xi|\le\eta$. This region corresponds to the upper triangular region between the two diagonals of the square. We refer to this region as the *north sector* of the computational domain. The mapping in this sector is based on the idea that we can map a horizontal line segment between points $(-d,d)$ and (d,d) to a circular arc of radius $R(d)$, passing through points $(-D(d),D(d))$ and $(D(d),D(d))$. The center of the circle on which this arc lies is given by $(x_0,y_0)=(0,D(d)-\sqrt{R(d)^2-D(d)^2})$. Using this idea, a general point (ξ,η) in the north sector is mapped to the unit disk by the mappings $X_n(\xi,\eta)$ and $Y_n(\xi,\eta)$ given by

$$\begin{aligned} X_n(\xi,\eta) &= D(\eta)\frac{\xi}{\eta}, \\ Y_n(\xi,\eta) &= D(\eta)-\sqrt{R(\eta)^2-D(\eta)^2}+\sqrt{R(\eta)^2-X_n(\xi,\eta)^2}, \qquad |\xi|\le\eta. \end{aligned} \tag{1}$$

The mapping to other sectors is easily obtained by negating and/or swapping the arguments (ξ,η) or functions (X_n,Y_n). For example, the mapping in the west sector is given by $X_w(\xi,\eta) = Y_n(\eta,-\xi)$ and $Y_w(\xi,\eta) = -X_n(\eta,-\xi)$. Matlab scripts for this and other mappings are given in [2].

There are many choices for $R(d)$ and $D(d)$ that lead to reasonable grids. For example, one choice that is well suited for the unit disk is

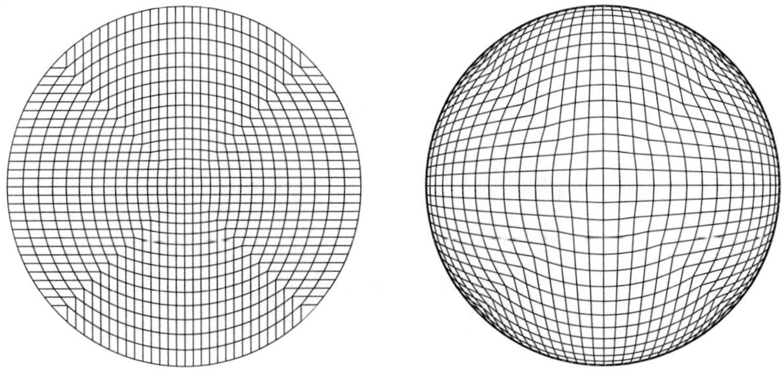

Fig. 1. Grids for circular domains. The grid on the *left* can be used for calculations in the disk. The grid on the *right* is useful for constructing the sphere grid described in Sec. 2.2

$$\begin{aligned} D(d) &= d/\sqrt{2} \\ R(d) &= 1 \end{aligned} \tag{2}$$

This choice leads to the grid on the left in Fig. 1. One key advantage of this grid for explicit computations is that the ratio of the largest to the smallest grid cells is about 2, and so there is no need to take artificially small times steps because of the presence of a few small cells. In [2], we use this grid to simulate a blast wave in a circular domain and observed no artifacts attributable to the nonsmoothness of the grid.

If one wants increased refinement near the boundary, either to model a boundary layer or, as we will see, to construct the sphere mapping, it may be useful to define $D(d)$ so that azimuthal grid lines are compressed near the outer edge. An example of a function $D(d)$ which does this is

$$D(d) = d(2-d)/\sqrt{2} \tag{3}$$

The plot at the right in Fig. 1 shows a grid generated using this $D(d)$, along with $R(d) = 1$.

2.2 A Mapping for the Unit Sphere

The above mapping for the unit disk can be used directly to create a grid for the unit sphere. In general, if we are given a mapping $X(\xi, \eta)$ and $Y(\xi, \eta)$ from the square $[-1, 1] \times [-1, 1]$ to the unit circle, we can define mapping functions $X_s(\xi, \eta)$, $Y_s(\xi, \eta)$, and $Z_s(\xi, \eta)$ from the rectangular region $[-3, 1] \times [-1, 1]$ to the sphere as

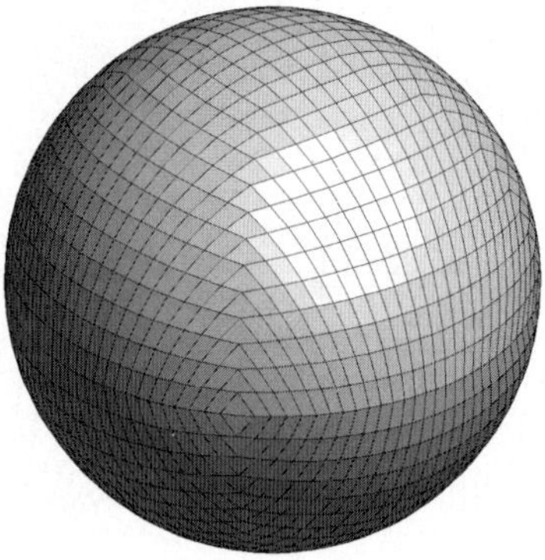

Fig. 2. Finite-volume sphere grid based on a single logically Cartesian grid

$$\begin{aligned} X_s(\xi,\eta) &= \begin{cases} X(-(\xi+2),\eta) & \text{if} \quad |\xi+2| \le 1 \\ X(\xi,\eta) & \text{if} \quad |\xi| \le 1 \end{cases} \\ Y_s(\xi,\eta) &= Y(\xi,\eta) \\ Z_s(\xi,\eta) &= \begin{cases} -\sqrt{1-(X(-(\xi+2),\eta)^2+Y(\xi,\eta)^2)} & \text{if} \quad |\xi+2| \le 1 \\ \sqrt{1-(X(\xi,\eta)^2+Y(\xi,\eta)^2)} & \text{if} \quad |\xi| \le 1. \end{cases} \end{aligned} \tag{4}$$

Using functions $X(\xi,\eta)$ and $Y(\xi,\eta)$ based on $D(d)$ as defined in (2) leads to a mapping with extremely elongated cells near the equator. Using (3) leads to a sphere mesh with cells which are more equi-distributed in size. The results are shown in Fig. 2.

3 Numerical Results

In this section, we present the results obtained using our mappings to solve hyperbolic problems. The algorithms we use are the wave propagation algorithms described in [4]. These algorithms are finite volume Godunov-type methods based on solving Riemann problems at the mapped cell interfaces. One reason for choosing these algorithms is that it is easy to take advantage of the rotational invariance that is often present in the equations of practical and physical interest. In particular, we can reuse existing one-dimensional Riemann solvers to solve Riemann problems at mapped cell interfaces aligned along any direction. Moreover, our approach does not require that we provide metric terms analytically, but only that we approximate these terms using corners of mapped mesh cells. We do not present the details of these solvers here, but rather refer the interested reader to [2] and [4].

3.1 Advection on the Sphere

Here we illustrate the use of the unit sphere mesh described in Sec. 2.2 for the approximation of advective transport problems. We discretize the equation

$$q_t + \mathbf{u} \cdot \nabla q = 0, \tag{5}$$

where $\mathbf{u}$ is a divergence-free velocity field on the sphere and $q(\mathbf{x}, t)$ is a concentration that depends on space and time. A particularly simple flow situation is solid body rotation. In this case the exact solution after N rotations is equal to the initial conditions and it is therefore easy to perform convergence studies.

We use the wave propagation algorithm for quadrilateral grids on a sphere as described in [2]. This requires appropriately scaled flow speeds normal to each grid cell interface. In general, we can obtain a divergence-free velocity field at the centers of cell interfaces by differencing a stream function ψ at cell vertices adjacent to the cell interface. For solid body rotation, the stream function we use is

$$\psi(\mathbf{x}) = 2\pi(\mathbf{x} \cdot \mathbf{a}_{rot}), \tag{6}$$

where $\mathbf{x} \in R^3$ is a position vector (e.g., point) on the surface of the sphere and $\mathbf{a}_{rot} \in R^3$ is a normalized vector in the direction of the specified axis of rotation.

To perform a convergence study we initialize the concentration field q with a smooth function given by

$$q(\mathbf{x}, 0) = 2\exp(-10d^2),$$

where $d = \arccos(\mathbf{x} \cdot \mathbf{x}_0)$ is the geodesic distance (i.e., distance along the surface of the sphere) between the points $\mathbf{x}$ and $\mathbf{x}_0$. For this example, we choose $x_0 = (1, 0, 0)$.

We compare the final results after one rotation with our initial conditions. Table 1 shows the error in the L_1-norm and the experimental order of convergence (EOC) for rotation about the y-axis ($\mathbf{a}_{rot} = (0, 1, 0)$) and rotation about the z-axis ($\mathbf{a}_{rot} = (0, 0, 1)$). For this smooth solution we obtain second-order convergence rates for rotation about both these axes.

Table 1. Convergence study for smooth advection test, $CFL = 0.9$, no limiters

	Axis of rotation: y-axis		Axis of rotation: z-axis			
	New grid		New grid		Lat-long grid	
Grid size	L_1-error	EOC	L_1-error	EOC	L_1-error	EOC
100×50	0.231495		0.098360		0.100258	
200×100	0.063987	1.85	0.025643	1.94	0.027075	1.89
400×200	0.015991	2.00	0.006438	1.99	0.006833	1.99
800×400	0.004004	2.00	0.001607	2.00	0.001693	2.01

For rotation about the z-axis, we also show the L_1-error and the EOC for calculations on a latitude–longitude grid with the same number of grid cells. On this latitude–longitude grid, the grid cells near the equator are larger than the grid cells on our new grid, and in fact, our new grid requires about twice as many times steps as the lat-long grid. The reduced resolution near the equator for the smooth lat-long grid may explain why this grid does not give better results than our new grid for flow around the equator, a flow field for which the lat-long grid is ideally suited. For rotation about the y-axis (e.g., over the poles) the situation is reversed, and the CFL restriction places severe time step restrictions on the lat-long grid.

Next we illustrate the performance of the method for advective transport of a discontinuous profile defined as

$$q(\mathbf{x},0) = \begin{cases} 1 & \text{if} \quad \sin(3\theta) > 0 \\ 0 & \quad\ \ \text{otherwise.} \end{cases}$$

Figure 3 shows plots of the solution after one rotation along the equator.

Adaptive Mesh Refinement on the Sphere Grid

One advantage of having the sphere grid represented by a single logically rectangular grid (as opposed to the multiblock structures required, for example, by [6]) is that we can easily use adaptive mesh refinement on our grid. To simplify the communication between boundaries, we double the size of the computational grid in the η-direction, creating a second copy of the sphere reflected across $\eta = -1$. Then the boundary conditions are simply periodic in both ξ and η, a case that is already handled in AMRCLAW [1]. This could be avoided with some rewriting of the refinement routines to implement the appropriate boundary conditions on the original rectangular domain, which are periodic

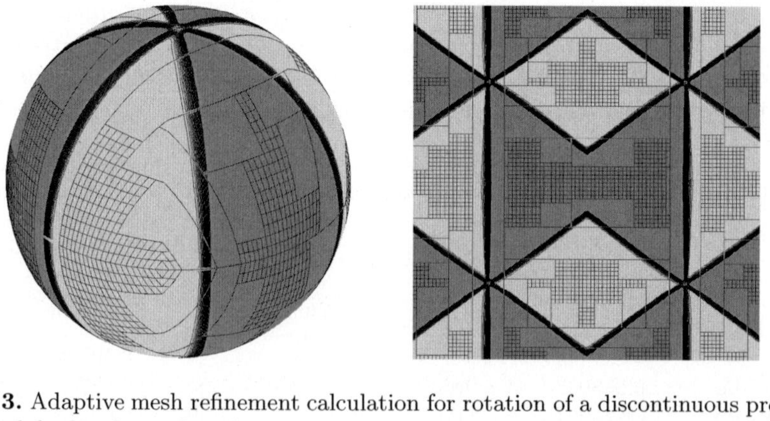

Fig. 3. Adaptive mesh refinement calculation for rotation of a discontinuous profile. The it left plot shows the solution on the sphere after one rotation, and the *right plot* shows the solution in the computational domain used for the AMRCLAW calculation

in ξ but along $\eta = -1$ and $\eta = +1$ are given by $q(\xi, \pm 1) = q(-2-\xi, \pm 1)$ for $-3 \leq \xi \leq 1$. For this example, we use the AMRCLAW algorithm with two levels of refinement. On the coarse grid the sphere is discretized with 100×50 grid cells. On the fine mesh a refinement factor of 4 is used in each direction.

3.2 Shallow Water Equations on the Sphere

The numerical solution to the shallow water equations on the sphere is of great interest to the global atmospheric and oceanic communities. Here, we solve the shallow water equations with no bottom topography on nonrotating sphere. The equations we solve are given by

$$\partial_t \mathbf{q} + \nabla \cdot \mathbf{f}(\mathbf{q}) = \mathbf{s}(\mathbf{x}, \mathbf{q}) \tag{7}$$

where $\mathbf{q} = (h, hu, hv, hw)$ is a vector of conserved quantities, and the flux function $\mathbf{f}(\mathbf{q})$ is given by

$$\mathbf{f}(\mathbf{q}) = \begin{pmatrix} hu & hv & hw \\ hu^2 + \frac{1}{2}gh^2 & huv & huw \\ huv & hv^2 + \frac{1}{2}gh^2 & hvw \\ huw & hvw & hw^2 + \frac{1}{2}gh^2 \end{pmatrix} \tag{8}$$

The source term $\mathbf{s}(\mathbf{x}, \mathbf{q})$, which acts only on the momentum equations, has the form

$$\mathbf{s}(\mathbf{x}, \mathbf{q}) = \begin{pmatrix} 0 \\ (\mathbf{x} \cdot \nabla \cdot \tilde{\mathbf{f}})\mathbf{x} \end{pmatrix} \tag{9}$$

where $\mathbf{x}$ is the position vector on the sphere. The flux vector $\tilde{\mathbf{f}} \in R^3$ consists of the second through fourth components of the flux vector $\mathbf{f}(\mathbf{q})$. The source term is included to ensure that the fluid velocity remains on the surface of the sphere. This approach was also taken by [3].

The height field is initialized on the sphere using the smooth function

$$h(\mathbf{x}) = 1 + 2\exp(-40(1 - (\mathbf{x} \cdot \mathbf{x}_0))^2). \tag{10}$$

This initial condition corresponds to a smooth, circular hump of fluid centered at the point $\mathbf{x}_0$ on the sphere. To solve the shallow water equations, we again use the algorithm described in [2]. As described there, we maintain conservation by subtracting out nonconservative terms after each time step.

In Fig. 4, we show the results at time $t = 0.9$, on an adaptively refined mesh with three levels of refinement used. From these results, we see that the solution appears to remain symmetric with respect to the axis of symmetry.

4 Conclusion

The mappings and the code used for the examples shown in this chapter are available at the Web site

`http://www.amath.washington.edu/~rjl/pubs/hyp2006`

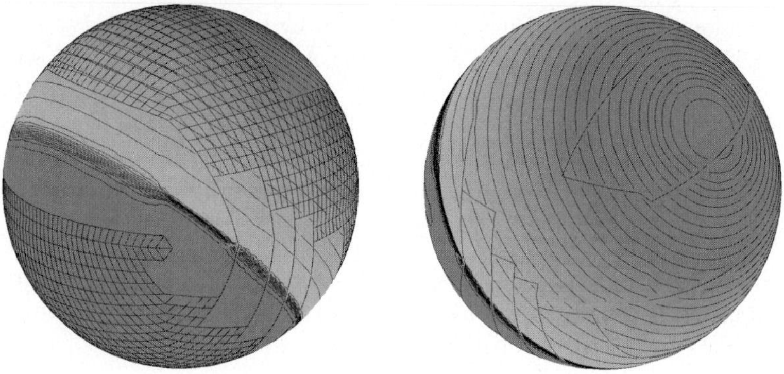

Fig. 4. Two views of the height field solution to the shallow water wave equation on the sphere at time $t = 0.9$. In the left plot, the mesh for the level 2 grid (out of 3 AMR grids) is shown

Acknowledgments

This work was supported in part by DOE grant DE-FC02-01ER25474, NSF grant DMS-0106511 and German Science Foundation (DFG) grant SFB611. The first author also gratefully acknowledges the support of the Laboratoire d'Études des Transferts et de Méchanique des Fluides at the C.E.A.

References

1. M.J. Berger and R.J. LeVeque. Adaptive mesh refinement using wave-propagation algorithms for hyperbolic systems. *SIAM J. Numer. Anal.*, 35:2298–2316, 1998.
2. D.A. Calhoun, C. Helzel, and R.J. LeVeque. Logically rectangular grids and finite volume methods for PDEs in circular and spherical domains. *submitted*, 2006. `http://www.amath.washington.edu/~rjl/pubs/circles`.
3. F.X. Giraldo. A spectral element shallow water model on spherical geodesic grids. *Int. J. Numer. Meth. Fluids*, 35:869–901, 2001.
4. R.J. LeVeque. *Finite Volume Methods for Hyperbolic Problems.* Cambridge University Press, 2002.
5. J.A. Rossmanith. A wave propagation method for hyperbolic systems on the sphere. *J. Comput. Phys.*, 213:629–658, 2006.
6. R. Sadourny. Conservative finite-difference approximations of the primitive equations on quasi-uniform spherical grids. *Mont. Weather Rev.*, 100:211–224, 1972.

Capturing Infinitely Sharp Discrete Shock Profiles with the Godunov Scheme

C. Chalons and F. Coquel

Summary. We show how to generate infinitely sharp discrete shock profiles with the Godunov method. This work is a first step toward the computation of nonclassical solutions associated with systems having at least one nongenuinely nonlinear nor linearly degenerate characteristic field.

1 Introduction and Motivations

In this chapter, we consider a nonlinear hyperbolic system of N conservation laws in one space dimension

$$\begin{cases} \partial_t \mathbf{v} + \partial_x \mathbf{f}(\mathbf{v}) = 0, \\ \mathbf{v}(x,0) = \mathbf{v}_0(x), \end{cases} \quad (x,t) \in \mathbb{R} \times \mathbb{R}^{+*}, \;\; \mathbf{v}(x,t) \in \mathbb{R}^N. \tag{1}$$

It is well known that this problem generally does not admit smooth solutions for large times so that weak solutions in the sense of distributions are considered. Due to the presence of discontinuities, these are generally not uniquely determined by $\mathbf{v}_0$ and the validity of an entropy criterion is added for the admissibility of discontinuities. More precisely, (1) is supplemented with the following entropy inequality:

$$\partial_t U(\mathbf{v}) + \partial_x F(\mathbf{v}) \leq 0, \tag{2}$$

to be satisfied in the sense of distributions. In (2), (U, F) is assumed to be an entropy–entropy flux pair (see [2]).

When (1) is strictly hyperbolic and admits only genuinely nonlinear (GNL) or linearly degenerate (LD) characteristic fields, existence and uniqueness of an entropy solution are proved for the Cauchy problem (1) and (2) (see for instance [1] for a review). From a numerical point of view, the celebrated Godunov method is an example of conservative and entropic numerical strategy that provides good numerical approximations. In fact, the method converges (if it does) toward the unique entropy solution under consideration by the Lax–Wendroff theorem (see [2] for details).

The situation becomes more complicated when one characteristic field (at least) is neither GNL nor LD. In this context, admissible weak solutions of (1) and (2) are no longer unique and an additional selection criterion must be introduced for shock discontinuities. This criterion is often called *kinetic criterion* and may take various forms (see [3]). For instance, the selection of admissible shock waves of (1) and (2) can be determined by the traveling waves associated with an augmented system of the following form:

$$\partial_t \mathbf{v} + \partial_x \mathbf{f}(\mathbf{v}) = \partial_x \mathcal{R}(\varepsilon \partial_x \mathbf{v}, \delta \varepsilon^2 \partial_{xx} \mathbf{v}) \tag{3}$$

which incorporates the effects of small scales like viscosity and capillarity. In (3), the rescaling parameter ε is expected to be very small, and δ represents the strength of the capillarity. This definition of admissible shock turns out to be very sensitive with respect to δ when one characteristic field fails to be either GNL or LD: two distinct values of δ generically give birth to two different families of admissible discontinuities in the limit system (1) and (2). The consequence of this sensitiveness is that the numerical approximation of the weak solutions of (1) and (2) becomes a particularly challenging issue. For instance, the Godunov method itself fails in providing good numerical approximations. The reason of this failure is that the artificial diffusion terms induced by most of the numerical methods generally disagree with the regularization operator $\mathcal{R}$ (i.e., with the prescribed value of δ) and eventually corrupt the discrete shocks. By contrast, the Glimm random choice method stays free from artificial numerical diffusion and converges to the correct solution. In particular, it provides sharp discrete shock profiles.

The difficulty in approximating the solutions of such systems being related to the artificial numerical diffusion, our main purpose in this chapter is to propose a numerical strategy based on the Godunov method which is free of numerical diffusion across the shock waves, in the classical framework of GNL or LD fields. From now on, we then assume that (1) and (2) are strictly hyperbolic with either GNL or LD characteristic fields. We claim that, with such an algorithm providing infinitely sharp discrete shock profiles, we will be in a better position to tackle the general case in a forthcoming study.

2 The Numerical Approximation

Let us first introduce a time step Δt and a space step Δx that we assume to be constant for simplicity in the forthcoming developments. We set $\lambda = \frac{\Delta t}{\Delta x}$ and define the mesh interfaces $x_{j+1/2} = j\Delta x$ for $j \in \mathbb{Z}$, and the intermediate times $t^n = n\Delta t$ for $n \in \mathbb{N}$. In the sequel, $\mathbf{v}_j^n$ denotes the approximated value at time t^n and on the cell $\mathcal{C}_j = [x_{j-1/2}, x_{j+1/2}$ of the solution of (1) and (2). Therefore, a piecewise constant approximated solution $x \to \mathbf{v}_\lambda(x, t^n)$ at time t^n is given by $\mathbf{v}_\lambda(x, t^n) = \mathbf{v}_j^n$ for all $x \in \mathcal{C}_j,\ \ j \in \mathbb{Z},\ \ n \in \mathbb{N}$. When $n = 0$, we set for instance

$$\mathbf{v}_j^0 = \frac{1}{\Delta x} \int_{x_{j-1/2}}^{x_{j+1/2}} \mathbf{v}_0(x)dx, \ \forall j \in \mathbb{Z}.$$

We follow by briefly recalling the two steps of the celebrated Godunov method.
Step 1: Evolution in time
In this first step, one solves the following Cauchy problem

$$\begin{cases} \partial_t \mathbf{v} + \partial_x \mathbf{f}(\mathbf{v}) = 0, & x \in \mathbb{R}, \\ \mathbf{v}(x,0) = \mathbf{v}_\lambda(x,t^n), \end{cases} \tag{4}$$

for times $t \in [0, \Delta t]$. Recall that $x \to \mathbf{v}_\lambda(x,t^n)$ is a piecewise constant. Then, under the following usual CFL restriction involving the characteristic speeds $\lambda_i, i = 1, \ldots, N$ of (1)

$$\frac{\Delta t}{\Delta x} \max_{\mathbf{v}} \{|\lambda_i(\mathbf{v})|, \ i = 1, \ldots, N\} \leq \frac{1}{2}, \tag{5}$$

for all the $\mathbf{v}$ under consideration, the solution of (4) is known by glueing together the solutions of the Riemann problems set at each interface (see Fig. 1). More precisely,

$$\mathbf{v}(x,t) = \mathbf{v_r}(\frac{x - x_{j+1/2}}{t}; \mathbf{v}_j^n, \mathbf{v}_{j+1}^n) \text{ for all } (x,t) \in [x_j, x_{j+1}] \times [0, \Delta t], \tag{6}$$

where $(x,t) \to \mathbf{v_r}(\frac{x}{t}; \mathbf{v}^l, \mathbf{v}^r)$ denotes the self-similar solution of the Riemann problem

$$\begin{cases} \partial_t \mathbf{v} + \partial_x \mathbf{f}(\mathbf{v}) = 0, \ \ x \in \mathbb{R}, \ \ t \in \mathbb{R}^{+,\star} \\ \mathbf{v}(x,0) = \begin{cases} \mathbf{v}^l \text{ if } x < 0, \\ \mathbf{v}^r \text{ if } x > 0, \end{cases} \end{cases} \tag{7}$$

whatever $\mathbf{v}^l$ and $\mathbf{v}^r$ are in the phase space.
Step 2: Projection
In this second step, we get back a piecewise constant approximated solution on each cell $\mathcal{C}_j$ at time t^{n+1} by averaging the solution $x \to \mathbf{v}(x, \Delta t)$

$$\mathbf{v}_j^{n+1} = \frac{1}{\Delta x} \int_{x_{j-1/2}}^{x_{j+1/2}} \mathbf{v}(x, \Delta t)dx, \ \ j \in \mathbb{Z}. \tag{8}$$

Using the Green's formula, we get the usual conservative update formula

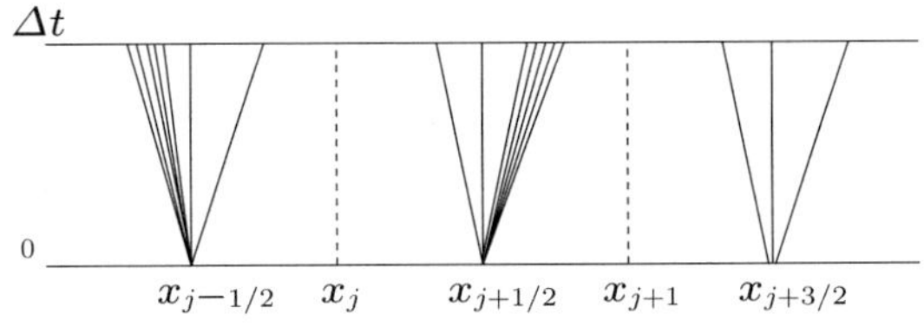

Fig. 1. An example of Riemann solutions at each interface

$$\mathbf{v}_j^{n+1} = \mathbf{v}_j^n - \lambda(\mathbf{g}_{j+1/2} - \mathbf{g}_{j-1/2}) \quad \text{with} \quad \mathbf{g}_{j+1/2} = \mathbf{f}(\mathbf{v_r}(0^{\pm}; \mathbf{v}_j^n, \mathbf{v}_{j+1}^n)). \quad (9)$$

As already stated in the introduction, this method is shown to be entropic (by Jensen's inequality) and generally gives numerical solutions in good agreement with exact ones. However, the associated discontinuities are generally smeared at the discrete level and even if this point is natural, it is not satisfactory with respect to our objective in this chapter. Actually, there exists a very particular situation for which the Godunov method does not produce smeared shock profiles, namely when considering an isolated stationary discontinuity. Indeed, if we consider a Riemann problem (7) such that $\mathbf{v}^l$ and $\mathbf{v}^r$ can be joined by an admissible stationary discontinuity, Godunov method is by construction exact and then keeps sharp the discontinuity. This property is in fact the starting point in the design of our numerical strategy for obtaining sharp discrete shock (moving or not) profiles. We are going to make artificially stationary in a first step all the shocks arising in the Riemann solutions set at interfaces $x_{j+1/2}$, the dynamics being taken into account in a second step. We hope in this way to get back sharp numerical shocks provided that a relevant method is used in the second step.

Before describing our numerical strategy in details, let us first be more precise on the proposed two-step decomposition. For that, let be given $\mathbf{v}_j^n$ and $\mathbf{v}_{j+1}^n$ such that the corresponding Riemann solution $(x,t) \to \mathbf{v_r}(\frac{x-x_{j+1/2}}{t}; \mathbf{v}_j^n, \mathbf{v}_{j+1}^n)$ contains an admissible shock. We denote $\sigma_{j+1/2}$ its speed of propagation and $\mathbf{v}_{j+1/2}^{n-}$ and $\mathbf{v}_{j+1/2}^{n+}$ its left and right states (see Fig. 2). On the interval $[x_j, x_{j+1}]$, we first rewrite equivalently $\partial_t \mathbf{v} + \partial_x \mathbf{f}(\mathbf{v}) = 0$ as $\partial_t \mathbf{v} + \partial_x \mathbf{f}(\mathbf{v}) - \sigma_{j+1/2}\partial_x \mathbf{v} + \sigma_{j+1/2}\partial_x \mathbf{v} = 0$ and we then propose to solve it using a splitting strategy:

First step ($t^n \to t^{n+1-}$). This step consists in solving

$$\begin{cases} \partial_t \mathbf{v} + \partial_x \mathbf{f}(\mathbf{v}) - \sigma_{j+1/2}\partial_x \mathbf{v} = 0 \\ \mathbf{v}(x,0) = \begin{cases} \mathbf{v}_j^n & \text{if } x < 0, \\ \mathbf{v}_{j+1}^n & \text{if } x > 0, \end{cases} \end{cases} \quad (10)$$

still supplemented with the entropy inequality (2). It is clear at this stage that the Riemann solution $(x,t) \to \tilde{\mathbf{v}}_{\mathbf{r}}(\frac{x-x_{j+1/2}}{t}; \mathbf{v}_j^n, \mathbf{v}_{j+1}^n)$ of (10) is obtained by simply shifting $(x,t) \to \mathbf{v_r}(\frac{x-x_{j+1/2}}{t}; \mathbf{v}_j^n, \mathbf{v}_{j+1}^n)$ so as to get

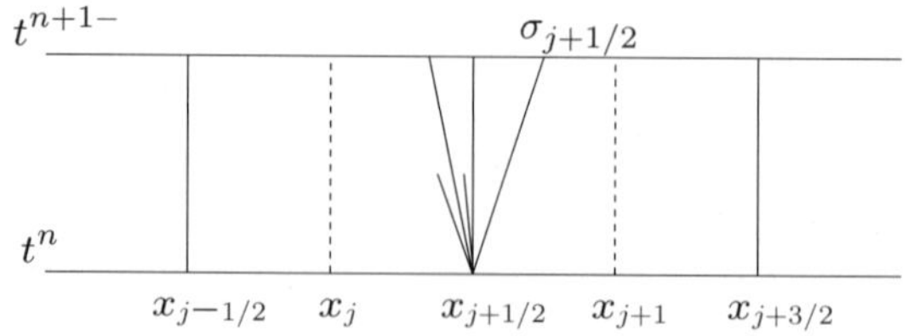

Fig. 2. An example of Riemann solution containing a shock wave

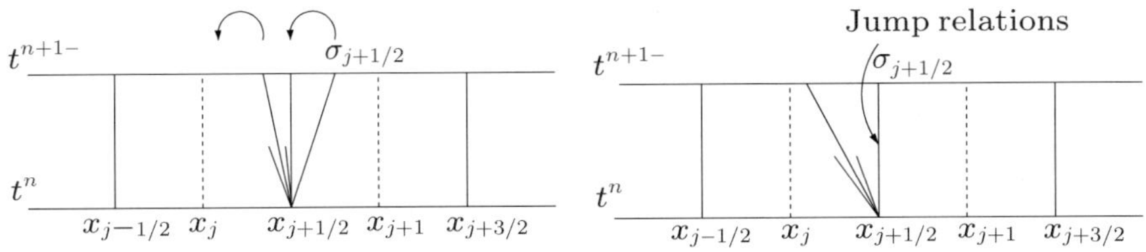

Fig. 3. Rotation on a Riemann solution containing a shock wave

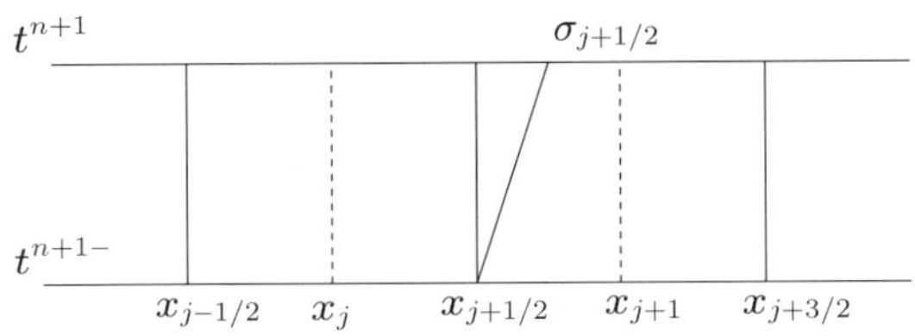

Fig. 4. Accounts for the dynamics of the shock wave in the splitting strategy

$$\tilde{\mathbf{v}}_{\mathbf{r}}(\frac{x - x_{j+1/2}}{t}; \mathbf{v}_j^n, \mathbf{v}_{j+1}^n) = \mathbf{v}_{\mathbf{r}}(\frac{x - x_{j+1/2} + \sigma_{j+1/2}t}{t}; \mathbf{v}_j^n, \mathbf{v}_{j+1}^n),$$

see Fig. 3. Observe that the discontinuity associated with the shock wave under consideration is now located along the $x = x_{j+1/2}$-axis. In other words, the shock has been made artificially stationary. When using the Godunov method in this first step, the generated discrete shock profiles are thus expected to be sharp.

Second step ($t^{n+1-} \to t^{n+1}$). This step takes into account the dynamics of the shock wave left stationary in the first step. It amounts to solve the following transport equation:

$$\partial_t \mathbf{v} + \sigma_{j+1/2}\partial_x \mathbf{v} = 0, \tag{11}$$

see Fig. 4. The solution obtained at the end of the first step will serve as a natural initial data for (11). To keep sharp the numerical profiles that are expected to be generated by the first step, we will make use of a random sampling strategy. Let us now describe the full strategy with details.

First step ($t^n \to t^{n+1-}$): the Godunov method

At each interface $x_{j+1/2}$, let us define $\sigma_{j+1/2}$ and $\mathbf{v}_{j+1/2}^{n\pm}$ as follows:

- If there is at least one shock wave in the Riemann solution $(x,t) \to \mathbf{v}_{\mathbf{r}}(\frac{x-x_{j+1/2}}{t}; \mathbf{v}_j^n, \mathbf{v}_{j+1}^n)$, then $\sigma_{j+1/2}$, $\mathbf{v}_{j+1/2}^{n-}$, and $\mathbf{v}_{j+1/2}^{n+}$ are, respectively, the speed of propagation and the corresponding left and right states of the shock with the larger amplitude for a given norm.
- Otherwise, we set $\sigma_{j+1/2} = 0$ and $\mathbf{v}_{j+1/2}^{n\pm} = \mathbf{v}_{\mathbf{r}}(0^{\pm}; \mathbf{v}_j^n, \mathbf{v}_{j+1}^n)$.

Then, under the CFL restriction

$$\frac{\Delta t}{\Delta x} \max_{\mathbf{v}}\{|\lambda_i(\mathbf{v})|,\ i = 1, \ldots, N\} \leq \frac{1}{4}, \tag{12}$$

for all the $\mathbf{v}$ under consideration, which is more restrictive than (5) due to the shift, we can define the function $(x,t) \to \tilde{\mathbf{v}}(x,t)$ for all $x \in \mathbb{R}$ and $t \in [0,\Delta t]$ as the juxtaposition of the Riemann solutions $(x,t) \to \tilde{\mathbf{v}}_{\mathbf{r}}(\frac{x-x_{j+1/2}}{t};\mathbf{v}_j^n,\mathbf{v}_{j+1}^n)$ defined on each interval $[x_j,x_{j+1}]$. By averaging on each cell $\mathcal{C}_j$ this solution at time Δt as in the usual Godunov method, we arrive at the following definition:

$$\mathbf{v}_j^{n+1-} = \frac{1}{\Delta x}\int_{x_{j-1/2}}^{x_{j+1/2}} \tilde{\mathbf{v}}(x,\Delta t)dx, \quad j \in \mathbb{Z}.$$

Invoking again the Green's formula, a straightforward calculation leads now to

$$\mathbf{v}_j^{n+1-} = \mathbf{v}_j^n - \lambda(\tilde{\mathbf{g}}_{j+1/2} - \tilde{\mathbf{g}}_{j-1/2}) - \lambda(\sigma_{j+1/2} - \sigma_{j-1/2})\mathbf{v}_j^n$$

with $\tilde{\mathbf{g}}_{j+1/2} = \mathbf{f}(\mathbf{v}_{j+1/2}^{n\pm}) - \sigma_{j+1/2}\mathbf{v}_{j+1/2}^{n\pm}$. Note that $\mathbf{f}(\mathbf{v}_{j+1/2}^{n-}) - \sigma_{j+1/2}\mathbf{v}_{j+1/2}^{n-} = \mathbf{f}(\mathbf{v}_{j+1/2}^{n+}) - \sigma_{j+1/2}\mathbf{v}_{j+1/2}^{n+}$ by the jump relations across the shock wave, while if $\sigma_{j+1/2} = 0$ these numerical fluxes coincide with those of the usual Godunov method.

Second step ($t^{n+1-} \to t^{n+1}$): a sampling strategy

In this step, we solve locally at each interface $x_{j+1/2}$ the transport equation (11) whose speed is $\sigma_{j+1/2}$. As an initial data, we consider the piecewise constant solution provided at time t^{n+1-} by the first step. To define the new approximation $\mathbf{v}_j^{n+1}$ at time $t^{n+1} = t^n + \Delta t$, we then propose to pick up randomly on the cell $\mathcal{C}_j$ a value at time Δt in the juxtaposition of the solutions of the transport equations. This choice is natural to avoid the appearance of new values in the shock profiles generated by the first step. More precisely, given a well-distributed random sequence (a_n) in $(0,1)$, it amounts to set:

$$\mathbf{v}_j^{n+1} = \begin{cases} \mathbf{v}_{j-1}^{n+1-} & \text{if } a_{n+1} \in [0, \lambda\sigma_{j-1/2}^+[, \\ \mathbf{v}_j^{n+1-} & \text{if } a_{n+1} \in [\lambda\sigma_{j-1/2}^+, 1 + \lambda\sigma_{j+1/2}^-[, \\ \mathbf{v}_{j+1}^{n+1-} & \text{if } a_{n+1} \in [1 + \lambda\sigma_{j+1/2}^-, 1[, \end{cases} \tag{13}$$

with $\sigma_{j+1/2}^+ = \max(\sigma_{j+1/2}, 0)$ and $\sigma_{j+1/2}^- = \min(\sigma_{j+1/2}, 0)$ (see also Fig. 5). In practice, we will use the celebrated van der Corput random sequence.

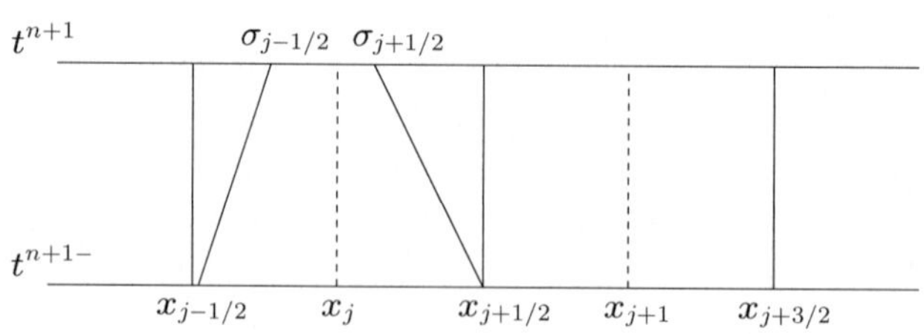

Fig. 5. Solutions arising in the second step

3 Numerical Experiments

In this section, we give some numerical evidences to illustrate the relevance of our strategy. To that purpose, we first consider without restriction the p-system in Lagrangian coordinates

$$\begin{cases} \partial_t \tau - \partial_x u = 0 \ \ (x,t) \in \mathbb{R} \times \mathbb{R}^{+*}, \\ \partial_t u + \partial_x p(\tau) = 0, \end{cases}$$

where $\tau > 0$ is the inverse of a density, u is the velocity, and $p > 0$ is the pressure of the fluid. We choose for instance $p(\tau) = \frac{1}{\tau^2}$ so that the system is strictly hyperbolic with characteristic speeds $\lambda_2 = -\lambda_1 = \sqrt{-p'(\tau)}$. We consider two Riemann problems (7) associated with $\mathbf{v}^l = (0.4, 0), \mathbf{v}^r = (1, 0)$ (test 1) and $\mathbf{v}^l = (0.8, 0), \mathbf{v}^r = (1, -1)$ (test 2) leading to solutions, respectively, made of a rarefaction wave followed by a 2-shock, and a 1-shock followed by a 2-shock. The covolumes τ are shown on Fig. 6. We observe that our strategy provides a very good approximation with in addition infinitely sharp discrete shock profiles. We follow by considering the system of gas dynamics in Lagrangian coordinates

$$\begin{cases} \partial_t \tau - \partial_x u = 0 \ \ (x,t) \in \mathbb{R} \times \mathbb{R}^{+*}, \\ \partial_t u + \partial_x p = 0, \\ \partial_t E + \partial_x pu = 0, \end{cases}$$

where $E = \frac{1}{2}u^2 + \frac{p\tau}{\gamma - 1}$, $\gamma = 2$ is the total energy. This system is strictly hyperbolic with characteristic speeds $\lambda_2 = 0$ and $\lambda_3 = -\lambda_1 = \sqrt{\gamma \frac{p}{\tau}}$. We consider two Riemann problems (7) associated with $\mathbf{v}^l = (0.5, 2, 2.625), \mathbf{v}^r = (2, 1, 1)$ (test 3) and $\mathbf{v}^l = (0.2, 1, 0.7), \mathbf{v}^r = (0.176, 0.875, 0.67)$ (test 4) leading to solutions, respectively, made of a contact discontinuity followed by a 3-shock, and a 1-shock followed by a contact discontinuity and a rarefaction wave. Again, we observe on Fig. 7 plotting the covolumes that our method generates infinitely sharp discrete shock profiles. With these two test cases, we also show that making stationary a shock wave in a system that already contains a stationary discontinuity (here a contact discontinuity) does not rise difficulties. To

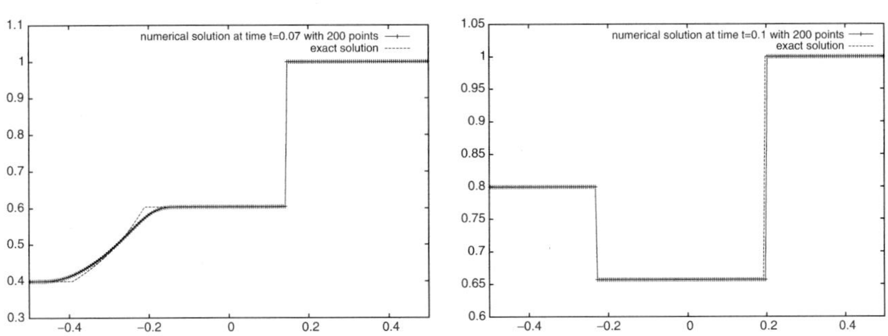

Fig. 6. τ: test 1 (*left*) and test 2 (*right*)

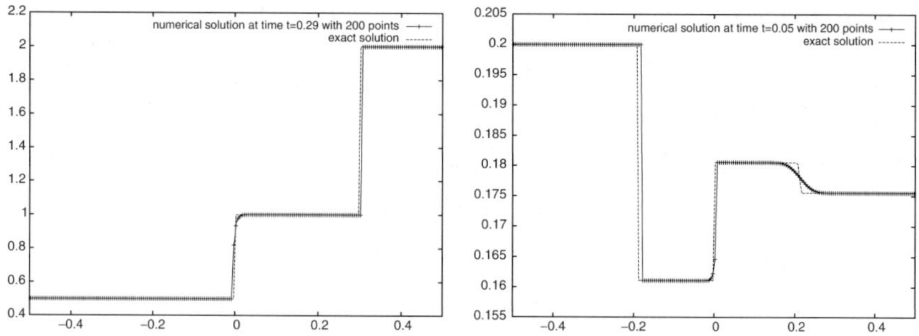

Fig. 7. τ: test 3 (*left*) and test 4 (*right*)

further validate the method, it is important to note that, when the Riemann initial data consist of two states that can be joined by an admissible shock, the proposed method simply reduces to the Glimm scheme and then converges to the expected solutions with an infinitely sharp shock profile.

As a conclusion, let us mention that it is actually possible to get rid of the exact Riemann solver used in our method while still keeping on generating infinitely sharp discrete shock profiles. Our approach is based on approximate Riemann solvers and will be described in a longer chapter. We will eventually apply the strategy for the computation of nonclassical solutions associated with non-GNL nor LD fields, which is the very motivation of this study.

References

1. Bressan, A.: The semigroup approach to systems of conservation laws. Mat. Contemp., **10**, 21–74 (1996)
2. Godlewsky, E., Raviart, P.-A., Numerical approximation of hyperbolic systems of conservation laws. Springer (1995)
3. LeFloch, P.-G, Hyperbolic systems of conservation laws: the theory of classical and nonclassical shock waves. Birkhäuser (2002)

Propagation of Diffusing Pollutant by a Hybrid Eulerian–Lagrangian Method

A. Chertock, E. Kashdan, and A. Kurganov

We present a hybrid numerical method for computing the propagation of a diffusing passive pollutant in shallow water. The flow is modeled by the Saint-Venant system of shallow water equations and the pollutant propagation is described by a convection–diffusion equation.

In this chapter, we extend the hybrid finite-volume-particle (FVP) method, which was originally introduced in [CK04, CKP06] for the model of inviscid pollutant propagation, to the case of a diffusing pollutant. The idea behind the FVP method is to use different schemes for the flow and pollution computations: the shallow water equations are numerically integrated using a finite-volume scheme, while the transport equation for the propagation of passive pollutant is solved by a deterministic particle method. When the pollutant diffuses, the second step of the FVP method has to be modified. We propose a new hybrid Eulerian–Lagrangian method, in which the convection term is treated by the method of characteristics (which, in the context of scalar transport equations, is very similar to the deterministic particle method), while the diffusion is resolved using the fast explicit operator splitting method recently developed in [CKP].

1 Introduction

Prediction of a pollution propagation is an important problem in many industrial and environmental projects. Different mathematical models are used to describe this phenomenon and to obtain an accurate location and concentration of pollutant.

In this chapter, we consider the transport of a passive pollutant by a flow modeled by the Saint-Venant system of shallow water equations [Sai1871]. In the one-dimensional (1-D) case, the system reads:

$$\begin{cases} h_t + (hu)_x = S, \\ (hu)_t + \left(hu^2 + \dfrac{gh^2}{2}\right)_x = -ghB_x. \end{cases} \tag{1}$$

Here, h and u are, respectively, the depth and the velocity of the water, g is the gravity constant, and S is a source term. The function B represents the bottom topography.

The propagation of the pollutant is modeled by the convection–diffusion equation

$$(hT)_t + (uhT)_x = T_S S + \nu h T_{xx}, \tag{2}$$

where T is the pollutant concentration, the coefficient T_S represents a concentration of the pollutant at the source, and ν is the viscosity coefficient.

For simplicity, we will assume that the pollution source has already been turned off, i.e., we will only consider the $S \equiv 0$ case (numerical treatment of the source term was discussed in detail in [CK04, CKP06]). Under this assumption, (2) and (1) are coupled only through the velocity u. This suggests the following strategy for developing numerical methods for the system (1) and (2): first, solve the Saint-Venant system (1), and then substitute the obtained velocity field u into (2), which can be whereupon viewed as a linear convection–diffusion equation with possibly discontinuous coefficients. Since (2) and (1) can be solved separately, they can be solved by two different methods: one method should be designed for the hyperbolic system of balance laws (1), while the another method should be capable to accurately solve the convection–diffusion equation (2). This simple hybrid strategy was realized in [CK04, CKP06], where the FVP method was introduced: the Saint-Venant system (1) was solved by a shock-capturing finite-volume method (the central-upwind scheme from [KL02, KNP01, KT00]), while the inviscid transport equation, (2) with $\nu = 0$, was accurately solved by the deterministic particle method (see, e.g., [Rav85] for a comprehensive description of particle methods for transport equations).

The main advantage of the FVP hybrid strategy is its flexibility. There is a wide variety of reliable finite-volume methods for the Saint-Venant system (see, e.g., [ABBKP04, AB03, GHS03, JW05, KL02, KP, NPPN06, XS06] for just a few examples of recently proposed methods). One may select one's favorite method for the first part of the hybrid algorithm. When the system (1) is solved for h and hu, a global approximation of u at each time level can be computed by dividing a piecewise polynomial approximation of hu by a piecewise polynomial approximation of h. This gives one a velocity coefficient in (2), which can be thus treated as a linear equation. Since particle methods are specifically designed as a nondiffusive numerical methods for transport equations, they allow to very accurately resolve contact discontinuities that typically appear in the pollutant concentration field, as it was demonstrated in [CK04, CKP06].

In this work, we extend the FVP method to the case of diffusing pollutant ($\nu \neq 0$). The first part of our hybrid algorithm (an Eulerian finite-volume method) is not affected by the presence of diffusion in (2), therefore, only its second part is to be modified. There are several ways to implement particle methods for convection–diffusion equations (see, e.g., [CHO73, CK00, DM91, DM90, LM99, RUS90]). These methods can be, in principle, applied to (2), but will require either smearing the particle approximation, which may introduce an extensive amount of numerical diffusion (especially in the convection-dominated case), or computing the numerical derivatives on a nonuniform mesh, formed by the particle locations, which may be inaccurate (especially in the two-dimensional (2-D) case).

To overcome these difficulties, we propose the following Lagrangian strategy for solving (2). First, we rewrite (2) in an equivalent nonconservative form:

$$T_t + uT_x = \nu T_{xx}. \tag{3}$$

We then use the Strang operator splitting [Str68] and solve the convection equation

$$T_t + uT_x = 0, \tag{4}$$

and the diffusion equation

$$T_t = \nu T_{xx}, \tag{5}$$

separately. As it has been shown in [CK04], the particle method applied to (2) with $\nu = 0$ is basically equivalent to the method of characteristics applied to (4). The linear heat equation (5) is very simple and can be solved exactly by convolving the initial condition with the heat kernel, as it has been done in [CKP]. Since we need to obtain point values of T after each parabolic step of the operator splitting method, a certain quadrature rule should be applied to the integral form of the solution, and this may be rather computationally expensive, especially in the 2-D case. However, in the convection-dominated case (i.e., when $\nu << 1$), splitting increments $\Delta t_{\rm spl}$ can be made very large (since the splitting error is proportional to $\nu^3(\Delta t_{\rm spl})^2$) and we can obtain a very accurate approximate solution by applying only a small number of operator splitting steps.

2 Description of the Method

In this section, we provide a description of our 1-D hybrid Eulerian–Lagrangian method for the system (1) and (3). Similarly, to the FVP method, the new hybrid method consists of two parts. We first solve the Saint-Venant system (1) by the semidiscrete second-order well-balanced positivity preserving central-upwind scheme, recently proposed in [KP], with time evolution carried out by the third-order strong stability preserving (SSP) Runge–Kutta solver [GST01]. We would like to emphasize that this part of our hybrid Eulerian–Lagrangian algorithm can be replaced with one's favorite finite-volume method.

We now focus on the second part of out hybrid algorithm: the Lagrangian version of the fast explicit operator splitting method for the convection–diffusion equation (3). In the following, we will assume that an approximate velocity $u(x,\cdot)$ is available at every time level. We split the convection–diffusion equation (3) into the convection equation (4) and the diffusion one (5). We denote the solution operators, associated with these equations by $S_{\mathcal{C}}$ and $S_{\mathcal{D}}$, respectively. Then, according to [Str68], the solution of the convection–diffusion equation (3) satisfies

$$T(t+\Delta t_{\rm spl}) = S_{\mathcal{C}}(\Delta t_{\rm spl}/2)S_{\mathcal{D}}(\Delta t_{\rm spl})S_{\mathcal{C}}(\Delta t_{\rm spl}/2) + E(\Delta t_{\rm spl}), \tag{6}$$

where $\Delta t_{\rm spl}$ is a splitting time increment, and the one time-step splitting error, $E(\Delta t_{\rm spl})$, is proportional to $\nu^3(\Delta t_{\rm spl})^3$ (an accumulative splitting error is then proportional to $\nu^3(\Delta t_{\rm spl})^2$). Note that a success of the fast explicit operator splitting method hinges on smallness of the viscosity coefficient ν, since then the splitting error is small even for sufficiently large values of $\Delta t_{\rm spl}$ and one can obtain a very accurate solution with only few splitting steps (see [CKP] for details).

In practice, the exact solution operators $S_{\mathcal{C}}$ and $S_{\mathcal{D}}$ are to be replaced with approximate ones. We use the (Lagrangian) method of characteristics to solve (4). To this end, we assume that a set of characteristic points, $\{x_i(t)\}$, and the corresponding pollution concentration values, $\{T(x_i(t),t)\}$, are available at a certain time level t. We then perform the first convection substep of (6) by evolving the characteristic points according to the following system of ODEs

$$\frac{dx_i(t)}{dt} = u(x_i(t),t),$$

which is numerically solved by the same SSP Runge–Kutta method used for time evolution of the shallow water solution. At the end of this substep, the solution will be realized as a set of point values $\{T_i^* := T(x_i^*, t+\Delta t_{\rm spl}/2) \equiv T(x_i(t),t)\}$, where $x_i^* := x_i(t+\Delta t_{\rm spl}/2)$. The diffusion substep is then carried out by evolving this intermediate solution exactly according to

$$S_{\mathcal{D}}(\Delta t_{\rm spl})T(x,\cdot) = T(x,\cdot) + \int\limits_{-\infty}^{\infty} G(x-\xi,\nu\Delta t_{\rm spl})(T(\xi,\cdot)-T(x,\cdot))\,d\xi, \tag{7}$$

where $G(x,t) := (4\pi t)^{-1/2}\exp(-x^2/4t)$ is the heat kernel. Since after the first convection substep only a discrete set of values $\{T_i^*\}$ is available, we apply the trapezoidal rule to (7) and end up with the new set of point values of T at the same locations $\{x_i^*\}$

$$T_i^{**} = T_i^* + \sum_j G(x_i^* - x_j^*, \nu\Delta t_{\rm spl})(T_j^* - T_i^*). \tag{8}$$

To complete one splitting step, we perform another convection substep and obtain the new set of characteristic points, $\{x_i(t+\Delta t_{\rm spl})\}$, with the same set of solution values there: $\{T(x_i(t+\Delta t_{\rm spl}), t+\Delta t_{\rm spl}) \equiv T_i^{**}\}$.

Remark. The 2-D extension of our hybrid Eulerian–Lagrangian method is rather straightforward. We would only like to comment here on the 2-D extension of the trapezoidal quadrature (8), which has been performed by arranging a nonuniform set of characteristic points $\{(x_i^*, y_i^*)\}$ into the Delaunay triangulation by implementing the algorithm based on the open source mesh generation package GeomPack, available at `www.geompack.org` (see also `www.csit.fsu.edu/~burkardt/f_src/geompack/geompack.html`). This determines the continuous piecewise linear approximation of the integrand in the 2-D analog of (7), which, in turn, can be integrated exactly: the integral over each triangle is equal to the area of the triangle multiplied by the average of the integrand values at the triangle vertices.

3 Numerical Examples

In this section, we illustrate the performance of the proposed hybrid Eulerian–Lagrangian method on two numerical examples. We compare the obtained results with those obtained by a "purely" Eulerian finite-volume method (we have used the second-order central-upwind scheme with the MinMod1.5 limiter [KL02, KNP01, KP]) applied to the Saint-Venant system (1) and the convection–diffusion equation (2) separately, as it has been suggested in [CKP06] as the way to improve the resolution of contact waves in T captured by the central-upwind scheme. In all our numerical experiments, the gravitation constant was taken $g = 1$ and viscosity coefficient $\nu = 10^{-5}$.

3.1 One-Dimensional Example

We start with the 1-D example taken from [CKP06]. In this example, we assume that the initial water level and the initial discharge are constant: $h(x,0) + B(x) \equiv 1$, $h(x,0)u(x,0) = 0.1$, the bottom topography is given by

$$B(x) = \begin{cases} 0.25(\cos(10\pi(x-0.5)) + 1, & \text{if } 0.4 \le x \le 0.6, \\ 0, & \text{otherwise,} \end{cases}$$

and the initially polluted area is $[0.4, 0.5]$:

$$T(x,0) = \begin{cases} 1, & \text{if } 0.4 \le x \le 0.5, \\ 0, & \text{otherwise.} \end{cases}$$

The pollution spot propagates to the right, and we numerically track its evolution. The pollutant concentration at times $t = 0$, 2, and 4, computed by the hybrid Eulerian–Lagrangian and the finite-volume methods, is shown in Fig. 1 (left), where both Δx for the central-upwind scheme and the distance between the initially uniformly distributed characteristic points is taken 1/200. One can clearly see the superiority of the results obtained by our hybrid method. In Fig. 1 (right), we refine the mesh employed by the finite-volume method.

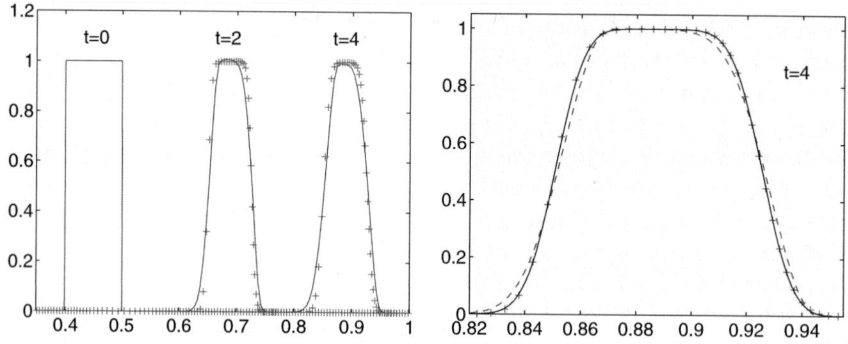

Fig. 1. Propagation of the 1-D pollution spot by the hybrid Eulerian–Lagrangian method ("pluses") and the second-order finite-volume method (*solid line*) at different times (*left figure*). In the *right figure*: comparison of the hybrid solution with the finite-volume ones computed on finer grids with $\Delta x = 1/400$ (*dashed line*) and $\Delta x = 1/1{,}600$ (*solid line*)

The solutions are shown at time $t = 4$. One can observe that the resolution achieved by the hybrid Eulerian–Lagrangian method with $\Delta x = 1/200$ is comparable with the one by the finite-volume method with $\Delta x = 1/1{,}600$. The difference becomes even more prominent in the 2-D case, considered in the next numerical example.

3.2 Two-Dimensional Example

Here, the initial pollutant spot is transferred with the flow over the exponentially shaped bump $B(x, y) = 0.25 \exp(-10x^2 - 5y^2)$. A 2-D version of the system (1) and (3) is solved subject to the initial conditions

$$h(x, y, 0) + B(x, y) = 1, \;\; h(x, y, 0)u(x, y, 0) = 0.2, \;\; h(x, y, 0)v(x, y, 0) = 0.05,$$

where h is, as before, the water depth, and u and v are the x- and y-components of the velocity vector. The initial concentration of the pollutant is

$$T(x, y, 0) = \begin{cases} 1, & -0.75 \le x \le -0.25, \; -0.25 \le y \le 0.25, \\ 0, & \text{otherwise.} \end{cases}$$

In Fig. 2, we show the pollutant concentration T at time $t = 4$, computed by our hybrid Eulerian–Lagrangian method (left) and the second-order central-upwind scheme (right) with $\Delta x = \Delta y = 1/50$ for both the central-upwind scheme and the initial uniform distribution of the characteristic points. One can clearly observe a much better resolution achieved by the hybrid method. As in the 1-D example, we refine the computational grid and apply the second-order central-upwind scheme with $\Delta x = \Delta y = 1/100$ (Fig. 3, left) and $\Delta x = \Delta y = 1/200$ (Fig. 3, right). As one can see there, our hybrid method still outperforms the finite-volume one.

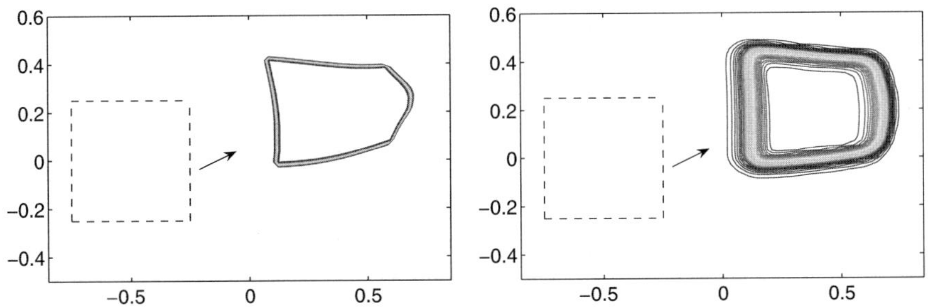

Fig. 2. Propagation of the 2-D pollution spot by the hybrid Eulerian–Lagrangian method (*left*) and the second-order finite-volume method (*right*). In both computations $\Delta x = \Delta y = 1/50$. The *dashed line* represents the boundary of the initially polluted domain

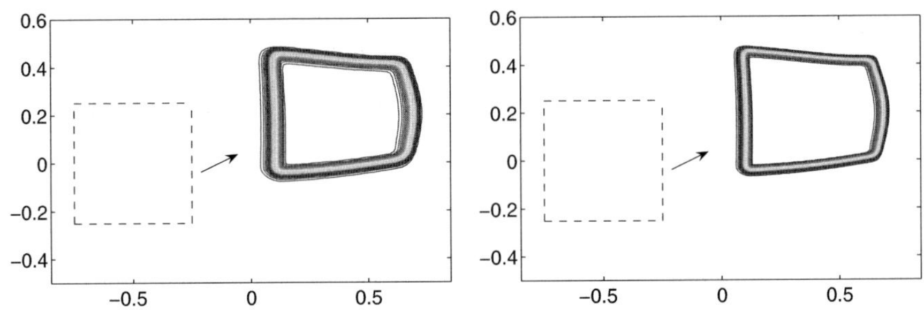

Fig. 3. Propagation of the 2-D pollution spot by the second-order finite-volume method with $\Delta x = \Delta y = 1/100$ (*left*) and $\Delta x = \Delta y = 1/200$ (*right*). The *dashed line* represents the boundary of the initially polluted domain

Acknowledgments

The work of A. Chertock was supported in part by the NSF Grant No. DMS-0410023. The research of A. Kurganov was supported in part by the NSF Grant Nos. DMS-0310585 and DMS-0610430.

References

[ABBKP04] Audusse, E., Bouchut, F., Bristeau, M.-O., Klein, R., Perthame, B.: A fast and stable well-balanced scheme with hydrostatic reconstruction for shallow water flows. SIAM J. Sci. Comput., **25**, 2050–2065 (2004).

[AB03] Audusse, E., Bristeau, M.-O.: Transport of pollutant in shallow water. A two time steps kinetic method. M2AN Math. Model. Numer. Anal., **37**, 389–416 (2003)

[CK04] Chertock, A., Kurganov, A.: On a hybrid finite-volume-particle method. M2AN Math. Model. Numer. Anal., **38**, 1071–1091 (2004)

[CKP06] Chertock, A., Kurganov, A., Petrova, G.: Finite-volume-particle methods for models of transport pollutant in shallow water. J. Sci. Comp, **27**, 323–333 (2006)

[CKP] Chertock, A., Kurganov, A., Petrova, G.: Fast explicit operator splitting method for convection-diffusion equation. To appear in Internat. J. Numer. Methods Fluids

[CHO73] Chorin, A.J.: Numerical study of slightly viscous flow. J. Fluid Mech., **57**, 785–796 (1973)

[CK00] Cottet, G.-H., Koumoutsakos, P.D.: Vortex Methods, Cambridge University Press, Cambridge (2000)

[DM91] Degond, P., Mas-Gallic, S.: The weighted particle method for convection-diffusion equations. Part 1 and Part 2. Math. Comp., **53**, 485–525 (1991)

[DM90] Degond, P., Mustieles, F.J.: A deterministic approximation of diffusion equations using particles. SIAM J. Sci. Stat. Comp., **11**, 293–310 (1990)

[GHS03] Gallouët, T., Hèrard, J.-M., Seguin, N.: Some approximate Godunov schemes to compute shallow-water equations with topography. Comput. Fluids, **32**, 479–513 (2003)

[GST01] Gottlieb, S., Shu, C.-W., Tadmor, E.: High order time discretization methods with the strong stability property. SIAM Rev., **43**, 89–112 (2001)

[JW05] Jin, S., Wen, X.: Two interface-type numerical methods for computing hyperbolic systems with geometrical source terms having concentrations. SIAM J. Sci. Comput., **26**, 2079–2101 (2005)

[KL02] Kurganov, A., Levy, D.: Central-upwind schemes for the Saint-Venant system. M2AN Math. Model. Numer. Anal., **36**, pp. 397–425 (2002)

[KNP01] Kurganov, A., Noelle, S., Petrova, G.: Semi-discrete central-upwind scheme for hyperbolic conservation laws and Hamilton-Jacobi equations. SIAM J. Sci. Comput., **23**, 707–740 (2001)

[KP] Kurganov, A., Petrova, G.: A second-order well-balanced positivity preserving central-upwind scheme for the Saint-Venant system. submitted to Commun. Math. Sci.

[KT00] Kurganov, A., Tadmor, E.: New high-resolution central schemes for nonlinear conservation laws and convection-diffusion equations. J. Comput. Phys., **160**, 241–282 (2000)

[LM99] Lacombe, G., Mas-Gallic, S.: Presentation and analysis of a diffusion-velocity method. ESAIM: Proceedings, **7**, 225–233 (1999)

[NPPN06] Noelle, S., Pankratz, N., Puppo, G., Natvig, J.: Well-balanced finite volume schemes of arbitrary order of accuracy for shallow water flows. J. Comput. Phys., **213**, 474–499 (2006)

[Rav85] Raviart, P.A.: An analysis of particle methods. In: Numerical methods in fluid dynamics (Como, 1983). Lecture Notes in Math., 1127, Springer, Berlin (1985)

[RUS90] Russo, G.: Deterministic diffusion of particles. Comm. Pure Appl. Math., **43**, 697–733 (1990)

[Sai1871] de Saint-Venant, A.J.C.: Thèorie du mouvement non-permanent des eaux, avec application aux crues des rivière at à l'introduction des marèes dans leur lit. C.R. Acad. Sci. Paris, **73**, 147–154 (1871)

[Str68] Strang, G.: On the construction and comparison of difference schemes. SIAM J. Numer. Anal., **5**, 506–517 (1968)

[XS06] Xing, Y., Shu, C.-W.: A new approach of high order well-balanced finite volume WENO schemes and discontinuous Galerkin methods for a class of hyperbolic systems with source terms, Commun. Comput. Phys., **1**, 100–134 (2006)

Nonlocal Conservation Laws with Memory

C. Christoforou

Systems of conservation laws in one-space dimension are equations of the form

$$\partial_t U(t,x) + \partial_x S(t,x) = 0, \tag{1}$$

where $x \in \mathbb{R}$ and U, $S \in \mathbb{R}^n$. Such equations are called *conservation laws* because in classical physics, conservation of mass, momentum, and energy of media is governed by equations of this type.

In this context, $U(t,x)$ determines the state of the medium at the point (t,x), and $S(t,x)$ is the flux of U. In general, materials are identified by constitutive relations which describe how the flux is determined by the state vector U. The medium is called *elastic* if $S(t,x)$ is determined by the value of state U at (t,x), i.e., (1) reduces to

$$\partial_t U(t,x) + \partial_x F(U(t,x)) = 0, \tag{2}$$

where $F : \mathbb{R}^n \to \mathbb{R}^n$ is a given smooth function. The Cauchy problem to (2) has been studied in great detail. When F is nonlinear, the profile of solutions starting out from smooth initial data gets progressively steeper and eventually breaks in finite time, by developing jump discontinuities which propagate on as shock waves. This raises the need to deal with weak solutions. Expositions of the current state of the theory may be found in the books [B, D, Se, Sm].

The model (2) fails in the presence of viscosity and relaxation phenomena. In such circumstances, the flux $S(x,t)$ depends also on the past history $U(x,\tau)$ for $\tau < t$ and we say that the material has memory. An important class of media of this type are materials with fading memory for which the constitutive relations yield

$$U_t + F(U(x,t))_x + \int_0^t k(t-\tau) G(U(x,\tau))_x \, d\tau = 0 \tag{3}$$

where F, $G : \mathbb{R}^n \to \mathbb{R}^n$ are smooth functions and $k : [0,\infty) \to \mathbb{M}_{n\times n}$ is a smooth kernel integrable over $[0,\infty)$. Extensive studies have been conducted to contrast the behavior of elastic media with the behavior of media with fading memory. It is known that, when the kernel satisfies appropriate conditions motivated by physical considerations, the influence of the memory term in (3) reflects a damping effect. Consequently, global classical solutions exist for "small" and sufficiently smooth initial data, in contrast to elastic media that classical solutions in general break down in finite time even when the initial data are small (cf. [DN, M1, M2, S, Sl.]). However, if the initial data are "large," then singularities develop in finite time (see [D2, MN]). For a general exposition on this subject, see [RHN]. Global weak solutions have been constructed for special equations of the form (3) mainly in L^∞ by the method of compensated compactness (see [D1, CD, NRT]).

Here, we study the existence of weak solutions of bounded variation (BV) to nonlocal conservation laws with fading memory (3). To be more precise, in Sect. 1 we first analyze the scalar conservation law thoroughly and present the results in [CC] obtained jointly with G.-Q. Chen. In particular, we show that there is a unique global weak solution and also establish the limit to the local conservation law when there is a relaxation kernel. Next, Sect. 2 outlines my results of [C2] that are the extension of the project [CC] to systems.

1 A Nonlocal Conservation Law with Fading Memory

Consider the scalar equation

$$\begin{cases} u_t + (f(u))_x + \int_0^t k(t-\tau)(f(u(\tau)))_x \, d\tau = 0 \\ u(0,x) = u_0(x) \end{cases}, \tag{4}$$

with $u_0 \in BV$. The main motivation is that (4) can be viewed as a linear Volterra equation under assumption on k. This was first observed by MacCamy [M1, M2] and later employed in Dafermos [D, D2] and Nohel–Rogers–Tzavaras [NRT] (also see [CD]). In this way, it is easy to extract the damping character of the memory term. Let $r(t)$ be the resolvent kernel associated with k:

$$r + k * r = -k. \tag{5}$$

Then we can write (4) as

$$-f(u)_x = u_t + \int_0^t r(t-\tau)\, u'(\tau)\, d\tau.$$

Integrating by parts yields

$$u_t + f(u)_x + r(0)u = r(t)u_0 - \int_0^t r'(t-\tau)u(\tau)\, d\tau, \tag{6}$$

which is equivalent to (4). Then under appropriate conditions on $r(t)$, we establish a global weak solution by the vanishing viscosity method.

Theorem 1. *Consider the Cauchy problem* (4)*. Let* r *be the resolvent kernel associated with* k*, i.e.,* $r + r * k = -k$*. Assume* r *is nonnegative, nonincreasing in* $L^1(\mathbb{R}_+)$*, then for each* $\varepsilon > 0$*, the Cauchy problem*

$$u_t^\varepsilon + f(u^\varepsilon)_x + \int_0^t k(t-\tau) f(u^\varepsilon(\tau))_x \, d\tau = \varepsilon \left(u^\varepsilon + \int_0^t k(t-\tau) u^\varepsilon(\tau) \, d\tau \right)_{xx}$$

has a unique solution u^ε *defined globally with a uniform* BV *bound and as* $\varepsilon \to 0$*,* u^ε *converges in* L^1_{loc} *to an entropy solution* $u \in BV$ *to* (4) *which satisfies*

$$TV\{u(t)\} + \int_0^t r(t-\tau) TV\{u(\tau)\} \, d\tau \le L\, M(u_0) \tag{7}$$

and

$$\|u(t) - u(s)\|_{L^1} \le C\, M(u_0)\, |t-s|, \qquad \|u\|_{L^\infty(\mathbb{R}^2_+)} \le \|u_0\|_{L^\infty(\mathbb{R})},$$

where $L = 1 + \|r\|_{L^1(\mathbb{R}_+)}$*,* $M(u_0) = TV\{u_0\} + 2\|u_0\|_{L^\infty(\mathbb{R})}$*, and* C *is a constant independent of* ε *and* $TV\{u_0\}$*.*

The vanishing viscosity approximation u^ε is equivalent to the following artificial viscosity approximation to (6):

$$u_t^\varepsilon + f(u^\varepsilon)_x + r(0) u^\varepsilon = r(t) u_0 - \int_0^t r'(t-\tau) u^\varepsilon(\tau) \, d\tau + \varepsilon\, u_{xx}^\varepsilon. \tag{8}$$

To prove the above theorem, the techniques are similar to those by Vol'pert [V] and Kruzkov [K]. The following result establishes the uniqueness and stability in L^1 of solutions to (4).

Theorem 2. *Let the resolvent kernel* $r(t)$ *associated with* k *be a nonnegative and nonincreasing function in* $L^1(\mathbb{R}_+)$*. Let* $u, v \in BV(\mathbb{R}^2_+)$ *be entropy solutions to* (4) *with initial data* $u_0, v_0 \in BV(\mathbb{R})$*, respectively. Then*

$$\|u(t) - v(t)\|_{L^1(\mathbb{R})} + \int_0^t r(t-\tau) \|u(\tau) - v(\tau)\|_{L^1(\mathbb{R})} \, d\tau \le L\, \|u_0 - v_0\|_{L^1(\mathbb{R})}. \tag{9}$$

That is, any entropy solution in BV *to* (4) *is unique and stable in* L^1*. As a consequence, if* u_0 *is only in* L^∞*, not necessarily in* $BV(\mathbb{R})$*, there exists a global entropy solution* $u \in L^\infty$ *to* (4)*.*

Having established the above results, we then analyze the case when the kernel in the scalar equation (4) is a relaxation kernel k_ν that depends on a small parameter $\nu > 0$ so that $k_\nu(t) \rightharpoonup (\alpha - 1)\,\delta(t)$ weakly as measures when $\nu \to 0+$, where $\delta(t)$ denotes the Dirac mass centered at the origin.

Theorem 3. *Consider the Cauchy problem*

$$\begin{cases} u^\nu_t + f(u^\nu)_x + \displaystyle\int_0^t k_\nu(t-\tau)(f(u^\nu(\tau)))_x \, d\tau = 0 \\ u^\nu(0,x) = u_0(x) \end{cases} \tag{10}$$

with $u_0 \in BV$. *Let the resolvent kernel* $r_\nu(t)$ *associated with* k_ν *be a nonnegative and nonincreasing function uniformly bounded in* $L^1(\mathbb{R}_+)$. *Then the entropy solutions* u^ν *to* (10) *are uniformly bounded in* L^∞ *and stable in* L^1. *Furthermore, if* $k_\nu(t) \rightharpoonup (\alpha-1)\,\delta(t)$ *weakly as measures when* $\nu \to 0+$, *then* u^ν *converges in* L^1_{loc} *to an admissible weak solution* u *of the local conservation law*

$$\begin{cases} u_t + \alpha(f(u))_x = 0 \\ u(0,x) = u_0(x). \end{cases} \tag{11}$$

There is a rich family of kernels k_ν that satisfy the assumptions of the theorems. For a set of kernels that satisfy these assumptions, see [CC]. The prototypical example is $k_\nu(t) = \frac{(\alpha-1)}{\nu} e^{-\frac{t}{\nu}}$, for $0 < \alpha < 1$, in which case (10) is equivalent to a 2×2 system with relaxation

$$\begin{cases} u_t + (f(u) + v)_x = 0 \\ v_t = \dfrac{(1-\alpha)f(u) - v}{\nu} \\ (u(0,x), v(0,x)) = (u_0, v_0) \end{cases}, \tag{12}$$

which has been studied extensively. Thus, our result that $u^\nu \to u$ as $\nu \downarrow 0$ reduces to this special case to the zero relaxation limit considered systematically in Chen–Levermore–Liu [CLL]. The restriction $0 < \alpha < 1$ is the corresponding subcharacteristic condition. See also [LN, STW, Yo] for the model.

2 Systems of Conservation Laws with Fading Memory

In this section, we outline the results of [C2] for the general case of systems. As already mentioned, conservation laws with fading memory (3) can be viewed as a linear Volterra equation under suitable choice of F and G. Following Dafermos suggestion in [D3], we consider hyperbolic conservation laws with fading memory in one-space dimension (3) that are equivalent to:

$$\begin{cases} U_t + A(U)U_x + g(U) = H(t)\,\bar{U} - \displaystyle\int_0^t K(t-\tau)U(\tau)\,d\tau \\ U(0,x) = \bar{U}(x) \end{cases}, \tag{13}$$

where $x \in \mathbb{R}$, $U(t,x) \in \mathbb{R}^n$, $A(U) \in M_{n\times n}$, $g : \mathbb{R}^n \to \mathbb{R}^n$, and H, $K : [0,+\infty) \to M_{n\times n}$. We assume that the system is strictly hyperbolic, i.e., $A(U)$ has n real distinct eigenvalues

$$\lambda_1(U) < \lambda_2(U) < \ldots < \lambda_n(U), \tag{14}$$

and therefore n linearly independent eigenvectors $r_i(U)$, $i = 1, \dots, n$. Note the similarity of the above equation with the one in the scalar case (6).

The goal is to establish entropy weak solutions of *bounded variation* (BV) to system (13), hence, to (3). To achieve this, we employ the vanishing viscosity method, namely we consider solutions U^ε to the viscous conservation laws with fading memory

$$\begin{cases} U_t^\varepsilon + A(U^\varepsilon)U_x^\varepsilon + g(U^\varepsilon) = H(t)\,\bar{U} - \displaystyle\int_0^t K(t-\tau)U^\varepsilon(\tau)\,d\tau + \varepsilon U_{xx}^\varepsilon \\ U^\varepsilon(0,x) = \bar{U}(x), \end{cases} \tag{15}$$

where $\varepsilon > 0$ is the viscosity parameter and obtain globally defined viscous approximations U^ε that are uniformly stable and converge in L^1_{loc} to the entropy weak solution U to (13). The vanishing viscosity method has been studied extensively. The scalar conservation law ($g \equiv 0$, $H \equiv 0$, $K \equiv 0$) was treated by Oleinik [O] in one-space dimension and by Kruzkov [K] in several space dimensions. The case of systems of conservation laws, which had been open for a long time, has been recently solved by Bianchini and Bressan in the fundamental paper [BiB] and later extended by Christoforou [C, C1] to systems of balance laws ($g \neq 0$ and $H \equiv 0$, $K \equiv 0$).

Here, to establish the convergence of the vanishing viscosity approximations, we need to impose "dissipative" conditions on the hyperbolic system (13). Namely, let U^* be a constant equilibrium to (13), then the assumptions are the following.

Assumptions $(\star)$

(1) $\tilde{B}(U^*) \doteq [R(U^*)]^{-1}Dg(U^*)R(U^*)$ is *strictly column diagonally dominant*, i.e., there exists a positive constant $\beta > 0$ such that

$$\tilde{B}_{ii}(u^*) - \sum_{j\neq i} |\tilde{B}_{ji}(U^*)| \geq \beta > 0 \qquad\qquad i = 1, \dots, n.$$

(2) $\tilde{K}(s) \doteq R(U^*)^{-1}K(s)R(U^*) \in L^1[0,+\infty)$ is *absolutely dominated by* $\tilde{B}$, i.e., there exists a positive constant $\kappa \geq 0$, such that for each $i = 1, \dots, n$

$$\sum_{j=1}^n \int_0^{+\infty} |\tilde{K}_{ji}(s)|\,ds < \kappa, \qquad \text{and} \qquad 0 \leq \kappa < \beta.$$

(3) $H(\cdot) \in L^1[0,+\infty)$.

The principal result of [C2] is the following.

Theorem 4. *Consider the Cauchy problem* (15). *Assume that the system is strictly hyperbolic and Assumptions* $(\star)$ *hold. Then there exists a constant* $\delta_0 > 0$ *such that if* $\bar{U} - U^* \in L^1$ *and* $TV\{\bar{U}\} < \delta_0$, *then for each* $\varepsilon > 0$, *the Cauchy problem* (15) *has a unique solution* U^ε, *defined for all* $t \geq 0$, *that satisfies*

$$TV\{U^{\varepsilon}(t,\cdot)\} + \int_0^t TV\{U^{\varepsilon}(s,\cdot)\}\,ds \le C\,TV\{\bar{U}\}, \tag{16}$$

where C is a positive constant that is independent of t and ε. Moreover, the solutions are stable in L^1, i.e., if V^{ε} is another solution of (15) *with initial data $\bar{V}$, then*

$$\|U^{\varepsilon}(t) - V^{\varepsilon}(t)\|_{L^1} + \int_0^t \|U^{\varepsilon}(\tau) - V^{\varepsilon}(\tau)\|_{L^1} d\tau \le L\,\|\bar{U} - \bar{V}\|_{L^1}. \tag{17}$$

Furthermore, the continuous dependence property with respect to time holds, i.e.,

$$\|U^{\varepsilon}(t) - U^{\varepsilon}(s)\|_{L^1} \le L'(|t-s| + \sqrt{\varepsilon}|\sqrt{t} - \sqrt{s}|), \tag{18}$$

for $t,\ s > 0$. Finally, as $\varepsilon \downarrow 0+$, U^{ε} converges in L^1_{loc} to a function U, which is the admissible weak solution U of (13).

The proof follows the fundamental ideas of Bianchini and Bressan [BiB] and the additional techniques of Christoforou [C] to treat the source terms. Because of the damping effect, we are actually able to show that the total variation as a function of time is integrable over $[0,\infty)$, hence the weak solution constructed by this method converges to the equilibrium as $t \to \infty$. Thus, we validate the results of the classical theory in the general context of weak solutions.

It should be noted that in [C], when $H = K = 0$, the total variation is exponentially decaying in time. Thus, the above result is an expected generalization of the result in [C].

The results mentioned in Sect. 1 are joint work with Professor Gui-Qiang Chen. The proof and more details can be found in [CC]. The results stated in Sect. 2 can be found in [C2].

Acknowledgments

The author would like to thank the organizers of the HYP2006 conference for the invitation and the hospitality in Lyon.

References

[BiB] Bianchini, S., Bresssan, A.: Vanishing viscosity of nonlinear hyperbolic systems, Ann. of Math. **161**, (1) 223–342 (2005)

[B] Bressan, A.: *Hyperbolic Systems of Conservation Laws. The One Dimensional Cauchy Problem.* Oxford University Press, (2000)

[CC] Chen, G.-Q., Christoforou, C.: Solutions for a nonlocal conservation law with fading memory, Proc. AMS, to appear

[CD] Chen, G.-Q., Dafermos, C.M.: Global solutions in L^∞ for a system of conservation laws of viscoelastic materials with memory, J. Diff. Eqs, **10**, (4), 369–383 (1997)

[CLL] Chen, G.-Q, Levermore, D., Liu, T.-P.: Hyperbolic conservation laws with stiff relaxation terms and entropy, Comm. Pure Appl. Math. **47**, 787–830 (1994)

[C] Christoforou, C.: Hyperbolic systems of balance laws via vanishing viscosity, J. Diff. Eqs, **221**/2, 470–541 (2006)

[C1] Christoforou, C.: Uniqueness and sharp estimates on solutions to hyperbolic systems with dissipative source, Comm. PDE, to appear

[C2] Christoforou, C.: Systems of conservation laws with memory, preprint

[D] Dafermos, C.M.: *Hyperbolic Conservation Laws in Continuum Physics*, Springer-Verlag, Berlin (2005)

[D1] Dafermos, C.M.: Solutions in L^∞ for a conservation law with memory, Analyse Mathematique et Applications, Gauthier-Viliars, Paris (1988)

[D2] Dafermos, C.M.: Development of singularities in the motion of materials with fading memory, Arch. Rational Mech. Anal. **91**, 193–205 (1986)

[D3] Dafermos, C.M.: Hyperbolic conservation laws with memory, Differential equations (Xanthi), 157–166 (1987)

[DN] Dafermos, C.M., Nohel, J.A.: Energy methods for nonlinear hyperbolic volterra integrodifferential equations, Comm. PDE **4**(3), 219–278 (1979)

[K] Kruzkov, S.: First-order quasilinear equations with several space variables, Mat. Sbornik **123** (1970), 228–255. English translation: Math. USSR Sbornik **10**, 217–273 (1970)

[LN] Luo, T., Natalini, R.: *BV* solutions and relaxation limit for a model in viscoelasticity, Proc. Roy. Soc. Edinburgh, **128**A, 775–795 (1998)

[M1] MacCamy, R.C.: An integrodifferential equation with applications in heat flow, Q. Appl. Math. **35**, 1–19 (1977)

[M2] MacCamy, R.C.: A model for one-dimensional nonlinear viscoelasticity. Quart. Appl. Math **35**, 21–33 (1977)

[MN] Malek-Madani, R., Nohel, J.A.: Formation of singularities for a conservation law with memory, SIAM J. Math. Anal. **16**, 530–540 (1985)

[NRT] Nohel, J.A., Rogers, R.C., Tzavaras, A.E.: Weak solutions for a nonlinear system in viscoelasticity. Comm. PDE **13** no. 1, 97–127 (1988)

[O] Oleinik, O.A.: Discontinuous solutions of non-linear differential equations. Usp. Mat. Nauk **12**, 3–73 (1957); AMS Translations, Ser. II, **26**, 95–172 (in English)

[RHN] Renardy, M., Hrusa W., Nohel, J.A.: *Mathematical Problems in Viscoelasticity*, Longman, New York, (1987)

[STW] Shen, W., Tveito, A., Winther, R.: On the zero relaxation limit for a system modeling the motions of a viscoelastic solid, SIAM J. Math. Anal. **30**, 1115–1135 (1999)

[Se] Serre, D.: *Systems of Conservation Laws, I,II*, Cambridge University Press, Cambridge, (1999)

[Sl.] Slemrod, M.: Instability of steady shearing flows in a nonlinear viscoelastic fluid, ARMA, **68**, no. 3, 211–225 (1978)

[Sm] Smoller, J.: *Shock Waves and Reaction-diffusion Equations*, Springer, New York, (1994)

[S] Staffans, O.J.: On a nonlinear hyperbolic Volterra Equation, SIAM J. Math. Anal., **11** , no. 5, 793–812 (1980)

[V] Vol'pert, A.I.: *BV* Space and quasilinear equations (Russian), Mat. Sb. (N.S.) **73**(115), 255–302 (1967)

[Yo] W.-A. Yong, A difference scheme for a stiff system of conservation laws, Proc. Roy. Soc. Edinburgh, **1**28A, 1403–1414 (1998)

Global Weak Solutions for a Shallow Water Equation

G.M. Coclite, H. Holden, and K.H. Karlsen

1 Introduction and Statement of the Main Result

Consider the equation

$$\partial_t u - \alpha^2 \partial_{txx}^3 u + 2\omega \partial_x u + 3u \partial_x u + \gamma \partial_{xxx}^3 u = \alpha^2 (2\partial_x u \partial_{xx}^2 u + u \partial_{xxx}^3 u), \quad (1)$$

in the space–time variables $t > 0$, $x \in \mathbf{R}$, augmented with the initial condition

$$u(0, x) = u_0(x), \qquad x \in \mathbf{R}, \quad (2)$$

where α, γ, ω are given real constants. Equation (1) was first introduced as a model describing propagation of unidirectional gravitational waves in a shallow water approximation over a flat bottom, with u representing the fluid velocity [DGH01].

For $\alpha = 0$ and for $\alpha^2 = 1$, $\gamma = 0$ we obtain the Korteweg–de Vries and the Camassa–Holm [CH93, J02] equations, respectively. Both of them describe unidirectional shallow water waves. Moreover, all these three equations have a bi-Hamiltonian structure, they are completely integrable, they have infinitely many conserved quantities.

From a mathematical point of view the Camassa–Holm equation is well studied, see [CHK05.1] for an extensive list of references. In particular, we recall that existence and uniqueness results for global weak solutions have been proved by Coclite et al. [CHK05.1], Constantin and Escher [CE98], Constantin and Molinet [CM00], and Xin and Zhang [XZ00, XZ02], see also Danchin [D01, D03] as well as others.

In [L06] the author proved the existence of global solutions to (1)–(2) under the assumptions

$$\gamma = -2\omega\alpha^2, \qquad u_0 \in H^s(\mathbf{R}), \quad s > \frac{3}{2}. \quad (3)$$

In this chapter we prove existence of global solutions under the assumptions

$$\alpha,\,\omega,\,\gamma \in \mathbf{R}, \qquad \alpha \neq 0, \qquad u_0 \in H^1(\mathbf{R}). \tag{4}$$

At least formally, (1) is equivalent to the elliptic–hyperbolic system

$$\begin{cases} \partial_t u + u\partial_x u - \frac{\gamma}{\alpha^2}\partial_x u + \partial_x P = 0, \\ -\alpha^2 \partial_{xx}^2 P + P = \frac{\alpha^2}{2}(\partial_x u)^2 + \left(2\omega + \frac{\gamma}{\alpha^2}\right) u + u^2. \end{cases} \tag{5}$$

We use the following definition of weak and admissible solutions.

Definition 1. *We call* $u\colon [0,\infty) \times \mathbf{R} \to \mathbf{R}$ *a weak solution of the Cauchy problem* (1)–(2) *if*

(a) $u \in C([0,\infty) \times \mathbf{R})$*;*
(b) $u \in L^\infty\big((0,\infty); H^1(\mathbf{R})\big)$*;*
(c) u *satisfies* (5) *in the sense of distributions;*
(d) $u(0,x) = u_0(x)$*, for every* $x \in \mathbf{R}$*.*

If, in addition, there exists a positive constant K *depending only on* $\|u_0\|_{H^1(\mathbf{R})}$*,* $\alpha,\,\omega,\gamma$*, such that*

$$\partial_x u(t,x) \le \frac{2}{t} + K, \qquad (t,x) \in (0,\infty) \times \mathbf{R}, \tag{6}$$

then we say that u *is an admissible solution of* (1)–(2)*.*

Our results are collected in the following theorem:

Theorem 1. *Assume that* (4) *holds. Then there exists an admissible weak solution* $u : [0,\infty) \times \mathbf{R} \to \mathbf{R}$ *to the Cauchy problem* (1)–(2)*. Moreover,*

$$\partial_x u \in L^p_{\mathrm{loc}}((0,\infty) \times \mathbf{R}), \quad \textit{for each } 1 \le p < 3. \tag{7}$$

Since the proof is similar to the one in [CHK05.1], we simply sketch it.

2 Viscous Approximations: Existence and A Priori Estimates

We prove existence of a weak solution to the Cauchy problem (1)–(2) by proving compactness of a sequence of smooth solutions $\{u_\varepsilon\}_{\varepsilon>0}$ solving the following viscous problems (see [CHK05.2])

$$\begin{cases} \partial_t u_\varepsilon + u_\varepsilon \partial_x u_\varepsilon - \frac{\gamma}{\alpha^2}\partial_x u_\varepsilon + \partial_x P_\varepsilon = \varepsilon \partial_{xx}^2 u_\varepsilon, \\ -\alpha^2 \partial_{xx}^2 P_\varepsilon + P_\varepsilon = \frac{\alpha^2}{2}(\partial_x u_\varepsilon)^2 + \left(2\omega + \frac{\gamma}{\alpha^2}\right) u_\varepsilon + u_\varepsilon^2, \\ u_\varepsilon(0,x) = u_{\varepsilon,0}(x), \end{cases} \tag{8}$$

that is equivalent to the following fourth-order equation

$$\begin{cases} \partial_t u_\varepsilon - \alpha^2 \partial_{txx}^3 u_\varepsilon + 2\omega\partial_x u_\varepsilon + 3u_\varepsilon \partial_x u_\varepsilon + \gamma \partial_{xxx}^3 u_\varepsilon \\ \qquad = \alpha^2 (2\partial_x u_\varepsilon \partial_{xx}^2 u_\varepsilon + u_\varepsilon \partial_{xxx}^3 u_\varepsilon) + \varepsilon \partial_{xx}^2 u_\varepsilon - \varepsilon\alpha^2 \partial_{xxxx}^4 u_\varepsilon, \\ u_\varepsilon(0,x) = u_{\varepsilon,0}(x). \end{cases} \tag{9}$$

Formally, sending $\varepsilon \longrightarrow 0$ in (9), (8) yields (1), (5), respectively.

We shall assume for each $\varepsilon > 0$ that

$$\begin{aligned} &u_{\varepsilon,0} \in H^2(\mathbf{R}), \qquad u_{\varepsilon,0} \longrightarrow u_0 \text{ in } H^1(\mathbf{R}), \\ &\|u_{\varepsilon,0}\|^2_{L^2(\mathbf{R})} + \alpha^2 \|\partial_x u_{\varepsilon,0}\|^2_{L^2(\mathbf{R})} \le \|u_0\|^2_{L^2(\mathbf{R})} + \alpha^2 \|\partial_x u_0\|^2_{L^2(\mathbf{R})}. \end{aligned} \tag{10}$$

The starting point of our analysis is the following well-posedness result for (8) (see [CHK05.2], Theorem 2.3).

Lemma 1. *Assume* (4), (10), *let* $\varepsilon > 0$. *There exists a unique smooth solution* $u_\varepsilon \in C\big([0,\infty); H^2(\mathbf{R})\big)$ *to the Cauchy problem* (8).

The next step in our analysis is to derive the following a priori estimates:

Lemma 2. *Assume* (4) *and* (10). *Then the following estimates hold:*

(a) (Energy conservation) for each $t \ge 0$

$$\begin{aligned} \|u_\varepsilon\,(t,\cdot)\|^2_{L^2(\mathbf{R})} &+ \alpha^2 \|\partial_x u_\varepsilon(t,\cdot)\|^2_{L^2(\mathbf{R})} \\ &+ 2\varepsilon \int_0^t \Big(\|\partial_x u_\varepsilon(s,\cdot)\|^2_{L^2(\mathbf{R})} + \alpha^2 \|\partial_{xx}^2 u_\varepsilon(s,\cdot)\|^2_{L^2(\mathbf{R})} \Big)\, ds \\ &\qquad = \|u_{\varepsilon,0}\|^2_{L^2(\mathbf{R})} + \alpha^2 \|\partial_x u_{\varepsilon,0}\|^2_{L^2(\mathbf{R})}; \end{aligned} \tag{11}$$

(b) (Oleĭnik type estimate [O63]) for any $t > 0$ *and* $x \in \mathbf{R}$

$$\partial_x u_\varepsilon(t,x) \le \frac{2}{t} + K, \tag{12}$$

where K *is a positive constant depending on* α, ω, γ, $\|u_0\|_{H^1(\mathbf{R})}$ *but independent on* ε;

(c) (Higher integrability estimate) for every $0 \le \beta < 1$, $T > 0$, *and* $a, b \in \mathbf{R}$, $a < b$, *there exists a positive constant* C_T *depending only on* α, β, ω, γ, $\|u_0\|_{H^1(\mathbf{R})}$ *but independent on* ε, *such that*

$$\int_0^T \int_a^b \left|\partial_x u_\varepsilon(t,x)\right|^{2+\beta} dt dx \le C_T. \tag{13}$$

Remark 1. Due to [LL01], Theorem 8.5, (10) and (11), we have for each $t \ge 0$

$$\|u_\varepsilon(t,\cdot)\|_{L^\infty(\mathbf{R})} \le \frac{1}{\sqrt{2}} \|u_\varepsilon(t,\cdot)\|_{H^1(\mathbf{R})} \le \frac{\max(|\alpha|,1)}{\min(|\alpha|,1)\sqrt{2}} \|u_0\|_{H^1(\mathbf{R})}. \tag{14}$$

In particular, the family $\{u_\varepsilon\}_{\varepsilon>0}$ is uniformly bounded in $L^\infty((0,\infty); H^1(\mathbf{R}))$.

Proof (Proof of Lemma 2). We begin with (a). Multiplying (9) by u_ε, integrating on $\mathbf{R}$, and integrating by parts we get

$$\frac{1}{2}\frac{d}{dt}\int_{\mathbf{R}}\left(u_\varepsilon^2+\alpha^2(\partial_x u_\varepsilon)^2\right)dx+\varepsilon\int_{\mathbf{R}}\left((\partial_x u_\varepsilon)^2+\alpha^2(\partial_{xx}^2 u_\varepsilon)^2\right)dx=0,$$

from which (11) follows.

We continue by proving (b). Introduce the notation $q_\varepsilon := \partial_x u_\varepsilon$. From (8) we get the following equation for q_ε

$$\partial_t q_\varepsilon + u_\varepsilon \partial_x q_\varepsilon + \frac{q_\varepsilon^2}{2} - \frac{\gamma}{\alpha^2}\partial_x q_\varepsilon + \frac{P_\varepsilon}{\alpha^2} - \frac{2\omega\alpha^2+\gamma}{\alpha^4}u_\varepsilon - \frac{u_\varepsilon^2}{\alpha^2} = \varepsilon\partial_{xx}^2 q_\varepsilon. \tag{15}$$

It follows from (11) and (14) (see [CHK05.1], Proof of Lemma 3.1) that

$$\left\|\frac{P_\varepsilon}{\alpha^2} - \frac{2\omega\alpha^2+\gamma}{\alpha^4}u_\varepsilon - \frac{u_\varepsilon^2}{\alpha^2}\right\|_{L^\infty((0,\infty)\times\mathbf{R})} \le L, \tag{16}$$

for some constant $L > 0$ independent of ε. Then, from (15),

$$\partial_t q_\varepsilon + u_\varepsilon \partial_x q_\varepsilon + \frac{q_\varepsilon^2}{2} - \frac{\gamma}{\alpha^2}\partial_x q_\varepsilon - \varepsilon\partial_{xx}^2 q_\varepsilon \le L.$$

Employing the comparison principle for parabolic equations, we get $q_\varepsilon(t,x) \le h(t)$, $t > 0$, $x \in \mathbf{R}$ where h solves $\frac{d}{dt}h + \frac{1}{2}h^2 = L$. Since the map $H(t) := \frac{2}{t} + \sqrt{2L}$, $t > 0$, is a super solution of that ordinary differential equation, we get $h(t) \le H(t)$ for all $0 < t \le T$. Therefore, (12) is proved.

Finally, we consider (c). The argument is very similar to the one of [CHK05.1], Lemma 4.1. Pick a cut-off function $\chi \in C^\infty(\mathbf{R})$ such that

$$0 \le \chi \le 1, \qquad \chi(x) = \begin{cases} 1, & \text{if } x \in [a,b], \\ 0, & \text{if } x \in (-\infty, a-1] \cup [b+1,\infty), \end{cases} \tag{17}$$

consider the map $\theta(\xi) := \xi\left(|\xi|+1\right)^\alpha$, $\xi \in \mathbf{R}$, then multiply (15) by $\chi\theta'(q_\varepsilon)$, integrate over $(0,T)\times\mathbf{R}$ and use (11), (14). □

3 Compactness

Due to (14) and to the elliptic structure of the second equation in (9) we have the following [CHK05.1], Lemma 5.1:

Lemma 3. *The family $\{P_\varepsilon\}_{\varepsilon>0}$ is uniformly bounded in $L^\infty([0,\infty);W^{1,\infty}(\mathbf{R}))$ and $L^\infty([0,\infty);H^1(\mathbf{R}))$.*

Lemma 4. *There exists a sequence $\{\varepsilon_j\}_{j\in\mathbf{N}}$ tending to zero and a function $u \in L^\infty([0,\infty);H^1(\mathbf{R})) \cap H^1([0,\infty]\times\mathbf{R})$ such that $u_{\varepsilon_j} \rightharpoonup u$ weakly in $H^1_{\text{loc}}([0,T]\times\mathbf{R})$, for each $T \ge 0$, $u_{\varepsilon_j} \to u$ strongly in $L^\infty_{\text{loc}}([0,\infty)\times\mathbf{R})$.*

Proof. Fix $T > 0$. Observe that, from (9), $\partial_t u_\varepsilon = -u_\varepsilon \partial_x u_\varepsilon + \frac{\gamma}{\alpha^2}\partial_x u_\varepsilon - \partial_x P_\varepsilon + \varepsilon\partial_{xx}^2 u_\varepsilon$ hence, by (11), (14), Lemma 3, and the Hölder inequality, $\{u_\varepsilon\}_{\varepsilon>0}$ is uniformly bounded in $H^1([0,T]\times\mathbf{R}) \cap L^\infty([0,T);H^1(\mathbf{R}))$, and the first part of the statement follows. Finally, since $H^1(\mathbf{R}) \subset\subset L^\infty_{\text{loc}}(\mathbf{R}) \subset L^2_{\text{loc}}(\mathbf{R})$, the second part of the statement is consequence of [S87], Theorem 5. □

Using the ellipticity of the second equation in (9), the estimates in (14) and Lemma 3 we can prove the following result, cf. [CHK05.1], Lemma 5.3.

Lemma 5. *The family $\{P_\varepsilon\}_{\varepsilon>0}$ is uniformly bounded in $W^{1,1}_{\text{loc}}([0,T)\times\mathbf{R})$ for any $T > 0$. In particular, there exists a sequence $\{\varepsilon_j\}_{j\in\mathbf{N}}$ tending to zero and a function $P \in L^\infty([0,T);W^{1,\infty}(\mathbf{R}))$ such that $P_{\varepsilon_j} \to P$ strongly in $L^p_{\text{loc}}([0,\infty)\times\mathbf{R})$ for each $1 \le p < \infty$.*

The next lemma is a direct consequence of Lemma 3 and (13).

Lemma 6. *There exist a sequence $\{\varepsilon_j\}_{j\in\mathbf{N}}$ tending to zero and two functions $q \in L^p_{\text{loc}}([0,\infty)\times\mathbf{R})$, $\overline{q^2} \in L^r_{\text{loc}}([0,\infty)\times\mathbf{R})$ such that $q_{\varepsilon_j} \rightharpoonup q$ in $L^p_{\text{loc}}([0,\infty)\times\mathbf{R})$, $q_{\varepsilon_j} \overset{\star}{\rightharpoonup} q$ in $L^\infty_{\text{loc}}([0,\infty);L^2(\mathbf{R}))$, $q^2_{\varepsilon_j} \rightharpoonup \overline{q^2}$ in $L^r_{\text{loc}}([0,\infty)\times\mathbf{R})$, for each $1 < p < 3$ and $1 < r < 3/2$. Moreover, $q^2(t,x) \le \overline{q^2}(t,x)$ for almost every $(t,x) \in [0,\infty)\times\mathbf{R}$ and $\partial_x u = q$ in the sense of distributions on $[0,\infty)\times\mathbf{R}$.*

In the remaining part of the chapter, for notational convenience, we replace the sequences $\{u_{\varepsilon_j}\}_{j\in\mathbf{N}}$, $\{q_{\varepsilon_j}\}_{j\in\mathbf{N}}$, $\{P_{\varepsilon_j}\}_{j\in\mathbf{N}}$ by $\{u_\varepsilon\}_{\varepsilon>0}$, $\{q_\varepsilon\}_{\varepsilon>0}$, $\{P_\varepsilon\}_{\varepsilon>0}$, respectively.

In view of Lemma 6, we conclude that for any $\eta \in C^1(\mathbf{R})$ with η' bounded, Lipschitz continuous on $\mathbf{R}$ and any $1 \le p < 3$ we have

$$\eta(q_\varepsilon) \rightharpoonup \overline{\eta(q)} \text{ in } L^p_{\text{loc}}([0,\infty)\times\mathbf{R}), \quad \eta(q_\varepsilon) \overset{\star}{\rightharpoonup} \overline{\eta(q)} \text{ in } L^\infty_{\text{loc}}([0,\infty);L^2(\mathbf{R})). \tag{18}$$

Multiplying the equation in (15) by $\eta'(q_\varepsilon)$, we get

$$\begin{aligned}\partial_t\eta(q_\varepsilon) &+ \partial_x(u_\varepsilon\eta(q_\varepsilon)) - q_\varepsilon\eta(q_\varepsilon) + q_\varepsilon^2\eta'(q_\varepsilon) - \tfrac{\gamma}{\alpha^2}\partial_x\eta(q_\varepsilon)\\ &+ \tfrac{P_\varepsilon}{\alpha^2}\eta'(q_\varepsilon) - \tfrac{2\omega\alpha^2+\gamma}{\alpha^4}u_\varepsilon\eta'(q_\varepsilon) - \tfrac{u_\varepsilon^2}{\alpha^2} = \varepsilon\partial_{xx}^2\eta(q_\varepsilon) - \varepsilon\eta''(q_\varepsilon)(\partial_x q_\varepsilon)^2.\end{aligned} \tag{19}$$

Lemma 7. *For any convex $\eta \in C^1(\mathbf{R})$ with η' bounded, Lipschitz continuous on $\mathbf{R}$, we have*

$$\begin{aligned}\partial_t\overline{\eta(q)} &+ \partial_x(u\overline{\eta(q)}) - \overline{q\eta(q)} + \overline{q^2\eta'(q)}\\ &- \tfrac{\gamma}{\alpha^2}\partial_x\overline{\eta(q)} + \tfrac{P}{\alpha^2}\overline{\eta'(q)} - \tfrac{2\omega\alpha^2+\gamma}{\alpha^4}u\overline{\eta'(q)} - \tfrac{u^2}{\alpha^2}\overline{\eta'(q)} \le 0\end{aligned} \tag{20}$$

in the sense of distributions on $[0,\infty)\times\mathbf{R}$. Here $\overline{q\eta(q)}$, $\overline{q^2\eta'(q)}$, $\overline{\eta'(q)}$, and $\overline{\eta'(q)q}$ denote the weak limits of $q_\varepsilon\eta(q_\varepsilon)$, $q_\varepsilon^2\eta'(q_\varepsilon)$, $\eta'(q_\varepsilon)$, and $\eta'(q_\varepsilon)q_\varepsilon$ in $L^r_{\text{loc}}([0,\infty)\times\mathbf{R})$, $1 < r < 3/2$, respectively.

Proof. By convexity $-\varepsilon\eta''(q_\varepsilon) \le 0$. Hence, due to Lemmas 4 and 6, sending $\varepsilon \to 0$ in (19) yields (20). □

Using Lemmas 4–6, and sending $\varepsilon \to 0$ in (15) we have the following result.

Lemma 8. *There holds*

$$\partial_t q + u\partial_x q + \frac{q^2}{2} - \frac{\gamma}{\alpha^2}\partial_x q + \frac{P}{\alpha^2} - \frac{2\omega\alpha^2+\gamma}{\alpha^4}u - \frac{u^2}{\alpha^2} = 0, \tag{21}$$

in the sense of distributions on $[0,\infty)\times\mathbf{R}$.

The next lemma contains a renormalized formulation of (21).

Lemma 9 ([CHK05.1], Lemma 5.8). *For any* $\eta \in C^1(\mathbf{R})$ *with* $\eta' \in L^\infty(\mathbf{R})$,

$$\begin{aligned} &\partial_t\eta(q) + \partial_x(u\eta(q)) - q\eta(q) + \left(\tfrac{\overline{q^2}}{2} - q^2\right)\eta'(q) \\ &\qquad - \tfrac{\gamma}{\alpha^2}\partial_x\eta(q) + \tfrac{P}{\alpha^2}\eta'(q) - \tfrac{2\omega\alpha^2+\gamma}{\alpha^4}u\eta'(q) - \tfrac{u^2}{\alpha^2}\eta'(q) = 0, \end{aligned} \tag{22}$$

in the sense of distributions on $[0,\infty)\times\mathbf{R}$.

Following [CHK05.1], Sect. 6 and [XZ00], we improve the weak convergence of q_ε in Lemma 6 to strong convergence (and then we have an existence result for (1)–(2). The idea is to derive a "transport equation" for the evolution of the defect measure $(\overline{q^2} - q^2)(t,\cdot) \geq 0$, so that if it is zero initially then it will continue to be zero at all later times $t > 0$.

Lemma 10. *For each* $t \geq 0$

$$\begin{aligned} &\int_{\mathbf{R}} \left(\overline{(q_+)^2}(t,x) - (q_+)^2(t,x)\right) dx \\ &\qquad \leq 2\int_0^t \int_{\mathbf{R}} S(s,x)\left[\overline{q_+}(s,x) - q_+(s,x)\right] ds dx, \end{aligned} \tag{23}$$

where

$$S := \frac{u^2}{\alpha^2} - \frac{P}{\alpha^2} - \frac{2\omega\alpha^2+\gamma}{\alpha^4}u.$$

Proof. Let $T > 0$, $R > K$ (see (12)). Subtract (22) from (20) using the renormalization

$$\eta_R^+(\xi) := \begin{cases} R\xi - \frac{R^2}{2}, & \text{if } \xi > R, \\ \frac{\xi^2}{2}, & \text{if } 0 \leq \xi \leq R, \\ 0, & \text{if } \xi < 0, \end{cases} \qquad \eta_R^-(\xi) := \begin{cases} 0, & \text{if } \xi > 0, \\ \frac{\xi^2}{2}, & \text{if } -R \leq \xi \leq 0, \\ -R\xi - \frac{R^2}{2}, & \text{if } \xi < -R. \end{cases}$$

Arguing as in [CHK05.1], Lemma 6.4 we get

$$\frac{d}{dt}\int_{\mathbf{R}} \left(\overline{(q_+)^2} - (q_+)^2\right) dx \leq 2\int_{\mathbf{R}} S(t,x)\left[\overline{q_+} - q_+\right] dx,$$

for $4/(\gamma(R - K_T)) < t < T$. First we apply the Gronwall lemma to the previous inequality on the interval $(4/(\gamma(R - K_T)), T)$, then sending $R \to \infty$ and using (see [CHK05.1], Lemma 6.2)

$$\lim_{t\to 0+}\int_{\mathbf{R}} \left(\overline{\eta_R^+(q)}(t,x) - \eta_R^+(q(t,x))\right) dx = 0, \qquad R > 0. \tag{24}$$

□

Lemma 11. *For any $t \geq 0$ and any $R > 0$,*

$$\begin{aligned} \int_{\mathbf{R}} \left[\overline{\eta_R^-(q)}(t,x) - \eta_R^-(q)(t,x) \right] dx \\ \leq \tfrac{R^2}{2} \int_0^t \int_{\mathbf{R}} \overline{(R+q)\chi_{(-\infty,-R)}(q)} ds dx \\ - \tfrac{R^2}{2} \int_0^t \int_{\mathbf{R}} (R+q)\chi_{(-\infty,-R)}(q) ds dx \\ + \tfrac{R}{2} \int_0^t \int_{\mathbf{R}} \left(\overline{(q_+)^2} - (q_+)^2 \right) ds dx \\ + R \int_0^t \int_{\mathbf{R}} \left(\overline{\eta_R^-(q)} - \eta_R^-(q) \right) ds dx \\ + \int_0^t \int_{\mathbf{R}} S(s,x) \left[\overline{(\eta_R^-)'(q)} - (\eta_R^-)'(q) \right] ds dx. \end{aligned} \tag{25}$$

Proof. The argument is very similar to the one of [CHK05.1], Lemma 6.3. We begin by subtracting (22) from (20), using the renormalization η_R^-. Then we integrate on $\mathbf{R}$ and use the Gronwall lemma and (see [CHK05.1], Lemma 6.2)

$$\lim_{t \to 0+} \int_{\mathbf{R}} \left(\overline{\eta_R^-(q)}(t,x) - \eta_R^-(q(t,x)) \right) = 0, \qquad R > 0. \tag{26}$$

□

Lemma 12. *There holds $\overline{q^2} = q^2$ almost everywhere in $[0,\infty) \times \mathbf{R}$.*

Proof. We follow the argument of [CHK05.1], Lemma 6.6. We add (23) and (25). Using the concavity of $\xi \mapsto (R+\xi)\chi_{(-\infty,-R)}(\xi)$, the Gronwall lemma, (24), (26) and

$$\lim_{t \to 0+} \int_{\mathbf{R}} q^2(t,x) dx = \lim_{t \to 0+} \int_{\mathbf{R}} \overline{q^2}(t,x) dx = \int_{\mathbf{R}} (\partial_x u_0)^2 \, dx,$$

we conclude that

$$\int_{\mathbf{R}} \left(\frac{1}{2} \left[\overline{(q_+)^2} - (q_+)^2 \right] + \left[\overline{\eta_R^-(q)} - \eta_R^-(q) \right] \right) (t,x) dx = 0,$$

for each $0 < t < T$. By the Fatou lemma and Lemma 6, sending $R \to \infty$ yields

$$0 \leq \int_{\mathbf{R}} \left(\overline{q^2} - q^2 \right) (t,x) dx \leq 0, \qquad 0 < t < T,$$

and, since the argument holds for each $T > 0$, we are done. □

Finally, we are ready to conclude the proof of Theorem 1.

Proof (Proof of Theorem 1). The conditions $(a), (c), (d)$ of Definition 1 are satisfied, due to (10), (11) and Lemma 4. We have to verify (b). Due to Lemma 12, we have

$$q_\varepsilon \to q \qquad \text{strongly in } L^2_{\text{loc}}([0,\infty) \times \mathbf{R}). \tag{27}$$

Clearly Lemma 4 and (27) imply that u is a distributional solution of (5). Since (6) comes from (12), u is an admissible solution of (1)–(2). Finally, (7) is consequence of (13). □

Acknowledgements

The research of H. Holden is supported in part by the Research Council of Norway. The research of K. H. Karlsen is supported by an Outstanding Young Investigators Award from the Research Council of Norway.

References

[CH93] Camassa, R., Holm, D.: An integrable shallow water equation with peaked solitons. Phys. Rev. Lett. **71**, 1661–1664 (1993).

[CHK05.1] Coclite, G.M., Holden, H., Karlsen, K.H.: Global weak solutions to a generalized hyperelastic-rod wave equation. SIAM J. Math. Anal., **37**, 1044–1069 (2005).

[CHK05.2] Coclite, G.M., Holden, H., Karlsen, K.H.: Wellposedness for a parabolic-elliptic system. Discrete Contin. Dyn. Syst., **13**, 659–682 (2005).

[CE98] Constantin, A., Escher, J.: Global weak solutions for a shallow water equation. Indiana Univ. Math. J., **47**, 1527–1545 (1998).

[CM00] Constantin, A., Molinet, L.: Global weak solutions for a shallow water equation. Comm. Math. Phys., **211**, 45–61 (2000).

[D01] Danchin, R.: A few remarks on the Camassa–Holm equation. Differential Integral Equations, **14**, 953–988 (2001).

[D03] Danchin, R.: A note on well-posedness for Camassa–Holm equation. J. Differential Equations, **192**, 429–444 (2003).

[DGH01] Dullin, H., Gottwald, G., Holm, D.: An integrable shallow water equation with linear and nonlinear dispersion. Phy. Rev. Lett., **87**, 194501 (2001).

[J02] Johnson, R.S.: Camassa–Holm, Korteweg–de Vries and related models for water waves. J. Fluid Mech., **455**, 63–82 (2002).

[LL01] Lieb, E.H., Loss, M.: Analysis. American Mathematical Society, Providence, RI, second edition (2001).

[L06] Liu, Y.: Global existence and blow-up solutions for a nonlinear shallow water equation. Math. Ann., **335**, 717–735 (2006).

[O63] Oleĭnik, O.A.: Discontinuous solutions of non-linear differential equations. Amer. Math. Soc. Transl. Ser. 2, **26**, 95–172 (1963).

[S87] Simon, J.: Compact sets in the space $L^p(0,T;B)$. Ann. Mat. Pura Appl., **146**, 65–96 (1987).

[XZ00] Xin, Z., Zhang, P.: On the weak solutions to a shallow water equation. Comm. Pure Appl. Math., **53**, 1411–1433 (2000).

[XZ02] Xin, Z., Zhang, P.: On the uniqueness and large time behavior of the weak solutions to a shallow water equation. Comm. Partial Differential Equations, **27**, 1815–1844 (2002).

Structural Stability of Shock Solutions of Hyperbolic Systems in Nonconservation Form via Kinetic Relations

B. Audebert and F. Coquel

Summary. We introduce stability conditions for shock solutions of hyperbolic systems in nonconservation form. The recently proposed framework of kinetic relations for defining shock solutions is shown to yield a natural extension of the structural stability conditions due to Majda in the conservative setting: besides the mandatory geometric Lax conditions, a direct extension of the Majda determinant must not vanish. We study these conditions for validity within the frame of PDE systems for modelling shock-turbulence interactions. We prove that a mostly neglected nonconservative correction to the PDEs plays a major role in the stability.

1 Statement of the Problem

This work treats stability properties of discontinuous solutions of the inviscid limit (i.e. $\epsilon \to 0^+$) of nonlinear hyperbolic systems with viscous perturbation in the following nonconservation form

$$\partial_t \mathbf{u}^\epsilon + \mathcal{A}(\mathbf{u}^\epsilon)\partial_x \mathbf{u}^\epsilon = \mathcal{B}(\mathbf{u}^\epsilon, \epsilon\partial_x \mathbf{u}^\epsilon, \epsilon^2 \partial_{xx}\mathbf{u}^\epsilon), \quad x \in I\!R,\ t > 0, \tag{1}$$

where $\mathcal{B}(\mathbf{u}, 0, 0) = 0$ for all $\mathbf{u} \in I\!R^n$, $n > 1$. In this limit, discontinuous solutions naturally arise but cannot be understood in the usual weak sense. Several mathematical frameworks are now at hand to give meaning to such limit solutions. We refer to [4] for a brief review. The key issue is that shock solutions in the nonconservative setting are generically small scales dependent: they do depend on the underlying dissipative mechanisms modelled by $\mathcal{B}$ in (1). To conveniently account for the reported sensitiveness, we propose to handle the inviscid limit of (1) in the recent framework of kinetic relations for hyperbolic systems in nonconservation form [4]. Roughly speaking, this framework assumes the existence of a smooth change of variable $\mathbf{u} \in I\!R^n \to \mathbf{v}(\mathbf{u}) \in I\!R^n$ so that smooth solutions of (1) obey the following equivalent form

$$\partial_t \mathbf{v}(\mathbf{u}^\epsilon) + \partial_x \mathcal{F}(\mathbf{v}(\mathbf{u}^\epsilon)) = D_{\mathbf{u}}\mathbf{v}(\mathbf{u}^\epsilon) \times \mathcal{B}(\mathbf{u}^\epsilon, \epsilon\partial_x \mathbf{u}^\epsilon, \epsilon^2 \partial_{xx}\mathbf{u}^\epsilon). \tag{2}$$

Namely, the left-hand side of (2) now stands in conservation form while the viscous perturbation still writes in nonconservation form. Hence in the limit $\epsilon \to 0^+$ and besides the left-hand side now in classical form, the viscous perturbation does again contribute in a nontrivial way: $D_{\mathbf{u}}\mathbf{v}(\mathbf{u}^\epsilon) \times \mathcal{B}(\mathbf{u}^\epsilon, \epsilon\partial_x\mathbf{u}^\epsilon, \epsilon^2\partial_{xx}\mathbf{u}^\epsilon)$ does not go to zero in general but rather to a bounded Borel measure, we denote $\Upsilon(\mathcal{B})_{\mathbf{u}}$, concentrated on the shocks of the limit function $\mathbf{u} = \lim_{\epsilon\to 0^+} \mathbf{u}^\epsilon$. Provided that suitable estimates on the sequence $\mathbf{u}^\epsilon$ and its derivative are met, one thus expect the limit function $\mathbf{u}$ to be a solution of

$$\partial_t\mathbf{v}(\mathbf{u}) + \partial_x\mathcal{F}(\mathbf{v}(\mathbf{u})) = \Upsilon(\mathcal{B})_{\mathbf{u}}, \quad t > 0,\ x \in I\!R. \tag{3}$$

For simplicity, the required notion of solution is investigated here within the class of piecewise Lipschitz continuous functions. Analysis of the travelling wave solutions of (1) motivates the following definition for the inviscid limit (3)

Definition 1. *A piecewise Lipschitz continuous function* $\mathbf{u}$ *is said to be a weak solution of (3) if it solves in the zone of smoothness*

$$\partial_t\mathbf{u} + \mathcal{A}(\mathbf{u})\partial_x\mathbf{u} = 0, \tag{4}$$

while at points of jump, it obeys

$$-\sigma(\mathbf{v}(\mathbf{u}_+) - \mathbf{v}(\mathbf{u}_-)) + \mathcal{F}(\mathbf{v}(\mathbf{u}_+)) - \mathcal{F}(\mathbf{v}(\mathbf{u}_-)) = \mathcal{K}_{\mathcal{B}}(\mathbf{u}_-, \sigma), \tag{5}$$

where $\mathcal{K}_{\mathcal{B}}(\mathbf{u}_-, \sigma)$ *denotes the mass of the measure* $\Upsilon(\mathcal{B})_{\mathbf{u}}$ *concentrated on the jump.*

The vector-valued function $\mathcal{K}_{\mathcal{B}} : I\!R^n \times I\!R \to I\!R^n$ is called a kinetic relation and is naturally defined as a function of solely the left trace $\mathbf{u}_-$ and the velocity σ when referring to shock solutions as the limit as $\epsilon \to 0^+$ of the family of travelling wave solutions $\{\mathbf{w}_\epsilon\}_{\epsilon>0}$ of (1). Indeed, recall briefly [7] that prescribing a state $\mathbf{u}_- \in I\!R^n$ and a velocity $\sigma \in I\!R$, a travelling wave $\mathbf{w}_\epsilon$ is a smooth solution of (1) of the form $\mathbf{w}_\epsilon(\xi)$, $\xi = x - \sigma t$, such that there exists an exit state $\mathbf{u}_+ \in I\!R^n$ with the next asymptotic conditions

$$\lim_{\xi\to-\infty} \mathbf{w}_\epsilon(\xi) = \mathbf{u}_-, \qquad \lim_{\xi\to+\infty} \mathbf{w}_\epsilon(\xi) = \mathbf{u}_+. \tag{6}$$

Now for fixed $\epsilon > 0$, such a travelling wave $\mathbf{w}_\epsilon(\xi)$ can be recovered by rescaling $\mathbf{w}_\epsilon(\xi) = \mathbf{w}(\xi/\epsilon; \mathbf{u}_-, \sigma)$ where for the prescribed state $\mathbf{u}_-$ and velocity σ, the function $\mathbf{w}(., \mathbf{u}_-, \sigma)$ must solve

$$-\sigma d_\xi\mathbf{w} + \mathcal{A}(\mathbf{w})d_\xi\mathbf{w} = \mathcal{B}(\mathbf{w}, d_\xi\mathbf{w}, d_{\xi\xi}\mathbf{w}), \quad \lim_{\xi\to\pm\infty} \mathbf{w}(\xi) = \mathbf{u}_\pm, \tag{7}$$

for the exit state $\mathbf{u}_+$ introduced in (6). Provided that the solution $\mathbf{w}(., \mathbf{u}_-, \sigma)$ of (7) exists, then the family of travelling wave solutions $\{\mathbf{w}_\epsilon\}_{\epsilon>0}$ can be seen under standard assumptions to converge to the next step function

$$\mathbf{u}(x,t) = \mathbf{u}_- + (\mathbf{u}_+ - \mathbf{u}_-)H(x - \sigma t), \quad t > 0,\ x \in I\!R, \tag{8}$$

which by definition is said to be a shock solution of (3). To infer the required generalized jump relations (5), observe that the solution $\mathbf{w}(., \mathbf{u}_-, \sigma)$ of (7) also satisfies the next rescaled form of (2)

$$-\sigma d_\xi \mathbf{v}(\mathbf{w}) + d_\xi \mathcal{F}(\mathbf{v}(\mathbf{w})) = D_\mathbf{u}\mathbf{v}(\mathbf{w}) \times \mathcal{B}(\mathbf{w}, d_\xi \mathbf{w}, d_{\xi\xi}\mathbf{w}), \tag{9}$$

which integrated over $I\!R_\xi$ yields in view of the asymptotic conditions in (7)

$$-\sigma(\mathbf{v}(\mathbf{u}_+)-\mathbf{v}(\mathbf{u}_-))+\mathcal{F}(\mathbf{v}(\mathbf{u}_+))-\mathcal{F}(\mathbf{v}(\mathbf{u}_-)) = \int_{I\!R_\xi} D_\mathbf{u}\mathbf{v}(\mathbf{w})\ \mathcal{B}(\mathbf{w}, d_\xi\mathbf{w}, d_{\xi\xi}\mathbf{w})d\xi.$$

Since $\mathbf{w}(., \mathbf{u}_-, \sigma)$ is built when (properly) prescribing $\mathbf{u}_-$ and σ in (7), it is natural to define the so-called kinetic function by

$$\mathcal{K}_\mathcal{B}(\mathbf{u}_-, \sigma) = \int_{I\!R_\xi} D_\mathbf{u}\mathbf{v}(\mathbf{w})\ \mathcal{B}(\mathbf{w}, d_\xi\mathbf{w}, d_{\xi\xi}\mathbf{w})d\xi. \tag{10}$$

Let us quote that the notion of kinetic relations traces back to the distinct setting of undercompressive discontinuous solutions for systems of conservation laws where they are understood as entropy rates of production. Such an interpretation actually persists in the present nonconservative setting. Kinetic relations have been proved useful in [6] for numerical purposes and also by Benzoni [3] in the stability analysis of undercompressive discontinuities. In turn this fruitful notion also provides a natural extension of the Majda stability conditions to shock solutions of nonconservative hyperbolic system. For simplicity in this brief presentation, the next statement only addresses shock solutions of the first family (see [2] for the general case).

Proposition 1. *A 1-shock solution (8) is said to be structurally stable iff*

(i) The Lax conditions are satisfied:

$$\lambda_1(\mathbf{u}_+) < \sigma < \lambda_1(\mathbf{u}_-), \quad \sigma < \lambda_2(\mathbf{u}_+) \tag{11}$$

with $\lambda_i(\mathbf{u}), 1 \leq i \leq n$, the increasingly arranged eigenvalues of $\mathcal{A}(\mathbf{u})$,

(ii) The next Majda determinant does not vanish

$$\begin{aligned} \Delta = Det\Big(D_\mathbf{u}^{-1}\mathbf{v}(\mathbf{u}_+)\Big\{(\mathbf{v}(\mathbf{u}_+) - \mathbf{v}(\mathbf{u}_-)) + \partial_\sigma \mathcal{K}_\mathcal{B}(\mathbf{u}_-, \sigma)\Big\}, \\ r_2(\mathbf{u}_+), \cdots, r_n(\mathbf{u}_+)\Big) \neq 0, \end{aligned} \tag{12}$$

with $\{r_i(\mathbf{u})\}_{1\leq i\leq n}$ a basis of right eigenvector for $\mathcal{A}(\mathbf{u})$.

Observe that the condition (12) equivalently reads

$$\Big({}^T(D_\mathbf{u}^{-1}\mathbf{v}(\mathbf{u}_+))\ l_1(\mathbf{u}_+), (\mathbf{v}(\mathbf{u}_+) - \mathbf{v}(\mathbf{u}_-)) + \partial_\sigma \mathcal{K}_\mathcal{B}(\mathbf{u}_-, \sigma)\Big) \neq 0, \tag{13}$$

where $l_1(\mathbf{u})$ stands for a left eigenvector of $\mathcal{A}(\mathbf{u})$ for the eigenvalue $\lambda_1(\mathbf{u})$.

2 A Nonconservative Model for Shock Turbulence Interaction

2.1 The PDE System

The proposed stability conditions are well exemplified by the next nonconservative system motivated by the physics of compressible turbulent flows

$$\begin{cases} \partial_t \rho^\epsilon + \partial_x(\rho u)^\epsilon = 0, \\ \partial_t(\rho u)^\epsilon + \partial_x(\rho u^2 + p(\rho) + R_{11})^\epsilon = \epsilon \partial_x(\mu \partial_x u^\epsilon), \\ \partial_t(\rho v)^\epsilon + \partial_x(\rho u v + R_{12})^\epsilon = \epsilon \partial_x(\nu \partial_x v^\epsilon), \\ \partial_t R_{11}^\epsilon + \partial_x(R_{11}u)^\epsilon + 2R_{11}^\epsilon \partial_x u^\epsilon = -2\epsilon \mathcal{T} \frac{R_{11}^\epsilon}{(\rho^\epsilon)^2} p'(\rho^\epsilon)(\partial_x \rho^\epsilon)^2, \\ \partial_t R_{22}^\epsilon + \partial_x(R_{22}u)^\epsilon + 2R_{12}^\epsilon \partial_x v^\epsilon = 0, \\ \partial_t R_{12}^\epsilon + \partial_x(R_{12}u)^\epsilon + R_{11}^\epsilon \partial_x v^\epsilon + R_{12}^\epsilon \partial_x u^\epsilon = -\epsilon \mathcal{T} \frac{R_{12}^\epsilon}{(\rho^\epsilon)^2} p'(\rho^\epsilon)(\partial_x \rho^\epsilon)^2. \end{cases} \tag{14}$$

This PDE system governs plane wave solutions of the celebrated Favre-averaged form of the compressible Navier–Stokes equations in two space dimensions. Here, u and v are the normal and axial components of $< U > = \overline{\rho U}/\overline{\rho}$, the Favre density-weighted average of the velocity U. Then $R_{ij}, 1 \leq i,j \leq 2$, are the components of the symmetric Reynolds stress tensor $< U' \otimes U' >$ where U' denotes the fluctuation of U from $< U >$. For simplicity but without real restriction, we assume an isothermal pressure law $p(\rho) = T_0 \rho$, $T_0 > 0$. Of central importance here, $\mathcal{T} \geq 0$ is the so-called Ristorcelli timescale. $\mathcal{T}$ is usually set to zero in the literature but large enough values of this timescale are seen below to be mandatory for stability issues. The reader is referred to [2] for precise definitions about $\mathcal{T}$ and the proof of the forthcoming statements.

The system (14) is equipped with the next phase space

$$\Omega_{\mathbf{u}} = \Big\{ \mathbf{u} = (\rho, \rho u, \rho v, R_{11}, R_{12}, R_{22})^T \in I\!R^6 / \ \rho > 0, \ (\rho u, \rho v) \in I\!R^2, \\ R_{ii} \geq 0, \ i = 1,2, \ R_{11}R_{22} - R_{12}^2 \geq 0 \Big\}. \tag{15}$$

The positivity requirements put on the Reynolds stresses, modelling $< U' \otimes U' >$, are actually natural. They are seen to be met by smooth solutions of the Cauchy problem (14) for initial data $\mathbf{u}_0$ taking values in $\Omega_{\mathbf{u}}$ (see [2]). Observe from (15) that a state with $R_{11} = 0$ necessarily obeys $R_{12} = 0$. To focus on the important case $R_{12} \neq 0$, we define $\Omega_{\mathbf{u}}^+ = \{\mathbf{u} \in \Omega_{\mathbf{u}}, \ R_{11} > 0\}$. Solutions with initial data in $\Omega_{\mathbf{u}}^+$ stay there for all finite time. We first state the following lemma.

Lemma 1. *The first-order system in (14) is hyperbolic over the phase space $\Omega_{\mathbf{u}}$. The following eigenvalues are in order*

$$u - c(\mathbf{u}) < \ u - a(\mathbf{u}) \leq u \ (double) \ \leq \ u + a(\mathbf{u}) < \ u + c(\mathbf{u}), \tag{16}$$

with $0 \leq a^2(\mathbf{u}) = \frac{R_{11}}{\rho} < c^2(\mathbf{u}) = T_0 + 3a^2(\mathbf{u})$. *The two extreme fields are genuinely nonlinear while all the other intermediate ones are linearly degenerate.*

2.2 Travelling Wave Analysis

It turns [2] that solely the shock solutions coming with the extreme fields deserve a special analysis in the limit $\epsilon \to 0+$. Discontinuous solutions of the intermediate families are well defined in this inviscid limit. Due to Galilean invariance, it suffices to study the viscous profiles of the first family. Existence follows from

Proposition 2. *For a given state* $\mathbf{u}_- \in \Omega^+_{\mathbf{u}}$, *let the velocity* $\sigma \in I\!R$ *be prescribed so that* $u_- - c(\mathbf{u}_-) > \sigma$. *Then for all values of the Ristorcelli timescale* $\mathcal{T} \geq 0$, *there exists a travelling wave solution of (14) with speed* σ, *issuing from* $\mathbf{u}_-$ *and arriving at some state* $\mathbf{u}_+ \in \Omega^+_{\mathbf{u}}$. *In addition the compression ratio* $\tau^+ = \rho(\mathbf{u}_-)/\rho(\mathbf{u}_+)$ *stays always less than one.*

By contrast, the amplitude of $\mathcal{T}$ plays a major role in uniqueness in view of

Proposition 3. *(i) Assume* $\mathcal{T} = 0$. *A left state* $\mathbf{u}_-$ *being fixed, there exits* $\sigma^\star \in I\!R$ *with* $u_- - c(\mathbf{u}_-) > \sigma^\star$ *so that infinitely many distinct right states* $\mathbf{u}_+ \in \Omega^+_{\mathbf{u}}$ *can be connected by travelling wave solutions of (14) propagating with a given speed* σ, $\sigma^\star > \sigma$, *and issuing from the fixed state* $\mathbf{u}_-$. *Each of these solutions is overcompressive, namely*

$$u_+ - c(\mathbf{u}_+) < \sigma < u_- - c(\mathbf{u}_-), \quad u_+ - a(\mathbf{u}_+) < \sigma. \tag{17}$$

(ii) A state $\mathbf{u}_-$ *and a velocity* σ *being given with* $u - c(\mathbf{u}_-) > \sigma$, *then there exists* $\mathcal{T}^\star > 0$ *large enough, so that for all values of the timescale* $\mathcal{T} > \mathcal{T}^\star$ *uniqueness of the travelling wave solution is restored. The resulting unique end state* $\mathbf{u}_+$ *obeys the Lax conditions*

$$u_+ - c(\mathbf{u}_+) < \sigma < u_- - c(\mathbf{u}_-), \quad \sigma < u_+ - a(\mathbf{u}_+). \tag{18}$$

In the conservative setting, failure of uniqueness takes place for overcompressive viscous profiles but due to the conservation property, all the solutions propagating at a given speed σ must connect the same pair of end states $\mathbf{u}_-$ and $\mathbf{u}_+$ (see [8] concerning stability issues in the viscous versus inviscid regime). In the present nonconservative setting, overcompressive profiles with speed σ do connect distinct states $\mathbf{u}_+$ from the same state $\mathbf{u}_-$. With this respect and besides stability issues, shock solutions cannot even be uniquely defined from the travelling wave solutions.

2.3 Shock Solutions

From now on, we always assume large enough values of the Ristorcelli timescale $\mathcal{T}$ to guarantee uniqueness in the inviscid limit of the family of travelling wave solutions. In view of (18), we therefore deal with Lax shock solutions. Let us the address the characterization of the resulting family of Lax 1-shock solutions thanks to the framework of kinetic relations. Useful nontrivial additional laws satisfied by their viscous profiles are inferred from

Proposition 4. *Smooth solutions* $\mathbf{u} : I\!R^+ \times I\!R \to \Omega_{\mathbf{u}}^+$ *of (14) verify*

$$\partial_t\{\rho E_t\}(\mathbf{u}^\epsilon) + \partial_x(\{\rho E_t\}(\mathbf{u}^\epsilon)u^\epsilon + R_{12}^\epsilon v^\epsilon) = \epsilon\partial_x(\nu v^\epsilon \partial_x v^\epsilon) - \epsilon\nu(\partial_x v^\epsilon)^2 \quad (19)$$

when defining the tangential energy ρE_t *by:*

$$\{\rho E_t\}(\mathbf{u}) = \rho\frac{v^2}{2} + \frac{R_{12}^2}{2R_{11}}. \quad (20)$$

In addition, these smooth solutions obey:

$$\partial_t\{\rho\mathcal{W}\}(\mathbf{u}^\epsilon) + \partial_x\{\rho\mathcal{W}\}(\mathbf{u}^\epsilon)u^\epsilon = 0, \quad \{\rho\mathcal{W}\}(\mathbf{u}) = R_{22} - \frac{R_{12}^2}{R_{11}}, \quad (21)$$

together with:

$$\partial_t\{\rho\mathcal{I}\}(\mathbf{u}^\epsilon) + \partial_x\{\rho\mathcal{I}\}(\mathbf{u}^\epsilon)u^\epsilon = -2\epsilon\mathcal{T}T_0\frac{\mathcal{I}(\mathbf{u}^\epsilon)}{\rho^\epsilon}(\partial_x\rho^\epsilon)^2, \quad \mathcal{I}(\mathbf{u}) = R_{11}\tau^3. \quad (22)$$

Observe that the mapping $\mathbf{u} \to \mathbf{v}(\mathbf{u}) = \{\rho, \rho u, \rho v, \rho\mathcal{I}(\mathbf{u}), \rho\mathcal{W}(\mathbf{u}), \rho E_t(\mathbf{u})\}$ does not bring a relevant change of variable for general solutions of (14) since the sign of the tangential Reynolds stress R_{12} cannot be determined from neither the law (20) or (21). Additional nontrivial laws are at hand [2] but share the same property. Nevertheless, the proposed mapping $\mathbf{u} \to \mathbf{v}(\mathbf{u})$ makes sense in the class of travelling wave solutions of (14) in view of the following proposition.

Proposition 5. *Let be given* $\mathbf{u}_- \in \Omega_{\mathbf{u}}^+$ *with* $R_{12}^- \neq 0$ *and a speed* σ *with* $u - c$ $(\mathbf{u}_-) > \sigma$*. Assume large enough value of the timescale* $\mathcal{T} > 0$ *for uniqueness of the travelling wave solution* $\mathbf{u}(\xi, \mathbf{u}_-, \sigma)$*. Then states in this solution obey* $R_{12}^- R_{12}(\xi) > 0$ *for all* $\xi \in I\!R$ *with* $\lim_{\xi\to+\infty} R_{12}^- R_{12}(\xi) > 0$*. For a state* $\mathbf{u}_-$ *with* $R_{12}^- = 0$*, the tangential Reynolds stress* $R_{12}(\xi)$ *stays identically zero for all time.*

This result therefore allows for the next characterization of Lax 1-shock solutions via a complete set of jump conditions with kinetic relations.

Proposition 6. *From any given pair* $(\mathbf{u}_-, \sigma) \in \Omega_{\mathbf{u}}^+ \times I\!R$ *with* $u - c(\mathbf{u}_-) > \sigma$*, let us define the following kinetic functions:*

$$\begin{aligned} \mathcal{K}_{\mathcal{I}}(\mathbf{u}_-, \sigma) &= -\int_{I\!R}\left\{2\mathcal{T}\rho\mathcal{I}\frac{d_\xi p\ d_\xi\rho}{\rho^2}\right\}(\mathbf{u}(\xi, \mathbf{u}_-, \sigma))d\xi, \\ \mathcal{K}_{E_t}(\mathbf{u}_-, \sigma) &= -\int_{I\!R}\nu\left\{d_\xi v\right\}^2(\mathbf{u}(\xi, \mathbf{u}_-, \sigma))d\xi, \end{aligned} \quad (23)$$

where $\mathbf{u}(., \mathbf{u}_-, \sigma)$ *denotes the travelling wave solution with speed* σ *issuing from* $\mathbf{u}_-$*. Then the state* $\mathbf{u}_+ \in \Omega_{\mathbf{u}}^+$ *in the Lax 1-shock solution (18) is the unique solution of the following generalized jump conditions:*

$$
\begin{aligned}
&-\sigma[\rho] + [\rho u] = 0, \\
&-\sigma[\rho u] + [\rho u^2 + p(\rho) + R_{11}] = 0, \\
&-\sigma[\rho \mathcal{I}(\mathbf{u})] + [\rho \mathcal{I}(\mathbf{u}) u] = \mathcal{K}_{\mathcal{I}}(\mathbf{u}_-, \sigma), \\
&-\sigma[\rho v] + [\rho v u + R_{12}] = 0, \\
&-\sigma[\tfrac{\rho v^2}{2} + \tfrac{R_{12}^2}{2R_{11}}] + [(\tfrac{\rho v^2}{2} + \tfrac{R_{12}^2}{2R_{11}})u + R_{12} v] = \mathcal{K}_{E_t}(\mathbf{u}_-, \sigma), \\
&-\sigma[\rho \mathcal{W}(\mathbf{u})] + [\rho \mathcal{W}(\mathbf{u}) u] = 0,
\end{aligned}
\tag{24}
$$

where for definiteness $R_{12}^- R_{12}^+ > 0$ *if* $R_{12}^- \neq 0$ *and* $R_{12}^+ = 0$ *otherwise.*

To conclude this section, we stress that the travelling wave analysis shows [2] that the dimensionless forms $\overline{\mathcal{K}}_{\mathcal{I}}$ and $\overline{\mathcal{K}}_{E_t}$ of the kinetic functions $\mathcal{K}_{\mathcal{I}}(\mathbf{u}, \sigma)$, $\mathcal{K}_{E_t}(\mathbf{u}, \sigma) : \Omega_{\mathbf{u}}^+ \times I\!R \to I\!R^-$ in (23) actually solely depend on two reduced real numbers: namely the relative Mach number $M(\mathbf{u}, \sigma)$ and the turbulent Mach number $M_t(\mathbf{u})$, respectively, defined by

$$
M(\mathbf{u}, \sigma) = \frac{u - \sigma}{c(\mathbf{u})}, \quad M_t(\mathbf{u}) = \frac{a(\mathbf{u})}{c(\mathbf{u})}. \tag{25}
$$

When the pair $(\mathbf{u}_-, \sigma)$ runs over $\Omega_{\mathbf{u}}^+ \times I\!R$ with $u_- - c(\mathbf{u}_-) > \sigma$, $M(\mathbf{u}_-, \sigma)$ covers $[1, \infty)$ while $M_t(\mathbf{u}_-)$ stays in $(0, 1/\sqrt{3})$. The reported properties make feasible the tabulation of the two dimensionless kinetic functions $\overline{\mathcal{K}}_{\mathcal{I}}(M, M_t)$ and $\overline{\mathcal{K}}_{E_t}(M, M_t)$ for practical values of the Mach number $M \in [1, 10]$ from a suitable numerical approximation of the travelling wave solutions of (14) (see [1] for the details).

2.4 Structural Shock Stability Conditions

The Ristorcelli timescale $\mathcal{T} > 0$ is chosen large enough so as to give birth to 1-shock solutions satisfying the Lax conditions stated in (11) Proposition 1. The stability of the resulting Lax shock family is then the matter of the following proposition.

Proposition 7. *Let us define from the Lax* 1*-shock family of Proposition 6* $M = M(\mathbf{u}_-, \sigma) > 1$, $M_t = M_t(\mathbf{u}_-) > 0$ *and also* $M_+ = M(\mathbf{u}_+, \sigma) \in (0, 1]$. *Let us consider the compression ratio* $\tau_+ = \rho(\mathbf{u}_-)/\rho(\mathbf{u}_+)$ *with* $\tau_+ \in (0, 1]$ *from Proposition 2. Then this family of Lax shocks is structurally stable if and only if the following dimensionless form of the Majda determinant (12)*

$$
\overline{\Delta} = \tau_+ M \left((1 - \tau_+) \tau_+^3 \frac{M}{M_+^2} (1 + M_+) - M_t^2 \frac{\partial \overline{\mathcal{K}}_{\mathcal{I}}}{\partial M}(M, M_t) \right) \tag{26}
$$

stays positive for all values of $M > 1$ *and* $M_t \in (0, 1/\sqrt{3})$.

A careful study of the travelling wave solutions of (14) then shows that it is possible to define the Ristorcelli timescale as a smooth function $\mathcal{T}(M, M_t)$ of the reduced numbers M and M_t, with sufficiently large values to ensure Lax shock solutions while achieving the following monotonicity property (see [2] for the details)

$$M \in [1, \infty) \to \overline{\mathcal{K}}_{\mathcal{I}}(M, M_t) \in I\!R^- \text{is a strictly decreasing function.} \tag{27}$$

In other words the Majda determinant (26) stays always positive: namely, Lax 1-shock solutions with (27) turn out to be structurally stable. The reported property (27) is actually achieved in [1] when enforcing for validity the next monotonicity of the compression ratio τ_+ expressed as a function of (M, M_t)

$$\partial_M \tau_+(M, M_t) < 0, \quad \text{for all } M > 1. \tag{28}$$

Let us underline that the decreasing property (28) is sufficient [2] (see also [5] for another setting) to ensure uniqueness of the solution of the Riemann problem for the inviscid limit model (14). In addition, if the compression rate $\tau^+(M, M_t)$ departs too much from the estimate (28), uniqueness is easily seen to be lost (see [5] for instance). We refer the reader to Serre [8] for the relationships between structural stability of Lax shock solutions and uniqueness of solutions in the Riemann problem.

To conclude, we emphasize that a natural extension of the uniform Lopatinski condition for the linearized stability of planar two-dimensional shock solutions [8] can be also derived thanks to the framework of kinetic relations (see [1]). In particular, the kinetic functions $\mathcal{K}_{\mathcal{I}}$ and $\mathcal{K}_{E_t}$ entering (24) are again in order thanks to frame invariance properties satisfied by the PDE model but with properly modified arguments to account for multidimensional effects. In [1], a numerical method based on the argument principle is derived to investigate possible zeros of the resulting determinant for relative Mach numbers M in the range $(1, 10)$. It is seen that the Ristorcelli timescale $\mathcal{T}$ once again plays a decisive role in the stability issue: its amplitude has to be chosen even larger to prevent planar shocks from strong instabilities.

Acknowledgments

The first author has been partly financially supported by EDF-R&D in the completion of the present work. The second author acknowledges part financial support from ONERA.

References

1. B. Audebert, Contribution à l'Etude de l'interaction turbulence onde de choc, Thesis, Paris VI University, Paris (2006)
2. B. Audebert and F. Coquel, Role of the Ristorcelli time scale in the traveling wave solutions of compressible turbulent models, work in preparation
3. S. Benzoni, Stability of multi-dimensional phase transitions in a van der Waals fluid, Nonlinear Analysis T.M.A. 31, No. 1/2, 243–263, (1998)

4. C. Berthon, F. Coquel and P.G. LeFloch, Kinetic relations for defining shock solutions of non conservative hyperbolic systems, submitted
5. C. Chalons and F. Coquel, The Riemann problem for the multi-pressure Euler system, J. Hyperbolic Differ. Equ. 2, No 3, 745–782, (2005)
6. C. Chalons and F. Coquel, Navier-Stokes equations with several indepedant pressure laws and explicit predictor-corrector schemes, Numer. Math., vol. 101, No. 3, 451–478, (2005)
7. P.G. LeFloch, Entropy weak solutions to nonlinear hyperbolic systems in nonconservative form, Comm. Partial Differential Equations, vol. 13, (1988)
8. D. Serre, Systems of conservation laws, I and II, Cambridge U. Press, (1999)

A Hyperbolic Model of Multiphase Flow

D. Amadori and A. Corli

1 Introduction

We consider the following model for the flow of an inviscid fluid admitting liquid and vapor phases:

$$\begin{cases} v_t - u_x & = 0, \\ u_t + p(v, \lambda)_x = 0, \\ \lambda_t & = 0. \end{cases} \tag{1}$$

Here $t > 0$ and $x \in \mathbb{R}$; moreover $v > 0$ is the specific volume, u the velocity, λ the mass density fraction of vapor in the fluid. Then $\lambda \in [0, 1]$, with $\lambda = 0$ characterizes the liquid and $\lambda = 1$ the vapor phase; intermediate values of λ model mixtures of the two pure phases. The pressure is $p = p(v, \lambda)$; under natural assumptions the system is strictly hyperbolic. We refer to [4, 3] for more information on the model. System (1) has close connections to a system considered by Peng [6]. A comparison of the two models is done in [1].

We consider and prove here, as a preliminary study for a forthcoming paper, the basic features of system (1): wave curves, Riemann problem, wave interactions. The results improve some of those in [6]; the proofs are different. We refer to [1] for more details as well as, for instance, refined interaction estimates and a simple but complete proof of Glimm estimates.

2 Assumptions, Wave Curves and the Riemann Problem

We consider a pressure law of the form

$$p(v, \lambda) = \frac{A(\lambda)}{v}, \tag{2}$$

where $A(\lambda) = a^2(\lambda)$ is a smooth function defined on $[0, 1]$ satisfying for every $\lambda \in [0, 1]$

$$A(\lambda) > 0, \qquad A'(\lambda) > 0.$$

We denote $U = (v, u, \lambda) \in \Omega = (0, +\infty) \times \mathbb{R} \times [0, 1]$ and by $\tilde{U} = (v, u)$ the projection of U onto the plane vu. Under the above assumptions on the pressure the system (1) is strictly hyperbolic in the whole Ω: the eigenvalues are $\pm\sqrt{-p_v(v, \lambda)}$, both genuinely nonlinear, and 0, which is linearly degenerate.

The direct shock-rarefaction curves through $U_o = (v_o, u_o, \lambda_o)$ for (1) are

$$\Phi_1(v, U_o) = \begin{cases} \left(v, u_o + a(\lambda_o)\dfrac{v - v_o}{\sqrt{vv_o}}, \lambda_o\right) & v < v_o \text{ shock}, \\ \left(v, u_o + a(\lambda_o)\log\dfrac{v}{v_o}, \lambda_o\right) & v > v_o \text{ rarefaction}, \end{cases} \tag{3}$$

$$\Phi_2(\lambda, U_o) = \left(v_o \frac{a^2(\lambda)}{a^2(\lambda_o)}, u_o, \lambda\right) \quad \lambda \in [0, 1] \quad \text{contact discontinuity}, \tag{4}$$

$$\Phi_3(v, U_o) = \begin{cases} \left(v, u_o - a(\lambda_o)\log\dfrac{v}{v_o}, \lambda_o\right) & v < v_o \text{ rarefaction}, \\ \left(v, u_o - a(\lambda_o)\dfrac{v - v_o}{\sqrt{vv_o}}, \lambda_o\right) & v > v_o \text{ shock}. \end{cases} \tag{5}$$

Remark that the pressure is constant along contact discontinuities. The curves Φ_1, Φ_2 and Φ_3 are *plane* curves; as for states we denote by $\tilde{\Phi}_i$ the projection of these curves on the plane vu. We denote the u-component of the 1-(3)-shock-rarefaction curves by $\phi_1(v, U_o)$ (resp. $\phi_3(v, U_o)$) so that $\Phi_i(v, U_o) = (v, \phi_i(v, U_o), \lambda_o)$ for $i = 1, 3$.

Lemma 1. *Fix any $(v_o, u_o, \lambda_o) \in \Omega$. Then for $i = 1, 3$ and any $\alpha > 0$, $\lambda_1, \lambda_2 \in [0, 1]$, $u_1, u_2 \in \mathbb{R}$:*

$$\phi_i\left(\alpha v, (\alpha v_o, u_o, \lambda_o)\right) = \phi_i\left(v, (v_o, u_o, \lambda_o)\right) \tag{6}$$

$$\frac{\phi_i\left(v, (v_o, u_1, \lambda_1)\right) - u_1}{a(\lambda_1)} = \frac{\phi_i\left(v, (v_o, u_2, \lambda_2)\right) - u_2}{a(\lambda_2)}. \tag{7}$$

Moreover, for $i, j = 1, 3$, $i \neq j$, and $\bar{v} = \frac{v_o^2}{v}$,

$$\phi_i(v, U_o) = \phi_j(\bar{v}, U_o). \tag{8}$$

Remark that property (8) exchanges shocks of the first family with shocks of the third-one, and analogously for rarefactions. The proof of the lemma follows from the definition of the functions ϕ_i; equality (6) is a consequence of the property of congruence of shock curves by rigid motions for fixed λ [5].

Definition 1. *With the notation (3)–(5) we define the strength ε_i of a i-wave by*

$$\varepsilon_1 = \frac{1}{2}\log\left(\frac{v}{v_o}\right), \quad \varepsilon_2 = 2\,\frac{a(\lambda) - a(\lambda_o)}{a(\lambda) + a(\lambda_o)}, \quad \varepsilon_3 = \frac{1}{2}\log\left(\frac{v_o}{v}\right). \tag{9}$$

We define the function h as $h(\varepsilon) = 2\varepsilon$ if $\varepsilon \geq 0$ and $h(\varepsilon) = 2\sinh\varepsilon$ if $\varepsilon < 0$. Then $|h(\varepsilon)| \geq 2|\varepsilon|$ and $\phi_i(v, U_o) = u_o + a(\lambda_o) \cdot h(\varepsilon_i)$ for $i = 1, 3$.

We consider the Riemann problem, i.e., the initial-value problem for (1) under the piecewise constant initial conditions

$$(v, u, \lambda)(0, x) = \begin{cases} (v_\ell, u_\ell, \lambda_\ell) = U_\ell & \text{if } x < 0, \\ (v_r, u_r, \lambda_r) = U_r & \text{if } x > 0, \end{cases} \tag{10}$$

for states U_ℓ and U_r in Ω. We denote $A_r = A(\lambda_r)$, $A_\ell = A(\lambda_\ell)$, $A_{r\ell} = A(\lambda_r)/A(\lambda_\ell)$; analogous notations are used for the function a.

If $\lambda_r = \lambda_\ell$ then the solution to the Riemann problem is classical, [7]; otherwise we proceed as follows. For any pair of states U_ℓ, U_r in Ω we introduce

$$U_{\ell r}^* = \Phi_2(\lambda_r, U_\ell) = (A_{r\ell} v_\ell, u_\ell, \lambda_r).$$

The state $U_{\ell r}^*$ is the unique state with mass density fraction λ_r that can be connected to U_ℓ by a 2-contact discontinuity; both states have then the same pressure. It lies on the right, resp. on the left, of $\tilde{U}_\ell$ according to either $\lambda_\ell < \lambda_r$ or $\lambda_\ell > \lambda_r$. Consider for $v > 0$ the curves

$$\Phi_2\left(\lambda_r, \Phi_1(v, U_\ell)\right) = (A_{r\ell} v, \phi_1(v, U_\ell), \lambda_r), \tag{11}$$

$$\Phi_2\left(\lambda_r, \Phi_3(v, U_\ell)\right) = (A_{r\ell} v, \phi_3(v, U_\ell), \lambda_r). \tag{12}$$

The first curve is the composition of the 1-curve through U_ℓ with the 2-curve to λ_r and passes through the point $U_{\ell r}^*$ when $v = v_\ell$. Similarly, the second is the composition of the 3-curve through U_ℓ with the 2-curve to λ_r and passes through the point $U_{\ell r}^*$ when $v = v_\ell$. We change parametrization $v \to v/A_{r\ell}$ in (11), (12) and define for $i = 1, 3$

$$\Phi_{i2}(v, U_\ell, \lambda_r) = \left(v, \phi_i\left(\frac{v}{A_{r\ell}}, U_\ell\right), \lambda_r\right) = (v, \phi_{i2}(v, U_\ell, \lambda_r), \lambda_r). \tag{13}$$

The point $U_{\ell r}^*$ corresponds now to $v = A_{r\ell} v_\ell$. In an analogous way we define for $v > 0$ and $i = 1, 3$ the curves Φ_{2i}, ϕ_{2i}, by

$$\Phi_{2i}(v, U_\ell, \lambda_r) = \Phi_i\left(v, \Phi_2(\lambda_r, U_\ell)\right) = (v, \phi_i(v, U_{\ell r}^*), \lambda_r) = (v, \phi_{2i}(v, U_\ell, \lambda_r), \lambda_r).$$

The curves defined above are the composition of the 2-curve from U_ℓ to $U_{\ell r}^*$ with the i-curve through $U_{\ell r}^*$. Both Φ_{21} and Φ_{23} pass through the point $U_{\ell r}^*$ when $v = A_{r\ell} v_\ell$.

Lemma 2 (Commutation of curves). *For $i = 1, 3$, the $i2$- and the $2i$-curves from U_ℓ are related by*

$$\frac{\phi_{i2}(v, U_\ell, \lambda_r) - u_\ell}{a_\ell} = \frac{\phi_{2i}(v, U_\ell, \lambda_r) - u_\ell}{a_r}. \tag{14}$$

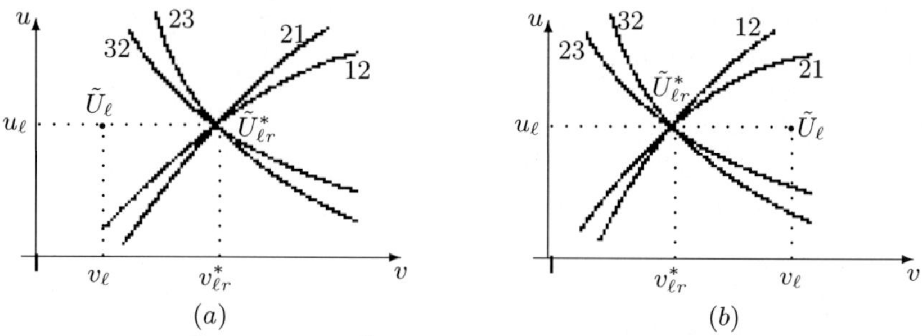

Fig. 1. The curves $\tilde\Phi_{i2}(v,U_\ell,\lambda_r)$, $\tilde\Phi_{2i}(v,U_\ell,\lambda_r)$, $i=1,3$. (**a**) $\lambda_r>\lambda_\ell$ and (**b**) $\lambda_r<\lambda_\ell$

Proof. Using first (6) and then (7), we have

$$\frac{\phi_{i2}(v,U_\ell,\lambda_r)-u_\ell}{a_\ell} = \frac{\phi_i\left(\frac{v}{A_{r\ell}},U_\ell\right)-u_\ell}{a_\ell} = \frac{\phi_i\left(v,(A_{r\ell}v_\ell,u_\ell,\lambda_\ell)\right)-u_\ell}{a_\ell}$$
$$= \frac{\phi_i\left(v,(A_{r\ell}v_\ell,u_\ell,\lambda_r)\right)-u_\ell}{a_r} = \frac{\phi_{2i}(v,U_\ell,\lambda_r)-u_\ell}{a_r}. \qquad \square$$

A consequence of the lemma is that if $\lambda_r\neq\lambda_\ell$ then the curves $\Phi_{i2}(v,U_\ell,\lambda_r)$ and $\Phi_{2i}(v,U_\ell,\lambda_r)$, $i=1,3$, meet only for $v=v^*_{\ell r}$. More precisely remark that $\phi'_{i2}(v,U_\ell,\lambda_r)=\frac{1}{a_{r\ell}}\phi'_{2i}(v,U_\ell,\lambda_r)$ for every $v>0$. Therefore if $\lambda_r>\lambda_\ell$ then $\phi'_{23}<\phi'_{32}<0<\phi'_{12}<\phi'_{21}$, while if $\lambda_r<\lambda_\ell$ then $\phi'_{32}<\phi'_{23}<0<\phi'_{21}<\phi'_{12}$. The mutual position of the four curves $\tilde\Phi_{ij}$ is as in Fig. 1.

Theorem 1. *For any pair of states U_ℓ, U_r in Ω the Riemann problem (1) (10) has a unique Ω-valued solution in the class of solutions consisting of simple Lax waves. If ε_i is the strength of the i-wave, $i=1,2,3$, then*

$$\varepsilon_3-\varepsilon_1=\frac{1}{2}\log\left(\frac{A_{r\ell}v_\ell}{v_r}\right), \tag{15}$$
$$a_\ell h(\varepsilon_1)+a_r h(\varepsilon_3)=u_r-u_\ell. \tag{16}$$

Moreover, let $\underline{v}>0$ be a fixed number. There exists a constant $C_1>0$ depending on $\underline{v}$ and $a(\lambda)$ such that if $v_l,v_r>\underline{v}$ then

$$|\varepsilon_1|+|\varepsilon_2|+|\varepsilon_3|\le C_1|U_\ell-U_r|. \tag{17}$$

Proof. The case $\lambda_r=\lambda_l$ is solved as in [7]. We consider the case $\lambda_r>\lambda_\ell$. Then $\tilde U^*_{\ell r}$ lies on the right of $\tilde U_\ell$: $v_\ell<v^*_{\ell r}=A_{r\ell}\cdot v_\ell$. The curves $\tilde\Phi_{12}(v,U_\ell,\lambda_r)$ and $\tilde\Phi_{23}(v,U_\ell,\lambda_r)$ divide the plane into four regions, see Fig. 2. The different patterns of the solution are classified as in the case of the p-system, [7], according to $\tilde U_r$ belongs to the regions RS, SS, SR, RR.

Assume that $\tilde U_r$ lies in region RS. Then by a continuity and transversality argument, [7], there exists a unique point $U_{mr}=(v_{mr},u_{mr},\lambda_r)$ on the curve

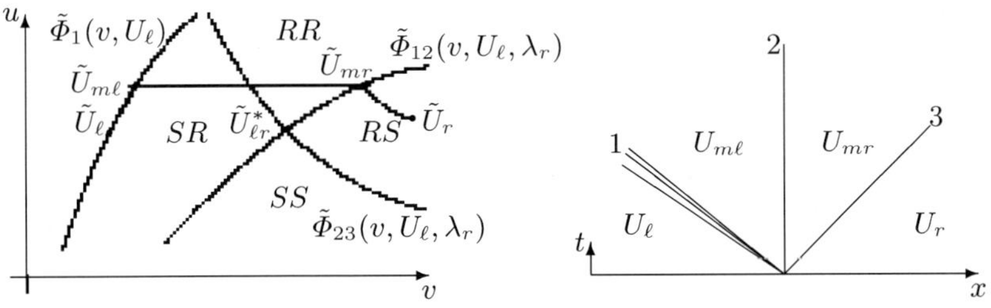

Fig. 2. Solution to the Riemann problem

$\Phi_{12}(v, U_\ell, \lambda_r)$ such that the 3-curve through U_{mr} passes by U_r. The curve $\phi_{12}(v, U_\ell, \lambda_r)$ is strictly monotone and surjective on $\mathbb{R}$; then we find a unique state $U_{m\ell} = (v_{m\ell}, u_{m\ell}, \lambda_\ell)$ on the 1-curve through U_ℓ, with $u_{m\ell} = u_{mr}$. The solution to the Riemann problem is then given by a 1-wave from U_ℓ to $U_{m\ell}$, a 2-wave to U_{mr} and a 3-wave to U_r. The other cases are treated analogously.

Formula (15) follows from the definition of strengths (9) as well as (16). Finally, let us prove (17). Concerning ε_2, remark that $|\varepsilon_2| \leq \frac{\max a'}{\min a} \cdot |\lambda_r - \lambda_\ell|$. If $\varepsilon_1 \varepsilon_3 \leq 0$, from (15) we get

$$
\begin{aligned}
|\varepsilon_1| + |\varepsilon_3| &= \frac{1}{2} |\log v_\ell - \log v_r| + |\log a_\ell - \log a_r| \\
&\leq \frac{1}{2\underline{v}} |v_r - v_\ell| + \frac{\max a'}{\min a} \cdot |\lambda_r - \lambda_\ell| .
\end{aligned}
$$

If $\varepsilon_1 \varepsilon_3 > 0$, from (16) we get $|u_r - u_\ell| = a_\ell \, |h(\varepsilon_1)| + a_r \, |h(\varepsilon_3)| \geq 2a_\ell \, |\varepsilon_1| + 2a_r \, |\varepsilon_3|$. Then $|\varepsilon_1| + |\varepsilon_3| \leq \frac{1}{2 \min a} \, |u_r - u_\ell|$. Then (17) follows for a suitable C_1. □

3 Interactions

We focus on interactions involving contact discontinuities, the interactions of 1- and 3-waves being treated as in [5, 7], see also [2].

Proposition 1. *Let λ_ℓ, λ_r be the side values of λ along a 2-wave. The interactions of* 1- *or* 3-*waves with the 2-wave give rise to the following pattern of solutions:*

interaction	*outcome*	
	$\lambda_\ell < \lambda_r$	$\lambda_\ell > \lambda_r$
$2 \times 1R$	$1R + 2 + 3R$	$1R + 2 + 3S$
$2 \times 1S$	$1S + 2 + 3S$	$1S + 2 + 3R$
$3R \times 2$	$1S + 2 + 3R$	$1R + 2 + 3R$
$3S \times 2$	$1R + 2 + 3S$	$1S + 2 + 3S$.

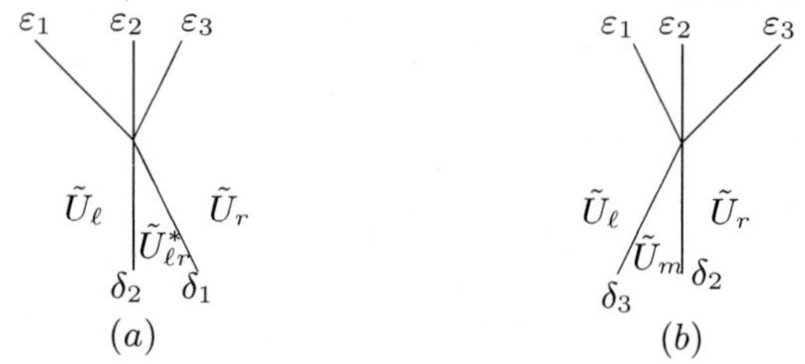

Fig. 3. Interactions: (**a**) from the right and (**b**) from the left

Proof. The interactions are solved in a geometric way by referring to Fig. 1. Thick lines split the plane according to different patterns of solution of the Riemann problem, thin lines describe the interaction.

Case 2×1. Consider the case of the interaction of the 2-wave with a 1-wave coming from the right, see Fig. 3. The state U_r lies on the curve $\Phi_{21}(v, U_\ell, \lambda_r)$. If the incoming wave is a rarefaction, then $v_r > v^*_{\ell r}$, if it is a shock, then U_r lies on the left branch of Φ_{21}: $v_r < v^*_{\ell r}$. Assume $\lambda_r > \lambda_\ell$. Then the curve $\tilde{\Phi}_{21}(v, U_\ell, \lambda_r)$ is contained either in the region RR if $v > v^*_{\ell r}$ or in SS if $v < v^*_{\ell r}$. The interaction is then solved according to Theorem 1 either by a rarefaction of the first family, a contact discontinuity and a rarefaction of the third family or by a shock of the first family, a contact discontinuity and a shock of the third family. If $\lambda_r < \lambda_\ell$ then the curve $\tilde{\Phi}_{21}(v, U_\ell, \lambda_r)$ is contained either in the region RS if $v > v^*_{\ell r}$ (incoming rarefaction) or in SR if $v < v^*_{\ell r}$ (incoming shock). The interaction pattern is solved again by Theorem 1.

Case 3×2. Consider now the case of the interaction of the 2-wave with a 3-wave coming from the left. The state U_r lies now on the curve $\Phi_{32}(v, U_\ell, \lambda_r)$. One can check that if the incoming wave is a rarefaction then $v_r < v^*_{\ell r}$, while if it is a shock then $v_r > v^*_{\ell r}$. If $\lambda_r > \lambda_\ell$ the curve $\tilde{\Phi}_{32}(v, U_\ell, \lambda_r)$ is contained either in the region SR if $v < v^*_{\ell r}$ or in RS if $v > v^*_{\ell r}$. The interaction is then solved either by a shock of the first family, a contact discontinuity and a rarefaction of the third family or by a rarefaction of the first family, a contact discontinuity and a shock of the third family. If $\lambda_r < \lambda_\ell$ then the curve $\tilde{\Phi}_{21}(v, U_\ell, \lambda_r)$ is contained either in the region RR if $v < v^*_{\ell r}$ (incoming rarefaction) or in SS if $v > v^*_{\ell r}$ (incoming shock). The interaction is solved consequently. □

Theorem 2. *Assume that a 1-wave of strength δ_1 or a 3-wave of strength δ_3 interacts with a 2-wave of strength δ_2. Then the strengths ε_i of the outgoing waves satisfy $\varepsilon_2 = \delta_2$ and, for $[\delta_2]_+ = \max\{\delta_2, 0\}$, $[\delta_2]_- = \max\{-\delta_2, 0\}$,*

$$|\varepsilon_i - \delta_i| = |\varepsilon_j| \le \frac{1}{2}|\delta_2| \cdot |\delta_i| \qquad i,j = 1,3,\ i \neq j\,. \tag{18}$$

$$|\varepsilon_1| + |\varepsilon_3| \le \begin{cases} |\delta_1| + |\delta_1|[\delta_2]_+ & \text{if 1 interacts,} \\ |\delta_3| + |\delta_3|[\delta_2]_- & \text{if 3 interacts.} \end{cases} \tag{19}$$

Proof. We prove (18) when $i = 1$, $j = 3$. Assume that a 1-wave of strength δ_1 interacts with a 2-wave of strength $\delta_2 = 2\frac{a_r - a_\ell}{a_r + a_\ell}$. From (15) and comparing the velocities before and after the interactions we have

$$\varepsilon_3 - \varepsilon_1 = -\delta_1, \tag{20}$$

$$a_\ell h(\varepsilon_1) + a_r h(\varepsilon_3) = a_r \cdot h(\delta_1). \tag{21}$$

Using (20), (21) we get

$$a_\ell h\left(\delta_1 + \varepsilon_3\right) + a_r h(\varepsilon_3) = a_r h(\delta_1). \tag{22}$$

Recall that $\varepsilon_1 = \delta_1 + \varepsilon_3$ and δ_1 have the same sign; observe that $\varepsilon_3\delta_1\delta_2 > 0$. We consider the four possible types of interaction, as listed in Proposition 1.

• $2 \times 1R$, $a_r > a_\ell$. The identity (22) gives $a_\ell\left(\delta_1 + \varepsilon_3\right) + a_r\varepsilon_3 = a_r\delta_1$ from which we obtain $(a_\ell + a_r)\varepsilon_3 = (a_r - a_\ell)\delta_1$. Then (18) follows.

• $2 \times 1R$, $a_r < a_\ell$. Here (22) gives $a_\ell\left(\delta_1 + \varepsilon_3\right) - a_r\delta_1 = -a_r \sinh\varepsilon_3 \geq -a_r\varepsilon_3$ from which we obtain $-(a_r + a_\ell)\varepsilon_3 = (a_r + a_\ell)|\varepsilon_3| \leq (a_\ell - a_r)\delta_1$.

• $2 \times 1S$, $a_r > a_\ell$. From (22) we have $a_\ell \sinh\left(|\delta_1| + |\varepsilon_3|\right) + a_r \sinh(|\varepsilon_3|) = a_r \sinh(|\delta_1|)$. Denote $x = |\delta_1|$, $y = |\varepsilon_3|$, $k = a_r/a_\ell > 1$. Then the previous identity is written for $x \geq 0$, $y \geq 0$ as $F(x,y) = \sinh\left(x + y\right) + k\sinh(y) - k\sinh(x) = 0$. We have $F(x,0) < 0$ and $F(0,y) > 0$ for $x > 0$, $y > 0$; moreover $\partial F/\partial y > 0$. Therefore the implicit equation above is solved globally by $y = y(x)$.

The estimate (18) writes now $y(x) \leq \frac{k-1}{k+1}x$. To prove this it is sufficient to show that $F(x, \frac{k-1}{k+1}x) \geq 0$. The Mac Laurin expansion of $F(x, \frac{k-1}{k+1}x)$ is

$$\sum_{n=0}^{\infty} \frac{x^{2n+1}}{(2n+1)!(k+1)^{2n+1}} \left[(2k)^{2n+1} + k(k-1)^{2n+1} - k(k+1)^{2n+1}\right].$$

Consider the term in brackets; we claim that for every $n \geq 0$ and $k > 1$

$$(2k)^{2n+1} + k(k-1)^{2n+1} - k(k+1)^{2n+1} \geq 0\,. \tag{23}$$

Now $(k+1)^{2n+1} - (k-1)^{2n+1} = [(k+1) - (k-1)]\sum_{j=0}^{2n}(k-1)^j(k+1)^{2n-j}$ and $(2k)^{2n+1} = 2k \cdot [(k+1) + (k-1)]^{2n}$. Then the left side of (23) equals

$$2k\Big\{\sum_{j=0}^{2n}\binom{2n}{j}(k-1)^j(k+1)^{2n-j} - \sum_{j=0}^{2n}(k-1)^j(k+1)^{2n-j}\Big\},$$

which is always positive. That proves the claim and hence (18).

• $2 \times 1S$, $a_r < a_\ell$. Here $0 < \varepsilon_3 < |\delta_1|$, and (22) gives $a_\ell \sinh\left(|\delta_1| - \varepsilon_3\right) - a_r\varepsilon_3 = a_r \sinh(|\delta_1|)$. As in the previous case set $x = |\delta_1|$, $y = \varepsilon_3$, $k = a_\ell/a_r > 1$ so that for $0 \leq y \leq x$ $F(x,y) = k\sinh\left(x - y\right) - y - \sinh(x) = 0$. Since $F(x,x) < 0$, $F(x,0) > 0$ for $x > 0$ and $\partial F/\partial y < 0$ we solve the above implicit equation globally with $y = y(x)$. In order to prove $y(x) \le \frac{k-1}{k+1}x$ we show that $F\left(x, \frac{k-1}{k+1}x\right) \le 0$. We have that $F\left(x, \frac{k-1}{k+1}x\right)$ equals

$$\sum_{n=0}^{\infty} \frac{x^{2n+1}}{(2n+1)!(k+1)^{(2n+1)}} \left[k2^{2n+1} - (k+1)^{2n+1}\right] - \frac{k-1}{k+1}x.$$

The first term in the sum in the right side is precisely $\frac{k-1}{k+1}x$, so we need to prove that for every $n \ge 1$, $k > 1$ we have $(k+1)^{2n+1} \ge k2^{2n+1}$, i.e., $(\frac{k+1}{2})^{2n+1} \ge k$. This follows by Bernoulli inequality: $(\frac{k+1}{2})^{2n+1} = (1+\frac{k-1}{2})^{2n+1} \ge 1+(2n+1)\frac{k-1}{2} \ge k$. Then (18) follows.

Finally, we prove (19), the first line. The case $\delta_2 < 0$ corresponds to $\lambda_\ell > \lambda_r$. From Proposition 1 the outcoming waves have different signs: $\varepsilon_1\varepsilon_3 < 0$. Hence we apply (20) to get $|\varepsilon_1| + |\varepsilon_3| = |\varepsilon_1 - \varepsilon_3| = |\delta_1|$ and we just get the equality in the first line of (19), for $\delta_2 < 0$. On the other hand, if $\delta_2 > 0$, the inequality simply follows by (18). In the other case, when a 3-wave interacts, one has $\varepsilon_1\varepsilon_3 < 0$ if $\lambda_\ell < \lambda_r$, that is $\delta_2 > 0$; the rest of the proof is as above. □

The inequalities (19) improve the inequality (3.3) in [6] in the case of two interacting wave fronts, one of them being of the second family. Under the notations of [6] we find a term $1/(a_r + a_\ell)$ instead of $1/\min\{a_r, a_\ell\}$. The proof differs from Peng's. Our estimates are sharp: in some cases (19) reduces to an identity.

We finally remark that, with the choice of the size of the wave-fronts in Definition 1, the total size of the strengths does not increase across any interaction of waves belonging only to the families 1 or 3 [1]. On the other hand, if a 2-wave is involved in the interaction, as in Theorem 2, the variation $|\varepsilon_1| + |\varepsilon_3| - |\delta_i|$ of the sizes of the strengths may be positive if and only if the incoming and the reflected waves are of the same type; this happens if and only if the colliding wave is moving toward a more liquid phase.

References

1. D. Amadori and A. Corli. On a model of multiphase flow. *Preprint*, 2006.
2. D. Amadori and G. Guerra. Global BV solutions and relaxation limit for a system of conservation laws. *Proc. Roy. Soc. Edinburgh Sect. A*, 131(1):1–26, 2001.
3. A. Corli and H. Fan. The Riemann problem for reversible reactive flows with metastability. *SIAM J. Appl. Math.*, 65(2):426–457, 2004/05.
4. H. Fan. On a model of the dynamics of liquid/vapor phase transitions. *SIAM J. Appl. Math.*, 60(4):1270–1301, 2000.
5. T. Nishida. Global solution for an initial boundary value problem of a quasilinear hyperbolic system. *Proc. Japan Acad.*, 44:642–646, 1968.
6. Y.-J. Peng. Solutions faibles globales pour un modèle d'écoulements diphasiques. *Ann. Scuola Norm. Sup. Pisa Cl. Sci. (4)*, 21(4):523–540, 1994.
7. J. Smoller. *Shock waves and reaction-diffusion equations.* Springer-Verlag, New York, 1994.

Nonlinear Stability of Compressible Vortex Sheets

J-F. Coulombel and P. Secchi

We present a result on the existence of two-dimensional contact discontinuities solutions to the compressible Euler equations. The problem is a free boundary nonlinear hyperbolic problem with two main difficulties: The free boundary is characteristic, and the so-called Lopatinskii condition holds only in a weak sense, which yields losses of derivatives in the energy estimates. A similar analysis is applied in the context of weakly stable shock waves and isothermal liquid–vapor phase transitions, and yields analogous existence results.

1 Introduction, Main Result

We are interested in the Cauchy problem for the compressible Euler equations in two space dimensions:

$$\begin{cases} \partial_t \rho + \nabla_x \cdot (\rho\, \mathbf{u}) = 0\,, \\ \partial_t (\rho\, \mathbf{u}) + \nabla_x \cdot (\rho\, \mathbf{u} \otimes \mathbf{u}) + \nabla_x\, p(\rho) = 0, \end{cases} \tag{1}$$

where $\rho > 0$ is the density of the fluid, $\mathbf{u} \in \mathbb{R}^2$ is the velocity field, p is the pressure law, t is the time variable and $x = (x_1, x_2) \in \mathbb{R}^2$ is the space variable. In what follows, p is a $\mathcal{C}^\infty$ function of ρ, defined on $]0, +\infty[$, and such that $p'(\rho) > 0$ for all ρ. In particular, the system (1) is a symmetrizable hyperbolic system of conservation laws and the (local in time) existence of smooth solutions follows from a general result by Kato [9]. Due to the classical blow-up in finite time (see [20] for an example), it is natural to look for weak solutions of the system (1). Since the functional setting for the global existence of such weak solutions is still a wide open problem, a first natural step is to construct local in time, piecewise smooth solutions to (1). This program was initiated in the pioneering work by Majda [13, 12] on shock waves. (We also mention the alternative approach by Blokhin [3] for gas dynamics.) A few years later, the construction of rarefaction waves was completed by Alinhac [1], while the construction of two shock waves and the construction of sonic waves was achieved by Métivier [15]. Let us also mention the work by Sablé-Tougeron [18] on several sonic fronts, and the recent work by Francheteau and Métivier [7] on the existence of weak shocks and the convergence to sonic waves. Here, we show how to construct the last basic family of waves, namely the contact discontinuities. In some sense, this is the last building block to solve the two-dimensional Cauchy problem with general piecewise smooth initial data.

A contact discontinuity solution to the Euler equations (1) is a function $(\rho, \mathbf{u})$ that is smooth on either side of a surface $\Gamma := \{x_2 = \varphi(t, x_1), t \in [0, T], x_1 \in \mathbb{R}\}$, and such that:

- The Euler equations (1) are satisfied on either side of Γ
- The Rankine–Hugoniot jump relations are satisfied in the following way on Γ

$$\begin{aligned} &\partial_t \varphi = \mathbf{u}^+ \cdot \nu = \mathbf{u}^- \cdot \nu, \quad \text{with } \nu = (-\partial_{x_1}\varphi, 1), \\ &\rho^+ = \rho^- . \end{aligned}$$

As usual, $\rho^\pm, \mathbf{u}^\pm$ denote the traces of $\rho, \mathbf{u}$ taken on either side of Γ.

The discontinuity surface Γ is part of the unknowns, and we are thus dealing with a free boundary problem. Moreover, this free boundary is characteristic with respect to either side, because of the Rankine–Hugoniot conditions. We also observe that the tangential velocity (with respect to the interface Γ) is the only quantity that experiments a jump across Γ, so contact discontinuities are vortex sheets.

Because the boundary Γ is part of the unknowns, it is convenient to introduce a change of variables to work in a fixed domain. Namely, we introduce two real-valued functions $\Phi^\pm$ that are smooth on the domain $\{t \in [0, T], x_1 \in \mathbb{R}, x_2 \geq 0\}$, and such that there exists a positive constant κ verifying $\pm\partial_{x_2}\Phi^\pm \geq \kappa$. We also require that there holds $\Phi^\pm(t, x_1, 0) = \varphi(t, x_1)$, where φ is the function that defines the interface Γ. Then we define the new unknown functions:

$$U^\pm(t, x_1, x_2) = (\rho, \mathbf{u})(t, x_1, \Phi^\pm(t, x_1, x_2)),$$

and we rewrite the Euler equations together with the jump relations for the functions $U^\pm, \Phi^\pm$. After fixing the front, the system is now set in $\{x_2 > 0\}$, and the Rankine–Hugoniot jump relations become boundary conditions on $\{x_2 = 0\}$.

There are many possible choices for the functions $\Phi^\pm$. An appropriate choice, that is justified with great details in [5, 6], is to choose $\Phi^\pm$ as the solution of some eikonal equations on the whole domain $\{t \in [0, T], x_1 \in \mathbb{R}, x_2 \geq 0\}$. This is made possible because the common trace φ has to solve simultaneously two eikonal equations on the boundary $\{t \in [0, T], x_1 \in \mathbb{R}, x_2 = 0\}$. We refer the reader to [5, 6] for more details.

Our goal is to construct local in time contact discontinuities solutions that are small perturbations of piecewise constant solutions. More precisely, up to

Galilean changes of frame, all piecewise constant contact discontinuities have the following form in the new straightened variables:

$$U^{\pm} \equiv (\overline{\rho}, \pm\overline{v}, 0)\,, \quad \Phi^{\pm} = \pm x_2\,, \quad \varphi \equiv 0, \tag{2}$$

where $\overline{\rho} > 0$ is a fixed density, $\overline{v} > 0$ is a fixed tangential velocity, while the normal velocity vanishes. We shall denote $\overline{U}^{\pm}, \overline{\Phi}^{\pm}$ this reference (stationary) solution. Our main result states the nonlinear stability of such a solution under the so-called "supersonic" condition:

Theorem 1. *Let $T > 0$, and let $\mu \in \mathbb{N}$, with $\mu \geq 6$. Assume that the stationary solution defined by (2) satisfies the "supersonic" condition:*

$$\overline{v} > \sqrt{2\,p'(\overline{\rho})}. \tag{3}$$

Assume that the initial data $(U_0^{\pm}, \varphi_0)$ have the form

$$U_0^{\pm} = \overline{U}^{\pm} + \dot{U}_0^{\pm},$$

with $\dot{U}_0^{\pm} \in H^{\mu+15/2}(\mathbb{R}^2_+)$, $\varphi_0 \in H^{\mu+8}(\mathbb{R})$, and that they satisfy some appropriate compatibility conditions. Assume also that $(\dot{U}_0^{\pm}, \varphi_0)$ have a compact support. Then, there exists $\delta > 0$ such that, if $\|\dot{U}_0^{\pm}\|_{H^{\mu+15/2}(\mathbb{R}^2_+)} + \|\varphi_0\|_{H^{\mu+8}(\mathbb{R})} \leq \delta$, then there exists a contact discontinuity solution $U^{\pm} = \overline{U}^{\pm} + \dot{U}^{\pm}, \Phi^{\pm} = \pm x_2 + \dot{\Phi}^{\pm}, \varphi$ on the time interval $[0, T]$. This solution satisfies $(\dot{U}^{\pm}, \dot{\Phi}^{\pm}) \in H^{\mu}(]0, T[\times\mathbb{R}^2_+)$, and $\varphi \in H^{\mu+1}(]0, T[\times\mathbb{R})$.

Observe that the initial datum is φ_0, and not $\Phi_0^{\pm}$. As a matter of fact, the physically relevant quantity is the front Γ (and thus φ) in the original space variables. The initial data $\Phi_0^{\pm}$ are determined by lifting φ_0, which can be done quite arbitrarily.

In the following section, we give an outline of the methods that were used in the proof. The reader is referred to the original papers [5, 6] for the complete details.

2 Main Steps of the Proof

The general strategy is the same as in the preceding works [13, 12, 1, 15] etc. We obtain the desired solution as the limit of a sequence of approximate solution that is constructed by solving linearized problems. As will be detailed later on, the linearized problems can be solved only up to the price of a loss of derivatives. We are thus led naturally to using a Nash–Moser procedure to overcome this loss of regularity. The iteration scheme is thus a Newton type iteration that is combined with a smoothing procedure.

We first describe the derivation of energy estimates for the linearized problem. Then we shall detail how these energy estimates yield the solvability of the linearized problem. In the end, we shall describe the iteration procedure.

2.1 A Priori Estimates for the Linearized Equations

Energy estimates for hyperbolic initial boundary value problems are usually obtained by means of symmetrizers. When the boundary conditions are dissipative, the symmetrizer is the identity and the energy estimate is obtained by integrating by parts. However, in the study of shock waves or contact discontinuities, the boundary conditions are not dissipative, and in this case, the construction of a symmetrizer is a more intricate problem. For noncharacteristic strictly hyperbolic problems, under the so-called uniform Lopatinskii condition, the construction of a microlocal symmetrizer is due to Kreiss [10]. It was later extended to uniformly characteristic boundaries by Majda and Osher [14], who also noted that strict hyperbolicity could be replaced by the now well-known "block structure condition." In both works, the construction heavily relies on the uniform Lopatinskii condition. For vortex sheets, the verification of the Lopatinskii condition dates back at least to Miles [17] (we also refer to [19, chapter 14] for a more concise calculation), and the following result holds:

- In three space dimensions, the Lopatinskii condition is never satisfied. Vortex sheets are violently unstable (the instability is the analogue of the Kelvin–Helmholtz instability for incompressible fluids).
- In two space dimensions, when $|\overline{v}| < \sqrt{2\,p'(\overline{\rho})}$, the vortex sheet is violently unstable.
- In two space dimensions, when $|\overline{v}| > \sqrt{2\,p'(\overline{\rho})}$, the vortex sheet is weakly stable. The Lopatinskii condition holds, but not uniformly.

This already explains why our existence result requires the "supersonic" condition (3).

Because the Lopatinskii condition is not satisfied, we can not use Majda and Osher's result. Our strategy is the following. We first study the constant coefficient problem and construct some degenerate Kreiss' symmetrizers that take into account the failure of the uniform Lopatinskii condition. In the end, we are able to prove an a priori estimate that exhibits a loss of one tangential derivative from the source terms to the solution. The construction relies on a detailed study of the degeneracy of the Lopatinskii condition. More precisely, we show that the Lopatinskii determinant has simple roots that belong to the "hyperbolic" region of the cotangent bundle of the boundary. After proving such an estimate for the constant coefficients linearized problem, we turn to the variable coefficients problem. In view of the nonlinear problem we want to solve, it is crucial to work with coefficients that have limited smoothness. To apply the preliminary symbolic construction, we use a suitable paradifferential calculus (we refer to [16] for an introduction). However, we need to have a constant rank boundary matrix, so the linearization state has to verify some appropriate constraints. Up to microlocalization errors, the paralinearization reduces more or less the analysis of the variable coefficients problem to the constant coefficient case. The final step in the derivation of the a priori estimate is the absorption of the microlocalization errors, and we perform this final step by showing that the solution can be controlled everywhere but on bicharacteristic curves emanating from the boundary, while the errors are microlocalized far from these curves. The details of the analysis can be found in [5].

2.2 Solvability of the Linearized Equations

For hyperbolic initial boundary value problems that satisfy a maximal a priori estimate (that is, an a priori estimate with no loss of derivative), well-posedness can be proved by following the arguments of [8] or [11]. One first constructs a weak solution by duality and then shows that weak solutions are strong solutions by commuting the equation with a tangential mollifier. When losses of derivatives occur, this analysis has to be modified because after commuting the equation with the mollifier, the errors can not be neglected anymore. As a matter of fact, the "weak=strong" argument can be modified into a "weak=semi-strong" argument, which still shows wellposedness under the assumption that the equations and a dual problem satisfy an estimate with a loss of one tangential derivative. To achieve this argument, the choice of the mollifier is crucial, and we refer to [4] for a deeper discussion of this topic.

In the context of contact discontinuities, the linearized equations satisfy an a priori estimate with a loss of one tangential derivative. To apply the wellposedness result of [4], it is sufficient to construct a dual problem that shares the same a priori estimate. This can be done "by hand," and in [6], we construct an explicit dual problem for which the Lopatinskii condition degenerates in exactly the same way as for the original problem. Following the analysis of the original problem, we end up with an a priori estimate for the dual problem, and the result of [4] implies wellposedness in the Hadamard's sense (at least for zero initial data).

2.3 The Iteration Procedure

To solve the nonlinear equations, we adopt a Nash–Moser iteration procedure. The convergence of such iterations relies on a suitable tame estimate in Sobolev spaces of arbitrarily large index. Consequently, after proving well-posedness in L^2 for the linearized equations, we show wellposedness in H^m, with m arbitrarily large. This is done in a classical way by first estimating tangential derivatives (one commutes the linearized equations with tangential derivatives, and applies Gagliardo–Nirenberg's estimates), and then by estimating normal derivatives. We need to pay special attention when commuting the equations with tangential derivatives since the commutators can not be neglected (because of the loss of regularity in the estimates). Moreover, the estimate of the normal derivatives is delicate because the boundary is characteristic. To recover all normal derivatives, we use the equations and we also compute the vorticity equation, which gives the estimate of the missing normal derivatives.

After proving the tame estimate in H^m, we are led to the final step of the analysis, that is to proving convergence of the iteration scheme. Starting from compatible initial data, we first construct an approximate solution that satisfies the equations in the sense of Taylor's expansions at $t = 0$. Then we adopt a Nash–Moser method that consists in solving linearized equations where both coefficients and source terms are smoothed. Here we face a technical difficulty because the estimates for the linearized equations are linked to some constraints that need to be satisfied by the coefficients (see the preceding paragraphs). We thus adapt the Nash–Moser procedure and perform an intermediate step where we enforce these constraints. Naturally, this step leads to new errors that need to be controlled in the convergence analysis. The definition of the source terms, as well as a complete proof of convergence can be found in [6].

3 Concluding Remarks

Theorem 1 shows that the weak Lopatinskii condition can be sufficient to ensure nonlinear wellposedness (up to the price of a loss of regularity from the initial data to the solution). As a matter of fact, uniqueness of contact discontinuities can be proved by using the estimates that were proved for the linearized equations. This however requires special attention because the loss of derivatives precludes the classical Gronwall's argument.

We point out that the analysis that we have developed in the study of contact discontinuities can also be applied to prove the existence of weakly stable shocks for the isentropic and nonisentropic Euler equations. The linear spectral stability of planar shocks was extensively studied in the literature of the 1950s and the 1960s. The precise stability/instability criteria can be found in [13] or [19, chapter 14]. In both isentropic and nonisentropic gas dynamics, weakly stable shocks correspond to Lopatinskii determinants that have simple roots in the "hyperbolic" region of the cotangent bundle of the boundary. The linearized equations thus satisfy energy estimates with a loss of one tangential derivative, and the solvability of the nonlinear equations follows by simply adapting the iteration scheme that was developed above. In this case, the analysis is simplified by the noncharacteristic nature of the front, which allows to choose explicit functions $\Phi^{\pm}$ to lift the front and to recover easily normal derivatives in the energy estimates.

Another application of our method is the study of dynamic liquid–vapor phase transitions. The linear spectral stability of such discontinuities was performed by Benzoni–Gavage [2]. We simply observe that these discontinuities are noncharacteristic, but they are undercompressive. As such, they require an additional jump relation. The physical motivation for this new jump relation is explained in [2]. The result of the spectral analysis is that dynamic liquid–vapor phase transitions are weakly stable, but this time the weak stability is associated with surface waves of finite energy (these waves are analogous to Rayleigh's waves in elastodynamics). The linearized equations thus satisfy an energy estimate that exhibits a slightly weaker loss of regularity. These estimates still imply nonlinear wellposedness, that is local in time existence of liquid–vapor phase transitions. This answers the question raised at the end of [2].

Eventually, we refer to the contribution by Yuri Trakhinin in this volume for a study of contact discontinuities in magnetohydrodynamics.

References

1. S. Alinhac. Existence d'ondes de raréfaction pour des systèmes quasi-linéaires hyperboliques multidimensionnels. *Comm. Partial Differential Equations*, 14(2): 173–230, 1989.
2. S. Benzoni-Gavage. Stability of multi-dimensional phase transitions in a van der Waals fluid. *Nonlinear Anal.*, 31(1-2):243–263, 1998.
3. A.M. Blokhin. Estimation of the energy integral of a mixed problem for gas dynamics equations with boundary conditions on the shock wave. *Sibirsk. Mat. Zh.*, 22(4):23–51, 229, 1981.
4. J.F. Coulombel. Well-posedness of hyperbolic initial boundary value problems. *J. Math. Pures Appl.*, 84(6):786–818, 2005.
5. J.F. Coulombel, P. Secchi. The stability of compressible vortex sheets in two space dimensions. *Indiana Univ. Math. J.*, 53(4):941–1012, 2004.
6. J.F. Coulombel, P. Secchi. Nonlinear stability of compressible vortex sheets in two space dimensions. *Preprint*, 2005.
7. J. Francheteau, G. Métivier. Existence de chocs faibles pour des systèmes quasi-linéaires hyperboliques multidimensionnels. *Astérisque*, 268:1–198, 2000.
8. K.O. Friedrichs. Symmetric positive linear differential equations. *Comm. Pure Appl. Math.*, 11:333–418, 1958.
9. T. Kato. The Cauchy problem for quasi-linear symmetric hyperbolic systems. *Arch. Rational Mech. Anal.*, 58(3):181–205, 1975.
10. H.O. Kreiss. Initial boundary value problems for hyperbolic systems. *Comm. Pure Appl. Math.*, 23:277–298, 1970.
11. P.D. Lax, R.S. Phillips. Local boundary conditions for dissipative symmetric linear differential operators. *Comm. Pure Appl. Math.*, 13:427–455, 1960.
12. A. Majda. The existence of multi-dimensional shock fronts. *Memoirs Amer. Math. Soc.*, 281, 1983.
13. A. Majda. The stability of multi-dimensional shock fronts. *Memoirs Amer. Math. Soc.*, 275, 1983.
14. A. Majda, S. Osher. Initial-boundary value problems for hyperbolic equations with uniformly characteristic boundary. *Comm. Pure Appl. Math.*, 28(5): 607–675, 1975.
15. G. Métivier. Ondes soniques. *J. Math. Pures Appl. (9)*, 70(2):197–268, 1991.
16. G. Métivier. Stability of multidimensional shocks. In *Advances in the theory of shock waves*, pages 25–103. Birkhäuser, 2001.

17. J.W. Miles. On the disturbed motion of a plane vortex sheet. *J. Fluid Mech.*, 4:538–552, 1958.
18. M. Sablé-Tougeron. Ondes de gradients multidimensionnelles. *Mem. Amer. Math. Soc.*, 106(511):viii+93, 1993.
19. D. Serre. *Systems of conservation laws. 2.* Cambridge University Press, 2000.
20. T.C. Sideris. Formation of singularities in three-dimensional compressible fluids. *Comm. Math. Phys.*, 101(4):475–485, 1985.

Regularity and Compactness for the DiPerna–Lions Flow

G. Crippa and C. De Lellis

1 Flow of Nonsmooth Vector Fields: The Regular Lagrangian Flow

When $b : [0,T] \times \mathbb{R}^d \to \mathbb{R}^d$ is a bounded smooth vector field, the *flow of* b is the smooth map $X : [0,T] \times \mathbb{R}^d \to \mathbb{R}^d$ such that

$$\begin{cases} \dfrac{dX}{dt}(t,x) &= b(t, X(t,x)) \,, \quad t \in [0,T] \\ \\ X(0,x) &= x \,. \end{cases} \tag{1}$$

Existence and uniqueness of the flow are guaranteed by the classical Cauchy–Lipschitz theorem. The study of (1) out of the smooth context is of great importance (for instance, in view of the possible applications to conservation laws or to the theory of the motion of fluids) and has been studied by several authors. What can be said about the well-posedness of (1) when b is only in some class of weak differentiability? We remark from the beginning that no generic uniqueness results (i.e., for a.e. initial datum x) are presently available.

This question can be, in some sense, "relaxed" (and this relaxed problem can be solved, for example, in the Sobolev or BV framework): we look for a *canonical selection principle*, i.e., a strategy that "selects," for a.e. initial datum x, a solution $X(\cdot, x)$ in such a way that this selection is stable with respect to smooth approximations of b. This in some sense amounts to redefine our notion of solution: we add some conditions that select a "relevant" solution of our equation. This is encoded in the following definition: we consider only the flows such that there are no concentrations of the trajectories. We will denote by $\mathcal{L}^d$ the d-dimensional Lebesgue measure in $\mathbb{R}^d$.

Definition 1 (Regular Lagrangian flow). *Let $b \in L^1_{\mathrm{loc}}([0,T] \times \mathbb{R}^d; \mathbb{R}^d)$. We say that a map $X : [0,T] \times \mathbb{R}^d \to \mathbb{R}^d$ is a* regular Lagrangian flow *for the vector field b if*

(i) for a.e. $x \in \mathbb{R}^d$ *the map* $t \mapsto X(t,x)$ *is an absolutely continuous integral solution of* $\dot\gamma(t) = b(t,\gamma(t))$ *for* t *in* $[0,T]$*, with* $\gamma(0) = x$*;*
(ii) there exists a constant L *independent of* t *such that*

$$\mathcal{L}^d\big(X(t,\cdot)^{-1}(A)\big) \le L\mathcal{L}^d(A) \qquad \text{for every Borel set } A \subset \mathbb{R}^d. \tag{2}$$

The constant L *in (ii) will be called the* compressibility constant *of* X*.*

2 The Link with the Transport Equation

Existence, uniqueness, and stability of regular Lagrangian flows have been proved in [DPL89] by DiPerna and Lions for Sobolev vector fields with bounded divergence. In a recent groundbreaking paper (see [Amb04]) this result has been extended by Ambrosio to BV coefficients bounded from below divergence.

The arguments of the DiPerna–Lions theory are quite indirect and they exploit (via the theory of characteristics) the connection between (1) and the Cauchy problem for the *transport equation*

$$\begin{cases} \partial_t u(t,x) + b(t,x)\cdot\nabla_x u(t,x) \;=\; 0 \\[2ex] u(0,\cdot) \;=\; \bar u\,. \end{cases} \tag{3}$$

Assuming that the divergence of b is in L^1 we can define bounded distributional solutions of (3) using the identity $b\cdot\nabla_x u = \nabla_x\cdot(bu) - u\nabla_x\cdot b$. Following DiPerna and Lions we say that a distributional solution $u \in L^\infty([0,T]\times\mathbb{R}^d)$ of (3) is a *renormalized solution* if

$$\begin{cases} \partial_t[\beta(u(t,x))] + b(t,x)\cdot\nabla_x[\beta(u(t,x))] \;=\; 0 \\[2ex] [\beta(u)](0,\cdot) \;=\; \beta(\bar u) \end{cases} \tag{4}$$

holds in the sense of distributions for every function $\beta \in C^1(\mathbb{R};\mathbb{R})$. In their seminal paper DiPerna and Lions showed that, if the vector field b has Sobolev regularity with respect to the space variable, then every bounded solution is renormalized. Ambrosio [Amb04] extended this result to BV vector fields with divergence in L^1. Under suitable compressibility assumptions (for instance $\nabla_x \cdot b \in L^\infty$), the renormalization property gives *uniqueness* and *stability* for (3) (the existence follows in a quite straightforward way from standard approximation procedures).

In turn, this uniqueness and stability property for (3) can be used to show existence, uniqueness, and stability of regular Lagrangian flows (we refer to [DPL89] for the original proofs and to [Amb04] for a different derivation of the same conclusions).

3 Properties of the Regular Lagrangian Flow

Having defined our notion of solution of (1) and having shown well-posedness for this problem under suitable regularity assumptions on the vector field, it is interesting to investigate some further properties of this solution. In particular we are interested to

- The *regularity* of the regular Lagrangian flow with respect to the initial datum: this amounts to the study of the map $x \mapsto X(t,x)$;
- The *compactness* of regular Lagrangian flows corresponding to vector fields satisfying natural uniform bounds.

The differentiability of the regular Lagrangian flow with respect to x has been first studied by Le Bris and Lions in [LBL04]. For vector fields with Sobolev regularity, they are able to show (using an extension of the theory of renormalized solutions) the existence of measurable maps $W_t : \mathbb{R}^d \times \mathbb{R}^d \to \mathbb{R}^d$ such that

$$\frac{X(t,x+\varepsilon y)-X(t,x)-\varepsilon W_t(x,y)}{\varepsilon} \to 0 \quad \text{locally in measure in } \mathbb{R}^d_x \times \mathbb{R}^d_y. \tag{5}$$

We recall that a sequence of Borel maps $\{f_n\}$ is said to be *locally convergent in measure* to f in $\mathbb{R}^k$ if

$$\lim_{n\to\infty} \mathcal{L}^k\big(\{x \in B_R(0) \ : \ |f_n(x) - f(x)| > \delta\}\big) = 0$$

for every $R > 0$ and every $\delta > 0$. If the sequence $\{f_n\}$ is locally equibounded in L^∞, then the local convergence in measure is equivalent to the strong convergence in $L^1_{\rm loc}$.

However, it turns out (see [AM05]) that the differentiability property expressed in (5) does not imply the classical approximate differentiability. We recall that a map $f : \mathbb{R}^k \to \mathbb{R}^m$ is said to be *approximately differentiable at* $x \in \mathbb{R}^k$ if there exists a linear map $L(x) : \mathbb{R}^k \to \mathbb{R}^m$ such that

$$\frac{f(x+\varepsilon y) - f(x) - \varepsilon L(x)y}{\varepsilon} \to 0 \quad \text{locally in measure in } \mathbb{R}^d_y.$$

Notice also that this concept has a pointwise meaning, while the one in (5) is global. Moreover, it is possible to show that the map f is approximately differentiable a.e. in $\Omega \subset \mathbb{R}^k$ if and only if the following Lusin-type approximation with Lipschitz maps holds: for every $\varepsilon > 0$ it is possible to find a set $\Omega' \subset \Omega$ with $\mathcal{L}^k(\Omega \setminus \Omega') \leq \varepsilon$ such that $f|_{\Omega'}$ is Lipschitz.

Approximate differentiability for regular Lagrangian flows relative to $W^{1,p}$ vector fields, with $p > 1$, has been first proved by Ambrosio, Lecumberry, and Maniglia in [ALM05]. The need for considering only the case $p > 1$ comes from the fact that some tools from the theory of maximal functions are used, as will be explained in the next section. In [ALM05] the strategy is no more an extension of the theory of renormalized solutions: the authors introduce some new estimates along the flow, inspired by the remark that, at a formal level, we can control the time derivative of $\log(|\nabla X(t,x)|)$ with $|\nabla b|(t,X(t,x))$. The strategy of [ALM05] allows to make this remark rigorous: it is possible to consider some integral quantities that contain a discretization of the space gradient of the flow and prove some estimates along the flow, in fact using the PDE formulation of the problem presented in the previous section. Then, the application of Egorov theorem allows the passage from integral estimates to pointwise estimates on big sets, and from this it is possible to recover Lipschitz regularity on big sets, and eventually one gets the approximate differentiability. However, the application of Egorov theorem implies loss of quantitative informations: this strategy does not allow a control of the Lipschitz constant in terms of the size of the "neglected" set.

4 Quantitative Estimates for $W^{1,p}$ Vector Fields

Starting from the result of Ambrosio, Lecumberry, and Maniglia [ALM05], the main point of [CDL06] is a modification of the estimates in such a way that quantitative informations are not lost. Let X be a regular Lagrangian flow relative to a bounded vector field $b \in L^1(W^{1,p})$ for some $p > 1$. For every $R > 0$ we define the quantity

$$A_p(R,X) := \left\| \sup_{0\le t\le T} \sup_{0<r<R} \fint_{B_r(x)} \log\left(\frac{|X(t,x)-X(t,y)|}{r}+1\right) dy \right\|_{L^p_x(B_R(0))}.$$

The strategy of the proof is based on two main steps.

4.1 A Priori Estimate

It is possible to give an a priori estimate for the functional $A_p(R,X)$ in terms of the $L^1_t(L^p_x)$ norm of Db and the compressibility constant of the flow. Trying to estimate the quantity

$$\frac{d}{dt} \fint_{B_r(x)} \log\left(\frac{|X(t,x)-X(t,y)|}{r}+1\right) dy \tag{6}$$

we get some difference quotients of the vector field computed along the flow. Here comes into play the theory of maximal functions.

We recall that, for $f \in L^1_{\mathrm{loc}}(\mathbb{R}^k;\mathbb{R}^m)$, we can define the *maximal function* of f as

$$Mf(x) := \sup_{r>0} \fint_{B_r(x)} |f(y)|\, dy\,.$$

It is well-known (see for example [Ste70]) that for every $p > 1$ the strong estimate

$$\|Mf\|_{L^p(\mathbb{R}^k)} \le C_{k,p}\|f\|_{L^p(\mathbb{R}^k)} \tag{7}$$

holds, while this is not true in the limit case $p = 1$. Moreover, if f has Sobolev regularity, we can estimate the increments using the maximal function of the derivative: there exists a negligible set $N \subset \mathbb{R}^k$ such that for every x, $y \in \mathbb{R}^k \setminus N$ we have

$$|f(x) - f(y)| \leq C_k |x - y| \big(MDf(x) + MDf(y) \big) .$$

Going back to (6), we see that it is possible to estimate the difference quotients that appear in the time differentiation using the maximal function of Db, computed along the flow. After we integrate with respect to time, we pass to the supremums and eventually we take the L^p norm in order to reconstruct the quantity $A_p(R, X)$. Then, changing variable (and for this we just pay a factor given by the compressibility constant L) and using the strong estimate (7) in order to express the bound in term of Db, we finally get the a priori quantitative estimate

$$A_p(R, X) \leq C\big(R, L, \|Db\|_{L^1(L^p)}\big) . \tag{8}$$

4.2 Quantitative Lipschitz Property

From the bound (8) we can obtain a *quantitative Lusin-type Lipschitz approximation* of the regular Lagrangian flow. This means that we are able to estimate the growth of the Lipschitz constant in terms of the size of the neglected set. For every fixed $\varepsilon > 0$ and every $R > 0$ we apply Chebyshev inequality to get a constant

$$M = M(\varepsilon) = \frac{A_p(R, X)}{\varepsilon^{1/p}}$$

and a set $K \subset B_R(0)$ with $\mathcal{L}^d(B_R(0) \setminus K) \leq \varepsilon$ such that for every $x \in K$

$$\sup_{0 \leq t \leq T} \sup_{0 < r < 2R} \fint_{B_r(x)} \log \left(\frac{|X(t,x) - X(t,y)|}{r} + 1 \right) dy \leq M.$$

From this it easily follows that

$$|X(t,x) - X(t,y)| \leq \exp\left(\frac{c_d A_p(R, X)}{\varepsilon^{1/p}} \right) |x - y|,$$

i.e., we obtain the following *explicit* control of the Lipschitz constant:

$$\mathrm{Lip}\,(X(t,\cdot)|_K) \sim \exp\left(C\varepsilon^{-1/p} \right). \tag{9}$$

5 Corollaries and Remarks

5.1 Immediate Consequences

Approximate Differentiability

Recalling the equivalence stated immediately after the definition of approximate differentiability, it follows that the regular Lagrangian flow is approximately differentiable with respect to the space variable a.e. in $\mathbb{R}^d$. We notice that the quantitative result expressed in (9) is not strictly necessary for this first consequence.

Compactness

The quantitative version of the Lusin-type Lipschitz approximation can be used to show the precompactness in $L^1_{\rm loc}$ for the regular Lagrangian flows $\{X_n\}$ generated by a sequence $\{b_n\}$ of vector fields equibounded in L^∞ and in $L^1(W^{1,p})$ (for $p > 1$), under the assumption that the compressibility constants of the regular Lagrangian flows are equibounded. We illustrate here the main idea to get this result.

On every ball $B_R(0)$, the regular Lagrangian flows are equibounded. We fix $\varepsilon > 0$. For every n we apply (9) to find a set K_n with $\mathcal{L}^d(B_R(0) \setminus K_n) \leq \varepsilon$ such that the Lipschitz constants of the maps $X_n|_{K_n}$ are equibounded. Then we can extend every map $X_n|_{K_n}$ to a map $\widetilde{X}_n$ defined on all $B_R(0)$ in such a way that the sequence $\{\widetilde{X}_n\}$ is equibounded and equicontinuous over $B_R(0)$. Hence we can apply Ascoli–Arzelà theorem to this sequence, getting strong compactness in L^∞. But since every map $\widetilde{X}_n$ coincides with the regular Lagrangian flow X_n out of a small set, it is simple to check that this implies strong compactness in L^1 for the regular Lagrangian flows $\{X_n\}$.

A merit of this approach is the fact that the compactness result holds under an assumption of equiboundedness of the compressibility constants: we do not need a uniform bound on the divergence of the vector fields (this would be in general a stronger condition). Some compactness results under uniform bounds on the divergence are already present in the literature: see for example [DPL89].

We finally remark that an extension of our strategy to the case $p = 1$ would give a positive answer to a conjecture proposed by Bressan in [Bre03B]. See also [Bre03A] for a related conjecture on mixing flows.

5.2 Quantitative Stability

With similar techniques it is possible to show a result of quantitative stability for regular Lagrangian flows relative to $W^{1,p}$ vector fields (here we need again the assumption $p > 1$). The stability results in [DPL89] and [Amb04] were obtained using some abstract compactness arguments, hence they do not give a rate of convergence. It is indeed possible to show that, for the regular Lagrangian flows X_1 and X_2 relative to bounded vector fields b_1 and b_2 belonging to $L^1(W^{1,p})$, the following estimate holds:

$$\|X_1(T,\cdot) - X_2(T,\cdot)\|_{L^1(B_r)} \leq C \Big| \log \big(\|b_1 - b_2\|_{L^1([0,T]\times B_R)} \big) \Big|^{-1}.$$

The constant C and R depend on the usual uniform bounds on the vector fields. We remark that this estimate also gives a new direct proof of the uniqueness of the regular Lagrangian flow.

5.3 A Regularity Result for the PDE

Using again (9) it is possible to show a result relative to solutions to the transport equation (3). We can prove that for bounded vector fields belonging to $L^1(W^{1,p})$ (as usual with $p > 1$) and with bounded divergence, solutions of (3) propagate a mild regularity, the same one of the corresponding regular Lagrangian flow expressed by (9).

6 An A Priori Estimates Approach

The estimates we have presented give a new possible approach to the theory of regular Lagrangian flows. In particular, we can develop (in the $W^{1,p}$ context with $p > 1$) a theory of ODEs completely independent from the associated PDEs theory. The general scheme is the following:

- The compactness we have illustrated can be used to show *existence* of the regular Lagrangian flow, via regularization, for vector fields with bounded divergence;
- The *uniqueness* comes together with the *stability*, which is recovered in a new quantitative fashion;
- In addition to these results, we can show the quantitative *regularity* expressed in (9), which implies the *approximate differentiability*;
- Finally, a new *compactness* result is obtained.

All this results are obtained at the Lagrangian level, with no mention to the transport equation theory.

References

[Amb04] Ambrosio, L.: Transport equation and Cauchy problem for BV vector fields. Invent. Math., **158**, 227–260 (2004).

[AC05] Ambrosio, L., Crippa, G.: Existence, uniqueness, stability and differentiability properties of the flow associated to weakly differentiable vector fields. Preprint available at http://cvgmt.sns.it (2005).

[ALM05] Ambrosio, L., Lecumberry, M., Maniglia, S.: Lipschitz regularity and approximate differentiability of the DiPerna–Lions flow. Rend. Sem. Mat. Univ. Padova, **114**, 29–50 (2005).

[AM05] Ambrosio, L., Malý, J.: Very weak notions of differentiability. Preprint available at http://cvgmt.sns.it (2005).

[Bre03A] Bressan, A.: A lemma and a conjecture on the cost of rearrangements. Rend. Sem. Mat. Univ. Padova, **110**, 97–102 (2003).

[Bre03B] Bressan, A.: An ill posed Cauchy problem for a hyperbolic system in two space dimensions. Rend. Sem. Mat. Univ. Padova, **110**, 103–117 (2003).

[CDL06] Crippa, G., De Lellis, C.: Estimates and regularity results for the DiPerna–Lions flow. Accepted by J. Reine Angew. Math.; preprint available at http://cvgmt.sns.it (2006).

[DPL89] DiPerna, R.J., Lions, P.-L.: Ordinary differential equations, transport theory and Sobolev spaces. Invent. Math., **98**, 511–547 (1989).

[Fed69] Federer, H.: Geometric measure theory. Springer, New York (1969)

[LBL04] Le Bris, C., Lions, P.-L.: Renormalized solutions of some transport equations with partially $W^{1,1}$ velocities and applications. Ann. Mat. Pura Appl., **183**, 97–130 (2004).

[Ste70] Stein, E.M.: Singular integrals and differentiability properties of functions. Princeton University Press, Princeton N. J. (1970)

A Note on L^1 Stability of Traveling Waves for a One-Dimensional BGK Model

C.M. Cuesta

Summary. We prove L^1 nonlinear stability of traveling waves for one-dimensional kinetic BGK models, regarded as relaxation models for scalar conservation laws with genuinely nonlinear fluxes. The proof relies on the L^1-contraction property and the monotonicity of the waves.

1 Introduction

We study L^1 nonlinear stability of traveling waves for the one-dimensional Bathnagar–Gross–Krook (BGK) type equation

$$\partial_t f + v\partial_x f = M(\rho_f, v) - f, \quad \text{with } t > 0\,,\; x \in \mathbb{R}\,,\; v \in \Omega \subset \mathbb{R}. \tag{1}$$

Here $f(t,x,v)$ is a time-dependent phase space density with time t, position x, and velocity v. The function $\rho_f(t,x)$ in (1) is the macroscopic density corresponding to the distribution f, i.e., the zeroth order velocity moment

$$\rho_f(t,x) = \int_\Omega f(t,x,v)\,dv. \tag{2}$$

The "Maxwellian" $M(\rho,v)$ is an equilibrium distribution satisfying the moment conditions

$$\int_\Omega M(\rho,v)\,dv = \rho, \quad \text{and} \quad \int_\Omega v\,M(\rho,v)\,dv = a(\rho), \tag{3}$$

for a macroscopic flux function $a(\rho)$. In addition, we assume that M is a strictly increasing function of ρ or, for smooth M, that

$$\partial_\rho M(\rho,v) > 0. \tag{4}$$

Formally, (3) ensures that the macroscopic limit equation (obtained using the scaling $(t,x) \to (t/\varepsilon, x/\varepsilon)$ and taking $\varepsilon \to 0$) of (1) is the scalar conservation law

$$\partial_t \rho_f + \partial_x a(\rho_f) = 0. \tag{5}$$

For smooth M, a Chapman–Enskog expansion gives up to $O(\varepsilon)$-terms a viscous regularization of (5), namely

$$\partial_t \rho_f + \partial_x a(\rho_f) = \varepsilon \partial_x (D(\rho_f)\partial_x \rho_f),$$

with $D(\rho_f) = \int v^2 \partial_\rho M(\rho_f, v)\, dv > 0$. It is therefore natural to expect that traveling waves of (1) share many features with viscous shock waves.

We also mention that (1) can be seen as a nonclosed hyperbolic system with relaxation, after applying the macroscopic scaling (see e.g. [8] for a review):

$$\begin{cases} \partial_t \rho_f + \partial_x m_f = 0, \\ \partial_t m_f + \partial_x \int_\Omega v^2 f\, dv = \frac{1}{\varepsilon}(a(\rho_f) - m_f), \end{cases}$$

where $m_f := \int_\Omega v f\, dv$. In this context, (4) resembles the subcharacteristic condition in relaxation systems, and guarantees the TVD property; cf. [1].

Examples of smooth Maxwellians satisfying (3) as well as (4) are given in [2] for $\Omega = \mathbb{R}$. Another well known kinetic model for scalar conservation laws is the Perthame–Tadmor model; see [9]:

$$\partial_t f + a'(v)\partial_x f = M(\rho_f, v) - f, \tag{6}$$

with

$$M(\rho, v) = \begin{cases} 1 & \text{if } 0 < v < \rho, \\ -1 & \text{if } \rho < v < 0, \\ 0 & \text{otherwise.} \end{cases} \tag{7}$$

Although here M is not continuous, it is increasing with respect to ρ.

Traveling waves of (1) are solutions that depend only on the variable $\xi = x - st$, where s is the wave speed:

$$(v - s)\partial_\xi f = M(\rho_f, v) - f\,, \qquad \xi \in \mathbb{R},\ v \in \Omega, \tag{8}$$

and that satisfy

$$\lim_{\xi \to \pm\infty} f(\xi, v) = M(\rho_\pm, v)\,, \qquad v \in \Omega, \tag{9}$$

for two constant values ρ_+ and ρ_-. As for (5), the wave speed is given by the Rankine–Hugoniot condition:

$$s = \frac{a(\rho_+) - a(\rho_-)}{\rho_+ - \rho_-}.$$

Existence of big amplitude traveling waves for (6)–(7) is proved in Golse [4]. The author and C. Schmeiser have proved existence of small [3] and big [2] amplitude traveling waves.

$$\frac{a(\rho) - a(\rho_-)}{\rho - \rho_-} - s > 0 \quad \text{for all } \rho \in (\min\{\rho_+, \rho_-\}, \max\{\rho_+, \rho_-\}) \tag{10}$$

is as a necessary condition for existence.

Local stability of the small amplitude traveling waves is also proved in [3]. The existence and stability of small waves hold for continuous as well as discrete velocity models, in contrast to the existence of big waves, only available for $\Omega = \mathbb{R}$. The stability result strongly relies on the smallness assumption of the wave amplitude, allowing a perturbative treatment. Also this stability result is local; the initial perturbation has to be small enough, and we do not have asymptotic stability. The purpose of this paper is to point out that L^1 nonlinear asymptotic stability of (small and big amplitude) traveling waves holds under rather weak assumptions. The key ingredient is the L^1 contraction property, and the proof is similar to that for the Xin–Jin relaxation model in one space dimension [6].

For simplicity we restrict our attention to the continuous velocity case mentioned above. The stability result is easily verified for other velocity spaces if existence of traveling waves holds, for example, it can be shown that small waves of a discrete velocity model are L^1 nonlinearly stable. As a final remark we mention that we also obtain monotonicity of (big amplitude) traveling waves. This is used in the stability proof, but also improves the existence results for big waves (see Sect. 3). The results are easily checked for (6)–(7) with $a(\rho) = \rho/2$ and can easily be generalized for a genuinely nonlinear flux function.

We start in Sect. 2 with some basic properties of (1). The stability analysis follows in Sect. 3.

2 Preliminaries

In this section we describe some properties of (1). The following existence theorem holds (see [7] for a discrete velocity model and [5] for the Perthame–Tadmor model):

Theorem 1. *Let $\rho_{init} \in L^1_x(\mathbb{R}) \cap L^\infty_x(\mathbb{R})$ and let $f_{init}(x,v) = M(\rho_{init}(x), v)$, then there exists a unique solution $f \in C(0,T; L^1(\mathbb{R}^2) \cap L^\infty_x(\mathbb{R}) \times L^\infty_{v,loc}(\Omega))$.*

The following comparison principle holds (see e.g. [1]):

Proposition 1. *Let the assumptions of Theorem 1 hold. If $\rho_{init} \in L^1_x(\mathbb{R}) \cap L^\infty_x(\mathbb{R})$ and there exist two constants $k_\pm$ such that*

$$k_- \le \rho_{init} \le k_+,$$

then the solution of (1) with initial condition $f_{init} = M(\rho_{init}, v)$ satisfies

$$M(k_-, v) \le f \le M(k_+, v).$$

Moreover, if f_{init} and g_{init} are two initial conditions satisfying

$$f_{init} \geq g_{init}$$

then the corresponding solutions f and g of (1) satisfy

$$f \geq g \quad \textit{for all} \quad t > 0.$$

Conservation of mass is easily checked. The L^1-contraction property holds; see e.g. [7] and [1].

Proposition 2. *If f and g are solutions of (1) with initial conditions such that $f_{init} - g_{init} \in L^1_{x,v}$, then*

$$\frac{d}{dt}\int_{\mathbb{R}}\int_{\Omega} |f-g|\,dv\,dx = \int_{\mathbb{R}}\int_{\Omega} (|M(\rho_f,v) - M(\rho_g,v)| - |f-g|)\,dv\,dx \leq 0. \tag{11}$$

Also, the inequality in (11) is an identity iff sgn$(f-g)$ is independent of v.

3 Stability

Before proving stability we recall the existence result given in [2], and prove monotonicity of traveling waves.

Theorem 2 (Existence of traveling waves). *Let $\Omega = \mathbb{R}$ and $M(\rho,v)$, satisfying (3) and (4), be continuous with respect to $v \in \Omega$ and such that*

$$\int_{\Omega} |v^2 M(\rho,v)|dv < \infty$$

for every $\rho \in \mathbb{R}$. Then, under the assumption (10), there exists a solution $f(\xi,v) \in C_\xi(\mathbb{R}; L^\infty_{loc,v}(\Omega))$ of (8) that satisfies (9) in the following sense: there exist sequences $\xi_n \to \infty$ and $\eta_n \to -\infty$ such that

$$f(\xi_n,v) \to M(\rho_+,v), \qquad f(\eta_n,v) \to M(\rho_-,v), \quad v\text{-a.e.}$$

For a proof the reader is referred to [4] and [2].

The following Corollary is a direct consequence of Proposition 2.

Corollary 1. *If f and g are two traveling waves such that $f-g \in L^1_{\xi,v}(\mathbb{R}\times\Omega)$ then*

$$\int_{\mathbb{R}}\int_{\Omega} (|M(\rho_f,v) - M(\rho_g,v)| - |f-g|)\,dv\,dx = 0,$$

and sgn$(f-g)$ is independent of v.

Proposition 3 (Monotonicity of traveling waves). *Under the assumptions of Theorem 2, a traveling wave f satisfies*

$$\partial_\xi f \begin{cases} > 0 & \text{if } a''(\rho) < 0 \quad \text{for } \rho \in (\rho_-, \rho_+) \\ < 0 & \text{if } a''(\rho) > 0 \quad \text{for } \rho \in (\rho_+, \rho_-) \end{cases} \qquad v\text{-a.e.}$$

In particular, ρ_f is strictly increasing (resp. decreasing) for a concave (resp. convex) flux.

Proof. If f is a solution of (8)–(9), let $g = f - M(\rho_-, v)$. Multiplying the g-equation by $(v - s)$ and integrating with respect to v yields

$$\partial_\xi \int_\Omega (v-s)^2 g \, dv = a(\rho_g + \rho_-) - a(\rho_-) - s\rho_g - \int (v-s) g \, dv. \tag{12}$$

Integrating (8) with respect to v implies that $\int (v-s) g \, dv = const.$, and further, after taking the limit along the sequences ξ_n and η_n of Theorem 2, $\int (v-s) g \, dv = 0$. Now, rewriting the left-hand side of (12) by (10) we get

$$\lim_{h \to 0} \frac{1}{h} \int_\Omega (v-s)^2 (g(\xi + h, v) - g(\xi, v)) \, dv \begin{cases} > 0 \text{ if } a''(\rho) < 0 \text{ for } \rho \in (\rho_-, \rho_+) \\ < 0 \text{ if } a''(\rho) > 0 \text{ for } \rho \in (\rho_+, \rho_-) \end{cases}. \tag{13}$$

Clearly, if g is a traveling wave then $g_h(\xi) = g(\xi + h)$ is also a traveling wave for any $h \in \mathbb{R}$, then Corollary 1 implies that $\operatorname{sgn}(g(\xi + h, v) - g(\xi, v))$ does not depend on v, and in particular the sign of $\partial_\xi g$ (and therefore of $\partial_\xi \rho_g = \partial_\xi \rho_f$) is the same as that of (13). □

As a direct consequence the far-field conditions are satisfied in the limit $\xi \to \pm\infty$, namely

Corollary 2. *Under the assumptions of Theorem 2 traveling waves satisfy*

$$\lim_{\xi \to \pm\infty} f(\xi, v) = M(\rho_\pm, v) \quad v\text{-a.e.}$$

In order to prove stability we use Proposition 2 for a perturbation $f - g$ of a traveling wave g. We shall prove that f has limits in L^1 as $t \to \infty$, and that, essentially, for these limits the right hand side of (11) vanishes. The key step is provided by the following lemma, which shows that when (11) is an identity then f is a translate of the traveling wave g.

Lemma 1. *Let f be a solution of the equation and g a traveling wave such that $sgn(f - g)$ is independent of v, then there exists a $k \in \mathbb{R}$ for which $f(t, \xi, v) = g(\xi + k, v)$ v-a.e.*

Proof. Because of the strict monotonicity of g with respect to ξ we can define a.e. in v a function ϕ such that

$$\phi(y, v) = \xi \quad \text{where} \quad g(\xi, v) = y,$$

and, as in [10], considering the shift $k(t,\xi,v) := \phi(f(t,\xi,v)) - \xi$ we have that

$$f(t,\xi,v) = g(\xi + k(t,\xi,v),v).$$

Let us prove that k is constant. That $\mathrm{sgn}(f-g)$ does not depend on v implies that k does not depend on v either. Let now $\eta := \xi + k(t,\xi)$, then $g(\eta,v) = g(\xi + k(t,\xi),v)$ and satisfies

$$\partial_\eta g\,\partial_t k + (v-s)\partial_\eta g\,(1+\partial_\xi k) = M(\rho_g(\eta),v) - g(\eta,v)\,,$$

but $g(\eta,v)$ satisfies (8) (with derivatives in η) and also $\partial_\eta g \neq 0$ (by Proposition 3), therefore

$$\partial_t k + (v-s)\partial_\xi k = 0. \quad \square$$

The main result of this paper is the following theorem

Theorem 3 (L^1 stability). *Let the assumptions of Theorem 2 hold. Let f_{init} be an initial condition such that there exists a traveling wave g_0 and a $\bar{\xi} \in \mathbb{R}$, such that*

$$g_0(\xi,v) \le f_{init}(\xi,v) \le g_1(\xi,v) := g_0(\xi+\bar{\xi},v)$$
$$\textit{and} \quad f - g_j \in L^1_{\xi,v}(\mathbb{R}\times\Omega) \quad \textit{for } j=1,2.$$

Then, if f is the corresponding solution of (1), there exists a unique $\theta \in (0,1)$ and a traveling wave $g_\theta(\xi,v) := g_1(\xi+\theta\bar{\xi},v)$ such that

$$\lim_{t\to\infty} \|f(t,\xi,v) - g_\theta(\xi,v)\|_{L^1_{\xi,v}} = 0.$$

Moreover, θ is given by

$$\int_{\mathbb{R}}\int_{\Omega} (f(t,\xi,v) - g_1(\xi,v))\,dv\,d\xi = (\rho_+ - \rho_-)\bar{\xi}\,\theta.$$

The proof follows along the lines of that in [6] for the Xin–Jin model (which can be written as a BGK model with two velocities). We do not give the proof in full detail.

Proof of Theorem 3. Let g denote a translate of g_0 and g_1 with $g_0 < g < g_1$ and let f be the solution with initial data f_{init}. We know by Proposition 2 that $\|f-g\|_{L^1_{\xi,v}}$ is a decreasing function of t, thus is uniformly bounded in t, and that

$$\lim_{t\to\infty}\int_{\mathbb{R}}\int_{\Omega} |f-g|\,dv\,d\xi = \lim_{t\to\infty}\int_{\mathbb{R}}\int_{\Omega} |M(\rho_f,v) - M(\rho_g,v)|\,dv\,d\xi.$$

Also $f-g$ is bounded by an $L^1_{\xi,v}(\mathbb{R}\times\Omega)$ function; by Propositions 1 and 3

$$g_0 - g \le f - g \le g_1 - g.$$

And $g_j - g \in L^1_{\xi,v}(\mathbb{R} \times \Omega)$ by Propositions 2 and 1. To get compactness of the family $\{f - g\}_{t<0}$ we need L^1_ξ equicontinuity: this is provided by using the L^1 contraction applied to $f(t, \xi + h, v) - f(t, \xi, v)$ and to $g(\xi + h, v) - g(\xi, v)$, to get that $\|\partial_\xi f - \partial_\xi g\|_{L^1_{\xi,v}}$ is uniformly bounded (see e.g. [1]).

Let f_∞ be an L^1 limit of f, i.e., there exists a sequence $t_n \to \infty$ such that

$$f \to f_\infty \quad \text{as } n \to \infty \text{ strongly in } L^1_\xi(\mathbb{R}; L^1_v(\Omega)).$$

Observe that the L^1-limits are preserved by the evolution of (1), if $\bar{f}$ is the solution with initial data f_∞, then there exists a sequence $t_k \to \infty$ such that $f \to \bar{f}$ strongly in $L^1_\xi(\mathbb{R}; L^1_v(\Omega))$ as $k \to \infty$.

Using Proposition 2, it is easy to check that

$$\lim_{t\to\infty} \int_{\mathbb{R}}\int_{\Omega} |\bar{f} - g|\, dv\, d\xi = \lim_{t\to\infty} \int_{\mathbb{R}}\int_{\Omega} |f - g|\, dv\, d\xi = \int_{\mathbb{R}}\int_{\Omega} |f_\infty - g|\, dv\, d\xi,$$

implying that

$$\int_{\mathbb{R}}\int_{\Omega} |\bar{f} - g|\, dv\, d\xi = const. \quad \text{for all } t > 0,$$

thus also

$$\int_{\mathbb{R}}\int_{\Omega} |\bar{f} - g|\, dv\, d\xi = \int_{\mathbb{R}}\int_{\Omega} |M(\rho_{\bar{f}}, v) - M(\rho_g, v)|\, d\xi\, dv$$

and $\text{sgn}(\bar{f} - g)$ does not depend on v. Now, by Lemma 1, there exists a $k \in \mathbb{R}$ such that $\bar{f}(t, \xi, v) = g(\xi + k, v)$. The same argument can be applied to all possible limits of f in L^1, and we conclude that there exists a k such that

$$\|f(t, \xi, v) - g(\xi + k, v)\|_{L^1_{\xi,v}} \to 0 \quad \text{as } t \to \infty.$$

Again by Lemma 1 and Proposition 1 there exists a $\theta \in (0, 1)$ such that $g(\xi + k, v) = g_1(\xi + \theta\bar{\xi}, v)$, that, using the conservation of mass, is given by

$$\int_{\mathbb{R}}\int_{\Omega} (f(t, \xi, v) - g_1(\xi, v))\, dv\, d\xi = \int_{\mathbb{R}} (\rho_{g_1}(\xi + \theta\bar{\xi}) - \rho_{g_1}(\xi))\, d\xi = \theta\bar{\xi}(\rho_+ - \rho_-).$$

□

Acknowledgements

The author gratefully acknowledges C. Schmeiser for helpful comments and the financial support provided by the Austrian Science Fund, project nr. P18367.

References

1. F. Bouchut. Construction of BGK models with a family of kinetic entropies for a given system of conservation laws. *J. Statist. Phys.*, 95:113–170, 1999.
2. C.M. Cuesta and C. Schmeiser. Kinetic profiles for shock waves of scalar conservation laws. To appear in Transport Theory and Statistical Physics.
3. C.M. Cuesta and C. Schmeiser. Weak shocks for a one-dimensional BGK kinetic model for conservation laws. *SIAM Journal on Mathematical Analisys.*, 38:637–656, 2006.
4. F. Golse. Shock profiles for the Perthame-Tadmor kinetic model. *Comm. Partial Differential Equations*, 23:1857–1874, 1998.
5. P.-L. Lions, B. Perthame, and E. Tadmor. A kinetic formulation of multidimensional scalar conservation laws and related equations. *J. Amer. Math. Soc.*, 7(1):169–191, 1994.
6. C. Mascia and R. Natalini. L^1 nonlinear stability of traveling waves for a hyperbolic system with relaxation. *J. Differential Equations*, 132(2):275–292, 1996.
7. R. Natalini. A discrete kinetic approximation of entropy solutions to multidimensional scalar conservation laws. *J. Differential Equations*, 148:292–317, 1998.
8. R. Natalini. Recent results on hyperbolic relaxation problems. In *Analysis of systems of conservation laws (Aachen, 1997)*, volume 99 of *Chapman & Hall/CRC Monogr. Surv. Pure Appl. Math.*, pages 128–198. Chapman & Hall/CRC, Boca Raton, FL, 1999.
9. B. Perthame and E. Tadmor. A kinetic equation with kinetic entropy functions for scalar conservation laws. *Comm. Math. Phys.*, 136:501–517, 1991.
10. D. Serre. Stabilité des ondes de choc de viscosité qui peuvent être caractéristiques. In *Nonlinear partial differential equations and their applications. Collège de France Seminar, Vol. XIV (Paris, 1997/1998)*, volume 31 of *Stud. Math. Appl.*, pages 648–654. North-Holland, Amsterdam, 2002.

The Weak Rankine Hugoniot Inequality

B. Després

1 Introduction

Consider a system of conservation law with a stiff right hand side

$$\partial_t U + \partial_x f(U) = \frac{1}{\lambda} R(U). \tag{1}$$

We assume essentially three hypotheses [6, 7]. *First hypothesis:* The left hand side of system (1) is endowed with an entropy–entropy flux pair $U \mapsto (\mathrm{e}(U), F(U)) \in \mathbb{R} \times \mathbb{R}$: if $R \equiv 0$ weak solutions satisfy $\partial_t \mathrm{e}(U) + \partial_x F(U) \le 0$. *Second hypothesis:* The stiff right hand side is compatible with the entropy principle. *Third hypothesis:* This hypothesis is of geometrical nature and is the most important one in our context. We assume the constraint is convex with respect to the primitive, adjoint or entropy variable $V = \nabla_U \mathrm{e}$. It implies that

$$R(0) = 0 \Longleftrightarrow V \in K,$$

where K is a closed convex set. So it is possible to reinterpret the system (1) as a system of conservation laws plus a convex primitive constraint. We address the question of discontinuous solutions for the limit $\lambda \to 0^+$. One has the following result [6, 7].

Lemma 1. *Consider a system satisfying the three above hypotheses. For all discontinuous limits of viscous solutions, one has the* **Weak Rankine Hugoniot Relation** *(or WRH inequality)*

$$-\sigma(\mathrm{e}_R - \mathrm{e}_L) + F_R - F_L - \left\langle -\sigma(U_R - U_L) + f_R - f_L, \overline{V} \right\rangle \le 0, \tag{2}$$

for all test vector $\overline{V} \in K$.

The velocity of the discontinuity is $\sigma \in \mathbb{R}$. The right (resp. left) state is referred to by the subscript R (resp. L). The WRH inequality (2) is a special polar transform. The proof is simple using ideas coming from the theory of moment and the geometrical definition of the constraint. The convexity of K is critical. If $K = \mathbb{R}^n$ then WRH inequality implies strong RH

$$-\sigma(U_R - U_L) + f_R - f_L = 0$$

and the entropy inequality

$$-\sigma(\mathrm{e}_R - \mathrm{e}_L) + F_R - F_L \leq 0.$$

This is the reason of the denomination **WRH**. Somehow this approach is an extension of the classical theory of shocks for conservation laws (Lax, Godunov, Friedrichs, ...). Determination of all cases such that WRH is enough for the construction of the Riemann problem in such a nonconservative framework is an open problem. The examples discussed in this work (the convex K shall be a disk, a hyper-plane, or a domain bounded by a parabola) give hints that this issue is connected to the geometry of the constraint in the adjoint space.

The outline of the paper is as follows. First we consider the linear wave system plus a convex constraint. On this example it is easy to prove the WRH. Then we discuss shocks and construct a solution to the Riemann problem. The second example comes from the method of moment, which is a special case of WRH. Finally we turn to the description of WRH for the system of St Venant with mass loss.

2 A Simple Example

Consider the viscous linear wave system with a stiff right hand side

$$\begin{cases} \partial_t u_{\lambda,\nu} + \partial_x v_{\lambda,\nu} = -\frac{1}{\lambda}(u_{\lambda,\nu} - u^*(u_{\lambda,\nu}, v_{\lambda,\nu})) + \nu\partial_{xx}u_{\lambda,\nu}, \\ \partial_t v_{\lambda,\nu} + \partial_x u_{\lambda,\nu} = -\frac{1}{\lambda}(v_{\lambda,\nu} - v^*(u_{\lambda,\nu}, v_{\lambda,\nu})) + \nu\partial_{xx}v_{\lambda,\nu}. \end{cases} \quad (3)$$

Here $\lambda \to 0^+$ is the relaxation parameter and $\nu \to 0^+$ is the viscosity parameter. The definition of right hand side is

$$\begin{cases} \text{if } u^2 + v^2 > 1, \ (u^*, v^*) = \frac{(u,v)}{\sqrt{u^2+v^2}}, \\ \text{if } u^2 + v^2 \leq 1, \ (u^*, v^*) = (u, v). \end{cases}$$

At the limit $\lambda \to 0^+$, the solution is constrained in the unit disk. The case that is new with respect to the standard theory [12] is the constrained one (4).

2.1 Limit Model for Smooth (Strong) Solution $\nu = 0$

The *limit constrained model* ($\lambda \to 0^+$) for smooth solutions corresponds to a saturation of the constraint

$$u^2 + v^2 \equiv 1. \quad (4)$$

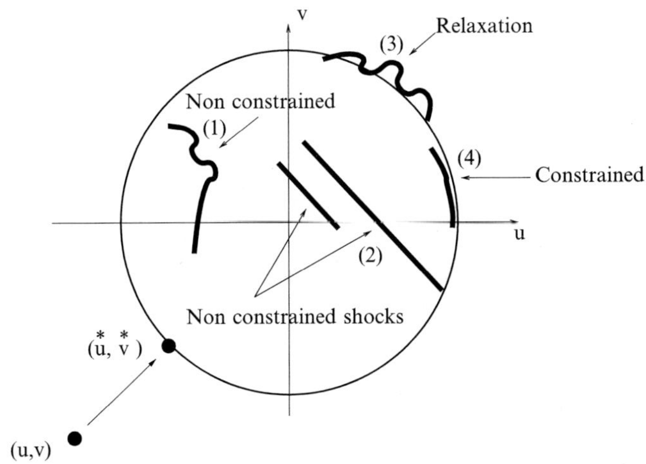

Fig. 1. Convexity of the set of constraints. Four types of solution

For smooth solutions the viscosity ν plays no role; therefore, we assume $\nu \equiv 0$ in (3). On the other hand the stiff right hand side may have a nonzero limit. Let us rewrite the system as

$$\begin{cases} \partial_t u_\lambda + \partial_x v_\lambda = -\frac{1}{\lambda} Q_\lambda u_\lambda, \\ \partial_t v_\lambda + \partial_x u_\lambda = -\frac{1}{\lambda} Q_\lambda v_\lambda, \end{cases}$$

for a convenient $Q_\lambda \in \mathbb{R}$. This is possible since the stiff right hand side of (3) is parallel to the vector (u, v). One has the identity

$$-\frac{1}{\lambda} Q_\lambda \left(u_\lambda^2 + v_\lambda^2\right) = \partial_t \left(\frac{u_\lambda^2 + v_\lambda^2}{2}\right) + \partial_x \left(u_\lambda v_\lambda\right).$$

Assume the strong convergences: $u_\lambda \to u$, $v_\lambda \to v$, and $u_\lambda^2 + v_\lambda^2 \to 1$ for all t and x. It means the constraint is saturated at the limit (4), see Fig. 1. Then formally $\partial_t(\frac{u_\lambda^2+v_\lambda^2}{2}) \to \partial_t 1 = 0$. So $-\frac{1}{\lambda} Q_\lambda \to \partial_x (uv)$. Therefore, the limit model is

$$\begin{cases} \partial_t u + \partial_x v = u \partial_x (uv), \\ \partial_t v + \partial_x u = v \partial_x (uv). \end{cases} \tag{5}$$

This reduced constrained system is *nonconservative*, i.e., nondivergent. The Jacobian matrix is

$$A = \begin{pmatrix} -uv & 1-u^2 \\ 1-v^2 & -uv \end{pmatrix}.$$

The wave velocities are

$$\lambda = 0, \quad \lambda_2 = -2uv \in [-1, 1] \text{ since } u^2 + v^2 = 1. \tag{6}$$

The limit model is hyperbolic but nondivergent. It is therefore of fundamental interest to determine if shocks are possible, or not, for such nonconservative systems. This theoretical problem has been adressed in many works [9, 5] in the context of the mathematical theory of plasticity systems [11, 4, 6, 7].

2.2 A New Inequality for Discontinuous (Weak) Solutions

This section is devoted to the derivation of a weak inequality, which is characteristic of weak solutions for the nonconservative system (5).

Definition 1. *A discontinuous solution (u, v) of the nonconservative system (5) is a limit $\lambda, \nu \to 0$ of $(u_{\lambda,\nu}, v_{\lambda,\nu})$ with*

$$\begin{cases} \partial_t u_{\lambda,\nu} + \partial_x v_{\lambda,\nu} = -\frac{1}{\lambda}(u_{\lambda,\nu} - u^*(u_{\lambda,\nu}, v_{\lambda,\nu})) + \nu \partial_{xx} u_{\lambda,\nu}, \\ \partial_t v_{\lambda,\nu} + \partial_x u_{\lambda,\nu} = -\frac{1}{\lambda}(v_{\lambda,\nu} - v^*(u_{\lambda,\nu}, v_{\lambda,\nu})) + \nu \partial_{xx} v_{\lambda,\nu}. \end{cases} \tag{7}$$

The next stage consists in the characterization of the limit.

Lemma 2. *Assume that a sequence of viscous solutions converges to a discontinuous solution of the nonconservative system (5). The discontinuous solution is defined by a left state (u_L, v_L), a right state (u_R, v_R), and a shock velocity σ. Then one has the Weak Rankine Hugoniot inequality: $\forall(\overline{u}, \overline{v})$ such that $\overline{u}^2 + \overline{v}^2 \leq 1$*

$$-\sigma \left[\frac{u^2 + v^2}{2} \right] + [uv] - (\overline{u}(-\sigma[u] + [v]) + \overline{v}(-\sigma[v] + [u])) \leq 0.$$

Take the scalar product of (7) by $(u_{\lambda,\nu} - \overline{u},\ v_{\lambda,\nu} - \overline{v})$. Here $(\overline{u}, \overline{v})$ is any test vector such that $\overline{u}^2 + \overline{v}^2 \leq 1$. Since the contribution of the stiff right hand side is nonpositive it gives

$$\begin{aligned} &\left(\partial_t \frac{u_{\lambda,\nu}^2 + v_{\lambda,\nu}^2}{2} + \partial_x u_{\lambda,\nu} v_{\lambda,\nu} \right) - \overline{u}\left(\partial_t u_{\lambda,\nu} + \partial_x v_{\lambda,\nu}\right) - \overline{v}\left(\partial_t v_{\lambda,\nu} + \partial_x u_{\lambda,\nu}\right) \\ &\leq \nu \partial_{xx} \left(\frac{u_{\lambda,\nu}^2 + v_{\lambda,\nu}^2}{2} - \overline{u} u_{\lambda,\nu} - \overline{v} v_{\lambda,\nu} \right) - \nu \left[(\partial_x u_{\lambda,\nu})^2 + (\partial_x v_{\lambda,\nu})^2 \right] \\ &\leq \nu \partial_{xx} \left(\frac{u_{\lambda,\nu}^2 + v_{\lambda,\nu}^2}{2} - \overline{u} u_{\lambda,\nu} - \overline{v} v_{\lambda,\nu} \right) \to 0 \text{ in the weak sense.} \end{aligned}$$

Therefore, one has at the limit the weak inequality

$$\left(\partial_t \frac{u^2 + v^2}{2} + \partial_x uv \right) - \overline{u}\left(\partial_t u + \partial_x v\right) - \overline{v}\left(\partial_t v + \partial_x u\right) \leq 0$$

true after integration by part against smooth test functions. This weak inequality is in divergent form even if the underlying system is not.

2.3 Weak Rankine Hugoniot Relation (or Inequality)

If a discontinuous (shock) weak limit exists, then necessary

$$-\sigma\left[\frac{u^2+v^2}{2}\right]+[uv]-(\overline{u}(-\sigma[u]+[v])+\overline{v}(-\sigma[v]+[u]))\leq 0. \tag{8}$$

This inequality if precise the WRH for the entropy–entropy flux pair e $=\frac{1}{2}u^2+\frac{1}{2}v^2$ and $F=uv$.

Let us discuss inequality (8). Since $(\overline{u},\overline{v})$ is arbitrary in $\overline{u}^2+\overline{v}^2\leq 1$, the inequality is equivalent to

$$\sqrt{a^2+b^2}\leq(ac+bd)\,,$$

where $a=-\sigma[u]+[v]$, $b=-\sigma[v]+[u]$, $c=-\frac{u_R+u_L}{2}$, $d=-\frac{v_R+v_L}{2}$, and $c^2+d^2\leq 1$. Any nontrivial solution $R\neq L$ is such that $c^2+d^2<1$. Therefore, any nontrivial discontinuous solution is such that

$$a=b=0\Longleftrightarrow\begin{cases}-\sigma[u]+[v]=0\\-\sigma[v]+[u]=0.\end{cases}$$

These equalities are the standard Rankine Hugoniot relations for the standard nonconstrained problem. For this problem all discontinuous solutions are standard discontinuous solutions for the standard nonconstrained problem.

2.4 Solution of the Riemann Problem

Consider the Riemann problem between a left state and a right state. For example, take

$$u_L=0.4,\ v_L=\sqrt{1-u_L^2}),\quad u_R=u_L, v_R=-v_L.$$

It is possible to construct a solution of the Riemann problem (Fig. 2). We first consider a standard shock with velocity $\sigma^-=-1$ (resp. $\sigma^+=1$) between the left (resp. right) state and the intersection of the straight line $u+v=u_L+v_L$ (resp. $u-v=u_L-v_L$) and the circle $u^2+v^2=1$. It gives two points on the circle. Then we connect these two points with a smooth wave fan. The construction is correct since $\sigma=\pm 1$ and $-1\leq\lambda\leq 1$ (see (6)). In this case WRH is *necessary and sufficient* for the solution of Riemann problem.

2.5 Numerical Results

Numerical solutions are particular viscous solutions. One can prove that under certain conditions, the analysis extends to discrete solutions. The computation is done in two steps. First step we solve the homogeneous linear wave equation

$$\partial_t u+\partial_x v=0,\quad \partial_t v+\partial_x u=0.$$

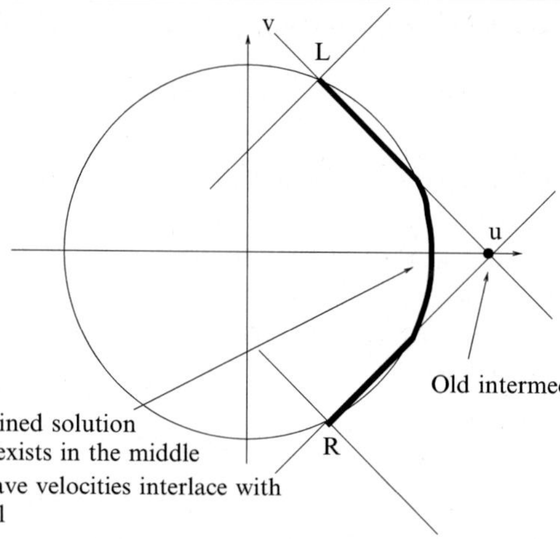

Fig. 2. Solution of the Riemann problem

Second step we solve the stiff right hand side

$$\text{if } u^2 + v^2 \leq 1 \text{ do nothing,} \quad \text{else } (u, v) = \frac{(u, v)}{\sqrt{u^2 + v^2}}.$$

We only mention that numerical results (not presented due to lack of place) are in agreement with the WRH.

3 Moment Models

There are different approaches to the method of moments. We refer mainly to [1, 2, 10]. The method of moment can be summed up in very few words.

Let $U \in \mathbb{R}^n$ be an unknown vector such that $\langle U, Z_i \rangle = \alpha_i$. Here $Z_i \in \mathbb{R}^n$ are *given* vectors and α_i are given real numbers. U is under-determined. The method of moments amounts to the definition of U, by taking the minimizer of the entropy. The minimizer U is the critical point $\nabla_U L = 0$ of the Lagrangian

$$L = \mathrm{e}(U) - \sum_{i=1}^{p<n} \lambda_i \left(\langle U, Z_i \rangle - \alpha_i \right)$$

whose solution is $V - \sum_i \lambda_i Z_i = 0$. It means the primitive variable belongs to a linear sub-space. Other formulation: K is flat. Therefore, we have proved

Lemma 3. *Shocks of moment models are a particular case of WRH with a linearly degenerate convex of constraints K (an hyper-plane).*

So the WRH is equivalent to

$$-\sigma(\mathrm{e}_R - \mathrm{e}_L) + F_R - F_L - \left\langle -\sigma(U_R - U_L) + f_R - f_L, \overline{V} \right\rangle \leq 0$$

for all $\overline{V} \in$ Span$(Z_1, \cdots, Z_p)$. Equivalently $-\sigma(\mathrm{e}_R - \mathrm{e}_L) + F_R - F_L \leq 0$ and

$$\langle -\sigma(U_R - U_L) + f_R - f_L, Z_i \rangle = 0, \quad i = 1, \cdots, p.$$

In this case WRH is *necessary and sufficient* to determine admissible shocks.

4 An Open Problem

Consider the nonconservative system of St-Venant with mass loss [3]

$$\begin{cases} \partial_t h + \partial_x hu = Q, \\ \partial_t hu + \partial_x \left(hu^2 + \frac{g}{2} h^2 \right) = Qu, \, g > 0, \end{cases} \tag{9}$$

with the constraint $h \geq 0$ (since this is not really a constraint, we shall not take it into account in the analysis) and much more important with the mass loss constraint

$$h \leq 1. \tag{10}$$

If $h < 1$ then $Q = 0$; if $h = 1$ then $Q = \partial_x hu = \partial_x u$. The entropy–entropy flux pair that we consider is $\mathrm{e} = \frac{g}{2}h^2 + \frac{1}{2}hu^2$, $F = \mathrm{e}u + \frac{g}{2}h^2 u$. Therefore, the primitive variable is $V = (V_1, V_2)$ with $V_1 = gh - \frac{1}{2}u^2$ and $V_2 = u$. The mass-loss constraint is convex with respect to V

$$V_1 + \frac{1}{2} V_2^2 \leq g.$$

So K is bounded by a parabola. Let us assume that any discontinuous solution of (9) must be a viscous limit in a certain sense compatible with the general hypotheses necessary for the derivation of the WRH. It is possible to show that the right hand side of (9) is a limit of a stiff right hand side compatible with the entropy principle.

Let us focus on the algebraic consequences of the WRH. Consider a discontinuous solution. Because of the Galilean invariance of the model, the WRH in the shock reference frame ($\sigma = 0$) is

$$[\frac{1}{2}u^3 + gu^2] + \max_{V_1 + \frac{1}{2}V_2^2 \leq g} \left(-[u]V_1 + [u^2 + \frac{1}{2}gu]V_2 \right) \leq 0.$$

The maximum is reached at the boundary of the domain of constraints

$$[\frac{1}{2}u^3 + \frac{1}{2}gu^2] + \max_{V_2} \left(-[u](g - \frac{1}{2}V_2^2) - [u^2]V_2 \right) \leq 0.$$

Therefore $[u] = u_R - u_L \leq 0$ and $V_2 = \frac{[u^2]}{[u]} = u_L + u_R$. After simplification one gets $u_L u_R \leq 0$. In the labs reference frame the condition is

$$u_L \geq \sigma \geq u_R.$$

This is a natural, physical, and necessary condition for mass loss. But there is not enough information to determine uniquely the states on both sides even if we assume that $h_R = h_L = 1$, which is the interesting case, because the values of u_L and u_R are not given. In particular $u_L = -u_R$ is not mandatory even if such a symmetry seems very physical. In this case WRH is *necessary* but is *not sufficient* to characterize discontinuous solutions of the nonconservative system (9).

5 Conclusion

We have presented on examples of the Weak Rankine Hugoniot Relation for discontinuous solutions of certain systems of PDEs with stiff right hand side

$$-\sigma(\mathrm{e}_R - \mathrm{e}_L) + F_R - F_L \left\langle -\sigma(U_R - U_L) + f_R - f_L, \overline{V} \right\rangle \leq 0.$$

This right hand side models a geometric constraint $V \in K$, where K is a closed convex set. Somehow this approach is an extension of the classical theory of shocks for conservation laws (Lax, Godunov, Friedrichs, ...). The WRH inequality is a special polar transform. The method of moments is a particular case. Determination of all cases such that WRH is enough for the construction of the Riemann problem in such a nonconservative framework is an open problem. The examples discussed in this work (the convex K is a disk, an hyper-plane, or a domain bounded by a parabola) give hints that this issue is connected to the geometry of the constraint in the adjoint space.

References

1. G. Boillat and T. Ruggeri, Hyperbolic principal subsystems: entropy convexity and subcharacteristic conditions. Arch. Rational Mech. Anal. 137 (1997), no. 4, 305–320.
2. G.Q. Chen, C.D. Levermore et T.P. Liu, Hyperbolic conservation laws with stiff relaxation and entropy, Communications in Pure an Applied Mathematics, 47 787–830, 1994.
3. F. Berthelin et F. Bouchut, Weak solutions for a hyperbolic system with unilateral constraint and mass loss, Ann. IHP Analyse non linéaire, 20, pp 975–997, 2003.
4. J.F. Colombeau and A.Y. Leroux, Multiplications of distributions in elasticity and hydrodynamics. J. Math. Phys. 29 (1988), no. 2, 315–319.
5. Dal Maso, Le Floch et F. Murat, Definition and weak stability of non conservative products, J. Math. Pures Appl., 74 (1995), 458–483.

6. B. Després, Weak Rankine Hugoniot relation for conservation laws with convex primitive constraints, CRAS Paris, Serie I, 342, p 73–78, 2006.
7. B. Després, A geometrical approach to non conservative shocks and elastoplastic shocks, to appear in Arch. Rat. Mech. Anal.
8. P.D. Lax, Hyperbolic systems of conservation laws and the theory of shock waves, SIAM, Philadelphia, 1973.
9. P. Le Floch, Entropy weak solutions to nonlinear hyperbolic systems under non conservative form, Comm. In P.D.E., 13 (6), 669–727, 1988.
10. I. Müller and T. Ruggeri, Rational Extended Thermodynamics, Springer, 1998.
11. M. Rascle, Elasto-Plasticity as a Zero-Relaxation Limit of Elastic Visco-Plasticity, Transp. Theory and Stat. Physics, 25 (3–5), 477–489, 1996.
12. D. Serre, Systems of conservation, Cambridge University Press, I (1999) and II (2000). Diderot, France, 1996.

Numerical Investigations Concerning the Strategy of Control of the Spatial Order of Approximation Along a Fitted Gas–Liquid Interface

C. Dickopp and J. Ballmann

Summary. With the motivation to simulate a collapsing gas or vapor bubble in a liquid, a finite element method using bilinear or piecewise constant shape and test functions for the spatial discretization and several Runge–Kutta methods for the time integration has been developed. The method approximates the solution of the Euler or Navier–Stokes equations in a Lagrangian formulation on an unstructured, moving grid in two space dimensions to fit the bubble–liquid interface. The nonlinear stability of the method is achieved by an improvement of the FCT-strategy for the control of the spatial order of approximation.

A series of numerical investigations indicates a strong dependence between the numerical solution of the model problem of a collapsing single bubble and the strategy of control of the spatial order of approximation along the bubble wall.

1 Introduction and Motivation

Cavitation damaging of solid surfaces is an important technical problem in liquids that is caused by the collapses of bubbles generated upstream in the flow. With the motivation to investigate the still unclarified mechanism of such damaging, the model problem of a single bubble that collapses near a solid surface is simulated numerically.

To overcome the restrictions of potential flow used in many publications, for example [PC] or [B], for the simulation of collapsing bubbles, a mathematical model based on the Euler or Navier–Stokes equations [DB98] can be used. One possible treatment of the bubbles wall within such an approach consists in the fitting of this phase interface using an arbitrary Euler–Lagrangian formulation of the transport equations.

Within such a so-called tracer or string approach stability problems arise near the gas–liquid interface. Also the redistribution, insert, or cancelation of the indicator nodes along the interface may be required in certain situations. A summary of such difficulties of tracking techniques using a Lagrangian description of a moving interface can be found inside the second chapter of Sethian's monograph about level set methods ([S]). Sethian motivates the introduction of level set formulation by the stated problems. Furthermore, a combination with the so-called ghost cell method can be used that solves in a first step two single-phase problems for both fluids and then combines these solutions along the interface whose course is indicated by the zero values of the level set function [OF]. The basic idea is to use values for the entropy of the ghost fluid along the interface instead of the incorporation of the jump conditions into the algorithm, for example, by a Riemann solver.

By contrast our finite element formulation takes the jump conditions for this kind of contact discontinuity into account by additional boundary integral terms along the interface according to the formulation of the problem in the sense of continuum mechanics. In the present paper concepts for the stabilization of this Lagrangian method along the interface by the control of the spatial order of approximation are discussed.

2 Physical Models and Mathematical Formulation

The mathematical description of the flow is based on the compressible Navier–Stokes equations in combination with different equations of state for the two fluids – the gas inside the bubble and the liquid around it. The assumed rotational symmetry with respect to the z-axis causes a geometrical source term $\frac{1}{r}\mathbf{Q}$. As turbulence models Wilcox' $k-\omega$-model as well as Menter's shear stress model were implemented and lead to additional source terms $\mathbf{Q}^{\mathbf{turb}}$ within the extended system of transport equations in an arbitrary Euler–Lagrangian formulation using the velocity $\mathbf{v}^{\mathbf{grid}}$ of the moving grid:

$$\mathbf{U^{turb}}_{,t} + \mathbf{F^{turb}}_r(\mathbf{U^{turb}})_{,r} + \mathbf{F^{turb}}_z(\mathbf{U^{turb}})_{,z} + \tfrac{1}{r}\mathbf{Q}(\mathbf{U^{turb}}) + \mathbf{Q^{turb}}(\mathbf{U^{turb}}) = \mathbf{0}. \tag{1}$$

$\mathbf{U^{turb}}$ denotes the vector of conservative variables, extended by the turbulent kinetic energy density ρk and the density of the turbulent dissipation rate $\rho\omega$:

$$\mathbf{U^{turb}} = (\rho, \rho v_r, \rho v_z, \rho e, \rho k, \rho\omega)^T, \tag{2}$$

and index behind comma means partial differentiation with respect to the appropriate variable. The flux functions as well as the thermal and calorical perfect gas law, Stokes' description of the molecular stress tensor, Reynolds' formulation for the turbulent stresses using Boussinesq's ansatz, Fourier's heat flux and the equations of the two implemented turbulence models, as well as the material laws to close the system of equations are well known and are not noted down here because of the limitation of place. $\mathbf{Q^{turb}}$ describes the production and the dissipation of k and ω, respectively.

While thermal and calorical perfect laws are assumed for the gas inside the bubble the behavior of a compressible liquid is described by Tait's equation of state in a barotropic way:

$$\frac{p+B}{p_0+B} = \left(\frac{\rho}{\rho_0}\right)^n \tag{3}$$

with the following values of the constants for pure water: $p_0 = 7,025\,\mathrm{Pa}$, $B = 3.01 \times 10^8\,\mathrm{Pa}$, $\rho_0 = 998.2\,\frac{\mathrm{kg}}{\mathrm{m}^3}$, $n = 7.15$.

3 Numerical Method

The solution of this system of equations is numerically approximated by a finite element approach that can be derived from a weighted residual formulation for the previously noted conservation form of the Navier–Stokes equations:

$$< D(\mathbf{U}^{\mathbf{turb}}(r,z,t)), w_m > = < \varepsilon, w_m > = 0 \quad \forall (r,z) \in \Omega, \qquad m \in M, \tag{4}$$

using a residual vector ε, a set of weighting functions $w_m, m \in M$ as well as the functional D for the governing equations

$$D(\mathbf{U}^{\mathbf{turb}}) = \mathbf{U}^{\mathbf{turb}}{}_{,t} + \mathbf{F}^{\mathbf{turb}}{}_r(\mathbf{U}^{\mathbf{turb}})_{,r} + \mathbf{F}^{\mathbf{turb}}{}_z(\mathbf{U}^{\mathbf{turb}})_{,z}$$
$$+ \frac{1}{r}\mathbf{Q}(\mathbf{U}^{\mathbf{turb}}) + \mathbf{Q}^{\mathbf{turb}}(\mathbf{U}^{\mathbf{turb}}). \tag{5}$$

Here $<,>$ denotes the scalar product that is defined for two continuous functions $a, b \in C^0(\Omega)$ by $< a, b > := \int_\Omega ab\, d\Omega$. The shape functions Φ_N are set up for the conservative variables as well as for the turbulent variables ρk and $\rho\omega$ as bilinear or piecewise constant functions, for example, $\rho = \Phi_N \rho_N$.

Also bilinear or piecewise constant weight functions w_m can be chosen, resulting in a Petrov–Galerkin approach. By this way a conform or nonconform discretization is obtained depending on the chosen shape and test functions.

To approximate the derivative $\mathbf{U}_{,t}$ explicit Runge–Kutta methods or Taylor expansion methods are implemented.

The method uses an unstructured, optionally moving grid composed by triangular or quadrangular elements. The time steps are chosen according to derived criteria for linear stability [DB04] that takes also the discretization of the source terms, the viscous terms as well as the additional boundary integral terms for the transition conditions along the gas–liquid interface into account. The analysis is based on the demand that an assumed perturbation of the solution with the wavelength of the computational grid or multiples of it may not be amplified by the numerical scheme. Further, the derived upper bound on the CFL-number depends on the chosen, explicit Runge–Kutta or Taylor scheme for the time integration.

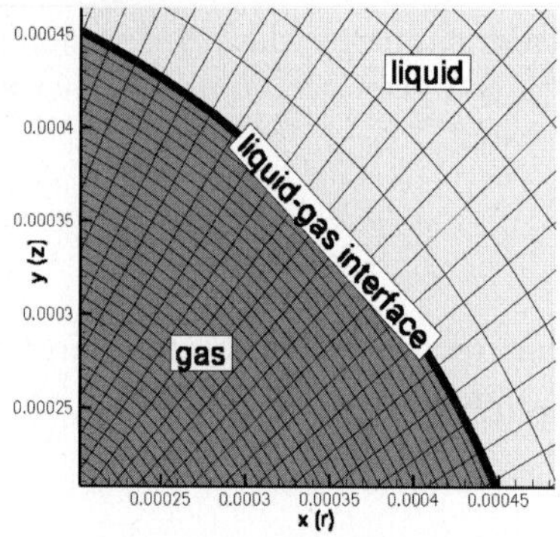

Fig. 1. Part of the computational mesh around the gas–liquid interface

A remeshing is needed within every time step to fit the course of the interface as a grid line.

Figure 1 shows a part of the computational grid to explain the setup of the method: the course of the interface divides the domain and the grid into two parts with identical vertices for both phases. By the incorporation of an additional boundary integral term for the jump conditions along the gas–liquid interface into the finite element formulation the flow calculations for both phases are strongly coupled in the sense of a tracer/string approach.

4 Nonlinear Stability

The basis for the nonlinear stabilization is formed by a theorem of Godunov, which states that discontinuous solutions can only be resolved in a stable manner by a method with first order of spatial approximation. On the other hand the numerical dissipation of a method of spatial order one is too strong to be used in the whole region. So the FCT-concept developed by Boris, Book, and Hain [BB], [BBH], mainly in the context of finite difference methods, combines the solutions of a low order scheme and a high order method to avoid oscillations caused by the nonlinearity of the system of equations. Löhner et al. [LMPV] transferred the basic idea of a limiter whose values are calculated by local comparisons of both solutions to finite element schemes for compressible flow problems.

Goodman and LeVeque have proven that a multi-dimensional method which fulfills a multidimensional extension of the TVD property has at most the spatial order 1 [LV]. Because the dissipation of general first order methods is too high, this theorem gave the motivation to introduce the ENO concept as a weak form of the TVD property.

To specify an ENO-reconstruction for a Godunov-type method a local selection process between different possible interpolations in the environment of a regarded node is used. Our intention was to improve the FCT-concept by a transfer of the idea behind ENO-reconstructions to control the spatial order for the used finite element method.

Although our concept of a common Petrov–Galerkin formulation within every step of the time integration scheme allows the choice of piecewise constant or bilinear shape and test functions, we restrict the last time integration step to a classical Bubunov–Galerkin scheme using bilinear shape and test functions. So the last step of the explicit time integration method takes the form

$$\mathbf{M}^{(\mathbf{k})} \Delta(\mathbf{U}^{\mathbf{h}})^{n+1} = \Delta t RHS. \tag{6}$$

The use of the consistent mass matrix $\mathbf{M}^{(\mathbf{k})}$ creates the second order of this approximation in space. To obtain a first order method this matrix is replaced by the lumped mass matrix $\mathbf{M_L}$ with the entries $(\mathbf{M_L})_{ij} = \begin{pmatrix} \sum_l (\mathbf{M}^{(\mathbf{k})})_{il} = \sum_l (\mathbf{M}^{(\mathbf{k})})_{li} \text{ for } i = j, \\ 0 \qquad\qquad\qquad \text{for } i \neq j \end{pmatrix}$. An additional viscosity term with the coefficient c_d improves the stability of the first order scheme:

$$\mathbf{M_L} \Delta(\mathbf{U^l})^{n+1} = \Delta t RHS + c_d \left(\mathbf{M}^{(\mathbf{k})} - \mathbf{M_L} \right) \mathbf{U}^n. \tag{7}$$

The task to control the spatial order of approximation can be expressed as the search for a limiter function Θ_e, $0 \leq \Theta_e \leq 1$, on each element e that combines the increments $\Delta \mathbf{U}^l$ and $\Delta \mathbf{U}^h$ from both methods in a linear way:

$$\Delta \mathbf{U} = \Delta \mathbf{U}^l + \sum_e \Theta_e \left(\Delta(\mathbf{U}^h_e)^{n+1} - \Delta(\mathbf{U}^l_e)^{n+1} \right). \tag{8}$$

The FCT strategy ensures preserving of monotonicity of the method by the calculation of the limiter values in such a way that the combined solution does not over- or undershoot the first order solution in the environment of each node.

Based on the truncation error for the one-dimensional Lax–Wendroff method, which is similar to the used finite element formulation, we take the initial guess for the diffusion coefficient $c_d = 3\,\eta\,(1-\eta)$ as a function of the local Courant number η. Then we try to improve c_d by scanning the whole possible interval $[0,1]$ with steps of about 0.01 with the intention that the first order solution becomes smoother. For this purpose a local measure of smoothness of the solution is needed that is evaluated for every choice of c_d and at every grid node no. N. In analogy to the measure of smoothness for linear reconstructions within Godunov-type methods (see for example [K], p. 246) that represents the steepness of an averaged gradient on the cell we introduce a similar measure using the bilinear shape functions:

type of the element e	shape function	measure σ_e
Triangle	$a + b\xi + c\eta$	$\|b\| + \|c\|$
Quadrangle	$a + b\xi + c\eta + d\xi\eta$	$\|b\| + \|c\| + \|d\|$

The coefficients a, b, c, and d are determined by the values q_i, $i = 1, 2, 3, 4$ of a regarded quantity q at the nodes:

Triangle		Quadrangle	
	$a = q_1$		$a = q_1$
	$b = q_2 - q_1$		$b = q_2 - q_1$
	$c = q_3 - q_1$		$c = q_3 - q_1$
	$d = 0$		$d = q_4 - q_3 - q_2 + q_1.$

Our experience is that

$$q = E + \frac{1}{2}\rho|v|^2 + \rho c_\infty^2 \tag{9}$$

is a convenient quantity to control the spatial order of approximation. c_∞ denotes the speed of sound of the gas or liquid resp. with respect to a reference state and serves as weighting parameter to detect density oscillations.

To obtain a measurement quantity for the nodes a summation over all elements to which a regarded node belongs is done with respect to the definition of σ_e by the two tables before and eq. (9):

$$\tilde{\sigma}_N = \sum_{Elements\ e\ to\ which\ N\ belongs} \sigma_e. \tag{10}$$

Then, at every node N of the grid the coefficient c_d is selected such that it leads to the smallest value of $\tilde{\sigma}_N$.

The limiter value Θ_N for a node no. N is determined by a similar selection process but that scans only the interval $[min_{neighbour\ i}\phi_i, max_{neighbour\ i}\phi_i]$ where ϕ_N represents the already mentioned limiters at the nodes within the FCT algorithm. We call this improvement of the FCT strategy by the two selection processes the ENO-version of FCT. The ENO-version of FCT leads to significant better results than the original FCT limiter for a physically multidimensional problem like the spherical bubble collapse.

For one-dimensional problems, for example shock tube problems, both limiters yield almost identical results.

5 Control of the Spatial Order Along the Interface

With the goal to simulate a single collapsing bubble, our special interest is dedicated to the numerical treatment of the bubbles wall. So we investigated eight versions of the nonlinear stabilizer along this fitted phase interface by a series of numerical tests.

In all cases the same model problem was considered: A gas-filled bubble surrounded by a compressible liquid is simulated near a solid surface. Initially both fluids are assumed at rest with different homogeneous state conditions such that the pressure inside the gas is lower than that in the surrounding liquid. So a rarefaction wave in the liquid is caused during a first time segment of such a simulation. Until an emitted rarefaction wave reaches the surface of the adjacent solid symmetric solutions have to be expected.

As summarized in the following table the three bold marked strategies fulfill both posed criteria: the rotational symmetry and the jump conditions. These three methods were used for longer time simulations of a collapsing bubble. The presentation of the corresponding results exceeds the scope of this paper and is planned for the future.

Method no.	Description of the method along the phase interface	Geometr. symmetry preserved?	Jump conditions fulfilled?
1	Same 2D-ENO-FCT as in the other parts of the domain	Disturbances	No
2	2D-ENO-FCT without artificial viscosity	Yes	No
3	**Pure second order method**	Yes	Yes
4	First order method including an artificial diffusion term	Yes	Too low Velocities
5	Use of an additional 1D-version of the ENO-FCT-algorithm in tangential direction and second order in the normal direction	Yes	Too low Velocities
6	**2D-ENO-FCT without artificial viscosity and 1D-ENO-FCT without artificial viscosity**	Yes	Yes
7	Use of 2D-ENO-FCT with high artificial viscosity with a correction of the momentum in the sense of an equilibrium of forces at the interface	Yes	No
8	**As before using an additional iteration towards an equilibrium of forces across the phase interface**	Yes	Yes

The first and obvious possibility consists in the use of the same ENO-FCT-strategy as in the other parts in the computational domain. A high jump of the pressure at the interface shows that the obtained solution does not fulfill the jump conditions posed there. Further little asymmetric effects can be recognized. So there is a need to change the control of the spatial order along the phase interface.

To clarify the reasons for these difficulties in further tests either always the second order scheme or always the first order scheme were applied along the bubbles wall together with a version of ENO-FCT without the term of artificial viscosity within the first order scheme. The asymmetric effect vanishes if the artificial viscosity is switched off so that this term seems to be at least one reason for the mentioned difficulties. The use of the pure second order scheme along the interface seems to be promising at a first glance but it leads to unstable solutions for long time simulations, resulting in a disturbed course of the interface. On the other hand too low velocities are calculated if only the first order method is applied at nodes of the interface.

Also an often used approach at fitted phase interfaces was implemented and tested: the equations at both sides of the interface are solved in an iterative manner until an equilibrium of forces is reached across the interface in the sense of the physical conditions there. The application of such an iteration leads of course to solutions that fulfill the jump conditions for contact discontinuities but the pressure levels reach unphysical values for long time simulation – too much energy is pumped into the bubble by the iteration process.

References

[B] Best, J.P.: The formation of toroidal bubbles upon the collapse of transient cavities. J. Fluid. Mech., **251**, 79–107 (1993)

[BB] Boris, J.P., Book, D.L.: Flux-Corrected Transport I., SHASTA, a fluid transport algorithm that works. J. Comp. Phys., **11**, 38–69 (1973)

[BBH] Book, D.L., Boris, J.P., Hain, K.: Flux-Corrected Transport II., Generalisation of the method. J. Comp. Phys., **18**, 248–283 (1975)

[DB98] Dickopp, C., Ballmann, J.: Numerical Simulation of Cavitation Phenomena in Accelerated Liquids, In: Mayinger, F., Giernoth, B. (ed) Transient Phenomena in Multiphase and Multicomponent Systems. WILEY-VCH (1998)

[DB04] Dickopp, C., Ballmann, J.: Two-Phase Flow Through Injection Nozzles, In: F. Asakura, H. Aiso, S. Kawashima, A. Matsumura, S. Nishibata, K. Nishihara (ed) Hyperbolic Problems - Theory, Numerics and Applications. Tenth International Conference in Osaka, September 2004, Vol. I, Yokohama Publishers, Japan (2006)

[K] Kroener, D.: Numerical Schemes for Conservation Laws. John Wiley & Sons and Teubner (1997)

[LV] LeVeque, R.J.: Finite Volume Methods for Hyperbolic Problems. Cambridge University Press, Cambridge (2002)

[LMPV] Löhner, R., Morgan, K., Peraire, J., Valdati, M.: Finite element flux-corrected transport (FEM-FCT) for the Euler- and Navier-Stokes-equations. Finite Elements in Fluids, **7**, 105–121 (1987)

[OF] Osher, S., Fedkiv, R.: Level Set Methods and Dynamic Implicit Surfaces. Springer-Verlag New York (2003)

[PC] Plesset, M.S., Chapman, R.B.: Collapse of an initially spherical vapour bubble in the neighbourhood of a solid boundary. J. Fluid. Mech., **47**, 283–290 (1971)

[S] Sethian, J.A.: Level Set Methods. Cambridge University Press, Cambridge (1996)

Domain Decomposition Techniques and Hybrid Multiscale Methods for Kinetic Equations

G. Dimarco and L. Pareschi

Summary. In this note we consider the development of a domain decomposition scheme directly obtained from the multiscale hybrid scheme described in [7]. The basic idea is to couple macroscopic and microscopic models in all cases in which the macroscopic model does not provide correct results. We will show that it is possible to view a Boltzmann–Euler domain decomposition method as a subset of the hybrid scheme [7], if we impose the value of the relaxation parameter equal to zero in some regions of the computational domain. Applications to the two-dimensional BGK equation is presented to show the performance of the method.

1 Introduction

In this note we afford the problem of developing efficient numerical methods for multiscale phenomena in Rarefied Gas Dynamic (RGD). In some fluid dynamic simulations the Navier–Stokes or the Euler equations do not give a satisfactory descriptions of the physical system and a kinetic description through the Boltzmann equation is often required. For regions not too far from thermodynamic equilibrium the BGK approximation is known to be accurate in describing the physics of the problem. However, the computational cost of a direct discretization of the BGK equation is still high and the use of particle simulation methods is preferable. The price to pay for the reduction of computational cost is the presence of fluctuations in the solution. As a consequence particle methods loose their competitiveness in continuum regions where a finite volume solver for the Euler or Navier–Stokes equations can be used. Domain decomposition techniques are then used in order to better treat these difficulties and to design suitable numerical schemes. In many situation in fact we do not need to solve the kinetic equations in the whole computational domain instead it is sufficient to solve the hydrodynamical equations except in small zones where departure from thermodynamical equilibrium like shock waves are present. To this aim we have developed a numerical scheme derived from the multiscale hybrid scheme [7], which can be used to obtain a decomposition of the domain, artificially imposing the value of the Knudsen number ε equal to zero, where the thermodynamic profiles provided by Euler equations are sufficiently accurate.

2 The BGK Equation

We will consider the BGK equation

$$\partial_t f + v \cdot \nabla_x f = \frac{1}{\varepsilon}(M_f - f), \tag{1}$$

with the initial condition

$$f(x, v, t = 0) = f_0(x, v), \tag{2}$$

where $f = f(x, v, t)$ is a nonnegative function describing the time evolution of the distribution of particles which move with velocity $v \in \mathbb{R}^3$ in the position $x \in \Omega \subset \mathbb{R}^3$ at time $t > 0$. The parameter $\varepsilon > 0$ is the Knudsen number and is proportional to the mean free path between collision. In the BGK equation the collision are modeled with a relaxation toward the Maxwellian equilibrium M_f. The local Maxwellian function is defined by

$$M_f(\varrho, u, T)(v) = \frac{\varrho}{(2\pi T)^{3/2}} \exp\left(\frac{-|u - v|^2}{2T}\right), \tag{3}$$

where ϱ, u, T are the density, mean velocity, and temperature of the gas

$$\varrho = \int_{\mathbb{R}^3} f dv, \; u = \int_{\mathbb{R}^3} v f dv, \; T = \frac{1}{3\varrho} \int_{\mathbb{R}^3} [v - u]^2 f dv. \tag{4}$$

In addition we define the energy E as

$$E = \frac{1}{2} \int_{\mathbb{R}^3} v^2 f dv, \tag{5}$$

and the kinetic entropy H_f by

$$H_f = \int_{\mathbb{R}^3} f \log f dv. \tag{6}$$

It is easy to show the entropy dissipation inequality

$$\partial_t \int_{\mathbb{R}^3} f \log f dv + \nabla_x \int_{\mathbb{R}^3} v f \log f dv \leq 0. \tag{7}$$

Now, if we consider the BGK equation (1) and multiply it for the collision invariants 1, v, $\frac{1}{2}|v^2|$ by integrating in v we obtain the evolution of the first three moments of the distribution function f

$$\partial_t \int_{\mathbb{R}^3} f\phi(v)dv + \nabla_x \int_{\mathbb{R}^3} vf\phi(v)dv = 0, \;\; \phi(v) = 1, v_1, v_2, v_3, |v|^2. \tag{8}$$

These equations are the corresponding conservations laws for mass, momentum, and energy. Unfortunately the differential system of (8) is not closed, since it involves higher order moments of the distribution function. As $\varepsilon \to 0$, from (1) we notice that $f \to M_f$. In this case the higher order moments of the distribution function can be computed as functions of ϱ, u, and T and we obtain the closed system of compressible Euler equations

$$\begin{aligned}
&\frac{\partial \varrho}{\partial t} + \nabla_x \cdot (\varrho u) = 0,\\
&\frac{\partial \varrho u}{\partial t} + \nabla_x \cdot (\varrho u \otimes u + p) = 0,\\
&\frac{\partial E}{\partial t} + \nabla_x \cdot (Eu + pu) = 0,\\
&p = \varrho T, \;\; E = \frac{3}{2}\varrho T + \frac{1}{2}\varrho |u|^2.
\end{aligned} \tag{9}$$

3 Fluid Solver Independent Hybrid Method

In this section we introduce the fluid solver independent (FSI) hybrid method, that will be used to compute our solution in the whole domain (see [7]). The domain decomposition will be achieved simply imposing the value of the Knudsen number equal to zero in some part of the domain. The FSI hybrid method is able to take advantage from the equilibrium part of the solution through a general fluid solver for the Euler equations instead of a kinetic scheme as in [6]. This represents an advantage in terms of computational flexibility and cost. In this way the resulting scheme is faster and more accurate then a conventional Monte Carlo (MC) solver. In fact, since only the perturbation from equilibrium is solved by a MC scheme we have a reduction of the number of particles in the computational domain with respect to a full MC. In order to explain the algorithm we introduce the projection operator P. The projection operator computes from the microscopic variables f or M_f the macroscopic variables $U = (\varrho, u, T)$, thus $P(f) = U$ and $P(M_f) = U$, since the local Maxwellian has the same moments of the distribution function. If we split the BGK equation (1) into a relaxation step and in a transport step, from the exact solution of first step we have

$$f(x,v,t) = e^{-\Delta t/\varepsilon} f(x,v,t) + (1 - e^{-\Delta t/\varepsilon}) M_f(x,v,t). \tag{10}$$

Thus we can write

$$U(x,t) = (1 - \beta(x)) P(f(x,v,t)) + \beta(x) P(M_f(x,v,t)), \tag{11}$$

where $\beta(x)$ indicate the value of the equilibrium fraction at each time step in each cell. Now we set $f^p(x,v,t) = (1 - \beta(x)) P(f(x,v,t))$ and solve the transport step for f^p with Monte Carlo methods while for $\beta(x)P(M_f(x,v,t))$ we use any deterministic macroscopic scheme. Unfortunately the knowledge of f^p is linked to the solution of the entire distribution function f. Let us now introduce the relaxation and transport operators R and T. The solution of the microscopic problem for one time step can be written in the following way

$$R(f^{MC}(x,v,t)) = (1-\beta(x))f^{MC}(x,v,t) + \beta(x)M_f^{MC}(x,v,t), \tag{12}$$

$$T(R(f^{MC}(x,v,t))) = T((1-\beta(x))f^{MC}(x,v,t)) + T(\beta(x)M_f^{MC}(x,v,t)), \tag{13}$$

where the apex MC , indicate that the solution is computed with MC scheme. After transport we loose the equilibrium structure of the solution, we define the quantity transported in the following way

$$\tilde{f}^p(x,v,t) = T((1-\beta(x))f^{MC}(x,v,t)), \tag{14}$$

$$\tilde{M}_f(x,v,t) = T(\beta(x)M_f^{MC}(x,v,t)). \tag{15}$$

On the other hand the solution of the macroscopic equations can be performed as

$$P(\beta(x)M_f^d(x,v,t)) = U^d(x,t), \tag{16}$$

$$T(U^d(x,t)) = \tilde{U}^d(x,t). \tag{17}$$

The final hybrid solution at each time step can be recovered by

$$U^h(x,t) = P(\tilde{f}^p(x,v,t)) + \tilde{U}^d(x,t). \tag{18}$$

At the next time step from relaxation we obtain

$$\begin{aligned} f(x,v,t+\Delta t) = (1-\beta(x))(\tilde{f}^p(x,v,t) + \tilde{M}_f(x,v,t)) \\ + \beta(x)M_f(x,v,t+\Delta t). \end{aligned} \tag{19}$$

The term $\tilde{f}^p + \tilde{M}_f$ represent the entire distribution function computed with MC algorithm, however, we need only $(1-\beta(x))$ of the latter quantity, thus instead of sampling from local Maxwellian a fraction of particle $\beta(x)U^h(x)$ we sample only $(1-\beta(x))\beta(x)U^h(x)$. This permits to avoid the effect of discarding and resampling particles at each time step.

The macroscopic scheme, for the compressible Euler equation, used in the computation is a MUSCL type (see [10, 7] for details), note however, that any numerical scheme is possible for the fluid equations to our purpose. The method deserves some remarks

Remark 1.

- We could try to compute the degree of equilibrium after transport of the distribution function, in that way it is possible to diminish particles and fluctuations similarly to [6]. Note however that this will cause an increase in the computational cost since we need the reconstruction of the distribution function f from samples. Moreover the evaluation of the equilibrium fraction will be rather difficult due to the probabilistic character of the component $\tilde{M}_f$ with respect to [5] and [6].

- It is possible to think to other strategies to compute the degree of equilibrium after transport. For example a strategy based on a best fitting of the higher order moments. Numerical results provided by this method are displayed in the right column of Fig. 1.

3.1 Domain Decomposition Method

We could view a domain decomposition technique as a subset of the FSI hybrid scheme just described. The method consist simply in setting artificially the value of the Knudsen number equal to zero, where the Euler profiles are sufficiently accurate. Thus the scheme becomes a coupling of a Euler solver (by MUSCL scheme) in one part of the domain and a BGK solver in the rest of the domain (by FSI method). In fact if $\varepsilon \equiv 0$ from (10) we get

$$f(x, v, t) = M_f(x, v, t). \tag{20}$$

Thus after transport since we project the entire solution toward the equilibrium, we are no more interested to the form of the distribution function but only by their moments. As a consequence $P(\tilde{f}^p) \equiv 0$ and the hybrid solution becomes $U^h = \tilde{U}^d$. In order to divide the domain we need to decide some criteria which permit to detect the zones in thermodynamical equilibrium with respect to the others, this is still an open problem and it will not be addressed in the present work, we quote for instance [8] in which the question is considered in detail. Two consideration about the boundary conditions are necessary:

- The deterministic scheme we use to solve the problem in regions where $\varepsilon = 0$, need a value which is supplied by the hybrid scheme. That value contains some statistical fluctuation that could be large, thus the full deterministic model could contain some boundary error given by the fluctuation which propagates in the rest of the domain. In order to avoid the problem a technique could be to make ε a smooth function of x which gently vary from some fixed value toward zero. For instance we could set ε as

$$\varepsilon(x) = \begin{cases} \varepsilon_F, & \text{for } x \le a \\ 0, & \text{for } x \ge b \\ \dfrac{x-b}{a-b}\varepsilon_F, & \text{for } x \in [a, b], \end{cases} \tag{21}$$

  where b represent the boundary of the equilibrium zones, a represent the boundary of the nonequilibrium zones in which we use the full hybrid scheme, while ε_F represent the value of the Knudsen number for the hybrid scheme.

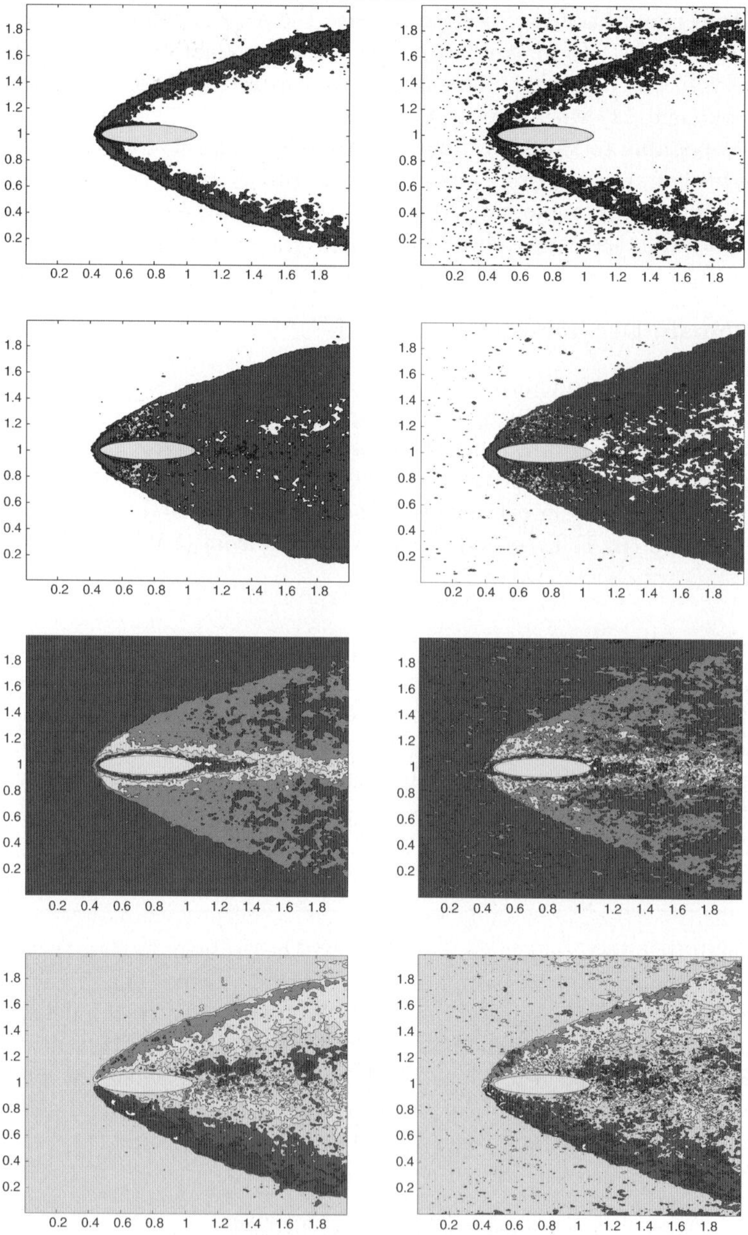

Fig. 1. Two-dimensional flow: domain decomposition hybrid scheme (*left*) Monte Carlo (*right*). Density (*top*), temperature (*middle*), velocity x-direction (*middle*) and velocity y-direction (*bottom*)

- We need also the boundary value for the MC part of the FSI where $\varepsilon \neq 0$, thus at the boundary between $\varepsilon \equiv 0$ and $\varepsilon \neq 0$ we need to sample some particles from the Maxwellian. The number of samples we need is given by $(1 - \beta_i)\beta_i U_i^h$, where i indicate the cell, divided for the mass of the particles. However the particles velocity have to be sampled from the local Maxwellian in the cell $(i - 1)$, if we are considering the left boundary, or from $i + 1$ if the right boundary is considered. Numerical results provided by this method are displayed in the left column of Fig. 1.

4 Numerical Test

In this section we compare the performance of MC and of DD-FSI (Domain Decomposition-FSI). We consider an ellipse embedded in a flow with the following characteristic

$$\varrho = 1, \quad T = 5, \quad M = 10, \tag{22}$$

where M indicates the Mach number, we choose full accommodation boundary condition with temperature $T_E = 10$. The Knudsen number is $\varepsilon = 10^{-4}$ that correspond to $\beta = 0.75$ if $((x-2)/1.8)^2 + (y-1)^2 < 0.8$ while $\varepsilon = 0$ if $((x-2)/1.9)^2 + (y-1)^2 > 0.8$ that correspond of course to $\beta = 1$. The value of the Knudsen number shift from the two reported value while we move from one region toward the other as shown in Fig. 2. The number of space cell are 200×200, the number of particles are 40 per cell. Since we are computing a stationary solution we could after some fixed time strongly diminish the fluctuation by averaging in time the solution, however we want to stress the difference of fluctuations and computational time of our hybrid scheme with respect to Monte Carlo, for this reason the solution is not averaged. What we could see is the better accuracy and performance we obtain with respect

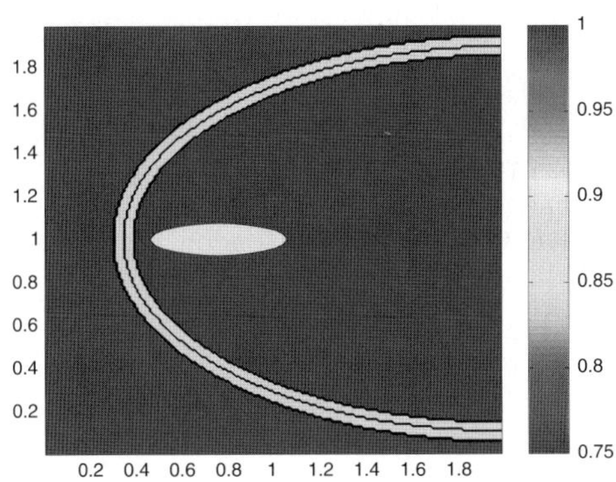

Fig. 2. Degree of equilibrium in different regions of the domain (β)

to Monte Carlo scheme. In the test presented we have a computational gain of about 40% which clearly increases in simulations in which we approach the hydrodynamic limit. In fact if $\varepsilon \to 0$ particles and fluctuations completely disappear and the time we need to perform the solution is the same of a solver for the compressible gas dynamic equations.

References

1. J.F. Bourgat, P. LeTallec, B. Perthame, and Y. Qiu, *Coupling Boltzmann and Euler equations without overlapping*, in Domain Decomposition Methods in Science and Engineering, Contemp. Math. 157, AMS, Providence, RI, (1994), pp. 377–398.
2. R.E. Caflisch and E. Luo, *A Uniform Hybrid Monte Carlo Method for Simulation of Rarefied Gas Dynamics*, preprint (2004).
3. N. Crouseilles, P. Degond, M. Lemou, *A hybrid kinetic-fluid model for solving the gas-dynamics Boltzmann BGK equation*, Journal of Computational Physics 199 (2004) 776–808.
4. P. Degond, S. Jin, and L. Mieussens, *A Smooth Transition Between Kinetic and Hydrodynamic Equations*, Journal of Computational Physics 209 (2005) 665–694.
5. G. Dimarco and L. Pareschi, *Hybrid multiscale methods I. Hyperbolic Relaxation Problems*, Comm. Math. Sci., 1 (2006), pp. 155–177.
6. G. Dimarco and L. Pareschi, *Hybrid multiscale methods II. Kinetic Equations*, Submitted to SIAM Journal of Multiscale Modeling and Simulation.
7. G. Dimarco and L. Pareschi, *A Fluid Solver Indipendent Hybrid Method for Multiscale Kinetic equations*, Preprint 2007.
8. P. Degond, G. Dimarco, L. Mieussens, *A Moving Interface Method For Dynamic Kinetic-Fluid Coupling*, Submitted to Journal of Computational Physics.
9. W.E. and B. Engquist, *The heterogeneous multiscale methods*, Comm. Math. Sci., 1, (2003), pp. 87–133.
10. S. Jin and Z.P. Xin, *Relaxation schemes for systems of conservation laws in arbitrary space dimensions*, Comm. Pure Appl. Math., 48 (1995), pp. 235–276.
11. P. LeTallec and F. Mallinger, *Coupling Boltzmann and Navier-Stokes by half fluxes* Journal of Computational Physics, 136 (1997) pp. 51–67.
12. L. Pareschi and R.E. Caflisch, *Towards an hybrid method for rarefied gas dynamics*, IMA Vol. App. Math. 135 (2004), pp. 57–73.
13. L. Pareschi, *Hybrid multiscale methods for hyperbolic and kinetic problems*, ESAIM: PROCEEDINGS, Vol. 15, T. Goudon, E. Sonnendrucker & D. Talay, Editors, pp. 87–120, (2005).

A Shock Sensor-Based Second-Order Blended (Bx) Upwind Residual Distribution Scheme for Steady and Unsteady Compressible Flow

J. Dobeš and H. Deconinck

1 Introduction

The class of compact residual distribution (RD) schemes has become an appealing alternative for solving fluid flows compared to other approaches, as, e.g., finite volume (FV) and finite element methods (FEM). RD schemes combine the second order of accuracy with nonoscillatory properties on a very compact stencil. They are routinely used for steady state calculations of a complex flow problems [DSA00].

Among linear RD schemes, probably two of the best-known schemes are the LDA and the N scheme. The LDA scheme is second-order accurate scheme and it is successfully used for subsonic computations. The N scheme is a first order, positive scheme. An example of state of the art second order and nonoscillatory schemes are the nonlinear blended (B) and N-modified schemes [DSA00, AM04]. Despite their robustness, both schemes suffer from rather poor convergence to the steady-state solution.

The aim of the chapter is to present another blending technique, which uses a shock capturing operator for blending the LDA and the N scheme. In the shock regions we use locally the N scheme for its monotone shock capturing, while in all other regions we use the LDA scheme, because of its high accuracy. Advantage is, that the LDA scheme contains enough dissipation to resolve smooth flows and contact discontinuities in a stable manner. The shock capturing operator is scaled such that in the smooth regions the N scheme introduces an error of higher order. Since our definition of the blending coefficient is smooth, the full Jacobian can be taken for the implicit method, which noticeably speeds up the convergence rate.

Unfortunately, a straightforward extension of the RD schemes for unsteady computations gives only first-order accuracy [AM03]. The space and time derivatives cannot be discretized separately as they are coupled by a FEM type mass matrix. For time dependent problems we present the version of the blended scheme, which uses the spatial LDA scheme with consistent mass matrix and as a nonoscillatory scheme the N scheme with the lumped mass matrix. The shock capturing operator is modified by inclusion of the time derivative of the pressure to account for unsteady shocks.

2 Residual Distribution Schemes for Steady Problem

In this section we are interested in a steady solution u of the Euler equations in conservative form

$$\frac{\partial u}{\partial t} + \boldsymbol{\nabla} \cdot \mathbf{f} = 0, \tag{1}$$

where $\mathbf{f}$ is the vector of fluxes. The system is closed by the equation for perfect gas with the specific heat ratio $\gamma = 1.4$.

The solution at time level $n+1$ is computed by the relation

$$U_i^{n+1} = U_i^n - k_i \sum_{T \in i} \phi_i^T, \tag{2}$$

where U_i is the solution in node i, k_i is a constant depending on the stability of the iteration procedure, and ϕ_i^T is the contribution from the element T (sharing node i) to the node i, given by the particular scheme. In the case of the LDA scheme, the nodal contribution is given by

$$\phi_i^{\mathrm{LDA}} = -K_i^+ N \phi^T, \quad N = \left(\sum_{j \in T} K_j^+ \right)^{-1}, \quad K_i = \frac{\partial \mathbf{F}}{\partial U} \cdot \frac{\mathbf{n}_i}{d}, \tag{3}$$

where K_i^+ is the positive-eigenvalue part of the upwind matrix K_i, evaluated in a mean state given by linearization [DRS93], $\mathbf{n}_i$ is the inward normal to the face opposite to the node i scaled by its surface, and ϕ^T is the element residual given by $\phi^T = \sum_{j \in T} K_j \bar{U}_j$, where $\bar{U}_j$ are nodal values evaluated according to linearization [DRS93]. The number of spatial dimensions is d ($d = 2$ for 2D). The summation is performed over all the nodes j contained in element T. The N scheme is given by

$$\phi_i^{\mathrm{N}} = K_i^+ (\bar{U}_i - U_{\mathrm{in}}), \qquad U_{\mathrm{in}} = N \left(\sum_{j \in T} K_j^- \bar{U}_j \right). \tag{4}$$

The original blended B scheme [DSA00, AM04] constructed from the N and the LDA scheme, is given by

$$\phi_i^{\mathrm{B}} = \theta \phi_i^{\mathrm{N}} + (1 - \theta) \phi_i^{\mathrm{LDA}}, \qquad \theta = \frac{|\phi^T|}{\sum_{j \in T} |\phi_j^N|}, \tag{5}$$

where θ is computed for each component of the vector of unknowns U_i.

3 New Formulation of Blended (Bx) Scheme

We first construct an element-wise shock capturing sensor given by

$$sc = \left(\frac{\nabla p \cdot \mathbf{v}}{\delta_{pv}}\right)^+ \approx \left(\frac{\int_T \nabla p \, \mathrm{d}x}{\delta_{pv}|T|} \cdot \mathbf{v}\right)^+, \tag{6}$$

where $\mathbf{v}$ is the approximation of the velocity vector in the element, p is the static pressure, and $|T|$ is the area of the element. The scaling parameter $\delta_{pv} \approx (p_{\max} - p_{\min}) \cdot \bar{v}$ is a global pressure variation multiplied by the magnitude of the mean velocity in the domain. The sensor sc is positive in a shock and compression, zero in expansion, and of order $sc = \mathcal{O}(1)$ in the smooth regions. One of the important properties of the scheme is its second order of accuracy. For 2D, in (5) the left-hand side has to give $\mathcal{O}(h^3)$ [Ric05], where h is the diameter of a circle with the same area as the element. The contribution from the N scheme gives $\mathcal{O}(h^2)$ and from the LDA scheme $\mathcal{O}(h^3)$ in 2D. Hence, the blending factor θ has to be of order $\mathcal{O}(h)$ in smooth parts. Multiplication of the shock sensor sc by h does not lead to sufficient damping in the shock regions. If multiplied by $\sqrt{h}$, then the amount of the numerical viscosity is correct, but the scheme is only $\mathcal{O}(h^{1.5})$ accurate. The solution is to take a blending factor as

$$\theta = \left(sc \cdot \sqrt{h}\right)^2 = sc^2 \cdot h, \tag{7}$$

which gives the right amount of artificial viscosity together with second order of accuracy in smooth regions. The nodal contribution of the Bx scheme is then given by (5) with the blending coefficient given by (6) and (7).

4 Numerical Results for Steady Problems

First, we present numerical results for transonic flow in the so-called Ni's channel [Ni81], with 10% circular bump. The flow is defined by the isentropic Mach number $M_{2\text{is}} = 0.675$. We use an unstructured grid consisting of 2,762 nodes and 5,281 elements, with 31 nodes along the bump, utilizing Weatherill triangulation. The Euler backward scheme is used for the time integration with numerical approximation of the Jacobian. For the N-modified and the B scheme we had to use the Jacobian of the N scheme. We start with the CFL number of 100 and every iteration we multiply it by the factor 1.2 until it reaches 10^6. In Fig. 1 isolines of the Mach number for the present Bx scheme are shown. One can observe a supersonic pocket on the bump. The isolines do not exhibit wiggles and they run straight into the shock, which corresponds to a not very dissipative scheme. The comparison of the Mach number distribution along the bottom wall is shown in Fig. 2 on a zoom to the beginning and the end of the shock. Before the shock, the Bx scheme follows the LDA scheme and creates a small overshoot on the Mach number distribution. After

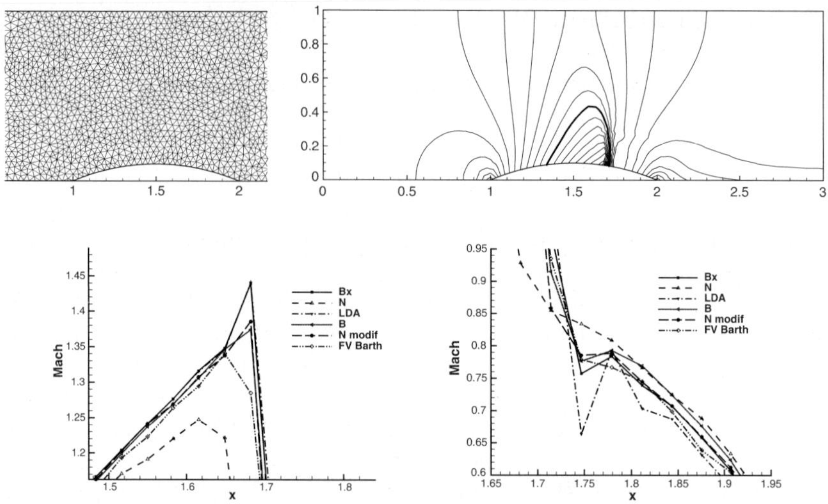

Fig. 1. Ni's channel. *Top*: Part of the Weatherill mesh and the Mach number isolines for the Bx scheme, $\Delta M = 0.05$, the bold line is $M = 1$. *Bottom*: distribution of the Mach number along the wall, zoom before and after the shock

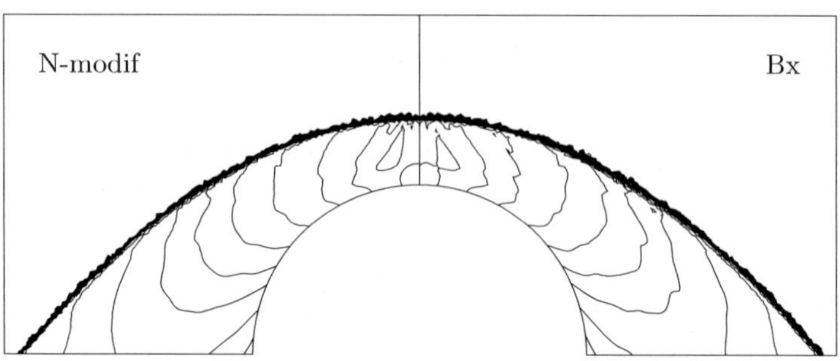

Fig. 2. Mach 20 bow shock. *Top*: isolines of the Mach number, the N-modified scheme is on the *left*, the Bx scheme on the *right*

the shock the Bx scheme behaves very similar as both the N-limited and the B scheme in terms of capturing after-shock singularity. For comparison, the solution for the cell-vertex Finite Volume method with Barth limiter [BJ89] and Roe's Riemann solver is shown. The FV scheme is clearly more dissipative than the nonlinear RD schemes, as can be observed on the last point before the shock and mainly in the more diffusive capture of the after-shock singularity. The new formulation gives a convergence rate very similar to the linear schemes, while all the other nonlinear schemes stall after a few orders of magnitude.

The next test examines robustness of the new scheme on a Mach 20 bow shock in front of the cylinder. The solution was computed using Bx, N, and

N-modified scheme on the mesh consisting of 10,531 nodes and 20,632 elements. In Fig. 2 the isolines of Mach number are shown, the left part of the figure is the N-modified scheme, while the right one presents the Bx scheme. One can notice, that the Bx scheme better captures irregularities of the solution, resulting from the interaction of the shock wave with the Weatherill type of mesh. No oscillation could be observed on the cut along the symmetry line (not shown). Convergence properties of the scheme are considerably worse than for the transonic flows, as is known also from other methods. On the other hand, one can employ a convergence fix, well known from the FV framework. After a certain number of iterations n_0, when the solution is fully developed, *we don't decrease anymore the blending factor* in the subsequent iterations, i.e.,

$$\theta^n = \max(\theta^n, \theta^{n-1}), \qquad \text{for all } n \geq n_0. \tag{8}$$

In this case we have chosen $n_0 = 4,000$. The scheme recovers convergence to machine accuracy with a possibly slight price of more dissipative solution.

As the last steady case, we present a subcritical flow around the cylinder [Pai95], with $M_\infty = 0.38$. This test examines behavior of the method in smooth flow regions. Our mesh consists of approximately equilateral triangles with 80 elements along the wall and 40 rows of triangles toward the free stream boundary. This gives a position of the farfield boundary approximately 10 diameters away from the cylinder. The exact solution is perfectly symmetric both with respect to x and y axis. The LDA and present Bx schemes give very similar almost symmetric results, see Fig. 3, with a small deviation behind the cylinder on the axis of symmetry. The N-modified scheme did not converge well and wake-like structure appear behind the cylinder. However, unlike the other results in [Pai95], no separation zone develops there. The results suggest that the Bx scheme contains very little unwanted artificial dissipation and it can capture smooth flow regions very well.

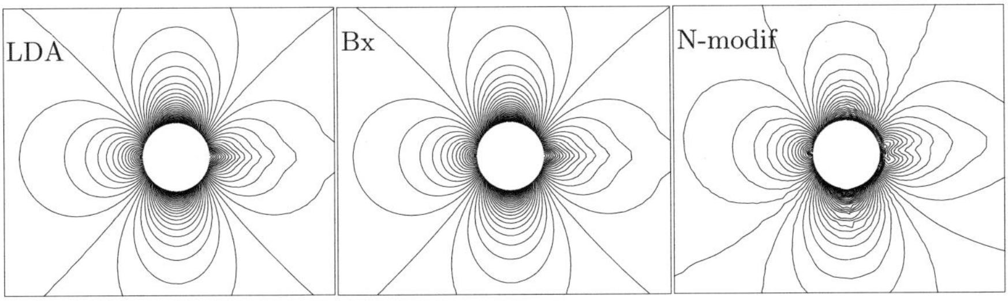

Fig. 3. Subcritical flow past the cylinder. Mach number isolines, $\Delta M = 0.02$

5 Blended (Bx) Scheme for Unsteady Problems

As a basis for the blended approach for unsteady problems we use the spatial LDA scheme with the consistent mass matrix given by

$$\phi_i^{\mathrm{STLDA_m}} = \Delta t \frac{|T|}{d+1}\left(-K_i^+ N + \frac{2}{d+2} - \frac{1}{d+1}\right)\frac{\partial U_i}{\partial t}$$

$$+ \Delta t \frac{|T|}{d+1} \sum_{\substack{j\in T\\ j\neq i}} \left[\left(-K_j^+ N + \frac{1}{d+2} - \frac{1}{d+1}\right)\frac{\partial U_j}{\partial t}\right] - \left[\Delta t K_i^+ N \sum_{j\in T} K_j \bar{U}_j\right]^{\alpha}. \tag{9}$$

In (2) index n has to be replaced by the pseudotime index τ and the steady solution in pseudotime is obtained for the solution in physical time at level $n+1$ by the scheme (2). For the detailed derivation see, e.g., [AM03]. We have chosen three points backward time integration for its superior stability. A space residual is then evaluated at time $\alpha = n+1$ and the time derivatives are discretized as $\partial U_i/\partial t = (3U_i^{n+1} - 4U_i^n + U_i^{n-1})/(2\Delta t)$. We choose the N scheme with the lumped mass matrix with the same time integration as the nonoscillatory scheme. It is given by

$$\phi_i^{\mathrm{STN_l}} = \frac{|T|}{d+1}\left(\frac{3}{2}U_i^{n+1} - 2U_i^n + \frac{1}{2}U_i^{n-1}\right) + \Delta t K_i^+(\bar{U}_i - U_{\mathrm{in}}). \tag{10}$$

The shock sensor is constructed using total derivative of the pressure

$$sc = \frac{1}{\delta_{pv}}\left(\frac{\partial p}{\partial t} + \mathbf{v}\cdot\boldsymbol{\nabla} p\right)^+. \tag{11}$$

6 Numerical Results for Unsteady Problems

We have chosen a smooth convection of a vortex as the first unsteady test case. The problem is solved on the square domain $[-0.5, 0.5]\times[-0.5, 0.5]$ filled with Weatherill triangulation with 41 points along each side. The flow velocity is given by the mean stream velocity $\mathbf{v}_m = (6, 0)$ and the circumferential perturbation

$$(v_x, v_y)_\theta = (-y, x)\cdot\omega, \quad \omega = 15\cdot(\cos 4\pi r + 1), \quad r = \sqrt{x^2+y^2} \tag{12}$$

for $r < 0.25$, $(v_x, v_y)_\theta = \mathbf{0}$ elsewhere. Density is chosen constant $\rho = 1.4$ and the pressure from the balance in the radial direction $p = p_m + \Delta p$, where

$$\Delta p = \frac{15^2\rho}{(4\pi)^2}\left(2\cos(4\pi r) + 8\pi r\sin(4\pi r) + \frac{\cos(8\pi r)}{8} + \frac{4\pi r\sin(8\pi r)}{4} + 12\pi^2 r^2\right) + C. \tag{13}$$

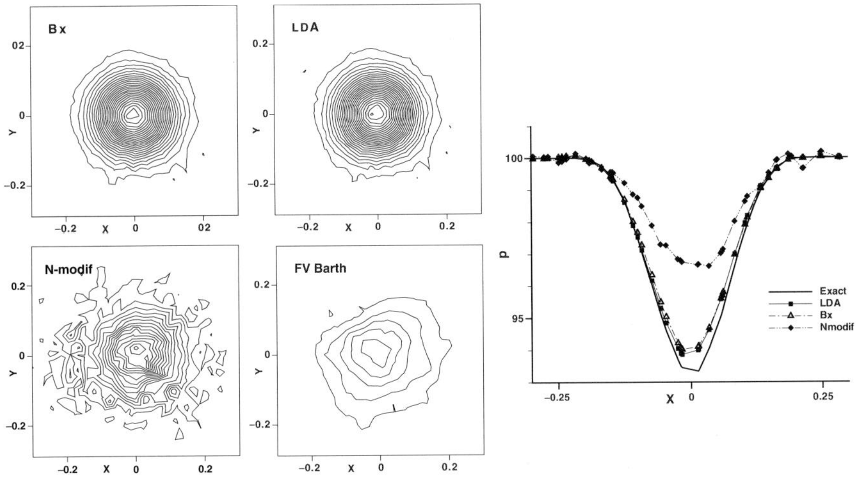

Scheme:	Bx	LDA	N-limited	FV nolim	FV Barth	Exact
$p_{\max}$	100.11	100.12	100.42	100.04	100.11	100
$p_{\min}$	94.00	93.84	96.27	94.35	98.76	93.213

Fig. 4. Advection of the vortex, $t = 1/6$. Isolines of the pressure for different schemes. Distribution of the pressure in the core of the vortex for different schemes. The maximal and the minimal pressure in the vortex core.

The constant C is such that $p|_{r=0.25} = p_m = 100$. The free stream values are prescribed on the boundaries $y = \pm 0.5$ and periodic boundaries are used for $x = \pm 0.5$. The simulation stops after one period, i.e., $t_{\max} = 1/6$. The computation was performed with $CFL = 1$. In Fig. 4 the isolines of the pressure for the different schemes are shown. Distribution of the pressure in the core of the vortex shows, that the Bx scheme performs essentially as the LDA scheme for the smooth regions. The performance of the N-limited scheme is noticeably worse. We have performed the same test with several other schemes, see Fig. 4. The finite volume method is cell centered with the three points backward time integration scheme formulated in dual time and with, or without Barth's limiter.

As a test to examine the scheme on a complex flow features, we present results for a 2D Riemann problem [AM03, Ric05]. In Fig. 5 comparison of the Bx scheme with the N-modified is presented on meshes with spatial resolution 1/200 and 1/400. The N-modified scheme captures the instabilities well, however it clearly gives lower resolution. Both the computations with the Bx scheme show much higher resolution. It can be observed on pronounced Kelvin–Helmholtz instabilities along the slip lines, where the Bx scheme gives a richer structure, which is seen from the growth of the rollers along the instability.

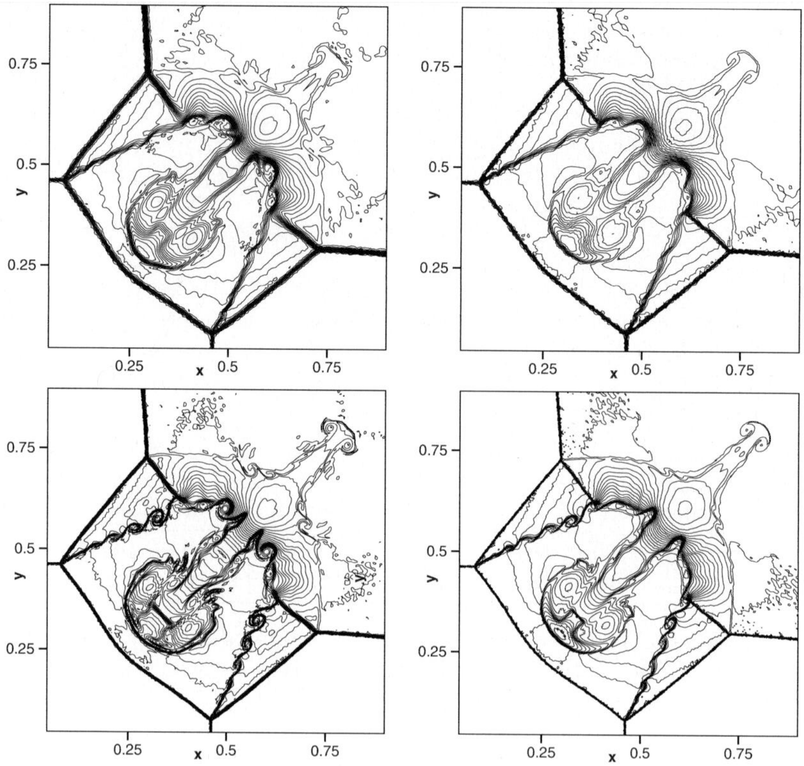

Fig. 5. 2D Riemann problem. Density contours $\Delta\rho = 0.05$. *Top*: $h = 1/200$, bottom $h = 1/400$. *Left*: Bx scheme. *Right*: N-modified scheme

Acknowledgments

This research was supported by Research plan of MSMT no. 6840770003. Dr. Mario Ricchiuto is acknowledged for a number important suggestions on the topic.

References

[AM03] Abgrall, R., and Mezine, M.: Construction of second order accurate monotone and stable residual distribution schemes for unsteady flow problems. *JCP*, **188**, 16–55 (2003).

[AM04] Abgrall, R., and Mezine, M.: Construction of second-order accurate monotone and stable residual distribution schemes for steady problems. *JCP*, **195**(2), 474–507 (2004)

[BJ89] Barth, T.J., Jesperson, D.C.: The design and application of upwind schemes on unstructured meshes. AIAA Paper 89–0366 (1989)

[DRS93] Deconinck, H., Roe, P., and Struijs, R.: A multidimensional generalization of Roe's flux difference splitter for the Euler equations. *C&F*, **22**, 215–222 (1993)

[DSA00] Deconinck, H., Sermeus, K., Abgrall, R.: Status of multidimensional upwind residual distribution schemes and applications in aeronautics. AAIA Paper 2000–2328 (2000)

[DD05] Dobeš, J., Deconinck, H.: Second order blended multidimensional residual distribution scheme for steady and unsteady computations. *Journal of Computational and Applied Mathematics (JCAM).* Special issue ACOMEN 2005. Accepted. (2005)

[Ni81] Ni, R.H.: A multiple grid scheme for solving Euler equations. *AIAA J.*, **20**(1) (1981)

[Pai95] Paillère, H.: *Multidimensional Upwind Residual Distribution Schemes for the Euler and Navier-Stokes Equations on Unstructured Grids.* PhD thesis, ULB/VKI (1995)

[Ric05] Ricchiuto, M.: *Construction and Analysis of Compact Residual Discretizations for Conservation Laws on Unstructured Meshes.* PhD thesis, ULB/VKI (2005)

Artificial Compressibility Approximation for the Incompressible Navier–Stokes Equations on Unbounded Domain

D. Donatelli and P. Marcati

1 Introduction

Let us consider the incompressible Navier–Stokes equation

$$\begin{cases} \partial_t u + \operatorname{div}(u \otimes u) - \Delta u = \nabla p \\ \operatorname{div} u = 0 \\ u(x,0) = u_0, \end{cases} \tag{1}$$

where $(x,t) \in \mathbb{R}^3 \times [0,T]$, $u \in \mathbb{R}^3$ denotes the velocity vector field , $p \in \mathbb{R}$ the pressure of the fluid. This chapter deals with the approximation of the Leray weak solutions in $3-D$ of the system (1). For this purpose we introduce the so-called "artificial compressibility approximation," namely

$$\begin{cases} \partial_t u^\varepsilon + \nabla p^\varepsilon = \Delta u^\varepsilon - (u^\varepsilon \cdot \nabla)\, u^\varepsilon - \dfrac{1}{2}(\operatorname{div} u^\varepsilon) u^\varepsilon \\ \varepsilon \partial_t p^\varepsilon + \operatorname{div} u^\varepsilon = 0, \end{cases} \tag{2}$$

where $(x,t) \in \mathbb{R}^3 \times [0,T]$, $u^\varepsilon = u^\varepsilon(x,t) \in \mathbb{R}^3$ and $p^\varepsilon = p^\varepsilon(x,t) \in \mathbb{R}$. In the system (2), the first equation resembles the momentum equation with the difference that the term $\frac{1}{2}(\operatorname{div} u^\varepsilon)u^\varepsilon$ is added in order to avoid the paradox of increasing energy. The second equation of (2) is a "linearized" compressibility constraint. This approximation was introduced by Chorin [1, 2], Temam [21, 22] and Oskolkov [15], in the case of bounded domains, in order to deal with the difficulty induced by the incompressibility constraints in the numerical approximations to the Navier–Stokes equation. In particular they obtain the convergence of these approximations by using the classical Sobolev compactness embedding. The interest into the artificial compressibility methods started with the previously mentioned results of Chorin and Temam and was later on investigated by Ghidaglia and Temam [6]. Later developments of numerical investigations in the directions of projections methods have been carried out by [9, 10, 14, 16, 17, 19, 24, 25]. In this chapter we will take a different point of view. First of all, we analyze the convergence problem in the case of the whole space and then, we wish to exploit the dispersive estimate given by the hyperbolic structure of the system (1). In fact the system (1) will be discussed as a semilinear wave type equation for the pressure function and the dispersive estimates will be carried out by using the L^p-type estimates due to Strichartz [7, 11]. Our analysis can also be related to the convergence of the incompressible limit problem via a formal expansion (see for instance Temam [23], Chap. 3). In particular a similar wave equation structure has been exploited in various way by the paper of Desjardins and Grenier [3] and references therein. In order to perform our analysis we need to assign initial data for the velocity and for the pressure in the system (1). If we observe that the Navier–Stokes equations require only one initial condition on the velocity u, we conclude that our approximation will be consistent if the initial datum on the pressure will be eliminated by an "initial layer" phenomenon. So, taking into account that in the limit we have to deal with Leray solutions, we can deduce a natural behavior to be imposed on the initial data $(u_0^\varepsilon, p_0^\varepsilon)$, namely

$$\begin{aligned} u_0^\varepsilon = u^\varepsilon(\cdot,0) &\longrightarrow u_0 = u(\cdot,0) \text{ strongly in } L^2(\mathbb{R}^3) \\ \sqrt{\varepsilon}p_0^\varepsilon = \sqrt{\varepsilon}p^\varepsilon(\cdot,0) &\longrightarrow 0 \text{ strongly in } L^2(\mathbb{R}^3). \end{aligned} \tag{ID}$$

Let us now state our main result. The convergence of $\{u^\varepsilon\}$ will be described by analyzing the convergence of the associated Hodge decomposition.

Theorem 1. *Let $(u^\varepsilon, p^\varepsilon)$ be a sequence of weak solution in $\mathbb{R}^3$ of the system (2), with the initial data satisfying (ID). Then there exists $u \in L_t^\infty L_x^2 \cap L_t^2 \dot{H}_x^1$ such that $u^\varepsilon \rightharpoonup u$ weakly in $L_t^2 \dot{H}_x^1$. The gradient component Qu^ε of the vector field u^ε satisfies $Qu^\varepsilon \longrightarrow 0$ strongly in $L_t^2 L_x^p$, for any $p \in [4,6)$, while the divergence free component Pu^ε of the vector field u^ε satisfies $Pu^\varepsilon \longrightarrow Pu = u$ strongly in $L_t^2 L_{x,loc}^2$. Moreover $u = Pu$ is a Leray weak solution to the incompressible Navier–Stokes equation.*

This chapter is organized as follows. In Sect. 2 we recall the mathematical tools needed in the chapter and recall same basic definitions. In Sects. 3 and 4 we sketch how to recover the a priori estimates needed to get the strong convergence of the approximating sequences. For the details of the proofs we refer to [4]. Finally, in Sect. 5 we give some concluding remarks.

2 Preliminaries

For convenience of the reader we establish some notations and recall some basic facts that will be useful in the sequel.

We will denote by $\mathcal{D}(\mathbb{R}^3 \times \mathbb{R}_+)$ the space of test function $C_0^\infty(\mathbb{R}^3 \times \mathbb{R}_+)$, by $\mathcal{D}'(\mathbb{R}^3 \times \mathbb{R}_+)$ the space of Schwartz distributions and $\langle \cdot, \cdot \rangle$ the duality bracket between $\mathcal{D}'$ and $\mathcal{D}$ and by $\mathcal{M}_t X'$ the space $C_c^0([0,T]; X)'$. Moreover

$W^{k,p}(\mathbb{R}^3) = (I-\Delta)^{-\frac{k}{2}}L^p(\mathbb{R}^3)$ and $H^k(\mathbb{R}^3) = W^{k,2}(\mathbb{R}^3)$ denote the nonhomogeneous Sobolev spaces for any $1 \le p \le \infty$ and $k \in \mathbb{R}$. $\dot{W}^{k,p}(\mathbb{R}^3) = (I-\Delta)^{-\frac{k}{2}}L^p(\mathbb{R}^3)$ and $\dot{H}^k(\mathbb{R}^3) = W^{k,2}(\mathbb{R}^3)$ denote the homogeneous Sobolev spaces. The notations $L^p_tL^q_x$ and $L^p_tW^{k,q}_x$ will abbreviate, respectively, the spaces $L^p([0,T];L^q(\mathbb{R}^3))$, and $L^p([0,T];W^{k,q}(\mathbb{R}^3))$.

We shall denote by Q and P, respectively, the Leray's projectors Q on the space of gradients vector fields and P on the space of divergence – free vector fields. Namely

$$Q = \nabla\Delta^{-1}\,\mathrm{div} \qquad P = I - Q. \tag{3}$$

Let us recall that if w is a (weak) solution of the following wave equation in the space $[0,T]\times\mathbb{R}^3$

$$(\partial_{tt}+\Delta)w(t,x) = F(t,x) \quad w(0,\cdot) = f, \quad \partial_t w(0,\cdot) = g,$$

for some data f,g,F, and time $0 < T < \infty$, then w satisfies the following Strichartz estimates, (see [7, 11])

$$\|w\|_{L^4_{t,x}} + \|\partial_t w\|_{L^4_tW^{-1,4}_x} \lesssim \|f\|_{\dot{H}^{1/2}_x} + \|g\|_{\dot{H}^{-1/2}_x} + \|F\|_{L^1_tL^{3/2}_x}. \tag{4}$$

Beside the Strichartz estimate (4) in the case of $d = 3$ (see [20]), a more refined estimate can also be deduced by the bilinear estimates of Klainerman and Machedon [12], namely

$$\|w\|_{L^4_{t,x}} + \|\partial_t w\|_{L^4_tW^{-1,4}_x} \lesssim \|f\|_{\dot{H}^{1/2}_x} + \|g\|_{\dot{H}^{1/2}_x} + \|F\|_{L^1_tL^2_x}. \tag{5}$$

3 A Priori Estimates

In this section we wish to establish the priori estimates, independent on ε, for the solutions of the system (2) which are necessary to prove the Theorem 1. First of all we will recover the a priori estimates that come from the classical energy estimates related to the system (2), then we get stronger estimates by exploiting the structure of the system.

3.1 Energy Estimates

The next results concerns the energy type estimate for the system (2).

Theorem 2. *Let us consider the solution $(u^\varepsilon, p^\varepsilon)$ of the Cauchy problem for the system (2). Assume that the hypotheses (ID) hold, then one has*

$$\frac{1}{2}\|u^\varepsilon(\cdot,t)\|^2_{L^2_x} + \frac{\varepsilon}{2}\|p^\varepsilon(\cdot,t)\|^2_{L^2_x} + \int_0^t \|\nabla u^\varepsilon(\cdot,s)\|^2_{L^2_x}ds = \frac{1}{2}\|u^\varepsilon_0\|^2_{L^2_x} + \frac{\varepsilon}{2}\|p^\varepsilon_0\|^2_{L^2_x} \tag{6}$$

in particular we have

$$\begin{aligned}
&\sqrt{\varepsilon}p^\varepsilon && \text{is bounded in } L_t^\infty L_x^2, && (7)\\
&\varepsilon p_t^\varepsilon && \text{is relatively compact in } H_{t,x}^{-1}, && (8)\\
&\nabla u^\varepsilon && \text{is bounded in } L_{t,x}^2, && (9)\\
&u^\varepsilon && \text{is bounded in } L_t^\infty L_x^2 \cap L_t^2 L_x^6, && (10)\\
&(u^\varepsilon \cdot \nabla)u^\varepsilon,\ (\operatorname{div} u^\varepsilon)u^\varepsilon && \text{are bounded in } L_t^2 L_x^1 \cap L_t^1 L_x^{3/2}. && (11)
\end{aligned}$$

3.2 Pressure Wave Equation

In this section by using the Strichartz estimates (4), (5) we get a priori estimates on p^ε. As we will see p^ε satisfies a wave type equation. We differentiate with respect to time $(2)_2$, by using $(2)_1$ and by rescaling the time variable, the velocity and the pressure in the following way

$$\tau = t/\sqrt{\varepsilon}, \quad \tilde{u}(x,\tau) = u^\varepsilon(x, \sqrt{\varepsilon}\tau), \quad \tilde{p}(x,\tau) = p^\varepsilon(x, \sqrt{\varepsilon}\tau), \tag{12}$$

we get that $\tilde{p}$ satisfies the following wave equation

$$\partial_{\tau\tau}\tilde{p} - \Delta\tilde{p} + \Delta \operatorname{div}\tilde{u} - \operatorname{div}\left((\tilde{u}\cdot\nabla)\,\tilde{u} + \frac{1}{2}(\operatorname{div}\tilde{u})\tilde{u} \right) = 0. \tag{13}$$

Now we consider $\tilde{p} = \tilde{p}_1 + \tilde{p}_2$, where $\tilde{p}_1$ and $\tilde{p}_2$ solve the following wave equations:

$$\begin{cases} \partial_{\tau\tau}\tilde{p}_1 - \Delta\tilde{p}_1 = -\Delta \operatorname{div}\tilde{u} \\ \tilde{p}_1(x,0) = \partial_\tau\tilde{p}_1(x,0) = 0, \end{cases} \qquad \begin{cases} \partial_{\tau\tau}\tilde{p}_2 - \Delta\tilde{p}_2 = \operatorname{div}\left((\tilde{u}\cdot\nabla)\,\tilde{u} + \frac{1}{2}(\operatorname{div}\tilde{u})\tilde{u}\right) \\ \tilde{p}_2(x,0) = \tilde{p}(x,0) \quad \partial_\tau\tilde{p}_2(x,0) = \partial_\tau\tilde{p}(x,0). \end{cases} \tag{14}$$

Therefore we are able to prove Theorem 3.

Theorem 3. *Let us consider the solution $(u^\varepsilon, p^\varepsilon)$ of the Cauchy problem for the system (2). Assume that the hypotheses (ID) hold. Then we set the following estimate*

$$\begin{aligned}
\varepsilon^{3/8}\|p^\varepsilon\|_{L_t^4 W_t^{-2,4}} + \varepsilon^{7/8}\|\partial_t p^\varepsilon\|_{L_t^4 W_t^{-3,4}} &\lesssim \sqrt{\varepsilon}\|p_0^\varepsilon\|_{L_x^2} + \|\operatorname{div} u_0^\varepsilon\|_{H_x^{-1}} \\
&+ \|\,(u^\varepsilon\cdot\nabla)\,u^\varepsilon + \frac{1}{2}(\operatorname{div} u^\varepsilon)u^\varepsilon\|_{L_t^1 L_x^{3/2}} + \sqrt{T}\|\operatorname{div} u^\varepsilon\|_{L_t^2 L_x^2}.
\end{aligned} \tag{15}$$

Proof. Since $\tilde{p}_1$ and $\tilde{p}_2$ are solutions of the wave equations (14), we can apply the Strichartz estimates (4) and (5), with $(x,\tau) \in \mathbb{R}^3 \times (0, T/\sqrt{\varepsilon})$. By using the estimate (5) on $\Delta^{-1}\tilde{p}_1$ we get

$$\|\tilde{p}_1\|_{L_\tau^4 W_x^{-2,4}} + \|\partial_\tau\tilde{p}_1\|_{L_\tau^4 W_x^{-3,4}} \lesssim \frac{\sqrt{T}}{\varepsilon^{1/4}}\|\operatorname{div}\tilde{u}\|_{L_\tau^2 L_x^2}. \tag{16}$$

In the same way, by applying the estimate (4) to $\Delta^{-1/2}\tilde{p}_2$, we have that $\tilde{p}_2$ satisfies

$$\|\tilde{p}_2\|_{L^4_\tau W^{-1,4}_x} + \|\partial_\tau \tilde{p}_2\|_{L^4_\tau W^{-2,4}_x} \lesssim \|\tilde{p}(x,0)\|_{H^{-1/2}_x} + \|\partial_\tau \tilde{p}(x,0)\|_{H^{-3/2}_x} + \|\,(\tilde{u}\cdot\nabla)\,\tilde{u} + \frac{1}{2}(\operatorname{div}\tilde{u})\tilde{u}\|_{L^1_\tau L^{3/2}_x}. \tag{17}$$

Now by using (16), (17) and taking into account the scaling (12) we end up with (15). □

4 Strong Convergence

In this section we conclude the proof of the Theorem 1. In particular we will show that the gradient part of the velocity Qu^ε converges strongly to 0, while the incompressible component of the velocity field Pu^ε converges strongly to $Pu = u$, where u is the limit profile as $\varepsilon \downarrow 0$ of u^ε. For further details we refer to [4]. Now, we wish to show that the gradient part of the velocity field Qu^ε goes strongly to 0 as $\varepsilon \downarrow 0$. As we will see in Proposition 1, this will be a consequence of the estimate (15) and of the following auxiliary result.

Lemma 1. *Let us consider a smoothing kernel $\psi \in C_0^\infty(\mathbb{R}^d)$, such that $\psi \geq 0$, $\int_{\mathbb{R}^d} \psi dx = 1$, and define $\psi_\alpha(x) = \alpha^{-d}\psi(x/\alpha)$. Then for any $f \in \dot{H}^1(\mathbb{R}^d)$, one has*

$$\|f - f * \psi_\alpha\|_{L^p(\mathbb{R}^d)} \leq C_p \alpha^{1-\sigma}\|\nabla f\|_{L^2(\mathbb{R}^d)}, \tag{18}$$

where $p \in [2,\infty)$ if $d = 2$, $p \in [2,6]$, if $d = 3$ and $\sigma = d\left(\frac{1}{2} - \frac{1}{p}\right)$. Moreover the following Young type inequality hold

$$\|f * \psi_\alpha\|_{L^p(\mathbb{R}^d)} \leq C\alpha^{s-d\left(\frac{1}{q}-\frac{1}{p}\right)}\|f\|_{W^{-s,q}(\mathbb{R}^d)}, \tag{19}$$

for any $p, q \in [1,\infty]$, $q \leq p$, $s \geq 0$, $\alpha \in (0,1)$.

Proposition 1. *Let us consider the solution $(u^\varepsilon, p^\varepsilon)$ of the Cauchy problem for the system (2). Assume that the hypotheses (ID) hold. Then as $\varepsilon \downarrow 0$,*

$$Qu^\varepsilon \longrightarrow 0 \quad \textit{strongly in } L^2_t L^p_x \textit{ for any } p \in [4,6). \tag{20}$$

Proof. In order to prove the Proposition 1 we split Qu^ε as follows

$$\|Qu^\varepsilon\|_{L^2_t L^p_x} \leq \|Qu^\varepsilon - Qu^\varepsilon * \psi_\alpha\|_{L^2_t L^p_x} + \|Qu^\varepsilon * \psi_\alpha\|_{L^2_t L^p_x} = J_1 + J_2,$$

where ψ_α is the smoothing kernel defined in Lemma 1. Now we estimate separately J_1 and J_2. By applying the inequality (18) we get

$$J_1 \le \alpha^{1-3(\frac{1}{2}-\frac{1}{p})} \left(\int_0^T \|\nabla Q u^\varepsilon(t)\|_{L_x^2}^2 dt \right) \le \alpha^{1-3(\frac{1}{2}-\frac{1}{p})} \|\nabla u^\varepsilon\|_{L_t^2 L_x^2}. \tag{21}$$

Hence from the identity $Qu^\varepsilon = -\varepsilon^{1/8}\nabla\Delta^{-1}\varepsilon^{7/8}\partial_t p$ and by the inequality (19), we get J_2 satisfies the following estimate

$$J_2 \le \varepsilon^{1/8}\alpha^{-2-3(\frac{1}{4}-\frac{1}{p})} T^{1/4} \|\varepsilon^{7/8}\partial_t p\|_{L_t^4 W_x^{-3,4}}. \tag{22}$$

Therefore, summing up (21) and (22) and by using (9) and (15), we conclude for any $p \in [4,6)$ that

$$\|Qu^\varepsilon\|_{L_t^2 L_x^p} \le C\alpha^{1-3(\frac{1}{2}-\frac{1}{p})} + C_T \varepsilon^{1/8}\alpha^{-2-3(\frac{1}{4}-\frac{1}{p})}. \tag{23}$$

Finally we choose α in terms of ε in order that the two terms in the right-hand side of the previous inequality have the same order, namely $\alpha = \varepsilon^{1/18}$. Therefore we obtain

$$\|Qu^\varepsilon\|_{L_t^2 L_x^p} \le C_T \varepsilon^{\frac{6-p}{36p}}, \quad \text{for any } p \in [4,6). \quad \square$$

It remains to prove the strong compactness of the incompressible component of the velocity field. This property can be obtained by using the time regularity of Pu^ε proved in the following lemma and theorems in the spirit of [18].

Lemma 2. *Let us consider the solution $(u^\varepsilon, p^\varepsilon)$ of the Cauchy problem for the system* (2)*. Assume that the hypotheses (ID) hold. Then for all $h \in (0,1)$, we have*

$$\|Pu^\varepsilon(t+h) - Pu^\varepsilon(t)\|_{L^2([0,T]\times\mathbb{R}^3)} \le C_T h^{1/5}. \tag{24}$$

Proof. Let us set $z^\varepsilon = u^\varepsilon(t+h) - u^\varepsilon(t)$, we have

$$\begin{aligned} &\|Pu^\varepsilon(t+h) - Pu^\varepsilon(t)\|_{L^2([0,T]\times\mathbb{R}^3)}^2 \\ &= \int_0^T \int_{\mathbb{R}^3} dtdx (Pz^\varepsilon)\cdot(Pz^\varepsilon - Pz^\varepsilon * \psi_\alpha) \\ &\quad + \int_0^T \int_{\mathbb{R}^3} dtdx (Pz^\varepsilon)\cdot(Pz^\varepsilon * \psi_\alpha) = I_1 + I_2. \end{aligned} \tag{25}$$

By using (18) we can estimate I_1 in the following way

$$I_1 \lesssim \alpha T^{1/2} \|u^\varepsilon\|_{L_t^\infty L_x^2} \|\nabla u^\varepsilon\|_{L_{t,x}^2}. \tag{26}$$

Now we reformulate Pz^ε in integral form by using the equation $(2)_1$, then integrating by parts and by applying (19), with $p = \infty$ and $q = 2$, we deduce

$$I_2 \leq h\|\nabla u^\varepsilon\|^2_{L^2_{t,x}} + C\alpha^{-3/2}T^{1/2}h\|u^\varepsilon\|_{L^\infty_t L^2_x}\| \left(u^\varepsilon \cdot \nabla\right) u^\varepsilon - \frac{1}{2}(\operatorname{div} u^\varepsilon)u^\varepsilon\|_{L^2_t L^1_x}. \tag{27}$$

Summing up I_1, I_2, by taking into account (9)–(11), and by choosing $\alpha = h^{2/5}$, we end up with (24). □

5 Some Remarks

In this chapter and in [4] we have analyzed the artificial compressibility method in the case of the whole space, but our method can be extended to exterior domains, which will be done in a forthcoming chapter.

Moreover we can apply the same method to the case of the so-called Navier–Stokes Fourier system which governs the velocity field u and the temperature field $\theta = \theta(x,t) \in \mathbb{R}$ in an incompressible fluid, namely

$$\begin{cases} \partial_t u + \operatorname{div}(u \otimes u) + \nabla p = \mu \Delta u \\ \partial_t \theta + u \cdot \nabla\theta = \kappa\Delta\theta \\ \operatorname{div} u = 0 \\ u(x,0) = u_0, \;\; \theta(x,0) = \theta_0. \end{cases} \tag{28}$$

For the definition and existence of weak solution of (28) we refer to [13]. Approximation of the system (28) have been of by Golse and Saint-Raymond in [8] in the contest of hydrodynamical limit, namely they recover the system (28) from the Boltzmann equation. Here we introduce a system similar to (2), which is in this case

$$\begin{cases} \partial_t u^\varepsilon + \nabla p^\varepsilon = \mu\Delta u^\varepsilon - \left(u^\varepsilon \cdot \nabla\right) u^\varepsilon - \dfrac{1}{2}(\operatorname{div} u^\varepsilon)u^\varepsilon \\ \partial_t \theta^\varepsilon + u \cdot \nabla\theta^\varepsilon = \kappa\Delta\theta^\varepsilon - \dfrac{1}{2}(\operatorname{div} u^\varepsilon)\theta^\varepsilon \\ \varepsilon\partial_t p^\varepsilon + \operatorname{div} u^\varepsilon = 0. \end{cases} \tag{29}$$

The details on the analysis of the convergence of this system can be found in [5].

References

1. A.J. Chorin, Numerical solution of the Navier-Stokes equations, Math. Comp. **22**, 745–762, (1968).
2. A.J. Chorin, On the convergence of discrete approximations to the Navier-Stokes equations, Math. Comp. **23**, 341–353, (1969).

3. B. Desjardins and E. Grenier, Low Mach number limit of viscous compressible flows in the whole space, R. Soc. Lond. Proc. Ser. A Math. Phys. Eng. Sci. **455**, no. 1986, 2271–2279, (1999).
4. D. Donatelli and P. Marcati, A dispersive approach to the artificial compressibility approximations of the Navier-Stokes equations in 3D, J. Hyperbolic Differ. Equ. **3**, no. 3, 575–588, (2006).
5. D. Donatelli and P. Marcati, Work in progress.
6. J.-M. Ghidaglia and R. Temam, Long time behavior for partly dissipative equations: the slightly compressible 2D-Navier-Stokes equations, Asymptotic Anal. **1**, no. 1, 23–49, (1988).
7. J. Ginibre and G. Velo, Generalized Strichartz inequalities for the wave equation, J. Funct. Anal. **133**, no. 1, 50–68, (1995).
8. F. Golse and L. Saint-Raymond, Hydrodynamic limits for the Boltzmann equation. Riv. Mat. Univ. Parma **7**, 4**, 1–144, (2005).
9. J.L. Guermond, P. Minev, and J. Shen, An overview of projection methods for incompressible flows, Comp. Meth. Appl. Mech. and Eng., **195**, 6011–6045, (2006).
10. B.G. Kuznecov and Š. Smagulov, Approximation of the Navier-Stokes equations, Čisl. Metody Meh. Splošn. Sredy **6**, no. 2, 70–79, (1975).
11. M. Keel and T. Tao, Endpoint Strichartz estimates, Amer. J. Math. **120**, no. 5, 955–980, (1998).
12. S. Klainerman and M. Machedon, Space-time estimates for null forms and the local existence theorem, Comm. Pure Appl. Math. **46**, no. 9, 1221–1268, (1993).
13. P.-L. Lions, Mathematical topics in fluid dynamics, incompressible models, Claredon Press, Oxford Science Publications, (1996).
14. R.H. Nochetto and J.-H. Pyo, Error estimates for semi-discrete gauge methods for the Navier-Stokes equations, Math. Comp. **74**, no. 250, 521–542 (electronic), (2005).
15. A.P. Oskolkov, A certain quasilinear parabolic system with small parameter that approximates a system of Navier-Stokes equations, Zap. Naučn. Sem. Leningrad. Otdel. Mat. Inst. Steklov. (LOMI), **21**, 79–103, (1971).
16. A. Prohl, *Projection and quasi-compressibility methods for solving the incompressible Navier-Stokes equations*, Advances in Numerical Mathematics, B. G. Teubner, Stuttgart, 1997.
17. R. Rannacher, On Chorin's projection method for the incompressible Navier-Stokes equations, The Navier-Stokes equations II—theory and numerical methods (Oberwolfach, 1991), Lecture Notes in Math., vol. 1530, Springer, Berlin, pp. 167–183, 1992.
18. J. Simon, Compact sets in the space $L^p(0,T;B)$, Ann. Mat. Pura Appl. (4) **146**, 65–96, (1987).
19. Š. Smagulov, Parabolic approximation of Navier-Stokes equations,Chisl. Metody Mekh. Sploshn. Sredy, **10**, no. 1 Gaz. Dinamika, 137–149, (1979).
20. C.D. Sogge, Lectures on nonlinear wave equations, Monographs in Analysis, II, International Press, Boston, MA, 1995.
21. R. Temam, Sur l'approximation de la solution des équations de Navier-Stokes par la méthode des pas fractionnaires. I, Arch. Rational Mech. Anal. **32**, 135–153, (1969).
22. R. Temam, Sur l'approximation de la solution des équations de Navier-Stokes par la méthode des pas fractionnaires. II, Arch. Rational Mech. Anal. **33**, 377–385, (1969).

23. R. Temam, *Navier-Stokes equations. Theory and numerical analysis*, AMS Chelsea Publishing, Providence, RI, 2001, Reprint of the (1984) edition.
24. W.E. and J.-G. Liu, Gauge method for viscous incompressible flows, Commun. Math. Sci. **1**, no. 2, 317–332, (2003).
25. N.N. Yanenko, B.G. Kuzne tsov, and Sh. Smagulov, On the approximation of the Navier-Stokes equations for an incompressible fluid by evolutionary-type equations, Numerical methods in fluid dynamics, "Mir", Moscow, pp. 290–314, (1984).

Traveling-Wave Solutions for Hyperbolic Systems of Balance Laws

A. Dressel and W.-A. Yong

Summary. This report is concerned with the existence of traveling-wave solutions for hyperbolic systems of balance laws satisfying a stability condition and a Kawashima-like condition. We focus on the case where the traveling-wave equations have a singularity. The basic idea is to understand the singular equations as a three-scale multidimensional connection problem. Based on this understanding, we make two center manifold reductions to convert the problem to a one-dimensional problem, for which there is a well-known criterion for existence. The main technical issue is to show the effectiveness of the reductions under the aforesaid structural conditions. We also show how to generalize the results in [2] to more general singularities.

1 Introduction

The goal of this report is to show the existence of traveling-wave solutions for hyperbolic systems of balance laws:

$$U_t + \sum_{j=1}^{d} F_j(U)_{x_j} = Q(U). \tag{1}$$

Here U is the unknown n-vector valued function of $(x,t) \equiv (x_1, x_2, \cdots, x_d, t) \in \mathbf{R}^d \times [0, +\infty)$, taking values in an open subset $\mathcal{U}$ of $\mathbf{R}^n$ (called *state space*); $Q(U)$ and $F_j(U) (j = 1, 2, \cdots, d)$ are given n-vector valued smooth functions of $U \in \mathcal{U}$.

Important examples of such balance laws occur in inviscid gas dynamics with relaxation, kinetic theories, traffic flows, non-linear optics, the numerical solution of conservation laws with relaxation schemes, and so on (cf. [13]).

In the aforementioned applications, we can assume that, after eventual linear transformation, the first $n - r$ entries of the source term $Q(U)$ vanish and its last r functions of $U \in \mathcal{U}$ are linearly independent. Accordingly, we often write

$$U = \begin{pmatrix} u \\ v \end{pmatrix}, \quad F_j(U) = \begin{pmatrix} f_j(u,v) \\ g_j(u,v) \end{pmatrix}, \quad Q(U) = \begin{pmatrix} 0 \\ q(u,v) \end{pmatrix}. \tag{2}$$

Throughout this report, we assume that $q(u,v)=0$ uniquely determines v in terms of u, say $v=h(u)$.

We are interested in traveling-wave solutions to system (1) of the form

$$U(x,t)=\phi(\omega\cdot x-s_*t)$$

with $\omega=(\omega_1,\cdots,\omega_d)\in\mathbf{R}^d\backslash\{0\}, \omega\cdot x=\sum_{j=1}^d\omega_jx_j$, $s_*\in\mathbf{R}$ and $\phi(\pm\infty)=U_\pm$ appropriately given. As a solution to (1), $\phi=\phi(\xi)$ solves

$$-s_*\phi_\xi+\sum_{j=1}^d\omega_jF_j(\phi)_\xi=Q(\phi). \tag{3}$$

Our task is to show the existence of a trajectory $\phi=\phi(\xi)$ of (3) connecting U_- on the left to U_+ on the right.

Note that system (3) together with $\phi(\pm\infty)=U_\pm$ suggests $Q(U_\pm)=0$. According to our assumption following (2), such $U_\pm$ can be written as

$$U_\pm=\begin{pmatrix}u_\pm\\ h(u_\pm)\end{pmatrix}$$

with $u_\pm\in\mathbf{R}^{n-r}$. Since the first $n-r$ components of $Q(U)$ vanish identically, it is easy to see from the first $n-r$ equations in (3) that the quantities $u_\pm, s_*$ and ω satisfy the following relation

$$s_*(u_+-u_-)=\sum_{j=1}^d\omega_j\left[f_j(u_+,h(u_+))-f_j(u_-,h(u_-))\right]. \tag{4}$$

This relation particularly determines s_* as a function of (u_-,u_+,ω). Note that as u_+ tends to u_-, $s_*=s_*(u_-,u_+,\omega)$ converges to an eigenvalue, say $\lambda_p(u_-,\omega)$, of $(n-r)\times(n-r)$ matrix $\sum_j\omega_j\partial_u(f_j(u,h(u)))|_{u=u_-}$.

Due to (4), it is well known that the piecewise-constant function

$$u(x,t)=\begin{cases}u_-, & \omega\cdot x<s_*t\\ u_+, & \omega\cdot x>s_*t\end{cases} \tag{5}$$

is a weak solution of the following conservation laws

$$u_t+\sum_{j=1}^d f_j(u,h(u))_{x_j}=0. \tag{6}$$

This is the so-called *equilibrium system* corresponding to the balance laws (1). It governs the dynamics of (1) in the *zero relaxation limit* (see, e.g., [13]). The piecewise-constant solution (5) is called a *shock front* if (6) is (strictly) hyperbolic and the solution satisfies Liu's entropy condition [5].

The above discussion relates the traveling-wave solutions to shock waves (5). Therefore, the former are often called *shock profiles* and their existence is of physical interest [1, 8, 11]. The analogous existence problem was first posed by Gelfand [3] for viscosity approximations of hyperbolic conservation laws and was solved by Majda and Pego [7].

The present existence problem was also studied in [15] for the balance laws satisfying the second stability condition in [12]. A crucial technical assumption in [15] is that the matrix $A(U_-) := -\lambda_p(u_-, \omega)I_n + \sum_{j=1}^{d} \omega_j F_{jU}(U_-)$ is invertible. This assumption seems quite artificial and restricts the applicability of the result in [15].

In this report, we replace this technical assumption with a Kawashima-like condition [10] and have an existence result similar to that in [15]. The Kawashima condition is also crucial for other results for hyperbolic systems of balance laws (see, e.g., [14]). For 2×2 systems with $r = 1$, the above existence problem was previously solved by Liu [6].

In case the matrix A is not invertible, (3) have a singularity when (ϕ, s_*) is close to $(U_-, \lambda_p(u_-, \omega))$. The basic idea is to understand the singular equations as a three-scale multidimensional connection problem. Based on this understanding, we make two center manifold reductions to convert the problem to a one-dimensional problem, for which there is a well-known criterion for existence. The main technical issue is to show the effectiveness of the reductions under some well-established structural conditions on the hyperbolic systems.

The report is organized as follows. In Sect. 2 we formulate our main result [2] and give the main ideas of the proof. Therein, our assumptions are the stability condition, a Kawashima-like condition and that zero is an isolated eigenvalue of $A(U_-)$ (also referred to as the *case of simple degeneracy*). In Sect. 3 we give an ansatz how to deal with the case, where zero is an eigenvalue of $A(U_-)$ with multiplicity larger than one (also referred to as the *case of multiple degeneracy*).

2 Main Result for the Case of Simple Degeneracy

The aim of this section is to formulate the structural conditions, the main result and to sketch out the proof for the case of simple degeneracy.

To begin with, we refer to (2) and recall the assumption that $q(u, v) = 0$ if and only if $v = h(u)$. On the *equilibrium manifold*

$$\mathcal{E} := \left\{ U \in \mathcal{U} \mid Q(U) = 0 \right\} = \left\{ U = \begin{pmatrix} u \\ h(u) \end{pmatrix} \in \mathcal{U} \right\}$$

we fix

$$U_- \equiv \begin{pmatrix} u_- \\ h(u_-) \end{pmatrix} \in \mathcal{E}$$

throughout this report. Moreover, we fix $\omega = (\omega_1, \omega_2, \cdots, \omega_d) \in \mathbf{R}^d \setminus \{0\}$.

Our next assumption on the balance laws (1) is the second stability condition in [12]. In [2] we have shown that, after appropriate linear transformation of the phase space and scaling of (x, t) by a constant factor, system (1) has *normalized form*, i.e. the following three conditions are fulfilled:

(h1). $B := Q_U(U_-) = \mathrm{diag}(0, S) \quad$ with $S \in \mathbf{R}^{r\times r}$

(h2). $\sum_{j=1}^{d} \omega_j F_{jU}(U_-)$ is a symmetric matrix

(h3). $S + S^* \leq -I_r$

Throughout this report, the subscript U refers to the usual partial derivatives with respect to U, the superscript $*$ denotes the transpose operator acting on matrices or vectors, and I_k is the unit matrix of order k.

The second stability condition and thereby (h1)–(h3) was shown in [12] and [13] to be fulfilled by many systems of balance laws (1) from mathematical physics. On the other hand, it was proved in [12] that the stability condition implies the (strong) hyperbolicity of the equilibrium system (6): for each $\tilde{\omega} = (\tilde{\omega}_1, \cdots, \tilde{\omega}_d) \in \mathbf{R}^d$, the characteristic matrix

$$a(\tilde{\omega}) := \sum_{j=1}^{d} \tilde{\omega}_j \partial_u (f_j(u, h(u)))|_{u=u_-} \tag{7}$$

has only real eigenvalues with a complete set of eigenvectors.

As in [15], we further assume that the matrix $a(\omega)$ has an isolated eigenvalue λ_p with r_p being a corresponding eigenvector. Then we know from [5] that the Rankine–Hugoniot relation

$$s(u - u_-) = \sum_{j=1}^{d} \omega_j (f_j(u, h(u)) - f_j(u_-, h(u_-)))$$

defines a smooth curve $(u(\rho), s(\rho))$, for $\rho \in (-\delta, \delta)$ with δ sufficiently small, satisfying

$$u(0) = u_-, \qquad s(0) = \lambda_p \quad \text{and} \quad \partial_\rho u(\rho)|_{\rho=0} = r_p.$$

According to Liu [5], the piecewise-constant solution (5) is called a *shock-front* if there is a $\rho_* \in (-\delta, \delta)$ such that $(u_+, s_*) = (u(\rho_*), s(\rho_*))$ and

$$s_* \leq s(\rho)$$

for all ρ between 0 and ρ_*.

Denote by $\mathrm{Ker}(M)$ the null space of matrix M. Our last assumption is the following Kawashima-like condition:

(h4). $\qquad \mathrm{Ker}(Q_U(U_-)) \cap \mathrm{Ker}(A(U_-)) = \{0\} \qquad$ with

$$A(U) = -\lambda_p I_n + \sum_{j=1}^{d} \omega_j F_{jU}(U). \tag{8}$$

Now we are in a position to state our main result of this section for the multidimensional connection problem (3).

Theorem 1. *([2]) Suppose $Q \in C^3(\mathcal{U}, \mathbf{R}^n)$, $F_j \in C^4(\mathcal{U}, \mathbf{R}^n)$, the system (1) admits the structural conditions (h1)–(h4), and λ_p is an isolated eigenvalue of the characteristic matrix $a(\omega)$ defined in (7). In addition, assume zero is an isolated eigenvalue of $A(U_-)$ defined in (8).*

Then there exists $\bar{\delta} > 0$ such that for any ρ_* satisfying $|\rho_*| < \bar{\delta}$ and $s_* := s(\rho_*) < \lambda_p$, the connection problem (3) has a solution $\phi \in C^1(\mathbf{R}, \mathcal{U})$ with $\phi(+\infty) = U_+ := \begin{pmatrix} u(\rho_*) \\ h(u(\rho_*)) \end{pmatrix}$, which decays exponentially to $U_\pm$ as ξ goes to $\pm\infty$.

The proof of Theorem 1 can be sketched out as follows. Rewrite the traveling-wave equations (3) as

$$\begin{aligned} \sigma_\xi &= 0, \\ (A(\phi) - \sigma I_n)\phi_\xi &= Q(\phi) \end{aligned} \tag{9}$$

with boundary conditions

$$(\sigma, \phi)(\pm\infty) = (s_* - \lambda_p, U_\pm),$$

where $A(\phi)$ is defined in (8). Due to our assumptions, zero is an isolated eigenvalue of $A(U_-)$. Hence, for ϕ close to U_-, $A(\phi)$ has only one eigenvalue $\lambda(\phi)$ satisfying $\lim_{\phi \to U_-} \lambda(\phi) = 0$. Note that $\lambda(\cdot)$ is of class C^3. Thus, when (σ, ϕ) is close to $(0, U_-)$ and $(A(\phi) - \sigma I_n)$ is invertible, we can write

$$(A(\phi) - \sigma I_n)^{-1} = (\lambda(\phi) - \sigma)^{-1} R(\phi) + \tilde{A}(\sigma, \phi). \tag{10}$$

Here the matrix $R(\phi)$ is the projector onto the eigenspace associated with the eigenvalue $\lambda(\phi)$ of $A(\phi)$ and $\tilde{A}(\sigma, \phi)$ is the generalized inverse of the matrix $(A(\phi) - \sigma I_n)$ associated with the eigenvalues of $(A(\phi) - \sigma I_n)$ which are different from zero at $(\sigma, \phi) = (0, U_-)$.

Motivated by (10), we rescale system (9) via $(\lambda(\phi) - \sigma) d\tau = d\xi$ and obtain

$$\begin{aligned} \sigma_\tau &= 0, \\ \phi_\tau &= \left[R(\phi) + (\lambda(\phi) - \sigma)\tilde{A}(\sigma, \phi) \right] Q(\phi). \end{aligned} \tag{11}$$

Let $\Phi = \Phi(\sigma, \psi)$ parametrize a center manifold containing the fixed point $(0, U_-)$ for the dynamical system (11). For $H = \Phi_\psi(0, 0)$, $\mathcal{Q}(\sigma, \psi) = H^* Q(\Phi(\sigma, \psi))$ and

$$\mathcal{W}(\sigma,\psi) = H^* \left[\sum_{j=1}^{d} \omega_j \left(F_j(\Phi(\sigma,\psi)) - F_j(U_-) \right) - (\sigma + \lambda_p)(\Phi(\sigma,\psi) - U_-) \right],$$

we obtain after multiplication of H^* to the second equation in (11), and we obtain after rescaling the following system of ordinary differential equations:

$$\begin{aligned} \sigma_\xi &= 0, \\ \mathcal{W}(\sigma,\psi)_\xi &= \mathcal{Q}(\sigma,\psi). \end{aligned} \tag{12}$$

In [2] we have shown that the structural conditions imply the invertibility of $\mathcal{W}_\psi(0,0) = H^* A(U_-) H$. Hence, the connection problem (3) is reduced to applying the classical center manifold theorem to system (12) and a modification of the argument in [15].

3 Case of Multiple Degeneracy

In this section the characteristic matrix $A(U)$ is allowed to have more than one eigenvalue tending to 0 as U goes to U_-.

Recall that $R(\phi)$ denotes the projector onto the eigenspace associated with the eigenvalues of $A(\phi)$ tending to zero if ϕ goes to U_-. As in Sect. 2, let $\tilde{A}(\sigma,\phi)$ be the generalized inverse of $(A(\phi) - \sigma I_n)$ associated with the eigenvalues which are different from zero for $(\sigma,\phi) = (0, U_-)$. Note that, for any (σ,ϕ) close to $(0, U_-)$,

$$R^2(\sigma,\phi) = R(\phi), \quad \tilde{A}(\sigma,\phi)R(\phi) = 0. \tag{13}$$

Let $\mu = \mu(\sigma,\phi)$ be a real-valued function of class C^3 on some open subset $\mathcal{O}$ of the domain of definition of $R(\phi)$ with the following properties:

(A1). $\mu(0, U_-) = 0$

(A2). $\{\mu(\sigma,\phi) \neq 0\}$ is dense in $\mathcal{O}$

For any (σ,ϕ) close to $(0, U_-)$ set

$$L(\sigma,\phi) := R(\phi)[(A(\phi) - \sigma I_n)] - \mu(\sigma,\phi)R(\phi). \tag{14}$$

From this definition and the relations in (13) we easily obtain

$$R(\phi)L(\sigma,\phi) = L(\sigma,\phi), \quad \tilde{A}(\sigma,\phi)L(\sigma,\phi) = \tilde{A}(\sigma,\phi)R(\phi)L(\sigma,\phi) = 0. \tag{15}$$

Using the relations in (15) it is not hard to show that, as long as the matrix $(A(\phi) - \sigma I_n - L(\sigma,\phi))$ is invertible and (σ,ϕ) is close to $(0, U_-)$, we have

$$[A(\phi) - \sigma I_n - L(\sigma,\phi)]^{-1} = \mu(\sigma,\phi)^{-1}R(\phi) + \tilde{A}(\sigma,\phi). \tag{16}$$

The traveling-wave system (9) can be rewritten as

$$\begin{aligned} \sigma_\xi &= 0, \\ (A(\phi) - L(\sigma,\phi) - \sigma I_n)\phi_\xi + L(\sigma,\phi)\phi_\xi &= Q(\phi). \end{aligned} \tag{17}$$

After rescaling $\mu(\sigma,\phi)d\tau = d\xi$, multiplication of

$$\mu(\sigma,\phi)^{-1}R(\phi) + \tilde{A}(\sigma,\phi)$$

to the second equation in (17) from the left and taking into account relations (15) we obtain

$$[\mu(\sigma,\phi)^{-1}R(\phi) + \tilde{A}(\sigma,\phi)]L(\sigma,\phi) = \mu(\sigma,\phi)^{-1}L(\sigma,\phi) =: M(\sigma,\phi). \tag{18}$$

Assume that, on some open neighborhood $\mathcal{O} \ni (0, U_-)$, the matrix $M(\sigma,\phi)$ satisfies the following conditions:

(A3). $(\sigma,\phi) \mapsto M(\sigma,\phi)$ has a continuation of class $C^3(\mathcal{O}, \mathbf{R}^{n\times n})$

(A4). $(I_n + M(0,U_-))^{-1}$ exists

Due to (16) and (18) (with the rescaling $\xi \mapsto \tau$ to be taken into account), system (17) can be rewritten as

$$\begin{aligned} \sigma_\tau &= 0, \\ (I_n + M(\sigma,\phi))\phi_\tau &= \left[R(\phi) + \mu(\sigma,\phi)\tilde{A}(\sigma,\phi)\right] Q(\phi). \end{aligned} \tag{19}$$

Due to assumptions (A3) and (A4), for any (σ,ϕ) close to $(0,U_-)$, the matrix

$$\mathcal{L}(\sigma,\phi) = (I_n + M(\sigma,\phi))^{-1}[R(\phi) + \mu(\sigma,\phi)\tilde{A}(\sigma,\phi)]$$

is well defined. Hence, system (19) is equivalent to the following system of ordinary differential equations:

$$\begin{aligned} \sigma_\tau &= 0, \\ \phi_\tau &= \mathcal{L}(\sigma,\phi)Q(\phi). \end{aligned} \tag{20}$$

The idea now is, similarly as in Sect. 2, to construct a center manifold to system (20) containing the fixed point $(\sigma,\phi) = (0,U_-)$ and then, after rescaling $\tau \mapsto \xi$, to apply a modification of the argument in [15] in order to solve the connection problem (3).

References

1. G. Boillat & T. Ruggeri: On the shock structure problem for hyperbolic system of balance laws and convex entropy. *Contin. Mech. Thermodyn.* **10**, 285–292 (1998)

2. A. Dressel & W.-A. Yong: Existence of Traveling-Wave Solutions for Hyperbolic Systems of Balance Laws. *Arch. Rational Mech. Anal.* **182**, 49–75 (2006).
3. I.M. Gelfand: Some problems in the theory of quasilinear equations. *Uspehi Mat. Nauk* **14**, 87–158 (1959); English transl., *Amer. Math. Soc. Transl. (2)* **29**, (1963)
4. A. Kelley: The stable, center-stable, center, center- unstable and unstable manifolds. *J. Differ. Eqns.* **3**, 546–570 (1967)
5. T.-P. Liu: The entropy condition and the admissibility of shocks. *J. Math. Anal. Appl.* **53**, 78–88 (1976)
6. T.-P. Liu: Hyperbolic conservation laws with relaxation. *Comm. Math. Phys.* **108**, 153–175 (1987)
7. A. Majda & R.L. Pego: Stable viscosity matrices for systems of conservation laws, *J. Differ. Eqns.* **56**, 229–262 (1985)
8. I. Müller & T. Ruggeri: *Rational Extended Thermodynamics.* Springer, New York, 1998
9. R.L. Pego: Stable viscosities and shock profiles for systems of conservation laws. *Trans. Am. Math. Soc.*, **282**, 749–763 (1984)
10. Y. Shizuta & S. Kawashima: Systems of equations of hyperbolic-parabolic type with applications to the discrete Boltzmann equation. *Hokkaido Math. J.*, **14**, 249–275 (1985)
11. W. Weiss: Continuous shock structure in extended thermodynamics. *Phys. Rev. E* **52**, 5760–5763 (1995)
12. W.-A. Yong: Singular Perturbations of First-Order Hyperbolic Systems. Ph.D. Thesis, Universität Heidelberg, 1992.
13. W.-A. Yong: Basic aspects of hyperbolic relaxation systems. In: *Advances in the Theory of Shock Waves.* Freistühler, H., Szepessy, A. eds., Progress in Nonlinear Differential Equations and Their Applications, Vol. 47, Birkhäuser, Boston, 2001
14. W.-A. Yong: Entropy and global existence for hyperbolic balance laws. *Arch. Rational Mech. Anal.* **172**, 247–266 (2004).
15. W.-A. Yong & K. Zumbrun: Existence of relaxation shock profiles for hyperbolic conservation laws. *Siam J. Appl. Math.* **60**, 1565–1575 (2000).

A Hyperbolic-Elliptic Model for Coupled Well-Porous Media Flow

S. Evje and K.H. Karlsen

1 Introduction

A natural model for well-porous media flow is obtained by coupling a hyperbolic system of two conservation laws corresponding to the isothermal Euler equations with source terms, and an integral equation. In dimensionless form it is given by

$$\begin{aligned} \partial_t(\rho) + \partial_x(\rho u) &= \frac{1}{\eta}\rho q_V, \qquad \partial_t(\rho u) + \partial_x(\rho u^2) + \partial_x p(\rho) = 0, \qquad \eta > 0, \\ p_0 - p(x,t) &= \int_0^t \int_0^1 H^r(x, x', t-t') q_V(x', t')\, dx' dt', \end{aligned} \tag{1}$$

for $x \in [0,1]$. ρ, u, and $p(\rho)$ are, respectively, density, velocity, and pressure, whereas q_V represents volumetric flow rate. p_0 is initial pressure (assumed to be constant) and η is a small parameter. The kernel $H^r(x, x', t-t')$ is characteristic for the porous media under consideration.

In order to get some understanding of basic underlying mechanisms present in the model (1), we assume that the fluid is incompressible. We then get a scalar conservation law on the form

$$\partial_t u + \partial_x(u^2) = -\partial_x p, \qquad p_0 - p(x,t) = \varepsilon \int_{-\infty}^{+\infty} G^r(x, x') u_{x'}(x', t)\, dx', \tag{2}$$

$$\text{with} \qquad \varepsilon = \frac{\mu D}{4\rho k}, \qquad G^r(x, x') = \frac{r^2}{\sqrt{(x-x')^2 + r^2}}, \qquad r > 0, \tag{3}$$

where μ is fluid viscosity, k is permeability, D is a characteristic time, r the well radius (which typically is small relatively the size of the porous media). We may write (2) on the form

$$\partial_t u + \partial_x(u^2) = \varepsilon G^r_x * u_x = \varepsilon G^r_{xx} * u, \qquad \varepsilon, r > 0. \tag{4}$$

Various properties of the model (4) was studied in [EK06]. In particular, well-posedness was demonstrated, respectively, in a L^∞ and L^2 setting.

In this note we replace the kernel $G^r(x,x')$ with an approximation $\widetilde{G}^r(x,x')$ such that $\|G^r_x - \widetilde{G}^r_x\|_{L^1(\mathbb{R})} = O(r)$, and then derive various estimates for the *approximate* well-reservoir model

$$\partial_t u + \partial_x(u^2) = \varepsilon \widetilde{G}^r_x * u_x, \qquad \varepsilon, r > 0. \tag{5}$$

These estimates, which imply existence and uniqueness of entropy weak solutions, are sharper than those presented in [EK06]. This is essentially due to the fact that the approximate kernel function $\widetilde{G}^r$ leads to a source term possessing a dissipative nature. In that respect, the approximate well-reservoir model (5) bears a clear link to the so-called radiating gas model, studied by many researchers more lately [KN99, N00, LT01, LM03, S03, S04]. This model can be written on the form

$$\partial_t u + \partial_x(\frac{1}{2}u^2) = \partial_x p, \quad p(x,t) = \int_{-\infty}^{+\infty} H(x,x')u_{x'}(x',t)\,dx' = H * u_x, \tag{6}$$

with $H(x,x') = \frac{1}{2}e^{-|x-x'|}$. Alternatively, one can express the model on the form $\partial_t u + \partial_x(\frac{1}{2}u^2) = H_{xx} * u = [H-\delta] * u = H * u - u$, where δ represents the Dirac delta function.

We emphasize that the aim of this note is to demonstrate how the well-reservoir model (4) can naturally be related to the radiating gas model through the proposed approximate well reservoir (5). In particular, the results we present for this model is obtained by applying the same techniques as those used for the radiating gas model. Nevertheless, one should also note that there is a clear difference between (5) and (6) since the kernel H corresponding to the latter is associated with the differential operator $(1-\partial^2_{xx})$. Consequently, (6) can be written on the form $u_t + uu_x = -p_x$, where $-p_{xx} + p = -u_x$. This is explicitly used, for example in traveling wave analysis [KN99, N00, S03].

2 Mathematical Models

2.1 Porous Media Flow

Darcy's law and the continuity equation for flow in porous medium can be combined to give a transient pressure equation [B88]

$$c\phi\frac{\partial p}{\partial t} - \nabla\cdot\left[\frac{k}{\mu}\nabla p\right] = Q_{\text{vol}}(\mathbf{x},t). \tag{7}$$

Here p is pressure, ϕ porosity, μ viscosity, whereas $Q_{\text{vol}}(\mathbf{x},t)$ accounts for the mass flow between well and porous media. We assume that the porosity ϕ and compressibility c is constant. Furthermore, let $\mathbf{X}_w(s) = (x_w(s), y_w(s), z_w(s))$ with $s \in [0,1]$ (dimensionless) be a parametrization of the line Γ_w describing the well path. The source term $Q_{\text{vol}}(\mathbf{x},t)$ represents a delta function singularity along the well path Γ_w given by

$$Q_{\text{vol}}(\mathbf{x},t) = \int_{\Gamma_w} q_V(\alpha,t)\delta(\mathbf{x} - \mathbf{X}_w(\alpha))\,d\alpha, \tag{8}$$

where $\delta(\mathbf{x})$ is a three-dimensional Dirac function, $q_V(\alpha,t)$ the volumetric influx or efflux rate per unit wellbore length, and α denotes the arc-length function. In the following we restrict ourselves to a straight line well geometry of length L_w. We also assume that Ω is a cube of length L. In terms of dimensionless variables the pressure equation (7) takes the form (see [EK06] for details)

$$\frac{\partial p}{\partial t} - \left[\frac{\partial^2 p}{\partial x^2} + \frac{\partial^2 p}{\partial y^2} + \frac{\partial^2 p}{\partial z^2}\right] = Q_{\text{vol}}(\mathbf{x},t), \tag{9}$$

where $(\mathbf{x},t) = (x,y,z,t) \in \Omega \times [0,T]$. In the following we shall apply the integral formulation of (9).

$$p_0(\mathbf{x}) - p(\mathbf{x},t) = \int_0^t \int_0^1 G(\mathbf{x}, \mathbf{X}_w(s'), t-t')q_V(s',t')\,ds'dt', \tag{10}$$

where G is the free-space kernel $G(\mathbf{x},\mathbf{x}',t-t') = \frac{1}{[4\pi(t-t')]^{3/2}}\exp\left[-\frac{\|\mathbf{x}-\mathbf{x}'\|^2}{4(t-t')}\right]$. By setting $\mathbf{x} = \mathbf{X}_w(s)+\mathbf{r}_w$ for $s \in [0,1]$ in (10), we note that $q_V(s',t')$ satisfies the integral equation

$$\Delta p(\mathbf{X}_w(s) + \mathbf{r}_w, t) = \int_0^t \int_0^1 G(s,s',t-t')q_V(s',t')\,ds'dt'. \tag{11}$$

Here $\Delta p(\mathbf{X}_w(s) + \mathbf{r}_w, t) = p_0(\mathbf{X}_w(s) + \mathbf{r}_w) - p(\mathbf{X}_w(s) + \mathbf{r}_w, t)$ represents the change in pressure at the well boundary. Equation (11) is an integral equation of first kind, Fredholm in space and Volterra in time.

Assuming the fluid is incompressible the pressure is given by

$$-\nabla \cdot \left[\frac{k}{\mu}\nabla p\right] = Q_{\text{vol}}(\mathbf{x},t),$$

where Q_{vol} is given by (8). Following the approach as described above, we arrive at the following integral equation

$$\Delta p(\mathbf{X}_w(s) + \mathbf{r}_w, t) = \int_0^1 G(\mathbf{X}_w(s) + \mathbf{r}_w, \mathbf{X}_w(s'))q_V(s',t)\,ds', \tag{12}$$

where the kernel G is the Green's function associated with the pressure equation $-\Delta p = \delta(\mathbf{x} - \mathbf{X}_w)$. That is, $G(\mathbf{x},\mathbf{x}') = \frac{1}{4\pi\|\mathbf{x}-\mathbf{x}'\|}$.

2.2 Well Flow

A single-phase, compressible, isothermal, and unsteady well-flow model is given in the form

$$\begin{aligned} \partial_t(A\rho_w) + \partial_\alpha(A\rho_w u) &= \rho_w q_{\mathrm{V}} \\ \partial_t(A\rho_w u) + \partial_\alpha(A\rho_w u^2) + A\partial_\alpha p_w &= 0, \end{aligned} \tag{13}$$

where α is the arc-length variable associated with the well-path Γ_w. Here ρ_w is density, u velocity, $p_w = p(\rho_w)$ pressure, q_{V} is volumetric flux per unit wellbore length. Moreover, $A = \pi r_w^2$ is the pipe cross-sectional area for a well of radius r_w. In terms of the nondimensional variables the model (13) takes the form (see [EK06] for more details)

$$\begin{aligned} &\partial_t(\rho_w) + \partial_s(\rho_w u) = \frac{1}{\eta}\rho_w q_{\mathrm{V}}, \qquad \eta = \frac{L_w A\mu}{DLk\bar{p}}, \\ &\partial_t(\rho_w u) + \partial_s(\rho_w u^2) + h_0\partial_s p_w = 0, \quad p_w = p_w(\rho_w), \quad h_0 = \frac{\bar{p}}{\bar{\rho}\bar{u}^2}. \end{aligned} \tag{14}$$

Here $\bar{p}, \bar{\rho}, \bar{u}$ represent characteristic quantities and $\bar{u} = L_w/D$.

2.3 Coupled Well-Porous Media Flow

Equipped with the well model (14) and the porous media model (11) we now formulate a *coupled* well-porous media flow model by imposing the coupling condition $p_w(\rho_w(s,t)) = p(\mathbf{X}_w(s) + \mathbf{r}_w, t) := p(s,t)$. This results in a model on the form (skipping the index "w")

$$\begin{aligned} &\partial_t(\rho) + \partial_s(\rho u) = \frac{1}{\eta}\rho q_{\mathrm{V}}, \qquad \partial_t(\rho u) + \partial_s(\rho u^2) + \partial_s P(\rho) = 0, \\ &P_0 - P(s,t) = h_0 \int_0^t \int_0^1 G(\mathbf{X}_w(s) + \mathbf{r}_w, \mathbf{X}_w(s'), t - t') q_{\mathrm{V}}(s', t')\, ds' dt', \end{aligned}$$

where $P(\rho) = h_0 p(\rho)$ and $h_0 = \frac{\bar{p}}{\bar{\rho}\bar{u}^2}$. This model corresponds to the model problem (1).

2.4 A Simplified "Compressible Well-Incompressible Porous Media" Model

We may treat the reservoir fluid as an incompressible fluid. In view of (12) we then obtain a well-porous media model on the form

$$\begin{aligned} &\partial_t(\rho) + \partial_s(\rho u) = \frac{1}{\eta}\rho q_{\mathrm{V}}, \qquad \partial_t(\rho u) + \partial_s(\rho u^2) + \partial_s P(\rho) = 0, \\ &P_0 - P(s,t) = \int_0^1 H^r(s,s') q_{\mathrm{V}}(s',t)\, ds', \end{aligned} \tag{15}$$

where $H^r(s,s') = h_0 G(\mathbf{X}_w(s) + \mathbf{r}, \mathbf{X}_w(s'))$ for $G(\mathbf{x},\mathbf{x}') = \frac{1}{4\pi\|\mathbf{x}-\mathbf{x}'\|}$. We introduce the dimensionless radius $\bar{r} = \frac{r_w}{L_w}$ and arrive at the following expression for the kernel $H^r(s,s')$ with $\varepsilon_1 = \frac{h_0 L}{4\pi\bar{r}L_w}$ (see [EK06] for more details).

$$H^r(s,s') = h_0 G(\mathbf{X}_w(s) + \mathbf{r}, \mathbf{X}_w(s')) = \varepsilon_1 \left[\left((s-s')/\bar{r} \right)^2 + 1 \right]^{-1/2}. \tag{16}$$

2.5 A Simplified Incompressible Well-Porous Media Model

We take a step further and impose in (15) that the well fluid is incompressible, i.e., $\rho = 1$. This yields the following simplified model

$$\begin{aligned} &\partial_s(u) = \frac{1}{\eta} q_V, \qquad \partial_t(u) + \partial_s(u^2) + \partial_s P = 0, \\ &P_0 - P(s,t) = \int_0^1 H^r(s,s') q_V(s',t)\, ds', \end{aligned} \tag{17}$$

where $\eta = \frac{L_w \mu A}{Lk\bar{p}D}$. In view of (16), we introduce the function $G^r(s,s')$ defined by (3) and see that $H^r(s,s') = \frac{h_0 L}{4\pi \bar{r}^2 L_w} G^r(s,s')$. Inserting this in (17), we obtain the model problem (2) and (3), where we have replaced the finite domain $[0,1]$ by the real axis.

2.6 An Approximate Well-Porous Media Model Relevant for (4)

In this section we focus on a well-reservoir model which represents an approximation to the well-porous media model (4). More precisely, we replace the kernel $G^r(x,x')$ given in (3) by an *approximate* kernel $\widetilde{G}^r(x,x')$ defined in the following. First, we observe that

$$G_x^r(x) = \frac{-r^2 x}{\left(x^2 + r^2\right)^{3/2}}, \qquad G_{xx}^r(x) = \frac{r^2\left(\sqrt{2}x - r\right)\left(\sqrt{2}x + r\right)}{\left(x^2 + r^2\right)^{5/2}}.$$

We then introduce the approximation

$$\widetilde{G}_x^r(x) = \begin{cases} G_x^r(x) & \text{if } |x| > r/\sqrt{2}, \\ c(1 - 2H(x)) & \text{if } |x| \le r/\sqrt{2}, \end{cases} \quad c = G_x^r(-r/\sqrt{2}),$$

where $H(x)$ is the Heaviside function and $c = \frac{2}{3\sqrt{3}}$. We easily see that $\|G_x^r - \widetilde{G}_x^r\|_{L^1(\mathbb{R})} = O(r)$. Next, we define $\widetilde{G}^r(x) = \int_{-\infty}^x \widetilde{G}_x^r(s)\, ds$. Moreover, it follows that $\widetilde{G}^r \in C^2(\mathbb{R}/\{0\})$ since

$$\widetilde{G}_{xx}^r(x) = \begin{cases} G_{xx}^r(x) & \text{if } |x| > r/\sqrt{2}, \\ -2c\delta(x) & \text{if } |x| \le r/\sqrt{2}, \end{cases} \qquad c = G_x^r(-r/\sqrt{2}) = -G_x^r(r/\sqrt{2}),$$

where $\delta(x)$ is the Dirac mass centered at $x = 0$ In particular, we note that $\widetilde{G}_{xx}^r$ is continuous at $\pm r/\sqrt{2}$ since $G_{xx}^r(r/\sqrt{2}) = 0$. We now consider the corresponding well-porous media model defined by

$$\begin{aligned}\partial_t\widetilde{u}+\partial_x(\widetilde{u}^2)&=\varepsilon\widetilde{G}^r_x*\widetilde{u}_x=\varepsilon\widetilde{G}^r_{xx}*\widetilde{u}\\&=\varepsilon\Big(G^r_{xx}\chi_{|x-x'|>r/\sqrt{2}}*\widetilde{u}-2c\delta(x-x')\chi_{|x-x'|\le r/\sqrt{2}}*\widetilde{u}\Big)\\&=\varepsilon(G^r_{xx}\chi_{|x-x'|>r/\sqrt{2}}*\widetilde{u}-2c\widetilde{u}),\end{aligned}\tag{18}$$

where $\chi_E(x)=1$ for $x\in E$ and $\chi_E(x)=0$ for $x\notin E$.

3 A Well-Posedness Result for the Model (18)

Definition 1 (Entropy weak solution). *A function $u\in L^\infty((0,T)\times\mathbb{R})\cap C([0,T];L^1(\mathbb{R}))$ for any $T>0$, is called an entropy weak solution to (18) provided for any convex C^2 entropy $\eta:\mathbb{R}\to\mathbb{R}$ with corresponding entropy flux $q:\mathbb{R}\to\mathbb{R}$ defined by $q'(u)=2u\eta'(u)$ there holds the inequality*

$$\int_0^T\int_{\mathbb{R}}[\eta(u)\phi_t+q(u)\phi_x-\eta'(u)p_x\phi]dxdt+\int_{\mathbb{R}}\eta(u_0(x))\phi(x,0)dx\ge 0,$$

$\forall\phi\in C_0^\infty([0,T)\times\mathbb{R})$, $\phi\ge0$, where $p_0-p(x,t)=\varepsilon\int_{\mathbb{R}}\widetilde{G}^r_x(x,x')u(x',t)\,dx'$.

We rely on the standard approach and seek for a solution to (18) by letting μ go to zero in the viscous approximation

$$\partial_t\widetilde{u}+\partial_x(\widetilde{u}^2)=\varepsilon\widetilde{G}^r_{xx}*\widetilde{u}+\mu\partial^2_{xx}\widetilde{u},\qquad\mu>0.\tag{19}$$

Lemma 1. *Let u and $\bar{u}$ be solutions of (19) with initial data $u_0,\bar{u}_0\in L^1(\mathbb{R})\cap L^\infty(\mathbb{R})$. Then, for any $t>0$,*

$$\int_{\mathbb{R}}[u(x,t)-\bar{u}(x,t)]^+\,ds\le\int_{\mathbb{R}}[u_0(x)-\bar{u}_0(x)]^+\,ds,$$

$$\|u(\cdot,t)-\bar{u}(\cdot,t)\|_{L^1(\mathbb{R})}\le\|u_0(\cdot)-\bar{u}_0(\cdot)\|_{L^1(\mathbb{R})},\tag{20}$$

$$\text{If } u_0(x)\le\bar{u}_0(x)\text{ a.e. on }\mathbb{R},\text{ then }u(x,t)\le\bar{u}(x,t)\text{ a.e. on }\mathbb{R}\times[0,T],\tag{21}$$

$$-a\le\|u(\cdot,t)\|_\infty,\|\bar{u}(\cdot,t)\|_\infty\le+a,\qquad a=\max\{\|u_0\|_\infty,\|\bar{u}_0\|_\infty\}.\tag{22}$$

Proof. We know that (19) has smooth (classical) solutions. We define $\eta_\delta(\cdot)$ such that, pointwise we have $\eta_\delta(u-\bar{u})\to[u-\bar{u}]^+$, $\eta_\delta'(u-\bar{u})\to\mathrm{sgn}(u-\bar{u})^+$, $\eta_\delta''(u-\bar{u})[u^2-\bar{u}^2]\to0$, as $\delta\downarrow0$. In view of (19) we can find an equation for $u-\bar{u}$. Multiplying this equation by $\eta_\delta'(u-\bar{u})$, we get

$$\begin{aligned}&\int_{\mathbb{R}}\eta_\delta(u-\bar{u})dx\le\int_{\mathbb{R}}\eta_\delta(u_0-\bar{u}_0)dx\\&+\int_0^t\int_{\mathbb{R}}\eta_\delta''(u-\bar{u})(u^2-\bar{u}^2)(u-\bar{u})_xdxdt+\varepsilon\int_0^t\int_{\mathbb{R}}\eta_\delta'\widetilde{G}^r_{xx}*(u-\bar{u})dxd\tau.\end{aligned}$$

Taking the limit $\delta\to0$, we get

$$\int_{\mathbb{R}} [u(x,t) - \bar{u}(x,t)]^+ \, dx \leq \int_{\mathbb{R}} [u_0(x) - \bar{u}_0(x)]^+ \, dx + R, \tag{23}$$

where $R = \varepsilon \int_0^t \int_{\mathbb{R}} \mathrm{sgn}(u - \bar{u})^+ \widetilde{G}^r_{xx} * (u - \bar{u}) \, dx d\tau$. We must estimate R.

$$\begin{aligned}
&\int_0^t \int_{\mathbb{R}} \mathrm{sgn}(u - \bar{u})^+ [G^r_{xx} \chi_{|x-x'|>r/\sqrt{2}}] * (u - \bar{u}) \, dx d\tau \\
&\leq \int_0^t \int_{\mathbb{R}} \int_{\mathbb{R}} G^r_{xx} \chi_{|x-x'|>r/\sqrt{2}} \cdot [u - \bar{u}]^+ (x', t) \, dx' \, dx d\tau \\
&= \int_0^t \int_{\mathbb{R}} [u - \bar{u}]^+ (x', t) \int_{\mathbb{R}} G^r_{yy} \chi_{|y|>r/\sqrt{2}} \, dy \, dx' d\tau \\
&= 2 \int_0^t \int_{\mathbb{R}} [u - \bar{u}]^+ (x', t) \int_{-\infty}^{-r/\sqrt{2}} G^r_{yy} \, dy \, dx' d\tau = 2c \int_0^t \int_{\mathbb{R}} [u - \bar{u}]^+ (x', t) \, dx' d\tau,
\end{aligned}$$

where we have used the transformation $y = x - x'$ for a fixed x'. Consequently,

$$\int_0^t \int_{\mathbb{R}} \mathrm{sgn}(u - \bar{u})^+ \Big([G^r_{xx} \chi_{|x-x'|>r/\sqrt{2}}] * (u - \bar{u}) - 2c(u - \bar{u}) \Big) \, dx d\tau \leq 0,$$

and it follows from (23) that $\int_{\mathbb{R}} (u - \bar{u})^+ \, dx \leq \int_{\mathbb{R}} (u_0 - \bar{u}_0)^+ \, dx$. From this, (20) and (21) follow. To show that (22) holds, we multiply (18) by a regularization of $p|u|^{p-1}\mathrm{sgn}(u)$ and observe that

$$\begin{aligned}
&p|u|^{p-1}\mathrm{sgn}(u) u_t = \partial_t(|u|^p), \quad p|u|^{p-1}\mathrm{sgn}(u)(u^2)_x = \partial_x\Big(\frac{2p}{p+1}\mathrm{sgn}(u)|u|^{p+1}\Big), \\
&p|u|^{p-1}\mathrm{sgn}(u)\mu\partial^2_{xx} u = \mu(u_x p|u|^{p-1}\mathrm{sgn}(u))_x - \mu p(p-1)|u|^{p-2}(u_x)^2.
\end{aligned}$$

Consequently, $\int_{\mathbb{R}} |u|^p \, dx \leq \int_{\mathbb{R}} |u_0|^p \, dx + \varepsilon \int_{\mathbb{R}} p|u|^{p-1}\mathrm{sgn}(u)\widetilde{G}^r_{xx} * u \, dx$. Moreover,

$$\begin{aligned}
&\int_{\mathbb{R}} |u|^{p-1}\mathrm{sgn}(u)[G^r_{xx}\chi_{|x-x'|>r/\sqrt{2}}] * u \, dx \leq \|u\|_p^{p-1} \|G^r_{xx}\chi_{|x-x'|>r/\sqrt{2}} * u\|_p \\
&\leq \|u\|_p^{p-1} \|G^r_{xx}\chi_{|x-x'|>r/\sqrt{2}}\|_1 \|u\|_p \leq 2c\|u\|_p^p.
\end{aligned}$$

Therefore, we conclude that $\int_{\mathbb{R}} p|u|^{p-1}\mathrm{sgn}(u)\widetilde{G}^r_{xx} * u \, dx \leq 0$, which implies that $\|u\|_p \leq \|u_0\|_p$ for all $p \geq 1$. Thus, (22) follows. □

Lemma 2. *Let u^μ be the solution to* (19) *with $u_0 \in L^1(\mathbb{R}) \cap L^\infty(\mathbb{R})$ as initial data with $\int_{\mathbb{R}} |u_0(x+h) - u_0(x)| \, dx \leq \omega(|h|)$, for any $h \in \mathbb{R}$, for some nondecreasing function ω on $\mathbb{R}^+$ with $\omega(r) \downarrow 0$ as $r \downarrow 0$. Then there exists a constant C, depending only on $\|u_0\|_\infty$ such that, for any $t > 0$,*

$$\begin{aligned}
&\int_{\mathbb{R}} |u^\mu(x+h,t) - u^\mu(x,t)| \, dx \leq \omega(|h|), \quad \textit{for any } h \in \mathbb{R}, \\
&\int_{\mathbb{R}} |u^\mu(x,t+k) - u^\mu(x,t)| \, dx \leq C(k + k^{2/3} + \mu k^{1/3})\|u_0\|_1 + 4\omega(k^{1/3}),
\end{aligned}$$

for any $k > 0$.

This can be proved along the line of, for example [LM03]. In view of Lemma 2, it follows by classical arguments that the sequence u^μ is compact in L^1_{loc}. More precisely, Theorems 1 and 2 holds.

Theorem 1. *Let u^μ be the solution to* (19) *with $u_0 \in L^1(\mathbb{R}) \cap L^\infty(\mathbb{R})$ as initial datum. Then, as $\mu \downarrow 0$, for any $T > 0$, $u^\mu \to u$ strongly in $L^p_{loc}(\mathbb{R} \times [0,T])$ for $p < \infty$, and $u \in L^1(\mathbb{R} \times [0,T]) \cap L^\infty(\mathbb{R} \times [0,T])$ is an entropy solution to* (18) *in the sense of Definition 1.*

Finally, we also mention that L^1-stability of entropy weak solutions (thus, uniqueness) can be proved along the line of [LM03]. It is also of interest to study the difference between the solutions of (4) and (18). We have the following result.

Theorem 2. *Let u and $\widetilde{u}$ denote entropy weak solutions, respectively, of* (4) *and* (18) *with initial data $u_0 \in BV(\mathbb{R})$. Then, for any $t > 0$,*

$$\|u(\cdot,t) - \tilde{u}(\cdot,t)\|_{L^1(\mathbb{R})} \le \frac{2\sqrt{2}}{3\sqrt{3}} \varepsilon r t e^{2\varepsilon t} \|u_0\|_{BV(\mathbb{R})}.$$

Proof. Let $e = u - \widetilde{u}$, where u and $\widetilde{u}$ are viscous approximations to (4) and (18). Then $e_t + (u^2 - \widetilde{u}^2)_x = \varepsilon \widetilde{G}^r_x * e_x + \varepsilon[G^r_x - c(1-2H)]\chi_{|x-x'|<r/\sqrt{2}} * u_x + \mu e_{xx}$. Along the line of Lemma 1 we get

$$\begin{aligned}\int_{\mathbb{R}} e(x,t)^+ \, dx \le \int_{\mathbb{R}} e_0(x)^+ \, dx + \varepsilon \int_0^t \int_{\mathbb{R}} \operatorname{sgn}(e)^+ \widetilde{G}^r_{xx} * e \, dx d\tau \\ + \varepsilon \int_0^t \int_{\mathbb{R}} \operatorname{sgn}(u - \bar{u})^+ [G^r_x - c(1-2H)]\chi_{|x-x'|<r/\sqrt{2}} * u_x \, dx d\tau.\end{aligned}$$

Here $\int_0^t \int_{\mathbb{R}} \operatorname{sgn}(e)^+ \widetilde{G}^r_{xx} * e \, dx d\tau \le 0$, whereas the last term is estimated as follows

$$\begin{aligned}&\left| \int_0^t \int_{\mathbb{R}} \operatorname{sgn}(u - \bar{u})^+ [G^r_x - c(1-2H)]\chi_{|x-x'|<r/\sqrt{2}} * u_x \, dx d\tau \right| \\ &\le \left\| [G^r_x - c(1-2H)]\chi_{|x-x'|<r/\sqrt{2}} \right\|_{L^1(\mathbb{R})} \int_0^t \|u_x\|_{L^1(\mathbb{R})} d\tau \\ &\le \sqrt{2} c r t e^{2\varepsilon t} \|u_0\|_{BV(\mathbb{R})},\end{aligned}$$

where we have used that $\|[G^r_x - c(1-2H)]\chi_{|x-x'|<r/\sqrt{2}}\|_{L^1(\mathbb{R})} \le 2G^r_x(r/\sqrt{2})r/\sqrt{2} = \sqrt{2}cr$ and $\|u(\cdot,t)\|_{BV(\mathbb{R})} \le e^{2\varepsilon t}\|u_0\|_{BV(\mathbb{R})}$, taken from Lemma 4.2 in [EK06]. □

References

[B88] J. Bear: Dynamics of Fluids in Porous Media. Dover Publications, Inc, New York, 1988.

[EK06] S. Evje and K.H. Karlsen: Hyperbolic-elliptic models for well-reservoir flow. Network Heterogenous Media, Vol **4**, 441–471, 2006.

[KN99] S. Kawashima and S. Nishibata: Shock waves for a model system of the radiating gas. SIAM J. Math. Anal., **30** (1), 95–117, 1999.

[LM03] C. Lattanzio and P. Marcati: Global well-posedness and relaxation limits of a model for radiating gas. J. Diff. Equations, **190**, 439–465, 2003.

[LT01] H. Liu and E. Tadmor: Critical thresholds in a convolution model for nonlinear conservation laws. SIAM J. Math. Anal., **33** (4), 930–945, 2001.

[N00] S. Nishibata: Asymptotic behavior of solutions to a model system of radiating gas with discontinuities initial data. Math. Models Methods Appl. Sci., **10** (8), 1209–1231, 2000.

[S03] D. Serre: L^1-stability of constants in a model for radiating gases. Comm. Math. Sci., **1** (1), 197–205, 2003.

[S04] D. Serre: L^1-stability of nonlinear waves in scalar conservation laws. Evolutionary equations. Vol. I, 473–553, Handb. Differ. Equ. (C. Dafermos, E. Feireisl eds.), North-Holland, Amsterdam, 2004.

High-Resolution Finite Volume Methods for Extracorporeal Shock Wave Therapy

K. Fagnan, R.J. LeVeque, T.J. Matula, and B. MacConaghy

1 Introduction

Extracorporeal Shock Wave Therapy (ESWT) is a noninvasive technique for the treatment of a variety of musculoskeletal conditions such as delayed union of bone fractures, plantar fasciitis, and calcified tendonitis of the shoulder. Shock waves were first used medically in the lithotripsy procedure (ESWL) to pulverize hardened calcified deposits such as kidney stones. This technique was then extended to musculoskeletal conditions as a treatment for calcifications in the shoulder, as these deposits are similar to renal calculi [1]. Later ESWT was shown to improve bone regeneration in the treatment of nonunions, which are bone fractures that fail to heal over time. In electrohydraulic lithotripsy, a shock wave is generated in a liquid bath, focused through the use of an ellipsoid reflector, and then propagates into the body where it ideally strikes the area of interest. Clinicians have indicated that this is not always the case and treatments will sometimes cause unforseen damage in parts of the body. Current numerical models are limited to simplified situations because the structure of the wave is highly nonlinear and therefore difficult to model with traditional finite difference and finite element techniques. We utilize high-resolution finite volume methods to capture the discontinuous pressure wave and model the wave propagation in bone and tissue. This approach has been successfully applied to many problems in acoustic or elastic wave propagation in heterogeneous media [2].

We are currently modeling shock wave propagation and reflection with linear elasticity equations, using finite volume methods in which each grid cell has material parameters associated to that cell. We use a Godunov-type method, which utilizes Riemann solvers in the wave propagation algorithm to solve the system of partial differential equations. Results were obtained in two- and three-dimensions using the CLAWPACK [3] and CHOMBOCLAW [4] software packages. To provide validation of these results, our modeling effort is compared to experiments.

2 Background

2.1 Extracorporeal Shock Wave Therapy

The basic physics of the shock wave generation and propagation are well understood. A shock wave is a rapidly traveling pressure disturbance and is characterized by a sudden rise from ambient pressure to the maximum pressure in the system. As pressure increases, the velocity of the sound wave increases and therefore wave components at higher pressure move faster than those at lower pressure. Since the velocity of the wave components is dependent upon the pressure, the wave deforms and steepens into a shock.

Electrohydraulic lithotripters have a spark plug source that generates a spherical shock wave pulse in water, which can be modeled as an underwater explosion. The shock wave reflects off an ellipsoidal reflector, and propagates through the fluid into the body. Attenuation is limited through the use of a coupling gel between the membrane and skin. When the shock wave enters the body it will be reflected and dissipated depending upon the material parameters of the tissues that are in its path. Eventually the wave will collide with the intended target centered at the second focus of the ellipse, as is illustrated in Fig. 1b. An essential property of ESWL and ESWT is that, due to shock wave focusing, maximal energy should be deposited in the treatment area centered at the second focus of the ellipsoid.

A typical shock wave has a high peak pressure of approximately 50–80 MPa with a short life cycle of approximately 10 ns, followed by a negative pressure wave that lasts for a few microseconds [1]. The diagram in Fig. 1a illustrates a typical wave form and its physical parameters as it would appear at the second focus. The tensile region is of particular interest, as it generates cavitation bubbles. This effect is one of the primary sources of energy leading to stone fragmentation in ESWL and is thought to contribute to the angiogenic response observed during ESWT treatments [5],[6]. Cavitation is not

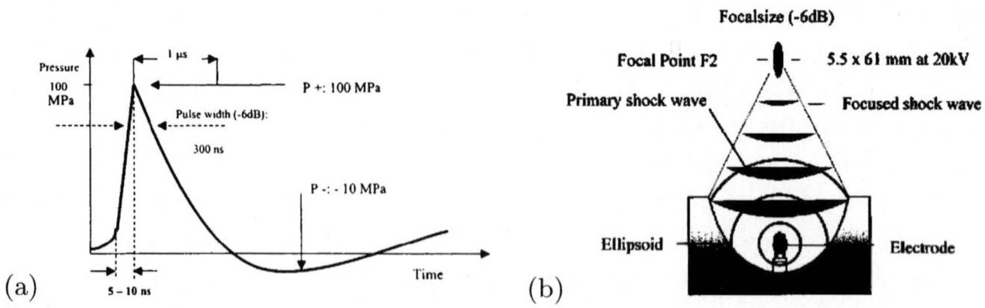

Fig. 1. Basic components of lithotripsy treatment. (**a**) The standard pressure wave profile at the second focus in lithotripsy and illustration of high pressure pulse followed by negative pressure or tensile region. (**b**) The basic setup for an electrohydraulic lithotripter. There is an ellipsoid reflector made of brass and a spark plug shock wave source at the F1 focus [1]

currently accounted for in our model. However, we have demonstrated with our results in Sect. 3 that it is possible to numerically calculate regions of maximal tension, which correspond to areas of cavitation in the laboratory experiments.

2.2 Mathematical Model

There are generally two types of waves that can propagate in an elastic solid, namely P-waves (pressure or primary waves) and S-waves (shear or secondary waves). In one-dimensional problems, these waves can be modeled by disjoint systems of equations. In higher dimensional problems, there is a coupling between P-waves and S-waves, and so the situation is more complex. We aim to model the behavior of these waves when propagating through soft tissue and bone in the body.

For the preliminary results presented here, we have used the 3D linear elasticity equations. We assume the soft tissue and bone are each isotropic materials and hence the equations have the form

$$\begin{aligned}
\sigma_t^{11} - (\lambda + 2\mu)u_x - \lambda v_y - \lambda w_z &= 0\\
\sigma_t^{22} - \lambda u_x - (\lambda + 2\mu)v_y - \lambda w_z &= 0\\
\sigma_t^{33} - \lambda u_x - \lambda v_y - (\lambda + 2\mu)w_z &= 0\\
\sigma_t^{12} - \mu(v_x + u_y) &= 0\\
\sigma_t^{23} - \mu(v_z + w_y) &= 0\\
\sigma_t^{31} - \mu(u_z + w_x) &= 0\\
\rho u_t - \sigma_x^{11} - \sigma_y^{12} - \sigma_z^{13} &= 0\\
\rho v_t - \sigma_x^{12} - \sigma_y^{22} - \sigma_z^{23} &= 0\\
\rho w_t - \sigma_x^{13} - \sigma_y^{23} - \sigma_z^{33} &= 0.
\end{aligned}$$

Here σ^{ij} represents the components of the stress tensor, $u = [u, v, w]$ are the components of velocity, and λ and μ are the Lame parameters. We use this nonconservative form of the elasticity equations to determine an analytical solution to the Riemann problem, which is the basis for the Reconstruct Evolve Average (REA) algorithm used in CLAWPACK [3].

The numerical tests presented here are compared to laboratory experiments where acrylic or acrylonitrile butadiene styrene (ABS) objects are immersed in water. We use the following values for the parameters:

$$\begin{array}{llll}
\text{water}: & \rho = 1,000\,\text{kg/m}^3, & \lambda = 2,190.4\,\text{MPa}, & \mu = 0\,\text{MPa},\\
\text{acrylic}: & \rho = 1,850\,\text{kg/m}^3, & \lambda = 9,305\,\text{MPa}, & \mu = 3,126.5\,\text{MPa},\\
\text{ABS}: & \rho = 1,210\,\text{kg/m}^3, & \lambda = 3,264\,\text{MPa}, & \mu = 1,032\,\text{MPa}.
\end{array}$$

Each grid cell is assigned values of ρ, λ, and μ depending on which material it is in, or to averaged values of these parameters if an interface cuts through the cell.

This isotropic linear model is only a starting point. Modeling the shock wave formation and propagation more accurately requires using a nonlinear formulation of the elasticity equations, and ideally one would also incorporate anisotropic effects for soft tissue and bone. We are currently working on extending the model, but even this linear model gives results that capture some of the main features of experiments, as illustrated in the next section.

3 Results

We have been able to make comparisons between two- and three-dimensional laboratory and numerical experiments through an interdisciplinary collaboration at the University of Washington. Some of these results will be summarized here. Our collaborators at The Applied Physics Lab (APL) at the University of Washington have performed several experiments with device similar to an HM3 Dornier lithotripsy device that we have used to verify our numerical model [7]. In our numerical experiments we have solved the linear elasticity equations using high-resolution finite volume methods with adaptive mesh refinement.

3.1 Stress Wave Experiment

Figure 2 shows the results of an experiment where the stress waves in an acrylic cylinder are imaged using a high-speed camera and polarized filters. Acrylic is used because it is birefringent, allowing us to observe interior stress waves using polarizing optics [8]. The focus for the treatment is at the right

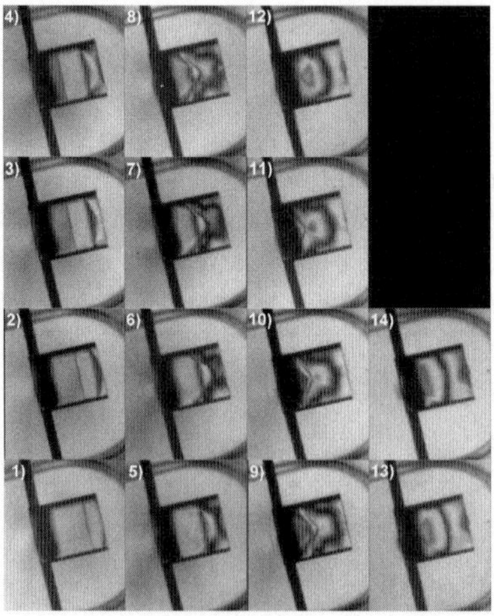

Fig. 2. Laboratory experiment where a shock wave is generated and propagates from right to left through an acrylic cylinder. The images are numbered to show the sequence in time. The stress waves are visualized using polarized film and images were taken with a high-speed camera (courtesy of B. MacConaghy)

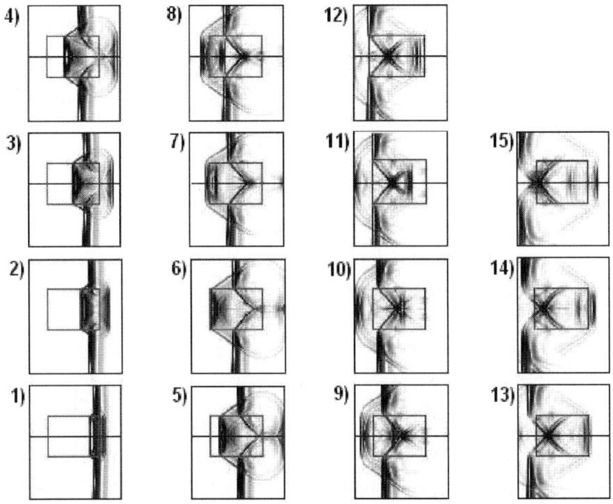

Fig. 3. A sequence of images from a numerical experiment for comparison with the laboratory result from Fig. 2. The pressure wave propagates through the acrylic cylinder and the conical-shaped, high-stress region develops in the simulation at a similar time and location to that of the lab experiment

edge of the cylinder and the shock wave propagates from right to left. There is a reflected wave resulting from interaction with the left edge of the cylinder. The photos show a conical shaped region where the stress is maximal. The location of the conical region agrees roughly with the fracture location of materials similar to kidney stones.

Figure 3 shows the numerical result of a pressure wave propagating through water and into an acrylic cylinder. We solve an axisymmetric version of the elasticity equations to generate these results. When we compare the numerical with the laboratory results in Fig. 2, it is clear that our model has captured both the conical region of maximal stress and the reflection of the wave off the left edge of the cylinder. This indicates that the pressure wave in the solid is behaving like a linear elastic wave.

3.2 Cavitation Field Experiments

The primary objective of these experiments was to investigate the behavior of the shock wave as it interacts with a 3D representation of a talus bone. The bone is composed of ABS and was created by a group in the Mechanical Engineering Department at the University of Washington [9]. The three-dimensional digital representation of the bone that is used in the numerical experiments (see Fig. 4) was provided by the same group. Therefore, we can make direct comparisons of our numerical results with the laboratory experiments.

The series of images in Fig. 5 shows the deflection of the trail of cavitation bubbles or tensile component of the shock wave, as a function of bone

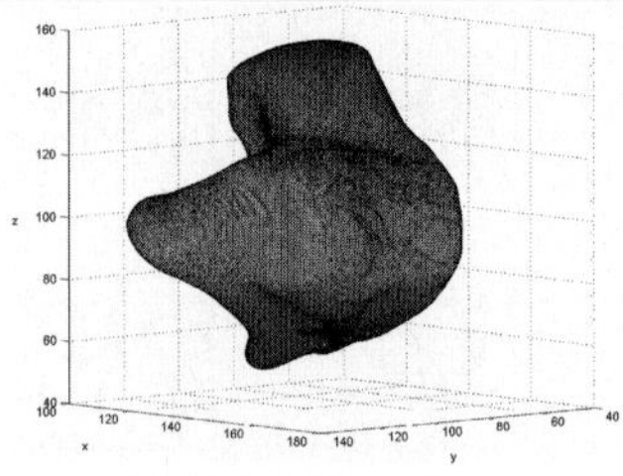

Fig. 4. A digital 3D representation of the talus used in the laboratory and numerical experiments [9]

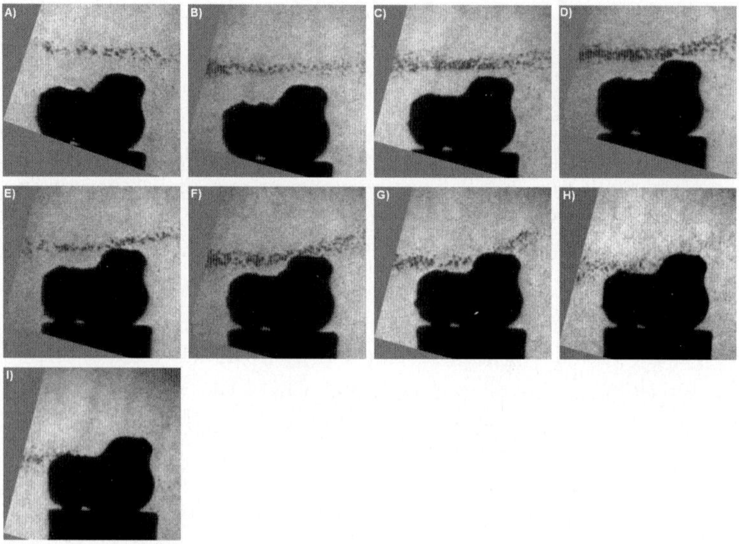

Fig. 5. Laboratory experiment where a shock wave is generated in the far left and the tensile component of the wave creates a cavitation field. The interaction of the cavitation field is imaged with a high-speed camera. The bone height changes from left to right where height corresponds to difference from the focus (**A**, −12; **B**, −9; **C**, −6; **D**, −3; **E**, 0; **F**, 3; **G**, 6; **H**, 9; **I**, 12) (courtesy of T. Matula)

placement within the field. As the location of the bone changes with respect to the focus, the deflection angle of the cavitation field is altered by the geometry of the bone. In some cases the cavitation field continues far past the desired treatment area and this could potentially cause tissue damage outside the focal region. The numerical results in Fig. 6 show the maximum tensile component of the pressure wave along $y = 0$, a two-dimensional slice of the full 3D calculation. Each image shows the deflection of the shock wave's tensile component with different bone locations. The dot is the location of F2 and the bone is shifted in the z-direction relative to the focus as in Fig. 5. The

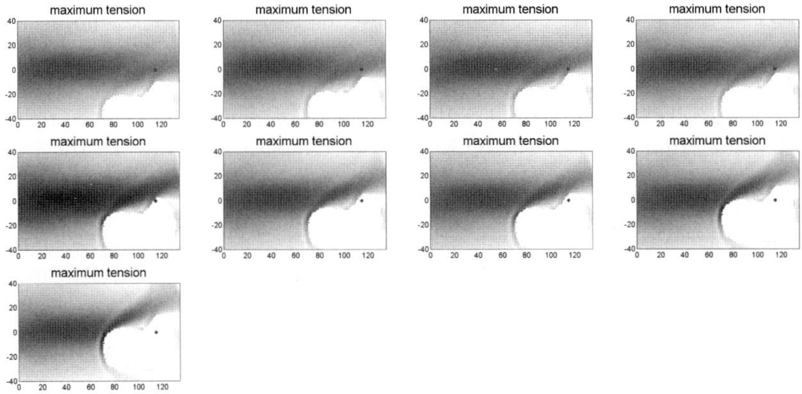

Fig. 6. Numerical simulation of laboratory experiment in Fig. 5. These results demonstrate a dependence of the deflection angle on the placement of the bone with respect to F2. The blue dot is F2 and in each photo the bone placement is changing in z with respect to F2 (**A**, −12; **B**, −9; **C**, −6; **D**, −3; **E**, 0; **F**, 3; **G**, 6; **H**, 9; **I**, 12). Although the pressure wave is more smeared out than in the experiment, there is more interaction with the front of the bone than in the laboratory experiment, the general behavior is correct

regions of maximum tension indicate where cavitation would occur, thus, this is the basis for comparison.

Because of the linear nature of the equations and computational limitations on the level of refinement, the pressure wave is more smeared out than in the laboratory experiment. As a result, the pressure wave interacts more with the front of the bone than the more focused laboratory shock wave. The deflection of the wave path due to the bone geometry is different from the laboratory result; however, the general behavior is similar in that there is an upward deflection of the cavitation path.

4 Conclusion

ESWT is a treatment for musculoskeletal conditions with its foundation in lithotripsy. Though clinical trials have indicated that this treatment is successful for a variety of conditions, the biological mechanisms are not well understood. We hope to model the propagation of the shock wave in bone and tissue so as to better understand this treatment. Our future goals include investigation on where energy is being deposited in the body and whether or not reflections from complicated geometry result in refocusing of the pressure wave.

We have used high-resolution finite volume methods and adaptive mesh refinement routines implemented in the CLAWPACK and CHOMBOCLOAW software packages to solve the equations of linear elasticity in the isotropic materials. Results have been compared to two laboratory experiments where it is crucial to properly capture wave reflection at interfaces. Our preliminary results show reasonable qualitative agreement. In future work we plan to make more quantitative comparisons and to extend the model with the inclusion of nonlinear and nonisotropic effects to more accurately capture the ESWT pressure wave and the complex nature of bone and tissue.

Acknowledgments

This work was supported in part by DOE grant DE-FC02-01ER25474, NSF grant DMS-0106511, NIH grant PO1-DK043881, and an ARCS Foundation Founders Fellowship.

References

1. Anna Toth-Kischkat John A. Ogden and Reiner Schultheiss. Principles of shock wave therapy. *Clinical Orthopaedics and Related Research*, 387:8–17, 2001.
2. R.J. LeVeque. High-resolution finite-volume methods for acoustic waves in periodic and random media. *J. Acoust. Soc. Am.*, 106:17–28, July, 1999.
3. R.J. LeVeque. Clawpack software. http://www.amath.washington.edu/∼claw.
4. D. Calhoun. Chomboclaw software and documentation. http://www.amath.washington.edu/∼ calhoun/demos/chomboclaw/.
5. Robin O. Cleveland and Oleg A. Sapozhinikov. Modeling elastic wave propagation in kidney stones with application to shock wave lithotripsy. *Journal of the Acoustical Society of America*, 118(4):2667–2676, 2005.
6. H.Y. Huang, C.J. Wang, and C.H. Pai. Shock wave enhances neovascularization at the tendon-bone junction. *Journal of Foot and Ankle Surgery*, 41:16–22, 2002.
7. Robin O. Cleveland et al. Design and characterization of a research electrohydraulic lithotripter patterned after the dornier hm3. *Review of Scientific Instruments*, 71(6):2514–2525, 2000.
8. Xufeng Xi and Pe Zhong. Dynamic photoelastic study of the transient stress field in solids during shock wave lithotripsy. *Journal of the Acoustical Society of America*, 9(3):1226–1239, 2001.
9. D. Stuarti, R.P. Ching, M. Ganter and Y. Hu. Mechanical engineering department. http://design.me.washington.edu/.

Asymptotic Properties of a Class of Weak Solutions to the Navier–Stokes–Fourier System

E. Feireisl

1 Introduction

1.1 Preliminaries

Many equations arising in continuum fluid dynamics do not, or at least are not known to, possess smooth solutions for general data. Therefore it is necessary to identify a larger class of "weak solutions" in order to develop a rigorous mathematical theory. Following the seminal paper of Leray [10] we introduce a class of weak solutions to the Navier–Stokes–Fourier system based on the concept of generalized derivatives (distributions). The main objective of the present paper is to illustrate the strength of the abstract theory reviewing several theoretical results on the asymptotic behavior of the weak solutions that are in good agreement with practical experiments as well as their numerical analysis. The main topics to be discussed include the following:

- A rigorous global-in-time existence theory for arbitrary large finite-energy data
- Asymptotic behavior of solutions for large time, convergence towards equilibrium states
- Scale analysis, in particular, the low Mach and Froude number limits and their relation to the so-called Oberbeck-Boussinesq system.

1.2 Weak Solutions

We shall assume that the state of a fluid at a given time $t \in (0,T)$ and a spatial position $x \in \Omega \subset R^3$ is characterized by the density $\varrho = \varrho(t,x)$, the velocity field $\mathbf{u} = \mathbf{u}(t,x)$, and the absolute temperature $\vartheta = \vartheta(t,x)$.

Pursuing the general strategy delineated above, we express the physical principle of *conservation of mass* through the integral identity

$$\int_0^T \int_\Omega \Big(\varrho \partial_t \varphi + \varrho \mathbf{u} \cdot \nabla \varphi \Big) \, \mathrm{d}x \, \mathrm{d}t = - \int_\Omega \varrho_0 \varphi(0, \cdot) \, \mathrm{d}x \tag{1}$$

to be satisfied for any test function $\varphi \in \mathcal{D}([0,T) \times \overline{\Omega})$. Note that (1) includes the satisfaction of the initial condition

$$\varrho(0,\cdot) = \varrho_0 \text{ in } \Omega \tag{2}$$

together with the impermeability boundary condition

$$\mathbf{u} \cdot \mathbf{n} = 0 \text{ on } \partial\Omega, \tag{3}$$

where the symbol $\mathbf{n}$ stands for the outer normal vector.

For future use, it is more convenient to write (1) in the renormalized form in the sense of DiPerna and Lions [2]:

$$\int_0^T \int_\Omega \Big(\varrho B(\varrho)\partial_t \varphi + \varrho B(\varrho)\mathbf{u}\cdot\nabla\varphi - b(\varrho)\nabla\cdot\mathbf{u}\varphi \Big) \, \mathrm{d}x \, \mathrm{d}t = - \int_\Omega \varrho_0 B(\varrho_0)\varphi(0,\cdot) \, \mathrm{d}x \tag{4}$$

for any $\varphi \in \mathcal{D}([0,T) \times \overline{\Omega})$, and any

$$b \in BC[0,\infty), \ B(\varrho) = B(1)\varrho + \int_1^\varrho \frac{b(z)}{z^2} \, \mathrm{d}z.$$

Similarly, *(angular) momentum balance* takes the form

$$\int_0^T \int_\Omega \Big(\varrho\mathbf{u} \cdot \partial_t \varphi + \varrho\mathbf{u} \otimes \mathbf{u} : \nabla\varphi + \frac{1}{\mathrm{Ma}^2} p \nabla \cdot \varphi \Big) \, \mathrm{d}x \, \mathrm{d}t \tag{5}$$

$$= \int_0^T \int_\Omega \Big(\mathbb{S} : \nabla\varphi - \frac{1}{\mathrm{Fr}^2} \varrho \nabla F \cdot \varphi \Big) \, \mathrm{d}x \, \mathrm{d}t - \int_\Omega \varrho_0 \mathbf{u}_0 \cdot \varphi \, \mathrm{d}x$$

for any $\varphi \in \mathcal{D}([0,T); \mathcal{D}(\Omega; R^3))$. Here, the symbol p denotes the pressure, $\mathbb{S}$ stands for the viscous stress tensor, and ∇F is the external force, to be determined below. The symbols Ma and Fr denote dimensionless parameters – the Mach and Froude number, respectively.

Similar to (1), formula (5) already "includes" the initial condition

$$\varrho\mathbf{u}(0,\cdot) = \varrho_0 \mathbf{u}_0 \text{ in } \Omega. \tag{6}$$

On the other hand, however, the class of test functions φ used in (5) is not rich enough in order to control the behavior of $\mathbf{u}$ on the boundary $\partial\Omega$. Instead, we shall assume integrability of $\nabla\mathbf{u}$ on the whole set $(0,T) \times \Omega$, in particular, the trace $\mathbf{u}|_{\partial\Omega}$ can be defined at least for a regular (Lipschitz) boundary (see e.g. Evans and Gariepy [3]). Accordingly, we are allowed to prescribe the no-slip boundary conditions

$$\mathbf{u}|_{\partial\Omega} = 0, \tag{7}$$

relevant for (linearly) viscous fluids considered in the present paper.

Mainly for technical reasons, the energy conservation principle will be replaced by *entropy balance* expressed through integral identity

$$\int_0^T \int_\Omega \Big(\varrho s \partial_t \varphi + \varrho s \mathbf{u} \cdot \nabla \varphi + \frac{\mathbf{q}}{\vartheta} \cdot \nabla \varphi \Big) \, \mathrm{d}x \, \mathrm{d}t + < \Phi, \varphi > = - \int_\Omega \varrho_0 s_0 \varphi(0, \cdot) \, \mathrm{d}x \tag{8}$$

for any $\varphi \in \mathcal{D}([0,T) \times \overline{\Omega})$, supplemented with *total energy conservation*

$$E(t) = \int_\Omega \Big(\mathrm{Ma}^2 \varrho |\mathbf{u}|^2 + \varrho e - \frac{\mathrm{Ma}^2}{\mathrm{Fr}^2} \varrho F \Big) \, \mathrm{d}x = E_0 \text{ for a.a. } t \in (0,T). \tag{9}$$

Here, the specific internal energy $e = e(\varrho, \vartheta)$ and the specific entropy $s = s(\varrho, \vartheta)$ are related to the pressure $p = p(\varrho, \vartheta)$ through *Gibbs' equation*

$$\vartheta D s(\varrho, \vartheta) = D e(\varrho, \vartheta) + p(\varrho, \vartheta) D \Big(\frac{1}{\varrho} \Big). \tag{10}$$

Accordingly, we set

$$s_0 = s(\varrho_0, \vartheta_0), E_0 = \int_\Omega \Big(\frac{\mathrm{Ma}^2}{2} \varrho_0 |\mathbf{u}_0|^2 + \varrho_0 e(\varrho_0, \vartheta_0) - \frac{\mathrm{Ma}^2}{\mathrm{Fr}^2} \varrho_0 F \Big) \, \mathrm{d}x. \tag{11}$$

Further in (8), the symbol Φ denotes the effective entropy production – a nonnegative measure on the set $[0,T] \times \overline{\Omega}$ satisfying

$$< \Phi, \varphi > \geq \int_0^T \int_\Omega \frac{1}{\vartheta} \Big(\mathrm{Ma}^2 \mathrm{S} : \nabla \mathbf{u} - \frac{\mathbf{q} \cdot \nabla \vartheta}{\vartheta} \Big) \varphi \, \mathrm{d}x \, \mathrm{d}t \tag{12}$$

for any nonnegative $\varphi \in C([0,T] \times \overline{\Omega})$, where $\mathbf{q}$ is the heat flux to be determined below.

If all quantities appearing in (1–12) are sufficiently smooth, it is a routine matter to check that Φ is absolutely continuous with respect to the Lebesgue measure, and

$$\Phi = \frac{1}{\vartheta} \Big(\mathrm{Ma}^2 \mathrm{S} : \nabla \mathbf{u} - \frac{\mathbf{q} \cdot \nabla \vartheta}{\vartheta} \Big).$$

In addition, we recover the initial condition

$$\varrho s(0, \cdot) = \varrho_0 s_0 \tag{13}$$

together with

$$\mathbf{q} \cdot \mathbf{n}|_{\partial \Omega} = 0 \tag{14}$$

(see Remark 2.3 in [7]).

1.3 Constitutive Relations

We shall assume that the fluid in question is a real monoatomic gas, for which

$$p(\varrho,\vartheta) = p_F(\varrho,\vartheta) + p_R(\vartheta), \tag{15}$$

where p_R is the radiation pressure,

$$p_R = \frac{a}{3}\vartheta^4,\ a > 0, \tag{16}$$

and

$$e(\varrho,\vartheta) = e_F(\varrho,\vartheta) + e_R(\varrho,\vartheta),\ e_R = a\frac{\vartheta^4}{\varrho}, \tag{17}$$

with

$$p_F(\varrho,\vartheta) = \frac{2}{3}\varrho e_F(\varrho,\vartheta) \tag{18}$$

(see [5]).

In addition, we impose the standard thermodynamics stability hypotheses

$$\frac{\partial p_F(\varrho,\vartheta)}{\partial \varrho} > 0,\ c_v = \frac{\partial e_F(\varrho,\vartheta)}{\partial \vartheta} > 0 \text{ for all } \varrho,\vartheta > 0. \tag{19}$$

Furthermore, we introduce a technical but still physically relevant stipulations

$$\overline{\lim_{\vartheta\to 0}}\, c_v(\varrho,\vartheta) < \infty \text{ for any fixed } \varrho,\ \lim_{\varrho\to 0}\frac{\partial p_F(\varrho,\vartheta)}{\partial\varrho} > 0 \text{ for any fixed } \vartheta. \tag{20}$$

Finally, we shall assume that the fluid is linearly viscous, with

$$\mathbb{S} = \mu\Big(\nabla\mathbf{u} + \nabla^t\mathbf{u} - \frac{2}{3}\nabla\cdot\mathbf{u}\mathbb{I}\Big) + \eta\nabla\cdot\mathbf{u}\mathbb{I}, \tag{21}$$

and that the heat flux is given by Fourier's law

$$\mathbf{q} = -\kappa\nabla\vartheta. \tag{22}$$

The transport coefficients $\mu = \mu(\vartheta)$, $\eta = \eta(\vartheta)$, $\kappa = \kappa(\vartheta)$ will be effective functions of the absolute temperature satisfying the growth restrictions

$$0 < c_1(1+\vartheta^\alpha) \le \mu(\vartheta),\ \eta(\vartheta) \ge 0,\ \mu(\vartheta),\eta(\vartheta) \le c_2(1+\vartheta^\alpha), \tag{23}$$

$$0 < c_1(1+\vartheta^3) \le \kappa(\vartheta) \le c_2(1+\vartheta^3) \tag{24}$$

(see [7]).

2 Large Data Existence Theory

Finding a proper notion of generalized solution for a given nonlinear problem is an important but also very subtle task. The main advantage of the weak formulation introduced in Sect. 1.2 is the fact that problem (1–24) is solvable on an arbitrary time interval $(0, T)$ and for any choice of the data ϱ_0, $\mathbf{u}_0$, ϑ_0 for which the initial energy E_0 is finite. More specifically, we report the following result.

Theorem 1. *Suppose that $\Omega \subset R^3$ is a bounded domain with a boundary of class $C^{2+\nu}$, $\nu > 0$. Let the initial data ϱ_0, $\mathbf{u}_0$, ϑ_0 be bounded measurable functions on Ω,*

$$\varrho_0(x) > 0,\ \vartheta_0(x) > \underline{\vartheta} > 0 \text{ for a.a. } x \in \Omega.$$

Furthermore, assume that

$$\frac{2}{5} < \alpha \leq 1,$$

where α is the exponent appearing in (23).

Then problem (1–24) admits a solution ϱ, $\mathbf{u}$, ϑ such that

$$\varrho \in L^\infty(0, T; L^{\frac{5}{3}}(\Omega)),\ \varrho \geq 0 \text{ a.a. on } \Omega,$$

$$\mathbf{u} \in L^2(0, T; W_0^{1,s}(\Omega; R^3)),\ s = \frac{8}{5 - \alpha},$$

$$\vartheta \in L^\infty(0, T; L^4(\Omega)) \cap L^2(0, T; W^{1,2}(\Omega)),\ \vartheta > 0,\ \log(\vartheta) \in L^2(0, T; W^{1,2}(\Omega)).$$

Theorem 1 can be proved by the method developed in [7], with the necessary modifications introduced in [4].

3 Long-Time Behavior of Solutions

3.1 Autonomous Case

A time varying solution to problem (1–24) can be viewed as a trajectory or curve in an infinite dimensional space. The main objective of the present section is to show that the piece of information available for the weak solutions, the existence of which is guaranteed by Theorem 1, is sufficient for studying their behavior for $t \to \infty$.

We start with the simplest case when the driving force is a gradient ∇F, with $F = F(x)$ independent of time. In such a situation, in accordance with the general Prigogine's principle, the system should approach the state that maximizes the entropy, which means, the static state

$$\varrho = \varrho_s(x),\ \mathbf{u} = 0,\ \vartheta = \overline{\vartheta} = \text{const} > 0,\ \nabla p(\varrho_s, \overline{\vartheta}) = \varrho_s \nabla F \text{ in } \Omega. \tag{25}$$

Given the two constants of motion:

$$\text{The total mass } m = \int_\Omega \varrho \, \mathrm{d}x \text{ and the total energy } E_0,$$

one can show that problem (25) admits a unique positive solution ϱ_s, $\overline{\vartheta}$ such that

$$\int_\Omega \varrho_s \, \mathrm{d}x = m, \ \int_\Omega \Big(\varrho_s e(\varrho_s, \overline{\vartheta}) - \varrho_s F \Big) \, \mathrm{d}x = E_0 \tag{26}$$

(see Březina [1]).

A complete description of the long-time behavior of solutions in the autonomous cases is provided by the following result (see Theorem 3.2 in [9]).

Theorem 2. *In addition to the hypotheses of Theorem 1 assume that* Ma $=$ Fr $= 1$, $F \in W^{1,\infty}(\Omega)$. *Let* $\{\varrho, \mathbf{u}, \vartheta\}$ *be a solution of problem (1–24) defined on the time interval* $(0, \infty)$.

Then

$$\varrho \mathbf{u}(t) \to 0 \text{ in } L^1(\Omega; R^3) \text{ as } t \to \infty,$$

$$\varrho(t) \to \varrho_s \text{ in } L^1(\Omega; R^3) \text{ as } t \to \infty,$$

and

$$\int_\Omega \varrho s(\varrho, \vartheta)(t) \, \mathrm{d}x \to \int_\Omega \varrho_s s(\varrho_s, \overline{\vartheta}) \, \mathrm{d}x \text{ as } t \to \infty,$$

where ϱ_s, $\overline{\vartheta}$ *is the unique positive solution of static problem (25) and (26).*

3.2 General Driving Force

In this section, we shall assume that system (1–24) is driven by a general force $\mathbf{f} = \mathbf{f}(t, x)$, specifically, setting Ma $=$ Fr $= 1$ we replace ∇F by f in (5) while (9) is converted to

$$\int_0^\infty E(t) \partial_t \psi \, \mathrm{d}t + \int_0^\infty \int_\Omega \varrho \mathbf{f} \cdot \mathbf{u} \, \mathrm{d}x \, \psi \, \mathrm{d}t = 0 \text{ for any } \psi \in \mathcal{D}(0, \infty), \tag{27}$$

where

$$E(t) = \int_\Omega \Big(\varrho |\mathbf{u}|^2 + \varrho e \Big) \, \mathrm{d}x.$$

We start with an auxiliary result stating that the global energy $E(t)$ is either bounded or blows up for $t \to \infty$ (the phenomenon called sometimes "grow-up").

Theorem 3. *In addition to the hypotheses of Theorem 1, assume* ∇F *being replaced by* $\mathbf{f}$ *in (5), and (27) instead of (9), where*

$$\mathbf{f} \in L^\infty(0, \infty; L^\infty(\Omega; R^3)).$$

Let $\{\varrho, \mathbf{u}, \vartheta\}$ *be a variational solution of problem (1–24) on the time interval* $I = (0, \infty)$.

Then either **(i)**

$$E(t) \to \infty \textit{ for } t \to \infty,$$

or **(ii)** *there is a constant* E_∞ *such that*

$$\operatorname*{ess\,sup}_{t>0} E(t) \leq E_\infty.$$

Furthermore, in the latter case, any sequence of times $\tau_n \to \infty$ *contains a subsequence such that*

$$\mathbf{f}_n \to \nabla F \textit{ weakly-(*) in } L^\infty(0, T; L^\infty(\Omega; R^3))$$

for any $T > 0$, *where we have set*

$$\mathbf{f}_n(t, x) = \mathbf{f}(t + \tau_n, x),$$

and where the limit function $F = F(x)$, $F \in W^{1,\infty}(\Omega)$ *may depend on the choice of* $\{\tau_n\}$.

The proof of Theorem 3 can be found in [9, Theorem 3.1]. One can read from Theorem 3 that the energy grows up to infinity for a "generic" set of driving forces. Indeed any small perturbation of the form

$$\mathbf{f} + \epsilon\mathbf{g}(t, x), \; 0 < \epsilon << 1,$$

where $\mathbf{g}$ is, say, a time periodic (nonconstant) function, causes the total energy of the system to diverge to infinity with growing time.

Another even more striking consequence of the previous results reads as follows (see Corollary 3.2 in [9]).

Theorem 4. *In addition to the hypotheses of Theorem 3, assume that the driving force* $\mathbf{f} \in BC(R; C(\overline{\Omega}; R^3))$ *is almost periodic with respect to* t.

Then either (i) $E(t) \to \infty$ *or (ii)* $\mathbf{f} = \nabla F$, $F = F(x)$, *and any solution to problem (1–24) tends to a single static state as established in Theorem 2.*

4 The Asymptotic Limit for Low Mach and Froude Numbers

4.1 The Oberbeck-Boussinesq Approximation

The Oberbeck-Boussinesq approximation is used when the density of a fluid is nearly constant but the density differences exist due to temperature changes, causing an imbalance of the hydrostatic equilibrium. Such a situation occurs with many convection problems where the temperature differences are introduced independent of the flow dynamics. The requirement that the Mach number tends to zero is needed, allowing the density ϱ to approach a constant $\overline{\varrho}$ except in the gravitational body force ∇F rescaled by a suitable choice of the Froude number Fr. Accordingly, the temperature differences are not caused by the flow but exist independent of the flow (see the expository paper by Zeytounian [14]).

To fix ideas and keep the presentation clear, we consider a very simple geometry of the underlying physical space, namely we take

$$\Omega = \Big([-\pi,\pi]|_{\{-\pi,\pi\}} \Big)^2 \times (0,\pi), \tag{28}$$

that is to say that from now on all quantities are assumed to be 2π-periodic with respect to x_1, x_2.

In this framework, the Oberbeck-Boussinesq approximation consists of equation of continuity,

$$\nabla \cdot \mathbf{U} = 0; \tag{29}$$

balance of momentum,

$$\overline{\varrho}\Big(\partial_t \mathbf{U} + \nabla \cdot (\mathbf{U} \otimes \mathbf{U}) \Big) + \nabla P = \nabla \cdot \ \mathrm{S} + r \nabla F, \ \nabla F = [0,0,-1]; \tag{30}$$

and balance of thermal energy,

$$\overline{\varrho}\, \overline{c}_p \Big(\partial_t \Theta + \nabla \cdot (\Theta \mathbf{U}) \Big) - \nabla \cdot (\overline{\kappa} \nabla \Theta) = 0, \tag{31}$$

where $\mathbf{U} = \mathbf{U}(t,x)$ denotes the velocity field, $\Theta = \Theta(t,x)$ is the temperature, and the symbol P stands for the normal stress (pressure).

Similar to (21), the viscous stress tensor S obeys Newton's rheological law

$$\mathrm{S} = \overline{\mu}\Big(\nabla \mathbf{U} + \nabla^t \mathbf{U} \Big), \tag{32}$$

where the viscosity coefficient $\overline{\mu}$ as well as the heat conductivity coefficient $\overline{\kappa}$, and the specific heat at constant pressure $\overline{c}_p$ are evaluated at constant density $\overline{\varrho}$ and constant temperature $\overline{\vartheta} = \frac{1}{|\Omega|} \int_\Omega \Theta \ \mathrm{d}x$. In accordance with the Boussinesq approximation, the temperature-dependent density $r = r(\Theta)$ appears only in the momentum equation and is assumed to vary with temperature as

$$r + \overline{\varrho}\, \overline{\alpha} (\Theta - \overline{\vartheta}) = 0, \tag{33}$$

where $\overline{\alpha}$ stands for the coefficient of thermal expansion (see Zeytounian [14] or Rajagopal et al. [12] for more details on the physical background of the problem).

Consistent with (28), we consider the periodic boundary conditions with respect to the spatial coordinates x_1, x_2, together with the conservative boundary conditions

$$\mathbf{U} \cdot \mathbf{n} = 0, \ (\mathrm{S}\mathbf{n}) \times \mathbf{n} = 0, \ \nabla \Theta \cdot \mathbf{n} = F_b \text{ on the lateral boundary} \{x_3 = 0, \pi\}. \tag{34}$$

Note that under these circumstances the total amount of thermal energy

$$E_{\rm th} = \int_\Omega \overline{\varrho}\ \overline{c}_p\ \Theta\ {\rm d}x = \overline{\varrho}\ \overline{c}_p\ \overline{\vartheta}|\Omega| \text{ is a constant of motion.}$$

4.2 The Asymptotic Limit

Following [6] we shall show that system (29–33) supplemented with the boundary conditions (34) can be viewed as an asymptotic limit of the full Navier–Stokes–Fourier system (1–24) provided both Mach and Froude numbers tend to zero. More specifically, introducing a small parameter $\epsilon > 0$, we set

$$\text{Ma} = \epsilon,\ \text{Fr} = \sqrt{\epsilon}. \tag{35}$$

in Navier–Stokes–Fourier system (1–24). Given the new geometry introduced by (28), the test functions φ in the momentum equation (5) are taken from the class

$$\varphi \in \mathcal{D}([0,T), \mathcal{D}(\overline{\Omega}; R^3)),\ \varphi \cdot \mathbf{n}|_{\partial\Omega} = 0, \tag{36}$$

with the no-slip boundary conditions (7) replaced by

$$\mathbf{u}\cdot\mathbf{n} = (\text{S}\mathbf{n}) \times \mathbf{n} = 0 \text{ on the lateral boundary} \{x_3 = 0, \pi\}. \tag{37}$$

Furthermore, we shall assume that the initial distribution of the density and temperature is a small perturbation of a spatially homogeneous equilibrium state $\{\overline{\varrho}, \overline{\vartheta}\}$:

$$\varrho_0 = \varrho_{\epsilon,0} = \overline{\varrho} + \epsilon\varrho^{(1)}_{\epsilon,0},\ \vartheta_0 = \vartheta_{\epsilon,0} = \overline{\vartheta} + \epsilon\vartheta^{(1)}_{\epsilon,0} \tag{38}$$

where

$$\overline{\varrho} = \frac{1}{|\Omega|}\int_\Omega \varrho_{\epsilon,0}\ {\rm d}x > 0,\ \overline{\vartheta} = \frac{1}{|\Omega|}\int_\Omega \vartheta_{\epsilon,0}\ {\rm d}x > 0, \tag{39}$$

and

$$\{\varrho^{(1)}_{\epsilon,0}\}_{\epsilon>0},\ \{\vartheta^{(1)}_{\epsilon,0}\}_{\epsilon>0} \text{ are bounded in } L^\infty(\Omega). \tag{40}$$

In addition, we take

$$\mathbf{u}_0 = \mathbf{u}_{\epsilon,0},\ \mathbf{u}_{\epsilon,0} \to \mathbf{V}_0 \text{ weakly-(*) in } L^\infty(\Omega; R^3). \tag{41}$$

The following result was proved in [6, Theorem 1.1].

Theorem 5. *In addition to the hypotheses of Theorem 1, assume that Ω is given by (28), which means, all x-dependent quantities are considered to be 2π-periodic with respect to x_1, x_2, (7) being replaced by (37). Furthermore, suppose that*

$$\text{Ma} = \epsilon,\ \text{Fr} = \sqrt{\epsilon},\ F = x_3,$$

$a = \epsilon$ in (16), and $\alpha = 1$ in (23).

Let $\{\varrho_\epsilon, \mathbf{u}_\epsilon, \vartheta_\epsilon\}_{\epsilon>0}$ be a family of weak solutions to Navier–Stokes–Fourier system (1–24) on $\Omega \times (0,T)$, with the initial data

$$\varrho_{\epsilon,0} = \overline{\varrho} + \epsilon \varrho^{(1)}_{\epsilon,0},\ \mathbf{u}_{\epsilon,0},\ \vartheta_{\epsilon,0} = \overline{\vartheta} + \epsilon \vartheta^{(1)}_{\epsilon,0}$$

satisfying (39), where

$$\varrho^{(1)}_{\epsilon,0} \to \varrho^{(1)}_0,\ \vartheta^{(1)}_{\epsilon,0} \to \vartheta^{(1)}_0 \text{ weakly-(*) in } L^\infty(\Omega),$$

$$\mathbf{u}_{\epsilon,0} \to \mathbf{V}_0 \text{ weakly-(*) in } L^\infty(\Omega; R^3).$$

Then, for $\epsilon \to 0$,

$$\varrho_\epsilon \to \overline{\varrho} \text{ in } C([0,T]; L^1(\Omega)) \cap L^\infty(0,T; L^{\frac{5}{3}}(\Omega)),$$

$$\vartheta_\epsilon \to \overline{\vartheta} \text{ in } L^2(0,T; W^{1,2}(\Omega)),$$

and, passing to a subsequence if necessary,

$$\mathbf{u}_\epsilon \to \mathbf{U} \text{ weakly in } L^2(0,T; W^{1,2}(\Omega; R^3)), \tag{42}$$

$$\varrho^{(1)}_\epsilon = \frac{\varrho_\epsilon - \overline{\varrho}}{\epsilon} \to \varrho^{(1)} \text{ weakly-(*) in } L^\infty(0,T; L^{\frac{5}{3}}(\Omega)),$$

$$\vartheta^{(1)}_\epsilon = \frac{\vartheta_\epsilon - \overline{\vartheta}}{\epsilon} \to \vartheta^{(1)} \text{ weakly in } L^2(0,T; W^{1,2}(\Omega)),$$

where the trio

$$\mathbf{U},\ r = \varrho^{(1)} + \overline{\varrho}\Big(\frac{1}{\partial_\varrho p_F(\overline{\varrho}, \overline{\vartheta})} - \frac{\overline{\vartheta}\,\overline{\alpha}^2}{\overline{c}_p}\Big)(x_3 - \frac{\pi}{2}),$$

$$\Theta = \overline{\vartheta} + \vartheta^{(1)} + \frac{\overline{\vartheta}\,\overline{\alpha}}{\overline{c}_p}(x_3 - \frac{\pi}{2})$$

represents a weak solution of Oberbeck-Boussinesq system (29 – 33), supplemented with the initial data

$$\mathbf{U}_0 = \mathbf{H}[\mathbf{V}_0],\ \Theta_0 = \overline{\vartheta} + \frac{\overline{c}_v}{\overline{c}_p}\vartheta^{(1)}_0 - \frac{2}{3}\frac{\overline{\vartheta}\,\overline{c}_v}{\overline{\varrho}\,\overline{c}_p}\varrho^{(1)}_0, \tag{43}$$

and the boundary conditions (34) with

$$F_b = \frac{\overline{\vartheta}\,\overline{\alpha}}{\overline{c}_p} \text{ on the lateral boundary } \{x_3 = 0, \pi\}.$$

*Here the symbol **H** stands for the Helmholtz projection onto the space of divergenceless vector fields.*

A similar result can be proved for a weakly stratified flow with $\mathrm{Fr} \approx 1$, in which situation, momentum equation (30) is free of any forcing, and the only coupling connecting equations (30) and (31) is through the convective term in (31) (see [8, Theorem 1.1]).

Relation (43) reveals a crucial aspect of the problem. If we desire to recover a nontrivial solution of Oberbeck-Boussinesq system in the asymptotic limit, the functions $\varrho_0^{(1)}$, $\vartheta_0^{(1)}$ must be nontrivial (nonzero) as well. In particular, we have to deal with the so-called *ill-prepared* data enhancing high frequency acoustic waves considered "harmless" in the asymptotic limit but still producing large amplitude velocity field oscillations in the original system (cf. the survey papers by Masmoudi [11] or Schochet [13]).

Acknowledgement

The work was supported by Grant IAA100190606 of GA ASCR under the general research programme of the Academy of Sciences of the Czech Republic, Institutional Research Plan AV0Z10190503.

References

1. J. Březina. On uniqueness of the static state for a general compressible fluid. *Nonlinear Anal.*, **64**:188–195, 2006.
2. R.J. DiPerna and P.-L. Lions. Ordinary differential equations, transport theory and Sobolev spaces. *Invent. Math.*, **98**:511–547, 1989.
3. L.C. Evans and R.F. Gariepy. *Measure theory and fine properties of functions.* CRC Press, Boca Raton, 1992.
4. E. Feireisl. Mathematics of viscous, compressible, and heat conducting fluids. *In Contemporary Mathematics 371, G.-Q. Chen, G. Gasper, J. Jerome Editors, Amer. Math. Soc.*, pages 133–151, 2005.
5. E. Feireisl. Stability of flows of real monoatomic gases. *Commun. Partial Differential Equations*, **31**:325–348, 2006.
6. E. Feireisl and Novotný. The Oberbeck-Boussinesq approximation as a singular limit of the full Navier-Stokes-Fourier system. *Trans. Amer. Math. Soc.*, 2006. Submitted.
7. E. Feireisl and A. Novotný. On a simple model of reacting compressible flows arising in astrophysics. *Proc. Royal Soc. Edinburgh*, **135A**:1169–1194, 2005.
8. E. Feireisl and A. Novotný. On the low Mach number limit for the full Navier-Stokes-Fourier system. *Arch. Rational Mech. Anal.*, **186**:77–107, 2007.
9. E. Feireisl and H. Petzeltová. On the long time behavior of solutions to the Navier-Stokes-Fourier system with a time dependent driving force. *J. Dynamics Differential Equations*, **19**:685–707, 2007.
10. J. Leray. Sur le mouvement d'un liquide visqueux emplissant l'espace. *Acta Math.*, **63**:193–248, 1934.
11. N. Masmoudi. Examples of singular limits in hydrodynamics. *In Handbook of Differential Equations, III, C. Dafermos, E. Feireisl Eds., Elsevier, Amsterdam,* 2006.

12. K.R. Rajagopal, M.Ružička, and A. R. Shrinivasa. On the Oberbeck-Boussinesq approximation. *Math. Models Meth. Appl. Sci.*, **6**:1157–1167, 1996.
13. S. Schochet. The mathematical theory of low Mach number flows. *M2ANMath. Model Numer. anal.*, **39**:441–458, 2005.
14. R.Kh. Zeytounian. Joseph Boussinesq and his approximation: a contemporary view. *C.R. Mecanique*, **331**:575–586, 2003.

A New Technique for the Numerical Solution of the Compressible Euler Equations with Arbitrary Mach Numbers

M. Feistauer and V. Kučera

Summary. This work is concerned with the numerical solution of an inviscid compressible flow. Our goal is to develop a sufficiently accurate and robust method allowing the solution of problems with a wide range of Mach numbers. The main ingredients are the discontinuous Galerkin finite element method (DGFEM), semi-implicit time stepping, special treatment of transparent boundary conditions, and a suitable stabilization procedure near discontinuities. Numerical tests show the robustness of the method.

1 Continuous Problem

For simplicity of the treatment we shall consider two-dimensional flow, but the method can be applied to 3D flow as well. The system of the Euler equations describing 2D inviscid flow can be written in the form

$$\frac{\partial \boldsymbol{w}}{\partial t} + \sum_{s=1}^{2} \frac{\partial \boldsymbol{f}_s(\boldsymbol{w})}{\partial x_s} = 0 \quad \text{in } Q_T = \Omega \times (0, T), \tag{1}$$

where $\Omega \subset I\!R^2$ is a bounded domain occupied by gas, $T > 0$ is the length of a time interval,

$$\boldsymbol{w} = (w_1, \ldots, w_4)^{\mathrm{T}} = (\rho,\, \rho v_1,\, \rho v_2,\, E)^{\mathrm{T}} \tag{2}$$

is the so-called state vector and

$$\boldsymbol{f}_s(\boldsymbol{w}) = (\rho v_s,\, \rho v_s v_1 + \delta_{s1} p,\, \rho v_s v_2 + \delta_{s2} p,\, (E + p)\, v_s)^{\mathrm{T}} \tag{3}$$

are the inviscid (Euler) fluxes of the quantity $\boldsymbol{w}$ in the directions x_s, $s = 1, 2$. We use the following notation: ρ – density, p – pressure, E – total energy, $\boldsymbol{v} = (v_1, v_2)$ – velocity, δ_{sk} – Kronecker symbol. The equation of state implies that

$$p = (\gamma - 1)\,(E - \rho|\boldsymbol{v}|^2/2). \tag{4}$$

Here $\gamma > 1$ is the Poisson adiabatic constant. The system (1)–(4) is *diagonally hyperbolic*. It is equipped with the initial condition

$$\boldsymbol{w}(\boldsymbol{x}, 0) = \boldsymbol{w}^0(x), \quad x \in \Omega, \tag{5}$$

and the boundary conditions, which are treated in Sect. 3. We define the matrix

$$\boldsymbol{P}(\boldsymbol{w}, \boldsymbol{n}) := \sum_{s=1}^{2} \boldsymbol{A}_s(\boldsymbol{w}) n_s, \tag{6}$$

where $\boldsymbol{n} = (n_1, n_2) \in I\!R^2$, $n_1^2 + n_2^2 = 1$ and

$$\boldsymbol{A}_s(\boldsymbol{w}) = \frac{D\boldsymbol{f}_s(\boldsymbol{w})}{D\boldsymbol{w}}, \; s = 1, 2, \tag{7}$$

are the Jacobi matrices of the mappings $\boldsymbol{f}_s$. It is possible to show that $\boldsymbol{f}_s$, $s = 1, 2$, are homogeneous mappings of order one, which implies that

$$\boldsymbol{f}_s(\boldsymbol{w}) = \boldsymbol{A}_s(\boldsymbol{w})\boldsymbol{w}, \; s = 1, 2. \tag{8}$$

2 Discrete Problem

Let Ω_h be a polygonal approximation of Ω. By $\mathcal{T}_h$ we denote a partition of Ω_h consisting of triangles or quadrilaterals $K_i \in \mathcal{T}_h$, $i \in I$ ($I \subset \{0, 1, 2, \ldots\}$ is a suitable index set). By Γ_{ij} we denote a common edge between two neighboring elements K_i and K_j or a face of K_i lying on $\partial\Omega_h$. We use a suitable numbering of elements and their faces and for each $i \in I$ introduce a set $S(i)$ so that

$$\partial K_i = \bigcup_{j \in S(i)} \Gamma_{ij}. \tag{9}$$

The symbol $\boldsymbol{n}_{ij} = ((n_{ij})_1, (n_{ij})_2)$ will denote the unit outer normal to ∂K_i on the face Γ_{ij}. By h_{K_i} and $|K_i|$ we shall denote the diameter and the area, respectively, of an element $K_i \in \mathcal{T}_h$.

The approximate solution will be sought at each time instant t as an element of the finite-dimensional space

$$S_h = S^{r,-1}(\Omega_h, \mathcal{T}_h) = \{v; v|_K \in P^r(K) \; \forall K \in \mathcal{T}_h\}^4,$$

where $r \geq 0$ is an integer and $P^r(K)$ denotes the space of all polynomials on K of degree $\leq r$. Functions $v \in S_h$ are in general discontinuous on interfaces Γ_{ij}. By $v|_{\Gamma_{ij}}$ and $v|_{\Gamma_{ji}}$ we denote the values of v on Γ_{ij} considered from the interior and the exterior of K_i, respectively.

To derive the discrete problem, we multiply (1) by a test function $\varphi \in S_h$, integrate over any element $K_i, i \in I$, apply Green's theorem and sum over all

$i \in I$. Then we approximate fluxes through the faces Γ_{ij} with the aid of a *numerical flux* $\boldsymbol{H} = \boldsymbol{H}(\boldsymbol{u}, \boldsymbol{w}, \boldsymbol{n})$ in the form

$$\int_{\Gamma_{ij}} \sum_{s=1}^{2} \boldsymbol{f}_s(\boldsymbol{w}(t))\,(n_{ij})_s \cdot \boldsymbol{\varphi}\,\mathrm{d}S \approx \int_{\Gamma_{ij}} \boldsymbol{H}(\boldsymbol{w}_h(t)|_{\Gamma_{ij}}, \boldsymbol{w}_h(t)|_{\Gamma_{ji}}, \boldsymbol{n}_{ij}) \cdot \boldsymbol{\varphi}\,\mathrm{d}S.$$

Further, we consider a partition $0 = t_0 < t_1 < t_2 \ldots$ of the time interval $(0, T)$ and set $\tau_k = t_{k+1} - t_k$. We use the notation $\boldsymbol{w}_h^k$ for the approximation of the solution at time t_k. Our goal is to construct a semi-implicit linearized scheme for obtaining the approximate solution. Using the homogeneity of fluxes $\boldsymbol{f}_s$ and the Vijayasundaram numerical flux and introducing a semi-implicit time discretization, similarly as in [DF04] we obtain the form

$$\begin{aligned}
& b_h(\boldsymbol{w}_h^k, \boldsymbol{w}_h^{k+1}, \boldsymbol{\varphi}_h) && (10)\\
&= -\sum_{K \in \mathcal{T}_h} \int_K \sum_{s=1}^{2} \boldsymbol{A}_s(\boldsymbol{w}_h^k(\boldsymbol{x}))\boldsymbol{w}_h^{k+1}(\boldsymbol{x}) \cdot \frac{\partial \boldsymbol{\varphi}_h(\boldsymbol{x})}{\partial x_s}\,\mathrm{d}\boldsymbol{x} && (11)\\
&+ \sum_{K_i \in \mathcal{T}_h} \sum_{j \in S(i)} \int_{\Gamma_{ij}} \left[\boldsymbol{P}^+\left(\langle \boldsymbol{w}_h^k \rangle_{ij}, \boldsymbol{n}_{ij}\right) \boldsymbol{w}_h^{k+1}|_{\Gamma_{ij}}\right. && (12)\\
&\left. + \boldsymbol{P}^-\left(\langle \boldsymbol{w}_h^k \rangle_{ij}, \boldsymbol{n}_{ij}\right) \boldsymbol{w}_h^{k+1}|_{\Gamma_{ji}}\right] \cdot \boldsymbol{\varphi}_h \mathrm{d}S.
\end{aligned}$$

Here $\langle \boldsymbol{w}_h^k \rangle_{ij} = (\boldsymbol{w}_h^k|_{\Gamma_{ij}} + \boldsymbol{w}_h^k|_{\Gamma_{ji}})/2$ and $\boldsymbol{P}^\pm = \boldsymbol{P}^\pm(\boldsymbol{w}, \boldsymbol{n})$ represents positive/negative part of the matrix $\boldsymbol{P}(\boldsymbol{w}, \boldsymbol{n})$. The form b_h is linear with respect to the second and third variable. Moreover, we denote $(\boldsymbol{u}, \boldsymbol{v})_h = \int_{\Omega_h} \boldsymbol{u} \cdot \boldsymbol{v}\, dx$ and introduce the following *semi-implicit scheme*: For each $k \geq 0$ find $\boldsymbol{w}_h^{k+1}$ such that

$$\begin{aligned}
&\text{(a)} \quad \boldsymbol{w}_h^{k+1} \in \boldsymbol{S}_h, && (13)\\
&\text{(b)} \quad \left(\frac{\boldsymbol{w}_h^{k+1} - \boldsymbol{w}_h^k}{\tau_k}, \boldsymbol{\varphi}_h\right)_h + b_h(\boldsymbol{w}_h^k, \boldsymbol{w}_h^{k+1}, \boldsymbol{\varphi}_h) = 0,\ \forall \boldsymbol{\varphi}_h \in \boldsymbol{S}_h,\ k = 0, 1, \ldots, \\
&\text{(c)} \quad \boldsymbol{w}_h^0 = \Pi_h \boldsymbol{w}^0 = \boldsymbol{S}_h - \text{interpolation of } \boldsymbol{w}^0.
\end{aligned}$$

It is obvious that the transition from the time level t_k to t_{k+1} is realized by the solution of a linear system of algebraic equations. This system is solved by the GMRES method with a block diagonal preconditioning. (It is also possible to apply the direct solver UMFPACK, [DD99]).

Scheme (13) is a first order accurate in time, but it is possible to construct a semi-implicit two-step second-order time discretization (see [DF04]).

3 Boundary Conditions

If $\Gamma_{ij} \subset \partial\Omega_h$, it is necessary to specify the boundary state $\boldsymbol{w}|_{\Gamma_{ji}}$ appearing in the definition of the form b_h. The appropriate treatment of boundary conditions plays a crucial role in the solution of low Mach number flows.

On a fixed impermeable wall we employ a standard approach using the condition $\boldsymbol{v} \cdot \boldsymbol{n} = 0$ and extrapolating the pressure. On the inlet and outlet it is necessary to use nonreflecting boundary conditions transparent for acoustic effects coming from inside of Ω. Therefore, *characteristics-based* boundary conditions are used.

Using the rotational invariance, we transform the Euler equations to the coordinates $\tilde{x}_1$, parallel with the normal direction $\boldsymbol{n}$ to the boundary, and $\tilde{x}_2$, tangential to the boundary, neglect the influence of the states on elements that are not adjacent to Γ_{ij} and linearize the resulting system around the state $\boldsymbol{q}_{ij} = \boldsymbol{Q}(\boldsymbol{n}_{ij})\boldsymbol{w}|_{\Gamma_{ij}}$, where

$$\boldsymbol{Q}(\boldsymbol{n}_{ij}) = \begin{pmatrix} 1, & 0, & 0, & 0 \\ 0, & (n_{ij})_1, & (n_{ij})_2, & 0 \\ 0, & -(n_{ij})_2, & (n_{ij})_1, & 0 \\ 0, & 0, & 0, & 1 \end{pmatrix} \tag{14}$$

is the rotational matrix. Then we obtain the linear system

$$\frac{\partial \boldsymbol{q}}{\partial t} + \boldsymbol{A}_1(\boldsymbol{q}_{ij})\frac{\partial \boldsymbol{q}}{\partial \tilde{x}_1} = 0, \tag{15}$$

for the vector-valued function $\boldsymbol{q} = \boldsymbol{Q}(\boldsymbol{n}_{ij})\boldsymbol{w}$, considered in the set $(-\infty, 0) \times (0, \infty)$ and equipped with the initial and boundary conditions

$$\begin{aligned} \boldsymbol{q}(\tilde{x}_1, 0) &= \boldsymbol{q}_{ij}, \quad \tilde{x}_1 < 0, \\ \boldsymbol{q}(0, t) &= \boldsymbol{q}_{ji}, \quad t > 0. \end{aligned} \tag{16}$$

The goal is to choose $\boldsymbol{q}_{ji}$ in such a way that this initial-boundary value problem is well posed, i. e., has a unique solution. The method of characteristics leads to the following process:

Let us put $\boldsymbol{q}^*_{ji} = \boldsymbol{Q}(\boldsymbol{n}_{ij})\boldsymbol{w}^*_{ji}$, where $\boldsymbol{w}^*_{ji}$ is a prescribed boundary state at the inlet or outlet. We calculate the eigenvectors $\boldsymbol{r}_s$ corresponding to the eigenvalues λ_s, $s = 1, \ldots, 4$, of the matrix $\boldsymbol{A}_1(\boldsymbol{q}_{ij})$, arrange them as columns in the matrix $\boldsymbol{T}$ and calculate $\boldsymbol{T}^{-1}$ (explicit formulae can be found in [FFS03], Sect. 3.1). Now we set

$$\boldsymbol{\alpha} = \boldsymbol{T}^{-1}\boldsymbol{q}_{ij}, \quad \boldsymbol{\beta} = \boldsymbol{T}^{-1}\boldsymbol{q}^*_{ji}. \tag{17}$$

and define the state $\boldsymbol{q}_{ji}$ by the relations

$$\boldsymbol{q}_{ji} := \sum_{s=1}^{4} \gamma_s \boldsymbol{r}_s, \quad \gamma_s = \begin{cases} \alpha_s, & \lambda_s \geq 0, \\ \beta_s, & \lambda_s < 0. \end{cases} \tag{18}$$

Finally, the sought boundary state $\boldsymbol{w}|_{\Gamma_{ji}}$ is defined as

$$\boldsymbol{w}|_{\Gamma_{ji}} = \boldsymbol{w}_{ji} = \boldsymbol{Q}^{-1}(\boldsymbol{n}_{ij})\boldsymbol{q}_{ji}. \tag{19}$$

4 Stabilization of Flow with Discontinuities

In the case of high speed flow, it is necessary to avoid spurious overshoots and undershoots in computed quantities near discontinuities (shock waves, contact discontinuities). This phenomenon does not occur in low Mach number regimes, but in transonic flow it causes instabilities in the numerical solution.

We apply the procedure motivated by [DFS03] and [JJS95]. First we introduce the discontinuity indicator $g(i)$ proposed in [DFS03] and defined by

$$g(i) = \int_{\partial K_i} [\rho_h^k]^2 \, \mathrm{d}S / (h_{K_i} |K_i|^{3/4}), \quad K_i \in \mathcal{T}_h. \tag{20}$$

By $[u]|_{\Gamma_{ij}} = u|_{\Gamma_{ij}} - u|_{\Gamma_{ji}}$, we denote the jump on Γ_{ij} of a function $u \in S_h$. Further, we define the discrete indicator

$$G(i) = 0 \text{ if } g(i) < 1 \text{ and } G(i) = 1 \text{ if } g(i) \geq 1, \quad K_i \in \mathcal{T}_h. \tag{21}$$

Now, to the left-hand side of (13), (b) we add the "artificial viscosity" forms

$$\beta_h(\boldsymbol{w}_h^k, \boldsymbol{w}_h^{k+1}, \boldsymbol{\varphi}) = \nu_1 \sum_{i \in I} h_{K_i} G^k(i) \int_{K_i} \nabla \boldsymbol{w}_h^{k+1} \cdot \nabla \boldsymbol{\varphi} \, \mathrm{d}\boldsymbol{x} \tag{22}$$

and

$$J_h(\boldsymbol{w}_h^k, \boldsymbol{w}_h^{k+1}, \boldsymbol{\varphi}) = \nu_2 \sum_{i \in I} \sum_{j \in s(i)} \frac{1}{2} (G^k(i) + G^k(j)) \int_{\Gamma_{ij}} [\boldsymbol{w}_h^{k+1}] \cdot [\boldsymbol{\varphi}] \, \mathrm{d}S, \tag{23}$$

where $\nu_1, \nu_2 \approx 1$.

Thus, the resulting scheme reads

$$\begin{aligned}
&\text{(a)} \quad \boldsymbol{w}_h^{k+1} \in \boldsymbol{S}_h, \\
&\text{(b)} \quad \left(\frac{\boldsymbol{w}_h^{k+1} - \boldsymbol{w}_h^k}{\tau_k}, \boldsymbol{\varphi}_h \right)_h + b_h(\boldsymbol{w}_h^k, \boldsymbol{w}_h^{k+1}, \boldsymbol{\varphi}_h) + \beta_h(\boldsymbol{w}_h^k, \boldsymbol{w}_h^{k+1}, \boldsymbol{\varphi}_h) \\
&\qquad + J_h(\boldsymbol{w}_h^k, \boldsymbol{w}_h^{k+1}, \boldsymbol{\varphi}_h) = 0, \ \forall \boldsymbol{\varphi}_h \in \boldsymbol{S}_h, \ k = 0, 1, \ldots, \\
&\text{(c)} \quad \boldsymbol{w}_h^0 = \Pi_h \boldsymbol{w}^0.
\end{aligned} \tag{24}$$

It is important that $G(i)$ vanishes in regions, where the solution is regular. Therefore, the scheme does not produce any nonphysical entropy in these regions (see Fig. 4).

We consider the CFL stability condition of the form

$$\tau_k \max_{K_i \in \mathcal{T}_h} \frac{1}{|K_i|} \left(\max_{j \in S(i)} |\Gamma_{ij}| \lambda^{\max}_{\boldsymbol{P}(\boldsymbol{w}_h^k|_{\Gamma_{ij}}, \boldsymbol{n}_{ij})} \right) \leq \text{CFL}. \tag{25}$$

Here CFL is a given constant and $\lambda^{\max}_{\boldsymbol{P}(\boldsymbol{w}_h^k|_{\Gamma_{ij}}, \boldsymbol{n}_{ij})}$ is the maximum over Γ_{ij} of the spectral radius of the matrix $\boldsymbol{P}(\boldsymbol{w}_h^k|_{\Gamma_{ij}}, \boldsymbol{n}_{ij})$.

Finally, an important ingredient of the method is the use of isoparametric elements near curved parts of the boundary $\partial\Omega$. (For details, see [FFS03], Sect. 4.6.8.)

5 Numerical Tests

To show the robustness of the described technique with respect to the Mach number, we present computational results obtained for some test problems. In all computations, quadratic elements were used.

We shall consider stationary flow past a negatively oriented Joukowski profile given by parameters $\Delta = 0.07, a = 0.5, h = 0.05$ (under the notation from [F93], Sect. 2.2.68) with zero angle of attack. The far field quantities are constant, which implies that the flow is irrotational and homoentropic. Using the complex function method from [F93], we obtain the exact solution of incompressible inviscid irrotational flow satisfying the Kutta–Joukowski trailing condition, provided the velocity circulation around the profile, related to the magnitude of the far field velocity, $\gamma_{\text{ref}} = 0.7158$. We assume that the far field Mach number of compressible flow $M_\infty = 0.0001$. The computational domain is of the form of a square with side of the length equal to 10 chords of the profile. The mesh (in the whole computational domain) was formed by 5,418 triangular elements and refined towards the profile. The steady state was obtained with the aid of time stabilization. The method (24) behaves as *unconditionally stable*.

Figure 1 shows streamlines of the computed compressible flow. We see that the flow past the trailing edge is smooth. Further, in Fig. 2, we compare the velocity distribution past the profile of incompressible and compressible flow, plotted in the direction from the leading edge to the trailing edge (ooo – exact solution of incompressible flow, —— – approximate solution of compressible flow). The agreement of both results is very good.

The maximum density variation in compressible flow is $1.04 \cdot 10^{-8}$. The computed velocity circulation related to the magnitude of the far field velocity is $\gamma_{\text{refcomp}} = 0.7205$, which gives the relative error 0.66% with respect to the

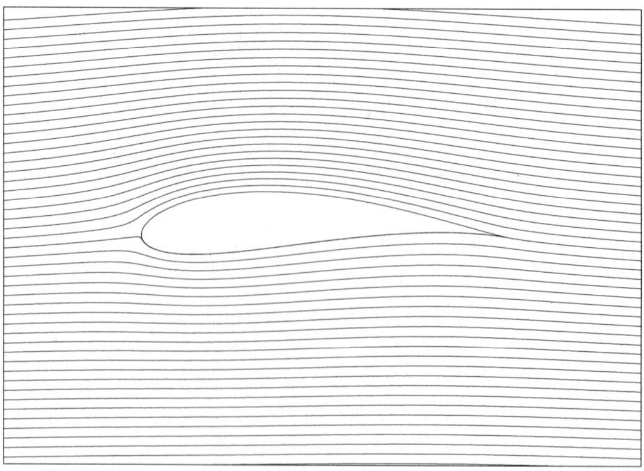

Fig. 1. Compressible low Mach flow past a Joukowski profile, approximate solution, streamlines

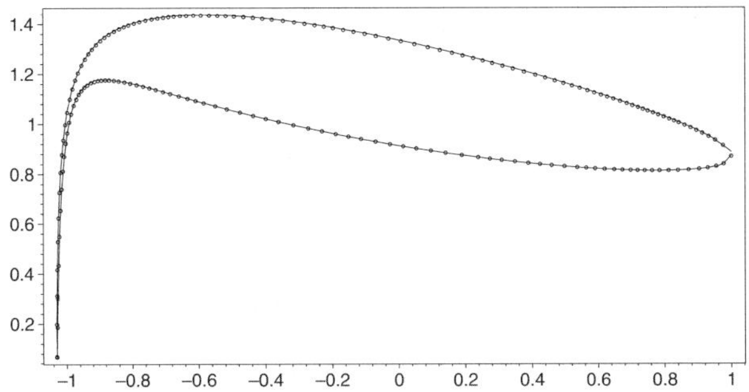

Fig. 2. Flow past a Joukowski profile, velocity distribution along the profile: ∘∘∘ – exact solution of incompressible flow, —— – approximate solution of compressible low Mach flow

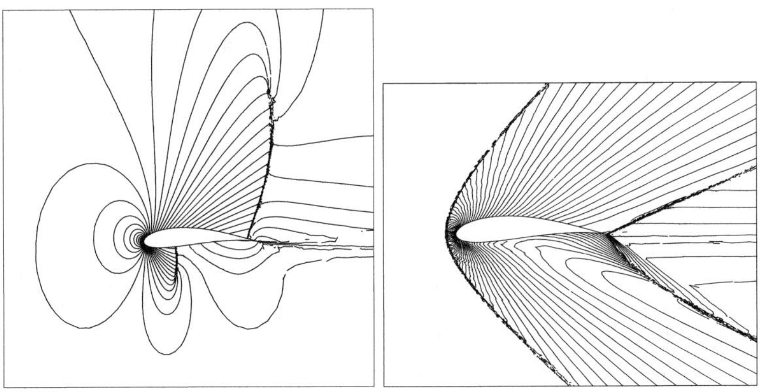

Fig. 3. Mach number isolines of the flow past a Joukowski profile with $M_\infty = 0.8$ (*left*) and $M_\infty = 2.0$ (*right*)

theoretical value γ_{ref} obtained for incompressible flow. The CFL number from the stability condition (25) was during the computational process successively increased from 1 (the start of the computation) to $6 \cdot 10^6$.

To demonstrate that the presented method allows the solution of high-speed flow with shock waves, we present the results obtained for the transonic and hypersonic flow with shock waves past the Joukowski profile. Figure 3 shows Mach number isolines of the transonic and hypersonic flow past the Joukowski profile for the far field Mach number $M_\infty = 0.8$ and $M_\infty = 2.0$, respectively. In these cases, the starting CFL number was chosen 0.08 due to the initial condition, which was constant in the whole computational domain. Then during the computational process the CFL number was successively increased up to 1,500 and 2,040 in the case of $M_\infty = 0.8$ and $M_\infty = 2.0$, respectively. The numbers of elements were 4,451 for $M_\infty = 0.8$ and 4,537 for $M_\infty = 2.0$. In both cases the constants from (22) and (23) had values $\nu_1 = \nu_2 = 0.1$. (For low Mach number flow the stabilization terms are not active.)

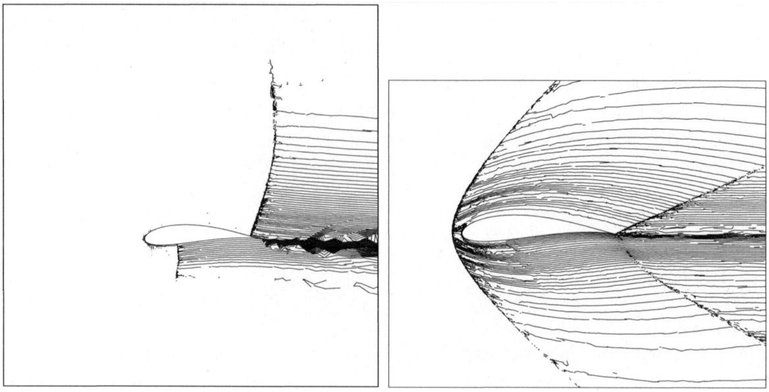

Fig. 4. Entropy isolines of the flow past a Joukowski profile with $M_\infty = 0.8$ (*left*) and $M_\infty = 2.0$ (*right*)

The maximum density variation was 0.94 and 2.61 in the case $M_\infty = 0.8$ and $M_\infty = 2.0$, respectively. In Fig. 4, entropy isolines are plotted showing that the entropy is produced only on shock waves.

Acknowledgements

This work is a part of the research project No. MSM 0021620839 of the Ministry of Education of the Czech Republic. The research of M. Feistauer was partly supported by the Grant No. 201/05/0005 of the Grant Agency of the Czech Republic and the research of V. Kučera was partly supported by the Nečas Center for Mathematical Modeling – Project LC06052 financed by the Ministry of Education of the Czech Republic. The authors acknowledge the support of these institutions.

References

[DD99] Davis, T.A., Duff, I.S.: A combined unifrontal/multifrontal method for unsymmetric sparse matrices. ACM Trans. Math. Softw., **25**, 1–19 (1999) (UMFPACK V4.0, http://www.cise.ufl.edu/research/sparse/umfpack)

[DF04] Dolejší, V., Feistauer, M.: A semi-implicit discontinuous Galerkin finite element method for the numerical solution of inviscid compressible flow. J. Comput. Phys., **198**, 727–746 (2004)

[DFS03] Dolejší, V., Feistauer, M., Schwab, C.: On some aspects of the discontinuous Galerkin finite element method for conservation laws. Math. Comput. Simul., **6**, 333–346 (2003)

[F93] Feistauer, M.: Mathematical Methods in Fluid Dynamics. Longman, Harlow, (1993)

[FFS03] Feistauer, M., Felcman, J., Straškraba, I.: Mathematical and Computational Methods for Compressible Flow. Clarendon Press, Oxford (2003)

[JJS95] Jaffre, J., Johnson, C., Szepessy, A.: Convergence of the discontinuous Galerkin finite elements method for hyperbolic conservation laws. Math. Models Methods Appl. Sci., **5**, 367–386 (1995)

Monokinetic Limits of the Vlasov-Poisson/Maxwell-Fokker-Planck System

L. Hsiao, F. Li, and S. Wang

1 Introduction

Kinetic equations are mathematical models used to describe the "dilute particle gases" at an intermediate scale between microscopic and macroscopic level. They appear in a variety of sciences such as plasma, astrophysics, aerospace engineering, nuclear engineering, particle–fluid interactions, semiconductor technology, social sciences, and biologies, for example see [Pe04].

Now the study on kinetic equations is one of the most active areas in applied mathematics. The quasineutral limit problem plays an important role in the theory of kinetic equations since it gives the relations between the microscopic mathematical models and the macroscopic ones.

In this note, we will discuss the monokinetic limits (the special case of quasineutral limits) for two kinds of equations including the Vlasov-Poisson-Fokker-Planck system and the Vlasov-Maxwell-Fokker-Planck system. It should be noted that for a monokinetic limit, the assumption $\sigma \to 0$ (see the theorems stated below) is necessary, since otherwise the Fokker-Planck scattering term would thermalize the distribution. We shall show that the solution of the former converges to the solution of incompressible Euler equations with or without damping, and the solution of the latter converges to the solution of so-called electron magnetohydrodynamics equations.

2 The Vlasov-Poisson-Fokker-Planck System

The Vlasov-Poisson-Fokker-Planck (VPFP) system takes the form

$$\partial_t f^\varepsilon + \xi \cdot \nabla_x f^\varepsilon - \nabla_x \Phi^\varepsilon \cdot \nabla_\xi f^\varepsilon = \frac{1}{\tau}\mathrm{div}_\xi\big(\xi f^\varepsilon + \sigma \nabla_\xi f^\varepsilon\big), \tag{1}$$

$$\int_{\mathbb{R}^N} f^\varepsilon(t,x,\xi)d\xi = 1 - \varepsilon \Delta \Phi^\varepsilon \tag{2}$$

with initial datum

$$f^\varepsilon(0,x,\xi) = f_0^\varepsilon(x,\xi) \geq 0. \tag{3}$$

Here $f(t,x,\xi)$ denotes the distribution function of particles, which expresses the probability of finding a particle at time $t \geq 0$ in a position $x \in [0,1]^N \equiv \mathbb{T}^N$, and with a velocity $\xi \in \mathbb{R}^N, N = 1,2,3$. The spatially periodic electric potential Φ^ε is coupled with $f^\varepsilon(t,x,\xi)$ through the Poisson equation (2). The parameter $\varepsilon > 0$ denotes the vacuum electric permittivity, $\sigma > 0$ denotes the thermal diffusion coefficient, and $\tau > 0$ denotes the relaxation time due to the collisions of particles with thermal bath.

We first recall some known results of VPFP system. Bouchut [Bo93] obtained the existence and uniqueness to smooth solutions of VPFP system. Poupaud and Soler [PS00] studied the parabolic limit of VPFP system and showed that the VPFP system converges to a parabolic equation. Goudon et al. [GNP05] studied the high-field limit and obtained that the above VPFP system converges to a nonlinear systems of PDEs related to Ohm's Law. Using the Hilbert expansion, Arnold et al. [ACG01] studied the low-field limit and obtained the convergence of VPFP system to the drift-diffusion equations.

We are interested in the monokinetic limits of VPFP system (1)–(2), and we discuss two cases: (i) $\varepsilon \to 0$, $\tau > 0$ is fixed, and $\sigma \to 0$, (ii) $\varepsilon \to 0$, $\tau \to +\infty$, and $\sigma > 0$ is fixed.

2.1 Convergence to Incompressible Euler Equations With Damping

In this section, we study the quasineutral limits of VPFP system for the first case, i.e., $\varepsilon \to 0$, $\tau > 0$ is fixed, and $\sigma \to 0$. We set $\tau = 1$ for convenience. Now the VPFP system reads

$$\partial_t f^\varepsilon + \xi \cdot \nabla_x f^\varepsilon - \nabla_x \Phi^\varepsilon \cdot \nabla_\xi f^\varepsilon = \mathrm{div}_\xi\big(\xi f^\varepsilon + \sigma \nabla_\xi f^\varepsilon\big), \tag{4}$$

$$\int_{\mathbb{R}^N} f^\varepsilon(t,x,\xi)d\xi = 1 - \varepsilon \Delta \Phi^\varepsilon \tag{5}$$

with initial datum

$$f^\varepsilon(0,x,\xi) = f_0^\varepsilon(x,\xi) \geq 0. \tag{6}$$

Denote

$$\rho^\varepsilon(t,x) = \int_{\mathbb{R}^N} f^\varepsilon(t,x,\xi)d\xi, \quad J^\varepsilon(t,x) = \rho^\varepsilon u^\varepsilon(t,x) = \int_{\mathbb{R}^N} \xi f^\varepsilon(t,x,\xi)d\xi. \tag{7}$$

Multiplying the equation (4) by 1 and ξ respectively, and integrating the results with respect to ξ, we have

$$\partial_t \rho^\varepsilon + \nabla \cdot J^\varepsilon = 0, \tag{8}$$

$$\begin{aligned}\partial_t J^\varepsilon + \nabla_x \cdot \int_{\mathbb{R}^N} (\xi \otimes \xi) f^\varepsilon d\xi + \nabla \Phi^\varepsilon \\ + \frac{\varepsilon}{2}\nabla(|\nabla\Phi^\varepsilon|^2) - \varepsilon \nabla \cdot (\nabla \Phi^\varepsilon \otimes \nabla \Phi^\varepsilon) + J^\varepsilon = 0,\end{aligned} \tag{9}$$

where we have used the identity

$$\nabla\Phi^\varepsilon(\Delta\Phi^\varepsilon) = \frac{1}{2}\nabla(|\nabla\Phi^\varepsilon|^2) - \nabla\cdot(\nabla\Phi^\varepsilon \otimes \nabla\Phi^\varepsilon)$$

and the symbol "$\otimes$," which denotes the tensor product of vectors.

We start with a purely formal analysis on the limit process as $\varepsilon \to 0$. First, it follows from the Poisson equation (5) that $\rho^\varepsilon \to 1$. For perfect cold electrons (the temperature vanishes), the probability measure (in ξ) $f^\varepsilon(t,x,\xi)$ is a delta function, which exactly means

$$f^\varepsilon(t,x,\xi) = \delta(\xi - J^\varepsilon(t,x)).$$

In this particular case, we obtain

$$\nabla\cdot J = 0, \quad \partial_t J + \nabla\cdot(J\otimes J) + \nabla\Phi + J = 0, \tag{10}$$

where J is the limit of J^ε (if it exists) in some sense, and Φ is a pressure function. Equations (10) are incompressible Euler equations with damping.

We will establish the above limit rigorously. Before stating our result, we give the following existence result on the incompressible Euler equations with damping. Consider the periodic boundary problem to the incompressible Euler equations with damping

$$\nabla\cdot u = 0, \quad \partial_t u + (u\cdot\nabla)u + \nabla p + u = 0,\ t>0,\ x\in\mathbb{T}^N, \tag{11}$$

with initial datum $u(0,x) = u_0(x) \in \mathcal{H}^s$, where the function space $\mathcal{H}^s$ is given by

$$\mathcal{H}^s = \{u \in H^s(\mathbb{T}^N) \mid \nabla\cdot u = 0\}. \tag{12}$$

Lemma 1 ([PW05]). *For each $u_0 \in \mathcal{H}^s (s > 1 + N/2)$, there exist a $T^* \in (0,\infty)$ and a unique solution $u \in L^\infty_{loc}([0,T^*),\mathcal{H}^s)$ satisfying, for any $T < T^*$, that*

$$\sup_{0\le t\le T}\left(\|u\|_{H^s} + \|\partial_t u\|_{H^{s-1}} + \|\nabla p\|_{H^s} + \|\partial_t\nabla p\|_{H^{s-1}}\right) \le C(T) \tag{13}$$

for some positive constant $C(T)$, depending only upon T.

Our result now reads as follows.

Theorem 1 ([HLWa]). *Let $0 < T < T^*$ and u_0 be a given vector in $\mathcal{H}^s(s > 1 + N/2)$, $\mathbb{Z}^N$ periodic in x. Assume that $f_0^\varepsilon(x,\xi) \ge 0$ is smooth, $\mathbb{Z}^N$ periodic in x with total mass 1, and decays fast as $\xi \to \infty$. In addition, we assume that there exists a constant $C_0 > 0$, independent of ε, such that*

$$\iint_{\mathbb{R}^N\times\mathbb{T}^N} (1 + |x|^2 + |\xi|^2 + |\ln f_0^\varepsilon(x,\xi)|) f_0^\varepsilon(x,\xi)\,dx\,d\xi \le C_0, \tag{14}$$

furthermore,

$$\int_{\mathbb{R}^N} f_0^\varepsilon(x,\xi)d\xi = 1 + o(\varepsilon^{1/2}) \quad \text{as} \quad \varepsilon \to 0 \tag{15}$$

in the strong sense of the space $H^{-1}(\mathbb{T}^N)$, *and*

$$\frac{1}{2}\iint_{\mathbb{R}^N\times\mathbb{T}^N} |\xi - u_0(x)|^2 f_0^\varepsilon(x,\xi)dxd\xi + \sigma\iint_{\mathbb{R}^N\times\mathbb{T}^N} f_0^\varepsilon(x,\xi)|\ln f_0^\varepsilon(x,\xi)|dxd\xi$$
$$+\frac{\varepsilon}{2}\int_{\mathbb{T}^N} |\nabla\Phi^\varepsilon(0,x)|^2dx \to 0 \quad \text{as} \quad \varepsilon,\sigma \to 0. \tag{16}$$

Let f^ε *be any nonnegative smooth solution of the problem* (4)–(6)*. Then, up to the extraction of a subsequence, the current* J^ε *converges weakly to the unique solution* $u(x,t)$ *of the incompressible Euler equations* (11) *with initial data* u_0*. Moreover, the divergence-free part of* J^ε *converges to* u *in* $L^\infty(0,T;L^2(\mathbb{T}^N))$.

Remark 1. The assumption on initial datum in Theorem 1 can be guaranteed, for example, by taking f_0^ε as

$$f_0^\varepsilon(x,\xi) = \frac{1}{(2\pi\varepsilon^\alpha)^{N/2}}\exp\Big\{-\frac{|\xi-u_0|^2}{2\varepsilon^\alpha}\Big\},$$

for some $\alpha > 0$.

The proof of Theorem 1 is based on compactness arguments and the so-called modulated energy method [Br00] (also known as relative-entropy method). The idea of this method is to modulate the energy of the system by test functions, and to obtain a stability inequality when these test functions are solutions of the limiting system.

2.2 Convergence to Incompressible Euler Equations Without Damping

In this section, we study the quasineutral limits of VPFP system for the second case, i.e. $\varepsilon \to 0$, σ is fixed, and $\tau \to +\infty$. For convenience, we set $\sigma = 1$ and then denote $\frac{1}{\tau} = \gamma$. Now the VPFP system reads

$$\partial_t f^\varepsilon + \xi\cdot\nabla_x f^\varepsilon - \nabla_x\Phi^\varepsilon\cdot\nabla_\xi f^\varepsilon = \gamma \mathrm{div}_\xi(\xi f^\varepsilon + \nabla_\xi f^\varepsilon), \tag{17}$$

$$\int_{\mathbb{R}^N} f^\varepsilon(t,x,\xi)d\xi = 1 - \varepsilon\Delta\Phi^\varepsilon \tag{18}$$

with initial datum

$$f^\varepsilon(0,x,\xi) = f_0^\varepsilon(x,\xi) \ge 0. \tag{19}$$

Multiplying the equation (17) by 1 and ξ respectively, and integrating the results with respect to ξ, we have

$$\partial_t \rho^\varepsilon + \nabla \cdot J^\varepsilon = 0, \tag{20}$$

$$\begin{aligned}\partial_t J^\varepsilon + \nabla_x \cdot \int_{\mathbb{R}^N} (\xi \otimes \xi) f^\varepsilon d\xi + \nabla \Phi^\varepsilon \\ + \frac{\varepsilon}{2} \nabla(|\nabla \Phi^\varepsilon|^2) - \varepsilon \nabla \cdot (\nabla \Phi^\varepsilon \otimes \nabla \Phi^\varepsilon) + \gamma J^\varepsilon = 0\end{aligned} \tag{21}$$

with $-\varepsilon \Delta \Phi^\varepsilon = \rho^\varepsilon - 1$. Here ρ^ε and J^ε are defined by (7).

The similar formal analysis on limit process as $\varepsilon, \gamma \to 0$ shows that

$$\nabla \cdot J = 0, \quad \partial_t J + \nabla \cdot (J \otimes J) + \nabla \Phi = 0, \tag{22}$$

for the case of perfectly cold electrons. (22) are nothing else but the Euler equations to the incompressible fluid, where J is the limit of J^ε (if it exists) in some sense, and Φ is a pressure function.

Consider the periodic boundary problem of Euler equations to the incompressible fluid without damping (the last term on the left hand side of (11) disappears), we have

Lemma 2 ([BKM84]). *For each $u_0 \in \mathcal{H}^s (s > 1 + N/2)$, there exist a $T^* \in (0, \infty] (T^* = +\infty$ if $N = 2)$ and a unique solution $u \in L^\infty_{loc}([0, T^*), H^s)$ satisfying, for any $T < T^*$, that*

$$\sup_{0 \le t \le T} \left(\|u\|_{H^s} + \|\partial_t u\|_{H^{s-1}} + \|\nabla p\|_{H^s} + \|\partial_t \nabla p\|_{H^{s-1}} \right) \le C(T) \tag{23}$$

for some positive constant $C(T)$, depending only upon T.

Now we state our result.

Theorem 2 ([HLW06]). *Let $0 < T < T^*$ and u_0 be a given vector in $\mathcal{H}^s (s > 1 + N/2)$, $\mathbb{Z}^N$ periodic in x. Assume that $f_0^\varepsilon(x, \xi) \ge 0$ is smooth, $\mathbb{Z}^N$ periodic in x, and decays fast as $\xi \to \infty$. In addition, we assume that there exists a constant $C_0 > 0$, independent of ε, such that*

$$\iint_{\mathbb{R}^N \times \mathbb{T}^N} (1 + |x|^2 + |\xi|^2) f_0^\varepsilon(x, \xi) dx d\xi \le C_0, \tag{24}$$

furthermore,

$$\int_{\mathbb{R}^N} f_0^\varepsilon(x, \xi) d\xi = 1 + o(\varepsilon^{1/2}) \quad \text{as} \quad \varepsilon \to 0 \tag{25}$$

in the strong sense of the space $H^{-1}(\mathbb{T}^N)$, and

$$\begin{aligned}\frac{1}{2} \iint_{\mathbb{R}^N \times \mathbb{T}^N} |\xi - u_0(x)|^2 f_0^\varepsilon(x, \xi) dx d\xi + \frac{\varepsilon}{2} \int_{\mathbb{T}^N} |\nabla \Phi^\varepsilon(0, x)|^2 dx \\ \to 0 \quad \text{as} \quad \varepsilon, \gamma \to 0.\end{aligned} \tag{26}$$

Let f^ε be any nonnegative smooth solution of the problem (17)–(19). Then, up to the extraction of a subsequence, the current J^ε converges weakly to the unique solution $u(x, t)$ of the Euler equations (11) (without the last term on the left-hand side) with initial data u_0. Moreover, the divergence-free part of J^ε converges to u in $L^\infty(0, T; L^2(\mathbb{T}^N))$.

Using the definition of dissipative solution to incompressible Euler equations due to Lions [Li96], we are able to prove that the divergence-free part of J^ε converges to the dissipative solution of the Euler equations to the incompressible fluid for the case of cold electrons. Namely, we have

Theorem 3 ([HLW06]). *Let* $0 < T < T^*$ *and* $J_0(x)$ *be a given vector which is divergence-free,* $\mathbb{Z}^N$ *periodic in* x*, and square integrable. Assume that* $f_0^\varepsilon(x,\xi) \geq 0$ *to be smooth,* $\mathbb{Z}^N$ *periodic in* x*, and* f_0^ε *decays fast as* $\xi \to \infty$*. In addition, we assume that* (24), (25) *hold, and*

$$\frac{1}{2}\iint_{\mathbb{R}^N\times\mathbb{T}^N} |\xi - v_0(x)|^2 f_0^\varepsilon(x,\xi)dxd\xi + \frac{\varepsilon}{2}\int_{\mathbb{T}^N} |\nabla\Phi^\varepsilon(0,x)|^2 dx$$
$$\to \frac{1}{2}\int_{\mathbb{T}^N} |J_0(x) - v_0(x)|^2 dx \quad \text{as} \quad \varepsilon,\gamma \to 0 \tag{27}$$

for any divergence-free and $\mathbb{Z}^N$ *periodic vector field* $v_0(x)$*. Let* f^ε *be any non-negative smooth solution to the problem* (17)–(19)*. Then, up to the extraction of a subsequence, the divergence-free part of* J^ε *converges in* $C^0([0,T],\mathcal{D}'(\mathbb{R}^N))$ *to a dissipative solution* $J \in C^0([0,T],L^2(\mathbb{T}^N) - w)$ *for the Euler equations* (22)*, in the sense of Lions, with initial datum* J_0*.*

Remark 2. In Theorem 2, we assume that the incompressible Euler equations has a smooth solution, however, for the Theorem 3, we just require the existence of dissipative solution to the Euler equations.

3 The Vlasov-Maxwell-Fokker-Planck System

The Vlasov-Maxwell-Fokker-Planck (VMFP) system is a mathematical model in the kinetic theory of plasmas. It provides a statistic description of plasma in the terms of charged particle density $f^\varepsilon(t,x,\xi)$ depending on the time $t \geq 0$, the position $x = (x_1,x_2,x_3) \in [0,1]^3 \equiv \mathbb{T}^3$, and the velocity $\xi = (\xi_1,\xi_2,\xi_3) \in \mathbb{R}^3$. The (rescaled) VMFP system takes the form

$$\partial_t f^\varepsilon + \xi\cdot\nabla_x f^\varepsilon + (E^\varepsilon + \alpha\xi\times B^\varepsilon)\cdot\nabla_\xi f^\varepsilon = \mathrm{div}_\xi\big(\beta\xi f^\varepsilon + \sigma\nabla_\xi f^\varepsilon\big), \tag{28}$$

$$\varepsilon^2\alpha\partial_t E^\varepsilon - \mathrm{curl}_x B^\varepsilon = -\alpha\int_{\mathbb{R}^3}\xi f^\varepsilon d\xi, \quad -\varepsilon^2 \mathrm{div}_x E^\varepsilon = 1 - \int_{\mathbb{R}^3} f^\varepsilon d\xi, \tag{29}$$

$$\alpha\partial_t B^\varepsilon + \mathrm{curl}_x E^\varepsilon = 0, \quad \mathrm{div}_x B^\varepsilon = 0, \tag{30}$$

where $\varepsilon, \alpha, \beta$, and σ are positive parameters. More explanation on the system and the parameters involved can be found in [DL89].

Before stating our result, we firs give a lemma, whose proof is similar to that of Lemma 5.1 in [PS04].

Lemma 3. *Let* $v_0, b_0 \in \mathcal{H}^s(s > 1 + 3/2)$ *satisfying*

$$\alpha v_0 - \mathrm{curl}_x b_0 = 0.$$

Then there exist a $T^* \in (0, \infty)$ *and a unique solution* $(v, b, e) \in C([0, T^*), \mathcal{H}^s)$, *which solves the problem*

$$\partial_t v + \mathrm{div}_x(v \otimes v) - e - \alpha v \times b + \beta v = 0, \tag{31}$$

$$\alpha \partial_t b + \mathrm{curl}_x e = 0, \quad \mathrm{div}_x v = 0, \tag{32}$$

$$\alpha v - \mathrm{curl}_x b = 0, \quad \mathrm{div}_x b = 0, \tag{33}$$

$$v(0, x) = v_0(x) \in \mathcal{H}^s, b(0, x) = b_0(x) \in \mathcal{H}^s. \tag{34}$$

Theorem 4 ([HLWb]). *Let* (v_0, b_0) *be two given vectors in* $\mathcal{H}^s(s > 2+3/2)$, $\mathbb{Z}^3$ *periodic in* x. *Consider the periodic boundary problem for the system* (28)–(30) *with initial data*

$$f^\varepsilon(0, x, \xi) = f_0^\varepsilon(x, \xi), \quad E^\varepsilon(0, x) = E_0^\varepsilon(x), \quad B^\varepsilon(0, x) = B_0^\varepsilon(x). \tag{35}$$

Assume that the initial data $f_0^\varepsilon(x, \xi) \geq 0, E_0^\varepsilon(x)$, *and* $B_0^\varepsilon(x)$ *are* $\mathbb{Z}^3$ *periodic in* x, *and satisfy the standard compatibility conditions*

$$-\varepsilon^2 \mathrm{div}_x E_0^\varepsilon = 1 - \int_{\mathbb{R}^3} f_0^\varepsilon d\xi, \quad \mathrm{div}_x B_0^\varepsilon = 0. \tag{36}$$

In addition, we assume that there exists a constant $C_0 > 0$, *independent of* ε, *such that*

$$\iint_{\mathbb{R}^3 \times \mathbb{T}^3} \left(1 + |x|^2 + |\xi|^2 + |\ln f_0^\varepsilon(x, \xi)|\right) f_0^\varepsilon(x, \xi) dx d\xi$$
$$+ \int_{\mathbb{T}^3} (\varepsilon |E_0^\varepsilon(x)|^2 + |B_0^\varepsilon(x)|^2) dx \leq C_0, \tag{37}$$

furthermore,

$$\frac{\beta}{2} \iint_{\mathbb{R}^3 \times \mathbb{T}^3} |\xi - v_0(x)|^2 f_0^\varepsilon(x, \xi) dx d\xi + \sigma \iint_{\mathbb{R}^3 \times \mathbb{T}^3} f_0^\varepsilon(x, \xi) |\ln f_0^\varepsilon(x, \xi)| dx d\xi$$
$$+ \frac{\beta \varepsilon^2}{2} \int_{\mathbb{T}^3} |E_0^\varepsilon(x)|^2 dx + \frac{\beta}{2} \int_{\mathbb{T}^3} |B_0^\varepsilon(x) - b_0(x)|^2 dx \to 0 \quad \text{as} \quad \varepsilon, \sigma \to 0. \tag{38}$$

Let $(f^\varepsilon, E^\varepsilon, B^\varepsilon)$ *be the smooth solutions of the VMFP system* (28)–(30) *with initial data* (35). *Then, for all* $0 < T < T^*$ *(defined in Lemma 1), the current* J^ε *converges weakly to* v *in* $L^\infty([0, T], \mathcal{D}'(\mathbb{T}^3))$, *the scaled electric field and the magnetic field converge strongly*

$$\varepsilon E^\varepsilon \to 0 \quad \textit{and} \quad B^\varepsilon \to b \quad \textit{in} \quad L^\infty([0, T], L^2(\mathbb{T}^3)) \quad \textit{as} \quad \varepsilon, \sigma \to 0,$$

where (v, b, e) *be the local strong solutions of the limiting equations* (31)–(34) *constructed in Lemma 3.*

Acknowledgments

The authors wish to express their gratitude to the referee for many constructive comments and suggestions. L. Hsiao is supported by NSFC (Grant 10431060). F. Li is supported by NSFC (Grant 10501047) and Nanjing University Talent Development Foundation. S. Wang is supported by NSFC (Grant 10471009) and by BSF (Grant 1052001).

References

[ACG01] Arnold, A., Carrillo, J.A., Gamba, I., Shu, C.-W.: Low and high field scaling limits for the Vlasov- and Wigner-Poisson-Fokker-Planck systems. Transport Theory Statist. Phys., **30**, 121–153 (2001)

[BKM84] Beale, J.T., Kato, T., Majda, A.: Remarks on the breakdown of smooth solutions for the 3-D Euler equations. Comm. Math. Phys., **94**, 61–66 (1984)

[Br00] Brenier, Y.: Convergence of the Vlasov-Poisson system to the incompressible Euler equations. Comm. Partial Differential Equations, **25**, 737–754 (2000)

[DL89] DiPerna, R.J., Lions, P.-L.: Global weak solutions of Vlasov-Maxwell systems. Comm. Pure Appl. Math., **42**, 729–757 (1989)

[Bo93] Bouchut, F.: Existence and uniqueness of a global smooth solution for the Vlasov-Poisson-Fokker-Planck system in three dimensions. J. Funct. Anal., **111**, 239–258 (1993)

[GNP05] Goudon, T., Nieto, J., Poupaud, F., Soler, J.: Multidimessional high-field limit of the electrostatic vlasov-poisson-fokker-planck system. J. Differential Equations, **213**, 418–442 (2005)

[HLW06] Hsiao, L., Li, F., Wang, S.: Convergence of the Vlasov-Poisson-Fokker-Planck system to the incompressible Euler equations. Sci. China Ser. A, **49**, 255–266 (2006)

[HLWa] Hsiao, L., Li, F., Wang, S.: The monokinetic limit of the Vlasov-Poisson-Fokker-Planck system, preprint.

[HLWb] Hsiao, L., Li, F., Wang, S.: Convergence of the Vlasov-Maxwell-Fokker-Planck system to the electron magnetohydrodynamics equations, preprint.

[Li96] Lions, P.-L.: Mathematical Topics in Fluid Mechanics, Vol. 1, Incompressible Models. Oxford University Press, New York (1996)

[PW05] Peng, Y.J., Wang, Y.G.: Convergence of compressible Ruler-Poisson equations to incompressible type Euler equations. Asymptot. Anal., **41**, 141–160 (2005)

[Pe04] Perthame, B.: Mathematical tools for kinetic equations. Bull. Amer. Math. Soc. (N.S.), **41**, 205–244 (2004)

[PS00] Poupaud, F., Soler, J.: Parabolic limit and stability of the Vlasov-Fokker-Planck system. Math. Models Methods Appl. Sci., **10**, 1027–1045 (2000)

[PS04] Puel M., Saint-Raymond L.: Quasineutral limit for the relativistic Vlasov-Maxwell system. Asymptot. Anal., **40**, 303–352 (2004)

High-Resolution Methods and Adaptive Refinement for Tsunami Propagation and Inundation

D.L. George and R.J. LeVeque

Summary. We describe the extension of high-resolution finite volume methods and adaptive refinement for the shallow water equations in the context of tsunami modeling. Godunov-type methods have been used extensively for modeling the shallow water equations in many contexts; however, tsunami modeling presents some unique challenges that must be overcome. We describe some of the specific difficulties associated with tsunami modeling, and summarize some numerical approaches that we have used to overcome these challenges. For instance, we have developed a well-balanced Riemann solver that is appropriate in the deep ocean regime as well as robust in near-shore and dry regions. Additionally, we have extended adaptive refinement algorithms to this application. We briefly describe some of the modifications necessary for using these adaptive methods for tsunami modeling.

1 Introduction

The shallow water equations are a commonly accepted governing approximation for tsunami propagation in the deep ocean as well as in near-shore regions – including inundation. Because of difficulties in the deep ocean associated with preserving the delicate steady state, the physically relevant form of the shallow water equations – a hyperbolic system for depth and momentum – can be problematic for transoceanic or global-scale tsunami modeling. Because of this, tsunami modelers have often used alternative forms of the shallow water equations which are valid for smooth solutions. However, these systems, and the numerical methods used to solve them, are rarely suited for modeling tsunami inundation since this regime can include steep gradients and shocks. We describe some numerical methods that are promising for modeling the physically relevant conservative form of the shallow water equations in all regimes of tsunami flow as well as adaptive algorithms that allow bridging the diverse scales at which the different regimes of tsunami flow occur.

The amplitude of tsunamis in the deep ocean is on the order of centimeters, meaning that tsunamis are a tiny perturbation to the steady state – a motionless body of water several kilometers deep. The steady state exists due to a nontrivial balance of momentum flux and a source term due to nonconstant bottom bathymetry. Modeling tsunami propagation accurately, therefore, demands resolving this perturbation against the background steady state. This requires well-balanced Riemann solvers (e.g., [7, 10, 13]) if Godunov methods are to be used for the conservative form of the equations. As tsunamis approach the shore, energy compression leads to more violent flow characteristics including shocks or turbulent bores in the inundation regime. Numerical schemes must be able to capture or track these shocks, and additionally must be robust and accurately given the appearance of dry states in the near-shore regime. A novel approximate Riemann solver that we have developed to handle these different challenges is briefly described in Sect. 3.

The diverse regimes of tsunami flow also occur at diverse spatial scales. In the deep ocean, tsunami wavelength is several hundred kilometers. Modeling transoceanic propagation obviously requires a large computing domain, yet, as tsunamis approach the shore the wavelength is compressed to several hundred meters in the shallow coastal waters. To accurately model coastal inundation, a much finer computational grid is required than can feasibly be used at the global scale. Grid cells of several hundred meters or significantly less are necessary if near-shore bathymetry features vary on a small scale. Additionally, because bathymetry focuses tsunami energy unpredictably, areas needing the most refinement cannot be easily determined a priori. Furthermore, tsunami waves, like those in other hyperbolic systems, are often highly localized at a given time but move throughout the computational domain. For these reasons, we believe that adaptive refinement is warranted for efficient global-scale tsunami calculations. We will briefly describe the extension of adaptive refinement algorithms, developed more generally for hyperbolic problems by Berger, Colella, Oliger, and LeVeque [4, 6, 5], to the specificities of tsunami modeling.

The methods and results described in this chapter are implemented in TsunamiClaw – an extension of the clawpack [11] software. We will describe the one-dimensional equations and algorithms for simplicity.

2 The Shallow Water Equations

The shallow water equations are a commonly used approximation for modeling tsunamis in all regimes – from the deep ocean to the inundation regime. We solve the physically relevant hyperbolic system

$$h_t + (hu)_x = 0, \tag{1}$$

$$(hu)_t + \left(hu^2 + \frac{1}{2}gh^2\right)_x = -ghb_x, \tag{2}$$

where $h(x,t)$ is the fluid depth, $u(x,t)$ the horizontal fluid velocity, and $b(x)$ the variable bottom bathymetry. We will use η to denote the fluid surface elevation

$$\eta(x,t) = h(x,t) + b(x). \tag{3}$$

The physically relevant steady state, against which tsunamis propagate in the deep ocean, comes from the nontrivial balance of the pressure flux and the source term due to bathymetry

$$\left(\frac{1}{2}gh^2\right)_x \approx -ghb_x. \tag{4}$$

It is well known (e.g., [10, 13]) that fractional step methods fail at preserving such steady states precisely. Therefore, considerable research has gone into developing well-balanced methods (e.g., [2, 3, 8, 10]) which preserve important discrete steady states, such as the "ocean at rest" steady state, where $\eta_x \equiv 0$ and $u \equiv 0$. When using adaptive refinement, it is equally important to maintain these steady states upon interpolating from coarse grids to fine grids, and averaging fine grids onto coarse grids.

3 Approximate Riemann Solvers for Wave Propagation Algorithms

The numerical method we use for this problem is based on the wave propagation algorithms described in [12]. These algorithms belong to the class of high-resolution Godunov-type finite volume methods that make use of Riemann problems at grid cell interfaces to solve hyperbolic systems of the form

$$q_t + f(q)_x = 0, \tag{5}$$

where $q \in \mathbb{R}^m$ and $f(q) \in \mathbb{R}^m$. A crucial component of the wave propagation algorithm is the determination of updating waves by a decomposition of the Riemann initial data into some set of appropriately chosen vectors

$$Q_i^n - Q_{i-1}^n = \sum_{p=1}^{M_w} \alpha_{i-1/2}^p r_{i-1/2}^p, \tag{6}$$

where $Q_i^n \in \mathbb{R}^m$ is the numerical solution in the ith grid cell at time t^n, M_w is the number of vectors used to approximate the Riemann solution, and $r^p \in \mathbb{R}^m$, $p = 1, \ldots, M_w$, are vectors chosen for the decomposition. Once the "waves" $\alpha_{i-1/2}^p r_{i-1/2}^p$ are determined, each wave propagates at a chosen speed $s_{i-1/2}^p$. Of course, one usually chooses $M_w = m$, to have a unique solution for the scalars $\alpha_{i-1/2}^p$, $p = 1, \ldots, M_w$. Typically $\{r_{i-1/2}^p, s_{i-1/2}^p\}$ is some approximation to the pth eigenpair of some locally linearized Jacobian: $\bar{A}(Q_{i-1}^n, Q_i^n) \approx f'(Q_{i-1}^n) \approx f'(Q_i^n)$.

A consistent and alternative method to (6) for determining waves of an approximate Riemann solution is to decompose the flux into a set of propagating vectors

$$f(Q_i^n) - f(Q_{i-1}^n) = \sum_{p=1}^{M_w} \beta_{i-1/2}^p r_{i-1/2}^p, \tag{7}$$

where again $\{r_{i-1/2}^p, s_{i-1/2}^p\}$ is some approximation to the pth eigenpair of some linearized Jacobian. In this case $\beta_{i-1/2}^p$ has the same dimension as $s_{i-1/2}^p \alpha_{i-1/2}^p$. This method, described more fully in [3], has the advantageous property of producing conservative Riemann solutions regardless of the form of the vectors $r_{i-1/2}^p$. If the eigenpairs of a Roe-averaged Jacobian are used to define $r_{i-1/2}^p$ and $s_{i-1/2}^p$ in (6) and (7), the two methods are the same.

For nonhomogeneous hyperbolic systems

$$q_t + f(q)_x = \psi(q, x), \tag{8}$$

it is consistent to include the effect of a source term directly into updating waves by performing the decomposition

$$f(Q_i^n) - f(Q_{i-1}^n) - \Psi_{i-1/2}\Delta x = \sum_{p=1}^{M_w} \beta_{i-1/2}^p r_{i-1/2}^p, \tag{9}$$

where $\Psi_{i-1/2}$ is some consistent approximation to the source term $\psi(q, x)$ at $x_{i-1/2}$, and $\Delta x = x_i - x_{i-1}$. See [3] for more details.

In [14], LeVeque and Pelanti explore the idea of performing a decomposition of the form

$$\begin{bmatrix} Q_i^n - Q_{i-1}^n \\ f(Q_i^n) - f(Q_{i-1}^n) \end{bmatrix} = \sum_{p=1}^{2m} \alpha_{i-1/2}^p \begin{bmatrix} w_{i-1/2}^p \\ \phi_{i-1/2}^p \end{bmatrix}, \tag{10}$$

using vectors $(w_{i-1/2}^p, \phi_{i-1/2}^p)^T \in \mathbb{R}^{2m}$, for $p = 1, \ldots, 2m$. Several options for $w_{i-1/2}^p$ and $\phi_{i-1/2}^p$ are explored, and connections to relaxation solvers are explored. For instance, it is possible to choose $(w_{i-1/2}^p, \phi_{i-1/2}^p)^T$ such that the methods (6), (7), and (10) are all the same.

For the shallow water equations (1), we have developed a solver based on a decomposition of the form

$$\begin{bmatrix} Q_i^n - Q_{i-1}^n \\ \varphi(Q_i^n) - \varphi(Q_{i-1}^n) \\ b_i - b_{i-1} \end{bmatrix} = \sum_{p=0}^{3} \alpha_{i-1/2}^p \tilde{r}_{i-1/2}^p, \tag{11}$$

where $\tilde{r}^p \in \mathbb{R}^4$, for $p = 0, \ldots, 3$, $q = (h, hu)^T$, $\varphi(q) = (hu^2 + \frac{1}{2}gh^2)$, and b_i is the bathymetry in the ith grid cell. The vectors $\tilde{r}_{i-1/2}^p$ and corresponding propagation speeds $\tilde{s}^p_{i-1/2}$ are chosen to be local approximations to the eigenvectors and eigenvalues of $\tilde{A}(\tilde{q}) \in \mathbb{R}^{4\times 4}$, where

$$\tilde{q}_t + \tilde{A}(\tilde{q})\tilde{q}_x = 0 \tag{12}$$

is an overdetermined system equivalent to the shallow water equations for smooth solutions $\tilde{q} = (h, hu, \varphi, b)^T$. By choosing certain averages in the local approximations $\tilde{r}^p_{i-1/2}$, the method has certain desirable properties. For instance, since $\tilde{s}^0_{i-1/2} = 0$, the solver preserves a large class of discrete steady state solutions as a stationary contact discontinuity at the cell interface $x_{i-1/2}$. Second, by defining the stationary eigenvector $\tilde{r}^0_{i-1/2}$ appropriately, the solver preserves depth nonnegativity of the approximate solution. Additionally, the solver can be shown to be equivalent to the Roe solver in the case of shock solutions over constant bathymetry. The Riemann solver is described in detail in [9].

4 Adaptive Mesh Refinement

To deal with the disparate spatial scales required to resolve a tsunami in the global propagation regime and the local inundation regime, we have extended adaptive mesh refinement routines (AMR) to this application. These algorithms (e.g., [4, 6, 5]) allow nesting of multiple rectangular Cartesian subgrids of several levels with integer refinement ratios within the computational domain. Because the subgrids evolve spatially and temporally, grids can essentially track moving features of the solution at various resolutions. For tsunami modeling, this allows transoceanic waves to be tracked on grids of suitable resolution without having to resolve unaffected regions of the ocean on fine grids. Additionally, since waves are compressed in the near-shore region, even finer subgrids appear as waves approach the shore – allowing inundation modeling on meter-scale grids.

As mentioned, modeling tsunami propagation requires resolving a small deviation from the background steady state. Since AMR must interpolate data on coarse grids to generate finer level grids, and since fine grids must be averaged to update underlying coarse grid cells, special care must be taken to preserve steady states during this process. For instance, the standard approach of using a linear interpolant within each coarse cell, based on the conserved variables in the surrounding cells, does not in general preserve the common steady state $\eta_x \equiv 0$, $hu \equiv 0$ on the new fine grid. Similarly, averaging the conserved variables in fine grid cells contained within coarse cells does not preserve the steady state on the coarser grid. A simple one-dimensional example of interpolation from a coarse grid with a refinement ratio of 2 is shown in Fig. 1. The steady state is not maintained. Given practical grid resolutions for computations of the deep ocean, the spurious waves generated by such interpolation could be orders of magnitude larger than the actual tsunami being modeled.

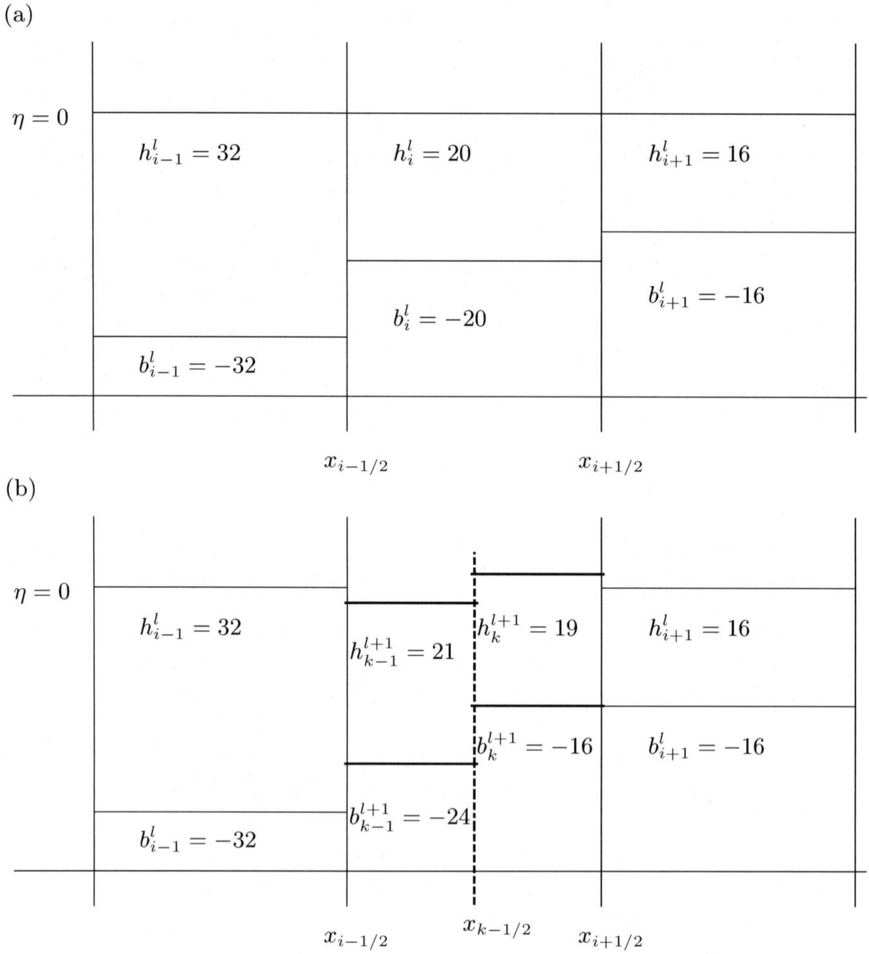

Fig. 1. Interpolating the water depth h from coarse grid cells to fine grid cells over nonlinear variable bathymetry. (**a**) A level l grid in one dimension for the common steady state problem of a motionless body of water with a flat surface $\eta = 0$. The numerical bathymetry is an average of the true bathymetry values in each grid cell. (**b**) Interpolation of h to a level $(l + 1)$ grid – refined by a factor of 2 – showing the two fine grid cells in the coarse cell $\mathcal{C}^l_i = [x_{i-1/2}, x_{i+1/2}]$. The bathymetry in the two fine grid cells reflects the average of the true bathymetry in those cells. Interpolating the depth h using minmod slopes destroys the steady state

A simple fix to the problem shown in Fig. 1 is to interpolate the surface elevation η rather than the conserved variable h. The water depths on the fine grid are then determined from the values of η by $h = \eta - b$. It is easy to show that this maintains conservation of mass upon interpolating so long as all of the depths remain positive. Since maintaining conservation of momentum

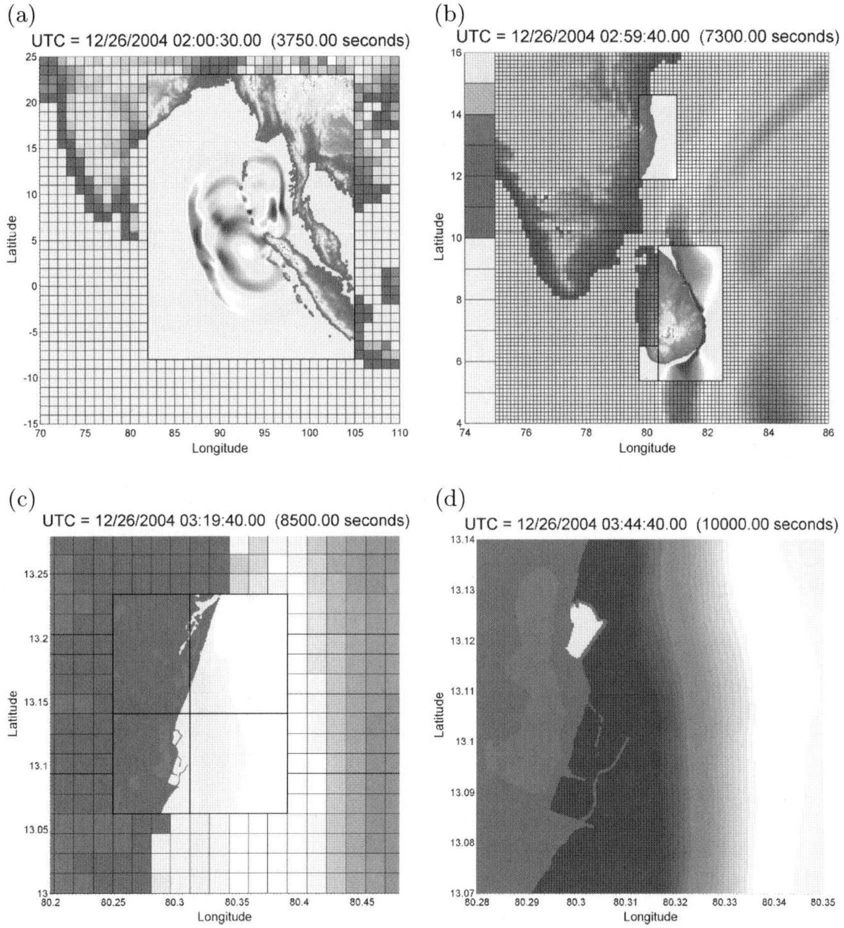

Fig. 2. TsunamiClaw simulation of the 2004 Indian Ocean Tsunami using adaptive refinement. The simulation had four levels of refinement with refinement ratios of 8, 8, and 64. This example features inundation modeling in Eastern India. Grid lines on the highest level of each figure are omitted for clarity. (**a**) The transoceanic tsunami waves are resolved on second level grids. Undisturbed areas of the Indian Ocean are maintained on the very coarse first level grid. (**b**) As the waves approach Sri Lanka and Eastern India, third level grids appear around the impacted shores. (**c**) The region around Madras, India is resolved on fourth level grids. This region is within the northernmost third level grids shown in the *upper right figure*. (**d**) Inundation of the Madras, India area is resolved on fine-scale fourth level grids. The commercial harbor to the south suffers greater inundation than the fishing harbor visible to the north. A color version is available at `http://www.amath.washington.edu/~rjl/pubs/hyp06tsunami`

requires interpolating hu rather than u, it was necessary for us to develop a limiting strategy for determining fine grid values of hu to prevent unbounded velocities u on the fine grid. Additional strategies of interpolation had to be developed to prevent spurious "shore waves" upon interpolating grids near the shoreline. These issues are described in detail in [9].

An example of TsunamiClaw simulation of the 2004 Indian Ocean Tsunami is shown in Fig. 2. The tsunami was generated dynamically at the start of the computation by using a spatial temporal model of the fault motion provided by the Seismolab at Caltech [1].

5 Conclusions

Tsunamis exhibit diverse flow regimes requiring a numerical method that can simultaneously resolve near steady state solutions for transoceanic propagation as well as converge to shock solutions representing turbulent bores. Additionally the numerical method must be robust to the appearance of dry states in the inundation regime. We have developed an approximate Riemann solver that can handle these multiple features. Additionally the regimes of tsunamis occur at diverse spatial scales, requiring some form of grid refinement. We have modified adaptive refinement algorithms for this application, so that transoceanic propagation and local inundation can be modeled in single global-scale computations.

Acknowledgments

This work was supported in part by NSF grants CMS-0245206, DMS-0106511, and DOE grant DE-FC02-01ER25474. The authors would also like to thank Marsha Berger and Harry Yeh.

References

1. C.J. Ammon et al. Rupture process of the 2004 Sumatra-Andaman earthquake. *Science*, 308:1133–1139, 2005.
2. E. Audeusse, F. Bouchut, M.O. Bristeau, R. Klein, and B. Perthame. A fast and stable well-balanced scheme with hydrostatic reconstruction for shallow water flows. *SIAM J. Sci. Comp.*, 25(6):2050–2065, 2004.
3. D. Bale, R.J. LeVeque, S. Mitran, and J.A. Rossmanith. A wave-propagation method for conservation laws and balance laws with spatially varying flux functions. *SIAM J. Sci. Comput.*, 24:955–978, 2002.
4. M.J. Berger and P. Colella. Local adaptive mesh refinement for shock hydrodynamics. *J. Comput. Phys.*, 82:64–84, 1989.
5. M.J. Berger and R.J. LeVeque. Adaptive mesh refinement using wave-propagation algorithms for hyperbolic systems. *SIAM J. Numer. Anal.*, 35:2298–2316, 1998.

6. M.J. Berger and J. Oliger. Adaptive mesh refinement for hyperbolic partial differential equations. *J. Comput. Phys.*, 53:484–512, 1984.
7. F. Bouchut. *Nonlinear Stability of Finite Volume Methods for Hyperbolic Conservation Laws and Well-Balanced Schemes for Sources.* Birkhäuser Verlag, 2004.
8. P. García-Navarro and M.E. Vázques-Cendón. On numerical treatment of the source terms in the shallow water equations. *Comput. Fluids*, 29:17–45, 2000.
9. D.L. George. *Finite volume methods and adaptive refinement for tsunami propagation and inundation.* PhD thesis, University of Washington, Seattle, WA, 2006.
10. J.M. Greenberg and A.Y. LeRoux. A well-balanced scheme for numerical processing of source terms in hyperbolic equations. *SIAM Journal of Numerical Analysis*, 33:1–16, 1996.
11. R.J. LeVeque. CLAWPACK software. `http://www.amath.washington.edu/~claw`.
12. R.J. LeVeque. Wave propagation algorithms for multi-dimensional hyperbolic systems. *Journal of Computational Physics*, 131:327–335, 1997.
13. R.J. LeVeque. Balancing source terms and flux gradients in high-resolution Godunov methods: The quasi-steady wave propagation algorithm. *Journal of Computational Physics*, 146:346–365, 1998.
14. R.J. LeVeque and M. Pelanti. A class of approximate Riemann solvers and their relation to relaxation schemes. *Journal of Computational Physics*, 172:572–591, 2001.

Young Measure Solutions of Some Nonlinear Mixed Type Equations

H.-P. Gittel

Summary. This contribution deals with measure-valued solutions to two types of nonlinear partial differential equations for which general existence results fail to exist. They are the potential equation for transonic flow and the associated unsteady problem (forward–backward diffusion equation). The solutions are constructed by an iteration scheme (Katchanov method) and additional time discretization (Rothe method) in the second case. The existence is proved in the sense of spatial gradient Young measures.

1 Transonic Flow

Let us study a standard model in transonic gas dynamics. The irrotational steady and isentropic flow of an inviscid gas in a bounded simply connected domain $\Omega \subseteq \mathbb{R}^N$ can be described by the continuity equation

$$\operatorname{div}\left(\rho\left(|\nabla u|^2\right)\nabla u\right) = 0 \qquad \textit{(full potential equation).} \tag{1}$$

Here, u denotes the velocity potential and ρ is the density of the gas related to $|\nabla u|^2$ by Bernoulli law. For a *polytropic (ideal) gas*, this function ρ is assumed to be

$$\rho(q) = \begin{cases} \rho_0\left(1 - \dfrac{q}{q_{max}}\right)^{1/(\kappa-1)} & \text{for } q \le q_1 \\ \rho_1 & \text{for } q > q_1 \end{cases} \tag{2}$$

with constants $\rho_0, q_{max} > 0$, $\kappa > 1$, $q_1 \in (q_{son}, q_{max})$, $q_{son} := \frac{\kappa-1}{\kappa+1} q_{max}$, and $\rho_1 := \rho(q_1)$. Note that the density is truncated for large values of the speed to keep the flow away from cavitation and to ensure the validity of the physical model (see [FN85, Git87]).

The type of the nonlinear partial differential operator in (1) is determined by the sign of the function $\rho(q) + 2q\rho'(q)$. It is a mixed one and depends on $|\nabla u|$:

elliptic	iff	$\lvert\nabla u\rvert^2 \in [0, q_{son}) \cup (q_1, \infty)$	*(subsonic region)*
parabolic	iff	$\lvert\nabla u\rvert^2 = q_{son}$	*(sonic region)*
hyperbolic	iff	$\lvert\nabla u\rvert^2 \in (q_{son}, q_1)$	*(supersonic region).*

Remark 1. In the case of *transonic flow*, we have to take into account that there exist subsonic regions as well as supersonic ones in Ω. The transitions between these regions are usually discontinuous, and shocks with jumps in ∇u occur. From the mathematical point of view, these facts cause many significant difficulties in existence, construction, approximation, and regularity of solutions to potential equation (1) (see, e.g., Morawetz [Mor04]).

To consider a boundary value problem for (1), we prescribe the mass flux of the flow orthogonal to the Lipschitz boundary $\partial\Omega$, i.e., we suppose the *boundary condition*

$$\rho\left(|\nabla u|^2\right) \frac{\partial u}{\partial \mathbf{n}} = g \quad \text{on } \partial\Omega\,. \tag{3}$$

Here, $g \in L^2(\partial\Omega)$ is a given function with $\int_{\partial\Omega} g\, do = 0$ and $\mathbf{n}$ is the exterior unit normal to $\partial\Omega$. The *weak formulation* of problem (1) and (3) in a subspace $\mathcal{V}$ of the Sobolev space $W^{1,2}(\Omega)$ reads

$$\int_\Omega \rho\left(|\nabla u|^2\right) \nabla u \nabla v\, dx = \int_{\partial\Omega} gv\, do \qquad \text{for all } v \in \mathcal{V}, \tag{4}$$

where

$$\mathcal{V} := \left\{ v \in W^{1,2}(\Omega) \,\middle|\, \int_\Omega v\, dx = 0 \right\} \quad \text{with} \quad \|v\|_{\mathcal{V}} := \left(\int_\Omega |\nabla v|^2\, dx \right)^{1/2}.$$

Let

$$J(v) := \frac{1}{2} \int_\Omega \left(\int_0^{|\nabla v|^2} \rho(\sigma)\, d\sigma \right) dx, \quad l(v) := \int_{\partial\Omega} gv\, do \qquad \text{for } v \in \mathcal{V}, \tag{5}$$

then relation (4) is the *Euler–Lagrange equation* to the variational problem

$$J(v) - l(v) \longrightarrow \min_{v \in \mathcal{V}}.$$

2 Iteration Scheme

With the notations introduced in (5) and the abbreviation

$$B(u; v, w) := \int_\Omega \rho\left(|\nabla u|^2\right) \nabla v \nabla w\, dx \qquad \text{for } u, v, w \in \mathcal{V}, \tag{6}$$

the weak problem (4) takes the abstract form

$$B(u; u, v) = l(v) \qquad \text{for all } v \in \mathcal{V}. \tag{7}$$

Now, the structure of the last relation implies a simple iteration procedure (*Katchanov method*):

Let $u_0 \in \mathcal{V}$ be arbitrary, $n = 0, 1, 2, \ldots$, and $u_{n+1} \in \mathcal{V}$:

$$B(u_n; u_{n+1}, v) = l(v) \qquad \text{for all } v \in \mathcal{V}. \tag{8}$$

Recall the properties of the symmetric, bounded, uniformly positive, and bilinear form $B(u; \cdot, \cdot)$ defined by (6) to deduce the following results of Feistauer and Nečas [FN85] and of Gittel [Git87].

Theorem 1. *The Katchanov sequence (u_n) from (8) satisfies the a priori estimates*

$$\begin{aligned} &\textbf{(a)} \quad \rho_1 \|u_n\|_{\mathcal{V}}^2 \le B(u_{n-1}; u_n, u_n) \le \|u_n\|_{\mathcal{V}} \|l\|_{\mathcal{V}^*}, \\ &\textbf{(b)} \quad \rho_1 \|u_{n+1} - u_n\|_{\mathcal{V}}^2 \le j_n - j_{n+1}, \end{aligned}$$

where $j_n := 2\,(J(u_n) - l(u_n))$ and $\mathcal{V}^*$ denotes the dual space of $\mathcal{V}$.

Corollary 1. *The functions u_n are uniformly bounded in $\mathcal{V}$ and*

$$B(u_n; u_n, v) = l(v) + R_n(v) \quad \textit{with} \quad \|R_n\|_{\mathcal{V}^*} \to 0.$$

Moreover, there exists a subsequence $(u_{n'})$ of (u_n):

$$u_{n'} \rightharpoonup u \textit{ in } \mathcal{V}, \;\; u_{n'} \to u \textit{ in } L^2(\Omega) \cap L^2(\partial\Omega), \;\; \textit{as } n \to \infty.$$

The question arises whether the limit function $u \in \mathcal{V}$ of a Katchanov sequence (u_n) is a solution of (7) (resp., (4)). Up to now, one can only show that u solves (4) provided additional assumptions (*entropy conditions*) are imposed on (u_n) (see [BF93, FN85, Git87]). But, each limit function u satisfies relation (4) in the more general sense of measures. Hence, the following considerations fit into the framework of DiPerna [DiP85] concerning systems of conservation laws.

3 Young Measure Solutions

We recall the definition of Young measures which were introduced by Young [You69] in optimal control theory.

Definition 1. *Let K be a compact subset of $\mathbb{R}^M$ and let* $\operatorname{Prob}(K)$ *be the collection of all nonnegative regular Borel measures μ with $\mu(K) = 1$. Then, a* Young measure *ν is a family of probability measures $\nu = (\nu_x)_{x\in\Omega}$:*

$$\nu : \Omega \to \operatorname{Prob}(K) \subset \left(C^0(K)\right)^*$$

if there is a sequence $(f_n) \subset L^\infty(\Omega)^M$ such that:

$$\begin{aligned} f_n(x) &\in K \quad \textit{for a.a. } x \in \Omega, \\ f_n &\overset{*}{\rightharpoonup} f \quad \textit{in } L^\infty(\Omega)^M, \\ F(f_n) &\overset{*}{\rightharpoonup} \int_K F(\lambda)\, d\nu(\lambda) =: \langle \nu, F\rangle \quad \textit{in } L^\infty(\Omega) \textit{ for all } F \in C^0(K). \end{aligned}$$

Applications of Young measures to nonlinear partial differential equations were studied by several authors and were first presented in a paper of Tartar [Tar79]. Here, a modified version of the above definition is used (see, e.g., Kinderlehrer and Pedregal [KP92, KP94]).

Definition 2. *A family of measures $\nu : \Omega \to \operatorname{Prob}(\mathbb{R}^N)$ is called a $W^{1,2}(\Omega)$-*gradient Young measure *generated by (u_n) if*

$$\begin{aligned} \|u_n\|_{W^{1,2}(\Omega)} &\le C \qquad \textit{for } n = 0, 1, 2, \ldots, \\ u_n &\rightharpoonup u \qquad \textit{in } W^{1,2}(\Omega), \\ F(\nabla u_n) &\rightharpoonup \langle \nu, F\rangle \quad \textit{in } L^r(\Omega) \end{aligned}$$

for all $F \in C^0(\mathbb{R}^N)$ satisfying the growth condition $|F(\lambda)| \le a + b|\lambda|^{2/r}$ with constants a, b, and $r > 1$.

Remark 2. Note that in Definition 2 $F(\nabla u_n) \rightharpoonup F(\nabla u)$ for all functions F in question iff ν_x equals to a Dirac measure or, more precisely, $\nu_x = \delta_{\nabla u(x)}$ for a.a. $x \in \Omega$. This is equivalent to $u_n \to u$ in $W^{1,2}(\Omega)$, as $n \to \infty$.

In this framework, the properties of the Katchanov sequence (u_n) studied in Sect. 2 immediately imply an existence theorem to the boundary value problem (4) (see [Git06]).

Theorem 2. *Let $g \in L^2(\partial\Omega)$, $\int_{\partial\Omega} g\, do = 0$. Then, there exists a Young measure solution (u, ν) to (4), i.e., a function $u \in \mathcal{V}$ and a $W^{1,2}(\Omega)$-gradient Young measure ν such that*

$$\int_\Omega \langle \nu, \rho\left(\lambda^2\right) \lambda\rangle \nabla v\, dx = \int_{\partial\Omega} g v\, do \quad \textit{for all } v \in \mathcal{V}.$$

Moreover,

$$\begin{aligned} \left\langle \nu_x, \rho\lambda^2\right\rangle &= \langle \nu_x, \rho\lambda\rangle \langle \nu_x, \lambda\rangle & &\textit{(independence relation)}, \\ \langle \nu_x, \lambda\rangle &= \nabla u(x) \quad \textit{for a.a. } x \in \Omega & &\textit{(coupling condition)}. \end{aligned} \tag{9}$$

By Remark 2, the function u in Theorem 2 would be a solution of (4) provided it could be proved that $\operatorname{supp}\nu_x$ reduces to a single point for a.a. $x \in \Omega$. But, except for particular cases, only a minor result is known (see [FN85, Git06]): Suppose that $\operatorname{supp}\nu_x \subset \{\lambda \in \mathbb{R}^N \mid |\lambda| < q_{son}\}$ (elliptic regime) for the associated Young measure ν. Then $\nu_x = \delta_{\nabla u(x)}$ and u is a weak solution of the boundary value problem (1) and (3). In general, the weak limit of $\left(\rho(|\nabla u_n|^2)\nabla u_n\right)$ may be unequal to $\rho(|\nabla u|^2)\nabla u$ and $\left\langle \nu_x, \rho\left(\lambda^2\right)\lambda\right\rangle$ measures the probability of values which are achieved at $x \in \Omega$ by this sequence in our approximation procedure.

Let us note some additional properties of the Young measure solution (u, ν) to (4) constructed above. They follow from the estimates in Theorem 1 and from Corollary 1, as $n \to \infty$:

$$\begin{aligned} \rho_1 \|u\|_{\mathcal{V}}^2 \le \int_\Omega \left\langle \nu, \rho\lambda^2 \right\rangle\, dx \le \|u\|_{\mathcal{V}} \|l\|_{\mathcal{V}^*}, \\ \| \left\langle \nu, \rho\lambda \right\rangle \|_{\mathcal{V}^*} \le \|l\|_{\mathcal{V}^*}. \end{aligned} \tag{10}$$

4 Forward–Backward Diffusion

The iteration procedure presented in Sect. 2 is robust enough to apply to the time-dependent analogon of (1) and (3). Let us consider the initial boundary value problem:

$$\begin{aligned} \frac{\partial u}{\partial t} &= \operatorname{div}\left(\rho\nabla u\right) \quad \text{in } (0,T)\times\Omega, \\ \rho\frac{\partial u}{\partial \mathbf{n}} &= g \qquad\qquad\quad \text{on } (0,T)\times\partial\Omega, \\ u(0,x) &= u_0(x) \qquad\quad \text{for } x\in\Omega. \end{aligned} \tag{11}$$

Remark 3. The differential equation in (11) describes a *forward–backward diffusion* since it behaves like the forward heat equation if $|\nabla u|^2 < q_{son}$ or $|\nabla u|^2 > q_1$ and like the backward one if $q_{son} < |\nabla u|^2 < q_1$ (see, e.g., Kawohl and Kutev [KK98]). It is sometimes called *anisotropic diffusion equation* and is used in image enhancement processes (see [PM90]). In these applications, diffusion functions ρ different from (2) are used, e.g., $\rho(q) = e^{-q}$, $\rho(q) = (1+q)^{-1}$. Truncate such a ρ for large values of q to get a modified function which has a positive lower bound and possesses properties analogous to those ones of (2). We want to point out that, in our approach, the threshold q_1 can be chosen arbitrary large.

For general data, problem (11) does not admit classical strong or weak solutions. Demoulini [Dem96] and Yin and Wang [YW02] proved the existence of Young measure solutions to such type of problems by another method. Asymptotic analysis in [Dem96], as $t \to \infty$, also yields a result for the steady-state problem, i.e., for a variant of (4).

To construct an approximation of problem (11), we substitute the time derivative with the corresponding backward difference quotient (*Rothe method*)

$$\frac{u(t,x) - u(t-h,x)}{h} = \operatorname{div}(\rho \nabla u)(t,x), \;\; h > 0.$$

Then, the weak problem in abstract form reads

$$\frac{1}{h}\int_\Omega (u-w)v\,dx + B(u;u,v) = l(v) \qquad \text{for all } v \in \mathcal{V} \tag{12}$$

where $h, t > 0$ are fixed, $w \in \mathcal{V}$ is given, and l and B are defined, respectively, by (5) and (6). If we use

$$\tilde{J}(v) := J(v) + \frac{1}{2h}\int_\Omega v^2\,dx \quad \text{and} \quad \tilde{l}(v) := l(v) + \frac{1}{2h}\int_\Omega wv\,dx \tag{13}$$

instead of J and l, respectively, the same procedure as for (4) yields a Young measure solution (u,ν) to (12) depending on h,t,w. Let $\left(u^{h,k},\nu^{h,k}\right)$ be this family of solutions (*Katchanov–Rothe sequence*) assigned to $t = kh$, $1 \le k \le \frac{T}{h}$, $w = u^{h,k-1}$, and $u^{h,0} = u_0$. Then

$$\|u^{h,k}\|^2_{L^2(\Omega)} + h\rho_1\|u^{h,k}\|^2_{\mathcal{V}} \le \frac{h}{\rho_1}\|l(kh,\cdot)\|^2_{\mathcal{V}^*} + \|u^{h,k-1}\|^2_{L^2(\Omega)}. \tag{14}$$

This inequality is a crucial ingredient in the derivation of the next results and follows from (10) where the modifications in (12) and (13) have to be observed. For $t \in [(k-1)h, kh]$, we define

$$\begin{array}{lll} u^h(t,x) & \text{by linear interpolation of} & u^{h,k}(x), \\ \nu^h_{t,x} & \text{by constant interpolation of} & \nu^{h,k}_x, \\ l^h(t,\cdot) & \text{by constant interpolation of} & l(kh,\cdot), \end{array}$$

and obtain the properties of u^h and ν^h (see [Git06]):

(i) $$\int_0^T \int_\Omega \left(\frac{\partial u^h}{\partial t} v + \langle \nu^h, \rho\lambda\rangle \nabla v\right) dt dx = \int_0^T l^h(v)\,dt$$
for all $v \in L^2((0,T),\mathcal{V})$,

(ii) $u^h(0,x) = u_0(x)$ for a.a. $x \in \Omega$,

(iii) $$\|u^h(T,.) - u_0\|_{L^2(\Omega)} + \|u^h\|_{L^2((0,T),\mathcal{V})} \le C\left(\|l^h\|_{L^2((0,T),\mathcal{V}^*)} + \sqrt{T}\|u_0\|_{\mathcal{V}}\right).$$

Remark 4. Estimate (iii) is a consequence of the nonlinearity and the uniform positivity of the diffusion function ρ and does not hold true for the usual linear backward heat equation.

At a first sight, the pair (u^h, ν^h) looks like a desired solution. But, it will not fulfill the fundamental relations corresponding to (9). This can be shown for the limit, as $h \to 0$, following the lines of the proofs in [Dem96]. Finally we get the existence theorem.

Theorem 3. *Let $g \in L^2((0,T) \times \partial\Omega)$, $\int_{\partial\Omega} g\,do = 0$ for a.a. $t \in (0,T)$ and $u_0 \in \mathcal{V}$. Then, there exists a Young measure solution (u,ν) to (11), i.e., a function $u \in L^2((0,T),\mathcal{V})$, $\frac{\partial u}{\partial t} \in L^2((0,T),\mathcal{V}^*)$, and a (spatial) $W^{1,2}(\Omega)$-gradient Young measure $(\nu_{t,.})$ such that*

$$\int_0^T \int_\Omega \left(\frac{\partial u}{\partial t} v + \langle \nu, \rho\lambda \rangle \nabla v \right) dtdx = \int_0^T \int_{\partial\Omega} gv\, dtdo$$

for all $v \in L^2((0,T),\mathcal{V})$ and

$$\begin{aligned} \|u(t,.) - u_0\|_{L^2(\Omega)} &\to 0, \quad \text{as } t \searrow 0, \\ \langle \nu_{t,x}, \rho\lambda^2 \rangle &= \langle \nu_{t,x}, \rho\lambda \rangle \langle \nu_{t,x}, \lambda \rangle, \\ \langle \nu_{t,x}, \lambda \rangle &= \nabla u(t,x) \quad \text{for a.a. } (t,x) \in (0,T) \times \Omega. \end{aligned}$$

A detailed analysis is necessary to derive further results concerning properties of a Young measure solution (u,ν) to (11) especially information on the size and the location of $\operatorname{supp} \nu_{t,x}$ (see [Dem96, Git06, YW02]).

Acknowledgment

We thank the reviewer for useful hints and comments.

References

[BF93] Berger, H., Feistauer, M.: Analysis of the finite element variational crimes in the numerical approximation of transonic flow. Math. Comp., **61**, 493–521 (1993)

[Dem96] Demoulini, S.: Young measure solutions for a nonlinear parabolic equation of forward-backward type. SIAM J. Math. Anal., **27**, 376–403 (1996)

[DiP85] DiPerna, R.J.: Compensated compactness and general systems of conservation laws. Trans. Amer. Math. Soc., **292**, 383–419 (1985)

[FN85] Feistauer, M., Nečas, J.: On the solvability of transonic potential flow problems. Z. Anal. Anw., **4**, 305–329 (1985)

[Git87] Gittel, H.-P.: Studies on transonic flow problems by nonlinear variational inequalities. Z. Anal. Anw., **6**, 449–458 (1987)

[Git06] Gittel, H.-P.: Young measure solutions of some nonlinear mixed type equations. (to appear)

[KK98] Kawohl, B., Kutev, N.: Maximum and comparison principle for one-dimensional anisotropic diffusion. Math. Ann., **311**, 107–123 (1998)

[KP92] Kinderlehrer, D., Pedregal, P.: Weak convergence of integrands and the Young measure representation. SIAM J. Math. Anal., **23**, 1–19 (1992)

[KP94] Kinderlehrer, D., Pedregal, P.: Remarks about the analysis of gradient Young measures. J. Geom. Anal., **4**, 59–90 (1994)

[Mor04] Morawetz, C.S.: Mixed equations and transonic flow. J. Hyperbolic Differ. Equ., **1**, 1–26 (2004)

[PM90] Perona, P., Malik, J.: Scale space and edge detection using anisotropic diffusion. IEEE Trans. Pattern Anal. Mach. Intell., **12**, 629–639 (1990)

[Tar79] Tartar, L.: Compensated compactness and applications to partial differential equations. In: Knops, R.J. (ed) Nonlinear Analysis and Mechanics: Heriot-Watt Symposium, Vol. IV. Pitman, Boston (1979)

[YW02] Yin, J., Wang, C.: Existence of Young measure solutions of a class of forward-backward diffusion equations. J. Math. Anal. Appl. **279**, 659–683 (2003)

[You69] Young, L.C.: Lectures on Calculus of Variations and Optimal Control Theory. Saunders, Philadelphia (1969)

Computing Phase Transitions Arising in Traffic Flow Modeling

C. Chalons and P. Goatin

Summary. A new version of Godunov's scheme is proposed to compute the solutions of a traffic flow model with phase transitions. The scheme is based on a modified averaging strategy and a sampling procedure.

1 Introduction

We propose a numerical scheme to compute the solutions of a traffic flow model with phase transitions. The model has been introduced by Colombo [7] to explain empirical flow-density relations. For low densities, the flow is *free* and is described by a scalar conservation law (Lighthill-Whitham [9] and Richards [10] (LWR) model). At high densities the flow is *congested* and is described by a 2×2 system. We get

$$
\begin{array}{ll}
\text{Free flow:} & \text{Congested flow:} \\
(\rho, q) \in \Omega_f, & (\rho, q) \in \Omega_c, \\
\partial_t \rho + \partial_x(\rho v) = 0, & \left\{ \begin{array}{l} \partial_t \rho + \partial_x(\rho v) = 0, \\ \partial_t q + \partial_x\left((q - Q)v\right) = 0, \end{array} \right. \\
q = \rho V, & \\
v = v_f(\rho) = V\left(1 - \frac{\rho}{R}\right). & v = v_c(\rho, q) = \left(1 - \frac{\rho}{R}\right)\frac{q}{\rho}.
\end{array}
\qquad (1)
$$

The conserved quantity $\rho \in [0, R]$ is the mean traffic density, and v is the mean traffic velocity. The parameter R is the positive maximal density, V is the maximal speed, and Q is a parameter of the road under consideration. The weighted linear momentum q is originally motivated by gas dynamics. It approximates the real flux ρv for ρ small compared to R.

The coupling is achieved by introducing a transition dynamics from free to congested flow.

The 2×2 system describing the congested flow turns out to be hyperbolic, the second characteristic field being linearly degenerate, while the first has an inflection point along the curve $q = Q$. Moreover, shock and rarefaction curves coincide, hence system (1), right, belongs to Temple class [11]. A detailed description of the Riemann solver, and analogies between solutions to (1) and real traffic features are given in [7] (see [8] for the well posedness of the Cauchy problem).

The domain Ω_f (respectively, Ω_c) is taken to be an invariant set for (1), left (respectively, right). The resulting domain is given by

$$\Omega_f = \{(\rho, q) \in [0, R] \times [0, +\infty[\, : v_f(\rho) \geq V_f, \, q = \rho \cdot V\},$$
$$\Omega_c = \left\{(\rho, q) \in [0, R] \times [0, +\infty[\, : v_c(\rho, q) \leq V_c, \, \tfrac{q-Q}{\rho} \in \left[\tfrac{Q_- - Q}{R}, \tfrac{Q_+ - Q}{R}\right]\right\},$$

where $V_f > V_c$ are the threshold speeds, i.e. above V_f the flow is free and below V_c the flow is congested. The parameters $Q_- \in]0, Q[$ and $Q_+ \in]Q, +\infty[$ depend on the environmental conditions and determine the width of the congested region.

The domain $\Omega = \Omega_f \cup \Omega_c$ turns out to be a disconnected set in $\mathbb{R}^2$, its two connected components representing the free and the congested phases. Due to the lack of convexity of the domain, the classical Godunov method is not applicable. In fact, in the presence of phase transitions, the projection step of the algorithm can give values which are not in the domain. Then the procedure is stopped. We are thus led to present a new version of the Godunov scheme, based on a modified averaging strategy and a sampling procedure. More precisely, we modify the mesh cells following the phase boundaries, so that the projection involves only values belonging to the same phase. To come back to the original cells, we complete the projection step with a Glimm-type sampling technique. The averaging procedure on modified cells has first been used (up to our knowledge) in [12] but in a different context and a slightly different form. However, the idea of going back to the initial cells by means of a sampling procedure is new and allows us to avoid dealing with moving meshes (as in [12]). Similar numerical techniques have recently been proposed by the first author for approximating nonclassical solutions arising in certain nonlinear hyperbolic equations (see [2], [1] and the references therein), and very recently by Chalons and Coquel in [3] for computing sharp discrete shock profiles.

Of course, the random choice method (Glimm's scheme) could be applied successfully in this case. Nevertheless, our method does not need to compute all the values in the Riemann solution, but only the values on both sides of the phase transition, and is then cheaper. Moreover, our algorithm coincides with the classical Godunov scheme, and hence it is conservative, away from phase transitions.

Numerical tests are showed to prove the validity of the method.

2 A New Version of the Godunov Scheme

We will use the following shorten form

$$\partial_t \mathbf{u} + \partial_x \mathbf{f}(\mathbf{u}) = 0, \qquad \mathbf{u} \in \Omega = \Omega_f \cup \Omega_c, \tag{2}$$

for the model (1), where

$$\begin{cases} \mathbf{u} = (\rho, q) \quad \text{and} \quad \mathbf{f}(\mathbf{u}) = (\rho v_f(\rho), q v_f(\rho)), & \text{if } (\rho, q) \in \Omega_f, \\ \mathbf{u} = (\rho, q) \quad \text{and} \quad \mathbf{f}(\mathbf{u}) = (\rho v_c(\rho, q), (q - Q) v_c(\rho, q)), & \text{if } (\rho, q) \in \Omega_c. \end{cases}$$

From now on, (2) will be supplemented with an initial datum, setting

$$\mathbf{u}(., t = 0) = \mathbf{u}_0 \in \Omega. \tag{3}$$

We introduce a space step Δx and a time step Δt, both assumed to be constant for simplicity. We set $\nu = \Delta t / \Delta x$. Then, we define the mesh interfaces $x_{j+1/2} = j\Delta x$ for $j \in \mathbb{Z}$ and the intermediate times $t^n = n\Delta t$ for $n \in \mathbb{N}$, and we seek at each time t^n an approximation $\mathbf{u}_j^n$ of the solution of (2)–(3) on the interval $[x_{j-1/2}, x_{j+1/2})$, $j \in \mathbb{Z}$. Therefore, a piecewise constant approximated solution $x \to \mathbf{u}_\nu(x, t^n)$ of the solution $\mathbf{u}$ is given by

$$\mathbf{u}_\nu(x, t^n) = \mathbf{u}_j^n \text{ for all } x \in C_j = [x_{j-1/2}; x_{j+1/2}), \ \ j \in \mathbb{Z}, \ \ n \in \mathbb{N}.$$

When $n = 0$, we set $x_j = 0.5 \cdot (x_{j-1/2} + x_{j+1/2})$ and $\mathbf{u}_j^0 = \mathbf{u}_0(x_j)$, for all $j \in \mathbb{Z}$. Note that the usual L^2-projection is not adapted in the present context since, depending on the proposed initial data, it could artificially introduce unphysical states which are not in the phase space at time $t = 0$ (recall that $\Omega = \Omega_f \cup \Omega_c$ is not convex).

Like the classical Godunov scheme, our method is composed of two steps: the first step in which the solution evolves in time according to the PDE model under consideration, and the second step of projection onto piecewise constant functions.

Step 1: Evolution in time
In this first step, one solves the following Cauchy problem

$$\begin{cases} \partial_t \mathbf{v} + \partial_x \mathbf{f}(\mathbf{v}) = 0, \ \ x \in \mathbb{R}, \\ \mathbf{v}(x, 0) = \mathbf{u}_\nu(x, t^n), \end{cases} \tag{4}$$

for times $t \in [0, \Delta t]$. Recall that $x \to \mathbf{u}_\nu(x, t^n)$ is piecewise constant. Then, under the usual CFL restriction

$$\frac{\Delta t}{\Delta x} \max_{\mathbf{v}} \{|\lambda_i(\mathbf{v})|, \ i = 1 \text{ if } \mathbf{v} \in \Omega_f, \ i = 1, 2 \text{ if } \mathbf{v} \in \Omega_c\} \leq \frac{1}{2}, \tag{5}$$

for all the $\mathbf{v}$ under consideration, the solution of (4) is known by gluing together the solutions of the Riemann problems set at each interface:

$$\mathbf{v}(x, t) = \mathbf{v_r}(\frac{x - x_{j+1/2}}{t}; \mathbf{u}_j^n, \mathbf{u}_{j+1}^n) \text{ for all } (x, t) \in [x_j, x_{j+1}] \times [0, \Delta t], \tag{6}$$

where $(x, t) \to \mathbf{v_r}(\frac{x}{t}; \mathbf{v}^l, \mathbf{v}^r)$ denotes the self-similar solution of the Riemann problem

$$\begin{cases} \partial_t \mathbf{v} + \partial_x \mathbf{f}(\mathbf{v}) = 0, & x \in \mathbb{R}, \quad t \in \mathbb{R}^{+,\star} \\ \mathbf{v}(x,0) = \begin{cases} \mathbf{v}^l & \text{if } x < 0, \\ \mathbf{v}^r & \text{if } x > 0, \end{cases} \end{cases}$$

whatever $\mathbf{v}^l$ and $\mathbf{v}^r$ are in the phase space $\Omega_f \cup \Omega_c$.

Step 2 (Modified): Projection ($t^n \rightarrow t^{n+1}$)
In the usual Godunov method and to get a piecewise constant approximated solution on each cell $\mathcal{C}_j$ at time t^{n+1}, the solution $x \rightarrow \mathbf{v}(x, \Delta t)$ given by (6) is simply averaged, as expressed by the following update formula:

$$\mathbf{u}_j^{n+1} = \frac{1}{\Delta x} \int_{x_{j-1/2}}^{x_{j+1/2}} \mathbf{v}(x, \Delta t) dt, \quad j \in \mathbb{Z}. \tag{7}$$

In the present context, this strategy may fail due to the lack of convexity of the domain $\Omega_f \cup \Omega_c$ in the (ρ, q)-plane and to the possible presence of phase transitions in the Riemann solutions. In this case, the state $\mathbf{u}_j^{n+1}$ resulting from the averaging procedure (7) can be outside $\Omega_f \cup \Omega_c$ for some $j \in \mathbb{Z}$ (even if the solution $\mathbf{u}_\nu(., t^n)$ belongs to the domain), so that the classical Godunov method stops. We then propose to average the solution $x \rightarrow \mathbf{v}(x, \Delta t)$ on (possibly) modified and nonuniform cells $\overline{\mathcal{C}}_j^n = [\overline{x}_{j-1/2}^n, \overline{x}_{j+1/2}^n)$ constructed as follows. Let $(\sigma_{j+1/2}^n = \sigma(\mathbf{u}_j^n, \mathbf{u}_{j+1}^n))_{j \in \mathbb{Z}}$ be a sequence of characteristic speeds of propagation at interfaces $(x_{j+1/2})_{j \in \mathbb{Z}}$ such that:

- If $\mathbf{u}_j^n$ and $\mathbf{u}_{j+1}^n$ are not in the same phase (free or congested), then $\sigma_{j+1/2}^n$ coincides with the speed of propagation of the phase transition in the Riemann solution $(x,t) \rightarrow \mathbf{v_r}(\frac{x}{t}; \mathbf{u}_j^n, \mathbf{u}_{j+1}^n)$.
- If $\mathbf{u}_j^n$ and $\mathbf{u}_{j+1}^n$ belong to the same phase, then $\sigma_{j+1/2}^n = 0$.

Then we define the new interfaces $\overline{x}_{j+1/2}^n$ at time t^{n+1} setting

$$\overline{x}_{j+1/2}^n = x_{j+1/2} + \sigma_{j+1/2}^n \Delta t, \quad j \in \mathbb{Z}. \tag{8}$$

We also introduce

$$\overline{\Delta x}_j^n = \overline{x}_{j+1/2}^n - \overline{x}_{j-1/2}^n, \quad j \in \mathbb{Z}.$$

In particular, on each modified cell $\overline{\mathcal{C}}_j^n = [\overline{x}_{j-1/2}^n, \overline{x}_{j+1/2}^n)$, the solution $x \rightarrow \mathbf{v}(x, \Delta t)$ given by (6) is fully either in the free phase or in the congested phase. Then, averaging this solution on cells $\overline{\mathcal{C}}_j^n$ provide us with a piecewise constant approximated solution $\overline{\mathbf{u}}_\nu(x, t^{n+1})$ on a nonuniform mesh defined by

$$\overline{\mathbf{u}}_\nu(x, t^{n+1}) = \overline{\mathbf{u}}_j^{n+1} \text{ for all } x \in \overline{\mathcal{C}}_j^n, \quad j \in \mathbb{Z}, \quad n \in \mathbb{N},$$

with

$$\overline{\mathbf{u}}_j^{n+1} = \frac{1}{\overline{\Delta x}_j^n} \int_{\overline{x}_{j-1/2}^n}^{\overline{x}_{j+1/2}^n} \mathbf{v}(x, \Delta t) dt, \quad j \in \mathbb{Z}.$$

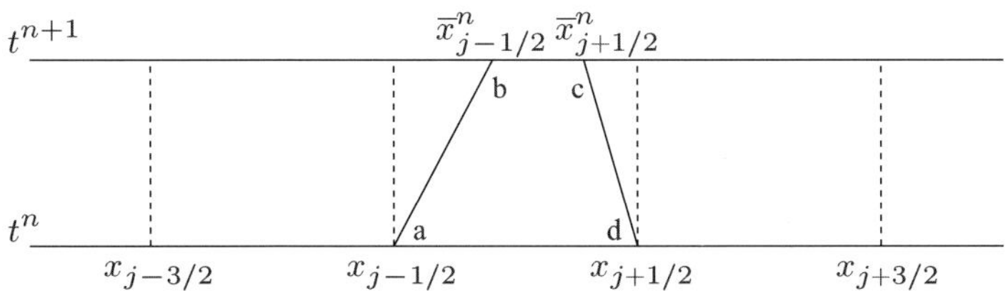

Fig. 1. An example of averaging element in the modified Godunov method

Let us underline that by the definition of the modified cells, we actually know to which phase each constant state of the solution $\overline{\mathbf{u}}_\nu(x, t^{n+1})$ belongs. In fact, note that both Ω_f and Ω_c are convex domains and then are stable under the process of an L^2-projection.

We can apply Green's formula on the domain $\overline{E} = (abcd)$ defined by:

$$\overline{E} = \{(x,t):\ t \in [0, \Delta t],\ x_{j-1/2} + \sigma^n_{j-1/2}\, t \leq x \leq x_{j+1/2} + \sigma^n_{j+1/2}\, t\}$$

(see Fig. 1). We get

$$\overline{\mathbf{u}}^{n+1}_j = \frac{\Delta x}{\overline{\Delta x}^n_j} \mathbf{u}^n_j - \frac{\Delta t}{\overline{\Delta x}^n_j} (\overline{\mathbf{f}}^{n,-}_{j+1/2} - \overline{\mathbf{f}}^{n,+}_{j-1/2}) \text{ for all } j \in \mathbb{Z}, \tag{9}$$

where the numerical fluxes are defined by

$$\overline{\mathbf{f}}^{n,\pm}_{j+1/2} = \mathbf{f}(\mathbf{v_r}(\sigma^{n,\pm}_{j+1/2}; \mathbf{u}^n_j, \mathbf{u}^n_{j+1})) - \sigma^n_{j+1/2} \mathbf{v_r}(\sigma^{n,\pm}_{j+1/2}; \mathbf{u}^n_j, \mathbf{u}^n_{j+1}) \text{ for all } j \in \mathbb{Z}, \tag{10}$$

using classical notations for the traces of the Riemann solutions at given points.

To go back to the (uniform) cells $\mathcal{C}_j$, $j \in \mathbb{Z}$, we now propose to pick up randomly on the cell $\mathcal{C}_j$ a value between $\overline{\mathbf{u}}^{n+1}_{j-1}$, $\overline{\mathbf{u}}^{n+1}_j$ and $\overline{\mathbf{u}}^{n+1}_{j+1}$, in agreement with their rate of presence in the cell. More precisely, given a well-distributed random sequence (a_n) within interval $(0, 1)$, it amounts to set:

$$\mathbf{u}^{n+1}_j = \begin{cases} \overline{\mathbf{u}}^{n+1}_{j-1} \text{ if } a_{n+1} \in (0, \frac{\Delta t}{\Delta x} \max(\sigma^n_{j-1/2}, 0)), \\ \overline{\mathbf{u}}^{n+1}_j \text{ if } a_{n+1} \in [\frac{\Delta t}{\Delta x} \max(\sigma^n_{j-1/2}, 0), 1 + \frac{\Delta t}{\Delta x} \min(\sigma^n_{j+1/2}, 0)), \\ \overline{\mathbf{u}}^{n+1}_{j+1} \text{ if } a_{n+1} \in [1 + \frac{\Delta t}{\Delta x} \min(\sigma^n_{j+1/2}, 0), 1), \end{cases} \tag{11}$$

for all $j \in \mathbb{Z}$. Following Collela [6], we consider the van der Corput random sequence (a_n) defined by $a_n = \sum_{k=0}^m i_k 2^{-(k+1)}$, where $n = \sum_{k=0}^m i_k 2^k$, $i_k = 0, 1$, denotes the binary expansion of the integers $n = 1, 2, \ldots$. This well-known sequence is often favorite since, when used in the context of Glimm's scheme, it leads to very good results in the smooth parts of the solutions (see for instance [6] and [5] for some illustrations).

We now propose to test our algorithm on three Riemann problems leading to solutions involving phase transitions. The parameters of the model are taken to be $R = 1$, $V = 2$, $V_f = 1$, $V_c = 0.85$, $Q = 0.5$, $Q^- = 0.25$, $Q^+ = 1.5$. The numerical solutions will be represented by the density and velocity profiles, and will be compared to the exact solutions. Solutions computed by means of a second-order extension of the method (see [4] for a full description) are also proposed.

For *Test A*, we consider $\rho^l = 0.7$, $\rho^l v^l = 0.3$ in the congested phase and $\rho^r = 0.3$ in the free phase, leading to a solution made of a rarefaction in the congested phase, followed by a phase transition to a free state, itself followed by a rarefaction wave in the free phase. The solutions are plotted in Fig. 2 at time $T_f = 0.5$. For this test case, we have used a mesh containing 500 points ($\Delta x = 0.002$).

We now address the case of phase transitions from a free state to a congested state. For *Test B*, we choose $\rho^l = 0.35$ in the free phase and $\rho^r = 0.6$, $\rho^r v^r = 0.25$, in the congested phase. The corresponding solution is a shock-like phase transition followed by a contact discontinuity. Figure 3 plots the solution at time $T_f = 0.6$ with $\Delta x = 0.002$.

For *Test C*, we take $\rho^l = 0.215$ in the free phase and $\rho^r = 0.7$, $\rho^r v^r = 0.2$, in the congested phase, leading a solution composed of a phase transition followed by a rarefaction wave, and a contact discontinuity propagating with a positive speed. In this case the congested state of the phase transition is very difficult to capture properly, due to the numerical diffusion of the scheme which

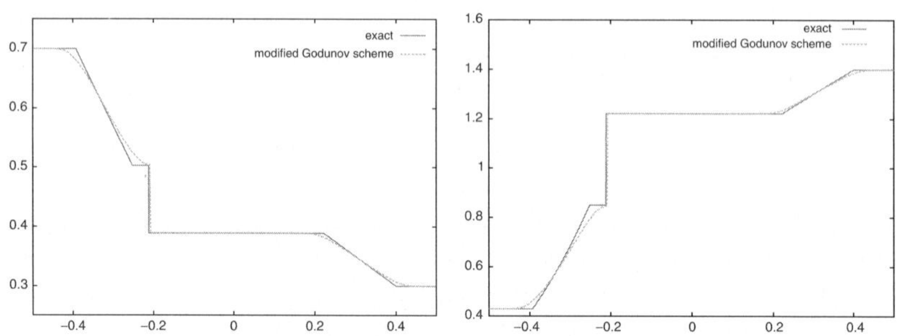

Fig. 2. *Test A*: ρ (*Left*) and v (*Right*)

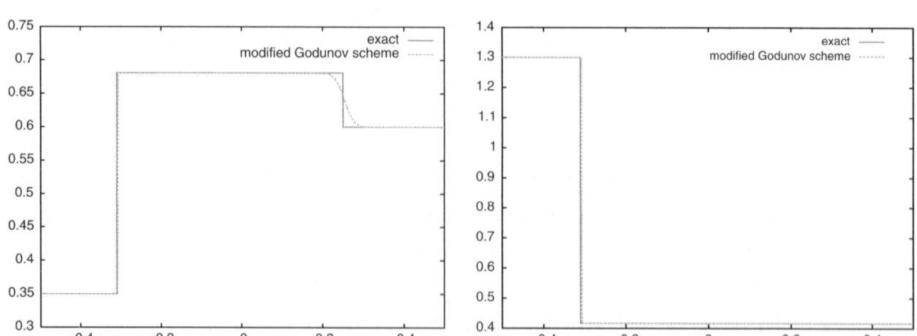

Fig. 3. *Test B*: ρ (*Left*) and v (*Right*)

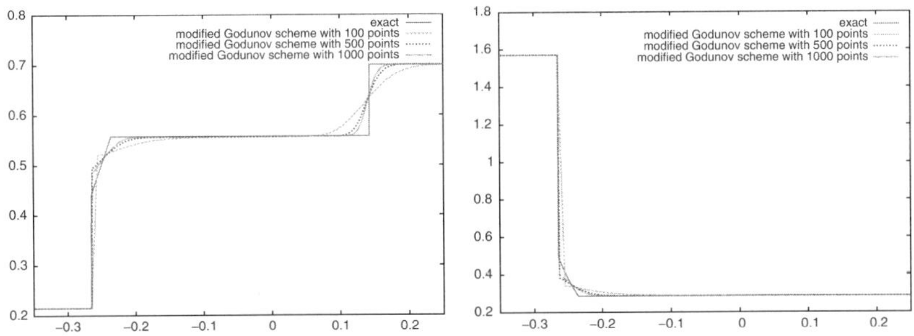

Fig. 4. *Test C*: ρ (*Left*) and v (*Right*)

is present in the rarefaction wave. Note that this state is always overestimated from the proposed averaging strategy. However we observe a good agreement between the numerical solution and the exact solution, and the numerical solution becomes better when the order of accuracy of the method is higher, as it is illustrated on Fig. 4 where we have taken $\Delta x = 0.01$, $\Delta x = 0.002$ and $\Delta x = 0.001$, $T_f = 0.8$.

3 Conservation Error

Due to the random sampling present in *Step 2 (Modified)*, our method does not strictly conserve the mass ρ. We then propose to measure the conservation errors on piecewise constant numerical solution ρ_ν defined as

$$\rho_\nu(x,t) = \rho_j^n \quad \text{if} \quad (x,t) \in [x_{j-1/2}, x_{j+1/2}) \times [t^n, t^{n+1}),$$

between times $t = 0$ and $t = T$, for some $T > 0$. We denote $[x_0, x_1]$ the computational domain and we proceed exactly as in [2]: we compare with 0 the function $E : T \in \mathbb{R}^+ \to E(T) \in \mathbb{R}$ with $E(T)$ defined by

$$\begin{aligned} \int_{x_0}^{x_1} \rho_\nu(x,T)dx \times E(T) &= \int_{x_0}^{x_1} \rho_\nu(x,T)dx - \int_{x_0}^{x_1} \rho_\nu(x,0)dx \\ &+ \int_0^T \{\rho v_c(\rho,q)\}_\nu(x_1,t)dt - \int_0^T \{\rho v_c(\rho,q)\}_\nu(x_0,t)dt. \end{aligned} \tag{12}$$

Recall that $q = \rho V$ in the free phase. $E(T)$ represents the relative conservation error of ρ at time T on the interval $[x_0, x_1]$. In the following table, we give for the *Tests A, B, C* the values of the L^1-norm $\frac{1}{T_f}||E||_{L^1(0,T_f)}$ of E, namely

$$\frac{1}{T_f}||E||_{L^1(0,T_f)} = \frac{1}{T_f}\int_0^{T_f} |E(T)|dT = \sum_{t^n=0}^{t^{n+1}=T_f} \frac{(t^{n+1}-t^n)}{T_f}|E(t^n)|,$$

where T_f is the final time of the corresponding simulations. We observe that the conservation errors are very small and decrease with the mesh size.

No. of points	*Test A* ($x_0 = -0.5$, $x_1 = 0.5$)	*Test B* ($x_0 = -0.3$, $x_1 = 0.3$)	*Test C* ($x_0 = -0.35$, $x_1 = 0.25$)
50	1.58%	2.02%	2.23%
100	0.81%	1.01%	1.11%
500	0.18%	0.17%	0.22%

References

1. C. Chalons. Numerical approximation of a macroscopic model of pedestrian flows. To appear in SIAM Journal of Scientific Computing.
2. C. Chalons. Transport-equilibrium schemes for computing nonclassical shocks. I. Scalar conservation laws. Submitted.
3. C. Chalons and F. Coquel. Capturing infinitely sharp discrete shock profiles with the godunov scheme. *Proceedings of the Eleventh International Conference on Hyperbolic Problems. Theory, Numerics, Applications*, 2007.
4. C. Chalons and P. Goatin. Godunov scheme and sampling technique for computing phase transitions in traffic flow modeling. Submitted.
5. C. Chalons and P.G. LeFloch. Computing undercompressive waves with the random choice scheme. nonclassical shock waves. *Interfaces and Free Boundaries*, 5:129–158, 2003.
6. P. Collela. Glimm's method for gas dynamics. *SIAM J. Sci. Stat. Comput.*, 3:76–110, 1982.
7. R.M. Colombo. Hyperbolic phase transitions in traffic flow. *SIAM J. Appl. Math.*, 63(2):708–721, 2002.
8. R.M. Colombo, P. Goatin, and F.S. Priuli. Global well posedness of a traffic flow model with phase transitions. *Nonlinear Anal. Ser. A.* To appear.
9. M.J. Lighthill and G.B. Whitham. On kinematic waves. II. A theory of traffic flow on long crowded roads. *Proc. Roy. Soc. London. Ser. A.*, 229:317–345, 1955.
10. P.I. Richards. Shock waves on the highway. *Operations Res.*, 4:42–51, 1956.
11. B. Temple. Systems of conservation laws with coinciding shock and rarefaction curves. *Contemp. Math.*, 17:143–151, 1983.
12. X. Zhong, T.Y. Hou, and P.G. LeFloch. Computational methods for propagating phase boundaries. *J. Comput. Phys.*, 124:192–216, 1996.

Dafermos Regularization for Interface Coupling of Conservation Laws

B. Boutin, F. Coquel, and E. Godlewski

Summary. We study the coupling of two conservation laws with different fluxes at the interface $x = 0$. The coupling condition yields (whenever possible) continuity of the solution at the interface. Thus the coupled model is not conservative in general. This gives rise to interesting questions such as nonuniqueness of self-similar solutions which we have chosen to analyze via a viscous regularization. Introducing a color function, we rewrite the problem in a conservative form involving a source term which is a Dirac measure. In turn, this leads to a nonconservative system which may be resonant. In this work, we analyze the regularization of this system by a viscous term following Dafermos's approach. We prove the existence of a viscous solution to the Cauchy problem with Riemann data and study the convergence to the solution of the coupled Riemann problem when the viscosity parameter goes to zero.

1 Introduction

We are interested in the study of the coupling of two different conservation laws at a fixed interface

$$\partial_t u + \partial_x f_L(u) = 0,\ x < 0,\ \partial_t u + \partial_x f_R(u) = 0,\ x > 0, \quad t > 0, \tag{1}$$

where $f_\alpha, \alpha = L, R$, are two smooth fluxes. The function u satisfies some initial condition

$$u(x, 0) = u_0(x), \quad x \in \mathrm{R}, \tag{2}$$

and a *coupling condition* at the interface $x = 0$. When u_0 is a Riemann data, as in the following sections,

$$u_0(x) = \begin{cases} u_L,\ x < 0, \\ u_R,\ x > 0, \end{cases} \tag{3}$$

where $u_L, u_R \in \mathrm{R}$ are two given constant states, we will say that u is a solution of a *coupled Riemann problem*.

The theoretical study of a new coupling condition (CC) was initiated in the scalar case [7]. This study results in expressing that two boundary value problems should be wellposed, and boils down to imposing (whenever possible) the continuity of the solution at the interface, without imposing the overall conservativity of the coupled model. We will call such a coupling a *state coupling.*

The coupling problem may be ill posed because the velocities may change sign at the interface and the nonuniqueness of self-similar solutions of the coupled Riemann problem was observed in [7]. We will consider only solutions u which are entropy solution in each half space. As opposed to the conservative approach, no *natural* global entropy criterium selecting a unique solution is associated to the nonconservative formulation, and using Dafermos's procedure [6] enables us to recover some of these solutions as self-similar zero-viscosity limits of a regularized system. In case of nonuniqueness, one can construct solutions of a coupled Riemann problem with arbitrary intermediate states but theses solutions are not allowed by this regularization procedure.

1.1 Conservative Approach: Flux Coupling

Let us introduce a color function a and consider the following conservative system

$$\begin{cases} \partial_t u + \partial_x f(u,a) = 0, & x \in \mathrm{R},\ t > 0, \\ \partial_t a = 0, \end{cases} \tag{4}$$

with

$$f(u,a) = a f_L(u) + (1-a) f_R(u) \tag{5}$$

and

$$a(x,0) = \begin{cases} 1, & x < 0 \\ 0, & x > 0. \end{cases} \tag{6}$$

For the conservative system (4), the Rankine–Hugoniot condition requires continuity of the flux at $x = 0$, i.e.,

$$f_L(u(0-,t)) = f_R(u(0+,t)) \tag{7}$$

which we may call *flux coupling.* Note that system (4), (5) is resonant (see [8]) when $\partial_u f(u,a) = a f_L'(u) + (1-a) f_R'(u)$ vanishes, which for $a \in [0,1]$ may occur only if f_L', f_R' do not have the same sign. Moreover, entropy conditions which select a unique solution can be naturally introduced in a number of situations (see [3, 1, 4]).

1.2 Nonconservative Approach: State Coupling

If (7) is a natural condition in a number of physical situations, this is not necessarily the case when the interface is artificially situated. Hence, we write instead

$$\partial_t u + \partial_x f(u,a) = \mathcal{M}, \quad x \in \mathrm{R},\ t > 0,\ \partial_t a = 0, \tag{8}$$

with (5) where $\mathcal{M}$ is a Dirac measure concentrated at the interface which involves $[f(u,a)] = f_R(u(0+,t)) - f_L(u(0-,t)) = (f_R(u) - f_L(u))(0,t)$, if one forces the continuity $u(0+,t) = u(0-,t)$ in the present case of *state coupling.* One notes that $f_R(u) - f_L(u) = -\partial_a(af_L(u) + (1-a)f_R(u))$ and moreover, since a is a Heaviside function, one gets the nonconservative formulation

$$\begin{cases} \partial_t u + (af_L'(u) + (1-a)f_R'(u))\partial_x u = 0, \\ \partial_t a = 0. \end{cases} \tag{9}$$

Note that the nonconservative product is well defined when u is continuous at $x = 0$. Let us set

$$\lambda(u,a) \equiv af_L'(u) + (1-a)f_R'(u). \tag{10}$$

System (9) is strictly hyperbolic if $\lambda \neq 0$, it has two characteristic fields, one which is genuinely nonlinear (GNL) associated to the eigenvalue λ, the other linearly degenerate (LD) associated to 0. As (5), it is resonant when (10) vanishes. Again this corresponds to the fact that f_L', f_R' may change sign. To define the nonconservative product $\lambda(u,a)\partial_x u$ even when u is not continuous, following [10], we use Dafermos regularization (see (12) below). We prove below the existence of self-similar solutions of (12) and the convergence of an extracted subsequence to a solution of the coupled Riemann problem for (1) which satisfies our coupling condition (see [5]). A closer study of the behavior of the limit profiles characterizes the possible boundary layers at the interface. Explicit conditions are obtained in the case of convex fluxes.

1.3 More General State Coupling

The above formalism enables us to treat more general cases where one forces the continuity of another variable (see [2] for a justification of transmitting variables other than the conservative ones in the state coupling). Indeed, consider some continuity condition associated with two monotone mappings simultaneously increasing (or decreasing) Φ_α

$$\Phi_R(u(0+,t)) = \Phi_L(u(0-,t)). \tag{11}$$

Introducing a change of variables $\Phi_\alpha(u_\alpha(\varphi)) = \varphi$, one wants $\varphi(0-) = \varphi(0+)$ whenever possible. In this aim, we define the function $g(\varphi,a) = f_R(u_R(\varphi)) - f_L(u_L(\varphi))$ and introduce the equation

$$\begin{cases} \partial_t\Big(au_L(\varphi) + (1-a)u_R(\varphi)\Big) + \partial_x\Big(af_L(u_L(\varphi)) + (1-a)f_R(u_R(\varphi))\Big) \\ \qquad\qquad\qquad\qquad + g(\varphi,a)\partial_x a = 0. \end{cases}$$

Setting for simplicity (in view of (10), this notation is a little abusing)

$$\lambda(\varphi,a) = \frac{af_L'(u_L)u_L' + (1-a)f_R'(u_R)u_R'}{au_L' + (1-a)u_R'}(\varphi)$$

we get the nonconservative system

$$\begin{cases} \partial_t \varphi + \lambda(\varphi, a)\partial_x \varphi = 0, \\ \partial_t a = 0. \end{cases}$$

The two systems (9) and (12) coincide for the identity mapping, i.e., when $\Phi_R(u) = \Phi_L(u) = u$. Again, system (12) has two eigenvalues, 0 and $\lambda(\varphi, a)$. We will now focus on (9) but a similar analysis can be developed for (12).

2 The Dafermos Regularization

To approximate solutions of the Riemann problem for the nonconservative system (9), following [6] we introduce for $\varepsilon > 0$ a system with a viscous regularization

$$\begin{cases} \partial_t u_\varepsilon + (a_\varepsilon f_L'(u_\varepsilon) + (1 - a_\varepsilon) f_R'(u_\varepsilon))\partial_x u_\varepsilon = t\varepsilon \partial_{xx} u_\varepsilon, \\ \partial_t a_\varepsilon = t\varepsilon^2 \partial_{xx} a_\varepsilon, \end{cases} \tag{12}$$

with Riemann data $u_\varepsilon(x, 0) = u_0(x)$ given by (3) (where u_L, u_R are given) and (6) for $a_\varepsilon(x, 0)$. In (12), two facts are noteworthy: one is the presence of the time variable t in the right-hand side viscous term and the other, the power 2 for ε in the second equation.

The term with t can be seen to correspond to a classical viscous regularization in variable $\xi = x/t, T = \ln t$ (see [11] for details). Having scale invariant solutions $u(x, t) = \tilde{u}(x/t)$, this regularization allows to study the approximation of self-similar solutions of (9). Indeed, Dafermos's conjecture for a system of conservation laws $u_t + f(u)_x = 0$ says that any Riemann solution is the limit (as $\varepsilon \to 0$) of self-similar solutions to the Dafermos regularized system $u_t + f(u)_x = \epsilon t u_{xx}$ (this is partially proven in [13], see also [12, 9]).

The different exponents for the small viscosity coefficient ε in (12) correspond to the fact that the system has two characteristic fields, one is GNL, associated to λ and the first equation, and one associated to 0 (thus LD) and the second equation. If ε is the size of the boundary layer for a GNL field, ε^2 is typical of a characteristic LD field.

We look for self-similar solutions of system (12) of the form $u_\varepsilon(x, t) = \tilde{u}_\varepsilon(x/t)$, $a_\varepsilon(x, t) = \tilde{a}_\varepsilon(x/t)$. Dropping the tilde for simplicity, they satisfy the ODE

$$\begin{cases} (-\xi + \lambda(u_\varepsilon, a_\varepsilon))d_\xi u_\varepsilon = \varepsilon d_{\xi\xi} u_\varepsilon \\ -\xi d_\xi a_\varepsilon = \varepsilon^2 d_{\xi\xi} a_\varepsilon \end{cases} \tag{13}$$

with boundary conditions

$$\begin{cases} \lim_{\xi \to -\infty} u_\varepsilon(\xi) = u_L, \ \lim_{\xi \to +\infty} u_\varepsilon(\xi) = u_R, \\ \lim_{\xi \to -\infty} a_\varepsilon(\xi) = 1, \ \lim_{\xi \to +\infty} a_\varepsilon(\xi) = 0. \end{cases} \tag{14}$$

To deal with a compact interval $[-M, M]$ for $\xi = x/t$, we use the fact that the propagation speeds of the limit problem are finite and bounded. Hence we may set for $M > \max|\lambda(u,a)|$

$$\begin{cases} u_\varepsilon(-M) = u_L, u_\varepsilon(+M) = u_R, \\ a_\varepsilon(-M) = 1, a_\varepsilon(+M) = 0. \end{cases}$$

One extends naturally $u_\varepsilon, a_\varepsilon$, respectively, by u_L, u_R and $1, 0$ outside $[-M, M]$. Let us define the interval $B_{LR} = [\min(u_L, u_R), \max(u_L, u_R)]$.

Proposition 1. *Assume $(u,a) \to \lambda(u,a)$ defined by (10) is bounded on $B_{LR} \times [0,1]$ and f_L', f_R' are Lipschitz on B_{LR}. There exists a solution $(u_\varepsilon(\xi), a_\varepsilon(\xi))$ of (13), (14) and a subsequence $(u_{\varepsilon'}(\xi), a_{\varepsilon'}(\xi))$ such that $u_{\varepsilon'}(\xi)$ converges as $\varepsilon' \to 0$ to $u \in \mathrm{L}^1_{loc}$ where $\overline{u}(x,t) \equiv u(x/t)$ is a weak self-similar solution of (1), (3) satisfying the entropy condition in each half space.*

Proof. The second equation in (13) can be integrated

$$a_\varepsilon(\xi) = \frac{\int_\xi^M e^{-s^2/2\varepsilon^2} ds}{\int_{-M}^M e^{-s^2/2\varepsilon^2} ds},$$

the resulting expression can be written in terms of the *error function* and the whole sequence converges to (6). The first equation in (13) is nonlinear and we use a fixed point argument. Define for fixed u a primitive of $\xi - \lambda(u(\xi), a_\varepsilon(\xi))$:

$$h(\xi; u, m) = \int_m^\xi (s - \lambda(u(s), a_\varepsilon(s)))ds \tag{15}$$

for some $m \in]-M, M[$ (which will be chosen in such a way that h remains nonnegative). Note that h also depends on ε through a_ε. Then, freezing u_ε, one has to solve the auxiliary linear problem

$$\varepsilon d_{\xi\xi} v_\varepsilon = -h(\xi; u_\varepsilon, \alpha) d_\xi v_\varepsilon, \tag{16}$$

which can be integrated twice

$$v_\varepsilon(\xi) = u_L + (u_R - u_L) \frac{\int_{-M}^\xi e^{-h(s)/\varepsilon} ds}{\int_{-M}^M e^{-h(s)/\varepsilon} ds} \tag{17}$$

(with shorthand notations for h). The right-hand side of (17) is in fact independent of m. Thus u_ε solution of (13) must be a fixed point of the mapping $T_\varepsilon : u_\varepsilon \to v_\varepsilon$, v_ε given by (17), defined on the set $\mathcal{C} = \{v \in \mathcal{C}^0([-M, M]), v(\xi) \in B_{LR}\}$. Schauder's theorem can be applied since T_ε is an operator on the closed-convex set $\mathcal{C}$ of $\mathcal{C}^0([-M, M])$ which transforms any (bounded) sequence v_n of $\mathcal{C}$ in a relatively compact set of $\mathcal{C}$. Moreover u_ε is BV since from (17), it satisfies the implicit representation formula

$$u_\varepsilon(\xi) = u_L + (u_R - u_L)\frac{\int_{-M}^{\xi} e^{-h(s;u_\varepsilon(s))/\varepsilon} ds}{\int_{-M}^{M} e^{-h(s;u_\varepsilon(s))/\varepsilon} ds}. \tag{18}$$

The embedding of $BV(-M, M)$ in L^1 is compact and we can extract a subsequence which converges to some $u \in \mathrm{L}^1$. It is not difficult to prove that u satisfies the ODE

$$-\xi d_\xi u + d_\xi f_\alpha(u) = 0, \ \alpha = L \text{ in } x < 0, \ \alpha = R \text{ in } x > 0,$$

together with the corresponding entropy inequalities, for $\eta \in \mathcal{C}^0(\mathrm{R})$ strictly convex

$$-\xi d_\xi \eta(u) + d_\xi q_\alpha(u) \leq 0 \text{ in } \mathcal{D}',$$

where q_α is the associated entropy flux, satisfying $q'_\alpha = \eta' f'_\alpha$. This is proved first in the intervals $(-M, 0)$, $(0, M)$ then in each half space (we refer to [5] for details). Thus $\overline{u}(x, t) = u(x/t)$ is a self-similar solution and a good candidate for a solution of the coupled Riemann problem if we can get more information on its behavior at $x = 0$. Note that the above proof does not guarantee uniqueness. □

3 Results at the Interface

We study the behavior of the limit solution and the existence of a possible boundary layer at the interface. For the *linear* case, $f'_L(.) = a_L, f'_R(.) = a_R$, we can study directly the limit u. It generally exhibits one discontinuity between u_L and u_R, propagating at speed a_L (if $a_L \leq 0$ and $a_R < -a_L$) or a_R (if $a_R \geq 0$ and $a_R > -a_L$). Then, u is continuous at $x = 0$, or at speed 0, i.e., stationary (if $a_L \geq 0 \geq a_R$) in which case $u(0-) = u_L, u(0+) = u_R$. It may have two discontinuities propagating at speed a_L and a_R only if $a_L = -a_R < 0$, in which case the intermediate state is $(u_L + u_R)/2$. No solution with arbitrary intermediate state is thus approximated.

The general case requires a zooming: we introduce a *stretched* variable $y = \xi/\varepsilon$ and define

$$\mathcal{U}_\varepsilon(y) = u_\varepsilon(\varepsilon y), \ \mathcal{A}_\varepsilon(y) = a_\varepsilon(\varepsilon y).$$

It is easily seen that $\mathcal{A}_\varepsilon$ converges to the smooth function

$$\mathcal{A}(y) = (1 - \mathrm{erf}(y/\sqrt{2}))/2,$$

satisfying $\mathcal{A}(-\infty) = 1, \mathcal{A}(+\infty) = 0$. Had we put ε in the second equation of (12), it would have resulted in a trivial limit profile for a ($a \equiv 1/2$) while, as we see, ε^2 provides a nontrivial smooth profile connecting $1 = a_L$ to $0 = a_R$. Now for $\mathcal{U}$, the characterization of the boundary layer, i.e., the position of $u(0-)$, $u(0+)$, is linked to the existence of nontrivial profiles for the limit $\mathcal{U}$ of $\mathcal{U}_\varepsilon$ since we might expect a profile connecting $\mathcal{U}(-\infty) = u(0-), \mathcal{U}(+\infty) = u(0+)$. The first equation (13) gives by rescaling

$$(-\varepsilon y + \lambda(\mathcal{U}_\varepsilon, \mathcal{A}_\varepsilon)(y))d_y\mathcal{U}_\varepsilon(y) = d_{yy}\mathcal{U}_\varepsilon(y).$$

Hence once we have proved the convergence of $\mathcal{U}_\varepsilon$, the limit $\mathcal{U}$ satisfies the ODE

$$\lambda(\mathcal{U}, \mathcal{A})(y)d_y\mathcal{U}(y) = d_{yy}\mathcal{U}(y).$$

The following result is not straightforward and requires a fine study of the integral for $y < 0$ and $y > 0$ and some technical lemmas, we refer to [5] for details.

Proposition 2. *Assume the hypothesis of Proposition 1. The functions $u, \mathcal{U}$ are both simultaneously increasing or decreasing and satisfy*

$$f_L(\mathcal{U}(-\infty)) = f_L(u(0-)),\ f_R(\mathcal{U}(+\infty)) = f_R(u(0+))$$

together with inequalities for the entropy fluxes

$$q_L(\mathcal{U}(-\infty)) \geq q_L(u(0-)),\ q_R(\mathcal{U}(+\infty)) \leq q_R(u(0+)).$$

Moreover, if f_L and f_R are assumed to be strictly convex, we have

$$\begin{cases} left: either\ \ \mathcal{U}(-\infty) = u(0-), \ right: either\ \ \mathcal{U}(+\infty) = u(0+), \\ \qquad\quad or \qquad \mathcal{U}(-\infty) < u(0-), \qquad\qquad or \qquad \mathcal{U}(+\infty) > u(0+), \\ f_L'(\mathcal{U}(-\infty)) < 0 < f_L'(u(0-)), \quad f_R'(\mathcal{U}(+\infty)) > 0 > f_R'(u(0+)). \end{cases} \tag{19}$$

We can get even more information when f_L, f_R are both strictly convex. A deeper study of all possible cases (respecting the above constraints) gives 35 cases, according to the signs of $f_L'(u(0-), f_R'(u(0+))$. The solutions we obtain are indeed solution of the *coupled Riemann problem* satisfying the coupling condition. Moreover, as in the linear case, if a solution involves an intermediate state, this state is not arbitrary. This occurs either with two shocks, one with negative $\sigma_L(u_L, \hat{u})$ and the other with positive speed $\sigma_R(u_R, \hat{u})$, then the intermediate state $\hat{u}$ solves

$$\sigma_R(u_R, \hat{u})(2f_R'(\hat{u}) - \sigma_R(u_R, \hat{u})) = \sigma_L(u_L, \hat{u})(2f_L'(\hat{u}) - \sigma_L(u_L, \hat{u})),$$

where $\sigma_\alpha(u, v) = (f_\alpha(u) - f_\alpha(v))/(u - v)$, or with two rarefactions on each side of the interface and $f_L'(\hat{u})^2 = f_R'(\hat{u})^2$. Thus we emphasize that some solutions of the coupled Riemann problem we might have constructed directly as in [7] are not attained by this regularization procedure.

For quadratic fluxes, $f_L(u) = u^2/2, f_R(u) = (u-c)^2/2$, example which was treated in [7], one can represent them in the $(u_L, u_R)-$plane. For example in the case $c > 0$, there are eight regions of uniqueness, two of nonuniqueness. Let us also mention that for $c < 0$, we may have up to four solutions when the velocities change sign, more precisely for initial states such that $\sigma_L(u_L, u_R) = (u_L + u_R)/2 < 0$, $\sigma_R(u_L, u_R) = (u_L + u_R)/2 - c > 0$, $f'_L(u_L) = u_L > 0$, $f'_R(u_R) = u_R - c < 0$, and such that $u_R - u_L - 2c < 0$ (the last inequality ensures $\sigma_L(u_L, \hat{u}) < 0$). Moreover some stationary discontinuities are not *stable*, in the sense that they are not obtained numerically.

The above study provides some insight on the existence of multiple solutions which can be attained by a regularization procedure. Some selection criteria such as continuity of the shock speed might be considered, i.e., for u_L fixed, and u_R varying, the possibility of selecting a solution which ensures the continuity of the shock speed.

This work falls within the scope of an ongoing joint research program on multiphase flows between CEA and University Pierre et Marie Curie-Paris6.

References

1. Adimurthi, Mishra, S., Gowda, G.D.V.: Optimal entropy solutions for conservation laws with discontinuous flux-functions. J. Hyperbolic Differ. Equ. **2**, No 4, 783–837 (2005)
2. Ambroso, A., Chalons, C., Coquel, F., Godlewski, E., Lagoutière, F., Raviart, P.-A., Seguin, N.: Coupling of general Lagrangian systems, submitted (2006)
3. Audusse, E., Perthame, B.: Uniqueness for a scalar conservation law with discontinuous flux via adapted entropies. Proc. Roy. Soc. Edinburgh A, **135** no. 2, 253–265 (2006)
4. Bachman F., Vovelle, J.: Existence and uniqueness of entropy solution of scalar conservation laws with a flux function involving discontinuous coefficients. Comm. Partial Differential Equations, **31** no. 1–3, 371–395 (2006)
5. Boutin, B.: Couplage de systèmes de lois de conservation scalaires par une régularisation à la Dafermos, Master dissertation, Paris (2005)
6. Dafermos, C.M.: Solution of the Riemann problem for a class of hyperbolic systems of conservation laws by the viscosity method. Arch. Ration. Mech. Anal., **52**, 1–9 (1973)
7. Godlewski E., Raviart, P.-A., The numerical interface coupling of nonlinear hyperbolic systems of conservation laws: I. The scalar case. Numer. Math., **97**, 81–130 (2004)
8. Isaacson, E., Temple, B.: Nonlinear resonance in systems of conservation laws. Siam J. Appl. Math., **52** no. 5, 1260–1278 (1992)
9. Joseph, K.T., LeFloch, P.G.: Boundary layers in weak solutions of hyperbolic conservation laws II. Self-similar vanishing diffusion limits. Comm. Pure Appl. Anal., **1** no. 1, 51–76 (2002)
10. LeFloch, P.G., Tzavaras, A.: Existence theory for the Riemann problem for non-conservative hyperbolic systems. C.R. Acad. Sci. Paris Sér. I Math., **23** no. 4, 347–352 (1996)
11. Lin, X.-B., Schecter, S.: Stability of selfsimilar solutions of the Dafermos regularization of a system of conservation laws. SIAM J. Math. Anal., **35** no. 4, 884–921 (2003)

12. Schecter, S., Szmolyan, P.: Composite waves in the Dafermos regularization. J. Dynam. Differential Equations, **16** no. 3, 847–867 (2004)
13. Tzavaras, A.E.: Wave interactions and variation estimates for self-similar zero-viscosity limits in systems of conservation laws. Arch. Rational Mech. Anal., **135** no. 1, 1–60 (1996)

Nonlocal Sources in Hyperbolic Balance Laws with Applications

R.M. Colombo and G. Guerra

1 Introduction and Main Result

We consider systems of conservation laws with nonlocal sources, i.e.,

$$\partial_t u + \partial_x f(u) = G(u), \tag{1}$$

where f is the flow of a nonlinear hyperbolic system of conservation laws and $G\colon \mathbf{L^1} \mapsto \mathbf{L^1}$ is a (possibly) *nonlocal* operator. As examples, we consider below the case $G(u) = g(u) + Q * u$ that enters a classical radiating gas model, see [22], as well as Rosenau regularization of Chapman–Enskog expansion of the Boltzmann equation, see [19, 20]. Here, by *nonlocal* we mean nonlocal in the space variable. Related results concerning the time variable, i.e., memory effects, are considered in [7].

We establish that (1) is well posed in $\mathbf{L^1}$, locally in time for general sources and globally for dissipative ones. The initial datum is assumed to have sufficiently small total variation.

To this aim, we require on (1) those assumptions that separately guarantee the well posedness of the convective part and of the source part

$$\partial_t u + \partial_x f(u) = 0 \tag{2}$$

$$\partial_t u = G(u). \tag{3}$$

These two equations generate two semigroups of solutions, say S and Σ. To obtain our results we exploit the techniques in [3, 9, 11], essentially based on the *fractional step* algorithm, see [9, 13, 15, 21]. Its core idea is to get a solution of the original equation as a limit of approximations obtained suitably merging S and Σ, i.e., for u in a suitable domain, $h \in \mathbb{N}$, and s positive sufficiently small, we define (see Fig. 1)

$$F_t^s u = S_{t-hs}(\Sigma_s \circ S_s)^h u \qquad \text{for } t \in [hs, (h+1)s[\ . \tag{4}$$

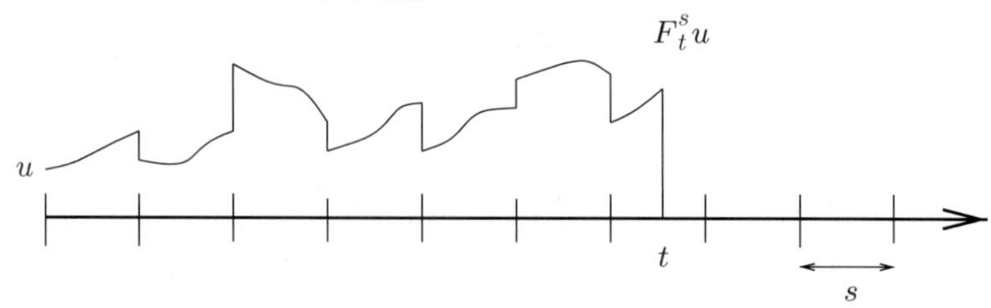

Fig. 1. Operator splitting: the vertical segments represent the application of Σ. The other continuous curves represent the application of S

In other words, in any interval $]hs, (h+1)s[$, we apply the semigroup S. In turn, at the times $t = hs$, Σ is applied. The semigroups S and Σ satisfy the two conditions:

- Stability: for all $s > 0$ and all u, v, $\|F_t^s u - F_t^s v\|_{\mathbf{L}^1} \le C\, \|u - v\|_{\mathbf{L}^1}$
- Commutativity: $\|\Sigma_\varepsilon S_\varepsilon u - S_\varepsilon \Sigma_\varepsilon u\|_{\mathbf{L}^1} \le C\, \varepsilon\, \omega(\varepsilon)$ where $\int_0^1 \frac{\omega(\varepsilon)}{\varepsilon}\, d\varepsilon < +\infty$

which, by [11], ensure the convergence of F^s to a limit semigroup F as $s \to 0$.

More precisely, let Ω be a open subset of $\mathbb{R}^n$ with $0 \in \Omega$. For all $\delta > 0$, define

$$\mathcal{U}_\delta = \left\{ u \in \mathbf{L}^1(\mathbb{R}; \Omega) \colon \mathrm{TV}(u) \le \delta \right\} .$$

As a general reference on conservation laws we refer to [6]. On the convective and on the source parts we assume throughout that

(F) $f \in \mathbf{C}^4(\Omega; \mathbb{R}^n)$, Df is strictly hyperbolic and each characteristic field is either genuinely nonlinear or linearly degenerate.

(G) For a positive δ_o, $G \colon \mathcal{U}_{\delta_o} \mapsto \mathbf{L}^1(\mathbb{R}, \mathbb{R}^n)$ is such that for suitable positive L_1, L_2, L_3

$$\begin{array}{ll} \forall\, u, w \in \mathcal{U}_{\delta_o} & \|G(u) - G(w)\|_{\mathbf{L}^1} \le L_1 \cdot \|u - w\|_{\mathbf{L}^1} \\ \forall\, u \in \mathcal{U}_{\delta_o} & \mathrm{TV}\,(G(u)) \le L_2 \cdot \mathrm{TV}(u) + L_3 . \end{array}$$

Note that **(F)**, respectively, **(G)**, ensures the local in time well posedness of (2), respectively, (3). A class of functions satisfying **(G)** is provided by the following result, see [10, Proposition 1.1]:

Proposition 1. *Let $g, h \colon \Omega \mapsto \mathbb{R}^n$ be locally Lipschitz and $Q \in \mathbf{L}^1(\mathbb{R}; \mathbb{R}^{n\times n})$. Then, the operator $G(u) = g(u) + Q * h(u)$ satisfies **(G)** with $L_3 = 0$.*

The main result is the following.

Theorem 1. *Let f satisfy **(F)** and G satisfy **(G)**. Then, there exist positive T, $\tilde\delta$, $\mathcal{L}$, closed domains $\mathcal{D}_t$ and processes*

$$F_t \colon \mathcal{D}_{T-t} \mapsto \mathcal{D}_T \qquad \forall\, t \in [0, T]$$

with the properties:

(1) *For* $t, s \in [0,T]$ *with* $t < s$, $\mathcal{U}_{\bar\delta} \subseteq \mathcal{D}_t \subseteq \mathcal{D}_s \subseteq \mathcal{U}_{\delta_o}$
(2) *For* u *in* $\mathcal{D}_T$, $F_0 u = u$; *for* $t, s \in [0,T]$ *with* $t+s \in [0,T]$, $F_s \mathcal{D}_t \subseteq \mathcal{D}_{t+s}$ *and for* $u \in \mathcal{D}_{T-t-s}$, $F_t F_s u = F_{t+s} u$
(3) *For* $\bar t \in [0,T]$ *and* $u \in \mathcal{D}_{\bar t}$, *the map* $t \mapsto F_t u$ *is a weak entropy solution to* (1) *for* $t \in [0, T-\bar t]$
(4) *If* S *is the SRS generated by* (2), *then for* $\bar t \in [0,T[$ *and* $u \in \mathcal{D}_{\bar t}$,

$$\lim_{t\to 0} \frac{1}{t} \left\| F_t u - \big(S_t u + t\, G(u)\big) \right\|_{\mathbf{L}^1} = 0$$

(5) *For* $t, s \in [0,T]$, $u, w \in \mathcal{D}_{T-t}$ *and* $s < t$, *then*

$$\begin{aligned} \|F_t u - F_t w\|_{\mathbf{L}^1} &\leq \mathcal{L} \cdot \|u-w\|_{\mathbf{L}^1} \\ \|F_t u - F_s u\|_{\mathbf{L}^1} &\leq \mathcal{L} \cdot \big(1 + \|u\|_{\mathbf{L}^1}\big) \cdot |t-s| \end{aligned} \tag{5}$$

(6) *For* $\bar t \in [0,T[$, $u \in \mathcal{D}_{\bar t}$ *and* $\tau \in [0, T-\bar t]$ *the map* $u(\tau) = F_\tau u$ *satisfies*

(6a) *For* $\xi \in \mathbb{R}$, $\displaystyle\lim_{\vartheta\to 0+} \frac{1}{\vartheta} \int_{\xi-\vartheta\hat\lambda}^{\xi+\vartheta\hat\lambda} \left\| \big(F_\vartheta u(\tau)\big)(x) - U^\sharp_{(u(\tau),\xi)}(\vartheta, x) \right\| dx = 0$

(6b) *There exists* $C > 0$ *such that for all* a, b, ξ *with* $-\infty \leq a < \xi < b \leq +\infty$

$$\begin{aligned} \overline{\lim_{\vartheta\to 0+}} \frac{1}{\vartheta} \int_{a+\vartheta\hat\lambda}^{b-\vartheta\hat\lambda} & \left\| \big(F_\vartheta u(\tau)\big)(x) - U^\flat_{(u(\tau),\xi)}(\vartheta, x) \right\| dx \leq \\ & C \cdot \Big(\mathrm{TV}\,\big(u(\tau);]a,b[\big)\Big)^2, \end{aligned}$$

where $U^\sharp_{(u(\tau),\xi)}$ *solves* (7) *and* $U^\flat_{(u(\tau),\xi)}$ *solves* (8)

(7) *For* $\bar t \in [0,T[$, *if a Lipschitz map* $u\colon [0, T-\bar t] \mapsto \mathcal{D}_T$ *is such that* $u(t) \in \mathcal{D}_{\bar t + t}$ *and for* $\tau \in [0, T-\bar t]$

(7a) *For* $\xi \in \mathbb{R}$, $\displaystyle\lim_{\vartheta\to 0+} \frac{1}{\vartheta} \int_{\xi-\vartheta\hat\lambda}^{\xi+\vartheta\hat\lambda} \left\| \big(F_\vartheta u(\tau)\big)(x) - U^\sharp_{(u(\tau),\xi)}(\vartheta, x) \right\| dx = 0$

(7b) *There exists a finite measure* μ_τ *such that for all* a, b, ξ *with* $-\infty \leq a < \xi < b \leq +\infty$

$$\limsup_{\vartheta\to 0+} \frac{1}{\vartheta} \int_{a+\vartheta\hat\lambda}^{b-\vartheta\hat\lambda} \left\| \big(F_\vartheta u(\tau)\big)(x) - U^\flat_{(u(\tau),\xi)}(\vartheta, x) \right\| dx \leq \Big(\mu_\tau\,\big(]a,b[\big)\Big)^2,$$

where $U^\sharp_{(u(\tau),\xi)}$ *solves* (7) *and* $U^\flat_{(u(\tau),\xi)}$ *solves* (8), *then* $u(t) = F_t u(0)$

Moreover, if f_1, f_2 *both satisfy* ***(F)*** *and* G_1, G_2 *both satisfy* ***(G)***, *then, denoting by* F^i *the process generated by* f_i *and* G_i, *for all* $t \in [0,T]$ *and* $u \in \mathcal{D}_0$

$$\begin{aligned} \left\| F^1_t u - F^2_t u \right\|_{\mathbf{L}^1} \leq\; & \mathcal{L} \cdot \|Df_1 - Df_2\|_{\mathbf{C^0}(\Omega, \mathbb{R}^{n\times n})} \cdot t \\ & + \mathcal{L} \cdot \|G_1 - G_2\|_{\mathbf{C^0}(\mathcal{U}_{\delta_o}; \mathbf{L}^1(\mathbb{R};\mathbb{R}^n))} \cdot t\,. \end{aligned} \tag{6}$$

For the definition and properties of the SRS, refer to [6]. Point 4 characterizes the tangent vector to $t \mapsto F_t u$ in the sense of [5, Sect. 5]. It is through 4 that the integral characterizations 6 and 7 are proved. Here, we need the following extension of [6, Sect. 9.2], [1, Sect. 5.2] and [4, Definition 15.1]. Let $v \in \mathcal{D}_T$ and $\xi \in \mathbb{R}$. The map $U^\sharp_{(v,\xi)}$ is the solution to the homogeneous Riemann problem

$$\begin{cases} \partial_t w + \partial_x f(w) = 0 \\ w(0,x) = \begin{cases} \lim\limits_{y\to\xi-} v(y) & \text{if} \quad x < \xi \\ \lim\limits_{y\to\xi+} v(y) & \text{if} \quad x > \xi \,. \end{cases} \end{cases} \tag{7}$$

The map $U^\flat_{(v,\xi)}$ is the broad solution (see [6, § 3.1]) to the Cauchy problem

$$\begin{cases} \partial_t w + Df\left(v(\xi)\right)\partial_x w = G\left(v\right) \\ w(0,x) = v(x). \end{cases} \tag{8}$$

The proof of Theorem 1 is in [10].

Remark that in (5) the term $\|u\|_{\mathbf{L}^1}$ is necessary, as the example $\partial_t u = u$ shows. Indeed, $\mathcal{U}_\delta$ is unbounded in $\mathbf{L}^1$. Moreover, note that the analogous estimate in [3, 8, 14] should be understood with a time Lipschitz constant dependent on the $\mathbf{L}^1$ norm of the initial datum.

We stress that the estimate (6b) is sharper than [1, formula (5.18)] thanks to the finite total variation of the source term, ensured by **(G)**.

2 Applications and Extensions

2.1 Euler System for a Radiating Gas

The following model for a radiating polytropic gas was considered in [22, Chap. XII, Sect. 6], see also [17, formula (1.2)]:

$$\begin{cases} \partial_t \rho + \partial_x \left(\rho\, v\right) = 0 \\ \partial_t \left(\rho\, v\right) + \partial_x \left(\rho\, v^2 + p\right) = 0 \\ \partial_t \left(\rho\, e + \frac{1}{2}\rho\, v^2\right) + \partial_x \left(v \left(\rho\, e + \frac{1}{2}\rho\, v^2 + p\right) + q\right) = 0 \\ -\partial^2_{xx} q + a\, q + b\, \partial_x \vartheta^4 = 0. \end{cases}$$

Here, as usual, ρ is the gas density, v its speed, e the internal energy, p the pressure, $\vartheta = e/c_v$ the temperature, and q is the radiative heat flux. The system is closed by means of the equation of state and specifying the values of the characteristic constants a and b.

Solving the latter equation in q we have $q = -\frac{b}{\sqrt{a}}\, Q_a * \left(\frac{d}{dx}\vartheta^4\right)$, where $Q_a(x) = \frac{1}{2}\exp\left(-\sqrt{a}\,|x|\right)$ and we are lead to consider the system

$$\begin{cases} \partial_t \rho + \partial_x (\rho v) = 0 \\ \partial_t (\rho v) + \partial_x (\rho v^2 + p) = 0 \\ \partial_t \left(\rho e + \frac{1}{2}\rho v^2\right) + \partial_x \left(v \left(\rho e + \frac{1}{2}\rho v^2 + p\right)\right) = b \left(-\vartheta^4 + \sqrt{a}\, Q_a * \vartheta^4\right). \end{cases} \tag{9}$$

It is well known that Euler system satisfies **(F)**. Condition **(G)** holds by Proposition 1. Hence, Theorem 1 applies and we obtain the local in time well posedness of (9). Note that this result also ensures the local Lipschitz dependence of the solutions to (9) from the parameters a and b.

2.2 Rosenau Regularization of the Chapman–Enskog Expansion

In his classical work [19], Rosenau proposed a system of balance laws that provides a regularized version of the Chapman–Enskog expansion for hydrodynamics in a linearized framework. The 1D version is the following:

$$\begin{cases} \partial_t \rho + \partial_x v = 0 \\ \partial_t v + \partial_x p = \mu_* * \partial^2_{xx} v \\ \partial_t \left(\frac{3}{2}\vartheta\right) + \partial_x v = \lambda_* * \partial^2_{xx}\vartheta, \end{cases}$$

where ρ is the fluid density, v is its speed, and ϑ is the temperature. μ_*, respectively, λ_*, is a convolution kernel related to viscosity, respectively, to thermal conductivity. This linear system motivated analytical results, see for instance [16, 18, 20], mostly related to the quasilinear scalar equation

$$\partial_t u + \partial_x \left(\frac{1}{2} u^2\right) = -u + Q * u$$

since the source term $-u + Q * u$ is equal to $Q * \partial^2_{xx} u$, provided $Q(x) = \frac{1}{2}\exp(-|x|)$. Therefore, it is natural to consider the following Euler system with Rosenau-type sources

$$\begin{cases} \partial_t \rho + \partial_x (\rho v) = 0 \\ \partial_t (\rho v) + \partial_x (\rho v^2 + p) = \mu_* * \partial^2_{xx} v \\ \partial_t \left(\rho e + \frac{1}{2}\rho v^2\right) + \partial_x \left(v \left(\rho e + \frac{1}{2}\rho v^2 + p\right) + q\right) = \lambda_* * \partial^2_{xx}\vartheta\,. \end{cases} \tag{10}$$

Rosenau kernels, see [19, formulae (4a) and (6)] read

$$\mu_*(x) = \frac{\mu}{2\,m\,\varepsilon}\exp(-|x|/\varepsilon) \quad \text{and} \quad \lambda_*(x) = \frac{\lambda}{2\,s\,\varepsilon}\exp(-|x|/\varepsilon)$$

for suitable positive parameters $\mu, \lambda, m, s, \varepsilon$. With the above choices, the sources in the last two equations in (10) can be rewritten as

$$\mu_* * \partial^2_{xx} v = \frac{1}{\varepsilon^2}\left(-\frac{\mu}{m} v + \mu_* * v\right) \quad \text{and} \quad \lambda_* * \partial^2_{xx}\vartheta = \frac{1}{\varepsilon^2}\left(-\frac{\lambda}{s}\vartheta + \lambda_* * \vartheta\right).$$

By Proposition 1, system (10) falls within the scope of Theorem 1. Thus, we prove the local in time well posedness of (10) as well as the local Lipschitz dependence of the solutions to (9) from the parameters $\mu, \lambda, m, s, \varepsilon$.

2.3 Local Inhomogeneous Source

Theorem 1 can be applied also in the standard case of a *local* source. Indeed, it is immediate to see that **(G)** is implied by the following conditions **(g1)** and **(g2)**.

Proposition 2. *Let $g\colon \mathbb{R} \times \Omega \mapsto \mathbb{R}^n$ be such that*

(g1) *There exists an $L_1 > 0$ such that for $u_1, u_2 \in \Omega$, $\|g(x,u_2) - g(x,u_1)\| \leq L_1 \cdot \|u_2 - u_1\|$.*

(g2) *There exists a finite measure μ on $\mathbb{R}$ such that for $u \in \Omega$ and $x_1, x_2 \in \mathbb{R}$ with $x_1 < x_2$, $\|g(x_2,u) - g(x_1,u)\| \leq \mu\left(]x_1,x_2]\right)$.*

*and assume that f satisfies **(F)**. Then, setting $\left(G(u)\right)(x) = g(x,u)$, Theorem 1 applies.*

Note that the integral estimates 6 and 7 in Theorem 1 ensure that the solution constructed here coincide with those in [1]. Similarly, the characterization 4 of the tangent vector imply that the present solutions coincide with those in [3].

2.4 The Nonautonomous Case

Theorem 1 can be extended to the nonautonomous balance law

$$\partial_t u + \partial_x f(u) = G(t,u)$$

provided f satisfies **(F)**, $G\colon [0,T_o] \times \mathbf{L^1} \mapsto \mathbf{L^1}$ satisfies

(G') For positive δ_o, T_o, the map $G\colon [0,T_o] \times \mathcal{U}_{\delta_o} \mapsto \mathbf{L^1}(\mathbb{R},\mathbb{R}^n)$ admits suitable positive L_1, L_2, L_3 such that for all $u, w \in \mathcal{U}_{\delta_o}$ and for all $t, s \in [0,T_o]$

$$\begin{aligned} \|G(t,u) - G(s,w)\|_{\mathbf{L^1}} &\leq L_1 \cdot \left(\|u-w\|_{\mathbf{L^1}} + |t-s|\right) \\ \mathrm{TV}\left(G(t,u)\right) &\leq L_2 \cdot \mathrm{TV}(u) + L_3 \,. \end{aligned}$$

Indeed, let $\hat\lambda$ be an upper bound for all moduli of characteristic speeds, i.e., $\hat\lambda > \sup_{\|u\| \leq \delta_o} \max_{i=1,\ldots,n} |\lambda_i(u)|$, and define

$$\tilde f(u,w) = \left(f(u), \hat\lambda\, w\right) \qquad \tilde G(u,w) = \left(G\left(\int_{\mathbb{R}} w, u\right), \chi_{[0,1]}\right).$$

Here, $\chi_{[0,1]}$ is the characteristic function of the real interval $[0,1]$. Then, $\tilde f$ satisfies **(F)** and $\tilde G$ satisfies **(G)**, so that Theorem 1 applies and the balance law $\partial_t(u,w) + \partial_x \tilde f(u,w) = \tilde G(u,w)$ generates the operator $\tilde F$. The Cauchy problem

$$\begin{cases} \partial_t u + \partial_x f(u) = G(t,u) \\ u(0,x) = u_o(x) \end{cases}$$

is solved by $t \mapsto F_t(u_o, 0)$, where $F_t(u_o, 0)$ is given by the first n component of $\tilde{F}_t(u_o, 0)$.

Again, the integral estimates 6 and 7 in Theorem 1 ensure that the solutions constructed here coincide with those in [14].

2.5 Dissipative Sources

Consider a source in (1) of the form

$$G(u) = g(u) + Q * u \tag{11}$$

and call R the matrix whose columns are the right eigenvectors of $Df(0)$. Recall, see [15], that an $n \times n$ matrix M is column diagonally dominant if there exists a $c > 0$ such that for $i = 1, \ldots n$

$$M_{ii} + \sum_{j=1,\, j \neq i}^{n} |M_{ji}| < -c, \tag{12}$$

see also [2] for a coordinate independent extension of diagonal dominance.

Theorem 2. *Let **(F)** hold. Fix $g \in \mathbf{C^2}(\Omega; \mathbb{R}^n)$, $Q \in \mathbf{L^1}(\mathbb{R}; \mathbb{R}^{n \times n})$ and define G as in (11). If the matrix $R^{-1}\, Dg(0)\, R$ is column diagonally dominant, then there exists a positive δ such that if*

$$\|Q\|_{\mathbf{L^1}} < \delta,$$

Theorem 1 applies globally in time with $\mathcal{D}_t = \mathcal{D}_0$ for all $t > 0$.

The constant δ above is of the form c/C, where c is as in (12) and C depends only on the $\mathbf{C^2}$ norm of g and $\mathbf{C^4}$ of f.

A proof of Theorem 2 can be obtained through a suitable mixing of the techniques in [3] with those in [10], see also [7]. A much shorter proof using the results in [11] is the subject of a forthcoming paper, see [12].

References

1. D. Amadori, L. Gosse, and G. Guerra. Global BV entropy solutions and uniqueness for hyperbolic systems of balance laws. *Arch. Ration. Mech. Anal.*, 162(4):327–366, 2002.
2. D. Amadori and G. Guerra. Global weak solutions for systems of balance laws. *Appl. Math. Lett.*, 12(6):123–127, 1999.
3. D. Amadori and G. Guerra. Uniqueness and continuous dependence for systems of balance laws with dissipation. *Nonlinear Anal.*, 49(7, Ser. A: Theory Methods):987–1014, 2002.
4. S. Bianchini and A. Bressan. Vanishing viscosity solutions of nonlinear hyperbolic systems. *Ann. of Math. (2)*, 161(1):223–342, 2005.

5. A. Bressan. On the Cauchy problem for systems of conservation laws. In *Actes du 29ème Congrès d'Analyse Numérique: CANum'97 (Larnas, 1997)*, pages 23–36 (electronic). Soc. Math. Appl. Indust., Paris, 1998.
6. A. Bressan. *Hyperbolic systems of conservation laws*, volume 20 of *Oxford Lecture Series in Mathematics and its Applications*. Oxford University Press, Oxford, 2000. The one-dimensional Cauchy problem.
7. G.-Q. Chen and C. Christoforou. Solutions for a nonlocal conservation law with fading memory. Preprint, 2006.
8. R.M. Colombo and A. Corli. On a class of hyperbolic balance laws. *J. Hyperbolic Differ. Equ.*, 1(4):725–745, 2004.
9. R.M. Colombo and A. Corli. A semilinear structure on semigroups in a metric space. *Semigroup Forum*, 68(3):419–444, 2004.
10. R.M. Colombo and G. Guerra. Hyperbolic balance laws with a non local source. To appear on *Comm. Partial Differential Equations*, 2006.
11. R.M. Colombo and G. Guerra. Differential equations in metric spaces with applications. In preparation, 2007.
12. R.M. Colombo and G. Guerra. Hyperbolic balance laws with a dissipative non local source. In preparation, 2007.
13. M. Crandall and A. Majda. The method of fractional steps for conservation laws. *Numer. Math.*, 34(3):285–314, 1980.
14. G. Crasta and B. Piccoli. Viscosity solutions and uniqueness for systems of inhomogeneous balance laws. *Discrete Contin. Dynam. Systems*, 3(4):477–502, 1997.
15. C.M. Dafermos and L. Hsiao. Hyperbolic systems and balance laws with inhomogeneity and dissipation. *Indiana Univ. Math. J.*, 31(4):471–491, 1982.
16. S. Jin and M. Slemrod. Remarks on the relaxation approximation of the Burnett equations. *Methods Appl. Anal.*, 8(4):539–544, 2001. IMS Conference on Differential Equations from Mechanics (Hong Kong, 1999).
17. C. Lattanzio and P. Marcati. Global well-posedness and relaxation limits of a model for radiating gas. *J. Differential Equations*, 190(2):439–465, 2003.
18. H. Liu and E. Tadmor. Critical thresholds in a convolution model for nonlinear conservation laws. *SIAM J. Math. Anal.*, 33(4):930–945 (electronic), 2001.
19. P. Rosenau. Extending hydrodynamics via the regularization of the Chapman-Enskog expansion. *Phys. Rev. A (3)*, 40(12):7193–7196, 1989.
20. S. Schochet and E. Tadmor. The regularized Chapman-Enskog expansion for scalar conservation laws. *Arch. Rational Mech. Anal.*, 119(2):95–107, 1992.
21. H.F. Trotter. On the product of semi-groups of operators. *Proc. Amer. Math. Soc.*, 10:545–551, 1959.
22. W. Vincenti and C. Kruger. *Introduction to Physical Gas Dynamics*. Wiley, 1965.

Comparison of Several Finite Difference Methods for Magnetohydrodynamics in 1D and 2D

P. Havlík and R. Liska

Summary. The comparison of several finite difference methods for ideal magnetohydrodynamics (MHD) is presented. Compared finite difference methods include composite schemes, central scheme, WENO, component wise CWENO, and public freely available packages Nirvana and Flash. 1D Cartesian tests concern smooth, Brio–Wu and intermediate shock formation problems. From 2D Cartesian tests we shortly present Orszag–Tang vortex problem and shock–cloud interaction problem. As we are interested in the generalization of schemes from Cartesian to cylindrical $r-z$ geometry, we include also generalization of composite and CWENO schemes to cylindrical geometry with their application to 2D conical z-pinch problem.

1 Introduction

For simplicity we use the system of ideal magnetohydrodynamics (MHD) equations in dimensionless units (with magnetic permeability $\mu = 1$)

$$\frac{\partial \varrho}{\partial t} + \nabla \cdot (\varrho \mathbf{v}) = 0, \tag{1}$$

$$\frac{\partial (\varrho \mathbf{v})}{\partial t} + \nabla \cdot \left(\varrho \mathbf{v}\mathbf{v}^{\mathrm{T}} + \mathrm{P}^{\star}\mathrm{I}^{3\times 3} - \mathbf{B}\mathbf{B}^{\mathrm{T}}\right) = 0, \tag{2}$$

$$\frac{\partial \mathbf{B}}{\partial t} + \nabla \cdot \left(\mathbf{v}\mathbf{B}^{\mathrm{T}} - \mathbf{B}\mathbf{v}^{\mathrm{T}}\right) = 0, \tag{3}$$

$$\frac{\partial E}{\partial t} + \nabla \cdot \left[\mathbf{v}\left(\mathrm{E} + \mathrm{P}^{\star}\right) - \mathbf{B}(\mathbf{v}\cdot\mathbf{B})\right] = 0, \tag{4}$$

where ϱ is mass density, $\mathbf{v}$ velocity, $\mathbf{B}$ magnetic induction, E total energy, and $P^{\star}$ is the sum of hydrodynamic and magnetic pressures $P^{\star} = p + \mathbf{B}^2/2$. Thermodynamical pressure p can be computed from equation of state for ideal gas $p = (\gamma - 1)\left[E - \varrho \mathbf{v}^2/2 - \mathbf{B}^2/2\right]$, where γ is ideal gas constant (ratio of specific heats). The symbol $\mathrm{I}^{3\times 3}$ denotes 3×3 unit matrix and ^{T} transposition. The MHD system (1)–(4) is coupled with constraint

$$\nabla \cdot \mathbf{B} = 0 \tag{5}$$

following directly from Maxwell equations. Using (3) one can verify, that the divergence of solenoidal magnetic field does not change in time, i.e., if the condition (5) is satisfied at initial time, it remains valid at any time later. Generally, this property does not hold for a numerical solutions of the system and special techniques have to be used for keeping numerical magnetic field solenoidal.

2 Cartesian Geometry

We first shortly describe numerical methods which have been used for comparisons and then present their performance on several 1D and 2D tests in Cartesian geometry. The numerical methods include composite, central, WENO, and component wise CWENO finite difference schemes together with the methods used in Nirvana (version 3, `http://nirvana-code.aip.de`) and Flash (version 2.5, `http://flash.uchicago.edu`) free available packages. All methods are of finite difference type and all results have been computed on rectangular uniform mesh. *Composite* scheme [9] performs one Lax–Friendrichs (LF) step after $n-1$ Lax–Wendroff (LW) steps and is denoted as LWLFn. The diffusive LF step serves as a consistent filter removing dispersive LW oscillations appearing behind shock waves. The LW scheme itself is second-order accurate, while the LF and LWLFn ones are only first. *Central* scheme (source code published in [2]) on the staggered grid uses limited piecewise polynomial reconstruction from cell averages. It requires neither Riemann solver nor eigendecomposition and also avoids dimensional splitting. *WENO* scheme [8] uses convex weighted combination of essentially nonoscillatory schemes on several stencils for space discretization and Runge–Kutta (RK) methods for time integrating. It is fifth-order accurate in space and we denote it as WENO3 in case of RK3 (TVD) time integration method and WENO5 in case of RK4 (non-TVD) method. WENO requires local eigenvector decomposition [10] which classifies it to be the slowest scheme from our choice. Avoiding eigendecomposition and applying WENO procedure directly to the conserved quantities we obtain *CWENO* (Component-wise WENO) scheme. After each time step of the above finite difference scheme the magnetic field is corrected to numerically satisfy the solenoidal condition (5) by the constrained transport method [5, 11]. Flash [6] package uses MUSCL-type limited gradient reconstruction method. Nirvana [12] employs second-order semidiscrete Godunov-type central method.

2.1 Smooth Periodic Problem in 1D

The first 1D problem originates in [7] for Euler equations however it provides an exact solution (periodic in ρ and constant in $\mathbf{v}, \mathbf{B}, p$) for the MHD system as well. We employ the particular exact solution $\varrho(x,t) = 1 + 0.2\sin(\pi(x-t))$, $v^x(x,t) = 1$, $B^y(x,t) = 1$, $p(x,t) = 1$ which we treat on interval $x \in [0,1]$

Table 1. Convergence for smooth periodic problem – $L_1(\varrho, 1600)$ errors and the numerical order of accuracy (NOA) $\log_2(L_1(\varrho, 800)/L_1(\varrho, 1600))$

Scheme	LWLF12	LW	Central	CWENO5	WENO3	WENO5	Flash	Nirvana3
$L_1(\varrho, 1600)$	2.5e-04	5.35e-07	1.01e-06	8.60e-14	1.21e-12	2.05e-14	2.51e-07	9.08e-07
NOA	1.0	2.0	2.4	5.0	3.1	5.0	2.1	2.1

Table 2. Convergence for Brio–Wu test problem – $L_1(\varrho, 1600)$ deviations from Flash solution using 6,400 cells and the NOA $\log_2(L_1(\varrho, 800)/L_1(\varrho, 1600))$

Scheme	LWLF12	Central	CWENO3	CWENO5	WENO3	WENO5	Flash	Nirvana3
$L_1(\varrho, 1600)$	2.6e-03	2.1e-03	1.4e-03	1.4e-03	1.0e-03	1.0e-03	6.2e-04	1.5e-03
NOA	0.7	1.0	1.0	1.0	1.0	1.0	1.2	1.0

till final time $t = 1$ with periodic boundary conditions and $\gamma = 1.4$. Table 1 shows for all numerical methods absolute L_1 errors of density on the grid with 1,600 cells (which we denote by $L_1(\varrho, 1600)$) together with the numerical order of accuracy (NOA) given by $\log_2(L_1(\varrho, 800)/L_1(\varrho, 1600))$. As expected: the composite scheme is first-order accurate; LW, central, Flash, and Nirvana are second order; WENO3 is third order; WENO5 and CWENO5 are fifth order.

2.2 Brio–Wu Problem in 1D

The Brio–Wu Riemann problem is classical MHD test problem [3] used in almost all papers numerically treating MHD equations. Here we present only similar results as for previous periodic problem. Table 2 shows $L_1(\varrho, 1600)$ deviations of numerical solution (on mesh with 1,600 cells) for density from a reference solution. As the reference solution we use Flash solution with 6,400 cells. NOA $\log_2(L_1(\varrho, 800)/L_1(\varrho, 1600))$ is presented in Table 2 too. As expected, all the numerical methods are first-order accurate for this problem involving discontinuous waves.

2.3 Intermediate Shock Formation in 1D

This test problem, originating in [8], starts from smooth initial conditions from which after some time shocks develop. Initially, $B^y(x) = \sin(2\pi x)/2$ for $x \in [0, 1]$ and all other conservative quantities are computed with using generalized Riemann invariants. All quantities are normalized to $\varrho = 1$, $\mathbf{v} = (0, 0, 0)^T$, $B^x = 1$, $B^z = 0$, and $p = 1$ at points, where $B^y = 0$. Periodic boundary conditions are applied. Table 3 presents $L_1(B^y, 1600)$ deviations of B^y numerical solutions with 1,600 cells from the reference WENO5 solution using 6,400 cells together with the NOA $\log_2(L_1(\varrho, 800)/L_1(\varrho, 1600))$ at three different times. At time $t = 0.25$ the solution is still smooth (see Fig. 1a) and

Table 3. Convergence for intermediate shock formation problem at three times $t = 0.25, 0.6, 1$ – $L_1(B^y, 1600)$ deviations (on $x \in [0, 1]$) from WENO5 solution using 6,400 cells and the NOA $\log_2(L_1(\varrho, 800)/L_1(\varrho, 1600))$ computed from deviations on $x \in [0, 1]$ and on $x \in [0.2, 0.4]$

Scheme	LWLF12	Central	CWENO3	CWENO5	WENO3	WENO5	Flash	Nirvana3
$t = 0.25, L_1(B^y, 1600)$	9.3e-05	2.8e-06	1.5e-10	8.1e-12	1.5e-10	9.7e-12	7.8e-07	1.8e-06
NOA on $x \in [0, 1]$	1.0	2.3	3.1	5.1	3.1	5.1	2.1	2.1
$t = 0.6, L_1(B^y, 1600)$	7.4e-04	4.4e-04	2.3e-04	2.3e-04	2.5e-04	2.5e-04	1.8e-04	3.2e-04
NOA on $x \in [0, 1]$	1.1	1.2	1.2	1.2	1.2	1.2	1.3	1.1
NOA on $[0.2, 0.4]$	1.0	2.4	4.2	4.3	4.9	5.0	1.9	2.0
$t = 1, L_1(B^y, 1600)$	1.0e-03	7.2e-04	6.9e-04	6.9e-04	3.5e-04	3.5e-04	1.9e-04	4.5e-04
NOA on $x \in [0, 1]$	0.9	1.0	0.6	0.6	1.1	1.1	1.1	1.0
NOA on $[0.2, 0.4]$	1.0	1.2	1.4	1.4	1.5	1.8	1.5	1.2

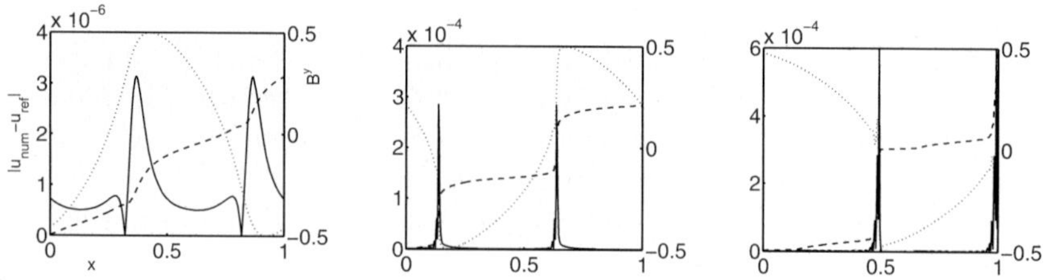

(a) $t = 0.25$ (b) $t = 0.6$ (c) $t = 1$

Fig. 1. Solution $B^y(x)$ by *dotted line*, local deviation $|u_{num}(x) - u_{ref}(x)|$ (of numerical solution from reference one) by *solid line* and cumulative deviation $\int_0^x |u_{num}(x') - u_{ref}(x')|dx'$ by *dashed line* for composite LWLF12 scheme (other schemes produce similar plots) with 400 cells at three times $t = 0.25, 0.6, 1$. Cumulative deviation at point $x = 1$ equals 9.7e-7 for $t = 0.25$, 6.8e-6 for $t = 0.6$, and 1.1e-5 for $t = 1$

the schemes have the NOA close to that one for smooth test presented in Table 1. In Fig. 1a the local deviation $|u_{num}(x) - u_{ref}(x)|$ of the numerical solution from the reference one is continuous and the cumulative deviation $\int_0^x |u_{num}(x') - u_{ref}(x')|dx'$ is smooth function of x showing that the L_1 deviation is distributed over the whole domain $x \in [0, 1]$. At time $t = 0.6$ two shocks are already formed around $x = 0.13$ and 0.63 (see Fig. 1b) and the NOA decreases rapidly toward one when L_1 deviations are evaluated on the whole computational domain $x \in [0, 1]$, while keeping high values when L_1 deviations are evaluated on $x \in [0.2, 0.4]$, where B^y still has smooth profile. In Fig. 1b the local deviation $|u_{num}(x) - u_{ref}(x)|$ is close to delta function at two shocks and the cumulative deviation $\int_0^x |u_{num}(x') - u_{ref}(x')|dx'$ has jumps at two shocks showing that most L_1 deviation is concentrated around the shocks. At time $t = 1.0$ the NOA is low even on subdomain $x \in [0.2, 0.4]$ where B^y is smooth, however the shock, which is now at $x = 0.5$ has already passed through this subdomain (see Fig. 1c).

2.4 Orszag–Tang Vortex Problem in 2D

The Orszag–Tang vortex problem is defined on a square $(x, z) \in [0, 2\pi] \times [0, 2\pi]$ with initial conditions $\varrho = \gamma^2, \mathbf{v} = (-\sin z, 0, \sin x)^T, \mathbf{B} = (-\sin z, 0, \sin(2x))^T$, $p = \gamma$ with $\gamma = 5/3$. Periodic boundary condition are applied in both directions. Figure 2a presents density contours at time $t = 3$ computed by Flash on the mesh with 400×400 cells. 1D cuts along line $z = \pi$ in density are shown for all schemes in Fig. 2b, c, where the differences between the schemes can be seen. For clarity, the results are split into two figures with identical axes and one reference (Flash) solution. Results can be compared with [2, 4, 8, 12].

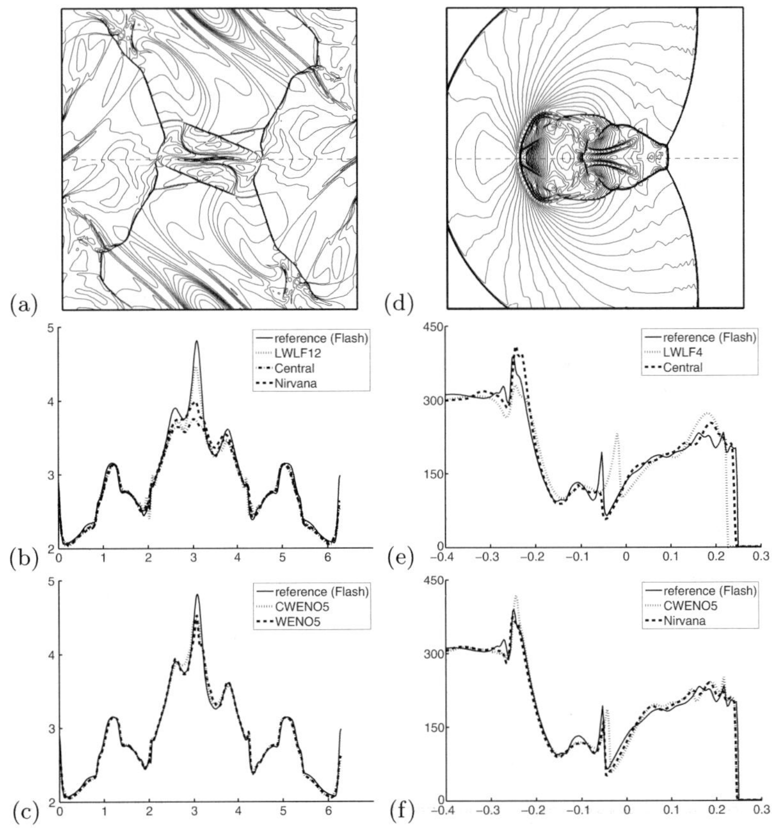

Fig. 2. Density contours for Orszag–Tang vortex by Flash (**a**); 1D slices in density along line $z = \pi$ for Orszag–Tang vortex (**b**), (**c**); pressure contours for shock–cloud interaction problem by Nirvana (**d**); 1D slices in pressure along line $z = 0$ for shock–cloud interaction problem (**e**), (**f**)

2.5 Shock–Cloud Interaction Problem in 2D

The initial conditions for the shock–cloud interaction problem [4, 11, 12] consist of two states in $(x, z) \in [-1/2, 1/2] \times [-1/2, 1/2]$ separated by shock discontinuity along line $x = 0.1$. The left state is defined by values $\varrho = 3.86859$, $\mathbf{v} = (0, 0, 0)^T$, $\mathbf{B} = (0, 2.1826182, -2.1826182)^T$, $p = 167.345$ while the right state is defined by values $\varrho = 1$, $\mathbf{v} = (-11.2536, 0, 0)^T$, $\mathbf{B} = (0, 0.56418958, 0.56418958)^T$, $p = 1$. At point $(0.3, 0)$ a spherical density clump with radius 0.15 and constant density $\varrho = 10.0$ and pressure $p = 1$ is located. On contour map of pressure in Fig. 2d we show result of Nirvana scheme computed on grid with 400×400 cells at time $t = 0.06$. Composite scheme has remarkably lower sharpness of shock wave, while Nirvana and Flash have both much better shock wave resolution, however the profile of density clump deformed by passing shock wave differs. The 2D profile from CWENO is closer to Nirvana. WENO scheme failed for this test during computation. 1D cuts of pressure in Fig. 2e, f along the line $z = 0$ show profiles of all methods. More differences can be seen in composite scheme results, which profiles are in some regions shifted, e.g., transmitted shock wave behind the clump is slower.

3 Cylindrical Geometry

In Cartesian geometry, there are no source terms and the MHD system (1)–(4) is in conservative form. On the other hand, in cylindrical geometry the divergence has the form $\nabla_{cyl} \equiv (1/r + \partial/\partial r, 1/r\,\partial/\partial\varphi, \partial/\partial z)$ which introduces geometrical source terms and the system is not conservative. Most of the geometric source terms can be included into the fluxes by multiplying the MHD system (except the equation for B^φ) by radius r. Then the source terms remain only in two equations for momentum conservation in r and φ directions. We have available only two methods - composite and CWENO schemes in 2D cylindrical $r - z$ geometry. Both LW and LF schemes in the composite use simple averaging to get the source terms on staggered mesh, just LW corrector uses sources from previous time step. CWENO scheme in cylindrical geometry has been developed as extension from Cartesian geometry. It employs the same weighted approximation procedure as for fluxes evaluation to get the source terms at the edges midpoints. The sources are averaged from the edges midpoints to get the source term inside the cell. Flash [6] supports MHD only in Cartesian geometry, we were quite surprised that with MHD setup in cylindrical geometry Flash happily computes in Cartesian geometry without any warning. Nirvana [12] does not include cylindrical MHD.

3.1 Conical z-Pinch in 2D

This test coming from [1] simulates compression of conical z-pinch by magnetic field. The problem is solved on rectangular area $(r, z) \in [0, 1.3] \times [0, 1]$ with initial conditions given by values $\varrho = 1$, $\mathbf{B} = (0, 0, 0)^T$ for $r \leq 1 + 0.3z$

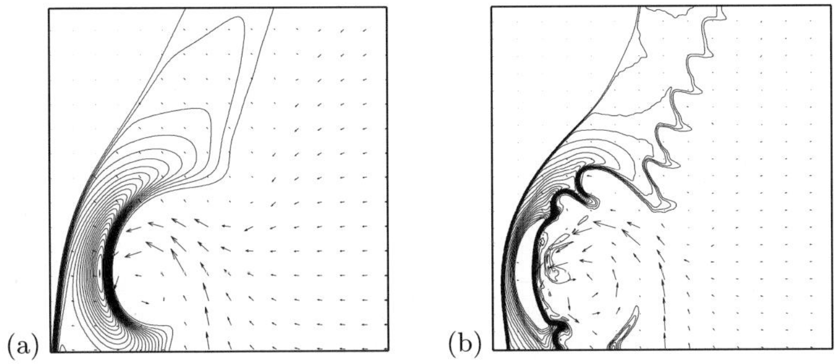

Fig. 3. B^y contours and velocity field by arrows of LWLF80 (**a**) and CWENO5 (**b**) for conical z-pinch in cylindrical geometry

holds and by values $\varrho = 10^{-4}$, $\mathbf{B} = (0, \sqrt{2}/r, 0)^T$ elsewhere. Velocity and pressure are same in the whole area $\mathbf{v} = (0, 0, 0)^T$ and $p = 10^{-4}$. Free boundary conditions on top at $z = 1$, bottom at $z = 0$ and free on right at $r = 1.3$ except Dirichlet boundary conditions $B^\varphi = \sqrt{2}/r$ for B^φ keeping the tangential magnetic induction. Figure 3 presents pressure contours and velocity fields by arrows obtained for this problem by composite (a) and CWENO (b) cylindrical schemes at time $t = 0.63$ on grid with 400×400 cells. Composite scheme is not able to resolve instabilities, seen in CWENO result. Similar instabilities appear also in [1].

4 Conclusion

Selected finite difference methods have been applied to a set of 1D and 2D test problems in Cartesian geometry and their numerical results have been compared. The results of composite schemes are the worst between others in the sense of resolution discontinuities. The most precise results in regions of smooth solution are typically obtained by WENO scheme, however it is very slow due to eigenvector decomposition and it fails in some cases as, e.g., for shock–cloud interaction problem. In most cases component-wise CWENO scheme produces results very close to WENO, however in some cases, as, e.g., for Brio–Wu problem, it produces very mild oscillations in regions of flat solution between the waves. In general it seems that the best results are obtained from Flash code, which is moreover remarkably fast. Nirvana produces also very good results. Composite and CWENO have been generalized to cylindrical $r - z$ geometry.

Acknowledgment

This research has been partly supported by the Czech Ministry of Education project MSM 6840770022, the Necas center for mathematical modeling project

LC06052 and Czech Technical University project CTU0621814. The software Flash used in this work was in part developed by the DOE-supported ASCI/ Alliance Center for Astrophysical Thermonuclear Flashes at the University of Chicago.

References

1. Aksenov, A.G.; Gerusov, A.V.: Comparative Analysis of Numerical Methods for 2-Dimensional High Compression MHD Flow Simulation, Plasma Physics Reports **21** (1): 11–19 Jan 1995
2. Balbas, J.; Tadmor, E.; Wu, C.-C.: Non-Oscillatory Central Schemes for One- and Two-Dimensional MHD Equations: I, J. of Comput. Phys. **201** (2004) 261–285
3. Brio, M.; Wu, C.C.: An Upwind Differencing Scheme for the Equations of Ideal Magnetohydrodynamics, J. of Comput. Phys. **75** (1998) 400–422
4. Dai, W.; Woodward, P.R.: A Simple Finite Difference Scheme for Multidimensional Magnetohydrodynamical Equations, J. of Comput. Phys. **142** (1998) 331–369
5. Evans, C.R.; Hawley, J.F.: Simulation of Magnetohydrodynamic Flows – A Constrained Transport Method, The Astrophysical J., 332:659–677, 1988
6. ASC FLASH Center: FLASH user's guide, version 2.3/2.5, University of Chicago, 2003/2005
7. Jiang, G.S.; Shu, C.W.: Efficient Implementation of Weighted ENO Schemes, J. Comp. Phys. **126** (1996) 202–228
8. Jiang, G.-S.; Wu, C.-C.: A High-Order WENO Finite Difference Scheme for the Equations of Ideal Magnetohydrodynamics, J. of Comput. Phys. **150** (1999) 561–594
9. Liska, R.; Wendroff, B.: Composite Schemes for Conservation Laws, SIAM J. Sci. Comput. **35**, no. 6 (1998) 2250–2271
10. Roe, P.L.; Balsara, D.S.: Notes on the Eigensystem of Magnetohydrodynamics, SIAM J. Appl. Math. **56**, no. 1 (1996) 57–61
11. Tóth, G.: The $\nabla \cdot \mathbf{B}$ Constraint in Shock-Capturing Magnetohydrodynamics Codes, J. of Comput. Phys. **161** (2000) 605–652
12. Ziegler, U.: A Central-constrained Transport Scheme for Ideal Magnetohydrodynamics, J. of Comput. Phys. **196** (2004) 393–416

On Global Large Solutions to 1-D Gas Dynamics

E.E. Endres and H.K. Jenssen

1 Introduction

We consider the 1-D Euler system (1)–(3) describing conservation of mass, momentum, and energy in compressible gas flow. For data with sufficiently small total variation Glimm's theorem [7] guarantees the existence of a global-in-time weak entropy admissible solution. The solution can be constructed by various methods: the Glimm scheme [7, 10], wave front-tracking [3], semidiscrete schemes [1], or vanishing viscosity [2]. The Euler system plays a distinguished role in the class of general conservation laws and much effort has been invested in extending Glimm's result to larger classes of data. For generic data the solution of the Euler equations is exceedingly complicated with a myriad of interactions resulting in complicated wave patterns.

The works by Nishida [12] and Nishida and Smoller [13] on the isothermal and isentropic equations, respectively, provide global existence for large BV data: any BV data in the case of isothermal flow, and large BV data with $(\gamma - 1) \times$ (total variation of data) sufficiently small, in the case of isentropic flow (γ being the adiabatic gas constant). These results were extended to the full Euler system by Liu [11] and Temple [18]. More recently Temple and Young [19] have established existence for the full Euler system up to an arbitrary time for data with large total variation and sufficiently small sup-norm.

Given the general form of Glimm's theorem it is natural to consider other systems of conservation laws and to investigate various large data regimes for these. However, recent examples show that no result similar to Glimm's theorem can hold at this level of generality. In particular [8, 9, 20] provides examples of special (unphysical) systems of conservation laws whose solutions may suffer blowup of amplitude and/or total variation. One is thus lead to ask if similar pathological behavior is possible in specific systems of physical interest. Such systems are typically equipped with a convex entropy and one could hope that their additional structure would prevent this type of breakdown. The most interesting system in this respect is the Euler system. Now, the blowup examples in [8, 20] were based on particular interaction patterns where repeated reflections created infinitely many waves in finite time. By judiciously choosing initial data one could generate solutions where wave strengths were amplified at each interaction, leading to finite time blowup. A related setup was considered by Nishida and Smoller in [14] who studied the single- and double-piston problems for the isentropic Euler equations. Under suitable conditions on the motion of the piston(s) they established global existence of entropy admissible weak solution for large data.

As a step toward a better understanding of large solutions for the full Euler system we consider solutions with similar wave patterns. For these solutions we then investigate if such solutions can produce growth, or even finite time blowup, of amplitudes or variation. Below we outline the construction and analysis of *one type* of solution for the ideal-polytropic Euler system where shocks undergo repeated reflections as they interact with two approaching contact waves. The particular pattern is motivated by the one for which blowup is known to occur in other systems. Our analysis indicates that no blowup occurs for solutions of this type. The particular wave pattern we consider here is simple enough to allow accurate estimates for the solution. To simplify the analysis we have imposed absorbing boundary conditions to the left and right of a central region (between the two approaching contacts). This amounts to disregarding interactions of shocks in the same family and renders the analysis tractable.

We also comment briefly on a particular scaling of the dependent variables in the Euler system which provides a large data result for certain types of "scaled up" data. The details of our analysis will appear in [6].

2 Euler Equations

Consider the 1-D Euler equations for compressible gases,

$$u_t + p_x = 0 \tag{1}$$

$$\tau_t - u_x = 0 \tag{2}$$

$$E_t + (pu)_x = 0, \tag{3}$$

where x is the Lagrangian space variable, u is velocity, p is pressure, τ is specific volume, $E = e + u^2/2$ is specific total energy, and $e(\tau, p)$ is internal energy. We consider ideal polytropic gases with the equation of state

$$e = \frac{p\tau}{\gamma - 1}, \tag{4}$$

where $\gamma > 1$ is the adiabatic constant, and the temperature T is given as

$$T = c_\nu^{-1} e, \tag{5}$$

where c_ν denotes the specific heat at constant volume. In this case shocks, contacts, and rarefaction waves can be calculated explicitly, [4, 15, 16, 17].

We denote backward and forward shocks by $S_\pm$, respectively. We let C_- denote contacts, where τ decreases from left to right across the discontinuity, while C_+ denote contacts where τ increases from left to right. As an example, by an S_+C_--interaction we mean an interaction where a forward shock meets a C_--contact from the left, etc.

2.1 Interactions and Interaction Pattern

The possible interactions for ideal polytropic gases have been analyzed in [4, 17, 15]. In particular, it can be shown that if an S_+C_--interaction results in a reflected S_--shock, a C_--contact, and a transmitted S_+-shock. Similarly, a C_+S_--interaction results in a transmitted S_--shock, a C_+-contact and a reflected S_--shock.

This allows us to setup an interaction pattern as follows. We select initial points $x_1 < x_2 < x_3$ and initial data,

$$(u(x,0), p(x,0), \tau(x,0)) = \begin{cases} (u_{l0}, p_{l0}, \tau_{l0}) & \text{if } x < x_1, \\ (u_{l0}, p_{l0}, \tau_{m0}) & \text{if } x_1 < x < x_2, \\ (u_{r0}, p_{r0}, \tau_{M0}) & \text{if } x_2 < x < x_3, \\ (u_{r0}, p_{r0}, \tau_{r0}) & \text{if } x_3 < x, \end{cases}$$

where the four states are connected, going from left to right, by a C_+-contact, an S_+-shock, and a C_--contact. See Fig. 1.

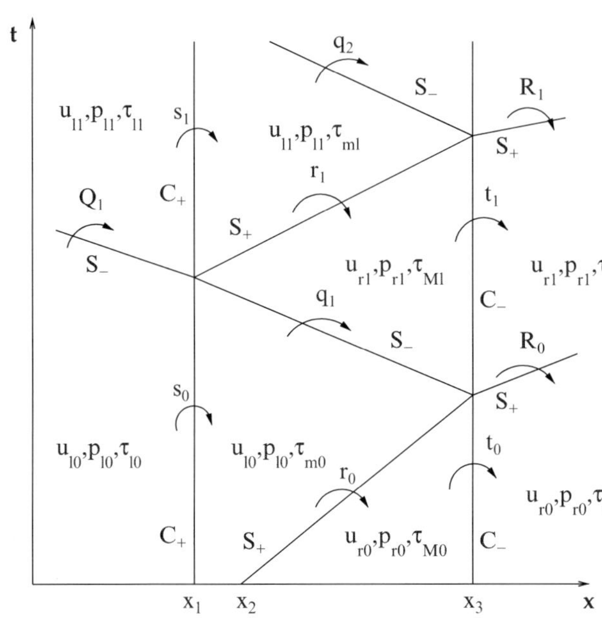

Fig. 1. Interaction pattern

The interaction of the S_+-wave with the C_--contact on the right results in outgoing S_-, C_-, and S_+ waves. In turn the reflected backward shock form the first interaction will later interact with the left C_+-contact. This second interaction yields outgoing S_-, C_+, and S_+ waves, and the same wave pattern repeats itself. We recall that velocity and pressure remain constant across contacts and we label the various states as in Fig. 1.

The transmitted shocks to the left of the central region between the two contacts will necessarily approach each other and interact in finite time. In turn the reflected waves generated in these interactions will result in a quite complicated wave pattern which appears to be too complicated to analyze explicitly. Similar interactions will occur to the right of the central region. As a first step we avoid dealing with these issues by imposing boundary conditions that absorb the transmitted shocks. This may be accomplished by letting the transmitted waves meet with gases whose densities and adiabatic constants are chosen such that no waves are reflected.

2.2 Properties of the Interaction Pattern

Consider the wave curves in (u, p, τ)-space. Given a left state (u_l, p_l, τ_l) the forward and backward shock curves emanating from (u_l, p_l, τ_l) are given by

$$S_\pm = \begin{cases} \tau = \frac{\tau_l(p_l+\mu^2 p)}{(p+\mu^2 p_l)}, \\ u = u_l \pm (p - p_l)\sqrt{\frac{(1-\mu^2)\tau_l}{p+\mu^2 p_l}}, \qquad p \lessgtr p_l, \end{cases} \tag{6}$$

where $\mu^2 = \frac{\gamma-1}{\gamma+1}$.

We notice that pressure increases along the S_--shock curve and that pressure decreases along the S_+-shock curve. This implies that within the three regions $x < x_1$, $x_1 < x < x_3$, and $x_3 < x$ the pressure increases in time. In particular, in the middle region we find,

$$p_{r0} < p_{l0} < p_{r1} < p_{l1} < p_{r2} < p_{l2} < \cdots . \tag{7}$$

From (6) we have that specific volume is a function of pressure which is monotonically decreasing. Together with (7) this implies that specific volume is decreasing in each of the three regions . That is,

$$\tau_{l0} > \tau_{l1} > \tau_{l2} > \cdots > 0 \tag{8}$$

$$\tau_{r0} > \tau_{r1} > \tau_{r2} > \cdots > 0 \tag{9}$$

$$\tau_{M0} > \tau_{m0} > \tau_{M1} > \tau_{m1} > \tau_{M2} > \tau_{m2} > \cdots > 0. \tag{10}$$

Thus, density is increasing and no vacuum formation takes place. A calculation shows that temperature is increasing in each region as time increases. Finally, the velocity is increasing in the right region and decreasing in the left region according to

$$u_{r0} < u_{r0} < u_{r0} < \cdots \tag{11}$$

$$u_{l0} > u_{l0} > u_{l0} > \cdots , \tag{12}$$

while in the middle we find that

$$u_{ri} < u_{lj} \qquad \forall\, i, j. \tag{13}$$

Therefore u_{li} and u_{ri} are monotonic bounded sequences and thus converge. For the wave speeds we have that the shock speeds corresponding to $S_{\pm}$ are given by

$$\sigma_{\pm} = \pm\sqrt{-\frac{(p_r - p_l)}{(\tau_r - \tau_l)}}, \tag{14}$$

where (p_l, τ_l) and (p_r, τ_r) are the values of pressure and specific volume to the immediate left and right of the shock, respectively. By a convexity argument it can be shown that the speeds of the shocks in each of the three regions are increasing in magnitude as time increases.

The properties above follow directly from the explicit expressions for the wave curves. In order to investigate whether magnification of waves can occur in the solution a more detailed analysis is required. At this point it is not clear if the interaction pattern exists for all time, or if the speeds of the shocks can approach infinity and result in infinitely many interactions in finite time. In either case we need to understand how the strengths of the waves develop. The following analysis shows that the solution is well behaved in the sense that shock speeds are bounded while wave strengths decay (quickly) in time. The crux of the matter is to carefully estimate the pressure and show that it is bounded for all times. We will first assume that pressure is bounded and see what this implies for the interaction pattern. The argument that the pressure is indeed bounded is then outlined in Sect. 2.4.

2.3 Properties Assuming Pressure is Bounded

We now assume that pressure is bounded in the three regions of the solution. From (7) it follows that there exists a $\bar{p}$ such that

$$\lim_{i\to\infty} p_{li} = \lim_{i\to\infty} p_{ri} = \bar{p}\,. \tag{15}$$

Thus the pressures in all three regions converge to the same value. From this and the fact that temperature is increasing in each region it follows that specific volume in each region converges to some strictly positive value. (This also shows that the temperature in each region converges to a finite value). An argument based on convexity then applies to show that the magnitude of the shock speeds are bounded. In particular this implies that there are only a finite number of interactions in finite time. A calculation shows that there exists a point $x_0 < x_1$ such that any interaction of transmitted shocks in the left region would occur to the left of x_0. Similarly, there is a point $x_4 > x_3$ such that any interaction of transmitted shocks in the right region would occur to the right of x_4. We can thus insert contacts with jumps in both specific volume and adiabatic constant such that no waves are reflected back toward the central region. This guarantees that the solutions we construct are indeed global-in-time solutions of the Euler equations, albeit with somewhat artificial boundary data.

2.4 Boundedness of Pressure

To show that pressure is bounded uniformly in time we will carefully estimate the solutions of the Riemann problems defined by S_+C_--interactions and C_+S_--interactions. Once the initial data are chosen as described above (that is, with one C_+-wave, one S_+-wave and one C_--wave), all ensuing wave strengths are determined through a system of coupled nonlinear algebraic equations. For these calculations it is convenient to use ratios of pressures and specific volumes to parameterize the wave curves. Referring to Fig. 1 we define the following ratios

$$r_i := \frac{p_{ri}}{p_{li}}, \qquad R_i := \frac{p_{ri}}{p_{r(i+1)}}, \tag{16}$$

$$q_{i+1} := \frac{p_{r(i+1)}}{p_{li}}, \qquad Q_{i+1} := \frac{p_{l(i+1)}}{p_{li}}, \tag{17}$$

$$s_i := \frac{\tau_{mi}}{\tau_{li}}, \qquad t_i := \frac{\tau_{ri}}{\tau_{Mi}}. \tag{18}$$

Here q_{i+1}, R_i and t_{i+1} are uniquely determined by r_i and t_i, while r_{i+1}, Q_{i+1} and s_{i+1} are uniquely determined by q_{i+1} and s_i. The equations for the ratios are not explicitly solvable in terms of formulas. However, a simple calculation shows that the ratios in the middle region satisfy

$$\frac{1}{r_1} > q_2 > \frac{1}{r_2} > q_3 > \cdots > 1, \tag{19}$$

$$t_1 < t_2 < t_3 < \cdots < 1, \tag{20}$$

$$s_1 > s_2 > s_3 > \cdots > 1. \tag{21}$$

Through a further argument (using properties that are specific to ideal polytropic gases) one can then compare the ratios r_i for example, to a sequence that is readily estimated. Specifically, defining

$$x_i := \frac{1}{r_i} - 1,$$

we show that for i sufficiently large $x_i < 1$ and therefore, without loss of generality, we assume $x_1 < 1$. We then proceed to show that there exist a $\delta > 0$ such that

$$p_{rn} < p_{r0}(1 + x_1) \prod_{k=1}^{2n} (1 + x_1^{(1+\delta)^k}). \tag{22}$$

Consequently the pressure in each region converges, since $\prod \left(1+\frac{1}{k^2}\right)$ converges and $x_1^{(1+\delta)^k} < \frac{1}{k^2}$ for k large.

3 Scaling Invariance of the Euler System

In the analysis above only the *ratios* of pressures and specific volumes were used. This reflects the general fact that the Euler system is formally invariant under the scaling

$$(\tau, u, p, E) \mapsto (\zeta\tau, \zeta u, \zeta p, \zeta^2 E) \qquad \zeta > 0.$$

While the four variables above are not independent, this scaling is consistent for ideal polytropic gases. More generally, the Euler system is invariant under the scaling $(\tau, u, p) \mapsto (\zeta\tau, \zeta u, \zeta p)$ provided the internal energy function $e(\tau, p)$ satisfies

$$e(\zeta\tau, \zeta p) = \zeta^2 e(\tau, p). \tag{23}$$

This clearly holds for ideal polytropic gases. A direct calculation shows that if (23) holds, then the eigenvalues λ_0, $\lambda_\pm$, as well as shock speeds, are invariant as well.

It follows that if $U(x,t) = \big(\tau(x,t),\, u(x,t),\, p(x,t)\big)$ is any solution of the Euler system (1)–(3) with initial data $U_0(x) = \big(\tau_0(x),\, u_0(x),\, p_0(x)\big)$, then $V(x) := \zeta U(x,t)$ ($\zeta > 0$) is again a solution with data $V_0 = \zeta U_0$. Furthermore, the wave patterns and interactions are identical in the two solutions. As a consequence we see that once we are given a solution we can obtain other solutions with arbitrarily large (or small) amplitudes by applying the scaling above.

As a final observation we note that it is natural to try and argue the other way as well. That is: given large BV data for the Euler system, we can scale these down (ζ small) to obtain BV data with small variation. Then apply Glimm's theorem to get a global-in-time solution, and finally scale up again (by $1/\zeta$). However, this argument is flawed as the amount of "allowable variation" of the data in Glimm's theorem depend on where the data are located in (τ, u, p)-space. In particular, as the neighborhood where the data lie shrinks toward the origin in (τ, u, p)-space under the scaling, so does the amount of variation allowed by Glimm's theorem.

References

1. Bianchini, S., BV solutions of the semidiscrete upwind scheme. Arch. Ration. Mech. Anal. **167** (2003), 1–81.

2. Bianchini, S., Bressan, A., Vanishing viscosity solutions of nonlinear hyperbolic systems, Ann. of Math. **161** (2005), 223–342.
3. Bressan, A., Hyperbolic systems of conservation laws. The one-dimensional Cauchy problem. Oxford University Press, 2000.
4. Chang, T., Hsiao, L., The Riemann problem and interaction of waves in gas dynamics. Pitman Monographs and Surveys in Pure and Applied Mathematics, 41. Longman Scientific & Technical, Harlow; copublished in the United States with John Wiley & Sons, Inc., New York, 1989.
5. Dafermos, C.M., Hyperbolic conservation laws in continuum physics. Second edition. Grundlehren der Mathematischen Wissenschaften, 325. Springer-Verlag, New York, 2005.
6. Endres, E.E., Jenssen, H.K., Large data solutions for 1D compressible Euler equations, work in progress.
7. Glimm, J., Solutions in the large for nonlinear hyperbolic systems of equations, Comm. Pure. Appl. Math. **18** (1965), 697–715.
8. Jenssen, H.K., Blowup for systems of conservation laws, SIAM J. Math. Anal. **31** (2000), 894–908.
9. Joly, J.L., Metivier, G., Rauch, J., A nonlinear instability for 3×3 systems of conservation laws, Comm. Math. Phys. **162** (1994), 47–59.
10. Liu, T.L., The Deterministic Version of the Glimm Scheme, Comm. Math. Phys. **57** (1977), 135–148.
11. Liu, T.L., Solutions in the large for the equations of nonisentropic gas dynamics, Indiana Univ. Math. J. **26** (1977), no. 1, 147–177.
12. Nishida, T., Global solution for an initial boundary value problem of a quasilinear hyperbolic system, Proc. Japan Acad. **44** (1968), 642–646.
13. Nishida, T., Smoller, J., Solutions in the large for some nonlinear hyperbolic conservation laws, Comm. Pure Appl. Math. **26** (1973), 183–200.
14. Nishida, T., Smoller, J., Mixed problems for nonlinear conservation laws, J. Differential Equations **23** (1977), 244–269.
15. Rozdestvenskii, B.L., Janenko, N.N., Systems of quasilinear equations and their applications to gas dynamics, Translations of Mathematical Monographs, vol. 55. American Mathematical Society, Providence, 1983.
16. Serre, D., Systems of conservation laws, 1–2, Cambridge University Press, Cambridge, 2000.
17. Smoller J., Shock waves and reaction-diffusion equations, Second edition. Grundlehren der Mathematischen Wissenschaften, 258. Springer-Verlag, New York, 1994.
18. Temple, B., Solutions in the large for the nonlinear hyperbolic conservation laws of gas dynamics, J. Differential Equations **41** (1981), 96–161.
19. Temple, B., Young, R., The large time stability of sound waves, Commun. Math. Phys. **179** (1996), 417–466.
20. Young, R., Exact solutions to degenerate conservation laws, SIAM J. Math. Anal. **30** (1999), 537–558.

A Carbuncle Free Roe-Type Solver for the Euler Equations

F. Kemm

Summary. Based on the idea of the HLLEM scheme, we propose a novel ansatz to cure the well known carbuncle instability. Instead of testing all neighboring cells for strong shocks we test the Riemann problem for contact and shear waves. As an indicator we suggest the residual of the Rankine–Hugoniot condition for the linear waves. By using known approaches, we can apply a well tempered amount of viscosity for the contact and shear waves of the Roe and HLLEM methods. The resulting Riemann solver approximates contact and shear waves exactly. However, according to the chosen value of a parameter, the Carbuncle phenomenon can be completely avoided.

1 Introduction

Since the famous paper of Quirk [11] appeared in 1994 a great research started on the instability known as *Carbuncle Phenomenon*. The origin of the name is the fact that in the simulation of strongly supersonic flows around a blunt body the middle part of the resulting bow shock degenerates to a carbuncle shaped structure. It was conjectured already by Quirk [11] that it is closely related to other instabilities such as the so called *odd-even-decoupling* encountered in straight shocks aligned with the grid. The research in this area was twofold. On the one hand, the stability of discrete shock profiles in one as well as in several space dimensions was investigated. On the other hand, there was a lot of effort to find cures for the failure of some schemes in numerical calculations. Unfortunately, the failure is only found in schemes giving high resolution of shear and entropy waves. This category includes, for example, the Godunov-, Roe-, Osher-, HLLC-, and HLLEM scheme. These schemes are preferable in calculations involving complex wave structures as well as boundary layers.

It was found that even in one space dimension there are some instabilities of discrete shock profiles: Slowly moving shocks produce small post-shock oscillations [11, 1, 7]. But also in the case of a steady shock, instabilities can be found depending on the value of the adiabatic coefficient γ [3]. The connection to two-dimensional instabilities is still not fully understood. The two-dimensional instabilites seem to be closely related to each other. In [12] also the connection to physical shock instabilities is investigated. In [4] the authors show by an ingenious numerical investigation that the odd–even-decoupling and the carbuncle follow the same mechanism. Dumbser et al. [5] present a method to test Riemann solvers for their tendency to the odd–even-decoupling.

To cure the carbuncle instability, several methods were suggested in the literature. First Quirk [11] discovered that adding some viscosity to the shear and entropy waves in the direction parallel to the shock front will stabilize the scheme. He then proposed a method of flux switching. The expression $\frac{|p_r - p_l|}{\min(p_r, p_l)}$ is evaluated for each cell interface and than compared to some threshold parameter. If a grid cell is found to be in the vicinity of a strong shock, it is marked. All Riemann problems involving marked cells are solved with HLLE instead of Roe. Related works are [9, 8, 10, 14].

All these methods have in common that the flux calculation does not only depend on two grid cells, but at least values of six cells to compute one intercell flux are needed. This considerably increases the computational effort. Therefore, we searched for an indicator that could be included in the Riemann solver itself.

The main idea of this paper is not to base the indicator on strong shocks in neighboring Riemann problems, but on shear and entropy waves in the Riemann problem itself. We found a useful criterion based on the residual in the Rankine–Hugoniot condition in the contact wave.

The paper is organised as follows: First we show how the instability arises and give the basic idea of the cure. In Sect. 3 we show how HLLEM can be understood as a reformulation of the Roe solver and how to apply the cure to this scheme. At the end we show some numerical experiments illustrating the effect of the new method and give some conclusions.

2 Arising of Carbuncle and the Idea of the Cure

The carbuncle instability mainly occurs in simulations of supersonic flows on structured grids. If a strong shock is aligned with the grid, and a Riemann solver with exact resolution of shear and entropy waves is employed, the shock profile becomes unstable. The worst case is found when the flow field is also aligned with the grid. If we go down to first order schemes the situation even deteriorates.

Besides genuinely two-dimensional effects the main source of the carbuncle phenomenon seems to be the unstable nature of the one-dimensional shock profile [3, 11, 1, 7].

In [3] the authors investigate steady shocks in one space dimension. They prove that the Godunov scheme for strong shocks can produce unstable shock profiles depending on the adiabatic coefficient γ. They report that in practice

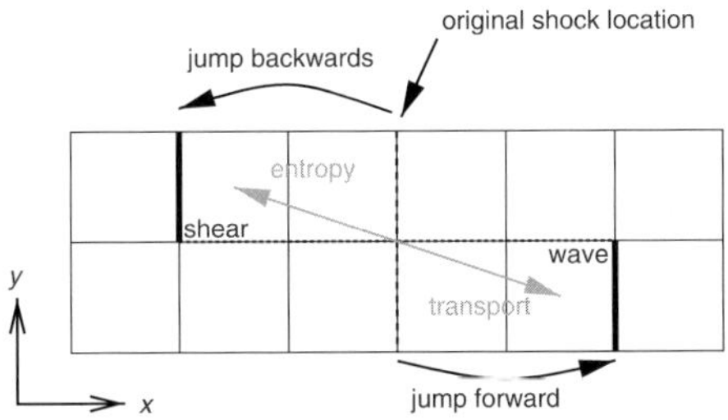

Fig. 1. Unstable two dimensional shock profile

this leads, after a transient regime, to a profile that is shifted by one or two grid cells.

Now we think of a two-dimensional situation like the one depicted in Fig. 1. In each of the two x-slices we have a one-dimensional unstable steady shock. Because of instability it may occur that one of them moves forward by one or two cells. It may also occur that the other one moves backwards by the same amount. For simplicity we think of the transition to be just a single jump. This would mean that on the dotted cell interfaces we have a strong shear wave and also have some entropy transport in an oblique direction. Both of them are artifacts introduced by the numerical scheme. If the Riemann solver used had enough diffusion on entropy and shear waves, we could avoid the instability and thus the carbuncle phenomenon.

In the usual cures all cell interfaces are checked for strong shocks. If a grid cell is in the vicinity of a strong shock, additional viscosity is added to entropy and shear waves. If we go back to the situation depicted in Fig. 1 we find that there are no shear nor entropy waves in the original problem. Thus we suspect that a Riemann solver that adds numerical diffusion to entropy and shear waves when there are none will also avoid the instability of the shock profile. Contrary to the usual approach, in this case there is no need for information from neighboring cell interfaces when computing the cell flux. The resulting Riemann solver is self contained.

Nevertheless, it is necessary to check for entropy and shear waves. In addition, information on the amount of numerical diffusion needed to prevent the carbuncle is desired. We apply the following criterion: The Rankine–Hugoniot condition for the middle wave in the Riemann problem. If $\hat{u}$ is the wave speed of this wave, $\mathbf{q}_r, \mathbf{q}_l$ are the right and left state and $f(.)$ is the physical flux function, then

$$f(\mathbf{q}_r) - f(\mathbf{q}_l) = \hat{u}(\mathbf{q}_r - \mathbf{q}_l) \tag{1}$$

implies that the Riemann problem consists of a single contact wave. Thus we take the Euclidean norm of the residual in (1) to be our indicator. As we will see in the next sections, this leads to the desired results.

3 Application in the Numerical Scheme

3.1 The Two-Dimensional Euler Equations

The two-dimensional Euler equations read as

$$\begin{aligned} \rho_t + \nabla \cdot (\rho \mathbf{v}) &= 0, \\ (\rho \mathbf{v})_t + \nabla \cdot (\rho \mathbf{v} \circ \mathbf{v}) + \nabla p &= 0, \\ e_t + \nabla \cdot ((e+p)\mathbf{v}) &= 0 \end{aligned}$$

with the density ρ, the velocity $\mathbf{v} = (u, v, w)^T$, the pressure p, and the energy e. Thus, the flux in x-direction is

$$f(\mathbf{q}) = \left(\rho u, \rho u^2 + p, \rho u v, (e+p)u\right)^T.$$

In this paper we restrict our analysis to the case of an ideal gas.

3.2 The Indicator Function

In the case of the two-dimensional Euler equations the state vector in conservative variables is $\mathbf{q} = (\rho, \rho u, \rho v, e)^T$. As mentioned above, we base our indicator function on the residual in the Rankine–Hugoniot condition for the contact waves. If $\tilde{u}$ is the Roe averaged speed of this wave, $\mathbf{q}_r, \mathbf{q}_l$ are the right and left states and $f(\cdot)$ is the physical flux function, then we have for the residual

$$\mathbf{R} = f(\mathbf{q}_r) - f(\mathbf{q}_l) - \tilde{u}(\mathbf{q}_r - \mathbf{q}_l).$$

Now we take our indicator θ to be the Euclidean norm of the residual related to the Roe averaged speed of sound:

$$\theta = \|\frac{\mathbf{R}}{\tilde{a}}\|_2.$$

We want a function ϕ, which is one, if θ is large, and is zero, if θ vanishes. So that we can smoothly switch between HLLE scheme and HLLEM (or Roe). Motivated by some heuristic considerations and numerical experiments we come up with the following ingredients:

If we now take some $\varepsilon > 0$, we could use

$$\phi(\theta) = \min\{1, \varepsilon\, \theta\}. \tag{2}$$

This defines the threshold value $1/\varepsilon$ above which the full viscosity of the HLLE scheme is applied. What we also should take into account is the fact that in the case of the flow direction being oblique to the shock front the carbuncle is much weaker. So we take another parameter $\alpha > 0$ and

$$\phi(\theta) = \min\{1, \varepsilon\, \theta \max\{0, 1 - M_u^{\alpha}\}\}, \qquad \alpha > 0, \tag{3}$$

where M_u is the directional Mach number perpendicular to the cell face. This is also preferably computed using the Roe mean values of the velocity and the speed of sound.

In addition, by numerical experiments, it seems to be advantageous to make ϕ concave:

$$\phi(\theta) = \min\{1, (\varepsilon\,\theta \max\{0, 1 - M_u^\alpha\})^\beta\}, \qquad 0 < \beta < 1. \tag{4}$$

Here we had to introduce an additional parameter β.

3.3 HLLEM: Roe as Correction of HLLE

The HLLEM solver originally was motivated by comparing the numerical viscosity of Roe- and HLLE-scheme [6]. Both fluxes can be written as

$$g(\mathbf{q}_r, \mathbf{q}_l) = \frac{1}{2}\big(f(\mathbf{q}_r) + f(\mathbf{q}_l)\big) - \frac{1}{2}V(\mathbf{q}_r - \mathbf{q}_l) \tag{5}$$

with some viscosity matrix V. For the Roe scheme we have

$$V_{\text{Roe}} = |\tilde{A}(\mathbf{q}_r, \mathbf{q}_l)|. \tag{6}$$

The HLLE-flux can be written as

$$g_{\text{HLL}}(\mathbf{q}_r, \mathbf{q}_l) = \frac{1}{2}\big(f(\mathbf{q}_r) + f(\mathbf{q}_l)\big) - \frac{1}{2}\frac{S_R + S_L}{S_R - S_L}\tilde{A}(\mathbf{q}_r - \mathbf{q}_l) + \frac{S_R S_L}{S_R - S_L}(\mathbf{q}_r - \mathbf{q}_l) \tag{7}$$

with $S_L \leq 0 \leq S_R$ being estimates for the leftmost and rightmost wave speeds. Using the Roe-matrix $\tilde{A}(\mathbf{q}_r, \mathbf{q}_l)$ we can find from this

$$V_{\text{HLL}} = \frac{S_R + S_L}{S_R - S_L}\tilde{A} - 2\frac{S_R S_L}{S_R - S_L}I. \tag{8}$$

From (6) and (8) it is easy to see that both viscosity matrices have the same eigenvectors. If we take S_L and S_R to be the corresponding wave speeds of the Roe matrix, the only remaining difference is the eigenvalue of the contact waves. We can express the Roe flux just by using the eigenvectors corresponding to the entropy and shear wave as a correction of the HLL-flux:

$$g_{\text{Roe}}(\mathbf{q}_r, \mathbf{q}_l) = g_{\text{HLL}}(\mathbf{q}_r, \mathbf{q}_l) - \frac{\tilde{u}^2 - \tilde{a}^2}{4\tilde{a}}\,\delta_{\tilde{q}}\left(\tilde{\mathbf{l}}_2(\mathbf{q}_r - \mathbf{q}_l)\,\tilde{\mathbf{r}}_2 + \tilde{\mathbf{l}}_3(\mathbf{q}_r - \mathbf{q}_l)\,\tilde{\mathbf{r}}_3\right) \tag{9}$$

with the antidiffusion coefficient $\delta_{\tilde{q}}$. If in the computation of g_{HLL} we allow S_L and S_R to differ from the corresponding Roe speeds, we end up with the HLLEM-flux. It is shown in [10] that, independent of the choice of S_L and S_R, HLLEM resolves single contact waves exactly as long as for the computation of the antidiffusion coefficient we stick to the Roe mean values of the quantities in consideration. According to [2] the best choice of wave speeds for HLLE, and thus for HLLEM, is

$$S_L = \min\{\tilde{u} - \tilde{a}, u_l - a_l, 0\} \qquad S_R = \max\{\tilde{u} + \tilde{a}, u_r + a_r, 0\}. \tag{10}$$

In this way, to every rarefaction wave a simple entropy fix is applied, even when the rarefaction is not sonic. Throughout this paper, especially in the numerical experiments, we use this setting.

For the cure of the carbuncle instability we just have to multiply the antidiffusion coefficient with $(1 - \phi(\theta))$. This means, we replace in (9) the original $\delta_{\tilde{q}}$ by the product $(1 - \phi(\theta)) \cdot \delta_{\tilde{q}}$. The additional effort in computation time compared to the original HLLEM scheme is minimal.

4 Numerical Results

In this section we present some numerical results. All of them are computed with the parameters in (3) set to $\varepsilon = 1/100$, $\alpha = \beta = 1/3$. The calculations are computed with Euler2d, a code of the group of C.-D. Munz at Stuttgart University.

4.1 Steady Shock

To test the behavior of our new solver in the case of steady shocks, we employ the setting proposed in [5]. To trigger the instability, we add some randomly generated numerical noise of order 10^{-6} to the initial data. Numerical results are displayed in Fig. 2.

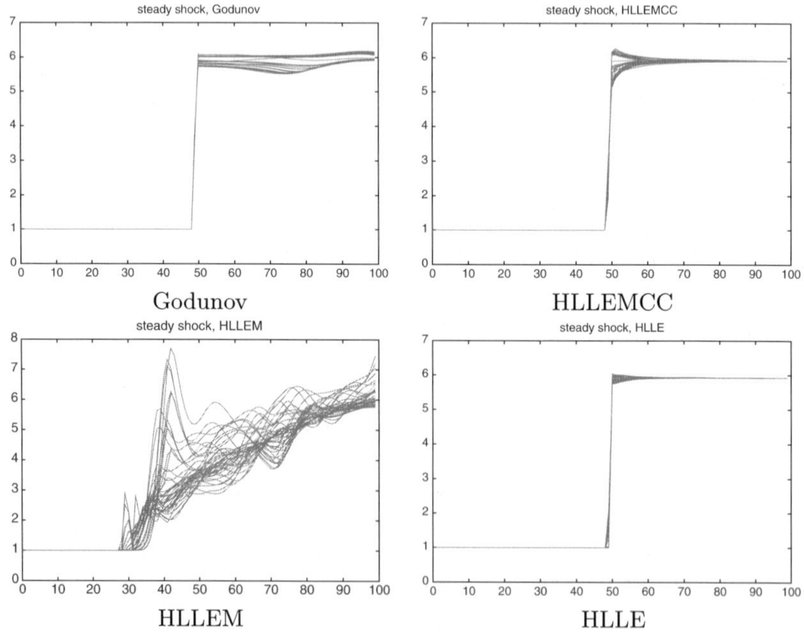

Fig. 2. Steady shock problem with different solvers

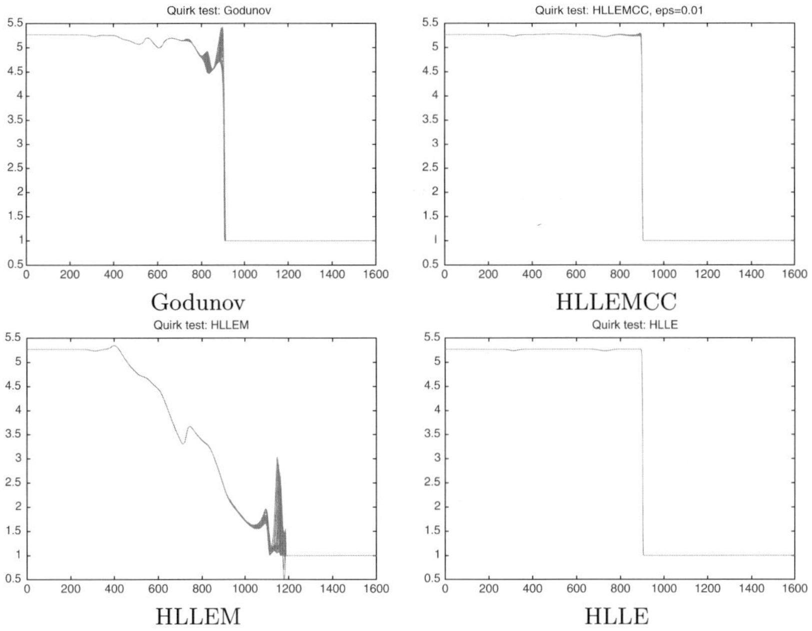

Fig. 3. Quirk test with different solvers

4.2 The Quirk Test Case

In [11] Quirk proposes an interesting test problem. A shock is running down a duct. With this configuration we can test for the stability of moving shock fronts. In the original setting of [11] the middle line of the mesh is perturbed. Because our test computations are done with a cartesian code, we instead superimpose some numerical noise of the order 10^{-3} to trigger the instability. The noise is generated using random numbers. Numerical results are displayed in Fig. 3.

5 Conclusions and outlook

In [13] Roe presents a new Riemann solver that faces the one-dimensional instability of profiles directly. The construction of the numerical flux function is based on entropy considerations. With this solver the carbuncle can be prevented on structured grids. If the approach we presented can be introduced in the new Roe solver it should be possible to avoid the carbuncle on all types of grids. Another question is how to adapt the new solvers in the context of wave propagation. Some of these questions as well as an investigation of the choice of parameters will be treated in a subsequent paper.

References

1. Mohit Arora and Philip L. Roe. On postshock oszillations due to shock capturing schemes in unsteady flows. *Journal of Computational Physics*, 130:25–40, 1997.
2. P. Batten, N. Clarke, C. Lambert, and D.M. Causon. On the choice of wavespeeds for the HLLC Riemann solver. *SIAM J. Sci. Comput.*, 18(6): 1553–1570, 1997.
3. Matthieu Bultelle, Magali Grassin, and Denis Serre. Unstable Godunov discrete profiles for steady waves. *SIAM Journal on Numerical Analysis*, 35(6):2272–2297, 1998.
4. Y. Chauvat, Jean-Marc Moschetta, and Jérḿie Gressier. Shock wave numerical structure and the carbuncle phenomenon. *International Journal for Numerical Methods in Fluids*, 47:903–909, 2005.
5. Michael Dumbser, Jean-Marc Moschetta, and Jérḿie Gressier. A matrix stability analysis of the carbuncle phenomenon. *Journal of Computational Physics*, 197:647–670, 2004.
6. Bernd Einfeldt. On Godunov-type methods for gas dynamics. *SIAM J. Numer. Anal.*, 25(2):294–318, 1988.
7. Shi Jin and Jian-Guo Liu. The effects of numerical viscosities i: Slowly moving shocks. *Journal of Computational Physics*, 126:373–389, 1996.
8. Sung-Soo Kim, Chongam Kim, Oh-Hyun Rho, and Seung Kyu Hong. Cures for the shock instability: Development of a shock-stable roe scheme. *Journal of Computational Physics*, 185:342–374, 2003.
9. Maurizio Pandolfi and Domenic D' Ambrosio. Numerical instabilities in upwind methods: Analysis and cures for the carbuncle phenomenon. *Journal of Computational Physics*, 166(2):271–301, 2001.
10. Soo Hyung Park and Jang Hyuk Kwon. On the dissipation mechanism of Godunov type schemes. *Journal of Computational Physics*, 188:524–542, 2003.
11. James J. Quirk. A contribution to the great Riemann solver debate. *International Journal for Numerical Methods in Fluids*, 18:555–574, 1994.
12. Jean-Christophe Robinet, Jérḿie Gressier, G. Casalis, and Jean-Marc Moschetta. Shock wave instability and the carbuncle phenomenon: Same intrinsic origin? *Journal of Fluid Mechanics*, 417:237–263, 2000.
13. Philip L. Roe. Affordable, entropy-consistent, flux functions. Oral talk at Eleventh International Conference on Hyperbolic Problems: Theory, Numerics, Applications, Lyon, 2006.
14. Richard Sanders, Eric Morano, and Marie-Claude Druguet. Multidimensional dissipation for upwind schemes: Stability and applications to gas dynamics. *Journal of Computational Physics*, 145(2):511–537, 1998.

WENOCLAW: A Higher Order Wave Propagation Method

D.I. Ketcheson and R.J. LeVeque

1 Introduction

Many important physical phenomena are governed by hyperbolic systems of conservation laws

$$\mathbf{q}_t + f(\mathbf{q})_x = 0, \tag{1}$$

for which a wide range of numerical methods have been developed. In this paper we present a numerical method for solution of (1) that is also applicable to general hyperbolic systems of the form

$$\mathbf{q}_t + A(\mathbf{q}, x, t)\mathbf{q}_x = 0. \tag{2}$$

In the nonlinear, nonconservative case, the method may be applied if the structure of the Riemann solution is understood. Examples of (1–2) include acoustics and elasticity in heterogeneous media. Wave-propagation methods of up to second order accuracy have been developed and applied to such systems ([4, 5]).

Many high order accurate methods for solution of (1) have been developed, including the essentially nonoscillatory (ENO) and weighted ENO (WENO) schemes. Such methods rely on calculating fluxes, and may not be applied to the more general system (2).

The method described in this work combines the notions of wave propagation and the method of lines, and can in principle be extended to arbitrarily high order accuracy by the use of high order accurate spatial reconstruction and a high order accurate ODE solver. In this work, we use Runge–Kutta methods.

The method is implemented in a software package, WENOCLAW, that is based on CLAWPACK [6] and makes use of Riemann solvers in the same form required for CLAWPACK. The software package and additional documentation are freely available at

`http://www.amath.washington.edu/~claw/`

2 One-Dimensional Discretization

2.1 Semidiscretization

Many numerical methods have been developed to approximately solve (1) based on Godunov's method. The method relies on solving Riemann problems, which consist of (2) with piecewise constant initial data

$$q(x,0) = q_0(x) = \begin{cases} q^- & \text{if } x < 0 \\ q^+ & \text{if } x > 0. \end{cases} \tag{3}$$

The conservation law is integrated over a cell to obtain

$$\frac{\partial Q_i}{\partial t} = -\frac{1}{\Delta x}\left(f(q^*_{i+\frac{1}{2}}) - f(q^*_{i-\frac{1}{2}})\right), \tag{4}$$

where $q^*_{i-\frac{1}{2}}$ is the solution to the Riemann problem along the ray $x = x_{i-\frac{1}{2}}$ (or equivalently, $f(q^*_{i-\frac{1}{2}})$ is the Godunov flux) and Q_i is the average of q over the ith grid cell. High order accuracy can be achieved using (4) as follows. First, reconstruct a piecewise polynomial approximation $\tilde{q}_i$ to q_i in each cell. In particular, obtain reconstructed values

$$q^+_{i-\frac{1}{2}} = \tilde{q}_i(x_{i-\frac{1}{2}}) = q^e(x_{i-\frac{1}{2}}) + \mathcal{O}(\Delta x^p) \tag{5}$$

$$q^-_{i+\frac{1}{2}} = \tilde{q}_i(x_{i+\frac{1}{2}}) = q^e(x_{i+\frac{1}{2}}) + \mathcal{O}(\Delta x^p), \tag{6}$$

which are approximations to q at the neighboring cell interfaces. Here q^e denotes the exact solution.

If the Godunov flux at $x_{i+\frac{1}{2}}$ is then obtained by solving a Riemann problem with left and right states $q^-_{i-\frac{1}{2}}$ and $q^+_{i-\frac{1}{2}}$, respectively, the resulting semidiscrete approximation (4) is accurate to order p.

Similar to the approach in ([4]), we now proceed to rewrite (4) in terms of fluctuations. Given a system of the form (2), we assume that the solution to the Riemann problem is a similarity solution consisting of a set of waves moving at constant speeds. For nonlinear systems where this may not be the case, we assume the use of an approximate Riemann solver that yields such a solution. This class of solvers includes linearized (such as Roe) solvers, as well as the simpler HLL and LLF solvers. In any case, we then have a decomposition of the jump in the Riemann problem into waves:

$$\Delta q = q_r - q_l = \sum_{p=1}^{M_w} \mathcal{W}^p, \tag{7}$$

where $\mathcal{W}^p$ is the jump across the pth wave, M_w is the number of waves, and each wave has an associated wave speed s^p.

We wish to express the flux difference in (4) in terms of fluctuations, $\mathcal{A}^{\pm}\Delta q$. The fluctuations may be defined in terms of the flux function:

$$\mathcal{A}^{-}\Delta q = f(q^*) - f(q^-) \tag{8}$$
$$\mathcal{A}^{+}\Delta q = f(q^+) - f(q^*), \tag{9}$$

where $\Delta q_{i-\frac{1}{2}} = q^+_{i-\frac{1}{2}} - q^-_{i-\frac{1}{2}}$. Notice that the fluctuations are a splitting of the flux difference:

$$f(q^+_{i-\frac{1}{2}}) - f(q^-_{i-\frac{1}{2}}) = \mathcal{A}^{+}\Delta q_{i-\frac{1}{2}} + \mathcal{A}^{-}\Delta q_{i-\frac{1}{2}}. \tag{10}$$

The sum on the right hand side of (10) is denoted by $\mathcal{A}\Delta q_{i-\frac{1}{2}}$ and referred to as a total fluctuation.

More generally, the fluctuations may be defined in terms of the waves:

$$\mathcal{A}^{+}\Delta q_{i-\frac{1}{2}} = \sum_{p=1}^{m} (s^p_{i-\frac{1}{2}})^{+}\mathcal{W}^p_{i-\frac{1}{2}} \tag{11}$$
$$\mathcal{A}^{-}\Delta q_{i+\frac{1}{2}} = \sum_{p=1}^{m} (s^p_{i+\frac{1}{2}})^{-}\mathcal{W}^p_{i+\frac{1}{2}}. \tag{12}$$

The fluctuation $\mathcal{A}^{-}\Delta q$ represents the net effect of all left-going waves on the solution, while $\mathcal{A}^{+}\Delta q$ represents the net effect of all right-going waves.

Using (8)–(9), the flux difference in (4) can be rewritten as

$$f(q^*_{i+\frac{1}{2}}) - f(q^*_{i-\frac{1}{2}}) = \mathcal{A}^{-}\Delta q_{i+\frac{1}{2}} + f(q^-_{i+\frac{1}{2}}) + \mathcal{A}^{+}\Delta q_{i-\frac{1}{2}} - f(q^+_{i-\frac{1}{2}}). \tag{13}$$

We now interpret $q^+_{i-\frac{1}{2}}$ and $q^-_{i+\frac{1}{2}}$ as the left and right states, respectively, for a Riemann problem within cell i. Then we have

$$f(q^-_{i+\frac{1}{2}}) - f(q^+_{i-\frac{1}{2}}) = \mathcal{A}\Delta q_i, \tag{14}$$

where $\Delta q_i = q^-_{i+\frac{1}{2}} - q^+_{i-\frac{1}{2}}$. Substitution of (14) into (13) gives

$$f(q^*_{i+\frac{1}{2}}) - f(q^*_{i-\frac{1}{2}}) = \mathcal{A}^{-}\Delta q_{i+\frac{1}{2}} + \mathcal{A}^{+}\Delta q_{i-\frac{1}{2}} + \mathcal{A}\Delta q_i. \tag{15}$$

Substitution into (4) yields the semidiscrete scheme

$$\frac{\partial Q_i}{\partial t} = -\frac{1}{\Delta x}(\mathcal{A}^{-}\Delta q_{i+\frac{1}{2}} + \mathcal{A}^{+}\Delta q_{i-\frac{1}{2}} + \mathcal{A}\Delta q_i). \tag{16}$$

Notice that the final term on the right hand side of (16) does not require the solution of a Riemann problem. It is clear from the derivation that this scheme reduces to the corresponding flux-differencing scheme when applied to (1). The advantage of the scheme over flux-differencing schemes lies in the ability to solve systems of the form (2). Since systems of this form generally cannot be rewritten in terms of a flux function, fluctuations are calculated in terms of the decomposition (11)–(12). Alternatively, the f-wave decomposition of [2] may be used to obtain the fluctuations.

Derivation of the scheme directly from (2), without the use of a flux function, is omitted here for the sake of brevity. The scheme (16) remains valid as long as the Jacobian A is constant (as a function of x) within each cell.

2.2 Reconstruction

In the previous section, we assumed that high order accurate point values of the solution were known. In fact, at the beginning of a time step, only cell-averaged quantities are known. We now discuss the problem of reconstructing point values.

Many details are omitted here; the reader is referred to [1] for details on WENO reconstruction and to [5] for details on TVD reconstruction.

Suppose we have a formula for obtaining the high order accurate approximations (5)–(6) for a scalar function $q(x)$. We will further assume that the formula can be written as

$$q^+_{i-\frac{1}{2}} = Q_i + \phi^+(\theta_{i+s_1-\frac{1}{2}}, ..., \theta_{i+s_2-\frac{1}{2}})\Delta q_{i-\frac{1}{2}} \tag{17}$$

$$q^-_{i-\frac{1}{2}} = Q_{i-1} + \phi^-(\theta_{i+s_1-\frac{1}{2}}, \ldots, \theta_{i+s_2-\frac{1}{2}})\Delta q_{i-\frac{1}{2}}, \tag{18}$$

where s_1, s_2 are parameters defining the stencil of the method and

$$\theta_{j-\frac{1}{2}} = \frac{\Delta q_{j-\frac{1}{2}}}{\Delta q_{i-\frac{1}{2}}}. \tag{19}$$

For instance, in the case of minmod reconstruction we have

$$\phi^{\pm} = \mp\frac{1}{2}\left(1 + \text{sgn}(\theta_{I-\frac{1}{2}})\right)\min(1, |\theta_{I-\frac{1}{2}}|), \tag{20}$$

where $I - \frac{1}{2}$ is the next interface upwind of $i - \frac{1}{2}$.

This scalar reconstruction may be applied to systems of equations in various ways. In the simplest approach, the scalar reconstruction is applied to each component of $\mathbf{q}$. This approach works well for some problems, but in other cases it is insufficient. In particular, it appears to become successively less satisfactory as the order of accuracy of the reconstruction is increased. See [7] for a detailed discussion with respect to central WENO schemes, for instance.

Accuracy can be improved by instead applying the reconstruction to the characteristic fields of $\mathbf{q}$. To do so, we reconstruct as follows. Let $A_{i-\frac{1}{2}} = A(q, x_{i-\frac{1}{2}})$. Let $r^p_{i-\frac{1}{2}}, l^p_{i-\frac{1}{2}}$ be the right and left eigenvectors of $A_{i-\frac{1}{2}}$. First each jump Δq is decomposed into characteristic components θ^p. Then $\mathbf{q}$ is reconstructed via

$$\mathbf{q}^{+}_{i-\frac{1}{2}} = Q_i + \sum_p \phi^{\pm}(\theta^p_{i+s_1-\frac{1}{2}}, ..., \theta^p_{i+s_2-\frac{1}{2}})\alpha^p_{i-\frac{1}{2}} r^p_{i-\frac{1}{2}}, \tag{21}$$

where $\alpha^p_{i-\frac{1}{2}} = l^p_{i-\frac{1}{2}} \cdot \Delta q_{i-\frac{1}{2}}$.

The only difference between the following methods is the manner in which the θ's are determined. If the system is nonlinear and/or A has explicit spatial dependence, the reconstruction must account for the variation in the characteristic structure over the stencil.

Wave-Slope Reconstruction

In this approach we first determine the waves entering each cell from the solution of the Riemann problems at each interface using the cell average states. This yields a set of waves $\mathcal{W}^p_{i-\frac{1}{2}}$ associated with each interface. To reconstruct $q^{\pm}_{i-\frac{1}{2}}$, we first project the pth wave at each interface in the stencil onto the pth wave at $x_{i-\frac{1}{2}}$:

$$\theta^p_{j-\frac{1}{2}} = \frac{\mathcal{W}^p_{j-\frac{1}{2}} \cdot \mathcal{W}^p_{i-\frac{1}{2}}}{\mathcal{W}^p_{i-\frac{1}{2}} \cdot \mathcal{W}^p_{i-\frac{1}{2}}}. \tag{22}$$

We then reinterpret each projected wave as an approximation to the slope of the corresponding characteristic field at that interface.

Typically the waves result from a linearized Riemann solution, and so $\mathcal{W}^p_{i-\frac{1}{2}} = (l^p_{i-\frac{1}{2}} \cdot \Delta q_{i-\frac{1}{2}}) r^p_{i-\frac{1}{2}}$. Then we can express $\theta^p_{j-\frac{1}{2}}$ as

$$\theta^p_{j-\frac{1}{2}} = \frac{\alpha^p_{j-\frac{1}{2}}\; r^p_{j-\frac{1}{2}} \cdot r^p_{i-\frac{1}{2}}}{\alpha^p_{i-\frac{1}{2}}\; r^p_{i-\frac{1}{2}} \cdot r^p_{i-\frac{1}{2}}}. \tag{23}$$

Characteristic-Wise Reconstruction

If the Jacobian A has rapidly varying spatial dependence, the previous methods may yield inaccurate results. In this case, the reconstruction is performed using

$$\theta^p_{j-\frac{1}{2}} = \frac{l^p_{i-\frac{1}{2}} \cdot \Delta q_{j-\frac{1}{2}}}{\alpha^p_{i-\frac{1}{2}}}. \tag{24}$$

Transmission-Based Reconstruction

In the reconstruction method of the previous section, waves from all characteristic fields at each interface in the stencil are decomposed and contribute to each characteristic field at interface $i - \frac{1}{2}$. For some systems, such as linear acoustics, it appears more reasonable to decompose only waves from the corresponding characteristic field. In the case of acoustics, this has the interpretation of taking waves approaching the interface $i+\frac{1}{2}$ and comparing the part of each that would be transmitted through that interface. In this case, the reconstruction is performed using

$$\theta^p_{j-\frac{1}{2}} = \frac{\alpha^p_{j-\frac{1}{2}}}{\alpha^p_{i-\frac{1}{2}}}(l^p_{i-\frac{1}{2}} \cdot r^p_{j-\frac{1}{2}}). \tag{25}$$

2.3 Time Stepping

Equation (16) may be integrated via an ODE solver. One-step methods are most convenient, and WENOCLAW is implemented using Runge–Kutta methods. Several strong stability preserving methods [3, 9] have been implemented, with order of accuracy ranging from two to five.

3 Extension to Two Dimensions

The semidiscrete scheme may be extended to two dimensions in a straightforward manner. The two-dimensional analog of (2) is

$$\mathbf{q}_t + A(\mathbf{q}, x, y)\mathbf{q}_x + B(\mathbf{q}, x, y)\mathbf{q}_y = 0. \tag{26}$$

It is possible to extend the semidiscrete wave propagation scheme using a simple dimension-by-dimension approach, meaning that the reconstruction at each face only uses data in a slice orthogonal to that face. The scheme is

$$\frac{\partial Q_{i,j}}{\partial t} = -\frac{1}{\Delta x}\left(\mathcal{A}\Delta q_{i,j} + \mathcal{A}^+\Delta q_{i-\frac{1}{2},j} + \mathcal{A}^-\Delta q_{i+\frac{1}{2},j}\right) \tag{27}$$

$$-\frac{1}{\Delta y}\left(\mathcal{B}\Delta q_{i,j} + \mathcal{B}^+\Delta q_{i,j-\frac{1}{2}} + \mathcal{B}^-\Delta q_{i,j+\frac{1}{2}}\right), \tag{28}$$

where the $\mathcal{B}^{\pm}\Delta q$ represent fluctuations in the y direction. Because it neglects certain cross-derivative terms, this scheme is formally of second order, regardless of the order of accuracy of the reconstruction and time stepping. However, in practice, the method yields solutions that are much better than traditional second order methods.

To formally achieve greater than second order accuracy, Gauss quadrature is used to integrate fluctuations over faces. The details of the genuinely multidimensional implementation are omitted here.

4 Sonic Crystal Bandgap Simulation

The semidiscrete wave propagation schemes we have described are especially suited for simulation of high-frequency waves in the presence of rapidly varying material parameters, as is the case for sound waves passing through a sonic crystal.

Small amplitude acoustic waves are described by the linear hyperbolic system

$$p_t + K(x)u_x = 0 \tag{29}$$

$$u_t + \frac{1}{\rho(x)}p_x = 0, \tag{30}$$

where p, u are pressure and velocity perturbations (respectively) relative to some ambient state. Note that this system is of the form (2), with

$$q = \begin{pmatrix} p \\ u \end{pmatrix}, \qquad A = \begin{pmatrix} 0 & K \\ 1/\rho & 0 \end{pmatrix}. \tag{31}$$

We consider a square array of square rods in air with a plane wave disturbance incident parallel to the x-axis. The array is seven periods deep, and periodic boundary conditions are applied in the y-direction. The lattice spacing is 10 cm and the rods have a cross-sectional side length of 4 cm, so that the filling fraction is 0.16. This crystal is similar to the one studied in [8], and it is expected that sound waves in the 1,200–1,800 Hz range will experience severe attenuation in passing through it, while longer wavelengths will not.

The results presented here were calculated using fifth order dimension-by-dimension WENO reconstruction with characteristic-wise limiting and a third order Runge–Kutta method.

Figure 1a shows a low frequency plane wave incident from the left. This wave has a frequency of about 800 Hz, well below the partial band gap. As expected, the wave passes through the crystal without significant attenuation. In Fig. 2a, the pressure is plotted along a line approximately midway between rows of rods.

Figures 1b and 2b show the same quantities for an incident plane wave with wavenumber $k = 29.22\,\mathrm{m}^{-1}$, $c = 344\,\mathrm{m\,s}^{-1}$. Notice that the wave is almost entirely reflected, resulting in a standing wave in front of the crystal.

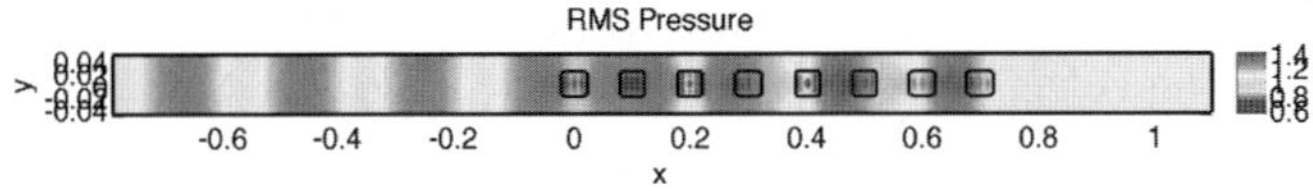

(a) 821 Hz

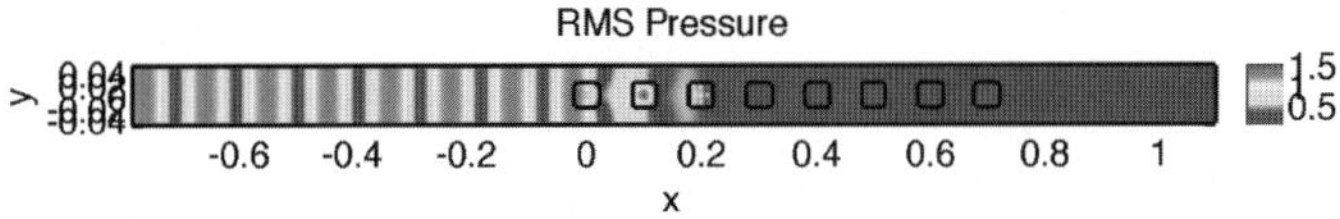

(b) 1,600 Hz

Fig. 1. RMS pressure in the sonic crystal

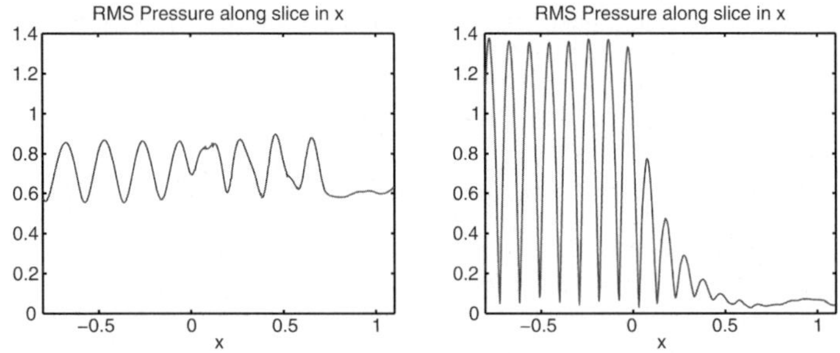

(a) 821 Hz (b) 1,600 Hz

Fig. 2. RMS pressure in the sonic crystal along a slice at $y = -0.05$

Acknowledgments

The authors thank Chi-Wang Shu and Yulong Xing for providing useful sample FORTRAN WENO codes. The first author was supported during this work by a U.S. Department of Homeland Security Fellowship and by a U.S. Department of Energy Computational Science Graduate Fellowship. This research was also supported in part by NSF grant DMS-0106511.

References

1. C.-W.S.B. Cockburn, C. Johnson and E. Tadmor, *ENO and WENO schemes for hyperbolic conservation laws*, in Lecture Notes in Mathematics, A. Quarteroni, ed., vol. 1697, Springer-Verlag, Berlin, pp. 325–432.
2. D.S. Bale, R.J. LeVeque, S. Mitran, and J.A. Rossmanith, *A wave propagation method for conservation laws and balance laws with spatially varying flux functions*, SIAM Journal of Scientific Computing, 24 (2002), pp. 955–978.
3. S. Gottlieb and C.-W. Shu, *Total variation diminishing runge-kutta schemes*, Mathematics of Computation, 67 (1998), pp. 73–85.
4. R.J. LeVeque, *Wave propagation algorithms for multidimensional hyperbolic systems*, Journal of Computational Physics, 131 (1997), pp. 327–353.
5. ——, *Finite Volume Methods for Hyperbolic Problems*, Cambridge University Press, 2002.
6. R.J. LeVeque, *CLAWPACK Version 4.2 User's Guide*, 2003. Available from www.amath.washington.edu/~claw/.
7. J. Qiu and C.-W. Shu, *On the construction, comparison, and local characteristic decomposition for high-order central WENO schemes*, Journal of Computational Physics, 183 (2002), pp. 187–209.
8. L. Sanchis, F. Cervera, J. Sanchez-Dehesa, J.V. Sanchez-Perez, C. Rubio, and R. Martinez-Sala, *Reflectance properties of two-dimensional sonic band-gap crystals*, J. Acoust. Soc. Am., 109 (2001), pp. 2598–2605.
9. R.J. Spiteri and S.J. Ruuth, *A new class of optimal high-order strong-stability-preserving time discretization methods*, SIAM Journal of Numerical Analysis, 40 (2002), pp. 469–491.

Unsteady Transonic Airfoil Flow Simulations using High-Order WENO Schemes

I. Klioutchnikov, J. Ballmann, H. Oliver, V. Hermes, and A. Alshabu

Summary. The flow over a supercritical airfoil in the transonic regime contains a lot of complex physical phenomena and interactions. High demands are made on the numerical model for an accurate numerical investigation of such a flow type. High-order ($N \geq 5$) WENO (Weighted Essentially Nonoscillatory) schemes are appropriate for the numerical simulation of transonic flows, as they have the capability to capture shocks, resolve vortices, and predict their interactions. A high-order finite difference WENO scheme is applied to simulate the unsteady flow over the supercritical BAC3-11 [1] airfoil. The numerical results are compared with experiments carried out in a transonic shock tube and found in good agreement with the experimental data. The influences of Ma and Re number are studied.

1 Numerical Method

The Navier–Stokes equations for unsteady two-dimensional compressible fluid flow are transformed from Cartesian (x, y) to general coordinates $\xi_r(x, y)$ with $r = 1, 2$ and read

$$\mathbf{U}_{,t} + \sum_{r=1}^{2} \frac{1}{J} \mathbf{F}_{r,\xi_r} = \sum_{r=1}^{2} \frac{1}{J} \mathbf{F}^{v}_{r,\xi_r}. \tag{1}$$

Here $\mathbf{U}$ is the solution vector of conserved variables, $\mathbf{F}_r$ and $\mathbf{F}^v_r$ represent the inviscid and viscous fluxes, respectively. J is the Jacobian of the transformation.

The **WENO** (Weighted Essentially Nonoscillatory) finite difference scheme [3], used for inviscid fluxes, is applied for the Navier–Stokes equations in general coordinates (1). The method is of arbitrary odd order $N = 3, 5, 7, 9, 11...$ in space and employs p sub-stencils with operators for spatial discretization of order $N = (2p - 1)$ of the scheme. For the time integration, an explicit Runge–Kutta TVD-scheme of third order accuracy is used that reads

$$\mathbf{U}^*_\bullet = \mathbf{U}^n_\bullet - \mathbf{Q}^n_\bullet, \qquad \mathbf{U}^{**}_\bullet = \mathbf{U}^*_\bullet - \frac{1}{4}\{-3\mathbf{Q}^n_\bullet + \mathbf{Q}^*_\bullet\},$$

$$\mathbf{U}_\bullet^{n+1} = \mathbf{U}_\bullet^{**} - \frac{1}{12}\{-\mathbf{Q}_\bullet^{n} - \mathbf{Q}_\bullet^{*} + 8\mathbf{Q}_\bullet^{**}\}, \qquad (2)$$

$$\mathbf{Q}_\bullet = \sum_{r=1}^{2} \lambda_r \{(\mathbf{F}_{r,i_r+1/2} - \mathbf{F}_{r,i_r-1/2}) - (\mathbf{F}^v_{r,i_r+1/2} - \mathbf{F}^v_{r,i_r-1/2})\}.$$

Here λ_r is defined $\lambda_r = \Delta t / J_\bullet \Delta\xi_r$. The upper indices n correspond to the time level. The lower indices are running in the spacial directions $r = 1, 2$ from $i_r = 1$ to $i_r = i_r max$. The symbol $\bullet$ corresponds to one grid point with the indices i_1, i_2. For monotonicity reasons of the spatial discretization the nonlinear Roe flux-difference splitting or the Lax–Friedrichs flux-vector splitting are available. Lax–Friedrichs flux-vector splitting is found to be more accurate and therefore applied in this work. The Lax–Friedrichs flux-vector splitting

$$\mathbf{F}_{i_r+1/2} = \mathbf{F}^+_{i_r+1/2} + \mathbf{F}^-_{i_r+1/2}$$

with $\mathbf{U} = (u_1, u_2, ..., u_M)$ (for 2D Navier–Stokes equations $M = 4$) reads

$$\begin{aligned} \mathbf{F}_{i_r+1/2} &= \phi_r \mathbf{F}_{i_r+1/2} + \sum_{m=1}^{M} [-\theta_N(\mathbf{R}^{-1}_{mri_r+1/2}\Delta\mathbf{F}^{m+}_{i_r-N/2+1}, ... \\ &..., \mathbf{R}^{-1}_{mri_r+1/2}\Delta\mathbf{F}^{m+}_{i_r-N/2+N-1}) + \theta_N(\mathbf{R}^{-1}_{mri_r+1/2}\Delta\mathbf{F}^{m-}_{i_r+N/2}, ... \\ &..., \mathbf{R}^{-1}_{mri_r+1/2}\Delta\mathbf{F}^{m-}_{i_r+N/2-N+2})]\mathbf{R}_{mri_r+1/2}, \end{aligned} \qquad (3)$$

$$\Delta\mathbf{F}^{m\pm}_{i_r+1/2} = \mathbf{F}^{m\pm}_{i_r+1} - \mathbf{F}^{m\pm}_{i_r}, \qquad \mathbf{F}^{m\pm}_{i_r} = 0.5(\mathbf{F}^m_{i_r} \pm \Lambda^{m(max)}_{i_r}\mathbf{U}_{i_r}),$$

where ϕ_r is the central difference operator of even order $(N-1)$. $\mathbf{R}^{-1}_{mri_r+1/2}$, $\mathbf{R}_{mri_r+1/2}$ are mth left and right eigenvectors of the Jacobian in general coordinates and $\Lambda^{m(max)}_{i_r}$ are the maximum local eigenvalues for each stencil. The nonlinear weight operators $\theta_N(a_1, a_2, ..., a_{N-1})$ are defined using smoothness estimators for every order N [4]. For the calculation of the viscous fluxes $\mathbf{F}^v_r$ a central high-order discretization of even order $(N+1)$ is used [2]. The numerical results shown here are obtained using the WENO method of order $N = 9$ in space and the third order Runge–Kutta scheme in time. Fictitious points are used outside the computational domain for the conformity of spatial discretization order.

Application Two-dimensional flow simulation is carried out on a C–H type grid with 1280×130 grid points on the RWTH Sun Fire cluster using 20 CPUs. No-slip boundary conditions are applied at solid walls with boundary layer related quantities $\Delta x^+ < 20$ and $\Delta y^+ < 1$. Calculations are performed with channel wall boundary condition on the one hand. In this case the computational domain is $-3 \le x \le 5$ and $-1.625 \le y \le 1.875$. The nondimensional coordinates x and y are scaled by the airfoil chord length c. The results are compared with experimental data. On the other hand, the study of the influence of *Mach* and *Reynolds* number on the flow field is accomplished numerically for unconfined flow field, because of the wind tunnel blocking for $Ma_\infty > 0.75$. In this case, the computational domain is $-20 \le x \le 20$ and $-20 \le y \le 20$. The laminar inflow Mach number Ma_∞ and Reynolds number based on the chord-length Re_c are varied in the following range:

$$Ma_\infty = 0.70 - 0.80 \qquad Re_c = 0.5 \times 10^6 - 5 \times 10^6.$$

The angle of attack α is kept constant $\alpha = 0°$.

2 Results

A numerical simulation with the inflow conditions $Ma_\infty = 0.75$, $Re_c = 1 \times 10^6$, and free stream boundary conditions is selected to describe the general flow field over a BAC3-11 airfoil. The local supersonic region on the suction side of the airfoil is for the chosen inflow conditions not yet well established. Nevertheless, small supersonic regions exist and are closed by λ-shaped shocks. The position, motion, and size of the family of λ-shocks depend on the *Ma* and *Re* number.

Vortical structures are observed on the upper and lower side of the airfoil (Fig. 1). The vortices are generated in the rear part ($x > 0.6$) of the upper side, where the pressure gradient becomes positive. On the lower side vortices are already observed in the front part of the airfoil, because of the special shape of the transonic airfoil. They move downstream and disappear in the region of the maximum airfoil thickness ($0.3 < x < 0.5$). In the region of the positive pressure gradient they reappear. The interaction of the vortices from the upper and lower side causes wake fluctuations.

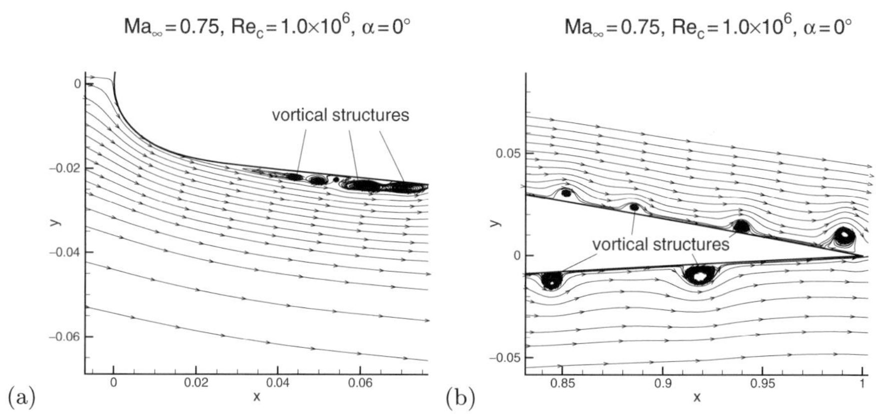

Fig. 1. BAC3-11 airfoil flow for $Ma_\infty = 0.75$ and $Re_c = 1 \times 10^6$. Streamlines with vortices around in the front (**a**) and rear (**b**) part of the airfoil

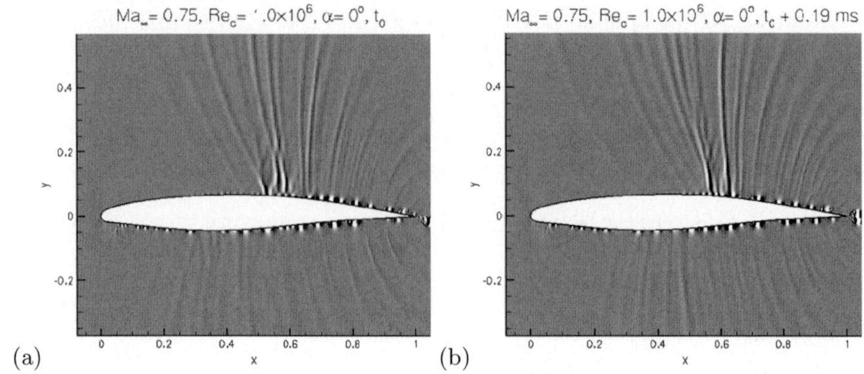

Fig. 2. BAC3-11 airfoil flow for $Ma_\infty = 0.75$ and $Re_c = 1 \times 10^6$. Instantaneous distributions of second derivative of density at different times: (**a**) t_0, (**b**) $t_0 + 0.19\,\mathrm{ms}$

Upstream traveling waves are observed in the simulated flow field (Fig. 2). The generation of these upstream moving waves is detected near the trailing edge of the airfoil. The mechanism of generation, interaction with the boundary layer, and influence on the transition process of these waves are not yet fully understood. The presented wave phenomena bears resemblance to so-called *Kutta*-waves [5], although the supersonic domain in the investigated *Ma* and *Re* number range is not yet fully established.

2.1 Shock Tube Flow

At the Shock Wave Laboratory of RWTH Aachen University, shock tube experiments at transonic *Mach* numbers have been performed. An experimental result of the BAC3-11 airfoil flow at $Ma_\infty = 0.71$ and $Re_c = 3.0 \times 10^6$ is chosen for comparison with the numerical simulation. In the calculation channel wall boundary conditions are used for the upper and lower wind tunnel walls. In Fig. 3a a shadowgraph of the experiments is presented. The numerically determined instantaneous distribution of the second derivative of the density, which corresponds to the experimental shadowgraph visualization, is depicted in Fig. 3c,d. The numerical results are in good agreement with the experimental data. The wave pattern is visualized in more detail (Fig. 3d) by increasing the resolution of the calculated second derivative of the density. Therefore, the depicted range of the derivative is reduced by factor 10 to stronger distinguish smaller gradients.

Time resolved pressure measurements have been also carried out in the experiment. Time averaged pressure coefficients of pressure sensors located at the upper airfoil side are additionally shown in the plot of the calculated pressure coefficient (Fig. 3b. The pressure history and its normalized power spectrum are presented in Fig. 4a for one sensor, which is located at $x = 0.61$. The dominant frequencies of the upstream moving waves are found to be 1

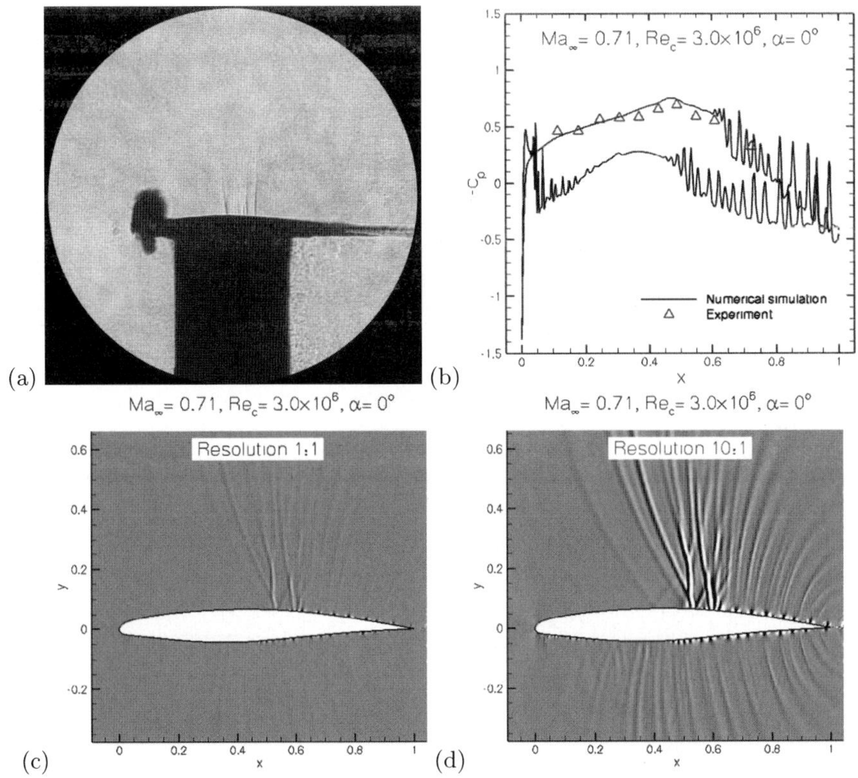

Fig. 3. BAC3-11 airfoil flow: (**a**) Shadowgraph, (**b**) Pressure coefficient, (**c**) and (**d**) instantaneous second derivative of density with different resolutions

and 1.5 kHz. The analysis of the power spectrum for the numerical simulation is performed inside ($y = 0.05$) and outside ($y = 0.07$) the boundary layer for $x = 0.7$ (Fig. 4b). The dominant frequencies outside the boundary layer are found to be 1 and 2 kHz. For low frequencies ($f < 3$ kHz) the power spectra are similar for positions outside and inside the boundary layer. The dominant frequencies inside the boundary layer are found in the range of 25–45 kHz. They are caused by the resolved downstream moving vortices. The high-frequency fluctuations are not reproduced by the experiment because the maximal measureable frequency for the experimental setup is limited to 5 kHz.

2.2 *Mach* Number Variation

The *Mach* number is varied in the range $Ma_\infty = 0.70$–0.80, whereas the *Reynolds* number is kept constant $Re_c = 1.0 \times 10^6$. A strong change in the airfoil flow is observed in the investigated *Mach* number range. The instantaneous *Mach* number distribution is depicted for different *Ma* numbers in

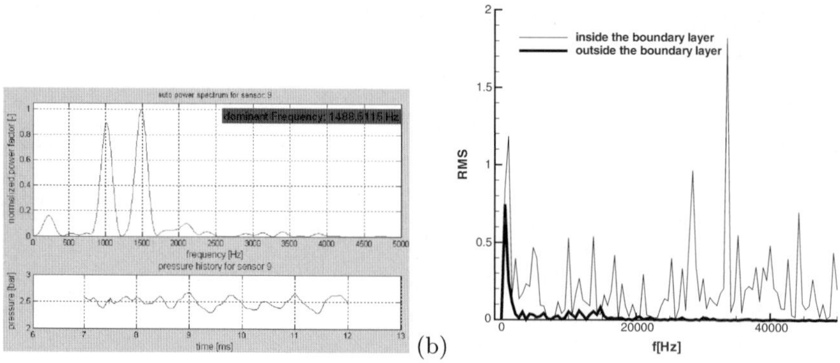

(a) (b)

Fig. 4. Pressure history and normalized power spectrum for one selected sensor (**a**) and power spectrum for the numerical result (**b**)

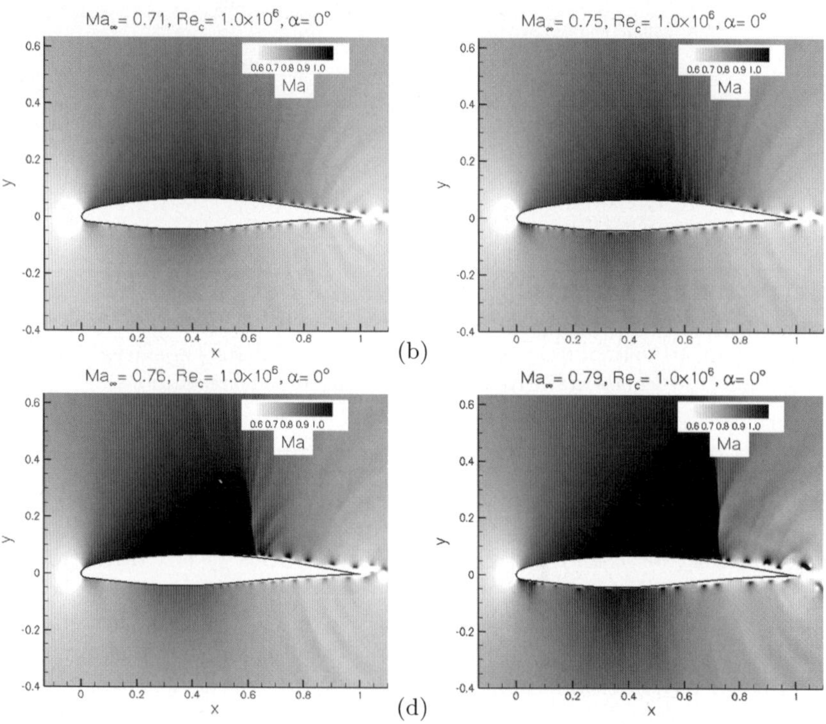

(a) (b) (c) (d)

Fig. 5. BAC3-11 airfoil flow. Instantaneous *Mach* number distributions for different inflow *Mach* numbers: (**a**) $Ma_\infty = 0.71$, (**b**) $Ma_\infty = 0.75$, (**c**) $Ma_\infty = 0.76$, (**d**) $Ma_\infty = 0.79$

Fig. 5. For $Ma_\infty = 0.70$ very small supersonic regions closed by λ-shocks are established on the upper airfoil side. The position of the λ-shock families is fluctuating and interactions with the upstream moving waves are observed. By increasing the inflow *Mach* number the size of the supersonic regions increase too. The λ-shocks become stronger. The position of the normal part of the λ-shock stabilizes, whereas the oblique part is still fluctuating. For $Ma_\infty > 0.76$ the small local supersonic regions merge to a big supersonic domain, which is closed by a strong normal shock.

2.3 *Reynolds* Number Variation

For the variation of the *Reynolds* number from $Re_c = 0.5 \times 10^6$ to $Re_c = 5.0 \times 10^6$ the *Mach* number is kept constant at $Ma_\infty = 0.75$. The results of the calculation with $Re_c = 2.1 \times 10^6$ are used as initial flow field conditions for all computations. Instantaneous *Mach* number distributions and pressure coefficients are depicted for different *Reynolds* numbers in Fig. 6. The pressure coefficient distribution is given at different times to better show the change in the flow field. The curves for t_0 correspond to the initial flow field at $Re_c = 2.1 \times 10^6$. By decreasing the *Reynolds* number to $Re_c = 0.5 \times 10^6$ the movement of the λ-shocks becomes stronger and the first λ-shock moves to the airfoil leading edge.

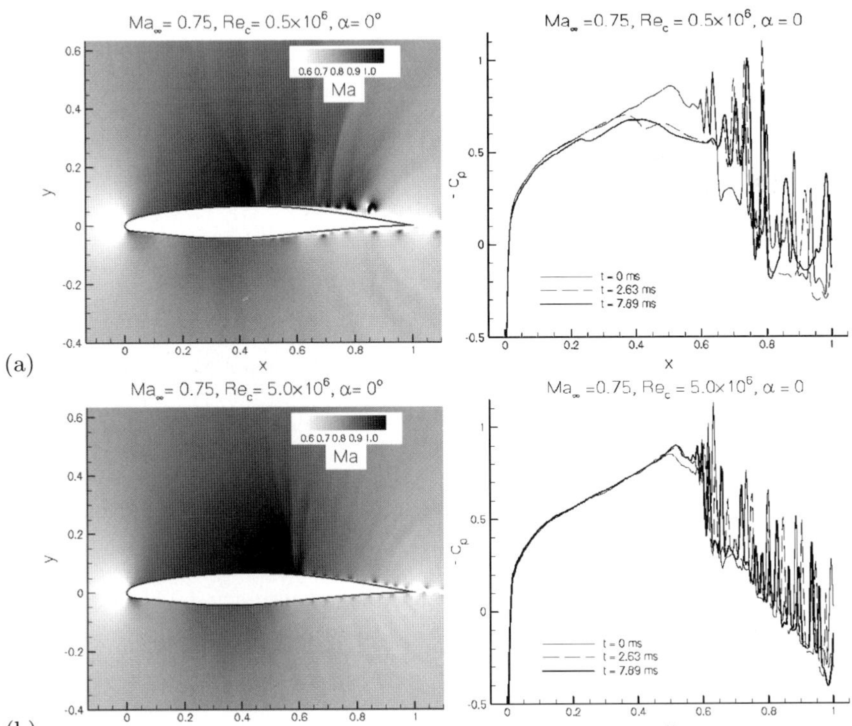

Fig. 6. BAC3-11 airfoil flow. Instantaneous *Mach* number distributions and related pressure coefficients for (**a**) $Re_c = 0.5 \times 10^6$ and (**b**) $Re_c = 5.0 \times 10^6$

By increasing the *Reynolds* number to $Re_c = 5 \times 10^6$, a big supersonic domain is observed which is closed by a strong normal shock. Λ-shocks are typical for laminar boundary layer shock interactions, whereas normal shocks are more typical for turbulent boundary layers. The study of the transition process by 3D numerical simulation and experimental investigations are in progress.

3 Conclusions

The simulation of the unsteady transonic BAC3-11 airfoil flow based on a high order WENO finite difference schemes is performed. The high order WENO schemes are found appropriate to reproduce transonic flow features. The high order accuracy in space enables to resolve well vortices and the dynamics of weak shock-structures in the transonic flow region. Results of calculations with channel wall boundary conditions are compared to experimental results and found to be in good accordance. *Mach* and *Reynolds* number variations have been made. The sensitivity of the flow field on the inflow conditions is found to be high in the investigated *Mach* and *Reynolds* number range.

Acknowledgments

The support of Deutsche Forschungsgemeinschaft within the Collaborative Research Center SFB 401 "Flow Modulation and Fluid-Structure Interaction" is greatfully acknowledged.

References

1. Moir IRM (1994) Measurements on a two-dimensional airfoil with high lift devices, AGARD-AR-303
2. Klioutchnikov I, Ballmann J (2004) Direct Numerical Simulation of Transonic Flow about a Supercritical Airfoil. DLES V, 2003, Kluwer Academic Publishers, 9, 223–230
3. Jiang GS, Shu CW (1996) Efficient implementation of weighted ENO schemes. J. of Comp. Physics, 126, 1, 202–228
4. Balsara D, Shu CW (2000) Monotonicity preserving weighted essentially non-oscillatory schemes with increasingly high order of accuracy. J. of Comp. Physics, 160, 2, 405–452
5. Lee BHK, Murty H, Jiang H (1994) Role of Kutta Waves on Oscillatory Shock Motion on an Airfoil. AIAA J., 32, 4, 789–796
6. Olivier H, Reichel T, Zechner M (2003) Flow visualisation and pressure measurements on an airfoil in high Reynolds number transonic flow. AIAA J., 41, 8, 1405–1412

The Predictor–Corrector Method for Solving of Magnetohydrodynamic Problems

T. Kozlinskaya and V. Kovenya

In the present work for solution of magnetohydrodynamic problems, the finite difference method of predictor–corrector type is chosen as basic. This method allows fulfilling above requirements on the algorithm constructing. At a predictor step, the initial equations are approximated in nondivergent form and the special splitting scheme on physical processes and spatial variables is proposed. The splitting of operators at fractional steps includes two steps: the splitting on spatial directions as in most schemes of approximate factorization and the special splitting of each one-dimensional problem so that the scheme is unconditionally stable but realized by scalar sweeps at the fractional steps on the one hand and on the other hand has minimal quantity of additional members arising from splitting. Proposed special splitting form of one-dimensional operators allows to minimize the splitting effect that makes these schemes to be similar to nonfactorizing schemes. At a corrector step, the equations are approximated on divergent form to satisfy the scheme conservative property. For linearized equations, it can be shown that proposed scheme is unconditionally stable in two-dimensional case. Efficiency of proposed algorithm is shown on solving the problem about high-temperature plasma spread in the magnetic field. Influence of the magnetic field on plasma configuration, and change of plasma shape in time are studied and plasma spread speed is evaluated.

1 The Initial Equations

We shall consider plasma motion in a nonuniform magnetic field in one-fluid magnetohydrodynamic approach under the assumption that effects of viscosity and heat conductivity are negligible small. Then the system of magnetohydrodynamic equations includes the equations of continuity, motion, full energy, and of a magnetic field [1]. In cylindrical coordinate system

$(r,\ z)$ in axisymmetric approach, the equations of magnetohydrodynamics in a dimensionless form can be written in the following form:

$$\begin{aligned}
&\frac{\partial n}{\partial t} + \operatorname{div}(n\vec{v}) = 0, \\
&\frac{\partial n\vec{v}}{\partial t} + \operatorname{div}(n\vec{v}\cdot\vec{v}) + k_0\nabla p = M_A^2\left[\operatorname{curl}\vec{B}\cdot\vec{B}\right], \\
&\frac{dp}{dt} + (\vec{v}\cdot\nabla)p + \gamma p \operatorname{div}\vec{v} = 0, \\
&\frac{\partial \vec{B}}{\partial t} = \operatorname{curl}\left[\vec{v}\cdot\vec{B}\right]
\end{aligned} \tag{1}$$

For convenience let us rewrite the initial equations (1) in the vector form:

$$\frac{\partial U}{\partial t} + W_r + W_z = 0, \tag{2}$$

$$W_r = \frac{1}{r}\begin{pmatrix}
\frac{\partial}{\partial r}(rnv_r) \\
\frac{\partial}{\partial r}\left(rnv_r^2\right) + k_0 r\frac{\partial p}{\partial r} + M_A^2 B_z\frac{\partial B_z}{\partial r} \\
\frac{\partial}{\partial r}(rnv_r v_z) - M_A^2 B_r\frac{\partial B_z}{\partial r} \\
\frac{\partial}{\partial r}(rv_r p) + lp\frac{\partial}{\partial r}rv_r \\
0 \\
\frac{\partial}{\partial r}r\left(v_z B_r - v_r B_z\right)
\end{pmatrix},$$

$$W_z = \begin{pmatrix}
\frac{\partial}{\partial z}(nv_z) \\
\frac{\partial}{\partial z}(nv_r v_z) - M_A^2 B_z\frac{\partial B_r}{\partial z} \\
\frac{\partial}{\partial z}\left(nv_z^2\right) + k_0\frac{\partial p}{\partial z} + M_A^2 B_r\frac{\partial B_r}{\partial z} \\
\frac{\partial}{\partial z}(v_z p) + lp\frac{\partial v_z}{\partial z} \\
\frac{\partial}{\partial z}\left(v_r B_z - v_z B_r\right) \\
0
\end{pmatrix},$$

where $U = (n,\ nv_r,\ nv_z,\ p,\ B_r,\ B_z)^T$, $l = \gamma - 1$, $M_A = B_0/\sqrt{4\pi n_0 U_0^2 m_i}$ is Alfven number of Mach, and $k_0 = T_0/(U_0^2 m_i)$, m_i is an ion weight. The system of (1) is closed by the equation of state in the form of

$$p = nT. \tag{3}$$

At transformations of (1), the condition

$$\operatorname{div} B = \frac{1}{r}\frac{\partial}{\partial r}(rB_r) + \frac{\partial B_z}{\partial z} = 0 \tag{4}$$

is used. At transformation to a dimensionless form, spatial coordinates are normalized by typical length L, time by L/U_0, and density, temperature, speed, and a magnetic field by their not indignant values n_0, T_0, U_0, and B_0. At construction of difference scheme at a predictor step, the equations in nondivergent form are used. Choosing as required functions a vector $f = (n, v_r, v_z, p, B_r, B_z)$, we rewrite (1) in the form of

$$\frac{\partial f}{\partial t} + C_r f + C_z f = 0. \tag{5}$$

Here

$$C_r = \begin{pmatrix} v_r\frac{\partial}{\partial r} & \frac{n}{r}\frac{\partial}{\partial r}r & 0 & 0 & 0 & 0 \\ 0 & v_r\frac{\partial}{\partial r} & 0 & \frac{k_0}{n}\frac{\partial}{\partial r} & 0 & M_A^2\frac{B_z}{n}\frac{\partial}{\partial r} \\ 0 & 0 & v_r\frac{\partial}{\partial r} & 0 & 0 & -M_A^2\frac{B_r}{n}\frac{\partial}{\partial r} \\ 0 & \frac{\gamma p}{r}\frac{\partial}{\partial r}r & 0 & v_r\frac{\partial}{\partial r} & 0 & 0 \\ 0 & 0 & 0 & 0 & \frac{v_r}{r}\frac{\partial}{\partial r}r & 0 \\ 0 & \frac{B_z}{r}\frac{\partial}{\partial r}r & -\frac{B_r}{r}\frac{\partial}{\partial r}r & 0 & 0 & v_r\frac{\partial}{\partial r} \end{pmatrix},$$

$$C_z = \begin{pmatrix} v_z & 0 & n & 0 & 0 & 0 \\ 0 & v_z & 0 & 0 & -M_A^2 B_z/n & 0 \\ 0 & 0 & v_z & k_0/n & M_A^2 B_r/n & 0 \\ 0 & 0 & \gamma p & v_z & 0 & 0 \\ 0 & -B_z & B_r & 0 & v_z & 0 \\ 0 & 0 & 0 & 0 & 0 & v_z \end{pmatrix}\frac{\partial}{\partial z} = C\frac{\partial}{\partial z}.$$

For obtaining matrix operators C_r and C_z, the condition (4) is used. The purpose of work is construction of the efficient schemes which, at the problem solution, require the minimal number of arithmetic realization operations. It can be reached on the basis of predictor–corrector method where at a predictor step the method of splitting possessing a sufficient stability factor is used, and at a corrector step a method accuracy raises and conservatism of scheme is restored. It is known that splitting of operators leads to occurrence of additional members in the difference scheme that, as a rule, leads to decreasing of calculation accuracy [2]. Therefore at construction of the scheme, we choose such splittings which lead to the minimal number of these members but keep properties of scheme unconditional stability and efficiency, i.e., scalar solvability of difference equations on fractional steps.

2 The Difference Scheme for One-Dimensional Problems

We consider algorithm of difference scheme construction for one-dimensional nonstationary problems. Let

$$\begin{aligned} &\frac{\partial f}{\partial t} + C\frac{\partial f}{\partial z} = 0, \\ &\frac{\partial U}{\partial t} + \frac{\partial W}{\partial z} = 0 \end{aligned} \tag{6}$$

is a system of the one-dimensional magnetohydrodynamic equations in the form of (2) and (5), vectors of unknown functions f and U are the same as in a two-dimensional case, and $\partial W/\partial z = W_z$, $C\partial/\partial z = C_z$. We calculate an approximate solution of problem (6) in the domain $Q = \{0 \le t \le T, 0 \le z \le Z\}$, using a uniform grid with steps τ and h. We approximate the first derivatives $\partial/\partial z$ by the difference operator with order $O(h^k)$.

The difference scheme of predictor–corrector type

$$\begin{aligned} &\frac{f^{n+1/2} - f^n}{\tau\alpha} + C^n \Lambda f^{n+1/2} = 0, \\ &\frac{U^{n+1} - U^n}{\tau} + \Lambda W^{n+1/2} = 0 \end{aligned} \tag{7}$$

approximates the initial equations (6) with order $O(\tau^m + h^k)$ where $m = 2$ at $\alpha = 0.5 + O(\tau)$, and at the predictor step it is realized by vector sweeps as follows from a type of the operator C. At $\alpha \ge 0.5$ it is unconditionally stable. For obtaining the schemes realized by scalar sweeps, we present the operator C in the form of splitting

$$C = C^1 + C^2, \tag{8}$$

where we specify the splitting in the following form:

$$C^1 = \begin{pmatrix} 0 & 0 & 0 & 0 & 0 & 0 \\ 0 & v_z & 0 & 0 & 0 & 0 \\ 0 & 0 & v_z & k_1 & k_2 B_r & 0 \\ 0 & 0 & \gamma p & 0 & 0 & 0 \\ 0 & 0 & B_r & 0 & 0 & 0 \\ 0 & 0 & 0 & 0 & 0 & 0 \end{pmatrix}, \quad C^2 = \begin{pmatrix} v_z & 0 & n & 0 & 0 & 0 \\ 0 & 0 & 0 & 0 & -k_2 B_z & 0 \\ 0 & 0 & 0 & 0 & 0 & 0 \\ 0 & 0 & 0 & v_z & 0 & 0 \\ 0 & -B_z & 0 & 0 & v_z & 0 \\ 0 & 0 & 0 & 0 & 0 & v_z \end{pmatrix}.$$

Here $k_1 = k_0/n$ and $k_2 = M_A^2/n$.

The difference scheme

$$\begin{aligned}
&\frac{f^{n+1/4} - f^n}{\tau\alpha} + C^1 \Lambda f^{n+1/4} = 0,\\
&\frac{f^{n+1/2} - f^{n+1/4}}{\tau\alpha} + C^2 \Lambda f^{n+1/2} = 0,\\
&\frac{U^{n+1} - U^n}{\tau} + \Lambda W^{n+1/2} = 0
\end{aligned} \tag{9}$$

approximates the initial equations (6) with the same order, as the base scheme, but unlike (7) it is realized by scalar sweeps. Introduction of splitting (8) results unlike nonfactorized schemes (7) in additional members only in the equations of motion because the matrix $C^1\Lambda C^2\Lambda$ contains four nonzero elements. All other forms of splitting (8) lead to greater number of additional members.

The difference scheme (9) for the linear equations obtained from (6) by "freezing" of factors is absolutely stable [3].

The scheme (9) is generalization of the scheme [4] offered for solving of gas dynamics equations.

3 The Scheme for a Multidimensional Case

The ideology of splitting at construction of difference schemes for the one-dimensional equations can be easily generalized for a many-dimensional case at combined splitting of the equations in spatial directions and splitting of each one-dimensional operator. Let τ, h_r, h_z are grid steps on time and space correspondingly. We enter splitting of every operators C_r and C_z at the sum of operators so that corresponding difference scheme is realized by scalar sweeps, absolutely stable, and contained the minimal number of additional members. Then by analogy with the scheme (9), the difference scheme for solving of two-dimensional problems can be written down in the form of

$$\begin{aligned}
&\frac{f^{n+1/8} - f^n}{\tau\alpha} + C^1_{h,r} f^{n+1/8} = 0,\\
&\frac{f^{n+1/4} - f^{n+1/8}}{\tau\alpha} + C^2_{h,r} f^{n+1/4} = 0,\\
&\frac{f^{n+3/8} - f^{n+1/4}}{\tau\alpha} + C^1_{h,z} f^{n+3/8} = 0,\\
&\frac{f^{n+1/2} - f^{n+3/8}}{\tau\alpha} + C^2_{h,z} f^{n+1/2} = 0,\\
&\frac{U^{n+1} - U^n}{\tau} + W^{n+1/2}_{h,r} + W^{n+1/2}_{h,z} - F^{l+1/2} = 0,
\end{aligned} \tag{10}$$

where $C_{h,r} = C^1_{h,r} + C^2_{h,r}$, $C_{h,z} = C^1_{h,z} + C^2_{h,z}$, $W_{h,r}$, and $W_{h,z}$ are the difference operators approximating corresponding differential operators with order

$O(h_i^k)$, $i = r, z$. Here the splittings are like in the one-dimensional case [3]. The difference scheme (10) approximates the initial equations with order $O(\tau^m + h^k)$, where $h = \max(h_r,\ h_z)$, $m = 2$ at $\alpha = 0,5 + O(\tau)$ and $m = 1$ otherwise, also it is absolutely stable at $\alpha \geq 0,5$. As it should be from a kind of the split operators, as in the one-dimensional case, it is realized at fractional steps by scalar sweeps, and at the corrector step explicitly.

4 Numerical Calculations

For offered difference scheme, calculations of a plasma cloud spread ($\gamma = 5/3$) are realized. Initially in the area of $\sqrt{r^2 + z^2} \leq 0.1$, the pressure jump exceeding background pressure on three orders (by virtue of symmetry of a problem, only the quarter of area is used for calculations) is set:

$$(n,\ v_r,\ v_z,\ p,\ B_r,\ B_z)^T_{t=0} = (n_0,\ 0,\ 0,\ 1000,\ 0,\ 1)^T \text{ for } \sqrt{r^2 + z^2} \leq 0.1,$$
$$\text{else } (n,\ v_r,\ v_z,\ p,\ B_r,\ B_z)^T_{t=0} = (1,\ 0,\ 0,\ 1,\ 0,\ 1)^T.$$

Cloud spread has been studied in time at various values of a magnetic field, density, and temperatures under following boundary conditions:

$$z = 0,\ 0 \leq r \leq 1:\ \partial n/\partial z = \partial v_r/\partial z = v_z = \partial p/\partial z = \partial B_r/\partial z = \partial B_z/\partial z = 0,$$
$$z = 1,\ 0 \leq r \leq 1:\ n = 1,\ v_r = 0,\ v_z = 0,\ p = 1,\ B_r = 0,\ B_z = 1,$$
$$r = 0,\ 0 \leq z \leq 1:\ \partial n/\partial z = v_r = \partial v_z/\partial z = \partial p/\partial z = \partial B_r/\partial z = \partial B_z/\partial z = 0,$$
$$r = 1,\ 0 \leq z \leq 1:\ n = 1,\ v_r = 0,\ v_z = 0,\ p = 1,\ B_r = 0,\ B_z = 1.$$

The numerical solving has been found by the predictor–corrector scheme (10) with the first order at $\alpha = 0.505$ on the calculated grids containing 100×100 and 200×200 units.

On Fig. 1, distributions of density and pressure at the time moment $t = 0.3$ are resulted for various values n_0.

Due to an initial gradient of pressure the cloud starts to extend. Two waves of density are formed. Inside of a cloud, the area with the lowered density of the order of background plasma density and pressure on some orders below background plasma pressure is formed. It is less n_0, plasma spread is more quickly and the density in the center of area decreases more quickly. In the subsequent calculations, influence of a magnetic field on distribution of a plasma cloud is studied at various values M_A. The magnetic field is directed along an axis z and as consequence, the gradient of magnetic pressure is directed along radius that should lead to compression of a cloud in a radial direction.

On Fig. 2, distributions of density and pressure for various values n_0 at the time moment $t = 0.15$ for $M_A^2 = 5$ are presented. It is seen that it is more n_0, spread of plasma and decrease of plasma density in the center of area are more slowly.

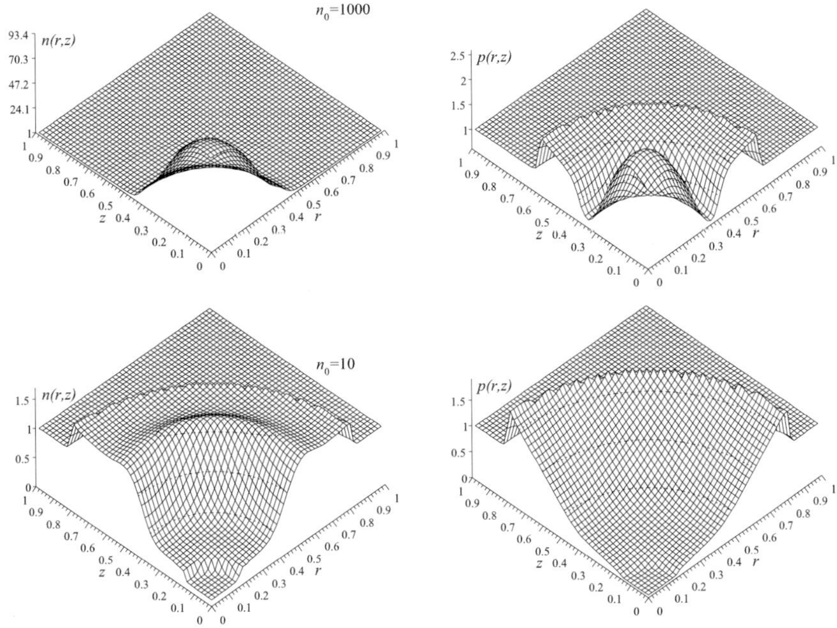

Fig. 1. Plasma spread at $M_A = 0$

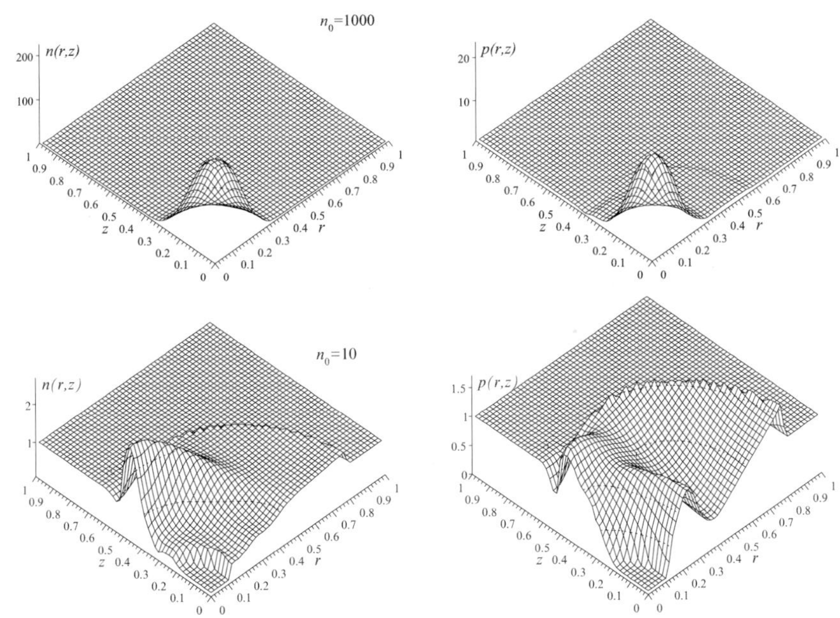

Fig. 2. Plasma spread in magnetic field

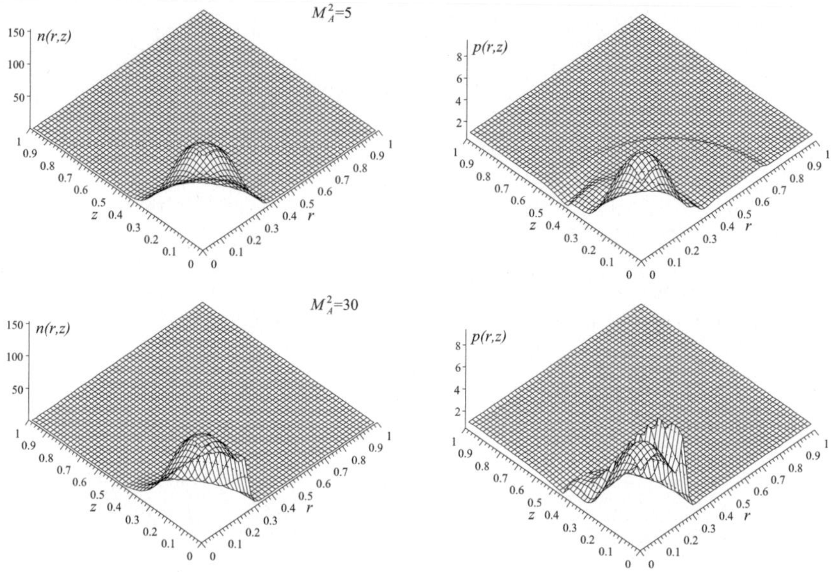

Fig. 3. Plasma spread at various M_A

On Fig. 3, distributions of density and pressure at the moment of time $t = 0.2$ are resulted at $n_0 = 1000$ for $M_A^2 = 5, 30$ accordingly. With increase of a magnetic field (increase M_A) spread of plasma in a radial direction becomes less, than in longitudinal, at $M_A^2 = 30$ it practically stops, the cloud is divided into two parts which spread along an axis z. Thus amplitudes of density and pressure strongly increase. Spread in a direction of an axis z does not depend on a magnetic field.

Results of the realized calculations correctly transfer physics of the studied phenomenon, qualitatively and quantitatively coincide with obtained in [5]. Unlike [5] offered the difference scheme is absolutely stable, allows to realize calculations with larger steps on time.

The realized calculations allow concluding about efficiency of the offered algorithm and an opportunity of its application for solving more difficult problems subject to real effects.

Acknowledgments

The work was partially supported by the Russian Foundation for Basis Research (Grant No. 05-01-00146a) and by the Siberian Branch of the Russian Academy of Sciences (Integration Project No. 116).

References

[1] Braginskiy, S.I.: Transport phenomena in plasma. Problems in plasma theory. Gosatomizdat, Moscow (1963)

[2] Kovenya, V.M.: Difference methods for solving multidimensional problems. Izd. Novosib. univ., Novosibirsk (2004)

[3] Kovenya, V.M., Kozlinskaya, T.V.: The method of predictor-corrector type for solving of magnetohydrodynamic problems. Vychisl. technol., Vol. 11, p. 1, 84–93 (2006)

[4] Kovenya, V.M., Kozlinskaya, T.V.: About calculation algorithm for plasma heating by electron beam. Vychisl. technol., Vol. 9, no 6, 59–67 (2004)

[5] Astrelin, V.T., Buredakov, A.V., Guber, N.A., Kovenya, V.M.: Simulation of non-uniform plasma motion and heating. PMTF, Vol. 42, no 6, 125–140 (2001)

A Central-Upwind Scheme for Nonlinear Water Waves Generated by Submarine Landslides

A. Kurganov and G. Petrova

We study a simple one-dimensional (1-D) toy model for landslides-generated nonlinear water waves. The landslide is modeled as a rigid bump translating down the side of the bottom while the water motion is modeled by the Saint-Venant system of shallow water equations. The resulting system is numerically solved using a well-balanced positivity preserving central-upwind scheme. The obtained numerical results are in good agreement with both the two-dimensional (2-D) incompressible flow numerical simulations and the experimental data.

1 Introduction

Landslides-generated water waves are of great interest to both ocean scientists and coastal and dams-keeping engineers. These landslides are natural phenomena that occur under certain conditions such as erosion, earthquakes, storms, heavy rainfalls, or water level fluctuations. In this chapter, we study submarine landslides, which are capable of generating several types of long waves, including very powerful and destructive tsunami waves.

Modeling of landslides-generated water waves is based on two components: the description of the motion of the boundaries of the fluid domain and the equations governing the water motion. In this chapter, we consider a simple 1-D toy model for landslides-generated water waves. The landslide is modeled by a rigid bump that loses its equilibrium and begins to translate down the side of the bottom (see, e.g., [GW05, Hei92, HPH01]). The water motion is described by the Saint-Venant system of shallow water equations [SV1871], which is commonly accepted approximation governing long wave propagation in deep ocean as well as in near-shore regions, including inundation. In the 1-D case, the studied system is

$$\begin{cases} h_t + (hu)_x = 0, \\ (hu)_t + \left(hu^2 + \frac{1}{2}gh^2\right)_x = -ghB_x, \end{cases} \tag{1}$$

where h is the fluid depth above the bottom, u is the velocity, g is the gravitational constant, and B represents the bottom elevation. Here, the only difference from the original Saint-Venant system is that the bottom topography function $B = B(x,t)$ is considered as a function not only of the space variable x but also of the time t.

In the past few decades, a wide variety of reliable finite-volume methods for the Saint-Venant system has been developed. We refer the reader to [ABBLP04, JW05, KL02, KP, NPPN06, XS06] for just several recently proposed methods. In the context of tsunami wave propagation, the shallow water equations have been recently numerically solved in [GL06].

To accurately capture landslides-generated water waves, a numerical method should possess two very important properties. First, it should be well-balanced, i.e., it should exactly preserve stationary steady states ($h + B \equiv$ const, $u \equiv 0$, $B_t \equiv 0$). Otherwise, even tiny perturbations of steady-state solutions may lead to "numerical storm," which would make the resolution of long water waves impossible due to their extremely small amplitude during the deep ocean propagation stage. The second important property of a good numerical method is the ability to preserve positivity of the water depth h. This feature is crucial for capturing water waves in near-shore regions and especially for resolving the inundation stage, when initially dry ($h = 0$) coastal areas flood after the wave hits the coastal line and then slowly drains out.

In this chapter, we numerically solve the model (1) by the second-order central-upwind scheme from [KP]. This scheme is both well-balanced and positivity preserving. It is also capable of treating discontinuous bottom topographies. Furthermore, its extension to the case of time-dependent B is straightforward. We describe our method in Sect. 2 and present the obtained numerical results in Sect. 3.

2 Description of the Central-Upwind Scheme

In this section, we describe a second-order semidiscrete well-balanced positivity preserving central-upwind scheme for the system (1). This scheme belongs to the class of high-resolution nonoscillatory Godunov-type central schemes first introduced in [NT90]. Compared to upwind finite-volume methods, the central schemes enjoy the advantages of simplicity, robustness, and universality. However, the numerical dissipation of the staggered central schemes, for example the one from [NT90], is relatively large, which is especially prominent when the system is integrated for large times. In [KL07, KNP01, KT00], a new class of the so-called semidiscrete central-upwind schemes has been introduced. These schemes enjoy all major advantages of Riemann-problem-solver-free central schemes, while their dissipation is lowered by incorporating some upwind information about local speeds of propagation into the central setting. Central-upwind schemes have been applied to the Saint-Venant system in [KL02, KP].

For simplicity, we introduce a uniform grid $x_\alpha := \alpha\Delta x$, where Δx is a small spatial scale, and we denote by I_j the finite-volume cells $I_j := [x_{j-\frac{1}{2}}, x_{j+\frac{1}{2}}]$.

We assume that at a certain time level t the computed solution is already available. We then replace the bottom topography function $B(\cdot,t)$ with its continuous piecewise linear approximation $\widetilde{B}(\cdot,t)$, consisting of the linear pieces that connect the points $(x_{j+\frac{1}{2}}, B_{j+\frac{1}{2}})$:

$$\widetilde{B}(x,t) = B_{j-\frac{1}{2}} + \left(B_{j+\frac{1}{2}} - B_{j-\frac{1}{2}}\right) \cdot \frac{x - x_{j-\frac{1}{2}}}{\Delta x}, \quad x_{j-\frac{1}{2}} \le x \le x_{j+\frac{1}{2}}, \tag{2}$$

where $B_{j+\frac{1}{2}} := (B(x_{j+\frac{1}{2}}+0,t) + B(x_{j+\frac{1}{2}}-0,t))/2$, which reduces to $B_{j+\frac{1}{2}} = B(x_{j+\frac{1}{2}},t)$ if $B(\cdot,t)$ is continuous at $x = x_{j+\frac{1}{2}}$. Note that $\{B_{j+\frac{1}{2}}\}$, like all other quantities reconstructed at time level t, depend on t, but we suppress this dependence to simplify the notation.

A central-upwind semidiscretization of (1) is the following system of ODEs

$$\frac{d}{dt}\overline{\mathbf{U}}_j(t) = -\frac{\mathbf{H}_{j+\frac{1}{2}}(t) - \mathbf{H}_{j-\frac{1}{2}}(t)}{\Delta x} + \overline{\mathbf{S}}_j(t), \tag{3}$$

where $\overline{\mathbf{U}}_j(t)$ are approximations of the cell averages of the solution:

$$\overline{\mathbf{U}}_j(t) \approx \frac{1}{\Delta x}\int_{I_j} \mathbf{U}(x,t)\,dx, \quad \mathbf{U} := (h,q)^T, \ q := hu,$$

$\overline{\mathbf{S}}_j$ is an appropriate discretization of the cell averages of the source term $\mathbf{S} := (0, -ghB_x)^T$:

$$\overline{\mathbf{S}}_j(t) = \left(0, -g\overline{h}_j \frac{B_{j+\frac{1}{2}} - B_{j-\frac{1}{2}}}{\Delta x}\right)^T \approx \frac{1}{\Delta x}\int_{I_j} \mathbf{S}(\mathbf{U}(x,t), B(x,t))\,dx,$$

and the central-upwind numerical fluxes $\mathbf{H}_{j+\frac{1}{2}}$ are given by:

$$\mathbf{H}_{j+\frac{1}{2}} = \frac{a^+_{j+\frac{1}{2}}\mathbf{F}(\mathbf{U}^-_{j+\frac{1}{2}}) - a^-_{j+\frac{1}{2}}\mathbf{F}(\mathbf{U}^+_{j+\frac{1}{2}})}{a^+_{j+\frac{1}{2}} - a^-_{j+\frac{1}{2}}} + \frac{a^+_{j+\frac{1}{2}}a^-_{j+\frac{1}{2}}}{a^+_{j+\frac{1}{2}} - a^-_{j+\frac{1}{2}}}\left[\mathbf{U}^+_{j+\frac{1}{2}} - \mathbf{U}^-_{j+\frac{1}{2}}\right]. \tag{4}$$

Here, $\mathbf{F}(\mathbf{U}) := (q, q^2/h + gh^2/2)^T$ and $\mathbf{U}^\pm_{j+\frac{1}{2}} = (h^\pm_{j+\frac{1}{2}}, q^\pm_{j+\frac{1}{2}})$ are the right-/left-sided values of the piecewise linear reconstruction $\widetilde{\mathbf{U}} = (\widetilde{h}, \widetilde{q})$ at $x = x_{j+\frac{1}{2}}$. For the discharge q, we have

$$\widetilde{q}(x) := \overline{q}_j + (q_x)_j(x - x_j), \quad x_{j-\frac{1}{2}} < x < x_{j+\frac{1}{2}}, \tag{5}$$

where the numerical derivative $(q_x)_j$ is (at least) first-order componentwise approximation of $q_x(x_j,t)$, computed using a nonlinear limiter needed to ensure a nonoscillatory nature of the reconstruction (5). We use the generalized minmod limiter (see, e.g., [KT00, KNP01, LN03, NT90])

$$(q_x)_j = \text{minmod}\left(\theta\frac{\overline{q}_j - \overline{q}_{j-1}}{\Delta x}, \frac{\overline{q}_{j+1} - \overline{q}_{j-1}}{2\Delta x}, \theta\frac{\overline{q}_{j+1} - \overline{q}_j}{\Delta x}\right), \quad \theta \in [1,2], \tag{6}$$

where the minmod function is defined as

$$\text{minmod}(z_1, z_2, \ldots) := \begin{cases} \min_j\{z_j\}, & \text{if } z_j > 0 \ \forall j, \\ \max_j\{z_j\}, & \text{if } z_j < 0 \ \forall j, \\ 0, & \text{otherwise.} \end{cases}$$

The parameter θ can be used to control the amount of numerical diffusion present in the resulting scheme. In our numerical experiments, we have used $\theta = 1$.

As suggested in [KP], the well-balanced and positivity preserving property of the scheme is guaranteed if, instead of reconstructing the water depth h, we perform a reconstruction for the fluid level $w := h + B$, and then use it to compute $h^{\pm}_{j+\frac{1}{2}} := w^{\pm}_{j+\frac{1}{2}} - B_{j+\frac{1}{2}}$. To this end, we first construct $\widetilde{w}$ in the same way as $\widetilde{q}$ was constructed in (5) and (6), with $\overline{w}_j := \overline{h}_j + (B_{j-\frac{1}{2}} + B_{j+\frac{1}{2}})/2$. Then, we correct the obtained $\widetilde{w}$ in the following two cases (see [KP] for details):

$$\begin{aligned} &\text{if } w^-_{j+\frac{1}{2}} < B_{j+\frac{1}{2}}, \quad \text{then take } w^-_{j+\frac{1}{2}} := B_{j+\frac{1}{2}}, \quad w^+_{j-\frac{1}{2}} := 2\overline{w}_j - B_{j+\frac{1}{2}}, \\ &\text{if } w^+_{j-\frac{1}{2}} < B_{j-\frac{1}{2}}, \quad \text{then take } w^-_{j+\frac{1}{2}} := 2\overline{w}_j - B_{j-\frac{1}{2}}, \quad w^+_{j-\frac{1}{2}} := B_{j-\frac{1}{2}}. \end{aligned}$$

This correction procedure guarantees that the resulting reconstruction $\widetilde{w}$ will stay above the piecewise linear approximant (2) of the bottom topography function. Therefore, the reconstructed values $h^{\pm}_{j+\frac{1}{2}}$ will be nonnegative.

To be able to accurately and efficiently treat (almost) dry regions, we also need to make a modification in the calculation of the velocities $u^{\pm}_{j+\frac{1}{2}}$ so that division by very small values of h is avoided. To achieve this goal, we follow [KP] and set

$$u^{\pm}_{j+\frac{1}{2}} := \frac{\sqrt{2}\, h^{\pm}_{j+\frac{1}{2}}\, q^{\pm}_{j+\frac{1}{2}}}{\sqrt{\left(h^{\pm}_{j+\frac{1}{2}}\right)^4 + \max\left(\left(h^{\pm}_{j+\frac{1}{2}}\right)^4, \varepsilon\right)}}, \tag{7}$$

where ε is a small a priori chosen positive number (in our numerical experiments, we have taken $\varepsilon = (\Delta x)^4$). Note that formula (7) can be replaced by an alternative desingularization strategy, but we selected this since (7) reduces to $u^{\pm}_{j+\frac{1}{2}} = q^{\pm}_{j+\frac{1}{2}}/h^{\pm}_{j+\frac{1}{2}}$ for large values of h. Next, we recompute the discharge q using

$$q^{\pm}_{j+\frac{1}{2}} := h^{\pm}_{j+\frac{1}{2}} \cdot u^{\pm}_{j+\frac{1}{2}},$$

where $u^{\pm}_{j+\frac{1}{2}}$ are given by (7). Equipped with the values of $h^{\pm}_{j+\frac{1}{2}}$ and $u^{\pm}_{j+\frac{1}{2}}$, we compute the one-sided local speeds of propagation $a^{\pm}_{j+\frac{1}{2}}$, needed in (4):

$$a^+_{j+\frac{1}{2}} = \max\left\{u^+_{j+\frac{1}{2}} + \sqrt{gh^+_{j+\frac{1}{2}}}\,,\, u^-_{j+\frac{1}{2}} + \sqrt{gh^-_{j+\frac{1}{2}}}\,,\, 0\right\},$$
$$a^-_{j+\frac{1}{2}} = \min\left\{u^+_{j+\frac{1}{2}} - \sqrt{gh^+_{j+\frac{1}{2}}}\,,\, u^-_{j+\frac{1}{2}} - \sqrt{gh^-_{j+\frac{1}{2}}}\,,\, 0\right\}.$$

Finally, the fully discrete central-upwind scheme is obtained by solving the system of ODEs (3) by a stable ODE solver of an appropriate order. In our numerical experiments, we have used the third-order strong stability preserving Runge–Kutta ODE solver from [GST01].

3 Numerical Results

In this section, we numerically solve a test problem from [Hei92], where both numerical and laboratory experiments were presented. The laboratory experiments were carried out in a $20 \times 0.55 \times 1.50$ m channel, where the water waves were generated by letting a triangular box slide down an inclined plane with a 45° slope. The shore is modeled by a second plane with a 15° slope. The box, a right isosceles triangle with 0.5 m sides, is initially 1 cm below the undisturbed free surface, and thus the bottom function at time $t = 0$ is given by (see Fig. 1)

$$B(x,0) = \begin{cases} -\alpha x, & x \le 0, \\ -x, & 0 \le x \le 0.01, \\ -0.01, & 0.01 \le x \le 0.51, \\ -x, & 0.51 < x \le 1, \\ 0, & x \ge 1, \end{cases} \tag{8}$$

where the shore slope is $\alpha = \tan(\pi/12)$, and the initial condition is

$$h(x,0) = \max\{1 - B(x,0), 0\}, \quad q(x,0) \equiv 0. \tag{9}$$

In [Hei92], the water motion is described by the 2-D incompressible Euler equations with free surface, which are numerically solved using the NASA-VOF2D code [TCMH85]. The numerical results reported in [Hei92] are comparable with the laboratory experiments.

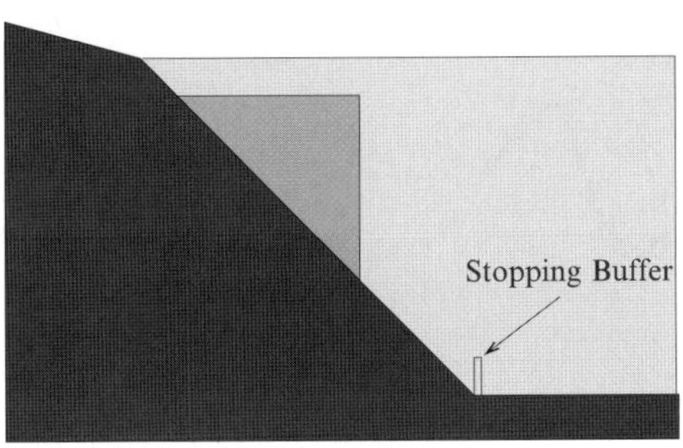

Fig. 1. Initial bottom topography

Here, we obtain similar results, using the much simpler shallow water model (1) with an added friction term:

$$\begin{cases} h_t + q_x = 0, \\ q_t + \left(\dfrac{q^2}{h} + \dfrac{1}{2}gh^2\right)_x = -ghB_x - \kappa(h)u. \end{cases} \tag{10}$$

In our numerical experiments, we have taken the friction coefficient $\kappa(h) = 0.1/(1 + 10h)$.

We numerically solve the system (10) subject to the initial condition (9) and the initial bottom topography (8). Following [GW05], we describe the motion of the triangular box by the displacement of its center of mass from its initial position:

$$S(t) = \frac{\sqrt{2\pi(\gamma+1)}}{4} \ln\left(\cosh\left(\frac{t\sqrt{2g(\gamma-1)}}{(\gamma+1)\sqrt{\pi}}\right)\right). \tag{11}$$

Here, γ is the ratio of the bulk and water densities. In our numerical experiments, $\gamma = 1.925$ and the gravitational constant $g = 9.812$.

Equations (11) and (8) give a closed formula for the function $B(x,t)$ participating in the studied system (10), which is numerically solved by the central-upwind scheme. The cell average of the friction term in (10) is approximated using the midpoint rule, which affects neither the well-balanced nor the positivity preserving properties of our scheme.

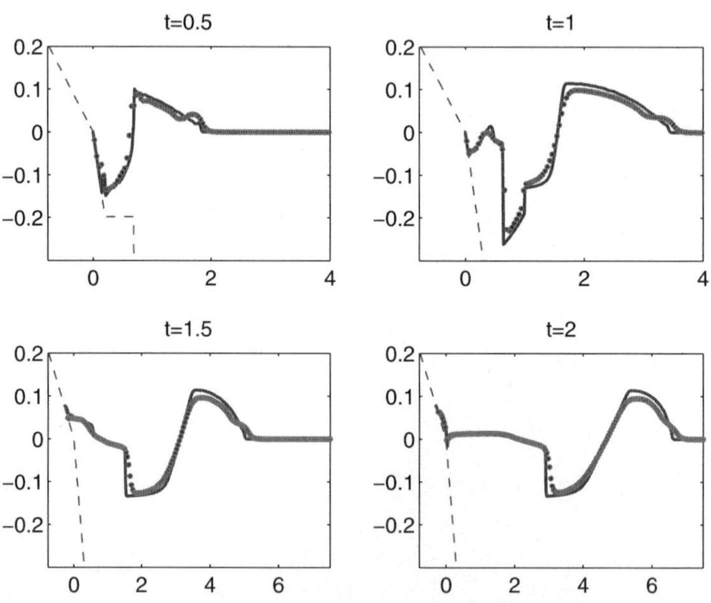

Fig. 2. Snapshots of the water surface $h + B$, computed on coarser (*dots*) and finer (*solid line*) grids. The *dashed line* represents the bottom

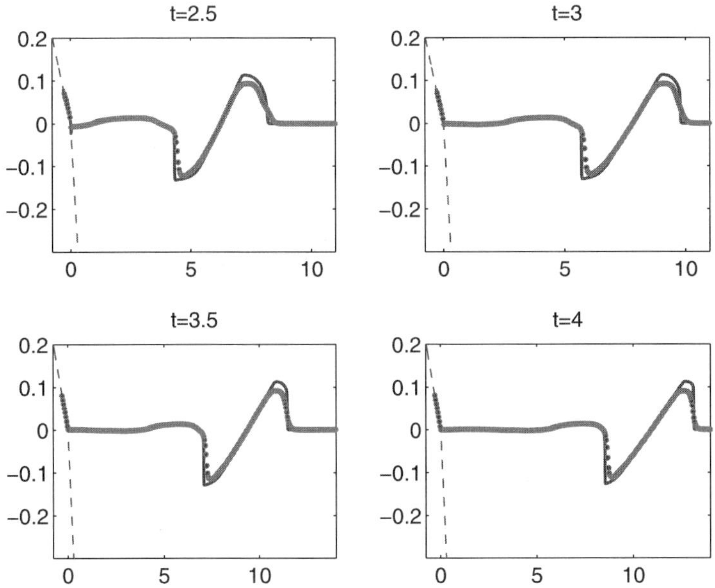

Fig. 3. The same as Fig. 2, but at later times

The water surface, computed on both coarse (with $\Delta x = 1/25$) and fine (with $\Delta x = 1/200$) uniform grids, is shown in Figs. 2 and 3, where snapshots of the solution are taken at times $t = 0.5$, 1, 1.5, 2, 2.5, 3, 3.5, and 4. Clearly, the downward motion of the triangular box generates a nonlinear water wave. While the left counterpart of this wave hits the shore (see Fig. 2, $t = 1.5$ and 2), the right counterpart gradually forms a stable "tsunami-like" shape that travels across the channel (see Fig. 3, $t = 2.5$, 3, 3.5, and 4). Our results are in good agreement with the ones reported in [Hei92].

Acknowledgments

The research of A. Kurganov was supported in part by the NSF Grant Nos. DMS-0310585 and DMS-0610430. The research of G. Petrova was supported in part by the NSF Grant No. DMS-0505501.

References

[ABBLP04] Audusse, E., Bouchut, F., Bristeau, M.-O., Klein, R., Perthame, B.: A fast and stable well-balanced scheme with hydrostatic reconstruction for shallow water flows. SIAM J. Sci. Comput., **25**, 2050–2065 (2004).

[GL06] George, D.L., LeVeque, R.J.: Finite volume methods and adaptive refinement for global tsunami propagation and local inundation. Science of Tsunami Hazards, **24**, 319–328 (2006).

[GST01] Gottlieb, S., Shu, C.-W., Tadmor, E.: High order time discretization methods with the strong stability property. SIAM Rev., **43**, 89–112 (2001)

[GW05] Grilli, S., Watts, P.: Tsunami generation by submarine mass failure. I: Modeling, experimental validation, and sensitivity analyses. J. Waterway Port Coast. Oc. Eng., **131**, 283–297 (2005)

[Hei92] Heinrich, Ph.: Nonlinear water waves generated by submarine and areal landslides. J. Waterway Port Coast. Oc. Eng., **118**, 249–266 (1992)

[HPH01] Heinrich, Ph., Piatanesi, A., Hebert, H.: Numerical modelling of tsunami generation and propagation from submarine slumps: the 1998 Papua New Guinea event. Geophys. J. Int, **145**, 97–111 (2001)

[JW05] Jin, S., Wen, X.: Two interface-type numerical methods for computing hyperbolic systems with geometrical source terms having concentrations. SIAM J. Sci. Comput., **26**, 2079–2101 (2005)

[KL02] Kurganov, A., Levy, D.: Central-upwind schemes for the Saint-Venant system. M2AN Math. Model. Numer. Anal., **36**, 397–425 (2002)

[KL07] Kurganov, A., Lin, C.-T.: On the reduction of numerical dissipation in central-upwind schemes. Commun. Comput. Phys., **2**, 141–163 (2007)

[KNP01] Kurganov, A., Noelle, S., Petrova, G.: Semi-discrete central-upwind scheme for hyperbolic conservation laws and Hamilton-Jacobi equations. SIAM J. Sci. Comput., **23**, 707–740 (2001)

[KP] Kurganov, A., Petrova, G.: A second-order well-balanced positivity preserving central-upwind scheme for the Saint-Venant system. submitted to Commun. Math. Sci.

[KT00] Kurganov, A., Tadmor, E.: New high-resolution central schemes for nonlinear conservation laws and convection-diffusion equations. J. Comput. Phys., **160**, 241–282 (2000)

[LN03] Lie, K.-A., Noelle, S.: On the artificial compression method for second-order nonoscillatory central difference schemes for systems of conservation laws. SIAM J. Sci. Comput., **24**, 1157–1174 (2003)

[NT90] Nessyahu, H., Tadmor, E.: Non-oscillatory central differencing for hyperbolic conservation laws. J. Comput. Phys., **87**, 408–463 (1990)

[NPPN06] Noelle, S., Pankratz, N., Puppo, G., Natvig, J.: Well-balanced finite volume schemes of arbitrary order of accuracy for shallow water flows. J. Comput. Phys., **213**, 474–499 (2006)

[ORM00] Ortiz, M., Gomez-Reyes, E., Velez-Munoz, H.: A fast preliminary estimation model for transoceanic tsunami propagation. Geofisica Internatcional, **39**, 207–220 (2000)

[SV1871] de Saint-Venant, A. J. C.: Thèorie du mouvement non-permanent des eaux, avec application aux crues des rivière at à l'introduction des marèes dans leur lit. C.R. Acad. Sci. Paris, **73**, 147–154 (1871)

[TCMH85] Torrey, M.D., Cloutman, L.D., Mjolness, R.C., Hirt, C.W.: NASA-VOF2D: A computer program for incompressible flows with free surfaces. Report LA-10612-MS, Los Alamos Nat. Lab., Los Alamos, N.M. (1985)

[XS06] Xing, Y., Shu, C.-W.: A new approach of high order well-balanced finite volume WENO schemes and discontinuous Galerkin methods for a class of hyperbolic systems with source terms, Commun. Comput. Phys., **1**, 100–134 (2006)

An A Posteriori Error Estimate for Glimm's Scheme

M. Laforest

We present an a posteriori error bound for Glimm's approximate solutions [2] to nonlinear scalar conservation laws containing only shock waves, that is

$$\begin{aligned} v_t + f(v)_x &= 0, \\ v(\cdot, 0) &= v_0(\cdot) \in L^\infty(\mathbb{R}) \quad \text{and decreasing.} \end{aligned} \tag{1}$$

Using Liu's wave-tracing method [9], we show that the L^1 norm of the error is bounded by a sum of residuals containing independent contributions from each wave in the approximate solution but allowing for error cancelation among waves. The proof can also be viewed as an explicit form of a construction of Hoff and Smoller [5]. This chapter contains an abbreviated description of the proof found in [8] and new numerical evidence that the error estimate should continue to hold for arbitrary L^∞ initial data.

The objective of this chapter is to present new error estimators for approximate solutions that may eventually be used to build efficient adaptive schemes. For conservation laws (1), adaptive schemes are important because it is particularly difficult to accurately solve the problem in the presence of smooth and discontinuous waves. These waves are an intrinsic part of the physical process modeled by conservation laws, namely bottlenecks in vehicular traffic or waves in channels.

As an a posteriori error estimate, this result is of interest since it shows explicitly that the errors are created, propagated, and canceled at the level of waves. This estimate contrasts with those based on Kruskov's stability theory, such as [6, 4], that do not account for such processes or with those based on the adjoint formulation which are currently limited to linear (or linearized) problems, see [1] for a survey. As a stability result, this approach might be useful to treat problems requiring an analysis in $\mathrm{BV}_{\mathrm{local}}$ such as when the initial data is periodic or in L^∞.

In Sect. 1 we review Glimm's scheme and introduce the local error estimator. In Sect. 2, we present the main results and show how certain local estimates are used to demonstrate the global estimate. Section 3 contains numerical results that show that the estimates are, in some sense, optimal and that they probably continue to hold for arbitrary initial data.

1 Glimm's Scheme

Glimm's scheme, coupled with Liu's wave-tracing method, plays a central role in the qualitative study of hyperbolic systems of conservation laws. It is well known that if f is strictly convex, then there exists a *unique* weak solution $v : \mathbb{R} \times \mathbb{R}^+ \to \mathbb{R}$ of (1) satisfying a physical entropy condition.

We now describe Glimm's scheme. Assume $v_0 \in L^\infty(\mathbb{R})$ is decreasing and pick a discretization $\Delta x, \Delta t$ satisfying the *CFL condition*, $\sup_v |f'(v)| \leq \Delta x/\Delta t$, where the sup is over the range of v. Begin by approximating v_0 by a piecewise constant and decreasing $w(\cdot, 0)$ such that (1) $w(\cdot, 0)$ is piecewise constant over $I_m \doteq [(m-1)\Delta x, (m+1)\Delta x]$, m even, and (2) $\|v_0(\cdot) - w(\cdot, 0)\|_{L^1} = \mathcal{O}(\Delta x)$. Writing $t_n \doteq n\Delta x$, the entropy solution over $\mathbb{R} \times [t_0, t_1)$ is a sequence of noninteracting discontinuities, each separated by different constant left- and right-hand states u^l, u^r, and traveling at the *Rankine–Hugoniot* speed

$$S(u^l, u^r) \doteq \frac{f(u^l) - f(u^r)}{u^l - u^r}. \tag{2}$$

In order to make $w(\cdot, t_1)$ into a piecewise constant function over the grid I_m with m odd, we pick a random number $\theta_1 \in [-1, 1]$ and set

$$w(\cdot, t_1)|_{I_m} \doteq \lim_{t \to t_1 -} w\big((m+1)\Delta x + \theta_1 \Delta x, t\big).$$

***The procedure can be repeated indefinitely using a random number θ_{n+1} to propagate a piecewise constant approximation $w(\cdot, t_n)$ to a piecewise constant approximation $w(\cdot, t_{n+1})$. Glimm showed that for almost all random sequences $\{\theta_n\}$, w converges to v as $\Delta x \to 0$.

We summarize Liu's so-called *wave-tracing* description [9], as it applies to scalar conservation laws with decreasing v_0, which describes w as a linear superposition of discrete *waves* propagating and interacting nonlinearly.

Theorem 1. *[9] Given a random sequence $\{\theta_n\}_{n\in\mathbb{N}}$, and a region $\mathbb{R} \times [0, T]$, Glimm's approximate solution w to* (1) *with decreasing $w(\cdot, 0)$ can be described as a family of waves $\mathcal{W}$, where each wave $\alpha \in \mathcal{W}$ has the following characteristics:*

(1) Two constant left- and right-hand states w^l_α and w^r_α
(2) A strength $\sigma_\alpha \doteq w^r_\alpha - w^l_\alpha$
(3) A position $x_\alpha(t) \in \mathbb{R}$ that satisfies $x_\alpha(t_{n+1}) - x_\alpha(t_n) = \pm\Delta x$

The approximate solution is then constructed as

$$w(x, t_n) = w(-\infty, 0) + \sum_{\{\alpha | x_\alpha(t_n) \leq x\}} \sigma_\alpha. \tag{3}$$

We briefly explain how the waves are defined. To each discontinuity in the initial data, we associate one unique (shock) wave α. We let $x_\alpha(0)$ be the location of that discontinuity at time t_0, and set

$$w^l_\alpha \doteq \lim_{x\to x_\alpha(0)-} w(x,0), \qquad w^r_\alpha \doteq \lim_{x\to x_\alpha(0)+} w(x,0). \tag{4}$$

If we know it is position at time t_n, then we can compute

$$w^-_\alpha(t_n) \doteq \lim_{x\to x_\alpha(t_n)-} w(x,t_n), \qquad w^+_\alpha(t_n) \doteq \lim_{x\to x_\alpha(t_n)+} w(x,t_n). \tag{5}$$

and thereby define

$$x_\alpha(t) = x_\alpha(t_n) + (t-t_n)S\big(w^-_\alpha(t_n), w^+_\alpha(t_n)\big), \tag{6}$$

$$x_\alpha(t_{n+1}) = x_\alpha(t_n) + \Delta x \,\mathrm{sign}\,\big(S\big(w^-_\alpha(t_n), w^+_\alpha(t_n)\big)\Delta t - \theta_{n+1}\Delta x\big). \tag{7}$$

The family of waves $\mathcal{W}$ has a natural ordering. A wave α is said to be smaller than a wave β, written $\alpha < \beta$, if $x_\alpha(0) < x_\beta(0)$, or when equality occurs, if $w^l_\alpha > w^l_\beta$.

Let u be the entropy solution of (1) with initial data $w(\cdot,0)$. Then at each time t, we can assign a position $y_\alpha(t)$ to the wave α in u by setting $y_\alpha(0) = x_\alpha(0)$ and, at each time t, making it equal to the position of the discontinuity in u at time t with which the initial discontinuity interacted.

If w is an approximate solution generated by Glimm's scheme then the residual is, in the sense of distributions, $w_t + f(w)_x$. Detailed arguments, found in [7, 8], show that there exists a well-defined notion of residuals that can be computed a posteriori and assigned uniquely to each wave in Liu's decomposition.

Definition 1. *Given a wave α at time t_n and $s = S(w^-_k(t_n-), w^+_\alpha(t_n-))$, then the residual is*

$$R(\alpha, t_n) \doteq \sigma_\alpha\big(\Delta x\,\mathrm{sign}(s\Delta t - \theta_n \Delta x) - s\Delta t\big). \tag{8}$$

2 Main Result

In this section we present a brief proof of Theorem 2. The proof depends on Lemmas 1–4 whose proofs can be found in [8].

Theorem 2. *Consider decreasing initial data $v_0 \in L^\infty(\mathbb{R})$ to (1). Suppose the approximation $w(\cdot,0)$ to the initial data contains only shocks and $v_0 - w(\cdot,0) \in L^1(\mathbb{R})$. Then for any Δt satisfying the CFL condition, any time t, and any sequence $\{\theta_k\}_k$ of numbers in $[-1,1]$, we have that the approximate solution w obtained with Glimm's scheme satisfies*

$$\big\|v(\cdot,t) - w(\cdot,t)\big\|_{L^1(\mathbb{R})} \le \big\|v_0(\cdot) - w(\cdot,0)\big\|_{L^1(\mathbb{R})} + \sum_{\alpha\in\mathcal{W}} \left| \sum_{k=1}^{\lfloor t/\Delta t\rfloor} R(\alpha, t_k)\right|. \tag{9}$$

This estimate contrasts with previous error bounds for finite difference schemes, such as those of Kröner and Ohlberger [6] or Gosse and Makridakis [4], that are of the form $\sum_\alpha \sum_k |R(\alpha, t_k)|$. We emphasize that this error bound is free of unknown constants, valid for all random sequences $\{\theta_k\}$, and accounts for error propagation and cancelation among waves.

Rather than compare the waves in w with those in v, we first compare them to those in the entropy solution u defined by the initial data $w(\cdot, 0)$.

Definition 2. *If the waves in $\mathcal{W}$ are ordered $\mathcal{W} = [\alpha(1), \ldots, \alpha(N)]$, then we call $X(t) = [x_{\alpha(i)}(t)]_{i=1}^N = [x_{\alpha(1)}(t), \ldots, x_{\alpha(N)}(t)]^T$ and $Y(t) = [y_{\alpha(i)}(t)]_{i=1}^N$, the trajectories of shock waves in, respectively, w and u.*

Given two sets of trajectories $Z(t) = [z_\alpha(t)]_{\alpha \in \mathcal{W}}$ and $\widetilde{Z}(t) = [\widetilde{z}_\alpha(t)]_{\alpha \in \mathcal{W}}$, we define the discrepancy as $d(\widetilde{Z}(t), Z(t)) \doteq \big[|\sigma_\alpha| \cdot |\widetilde{z}_\alpha(t) - z_\alpha(t)|\big]_{\alpha \in \mathcal{W}}$.

Definition 3. *We define the continuous trajectories $X^{(0)}(t) = [x_\alpha^{(0)}(t)]_{\alpha \in \mathcal{W}}$ by requiring that the position $x_\alpha^{(0)}(t)$ of each wave α satisfy $x_\alpha^{(0)}(0) = x_\alpha(0)$ and $\dot{x}_\alpha^{(0)}(t) = S(w_\alpha^-(t), w_\alpha^+(t))$, for all positive t.*

Lemma 1. *If $u_0(\cdot) = w(\cdot, 0)$ is decreasing and $\mathbf{1}^T = [1, 1, \ldots, 1]$, then*

$$\|u(\cdot, t) - w(\cdot, t)\|_{L^1(\mathbb{R})} = \mathbf{1}^T d(Y(t), X(t)), \qquad \forall t > 0. \tag{10}$$

Lemma 2. *If $\alpha \in \mathcal{W}$ is a shock wave then for all $t \geq 0$, the α-th component of $d(X^{(0)}(t), X(t))$ satisfies*

$$|\sigma_\alpha| \cdot |x_\alpha^{(0)}(t) - x_\alpha(t)| = \left| \sum_{k=1}^{\lfloor t/\Delta t \rfloor} R(\alpha, t_k) \right|. \tag{11}$$

The first lemma relates the L^1 norm to discrepancies between Y and X while the second lemma relates the residuals to discrepancies between $X^{(0)}$ and X. The problem is now to find a sequence of corrections that, when applied to $X^{(0)}$, provide the exact trajectories Y yet preserve the total quantity of residuals $\mathbf{1}^T R(t) = \mathbf{1}^T d(X^{(0)}(t), X(t))$.

Definition 4. *Consider a consecutive set of waves $\mathcal{F} = [\alpha(1), \ldots, \alpha(n)]$. The free trajectories of $\mathcal{F}$ are the trajectories $F(t) = [f_{\alpha(i)}(t)]_{i=1}^n$ of the discontinuities in the solution $V(x, t)$ to the conservation law (1) with initial data*

$$V(x, 0) = \begin{cases} w_{\alpha(1)}^l & \text{if } x \leq x_{\alpha(1)}(0), \\ w_{\alpha(i)}^r & \text{if } x_{\alpha(i)}(0) < x \leq x_{\alpha(i+1)}(0) \text{ and } i \in \{1, \ldots, n-1\}, \\ w_{\alpha(n)}^r & \text{if } x_{\alpha(n)}(0) < x. \end{cases} \tag{12}$$

We will say that an interaction occurred in F at time t_* if it occurred in the entropy solution V at time t_*.

Definition 5. *A matrix C is conservative if $\mathbf{1}^T C = \mathbf{1}^T$.*

Definition 6 provides, for each fixed time t_*, the two trajectories $X^{(1)}$ and $X^{(2)}$. The Lemmas 3 and 4 then indicate how to relate Y and $X^{(2)}$ and finally to $X^{(1)}$.

Definition 6. *For a fixed time t_*, let $\mathcal{F}^{(1)}, \ldots, \mathcal{F}^{(m)}$ be the set of waves forming the m discontinuities in w at time t_*. For each set $\mathcal{F}^{(i)}$, suppose the free trajectories $F^{(i)}(t)$ associated to $\mathcal{F}^{(i)}$ posses n_i discontinuities at time $t = t_*$ and suppose these discontinuities form the sets $\mathcal{S}_{(i,1)}, \ldots, \mathcal{S}_{(i,n_i)}$. Let F be the trajectories defined by*

$$F(t) = \left[F^{(1)}(t)^T, \cdots, F^{(m)}(t)^T\right]^T, \tag{13}$$

and construct the trajectories $X^{(1)}$ satisfying

$$\begin{aligned} X^{(1)}(t) &= F(t), && \forall t \in [0, t_*], \\ \dot{X}^{(1)}(t) &= \dot{X}(t), && \forall t > t_*. \end{aligned} \tag{14}$$

Let T_* be the next time at which either an interaction occurs in w, or in one of the free trajectories $F^{(1)}, \ldots, F^{(m)}$. Crossings when $x_\alpha^{(1)}(t) = x_\beta^{(1)}(t)$ but $\alpha \in \mathcal{F}^{(i)}$ and $\beta \in \mathcal{F}^{(j)}$, $i \neq j$, are not considered interactions. Define the trajectories $X^{(2)}$ by

$$\begin{aligned} X^{(2)}(t) &= F(t), && \forall t \in [0, T_*], \\ \dot{X}^{(2)}(t) &= \dot{X}(t), && \forall t > T_*. \end{aligned} \tag{15}$$

Lemma 3. *For each fixed time t_*, there exists a conservative matrix $B(t)$ such that*

$$d\big(X^{(2)}(t), X(t)\big) \le B(t) \cdot d\big(X^{(1)}(t), X(t)\big), \qquad \forall t. \tag{16}$$

Lemma 4. *For each fixed time t_*, there exists a conservative matrix $C(t)$ such that*

$$d\big(Y(t), X(t)\big) \le C(t) \cdot d\big(X^{(2)}(t), X(t)\big), \qquad \forall t \le T_*. \tag{17}$$

We now sketch the proof of Theorem 2.

Proof of Theorem 2. The proof proceeds by induction. The induction hypothesis is that at each time t_*, there exists a conservative matrix $A(t)$ such that

$$d\big(X^{(1)}(t), X(t)\big) \le A(t) \cdot d\big(X^{(0)}(t), X(t)\big), \qquad \forall t. \tag{18}$$

We begin by showing that if the induction hypothesis holds at time t_*, then the theorem holds for all $t \in [t_*, T_*]$. Applying in the following order

Lemmas 1, 3, 4, the induction hypothesis (18), Definition 5, and Lemma 2, we obtain

$$\begin{aligned}\|u(\cdot,t)-w(\cdot,t)\|_{L^1(\mathbb{R})} &= \mathbf{1}^T d\big(Y(t),X(t)\big)\\ &\le \mathbf{1}^T C(t)\cdot d\big(X^{(2)}(t),X(t)\big)\\ &\le \mathbf{1}^T C(t)B(t)\cdot d\big(X^{(1)}(t),X(t)\big)\\ &\le \mathbf{1}^T C(t)B(t)A(t)\cdot d\big(X^{(0)}(t),X(t)\big)\\ &= \sum_{\alpha\in\mathcal{W}}\left|\sum_{k=1}^{\lfloor t/\Delta t\rfloor} R(\alpha,t_k)\right|. \end{aligned} \tag{19}$$

To prove estimate (9), it now suffices to apply the triangle inequality on $\|v(\cdot,t)-u(\cdot,t)+u(\cdot,t)-w(\cdot,t)\|_{L^1}$ and the fact that the evolution operator for (1) is contracting in L^1.

The induction hypothesis holds at time $t_* = 0$ because $X^{(1)}(0) = X^{(0)}(0) = X(0)$. We have thus reduced the proof of the theorem to demonstrating that if the induction hypothesis (18) holds at time t_*, then it must hold at time T_*. At time $T_* > t_*$, the time of the next interaction in w or among the free trajectories, two cases can occur.

Case 1. Two discontinuities, $\mathcal{S}_{(k,l)}$ and $\mathcal{S}_{(k,l+1)}$ from the free trajectories $F^{(k)}(t)$, meet at time T_*. Let the trajectories (14) and (15) defined with respect to time T_* be distinguished by a superscript tilde from those defined with respect to time t_*. Then $\widetilde{X}^{(1)} \equiv X^{(2)}$ and Lemma 3 imply the induction hypothesis at time T_*,

$$\begin{aligned} d\big(\widetilde{X}^{(1)}(t),X(t)\big) &\le B(t)\cdot d\big(X^{(1)}(t),X(t)\big)\\ &\le B(t)A(t)\cdot d\big(X^{(0)}(t),X(t)\big), \qquad \forall t. \end{aligned} \tag{20}$$

Case 2. Suppose the discontinuities $\mathcal{F}^{(k)}$ and $\mathcal{F}^{(k+1)}$ in w meet at time T_*. A more general version of Lemma 4, presented in [8], would apply directly to this case. To keep the presentation short, we briefly explain how a slight re-interpretation of Lemma 4 would suffice to prove the induction hypothesis.

Suppose that the two discontinuities in w at time t_*, $\mathcal{F}^{(k)}$ and $\mathcal{F}^{(k+1)}$, interacted to form a discontinuity $\widetilde{\mathcal{F}}^{(k)}$. The trajectories $\widetilde{X}^{(1)}(T_*) = X^{(2)}(T_*)$ for all waves in $\mathcal{W}\setminus\widetilde{\mathcal{F}}^{(k)}$ and therefore it suffices to find bounds for the discrepancies involving only the waves in $\widetilde{\mathcal{F}}^{(k)}$. By definition, the trajectories $\widetilde{X}^{(1)}(T_*)|_{\widetilde{\mathcal{F}}^{(k)}}$ correspond to the positions $\widetilde{Y}(T_*)$ of waves in an entropy solution with $\widetilde{\mathcal{F}}^{(k)}$ as initial data, say $\widetilde{u}$. The trajectories $X^{(2)}(T_*)|_{\widetilde{\mathcal{F}}^{(k)}}$ correspond to the positions of waves in two entropy solutions with, respectively, $\mathcal{F}^{(k)}$ and $\mathcal{F}^{(k+1)}$ as initial data. In this case, $X^{(2)}(T_*)|_{\widetilde{\mathcal{F}}^{(k)}}$ is a trajectory of the form (15) for a subdivision $\mathcal{F}^{(k)}$, $\mathcal{F}^{(k+1)}$ of a total set of waves $\widetilde{\mathcal{F}}^{(k)}$. We now reinterpret Lemma 4 by replacing the positions $Y(t)$ of waves associated to a solution u having initial data $\mathcal{W}$, with the positions $\widetilde{Y}(t)$ of the waves in a solution $\widetilde{u}$ having as initial data $\widetilde{\mathcal{F}}^{(k)}$. Lemma 4 then provides a conservative matrix $C(t)$ such that for all $t \leq \widetilde{t}_* = T_*$, we have

$$d\big(\widetilde{X}^{(1)}(t), X(t)\big)\big|_{\widetilde{\mathcal{F}}^{(k)}} \leq C(t) \cdot d\big(X^{(2)}(t), X(t)\big)\big|_{\widetilde{\mathcal{F}}^{(k)}}. \tag{21}$$

The definition of $C(t)$ can be extended in such a way that (21) holds for all $t > T_*$ by using Definition 3 and the fact that (14) implies $d/dt(\widetilde{X}^{(1)}) = d/dt(X^{(0)})$ for $t > T_*$.

Extending $C(t)$ trivially to all waves in $\mathcal{W}$ and combining this with Lemma 3 and the induction hypothesis at time t_*, we find

$$\begin{aligned} d\big(\widetilde{X}^{(1)}(t), X(t)\big) &\leq C(t) \cdot d\big(X^{(2)}(t), X(t)\big) \\ &\leq C(t)B(t) \cdot d\big(X^{(1)}(t), X(t)\big) \\ &\leq C(t)B(t)A(t) \cdot d\big(X^{(0)}(t), X(t)\big). \end{aligned}$$

Since $\widetilde{X}^{(0)} = X^{(0)}$, the induction hypothesis holds at time T^*. □

3 Numerical Results

We present numerical experiments comparing the effectivity of (9) as an estimator of the true error in L^1. The "true" error was computed by comparing the approximate solution w to a numerical approximation of u, the exact solution with $u(\cdot, 0) = w(\cdot, 0)$. We used the second-order slope-limiter method of Goodman and LeVeque [3] to obtain a sufficiently precise approximation of u.

The first experiment considers initial data formed exclusively of shock waves. Given the values $[a_i]_{i=1}^N = [0.0, -0.25, -0.5, -2.5, -2.75, -3.0, -3.5]$, where $N = 7$, we define

$$w(x, 0) = \begin{cases} a_i & \text{if } x \in [(i-1), i], \\ a_1 & \text{if } x < x_0, \\ a_N & \text{if } x > x_N. \end{cases} \tag{22}$$

In Table 1, the error estimator (9) and the L^1 norm are presented, both evaluated at time $t = 8.0$. In Fig. 1, we present the true error (solid) and the

Table 1. Errors w.r.t. Δx for (22)

Δx	$\|u - w\|_{L^1}$	$\sum_\alpha \lvert \sum_k R \rvert$
0.125	4.750	4.750
0.0625	0.0631	0.950
0.03125	5.969	5.969
0.015625	4.766	4.766

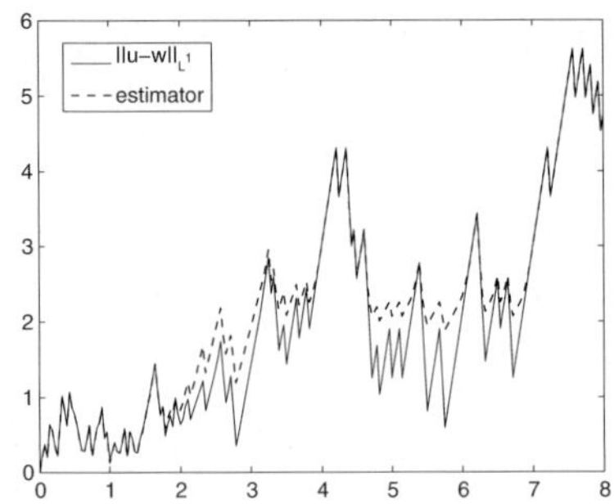

Fig. 1. Initial data (22)

Table 2. Errors w.r.t. Δx for (23)

Δx	$\|u-w\|_{L^1}$	$\sum_\alpha \vert \sum_k R\vert$
0.5	49.576	93.465
0.25	29.903	57.716
0.125	10.340	30.559
0.0625	5.744	35.428
0.03125	2.466	24.739

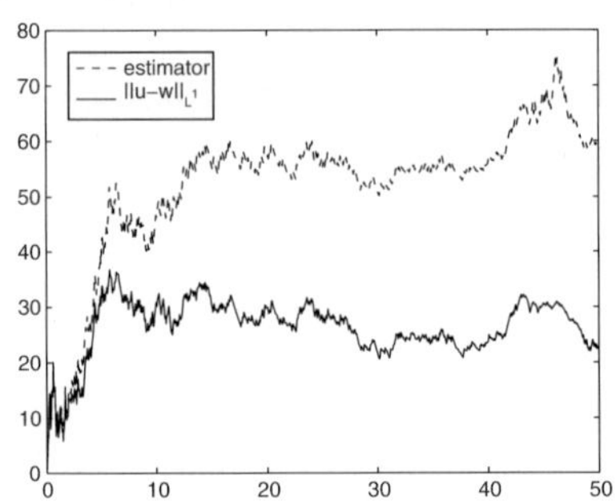

Fig. 2. Initial data (23)

error estimator (dash) as a function of time for the experiment with $\Delta x = 0.125$. Experiments with other values of Δx confirmed the exactness of the error estimator. The temporary overestimates of the error, as in Fig. 1, are to be expected but beyond the scope of this chapter.

We also present numerical experiments with initial data containing bock shock and rarefaction waves. Using the sequence of $N = 12$ values,

$$[a_i]_{i=1}^N = [0, -5, -3, -4, -1, -2, 3, 2, 4, 3, 5, 0], \tag{23}$$

we define initial data $w(\cdot, 0)$ in the same manner as (22). The solution converges to an N-wave profile and in Table 2 we compare the error and the error estimator at time $t = 50.0$, when $\|u\|_{L^\infty} \approx 1$. Figure 2 shows the comparison as a function of time for $\Delta x = 0.25$. This confirms numerically that the estimate (9) should continue to hold for arbitrary L^∞ initial data.

References

1. T.J. Barth and H. Deconinck, editors. *Error Estimation and Adaptive Discretization Methods in Computational Fluid Dynamics*, volume 25 of *Lecture Notes in Computational Sciences and Engineering*. Springer Verlag, Berlin, 2003.
2. J. Glimm. Solutions in the large for nonlinear hyperbolic systems of equations. *Comm. Pure Appl. Math.*, 18:697–715, 1965.
3. J.B. Goodman and R.J. LeVeque. A geometric approach to high resolution TVD schemes. *SIAM J. Numer. Anal.*, 25:268–284, 1988.
4. L. Gosse and C. Makridakis. Two a posteriori error estimates for one-dimensional scalar conservation laws. *SIAM J. Numer. Anal.*, 38(3):964–988, 2000.
5. D. Hoff and J. Smoller. Error bounds for Glimm difference approximations for scalar conservation laws. *Trans. Amer. Math. Soc.*, 289:611–642, 1985.
6. D. Kröner and M. Ohlberger. A posteriori error estimates for upwind finite volume schemes for nonlinear conservation laws in multi dimensions. *Math. Comput.*, 69(229):25–39, 1999.
7. M. Laforest. A posteriori error estimate for front-tracking: Systems of conservation laws. *SIAM J. Math. Anal.*, 35(5):1347–1370, 2004.
8. M. Laforest. Mechanisms for error propagation and cancellation in Glimm's scheme without rarefactions. 2006. submitted for publication to J. Hyp. Diff. Eq.
9. T.-P. Liu. The deterministic version of the Glimm scheme. *Comm. Math. Phys.*, 57:135–148, 1977.

Multiphase Flows in Mass Transfer in Porous Media

W. Lambert and D. Marchesin

1 Introduction

Mass interchange between different phases occurs typically in multiphase flows in porous media. Such flows are not represented by conservation laws or hyperbolic equations both because they have source terms and because they are not evolutionary in all the variables. In the absence of gravitational effects, such flows can be modeled by systems of $m+1$ equations:

$$\frac{\partial}{\partial t}\mathcal{G}(\mathcal{V}) + \frac{\partial}{\partial x}u\mathcal{F}(\mathcal{V}) = Q(\mathcal{V}), \tag{1}$$

where $\mathcal{V} \in \Omega \subset R^m$ and $u \in \mathcal{R}$. The first m equations represent mass balance for different chemical components in different phases and the last one represents conservation of energy. The variable u is called speed because this is its interpretation in many applications is a flow rate; the pair $(\mathcal{V}, u)$ in R^{n+1} is called *state variable.* $\mathcal{G}$ and $\mathcal{F}$ represent the vector-valued functions $\mathcal{G} = (\mathcal{G}_1, \mathcal{G}_2, \cdots, \mathcal{G}_{m+1})^T$: $\Omega \longrightarrow R^{m+1}$ and $\mathcal{F} = (\mathcal{F}_1, \mathcal{F}_2, \cdots, \mathcal{F}_{m+1})^T$: $\Omega \longrightarrow R^{m+1}$, where $u\mathcal{F}_i$ is the flux for the conserved quantity $\mathcal{G}_i$ and $\partial\mathcal{G}_i/\partial t$ is the corresponding accumulation term, for $i = 1, 2, \cdots, m+1$. On the right-hand side $Q = (Q_1, Q_2, \cdots, Q_{m+1})^T$: $\Omega \longrightarrow R^{m+1}$. Generically for a thermally insulated system the total energy is conserved, which usually can be represented by setting $Q_{m+1} = 0$. The functions $\mathcal{G}$, $\mathcal{F}$, and Q are continuous in the whole domain Ω, but they are only piecewise smooth.

The solution $\mathcal{V}(x,t)$ and $u(x,t)$ for $x \in R$ and $t \in R^+$ needs to be determined. The terms Q often generate fast variations in the solution, which may be regarded as distribution so that (1) must be taken in the weak sense. This equation has an important feature: the variable u does not appear in the accumulation term, but only in the flux term.

Generically, the equations (1) model flows that encompass particular physical situations, where one or several phases or chemical components are missing. There is a linear map E that reduces the balance system (1) in each particular physical situation to simpler systems of conservation laws with fewer equations of the form:

$$\frac{\partial}{\partial t}G(V) + \frac{\partial}{\partial x}uF(V) = 0, \tag{2}$$

where the source term Q is absent. In each physical situation, the corresponding set of variables V is a subset of the set of variables $\mathcal{V}$; F; and G are obtained from $\mathcal{F}$ and $\mathcal{G}$; they have $n+1$ components, $n \leq m$. Under suitable hypothesis, one can prove that the solution does not depend on the E, see [Lamb1].

Typically, the laws of thermodynamics play a central role in multiphase models (1). Each physical situation where the system (1) reduces to a system of type (2) is actually a *physical situation under local thermodynamical equilibrium or quasiequilibrium*; this equilibrium is enforced by thermodynamic relationships between the quantities in V. Generically, when physical changes occur in waves under strict thermodynamical equilibrium, the solutions of the systems (2) and (1) are the same; however, when there is only local thermodynamical equilibrium, the solution of (1) is an approximation of (2). In future work, we intend to understand how small deviations from thermodynamical equilibrium determines the discrepancy between the two solutions.

There are three groups of variables in each physical situation, the basic variables V, or "primary variables"; the "secondary variable" u is obtained from the primary variables; and the "trivial variables" are constant or they can be recovered from other variables in a simple way. Notice that the number of total variables always add up to $m+1$.

As an example we study flow of nitrogen and steam in a porous medium with water in [Lamb1], proposed originally and partially solved in [Brui]. The water, steam, nitrogen mass balance equation, and the energy balance equations are:

$$\frac{\partial}{\partial t}\varphi\rho_W s_w + \frac{\partial}{\partial x}u\rho_W f_w = q_{g\longrightarrow a,w}, \tag{3}$$

$$\frac{\partial}{\partial t}\varphi\rho_{gw} s_g + \frac{\partial}{\partial x}u\rho_{gw} f_g = -q_{g\longrightarrow a,w}, \tag{4}$$

$$\frac{\partial}{\partial t}\varphi\rho_{gn} s_g + \frac{\partial}{\partial x}u\rho_{gn} f_g = 0, \tag{5}$$

$$\frac{\partial}{\partial t}\left(\rho_r h_r + \varphi(\rho_W h_w s_w + (\rho_{gw}h_{gw} + \rho_{gn}h_{gn})s_g)\right) + \frac{\partial}{\partial x}u\left(\rho_W h_w f_w + (\rho_{gw}h_{gw} + \rho_{gn}h_{gn})f_g\right) = 0. \tag{6}$$

Here φ is the constant rock porosity; s_w and s_g are the water and steam saturations; T is the temperature; u is the Darcy velocity; f_w and f_g are the fractional flow functions for water and steam; the concentration of steam ρ_{gw} is steam mass per unit gas volume, and the concentration of nitrogen is ρ_{gn}, which depends on T; ρ_r, and ρ_W are the constant rock and liquid water densities. The rock, water, and steam enthalpies per unit mass are h_r, h_w, and

h_g; they depend on temperature. Finally $q_{g\longrightarrow a,w}$ is the water mass transfer or steam condensation term, see [Lamb1].

There are three physical situations: the single-phase gaseous situation, with steam and nitrogen; the two-phase situation, with steam, liquid water, and nitrogen; and the single-phase liquid situation, with water.

The accumulation and flux functions G and F depend on the physical situation; e.g., in the single phase gaseous situation they can be written in the form:

$$\frac{\partial}{\partial t}\varphi\frac{M_W p_{at}}{R}\psi_{gw}T^{-1} + \frac{\partial}{\partial x}u\frac{M_W p_{at}}{R}\psi_{gw}T^{-1} = 0,$$
$$\frac{\partial}{\partial t}\varphi\frac{M_N p_{at}}{R}(1-\psi_{gw})T^{-1} + \frac{\partial}{\partial x}u\frac{M_N p_{at}}{R}(1-\psi_{gw})T^{-1} = 0,$$
$$\frac{\partial}{\partial t}\varphi\left(\hat{H}_r + \psi_{gw}\frac{M_W p_{at}}{RT}h_{gW} + (1-\psi_{gw})\frac{M_N p_{at}}{RT}h_{gN}\right)$$
$$+\frac{\partial}{\partial x}u\left(\psi_{gw}M_W h_{gW} + (1-\psi_{gw})M_N h_{gN}\right)\frac{p_{at}}{RT} = 0,$$

where the constants can be found in [Lamb1]. Here the primary and trivial variables are $\{\psi_{gw}, T\}$ and $\{s_w = 0\}$. In the two-phase situation, the primary and trivial variables are $\{s_w, T\}$ and $\{\psi_{gw}(T)\}$. In the single-phase liquid situation, the primary and trivial variables are $\{T\}$ and $\{s_w = 1, \psi_{gw}(T)\}$; the Darcy speed u is a constant. Even though there is no gas in this physical situation, it is useful to define the gas composition by continuity, see [Lamb1].

In Fig. 1, we represent the space of primary and trivial variables in all physical situations.

2 The Riemann Problem

We are interested in the Riemann problem for (1), with initial data

$$\begin{cases} W_L = (V_L, u_L) & \text{if } x > 0, \\ W_R = (V_R, \cdot) & \text{if } x < 0, \end{cases} \tag{7}$$

where $W = (V, u)$, with V the primary and u the secondary variables. Since the system (1) has an infinite characteristic speed associated to u, only one of the speeds u_L or u_R is given (we have chosen u_L). Later, it will be clear that the other speed can be determined from the other equations in the system.

The general solution of the Riemann problem associated to the set of several equations (2) for different physical situations consists of a sequence of elementary waves, rarefactions and shocks, separated by constant states.

2.1 Characteristic Speeds

Since the cumulative and flux terms are not smooth between physical situations, we calculate the characteristic speeds only within each physical

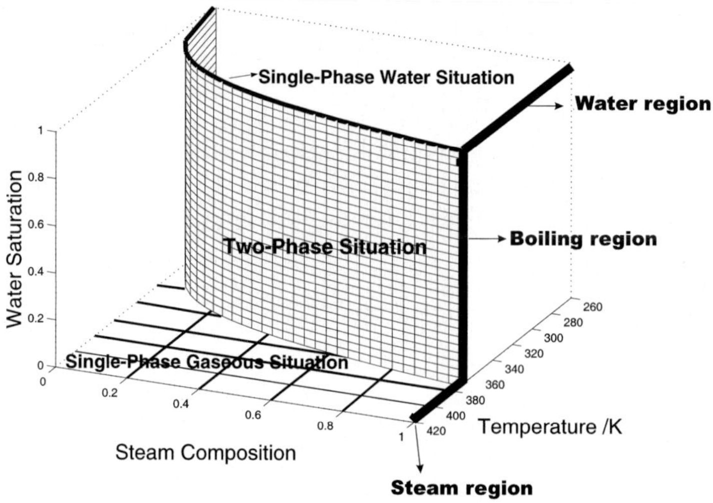

Fig. 1. State space, physical situations, and their domains in Example 1. In the single-phase gaseous situation, the primary and trivial variables are $\{\psi_{gw}, T\}$ and $\{s_w = 0\}$, respectively. In the two-phase situation, the primary and trivial variables are $\{s_w, T\}$ and $\{\psi_{gw}(T)\}$, respectively. In the single-phase liquid situation, the primary and trivial variables are $\{T\}$ and $\{s_w = 1, \psi_{gw}(T)\}$

situation, where the system (1) of balance equations reduces to a conservation form (2).

For a smooth wave we differentiate (2) obtaining:

$$B\frac{\partial}{\partial t}\begin{pmatrix} V \\ u \end{pmatrix} + A\frac{\partial}{\partial x}\begin{pmatrix} V \\ u \end{pmatrix} = 0; \tag{8}$$

the matrices B and A are the derivatives of $G(V)$ and $uF(V)$ with respect to $W = (V, u)$. Since $G(V)$ does not depend on u, the matrix $\partial G/\partial W$ has a zero column. It is useful to define $\mathcal{C} = \{1, 2, \cdots, n+1\}$. The following results are proved in [Lamb1]:

Lemma 1. *Assume that $u \neq 0$ and that $F_k(V) \neq 0$ for some fixed $k \in \mathcal{C}$ for all V. The eigenvalue, right and left eigenvectors for (8) have the form:*

$$\lambda = u\vartheta(V), \tag{9}$$

$$r = (g_1(V), \cdots, g_n(V), ug_{n+1}(V)) \text{ and } \ell = (\ell_1(V), \cdots, \ell_{n+1}(V)), \tag{10}$$

where ϑ, g_i, and ℓ_i (for all $i \in \mathcal{C}$) depend only on V. There are at most n characteristic pairs for this system of $n+1$ equations.

Theorem 1. *Assume that the eigenvector r associated to a certain family forms a local vector field. We calculate the primary variables V on the rarefaction waves in the (x,t) plane independently of u, i.e., first we obtain the primary variables in the classical way, and then we calculate u :*

$$u = u^- exp(\gamma(\xi)), \ \ \gamma(\xi) = \int_{\xi^-}^{\xi} g_{n+1}(V(\eta))d\eta, \tag{11}$$

where $\xi = x/t$, $\xi^- = \lambda(W^-)$ *and* u^- *is the "leftmost" value of* u *on the rarefaction wave, i.e.,* $u = u^-$ *for* $\xi = \xi^-$.

2.2 Shocks and the Rankine–Hugoniot Condition

The main feature of the class of (1) is the existence of separate regions, where the system of balance equations (1) reduces to different systems of conservation equations (2). To obtain the complete Riemann problem solution it is necessary to link these regions. Because the term Q often gives rise to fast changes between regions, shocks link such regions. Eliminating Q by means of E, the RH condition can be written as:

$$v^s \left(G^+(V^+) - G^-(V^-)\right) = u^+F^+(V^+) - u^-F^-(V^-), \tag{12}$$

where (V^-, u^-) are the states on the left side and (V^+, u^+) are states on the right side. Notice here that if the shock sides are in different physical situations, the accumulations and flux functions have different expressions on each side.

Writing (12) as a linear homogeneous system, we obtain:

$$\begin{pmatrix} [G_1] & -F_1^+ & F_1^- \\ [G_2] & -F_2^+ & F_2^- \\ \vdots & \vdots & \vdots \\ [G_n] & -F_n^+ & F_n^- \\ [G_{n+1}] & -F_{n+1}^+ & F_{n+1}^- \end{pmatrix} \begin{pmatrix} v^s \\ u^+ \\ u^- \end{pmatrix} = 0, \text{ or } \mathcal{M} \begin{pmatrix} v^s \\ u^+ \\ u^- \end{pmatrix} = 0, \tag{13}$$

where $[G_i] = G_i^+(V^+) - G_i^-(V^-)$, $F_i^+ = F_i^+(V^+)$ and $F_i^- = F_i^-(V^-)$.

The 3×3 minors of the matrix $\mathcal{M}$ are denoted by $\mathcal{M}_{pqs}$:

$$\mathcal{M}_{pqs} = \begin{pmatrix} [G_p] & -F_p^+ & F_p^- \\ [G_q] & -F_q^+ & F_q^- \\ [G_s] & -F_s^+ & F_s^- \end{pmatrix} \text{ for all distinct } p,\ q, \text{ and } s \text{ in } \mathcal{C}. \tag{14}$$

The homogeneous linear system (13) has a nontrivial solution, if and only if

$$\mathcal{H}_{pqs} = det\left(\mathcal{M}_{pqs}\right) = 0 \text{ for all distinct } p,\ q, \text{ and } s \text{ in } \mathcal{C}. \tag{15}$$

Definition 1. *The RH locus is given by implicit expressions in the primary variables. For* (V^-, u^-) *fixed in some physical situation, the RH locus in the projected space* V *consists of the* V^+ *that satisfy* $det(\mathcal{M}_{pqs}) = 0$ *for all distinct* p, q, *and* s *in* $\mathcal{C}$. *Notice that* V^+ *can lie in the same or in a distinct physical situation relatively to* V^-. *We denote the RH locus of a state* V^- *as* $\mathcal{RH}(V^-)$.

Notice that there are C^3_{n+1} equations satisfying (15); however, the following Lemma guarantees that we only need to solve a set of $(n-1)$ equations of type (15). Since we want to characterize the solution of (15), it is useful to define for each $i \in \mathcal{C}$, the vector

$$\mathcal{D}_i \equiv \mathcal{D}_i(V^-;V^+) \equiv \left([G]_i, -F_i^+, F_i^-\right). \tag{16}$$

Lemma 2. *Given V^- in some physical situation, assume that for $V^+ \in \mathcal{RH}(V^-)$ and in a neighborhood of V^+ there exists at least a pair of linearly independent vectors $\mathcal{D}_p$ and $\mathcal{D}_q$ of the form (16) with $p, q \in \mathcal{C}$. Assume also that for all $l \neq m \in \mathcal{C}$ and $V^+ \in \mathcal{B}^j_\delta(V^+) - X_{lm}(V^-;V^+)$ the vectors $\mathcal{D}_l$ and $\mathcal{D}_m$ are linearly independent. Then the RH locus is a 1-dimensional structure that solves the $(n-1)$ equations:*

$$det \begin{pmatrix} \mathcal{D}_p \\ \mathcal{D}_q \\ \mathcal{D}_r \end{pmatrix} = 0, \ \ \forall\, r \in \mathcal{C} - \{p, q\}, \tag{17}$$

where $X_{lm}(V^-;V^+)$ is a structure with dimension at most 1. Moreover, $X_{lm}(V^-;V^+)$ for $l \neq m \in \mathcal{C}$ belongs to the RH locus.

Let us assume that $\mathcal{RH}(V^-)$ was found. Now we want to determine u^+. Let $p, q \in \mathcal{C}$. Given V^- in some physical situation, if $\mathcal{D}_p$ and $\mathcal{D}_q$ are LI in a neighborhood of V^+, then:

$$det \begin{pmatrix} [G_p] & -F_p^+ \\ [G_q] & -F_q^+ \end{pmatrix} \neq 0. \tag{18}$$

In this case v^s and u^+ can be written in terms of u^- and $\mathcal{V}^-$ as:

$$v^s = u^- \frac{F_p^+ F_q^- - F_q^+ F_p^-}{F_q^+[G_p] - F_p^+[G_q]} \quad \text{and} \quad u^+ = u^- \frac{F_q^-[G_p] - F_p^-[G_q]}{F_q^+[G_p] - F_p^+[G_q]}. \tag{19}$$

Of course, we could have chosen to solve v^s and u^- in terms of u^+.

In Theorem 2, we summarize the connection between rarefaction and shock waves. Here we index the rarefaction curves in the Riemann solution by l for $l = 1, 2, \cdots, \rho_1$; the shocks are indexed by m for $m = 1, 2, \cdots \rho_2$:

Theorem 2. *Let u_L be positive. The primary variables V in the shock and rarefaction curves do not depend on the left speed $u_L > 0$. If a sequence of waves and states solve the Riemann problem in the primary variables, for a given $u_L > 0$, then it is also a solution for any other $u_L > 0$:*

$$\begin{cases} (\mathcal{V}_L, u_L) & if\ x < 0 \\ (\mathcal{V}_R, \cdot) & if\ x > 0. \end{cases} \tag{20}$$

Assume that for each m there are $p, q \in \mathcal{C}$ such that $F^+_{q,m}[G_{p,m}] - F^+_{p,m}[G_{q,m}] \neq 0$. Then u_R is given by:

$$u_R = u_L \prod_{l=1}^{\varrho_1} exp\left(\int_{\xi^{-,l}}^{\xi^{+,l}} g_{n+1}^l(V(\eta))d\eta\right) \prod_{m=1}^{\varrho_2} \frac{F_{q,m}^-[G_{p,m}] - F_{p,m}^-[G_{q,m}]}{F_{q,m}^+[G_{p,m}] - F_{p,m}^+[G_{q,m}]}, \quad (21)$$

where g_{n+1}^l is the $(n+1)$-th component of the eigenvector r^l, $\xi^{-,l}$ and $\xi^{+,l}$ are the leftmost and rightmost values of ξ^l associated to the l-th rarefaction wave.

In the theorem above, $G_{j,m}$ and $F_{j,m}$ represent the jth components ($j \in \mathcal{C}$) of G and F on the mth shock wave. Similarly $exp\left(\int_{\xi^{0,l}}^{\xi^l} g_{n+1}(V(\eta))d\eta\right)$ is computed along the lth-rarefaction curve, see (11).

In Theorem 2, instead of u_L we could have prescribed the right speed u_R and obtain u_L and the other speeds as function of V and u_R.

Theorem 3. *Assume that $u_L \neq 0$. Then the Riemann solution can be obtained in the primary variables V for each physical situation. The solution for the speed u can be obtained in terms of $\mathcal{V}$ and u_L by (21). The same is true for rarefaction and shock speeds.*

In [Lamb1], we extended two theorems used to obtain bifurcation regions in the state space. The first result is the Bethe–Wendroff theorem, [Wend]

Theorem 4. *Let $v^s(W^+;W^-)$ be the speed of the shock between different physical situations. Assume that $\ell^p(V^+) \cdot [G] \neq 0$. Then v^s has a critical point at W^+ (so $\hat{v}^+(V^-,V^+)$ has a critical point at V^+), if and only if:*

$$\hat{v}^+(V^-;V^+) = \lambda^{p,+}(V^+) \quad \textit{for the family } p, \quad (22)$$

where $\hat{v}^+(V^-;V^+) := v^s/u^-$ and $\lambda^{p,+}(V^+) = \lambda^p(V^+)/u^+$. In this case the Hugoniot locus in the space of primary variables is parallel to the $(r_1^p, r_2^p, \cdots, r_n^p)$ in V^+.

Another result is a generalization of the triple shock rule [Isa], modified to a quadruple shock rule. We refer the reader to [Lamb1].

Example of Riemann Solution for the Nitrogen Example

We consider the Riemann problem (20) with left state L in the single phase gaseous situation (*spg*) and right state R in the single phase gaseous situation (*spl*).

In the *spg* there are a thermal rarefaction wave, with characteristic speed λ_T and a compositional wave with speed λ_c, which is a contact discontinuity; the speeds satisfy $\lambda_c > \lambda_T > 0$.

In the two-phase situation (*tp*) there are an evaporation rarefaction where the saturation, temperature, and speed change, with characteristic speed λ_e; a condensation shock S_c with speed v_c; an isothermal wave, consisting of a Buckley–Leverett shock S_{BL} and rarefaction R_{BL}, with speeds v^{BL} and λ^{BL}.

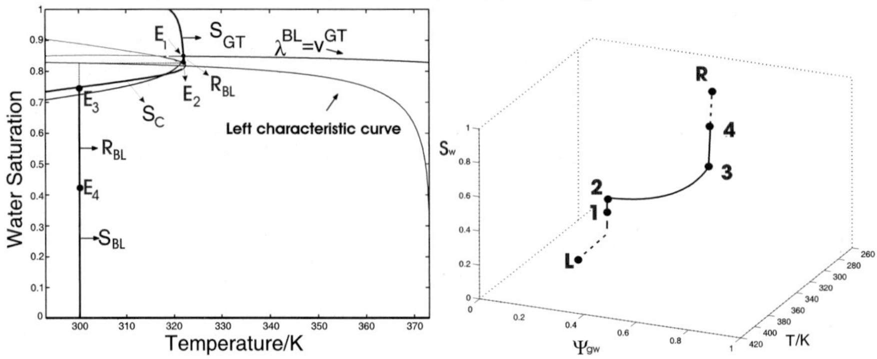

Fig. 2. (**a**) *Left*: Wave and state sequences that appear in the Riemann solution (**b**) *Right*: Riemann solution in phase space V; in the Riemann solution, we have removed the surface in phase space of Fig. 1

In Fig. 2 we show the Riemann solution, details are found in [Lamb1].

The class of equations studied in this work was motivated by practical processes of steam and gas injection in a porous medium. Such techniques can be used in oil industry to recover heavy oil or clean up sites. In [Lamb1], we provide the Riemann solution for steam and water injection and an example of Riemann solution for steam, water and nitrogen injection in a porous media.

Acknowledgments

We thank Prof. Johannes Bruining (TUDelft) for important discussions about the physical model. This work was supported in part by: CNPq scholarship 141573/2002-3, ANP/PRH-32 scholarship 2001.6787-0, CNPq grant 301532/2003-6 and FAPERJ grant E-26/152.163/2002.

References

[Brui] Bruining, J., Marchesin, D., Nitrogen and Steam Injection in a Porous Medium with Water, Transport in Porous Media, Vol. 62, No. 3, 251–281, 2006.

[Lamb1] Lambert, W., Doctoral Thesis: Riemann solution of balance system with phase change for thermal flow in porous media, Rio de Janeiro, 2006, http://www.preprint.impa.br.

[Isa] Isaacason, E., Marchesin, D., Plohr, B., and Temple, J.B., The Riemann Problem near a hyperbolic singularity: Classification of quadratic Riemann problems I, SIAM Journal of Appl. Math., 48, 1009–1032, 1988.

[Wend] Wendroff, B., The Riemann Problems for materials with non-convex equations of state II. J. Math. Anal. Appl. 38, 454–466, (1972).

Nonlinear Hyperbolic–Elliptic Coupled Systems Arising in Radiation Dynamics

C. Lattanzio, C. Mascia, and D. Serre

Summary. In this chapter we deal with a hyperbolic system of conservation laws with a nonlinear coupling with a scalar elliptic equation, modeling radiation dynamics. In particular, we study existence and regularity of *radiative shock waves* for that system, that is (possibly discontinuous) traveling wave-type solutions, both in the strictly convex and in the general case.

1 Introduction

The aim of this chapter is the study of traveling waves for the following hyperbolic–elliptic coupled system

$$\begin{cases} u_t + f(u)_x + Lq_x = 0 \\ -q_{xx} + Rq + G(u)_x = 0, \end{cases} \tag{1}$$

where $x \in \mathbb{R}$, $t > 0$, $u \in \mathbb{R}^n$, $q \in \mathbb{R}$, and $R > 0$, $L \in \mathbb{R}^n$ is a constant vector and $G : \mathbb{R}^n \to \mathbb{R}$ is smooth. Moreover, the flux function $f : \mathbb{R}^n \to \mathbb{R}^n$ is assumed to be smooth and such that the reduced system

$$u_t + f(u)_x = 0 \tag{2}$$

is strictly hyperbolic, i.e., $\nabla f(u)$ has n distinct real eigenvalues $\lambda_1(u) < \cdots < \lambda_n(u)$ with corresponding left and right eigenvectors $\ell_i(u)$ and $r_j(u)$, normalized so that $\ell_i(u) \cdot r_j(u) = \delta_{ij}$, for any i, j and for any state u under consideration. This kind of systems arises in the study of the dynamics of a gas in presence of radiation, which yields to the compressible Euler equations with an additional term in the flux of energy:

$$\begin{cases} \rho_t + (\rho u)_x = 0 \\ (\rho u)_t + (\rho u^2 + p)_x = 0 \\ \left\{\rho\left(e + \frac{u^2}{2}\right)\right\}_t + \left\{\rho u\left(e + \frac{u^2}{2}\right) + pu + q\right\}_x = 0 \\ -q_{xx} + aq + b(\theta^4)_x = 0. \end{cases} \tag{3}$$

Here we recall that a traveling wave solution, or *radiative shock wave*, of the hyperbolic–elliptic system (1) is a weak solution $(u, q)(x, t) := (U, Q)(x - st)$ such that

$$\lim_{\xi \to \pm\infty} (U, Q)(\xi) = (u_\pm, 0),$$

where $u_\pm \in \mathbb{R}^n$, $s \in \mathbb{R}$, define a shock wave for the reduced hyperbolic system (2), namely, it verifies the usual Rankine–Hugoniot condition

$$f(u_+) - f(u_-) = s(u_+ - u_-), \tag{4}$$

and the admissibility condition (Lax condition for GNL characteristic fields, Liu's E-condition otherwise). It is worth to observe that, being a weak solution, a radiative shock wave may be discontinuous at some point, where it shall verify the same Rankine–Hugoniot and admissibility conditions.

Motivated by the aforementioned application, the study of hyperbolic–elliptic coupled systems is been very active nowadays: see the papers of Kawashima and collaborators [KNN98, KN99b, KN99, KN01, IK02, KNN03, KT04] and the papers of of Liu, Schochet and Tadmor [ST92, LT01]. In the above papers, both the asymptotic behavior of small solutions and the existence of traveling waves for (1) is analyzed, in different particular cases. Concerning the study of traveling waves, the results are mainly confined to the scalar, strictly convex, linear case ($u \in \mathbb{R}$ and $G(u) = Gu$, for a constant $G \in \mathbb{R}$). In this context, the existence of *smooth* traveling waves has been shown in [ST92], while the complete analysis, including the case of *discontinuous* traveling waves for large shocks, has been carried out only for Burgers' type fluxes ($f(u) = u^2/2$), the so–called *Hamer model* [Ham71], in [KN99]. Finally, stability results can be found in [Ser03, Ser04a, Ser04b].

In the case of systems, this problem has been addressed only recently: in [LCG06], the existence of traveling waves is proved for the physical system (3), again in the smooth case, while [LMS06] treats the general system (1), including also discontinuous traveling waves, but for linear coupling, that is $G(u) = G \cdot u$, for a constant vector $G \in \mathbb{R}^n$. In this last paper, the authors proved the result via a reduction procedure: the dynamics of traveling waves is basically governed by a general scalar conservation law for the variable $w = G \cdot u$, coupled with the linear elliptic equation. Moreover, the scalar case is treated in full generality, including discontinuous traveling waves for large shocks and general (nonconvex) fluxes.

The aim of the present chapter is to generalize the results of [LMS06] in the case of the nonlinear coupling $G(u)$ in $(1)_2$, including in this way the physical model (3). For the sake of clarity, we confine ourselves to the strictly convex framework, but our arguments apply also to general nonlinear fluxes. Thus, we assume the kth characteristic field of (2) is genuinely nonlinear, that is

$$\nabla \lambda_k(u) \cdot r_k(u) \neq 0 \tag{5}$$

and we fix states $u_\pm \in \mathbb{R}^n$ and a speed $s \in \mathbb{R}$ such that the triple $(u_-, u_+; s)$ defines a k-Lax shock wave for (2), namely, (4) and the admissibility condition

$$\begin{aligned} \lambda_{k-1}(u_-) < s < \lambda_k(u_-) \\ \lambda_k(u_+) < s < \lambda_{k+1}(u_+) \end{aligned} \tag{6}$$

are satisfied. The nondegeneracy hypothesis we shall require states as follows

$$\ell_k(u_\pm) \cdot \nabla G(u_\pm) \otimes L\, r_k(u_\pm) > 0. \tag{7}$$

The meaning of condition (7) is clear as soon as we eliminate the q variable from system (1). Indeed, we end up with the following system for u:

$$u_t + f(u)_x = R^{-1}(\nabla G(u) \otimes L u_x)_x + R^{-1}\big(u_t + f(u)_x\big)_{xx},$$

which gives a "good" dissipation for $u \sim u_\pm$ along the kth characteristic field, provided (7) is satisfied.

The rest of this chapter is organized as follows. We recall the result of [LMS06] for the scalar case in Sect. 2, while the changes of the reduction procedure due to the nonlinear coupling are described in Sect. 3.

2 Results for the Scalar Linear Case

We recall in this section the result contained in [LMS06] concerning the existence of radiative shock waves for the following scalar conservation law with linear coupling with an elliptic equation:

$$\begin{cases} w_t + \hat{f}(w)_x + q_x = 0 \\ -q_{xx} + q + w_x = 0, \end{cases} \tag{8}$$

where $\hat{f}$ is a strictly convex function. Hence we are dealing with the 2×2 adimensionalized (via a rescaling) version of (7). It is clear from the arguments below that if we replace w in $(8)_2$ with a regular, invertible, nonlinear function $G(w)$, we do not have significant changes in the present discussion.

Let us consider a solution to (8) of the form $(w, q) = (w(x - st), q(x - st))$ such that

$$w(\pm\infty) = w_\pm, \qquad s = \frac{\hat{f}(w_+) - \hat{f}(w_-)}{w_+ - w_-}, \qquad w_+ < w_-.$$

We introduce the variable z as the opposite of the antiderivative of q, that is $z_x := -q$, with $z(\pm\infty) = z_\pm = w_\pm$. Thus, after integration of both equations, the system for the profiles $w(x - st)$ and $z(x - st)$ takes the form

$$\begin{cases} -s(w - w_\pm) + \hat{f}(w) - \hat{f}(w_\pm) = z' \\ w = z - z'' \end{cases} \tag{9}$$

or

$$z' = \hat{F}(z - z''; s), \tag{10}$$

where $\hat{F}(w;s) := \hat{f}(w) - \hat{f}(w_\pm) - s(w - w_\pm)$. Hence, our results for the profile $u(x-st)$ will be consequence of the existence and regularity results of solutions of the (singular) second-order equation (10) with $z(\pm\infty) = z_\pm$. The results for the profiles state as follows.

Theorem 1. *There exists a (unique up to space translations) C^1 profile z with $z(\pm\infty) = z_\pm = u_\pm$ such that the function $z(x-st)$ is solution of (10), where s is given by the Rankine–Hugoniot condition. The solution z is C^2 away from a single point, where z'' has at most a jump discontinuity.*

Moreover, there exists a (unique up to space translations) profile w with $w(\pm\infty) = w_\pm$ such that the function $w(x-st)$ is solution of (9), where s is given by the Rankine–Hugoniot condition. This profile is continuous away from a single point, where it has at most a jump discontinuity which verifies the Rankine–Hugoniot and the admissibility conditions of the scalar conservation law $w_t + \hat{f}(w)_x = 0$.

Remark 1. The results in [LMS06] include also the increase of regularity of the traveling waves, as the strength of the underlying shock decreases, both for the scalar and (via the reduction procedure) for the system case. Clearly, for regularity issues, the nonlinearity in $(1)_2$ does not play any role in the reduction procedure and therefore we do not discuss these results in the present chapter.

3 Reduction Procedure for Nonlinear Couplings

Let us consider the general system (1) and let us look for traveling waves, namely solution of the form $(u,q) = (u(x-st), q(x-st))$, $(u,q)(\pm\infty) = (u_\pm, 0)$, with $(u_-, u_+; s)$ that define a shock for (2). Introducing as before the new variable z via $z_x := -q$, $z(\pm\infty) = z_\pm = u_\pm$, after an integration, the system of ODEs satisfied by the profiles is given by

$$\begin{cases} F(u;s) = L\,z' \\ -z'' + Rz = G(u), \end{cases} \tag{11}$$

where, as before, $F(u;s) := f(u) - f(u_\pm) - s(u - u_\pm)$.

Given the vector L, let $P : \mathbb{R}^n \to \mathbb{R}^n$ and $Q : \mathbb{R}^n \to \mathbb{R}$ be linear applications such that

$$\ker P = \text{span}\,\{L\}, \qquad QL = 1.$$

Then system (11) becomes

$$\begin{cases} P\,F(u;s) = 0 \\ z' = Q\,F(u;s) \\ G(u) = Rz - z''. \end{cases} \tag{12}$$

Therefore, the dynamics of the profiles u and z reduces to a scalar, singular, second-order ODE for z of the form (10), provided the constraints $(12)_1$, together with condition (7) imply

$$Q\,F(u;s) = \hat{F}(w;s), \quad w := G(u), \tag{13}$$

in a neighborhood of $(u_\pm;\lambda_k(u_\pm))$, that is for sufficiently small shocks $(u_-,u_+;s)$.

Lemma 1. *Let f be C^1 and assume condition (7) holds. Then, there exist neighborhoods I of $\lambda_k(u_\pm)$ and $\mathcal{U}$ of $u_\pm$ such that for $s \in I$,*

$$P\,F(u;s) = 0 \iff u = \Psi(G(u)), \text{ for any } u \in \mathcal{U},$$

with $\Psi \in C^1(\mathbb{R};\mathcal{U})$, not depending on the choice of P such that $\ker P = span\{L\}$.

Proof. From the discussion in [LMS06] concerning the linear case, we already know that, in our hypotheses, there exist neighborhoods I of $\lambda_k(u_\pm)$ and Ω of $u_\pm$ such that for $s \in I$,

$$P\,F(u;s) = 0 \iff u = \Phi(\nabla G(u_-)\cdot u) =: \Phi(\omega),$$

for any $u \in \Omega$. Expanding $w = G(u)$ for u close to u_- we obtain

$$w = G(u_-) + \nabla G(u_-)\cdot(u-u_-) + o(|u-u_-|) = C + \omega + o(|u-u_-|)$$

with $C := G(u_-) - \nabla G(u_-)\cdot u_-$. Hence ω can be expressed as a function of w for u close to u_-, namely, there exitsts a neighborhood $\widetilde{\mathcal{U}}$ of u_- such that

$$\omega = \omega(w), \ u \in \widetilde{\mathcal{U}}.$$

Thus, for $s \in I$,

$$P\,F(u;s) = 0 \iff u = \Phi(\omega(w)) =: \Psi(w) \text{ for any } u \in \mathcal{U} := \Omega \cap \widetilde{\mathcal{U}}$$

and the proof is complete.

At this point, we consider the genuinely nonlinear case, namely conditions (5) and (6) are satisfied. As for linear couplings, we shall prove that in this framework, in the reduced scalar dynamics yielding the existence of our profile, that is $(12)_2$, the flux function $\hat{F}(\cdot;s)$ does not change convexity.

Lemma 2. *Let us assume condition (7) and (5) hold. Then the function $\hat{F}(\cdot;s) : \mathbb{R} \to \mathbb{R}$ defined in (13) is either strictly convex or strictly concave in a neighborhood of $w_\pm = G(u_\pm)$.*

Proof. Since $(u_-, u_+; s)$ form a k-Lax shock for (2), it suffices to prove

$$\frac{d^2}{dw^2}\hat{F}(G(u_\pm); s) \neq 0,$$

for sufficiently small shocks.

Since $P\,F(u; s) = 0$ and $Q\,F(u; s) = \hat{F}(G(u); s)$, for any $u \in \mathcal{U}$ we have

$$F(u; s) = \hat{F}(G(u); s)\,L.$$

Differentiating with respect to u, one has

$$\nabla f(u) - sI = \frac{d}{dw}\hat{F}(G(u); s)\,L \otimes \nabla G(u). \tag{14}$$

Applying $\ell_k(u)$ and $r_k(u)$, respectively, to the left and to the right of (14), we get

$$\lambda_k(u) - s = \frac{d}{dw}\hat{F}(G(u); s)\,\ell_k(u) \cdot L \otimes \nabla G(u)\,r_k(u). \tag{15}$$

Choosing $u = u_\pm$ and assuming $|u_+ - u_-|$ small enough, (7) implies

$$\frac{d}{dw}\hat{F}(G(u); s)\Big|_{u=u_\pm} = o(1), \qquad \text{as} \quad |u_- - u_+| \to 0. \tag{16}$$

Differentiating (15) with respect to u in the direction of $r_k(u)$, we obtain

$$\begin{aligned}\nabla\lambda_k(u) \cdot r_k(u) = {}& \frac{d}{dw}\hat{F}(G(u); s)\nabla(\ell_k(u) \cdot L \otimes \nabla G(u) r_k(u)) \cdot r_k(u)\\ &+ \frac{d^2}{dw^2}\hat{F}(G(u); s)\big(\ell_k(u) \cdot L\big)\big(\nabla G(u) \cdot r_k(u)\big)^2.\end{aligned}$$

Evaluating this relation at $u = u_\pm$ and taking into account (16), we obtain

$$\begin{aligned}\frac{d^2}{dw^2}\hat{F}(G(u_\pm); s) = {}& \frac{\nabla\lambda_k(u_\pm) \cdot r_k(u_\pm)}{\big(\ell_k(u_\pm) \cdot L\big)\big(\nabla G(u_\pm) \cdot r_k(u_\pm)\big)^2} + o(1),\\ &\text{as} \quad |u_- - u_+| \to 0\end{aligned}$$

and the conclusion follows from GNL condition (5) and assumption (7).

Now we state the main result of the section.

Theorem 2. *Let $u_- \in \mathbb{R}^n$ be such that the kth characteristic field of (2) is genuinely nonlinear at u_-, that is $\nabla\lambda_k(u_-) \cdot r_k(u_-) \neq 0$. Assume that*

$$(\ell_k(u_-) \cdot L)\,(\nabla G(u_-) \cdot r_k(u_-)) > 0.$$

Then there exists a sufficiently small neighborhood $\mathcal{U}$ of u_- such that for any $u_+ \in \mathcal{U}$, $s \in \mathbb{R}$ such that the triple $(u_-, u_+; s)$ defines a shock wave for (2), there exists a (unique up to shift) admissible radiative shock wave for (1).

We skip the proof of that theorem because it follows the same lines of the one in [LMS06] for the linear case, by replacing the linear function $G \cdot u$ with $G(u)$ and we prefer to illustrate it in full details for a specific example.

Remark 2. It is clear from the above arguments that the smallness of the shock required in Theorem 2 is due to the reduction procedure (Lemmas 1 and 2), which is basically connected with the coupling between the hyperbolic and the elliptic part of system (1). Hence, it is not connected with the smoothness of the resulting radiative profile, which is linked solely to the flux of the scalar reduced model. In other words, applying our procedure to concrete examples, it is possible to construct *discontinuous* radiative shock waves, where the reduction procedure can be performed *by hand*, without using the implicit function theorem. Here we show this property in the following toy model, given by the p-system coupled with a linear elliptic equation:

$$\begin{cases} u_t - v_x = 0 \\ v_t + p(u)_x + q_x = 0 \\ -q_{xx} + q + v_x = 0. \end{cases} \tag{17}$$

In this case, both the back and the forward characteristic fields are GNL and verify (7), provided $p'(u) < 0$ and $p''(u) > 0$ for any u under consideration. Moreover, (12) reduces to

$$\begin{cases} -v + v_\pm - s(u - u_\pm) = 0 \\ p(u) - p(u_\pm) - s(v - v_\pm) = z' \\ v = z - z''. \end{cases} \tag{18}$$

Hence, since $s \neq 0$ for both forward or backward Lax shocks, the reduction performed in Lemma 1 is indeed global in the state space:

$$u = \frac{-v + v_\pm}{s} + u_\pm \tag{19}$$

and the reduced dynamic is given by

$$z' = p\left(\frac{-v + v_\pm}{s} + u_\pm\right) - p(u_\pm) - s(v - v_\pm). \tag{20}$$

Hence, the above equation comes from a conservation law with flux

$$\hat{p}(v) = p\left(\frac{-v + v_\pm}{s} + u_\pm\right),$$

which is a strictly convex function, again without restrictions for v.

At this point, let us consider the connection between Rankine–Hugoniot and entropy conditions for the complete and the reduced model $v_t + \hat{p}(v)_x = 0$. To fix ideas, let us consider a radiative shock wave for an underlying backward

Lax shock of the p-system. Then the Rankine–Hugoniot condition for the scalar model reduces to

$$s(v_+ - v_-) = \hat{p}(v_+) - \hat{p}(v_-) = p(u_+) - p(u_-),$$

which is implied by the Rankine–Hugoniot condition for the original shock. Moreover, in view of the strict convexity of the flux, the admissibility condition for that discontinuity reduces to $v_- > v_+$, and this is implied by the Lax condition for a backward shock in the p-system, namely

$$-\sqrt{-p'(u_+)} < s < -\sqrt{-p'(u_-)} < 0 \iff u_- > u_+,$$

because

$$v = -su + v_+ + su_+ = -su + v_- + su_-$$

is an increasing function of u ($s < 0$ for backward shocks).

Hence, we can apply Theorem 1 and we have a radiative shock wave with (possibly) a discontinuity $(v_l, v_r; s)$ satisfying Rankine–Hugoniot and entropy conditions for the reduced model $v_t + \hat{p}(v)_x = 0$. Then, from (19) we obtain u_l and u_r verifying

$$\begin{aligned} s(u_r - u_l) &= -v_r + v_l \\ s(v_r - v_l) &= \hat{p}(v_r) - \hat{p}(v_l) = p(u_r) - p(u_l), \end{aligned}$$

that is, the Rankine–Hugoniot conditions for the states $U_l := (u_l, v_l)$ and $U_r := (u_r, v_r)$ in the original p-system.

Finally, the entropy condition $\hat{p}'(v_r) < s < \hat{p}'(v_l) < 0$ becomes

$$-\frac{1}{s}p'(u_r) < s < -\frac{1}{s}p'(u_l).$$

Multiplying the above relation by $s < 0$, we obtain

$$-p'(u_l) < s^2 < -p'(u_r),$$

which reduces to

$$-\sqrt{-p'(u_r)} < s < -\sqrt{p'(u_l)},$$

that is, the Lax condition for the backward shock $(U_l, U_r; s)$ of the p-system.

In conclusion, in the above example we are able to reduce the ODEs in (18), yielding to the existence of radiative shock waves for the complete system (17), to the scalar, singular ODE given by $(18)_3$ and (20), *without any smallness assumption* for the underlying shock wave of the p-system. Then, thanks to Theorem 1, we get (possibly discontinuous) radiative shock waves, by solving (20) for *arbitrary strong shocks*. Finally, coming back to the original system, we end up with the existence of (possibly discontinuous) radiative shock waves, for arbitrary strong shocks of the p-system.

References

[Ham71] Hamer, K.: Nonlinear effects on the propagation of sound waves in a radiating gas. Quart. J. Mech. Appl. Math., **24**, 155–168 (1971)

[IK02] Iguchi, T., Kawashima, S.: On space-time decay properties of solutions to hyperbolic-elliptic coupled systems. Hiroshima Math. J. **32**, no. 2, 229–308 (2002)

[KNN98] Kawashima, S., Nikkuni, Y., Nishibata, S.: The initial value problem for hyperbolic-elliptic coupled systems and applications to radiation hydrodynamics. In: Analysis of systems of conservation laws (Aachen, 1997), Chapman & Hall/CRC Monogr. Surv. Pure Appl. Math., vol. 99, Chapman & Hall/CRC, Boca Raton, FL, (1999)

[KNN03] Kawashima, S., Nikkuni, Y., Nishibata, S.: Large-time behavior of solutions to hyperbolic-elliptic coupled systems. Arch. Ration. Mech. Anal., **170**, no. 4, 297–329 (2003)

[KN99b] Kawashima, S., Nishibata, S.: Cauchy problem for a model system of the radiating gas: weak solutions with a jump and classical solutions. Math. Models Methods Appl. Sci., **9**, no. 1, 69–91 (1999)

[KN99] Kawashima, S., Nishibata, S.: Shock waves for a model system of a radiating gas. SIAM J. Math. Anal., **30**, 95–117 (1999)

[KN01] Kawashima, S., Nishibata, S.: A singular limit for hyperbolic-elliptic coupled systems in radiation hydrodynamics. Indiana Univ. Math. J., **50**, no. 1, 567–589 (2001)

[KT04] Kawashima, S., Tanaka, Y.: Stability of rarefaction waves for a model system of a radiating gas. Kyushu J. Math. **58**, no. 2, 211–250 (2004)

[LMS06] Lattanzio, C., Mascia, C., Serre, D.: Shock waves for radiative hyperbolic–elliptic systems. Indiana Univ. Math. J., to appear

[LCG06] Lin, C., Coulombel, J.-F., Goudon, T.: Shock profiles for non equilibrium radiating gases. Physica D, **218**, 83–94 (2006)

[LT01] Liu, H., Tadmor, E.: Critical thresholds in a convolution model for nonlinear conservation laws. SIAM J. Math. Anal. **33**, no. 4, 930–945 (2001)

[Nis00] Nishibata, S.: Asymptotic behavior of solutions to a model system of radiating gas with discontinuous initial data. Math. Models Methods Appl. Sci., **10**, no. 8, 1209–1231 (2000)

[ST92] Schochet, S., Tadmor, E.: The regularized Chapman–Enskog expansion for scalar conservation laws. Arch. Rational Mech. Anal., **119**, no. 2, 95–107 (1992)

[Ser03] Serre, D.: L^1–stability of constants in a model for radiating gases. Comm. Math. Sci. **1**, 197–205 (2003)

[Ser04a] Serre, D.: Hyperbolic–elliptic systems of conservation laws. Tech. report, `http://www.umpa.ens-lyon.fr/~serre/DPF/SerreCR.pdf` (2004)

[Ser04b] Serre, D.: L^1-stability of nonlinear waves in scalar conservation laws. In: Evolutionary equations. Vol. I, Handb. Differ. Equ., North-Holland, Amsterdam, (2004)

The Lagrangian Coordinates Applied to the LWR Model

L. Ludovic, C. Estelle, and L. Jorge

1 Introduction

Traditionally, the LWR model has been formulated in Eulerian coordinates (x,t) as a scalar conservation law [LW55]:

$$\partial_t k + \partial_x (kv) = 0. \tag{1}$$

This model is fully described by vehicle density k. The speed v and the flow q are related to k by a fundamental diagram (FD) $v = V(k)$ or $q = Q(k) = kV(k)$. This model is appealing because of its simplicity, parsimony, and its robustness to replicate basic traffic features. Its solution is usually computed with the Godunov scheme [God59], which is based on iterative solutions of Riemann problems [Dag94] [Leb96]. This scheme is known to introduce important numerical viscosity.

Alternatively, Newell [New93] proposed the use of cumulative count function $N(x,t)$ and conjectured that the LWR solution is the lower envelope of the multiple values of $N(x,t)$ that can be obtained from proper boundary and initial data. Recently, Daganzo [Dag05] proved Newell's conjecture using variational theory, which reduces the LWR model to the solution of the Hamilton–Jacobi equation:

$$\partial_t N = Q(-\partial_x N) \tag{2}$$

derived from $q = Q(k) = kV(k)$ since $k = -\partial_x N$ and $q = \partial_t N$.
Variational theory opens the door to powerful numerical methods based on shortest-path algorithms in (x,t) coordinates [Dag05b].

The aim of this paper is to go further into the use of the N-function by formulating the LWR model in the transformed coordinate system (N,t). These Lagrangian coordinates are fixed to a given fluid particle and move with it in space-time. In this new coordinate system, the purpose is no longer to determine the local density k but the position $X(n,t)$ of vehicle number n. Note that in the continuum, n is not necessarily an integer. In the remainder of the paper, capital N (respectively X) will stand for the $N(x,t)$ (respectively $X(n,t)$) function while n (respectively x) will define a value taken by this function.

This paper is organized as follow: Sect. 2 formulates the LWR model in Lagrangian coordinates as a conservation law and as a variational principle. Section 3 derives relevant numerical schemes. Finally, Sect. 4 presents a brief discussion.

2 The LWR Model

2.1 Lagrangian Conservation Law

The conservation law in Lagrangian coordinates was introduced by Courant and Friedrichs [CF48] in the case of gas dynamics. In the case of traffic flow, (1) becomes

$$\partial_t s + \partial_n v = 0, \tag{3}$$

where the spacing s corresponds to $1/k$.
The corresponding fundamental diagram V^* can be expressed as a concave function of s, $v = V^*(s) = V(1/k)$. Therefore, the LWR model in Lagrangian coordinates corresponds to the following hyperbolic equation in s:

$$\partial_t s + \partial_n V^*(s) = 0. \tag{4}$$

Wagner [Wag87] has proven the equivalence between (4) and (1) for weak solutions, even in vacuum cases where s is not defined; i.e., when $k = 0$.

2.2 Lagrangian Variational Principle

The reader should refer to [Dag05] and [Dag05b] for a complete description of this theory in Eulerian coordinates. For simplicity, we will study here homogeneous problems but inhomogeneous ones can be treated similarly as in [Dag05].

What follows shows that the transformation to Lagrangian coordinates preserves the nature of the problem; i.e., that a partial differential equation similar to (2) has to be solved, i.e.,

$$\partial_t X = V^*\left(-\partial_n X\right) \text{ derived from } v = V(1/k) = V^*(s). \tag{5}$$

To prove the existence of $\partial_t X$ and $\partial_n X$, we only need to show that X exists and is differentiable. To this end, we note that the function $X(n,t)$ can be obtained by inverting $N(x,t)$; i.e., solving for x in $n = N(x,t)$. Two cases may arise:

1. Non-vacuum ($k = \partial_x N \neq 0$): in this case, N is continuous and strictly decreasing in space. Hence, $N(x,t)$ is bijective and the inversion is possible.

Thus, $X(n,t)$ exists, is continuous, and strictly increasing in n. Furthermore, as N is differentiable except on shockwaves, X verifies the same property.

2. Vacuum ($\partial_x N = 0$ and $\partial_n X = +\infty$): this case corresponds to step-jumps in the X profile with respect to n, i.e., voids in traffic flow. Intuitively, these jumps should not be a problem because a void separates two independent LWR problems: the solution of one does not influence the solution of the other. Reassuringly, Wagner's results [Wag87] imply that the general problem remains well-posed even when N is not invertible.

Therefore, it is possible to formulate the Lagrangian variational principle analogously to the Eulerian one; i.e., it can be treated similarly as in [Dag05]. One just has to transpose variables using Table 1.

All the results proven in [Dag05] and [Dag05b] can thus be applied to the Lagrangian variational formulation of the LWR model. Notably, the value of X at a point P in the (n,t) plane, X_P, can be expressed as a least-cost path problem:

$$\begin{array}{l} X_P = \min\left(B_\wp + \Delta\left(\wp\right) : \forall \wp \in V \cap P_P\right), \text{ where} \\ \left| \begin{array}{l} V\text{: set of all valid paths} \\ P_P\text{: set of all path from the boundary condition to } P \\ B_\wp\text{: } X \text{ value at the beginning of the path} \\ \Delta\left(\wp\right)\text{: cost of path } \wp. \end{array} \right. \end{array} \tag{6}$$

Analogously to [Dag05], "waves" in (n,t) coordinates are characteristics where s is constant, they have velocity $u = \partial_s V^*(s)$ representing a passing rate. We define two types of passing rates: (a) u is a "possible passing rate" if there exists s such that $u = \partial_s V^*(s)$; (b) $\hat{u}$ is an "allowable passing rate" if $\min(\partial_s V^*(s)) \leq \hat{u} \leq \max(\partial_s V^*(s))$. "Valid paths" are continuous and piecewise differentiable paths $n(t)$ in the (n,t) plane whose slopes $n'(t)$ are allowable passing rates. 'Wave paths' are valid paths whose slopes are possible passing rates and are thus composed of a succession of waves.

The cost rate r on a wave path is given by $\mathbf{d}_t X$. The scalar r represents the speed of the Eulerian characteristic associated to the passing rate u.

$$r = \mathrm{d}_t X = \partial_t X + \partial_n X \partial_t n = v - su. \tag{7}$$

Table 1. Correspondence between Eulerian and Lagrangian variational principles

	Eulerian coordinates	Lagrangian coordinates
Unknown function	N	X
Main variable	$k = -\partial_x N$	$s = -\partial_n X$
Flux	$q = \partial_t N$	$v = \partial_t X$
FD	$Q(k)$	$V^*(s)$

As (5) holds and V^* is concave, one can express r only has a function $R(u)$ using the Legendre transformation as in [Dag05]:

$$r = R(u) = \sup_s \left(V^*(s) - su\right). \tag{8}$$

The cost on a Lagrangian valid path $\wp$ from B to P is thus

$$\Delta(\wp) = \int_{t_B}^{t_P} R(n'(t)). \tag{9}$$

3 Numerical Resolution

In this section, numerical scheme derived from each formulation of the LWR model in Lagrangian coordinates will be proposed. The equivalence between both numerical schemes will also be proven under specific assumptions. We will also show that the numerical schemes are exact in that latter case.

3.1 Godunov Scheme

In the Godunov scheme, the spacing s is approximated by a constant value between n and $n + \Delta n$ and is calculated at every time step Δt; see Fig. 1. Since the flux function V^* for (4) is increasing in s, the characteristic speed is always nonnegative (traffic anisotropy). The Godunov method reduces in this case to the upwind method:

$$s_i^{t+\Delta t} = s_i^t + \frac{\Delta t}{\Delta n}\left(V^*(s_i^t) - V^*(s_{i-1}^t)\right). \tag{10}$$

The Courant–Friedrichs–Lewy's (CFL) condition (11) defines the stability domain of (10). As the Godunov scheme is consistent and conservative, this also guarantees that it converges [Lev92]

$$\Delta n \geq \max_s |\partial_s \left(V^*(s)\right)| \Delta t. \tag{11}$$

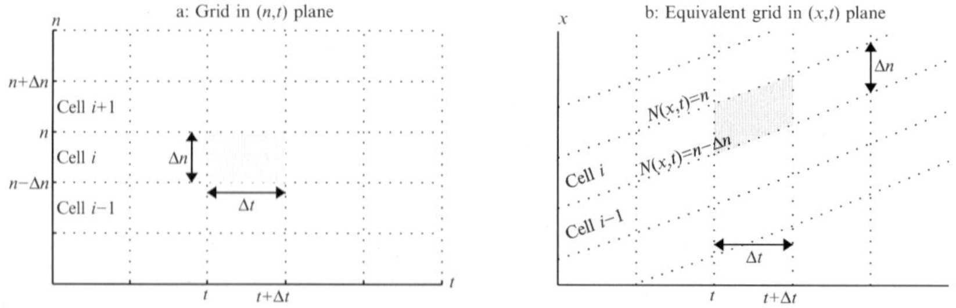

Fig. 1. Lagrangian grid

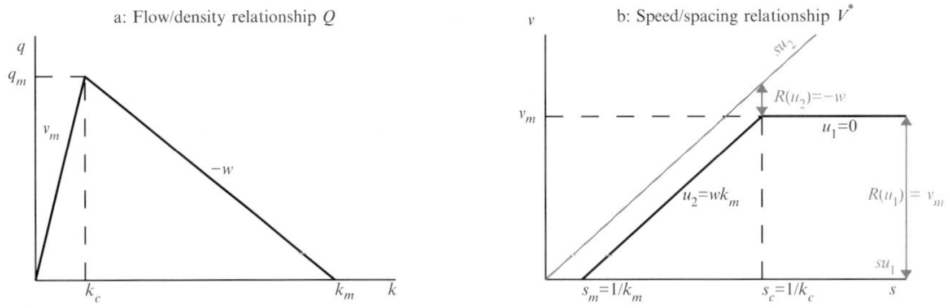

Fig. 2. Triangular fundamental diagram

Notice that nonnegative wave speeds imply that Lagrangian rarefaction waves never influence flux values at cell boundaries. The entropy condition is not necessary to define the flux value at cell boundaries. This condition is therefore naturally handled in the numerical scheme.

The Lagrangian Godunov scheme can also be expressed in terms of $X(n,t)$ by noting that the flux $V^*(s_i^t)$ at a boundary n of a cell i is given by (12)

$$\frac{X(n,t+\Delta t)-X(n,t)}{\Delta t}=V^*(s_i^t)=V^*\left(-\frac{X(n,t)-X(n-\Delta n,t)}{\Delta n}\right). \quad (12)$$

If we suppose now that Q is triangular as in Fig. 2a, V^* can be expressed as

$$V^*(s)=\min(v_m, wk_m s-w), \quad (13)$$

where v_m is the free-flow speed, w is the wave speed, and k_m is the maximal density; see Fig. 2b. After simplification (12) becomes

$$X(n,t+\Delta t)=\min\left(X(n,t)+v_m\Delta t,(1-\alpha)X(n,t)+\alpha X(n-\Delta n,t)-w\Delta t\right) \quad (14)$$

where $\alpha = wk_m\Delta t/\Delta n$.

If the CFL condition (11) is satisfied as an equality, then $\Delta n = wk_m\Delta t$ and $\alpha = 1$; see Fig. 2b. In this case, (14) reduces to

$$X(n,t+\Delta t)=\min(X(n,t)+v_m\Delta t, X(n-\Delta n,t)-w\Delta t). \quad (15)$$

When n is an integer and $\Delta n = 1$ then $X(n,t)$ corresponds to the position x_n^t of vehicle n at time t and $X(n-1,t)$ to the position x_{n-1}^t of its leader at the same time. Notice that (15) reduces to Newell's simplified car-following model [New02]:

$$x_n^{t+\Delta t}=\min\left(x_n^t+\frac{v_m}{wk_m}, x_{n-1}^t-\frac{1}{k_m}\right). \quad (16)$$

We will show that this scheme is exact using the Lagrangian variational principle.

3.2 Lagrangian Variational Principle

Daganzo [Dag05b] proposed efficient methods to solve the LWR model using the concept of "sufficient networks." A "network" is defined as directed graph of nodes and arcs in the considered plane (Eulerian or Lagrangian), where arcs are valid paths. A network is "sufficient" when the least-cost path through the network between every valid pair of nodes is an optimum path. A pair of nodes is said to be "valid" if a valid path exists between them. Notice that this valid path may not necessarily be included in the network. An "optimum path" between a valid pair of nodes is a least-cost valid path between these two points. Notice, again, that this optimum path may not be included in the network and this may introduce errors. In a sufficient network, the solution is exact at every node provided that the initial data is linear between two consecutive initial nodes.

Next, we will apply this method in Lagrangian coordinates to the variational principle (5) supposing that Q is triangular. In this case, waves have only two possible velocities: $u_1 = 0$ (free-flow) and $u_2 = wk_m$ (congestion). The resulting cost rates (8) are $R(u_1) = v_m$ and $R(u_2) = -w$; see Fig. 2b. It can be shown, following the derivation in [Dag05b], that any geometric network formed by two families of parallel equidistant lines with slopes u_1 and u_2 and separated by Δn_1 and Δn_2 is sufficient; see Fig. 3a. Therefore, with appropriate initial data, the solution at nodes is exact.

Since $u_1 = 0$ nodes are always lined-up along rows where n values are constant. Furthermore, if one sets $\Delta n_1 = \Delta n_2 = \Delta n$, nodes also line-up along "time-columns"; see Fig. 3b. This defines a rectangular lattice in the (n,t) plane with $\Delta t = \Delta n / wk_m$, which is very practical for computational implementation. Furthermore, with only two incoming arcs per node, the computation of the optimal value of X (given by (6)) at each node is straightforward:

$$\begin{aligned} X(n, t + \Delta t) &= \min\big(X(n,t) + \Delta_{(n,t)\to(n,t+\Delta t)}, X(n - \Delta n, t) \\ &\quad + \Delta_{(n-\Delta n,t)\to(n,t+\Delta t)}\big) \\ &= \min(X(n,t) + v_m \Delta t, X(n - \Delta n, t) - w\Delta t), \end{aligned} \tag{17}$$

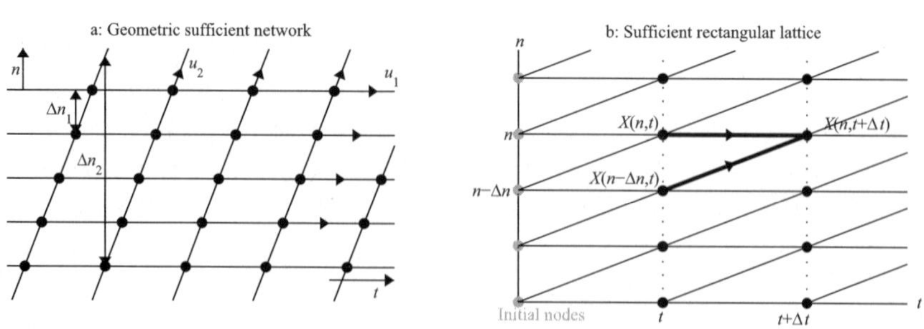

Fig. 3. Geometric networks associated to the Lagrangian variational principle

where $\Delta_{(n,t)\to(n',t')}$ is the cost of the arc between the two grid-points (n,t) and (n',t').

Notice that (17) and (15) are similar. Therefore, the Godunov and the variational schemes are equivalent when Q is triangular and the CFL condition (11) is satisfied as an equality; i.e., $\Delta n = wk_m \Delta t$. In this case, the Godunov scheme computes the exact solution and no numerical viscosity appears.

4 Discussion

From the authors' point of view, the main advantage of the Lagrangian approach is its exactitude when Q is triangular. This is important because a triangular Q is parsimonious while being an accurate representation of reality.

Table 2 presents a summary of the exactitude of the Lagrangian approaches compared to Eulerian ones. At a glance, the superiority of the Lagrangian approach is apparent as it remains exact in all cases contrary to the Godunov scheme when Q is piecewise linear with more than two wave speeds (PWL). We note that when Q is PWL the Eulerian variational numerical method introduces errors because one is forced to introduce horizontal arcs (see [Dag05b] Sect. 3.1), which are not necessarily wave paths. Interestingly, in the Lagrangian counterpart, horizontal arcs are also needed but correspond to wave paths. Therefore, no numerical errors are introduced. Another interesting advantage of the Lagrangian variational method is that it does not require $v_m/w = j$ to be an integer in order to have a rectangular lattice; in the Eulerian counterpart, this is not only necessary but one has to memorize for the j time-steps preceding the current simulation time.

Further research is needed to see how major extensions of the LWR model (moving bottlenecks, intersections, multipipe representation. . .) can be formulated in Lagrangian coordinates.

Table 2. Exactitude by approach ($\surd$, exact; X, nonexact)

	Lagrangian approach		Eulerian approach	
	Godunov scheme	Variational method	Godunov scheme	Variational method
Q triangular + rectangular lattice	$\surd$	$\surd$	X	$\surd$ only if v_m/w is an integer and memory is used
Q triangular	$\surd$	$\surd$	X	$\surd$ only for geometric networks
Q PWL	X	$\surd$	X	X

Acknowledgment

The authors thank Professor Carlos Daganzo for suggesting the use of variational theory in Lagrangian coordinates. This work has been partially supported by the French ACI-NIM (Nouvelles Interactions des Mathématiques) No 193 (2004).

References

[CF48] Courant, R., Friedrichs, K.O.: Supersonic Flows and Shock Waves. Pure Appl. Math, **1**, (1948)

[Dag05] Daganzo, C.F.: A variational formulation of kinematic waves: basic theory and complex boundary conditions. Transportation Research B, **39**(2), 187–196 (2005)

[Dag05b] Daganzo, C.F.: A variational formulation of kinematic waves: Solution methods. Transportation Research B, **39**(10), 934–950 (2005b)

[Dag94] Daganzo, C.F.: The cell transmission model: A dynamic representation of highway traffic consistent with the hydrodynamic theory. Transportation Research B, **28**(4), 269–287 (1994)

[DM05] Daganzo, C.F., Menendez, M.: A variational formulation of kinematic waves: bottlenecks properties and examples. In: Mahmassani, H.S. (ed) ISTTT. Pergamon, London, 345–364 (2005)

[God59] Godunov, S.K.: A difference scheme for numerical computation of discontinuous solutions of equations of fluid dynamics. Mat. Sb., **47**, 271–290 (1959).

[Leb96] Lebacque, J.P.: The Godunov scheme and what it means for first order traffic flow models. In: Lesort, J.B. (ed) ISTTT. Pergamon, London, 647–678 (1996)

[Lev92] Leveque, R.J.: Numerical methods for conservation laws. 2nd Edition. Bâle, Switzerland, Birkhäuser, 214 p. (1992)

[LW55] Lighthill, M.J., Whitham, J.B.: On kinematic waves II: A theory of traffic flow in long crowded roads. Proceedings of the Royal Society, **A229**, 317–345 (1955)

[New02] Newell, G.F.: A simplified car-following theory: a low-order model. Transportation Research B, **36**(3), 195–205 (2002)

[New93] Newell, G.F.: A simplified theory of kinematic waves in highway traffic, part I General Theory. Transportation Research B, **27**(4), 281–287 (1993)

[Ric56] Richards, P.I.: Shockwaves on the highway. Operations Research, **4**, 42–51 (1956)

[Wag87] Wagner, D.: Equivalence of the Euler and Lagrangian equations of gas dynamics for weak solutions. J. Diff. Eq., **68**, 118–136 (1987)

Hyperbolic Conservation Laws and Spacetimes with Limited Regularity

P.G. LeFloch

1 Introduction

Hyperbolic conservation laws posed on manifolds arise in many applications to geophysical flows and general relativity. Recent work by the author and his collaborators attempts to set the foundations for a study of weak solutions defined on Riemannian or Lorentzian manifolds and includes an investigation of the existence and qualitative behavior of solutions. The metric on the manifold may either be fixed (shallow water equations on the sphere, for instance) or be one of the unknowns of the theory (Einstein–Euler equations of general relativity). This work is especially concerned with solutions and manifolds with limited regularity. We review here results on three themes: (1) Shock wave theory for hyperbolic conservation laws on manifolds, developed jointly with M. Ben-Artzi (Jerusalem); (2) Existence of matter Gowdy-type spacetimes with bounded variation, developed jointly with J. Stewart (Cambridge). (3) Injectivity radius estimates for Lorentzian manifolds under curvature bounds, developed jointly with B.-L. Chen (Guang-Zhou).

2 Conservation Laws on a Riemannian Manifold

In the present section, $(\mathbf{M}^n, g)$ is a compact, oriented, n-dimensional Riemannian manifold. As usual, the tangent space at a point $x \in \mathbf{M}^n$ is denoted by $T_x\mathbf{M}^n$ and the tangent bundle by $T\mathbf{M}^n := \bigcup_{x\in\mathbf{M}^n} T_x\mathbf{M}^n$, while the cotangent bundle is denoted by $T^\star\mathbf{M}^n = T^\star_x\mathbf{M}^n$. The metric structure is determined by a positive-definite, two-covariant tensor field g.

A *flux* on the manifold $\mathbf{M}^n$ is a vector field $f = f_x(\bar{u})$ depending smoothly upon the parameter $\bar{u}$. The conservation law associated with f reads

$$\partial_t u + \mathrm{div}_g\,(f(u)) = 0, \tag{1}$$

where the unknown $u = u(t,x)$ is defined for $t \geq 0$ and $x \in \mathbf{M}^n$ and the divergence operator is applied to the vector field $x \mapsto f_x(u(t,x)) \in T_x\mathbf{M}^n$. We say that the flux is *geometry-compatible* if

$$\operatorname{div}_g f_x(\bar u) = 0, \qquad \bar u \in \mathbb{R},\ x \in \mathbf{M}^n. \tag{2}$$

We propose to single out this class of conservation laws as an important case of interest, which leads to robust L^p estimates that do not depend on the geometry of the manifold. The equations arising in continuum physics do satisfy this condition.

Equation (1) is a geometric partial differential equation that depends on the geometry of the manifold only. All estimates must take a coordinate-independent form; in the proofs, however, it is often convenient to introduce particular coordinate charts. We are interested in solutions $u \in L^\infty(\mathbb{R}_+ \times \mathbf{M}^n)$ assuming a prescribed initial condition $u_0 \in L^\infty(\mathbf{M}^n)$:

$$u(0,x) = u_0(x), \quad x \in \mathbf{M}^n. \tag{3}$$

We extend Kruzkov theory of the Euclidian space $\mathbb{R}^n$ (see [8]) to the case of a Riemannian manifold, as follows.

Let $f = f_x(\bar u)$ be a geometry-compatible flux on the Riemannian manifold (M,g). A *convex entropy/entropy-flux pair* is a pair (U,F) where $U : \mathbb{R} \to \mathbb{R}$ is convex and $F = F_x(\bar u)$ is the vector field defined by

$$F_x(\bar u) := \int^{\bar u} \partial_{u'} U(u')\, \partial_{u'} f_x(u')\, du', \qquad \bar u \in \mathbb{R},\ x \in \mathbf{M}^n.$$

Given $u_0 \in L^\infty(\mathbf{M}^n)$, a function $u \in L^\infty\big(\mathbb{R}_+, L^\infty(\mathbf{M}^n)\big)$ is called an *entropy solution* to the initial value problem (1), (3) if the following entropy inequalities hold

$$\begin{aligned}\int\!\!\int_{\mathbb{R}_+\times\mathbf{M}^n} \Big(U(u(t,x))\, \partial_t\theta(t,x) + g_x\big(F_x(u(t,x)), \operatorname{grad}_g \theta(t,x)\big)\Big)\, dV_g(x)dt \\ + \int_{\mathbf{M}^n} U(u_0(x))\, \theta(0,x)\, dV_g(x) \geq 0,\end{aligned}$$

for every convex entropy/entropy flux pair (U,F) and all smooth function $\theta = \theta(t,x) \geq 0$ compactly supported in $[0,\infty) \times \mathbf{M}^n$.

Theorem 1 (Well-posedness theory on a Riemannian manifold. I). *Let $f = f_x(\bar u)$ be a geometry-compatible flux on a Riemannian manifold $(\mathbf{M}^n, g)$. Given $u_0 \in L^\infty(\mathbf{M}^n)$, there exists a unique entropy solution $u \in L^\infty(\mathbb{R}_+ \times \mathbf{M}^n)$ to the problem (1)–(3). Moreover, for each $1 \leq p \leq \infty$,*

$$\|u(t)\|_{L^p(\mathbf{M}^n; dV_g)} \leq \|u_0\|_{L^p(\mathbf{M}^n; dV_g)}, \qquad t \in \mathbb{R}_+,$$

and, given two entropy solutions u, v associated with initial data u_0, v_0,

$$\|v(t) - u(t)\|_{L^1(\mathbf{M}^n; dV_g)} \leq \|v_0 - u_0\|_{L^1(\mathbf{M}^n; dV_g)}, \qquad t \in \mathbb{R}_+.$$

The framework proposed here allows us to construct entropy solutions on a Riemannian manifold via the vanishing diffusion method or the finite volume method [1, 4]. Following DiPerna [6] we can introduce the (larger) class of entropy measure-valued solutions $(t, x) \in \mathbb{R}_+ \times \mathbf{M}^n \mapsto \nu_{t,x}$.

Theorem 2 (Well-posedness theory on a Riemannian manifold. II). *Let $f = f_x(\bar{u})$ be a geometry-compatible flux on a Riemannian manifold $(\mathbf{M}^n, g)$. Let ν be an entropy measure-valued solution to (1)–(3) for some $u_0 \in L^\infty(\mathbf{M}^n)$. Then, for almost every (t, x), $\nu_{t,x} = \delta_{u(t,x)}$, where $u \in L^\infty(\mathbb{R}_+ \times \mathbf{M}^n)$ is the unique entropy solution to the problem.*

Finally we can relax the geometry compatibility condition and consider a general conservation law associated with an arbitrary flux f. More general conservation laws solely enjoy the L^1 contraction property and leads to a unique contractive semigroup of entropy solutions.

Theorem 3 (Well-posedness theory on a Riemannian manifold. III). *Let $f = f_x(\bar{u})$ be an arbitrary (not necessarily divergence-free) flux on $(\mathbf{M}^n, g)$, satisfying the linear growth condition*

$$|f_x(\bar{u})|_g \lesssim 1 + |\bar{u}|, \qquad \bar{u} \in \mathbb{R},\ x \in \mathbf{M}^n.$$

Then there exists a unique contractive, semigroup of entropy solutions $u_0 \in L^1(\mathbf{M}^n) \mapsto u(t) := S_t u_0 \in L^1(\mathbf{M}^n)$ to the initial value problem (1), (3).

For the proofs we refer to [4]. See [1] for the convergence of the finite volume schemes on a manifold. Earlier material can be found in Panov [12] (n-dimensional manifold) and in LeFloch and Nedelec [10] (Lax formula for general metrics including the case of spherical symmetry).

3 Conservation Laws on a Lorentzian Manifold

Motivated by the application to general relativity, we can extend the theory to a Lorentzian manifold. Let $(\mathbf{M}^{n+1}, g)$ be a time-oriented, $(n+1)$-dimensional Lorentzian manifold, g being a metric tensor with signature $(-, +, \dots, +)$. Tangent vectors X can be separated into time-like vectors ($g(X, X) < 0$), null vectors ($g(X, X) = 0$), and space-like vectors ($g(X, X) > 0$). The null cone separates time-like vectors into future-oriented and past-oriented ones. Let ∇ be the Levi–Cevita connection associated with the Lorentzian metric g.

A *flux* on the manifold $\mathbf{M}^{n+1}$ is a vector field $x \mapsto f_x(\bar{u}) \in T_x\mathbf{M}^{n+1}$, depending on a parameter $\bar{u} \in \mathbb{R}$. The conservation law on $(\mathbf{M}^{n+1}, g)$ associated with f is

$$\mathrm{div}_g\,\big(f(u)\big) = 0, \qquad u : \mathbf{M}^{n+1} \to \mathbb{R}. \tag{4}$$

It is said to be *geometry compatible* if

$$\mathrm{div}_g\, f_x(\bar{u}) = 0, \quad \bar{u} \in \mathbb{R},\ x \in \mathbf{M}^{n+1}. \tag{5}$$

Furthermore, f is said to be a *time-like flux* if $g_x\big(\partial_u f_x(\bar u), \partial_u f_x(\bar u)\big) < 0$, $x \in \mathbf{M}^{n+1}$, $\bar u \in \mathbb{R}$.

Note that our terminology here differs from the one in the Riemannian case, where the conservative variable was singled out. We are interested in the initial-value problem associated with (4). We fix a space-like hypersurface $\mathcal{H}_0 \subset \mathbf{M}^{n+1}$ and a measurable and bounded function u_0 defined on $\mathcal{H}_0$. Then, we search for $u = u(x) \in L^\infty(\mathbf{M}^{n+1})$ satisfying (4) in the distributional sense and such that the (weak) trace of u on $\mathcal{H}_0$ coincides with u_0:

$$u_{|\mathcal{H}_0} = u_0. \tag{6}$$

It is natural to require that the vectors $\partial_u f_x(\bar u)$ are time-like and future-oriented.

We assume that the manifold $\mathbf{M}^{n+1}$ is *globally hyperbolic,* in the sense that there exists a foliation of $\mathbf{M}^{n+1}$ by space-like, compact, oriented hypersurfaces $\mathcal{H}_t$ $(t \in \mathbb{R})$: $\mathbf{M}^{n+1} = \bigcup_{t\in\mathbb{R}} \mathcal{H}_t$. Any hypersurface $\mathcal{H}_{t_0}$ is referred to as a Cauchy surface in $\mathbf{M}^{n+1}$, while the family $\mathcal{H}_t$ $(t \in \mathbb{R})$ is called an admissible foliation associated with $\mathcal{H}_{t_0}$. The future of the given hypersurface will be denoted by $\mathbf{M}^{n+1}_+ := \bigcup_{t\geq 0} \mathcal{H}_t$. Finally, we denote by n^t the future-oriented, normal vector field to each $\mathcal{H}_t$, and by g^t the induced metric. Finally, along $\mathcal{H}_t$, we denote by X^t the normal component of a vector field X, thus $X^t := g(X, n^t)$.

A flux $F = F_x(\bar u)$ is called a *convex entropy flux* associated with the conservation law (4) if there exists a convex $U : \mathbb{R} \to \mathbb{R}$ such that

$$F_x(\bar u) = \int_0^{\bar u} \partial_u U(u')\, \partial_u f_x(u')\, du', \qquad x \in \mathbf{M}^{n+1},\ \bar u \in \mathbb{R}.$$

A measurable and bounded function $u = u(x)$ is called an *entropy solution* of the geometry-compatible conservation law (4)–(5) if

$$\int_{\mathbf{M}^{n+1}_+} g(F(u), \operatorname{grad}_g {}_g\theta)\, dV_g + \int_{\mathcal{H}_0} g_0(F(u_0), n_0)\, \theta_{\mathcal{H}_0}\, dV_{g_0} \geq 0.$$

for all convex entropy flux $F = F_x(\bar u)$ and all smooth $\theta \geq 0$ compactly supported in $\mathbf{M}^{n+1}_+$.

Theorem 4 (Well-posedness theory on a Lorentzian manifold). *Consider a geometry-compatible conservation law (4)–(5) posed on a globally hyperbolic Lorentzian manifold $\mathbf{M}^{n+1}$. Let $\mathcal{H}_0$ be a Cauchy surface in $\mathbf{M}^{n+1}$, and $u_0 : \mathcal{H}_0 \to \mathbb{R}$ be measurable and bounded. Then, the initial-value problem (4)–(6) admits a unique entropy solution $u = u(x) \in L^\infty(\mathbf{M}^{n+1})$. For every admissible foliation $\mathcal{H}_t$, the trace $u_{\mathcal{H}_t}$ exists and belong to $L^1(\mathcal{H}_t)$, and $\|F^t(u_{\mathcal{H}_t}\|_{L^1(\mathcal{H}_t)}$ is nonincreasing in time, for any convex entropy flux F. Moreover, given any two entropy solutions u, v,*

$$\|f^t(u_{\mathcal{H}_t}) - f^t(v_{|\mathcal{H}_t})\|_{L^1(\mathcal{H}_t)}$$

is nonincreasing in time.

We emphasize that, in the Lorentzian case, no time-translation property is available in general, contrary to the Riemannian case. Hence, no time-regularity is implied by the L^1 contraction property.

4 Existence of Matter Gowdy-Type Spacetimes with Bounded Variation

Vacuum Gowdy spacetimes are inhomogeneous spacetimes admitting two commuting spatial Killing vector fields. The existence of *vacuum* spacetimes with Gowdy symmetry is well-known and the long-time asymptotics of solutions have been found to be particularly complex. In comparison, much less emphasis has been put on *matter* spacetimes. Recently, LeFloch et al. [3, 11] initiated a rigorous mathematical treatment of the coupled Einstein–Euler system on Gowdy spacetimes. The unknowns of the theory are the density and velocity of the fluid together with the components of the metric tensor. The existence for the Cauchy problem in the class of solutions with (arbitrary large) bounded total variation is proven by a generalization of the Glimm scheme. Our theory allows for the formation of shock waves in the fluid and singularities in the geometry. The first results on shock waves and the Glimm scheme on special and general relativity are due to Smoller and Temple [14] (flat Minkowski spacetime) and Groah and Temple [7] (spherically symmetric spacetimes). The novelty in [3, 11] is the generalization to a model allowing for both gravitational waves and shock waves.

The metric is given in the polarized Gowdy symmetric form

$$ds^2 = e^{2a}\,(-dt^2 + dx^2) + e^{2b}\,(e^{2c}\,dy^2 + e^{-2c}\,dz^2), \tag{7}$$

where the variables a, b, c depend on the time variable t and the space variable x, only. We consider *Einstein field equations* $G^{\alpha\beta} = \kappa T^{\alpha\beta}$ for perfect fluids with energy density $\mu > 0$ and pressure $p = \mu c_s^2$. Here, the sound speed c_s is a constant with $0 < c_s < 1$ and $G^{\alpha\beta}$ denotes the Einstein tensor, while κ is a normalization constant.

The four-velocity vector u^α of the fluid is time-like and is normalized to be of unit length and we define the scalar velocity v and relativistic factor $\xi = \xi(v)$ by $(u^\alpha) = e^{-a}\,\xi\,(1, v, 0, 0)$ and $\xi = (1 - v^2)^{-1/2}$. The matter is described by the energy–momentum tensor $T^{\alpha\beta} = (\mu + p)\,u^\alpha u^\beta + p\,g^{\alpha\beta}$, from which we extract the fields τ, S and Σ:

$$\begin{aligned}
T^{00} &= e^{-2a}\big((\mu + p)\,\xi^2 - p\big) =: e^{-2a}\tau,\\
T^{01} &= e^{-2a}(\mu + p)\,\xi^2 v =: e^{-2a} S,\\
T^{11} &= e^{-2a}\big((\mu + p)\xi^2 v^2 + p\big) =: e^{-2a}\,\Sigma.
\end{aligned}$$

After very tedious calculations we arrive at the *constraint equations*

$$\begin{aligned} 2\,a_t\,b_t + 2\,a_x\,b_x + b_t^2 - 2\,b_{xx} - 3\,b_x^2 - c_t^2 - c_x^2 &= \kappa\,e^{2a}\,\tau, \\ -2\,a_t\,b_x - 2\,a_x\,b_t + 2\,b_{tx} + 2\,b_t\,b_x + 2\,c_t\,c_x &= \kappa\,e^{2a}\,S, \end{aligned}$$

and the *evolution equations*

$$\begin{aligned} a_{tt} - a_{xx} &= b_t^2 - b_x^2 - c_t^2 + c_x^2 + \frac{\kappa}{2}\,e^{2a}\,(-\tau + \Sigma - 2\,p), \\ b_{tt} - b_{xx} &= -2b_t^2 + 2b_x^2 + \frac{\kappa}{2}\,e^{2a}\,(\tau - \Sigma), \\ c_{tt} - c_{xx} &= -2\,b_t\,c_t + 2\,b_x\,c_x. \end{aligned}$$

The evolution equations for the fluid, $\nabla_\beta T^{\alpha\beta} = 0$, are the *Euler equations*

$$\tau_t + S_x = T_1, \qquad S_t + \Sigma_x = T_2,$$

in which the source terms T_1, T_2 are nonlinear in first-order derivatives of the metric and fluid variables.

We propose to reformulate the Einstein–Euler equations in the form of a nonlinear hyperbolic system of balance laws with integral source-term, in the variables (μ, v) and $w := \big(a_t, a_x, \beta_t, \beta_x, c_t, c_x\big)$, where $\beta = e^{2b}$. It is convenient to also set $\alpha = e^{2a}$. The functions α, b (and a, β) are determined by

$$\begin{aligned} \alpha(t,x) &= e^{2a(t,x)}, & a(t,x) &= \textstyle\int_{-\infty}^{x} w_2(t,y)\,dy, \\ b(t,x) &= \tfrac{1}{2}\,\ln\beta(t,x), & \beta(t,x) &= 1 + \textstyle\int_{-\infty}^{x} w_4(t,y)\,dy. \end{aligned}$$

Obviously, we are interested in solutions such that β remains positive.

The equations under consideration consist of three sets of two equations associated with the propagation speeds ± 1, the speed of light (after normalization). The principal part of the fluid equations are the standard relativistic fluid equations in a Minkowski background, with wave speeds $\lambda_\pm = (v \pm c_s)/(1 \pm v\,c_s)$. To formulate the initial-value problem it is natural to prescribe the values of μ, v, w on the initial hypersurface at $t = 0$, denoted by (μ^0, v^0, w^0).

Our main result is as follows:

Theorem 5 (Existence of Gowdy spacetimes with compressible matter). *Consider the (μ, v, w)-formulation of the Einstein–Euler equations on a polarized Gowdy spacetime with plane-symmetry. Let the initial data (μ^0, v^0, w^0) be of bounded total variation, $TV(\mu^0, v^0, w^0) < \infty$, satisfying the constraints, and suppose that the corresponding functions α^0, b^0 are measurable and bounded, $\sup|(\alpha^0, b^0)| < \infty$. Then the Cauchy problem admits a weak solution μ, v, w such that for some increasing $C(t)$*

$$TV(\mu, v, w)(t) + \sup|(\alpha, b)(t, \cdot)| \leqslant C(t), \qquad t \geq 0,$$

and are defined up to a maximal time $T \leqslant \infty$. If $T < \infty$ then either the geometry variables α, b blow up: $\lim_{t\to T}\big(\sup_{\mathbb{R}}|\alpha(t,\cdot)| + |b(t,\cdot)|\big) = \infty$, or the energy density blows up: $\lim_{t\to T}\sup_{\mathbb{R}}|\mu(t,\cdot)| = \infty$.

Hence, the solution exists until either a singularity occurs in the geometry (e.g., the area β of the two-dimensional space-like orbits of the symmetry group vanishes) or the matter collapses to a point. To our knowledge this is the first *global* existence result for the Euler–Einstein equations.

If a shock wave forms in the fluid, then μ, v will be discontinuous and, as a consequence, w_{3x} and w_{4x} might also be discontinuous. In fact, Theorem 5 allows not only such discontinuities in second-order derivatives of the geometry components (i.e., at the level of the curvature), but also discontinuities in the first-order derivatives that propagate at the speed of light. The latter correspond to *Dirac distributions* in the curvature of the metric.

5 Lower Bounds on the Injectivity Radius of Lorentzian Manifolds

Motivated by the application to spacetimes of general relativity and by earlier results established by Anderson [2] and Klainerman and Rodnianski [13], we investigate in [5] the geometry and regularity of $(n+1)$-dimensional Lorentzian manifolds (M, g). Under curvature and volume bounds we establish new injectivity radius estimates, which are valid either in arbitrary directions or in null cones. Our estimates are purely local and are formulated via the "reference" Riemannian metric $\widehat{g}_T$ associated with an arbitrary future-oriented time-like vector field T.

Our proofs are based on suitable generalizations of arguments from Riemannian geometry and rely on the observation that geodesics in the Euclidian and Minkowski spaces coincide, so that estimates for the reference Riemannian metric can be carried over to the Lorentzian metric. Our estimates should be useful to investigate the qualitative behavior of spacetimes satisfying Einstein field equations.

We state here one typical result from [5] encompassing a large class of Lorentzian manifolds. Fix a point $p \in M$, and let us assume that a domain $\Omega \subset M$ containing p is foliated by spacelike hypersurfaces Σ_t with normal T, say $\Omega = \bigcup_{t\in[-1,1]} \Sigma_t$. Assume also that the geodesic ball $B_{\Sigma_0}(p,1) \subset \Sigma_0$ is compactly contained in Σ_0. Consider the following assumptions where K_0, K_1, K_2, and v_0 are positive constants:

$$K_0 \leq -\left|\frac{\partial}{\partial t}\right|_g^2 \leq 1/K_0 \quad \text{in } \Omega, \tag{8}$$

$$|\mathcal{L}_T g|_{\widehat{g}_T} \leq K_1 \quad \text{in } \Omega, \tag{9}$$

$$|\mathrm{Rm}_g|_{\widehat{g}_T} \leq K_2 \quad \text{in } \Omega, \tag{10}$$

$$\mathrm{vol}_g(B_{\Sigma_0}(p,1)) \geq v_0, \tag{11}$$

We prove in [5]:

Theorem 6 (Injectivity radius estimate for Lorentzian manifolds). *Let* (M, g) *be a Lorentzian manifold satisfying* (8)–(11) *at a point* $p \in M$. *Then, there exists a positive constant* i_0 *depending on the foliation bounds* K_0, K_1, *curvature bound* K_2, *volume bound* v_0, *and dimension* n *so that the injectivity radius at* p *is bounded below by* i_0, *that is* $Inj(M, g, p) \geq i_0$.

Acknowledgments

The author was partially supported by the A.N.R. grant 06-2-134423 entitled "Mathematical Methods in General Relativity" (MATH-GR).

References

1. Amorim, P., Ben-Artzi, M., LeFloch, P.G.: Hyperbolic conservation laws on manifolds. Total variation estimates and the finite volume method, Meth. Appli. Anal. **12** (2005), 291–324.
2. Anderson, M.T.: Regularity for Lorentz metrics under curvature bounds, Jour. Math. Phys. **44** (2003), 2994–3012.
3. Barnes, A.P., LeFloch, P.G., Schmidt, B.G., Stewart, J.M.: The Glimm scheme for perfect fluids on plane-symmetric Gowdy spacetimes, Class. Quantum Grav. **21** (2004), 5043–5074.
4. Ben-Artzi, M., LeFloch, P.G.: The well-posedness theory for geometry compatible hyperbolic conservation laws on manifolds, Ann. Inst. H. Poincaré, Nonlinear Anal., Arxiv preprint math.AP/0612846, 2006.
5. Chen B.-L., LeFloch, P.G.: Injectivity radius estimates for Lorentzian manifolds, Arxiv preprint math.AP/0612860, 2006.
6. DiPerna, R.J.: Measure-valued solutions to conservation laws, Arch. Rational Mech. Anal. **88** (1985), 223–270.
7. Groah, J.M., Temple, B.: Shock wave solutions of the Einstein equations with perfect fluid sources; existence and consistency by a locally inertial Glimm scheme, Mem. Amer. Math. Soc. **172** (2004).
8. Kruzkov, S.: First-order quasilinear equations with several space variables, Math. USSR Sb. **10** (1970), 217–243.
9. LeFloch, P.G.: Hyperbolic systems of conservation laws, Lectures in Mathematics, ETH Zürich, Birkhäuser, 2002.
10. LeFloch, P.G., Nedelec, J.-C.: Explicit formula for weighted scalar nonlinear conservation laws, Trans. Amer. Math. Soc. **308** (1988), 667–683.
11. LeFloch, P.G., Stewart, J.M.: Shock and gravitational waves in matter spacetimes with Gowdy symmetry, Portugal. Math. **62** (2005), 349–370.
12. Panov, E.Y.: On the Cauchy problem for a first-order quasilinear equation on a manifold, Differential Equations **33** (1997), 257–266.
13. Klainerman, S., Rodnianski, I.: On the radius of injectivity of null hypersurface, J. Amer. Math. Soc., Arxiv preprint math.DG/0603010, 2006.
14. Smoller, J.A., Temple, B.: Global solutions of the relativistic Euler equations, Commun. Math. Phys. **156** (1993), 67–99.

Arbitrary Lagrangian–Eulerian (ALE) Method in Cylindrical Coordinates for Laser Plasma Simulations

M. Kucharik, R. Liska, R. Loubere, and M. Shashkov

Summary. The Cartesian arbitrary Lagrangian–Eulerian (ALE) method is generalized to cylindrical coordinates and implemented on logically rectangular quadrilateral mesh. For laser plasma applicationsg the code is extended to model also laser absorption and heat conductivity. The code is used for simulation of high velocity impact problem for which pure Lagrangian method fails due to severe mesh distortion.

1 Introduction

For solving of compressible fluid equations, one can use two different approaches – the Eulerian and Lagrangian methods. In the Eulerian approach, the fluid flows through a static computational mesh, and in the Lagrangian approach, the computational mesh moves with the fluid; there is no mass flux between computational cells. The advantage of the Lagrangian approach is obvious – it can simulate situations when the investigated fluid considerably changes its volume (like compression or expansion) or its shape and when moving boundary conditions are treated naturally. This is exactly the situation in laser–plasma simulations, where the computational domain dramatically changes during target compression or corona expansion, and thus the Lagrangian approach is convenient. Unfortunately, the moving computational mesh can degenerate or even tangle during the simulation, it can lose its regularity, and the Lagrangian computation cannot continue as its assumptions on the mesh quality do not remain valid.

The Arbitrary Lagrangian–Eulerian (ALE) [1] method combines positives of both approaches to treat this problem. The computational mesh moves as in the Lagrangian approach, and so one can treat changing computational domains. When mesh quality decreases, the Eulerian part of the ALE method comes into play – the mesh is rezoned (smoothed) and the conservative quantities are conservatively remapped to a new, smoothed mesh. Then, the Lagrangian step can continue till the next mesh smoothing step.

In [2], we presented the ALE method in Cartesian geometry, working on quadrilateral 2D computational meshes. Here we extend this ALE method to the cylindrical $r-z$ geometry needed for modeling of laser plasma interactions. The laser plasma is modeled by the Lagrangian hydrodynamical equations

$$\frac{d\,\rho}{d\,t} + \rho\,\mathrm{div}\,\mathbf{u} = 0, \quad \rho\frac{d\,\mathbf{u}}{d\,t} + \mathbf{grad}\,p = 0$$
$$\rho\frac{d\,\epsilon}{d\,t} + p\,\mathrm{div}\,\mathbf{u} = -\mathrm{div}(\mathbf{I}) + \mathrm{div}(\kappa\,\mathbf{grad}\,T), \tag{1}$$

where ρ is density, $\mathbf{u}$ is velocity, p is pressure, ϵ is specific internal energy, T is temperature, κ is heat conductivity, and $\mathbf{I}$ is vector of laser intensity. The total Lagrangian time derivatives in this system include convective terms $d/d\,t = \partial/\partial\,t + \mathbf{u}\cdot\mathbf{grad}$. The movement of each node of the Lagrangian mesh is defined by an ordinary differential equation $d\,\mathbf{x}(t)/d\,t = \mathbf{u}$. The system is closed by the equation of state defining relations $p = p(\epsilon,\rho), T = T(\epsilon,\rho)$ between thermodynamical quantities ϵ, ρ, p, T. This system is numerically treated by splitting it into hyperbolic and parabolic (heat conductivity) parts.

2 Lagrangian Step

Lagrangian hyperbolic hydrodynamical equations are numerically treated by a compatible staggered method [3], which defines the scalar quantities (ρ, p, ϵ, T) inside grid cells and the vector quantities $(\mathbf{x}, \mathbf{u})$ in grid nodes. There are two versions of the compatible method in the cylindrical $r-z$ geometry [3]: Area Weighted (AW) and Control Volume (CV) methods. The AW method cannot be used in the ALE framework as it does not define well the cylindrical volumes needed during remapping and without these volumes we are not able to construct an AW compatible remapping.

The CV method defines the quadrilateral cell volume (based on Green's theorem)

$$V_c = \int_c 1\,r\,dr\,dz = \frac{1}{6}\sum_{l=1}^{4}(z_{l+1} - z_l)\left(r_l^2 + r_{l+1}^2 + r_l\,r_{l+1}\right), \tag{2}$$

where $(r_l, z_l), l = 1, \cdots, 4$ (with cyclic extension) are the coordinates of four nodes defining the cell c. However, the cylindrical CV method has to be modified to be compatible with ALE remapping. Originally the CV method defines cell center as the average of the cell nodes. The average is replaced by the mass center of the cell

$$r_c = \frac{1}{V_c}\int_c r^2\,dr\,dz, \quad z_c = \frac{1}{V_c}\int_c z\,r\,dr\,dz,$$

so that the remapping can be conservative. The compatible method is based on the zonal, subzonal, and viscosity forces in each grid node. The cell integrals appearing in the zonal and subzonal forces in cylindrical geometry contain additional factor r compared to the Cartesian geometry case. Using Green's theorem the cell integrals are transformed into the integrals over the cell edges, which are evaluated exactly (see further).

3 Rezoning and Remapping

The rezoning stage of the ALE method performs mesh smoothing. Most often we employ the simple Winslow smoothing technique [4] for rezoning, like in Cartesian geometry, even though more advanced methods as the reference Jacobian approach exist. Special boundary conditions in cylindrical geometry are applied during rezoning on the z axis so that the boundary nodes movement is constrained to this axis.

The remapping stage of the ALE method includes a conservative interpolation of the discrete conserved quantities from an old mesh to a new smoother one. The remapping stage consists of three steps: reconstruction, integration, and repair. First the remapped conservative function g (as e.g. density ρ) is reconstructed from the discrete values by a piecewise linear function on each old cell, typically with the Barth–Jespersen limiter. Then the reconstructed piecewise linear function is integrated over each new cell $\tilde{c}$ (objects related to the new mesh are accented by a tilde) to get the total value $G_{\tilde{c}} = \int_{\tilde{c}} g\, r dr\, dz$ of the conserved quantity (e.g. mass of the cell) inside the new cell, which defines the remapped density of conserved quantity $\tilde{g}_{\tilde{c}} = G_{\tilde{c}}/V_{\tilde{c}}$.

The natural exact integration of the piecewise linear function over the new cell requires computing intersections of the new cell with all neighboring old cells, see Fig. 1a where the new cell $\tilde{c}_{i,j} = \tilde{P}_{i,j}, \tilde{P}_{i+1,j}, \tilde{P}_{i+1,j+1}, \tilde{P}_{i,j+1}$ intersects with nine (3×3 patch) old cells $c_{k,l}, k = i-1, i, i+1, l = j-1, j, j+1$.

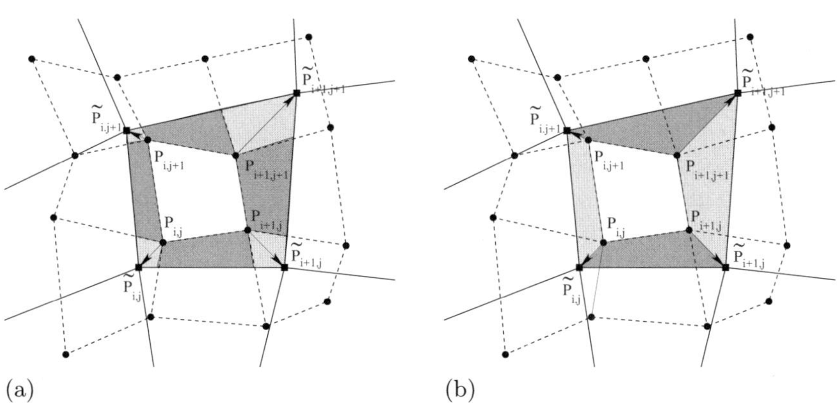

Fig. 1. Old (*dashed*) and new (*solid segments*) mesh with intersection regions for the exact integration (**a**) and swept regions for the approximate integration (**b**)

The linear reconstruction, given by the old cell c'

$$g(r,z) = g_{c'} + \left(\frac{\partial g}{\partial r}\right)_{c'} (r - r_{c'}) + \left(\frac{\partial g}{\partial z}\right)_{c'} (z - z_{c'}),$$

inside each such intersection $I_{c'}^{\tilde{c}}$ results in the contribution

$$\begin{aligned} G_{I_{c'}^{\tilde{c}}} = g_{c'} \int_{I_{c'}^{\tilde{c}}} r\,dr\,dz &+ \left(\frac{\partial g}{\partial r}\right)_{c'} \left(\int_{I_{c'}^{\tilde{c}}} r^2\,dr\,dz - r_{c'} \int_{I_{c'}^{\tilde{c}}} r\,dr\,dz\right) \\ &+ \left(\frac{\partial g}{\partial z}\right)_{c'} \left(\int_{I_{c'}^{\tilde{c}}} z\,r\,dr\,dz - z_{c'} \int_{I_{c'}^{\tilde{c}}} r\,dr\,dz\right) \end{aligned} \tag{3}$$

to the whole integral $G_{\tilde{c}}$. The integrals in this contribution over the polygonal intersection are transformed using Green's theorem into integrals over the edges of the polygon. The exact integration is computationally rather expensive because it requires finding all cell intersections.

The approximate integration over swept regions (see [5] for Cartesian geometry case), which are the regions swept by the cell edges moving from old mesh to the new position in the new mesh (see Fig. 1b), is much faster. The contribution from each of the four swept regions has similar form as (3) with the intersection $I_{c'}^{\tilde{c}}$ replaced by the swept region. Green's theorem again transforms integrals over polygons into integrals over the edges of the polygon. In all the cases the integration from the Lagrangian step and from the remapping, the integrals of low degree polynomials in r, z over the given polygon P can be exactly evaluated as e.g. (2) or

$$\begin{aligned} \int_P r^2\,dr\,dz &= \frac{1}{12} \sum_{e \in \partial P} (z_2 - z_1)\left(r_1^2 + r_2^2\right)(r_1 + r_2) \\ \int_P z\,r\,dr\,dz &= \frac{1}{24} \sum_{e \in \partial P} (z_2 - z_1)\left(r_1^2\,(3\,z_1 + z_2) + r_2^2\,(3\,z_2 + z_1)\right. \\ &\qquad \left. + \,2\,r_1\,r_2\,(z_1 + z_2)\right), \end{aligned}$$

where the edge e connects the polygon P vertexes (r_1, z_1) and (r_2, z_2). In the swept regions method the integrals of the reconstructed function over the swept regions can be interpreted as fluxes through the mesh edges and the remapping formula can be written in a conservative flux form.

The last step of the remapping phase is repair [6], which conservatively redistributes conserved quantities in such a way that the remapping does not introduce any new local extrema.

4 Laser Absorption and Heat Conductivity

For laser plasma simulations, several more issues have to be included to get realistic results. The first is the model of laser absorption that appears as a source term $\operatorname{div}(\mathbf{I})$ in the energy equation (1). We assume that the absorption appears only on a critical surface, which is the isosurface with a critical density. The laser penetrates through the subcritical density where its intensity is given by laser radial and temporal profiles. Behind the critical surface the laser intensity is zero. The laser intensity is projected on the edge normals at the edge midpoints and div(**I**) approximated by the standard formula (derived from Green's theorem) is included as a source in the hyperbolic part of the energy equation (1) in Lagrangian step.

The simulation of laser absorption is considerably influenced by plasma heat conductivity, which is included as a parabolic term in the energy equation (1). The parabolic part of the energy equation is treated separately by splitting by an implicit mimetic finite difference method [7] generalized to cylindrical geometry. The mimetic method works well on poor quality meshes appearing quite often in the Lagrangian simulations and it can resolve the heat waves caused by nonlinear heat conductivity $\kappa(T)$, which is typically proportional to $T^{5/2}$ for plasma.

5 Simulations of Flyer Targets

To show the ability of our cylindrical code to perform complex modeling of the laser–plasma interactions, we simulated a set of problems based on the real experiments [8] performed at the PALS laser facility in Prague. Here we shortly present one of these simulations. A thin aluminum disc is irradiated by an intense laser beam and ablatively accelerated up to a very high speed. The disc flyer impacts the massive aluminum target creating a crater. The experimental setup is shown in Fig. 2a. The aluminum disc flyer has radius $r = 150\,\mu\mathrm{m}$ and thickness $d = 11\,\mu\mathrm{m}$ and is originally placed at distance $L = 200\,\mu\mathrm{m}$ from the massive aluminum target. The laser operates at 3-rd harmonics with a total energy 240 J, the laser pulse length is 400 ps and radius of focus (laser spot on target) is $r_f = 125\,\mu\mathrm{m}$.

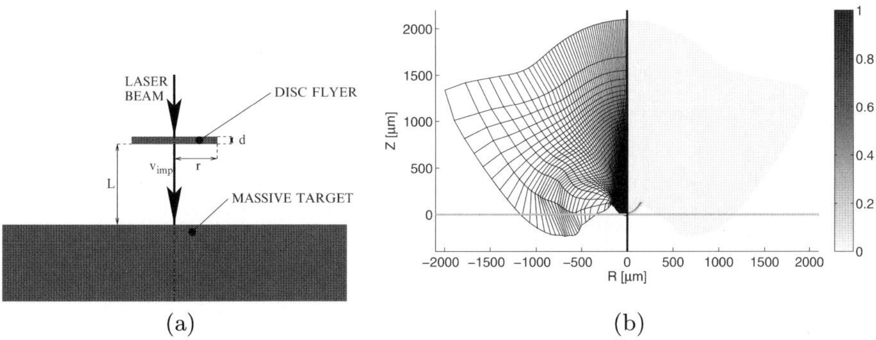

Fig. 2. Experimental setup (**a**); mesh and density colormap of the accelerated disc flyer hitting the massive target (**b**) where the horizontal line at $z = 0$ marks the upper surface of the target

We split the simulation into two parts. In the first part the disc flyer is ablatively accelerated by the laser beam and in the second part the impact of the disc flyer into the massive target is modeled. The laser absorption heats the upper disc surface, which evaporates and expands upwards very rapidly, accelerating the disc downwards due to momentum conservation. The first part of the simulation is stopped at time $t = 1.3\,\text{ns}$ when the disc reaches the massive target as shown in Fig. 2b and zoomed at Fig. 3a. The average velocity of the disc flyer impacting the target is $134\,\text{km}\,\text{s}^{-1}$. Note in Fig. 2b that the size of the computing domain has expanded from $d = 11$ to 2,000 µm in z direction and from $r = 150$ µm to more than 1,500 µm in r direction showing the need to use the Lagrangian moving mesh formulation. Similar expansion rate appears also during the second part of simulation, the huge area of reflected material moving upwards is cut off in Fig. 4 in order to be able to distinguish the features of the crater.

The initial conditions for the impact simulation, shown in Fig. 3b, are obtained by a conservative interpolation from the results of the ablative disc

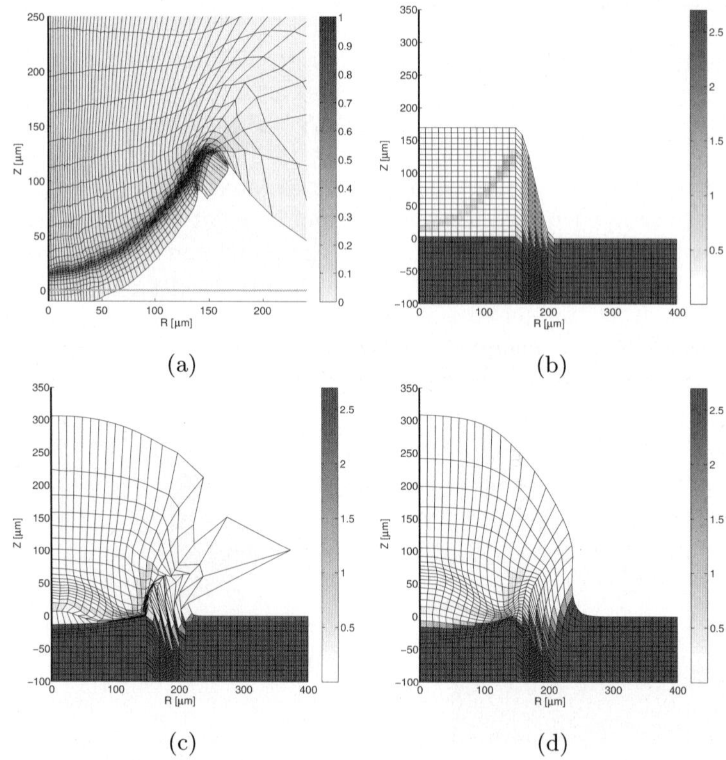

Fig. 3. Density colormaps of the accelerated disc flyer hitting the massive target (**a**) (zoomed Fig. 2b); the initial conditions and the mesh for the impact simulation (**b**); impact simulation at $t = 0.5\,\text{ns}$ by pure Lagrangian (**c**); and ALE method (**d**)

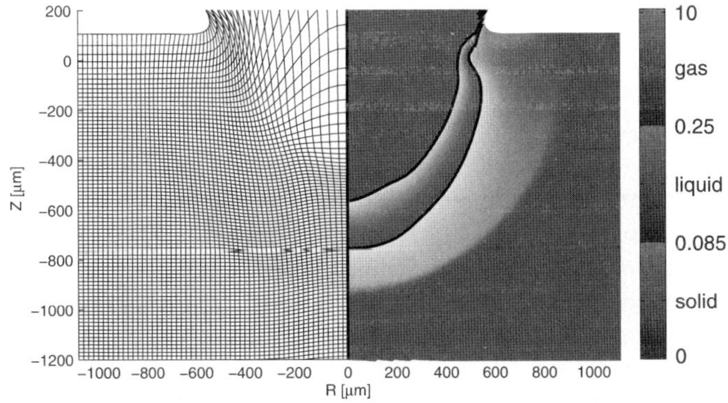

Fig. 4. Computational mesh (only every second edge in each logical direction is visible) and temperature (in eV) colormap at time $t = 80$ ns after the impact. Solid, liquid, and gas phases (ordered on the z axis at $R = 0$ from bottom to top) are separated by isolines at temperature of melting and evaporation of aluminum

acceleration, shown in Fig. 3a, to the initial mesh constructed for impact simulation. The impact simulation by pure Lagrangian method fails soon after time 0.5 ns due to a seriously distorted computational mesh, shown in Fig. 3c, while the ALE method keeps the mesh reasonably smooth as seen in Fig. 3d. The impact simulation is possible only by the ALE method. After the impact the large kinetic energy of the disc flyer is transformed into an internal energy, which melts and evaporates the target material. The circular shock wave visible in the solid phase region in Fig. 4 propagates into the massive target. The computational mesh and temperature colormap is presented in Fig. 4 at time $t = 80$ ns. The temperature colormap is separated by two temperature isolines, corresponding to aluminum melting and evaporating temperature, into three regions (from bottom to top on z axis) of solid, liquid, and gas phases. The crater is formed by the impact in the target. We interpret the crater boundary by the gas–liquid interface. After time $t = 80$ ns this interface does not move any further into the target, so that the upper temperature isoline in Fig. 4 presents the final shape of the simulated crater. The simulated crater formations are reasonably similar to the experimental data. More physics related details can be found in [9, 10].

6 Conclusion

All parts of the ALE method have been generalized from Cartesian to cylindrical geometry and implemented into the 2D laser plasma simulation code, including also heat conductivity and laser absorption. The code has been applied to the simulations of the disc flyer targets originated from PALS experiments. The pure Lagrangian method without ALE extension is unable to simulate such high velocity impact problems due to severe mesh distortion appearing very soon after the impact, while ALE gives results that compare reasonably well with the experimental data.

Acknowledgment

This work was carried out under the auspices of the National Nuclear Security Administration of the U.S. Department of Energy at Los Alamos National Laboratory under Contract No. DE-AC52-06NA25396. This research has been partly supported by the Czech Ministry of Education project MSM 6840770022, research center LC 528 and the Czech Science Foundation project GACR 202/03/H162.

References

1. C.W. Hirt, A.A. Amsden, and J.L. Cook. An arbitrary Lagrangian-Eulerian computing method for all flow speeds. *J. Comp. Phys.*, 14:227–253, 1974. Reprinted at *J. Comp. Phys.*, 135:203–216, 1997.
2. M. Kuchařík, R. Liska, and M. Shashkov. Conservative remapping and ALE methods for plasma physics. In F. Asakura, H. Aiso, S. Kawashima, A. Matsumura, S. Nishibata, and K. Nishihara, editors, *Hyperbolic Problems: Theory, Numerics and Applications II*, pages 133–140. Osaka University, Yokohama Publishers, 2006.
3. E.J. Caramana, D.E. Burton, M.J. Shashkov, and P.P. Whalen. The construction of compatible hydrodynamics algorithms utilizing conservation of total energy. *J. Comp. Phys.*, 146(1):227–262, 1998.
4. A.M. Winslow. Equipotential zoning of two-dimensional meshes. Technical Report UCRL-7312, Lawrence Livermore National Laboratory, 1963.
5. M. Kuchařík, M. Shashkov, and B. Wendroff. An efficient linearity-and-bound-preserving remapping method. *J. Comp. Phys.*, 188(2):462–471, 2003.
6. Mikhail Shashkov and Burton Wendroff. The repair paradigm and application to conservation laws. *J. Comp. Phys.*, 198(1):265–277, 2004.
7. M. Shashkov and S. Steinberg. Solving diffusion equation with rough coefficients in rough grids. *J. Comp. Phys.*, 129:383–405, 1996.
8. S. Yu. Guskov and et al. Investigation of shock wave loading and crater creation by means of single and double targets in PALS-laser experiment. *J. of Russian Laser Research*, 26:228–244, 2005.
9. M. Kuchařík, J. Limpouch, and R. Liska. Cylindrical 2D ALE simulations of laser interactions with flyer targets. *Czech. J. Phys.*, 56:B522–B527, 2006.
10. M. Kuchařík, J. Limpouch, and R. Liska. Laser plasma simulations by arbitrary Lagrangian Eulerian method. *J. de Physique IV*, 133:167–169, 2006.

Numerical Aspects of Parabolic Regularization for Resonant Hyperbolic Balance Laws

M. Kraft and M. Lukáčová-Medviďová

1 FVEG Schemes and Hyperbolic Balance Laws

Many problems arising in geophysics and engineering yield hyperbolic balance laws. Let us mention, for example, the compressible duct flow, multiphase flow or shallow water flow with variable bottom topography. All of them can be written in the following form

$$\boldsymbol{u}_t + \sum_{i=1}^{d} (\boldsymbol{f}(\boldsymbol{u}))_{x_i} = \boldsymbol{b}(\boldsymbol{u}), \tag{1}$$

where d is the space dimension. This system belongs to the class of *nonconservative systems*, i.e., systems which cannot be written in the divergence form. The nonconservative products cannot be defined in the distributional way and new tools have to used for theoretical investigations, see Dal Maso et al. [DLM95]. Additionally, the system is *nonstrictly hyperbolic* with *resonant behavior*, i.e., we have for the extended system with an additionally equation $\boldsymbol{b}(\boldsymbol{u})_t = 0$, cf. (9), some coinciding eigenvalues and linearly dependent eigenvectors.

As showed by several authors one can construct nonunique solutions of the one-dimensional Riemann problem, which locally satisfy entropy criterion, see, e.g., Chinnaya et al. [CLS04] and Andrianov and Warnecke [AW04a, AW04b]. For further theoretical results on resonant systems see also Goatin and LeFloch [GL04].

Despite these theoretical difficulties several numerical schemes have been proposed to solve the problem (1). One possibility is to use the operator-splitting technique which yields two subsystems

$$\boldsymbol{u}_t + \sum_{i=1}^{d} (\boldsymbol{f}(\boldsymbol{u}))_{x_i} = 0 \tag{2}$$

and

$$\boldsymbol{u}_t = \boldsymbol{b}(\boldsymbol{u}). \tag{3}$$

They can be solved by some finite volume scheme and a suitable ODE-solver. Anyway, in the case where the desired solution is close to the equilibrium state, i.e.

$$\sum_{i=1}^{d}(\boldsymbol{f}(\boldsymbol{u}))_{x_i} = \boldsymbol{b}(\boldsymbol{u}), \tag{4}$$

the operator-splitting approach is not appropriate. In fact, using two different types of discretizations for (2) and (3) will produce unbalanced higher order errors. As a result the balance state (4) can be obtained only for very fine discretizations, which is not efficient. In the recent literature one can find the so-called well-balanced schemes, in which the balance between the gradient of fluxes and the source terms is dictated by the construction of the scheme, see, e.g., [GL96, ABBKP04, BKLL04, NPPN06].

In [LV05] and [LNK06] the well-balanced finite volume evolution Galerkin (FVEG) scheme has been proposed for the shallow water equations with source terms. The FVEG scheme is a two-step predictor–corrector method. In the corrector step the classical finite volume update is done. To evaluate fluxes on cell interfaces the so-called approximate evolution operator $E_{\Delta t/2}$ is used. Such an evolution operator is based on the theory of bicharacteristics and takes all of the infinitely many directions of wave propagation into account. Derivation of evolution operators can be found in [LMW00, LS03, LSW02, LMW04] and [LNK06]. Thus, in the predictor step the values on cell interfaces are evolved at certain integrations points at the intermediate time $t_{n+1/2} := t_n + \Delta t/2$. Let us denote by $E_{\Delta t/2}\boldsymbol{U}^n$ the predicted value of the approximate solution at time $t_{n+1/2}$. Then the fluxes at the cell interfaces can be approximated by

$$\bar{\boldsymbol{f}}_k^{n+1/2} := \sum_j \omega_j \boldsymbol{f}_k(E_{\Delta t/2}\boldsymbol{U}^n(\boldsymbol{x}^j(\mathcal{E}))), \tag{5}$$

where the ω_j denotes the weights of the integration rule and $\mathcal{E}$ is the cell interface.

Let us consider, for simplicity, a regular rectangular mesh consisting of mesh cells $\Omega_{ij} = [x_i - \hbar/2, x_i + \hbar/2] \times [y_j - \hbar/2, y_j + \hbar/2]$, $i, j \in Z$, $\hbar$ is a mesh size. Further, let us denote the cell averages at time t_n by $\boldsymbol{U}_{ij}^n$ and the central difference operator in x_k-direction by $\delta_{x_k}^{ij}$, i.e., $\delta_{x_1}^{ij} u = u_{i+1/2,j} - u_{i-1/2,j}$. Then the finite volume update reads

$$\boldsymbol{U}_{ij}^{n+1} = \boldsymbol{U}_{ij}^n - \lambda \sum_{k=1}^{2} \delta_{x_k}^{ij} \bar{\boldsymbol{f}}_k^{n+1/2} + \lambda \boldsymbol{B}_{ij}^{n+1/2}, \tag{6}$$

where $\lambda = \Delta t/\hbar$ and $\boldsymbol{B}_{ij}$ stands for the approximation of the source term multiplied by the mesh size, $\hbar\boldsymbol{b}$. Note that in order to obtain a well-balanced scheme a care has to be taken in order to approximate the source term. In our scheme we are using the cell interface approach, see [J01, LV05, LNK06].

To solve the problems in numerical solutions arising by nonuniqueness of the Riemann problem for resonant hyperbolic balance laws we propose to use a parabolic regularization by adding a suitable viscosity term. This is motivated by an original physical problem from which the hyperbolic problem arises. Note that the approach proposed in this chapter is general and can be applied to any numerical scheme (typically finite volume scheme) for the problem (1). For the nonlocal regularization of phase-transition problems see, e.g., Rohde [R04], Schofer [S06] and the references therein.

2 Shallow Water Equations with Bottom Topography

Let us consider the shallow water equations with variable bottom topography. Such a system arises in many geophysical problems, for example in oceanography, river flow or atmospheric flows

$$\begin{aligned} h_t + (hu)_x + (hv)_y &= 0, \\ (hu)_t + (hu^2 + gh^2/2)_x + (huv)_y &= -ghb_x, \\ (hv)_t + (huv)_x + (hv^2 + gh^2/2)_y &= -ghb_y. \end{aligned} \tag{7}$$

Here h denotes the depth of the shallow water, $(u, v)^T$ is the velocity vector, b represents the bottom topography and g is the gravitational constant.

In [LNK06] the following approximate evolution operator for (7) has been derived

$$\begin{aligned} h\,(P) &= -b(P) \\ &\quad + \frac{1}{2\pi} \int_0^{2\pi} (h\,(Q) + b(Q)) - \frac{\tilde{c}}{g}\,(u\,(Q)\,\mathrm{sgn}(\cos\theta) + v\,(Q)\,\mathrm{sgn}(\sin\theta))\,\mathrm{d}\theta \\ u\,(P) &= \frac{1}{2\pi} \int_0^{2\pi} -\frac{g}{\tilde{c}}(h(Q) + b(Q))\mathrm{sgn}(\cos\theta)\mathrm{d}\theta \\ &\quad + \frac{1}{2\pi} \int_0^{2\pi} u\,(Q) \left(\cos^2\theta + \frac{1}{2}\right) + v\,(Q) \sin\theta\cos\theta\mathrm{d}\theta \end{aligned} \tag{8}$$

with an analogous equation for the velocity v. Here $P = (x, y, t_n + \Delta t/2)$ is the apex of the so-called bicharacteristic cone, i.e., an integration point for cell interface fluxes. The values $\tilde{c}$, $\tilde{u}$, $\tilde{v}$ are suitable linearizations around P and $Q = (x - \tilde{u}\Delta t/2 + \tilde{c}\Delta t/2\cos\theta, y - \tilde{v}\Delta t/2 + \tilde{c}\Delta t/2\sin\theta, t_n)$. This operator, denoted by $E_{\Delta t/2}^{const}$, is used to evolve piecewise constant approximate solutions. An analogous operator for bilinear data has been derived in [LNK06].

2.1 Transcritical States and Parabolic Regularization

We now turn our attention to transcritical flows, i.e., flows that change from $Fr > 1$ to $Fr < 1$ or vice versa, where $Fr = \|(u, v)^T\|_2/c$ denotes the Froude number and $c = \sqrt{gh}$ is the wave celerity. For better understanding let us consider one-dimensional shallow water equations and rewrite them in an extended form such that the bottom topography is formally a part of the variables

$$\begin{aligned} b_t &= 0, \\ h_t + (hu)_x &= 0, \\ (hu)_t + (hu^2 + gh^2/2)_x + ghb_x &= 0. \end{aligned} \tag{9}$$

We can rewrite these equations in a quasilinear form

$$\boldsymbol{w}_t + \boldsymbol{A}(\boldsymbol{w})\boldsymbol{w}_x = 0, \tag{10}$$

where

$$\boldsymbol{w} = \begin{pmatrix} b \\ h \\ hu \end{pmatrix}, \qquad \boldsymbol{A}(\boldsymbol{w}) = \begin{pmatrix} 0 & 0 & 0 \\ 0 & 0 & 1 \\ c^2 & c^2 - u^2 & 2u \end{pmatrix}. \tag{11}$$

The eigenvalues of the matrix $\boldsymbol{A}$ are $\lambda_1 = 0$, $\lambda_2 = u - c$ and $\lambda_3 = u + c$. The eigenvectors written in a matrix form give

$$\boldsymbol{R} = \begin{pmatrix} c^2 - u^2 & 0 & 0 \\ -c^2 & 1 & 1 \\ 0 & u - c & u + c \end{pmatrix} \overset{c=u}{\Longrightarrow} \boldsymbol{R} = \begin{pmatrix} 0 & 0 & 0 \\ -u^2 & 1 & 1 \\ 0 & 0 & 2u \end{pmatrix}. \tag{12}$$

So in the case $c = \pm u$ the matrix $\boldsymbol{R}$ is singular and the system is parabolic degenerate locally. This is called in the literature the resonant case, see, e.g., [GL04]. It has been pointed out by LeVeque [LeV98] that some schemes relying closely on the hyperbolic structure of the problem may show some deficiencies for this type of transcritical states. LeVeque showed, for example, that the wave propagation algorithm was not able to approximate correctly the steady transcritical shock. In Sect. 2, we will present results obtained by the FVEG scheme, which also yields some oscillations on transcritical shocks. This is a typical behavior for resonant systems. However, it has been pointed out by LeFloch [L99] that these shocks are sensitive on regularization. Thus we use a parabolic regularization and add a viscous term to the momentum equation. In [GP00] asymptotic derivation of the viscous shallow water equations was done. Note that in [GP00] no bottom topography has been considered. In an analogous way we propose the following form of the viscous term in the momentum equation for the shallow water system with a bottom topography

$$(hu)_t + (hu^2 + gh^2/2)_x = -ghb_x + 4\mu(hu_x)_x, \tag{13}$$

where μ is a viscosity parameter. Our aim is to choose μ in a numerical scheme in such a way, that it vanishes as the mesh is refined.

2.2 Numerical Experiments

Transcritical Shock

Let us consider the one-dimensional transcritical flow problem firstly proposed by LeVeque in [LeV98]. The computational domain is $[0, 1]$ and the bottom topography is given by

$$b(x) := \begin{cases} 0.25(\cos(\pi(x-0.5)/0.1)+1) & \text{if } |x-0.5|<0.1 \\ 0 & \text{otherwise.} \end{cases} \tag{14}$$

The initial data for the water depth h are $h(x,0) = 1 - b(x)$ and for the velocity $u(x,0) = 0.3$. The gravitational constant is set to be $g = 1$. The numerical solution has been computed up to the final time $T = 5$, where a steady state is already formed. The interest of this example is a steady transcritical shock (hydraulic jump) at which the FVEG scheme produces oscillatory behavior which does not disappear as the grid size is refined.

In all our examples a second-order accurate scheme is used and the minmod limiter is applied. We have used extrapolation boundary conditions at $x = 0$ and $x = 1$. Figure 1 shows the water depth h with bottom topography (left) and the Froude number Fr (right) calculated by the FVEG scheme without a parabolic regularization. From left to right the flow turns continuously from subcritical (fluvial) to supercritical (torrential) and then through a transcritical shock it goes back to subcritical. As can be noticed the oscillations arise under the great change in the bottom topography and a sudden change of the flow regime from the supercritical to subcritical flow.

Recall that we have denoted by u_i^n the cell average of u on the ith cell at time t_n. Define the following finite difference operators

$$\delta_x u_i := u_{i+1/2} - u_{i-1/2}, \qquad \mu_x h_i := (h_{i+1/2} + h_{i-1/2})/2.$$

We approximate the viscous term by the central finite differences at the time t_n as follows

$$4\mu(hu_x)_x \approx 4\mu\delta_x(\mu_x h_i^n \, \delta_x u_i^n)/\hbar^2. \tag{15}$$

The viscosity parameter μ needs to be chosen as small as possible but big enough to damp the oscillations completely. Numerical experiments indicated that μ should be of the form

$$\mu = \alpha\hbar, \tag{16}$$

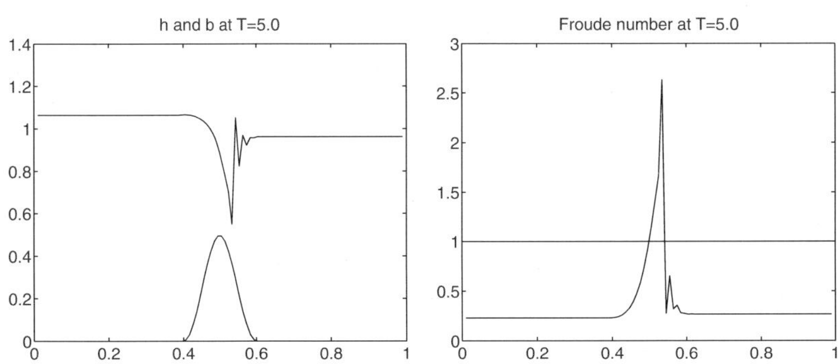

Fig. 1. Transcritical shock without parabolic regularization, 100 mesh cells

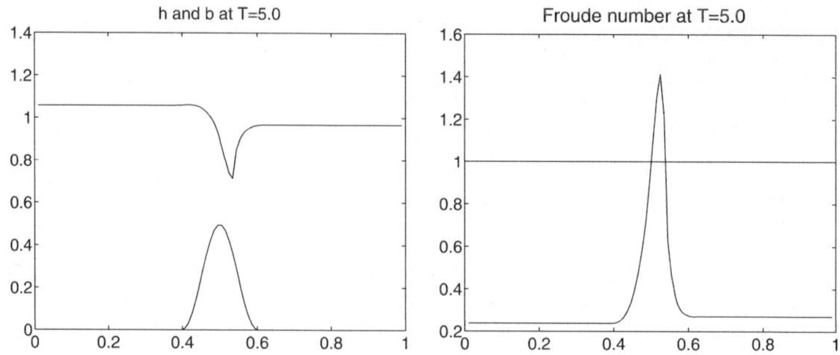

Fig. 2. Transcritical shock with parabolic regularization, 100 mesh cells

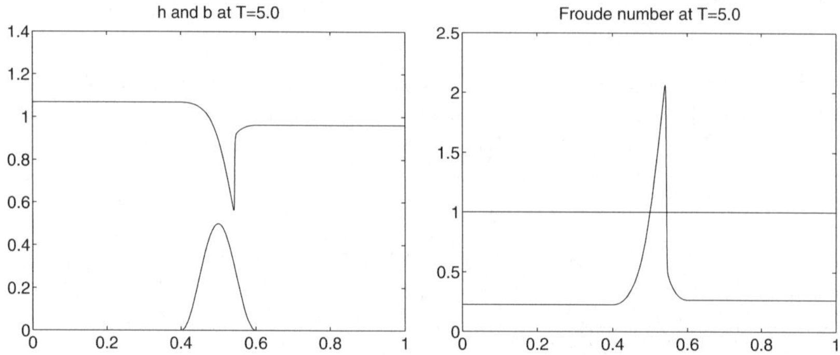

Fig. 3. Transcritical shock with parabolic regularization, 1,000 mesh cells

where α is a constant, $\alpha = \mathcal{O}(1)$, $\hbar$ is a mesh size. For this choice of α the regularization acts in a uniform manner for different mesh sizes. For our experiments we set $\alpha = 0.1$. Furthermore, the regularization term is a first-order term and thus one wants to apply the regularization only if necessary. A suitable switch has to be found to localize the resonant phenomenon. In the examples presented in this chapter the regularization has been applied over the whole computational domain.

In Figure 2 the water depth h is plotted for a mesh with 100 mesh cells. We can notice a stable steady transcritical shock without any oscillations. The shock is smeared slightly due the added numerical diffusion. For finer meshes the schock resolution is sharper, see Fig. 3, where the results on a mesh with $1,000$ cells are shown.

Transcritical Steady State Without Shock

Now, we consider the same problem as before but assume the following initial data for the velocity $u(x, 0) = 0.6$. We set the final time to $T = 10$. In this case the steady state is again transcritical but smooth. In Fig. 4 the bottom

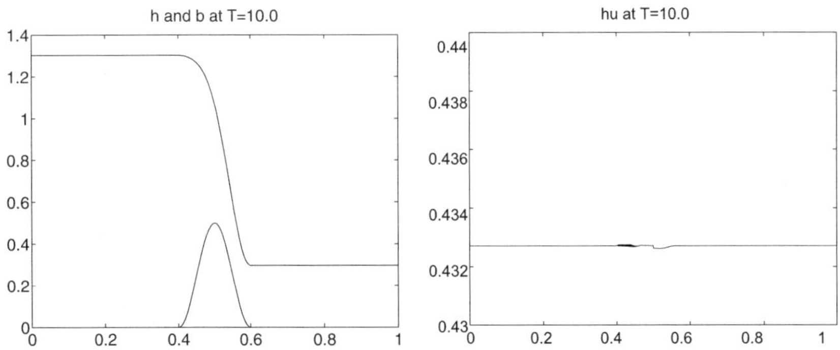

Fig. 4. Transcritical flow without parabolic regularization, 1,000 mesh cells

topography and water depth are depicted (left) as well as the momentum hu (right). Note, that no regularization has been necessary now.

Acknowledgments

We want to thank Christian Rohde, University of Stuttgart, for a fruitful discussion on regularization aspects of conservation laws. The second author has been partially supported by the grant LU 1194/4-1 of the Deutsche Forschungsgemeinschaft as well as by the DAAD. She gratefully acknowledges these supports.

References

[ABBKP04] Audusse E., Bouchut F., Bristeau M.-O., Klein R., Perthame B.: A fast and stable well-balanced scheme with hydrostatic reconstruction for shallow water flows, to appear in SIAM J. Sci. Comp., (2004)

[BKLL04] Botta N., Klein R., Langenberg S., Lützenkirchen S.: Well balanced finite volume methods for nearly hydrostatic flows, J. Comp. Phys. **196(2)**, 539–565 (2004)

[CLS04] Chinnayya A., LeRoux A.-Y., Seguin N.: A well-balanced numerical scheme for the approximation of the shallow-water equations with topography: the resonance phenomenon, International Journal on Finite Volume (electronic), **1(1)**, 1–33 (2004)

[AW04a] Andrianov N., Warnecke G.: On the solution to the Riemann problem for the compressible duct flow, SIAM J. Appl. Math. **64(3)**, 878–901 (2004)

[AW04b] Andrianov N., Warnecke G.: The Riemann problem for the Baur-Nunziato model of two-phase flows, J. Comp. Phys. **195**, 434–464 (2004)

[DLM95] Dal Maso G., LeFloch P.G., Murat F.: Definition and weak stability of non conservative products, J. Math. Pures Appl. **74**, 483–548 (1995)

[GL04] Goatin P., LeFloch P.G.: The Riemann problem for a class of resonant hyperbolic systems of balance laws, Ann. Inst. Henri Poincare, Anal. Non Lineaire, **21(6)**, 881–902 (2004)

[GP00] Gerbeau J.-F., Perthame B.: Derivation of viscous Saint-Venant system for laminar shallow water; numerical validation, Rapport de recherche **4084**, INRIA (2000)

[GL96] Greenberg J.M., LeRoux A.-Y.: A well-balanced scheme for numerical processing of source terms in hyperbolic equations, SIAM J. Numer. Anal., **33(1–16)** (1996)

[J01] Jin S.: A steady-state capturing method for hyperbolic systems with geometrical source terms, M^{2}AN, Math. Model. Numer. Anal., **35(4)**, 631–646 (2001)

[L99] LeFloch: An Introduction to Nonclassical Shocks of Systems of Conservative Laws. In: Kröner et al. (eds.), An Introduction to Recent Development in Theory and Numerics for Conservation Laws, 28–72, Springer (1999)

[LeV98] LeVeque R.J.: Balancing source terms and flux gradients in high-resolution Godunov methods: The quasi-steady wave propagation algorithm, J. Comp. Phys. **146**, 346–365 (1998)

[LMW00] Lukáčová-Medviďová M., Morton K.W., Warnecke G.: Evolution Galerkin methods for hyperbolic systems in two space dimensions, MathComp., **69**, 1355–1384 (2000)

[LMW04] Lukáčová-Medviďová M., Morton K.W., Warnecke G.: Finite volume evolution Galerkin (FVEG) methods for hyperbolic problems, SIAM J. Sci. Comput., **26(1)**, 1–30 (2004)

[LNK06] Lukáčová-Medviďová M., Noelle S., Kraft M.: Well-balanced Finite Volume Evolution Galerkin Methods for the Shallow Water Equations, in print J. Comp. Phys. (2006)

[LSW02] Lukáčová-Medviďová M., Saibertová J., Warnecke G.: Finite volume evolution Galerkin methods for nonlinear hyperbolic systems. J. Comp. Phys., **183**, 533–562 (2002)

[LS03] Lukáčová-Medviďová M., Saibertová J.: Genuinly multidimensional Evolution Galerkin schemes for the shallow water equations. In: F. Brezzi et al. (eds.) Numerical Mathematics and Advanced Applications ENUMATH 2001, Springer, 145–154 (2003)

[LV05] Lukáčová-Medviďová M., Vlk Z.: Well-balanced Finite Volume Evolution Galerkin Methods for the Shallow Water Equations with Source Terms, Int. J. Num. Fluids **47(10–11)**, 1165–1171 (2005)

[NPPN06] Noelle S., Pankratz N., Puppo G., Natvig J.: Well-balanced finite volume schemes of arbitrary order of accuracy for shallow water flows, J. Comput. Phys. **213**, 474–499 (2006)

[R04] Rohde Ch.: Approximation of Solutions of Conservation Laws by Non-Local Regularization and Discretization, Habilitation Thesis, University of Freiburg (2004).

[S06] Schofer D.: Approximation von komplexen Riemannlösungen für gemischt hyperbolisch-elliptische Systeme, Diploma Thesis, University of Freiburg (2006)

Three-Dimensional Adaptive Central Schemes on Unstructured Staggered Grids

A. Madrane

Summary. We present a new formulation of three-dimensional central finite volume methods on unstructured staggered grids for solving systems of hyperbolic equations. Based on the Lax–Friedrichs and Nessyahu–Tadmor one-dimensional central finite difference schemes, the numerical methods we propose involve a staggered grids to avoid solving Riemann problems at cell interfaces. The cells are barycentric, while those of the staggered grid are diamond shaped. To reduce artificial viscosity, we start with an adaptively refined primal grid in three-dimensional, where the theoretical a posteriori result of the first-order scheme is used to derive appropriate refinement indicators. We apply those methods and solve Euler equations. Our numerical results are in good agreement with corresponding results appearing in the literature.

1 Introduction

Staggered central finite volume schemes have been introduced by Nessyahu and Tadmor in 1990 [NT90]. The main advantage of these schemes is that no information about solutions to local Riemann problems is needed. Using staggered grids one can replace the upwind fluxes by central differences. The price one has to pay is the occurrence of excessive numerical viscosity since the resulting scheme can be interpreted as a Lax–Friedrichs scheme. Therefore, a MUSCL-type higher order scheme in one spatial dimension was proposed in [NT90]. Later in [AVM97] and [AMS01] the central schemes have been generalized to multidimensional schemes on unstructured grids. The present chapter focusses on the derivation of high-order central schemes on unstructured staggered grids. Furthermore we extend an adaptation strategy of [KO02] for first- and second-order methods where the theoretical a posteriori result of first-order schemes is used to derive appropriate refinement indicators.

2 Mathematical Modeling

2.1 Governing Equations

Let $\Omega \subset R^3$ be the domain of interest of the flow with boundary Γ. We write $\Gamma = \Gamma_B \cup \Gamma_\infty$, where Γ_B denotes the part of the body boundary which is relevant for the computational domain and Γ_∞ is the (upwind) far-field boundary. Three-dimensional compressible inviscid flows are described by Euler's equations, written in their conservation form

$$\frac{\partial U}{\partial t} + \nabla \cdot \vec{\mathcal{F}}(U) = 0, \tag{1}$$

where

$$U = (\rho, \rho u, \rho v, \rho w, E)^T, \quad \vec{\mathcal{F}}(U) = (F(U), G(U), H(U))^T.$$

Here $\vec{\mathcal{F}}(U)$ denotes the convective flux, ρ is the density, $\vec{V} = (u, v, w)^T$ is the velocity vector, $E = \rho e = \rho\epsilon + \frac{1}{2}\rho(u^2 + v^2 + w^2)$ is the total energy per unit volume, and p is the pressure of the fluid.

3 Space and Time Discretization

3.1 Definitions

We assume that Ω is a bounded polyhedral domain of R^3 and we start from an arbitrary FEM tetrahedral grid $\mathcal{T}_h$, where h is the maximal length of the edges in $\mathcal{T}_h$.

A dual finite volume partition is derived from the construction of median planes, that is, for every vertex i of $\mathcal{T}_h$, a cell C_i is defined around i as follows.

Every tetrahedron having i as a vertex is subdivided into 24 sub-tetrahedra by planes containing an edge and the midpoint of the opposite edge; then the cell C_i is the union of sub-tetrahedra having i as a vertex (see Fig. 1). In particular, the boundary ∂C_i of C_i is the union of $\partial C_{ij} = \partial C_i \cap \partial C_j$ that can be defined as the union of triangles (see Fig. 1). As for two-dimensional extensions [AVM97], the present three-dimensional extension also uses a dual grid, with dual cells L_{ij} associated with the edges of $\mathcal{T}_h$. The dual ("diamond") cell L_{ij} is composed of the sub-tetrahedra (defined above) sharing edge $[i, j]$ (see Fig. 1). The following notation will be needed.

Notation 1 *Let i, j, k, l be the four nodes defining a tetrahedron τ, $\tau \in \mathcal{T}_h$. Then*

- *T_{ij} denotes the set of all tetrahedra which share edge $[i, j]$ as a common edge*
- *$K(i)$ is the set of nodes (vertices) which are neighbors of node i*

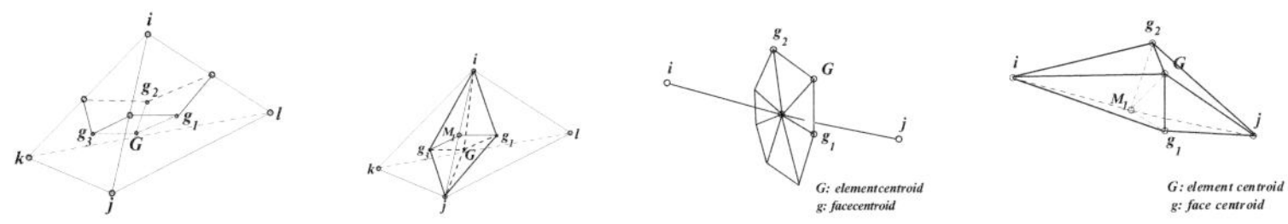

Fig. 1. Barycentric cell C_i and diamond cell L_{ij} and Sketch of $\partial C_{ij} = \partial C_i \cap \partial C_j$ and $\partial L_{ij} = \partial L_{ij} \cap \tau$

- $C_i = \bigcup_{j \in K(i)} (C_i \cap L_{ij})$
- $\partial C_i = \bigcup_{j \in K(i)} \{\partial C_i \cap \partial C_j\} \cup \{\partial C_i \cap \Gamma_B\} \cup \{\partial C_i \cap \Gamma_\infty\},$
- $L_{ij} = \bigcup_{\tau \in T_{ij}} (L_{ij} \cap C_i \cap \tau) \cup (L_{ij} \cap C_i \cap \tau)$
- $\partial L_{ij} = \bigcup_{\tau \in T_{ij}} (\partial L_{ij} \cap \tau) \cup (\partial L_{ij} \cap \Gamma_\infty) \cup (\partial L_{ij} \cap \Gamma_B),$
- $\mathbf{n}_{ij} = (n_{ij_x}, n_{ij_y}, n_{ij_z})$ *is the unit outward normal vector to* ∂L_{ij}
- $\boldsymbol{\nu}_i = (\nu_{i_x}, \nu_{i_y}, \nu_{i_z})$ *is the unit outward normal vector to* ∂C_i

Let m_{ij} denote the midpoint of edge $[i, j]$, also written as M_1 in Fig. 1, and let $U_i^n \cong U(a_i, t^n)$ and $U_{ij}^{n+1} \cong U(m_{ij}, t^{n+1})$ denote the nodal (resp. cell average) values in the first and second grid at time $t = t^n$ and $t = t^{n+1}$, respectively, (n even). The union of all the barycentric cells constitutes a partition of the computational domain Ω_h and the same holds for diamond cells. Now we can define the two steps of our high-order accurate (staggered, Lax–Friedrichs type) finite volume method.

3.2 High-Order Accurate Approximations

To obtain second-order accuracy, we introduce cell-wise piecewise linear interpolation.

First step. We integrate (1) on $L_{ij} \times [t^n, t^{n+1}]$, assuming we have obtained, from the cell average values U_i^n, piecewise linear reconstructions given by

$$U_h(\mathbf{x}, t^n)\big|_{C_i} = \mathcal{L}_i(\mathbf{x}, t^n) = U_i^n + \nabla U_i^n \cdot (\mathbf{x} - \mathbf{x}_i), \qquad \forall \mathbf{x} \in C_i,\ \mathbf{x} \in R^3. \quad (2)$$

For the integration with respect to time, to ensure "nearly" second-order accuracy, we adopt a "quasimidpoint formula" time discretization, where the convective flux is computed at the intermediate time $t^{n+1/2}$, thus requiring the computation of *predicted* values $U_h(\mathbf{x}, t^{n+1/2})$ given ∂L_{ij}.

Predictor's first step. On each face of the cell L_{ij}, using Euler's equations, we define a predicted vector

$$U^{n+1/2}_{i,g_1,G} = U^n_{i,g_1,G} - \frac{\Delta t}{2} \vec{\mathcal{F}}'(U^n_{i,g_1,G}) \cdot \nabla U^n_i, \tag{3}$$

where, by (2), the value of U^n_h is taken equal to

$$U_h(\mathbf{x}, t^n) \cong U^n_i + \nabla U^n_i \cdot (\mathbf{x}_{i,g_1,G} - \mathbf{x}_i) \equiv U^n_{i,g_1,G} \tag{4}$$

along the face ig_1G of the diamond cell L_{ij}.

Corrector's first step. By (3) the corrector can be written as follows:

$$\begin{aligned} &\mathrm{Vol}(L_{ij})U^{n+1}_{ij} - \sum_{\tau \in T_{ij}} \left[\int_{L_{ij} \cap C_i \cap \tau} \mathcal{L}(\mathbf{x}, t^n)\, d\mathbf{x} + \int_{L_{ij} \cap C_j \cap \tau} \mathcal{L}(\mathbf{x}, t^n)\, d\mathbf{x} \right] \\ &+ \Delta t \sum_{\tau \in T_{ij}} \left[\int_{\partial L^1_{ij} \cap \tau} + \int_{\partial L^2_{ij} \cap \tau} + \int_{\partial L^3_{ij} \cap \tau} + \int_{\partial L^4_{ij} \cap \tau} \right] \vec{\mathcal{F}}(U^{n+1/2}_h) \cdot \mathbf{n}_{ij}\, dA \\ &+ \Delta t \int_{\partial L_{ij} \cap \Gamma_B} \vec{\mathcal{F}}(U^n_h) \cdot \mathbf{n}\, dA + \Delta t \int_{\partial L_{ij} \cap \Gamma_\infty} \vec{\mathcal{F}}(U^n_h) \cdot \mathbf{n}\, dA = 0. \end{aligned} \tag{5}$$

Where the volume and the boundary integrals are approximated by the midpoint rule.

Second step. To obtain the second step of the time discretization, we integrate (1) on the cell $C_i \times [t^{n+1}, t^{n+2}]$, assuming that, from the diamond cell average values U^{n+1}_{ij} computed in the first time step, we have obtained piecewise linear reconstructions given by

$$U_h(\mathbf{x}, t^{n+1})|_{L_{ij}} = \mathcal{L}_{ij}(\mathbf{x}, t^{n+1}) = U^{n+1}_{ij} + \nabla U^{n+1}_{ij} \cdot (\mathbf{x} - \mathbf{x}_{ij}). \tag{6}$$

Predictor's second step. Proceeding as in the first step, we obtain the predictor's second step:

$$U^{n+3/2}_{M_1 g_1 G} = U^{n+1}_{M_1 g_1 G} - \frac{\Delta t}{2} \vec{\mathcal{F}}'(U^{n+1}_{M_1 g_1 G}) \cdot \nabla U^{n+1}_{ij}, \tag{7}$$

where

$$U_h(\mathbf{x}, t^{n+1}) \cong U^{n+1}_{ij} + \nabla U^{n+1}_{ij} \cdot (\mathbf{x}_{M_1,g_1,G} - \mathbf{x}_{M_1}) \equiv U^{n+1}_{M_1,g_1,G} \tag{8}$$

defines an approximation to the value of U on the boundary element $[M_1, g_1, G]$ of cell C_i.

Corrector's second step. The second step is

$$\begin{aligned} &\mathrm{Vol}(C_i)U^{n+2}_i - \sum_{j \in K(i)} \int_{C_i \cap L_{ij}} \mathcal{L}_{ij}(\mathbf{x}, t^{n+1})\, d\mathbf{x} \\ &+ \Delta t \sum_{j \in K(i)} \int_{\partial C_i \cap \partial C_j} \vec{\mathcal{F}}(U(\mathbf{x}, t^{n+3/2})) \cdot \boldsymbol{\nu}_\mathbf{i}\, dA \\ &+ \Delta t \int_{\partial C_i \cap \Gamma_B} \vec{\mathcal{F}}(U^{n+1}_h) \cdot \boldsymbol{\nu}\, dA + \Delta t \int_{\partial C_i \cap \Gamma_\infty} \vec{\mathcal{F}}(U^{n+1}_h) \cdot \boldsymbol{\nu}\, dA = 0, \end{aligned} \tag{9}$$

where the volume and the boundary integrals are computed as above.

Approximation of the Slopes and Limitation

Numerical experiments have led us to choose Green–Gauss' method [BJ89] for the gradients used in the reconstruction for the cells C_i and a least squares weighted procedure [HB98] for the cells L_{ij}. For the limitation we use Venkatakrishnan's limiter see [Ven95].

4 Mesh Adaptation Algorithm

4.1 General Description

The theory behind the mesh adaptation technique for central schemes on unstructured staggered grids has been developed in [KO02]. We introduce the following three main steps in this technique.

- First, a strategy to determine where a modification is needed in the field of the grid, e.g., by means of an (a posteriori) error estimate.
- Second, a rule that selects the elements or edges in $\mathcal{T}_h$ (marking strategy).
- Third, a rule that refines the elements in $\mathcal{T}_h$ (refinement strategy).

A Posteriori Error Estimate

For stationary problems, following the theory of [KO02], for each edge $e_{ij} \in \mathcal{T}_h$ we have the error estimate $\eta_{e_{ij}}$:

$$\|u - u_h\|_{L^1(e_{ij})} \leq \eta(u_{e_{ij}}) = aQ + b\sqrt{Q}, \tag{10}$$

where $a = 2 + 2\omega$, $\omega = 0.5$, $b = 4 + 2^d$, $d = 3$,

$$\begin{aligned} Q = \frac{1}{2} \sum_{e_{ij}=1}^{ne} h_{e_{ij}} \mathrm{Vol}(L_{e_{ij}}) \sum_{\tau \in T_{ij}} & \frac{\mathrm{Vol}(L_{e_{ij}} \cap C_i \cap \tau)}{\mathrm{Vol}(L_{e_{ij}})} \frac{\mathrm{Vol}(L_{e_{ij}} \cap C_j \cap \tau)}{\mathrm{Vol}(L_{e_{ij}})} |u_i - u_j| \\ + \sum_{e_{ij}=1}^{ne} \mathrm{Vol}(L_{e_{ij}}) & \left| u_{e_{ij}} - \sum_{j \in K(i)} \frac{\mathrm{Vol}(L_{e_{ij}} \cap C_i)}{\mathrm{Vol}(L_{e_{ij}})} u_{e_{ij}} \right| \\ + 6 \sum_{e_{ij}=1}^{ne} h_{e_{ij}} & \sum_{\tau \in T_{ij}} \mathrm{Area}(\partial C_i \cap \partial C_j \cap \tau) |u_i - u_j|, \end{aligned}$$

where $h_{e_{ij}} = \sqrt{(x_i - x_j)^2 + (y_i - y_j)^2 + (z_i - z_j)^2}$ and ne is the number of edges of the original finite element triangulation $\mathcal{T}_h$.

Marking Strategy

In this subsection, we introduce the maximum strategy to determine the set $\widetilde{\mathcal{T}_h}$ in the general adaptive algorithm.

Algorithm 1 (Maximum strategy) *This algorithm determines the set* $\widetilde{\mathcal{T}_h}$.

(a) *Given: a partition* $\mathcal{T}_h$*, error estimates* $\eta_{e_{ij}}$ *for the edges* $e_{ij} \in \mathcal{T}_h$*, and a threshold* $\theta \in (0,1)$.
Sought: a subset $\widetilde{\mathcal{T}_h}$ *of marked edges that should be refined.*
(b) *Compute* $\eta_{\mathcal{T}_h,\max} = \max_{e_{ij} \in \mathcal{T}_h} \eta_{e_{ij}}$.
(c) *If* $\eta_{e_{ij}} \geq \theta \eta_{\mathcal{T}_h,\max}$, *then mark the edge* e_{ij} *for refinement and put it into the set* $\widetilde{\mathcal{T}_h} = \big\{e_{ij} \in \mathcal{T}_h | \eta_{e_{ij}} \geq \theta \eta_{\mathcal{T}_h,\max}, \theta \in (0,1)\big\}$.

Refinement Strategy

The set of marked edges is examined, tetrahedron by tetrahedron, and additional edges are marked in an attempt to maintain the grid quality and to get a conforming mesh (see Fig. 2). The final set of marked edges results in tetrahedra with one edge, three edges on one face, or all six edges. A tetrahedron with all six marked edges is shown in Fig. 2. The mesh is then refined by inserting new nodes on the midpoints of the marked edges and reconnecting these nodes into new tetrahedra and boundary faces. For the last configuration, cutting off tetrahedra on all four corners leaves an octahedron which can be split into four tetrahedra by adding an inner edge connecting two diagonally opposite corners of the octahedron. To minimize distortions of the created tetrahedra, the shortest of the three possible inner diagonals should be chosen.

Tetrahedral Mesh Improvement Using Face and Edge Swapping

The edge and face swapping techniques effectively improve shape measures. The swapping algorithm minimizes a shape function, such as the aspect ratio,

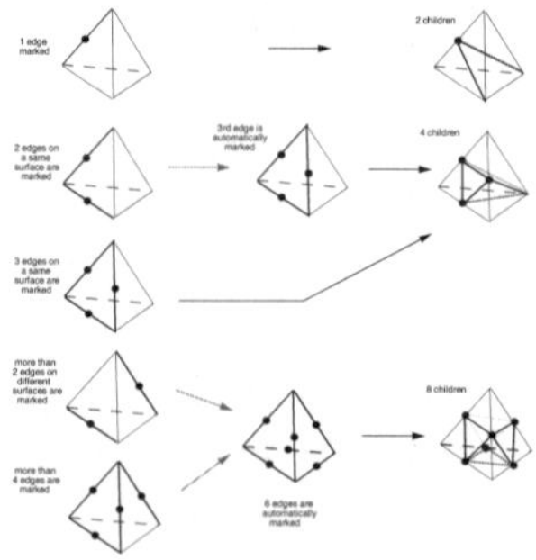

Fig. 2. Refinement strategies for a tetrahedron

$AR = \frac{1}{3}\frac{\rho_{out}}{\rho_{in}}$, where ρ_{out} is the tetrahedral circumsphere radius, ρ_{in} is the tetrahedral in-sphere radius, for tetrahedra [LJ94] as follows. The value of the aspect ratio varies from 1, for an ideal element, to ∞, for badly shaped elements. Reconnections of tetrahedra with undesirable shape measures are investigated and new local configurations for tetrahedra are selected with more desirable shape measures. Edges on boundary faces can also be swapped. Details of the way in which the face swapping can be implemented in practice can be found in [Joe95].

Boundary Modification

The inserted boundary nodes may not be located on the surface geometry of the model to be simulated because they were inserted at the midpoints of existing edges. To address this issue, we have implemented a boundary curvature correction based on Hermite interpolation [Loh96].

5 Numerical Experiments

5.1 Transonic NACA0012

A NACA0012 wing configuration has been employed to demonstrate the transonic shock capturing capability of the present adaptive grid solution method. The flow condition is at a free stream Mach number of 0.85 and incidence angle of 3.5°. A reasonably coarse grid with a chordwise nearly uniform point distribution was generated to serve as the initial grid for adaptation. The grid, contains 76,125 points and 395,203 tetrahedral cells. An inviscid flow computation on this grid reveals the presence of a weak shock wave on the upper surface of the wing. The Mach and Cp contours are also illustrated in Fig. 3. As expected, the shock wave is diffused due to the grid coarseness and excessive numerical viscosity. Using the local remeshing procedures described earlier, four levels of adaptive refinement were performed in this case. The final grid

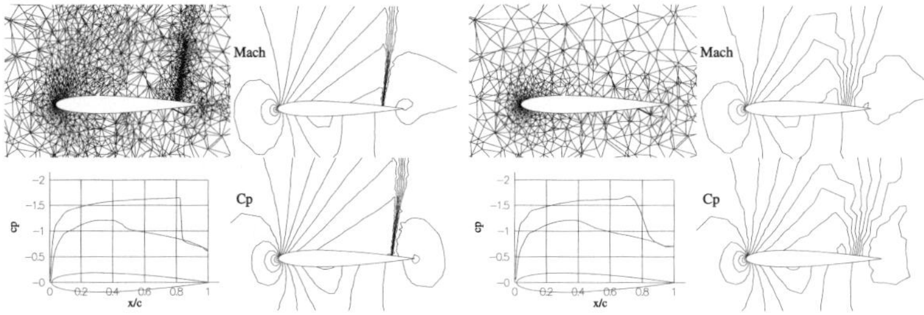

Fig. 3. Mach and Cp contours at $M_\infty = 0.85$ and $\alpha = 3.5°$ for the initial grid and adapted grid

contains 305,458 points and 1,675,668 tetrahedra. A threshold value $\theta = 0.5$ is used. Figure 3 shows the adapted grid and the corresponding Mach and C_p contours. As evidenced, the grid is efficiently refined at the shock location, which shows a sharp shock definition. Figure 3 illustrates the chordwise distributions of the surface pressure coefficient C_p for the initial coarse and adapted grids. As expected, there are significant differences between the adapted and the initial grid results. From the C_p distribution, it appears that the shock location of adapted grids is well captured compared to that of coarse grids. Furthermore, this example emphasizes the advantage of grid adaptation in providing more accurate flow solutions economically.

References

[AVM97] P. Arminjon, M.C. Viallon and A. Madrane, *A finite volume extension of the Lax–Friedrichs and Nessyahu–Tadmor schemes for conservation laws on unstructured grids*, revised version with numerical applications, Int. J. of Comp. Fluid Dynamics 9, No. 1 (1997), pp. 1–22.

[AMS01] P. Arminjon, A. Madrane and A. St-Cyr, *Numerical simulation of 3D flows with a non-oscillatory central scheme on staggered unstructured tetrahedral grids*, in H. Freistuehler, G. Warnecke, Birkhauser, Eds., Proceedings of the Eighth International Conference on Hyperbolic Problems, Int. Series of Num Math. 140 (2001), pp. 59–68.

[BJ89] T.J. Barth and D.C. Jespersen, (1989), *The design and application of upwind schemes on unstructured meshes*, AIAA Paper No. 89-0366, 27th Aerospace Sciences Meeting, January 9–12, 1989, Reno, Nevada.

[HB98] A. Haselbacher and J. Blazek, *On the accurate and efficient discretisation of the Navier–Stokes equations on mixed grids.* Computational Fluid Dynamics Conference, 14th, Norfolk, VA, June 28–July 1, 1999, Collection of Technical Papers. Vol. 2 (A99-33467 08-34).

[Joe95] B. Joe, *Construction of three-dimensional improved quality triangulations using local transformations*, SIAM J. Sci. Comput., 16, 1292–1307, 1995.

[KO02] M. Küther and M. Ohlberger, *Adaptive second-order central schemes on unstructured staggered grids*, in T.Y. Hou and E. Tadmor, Eds., Proceedings of the Ninth International Conference on Hyperbolic Problems held at Caltech, Pasadena CA, March 25–29, 2002, pp. 675–684, Springer Berlin/Heidelberg/New York, 2003.

[LJ94] A. Liu and B. Joe, *Relationship between tetrahedron shape measures*, BIT, 34 (1994), pp. 268–287.

[Loh96] R. Löhner, *Regridding Surface Triangulations*, J.C.P. 126, 1–10, 1996.

[NT90] H. Nessyahu and E. Tadmor, *Non-oscillatory central differencing for hyperbolic conservation laws*, J. Comp. Phys., 87, No. 2 (1990), pp. 408–463.

[Ven95] V. Venkatakrishnan, *Convergence to steady state solutions of the Euler equations on unstructured grids with limiters*, J. Comp. Phys. 118 (1995), pp. 120–130.

High Amplitude Solutions for Small Data in Pairs of Conservation Laws that Change Type

V. Matos and D. Marchesin

Summary. We prove that high amplitude Riemann solutions arise from Riemann data with arbitrarily small amplitude in the hyperbolic region near the point where the rarefaction curves are tangent to the elliptic region. These solutions arise in a quadratic system of conservation laws with a compact elliptic region. The second-order terms in the fluxes correspond to type IV in Schaeffer and Shearer classification. For such Riemann data there is no small amplitude solution. Whitney perturbations of the fluxes do not change the result. Thus, we understand one possible consequence for violating strict hyperbolicity in the neighborhood of the Riemann data points. The violation of this hypothesis in Lax's renowned theorem may yield large amplitude solutions for small data.

1 Introduction

A famous theorem of Lax states that systems of n conservation laws with small data have Riemann solution consisting of n small waves, rarefactions or shocks, separated by constant states, under certain hypotheses. What happens if the hypotheses are violated? Liu [2] showed in 1974 that if the hypothesis of genuine nonlinearity is violated, the rarefactions and shocks can join. Still, they form n groups separated by $n-1$ constant states.

In this work, we find an example of two conservation laws for which the Riemannn solution consists of two shocks with $O(1)$ amplitude no matter how small the data are, provided it is close to a special point on the locus where the characteristic speeds coincide. At this special point the characteristic direction is tangent to the the boundary of the elliptic region.

Though our example occurs in a system with quadratic flux functions, such a point exists generically for systems that change from hyperbolic to elliptic type. In [3] it is established that this point persists under C^3 Whitney perturbations of the flux functions. Thus the existence of large Riemann solutions for small data is generic.

In Sect. 2 we review some results for systems of two conservation laws in one space dimension. In Sect. 3 we present our results. Proofs are in Sect. 4. In Sect. 5 we present some remarks about our result.

2 Background

In this section we review some results for systems of two conservation laws in one space dimension. These are partial differential equations of the form

$$U_t + F(U)_x = 0, \tag{1}$$

where $U(x,t) = (u,v)^T \in \mathbb{R}^2$ for $x \in \mathbb{R}$ and $t \geq 0$, $F \in C^2\left(\mathbb{R}^2, \mathbb{R}^2\right)$.

A Riemann problem (RP) is an initial value problem with constant states on the left- and right-hand sides of the origin, called U_L and U_R, that is,

$$U(x,0) = U_L \text{ if } x < 0,\ U(x,0) = U_R \text{ if } x > 0. \tag{2}$$

Definition 1. *The set of U in $\mathbb{R}^2$ where $DF(U)$ has:*

(1) Two distinct real eigenvalues is called the strictly hyperbolic region, SHR

(2) Two distinct complex conjugate eigenvalues is called the elliptic region

(3) One double real eigenvalue is called the coincidence locus

In the SHR the characteristic speeds of $DF(U)$ are ordered so that the lowest is called 1-speed, $\lambda_1(U)$, and the highest is called 2-speed, $\lambda_2(U)$. The corresponding eigenvectors are $r_1(U)$ and $r_2(U)$.

It is well known that solutions of (1) and (2) can have shocks, i.e., discontinuities. Each shock satisfies two Rankine–Hugoniot (RH) conditions

$$F(U_+) - F(U_-) - s(U_+ - U_-) = 0, \tag{3}$$

where U_- and U_+ are, respectively, the left and right states of the shock and s is its speed. We denote the shock by the triplet (U_-, U_+, s); we may use $s(U_-, U_+)$ or just s for the shock speed. The state $U_+ = U_-$ is always a trivial zero of the RH condition.

Based on Lax we define:

Definition 2. *Generic shocks between states in the SHR:*

(1) 1-Shocks: $\lambda_1(U_+) < s < \lambda_1(U_-)$ and $s < \lambda_2(U_+)$ (1S in the figures)

(2) 2-Shocks: $\lambda_2(U_+) < s < \lambda_2(U_-)$ and $\lambda_1(U_-) < s$ (2S in the figures)

Definition 3. *Other shocks are important in our problems, namely:*

(1) Over-compressive: $\lambda_2(U_+) < s < \lambda_1(U_-)$ (C in the figures)

(2) Left-characteristic 1-shocks: $s = \lambda_1(U_-)$ and $\lambda_1(U_+) < s < \lambda_2(U_+)$

(3) Left-characteristic over-compressive: $s = \lambda_1(U_-)$ and $\lambda_2(U_+) < s$

Every discontinuity within the classification of Definition 2 or 3 is called an *admissible shock*. Otherwise it is called inadmissible.

Definition 4. *The RH set for a fixed U_- is a one-dimensional set in U-space:*

$$\mathcal{H}(U_-) = \{U \in \mathbb{R}^2 : \exists s \in \mathbb{R} \text{ such that (3) holds}\}. \tag{4}$$

Each point of the RH set $\mathcal{H}$ is classified as inadmissible or admissible and the latter follow the classification according to Definitions 2 and 3. Typically, there are connected parts in $\mathcal{H}$ consisting of 1-shocks, of 2-shocks and of over-compressive shocks, or OC shocks. Similarly, there are isolated points in $\mathcal{H}$ representing left-characteristic shocks. Typically, for left-characteristic shocks U_- is a double rather than a single zero of the RH condition: this can be verified by examining the Jacobian of the RH (3) at U_-.

We state a loose version of Lax's classical theorem for two equations in a neighborhood N with $\bar{N}$ in the SHR, such that $\nabla\lambda_i \cdot r_i \neq 0$, $i = 1, 2$.

Theorem 1. *Given U_L and U_R in N, there exist two transverse curves, tangent to r_1 at U_L and to r_2 at U_R, and a U_M such that the curve segment from U_L to U_M along the slow speed curve followed by the curve segment from U_M to U_R along the fast speed curve parameterize the unique solution of the RP with data U_L, U_R. These curves represent shocks and rarefactions.*

Corollary 1. *Let U_M be the middle point of the solution of the RP with data U_L, U_R. Then $|U_M - U_L| \searrow 0$ as $|U_R - U_L| \searrow 0$.*

3 The Local RP with Nonlocal Solution

We study a type IV model in Schaeffer and Shearer's classification with flux

$$F\begin{pmatrix} u \\ v \end{pmatrix} = \frac{1}{2}\begin{pmatrix} 3u^2 + v^2 \\ 2uv \end{pmatrix} + \begin{pmatrix} 2v \\ 0 \end{pmatrix}. \tag{5}$$

We remark that the linear factor on the right-hand side of (5) could be multiplied by any positive number, and the results would be similar except for scaling. Also, the quadratic term on the right-hand side arises from setting $a = 3$ and $b = 0$ in the classification given in [4]. We expect that other type IV models with nearby parameters lead to similar results. The eigenvalues of DF are $\lambda_1 = 2u - \sqrt{u^2 + (v+1)^2 - 1}$ and $\lambda_2 = 2u + \sqrt{u^2 + (v+1)^2 - 1}$. Notice that $\lambda_1 = \lambda_2$ along the circle $u^2 + (v+1)^2 = 1$, the coincidence locus. The interior of this circle is the elliptic region in this model.

We show that nonlocal solutions arise from RPs with arbitrarily small data in the SHR near the origin. This result is stated in the following theorems.

Theorem 2. *Let $\mathcal{O}$ be $(0,0)$. There is an open set B in the SHR ($\mathcal{O} \in \partial B$) with the following property. For any small $\beta > 0$, for any $U_R \in B$ with $|U_R - \mathcal{O}| < \beta$ the solution of the Riemann problem $U_L = \mathcal{O}$, U_R has amplitude close to 4.*

This behavior can be extended for U_L, U_R in open sets near $\mathcal{O}$ in the SHR. Let $T(\beta)$ be the following family of open triangles in the SHR (the choice $\frac{\beta^2}{9}$ is explained in the proof of Lemma 1):

$$T(\beta) = \left\{(u,v) \in \mathbb{R}^2 : 0 < v < \tfrac{\beta^2}{9} \text{ and } -v < u < v\right\}. \tag{6}$$

Theorem 3. *Let be $\beta \gtrsim 0$. For every $U_L \in T(\beta)$ there is a nonempty open sector $A(U_L, \beta)$ closer than β from U_L with the following properties: (1) The sector $A(U_L, \beta)$ lies in the SHR; (2) For all points U_R in $A(U_L, \beta)$ the solution of the RP with data (U_L, U_R) has amplitude larger than 4.*

We remark that both $T(\beta)$ and $A(U_L, \beta)$ approach $\mathcal{O}$ as β vanishes.

4 Sketch of Proof of the Theorems

We are concerned with solutions of (1), (2), and (5) consisting of a sequence of two shocks. The first shock between the states U_L and U_M must be not faster than the second one between U_M and U_R.

Substituting (5) in the RH relation (3) yields

$$-s(u_+ - u_-) + 3(u_+^2 - u_-^2)/2 + (v_+^2 - v_-^2)/2 + 2(v_+ - v_-) = 0, \tag{7a}$$

$$-s(v_+ - v_-) + u_+v_+ - u_-v_- = 0. \tag{7b}$$

Fixing (u_-, v_-), the curves in the variables (u_+, v_+) are conic sections, so there are 0, 2, or 4 intersections or zeros counting multiplicity. Since $U_+ = U_-$ is always a solution of (7), there are 2 or 4 zeros.

For $U_- = \mathcal{O} = (0,0)$, (7) leads to $Q \equiv \frac{3}{2}\left(u_+ - \frac{s}{3}\right)^2 + \frac{1}{2}(v_+ + 2)^2 = 2 + \frac{s^2}{6}$ and $(u_+ - s)\, v_+ = 0$. The RH locus $\mathcal{H}(\mathcal{O})$ defined in (4) consists of the horizontal axis $v_+ = 0$ together with the circle $u_+^2 + (v_+ + 2)^2 = 4$. On the horizontal axis the shock velocity is given by $s = \frac{3}{2}u_+$. On the circle, $s = u_+$, so that $s < \lambda_1(U_+)$ if and only if $u_+ > 0$ and $-2 < v_+ < 0$; also $s > \lambda_2(U_+)$ if and only if $u_+ < 0$ and $-2 < v_+ < 0$.

Now we can classify the points in $\mathcal{H}(\mathcal{O})$ according to the Definitions 2 and 3 as shown in Fig. 1. The points D_1, D_2, and D_3 will be used later.

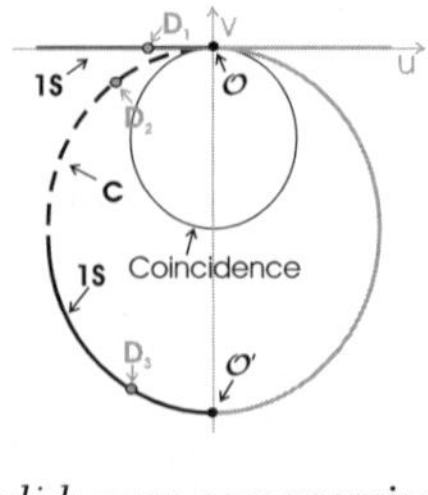

Fig. 1. $\mathcal{H}(\mathcal{O})$: 1-shocks (1S), *solid*; over-compressive (C), *dashed*; inadmissible, *gray*

For $s = 0$ there are just two distinct zeros of the RH condition, $\mathcal{O}$ and $\mathcal{O}' = (0, -4)$ (Fig. 2). Notice that the 1-shock $(\mathcal{O}, \mathcal{O}', 0)$ is doubly left characteristic, i.e., $s(\mathcal{O}, \mathcal{O}') = \lambda_1(\mathcal{O}) = 0 = \lambda_2(\mathcal{O})$, and $\mathcal{O}$ is a triple zero. We will see that the zero $\mathcal{O}'$ plays an important role.

For $U_- = \mathcal{O}$ and shock speed $s < \lambda_1(\mathcal{O})$ there are four distinct zeros of the RH condition, (Fig. 3). We see that $\mathcal{O}$ splits into three zeros, $\mathcal{O}$, D_1, and D_2, while $\mathcal{O}'$ moves to D_3; D_1 and D_3 lie on the 1-shock part $1S$ of $\mathcal{H}(\mathcal{O})$ while D_2 lies on the part C near $\mathcal{O}$ (Fig. 1). These four zeros define several shocks (Fig. 4): the 1-shocks $(\mathcal{O}, D_1, s)$ and $(\mathcal{O}, D_3, s)$, the OC shock $(\mathcal{O}, D_2, s)$ and the 2-shocks (D_1, D_2, s) and (D_3, D_2, s); D_2 belongs to both the 2-shock parts of $\mathcal{H}(D_1)$ and $\mathcal{H}(D_3)$. Therefore the RP with $U_L = \mathcal{O}$ and $U_R = D_2$ has multiple solutions in phase space that coincide in physical space.

We remove this degeneracy of the Riemann solution. Let U_R be a point out of C on the 2-shock part of $\mathcal{H}(D_1)$ near D_2 (Fig. 5a). If U_R lies above D_2 then $(\mathcal{O}, D_1, s)$ followed by (D_1, U_R, s_u) is compatible, i.e., the second shock is not slower than the first; actually the second shock is faster than the first,

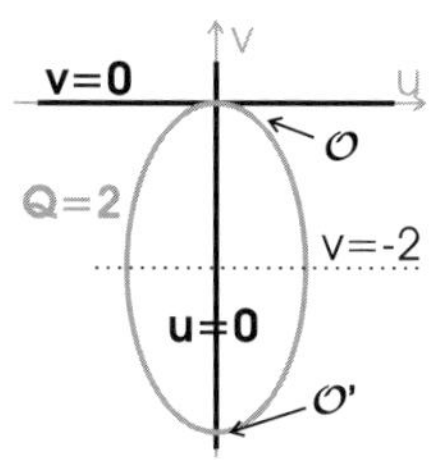

Fig. 2. Quadratic curves for $U_- = \mathcal{O}$ and $s = 0$

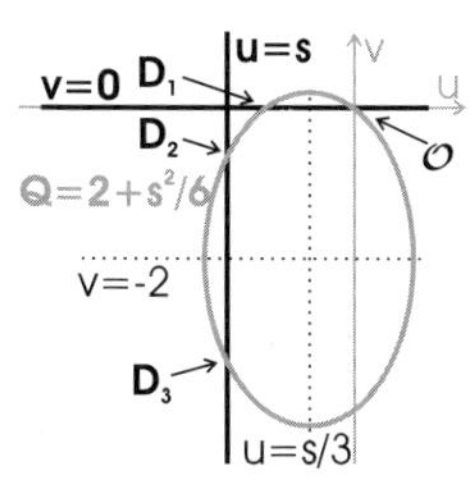

Fig. 3. Quadratic curves for $U_- = \mathcal{O}$ and $s < 0$

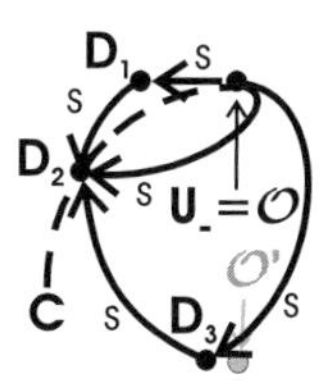

Fig. 4. The multiple solutions: $U_L = \mathcal{O}$, $U_R = D_2$

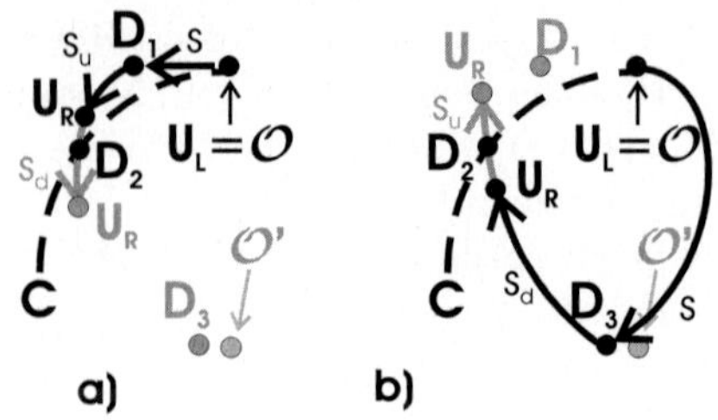

Fig. 5. Shocks for $U_L = \mathcal{O}$ and U_R near D_2. Compatible sequences in *black*; incompatible in *gray*

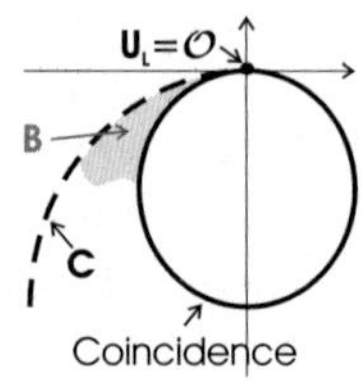

Fig. 6. The open set B in Theorem 2

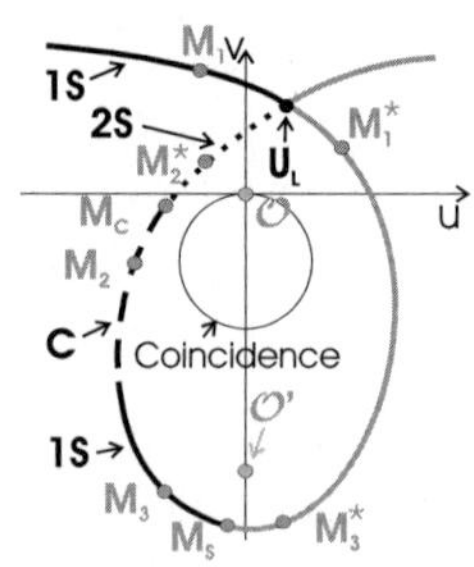

Fig. 7. $\mathcal{H}(U_L)$ for $U_L \in T(\beta)$; 1-shocks, *solid*; over-compressive, *dashed*; 2-shocks, *dotted*; inadmissible, *gray*

$s < s_u$. On the other hand, if U_R lies below D_2 then $(\mathcal{O}, D_1, s)$ followed by (D_1, U_R, s_d) is incompatible ($s_d < s$).

Let U_R be a point out of C on the 2-shock part of $\mathcal{H}(D_3)$ near D_2 (Fig. 5b). If U_R lies below D_2 then $(\mathcal{O}, D_3, s)$ followed by (D_3, U_R, s_d) is compatible ($s < s_d$). On the other hand, if U_R lies above D_2 then $(\mathcal{O}, D_3, s)$ followed by (D_3, U_R, s_u) is incompatible ($s_u < s$). In summary, the solution of the RP with data $\mathcal{O}$, U_R has D_1 as middle state if U_R lies above C; and D_3 as middle state if U_R lies below C. Therefore, the RP does not have a local solution, i.e., in the latter case there is no small amplitude solution. Because C starts at $\mathcal{O}$ we can choose D_2 as close to $\mathcal{O}$ as we wish, so there are RPs with data $U_L = \mathcal{O}$, U_R arbitrarily close to $\mathcal{O}$ with nonlocal solutions. The open set B (Fig. 6) lies between C and the coincidence curve. The proof is complete. □

We show that this behavior actually occurs also for U_L lying in triangles above $\mathcal{O}$. Let $T(\beta)$ be the family of open triangles defined in (6). For U_L in $T(\beta)$ the RH curve is shown in Fig. 7; the points M_i will be defined later.

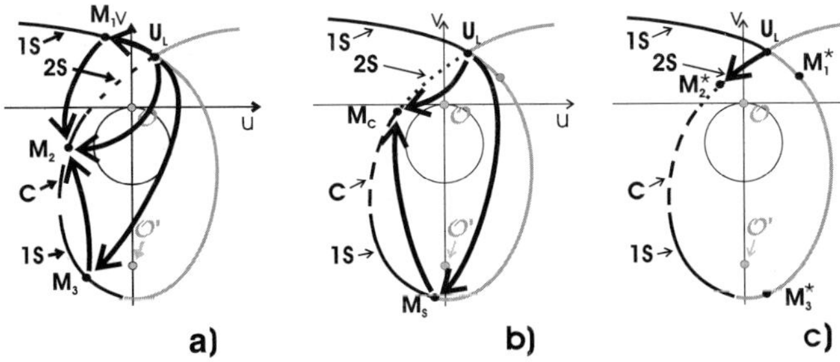

Fig. 8. Thick admissible shocks for $U_L \in T(\beta)$ and (**a**) $s < s_0$; (**b**) $s = s_0$; and (**c**) $s > s_0$

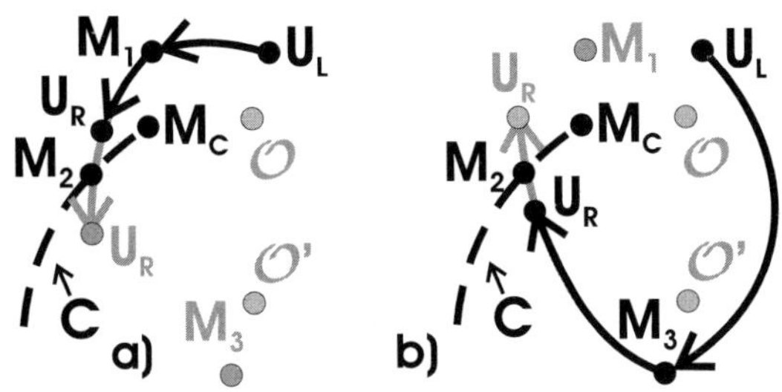

Fig. 9. Upper and lower sequences of shocks for $U_L \in T(\beta)$ and U_R out the compressive part but near M_2. Compatible sequences in *black*; incompatible sequences in *gray*

As U_L now lies in the SHR it has two characteristic speeds, and we set $s_0 = \lambda_1(U_L)$. The Lax theorem guarantees that the OC part does not reach U_L.

Let $U_- = U_L \in T(\beta)$. For $s < s_0$ there are four zeros of the RH condition, namely, U_L, M_1, M_2, and M_3 (Fig. 8a). They define the 1-shocks (U_L, M_1, s) and (U_L, M_3, s), the OC shock (U_L, M_2, s), and the 2-shocks (M_1, M_2, s) and (M_3, M_2, s). Therefore the RP with U_L and $U_R = M_2$ has multiple solutions in phase space that coincide in physical space.

By increasing the speed back to s_0 there are only three distinct zeros of the RH condition (Fig. 8b). This is so because the zeros M_1 and U_L collapse into each other but M_2 stays away from U_L and moves to the boundary between the 2-shock and the OC parts; M_3 moves to the boundary of the 1-shock part. In this case we rename M_2 and M_3 as M_C and M_S, respectively. The zeros define the left-characteristic 1-shock (U_L, M_S, s_0), the left-characteristic OC shock (U_L, M_C, s_0) and the 2-shock (M_C, M_S, s_0).

For $s > s_0$ there are again four distinct zeros of the RH condition, U_L, M_1^*, M_2^*, and M_3^* (Fig. 8c). However, there is just one admissible shock starting at U_L, namely the 2-shock from U_L to M_2^*.

We are interested in solutions of the RP with $U_L \in T(\beta)$ and right states near the part C of $\mathcal{H}(U_L)$, i.e., near a point M_2.

If U_R lies above C (Fig. 9a), the solution is a 1-shock from U_L to M_1 followed by a faster 2-shock from M_1 to U_R. We remark that the 1-shock from U_L to M_3 and the subsequent 2-shock from M_3 to U_R have incompatible speeds (Fig. 9b). Therefore, the RP with data U_L, U_R, with U_R above C has a unique local solution with middle state M_1, as in Lax Theorem.

On other hand if U_R lies below C (Fig. 9b) the solution is a 1-shock from U_L to M_3 followed by a faster 2-shock from M_3 to U_R. We will show that M_3 stays away from U_L, therefore, this RP has only a large amplitude solution. We remark that the 1-shock from U_L to M_1 followed by the 2-shock from M_1 to U_R have incompatible speeds.

We determine the end of the OC part, i.e., the point M_C separating the OC from the 2-shocks. We also locate M_S, the frontier of the 1-shocks.

Lemma 1. *For $U_L \in T(\beta)$ with small β we have $|U_L - M_C| < \beta$ and $|U_L - M_S| > 4$.*

Proof. Consider $M_C \equiv (u_C, v_C)$ and $M_S \equiv (u_S, v_S)$. For $U_L = (\alpha v_L, v_L) \in T(\beta)$, with $-1 < \alpha < 1$ and $0 < v_L < \beta^2/9$, calculations using (7) with $s = \lambda_1(U_L)$ lead to $v_C = -v_L - 2 + b$, $v_S = -v_L - 2 - b$, $u_i = 2\alpha v_L - a - (\alpha v_L^2 + a v_L)/v_i$ for $i = C, S$, with $a = \sqrt{2v_L + (1+\alpha^2)v_L^2}$, $b = \sqrt{4 + (6\alpha a - 2)v_L - 6(\alpha^2 - 2)v_L^2}$. For small positive β, a and b are real and both M_C and M_S lie in the SHR. Expanding the distances from U_L to M_S and M_C in v_L we have $|U_L, M_C| \simeq (5\sqrt{2v_L} - \alpha v_L)/3$ and $|U_L, M_S| \simeq 4 + 7v_L/4$, with error $O\left(v_L^{3/2}\right)$, so for small $v_L > 0$ we have $|U_L, M_C| < 3\sqrt{v_L} < \beta$ and $|U_L, M_S| > 4$. □

Consider the Riemann solution for U_R lying in the region below the part C left of M_C (Fig. 10). The 1-shock from U_L to M_3 near M_S has speed s slightly lower than s_0; the 2-shock from M_3 to U_R near M_2 and M_C has speed higher than s. By continuity $|U_L, M_3| > 4$ and $|U_L, U_R| < \beta$.

The sets $\mathcal{H}(U_L)$ and $\mathcal{H}(M_S)$ are transverse at M_C because neither U_L, M_C and M_S are collinear nor $\lambda_1(U_L)$ equals any characteristic speed of M_C. So we define a sector $A(U_L, \beta)$ (Fig. 10), with vertices on M_C and angle given

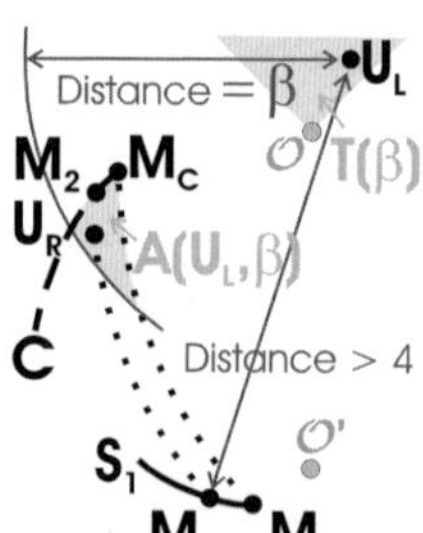

Fig. 10. The set $A(U_L, \beta)$

by the tangents of $\mathcal{H}(U_L)$ and $\mathcal{H}(M_S)$ at M_C and distance to U_L that is smaller than β. The proof is complete. □

5 Remarks

For RPs with a type IV umbilic point, it is shown in [1] that high amplitude solutions do not appear; this is so despite the fact that taking U_L near the umbilic point both the RH set and the solution of the RPs are analogous to our case for $U_L \in T(\beta)$ (Figs. 7 and 9). But, for U_L equal to umbilic point and characteristic speed the situation is different from $U_L = \mathcal{O}$ and $s = 0$, because in the umbilic case there is just one (quadruple) zero of the RH condition. Thus, there is no point playing the role of the zero $\mathcal{O}'$ (Fig. 1), which is fundamental for the high amplitude solutions in our example.

Acknowledgments

V. Matos had financial support from the Fundação para a Ciência e a Tecnologia (FCT), Portugal, through programs POCTI and POSI of Quadro Comunitário de Apoio III (2000–2006) with national and EU (FEDER) funding and from PCI of the MCT, Brazil, under Grant BEV 170207/04-8. This work was also supported in part by CNPq under Grant 301532/2003-06 and by FAPERJ under Grant E-26/150.163/2002.

References

1. Isaacson, E., Marchesin, D., Plohr, and Temple, B.: The Riemann Problem near a Hyperbolic Singularity: the Classification of Solutions of Quadratic Riemann Problems I, SIAM J. Appl. Math., Vol. 48, October, 1009–1032 (1988).
2. Liu, T.P.: The Riemann Problem for General 2×2 Conservation Laws, J. Trans. Amer. Math. Soc., Vol. 199, 89–112, (1974).
3. Marchesin, D., and Palmeira, F.: Topology of Elementary Waves for Mixed-Type Systems of Conservation Laws, J. Dyn. Diff. Eq., Vol. 6, 427–446 (1994).
4. Schaeffer, D., and Shearer, M., The Classification of 2x2 Systems of Non-strictly Hyperbolic Conservation Laws, with Application to Oil Recovery, with Appendix by D. Marchesin, P.J. Paes-Leme, Comm. Pure Appl. Math., Vol. 15, 141–178 (1987).

Asymptotic Behavior of Riemann Problem with Structure for Hyperbolic Dissipative Systems

A. Mentrelli and T. Ruggeri

Summary. We test for a 2×2 hyperbolic dissipative system, by numerical experiments, the conjecture according to which the solutions of Riemann problem and Riemann problem with structure converge, for large time, to a combination of shock structures (with or without subshocks) and rarefactions of the equilibrium subsystem.

1 Introduction

When modeling physical phenomena, systems of equations of both conservation and balance laws are often encountered. Nonequilibrium theories, like Extended Thermodynamics [MR98], are good examples. The behavior of smooth solutions of this kind of systems has only been successfully investigated for particular initial data (weakly perturbed constant states [HN03, Yon04, RS04, BHN05]) and no general theorems for the prediction of the behavior of the Riemann problem and Riemann problem with structure (when initial data are made of two constant equilibrium states smoothly connected, see Fig. 1) are available. With this regard, Brini and Ruggeri – following Liu [Liu96] – have recently proposed a conjecture according to which the solutions of both Riemann problems converge, for large time, to a combination of shock structures and rarefactions of the equilibrium subsystem [BR06a, BR06b]. Here, the conjecture is tested, in the case of the Riemann problem with structure, for a 2×2 dissipative hyperbolic system that has many features in common with physically relevant systems but is simple enough to allow an analytical approach to the calculation of the shock structure and rarefaction of the equilibrium subsystem. We have thus tested the conjecture by comparison of numerical and predicted analytical solution.

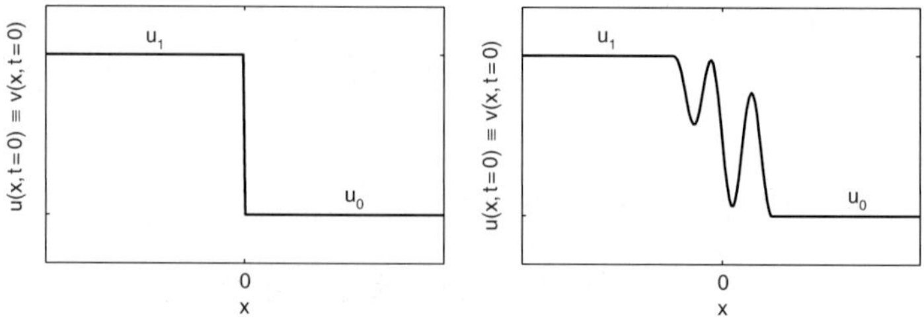

Fig. 1. Riemann problem (*left*) and Riemann problem with structure (*right*)

2 Dissipative Equilibrium Systems

When modeling physical phenomena, systems of balance laws are in the form

$$\frac{\partial \mathbf{F}^0(\mathbf{u})}{\partial t} + \frac{\partial \mathbf{F}^i(\mathbf{u})}{\partial x^i} = \mathbf{f}(\mathbf{u}) \tag{1}$$

where $\mathbf{F}^0(\mathbf{u}), \mathbf{F}^i(\mathbf{u}), \mathbf{f}(\mathbf{u}) \in \mathbb{R}^N$ are, respectively, the densities, the fluxes, and the productions ($i = 1, 2, 3$), while $\mathbf{u} \equiv \mathbf{u}(t, x^i)$ is the field. In physical cases every field $\mathbf{u}$ satisfying (1) satisfies also the supplementary law:

$$\frac{\partial h^0}{\partial t} + \frac{\partial h^i}{\partial x^i} = \Sigma \leq 0, \tag{2}$$

where in the thermodynamical case $-h^0$, $-h^i$ and Σ are the entropy density, flux, and production, respectively. If h^0 is a convex function of $\mathbf{u}$ in a general domain $D \subseteq \mathbb{R}^N$, it is known that a necessary and sufficient condition for the system (1) to admit a supplementary law (2) is the existence of a *main field* $\mathbf{u}'$ and a set of "potentials" $h'^0 \equiv h'^0(\mathbf{u}')$ and $h'^i \equiv h'^i(\mathbf{u}')$ such that [Boi74, Boi96, RS81]:

$$\mathbf{F}^0 = \frac{\partial h'^0}{\partial \mathbf{u}'}, \qquad \mathbf{F}^i = \frac{\partial h'^i}{\partial \mathbf{u}'}, \qquad \mathbf{u}' \cdot \mathbf{f} = \Sigma \leq 0.$$

Under these circumstances, the system (1) may be written as follows

$$\frac{\partial}{\partial t}\left(\frac{\partial h'^0}{\partial \mathbf{u}'}\right) + \frac{\partial}{\partial x^i}\left(\frac{\partial h'^i}{\partial \mathbf{u}'}\right) = \mathbf{f} \tag{3}$$

and it becomes a very special symmetric hyperbolic system where

$$h^0 = \mathbf{u}' \cdot \frac{\partial h'^0}{\partial \mathbf{u}'} - h'^0, \qquad h^i = \mathbf{u}' \cdot \frac{\partial h'^i}{\partial \mathbf{u}'} - h'^i.$$

If the system (3) is made up of M conservation laws $(0 < M < N)$ and $(N - M)$ balance laws, it may be written as [BR97]

$$\begin{aligned} \frac{\partial}{\partial t}\left(\frac{\partial h'^0}{\partial \mathbf{v}'}\right) + \frac{\partial}{\partial x^i}\left(\frac{\partial h'^i}{\partial \mathbf{v}'}\right) &= \mathbf{0} \\ \frac{\partial}{\partial t}\left(\frac{\partial h'^0}{\partial \mathbf{w}'}\right) + \frac{\partial}{\partial x^i}\left(\frac{\partial h'^i}{\partial \mathbf{w}'}\right) &= \mathbf{g}\left(\mathbf{v}', \mathbf{w}'\right) \end{aligned} \tag{4}$$

where $\mathbf{f} = (\mathbf{0}, \mathbf{g})$ with $\mathbf{g} \in \mathbb{R}^{N-M}$ and $\mathbf{v}' \in \mathbb{R}^M$, $\mathbf{w}' \in \mathbb{R}^{N-M}$ are such that $\mathbf{u}' \equiv (\mathbf{v}', \mathbf{w}')$. Assuming that the equilibrium state is the one for which Σ vanishes and attains the maximum value, Boillat and Ruggeri proved [BR97] that at equilibrium $\mathbf{w}'\Big|_E = 0$ and the principal equilibrium subsystems related to (4) is

$$\frac{\partial}{\partial t}\left(\frac{\partial h'^0}{\partial \mathbf{v}'}(\mathbf{v}', \mathbf{0})\right) + \frac{\partial}{\partial x^i}\left(\frac{\partial h'^i}{\partial \mathbf{v}'}(\mathbf{v}', \mathbf{0})\right) = \mathbf{0}. \tag{5}$$

Moreover $\mathbf{g} = -\mathbf{B}\mathbf{w}'$, being $\mathbf{B} \equiv \mathbf{B}\left(\mathbf{v}', \mathbf{w}'\right)$ a positive definite matrix at equilibrium, and the subcharacteristic conditions for (4) and (5) are automatically satisfied. These systems have been studied for initial data that are perturbations of constant states and it was proven that under the Shizuta-Kawashima condition [SK85, Kaw87] smooth solutions exist for all times, for small initial data, and constant states are stable [HN03, Yon04, RS04, BHN05].

3 A 2 × 2 Hyperbolic Dissipative System

We study the following hyperbolic dissipative toy system (τ is a relaxation parameter)

$$\begin{aligned} (u+v)_t + \left(v^2 + 2uv\right)_x &= 0 \\ u_t + \left(v^2\right)_x &= \frac{1}{\tau}(v - u) \end{aligned} \tag{6}$$

that has the following characteristics eigenvalues:

$$\lambda^{(1)} = u - \sqrt{u^2 + 4v^2}, \qquad \lambda^{(2)} = u + \sqrt{u^2 + 4v^2}. \tag{7}$$

The equilibrium subsystem of (6), and its eigenvalue μ, are

$$u_t + 3uu_x = 0, \qquad \mu = 3u. \tag{8}$$

4 Exact Solutions

We present the exact solutions of the system (6) and of its equilibrium subsystem (8) that, following the Brini–Ruggeri conjecture, are the asymptotic limit of the Riemann problem and Riemann problem with structure. If the constant equilibrium states characterizing the initial data are $\mathbf{u}_1 \equiv (u_1, v_1 = u_1)^T$ and $\mathbf{u}_0 \equiv (u_0, v_0 = u_0)^T$, three different cases are possible, depending on the value of the ratio $\gamma = u_1/u_0$. The first case (*case a*) arises when $1 < \gamma \leq \gamma_*$, with $\gamma_* = \left(2\sqrt{5} - 1\right)/3$, and it is characterized by a continuous shock structure solution of the system (6). The second case (*case b*) arises when $\gamma > \gamma_*$ and it corresponds to a shock structure solution with subshock of the system (6). Finally, the third case (*case c*), arising when $\gamma < 1$, is characterized by the rarefaction of the equilibrium subsystem (8). In the next sections, we briefly present these exact solutions (more details are available in [MR06]) and then, in Sect. 5, we verify by means of numerical experiments that the solution of (6) and (8) for the Riemann problem with structure converges for large time to one of these three solutions, depending on the value of γ.

4.1 Continuous Shock Structure Solution (*case a*)

We recall that a smooth shock structure solution may exist only for shock velocity s not greater than the maximum characteristic speed evaluated in the equilibrium unperturbed state [BR98]. Taking into account (7), this condition turns into $\gamma \leq \gamma_*$. The growth of the entropy gives another limitation on the constant states u_0 and u_1 for a continuous shock structure to exist. This condition implies, in the present case, $u_1 > u_0$, i.e. $\gamma > 1$.

Introducing the dimensionless variables $\hat{s} = s/u_0, U = u/u_0, V = v/u_0, z = (x - st)/(\tau u_0)$, where $s = 3(u_0 + u_1)/2$, we can see that the system (6) admits a shock structure solution given by

$$U = \frac{6\gamma - 3\left(1+\gamma\right)V + 2V^2}{3\left(1+\gamma\right) - 4V}, \tag{9}$$

$$-\frac{3\left(1+\gamma\right)}{2}\log\left|3\left(1+\gamma\right)+4V\right| + \frac{19\gamma^2 - 6\gamma - 9}{12\left(\gamma-1\right)}\log\left|\gamma - V\right| + \frac{9\gamma^2 + 6\gamma - 19}{12\left(\gamma - 1\right)}\log\left|V-1\right| + \frac{4}{3}V = z + C_1. \tag{10}$$

The presence of the constant C_1 is in agreement with the fact that the shock structure solutions are defined except for a translation.

4.2 Shock Structure Solution with Subshock (*case b*)

When $\gamma > \gamma_*$, a C^1 shock structure solution cannot exist. In this case a subshock appears and an analytical representation of the solution may still be given. In this case, we may obtain the values of the braking point (u_*, v_*), where the shock appears, by means of the Rankine–Hugoniot relation:

$$U_* = \frac{9\gamma^2 + 18\gamma + 5 - 2\sqrt{2\,(9\gamma^2 + 6\gamma - 1)}}{12\,(1+\gamma)},$$

$$V_* = \frac{-2 + \sqrt{2\,(9\gamma^2 + 6\gamma - 1)}}{4}.$$

Therefore, if $\gamma > \gamma_*$, there is a shock structure with subshock that is made up by a continuous solution connecting the states (u_1, u_1) and (u_*, v_*), and a shock connecting (u_*, v_*) and (u_0, u_0). The analytical representation of the solution connecting the states (u_1, u_1) and (u_*, v_*) is given in the dimensionless variables, for $\gamma \neq 3$, by the equations (9) and (10).

4.3 Rarefaction Solution of the Equilibrium Subsystem (*case c*)

The equilibrium subsystem (8) when $\gamma < 1$ admits the rarefaction solution

$$u(x,t) = v(x,t) = \begin{cases} u_1 & \frac{x}{t} \leq 3u_1 \\ \frac{x}{3t} & 3u_1 < \frac{x}{t} < 3u_0 \\ u_0 & \frac{x}{t} \geq 3u_0. \end{cases} \tag{11}$$

5 Results of the Numerical Simulations

To test the Brini–Ruggeri conjecture, solutions of the system (6) with $\tau = 10^{-2}$ have been calculated under initial data of both Riemann and Riemann with structure type by means of a numerical code based on the algorithms presented in [LRR00]. Simulations have been carried out for different values of u_1 and u_0, giving rise to different values of γ, in order to investigate the behavior of the system for the three main cases previously discussed in Sect. 4. In the following, numerical results of the Riemann problem with structure are presented and compared to the expected asymptotic solutions, obtained analytically, in order to test the conjecture.

The results of the simulation of a Riemann problem with structure with $u_1 = 1.1$ and $u_0 = 1$ (*case a*) are presented in Fig. 2, where the behavior of the solution in terms of U and V as functions of z for different times t is shown. It is easy to check that the behavior of both U and V seems to

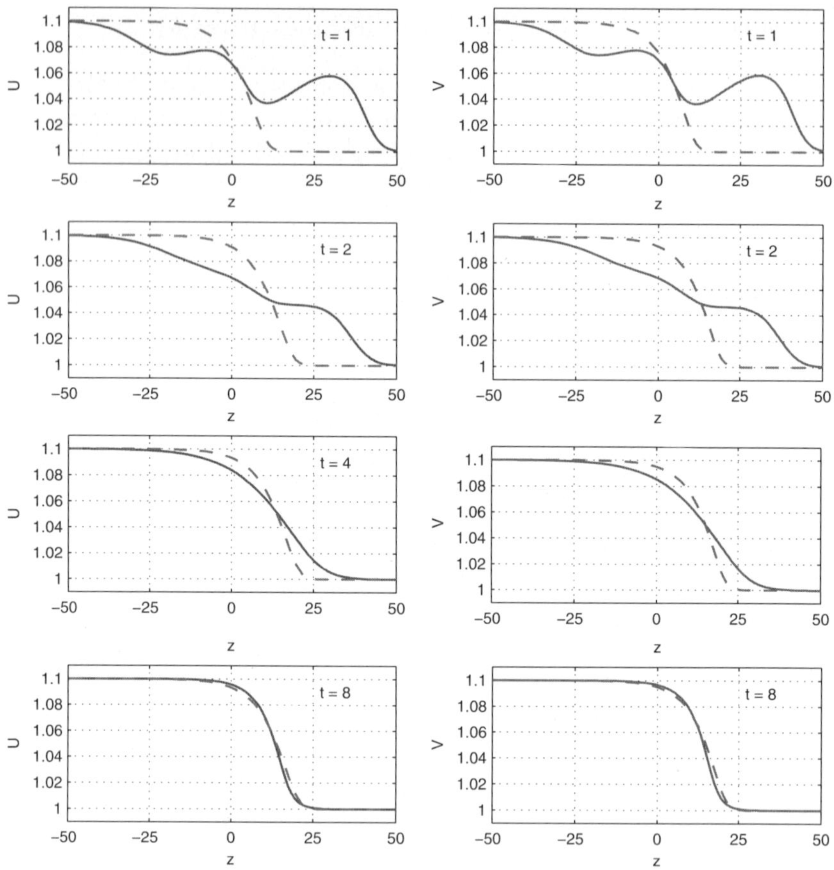

Fig. 2. *Case a*: numerical solution (*solid line*) and predicted asymptotic solution (*dashed line*) at different times t

asymptotically get closer and closer to the expected results. The behavior of the solution for the case with $u_1 = 2$ and $u_0 = 1$ (*case b*) are shown in Fig. 3. In this case, according to the conjecture, the solution of the system is expected to converge to a shock structure solution with a subshock, described in Sect. 4.2. As expected, a sharp discontinuity is met in the numerical results and the position of the point where the continuity is lost fits very well the expectations. Even in this case, the numerical solution is plotted together with the expected one, evaluated as described in Sect. 4.2. Finally, in the case with $u_1 = 1$ and $u_0 = 2$ (*case c*), we can see from Fig. 4 that the solution of the Riemann problem with structure converge to the equilibrium rarefaction given by (11).

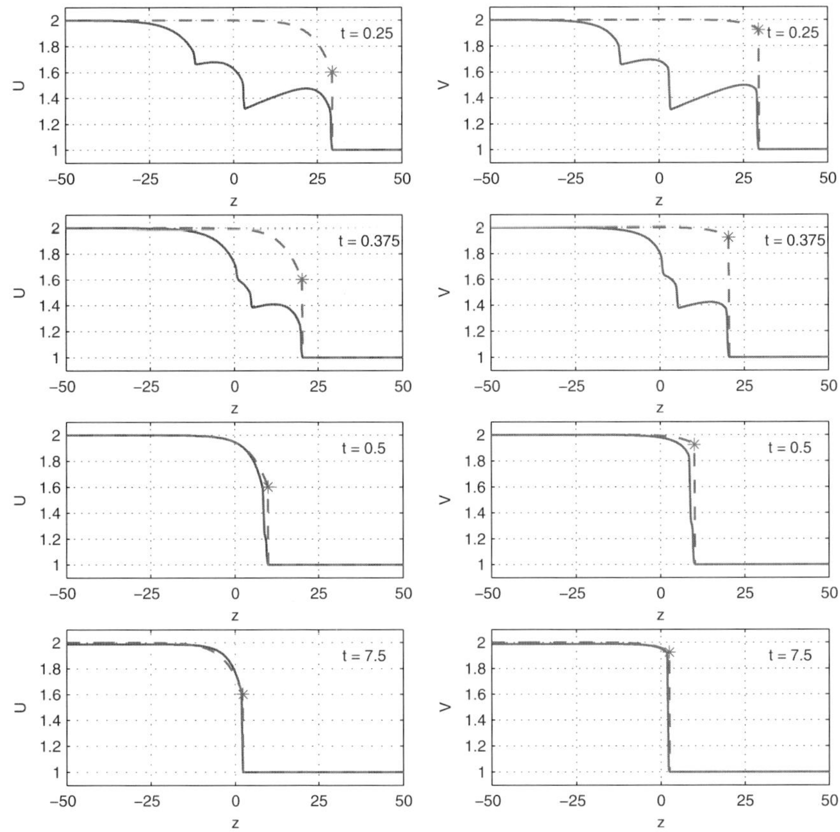

Fig. 3. *Case b*: numerical solution (*solid line*) and predicted asymptotic solution (*dashed line*) at different times t. The symbol $*$ indicates the point (U_*, V_*) where the continuity of the analytical solution is lost

A discussion of the results concerning the Riemann problem, as well as a deeper investigation on the Riemann problem with structure, are available in [MR06] and in a forthcoming paper where both Riemann problems are investigated in the case of more physically relevant dissipative hyperbolic systems.

Acknowledgments

This paper was supported by MIUR "Programma di Ricerca di Interesse Nazionale" (PRIN) *Non-linear propagation and stability in thermodynamical processes of continuous media* (Coordinator: T. Ruggeri) and

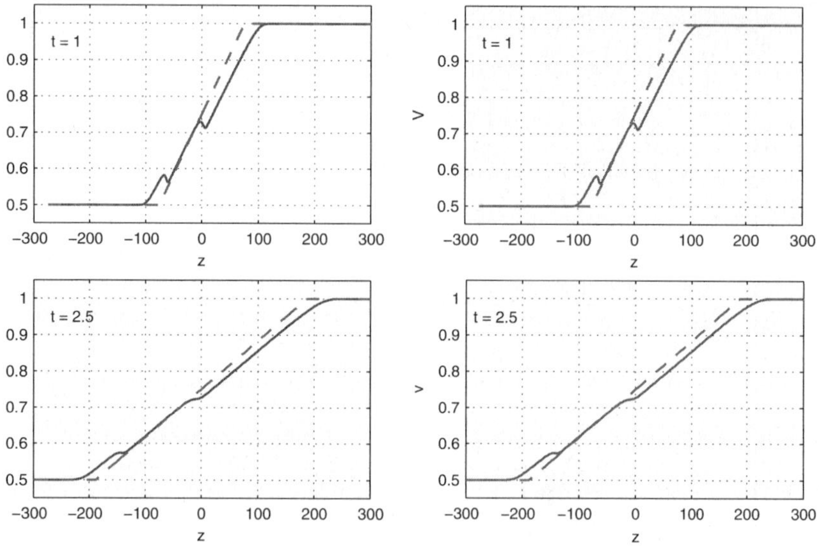

Fig. 4. *Case c*: numerical solution (*solid line*) and predicted asymptotic solution (*dashed line*) at different times t

by GNFM/INdAM "Progetto Giovani Ricercatori" *Asymptotic behavior of Riemann problem for hyperbolic dissipative systems* (Coordinator: A. Mentrelli).

References

[BHN05] Bianchini, S., Hanouzet, B., Natalini, R.: Asymptotic behavior of smooth solutions for partially dissipative hyperbolic systems with a convex entropy. IAC Report, **79** (2005)

[Boi74] Boillat, G.: Sur l'existence et la recherche d'équations de conservation supplémentaires pour les systèmes hyperboliques. C.R. Acad. Sc. Paris **278A** (1974)

[Boi96] Boillat, G.: Non linear hyperbolic fields and waves. In: Ruggeri, T. (ed) Recent Mathematical Methods in Nonlinear Wave Propagation, Lecture Notes in Mathematics, **1640**, 1–47, Springer-Verlag (1996)

[BR97] Boillat, G., Ruggeri, T.: Hyperbolic principal subsystems: Entropy convexity and subcharacteristic conditions. Arch. Rat. Mech. Anal., **137**, 305–320 (1997)

[BR98] Boillat, G., Ruggeri, T.: On the shock structure problem for hyperbolic system of balance laws and convex entropy. Continuum Mech. Thermodyn., **10**, 285–292 (1998)

[BR06a] Brini, F., Ruggeri, T.: On the Riemann problem in Extended Thermodynamics. Proceedings of the 10th Int. Conf. on Hyperbolic Problems (HYP2004); Osaka, September 13–17, 2004; Yokohama Pub. Inc., **I**, 319–326 (2006)

[BR06b] Brini, F., Ruggeri, T.: On the Riemann problem with structure in Extended Thermodynamics. Suppl. Rend. Circ. Mat. Palermo "Non Linear Hyperbolic Fields and Waves, A tribute to Guy Boillat", Serie II, **78**, 21–33 (2006)

[HN03] Hanouzet, B., Natalini, R.: Global existence of smooth solutions for partially dissipative hyperbolic systems with a convex entropy. Arch. Rat. Mech. Anal., **169**, 89–117 (2003)

[Kaw87] Kawashima, S.: Large-time behavior of solutions to hyperbolic-parabolic systems of conservation laws and applications. Proc. Roy. Soc. Edimburgh, **106A**, 169–194 (1987)

[Liu96] Liu, T.-P.: Nonlinear hyperbolic-dissipative partial differential equations. In: Ruggeri, T. (ed) Recent Mathematical Methods in Nonlinear Wave Propagation, Lecture Notes in Mathematics, **1640**, 103–136, Springer-Verlag (1996)

[LRR00] Liotta, S.F., Romano, V., Russo, G.: Central scheme for balance laws of relaxation type. SIAM J. Numer. Anal., **38**, 1337–1356 (2000)

[MR98] Müller, I., Ruggeri, T.: Rational Extended Thermodynamics. Springer Tracts in Natual Philosophy **37**, Second Edition, Springer Verlag (1998)

[MR06] Mentrelli, A., Ruggeri, T.: Asymptotic behavior of Riemann and Riemann with structure problems for a 2×2 hyperbolic dissipative system. Suppl. Rend. Circ. Mat. Palermo "Non Linear Hyperbolic Fields and Waves, A tribute to Guy Boillat", Serie II, **78**, 201–226 (2006)

[RS81] Ruggeri, T., Strumia, A.: Main field and convex covariant density for quasi-linear hyperbolic systems. Relativistic fluid dynamics. Ann. Inst. H. Poincaré, **34**, 65–84 (1981)

[RS04] Ruggeri, T., Serre, D.: Stability of constant equilibrium state for dissipative balance laws system with a convex entropy. Quarterly of Applied Math., **62** (1), 163–179 (2004)

[SK85] Shizuta, Y., Kawashima, S.: Systems of equations of hyperbolic-parabolic type with applications to the discrete Boltzmann equation. Hokkaido Math. J., **14**, 249–275 (1985)

[Yon04] Yong, W.-A.: Entropy and global existence for hyperbolic balance laws. Arch. Rat. Mech. Anal., **172** (2), 247–266 (2004)

Maximal Entropy Solutions for a Scalar Conservation Law with Discontinuous Flux

S. Mishra

1 Introduction

We are interested in the following one-dimensional single conservation law,

$$\begin{aligned} u_t + (f(k(x), u))_x &= 0 \\ u(x, 0) &= u_0(x), \end{aligned} \tag{1}$$

with u as the unknown, f as the flux, and k as a spatially dependent coefficient, which may be discontinuous. The simplest form of (1) is the so-called 2−flux case given by

$$u_t + (H(x)f(u) + (1 - H(x))g(u))_x = 0, \tag{2}$$

where H is the Heaviside function. For the rest of this paper, we will consider (2) for simplicity. All results obtained for it can be easily transfered to (1). See [4] for details.

Equations of the type (1) arise in a wide variety of applications like two-phase flows in a heterogeneous porous medium, the simulation of the Clarifier-Thickener unit used in waste water treatment plants and in traffic flow on highways with changing surface conditions. A detailed account of applications is provided in [10].

The main issues in the study of (1) is to formulate a suitable concept of entropy solutions for these equations and to investigate the existence and stability of entropy solutions. Another crucial issue is to derive suitable numerical schemes to approximate the entropy solutions. A lot of papers in recent years have addressed these questions in great detail, see [6],[5],[1],[7],[11],[12] and the references therein to name just a few.

One interesting feature of (1) was the discovery that it can give rise to different classes of meaningful entropy solutions in different physical regimes. In particular, the semigroup identified in [7] in the context of a vanishing viscosity limit was different from the one identified in [1], which corresponded to the physically meaningful solutions for two-phase flows in heterogeneous porous media. This surprising feature was formalized for (2) with convex fluxes in [2].

In [2], the authors were able to identify infinitely many L^1 stable semigroups of entropy solutions for (1). Each of these semigroups was characterized in terms of an interface connection. Existence was proved by showing convergence for a Godunov type scheme and stability was based on a doubling of variables argument. The authors extended these results to general fluxes in [3] and to the more general equation (1) in [4].

Hence, the task of formulating a suitable concept of entropy solution for a given physical regime reduced to choosing a suitable connection. In [2], the authors proposed a concept of *optimal* entropy solutions for (2) based on optimizing a suitable functional at the interface. This functional was a measure of the total variation of the solution (measured in terms of a singular mapping) and the resulting optimizing connection turned out to be unique and coincided with the physically relevant nonundercompressive solution for two-phase flows in porous media. Yet, one undesirable feature of the interface functional was that it depended on the variation of the solution and did not rely on any entropy criteria. This approach is also difficult to extend for more general flux geometries. See [3] for a discussion.

We propose a new approach to address the above problem in this paper. Our approach is based on defining a new functional depending on the entropy dissipated across the interface and we demand that not only is entropy dissipated across the interface, but we also choose the connection that dissipates the maximum entropy across the interface. For convex fluxes, this leads to a unique choice of the maximizer that corresponds to the *maximal* entropy solutions for (2). The maximal entropy solutions are easy to extend to more general fluxes and coincide with the nonundercompressive solution that is physically meaningful in the context of two-phase flows in a heterogeneous porous medium. In fact, we believe that a traveling wave analysis will lead to the same entropy criteria, although the calculations are beyond the scope of this paper.

The situation above is similar to the one encountered while dealing with nonclassical shocks resulting from vanishing diffusion-dissipation limits of conservation laws. See [8] for more details. In [8], there are infinitely many stable solutions and the physically relevant solution is selected by a maximization of the entropy dissipation measure. This motivates us to use a similar approach for (1).

Another novel feature of this paper is to derive a suitable modification of the Enquist–Osher scheme for approximating any semigroup of entropy solutions for (2). In [2], it was mentioned that we need Riemann solvers to be able to approximate these infinite classes of solutions but the modifications of the Enquist–Osher flux suggests that we can work with finite difference schemes in approximating these infinite semigroups of solutions. The rest of this paper is organized as follows. In Sect. (2),we state the hypothesis of the fluxes and identify the infinite classes of entropy solutions for (2). The Enquist–Osher flux is proposed and shown to converge in Sect. (3) and the concept of maximal entropy solutions is described in Sect. (4). We conclude with some numerical experiments in Sect. (5).

2 AB-Entropy Solutions

Let $-\infty < s \leq S < \infty$. We consider the fluxes f, g with the following hypothesis,

H_1 . $f, g \in \mathrm{Lip}[s, S]$.
H_2 . $f(s) = g(s), f(S) = g(S)$.
H_3 . g, f have at most one minimum (resp. maximum) and no maxima (resp. minima) in $[s, S]$.

The above hypothesis are standard for fluxes that arise in two-phase flows across changing rock types. See Fig. 1 for a typical example of the fluxes. We use the standard definition of weak solutions of (2) and the standard Kruzkhov type interior entropy conditions. See Sect. 2 of [2], Definitions 2.1 and 2.3. We know that the interior entropy conditions are not enough to guarantee uniqueness. Hence as in [2], we proceed to define the following.

Definition 1. *Connection : In case the fluxes f and g satisfy the hypothesis (H_1, H_2, H_3) with θ_f be the unique minimum of f and θ_g being the unique minimum of g, then the pair $(A, B) \in [s, S]$ is said to be a connection if satisfies the following,*
1. $g(A) = f(B)$.
2. $A \leq \theta_g$, and $\theta_f \geq B$.
It is said to be undercompressive if $A < \theta_g$ and $B > \theta_f$.

Some examples of connections are shown in Fig. 1.

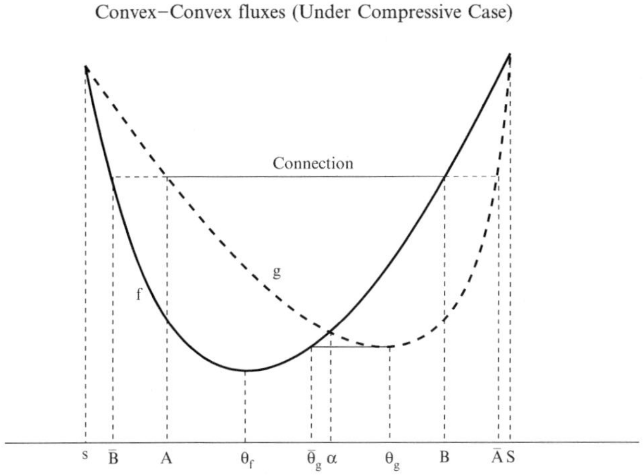

Fig. 1. Shape of fluxes f, g and connections

Definition 2. *Interface entropy functional I_{AB}: For any time t, we assume that the traces at the interface $u^-(t) = u(0-,t)$ and $u^+(t) = u(0+,t)$ exist. Then, we define the following Interface entropy functional relative to the connection (A,B) as*

$$I_{AB}(t) = sign\ (u^-(t) - A)(g(u^-(t)) - g(A)) - sign\ (u^+(t) - B)(f(u^+(t)) - f(B)) \tag{3}$$

Now, we are in position to define the following.

Definition 3. *Interface Entropy Condition: We have the following interface entropy condition relative to a connection (A,B),*

$$I_{AB}(t) \geq 0 \quad a.e \quad t \tag{4}$$

Note that this condition is defined with respect to each connection (A,B). Now for every choice of connection (A,B), we are in position to define the AB-entropy solution as the following.

Definition 4. *AB-Entropy solution: A function $u \in L^\infty((-\infty,\infty) \times [0,\infty)$ is defined as the entropy solution of (2) relative to the connection (A,B) if the following holds,*
1. u is a weak solution of (2).
2. u satisfies the interior entropy conditions, and
3. u satisfies the interface entropy condition (4) relative to the connection (A,B).

We state the following existence and stability theorem from [2] (without proof).

Theorem 1. *Let the fluxes satisfy the hypothesis H_1, H_2, and H_3 and the let the initial data $u_0 \in L^\infty(s,S) \cap BV(s,S)$, then for each choice of connection (A,B), the corresponding AB-entropy solutions exist and are unique. Furthermore, let u, v be the AB-entropy solutions corresponding to data u_0, v_0, then the following estimate holds,*

$$\int_{-\infty}^{\infty} |u(x,t) - v(x,t)| dx \leq \int_{-\infty}^{\infty} |u_0(x) - v_0(x)| dx \tag{5}$$

The proof of the above theorem is presented in detail in [2]. Existence is proved by showing convergence for Godunov-type schemes and stability follows by a Kruzkhov-type doubling of variables argument. In the next section, we provide an alternative existence proof by showing convergence for a Enquist–Osher type scheme.

3 Enquist–Osher Type Scheme

We start with a few definitions

Definition 5. *Let $h \in Lip[s,S]$ be a given function. Then, the standard Enquist–Osher flux corresponding to h is given by*

$$H(a,b) = \frac{1}{2}(h(a) + h(b) - \int_a^b |h'(\xi)| d\xi). \tag{6}$$

For the special case where h has at most one minimum $\theta \in [s,S]$ and no maxima, (6) has a very simple form given by

$$H(a,b) = h(\max(a,\theta)) + h(\min(\theta,b)) - h(\theta). \tag{7}$$

Next we need to define a flux at the interface below.

Definition 6. *Interface Enquist–Osher Flux: Let (A,B) be a connection for the fluxes f and g satisfying the hypothesis (H_1, H_2, H_3), then the interface Enquist–Osher flux F_{AB} is defined as*

$$F_{AB}(a,b) = g(\max(a,A)) + f(\min(b,B)) - g(A). \tag{8}$$

Note that the above interface flux is a simple modification of the standard Enquist–Osher flux to this situation. If we take $(A,B) = (\theta_g, \overline{\theta_f})$, then when $f = g$, (8) reduces to the standard Enquist–Osher flux thus preserving the essential consistency.

We discritize in space on a uniform grid with mesh parameters $h > 0$ and define the space grid points x_i as follows.

$$x_i = \left(\frac{2i-1}{2}\right) h \quad \text{for } i \geq 1, \quad x_i = \left(\frac{2i+1}{2}\right) h \quad \text{for } i \leq -1.$$

Similarly we discretize uniformly in time with the time step Δt and denote the discrete time levels as t^n. The CFL condition is given by

$$2\frac{\Delta t}{h} M \leq 1,$$

where $M = \max\{Lip(f), Lip(g)\}$. The Enquist–Osher scheme is then given by

$$\begin{aligned}
u_i^{n+1} &= u_i^n - \lambda(F(u_i^n, u_{i+1}^n) - F(u_{i-1}^n, u_i^n)) \qquad && \text{if } i \geq 2 \\
u_1^{n+1} &= u_1^n - \lambda(F(u_1^n, u_2^n) - F_{AB}(u_{-1}^n, u_1^n)) && \\
u_{-1}^{n+1} &= u_{-1}^n - \lambda(F_{AB}(u_{-1}^n, u_1^n) - G(u_{-2}^n, u_{-1}^n)), && \\
u_i^{n+1} &= u_i^n - \lambda(G(u_i^n, u_{i+1}^n) - G(u_{i-1}^n, u_i^n)) && \text{if } i \leq -2,
\end{aligned} \tag{9}$$

where F, G are the standard Enquist–Osher fluxes corresponding to f and g respectively and F_{AB} is the interface flux. We also need the approximating functions

$$u_h(x,t) = u_i^n \text{ for } (x,t) \in [x_{i-/2}, x_{i+1/2}) \times [n\Delta t, (n+1)\Delta t), \quad i \neq 0. \tag{10}$$

We have the following convergence theorem.

Theorem 2. *For $h > 0$, let u_h be the corresponding calculated solution given by (9). Then there exists a subsequence $h_k \to 0$ such that u_{h_k} converges a.e. to a weak solution u of (2) satisfying interior entropy condition. Suppose the discontinuities of every limit function u of $\{u_h\}$ is a discrete set of Lipschitz curves; then $u_h \to u$ in $L^\infty_{loc}((0,\infty), L^1_{loc}(-\infty,\infty))$ as $h \to 0$, and u satisfies the interface entropy condition (4) relative to the connection (A, B).*

Sketch of the Proof: The proof of convergence is based on a L^∞ estimate $s \le u_i^n \le S$, which follows from a invariant region principle. The key step is to use the singular mapping method to control the variation of the approximate solutions. This is presented in detail in the paper [9] for the Enquist–Osher scheme. An easy modification of the arguments in [9] apply in this case. The details of the calculation are not presented here.

4 Maximal Entropy Solutions

Theorem (1) gives the existence of infinitely many stable semigroups of solutions for (2). The problem of choosing one of these infinite classes is hard and depends on the physical regime being considered. As pointed out in the introduction, we will be presenting a new concept of solutions – the so called maximal entropy solutions. The key step is that in the proof of stability (hence uniqueness), we need a sign on the interface entropy functional I_{AB} (see (3)). This functional measures the entropy dissipated across the interface for each choice of connection and we want it to be nonnegative for each admissible connection in order to obtain uniqueness. It is then reasonable to demand that instead of just asking for nonnegativity, we want to maximize the maximum entropy dissipated across the interface for each connection among all admissible connections. This gives rise to the concept of a maximal connection defined below. First we need to define the admissible class of trace values,

Definition 7. *Admissible traces: The pair $u^-, u^+ \in [s, S]$ are defined as admissible traces for (2) with respect to the connection (A, B) if the following holds*

1. $g(u^-) = f(u^+)$.
2. $I_{AB}(u^-, u^+) \ge 0$.

The set of admissible traces with respect to the connection (A, B) is denoted by T_{AB}.

Definition 8. *Maximal connection: A connection (A_0, B_0) is defined to be a maximal entropy connection if the following holds*

$$\max_{u^-,u^+ \in T_{A_0B_0}} I_{A_0B_0}(u^-, u^+) = \max_{A,B} \max_{u^-,u^+ \in T_{AB}} I_{AB}(u^-, u^+). \tag{11}$$

The corresponding A_0B_0-entropy solution is defined as the maximal entropy solution of (2).

Therefore, for each choice of admissible connection we measure the maximum entropy dissipated at the interface and then maximize this quantity across all admissible connections. The characterization of the maximal entropy solutions is done in the following theorem.

Theorem 3. *Let f, g satisfy the hypothesis H_1, H_2, and H_3, For simplicity, assume that $g(\theta_g) \geq f(\theta_f)$, then there exists a unique maximal connection given by*

$$(A_0, B_0) = (\theta_g, \overline{\theta_f}),$$

where $\overline{\theta_f} > \theta_f$ such that $g(\theta_g) = f(\overline{\theta_f})$.

Note that the maximal entropy solution coincides with the physically relevant vanishing capillarity solution for two-phase flows in a heterogeneous porous medium.

Proof. First we characterize the admissible traces in the claim below.
Claim: If $u^-, u^+ \in T_{AB}$, then $g(u^-) \geq g(A)$.

To prove the above claim, we proceed by contradiction. Let $g(u^-) < g(A)$. Then as $(u^-, u^+) \in T_{AB}$, we have that

$$I_{AB}(u^-, u^+) = (\text{sign}(u^- - A) - \text{sign}(u^+ - B))(g(u^-) - g(A)) \geq 0.$$

This implies that $u^{\leq} A$ and $u^+ > B$. But as g is decreasing in (s, A), this implies that $g(u^-) > g(A)$, which is a contradiction, thus proving the claim.

It is easy to check that for each connection (A, B) and for the given flux geometry, we have

$$\max_{(u^-, u^+ \in T_{AB})} I_{AB}(u^-, u^+) = \max\{g(s), g(S)\} - g(A).$$

Now, when we maximize over all connections (A, B), we optimize the above quantity when $A = \theta_g$. Hence $(\theta_g, \overline{\theta_f})$ is the maximal entropy connection in this case. □

5 Numerical Experiments

We present a simple numerical experiment with the following fluxes and data

$$f(u) = \frac{2u(u-1)}{u+1}, \; g(u) = \frac{2u(u-1)}{2-u}$$
$$u(x, 0) \equiv 0.5.$$

The above fluxes represent flows across two rock-types for two phase separated at $x = 0$. The initial mixture of phases is redistributed by gravity. We show the solutions obtained with an Enquist–Osher scheme (9) for the maximal entropy connection in Fig. 2. As expected, the schemes gives good resolution and converges to the maximal solution.

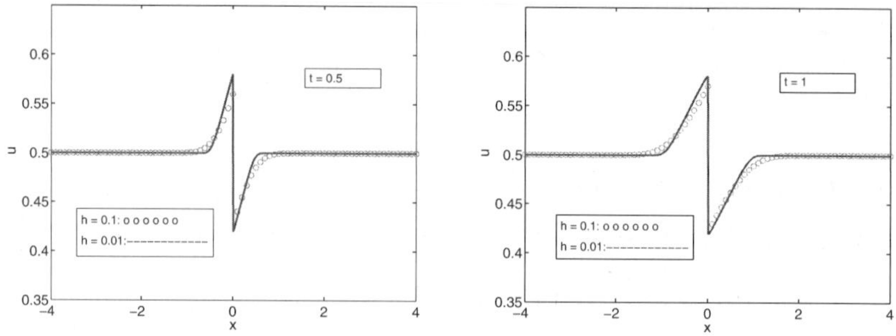

Fig. 2. Approximate solutions by the Enquist–Osher scheme for the maximal entropy solutions with $h = 0.1$ and $h = 0.01$ at times $t = 0.5$ (*left*) and $t = 1$ (*right*)

References

1. Adimurthi, J. Jaffre and G.D. Veerappa Gowda. Godunov type methods for Scalar Conservation Laws with Flux function discontinuous in the space variable. *SIAM J. Numer. Anal.*, 42(1): 179–208, 2004.
2. Adimurthi, Siddhartha Mishra and G.D. Veerappa Gowda. Optimal entropy solutions for conservation laws with discontinuous flux. *Journal of Hyp. Diff. Eqns.*, 2(4): 787–838, 2005.
3. Adimurthi, Siddhartha Mishra and G.D. Veerappa Gowda. Conservation laws with flux function discontinuous in the space variable - III, The general case. *Preprint*, 2005.
4. Adimurthi, Siddhartha Mishra and G.D. Veerappa Gowda. Convergence of Godunov type methods for conservation laws with spatially varying discontinuous flux functions. *Math. Comput.*, 76: 1219–1242, 2007.
5. R. Bürger, K.H. Karlsen, N.H. Risebro and J.D. Towers. Well-posedness in BV_t and convergence of a difference scheme for continuous sedimentation in ideal clarifier thickener units. *Numer. Math.*, 97(1): 25–65, 2004.
6. T. Gimse and N.H. Risebro. Solution of Cauchy problem for a conservation law with discontinuous flux function. *SIAM J. Math. Anal*, 23(3): 635–648, 1992.
7. K.H. Karlsen, N.H. Risebro and J.D. Towers. L^1 stability for entropy solution of nonlinear degenerate parabolic convection-diffusion equations with discontinuous coefficients. *Skr. K. Nor. Vidensk. Selsk.*, 3, 2003, 49 pages.
8. P. LeFloch. Hyperbolic systems of conservation laws: The theory of classical and non-classical shock waves. Birkhauser, Basel, 2002.
9. S. Mishra. Convergence of upwind finite difference schemes for a scalar conservation law with indefinite discontinuities in the flux function. *SIAM Jl. Num. Anal*, 43(2): 559–577, 2005.
10. S. Mishra. Analysis and Numerical approximation of conservation laws with discontinuous coefficients *PhD Thesis,* Indian Institute of Science, Bangalore, 2005.
11. J.D. Towers. Convergence of a difference scheme for conservation laws with a discontinuous flux. *SIAM J. Numer. Anal.*, 38(2): 681–698, 2000.
12. J.D. Towers. A difference scheme for conservation laws with a discontinuous flux-the nonconvex case. *SIAM J. Numer. Anal.*, 39(4): 1197–1218, 2001.

Semidiscrete Entropy Satisfying Approximate Riemann Solvers and Application to the Suliciu Relaxation Approximation

T. Morales and F. Bouchut

Summary. We establish conditions for an approximate simple Riemann solver to satisfy a semidiscrete entropy inequality. Classically, a discrete entropy inequality allows to analyze the stability of a numerical scheme for a conservative system. A semidiscrete entropy inequality gives a simpler and less restrictive approach than a fully discrete entropy inequality and leads to the definition of less restrictive conditions for numerical schemes to satisfy. First, conditions are established in an abstract framework for simple Riemann solvers to satisfy a semidiscrete entropy inequality and then the results are applied, as a particular case, to the Suliciu system. The Suliciu relaxation system is attached to the resolution of the isentropic gas dynamics system and can also handle full gas dynamics. It allows to define a simple approximate Riemann solver for gas dynamics. Conditions have already been established for the scheme to be entropy satisfying. Our approach allows to relax the conditions established in the fully discrete case and leads to the definition of a numerical scheme for gas dynamics that satisfies a semidiscrete entropy inequality while allowing exact resolution of shocks.

1 Introduction

A good solver for a conservation law has to be at the same time accurate, stable, and sufficiently simple. A way to analyze the stability is to verify some entropy inequality, which is the approach introduced in [2]. A desirable property for the scheme is to solve shocks exactly.

Consider, as a particular case, the isentropic gas dynamics system. The Suliciu relaxation system, described in [4], [1], and [2], allows the definition of a discrete entropy satisfying scheme for the isentropic gas dynamics system, as it is shown in [2]. But under this approach, the shocks are not preserved. On the other hand, some schemes have been presented in [5], and [3] that preserve shocks but are not entropy satisfying.

We aim the definition of a numerical scheme for the isentropic gas dynamics that verifies, at the same time, some entropy inequality and gives the correct speed for shocks. This will be done using the concept of semidiscrete entropy satisfying scheme, which is less restrictive than fully discrete entropy satisfying. Doing so, we will be able to impose other properties to the scheme such as the preservation of shocks.

We will first introduce the main concepts and results related to entropy inequalities in an theoretical framework. Then, we will focus in the particular case of the isentropic gas dynamics system and the Suliciu solver.

Everything can be extended to other systems and, in particular, to the full gas dynamics.

2 Entropy Inequalities

We consider a system of conservation laws

$$\partial_t U + \partial_x(F(U)) = 0, \tag{1}$$

and we define a numerical scheme for this system

$$U_i^{n+1} - U_i^n + \frac{\Delta t}{\Delta x_i}(F(U_i^n, U_{i+1}^n) - F(U_{i-1}^n, U_i^n)) = 0. \tag{2}$$

An entropy for system (1) is a convex function $\eta(U)$ with real values such that there exists another real-valued function $G(U)$, called the entropy flux, satisfying $G'(U) = \eta'(U)F'(U)$.

Consider a scheme (2) and suppose a numerical entropy flux function $G(U_l, U_r)$ that is consistent with the exact entropy flux, in the sense that $G(U, U) = G(U)$.

Definition 1. *The scheme (2) satisfies a discrete entropy inequality associated to the convex entropy η and the numerical entropy flux G if, under some CFL condition, the discrete values computed by (2) satisfy*

$$\eta(U_i^{n+1}) - \eta(U_i^n) + \frac{\Delta t}{\Delta x_i}(G(U_i^n, U_{i+1}^n) - G(U_{i-1}^n, U_i^n)) \leq 0. \tag{3}$$

Definition 2. *The scheme (2) satisfies a semidiscrete entropy inequality associated to the convex entropy η and the numerical entropy flux G if*

$$G(U_r) + \eta'(U_r)(F(U_l, U_r) - F(U_r)) \leq G(U_l, U_r), \tag{4}$$

$$G(U_l, U_r) \leq G(U_l) + \eta'(U_l)(F(U_l, U_r) - F(U_l)). \tag{5}$$

Lemma 1. *If the scheme is discrete entropy satisfying (Definition 1) then it is semidiscrete entropy satisfying (Definition 2).*

For further details refer to [2].

3 Riemann Solvers

Definition 3. *An approximate Riemann solver for (1) is a vector function $R(x/t, U_l, U_r)$, which is an approximation of the solution to the Riemann problem with data U_l, U_r, in the sense that it must satisfy the consistency relation*

$$R(x/t, U, U) = U, \tag{6}$$

and the conservative identity $F_l(U_l, U_r) = F_r(U_l, U_r)$, where

$$F_l(U_l, U_r) = F(U_l) - \int_{-\infty}^{0} (R(v, U_l, U_r) - U_l)\, dv, \tag{7}$$

$$F_r(U_l, U_r) = F(U_r) + \int_{0}^{\infty} (R(v, U_l, U_r) - U_r)\, dv. \tag{8}$$

Definition 4. *An approximate Riemann solver for (1) is said to preserve the entropic shocks of the system if $R(x/t, U_l, U_r)$ is the exact solution to the Riemann problem with initial data U_l, U_r, whenever U_l, U_r can be connected by an entropic shock.*

Lemma 2. *Given an approximate Riemann solver R, we define the left and right entropy fluxes*

$$G_l(U_l, U_r) = G(U_l) - \int_{-\infty}^{0} (\eta(R(v, U_l, U_r)) - \eta(U_l))\, dv, \tag{9}$$

$$G_r(U_l, U_r) = G(U_r) + \int_{0}^{\infty} (\eta(R(v, U_l, U_r)) - \eta(U_r))\, dv. \tag{10}$$

Suppose that

$$G_r(U_l, U_r) - G_l(U_l, U_r) \leq 0, \tag{11}$$

then, under a CFL condition 1/2, the scheme associated to the approximate Riemann solver R satisfies (3) for any $G_r(U_l, U_r) \leq G(U_l, U_r) \leq G_l(U_l, U_r)$.

The proof of this lemma can be found in [2].

Lemma 3. *Given an approximate Riemann solver R, we define the semidiscrete left and right entropy fluxes*

$$G_l^s(U_l, U_r) = G(U_l) - \eta'(U_l) \int_{-\infty}^{0} (R(v, U_l, U_r) - U_l)\, dv, \tag{12}$$

$$G_r^s(U_l, U_r) = G(U_r) + \eta'(U_r) \int_{0}^{\infty} (R(v, U_l, U_r) - U_r)\, dv, \tag{13}$$

which are linearizations of (9) and (10).
The scheme associated to R is semidiscrete entropy satisfying with respect to the convex entropy η if, and only if,

$$G_r^s(U_l, U_r) - G_l^s(U_l, U_r) \leq 0. \tag{14}$$

Proof. One can take any G verifying $G_r^s(U_l, U_r) \leq G(U_l, U_r) \leq G_l^s(U_l, U_r)$. Then, using (7) and (8), we recover (4) and (5).

Definition 5. *We say that a Riemann solver R is a simple solver if there exists a finite number $m \geq 1$ of speeds $\sigma_0 = -\infty < \sigma_1 < \ldots < \sigma_m < \sigma_{m+1} = +\infty$, and intermediate states $U_0 = U_l, U_1, \ldots, U_{m-1}, U_m = U_r$ such that*

$$R(x/t, U_l, U_r) = U_i \quad \text{if } \sigma_i < x/t < \sigma_{i+1}. \tag{15}$$

Proposition 1. *Let R be a simple solver and k such that $\sigma_k \leq 0 < \sigma_{k+1}$. A sufficient condition for (14) to hold is that there exist quantities E_i, $i = 0, \ldots, m$ satisfying*

$$G(U_r) - G(U_l) = \sum_{i=0}^{m-1} \sigma_{i+1}(E_{i+1} - E_i), \tag{16}$$

$$E_m = \eta(U_r), \quad E_0 = \eta(U_l), \tag{17}$$

$$\begin{aligned} E_i &\geq \eta'(U_l)(U_i - U_l) + \eta(U_l), \quad i = 1, \ldots, k-1, k, \\ E_i &\geq \eta'(U_r)(U_i - U_r) + \eta(U_r), \quad i = k, k+1, \ldots, m-1. \end{aligned} \tag{18}$$

Proof. The idea of the proof is to use the inequalities (18) in (16) and the particular expression of (14) for a simple solver.

Remark 1. There is an analogue result for fully discrete entropy inequalities. The condition $E_i \geq \eta(U_i)$, $i = 1, \ldots, m-1$ is a sufficient condition for (11) to hold. As η is convex, this is a stronger condition than (18).

4 Modified Suliciu Relaxation System

4.1 Suliciu Solver

Consider the isentropic gas dynamics system

$$\begin{cases} \partial_t \rho + \partial_x(\rho u) = 0, \\ \partial_t(\rho u) + \partial_x(\rho u^2 + p(\rho)) = 0. \end{cases} \tag{19}$$

We will note (G, η) the entropy pair for this system,

$$\eta(U) = \rho u^2/2 + \rho e(\rho), \quad G(U) = (\rho u^2/2 + \rho e + p(\rho))u, \tag{20}$$

where $e'(\rho) = \frac{p(\rho)}{\rho^2}$.

We introduce now the Suliciu relaxation system in order to define an approximate Riemann solver for (19).

$$\begin{cases} \partial_t \rho + \partial_x(\rho u) = 0, \\ \partial_t(\rho u) + \partial_x(\rho u^2 + \pi) = 0, \\ \partial_t(\rho\pi/c^2) + \partial_x(\rho\pi u/c^2) + \partial_x u = 0, \\ \partial_t(\rho c) + \partial_x(\rho c u) = 0, \\ \partial_t(\rho u^2/2 + \rho e) + \partial_x((\rho u^2/2 + \rho e + \pi)u) = 0. \end{cases} \tag{21}$$

Consider a Riemann problem for (21) with initial data $(\rho_l, \rho_l u_l, \rho_l \pi_l/c_l^2, \rho_l c_l, E_l)$, $(\rho_r, \rho_r u_r, \rho_r \pi_r/c_r^2, \rho_r c_r, E_r)$, where $E = \rho u^2/2 + \rho e$.

All the eigenvalues of (21) are linearly degenerate, so that we can easily compute the exact solution to the Riemann problem. It has three wave speeds

$$\sigma_1 = u_l - c_l/\rho_l, \quad \sigma_2 = u_l^* = u_r^*, \quad \sigma_3 = u_r + c_r/\rho_r, \tag{22}$$

with two intermediate states that we shall index l^* and r^*. The states are obtained from the relations

$$\begin{aligned} c_l^* &= c_l, \quad c_r^* = c_r, \\ u_l^* &= u_r^*, \quad \pi_l^* = \pi_r^*, \\ (\pi + cu)_l^* &= (\pi + cu)_l, \quad (\pi - cu)_r^* = (\pi - cu)_r, \\ (1/\rho + \pi/c^2)_l^* &= (1/\rho + \pi/c^2)_l, \quad (1/\rho + \pi/c^2)_r^* = (1/\rho + \pi/c^2)_r, \\ (e - \pi/2c^2)_l^* &= (e - \pi/2c^2)_l, \quad (e - \pi/2c^2)_r^* = (e - \pi/2c^2)_r. \end{aligned} \tag{23}$$

Now, consider a Riemann problem for (19) with left and right data $(\rho, \rho u)_{l,r}$ and suppose that we have a suitable definition for the constants c_l, c_r. We solve the Riemann problem for (21) with left and right data $(\rho, \rho u, \rho p(\rho)/c^2, \rho c, \rho u^2/2 + \rho e(\rho))_{l,r}$. The first two variables ρ, ρu of the solution for this new problem allow to define an approximate solution for the Riemann problem for (19). In this way we can define an approximate simple solver for the isentropic gas dynamics system that we call the Suliciu solver.

4.2 Entropy Inequalities for the Suliciu Solver

As we said previously, the Suliciu solver depends on a suitable definitions for the constants c_l, c_r. In this section, we focus on the choice of this constants so that some entropy inequality is satisfied.

We shall note by index $0, 1, 2, 3$ the states l, l^*, r^*, r respectively.

We notice that $\sigma_i(E_i - E_{i-1}) = \xi_i - \xi_{i-1}$, where $E = \rho u^2/2 + \rho e$ and $\xi = (\rho u^2/2 + \rho e + \pi)u$. Recall that here e is an independent variable.

Thus, from (20), we get

$$G(U_r) - G(U_l) = \xi_3 - \xi_0 = \sum_{i=0}^{2} \sigma_{i+1}(E_{i+1} - E_i), \tag{24}$$

and we may apply proposition 1 to this decomposition.

In [2] it is shown that under the hypothesis $E_l^* \geq \eta(U_l^*)$, $E_r^* \geq \eta(U_r^*)$ one gets that the Suliciu solver is entropy satisfying. As η is a convex function, this hypothesis implies (18) as expected.

We will replace conditions (18) by

$$E_i \geq \eta(U) + \eta'(U)(U_i - U), \tag{25}$$

where U stands for either U_l or U_r, U_i stands for either U_l^* or U_r^*.

As $\eta(U) = \rho u^2/2 + \rho e(\rho)$, assuming $\rho_i \geq 0$, condition (25) simplifies to

$$e(\rho) - e_i + p(\rho)\left(\frac{1}{\rho} - \frac{1}{\rho_i}\right) - \frac{(u_i - u)^2}{2} \leq 0, \tag{26}$$

where (ρ, u) stands for either (ρ_l, u_l) or (ρ_r, u_r) and (ρ_i, u_i, e_i) stands for either (ρ_l^*, u_l^*, e_l^*) or (ρ_r^*, u_r^*, e_r^*).

Theorem 1. *Let*

$$c_l^2 = \frac{(p_r - p_l)^2}{(p_r - p_l)\left(\frac{1}{\rho_l} - \frac{1}{\rho_r}\right) - \left[-\left(2(e_l - e_r) + (p_r + p_l)\left(\frac{1}{\rho_l} - \frac{1}{\rho_r}\right)\right)\right]_+}, \tag{27}$$

$$c_r^2 = \frac{(p_r - p_l)^2}{(p_r - p_l)\left(\frac{1}{\rho_l} - \frac{1}{\rho_r}\right) - \left[2(e_l - e_r) + (p_r + p_l)\left(\frac{1}{\rho_l} - \frac{1}{\rho_r}\right)\right]_+}. \tag{28}$$

We set

$$\widetilde{c}_l = \max\left(c_l, \rho_l(u_l - u_r)_+, \sqrt{\rho_l(p_r - p_l)_+}\right), \tag{29}$$

$$\widetilde{c}_r = \max\left(c_r, \rho_r(u_l - u_r)_+, \sqrt{\rho_r(p_l - p_r)_+}\right). \tag{30}$$

The Suliciu solver defined in Sect. 4.1 with the definition given by $\widetilde{c}_l$, $\widetilde{c}_r$ for the parameter c is a simple Riemann solver for the isentropic gas dynamics system (19) that satisfy the following:

(i) It preserves the non negativity of ρ,
(ii) It satisfies a semidiscrete entropy inequality,
(iii) It preserves the entropic shocks for the isentropic gas dynamics system.

Remark 2. We recall that two different states U_l, U_r can be connected by a shock for the system (19) if, and only if, $(u_r - u_l)^2 = (p_r - p_l)\left(\frac{1}{\rho_l} - \frac{1}{\rho_r}\right)$.
An entropic 1-shock for (19) satisfies $u_l \geq u_r$, and $\rho_r \geq \rho_l$.
An entropic 2-shock for (19) satisfies $u_l \geq u_r$, and $\rho_r \leq \rho_l$.

The proof of this theorem is done via some lemmas.

Lemma 4. *Under the assumptions*

$$\frac{(p_r - p_l)^2}{2\left(e_l - e_r + p_r\left(\frac{1}{\rho_l} - \frac{1}{\rho_r}\right)\right)} \leq c_l^2, \quad \frac{(p_r - p_l)^2}{2\left(e_r - e_l + p_l\left(\frac{1}{\rho_r} - \frac{1}{\rho_l}\right)\right)} \leq c_r^2, \tag{31}$$

condition (26) is satisfied.

Proof. For $U = U_l$ and $U_i = U_l^*$ or $U = U_r$ and $U_i = U_r^*$, the condition is automatically satisfied, using the corresponding definitions.

In the case $U = U_r$ and $U_i = U_l^*$ (the case $U = U_l$ and $U_i = U_r^*$ being analogue), some manipulations lead to

$$\begin{aligned} &e(\rho_r) - e_l^* + p(\rho_r)\left(\frac{1}{\rho_r} - \frac{1}{\rho_l^*}\right) - \frac{(u_l^* - u_r)^2}{2} \\ &\leq e_r - e_l + p_r\left(\frac{1}{\rho_r} - \frac{1}{\rho_l}\right) + \frac{1}{2c_l^2}(p_r - p_l)^2, \end{aligned} \tag{32}$$

and the result follows.

Lemma 5. *Let $(\rho_l, \rho_l u_l)$,$(\rho_r, \rho_r u_r)$ two different states such that they can be connected by an entropic 1-shock (resp. 2-shock) for the system (19) with shock speed σ. Then the following conditions are equivalent.*

(i) The Suliciu relaxation system preserves the 1-shock (resp. preserves the 2-shock),
(ii) $c_l = \frac{p_r - p_l}{u_l - u_r}$ (resp. $c_r = \frac{p_l - p_r}{u_l - u_r}$).

Proof. It can be proved by using Remark 2 and imposing $\rho_l^* = \rho_r^* = \rho_r, \quad u_l^* = u_r^* = u_r, \quad u_l - c_l/\rho_l = \sigma$ or $\rho_l^* = \rho_r^* = \rho_l, \quad u_l^* = u_r^* = u_l, \quad u_r + c_r/\rho_r = \sigma$, for a 1 or 2 shock, respectively.

Lemma 6. *The quantities (27) and (28) are well defined for $\rho_l \neq \rho_r$ and verify $c_l, c_r \to \rho\sqrt{p'(\rho)}$ for $\rho_l, \rho_r \to \rho$.*
Moreover, (31) is satisfied and $c_l = \frac{p_r - p_l}{u_l - u_r}$, (resp. $c_r = \frac{p_l - p_r}{u_l - u_r}$,) when the left and right states are connected by an entropic 1-shock (resp. 2-shock) for (19).

Proof of Theorem 1.

The proof is straightforward from the lemmas.

We remark that take (29) and (30) in order to prove $\rho_l^*, \rho_r^* \geq 0$.

5 Numerical Tests

Numerical tests have been done in order to compare the semidiscrete entropy approach introduced here with discrete entropy approach introduced in [2] for the Suliciu solver. In general, the first one is a bit more accurate in shock regions while both behave similarly for rarefactions.

We show here just a really simple example. We consider a Riemann problem with left and right data $\rho_l = 1$, $u_l = 3.2125$, and $\rho_r = 2$, $u_r = 0.5$. The exact solution consists in a nonstationary shock and the simulation is shown in Fig. 1. We see that the semidiscrete entropic Suliciu solver is less diffusive than the discrete entropic one.

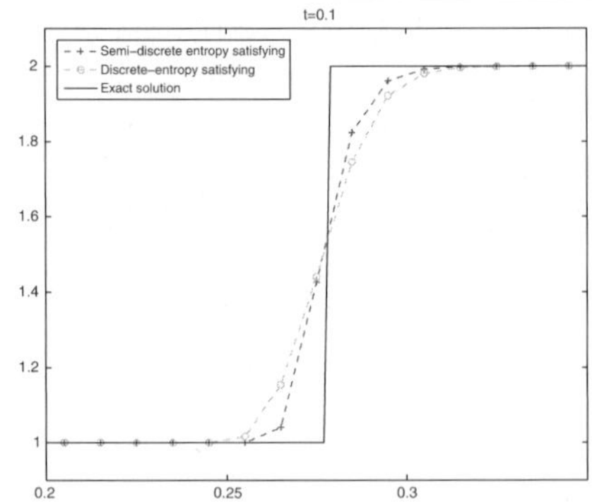

Fig. 1. Comparison of semidiscrete and discrete entropy satisfying Suliciu solver for a shock

6 Conclusions

Imposing a semidiscrete entropy inequality for a numerical scheme is less restrictive than a fully discrete entropy inequality. This gives more *freedom* in order to impose some other properties such as exact shock capturing.

Here we have shown the particular case of Suliciu solver applied to the isentropic gas dynamics. We get a scheme that is less diffusive in shock regions, solve exactly stationary shocks, and verify a semidiscrete entropy inequality.

The technique can be applied to other cases; in particular, it has been developed by the authors for the full gas dynamics system.

References

1. F. Bouchut. Entropy satisfying flux vector splittings and kinetic BGK models. *Numer. Math.*, 94(4):623–672, 2003.
2. F. Bouchut. *Nonlinear stability of finite volume methods for hyperbolic conservation laws and well-balanced schemes for sources.* Frontiers in Mathematics. Birkhäuser Verlag, Basel, 2004.
3. P. Colella and H.M. Glaz. Efficient solution algorithms for the Riemann problem for real gases. *J. Comput. Phys.*, 59(2):264–289, 1985.
4. I. Suliciu. On modelling phase transitions by means of rate-type constitutive equations. Shock wave structure. *Internat. J. Engrg. Sci.*, 28(8):829–841, 1990.
5. E.F. Toro. *Riemann solvers and numerical methods for fluid dynamics.* Springer-Verlag, Berlin, second edition, 1999. A practical introduction.

On the L^2-Well Posedness of an Initial Boundary Value Problem for the Linear Elasticity in Two and Three Space Dimensions

A. Morando and D. Serre

1 Introduction

We are concerned with the system of d-dimensional linear elasticity, where $d = 2, 3$; the system reads as follows

$$\begin{aligned} \partial_t F + \nabla z &= 0, \\ \partial_t z + \mathrm{div} T &= 0, \end{aligned} \tag{1}$$

where $F(x,t) \in \mathbf{M}_{d\times d}(\mathbf{R})$ and $z(x,t) \in \mathbf{R}^d$ (both functions of the position $x = (x_1, \ldots, x_d)$ and the time t) are the unknowns and

$$T := \lambda(F + F^T) + \mu(\mathrm{Tr} F) I_d, \tag{2}$$

with I_d the $d\times d$ identity matrix. The functions z and F represent, respectively, the opposite of the material velocity and the infinitesimal deformation tensor; T defined by (2) is the stress tensor; and the positive constants λ, μ are the so-called *Lamé coefficients* of the material. A thorough analysis of the elasticity model can be found in the books of Ciarlet [C88] and Dafermos [D00]. Since in (1) the skew-symmetric part of F decouples from the rest, we may restrict to the system describing the evolution of z and the symmetric part of F. The latter is a Friedrichs symmetrizable system of size $n = \frac{d(d+3)}{2}$; therefore by an appropriate linear change of unknowns, it can be written in the symmetric hyperbolic form

$$Lu := \frac{\partial u}{\partial t} + \sum_{\alpha=1}^{d} A^\alpha \frac{\partial u}{\partial x_\alpha} = 0. \tag{3}$$

In (3), u is the vector of the new unknowns: for $d = 2$ it is provided by

$$u := (2\sqrt{\lambda(\lambda+\mu)} F_{1,1}, c_P\sqrt{\lambda}(F_{1,2} + F_{2,1}), c_P^2 F_{2,2} + \mu F_{1,1}, c_P z_1, c_P z_2)^T,$$

where $c_P := \sqrt{2\lambda + \mu}$ is the *velocity of pressure waves*; a similar expression also appears in the three-dimensional model. The coefficients A^α of L are appropriate $n \times n$ real symmetric matrices; in particular the coefficient of the x_d-spatial derivative is a singular matrix carrying the form

$$A^d = \begin{pmatrix} \mathbf{0}_m & 0 \\ 0 & a^d \end{pmatrix}, \tag{4}$$

where a^d is an invertible matrix of size $n - m$, $\mathbf{0}_m$ is the null matrix of size m, and m is a positive integer ($\leq n$), depending as follows on the space dimension d:

$$\text{for } d = 2: \ m = 1; \qquad \text{for } d = 3: \ m = 3.$$

Let us denote by $\xi = (\xi_1, \ldots, \xi_{d-1}, \xi_d) = (\eta, \xi_d)$ the dual space variables; then the symbol $A(\xi) := \sum_{\alpha=1}^{d} A^\alpha \xi_\alpha$ of the linear operator L in (3) carries the following structure by block

$$A(\xi) = \begin{pmatrix} \mathbf{0}_m & a_{2,1}(\eta)^T \\ a_{2,1}(\eta) & a_2(\eta) + \xi_d a^d \end{pmatrix}, \quad \xi = (\eta, \xi_d); \tag{5}$$

where $a_{2,1}(\eta)$ and $a_2(\eta)$ are suitable real matrices, linearly depending on $\eta = (\xi_1, \ldots, \xi_{d-1})$.

2 The Initial Boundary Value Problem

We study the L^2-well posedness of an initial boundary value problem (ibvp) on the positive d-dimensional half-space $\mathbf{R}^d_+ = \{x \in \mathbf{R}^d : x_d > 0\}$; the boundary of $\mathbf{R}^d_+$ is identified with $\mathbf{R}^{d-1}$ and the tangential variables $(x_1, \ldots, x_{d-1})$ are denoted by y (thus we shortly write $x = (y, x_d)$). The ibvp reads as follows:

$$Lu(y, x_d, t) = f(y, x_d, t), \ y \in \mathbf{R}^{d-1}, x_d, t > 0, \tag{6}$$

$$Bu(y, 0, t) = g(y, t), \ y \in \mathbf{R}^{d-1}, t > 0, \tag{7}$$

$$u(y, x_d, 0) = a(y, x_d), \ y \in \mathbf{R}^{d-1}, x_d > 0. \tag{8}$$

The data f, g, a are given square-integrable functions; B is a given $p \times n$ real matrix, with rank $B = p = \frac{n-m}{2}$; note that p is just the number of the positive eigenvalues of A^d (i.e., the so-called number of *incoming characteristics* of L). Furthermore, since $\mathrm{Ker} A^d = \mathbf{R}^m \times \{0_{n-m}\}$, the problem (6)–(8) has *uniformly characteristic boundary* in the sense of Majda–Osher [MO75]. We assume that $\mathrm{Ker} A^d \subseteq \mathrm{Ker} B$, which yields $B = (0_{p \times m}, B_2)$ ($0_{p \times m}$ the $p \times m$ null matrix) with some $p \times (n - m)$ real matrix B_2 of full rank p; the aforementioned property, called *reflexivity* by Ohkubo [O81], is a rather natural assumption for characteristic ibvps; indeed the best control expected for the trace of weak solutions to such problems regards only the noncharacteristic part of the solutions. It is well known that an ibvp such as (6)–(8) is strongly L^2-well posed when the boundary operator B is *maximally strictly dissipative* for L: this means that the quadratic form $w \mapsto w^* A^d w$ is nonpositive on Ker$B$ and its restriction to Ker$B$ vanishes only on KerA^d; moreover, Ker$B$ must be maximal with respect to the preceding assumption. Here we are interested in the strong L^2-well posedness of (6)–(8), in the case when the boundary condition (7) satisfies the weaker but necessary *uniform Kreiss–Lopatinskii condition* (UKL). Recall that a characteristic ibvp such as (6)–(8) is said to fulfill the UKL condition, provided there exists a positive constant C such that

$$\|A^d V\| \leq C\|BV\|, \; V \in \mathcal{E}_-(\tau, \eta) \tag{9}$$

holds true for all pairs $(\tau, \eta) \in \mathbf{C} \times \mathbf{R}^{d-1}$ with $\mathrm{Re}\,\tau > 0$. For any (τ, η) as before, $\mathcal{E}_-(\tau, \eta)$ denotes the stable subspace of the system

$$(\tau I_n + iA(\eta, 0))V + A^d \frac{dV}{dx_d} = 0,$$

obtained by taking the Fourier–Laplace transform of (3) with respect to (y, t). We refer to [K70], [S00, Chap. 14], and [BGS06] for an exhaustive accounting of the UKL condition, in both the cases of noncharacteristic and characteristic boundary. Under the assumptions discussed above, we construct a symbolic symmetrizer of (6)–(8) in the sense of Kreiss [K70]; as a consequence, standard Kreiss' methods apply to derive the strong L^2-well posedness of our ibvp. Before stating (and sketching the proof) of our results, let us make a few comments. In the already quoted work [MO75], Majda and Osher prove the existence of a degenerate Kreiss symmetrizer for a variable coefficient symmetric hyperbolic ibvp, with uniformly characteristic boundary, when the linear differential operator L obeys additional structural hypotheses; among them, it is required (cf. [MO75, Assumption 1.1]) that the $m \times m$ upper-left block of the symbol $A(\eta, 0)$ of L has only simple eigenvalues, as long as $\eta \neq 0$: actually this assumption fails to be satisfied by the three-dimensional elasticity model, where the upper-left block of $A(\eta, 0)$ has size 3 and is identically zero (cf. (5)). Moreover, as we will see later, the Kreiss symmetrizer we construct explicitly for (6)–(8) is nondegenerate (in either two or three space dimensions), whereas the general construction given in [MO75] leads to a degenerate symbolic symmetrizer. In [BGS06], the existence of a nondegenerate Kreiss symmetrizer of any symmetric hyperbolic constant coefficient ibvp, with characteristic boundary, is proven assuming only that the matrix $a_2(\eta)$ (appearing in the lower-right block of the operator symbol $A(\eta, 0)$; cf. (5)) is identically zero; however, this requirement is violated by the linear elasticity model.

3 The Main Result

We prove the following result.

Theorem 1. *Let us consider the ibvp (6)–(8) in dimension* $d = 2, 3$*; assume that the matrix* B *satisfies the reflexivity and the UKL conditions. Then for* *every $T>0$ and all data $f\in L^2(\mathbf{R}^d_+\times(0,T))$, $g\in L^2(\mathbf{R}^{d-1}\times(0,T))$, and $a\in L^2(\mathbf{R}^d_+)$, there exists one and only one solution $u\in L^2(\mathbf{R}^d_+\times(0,T))$ to (6)–(8) such that:*

(i) $u\in C([0,T];L^2(\mathbf{R}^d_+))$.
(ii) $A^d u$ admits a trace $\gamma_0 A^d u$ on the boundary of $\mathbf{R}^d_+$, in the class $L^2(\mathbf{R}^{d-1}\times(0,T))$.

Moreover, for every positive number γ, the solution of (6)–(8) obeys the following a priori energy estimate

$$e^{-2\gamma T}\|u(T)\|^2_{L^2}+\|u\|^2_{\gamma,T}\le C\left(\|a\|^2_{L^2}+\int_0^T e^{-2\gamma t}\left(\frac{1}{\gamma}\|f(t)\|^2_{L^2}+\|g(t)\|^2_{L^2}\right)dt\right),$$

where the constant $C>0$ does not depend on f,g,a and γ,T. In the estimate above, $\|\cdot\|_{L^2}$ denotes both the norms in $L^2(\mathbf{R}^d_+)$ and $L^2(\mathbf{R}^{d-1})$; moreover, we have set

$$\|u\|^2_{\gamma,T}:=\int_0^T\int_{\mathbf{R}^{d-1}}e^{-2\gamma t}|\gamma_0 A^d u(y,t)|^2dydt+\gamma\int_0^T\int_{\mathbf{R}^d_+}e^{-2\gamma t}|u(x,t)|^2dxdt.$$

As was discussed earlier, the proof of Theorem 1 is a consequence of the existence of a *nondegenerate dissipative Kreiss symmetrizer* of (6)–(8); recall that a dissipative symmetrizer of our ibvp consists of a bounded function $(\tau,\eta)\mapsto K(\tau,\eta)\in\mathbf{M}_{n\times n}(\mathbf{C})$, defined in the frequency space $\Xi:=\{(\tau,\eta)\in\mathbf{C}\times\mathbf{R}^{d-1}:\ \mathrm{Re}\,\tau\ge 0\}$ and homogeneous of degree zero, satisfying the following properties:

I. For all $(\tau,\eta)\in\Xi$, $\mathcal{K}(\tau,\eta):=K(\tau,\eta)A^d$ is Hermitian.
II. $\mathcal{K}(\tau,\eta)$ must be nonpositive on $\mathrm{Ker}B$, uniformly in $(\tau,\eta)\in\Xi$, and its restriction to $\mathrm{Ker}B$ vanishes only on $\mathrm{Ker}A^d$.
III. If we set $\mathcal{P}(\tau,\eta):=K(\tau,\eta)(\tau I_n+iA(\eta,0))$, there exists a positive constant c_0 such that $\mathrm{Re}\,\mathcal{P}(\tau,\eta)\ge c_0(\mathrm{Re}\tau)I_n$ for all $(\tau,\eta)\in\Xi$.

Here below, we illustrate the main steps of the construction of a nondegenerate dissipative symmetrizer of (6)–(8), according to the preceding definition. Since the symmetrizer K is required to be homogeneous of degree zero in (τ,η), we may restrict to define K on the *unit hemisphere* of Ξ, namely the set of unitary frequencies $\Sigma:=\{(\tau,\eta):\ |\tau|^2+\|\eta\|^2=1,\ \mathrm{Re}\,\tau\ge 0\}$; since the latter is a compact subset of Ξ, by introducing a smooth partition of unity, we can further reduce to construct K microlocally in a neighborhood of each point of Σ. Agreeing with the terminology used in [BGS06], we have to argue separately on three different classes of frequencies in Σ:

(a) The "interior points" $(\tau,\eta)\in\Sigma$, where $\mathrm{Re}\,\tau>0$
(b) The "boundary points" $(\tau,\eta)\in\Sigma$, where $\mathrm{Re}\,\tau=0$ and $\tau\ne 0$
(c) The "central points" $(0,\eta)$, where $\|\eta\|=1$

As regards to the construction of K in a neighborhood of points lying to the classes (a) or (b), essentially one follows the arguments developed in [BGS06, Chap. 6] to build a symmetrizer of an ibvp, the symbol $A(\xi)$ of which fulfills Ohkubo's assumption $a_2(\eta) \equiv 0$. On the contrary, in the construction proposed by [BGS06], Ohkubo's assumption plays a fundamental role to build the symmetrizer near the frequency points of kind (c).

4 A Dissipative Symmetrizer Near the Central Points

This section is devoted to summarize the main ideas of the construction of a Kreiss symmetrizer to the elasticity problem (6)–(8), in the vicinity of a central point $(0, \eta_0)$ with $\|\eta_0\| = 1$. In view of the simple structure of A^d (cf. (4)), the requirement I in the definition of a dissipative symmetrizer yields that K need to display the form

$$K = \begin{pmatrix} k_{I,I} & 0 \\ k_{II,I} & K_2 \end{pmatrix}, \tag{10}$$

where $K_2 = K_2(\tau, \eta)$ is a $m \times m$ complex matrix such that $\mathcal{K}_2 := K_2 a^d$ is Hermitian. Condition II even implies that $\mathcal{K}_2$ must be positive definite (uniformly in (τ, η)) on $\operatorname{Ker} B_2$. For the time being, it will be convenient splitting the variable z in $\mathbf{C}^n$ as $z = (z', \zeta', \zeta'')^T$, where z' and ζ', ζ'' span $\mathbf{C}^m$ and $\mathbf{C}^p$, respectively (recall that $n - m = 2p = \operatorname{rank} A^d$; cf. (4)). The way of defining the matrices $k_{I,I}, k_{II,I}, K_2$ is suggested by deriving an explicit representation of the linear space $\operatorname{Ker} B_2$. Exploiting some algebraic properties of a wide class of constant coefficient symmetric hyperbolic ibvps (including the elasticity model), discussed in [BGS06], and combining them with an appropriate reformulation of the UKL condition near the central points, again proven in [BGS06, Chap. 6, Proposition 6.6], we get the desired representation of $\operatorname{Ker} B_2$; namely one can show that there exists a C^∞ mapping $\eta \mapsto D(\eta) \in M_{p \times p}(\mathbf{C})$, defined in some neighborhood $\mathcal{V}_0$ of η_0 and homogeneous of degree zero, such that

$$\operatorname{Ker} B_2 = \left\{ (D(\eta)\zeta'', \zeta'')^T \, : \, \zeta'' \in \mathbf{C}^p \right\}, \tag{11}$$

as long as η belongs to $\mathcal{V}_0$. The different components of the matrices $k_{I,I}$, $k_{II,I}$, and K_2 can be now defined in such a way to obtain the following formula

$$(z'')^* \mathcal{K}_2(\tau, \eta) z'' = \Phi\,(z''; \tau, \eta) - AQ\,(\zeta'') + R_\chi\,(z''; \tau, \eta), \tag{12}$$

where $z'' := (\zeta', \zeta'')^T$. In (12), Q is a positive definite quadratic form in $\mathbf{C}^p$ (independent of (τ, η)); therefore, one gets

$$Q(\zeta'') \geq \varepsilon_0 \|\zeta''\|^2, \ \forall \zeta'' \in \mathbf{C}^p, \tag{13}$$

for some positive constant $\varepsilon_0 > 0$. For each (τ, η) near $(0, \eta_0)$, $\Phi(\cdot; \tau, \eta)$ and $R_\chi(\cdot; \tau, \eta)$ are suitable quadratic forms in $\mathbf{C}^{2p}$, and A and χ are positive numbers to be chosen appropriately. Using that the norm of $z'' = (\zeta', \zeta'')^T$ is equivalent to that of ζ'' in $\operatorname{Ker} B_2$ (this is due to formula (11)) and estimating Φ and R_χ by

$$\begin{aligned} &\Phi(z''; \tau, \eta) \le C_1 \|z''\|^2, \ \forall z'' \in \mathbf{C}^{2p}, \\ &R_\chi(z''; \tau, \eta) \le C_2 \chi |\tau| \|z''\|^2, \ \forall z'' \in \mathbf{C}^{2p}, \end{aligned} \tag{14}$$

where the positive constants C_1, C_2 are independent of (τ, η), one gets for $\mathcal{K}_2$ the following upper bound

$$(z'')^* \mathcal{K}_2(\tau, \eta) z'' \le (C_1 - \varepsilon A + C_2 \chi |\tau|) \|z''\|^2, \ \forall z'' \in \operatorname{Ker} B_2. \tag{15}$$

Hence, taking $A > 0$ sufficiently large (and assuming that $\|\eta - \eta_0\|$ and $|\tau|$ are small enough), one finds that $\mathcal{K}_2$ is positive definite on $\operatorname{Ker} B_2$, uniformly in (τ, η), for any positive χ; this entails that condition II is satisfied by our symbolic symmetrizer. Eventually, a suitable choice of the value of χ, together with some further manipulations of additional parameters involved in Φ, implies that the constructed symbolic symmetrizer satisfies also condition III (cf. [MS05-1] and [MS05-2] for further details). In the case of dimension $d = 2$, the preceding arguments lead to define $K = K(\tau, \eta)$, in a neighborhood $\mathcal{W}_0 \subset \Sigma$ of $(0, \eta_0)$, by the following expression

$$K(\tau, \eta) = \begin{pmatrix} h + \chi\overline{\tau} & 0 & 0 & 0 & 0 \\ \gamma\eta & h + \chi\overline{\tau} & -\frac{\sqrt{\lambda}\rho\gamma}{\Theta c_P} & -\frac{\rho\gamma}{\Theta} & -\frac{\sqrt{\lambda}\rho\gamma}{\Theta c_P} \\ \gamma\eta & 0 & h\chi\overline{\tau} & -\frac{\rho\gamma}{\Theta} & -\frac{\mu\rho\gamma}{\sqrt{\lambda}\Theta c_P} \\ -iM\eta & -A & iN\eta & h + \chi\tau & 0 \\ \gamma\eta & -\frac{c_P}{\sqrt{\lambda}} iN\eta & -\frac{\sqrt{\lambda}c_P A}{\mu} & -\frac{\rho\gamma}{\Theta} & h + \chi\tau \end{pmatrix},$$

for $\tau = \gamma + i\rho$; here above M and N are two positive parameters to be fixed appropriately and $\Theta := \frac{2\sqrt{\lambda(\lambda+\mu)}}{c_P}$. A similar expression can be also used to define a symbolic symmetrizer of the three-dimensional elasticity problem, near any central point $(0, \eta_0) \in \Sigma$ (cf. [MS05-2]).

References

[BGS06] Benzoni-Gavage, S., Serre, D.: Multi-dimensional hyperbolic partial differential equations. First-order systems and applications. Oxford Mathematical Monographs. The Clarendon Press, Oxford University Press (2006).

[C88] Ciarlet, P.G.: Mathematical elasticity. Vol. 1. Three-dimensional elasticity. Studies in Mathematics and its Applications, 20. North-Holland Publishing Co. (1988).

[D00] Dafermos, C.: Hyperbolic Conservation Laws in Continuum Physics. Grundlehren der Mathematischen Wissenschaften, 325. Springer-Verlag (2000).

[K70] Kreiss, H.-O.: Initial boundary value problems for hyperbolic systems. Comm. Pure Appl. Math., **23**, 277–298 (1970).

[MO75] Majda, A., Osher, S.: Initial-boundary value problems for hyperbolic equations with uniformly characteristic boundary. Comm. Pure Appl. Math., **28**(5), 607–675 (1975).

[M00] Métivier, G.: The block structure condition for symmetric hyperbolic systems. Bull. London Math. Soc., **32**(6), 689–702 (2000).

[MS05-1] Morando, A., Serre, D.: A result of L^2-well posedness concerning the system of linear elasticity in 2D. Comm. Math. Sci., **3**(3), 317–334 (2005).

[MS05-2] Morando, A., Serre, D.: On the L^2-well posedness of an initial boundary value problem for the 3D linear elasticity. Comm. Math. Sci., **3**(4), 575–586 (2005).

[O81] Ohkubo, T.: Regularity of solutions to hyperbolic mixed problems with uniformly characteristic boundary. Hokkaido Math. J., **10**, 93–123 (1981).

[S00] Serre, D.: Systems of conservation laws. 2. Geometric structures, oscillations, and initial-boundary value problems. Cambridge University Press (2000). Translated from the 1996 French original by I.N. Sneddon.

Intersections Modeling with a Class of "Second-Order" Models for Vehicular Traffic Flow

M. Herty, S. Moutari, and M. Rascle

Summary. In a recent paper, Herty and Rascle (SIAM J. Appl. Math. 38:595–616, 2006) introduced some coupling conditions to model road junctions with the Aw–Rascle traffic flow model (Aw and Rascle, SIAM J. Appl. Math. 60:916–944, 2000). In this chapter, we study an extension of the work in Herty and Rascle (SIAM J. Appl. Math. 38:595–616, 2006). Here, in contrast to Herty and Rascle (SIAM J. Appl. Math. 38:595–616, 2006), we propose a general formulation for suitable coupling conditions at an intersection, *without* imposing any fixed mixture principle. The solution conserves mass and "pseudomomentum" and additionally *maximizes the total (mass) flux* at an intersection. This problem has been completely solved in the cases of $2 \longmapsto 1$ (merge) and $1 \longmapsto 2$ (diverge) junctions. Furthermore these two simple cases of merging and diverging junctions can be combined to model more complex junctions, like roundabouts.

1 Introduction

Traffic modeling on road networks has been intensively investigated under the framework of the celebrated LWR first-order model [LW55, Ric56]. For instance, see [CGP02, HK04, HR95] and the references therein. In this chapter, we are interested with a (class of) "second-order" model(s): the "Aw–Rascle" (AR) model. This model consists of a nonlinear, coupled system of conservation laws, introduced in [AR00] and independently in [Zha02]. Such class of models describes the behavior of traffic density and velocity where different cars can have a different response to local traffic situations, e.g., the model distinguishes trucks and cars. One of the major issues when modeling a traffic network is the definition of the coupling conditions at road intersections, i.e., the well definition of the problem and the existence of a unique solution (see [CGP02, HR95, HK04] for the discussion in the scalar case). In this work we introduce some new coupling conditions for the AR system. Contrary to the previous works [GP05] and [HR06], those new conditions conserve all moments of the system, and the derived conditions maximize the flux at the intersections without any further constraint. This chapter is organized as follows. In Sect. 2, we briefly present the AR model on a road network (see [GP05, HMR06, HR06] for detailed discussions). Section 3 provides the main results on the definition of the Riemann problem solution at a road junction. In Sect. 4, we summarize the discussion on the merge and diverge junctions, followed by some numerical simulations. Finally, in Sect. 5, we give some brief concluding remarks and the possible extensions of the approach.

2 The Model

First of all, we consider a road network as a finite directed graph $(\mathcal{I}, \mathcal{N})$, with $|\mathcal{I}| = \mathbf{I}$ and $|\mathcal{N}| = \mathbf{N}$. Each arc $i = 1, \ldots, \mathbf{I}$ corresponds to a road and each vertex $n \in \mathcal{N}$ to a junction. For a given junction n we denote by δ_n^- and δ_n^+, respectively, the set of incoming roads (indexed k) and the set of outgoing roads (indexed j) to n. Each road i is modeled by an interval $I_i := [a_i, b_i]$. We require the AR equations (1) to hold on each arc $i \in \mathcal{I}$ of the network

$$\partial_t \rho_i + \partial_x(\rho_i v_i) = 0 \tag{1a}$$

$$\partial_t(\rho_i w_i) + \partial_x(\rho_i v_i w_i) = 0 \tag{1b}$$

$$w_i = v_i + p_i(\rho_i) \tag{1c}$$

where, for each i, $\rho_i \mapsto p_i(\rho_i)$ is a known function ("traffic pressure") with the following properties

$$\forall \rho_i, \rho_i p_i''(\rho_i)) + 2p_i'(\rho_i) > 0 \text{ and, e.g., } p_i(\rho_i) \sim \rho_i^\gamma \text{ at } \rho_i = 0, \tag{2}$$

and where $\gamma > 0$. ρ_i and v_i, respectively, describe the density and velocity of traffic on road i.

For more details on this model, we refer the reader to [AR00, HR06].

We denote by α_{jk} the percentage of cars on road k willing to go (and actually going, see below) to road j. The corresponding matrix $A := (\alpha_{jk})_{j\in\delta^+, k\in\delta^-}$ is assumed to be *known*, see [CGP02, GP05, HR06]. By definition we have

$$\sum_{j\in\delta^+} \alpha_{jk} = 1 \ \forall k \in \delta^-. \tag{3}$$

Next, let

$$q_k(t) := \rho_k v_k(b_k-, t), \ q_j(t) := \rho_j v_j(a_j+, t)$$

denote the (initially unknown) total first component of the flux (i.e., the mass flux) on the incoming road k (resp., on the outgoing road j). Furthermore, let us introduce the (initially unknown) flux q_{jk} of cars actually going from road k to road j and let $\beta_{jk} := q_{jk}/q_j$, which is also initially unknown. Then, by the above definitions, we have

$$\alpha_{jk} = \frac{q_{jk}}{q_k} \text{ and } \sum_{k\in\delta^-} \beta_{jk} = 1.$$

Finally, we describe the construction of the demand and supply functions (on an arbitrary road) for a given level curve of $\{\mathbf{w}(U) = c\}$, $c \geq 0$. Recall, $\mathbf{w}(U) = v + p(\rho)$ and its level curve is a concave function in the $(\rho, \rho v)$ plane with a unique maximum. As in the case of first-order models, e.g., [Leb93, GP05], in the $(\rho, \rho v)$ plane the demand function $d(\rho; \mathbf{w}, c)$ is an extension of the *nondecreasing* part of this level curve $\{\mathbf{w}(U) = c\}$ for $\rho \geq 0$ and the supply function $s(\rho; \mathbf{w}, c)$ is an extension of the *nonincreasing* part of this curve $\{\mathbf{w}(U) = c\}$ and $\rho \geq 0$. We denote by $d_k := d(\rho_k; \mathbf{w}, c_k)$ the demand on an *incoming* road k and by $s_j := s(\rho_j, \mathbf{w}, c_j)$ the supply on an *outgoing* road j.

3 Solution to the Riemann Problem

In this section, we state some of the main results on the definition of the solution to the Riemann problem at a junction and we refer the reader to [HR06, GP05, HR95] for a complete discussion on the derivation of the necessary conditions (Figs. 1 and 2).

Proposition 1. *[HMR06] Let $U^- = (\rho^-, \rho^- v^-) \neq (0,0)$ be the initial value on an incoming road. Let the 1-curve through U^- be $\mathbf{w}(U) = v + p(\rho) = w^-$ with $w^- := \mathbf{w}(U^-)$. Then the "admissible" states $U^+ = (\rho^+, \rho^+ v^+)$ for the associated Riemann problem must belong to that curve. The maximal possible flux associated with any admissible state U^+ is $d(\rho^-; \mathbf{w}, w^-)$.*

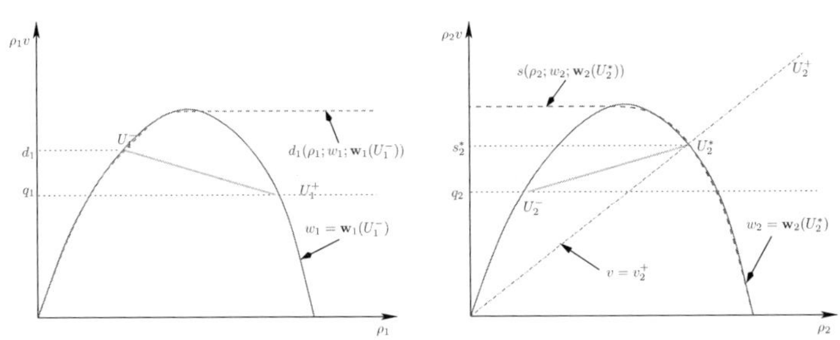

Fig. 1. (Half-)Riemann problem in the $(\rho, \rho v)$ plane on an incoming road (*left*) and an outgoing road (*right*)

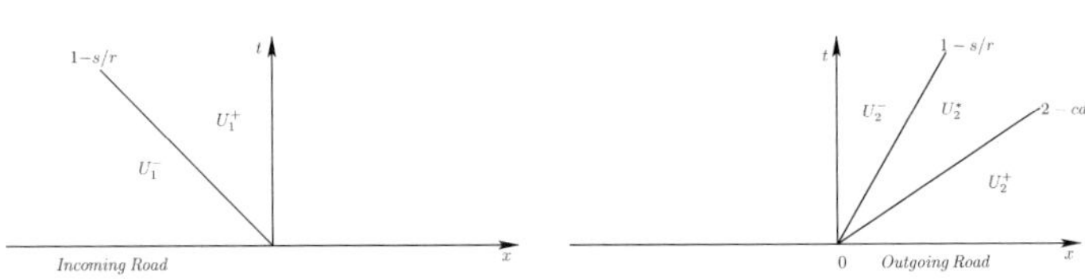

Fig. 2. (Half-)Riemann problem in the (x, t) plane on an incoming road (*left*) and an outgoing road (*right*)

Proposition 2. *[HMR06] Consider the initial state $U^+ \neq (0,0)$ on an outgoing road and the level curve of the first Riemann invariant $\{\mathbf{w}(U) = c\}$ with an arbitrary nonnegative constant c. Let $U^\dagger = (\rho^\dagger, \rho^\dagger v^\dagger)$ be the point of intersection, if it exists, of the two Riemann invariants $\{\mathbf{v}(U) = v^+\}$ and $\{\mathbf{w}(U) := v + p(\rho) = c\}$ with $\rho > 0$ and $v > 0$. Then the "admissible" states U^- for the associated Riemann problem must satisfy $\mathbf{w}(U^-) = c$ and $\rho^- v^- \geq 0$. Note that the set of admissible states U^- depends on the existence of the point $U^\dagger$. The maximal possible flux associated with any "admissible" state U^- is $s(\rho^\dagger; \mathbf{w}, c)$.*

Definition 1. *[HMR06] Consider a junction with m incoming and n outgoing roads, with constant initial data $U_{i,0} = (\rho_{i,0}, \rho_{i,0} v_{i,0})_{i \in \delta^- \cup \delta^+}$ under assumptions (3).*

We say that the family $\{U_i(x,t)\}_{i \in \delta^- \cup \delta^+}$ is an admissible solution of the associated Riemann problems if and only if it satisfies:

(C1) $\forall \; i \in \delta^- \cup \delta^+$, $U_i(x,t)$ is a weak entropy solution (in the sense of [HMR06, HR06]) of the network problem, where $p_i^\dagger \equiv p_i, \forall i \in \delta^-$.
On an outgoing road $j \in \delta^+$, the solution $U_j(x,t)$ is constructed as in [HR06]: in the triangle $\{(x,t); a_j < x < a_j + t v_{j,0}\}$, U_j is the homogenized solution defined below with $p_j^\dagger \equiv p_j^*$, whereas for $x > a_j + t v_{j,0}$, $p_j^\dagger \equiv p_j$.

(C2) The flux distribution satisfies the Rankine–Hugoniot conditions (see [HMR06]).

(C3) The sum of the incoming fluxes $\sum_{k \in \delta^-} \rho_k v_k(b_k -, t) = \sum_{k \in \delta^-} q_k$ (or equivalently the sum of the outgoing fluxes $\sum_{j \in \delta^+} \rho_j v_j(a_j +, t) = \sum_{j \in \delta^+} q_j$) is maximal subject to (C1) and (C2).

Next, we describe the homogenization to define p_j^*. For a motivation and a detailed discussion of the homogenization, see [BR03] and [HR06, Sect. 6]. We recall that for each $k \in \delta^-$, $p_k^\dagger \equiv p_k$ and $\mathbf{w}_k(U) = v + p_k(\rho)$ are well defined. First, we define the (initially unknown) homogenized value for each outgoing road $j \in \delta^+$

$$\bar{w}_j := \sum_{k \in \delta^-} \beta_{jk} \mathbf{w}_k(U_{k,0}). \tag{4}$$

Then, for each $j \in \delta^+$, $p_j^*(\cdot)$ is defined as in [HR06]. Namely, we first define the function

$$P_j(\tau) := p_j(1/\tau), \tag{5}$$

where $\tau = \frac{1}{\rho}$ is the specific volume, see [AKMR02, BR03].

Now, we consider the function:

$$v \longmapsto \tau := \sum_{k \in \delta^-} \frac{q_{jk}}{q_j} P_j^{-1}(\mathbf{w}_k(U_{k,0}) - v) = \sum_{k \in \delta^-} \beta_{jk} P_j^{-1}(\mathbf{w}_k(U_{k,0}) - v). \tag{6}$$

Then, we *choose* to define a new invertible function P_j^* by rewriting (6) under the form

$$\tau := \left(P_j^*\right)^{-1}(\bar{w}_j - v), \tag{7}$$

which we *only* use with the particular value $\bar{w}_j$ defined by (4), see [BR03] for more details.

Finally, we set

$$p_j^\dagger(\rho) := p_j^*(\rho) := P_j^*(1/\rho), \tag{8a}$$

$$\mathbf{w}_j^\dagger(U) := v + p_j^\dagger(\rho). \tag{8b}$$

This construction is perfectly well defined once the proportions $\beta_{jk} = q_{jk}/q_j$ are known. In [HR06] these proportions are assumed to be known a priori. Here, in contrast, we show that the proportions β_{jk} can be determined by solving a maximization problem stated below. Unfortunately, so far, the problem is only tractable for particular types of junctions. With all the previous remarks in mind, we conclude: there exists a unique solution $\{U_i\}_{i\in\delta^-\cup\delta^+}$ in the sense of Definition 1 if the following maximization problem

$$\max \sum_{j\in\delta^+} q_j \text{ subject to} \tag{9}$$

$$\forall k \in \delta^-,\ 0 \le q_k \le d_k(\rho_{k,0}; \mathbf{w}_k, \mathbf{w}_k(U_{k,0})), \tag{10}$$

$$\forall j \in \delta^+,\ 0 \le q_j \le s_j(\rho_{j,0}, \mathbf{w}_j^\dagger, w_j^*), \tag{11}$$

$$\forall k \in \delta^-, \forall j \in \delta^+,\ \beta_{jk} q_j = \alpha_{jk} q_k, \tag{12}$$

$$\forall j \in \delta^+,\ \sum_{k\in\delta^-} \beta_{jk} = 1, \tag{13}$$

$$\forall k \in \delta^-, \forall j \in \delta^+,\ 0 \le \beta_{jk} \le 1. \tag{14}$$

has a unique solution, with $\mathbf{w}_j^\dagger(U) \equiv v + p_j^\dagger(\rho)$ and $w_j^* \equiv \sum_{k\in\delta^-} \beta_{jk}\mathbf{w}_k(U_{k,0})$.

We now move to the two types of junctions considered here (merge and diverge).

4 Merge and Diverge Junctions

4.1 The Merge Junction

In this case, we have $k = 1, 2$ for incoming roads, $j = 3$ for the outgoing road. We pose $\beta_1 := \beta_{31}$, $\beta_2 := \beta_{32}$, $d_1 := d(\rho_{1,0}; \mathbf{w}_1, \mathbf{w}_1(U_{1,0}))$, and $d_2 := d(\rho_{2,0}; \mathbf{w}_2, \mathbf{w}_2(U_{2,0}))$. By assumption (3), $\alpha_{31} = \alpha_{32} = 1$.

The crucial point in solving (9) is to determine the supply s_j. We briefly describe the homogenization leading to s_j, before describing the solution.

Let $U := (\rho, \rho v)$. In this particular case, the homogenization process described above can be rewritten as follows.

First, on each incoming road $k = 1, 2$, the curve $\{\mathbf{w}_k(U) = \mathbf{w}_k(U_{k,0})\}$ becomes in Lagrangian coordinates

$$\{U(\tau, v);\quad v + P_k(\tau) = \mathbf{w}_k(U_{k,0})\}.$$

Now, on the outgoing road 3, the general equations (4)–(8) become

$$\bar{w} = \beta_1 \mathbf{w}_1(U_{1,0}) + (1 - \beta_1)\mathbf{w}_2(U_{2,0}) \tag{15}$$

and

$$\tau = \beta_1 P_1^{-1}(\mathbf{w}_1(U_{1,0}) - v) + (1 - \beta_1)P_2^{-1}(\mathbf{w}_2(U_{2,0}) - v), \tag{16}$$

where β_1 is still unknown.

As a prototype, we treat the case where $p_i(\rho) = \rho^\gamma$ (or $P_i(\tau) = 1/\tau^\gamma$) for $i = 1, 2, 3$, with $\gamma = 1$. Then, (16) becomes

$$\tau_3 = \frac{\beta_1}{w_1 - v} + \frac{(1 - \beta_1)}{w_2 - v}, \tag{17}$$

where w_k is the *constant*, $w_k := \mathbf{w}_k(U_{k,0})$, $k = 1, 2$.

This homogenized relation (17) implies

$$\rho_3 v = \frac{(w_2 - v)(w_1 - v)v}{\beta_1(w_2 - w_1) + w_1 - v}. \tag{18}$$

Combining (18) and (17), problem (9)–(14) is equivalent to the following maximization problem:

$$\max q_3 \text{ subject to} \tag{19a}$$

$$0 \le q_3 \le \frac{d_1}{\beta_1}; \tag{19b}$$

$$0 \le q_3 \le \frac{d_2}{(1 - \beta_1)}; \tag{19c}$$

$$0 \le q_3 \le s_3(U_{3,0}, \mathbf{w}_3^\dagger, \mathbf{w}_3^*); \tag{19d}$$

$$0 \le \beta_1 \le 1. \tag{19e}$$

We set $v_3 := v_{3,0}$. Then, for each given β_1, we denote by v_c the velocity corresponding to the maximal flux on the outgoing road, according to the supply, i.e., v_c is obtained by solving $\frac{d(\rho_3 v)}{dv} = 0$ for any *fixed* β_1. The supply $s_3(U_{3,0}, \mathbf{w}_3^\dagger, \mathbf{w}_3^*)$ is then:

$$s_3 = \begin{cases} \dfrac{(w_2 - v_3)(w_1 - v_3)v_3}{\beta_1(w_2 - w_1) + w_1 - v_3} & if \ \ v_3 \le v_c; \\ \dfrac{(w_2 - v_c)(w_1 - v_c)v_c}{\beta_1(w_2 - w_1) + w_1 - v_c} & if \ \ v_3 > v_c. \end{cases} \tag{20}$$

For any fixed v_3, we note that the function $\beta_1 \longmapsto s_3(v_3, \beta_1)$ is nondecreasing if $w_1 > w_2$, nonincreasing if $w_1 < w_2$, and constant if $w_1 = w_2$.

Some Examples of Optimal Solution When $w_1 > w_2$

To illustrate the computation of the optimal solution of the problem (19), we consider two generic cases (see [HMR06] for more details) (Fig. 3):

$$\text{Case 1} \quad \begin{cases} q_3 = \frac{d_1}{\beta_1}; \\ q_3 = \frac{d_2}{1-\beta_1}; \\ q_3 < s_3(v_3, \beta_1) \end{cases} \qquad \text{Case 2} \quad \begin{cases} q_3 < \frac{d_1}{\beta_1}; \\ q_3 < \frac{d_2}{1-\beta_1}; \\ q_3 = s_3(v_3, \beta_1); \end{cases}$$

Numerical Simulations

The numerical simulation uses a standard first-order relaxation scheme [JX95], with a fixed discretization size $\Delta x = 1/800$ and the time step is chosen according to the CFL condition. With $U = (\rho, \rho v)$, we set initial data $U_1^- = (3, 5)$, $U_2^- = (2, 3)$, and $U_3^+ = (3, 7)$ with the same function $p_i(\rho) = p(\rho)$ on all roads $i = 1, 2, 3$. The corresponding solution on the different roads is depicted in Fig. 4.

4.2 The Diverge Junction

Here, we follow the presentation of [HR06]. The results are also recovered by the presentation in [GP05]. In this case, $k = 1$ for the incoming road and

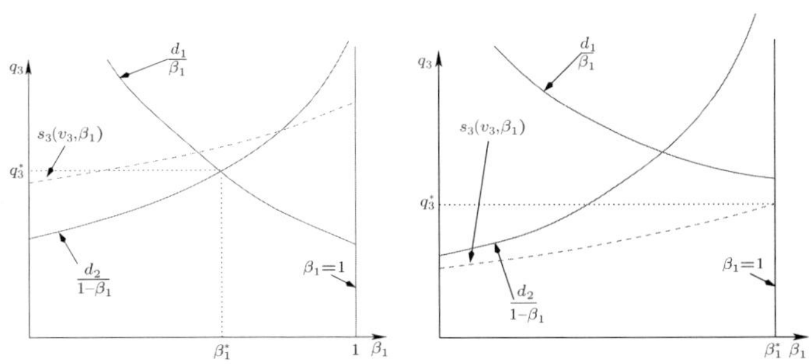

Fig. 3. Optimal solution (β_1^*, q_3^*) in the (β_1, q_3) plane: Case 1 (*left*) and Case 2 (*right*)

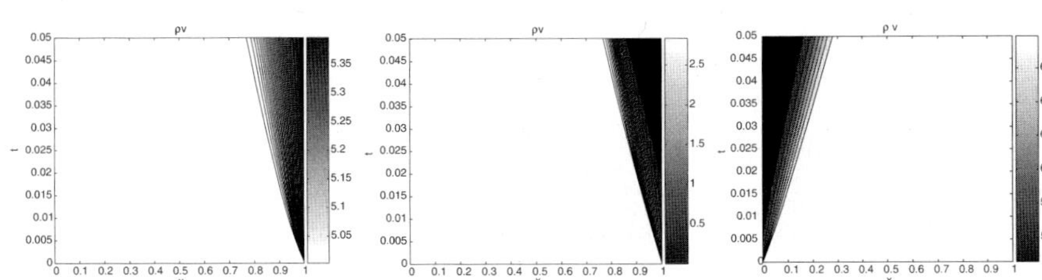

Fig. 4. Plots of the level curves of the flux ρv in the (x, t) plane on road 1 (*right*), road 2 (*middle*), and road 3 (*left*)

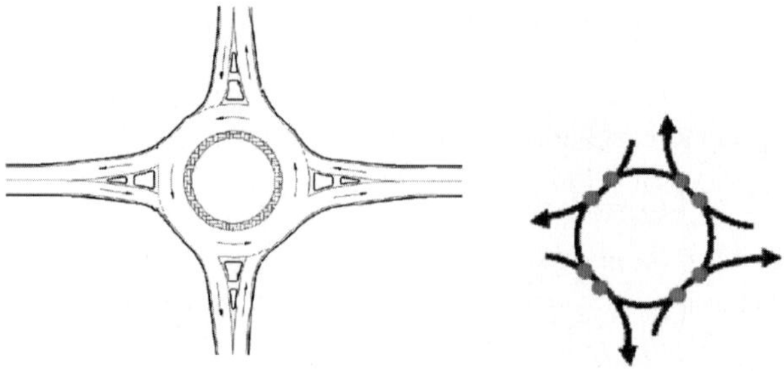

Fig. 5. Roundabouts for a 4–4 junction

$j = 2, 3$ for the outgoing roads. For notational convenience, we set $\alpha_{21} = \alpha$ and $\alpha_{31} = (1 - \alpha)$. Furthermore, we set $w_1 := \mathbf{w}_1(U_{1,0})$. Again, we simplify the general maximization problem (9). From (12), we obtain

$$\beta_{j1} = \frac{\alpha_{j1} q_1}{q_j}, \qquad j = 2, 3.$$

Since there is only one incoming road, $q_j = \alpha_{j1} q_1$, $j = 2, 3$ and therefore $\beta_{21} = \beta_{31} = 1$. Obviously, here, *no* homogenization is needed, since there is a single incoming road. Therefore the supply on the outgoing road is known and the problem (9) is nothing but a linear program.

5 Extension to a Roundabout and Concluding Remarks

Naturally due to the strong nonlinearities arising in particular in (11), the general maximization problem (9) is rather complex. However, most of the traffic intersections with n incoming and m outgoing roads can be seen and (in fact are designed) as roundabouts. For this type of junctions, see Fig. 5 (right), all the *conflict points* (i.e., points of intersections of roads) are either $2 \mapsto 1$ or $1 \mapsto 2$ junctions. In this two cases, we solved completely the problem and we have given an example of numerical results for the more interesting case of merge junction.

Acknowledgments

This work has been partially supported by the Kaiserslautern Excellence Cluster "Dependable Adaptive Systems and Mathematical Modeling," by the former HYKE No. HPRN-CT-2002-00282 and by the French ACI-NIM (Nouvelles Interactions des Mathématiques) No. 193 (2004).

References

[AKMR02] Aw, A., Klar, A., Materne, M., Rascle, M.: Derivation of continuum flow traffic models from microscopic follow the leader models, SIAM J. Appl. Math., **63**, 259–289, (2002).

[AR00] Aw, A., Rascle,: Resurection of second order models of traffic flow, SIAM J. Appl. Math., **60**, 916–944, (2000).

[BR03] Bagnerini, P., Rascle, M.: A multi-class homogenized hyperbolic model of traffic flow, SIAM J. Math. Anal., (4) **35**, 949–973, (2003).

[CGP02] Coclite, G., Garavello, M., Piccoli, B.: Traffic flow on road networks, SIAM J. Math. Anal., **36**, 1862–1894, (2005).

[Dafer00] Dafermos, C.M.: Hyperbolic Conservation Laws in Continuum Physics, Springer Verlag, Berlin, Heidelberg, New York, (2000).

[Dag95] Daganzo, C.F.: Requiem for second order fluid approximations of traffic flow, Trans. Res. B, **29**, 277–289, (1995).

[GP05] Garavello, M., Piccoli, B.: Traffic flow on a road network using the "Aw-Rascle" model, Comm. Partial Dif. Eq., To appear, (2005).

[HK04] Herty, M., Klar, A.: Modelling and optimization of traffic networks, SIAM J. Sci. Comp., **25**, 1066–1087, (2004).

[HMR06] Herty, M., Moutari, S., Rascle, M.: Optimization criteria for modelling intersections of vehicular traffic flow, Networks and Heterogeneous Media, **1**, 275–294, (2006).

[HR06] Herty, M., Rascle, M.: Coupling conditions for a class of "second-order" models for traffic flow, SIAM J. Appl. Math. **38**, 595–616, (2006).

[HR95] Holden, H., Risebro, N.H.: A mathematical model of traffic flow on a network of unidirectional roads, SIAM J. Math. Anal., **26**, 999–1045, (1995).

[JX95] Jin, S., Xin, Z.: The Relaxation Schemes for Systems of Conservation Laws in Arbitrary Space Dimensions, Comm. Pure Appl. Math., **48**, 235–255, (1995).

[Leb93] Lebacque, J.P.: Les modèles macroscopiques du traffic, Annales des Ponts., **67**, 24–45, (1993).

[LW55] Lighthill, M., Whitham, J.: On kinematic waves, Proc. Royal Soc. Edinburgh, A, **229**, 281–297, (1955).

[Ric56] Richards, P.I.: Shock waves on the highway, Operation Research, **4**, 42–51, (1956).

[Pay79] Payne, H.: FREFLO: A macroscopic simulation model for freeway traffic, Transportation Research Record, **722**, 68–89, (1979).

[Zha02] Zhang, H.M.: A non-equilibrium traffic model devoid of gas-like behaviour, Trans. Res. B, **36**, 275–298, (2002).

Some Contributions About an Implicit Discretization of a 1D Inviscid Model for River Flows

A. Bermúdez de Castro, R. Muñoz-Sola, C. Rodríguez, and M. Ángel Vilar

1 Statement of the Equations

We consider a 1D inviscid model to simulate river or channel flows with variable cross section. Let $\Omega(x,t)$ be the wet section of the river at point x and time t. We denote by $h(x,t)$ the height of water from the lowest point of this section and by $b(x)$ the height of this point with respect to a fixed reference. We put $\eta(x,t) = h(x,t) + b(x)$ (see Fig. 1).

By integrating the incompressible Euler equations on the wet section for each point x and time t and neglecting friction at the bottom, we can obtain the following equations (see [A79])

$$\frac{\partial a}{\partial t} + \frac{\partial Q}{\partial x} = q \text{ in } (0,L) \times (0,T), \tag{1}$$

$$\frac{\partial Q}{\partial t} + \frac{\partial (uQ)}{\partial x} + ga\frac{\partial \eta}{\partial x} = f \text{ in } (0,L) \times (0,T), \tag{2}$$

where $a(x,t)$ is the area of the wet section $\Omega(x,t)$, $Q(x,t)$ is the flow rate across this section, $u(x,t)$ is the averaged velocity ($u = Q/a$), L is the length of the channel, g is gravity acceleration, $q(x,t)$ is the flow rate per unit length of external discharges at point x and time t from effluents, and $f = \mid U \mid q\cos\theta$, where $U(x,t)$ is the velocity of the flow of external discharges and θ is the angle between the velocity vector of discharges and the river.

To get a closed system, we take into account the geometry of the cross section through the relationship $a = A(h,x)$. Function $A(\cdot,x)$ is strictly increasing and $A(0,x) = 0$ for every $x \in [0,L]$. By writing $h = B(a,x)$ and introducing the functions $E(h,x) = \int_0^h A(r,x)\,dr$, $F(a,x) = \int_0^{B(a,x)} \frac{\partial A}{\partial x}(r,x)\,dr$, and $G(a,x) = E(B(a,x),x)$, (2) can be rewritten as:

$$\frac{\partial Q}{\partial t} + \frac{\partial (uQ)}{\partial x} + g\frac{\partial}{\partial x}\Big(G(a,x)\Big) - gF(a,x) + gab'(x) = f. \tag{3}$$

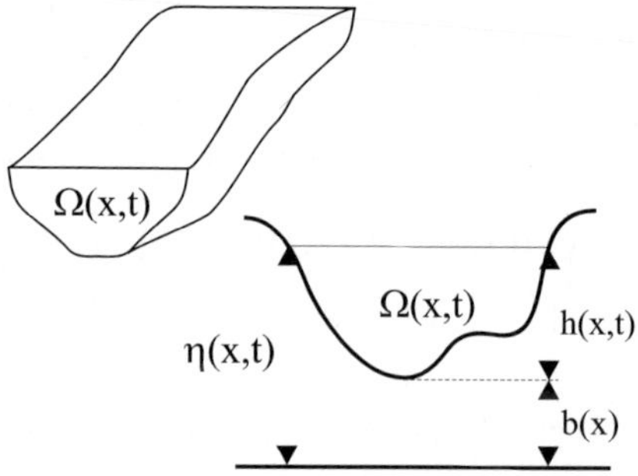

Fig. 1. Section of the river

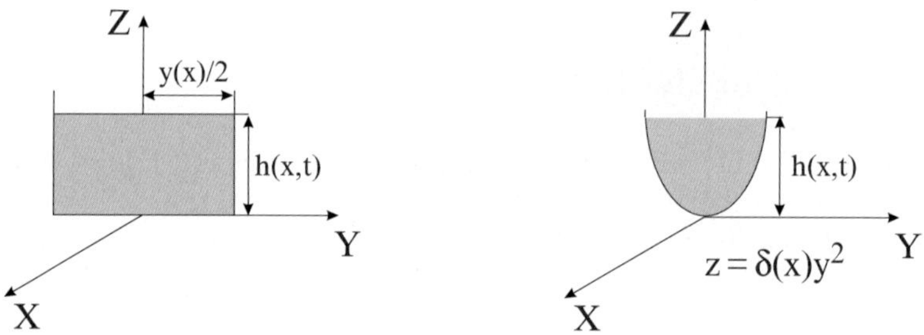

Fig. 2. Profiles of the river: rectangular and parabolic

Equations (1) and (3) constitute a hyperbolic system of balance laws with spatially dependent flux. This system is supplemented with boundary and initial conditions: $Q(0,t) = \gamma_0(t)$, $a(L,t) = \gamma_L(t) > 0$, $a(x,0) = a^0(x) > 0$, and $Q(x,0) = Q^0(x)$.

An existence result for a closely related model (including the bottom friction term), where the river longitudinally occupies the whole real line and the cross section is independent of x, can be seen in [KY97].

To enforce the physical condition $a \geq 0$, we introduce the maximal monotone operator

$$\widetilde{G}(a,x) = \begin{cases} G(a,x) & \text{if } a > 0, \\ (-\infty, 0] & \text{if } a = 0, \\ \mathcal{O}(1) & \text{if } a < 0. \end{cases}$$

and replace (3) by

$$\frac{\partial Q}{\partial t} + \frac{\partial (uQ)}{\partial x} + g\frac{\partial \zeta}{\partial x} - gF(a,x) + gab'(x) = f, \tag{4}$$

$$\zeta(x,t) \in \widetilde{G}(a(x,t),x) \quad \text{for a.e.} \quad (x,t) \in (0,L) \times (0,T). \tag{5}$$

As examples, we show the functions in our model for two specific sections: rectangular and parabolic. See Fig. 2 for notations.

For the rectangular profile, we have $A(h,x) = hy(x)$, $F(a,x) = \dfrac{y'(x)}{2y^2(x)}a^2$, and $G(a,x) = \dfrac{a^2}{2y(x)}$, and for the parabolic one, $A(h,x) = \dfrac{4}{3}\sqrt{\dfrac{h}{\delta(x)}}\,h$, $F(a,x) = -\frac{\delta'(x)}{5}\left(\frac{3}{4\delta(x)}\right)^{2/3} a^{5/3}$, and $G(a,x) = \frac{1}{5}\sqrt[3]{\frac{9}{2}\,\delta(x)}\,a^{\frac{5}{3}}$.

2 Derivation of the Time Semidiscretization Scheme

To make an implicit discretization of the convective term $\dfrac{\partial(uQ)}{\partial x}$, we use a method of characteristics (see [BPS83]). Let $X(x,t;\tau)$ and $J(x,t;\tau)$ be the solution of

$$\begin{cases} \dfrac{dX}{d\tau}(x,t;\tau) = u(X(x,t;\tau),\tau) \\ X(x,t;t) = x \end{cases}$$

$$\begin{cases} \dfrac{dJ}{d\tau}(x,t;\tau) = \dfrac{\partial u}{\partial x}(X(x,t;\tau),\tau)\; J(x,t;\tau) \\ J(x,t;t) = 1 \end{cases}$$

Let Δt be a time step, $t_n = n\Delta t$, $X^n(x) = X(x,t_{n+1};t_n)$, and $J^n(x) = J(x,t_{n+1};t_n)$. We denote $f^{n+1}(x) = f(x,t_{n+1})$ and $q^{n+1}(x) = q(x,t_{n+1})$.

Thus, (1), (4), and (5) are semidiscretized in time by means of the following scheme:

$$\frac{a^{n+1}(x) - a^n(x)}{\Delta t} + \frac{\partial Q^{n+1}(x)}{\partial x} = q^{n+1}(x) \text{ in } (0,L), \tag{6}$$

$$\begin{aligned} \frac{Q^{n+1}(x) - Q^n(X^n(x))J^n(x)}{\Delta t} &+ g\frac{\partial \zeta^{n+1}}{\partial x}(x) - gF(a^{n+1}(x),x) \\ &+ g\,a^{n+1}(x)b'(x) = f^{n+1}(x) \text{ in } (0,L), \end{aligned} \tag{7}$$

$$\zeta^{n+1}(x) \in \tilde{G}(a^{n+1}(x),x) \text{ a. e. in } (0,L), \tag{8}$$

By using (6), we eliminate a^{n+1} in (7) and (8). We then put the first of the resulting equations in weak form taking into account the boundary condition $a(L,t) = \gamma_L(t)$.

Let $\mathcal{V} = \{z \in W^{1,p}(0,L) : z(0) = 0\}$, $p \in [1,+\infty]$. We look for

$$Q^{n+1} \in \gamma_0(t_{n+1}) + \mathcal{V} \text{ and } \zeta^{n+1} \in L^{p'}(0,L)\ \left(\frac{1}{p'} + \frac{1}{p} = 1\right) \tag{9}$$

such that

$$\int_0^L \frac{Q^{n+1}(x) - Q^n(X^n(x))J^n(x)}{\Delta t} z(x)\, dx - g \int_0^L \zeta^{n+1}(x) \frac{\partial z}{\partial x}(x)\, dx$$
$$- g \int_0^L F\left(a^n(x) + \Delta t(q^{n+1}(x) - \frac{\partial Q^{n+1}}{\partial x}(x)), x\right) z(x)\, dx$$
$$+ g \int_0^L \left(a^n(x) + \Delta t(q^{n+1}(x) - \frac{\partial Q^{n+1}}{\partial x}(x))\right) b'(x)\, z(x)\, dx$$
$$= \int_0^L f^{n+1}(x) z(x)\, dx - gG(\gamma_L(t^{n+1}), L) z(L) \ \ \forall z \in \mathcal{V}, \quad (10)$$

$$\zeta^{n+1}(x) \in \tilde{G}\left(a^n(x) + \Delta t(q^{n+1}(x) - \frac{\partial Q^{n+1}}{\partial x}(x)), x\right) \text{ a. e. in } (0, L). \quad (11)$$

3 Existence of Solution of the Time Semidiscretized Problem

We first rewrite the time semidiscretized problem as a variational inequality. To do this, we assume that:

- $1 \le p < +\infty$, a^n, $q^{n+1} \in L^p(0, L)$, $b' \in L^{max(2,p')}(0, L)$, $f^{n+1} \in L^1(0, L)$, and $(Q^n \circ X^n)J^n \in L^1(0, L)$.
- $G : [0, +\infty) \times (0, L] \to \mathbb{R}$ is continuous and $G(s, x) \le cs^{p-1}$ $\forall s \ge 0$ $\forall x \in (0, L]$.
- $F : \mathbb{R} \times (0, L) \to \mathbb{R}$ is a Caratheodory function and satisfies $|F(s, x)| \le \hat{a}(x) + C|s|^p$ with $\hat{a} \in L^1(0, L)$, $\hat{a} \ge 0$, $C > 0$.
- $\gamma_0(t) = 0$ (for the sake of simplicity).

We remark that assumptions on function A imply that $G(0, x) = 0$ $\forall x \in (0, L]$ and $G(\cdot, x)$ is strictly increasing on $[0, \infty)$ $\forall x \in (0, L]$.

Let us introduce:

- The mapping $W : \mathcal{V} \to L^p(0, L)$ defined by $W(z) = a^n + \Delta t(q^{n+1} - \frac{\partial z}{\partial x})$.
- The closed convex set $\mathcal{K} = \{z \in \mathcal{V} / \ W(z) \ge 0\}$.
- The convex C^1 functional $\Phi_0 : L^p(0, L) \to \mathbb{R}$ defined by
$$\Phi_0(w) = \int_0^L \phi_0(w(x), x)\, dx, \text{ where } \phi_0(a, x) = \int_0^{|a|} G(r, x)\, dr.$$
- $\tilde{L} \in \mathcal{V}'$ and the operator $\tilde{A} : \mathcal{V} \to \mathcal{V}'$ defined, respectively, by

$$\langle \tilde{L}, z \rangle = \frac{1}{\Delta t} \int_0^L Q^n(X^n(x))J^n(x)\, z\, dx - g \int_0^L (a^n + \Delta t\, q^{n+1})\, b'(x)\, z\, dx$$
$$+ \int_0^L f^{n+1}\, z\, dx - gG(\gamma_L(t^{n+1}), L)\, z(L),$$

$$\langle \tilde{A}(v), z \rangle = \int_0^L v\, z\, dx - g\Delta t^2 \int_0^L b'(x) \frac{\partial v}{\partial x}\, z\, dx - g\Delta t \int_0^L F(W(v), x)\, z\, dx.$$

Lemma 1. *Problem (9)–(11) is equivalent to*

$$\begin{cases} \textit{Find } Q^{n+1} \in \mathcal{K} \textit{ such that} \\ < \tilde{A}(Q^{n+1}) + j'(Q^{n+1}), z - Q^{n+1} > \geq < \Delta t\, \tilde{L}, z - Q^{n+1} > \quad \forall z \in \mathcal{K}, \end{cases} \tag{12}$$

where $j(z) = g\Phi_0(W(z)) \ \forall z \in \mathcal{V}$.

The proof is based on the rule chain for subdifferentials (see [BMRV06]).

Theorem 1. *With the previous notations, let us assume* $p > 2$, $b' \in L^\infty(0, L)$, a^n, $q^{n+1} \in L^p(0, L)$, *and:*

- F *is a Caratheodory function that satisfies* $|F(s, x)| \leq \hat{a}(x) + C|s|^r$, *where* $\hat{a} \in L^{p'}(0, L)$, $\hat{a} \geq 0$, $C > 0$, $1 \leq r \leq p - 1$.
- $G : [0, +\infty) \times (0, L] \to \mathbb{R}$ *is continuous,* $G(\cdot, x)$ *is strictly increasing in* $[0, +\infty)$ $\forall x \in (0, L]$, $G(0, x) = 0$ $\forall x \in (0, L]$ *and there are constants* $c_1 > 0$, $c_2 > 0$, $c_3 \geq 0$ *such that*

$$c_2 s^{p-1} - c_3 \leq G(s, x) \leq c_1 s^{p-1} \ \ \forall s \in [0, +\infty) \ \forall x \in (0, L].$$

If $a^n \geq 0$ *and either (1)* $r < p - 1$ *or (2)* $r = p - 1$, *and* $F(a, x) \geq 0$ $\forall a \in [0, +\infty)$ *a.e.* $x \in (0, L)$, *then the semidiscretized problem (12) has at least one solution.*

We refer to [BMRV06] for the proof. It is done by checking that $\tilde{A} + j'$ is a Leray–Lions operator (hence bounded, continuous, and pseudomonotone) and coercive ([S97]), and using an existence theorem of solution for variational inequalities ([L69]).

4 Finite Element Discretization

Let τ_m be a partition of $[0, L]$ in m subintervals $I_k = [x_{k-1}, x_k]$ ($k = 1, \ldots, m$). We associate to τ_m the finite-dimensional spaces: $\mathcal{V}_m = \{Q_m \in C([0, L]) : Q_{m|_{I_k}} \in P_1 \ \forall k = 1, \ldots, m\}$, $\mathcal{V}_{m,0} = \{Q_m \in \mathcal{V}_m : Q_m(0) = 0\}$, and $\mathcal{W}_m = \{a_m : a_{m|_{I_k}} \in P_0 \ \forall k = 1, \ldots, m\}$. To alleviate notation, we shall suppress index m related to the spatial discretization, when dealing with discrete functions. Thus, the fully discretized problem is the following: given $q^{n+1} \in \mathcal{W}_m$ and $a^n \in \mathcal{W}_m$, find $Q^{n+1} \in \mathcal{V}_m$ and $a^{n+1} \in \mathcal{W}_m$ such that

$$a^{n+1} = a^n + \Delta t \left(q^{n+1} - \frac{\partial Q^{n+1}}{\partial x} \right), \tag{13}$$

$$\int_0^L \frac{Q^{n+1} - (Q^n \circ X^n)J^n}{\Delta t} z\, dx - \int_0^L \zeta^{n+1} \frac{\partial z}{\partial x}\, dx - g \int_0^L F(a^{n+1}) z\, dx$$

$$+ g \int_0^L a^{n+1} b'\, z\, dx = \int_0^L f^{n+1} z\, dx - gG(\gamma_L(t^{n+1}), L) z(L) \;\; \forall z \in \mathcal{V}_{m,0}, \quad (14)$$

$$\zeta^{n+1} \in \tilde{G}(a^{n+1}), \quad (15)$$

$$Q_0^{n+1} = \gamma_0(t^{n+1}). \quad (16)$$

5 An Algorithm for Solving the Fully Discrete Problem

Nonlinearities are due to F and G. Operator F is treated implicitly. Multi-valued operator G is handled with the Bermúdez–Moreno iterative algorithm (see [BM81]). This is based on the following fact: for $\omega > 0$, let $\tilde{G}^w = \tilde{G} - \omega I$ where I is the identity operator and $\tilde{G}^\omega_\lambda$ be the Yosida regularization of $\tilde{G}^\omega$. Let $p^{n+1} = \zeta^{n+1} - \omega a^{n+1}$, then

$$(15) \Leftrightarrow p^{n+1} \in \tilde{G}^w(a^{n+1}) \Leftrightarrow p^{n+1} = \tilde{G}^\omega_\lambda(a^{n+1} + \lambda p^{n+1}), \text{ for } \lambda\omega < 1 \text{ and } \lambda > 0$$

We rewrite (14) by substituting $\zeta^{n+1} = p^{n+1} + w(a^n + \Delta t(q^{n+1} - \dfrac{\partial Q^{n+1}}{\partial x}))$ and keep in the left-hand side only the terms which are linear in Q^{n+1}. The resulting equation suggests the following algorithm:

- $p^{n+1,0}$ is arbitrarily given.
- $a^{n+1,0} = a^n + \Delta t \left(q^{n+1} - \dfrac{\partial Q^n}{\partial x} \right)$.
- For $r \geq 0$, we calculate $Q^{n+1,r}$ as the solution of the linear system

$$\begin{aligned} &\int_0^L Q^{n+1,r} z\, dx + g\omega \Delta t^2 \int_0^L \frac{\partial Q^{n+1,r}}{\partial x} \frac{\partial z}{\partial x}\, dx - g\Delta t^2 \int_0^L \frac{\partial Q^{n+1,r}}{\partial x} b'\, z\, dx \\ &= \Delta t \int_0^L f^{n+1} z\, dx + \int_0^L (Q^n \circ X^n) J^n z\, dx + g\Delta t \int_0^L F(a^{n+1,r}) z\, dx \\ &\quad + g\Delta t \int_0^L p^{n+1,r} \frac{\partial z}{\partial x}\, dx + g\omega\Delta t \int_0^L a^n \frac{\partial z}{\partial x}\, dx + g\omega \Delta t^2 \int_0^L q^{n+1} \frac{\partial z}{\partial x}\, dx \\ &\quad - g\Delta t \int_0^L a^n\, b'\, z\, dx - g\Delta t^2 \int_0^L q^{n+1}\, b'\, z\, dx - g\Delta t G(\gamma_L(t^{n+1}), L) z(L) \\ &\quad \forall z \in \mathcal{V}_{m,0}. \end{aligned} \quad (17)$$

- $a^{n+1,r+1} = a^n + \Delta t \left(q^{n+1} - \dfrac{\partial Q^{n+1,r}}{\partial x} \right)$.
- $p^{n+1,r+1} = \tilde{G}^\omega_\lambda \left(a^{n+1,r+1} + \lambda p^{n+1,r} \right)$.

We discretize the integrals in (17) by using the trapezoidal rule.

6 Numerical Tests

Test 1: An academic test with parabolic section and explicit smooth solution. We take $L = 1{,}000$, $b(x) = 5 - \frac{x(x-500)}{10^5}$, and $h(x) = b(x) + 1$, and consider a parabolic channel $z = \delta(x)y^2$ with $\delta(x) = \dfrac{h(x)^3}{36}$ to have $a = 8$. We take also $\gamma_0(t) = \dfrac{1}{t+1} + 10$ and $q(x) = \dfrac{1}{10^3 + (x-400)^2}$. Then, from (1) we deduce that $Q(x,t) = \gamma_0(t) + \int_0^x q(s,t)ds$. We take f accordingly with (2) and impose the boundary and initial conditions satisfied by a and Q. Hence we have constructed a test problem with exact solution a and Q. The numerical results for wet area and velocity can be seen in Fig. 3. The errors in the $L^2(0,L)$ norm at time $t = 800\,\text{s}$, against the space step $\Delta x = 0.125, 0.25, 0.5, 1(\text{m})$, can

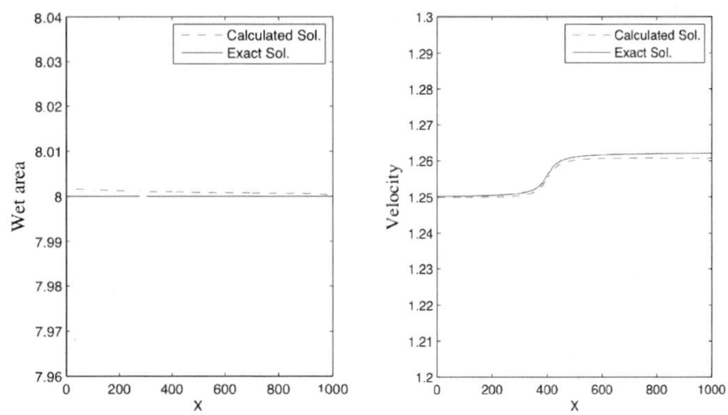

Fig. 3. Test 1: wet area and velocity for $t = 800\,\text{s}$

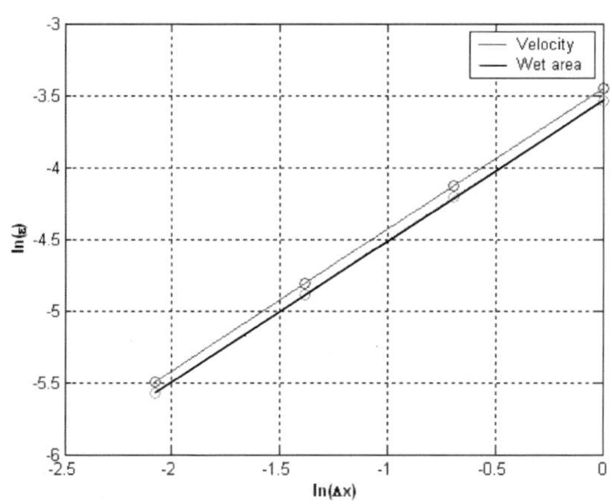

Fig. 4. Test 1: error in the $L^2(0,L)$ norm at time $t = 800\,\text{s}$, against Δx in log–log scale

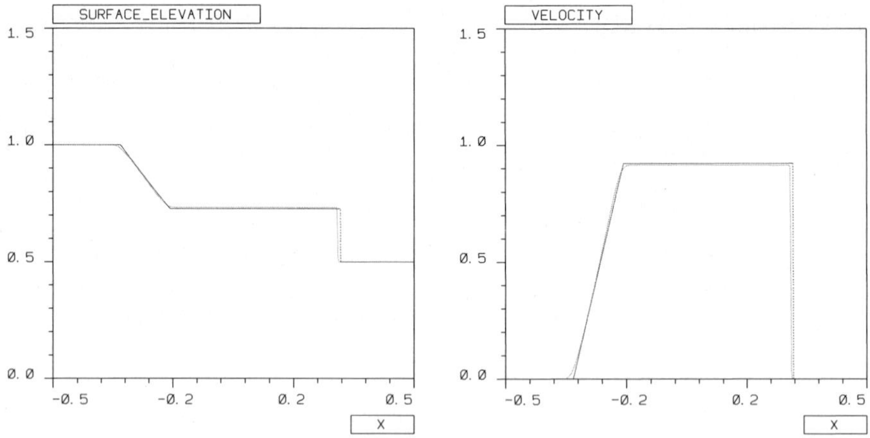

Fig. 5. Test 2: surface elevation $\eta(x,t) = h(x,t)$ and velocity $u(x,t)$ for $t = 0.1\,\mathrm{s}$. Exact solution (*dotted line*) vs. numerical solution (*continuous line*) computed with $\Delta x = 10^{-3}$ and $\Delta t = 10^{-4}$

be seen in Fig. 4 in the log–log scale. The time step Δt has been chosen to keep the ratio $\Delta t/\Delta x = 0.1(\mathrm{s\,m^{-1}})$.

Test 2: Dam-break problem for a flat channel with constant rectangular section. We consider a channel with constant rectangular section $y(x) = 1$, $x \in [-0.5, 0.5]$ and $b(x) = 0$ (Fig. 5). We take $f = 0$, $q = 0$, and the following boundary and initial conditions: $Q(-0.5,t) = 0$, $\eta(0.5,t) = 0.5$, $Q(x,0) = 0$, and

$$\eta(x,0) = \begin{cases} 1 & \text{if } -0.5 < x < 0, \\ 0.5 & \text{if } 0 < x < 0.5 \end{cases}.$$

Acknowledgments

Authors have been supported by Ministerio de Ciencia y Tecnologia (Spain) under Research Projects DPI2001-1613-C02-02 and BFM2003-00373 and also by PGIDIT02PXIC2070IPN, Xunta de Galicia (Spain).

References

[A79] Abbot, M.B.: Computational Hydraulics: Elements of the theory of free surface flows. Pitman (1979)

[BM81] Bermúdez, A., Moreno, C.: Duality methods for solving variational inequalities. Comp. and Maths. with Appls., **7**, 43–58 (1981)

[BMRV06] Bermúdez, A., Muñoz-Sola, R., Rodríguez, C., Vilar, M.A.: Theoretical and numerical study of an implicit discretization of a 1D inviscid model for river flows. Math. Models Methods Appl. Sci., **16**, 375–395 (2006)

[BPS83] Bercovier, M., Pironneau, O., Sastri, V.: Finite elements and characteristics for some parabolic-hyperbolic problems. Appl. Math. Modelling.,**7**, 89–96 (1983)

[KY97] Klingenberg, C., Yung-guang Lu.: Existence of solutions to hyperbolic conservation laws with a source. Commun. Math. Phys., **187**, 327–340 (1997)

[L69] Lions, J.L.: Quelques méthodes de résolution des problemes aux limites non linéaires. Dunod (1969)

[S97] Showalter, R.E.: Monotone operators in Banach space and nonlinear partial differential equations. American Mathematical Society. Series: Mathematical surveys and monographs, Vol. 49 (1997)

Remarks on the Nonhomogeneous Oseen Problem Arising from Modeling of the Fluid Around a Rotating Body

S. Kračmar, S. Nečasová, and P. Penel

1 Introduction

Over the past years, there has been a great impulse in studying the motion of a rigid body. This is a fascinating area of new problems for theory and numerical investigation. We would like to mention some historical attempt of this problem. The first systematic study on this subject initiated with the pioneering work of Kirchhoff [17], Lord Kelvin [28] regarding the motion of one or more bodies in a frictionless liquid. After that many mathematicians have furnished significant contributions to this fascinating field under different assumptions on the body and on the fluid. We wish to quote the work of Brenner [1] concerning the steady motion of one or more bodies in a linear viscous liquid in the Stokes approximation. Weinberger [30], [31] and Serre [25] regarding the fall of a body in an incompressible Navier–Stokes fluid under the action of gravity. Recently see e.g. Galdi and Silvestre [12], Farwig et al. [4], Farwig [2], [3], Hishida [20]–[22], Hieber et al. [16], Kračmar et al. [18], [19], Farwig et al. [5], [6], Farwig et al. [8].

2 Formulation of the Problem

In this paper we consider a three-dimensional rigid body rotating with angular velocity $\omega = \tilde{\omega}(0,0,1)^T$, $\tilde{\omega} \neq 0$ and assume that the complement is filled with a viscous incompressible fluid modeled by the Navier–Stokes equations. We consider the viscous flow either past a rotating body $K \subset\subset \mathbb{R}^3$ with axis of rotation ω and with the velocity $u_\infty = ke_3 \neq 0$ at infinity or around a rotating body K, which is moving in the direction of its axis of rotation. Given the coefficient of viscosity $\nu > 0$ and an external force $\tilde{f} = \tilde{f}(y,t)$, we are looking for the velocity $v = v(y,t)$ and the pressure $q = q(y,t)$ solving the nonlinear system

$$v_t - \nu \Delta v + v \cdot \nabla v + \nabla q = \tilde{f}, \tag{1}$$

$$\operatorname{div} v = 0, \tag{2}$$

$$v(y,t) = \omega \wedge y \text{ on } \partial\Omega(t), \tag{3}$$
$$v(y,t) \to u_\infty \neq 0, |y| \to \infty. \tag{4}$$

Here the time-dependent exterior domain $\Omega(t)$ is given – due to the rotation with angular velocity ω – by

$$\Omega(t) = O_\omega \Omega,$$

where $\Omega \subset \mathbb{R}^3$ is a fixed exterior domain and $O_\omega(t)$ denotes the orthogonal matrix

$$O_\omega(t) = \begin{pmatrix} \cos\tilde{\omega}t & -\sin\tilde{\omega}t & 0 \\ \sin\tilde{\omega}t & \cos\tilde{\omega}t & 0 \\ 0 & 0 & 1 \end{pmatrix}. \tag{5}$$

Introducing the change of variables

$$x = O_\omega(t)^T y \tag{6}$$

and the new functions

$$u(x,t) = O^T_{\omega(t)}(v(y,t) - u_\infty), \quad p(x,t) = q(y,t), \tag{7}$$

as well as the force term $f(x,t) = O_{\omega(t)}(t)^T \tilde{f}(y,t)$. After linearization in u at $u \equiv 0$ and considering the case $u_\infty || \omega$ to be considered here $O^T_\omega(t) u_\infty = ke_3$, for all $t > 0$ and thus considering only the stationary nonhomogeneous problem in the whole space problem we arrive at the modified Oseen system

$$-\nu\Delta u + k\,\partial_3 u - (\omega \wedge x)\cdot\nabla u + \omega \wedge u + \nabla p = f, \tag{8}$$
$$\operatorname{div} u = g, \tag{9}$$
$$u \to 0 \quad \text{as} \quad |x| \to \infty. \tag{10}$$

The linear system (8)–(10) has been analyzed in L^q–spaces, $1 < q < \infty$, in [2], [3] proving the priori estimate

$$\|\nu\nabla^2 u\|_q + \|\nabla p\|_q \leq c\|f\|_q, \tag{11}$$
$$\|k\partial_3 u\|_q + \|(\omega \wedge x)\cdot\nabla u + \omega \wedge u\|_q \leq c\left(1 + \frac{k^4}{\nu^2|\omega|^2}\right)\|f\|_q \tag{12}$$

with the constant $c > 0$ independent of ν, k, ω.
We introduce notation and then we will give a formulation of our problem. Given a domain $\Omega = \mathbb{R}^3$, the class $C_0^\infty(\Omega)$ consists of C^∞ functions with compact supports contained in Ω. By $L^q(\Omega)$ we denote the usual Lebesgue space with norm $\|\cdot\|_{q,\Omega}$. We define the homogeneous Sobolev spaces

$$\widehat{W}^{1,q}(\mathbb{R}^3) = \overline{C_0^\infty(\mathbb{R}^3)}^{\|\nabla\cdot\|_{q,\mathbb{R}^3}} = \{v \in L^q_{\mathrm{loc}}(\mathbb{R}^3);\ \nabla v \in L^q(\mathbb{R}^3)^3\}/\mathbb{R}, \tag{13}$$

and their dual space

$$\widehat{W}^{-1,q}(\mathbb{R}^3) = (\widehat{W}^{1,q/(q-1)}(\mathbb{R}^3))^*, \tag{14}$$

with norm $\|\cdot\|_{-1,q,\mathbb{R}^3}$.
Let us consider the problem (8)–(10).

Definition 1. *Let $1 < q < \infty$. Given $f \in \widehat{W}^{-1,q}(\mathbb{R}^3)^3$, $g \in L^q(\mathbb{R}^3)$, $(\omega \wedge x)g \in \widehat{W}^{-1,q}(\mathbb{R}^3)^3$ we call $\{u,p\} \in \widehat{W}_0^{1,q}(\mathbb{R}^3)^3 \times L^q(\mathbb{R}^3)$ weak solution to (8)–(10) if*

$$(i) \qquad \nabla \cdot u = g \quad in \quad L^q(\mathbb{R}^3),$$
$$(ii) \qquad (\omega \wedge x) \cdot \nabla u - \omega \wedge u \in \widehat{W}^{-1,q}(\mathbb{R}^3),$$

$\{u,p\}$ satisfies (15) in the sense of distributions, that is,

$$\begin{aligned} &< \nabla u, \nabla \varphi > - < (\omega \wedge x) \cdot \nabla u - \omega \wedge u, \varphi > \\ &+ k\Big\langle \frac{\partial u}{\partial x_3}, \varphi \Big\rangle - < p, \nabla \cdot \varphi > \; = \; < f, \varphi >, \\ &\qquad\qquad\qquad\qquad \varphi \in C_0^\infty(\mathbb{R}^3), \end{aligned} \tag{15}$$

where $< .,. >$ denotes the duality pairings.
$\{u,p\}$ satisfies (15) for all $\varphi \in \widehat{W}^{1,q/(q-1)}(\mathbb{R}^3)$.

Theorem 1. *Let $1 < q < \infty$ and suppose*

$$f \in \widehat{W}^{-1,q}(\mathbb{R}^3)^3, g \in L^q(\mathbb{R}^3), (\omega \wedge x)g, \nabla g \in W^{-1,q}(\mathbb{R}^3)^3 \tag{16}$$

then the problem (8)–(10) possesses a weak solution $\{u,p\} \in \widehat{W}^{1,q}(\mathbb{R}^3) \times L^q(\mathbb{R}^3)$

$$\begin{aligned} &\|\nabla u\|_q + \|p\|_q + \|(\omega \wedge x) \cdot \nabla u - \omega \wedge u\|_{-1,q} \\ &\le C(\|f\|_{-1,q} + \|g\|_q + \|(\omega \wedge x) g\|_{-1,q}) \end{aligned} \tag{17}$$

with some $C > 0$, depends on q. The solution is unique in $\widehat{W}^{1,q}(\mathbb{R}^3)^3$ up to a constant multiple of ω for u.

3 Proof of the Main Theorem

We give a sketch of the proof of Theorem 1, for more details see [19]. For a rapidly decreasing function $u \in \mathcal{S}(\mathbb{R}^n)$ let

$$\mathcal{F}u(\xi) = \hat{u}(\xi) = \frac{1}{(2\pi)^{n/2}} \int_{\mathbb{R}^n} e^{-ix\cdot\xi} u(x)\, dx,$$

with $\xi \in \mathbb{R}^n$, be the Fourier transform of u. Its inverse is denoted by $\mathcal{F}^{-1}$.

Because of the geometry of the problem it is reasonable to introduce cylindrical coordinates $(r, x_3, \theta) \in (0, \infty) \times \mathbb{R} \times [0, 2\pi)$. Then the term $(\omega \wedge x) \cdot \nabla u = -x_2 \partial_1 u + x_1 \partial_2 u$ may be rewritten in the form

$$(\omega \wedge x) \cdot \nabla u = \partial_\theta u$$

using the angular derivative ∂_θ applied to $u(r, x_3, \theta)$.

Since

$$\operatorname{div}((\omega \wedge x) \cdot \nabla u - \omega \wedge u) = (\omega \wedge x) \cdot \nabla \operatorname{div} u = \partial_\theta g$$

the pressure p will satisfy the equation

$$\Delta p = \operatorname{div} f + \Delta g + \partial_\theta g \text{ in } R^3.$$

With given p and ignoring (9) we arrive at the system

$$-\nu \Delta u + k \partial_3 u - \partial_\theta u + \omega \wedge u = f \quad \text{in} \quad \mathbb{R}^n, \tag{18}$$

where by f we mean $f - \nabla p$.

Now we will solve (18) explicitly using Fourier transforms and multiplier operators. Working first of all formally or in the space $\mathcal{S}'(\mathbb{R}^n)$ of tempered distributions we apply the Fourier transform $\mathcal{F} = \widehat{\ }$ to (8)–(10). With the Fourier variable $\xi = (\xi_1, \xi_2, \xi_3) \in \mathbb{R}^3$ and $s = |\xi|$ we get from (18).

$$(\nu s^2 + ik\xi_3)\widehat{u} - \partial_\varphi \widehat{u} + \tilde{\omega} e_3 \times \widehat{u} = \widehat{f}. \tag{19}$$

Theorem 2. *Let $1 < q < \infty$ and $f \in \widehat{W}^{-1,q}(\mathbb{R}^3)$. Then the equation*

$$\begin{aligned} &Lu \equiv \\ &-\Delta u + \frac{\partial u}{\partial x_3} - (\omega \wedge x) \cdot \nabla u + \omega \wedge u = f \ \text{ in } \ \mathbb{R}^3 \end{aligned} \tag{20}$$

possesses a weak solution $u \in \widehat{W}^{1,q}(\mathbb{R}^3)$ subject to the estimate

$$\begin{aligned} &\|\nabla u\|_{q,\mathbb{R}^3} + \|(\omega \wedge x) \cdot \nabla u \ - \omega \wedge u\|_{-1,q,\mathbb{R}^3} \\ &\le C \|f\|_{-1,q,\mathbb{R}^3}, \end{aligned} \tag{21}$$

with some $C > 0$, depends on q. The solution is unique in $\widehat{W}^{1,q}(\mathbb{R}^3)^3$ up to a constant multiple of ω for u.

In Fourier space, using cylindrical coordinates $(s, \varphi, \xi_3) \in \bar{\mathbb{R}}_+ \times [0, 2\pi] \times \mathbb{R}$, $s = \sqrt{\xi_1^2 + \xi_2^2}$, for $\xi = (\xi_1, \xi_2, \xi_3)^T$ as well and note that $\widehat{\partial_\varphi u} = \partial_\varphi \widehat{u}$, $\partial_\varphi u = (e_3 \times x) \cdot \nabla u$ and $\widehat{u}$ satisfies the equation

$$\frac{1}{\tilde{\omega}}(\nu |\xi|^2 + ik\xi_3)\widehat{u} - \partial_\varphi \widehat{u} + e_3 \wedge \hat{u} = \frac{1}{\tilde{\omega}} \widehat{f}$$

with respect to φ. Denoting $\widehat{v}(\varphi) = O_{e3}^T(\varphi)\widehat{u}(s, \varphi, \xi_3)$ then we are looking for the solution of the following problem:

$$\frac{1}{\tilde{\omega}}(\nu|\xi|^2 + ik\xi_3)\widehat{v} - \partial_\varphi \widehat{v} = \frac{1}{\tilde{\omega}} O_{e3}^T(\varphi)\widehat{f}. \tag{22}$$

After some calculation we obtain

$$\widehat{u}(\xi) \equiv \int_0^\infty e^{-\nu|\xi|^2 t} O_\omega^T(t)(\mathcal{F}f(O_\omega(t). - kte_3))(\xi)dt. \tag{23}$$

Finally note that $e^{-\nu|\xi|^2 t}$ is the Fourier transform of the heat kernel

$$E_t(x) = \frac{1}{(4\pi\nu t)^{3/2}} e^{-|x|^2/4\nu t}, \tag{24}$$

yielding

$$u(x) = \int_0^\infty E_t * O_\omega^T(t) f(O_\omega(t). - kte_3)(x)dt. \tag{25}$$

Note that $F = f - \nabla p$ is solenoidal so that the identity $i\xi \cdot \widehat{F} = 0$ implies that u is also solenoidal.
An essential step is to show

$$\|\nabla u\|_{q,\mathbb{R}^3} \le C\|G\|_{q,\mathbb{R}^3}$$

for the force of the form $f = \nabla \cdot G$ with $G \in C_0^\infty(\mathbb{R}^3)^9$, which is obtained by using Littlewood–Paley technique, see [19].

It follows from [7] that for all $f \in \widehat{W}^{-1,q}(\Omega)$, there is $G \in L^q(\Omega)$ such that

$$\nabla \cdot G = f \tag{26}$$

$$\|G\|_{q,\Omega} \le C\|f\|_{-1,q,\Omega} \tag{27}$$

with some $C > 0$.
As a result, the space $\{\nabla \cdot G; G \in C_0^\infty(\Omega)^n\}$ is dense in $\widehat{W}^{-1,q}(\Omega)$.

Let us derive the L^q estimate of the operator T defined by

$$\begin{aligned} TG(x) &= \nabla u(x) = -\textstyle\int_{\mathbb{R}^3} \nabla_x \nabla_y \Gamma(x, y) : G(y)dy, \\ \Gamma(x, y) &= \textstyle\int_0^\infty E_t O_w^T(t)dt. \end{aligned} \tag{28}$$

The following proposition indicates that fundamental solution does not define a classical Calderon–Zygmund integral operator and we need to use the Littlewood–Paley theory.

Proposition 1. *For $|x|$, $|y| \to \infty$, the fundamental solution $\Gamma(x, y)$ is not bounded by $C|x - y|^{-1}$. Precisely, there exists an $\alpha > 0$ such that for suitable x, $y \in R^3$ with $|x|$, $|y| \to \infty$*

$$\Gamma(x, y)| \ge \alpha \frac{log|x - y|}{|x - y}.$$

Proof. see [22] or [4]. □

Remark 1. We would like to mention very important property that the terms $\omega \wedge x\nabla u, \omega \wedge u$ cannot be estimated separately in general case but unless in a case that we require special type of compatibility condition on f

$$\frac{1}{2\pi}\int_0^{2\pi} O(\theta)^T f(r, x_3, \theta)d\theta = 0 \text{ for a.a. } r > 0, x_3 \in \mathbb{R}.$$

For more details see [4].

Proof of Theorem 1. As we explain before the pressure is formally obtained from the problem

$$p = -\nabla \cdot (-\Delta)^{-1}(f + \nabla g + (\omega \wedge x)g)$$

Since $(-\Delta)^{-1}$ can be justified as a bounded operator from $\widehat{W}^{-1,q}(\mathbb{R}^3)$ to $\widehat{W}^{1,q}(\mathbb{R}^3)$, see [15], we get

$$\|p\|_q \leq c\|f + \nabla g + (\omega \wedge g)\|_{-1,q},$$

which implies that

$$\|f - \nabla p\|_{-1,q} \leq c(\|f\|_{-1,q} + \|\nabla g + (\omega \wedge x)g\|_{-1,q}).$$

This completes the proof of Theorem 1. □

Acknowledgment

Š.N. and S.K. was supported by the Grant Agency of the Academy of Sciences No. IAA100190505. The research of S.K. was supported by the research plans of the Ministry of Education of the Czech Republic N. 6840770010 and the research of Š.N. was supported by the Academy of Sciences of the Czech Republic, Institutional Research Plan N. AV0Z10190503.

References

1. Brenner, H., *The Stokes Resistance of an Arbitrary Particle II*, Chem. Engng. Sci. **19**, (1959) 599–624
2. Farwig, R., *An L^q-analysis of viscous fluid flow past a rotating obstacle.* Tohoku Math. J. (2) **58**, 1, (2006), 129–147.
3. Farwig, R., *Estimates of lower order derivatives of viscous fluid flow past a rotating obstacle.* Banach Center Publications, Vol. 70, Warsaw 2005
4. Farwig, R., Hishida, T. and Müller, D., *L^q-Theory of a singular "winding" integral operator arising from fluid dynamics.* Pacific J. Math. 215 (2004), 297–312.
5. Farwig, R., Krbec, M., Nečasová, Š., *A weighted L^q approach to Oseen flow around a rotating body*, Preprint 2006

6. Farwig, R., Krbec, M., Nečasová, Š., *A weighted L^q approach to Stokes flow around a rotating body*, Preprint 2005, Univ. of Darmstadt
7. Farwig, R., Sohr, H., *Generalized resolvent estimates for the Stokes system in bounded and unbounded domains*, J. Math. Soc. Japan, 48, 4, 1994, 607–643
8. Farwig, R., Neustupa, J., *On the spectrum of a Stokes-type operator arising from flow around a rotating body.* Manuscripta mathematica, **122**, (2007), 419–437
9. Galdi, G.P., *An introduction to the mathematical theory of the Navier-Stokes equations: Linearised steady problems.* Springer Tracts in Natural Philosophy, Vol. 38, 2nd edition, Springer 1998
10. Galdi, G.P., *On the motion of a rigid body in a viscous liquid: a mathematical analysis with applications.* Handbook of Mathematical Fluid Dynamics, Vol. 1, Ed. by S. Friedlander, D. Serre, Elsevier 2002
11. Galdi, G.P., *Steady flow of a Navier-Stokes fluid around a rotating obstacle.* J. Elasticity 71 (2003), 1–31
12. Galdi, G.P., Silvestre A.L., *The Steady Motion of a Navier-Stokes Liquid Around a Rigid Body.* Archive for Rational Mechanics and Analysis, **184**, (2007), 371–400
13. Galdi, G.P., Vaidya, A., *Translational steady Fall of Symmetric Body in a Navier-Stokes Liquid with application to particle sedimentation*, Journal of Math. Fluid Mech., **3**, 2, (2001), 183–211
14. Galdi, G.P., Vaidya, A., Pokorný, M., Joseph, D., Feng, J., *Orientation of symmetric bodies falling in a second-order liquid at nonzero Reynolds number*, Math. Models Method Appl. Sci., **12**, (2002), 11, 1653–1690
15. Galdi, G., Simader, C.G., *Existence, uniqueness and L^q - estimates for the Stokes problem in exterior domains,* Arch. Rational Mech. Anal., **112**, (1990), 291–318
16. Geissert, M. Heck, H., Hieber, M., *L^p-theory of the Navier-Stokes flow in the exterior of a moving or rotating obstacle.* J. Reine Angew. Math., **596**, (2006), 45–62
17. Kirchoff, G., *Uber die Bewegung eines Rotationskorpers in einer Flussigkeit, Crelle*, **71**, (1869) 237–281
18. Kračmar, S., Nečasová, Š. and Penel, P., *Estimates of weak solutions in anisotropically weighted Sobolev spaces to the stationary rotating Oseen equations.* IASME Transactions 2 (2005), 854–861
19. Kračmar, S., Nečasová, Š. and Penel, P., *A L^q approach of the weak solution of the problem of Oseen flow around a rotating body*, Preprint 2006
20. Hishida, T., *An existence theorem for the Navier-Stokes flow in the exterior of a rotating obstacle*, Arch. Rational Mech. Anal. 150, (1999), 307–348
21. Hishida, T., *The Stokes operator with rotating effect in exterior domains*, Analysis 19, (1999), 51–67
22. Hishida, T., *L^q estimates of weak solutions to the stationary Stokes equations around a rotating body*, Hokkaido Univ. Preprint series in Math., No. 691, (2004)
23. Nečasová, Š., *On the problem of the Stokes flow and Oseen flow in $\mathbb{R}^3$ with Coriolis force arising from fluid dynamics.* IASME Transaction 2 (2005), 1262–1270.
24. Nečasová, Š., *Asymptotic properties of the steady fall of a body in viscous fluids.* Math. Meth. Appl. Sci. 27 (2004), 1969–1995
25. Serre, D., *Chute Libre d'un Solide dans un Fluide Visqueux Incompressible. Existence.* Jap. J. Appl. Math., 4, **(1)**, (1987) 99–110
26. Stein, E.M., *Singular integrals and differentiability properties of functions*, Princeton Univ. Press, Princeton 1970

27. Stein, E.M., *Harmonic Analysis: Real- Variable Methods, Orthogonality, and Oscillatory Integrals.* Princeton University Press, Princeton, N.J., 1993
28. Thomson, W. (Lord Kelvin), *Mathematical and Physical Papers,* Vol. **4**, Cambridge University Press, (1882)
29. von Wahl, W., *Vorlesungen über das Aussenraumproblem für die instationären Gleichungen von Navier- Stokes,* SFB 256 Nichtlineare partielle Differentialgleichungen, Vorlesungsreihe Nr. 11, Universität Bonn, (1989)
30. Weinberger, H.F., *On the steady fall of a body in a Navier-Stokes fluid,* Proc. Symp. Pure Mathematics, **23**, (1973), 421–440
31. Weinberger, H.F., *Variational Principles for a Body Falling in Steady Stokes Flow,* J. Fluid Mech., **52**, (1972), 321–344

Multi-D Bony Type Potential for the Boltzmann–Enskog Equation

S.-Y. Ha and S.E. Noh

1 Introduction

This chapter is devoted to the multidimensional Bony type potential to the Boltzmann–Enskog equation, which models a transport phenomena in moderately dense gases of hard sphere molecules. The Boltzmann–Enskog equation is an evolution equation for a velocity distribution function $f = f(x, \xi, t)$, which takes into account some geometric effects due to the overall dimensions of hard sphere molecules. In the absence of external forces, f satisfies an integro-differential equation:

$$\begin{aligned} \partial_t f + \xi \cdot \nabla_x f &= Q_E(f, f), \quad x, \xi \in R^3, \quad t > 0, \\ f(x, \xi, 0) &= f_0(x, \xi). \end{aligned} \tag{1}$$

Here $Q_E(f, f)$ denotes a collision operator due to binary collisions between particles, and whose specific form will be addressed later.

Let (ξ, ξ_*) and (ξ', ξ'_*) be pairs of precollisional and postcollisional velocities satisfying a collision transformation:

$$\xi' = \xi - [(\xi - \xi_*) \cdot \omega]\omega \quad \text{and} \quad \xi'_* = \xi_* + [(\xi - \xi_*) \cdot \omega]\omega, \quad \omega \in \mathbf{S}^2_+, \tag{2}$$

where $v \cdot w$ is the standard inner product between v and w in R^3 and $\mathbf{S}^2_+ = \{\omega \in \mathbf{S}^2 : (\xi - \xi_*) \cdot \omega \geq 0\}$. The Boltzmann–Enskog collision operator $Q_E(f, f)$ takes the form of

$$Q_E(f, f)(x, \xi, t) \equiv a^2 \int_{R^3 \times \mathbf{S}^2_+} [(\xi - \xi_*) \cdot \omega](f' f'_{*-} - f f_{*+}) d\omega d\xi_*, \tag{3}$$

where a is the diameter of a hard sphere molecule, and we have used abbreviated notations:

$$\begin{aligned} f' &\equiv f(x, \xi', t), \quad f'_{*-} \equiv f_*(x - a\omega, \xi'_*, t), \\ f &\equiv f(x, \xi, t), \quad f_{*+} \equiv f(x + a\omega, \xi_*, t). \end{aligned}$$

Throughout the paper, we use auxiliary functions as the functions f and $Q_E(f,f)$ evaluated along the particle trajectories:

$$f^\sharp(x,\xi,t) \equiv f(x+t\xi,\xi,t) \quad \text{and} \quad Q_E^\sharp(f,f)(x,\xi,t) \equiv Q_E(f,f)(x+t\xi,\xi,t).$$

We integrate (1) along the particle path $(x+s\xi,\xi,s)$ to find a mild form

$$f^\sharp(x,\xi,t) = f_0(x,\xi) + \int_0^t Q_E^\sharp(f,f)(x,\xi,s)ds, \qquad t \geq 0. \tag{4}$$

The definitions of mild and classical solutions are as follows.

Definition 1. *1. A nonnegative function $f(x,\xi,t) \in C([0,T);L^1_+(R^3 \times R^3))$ is a mild solution of* (1) *with a nonnegative initial datum f_0 if and only if for all $t \in [0,T)$ and a.e $(x,\xi) \in R^3 \times R^3$ f satisfies the integral equation* (4).

2. A function $f = f(x,\xi,t) \in C(R^3 \times R^3 \times [0,T))$ is a classical solution of (1) *with a nonnegative initial datum f_0 if and only if f is continuously differentiable with respect to (x,t) and f satisfies* (1) *pointwise.*

Local existence to the Enskog equation was first studied in [17], while global existence of classical and mild solutions when initial datum is a small perturbation of a vacuum was proved in [9, 20]. In contrast, for large initial data, global mild solutions and normalized solutions were obtained in [1, 10, 12, 18, 19] for Enskog type equations. On the other hand, asymptotic equivalence between the Enskog equation and the Boltzmann equation was investigated in [2, 3, 4]. For notational simplicity, we surpress from now on the t dependence and write

$$f(x,\xi) \equiv f(x,\xi,t) \qquad \text{and} \qquad Q_E(f,f)(x,\xi) \equiv Q_E(f,f)(x,\xi,t).$$

The main purpose of this paper is to present the multidimensional Bony type functional addressed in authors' recent paper [16]. The Bony type functional $\mathcal{B}(f(t))$ measures the future interactions between mass and momentum:

$$\mathcal{B}(f(t)) \equiv \int_{R^6} f(x,\xi)\Big[\int_{R^3 \times B_a(x)^c} \omega(x,y)\cdot(\xi_* - \xi)f(y,\xi_*)d\xi_* dy\Big]d\xi dx, \tag{5}$$

where $B_a(x)$ denotes an open ball with a center x and a radius a, and

$$\omega(x,y) = \frac{x-y}{|x-y|}, \qquad y \in B_a(x)^c.$$

Then it is easy to see that

$$|\mathcal{B}(f(t))| \leq 2\|f(t)\|_{L^1}^{3/2}\||\xi|^2 f(t)\|_{L^1} < \infty, \tag{6}$$

where we used a simplified notation

$$\|f(t)\|_{L^1} \equiv \|f(\cdot,\cdot,t)\|_{L^1(R^6)}.$$

The time-evolution estimate of the functional $\mathcal{B}(f(t))$ yields the following a priori estimate for the Boltzmann–Enskog equation.

Theorem 1. *Let f be a classical solution corresponding to initial data f_0 with finite energy and mass $\|f_0\|_{L^1} < \frac{\sqrt{2}}{4}$. Then we have the following estimate*

$$a^2 \int_0^\infty \int_{R^9 \times \mathbf{S}^2} [(\xi - \xi_*) \cdot \omega]^2 f(x, \xi) f(x + a\omega, \xi_*) d\omega d\xi d\xi_* dx dt$$
$$+ \int_0^\infty \int_{R^9 \times B_a(x)^c} \frac{|(\xi - \xi_*) \times (y - x)|^2}{|y - x|^3} f(x, \xi) f(y, \xi_*) d\xi d\xi_* dy dx dt \le C(f_0).$$

Here $C(f_0)$ is a positive constant independent of t.

Remark 1. For the Boltzmann equation with local conservation of mass, momentum, and energy, with a slight modification of Bony type functional (5), we can obtain the following a priori estimate without assuming smallness of the total mass (see [16] for details):

$$\int_0^\infty \int_{R^{12}} \frac{|(\xi - \xi_*) \times (y - x)|^2}{|y - x|^3} f(x, \xi) f(y, \xi_*) d\xi d\xi_* dy dx dt < \infty.$$

The rest of the paper is organized as follows. In Sect. 2 we discuss the explicit construction of multidimensional Bony type functional and its time-decay estimate. Finally, in Sect. 3 we discuss some perspective on search of Bony type functionals.

2 Multi-D Bony Type Functional

In this section, we briefly present the construction of the Bony type functional and its time-variation along the smooth solutions.

2.1 Construction of Potential

Let f be a classical solution to (1)–(3). Consider hard sphere test particles at $B_a(x)$ with a velocity ξ and their neighboring field particles located at $y \in B_a(x)^c \equiv R^3 - B_a(x)$ with a velocity ξ_* (see Fig. 1). In this case, local net interactions between mass and effective momentum flux due to the field particles are measured by

$$\mathcal{B}(x, \xi, t) \equiv f(x, \xi) \Big(\int_{B_a(x)^c \times R^3} \omega(x, y) \cdot (\xi_* - \xi) f(y, \xi_*) d\xi_* dy \Big). \tag{7}$$

As can be seen in Fig. 2, we note that

$$\mathcal{B}(x, \xi, t) \begin{cases} > 0, & \text{two particles are approaching,} \\ < 0, & \text{otherwise.} \end{cases}$$

Next, we define the multidimensional Bony type functional $\mathcal{B}$ measuring interactions between mass and effective momentum:

$$\mathcal{B}(f(t)) \equiv \int_{R^6} f(x, \xi) \Big[\int_{B_a(x)^c \times R^3} \omega(x, y) \cdot (\xi_* - \xi) f(y, \xi_*) dy d\xi_* \Big] d\xi dx. \tag{8}$$

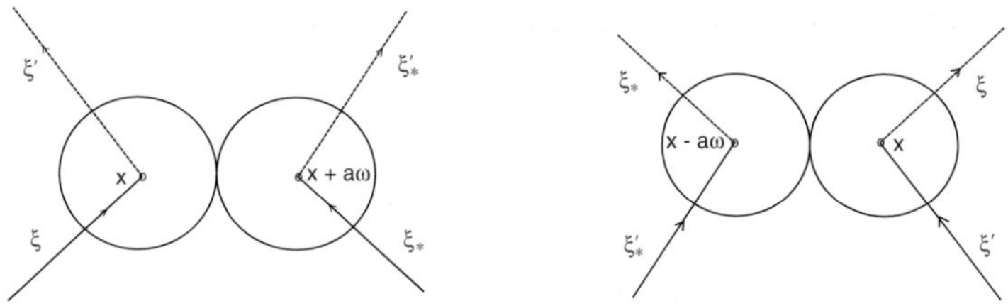

1: Forward collision. 2: Backward collision.

Fig. 1. Types of collisions

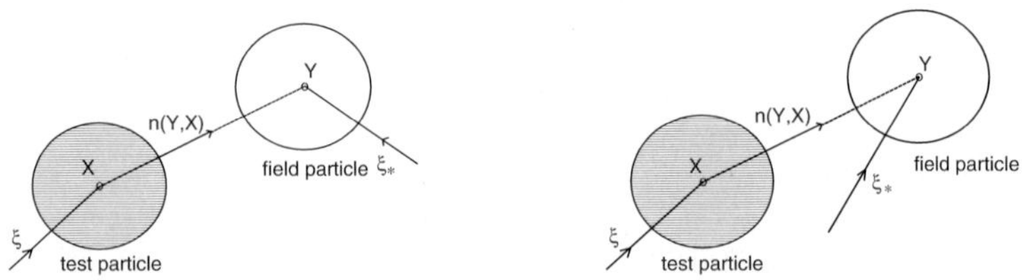

1: $\omega(x, y) \cdot (\xi_* - \xi) > 0$. 2: $\omega(x, y) \cdot (\xi_* - \xi) < 0$.

Fig. 2. Net interaction between mass and effective momentum; sign of the contribution

2.2 Time-Decay Estimate

We first define interaction production rates $\Lambda_i(f(t)), i = 1, 2$:

$$\Lambda_1(f(t)) \equiv \frac{a^2}{\sqrt{2}} \int_{R^9 \times \mathbf{S}^2} [(\xi_* - \xi) \cdot \omega]^2 f(x, \xi) f(x + a\omega, \xi_*) d\omega dx d\xi d\xi_*,$$

$$\Lambda_2(f(t)) \equiv \int_{R^9 \times B_a(x)^c} \frac{|(\xi - \xi_*) \times (y - x)|^2}{|y - x|^3} f(x, \xi) f(y, \xi_*) dy dx d\xi d\xi_*.$$

Lemma 1. *Let $f \in L^1(R^6)$ be rapidly decaying at infinity in phase space. Then for fixed $(x, t) \in R^3 \times R_+$, we have*

(1) $\displaystyle\int_{R^3} Q_E(f, f)(\xi) d\xi = 0.$

(2) $\displaystyle\int_{R^3} \xi Q_E(f, f)(\xi) d\xi = -a^2 \int_{R^6 \times \mathbf{S}^2_+} [(\xi - \xi_*) \cdot \omega]^2 \omega f f_{*+} d\omega d\xi_* d\xi.$

Lemma 2. *Let f be a classical solution to (1)–(3) with a finite energy and mass. Then we have*

$$\frac{d}{dt} \mathcal{B}(f(t)) \le (-1 + 2\sqrt{2}\|f_0\|) \Lambda_1(f(t)) - \Lambda_2(f(t)), \qquad t \ge 0.$$

Proof. Consider equations for $f(x,\xi)$ and $f(y,\xi_*)$:

$$\partial_t f(x,\xi) + \xi\cdot\nabla_x f(x,\xi) = Q_E(f,f)(x,\xi), \tag{9}$$
$$\partial_t f(y,\xi_*) + \xi_*\cdot\nabla_y f(y,\xi_*) = Q_E(f,f)(y,\xi_*). \tag{10}$$

Then $(\xi_* - \xi)\cdot\omega(x,y)[(9)\cdot f(y,\xi_*) + (10)\cdot f(x,\xi)]$ yields

$$\begin{aligned}
&\partial_t\Big((\xi_* - \xi)\cdot\omega(x,y) f(x,\xi) f(y,\xi_*)\Big)\\
&\quad = -\mathrm{div}_{(x,y)}\Big((\xi,\xi_*)(\xi_* - \xi)\cdot\omega(x,y) f(x,\xi) f(y,\xi_*)\Big)\\
&\qquad + \Big[\nabla_{(x,y)}\Big((\xi_* - \xi)\cdot\omega(x,y)\Big)\cdot(\xi,\xi_*)\Big] f(x,\xi) f(y,\xi_*)\\
&\qquad + (\xi_* - \xi)\cdot\omega(x,y)\big(Q_E(f,f)(x,\xi) f(y,\xi_*)\\
&\qquad + Q_E(f,f)(y,\xi_*) f(x,\xi)\big).
\end{aligned}$$

We now integrate the above equation over $R^9\times B_a(x)^c$ to get

$$\begin{aligned}
&\frac{d}{dt}\mathcal{B}(f(t))\\
&\quad = -\int_{R^9\times B_a(x)^c} \mathrm{div}_{(x,y)}\big((\xi,\xi_*)(\xi_* - \xi)\cdot\omega(x,y) f(x,\xi) f(y,\xi_*)\big) dy dx d\xi d\xi_*\\
&\qquad + \int_{R^9\times B_a(x)^c} \big[\nabla_{(x,y)}\big((\xi_* - \xi)\cdot\omega(x,y)\big)\cdot(\xi,\xi_*)\big] f(x,\xi) f(y,\xi_*) dy dx d\xi d\xi_*\\
&\qquad + \int_{R^9\times B_a(x)^c} (\xi_* - \xi)\cdot\omega(x,y)\\
&\qquad\quad \times\big(Q_E(f,f)(x,\xi) f(y,\xi_*) + Q_E(f,f)(y,\xi_*) f(x,\xi)\big) dy dx d\xi d\xi_*\\
&\quad \equiv \mathcal{J}_1 + \mathcal{J}_2 + \mathcal{J}_3.
\end{aligned} \tag{11}$$

We next estimate $\mathcal{J}_i$ separately.

Case 1. ($\mathcal{J}_3$): We use Lemma 1 and a straightforward calculations to get

$$|\mathcal{J}_3| \le 2\sqrt{2}\|f_0\|\Lambda_1(f(t))$$

Case 2. ($\mathcal{J}_1$): We apply divergence theorem to get

$$\mathcal{J}_1 = -\frac{a^2}{\sqrt{2}}\int_{R^9\times\mathbf{S}^2} [(\xi_* - \xi)\cdot\omega]^2 f(x,\xi) f(x + a\omega,\xi_*) d\omega dx d\xi d\xi_*.$$

Case 3. ($\mathcal{J}_2$): By direct calculation, we have

$$\mathcal{J}_2 = -\int_{R^9\times B_a(x)^c} \frac{|(\xi - \xi_*)\times(y - x)|^2}{|y - x|^3} f(x,\xi) f(y,\xi_*) dy dx d\xi d\xi_*.$$

In (4.5), we combine all estimates for $\mathcal{J}_i$ to find

$$\frac{d}{dt}\mathcal{D}(f(t)) \le (-1 + 2\sqrt{2}\|f_0\|)\Lambda_1(f(t)) - \Lambda_2(f(t)). \quad \square$$

Proof of Theorem 1. Note that $\|f_0\|_{L^1} < \frac{\sqrt{2}}{4}$, it follows from Lemma 4.1 that

$$\frac{d}{dt}\mathcal{B}(f(t)) \le -C_2\Lambda_1(f(t)) - \Lambda_2(f(t)), \tag{12}$$

where C_2 is a positive constant independent of time t. We now integrate (4.7) from $s = 0$ to $s = t$ to get

$$\mathcal{B}(t) + C_2\int_0^t \Lambda_1(f(s))ds + \int_0^t \Lambda_2(f(s))ds \le \mathcal{B}(0).$$

This yields

$$\begin{aligned}
&\frac{a^2C_2}{\sqrt{2}}\int_0^\infty \int_{R^9\times \mathbf{S}^2} [(\xi_* - \xi)\cdot\omega]^2 f(x,\xi)f(x+a\omega,\xi_*)d\omega d\xi d\xi_* dxdt \\
&\quad + \int_0^\infty \int_{R^9\times B_a(x)^c} \frac{|(\xi-\xi_*)\times(y-x)|^2}{|y-x|^3} f(x,\xi)f(y,\xi_*)d\xi d\xi_* dydxdt \\
&\quad \le |\mathcal{B}(0)| + |\mathcal{B}(f(t))| \\
&\quad \le 4\|f_0\|_{L^1}^{3/2}\||\xi|^2 f_0\|_{L^1} < \infty, \qquad \text{by (6).}
\end{aligned}$$

This completes the proof. □

3 Conclusion

In this work, we have presented a multi-D Bony type functional for the Boltzmann–Enskog equation. Using this functional, we also obtain a new a priori estimate for the full Boltzmann equation without smallness assumption on the initial data. However, the genuine version [5] of the Bony functional for the full Boltzmann equation is still missing, although it is available for the one-dimensional Boltmzann equation with some special collision kernel. Recently, Cercignani [7] has shown that DiPerna-Lions's renormalized solution for the one-dimensional Boltzmann equation is actually a weak solution in distribution sense. Hence once we can construct a multi-D Bony functional for the full Boltzmnan equation, we might show that renormalized solutions are the weak solutions in distribution sense.

Acknowledgment

The research of S.-Y. Ha is partially supported by BK21 and Research Institute of Mathematics at SNU, and the research of S.E. Noh is supported by BK21-Mathematical Division of SNU.

References

1. Arkeryd, L.: On the Enskog equation with large initial data. SIAM J. Math. Anal. **21**, 631–646 (1990)
2. Arkeryd, L., Cercignani, C.: Global existence in L^1 for the Enskog equation and convergence of the solutions to solutions of the Boltzmann equation. J. Stat. Phys. **59**, 845–867 (1990)
3. Arkeryd, L., Cercignani, C.: On the convergence of solutions of the Enskog equation to solutions of the Boltzmann equation. Comm. Partial Differential Equations **14**, 1071–1090 (1989)
4. Bellomo, N., Lachowicz, M.: On the asymptotic equivalence between the Enskog and the Boltzmann equaitons. J. Stat. Phys. **51**, 233–247 (1988)
5. Bony, J.M.: Solutions globales bornées pour les modèles discrete de l'équation de Boltzmann en dimension 1 d'espace, Actes Journees E.D.P.St. Jean de Monts, n° XVI, (1987)
6. Cercignani, C.: Global weak solutions of the Boltzmann equation. J. Stat. Phys. **118**, 333–342 (2005)
7. Cercignani, C.: Weak solutions of the Boltzmann equation and energy conservation. Appl. Math. Lett. **8**, 53–59 (1995). See also: Errata, Appl. Math. Lett. **8**, 95–99 (1995)
8. Cercignani, C.: A remarkable estimate for the solutions of the Boltzmann equation. Appl. Math. Lett. **5**, 59–62 (1992)
9. Cercignani, C.: Small data existence for the Enskog equation in L^1. J. Stat. Phys. **51**, 291–297 (1988)
10. Cercignani, C.: Existence of global solutions for the space inhomogeneous Enskog equation. Transport Theory Stat. Phys. **16**, 213–221 (1987)
11. Enskog, D.: Kinetiske theorie, svenska akad., **63** (1921). English translation in kinetic theory, S. Brush Ed., vol 3, Pergamon Press, New York, (1972). New York 1972.
12. Esteban, M.J., Perthame, B.: On the mod ified Enskog equation for elastic and inelastic collisions. Models with spin. Ann. Inst. H. Poincaré Anal. Non Linéaire **8**, 289–308 (1991)
13. Ha, S.-Y.: Lyapunov functionals for the Boltzmann-Enskog equation. Indiana Univ. Math J. **54**, 997–1014 (2005)
14. Ha, S.-Y.: Nonlinear functionals of the Boltzmann equation and uniform stability estimates. J. Differential Equations **215**, 178–205 (2005)
15. Ha, S.-Y.: L^1 stability of the Boltzmann equation for a hard sphere model. Arch. Rational Mech. Anal. **173**, 279–296 (2004)
16. Ha, S.-Y., Noh, S.E.: New a priori estimate for the Boltzmann-Enskog equation. Nonlinearity **19**, 1219–1232 (2006)
17. Lachowicz, M: On the local existence and uniqueness of solution of initial-value problem for the Enskog equation. Bull. Polish. Acad. Sci. **31**, 89–96 (1983)
18. Polewczak, J.: Global existence in L^1 for the generalized Enskog equation. J. Stat. Phys. **59**, 461–500 (1990)
19. Polewczak, J.: Global existence in L^1 for the modified nonlinear Enskog equation in R^3. J. Stat. Phys. **56**, 159–173 (1989)
20. Toscani, G., Bellomo, N.: The Boltzmann-Enskog equation in the whole space R^3. Some global existence, uniqueness and stability results. Comput. Math. Appl. **13**, 851–859 (1987)

Convergence of Well-Balanced Schemes for the Initial Boundary Value Problem for Scalar Conservation Laws in 1D

M. Nolte and D. Kröner

1 Introduction

In $\Omega :=]x_l, x_r[\subset \mathbb{R}$ we consider the one-dimensional initial boundary value problem

$$\begin{aligned} \partial_t u + \partial_x f(u) + z'(x)b(u) &= 0 \quad \text{in } \Omega\times]0,T[, \\ u(\cdot,0) &= u_0 \quad \text{in } \Omega, \\ u(x_l,\cdot) &= u_l \quad \text{on }]0,T[, \text{ if } f'(u) > 0, \\ u(x_r,\cdot) &= u_r \quad \text{on }]0,T[, \text{ if } f'(u) < 0, \end{aligned} \tag{1}$$

where we assume that the data satisfies

$$\begin{aligned} u_0 \in L^\infty(\Omega), \quad u_r, u_l \in L^\infty(]0,T[), \quad f \in C^2(\mathbb{R},\mathbb{R}), \\ z \in H^{1,\infty}(\Omega), \quad b \in C^1(\mathbb{R}), \quad b' \in L^\infty(\mathbb{R}). \end{aligned} \tag{2}$$

Furthermore, we define

$$D(s) := \int_0^s \frac{f'(\xi)}{b(\xi)}\, \mathrm{d}\xi \tag{3}$$

and assume that

$$D \in C^2(\mathbb{R}), \quad D(\mathbb{R}) = \mathbb{R}, \quad \inf_{\mathbb{R}} D' > 0. \tag{4}$$

Greenberg et al. [5] have developed and investigated a well-balanced scheme for the corresponding initial value problem. In particular, this scheme preserves equilibria at the discrete level. Numerical experiments (see, e.g., [2]) indicate that the well-balanced schemes are much more efficient than classical schemes. Convergence results and error estimates for the initial value problem can be found in [2] and [4].

Existence and uniqueness of the entropy solution for the initial boundary value problem have been obtained in [1] within the framework of BV-functions. In the case $z' \equiv 0$, Otto defined the entropy solution in the class of L^∞-functions in [8] (see also [6]) and Vovelle obtained convergence of numerical schemes towards this unique entropy solution in [9].

We are interested in a convergence result for a well-balanced scheme for (1) in $L^\infty(\Omega\times]0,T[)$. Therefore, we will adapt Otto's notion of entropy solution to take the source term into account. As in [9], using entropy process solutions, we then prove the convergence of numerical solutions from a well-balanced scheme toward the entropy solution of (1).

2 Entropy and Entropy Process Solutions

As in [8], we use a special class of entropy pairs, so-called boundary entropy pairs, to define the entropy solution. They are defined as follows:

Definition 1 (Boundary Entropy Pairs). *Let $\eta \in C^2(\mathbb{R})$ be convex, $q \in C^1(\mathbb{R})$ and $q' = \eta' f'$. Then (η, q) is called an* entropy pair *for the partial differential equation in (1).*

The entropy pair (η, q) is called a boundary entropy pair *if there exists a $w \in \mathbb{R}$ such that*

$$\eta(w) = \eta'(w) = q(w) = 0.$$

Note, that this definition of a boundary entropy pair is equivalent to the definition by Otto.

Using the boundary entropy pairs, we can define an entropy solution to the initial boundary value problem (1) as follows:

Definition 2 (Entropy Solution). *The function $u \in L^\infty(\Omega\times]0,T[)$ is called an entropy solution to (1), if the following inequality holds for all boundary entropy pairs (η, q) and all test functions $\varphi \in C_0^\infty(\overline{\Omega} \times [0,T[)$, $\varphi \geq 0$:*

$$\int_0^T \int_\Omega \eta(u)\partial_t\varphi + q(u)\partial_x\varphi - \eta'(u)z'(x)b(u)\varphi \,\mathrm{dx}\,\mathrm{dt} + \int_\Omega \eta\big(u_0(x)\big)\varphi(x,0)\,\mathrm{dx}$$
$$+ \operatorname*{Lip}_{[-C,C]}(f) \int_0^T \eta\big(u_r(t)\big)\varphi(x_r,t) + \eta\big(u_l(t)\big)\varphi(x_l,t)\,\mathrm{dt} \geq 0. \tag{5}$$

Here $C = \|u\|_{L^\infty(\Omega\times]0,T[)}$ and by $\mathrm{Lip}_{[-C,C]}(f)$ we denote the Lipschitz constant of $f_{|[-C,C]}$.

This definition directly corresponds to the one by Otto, if $z' \equiv 0$.

As we are interested in convergence of numerical solutions $u_{\Delta x}$ given by a well-balanced scheme, we will derive an equation similar to (5) for $u_{\Delta x}$. If the sequence $u_{\Delta x} \subset L^\infty(\Omega\times]0,T[)$ is bounded, we can use the following lemma to obtain a weak$*$ convergent subsequence:

Lemma 1 (Eymard et al. [3]). *Given a bounded sequence $(u_k)_{k\in\mathbb{N}} \subset L^\infty(\Omega\times]0,T[)$, there exists a subsequence $(u_{k'})_{k'\in\mathbb{N}}$ and a function $\mu \in L^\infty(\Omega\times]0,T[\times]0,1[)$ such that for any $g \in C^0(\mathbb{R})$ we have*

$$g(u_{k'}) \to \int_0^1 g(\mu(x,t,\alpha))\,\mathrm{d}\alpha \quad weak*.$$

Obviously, the weak$*$-limit μ cannot satisfy (5). Therefore, we make use of the so-called entropy process solutions introduced by Eymard et al. in [3]:

Definition 3 (Entropy Process Solution). *Let $\mu = \mu(x,t,\alpha)$ be an L^∞ $(\Omega\times]0,T[\times]0,1[)$-function, $C := \|\mu\|_{L^\infty(\Omega\times]0,T[\times]0,1[)}$ and for all boundary entropy pairs (η,q) and all $\varphi \in C_0^\infty(\overline{\Omega}\times[0,T[)$, $\varphi \geq 0$, let*

$$\int_0^T \int_\Omega \int_0^1 \eta(\mu)\partial_t\varphi + q(\mu)\partial_x\varphi - \eta'(\mu)z'(x)b(\mu)\varphi\,\mathrm{d}\alpha\,\mathrm{d}x\,\mathrm{d}t + \int_\Omega \eta(u_0(x))\varphi(x,0)\,\mathrm{d}x + \operatorname*{Lip}_{[-C,C]}(f)\int_0^T \eta(u_r(t))\varphi(x_r,t) + \eta(u_l(t))\varphi(x_l,t)\,\mathrm{d}t \geq 0.$$

Then μ is called an entropy process solution of (1).

Indeed, by the following theorem, there is at most one entropy process solution of (1). The proof will be omitted for the sake of a brief presentation.

Theorem 1 (Uniqueness of the Entropy Process Solution). *Let $\mu,\nu \in L^\infty(\Omega\times]0,T[\times]0,1[)$ be two entropy process solutions of (1) with respect to the inital data $u_0 \in L^\infty(\Omega)$ and to the boundary data $u_l, u_r \in L^\infty(]0,T[)$. Then $\mu = \nu$ a.e. on $\Omega\times]0,T[\times]0,1[$ and*

$$u(x,t) := \int_0^1 \mu(x,t,\alpha)\,\mathrm{d}\alpha$$

is an entropy solution of (1).

As a consequence, we obtain uniqueness of the entropy solution of (1).

3 The Well-Balanced Scheme

In this section we will describe a well-balanced scheme for the initial boundary value problem (1). The basic idea of the scheme goes back to Greenberg et al. (see [5]) and it was, in the case of the initial value problem, also studied in [2] using a kinetic formulation. Here, we are going to generalize this result to the initial boundary value problem (1).

First, we need some notation for the discretization. Let $\Delta x > 0$ such that $\frac{|\Omega|}{\Delta x} \in \mathbb{N}$, $j_l \in \mathbb{Z}$, and $x_j := x_l + (j - j_l + \frac{1}{2})\Delta x$. Denoting $C_j :=]x_{j-\frac{1}{2}}, x_{j+\frac{1}{2}}[$, we choose $J \subset \mathbb{Z}$ such that $\overline{\Omega} = \bigcup_{j\in J}\overline{C}_j$ and define $j_r := \max J$.

For the time discretization, let $\Delta t > 0$ such that $N_T := \frac{T}{\Delta t} \in \mathbb{N}$ and $t^n := n\Delta t$.

For any function $\varphi \in C^0(\mathbb{R}^2)$ we define $\varphi_j^n := \varphi(x_j, t^n)$ and the piecewise constant function

$$\overline{\varphi}(x,t) := \varphi_j^n \quad t \in [t^n, t^{n+1}), \ \ x \in C_j.$$

As to the discretization of the initial data, let u_j^0 be the average of u_0 over the cell C_j. To incorporate the boundary data, we use a ghost cell approach. Let

$$u_{j_l-1}^n := \frac{1}{\Delta t}\int_{t^n}^{t^{n+1}} u_l(t)\,\mathrm{dt}, \quad u_{j_r+1}^n := \frac{1}{\Delta t}\int_{t^n}^{t^{n+1}} u_r(t)\,\mathrm{dt}. \tag{6}$$

The standard form of a numerical scheme in conservation form for the partial differential equation in (1) is

$$u_j^{n+1} := u_j^n - \frac{\Delta t}{\Delta x}\left(g\left(u_j^n, u_{j+1}^n\right) - g\left(u_{j-1}^n, u_j^n\right)\right) - \Delta t z_j' u_j^n \quad \text{for } j \in J,$$

where g is a numerical flux and z_j' a discretization of z' in C_j, e.g., the cell average. This scheme is very inefficient, especially for the approximation of stationary solutions.

The main idea behind the well-balanced scheme is to preserve stationary solutions over time. By (3), a stationary solution v of (1), i.e. $\partial_t v \equiv 0$, also satisfies

$$\partial_x\left(D\left(v\left(x\right)\right) + z\left(x\right)\right) = 0.$$

This is equivalent to

$$D\left(v\right) + z\left(x\right) = c \in \mathbb{R}. \tag{7}$$

This equation will be the main building block of the scheme. Notice that under our assumptions (4), D is invertible.

Definition 4 (Well-Balanced Scheme). *For $j \in J$ let z_j be the cell average of z over the cell C_j. Moreover, let $z_{j_l-1} = z_{j_l}$ and $z_{j_r+1} = z_{j_r}$.*

Assume that $(u_j^n)_{j\in J}$ is already defined. The values u_j^{n+1} for the new time step of the well-balanced scheme are given by

$$u_j^{n+1} := u_j^n - \frac{\Delta t}{\Delta x}\left(g(u_j^n, u_{j+1,-}^n) - g(u_{j-1,+}^n, u_j^n)\right) \quad \textit{for } j \in J$$

where, due to (7), $u_{j+1,-}^n$ and $u_{j-1,+}^n$ are defined by

$$\begin{aligned} D(u_{j-1,+}) + z_j &= D(u_{j-1}) + z_{j-1}, \\ D(u_{j+1,-}) + z_j &= D(u_{j+1}) + z_{j+1}. \end{aligned} \tag{8}$$

Using the discrete data $(u_j^n)_{j\in J, 0\le n<N_T}$ we define the numerical solution $u_{\Delta x} \in L^\infty(\Omega \times [0,T[)$ by

$$u_{\Delta x}(x,t) := u_j^n \quad \text{for } t \in [t^n, t^{n+1}[, \ \ n < N_T, \ \ x \in C_j, \ \ j \in J.$$

Throughout this paper we will assume that g is the Engquist–Osher flux given by

$$g(u,v) = \int_0^u f'^{+}(\xi)\, \mathrm{d}\xi - \int_0^v f'^{-}(\xi)\, \mathrm{d}\xi + f(0) \quad \text{for } u,v \in \mathbb{R}. \tag{9}$$

Under these assumptions, we can prove that the numerical solution remains bounded.

Lemma 2 (Boundedness of the Numerical Solutions). *Let g be the Engquist–Osher flux and assume (2), (4). Furthermore, we define*

$$M := \max\{\|u_0\|_{L^\infty(\Omega)}, \|u_l\|_{L^\infty(]0,T[)}, \|u_r\|_{L^\infty(]0,T[)}\},$$
$$C_T^{\Delta x} := \left(M + |b(0)| + \Delta x \frac{\|z'\|_{L^\infty(\Omega)}}{\inf_{\mathbb{R}} D'}\right) e^{4T(\|b'\|_{L^\infty(\mathbb{R})}+1)\|z'\|_{L^\infty(\Omega)}}$$

and assume the CFL-condition

$$\operatorname*{Lip}_{[-C_T^{\Delta x}, C_T^{\Delta x}]}(f)\frac{\Delta t}{\Delta x} \le 1. \tag{10}$$

Then we obtain for all $n\,\Delta t \le T$

$$\sup_{j\in J} |u_j^n| \le C_T^{\Delta x}.$$

Moreover, $u_{j+1,-}^n$ and $u_{j-1,+}^n$, as defined in (8), are also bounded by $C_T^{\Delta x}$.

The proof is similar to Lemma 3.1 in [2]. However, the term containing Δx is missing there, though it is necessary.

One of the main tools for the convergence proof in Sect. 4 will be a cell entropy inequality, i.e., for any boundary entropy pair (η, q) we have to find a numerical entropy flux G, such that for $C > 0$ and for $\frac{\Delta t}{\Delta x}$ sufficiently small

$$\eta\Big(v - \frac{\Delta t}{\Delta x}\big[g(v,w) - g(u,v)\big]\Big) - \eta(v) + \frac{\Delta t}{\Delta x}\big(G(v,w) - G(u,v)\big) \le 0 \tag{11}$$

holds for all $u, v, w \in [-C, C]$.

As a slight generalization of a result in [2] we have the following lemma:

Lemma 3. *Let $|u|, |v|, \le C \in \mathbb{R}$ and let (η, q) be a boundary entropy pair, i.e., $\eta(w) = \eta'(w) = q(w) = 0$ for some $w \in \mathbb{R}$. Assuming the CFL-condition $\mathrm{Lip}_{[-C,C]}(f)\frac{\Delta t}{\Delta x} \le 1$, the function $G : \mathbb{R}\times\mathbb{R} \to \mathbb{R}$ given by*

$$G(u,v) := \int_w^u \eta'(\xi) f'^{+}(\xi)\,\mathrm{d}\xi - \int_w^v \eta'(\xi) f'^{-}(\xi)\,\mathrm{d}\xi \tag{12}$$

is a consistent numerical entropy flux, i.e., the cell entropy inequality (11) is satisfied.

4 Convergence of the Numerical Solutions

In this section we will present a convergence result for the numerical solutions obtained by the well-balanced scheme described in the previous section. We seek to prove the following theorem:

Theorem 2. *Let g be the Engquist–Osher flux defined in (9) and let (2), (4) be satisfied. Let $(u_{\Delta x})_{\Delta x>0}$ be the sequence of numerical solutions of (1) obtained from the well-balanced scheme defined in Definition 4 under the CFL-condition (10). Then, there is a function $u \in L^\infty(\Omega\times]0,T[)$ such that*

$$u_{\Delta x} \overset{\Delta x\to 0}{\longrightarrow} u \quad in\ L^p(\Omega_T),\ \ 1 \le p < \infty,$$

and u is the unique entropy solution of (1).

For the proof we need the following lemmata. We start with an estimate on the numerical entropy flux across the boundary of Ω:

Lemma 4. *Let (η, q) be a boundary entropy pair, let $w \in \mathbb{R}$ such that $\eta(w) = \eta'(w) = q(w) = 0$ and let G be defined by (12). If additionally we have $u, v, w \in [-C, C]$ for some $C > 0$, the following inequality holds true:*

$$-\operatorname*{Lip}_{[-C,C]}(f)\eta(v) \le G(u, v) \le \operatorname*{Lip}_{[-C,C]}(f)\eta(u).$$

Proof. Since $\eta \in C^2(\mathbb{R})$ is convex and $\eta'(w) = 0$, we have

$$\operatorname{sign}(u - w)\eta'(\xi) \ge 0 \quad \text{for all } \xi \in [u, w].$$

Therefore, as $f'^+ \ge 0$, we see that

$$\int_w^u \eta'(\xi)f'^+(\xi)\,\mathrm{d}\xi = \int_{[w,u]} \operatorname{sign}(u-w)\eta'(\xi)f'^+(\xi)\,\mathrm{d}\xi \ge 0.$$

On the other hand $\eta(w) = 0$ and we can estimate

$$\int_w^u \eta'(\xi)f'^-(\xi)\,\mathrm{d}\xi \le \|f'\|_{L^\infty([-C,C])} \int_{[w,u]} \operatorname{sign}(u-w)\eta'(\xi)\,\mathrm{d}\xi.$$

Putting the two inequalities together, we obtain the left inequlity. Using the same line of argumentation, the right inequality can be proved. □

Lemma 5. *Let g be the Engquist–Osher flux and assume (2), (4), and (11). Let $(u_j^n)_{0\le n\le N_T, j\in J}$ be the numerical solution of (1) in the sense of Definidvips tion 4 and assume $\frac{\Delta t}{\Delta x} \le \lambda$ for some λ. Let (η, q) be an entropy pair, $\eta(w) = \eta'(w) = q(w) = 0$ and $|u_j^n|, |u_{j+1,-}^n|, |u_{j-1,+}^n|, |w| \le C$ for some $C > 0$. Then, for all $\varphi \in C_0^2(\mathbb{R}^2) \cap C_0^\infty(\overline{\Omega} \times [0, T[)$, $\varphi \ge 0$, we have*

$$\begin{aligned}&\int_0^T \int_\Omega \eta(u_{\Delta x})\partial_t\varphi + q(u_{\Delta x})\partial_x\varphi - \eta'(u_{\Delta x})b(u_{\Delta x})z'(x)\varphi \,\mathrm{dx}\,\mathrm{dt}\\&+\int_\Omega \eta\big(u_0(x)\big)\varphi(x,0)\,\mathrm{dx} + \operatorname*{Lip}_{[-C,C]}(f)\int_0^T \eta\big(u_r(t)\big)\varphi(x_{j_r},t)\,\mathrm{dt}\\&\qquad+\operatorname*{Lip}_{[-C,C]}(f)\int_0^T \eta\big(u_l(t)\big)\varphi(x_{j_l},t)\,\mathrm{dt} \geq \mathcal{O}(\Delta x).\end{aligned}\tag{13}$$

Proof. Multiplying the cell entropy inequality (11) by $\Delta x\,\varphi_j^n$ and summing over $n < N_T$ and $j \in J$ we obtain

$$\begin{aligned}0 \geq \Delta x &\sum_{n=0}^{N_T-1}\sum_{j\in J}\left(\eta\left(u_j^{n+1}\right)-\eta\left(u_j^n\right)\right)\varphi_j^n\\&+\Delta t\sum_{n=0}^{N_T-1}\sum_{j\in J}\left(G\left(u_j^n,u_{j+1}^n\right)-G\left(u_{j-1}^n,u_j^n\right)\right)\varphi_j^n\\&+\Delta t\sum_{n=0}^{N_T-1}\sum_{j\in J}\left(G\left(u_j^n,u_{j+1,-}^n\right)-G\left(u_j^n,u_{j+1}^n\right)\right)\varphi_j^n\\&-\Delta t\sum_{n=0}^{N_T-1}\sum_{j\in J}\left(G\left(u_{j-1,+}^n,u_j^n\right)-G\left(u_{j-1}^n,u_j^n\right)\right)\varphi_j^n.\end{aligned}\tag{14}$$

The first two terms are treated by classical arguments and we will not repeat them here. The only difference is the incorporation of the boundary data, which is handled using Lemma 4.

Let us just sketch the argumentation for the third term (the fourth term is handled similarly). By the mean value theorem there is a $\zeta_{j+1}^n \in [u_{j+1}^n, u_{j+1,-}^n]$ such that

$$\begin{aligned}G(u_j^n,u_{j+1,-}^n)-G(u_j^n,u_{j+1}^n) &= -\int_{u_{j+1}^n}^{u_{j+1,-}^n}\eta'(\xi)f'^-(\xi)\,\mathrm{d}\xi\\&= -\eta'(\zeta_{j+1}^n)f'^-\left(\zeta_{j+1}^n\right)\left(u_{j+1,-}^n-u_{j+1}^n\right).\end{aligned}$$

Moreover, by (8) there is a $\vartheta_{j+1}^n \in [u_{j+1}^n, u_{j+1,-}^n]$ so that

$$z_{j+1}-z_j = D(u_{j+1,-}^n)-D(u_{j+1}^n) = D'(\vartheta_{j+1}^n)(u_{j+1,-}^n-u_{j+1}^n).$$

Since $z_{j_l}-z_{j_l-1} = z_{j_r+1}-z_{j_r} = 0$, we obtain

$$\begin{aligned}&\sum_{j\in J}\big(G(u_j^n,u_{j+1,-}^n)-G(u_j^n,u_{j+1}^n)\big)\varphi_j^n\\&\quad= -\Delta x\sum_{j\in J}\eta'(\zeta_j^n)\frac{f'^-(\zeta_j^n)}{D'(\vartheta_j^n)}\frac{z_j-z_{j-1}}{\Delta x}\varphi_{j-1}^n\\&\quad= -\Delta x\sum_{j\in J}\eta'(u_j^n)\frac{f'^-(u_j^n)}{D'(u_j^n)}\frac{z_j-z_{j-1}}{\Delta x}\varphi_j^n + \mathcal{O}(\Delta x).\end{aligned}$$

Bearing in mind that $b = \frac{f'}{D'}$, it is not hard to see that this yields the source term. □

Proof 3 (Proof of Theorem 2). For fixed $\delta > 0$ and $\Delta x \leq \delta$, Lemma 2 and the CFL-condition imply

$$\|u_{\Delta x}\|_{L^\infty(\Omega\times]0,T[)} \leq C_T^{\Delta x} \leq C_T^{\delta}$$

and the boundedness of the states $u_{j+1,-}^n$ and $u_{j-1,+}^n$, as defined in (8).

By verifying the assumptions of Lemma 5, we see that (13) holds for any boundary entropy pair (η, q) and any nonnegative $\varphi \in C^0(\infty(\overline{\Omega} \times [0, T[)$. Using Lemma 1, we obtain a subsequence converging to an entropy process solution $\mu \in L^\infty(\Omega\times]0, T[\times]0, 1[)$ of (1).

By Theorem 1, we find an entropy solution $u \in L^\infty(\Omega\times]0, T[)$ such that $\mu(\cdot, \cdot, \alpha) = u$ for almost all $\alpha \in]0, 1[$. Because of uniqueness, the entire sequence $(u_{\Delta x})_{\Delta x>0}$ must converge to u weak$*$ in $L^\infty(\Omega\times]0, T[)$.

Applying a result by Vovelle (see [9], Lemma 10), we obtain strong L^p-convergence, $1 \leq p < \infty$. □

References

1. C. Bardos, A.Y. Le Roux, and J.C. Nedelec, *First order quasilinear equations with boundary conditions*, Commun. Partial Differ. Equations **4** (1979), 1017–1034.
2. R. Botchorishvili, B. Perthame, and A. Vasseur, *Equilibrium schemes for scalar conservation laws with stiff sources*, Math. Comput. **72** (2003), 131–157.
3. R. Eymard, T. Gallouët, and R. Herbin, *Existence and uniqueness of the entropy solution to a nonlinear hyperbolic equation*, Chin. Ann. Math., Ser. B **16** (1995), 1–14.
4. L. Gosse, *A priori error estimate for a well-balanced scheme designed for inhomogenous scalar conservation laws*, C.R. Acad. Sci., Paris, Sér. I **327** (1998), 467–472.
5. J.M. Greenberg and A.Y. Le Roux, *A well-balanced scheme for the numerical processing of source terms in hyperbolic equations*, SIAM J. Numer. Anal. **33** (1996), 1–16.
6. J. Málek, J. Nečas, M. Rokyta, and M. Ružička, *Weak and measure-valued solutions to evolutionary PDEs*, Chapman & Hall, 1996.
7. M. Nolte and D. Kröner, *Well-balanced schemes for the initial boundary value problem for 1d scalar conservation laws*, Arxiv preprint math.NA/0608567, 2006.
8. F. Otto, *Initial-boundary value problem for a scaler conservation law*, C.R. Acad. Sci., Paris, Sér. I **322** (1996), 729–734.
9. J. Vovelle, *Convergence of finite volume monotone schemes for scalar conservation laws on bounded domains*, Num. Math. **90** (2002), 563–596.

Stability for Multidimensional Periodic Waves Near Zero Frequency

M. Oh and K. Zumbrun

1 Introduction

Consider a system of conservation laws

$$u_t + \sum_j f^j(u)_{x_j} = \sum_{j,\,k} \left(B^{jk}(u)u_{x_k}\right)_{x_j}, \tag{1}$$

$u \in \mathcal{U}(\text{open}) \in \mathbb{R}^n$, $f^j \in \mathbb{R}^n$, $B^{jk} \in \mathbb{R}^{n\times n}$, $x \in \mathbb{R}^d$, $d \geq 2$, and a periodic traveling wave solution

$$u = \bar{u}(x\cdot\nu - st), \tag{2}$$

of period X, satisfying the traveling wave ordinary differential equation

$$(\sum_{j,k} \nu_j\nu_k B^{jk}(\bar{u})\bar{u}')' = (\sum_j \nu_j f^j(\bar{u}))' - s\bar{u}', \tag{3}$$

with initial conditions

$$\bar{u}(0) = \bar{u}(X) =: u_0.$$

Integrating (3), we reduce to a first-order profile equation

$$\sum_{j,k} \nu_j\nu_k B^{jk}(\bar{u})\bar{u}' = \sum_j \nu_j f^j(\bar{u}) - s\bar{u} - q \tag{4}$$

encoding the conservative structure of the equations, where q is a constant of motion.

The purpose of this article is to extend to multiple dimensions the important observation of Serre, relating the linearized dispersion relation near zero to a multidimensional version of the homogenized system developed in [Se.1]. As an immediate corollary, similarly as in [OZ.1, Se.1] in the one-dimensional case, this yields as a necessary condition for multidimensional stability the hyperbolicity of the multidimensional homogenized system.

2 Preliminaries

We make only generic assumptions ensuring that the set of periodic traveling waves is a manifold of maximal dimension subject to the conservative properties of the equations (encoded in form (4)). Given $(a, s, \nu, q) \in \mathcal{U} \times \mathbb{R} \times S^{d-1} \times \mathbb{R}^n$, (4) admits a unique local solution $u(y; a, s, \nu, q)$ such that $u(0; a, s, \nu, q) = a$. Denote X the period, $\omega := 1/X$ the frequency and M and F^j the averages over the period:

$$M := \frac{1}{X}\int_0^X u(y)dy, \quad F^j := \frac{1}{X}\int_0^X \Big(f^j(u) - \sum_{k=1}^d B^{jk}(u)\omega\nu_k\partial_y u\Big)dy,$$

when u is a periodic solution of (4). Since these quantities are translation invariant, we consider the set P of periodic functions u that are solutions of (4) for some triple (s, ν, q), and construct the quotient set $\mathcal{P} := P/\mathcal{R}$ under the relation

$$(u\,\mathcal{R}\,v) \Longleftrightarrow (\exists h \in \mathbb{R};\ v = u(\cdot - h)).$$

We thus have class functions:

$$X = X(\dot{u}),\ \omega = \Omega(\dot{u}),\ s = S(\dot{u}),\ \nu = N(\dot{u}),\ q = Q(\dot{u}),\ M = M(\dot{u}),$$
$$F^j = F^j(\dot{u}),$$

where $\dot{u}$ is the equivalence class of translates of different periodic functions. Note that $\bar{u}$ is a nonconstant periodic solution. Without loss of generality, assume $S(\bar{u}) = 0$ and $N(\bar{u}) = e_1$, so that (4) takes the form

$$B^{11}(\bar{u})\bar{u}' = f^1(\bar{u}) - \bar{q}$$

for $\bar{q} = Q(\bar{u})$. Letting $\bar{X} = X(\bar{u})$ and $\bar{a} = \bar{u}(0) = u_0$, the map $(y, a, s, \nu, q) \mapsto u(y; a, s, \nu, q) - a$ is smooth and well defined in a neighborhood of $(\bar{X}; \bar{a}, 0, e_1, \bar{q})$, and it vanishes at this special point. Here and elsewhere, e_j denotes the jth standard Euclidean basis element. We assume:

(H0) f^j, $B^{jk} \in C^2$.
(H1) $\operatorname{Re}\sigma(\sum_{jk}\nu_j\nu_k B^{jk}) \geq \theta > 0$.
(H2) The map $H : \mathbb{R} \times \mathcal{U} \times \mathbb{R} \times S^{d-1} \times \mathbb{R}^n \to \mathbb{R}^n$ taking $(X; a, s, \nu, q) \mapsto u(X; a, s, \nu, q) - a$ is a submersion at point $(\bar{X}; \bar{a}, 0, e_1, \bar{q})$.

As a consequence of (H0), (H2), there is a smooth $n + d$-dimensional manifold $\mathcal{P}$ of periodic solutions $\dot{u}$ in the vicinity of $\bar{u}$, where d is the spatial dimension.

3 Slow Modulation Approximation

We carry out a multidimensional version of the slow modulation (WKB) expansion in [Se.1]. Rescale $(x, t) \mapsto (\epsilon x, \epsilon t)$ in (1) to obtain

$$u_t + \sum_j f^j(u)_{x_j} = \epsilon \sum_{j,\,k} \big(B^{jk}(u)u_{x_k}\big)_{x_j}. \tag{5}$$

Let

$$u^\epsilon(x,t) = u^0\left(x, t, \frac{\phi(x,t)}{\epsilon}\right) + \epsilon u^1\left(x, t, \frac{\phi(x,t)}{\epsilon}\right) + \cdots, \tag{6}$$

where $y \mapsto u^0(x,t,y)$ is a periodic function with $\partial_x \phi \neq 0$. We plug (6) into (5) and consider the equations obtained by equating coefficients at successive powers of ϵ.

At order ϵ^{-1}, we have

$$-s\partial_y u^0 + \sum_j \omega\nu_j \partial_y(f^j(u^0)) - \partial_y\left(\sum_{j,k} \omega^2 \nu_j \nu_k B^{jk}(u^0)\partial_y u^0\right) = 0,$$

with

$$s := -\frac{\partial_t \phi}{|\partial_x \phi|}, \quad \nu := \frac{\partial_x \phi}{|\partial_x \phi|}, \quad \omega := |\partial_x \phi|, \tag{7}$$

which may be recognized as the traveling-profile equation after rescaling $y \to \omega y$. That is, $u^0(y) = \bar{u}(\omega y)$ for a periodic profile of period $X = \omega^{-1}$, hence u^0 is periodic of *period one*, as described in [Se.1]. The quantities $\omega(x,t)$, $s(x,t)$, $\nu(x,t)$ are the local frequency, speed, and direction of the modulated wave.

At order ϵ^0, we have

$$\partial_t u^0 + \sum_{j=1}^d \partial_{x_j}\left(f^j(u^0) - \sum_{k=1}^d B^{jk}(u^0)\omega\nu_k \partial_y u^0\right) = \partial_y(\ldots).$$

Taking the average with respect to y, and rescaling with $y := \omega y$, we obtain

$$\partial_t M(u^0) + \sum_{j=1} \partial_{x_j} F^j(u^0) = 0, \tag{8}$$

where

$$F^j(u^0) = \frac{1}{X}\int_0^X \left(f^j(u^0) - \sum_{k=1}^d B^{jk}(u^0)\omega\nu_k \partial_y u^0\right) dy \tag{9}$$

is the averaged flux along orbit u^0 (now rescaled to actual period $\bar{X}$), with

$$\sum_j \nu_j F^j = (SM + Q)(u^0),$$

by the profile equation. In the Laplacian case $B^{jk} = \delta^j_k$, (9) simplifies to

$$F^j(u^0) = \frac{1}{X}\int_0^X (f^j(u^0) - \nu_j(u^0)\partial_y u^0) dy. \tag{10}$$

We have an additional d equations

$$\partial_t(\Omega N)(u^0) + \partial_x(\Omega S(u^0)) = 0 \tag{11}$$

from the Schwarz identity $\partial_t\partial_x\phi = \partial_x\partial_t\phi$, where d is the dimension of the spatial variable x. (Note: (Ω, N) may be regarded as polar coordinates for ΩN.)

Combining, we obtain finally the closed homogenized system

$$\partial_t(M, \Omega N) + \sum_j \partial_{x_j}(F^j, \Omega S e_j) = 0 \tag{12}$$

consisting of $n+d$ equations in $n+d$ unknowns. The problem of stability of $\bar{u}$ may heuristically be expected to be related to the linearized equations of (12) about the constant solution $\dot{u}(x,t) \equiv u^0$, $u^0 \sim \bar{u}$, provided that the WKB expansion is justifiable by stability considerations. This leads to the homogeneous degree $n+d$ linearized dispersion relation

$$\hat{\Delta}(\xi, \lambda) := \det\left(\lambda \frac{\partial(M, \Omega N)}{\partial \dot{u}}(\dot{u}) + \sum_j i\xi_j \frac{\partial(F^j, S\Omega e_j)}{\partial \dot{u}}(\dot{u})\right) = 0. \tag{13}$$

4 The Evans Function Calculations

Recall that $\bar{u} = \bar{u}(x_1)$ represents a stationary solution. Linearizing (1) about $\bar{u}(\cdot)$, we obtain

$$v_t = Lv := \sum (B^{jk} v_{x_k})_{x_j} - \sum (A^j v)_{x_j}, \tag{14}$$

where coefficients

$$B^{jk} := B^{jk}(\bar{u}), \quad A^j v := Df^j(\bar{u})v - (DB^{j1}(\bar{u})v)\bar{u}_{x_1} \tag{15}$$

are now periodic functions of x_1.

Taking the Fourier transform in the transverse coordinate $\tilde{x} = (x_2, \cdots, x_d)$, we obtain

$$\begin{aligned}\hat{v}_t = L_{\tilde{\xi}}\hat{v} = {} & (B^{11}\hat{v}_{x_1})_{x_1} - (A^1\hat{v})_{x_1} + i(\sum_{j\neq 1} B^{j1}\xi_j)\hat{v}_{x_1} \\ & + i(\sum_{k\neq 1} B^{1k}\xi_k\hat{v})_{x_1} - i\sum_{j\neq 1} A^j\xi_j\hat{v} - \sum_{j\neq 1, k\neq 1} B^{jk}\xi_k\xi_j\hat{v},\end{aligned} \tag{16}$$

where $\tilde{\xi} = (\xi_2, \cdots, \xi_d)$ is the transverse frequency vector. The Laplace transform in time t leads us to study the family of eigenvalue equations

$$\begin{aligned}0 = (L_{\tilde{\xi}} - \lambda)w = {} & (B^{11}w')' - (A^1 w)' + i\sum_{j\neq 1} B^{j1}\xi_j w' + i(\sum_{k\neq 1} B^{1k}\xi_k w)' \\ & - i\sum_{j\neq 1} A^j \xi_j w - \sum_{j\neq 1, k\neq 1} B^{jk}\xi_k\xi_j w - \lambda w,\end{aligned} \tag{17}$$

associated with operators $L_{\tilde{\xi}}$ and frequency $\lambda \in \mathbb{C}$, where prime denotes $\partial/\partial x_1$. Clearly, a necessary condition for stability of (1) is that (17) have no L^2 solutions w for $\tilde{\xi} \in \mathbb{R}^{d-1}$ and Re $\lambda > 0$. For solutions of (17) correspond to normal modes $\hat{v}(x,t) = e^{\lambda t} e^{i\tilde{\xi}\cdot\tilde{x}} w(x_1)$ of (14).

The difficulty of our problem is due to accumulation at the origin of the essential spectrum of the linearized operator L about the wave as in the one-dimensional case. Multidimensional stability concerns the behavior of the perturbation of the top eigenvalue, $\lambda = 0$ under small perturbations in $\tilde{\xi}$. For study this stability, we use Floquet's theory and an Evans function [G] which not only depends on λ but also on ξ_1 which corresponds to the phase shift and $\tilde{\xi}$. To define the Evans function, we choose a basis $\{w^1(x_1, \tilde{\xi}, \lambda), \dots, w^{2n}(x_1, \tilde{\xi}, \lambda)\}$ of the kernel of $L_{\tilde{\xi}} - \lambda$, which is analytic in $(\tilde{\xi}, \lambda)$ and is real when λ is real, for details see [OZ.1, Se.1]. Now we can define the Evans function by

$$D(\lambda, \xi_1, \tilde{\xi}) := \left| \begin{matrix} w^l(X, \tilde{\xi}, \lambda) - e^{iX\xi_1} w^l(0, \tilde{\xi}, \lambda) \\ (w^l)'(X, \tilde{\xi}, \lambda) - e^{iX\xi_1} (w^l)'(0, \tilde{\xi}, \lambda) \end{matrix} \right|_{1 \le l \le 2n}, \tag{18}$$

where $\xi_1 \in \mathbb{R}$. Note that $X\xi_1$ is exactly θ in [Se.1]. We remark that D is analytic everywhere, with associated analytic eigenfunction w^l for $1 \le l \le 2n$. A point λ is in the spectrum of $L_{\tilde{\xi}}$ if and only if $D(\lambda, \xi) = 0$ with $\xi = (\xi_1, \tilde{\xi})$. The zero set $(\xi, \lambda(\xi))$ determines the linearized dispersion relation for (1), with $\lambda(\xi)$ running over the spectrum of L as ξ runs over $\mathbb{R}^d$. In particular, the low-frequency expansion of $\lambda(\xi)$ near $(\xi, \lambda) = (0, 0)$ may be expected to determine long-time asymptotic behavior, provided that spectrum away from $\lambda = 0$ has strictly negative real part, and this in turn may be expected to derive from the lowest order terms of the Taylor expansion of D. A tedious, but fairly straightforward calculation following [OZ.1, Se.1] shows that

$$D(\xi, \lambda) = \Delta_1(\xi, \lambda) + \mathcal{O}(|\xi, \lambda|^{n+2}), \tag{19}$$

where Δ_1 is a homogeneous degree $n+1$ polynomial expressed as the determinant of a rather complicated $2n \times 2n$ matrix in (ξ, λ).

5 Main Results

Our main result is the following theorem relating these two expansions (13) and (19), generalizing the result of [Se.1] in the one-dimensional case. Define

$$\Delta(\xi, \lambda) := \lambda^{1-d} \hat{\Delta}(\xi, \lambda), \tag{20}$$

where $\hat{\Delta}$ is defined as in (13).

Theorem 1. *Under assumptions (H0)–(H2), $\Delta_1 = \Gamma_0 \Delta$, i.e.,*

$$D(\xi, \lambda) = \Gamma_0 \Delta(\xi, \lambda) + \mathcal{O}(|\xi, \lambda|^{n+2}) \tag{21}$$

$\Gamma_0 \neq 0$ constant, for $|\xi, \lambda|$ sufficiently small.

That is, up to an additional factor of λ^{d-1}, the dispersion relation (13) for the averaged system (12) indeed describes the low-frequency limit of the exact linearized dispersion relation

$$D(\xi, \lambda) = 0.$$

The discrepancy λ^{d-1} is an interesting and at first glance puzzling new phenomenon in the multidimensional case. However, it is easily explained by a closer look at the formal approximation procedure described in Sect. 3. For, in the derivation of (12), it was assumed that ΩN represent the gradient $\nabla_x \phi$ of a certain phase function $\phi(x,t)$ In one dimension, this is no restriction, since we may always take $\phi(x,t) := \int_0^x \omega(z)dz$. However, in multidimensions, it imposes the additional constraint

$$\text{curl } (\Omega N) \equiv 0, \tag{22}$$

which properly should be adjoined to the averaged system.

Taking the curl of (11), we obtain the simple equation

$$\partial_t \text{ curl } (\Omega N) = 0,$$

revealing at once that constraint (22) is compatible with the time-evolution of the system, and that the unconstrained system possesses $(d-1)$ spurious zero characteristices $\lambda(\xi) \equiv 0$, corresponding to the $(d-1)$ Fourier modes in the range of the curl operator $\hat{f} \to \xi$ curl $\hat{f}$, lying in $(\xi/|\xi|)^\perp$. *Thus,* $\Delta(\xi, \lambda) = \lambda^{1-d}\hat{\Delta}(\xi, \lambda) = 0$ *is exactly the linearized dispersion relation for the constrained averaged system (12), (22) relevant to time-asymptotic behavior.*

As an immediate consequence of Theorem 1, we obtain the following two corollaries, yielding a *necessary* condition for low-frequency multidimensional spectral stability strengthening the one-dimensional version obtained in [OZ.1, Se.1].

Corollary 1. *Assuming (H0)–(H2) and the nondegeneracy condition*

$$\det \left(\frac{\partial(M, \Omega N)}{\partial \dot{u}}(\bar{u}) \right) \neq 0, \tag{23}$$

then for λ, ξ sufficiently small, the zero-set of $D(\cdot,\cdot)$, corresponding to spectra of L, consists of $n+1$ characteristic surfaces:

$$\lambda_j(\xi) = -ia_j(\xi) + o(\xi), \quad j = 1, \ldots, n+1, \tag{24}$$

where $a_j(\xi)$ denote the eigenvalues of

$$A := \sum_j \xi_j \frac{\partial(F^j, S\Omega e_j)}{\partial(M, \Omega N)}, \tag{25}$$

excluding $(d-1)$ identically zero eigenvalues associated with modes not satisfying constraint (22).

Proof. Similarly as in as in the proof of the analogous Lemma 7.5 [ZS] in the shock wave case, assuming (23), we may easily deduce (24) from (21) using Rouché's Theorem. Defining

$$\mathcal{D}^{\rho,\hat{\xi}}(\hat{\lambda}) := \rho^{-(n+1)} D(\rho\hat{\xi}, \rho\hat{\lambda}), \tag{26}$$

for $(\rho, \hat{\xi}, \hat{\lambda}) \in \mathbb{R} \times S^{d-1} \times \mathbb{C}$, we obtain a d-parameter family of analytic maps, converging as $\rho \to 0$ to $\mathcal{D}^{0,\hat{\xi}} = \Delta(\hat{\xi}, \cdot)$. Under assumption (23), $\mathcal{D}^{0,\hat{\xi}} = \Delta(\hat{\xi}, \cdot) \sim \hat{\lambda}^{n+1}$ as $|\lambda| \to \infty$, hence, for ρ sufficiently small, $\mathcal{D}^{\rho,\hat{\xi}}$ has $n+1$ continuously varying roots $\lambda = \hat{a}_j(\hat{\xi}, \rho)$. Defining $a_j(\xi) := |\xi|\hat{a}_j(\xi/|\xi|, |\xi|)$, we obtain the result. □

Remark 1. Evidently, $a_j(\xi)$ are smooth in $|\xi|$ for fixed $\hat{\xi}$, but in general have a conical singularity at $\xi = 0$ when considered as a function of ξ, i.e., $\partial a_j/\partial \xi$ is discontinuous at $\xi = 0$.

Corollary 2. *Assuming (H0)–(H2) and the nondegeneracy condition (23), a necessary condition for low-frequency spectral stability of $\bar{u}$, defined as Re $\lambda \le 0$ for $D(\xi, \lambda) = 0$, $\xi \in \mathbb{R}^d$, and $|\xi, \lambda|$ sufficiently small, is that the averaged system (12) be "weakly hyperbolic" in the sense that it possesses a full set of real characteristics $\hat{\lambda}_j(\xi)$ for each $\xi \in \mathbb{R}^d$, i.e., the eigenvalues of*

$$\mathcal{A} = \sum_j \xi_j \frac{\partial(F^j, S\Omega e_j)}{\partial(M, \Omega N)}$$

are real.

Remark 2. Condition (23), or equivalently $(\partial/\partial\lambda)^{n+1} D(0,0) \neq 0$, is a necessary condition for one-dimensional linearized stability [OZ.2], while hyperbolicity is necessary for stability of the homogenized system linearized about a constant state. Thus, Corollaries 1 and 2 are analogous to results of [ZS] in the shock wave case, stating that, given one-dimensional stability, stability of the inviscid equations linearized about an ideal shock is necessary for multidimensional stability of a viscous shock wave.

Finally, we mention that, though the averaged system may in some cases be hyperbolic [OZ.1], so far, only unstable periodic traveling-wave solutions have been found for viscous conservation laws. However, essentially only the single, 2×2 model of van der Waals gas dynamics with viscosity–capillarity in one dimension has so far been considered in detail [OZ.1, Se.1, Se.2], and we see no obvious reason why a stable wave should not exist for other models. It would be extremely interesting to either find such an example, with the associated rich behavior described by the modulation equations, or show that it can in no case exist. As suggested by Serre (private communication), a useful starting point might be to consider whether the averaged system (12) might ever possess an entropy.

References

[G] R. Gardner, *On the structure of the spectra of periodic traveling waves*, J. Math. Pures Appl. 72 (1993), 415–439.

[GMWZ.1] Gues, O., Metivier, G., Williams, M., and Zumbrun, K., *Multidimensional viscous shocks I: degenerate symmetrizers and long time stability*, Journal of the Amer. Math. Soc. 18. (2005), 61–120.

[GMWZ.2] Guès, O., Métivier, G., Williams, M., and Zumbrun, K., *Multidimensional viscous shocks II: the small viscosity problem*, Comm. Pure Appl. Math. 57. (2004), 141–218.

[GMWZ.3] Guès, O., Métivier, G., Williams, M., and Zumbrun, K., *Existence and stability of multidimensional shock fronts in the vanishing viscosity limit*, Arch. Rat. Mech. Anal. 175. (2004), 151–244.

[MZ] Metivier, G. and Zumbrun, K., *Large viscous boundary layers for noncharacteristic nonlinear hyperbolic problems*, Mem. Amer. Math. Soc. 175 (2005), no. 826, vi+107 pp.

[OZ.1] M. Oh and K. Zumbrun, *Stability of periodic solutions of viscous conservation laws with viscosity-1. Analysis of the Evans function*, to appear, Arch. Rational Mech. Anal. (2002).

[OZ.2] M. Oh and K. Zumbrun, *Stability of periodic solutions of viscous conservation laws with viscosity- Pointwise bounds on the Green function*, to appear, Arch. Rational Mech. Anal. (2002).

[Se.1] D. Serre, *Spectral stability of periodic solutions of viscous conservation laws: Large wavelength analysis*, Comm. Partial Differential Equations 30 (2005), no. 1–3, 259–282.

[Se.2] D. Serre, *Entropie du mélange liquide–vapeur d'un fluide thermo–capillaire,* Archy. Rational Mech. Anal., No. 128 (1994) 33–73.

[Z-k] K. Zumbrun, *Multidimensional stability of planar viscous shock waves*, TMR Summer School Lectures: Kochel am See, May, 1999, Birkhauser's Series: Progress in Nonlinear Differential Equations and their Applications (2001), 207 pp.

[ZS] K. Zumbrun and D. Serre, *Viscous and inviscid stability of multidimensional planar shock fronts,* Indiana Univ. Math. J. 48 (1999) 937–992.

Existence of Strong Traces for Quasisolutions of Scalar Conservation Laws

E.Y. Panov

1 Introduction

Consider a scalar conservation law

$$\mathrm{div}_x \varphi(u) = 0, \tag{1}$$

where $u = u(x)$, $x \in \Omega$, $\Omega \subset \mathbb{R}^n$ is a domain with C^1-smooth boundary $S = \partial\Omega$; $\varphi(u) = (\varphi_1(u), \dots, \varphi_n(u)) \in C(\mathbb{R}, \mathbb{R}^n)$. Denote by $\bar{M}_{loc}(\Omega)$ the space of Borel measures on Ω having locally finite variation up to the boundary S.

Definition 1. *A function $u = u(x) \in L^\infty(\Omega)$ is called a quasisolution (quasi-s. for short) of (1) if $\forall k \in F$, where $F \subset \mathbb{R}$ is a dense set (in the sequel we suppose it being countable), there exists a Borel measure $\gamma_k \in \bar{M}_{loc}(\Omega)$ such that $\mathrm{div}\psi_k(u) = -\gamma_k$ in $\mathcal{D}'(\Omega)$, where $\psi_k(u) = \mathrm{sign}\,(u - k)(\varphi(u) - \varphi(k))$ is the Kruzhkov entropy flux.*

Remark that, by the same definition, we can introduce quasi-s. for more general equation $\mathrm{div}_x\varphi(x, u) = 0$.

One could show that generalized entropy solutions (in Kruzhkov sense) as well as generalized entropy sub and supersolutions of (1) are quasi-s. Observe also that if $a, b \in F$, $a < b$, and $u = u(x)$ is a quasi-s, then $v = \max(a, \min(u, b))$ is also a quasi-s. Our main result is the following

Theorem 1. *There exists a function $u_0(y) \in L^\infty(S)$ (the trace) such that*

$$\forall k \in \mathbb{R} \quad \psi_k(u(x)) \cdot \nu \to \psi_k(u_0(y)) \cdot \nu \quad \text{as } x \to y \in S \quad \text{in } L^1_{loc}. \tag{2}$$

Here $\nu = \nu(y)$ is the normal vector at $y \in S$, and $\cdot$ signifies a scalar multiplication. Besides, if for a.e. $y \in S$ functions $u \to \varphi(u) \cdot \nu(y)$ are not constant on nondegenerate intervals then $u(x) \to u_0(y)$ as $x \to y$ strongly.

In particular, the normal flux component $\varphi(u(x)) \cdot \nu$ has the strong trace.

The exact sense of trace relation (2) will be explained later, in Theorem 2. This relation becomes clear after passage to local coordinates at a neighborhood of a boundary point $y \in S$.

Theorem 1 generalizes the previous results of [3, 12], obtained under nondegeneracy conditions on the flux vector, as well as results of [9], where existence of the strong traces was proved for quasi-s. of the equation $u_t + \mathrm{div}_x \varphi(u) = 0$ at the initial hyperplane $t = 0$. Theorem 1 allows to formulate boundary value problems for (1) in the sense of Bardos et al. [1]. For instance the Dirichlet problem $u|_S = u_b$ can be understood in the sense of the inequality: $\forall k \in \mathbb{R}$

$$(\mathrm{sign}\,(u-k) + \mathrm{sign}\,(k-u_b))(\varphi(u) - \varphi(k)) \cdot \nu \geq 0 \quad \text{on } S,$$

which is well defined due to existence of strong traces of $\varphi(u) \cdot \nu$ and

$$\mathrm{sign}\,(u-k)(\varphi(u) - \varphi(k)) \cdot \nu \,.$$

1.1 Reformulation of the Problem

The assertion of Theorem 1 has local character. Thus, we can assume that for some $f(y) \in C^1(\mathbb{R}^{n-1})$

$$\Omega = \{\, x \in \mathbb{R}^n \mid x_1 > f(x_2, \ldots, x_n) \,\}.$$

Making the change $t = x_1 - f(x_2, \ldots, x_n)$, we transform (1) to the equation

$$(\varphi_0(x,u))_t + \mathrm{div}_x \tilde{\varphi}(u) = (\varphi_0(x,u))_t + \sum_{i=2}^{n} \varphi_i(u)_{x_i} = 0, \tag{3}$$

$u = u(t,x)$, $(t,x) \in \Pi = \mathbb{R}_+ \times \mathbb{R}^{n-1}$; here $\tilde{\varphi}(u) = (\varphi_2(u), \ldots, \varphi_n(u))$,

$$\varphi_0(x,u) = \varphi_1(u) - \sum_{i=2}^{n} \varphi_i(u) \frac{\partial f}{\partial x_i} = \varphi(u) \cdot \nu(x).$$

Clearly $u(x)$ is a quasi-s. of (1) if and only if $u(t,x)$ is a quasi-s. of (3). For (3) Theorem 1 is formulated as follows.

Theorem 2. *There exists a function $u_0(y) \in L^\infty(\mathbb{R}^{n-1})$ such that $\forall k \in \mathbb{R}$*

$$\operatorname*{ess\,lim}_{t \to 0} \psi_{0k}(x, u(t,x)) = \psi_{0k}(x, u_0(x)) \quad \text{in } L^1_{loc}(\mathbb{R}^{n-1}).$$

Here

$$\psi_{0k}(x,u) = \mathrm{sign}\,(u-k)(\varphi_0(x,u) - \varphi_0(x,k)) = \mathrm{sign}\,(u-k)(\varphi(u) - \varphi(k)) \cdot \nu(x),$$

$u \in \mathbb{R}$, $x = (x_2, \ldots, x_n) \in \mathbb{R}^{n-1}$. *Moreover, if for a.e. $x \in \mathbb{R}^{n-1}$ the functions $u \to \varphi(u) \cdot \nu(x)$ are not constant on nondegenerate intervals then*

$$\operatorname*{ess\,lim}_{t \to 0} u(t,x) = u_0(x) \quad \text{in } L^1_{loc}(\mathbb{R}^{n-1}),$$

i.e., the function $u(t,x)$ itself has the strong trace.

2 Preliminaries

2.1 Measure-Valued Functions

Recall (see [4, 10]) that a measure-valued function on a domain $\Omega \subset \mathbb{R}^N$ is a weakly measurable map $x \to \nu_x$ of the set Ω into the space of Borel probability measures having compact supports on $\mathbb{R}$. Weak measurability of ν_x means that for any continuous function $p(\lambda)$ the function

$$x \to \langle p(\lambda), \nu_x(\lambda)\rangle = \int p(\lambda) d\nu_x(\lambda)$$

is Lebesgue measurable on Ω. A measure-valued function ν_x is said to be bounded if there is a constant $M > 0$ such that $\operatorname{supp} \nu_x \subset [-M, M]$ for a.e. $x \in \Omega$.

Finally, measure-valued functions of the kind $\nu_x(\lambda) = \delta(\lambda - u(x))$, where $\delta(\lambda - u)$ is the Dirac measure at the point u, are called regular and they are identified with the corresponding functions $u(x)$. Thus, the set $\mathrm{MV}(\Omega)$ of bounded measure-valued functions on Ω includes the space $L^\infty(\Omega)$.

Measure-valued functions naturally arise as weak limits of bounded sequences in $L^\infty(\Omega)$ that is (see [10]) for any bounded sequence $u_m(x) \in L^\infty(\Omega)$ there exist a subsequence $u_r(x)$ weakly convergent to a measure-valued function $\nu_x \in \mathrm{MV}(\Omega)$ in the sense of the relation: $\forall p(\lambda) \in C(R)\ \ p(u_r(x)) \underset{r\to\infty}{\to} \langle p(\lambda), \nu_x(\lambda)\rangle$ $*$-weakly in $L^\infty(\Omega)$.

Measure-valued solutions to conservation laws are widely investigated (see, for instance, [4, 10]). We will need only the notion of a measure-valued isentropic solution (i.s.) of the equation

$$\varphi_0(u)_t + \operatorname{div}_x \tilde{\varphi}(u) = 0, \tag{4}$$

$(t, x) \in \Pi = \mathbb{R}_+ \times \mathbb{R}^{n-1}$, which is defined as a measure-valued function $\nu_{t,x} \in \mathrm{MV}(\Pi)$ such that for each $k \in \mathbb{R}$ $\langle \psi_{0k}(\lambda), \nu_{t,x}(\lambda)\rangle_t + \operatorname{div}_x \langle \psi_k(\lambda), \nu_{t,x}(\lambda)\rangle = 0$ in $\mathcal{D}'(\Pi)$. As is easily verified, $\nu_{t,x}$ is a measure-valued i.s. of every equation

$$(T_\eta \varphi_0(u))_t + \operatorname{div}_x T_\eta \varphi(u) = 0, \tag{5}$$

where T_η is a bounded linear operator on $C(\mathbb{R})$ defined by the relation $T_\eta g(u) = \eta(u) g(u) - \int_0^u g(s) d\eta(s)$, and $\eta(u) \in \mathrm{BV}_{loc}(\mathbb{R})$ is an arbitrary function with locally bounded variation. It turns out that for i.s. of (4) the statement of Theorem 2 holds. More precisely, we have the following

Theorem 3. *Let $\nu_{t,x}$ be a measure-valued i.s. of (4) such that for a.e. $(t, x) \in \Pi$ $\operatorname{supp} \nu_{t,x}$ contains in a segment where $\varphi_0(u)$ is constant. Then $\forall k \in \mathbb{R}$ the functions $w_k(t, x) = \langle \psi_{0k}(\lambda), \nu_{t,x}(\lambda)\rangle$ have the strong traces $v_k(x)$.*

To prove Theorem 3 we utilize the measure-valued version of the Kruzhkov method of doubling variables for two coinciding i.s. $\nu_{t,x}$ of (5), and derive the following relation

$$\iint \operatorname{sign}(\lambda-\mu)(T_\eta\varphi_0(\lambda)-T_\eta\varphi_0(\mu))d\nu_x(\lambda)d\nu_x(\mu)=0$$

for each $\eta(u)\in \mathrm{BV}_{loc}(\mathbb{R})$, where ν_x is the measure-valued trace for $\nu_{t,x}$. This relation implies that $\varphi_0(\lambda)$ must be constant on convex hulls of $\operatorname{supp}\nu_x$ for a.e. x. The latter implies in turn that $v_k(x)=\langle\psi_{0k}(\lambda),\nu_x(\lambda)\rangle$ are the strong traces for $w_k(t,x)$, as was to be proved.

2.2 Weak Traces

From the condition

$$(\psi_{0k}(x,u))_t+\operatorname{div}_x\tilde{\psi}_k(u)=-\gamma_k\in \bar{M}_{loc}(\Pi),\ \ k\in F$$

it follows (see Chen and Frid [2]) existence of the weak traces:

$$\operatorname*{ess\,lim}_{t\to 0}\psi_{0k}(x,u(t,x))=v_k(x)\ \ \text{weakly-}*\text{ in } L^\infty(\mathbb{R}^{n-1})$$

for all $k\in\mathbb{R}$ (we also take into account that $\psi_{0k}(x,u)$ depends continuously on k). More precisely, there exists a measure-valued trace function $\nu_x\in \mathrm{MV}(\mathbb{R}^n)$, such that as $t\to 0$, running over a set $E\subset\mathbb{R}_+$ of full measure, $\psi_{0k}(x,u(t,x))$ weakly converge to measure-valued functions $\psi_{0k}(x,\lambda)^*\nu_x(\lambda)$. We find the following simple criterion for existence of the strong traces.

Proposition 1. *The traces $v_k(x)$ are strong if and only if there exists a bounded function $u_0(x)$ such that $v_k(x)=\psi_{0k}(u_0(x))$ a.e. in $\mathbb{R}^{n-1}$ for all $k\in\mathbb{R}$.*

2.3 Reduction of the Dimension

Suppose one of the component, say $\varphi_n(u)\equiv\text{const}$. The following statement allows to use induction in space dimension in the proof of Theorem 2.

Proposition 2 ([9]). *For a.e. x_n the function $u(t,x')=u(t,x',x_n)$ is a quasi-s. of reduced equations (3)*

$$\varphi_0(x,u)_t+\sum_{i=2}^{n-1}\varphi_i(u)_{x_i}=0,\ \ (t,x')\in\Pi'=\mathbb{R}_+\times\mathbb{R}^{n-2}.$$

2.4 "Blow-Up" Procedure

Let $u(t,x)$ be a quasi-s. of (3), $y \in \mathbb{R}^{n-1}$, $\varepsilon = \varepsilon_m \to 0$. Introduce the sequences $u^\varepsilon(t,x;y) = u(\varepsilon t, y + \varepsilon x) \in L^\infty(\Pi)$, $v_k^\varepsilon(x;y) = v_k(y+\varepsilon x) \in L^\infty(\mathbb{R}^{n-1})$, and the sequences of measures $\gamma_k^\varepsilon = \varepsilon\gamma_k(\varepsilon t, y+\varepsilon x)$ (in the sense of distributions). Evidently,

$$(\psi_{0k}(y+\varepsilon x, u^\varepsilon))_t + \operatorname{div}_x \psi_k(u^\varepsilon) = -\gamma_k^\varepsilon; \tag{6}$$

$$\psi_{0k}(y+\varepsilon x, u^\varepsilon)|_{t=0} = v_k^\varepsilon(x;y) \quad \text{in the weak sense.} \tag{7}$$

Our interest to the sequence $u^\varepsilon(t,x;y)$ is stipulated by Theorem 4:

Theorem 4. *(i) There exist a sequence $\varepsilon = \varepsilon_m \to 0$ and a set of full measure $Y \subset \mathbb{R}^{n-1}$ such that for $y \in Y$*

$$v_k^\varepsilon(x;y) \to v_k(y) = \text{const} \quad in\ L^1_{loc}(\mathbb{R}^{n-1});\ \gamma_k^\varepsilon \to 0 \quad in\ \bar{M}_{loc}(\Pi);$$

(ii) If the strong traces $v_k(x)$ exist then there is a sequence $\varepsilon = \varepsilon_m \to 0$ such that for a.e. $y \in \mathbb{R}^{n-1}$ the sequence $\psi_{0k}(y, u^\varepsilon(t,x;y))$ is strongly (i.e., in $L^1_{loc}(\Pi)$) convergent. (iii) Conversely, if for a.e. $y \in \mathbb{R}^{n-1}$ there is a sequence $\varepsilon = \varepsilon_m \to 0$ such that the sequence $\psi_{0k}(y, u^\varepsilon(t,x;y))$ is strongly convergent then the traces v_k are strong.

The statement (i), (ii) are proved in [12] and [9]. We explain how to prove (iii). Let Y be the set of full measure from assertion (i), and $y \in Y$ be fixed. Then by (6), (7) and (i) in the limit as $m \to \infty$ the sequence $u^\varepsilon(t,x;y)$, $\varepsilon = \varepsilon_m$ converges weakly to a measure-valued i.s. $\nu_{t,x}$ of (4) with $\varphi_0(u) = \varphi_0(y,u)$, and the functions $w_k(t,x) = \langle \psi_{0k}(\lambda), \nu_{t,x}(\lambda)\rangle$ have weak traces $v_k(y)$. By the strong convergence of $\psi_{0k}(y, u^\varepsilon(t,x;y))$ we see that $\nu_{t,x}$ satisfies assumptions of Theorem 3 and by this theorem the traces $v_k(y)$ are strong. This implies that $v_k(y) = \psi_{0k}(u_0)$ for some constant $u_0 = u_0(y)$ (for instance, this constant can be taken as $\langle \lambda, \nu_x(\lambda)\rangle$, where ν_x is a measure-valued trace function for $\nu_{t,x}$, and x is an arbitrary "Lebesgue point" of ν_x). Thus, for all $y \in Y$ $v_k(y) = \psi_{0k}(u_0(y))$ and by Proposition 1 we conclude that traces v_k are strong.

3 *H*-Measures

Suppose $\Omega \subset \mathbb{R}^n$ and the sequence $u_m = u_m(x)$ is bounded in $L^\infty(\Omega)$, u_m converges weakly to a measure-valued function ν_x. We denote $U_m^p(x) = \delta_{u_m}((p,+\infty)) - \nu_x((p,+\infty))$. Let $F(u)(\xi)$, $\xi \in \mathbb{R}^n$ be the Fourier transform of $u(x) \in L^2(\mathbb{R}^n)$; $S = S^{n-1} = \{\, \xi \in \mathbb{R}^n \mid |\xi| = 1 \,\}$ be the unit sphere; $u \to \bar{u}$, $u \in \mathbb{C}$ be the complex conjugation.

Theorem 5 ([6, 7, 8]).

(i) There exists a set $P \subset \mathbb{R}$ *with at most countable complement such that* $U^p_m(x) \to 0$ *as* $m \to \infty$ *weakly-∗ in* $L^\infty(\Omega)$

(ii) There exists a family of complex-valued locally finite Borel measures $\{\mu^{pq}\}_{p,q\in P}$ *on* $\Omega \times S$ *and a subsequence* $U_r(x) = \{U^p_r(x)\}_{p\in P}$ *such that* $\forall \Phi_1(x), \Phi_2(x) \in C_0(\Omega)$, $\psi(\xi) \in C(S)$

$$\langle \mu^{pq}, \Phi_1(x)\overline{\Phi_2(x)}\psi(\xi)\rangle = \lim_{r\to\infty} \int_{\mathbb{R}^n} F(\Phi_1 U^p_r)(\xi)\overline{F(\Phi_2 U^q_r)(\xi)}\psi(\xi/|\xi|)d\xi$$

(iii) The map $(p,q) \to \mu^{pq}$ *is continuous as a map from* $P \times P$ *into the space* $M_{loc}(\Omega \times S)$ *of locally finite Borel measures on* $\Omega \times S$

Following Tartar [11], and Gerárd [5] we call the family $\{\mu^{pq}\}_{p,q\in P}$ the H-measure corresponding to the subsequence $u_r(x)$. We will need the following

Lemma 1. *Let* $\{\mu^{pq}\}_{p,q\in P}$ *be the* H*-measure corresponding to a bounded in* $L^\infty(\Omega)$ *sequence* $u_r(x)$, $r \in \mathbb{N}$. *Suppose* $u_r(x)$ *converges weakly to a measure-valued function* ν_x *and* $[a(x), b(x)]$ *is the convex hull of* $\operatorname{supp}\nu_x$. *Then for a.e.* $x \in \Omega$ $\mu^{pp} \neq 0$ *for all* $p \in (a(x), b(x)) \cap P$.

Now, let $\varphi(u) \in C(\mathbb{R}, \mathbb{R}^n)$ be a continuous vector function. Suppose that the sequence $u_r(x)$ satisfies the condition: (C) $\forall p \in \mathbb{R}$ *the sequence of distributions* $\mathcal{L}^p_r = \operatorname{div}_x[\operatorname{sign}(u_r - p)(\varphi(u_r) - \varphi(p))]$ *is precompact in* $H^{-1}_{loc}(\Omega)$.

Theorem 6 plays a key role in the proof of our main Theorem 2.

Theorem 6 ([7, 8]). *Let* $\{\mu^{pq}\}_{p,q\in P}$ *be the* H*-measure corresponding to the sequence* $u_r(x)$. *Suppose that condition (C) is satisfied and for some* $p \in P$ $\mu^{pp} \neq 0$. *Then there exists a segment* $I = [p-\delta, p+\delta]$ *and a vector* $\xi \in S$ *such that* $(\xi, \varphi(u)) = \text{const}$ *on* I.

In particular from Theorem 6 it follows that under the nondegeneracy condition $\forall \xi \in S$ *the function* $u \to (\xi, \varphi(u))$ *is not constant on nondegenerate intervals* $\mu^{pp} \equiv 0$ and, therefore, the sequence $u_r(x)$ is strongly convergent.

4 Sketch of the Proof of Theorem 2

We apply the induction in n. If $n = 1$ then we have $u = u(t)$, $\psi_{0k}(u)_t = -\gamma_k$, i.e., $\psi_{0k}(u(t))$ are BV-functions in a vicinity of $t = 0$. Therefore, $\psi_{0,k}(u(t)) \to v_k$ as $t \to 0$. Clearly $v_k = \psi_{0k}(u_0)$, where u_0 is some limit point of $u(t)$ as $t \to 0$.

Now suppose that Theorem 2 is true for space dimension less than n. Introduce the countable set $\mathcal{I}$ consisting of intervals $I = (a,b)$ such that $a, b \in F$ and for some $\xi \in \mathbb{R}^n$, $\xi \neq 0$ $\xi \cdot \varphi(u) = \text{const}$ on I. We denote $u_I(t,x) = \max(a, \min(u(t,x), b))$. Recall that u_I is also a quasi-s. of (3). Let us show that for $I \in \mathcal{I}$ u_I satisfies the strong trace property. Let $\xi \in \mathbb{R}^n$, $\xi \neq 0$ be such that $\xi \cdot \varphi(u) = \text{const}$ on I. Consider the following two cases:

Case 1: $\xi_1 = 0$. In this case $\xi' = (\xi_2, \dots, \xi_n) \neq 0$ and there exists a linear change $z = z(x)$, $x = (x_2, \dots, x_n) \in \mathbb{R}^{n-1}$ such that $z_n = \xi' \cdot x$. After this change (3) reduces to the form

$$\varphi_0(z,u)_t + \operatorname{div}_z \bar{\varphi}(u) = 0, \tag{8}$$

where $\bar{\varphi}_n = \xi' \cdot \tilde{\varphi}(u) = \text{const}$ on I. By Proposition 2 $u_I(t, z', z_n)$ is a quasi-s. of reduced equation for a.e. z_n. By induction hypothesis, the strong trace property is satisfied in $\Pi' = \mathbb{R}_+ \times \mathbb{R}^{n-2}$. This easily implies that it is valid for original equation in Π.

Case 2: $\xi_1 \neq 0$. Let $E \subset \mathbb{R}^{n-1}$ be the set of x such that $\xi' + \xi_1 \nabla f(x) = 0$. Since $\varphi_1(u) = \varphi_0(x,u) + \nabla f(x) \cdot \tilde{\varphi}(u)$ we have the relation

$$\xi_1 \varphi_0(x,u) + (\xi' + \xi_1 \nabla f(x)) \cdot \tilde{\varphi}(u) = \text{const} \quad \forall u \in I.$$

Thus, for $x \in E$ $\varphi_0(x,u) = \text{const}$ on I and therefore for $u_0 \in I$ $\psi_{0k}(x, u(t,x)) \to \psi_{0k}(x, u_0)$ as $t \to 0$ in $L^1_{loc}(E)$.

If $x \notin E$, when in a vicinity of this point we can make the change of variables $z_i = z_i(x)$, $i = 2, \dots, n-1$; $z_n = \xi_1(t + f(x)) + \xi' \cdot x$, which reduces (3) to the form (8) with $\bar{\varphi}_n = \xi_1 \varphi_0(x,u) + (\xi' + \xi_1 \nabla f(x)) \cdot \tilde{\varphi}(u) = \text{const}$. As in the case 1, we deduce from this representation the strong trace property in a vicinity of x. The above arguments show that the strong trace property is satisfied on the whole hyperplane $t = 0$.

By Theorem 3 we see that after extraction of a subsequence $\varepsilon = \varepsilon_m \to 0$ the sequence $\psi_{0k}(y, (u_I)^\varepsilon(t,x;y))$ is strictly convergent for all $I \in \mathcal{I}$ and $y \in Y$. We assume in addition that for the sequence ε_m and the set of full measure $Y \subset \mathbb{R}^{n-1}$ the assertion of Theorem 4(i) holds.

Now we fix $y \in Y$ and a subsequence of ε_m such that the sequence $u^\varepsilon(t,x;y)$ weakly converges to a measure-valued function $\nu_{t,x}$ and the corresponding H-measure μ^{pq} is defined. As follows from strong convergence of $\psi_{0k}(y, (u_I)^\varepsilon(t,x;y))$, for a.e. $(t,x) \in \Pi$ all the functions $\psi_{0k}(y, \lambda)$ are constant on $[a(t,x), b(t,x)] \cap I$ $\forall I \in \mathcal{I}$, where $[a(t,x), b(t,x)]$ is the convex hull of $\operatorname{supp} \nu_{t,x}$. Further, we verify that the sequence $u^\varepsilon(t,x;y)$ satisfies property (C) applied to the vector $(\varphi_0(y,u), \varphi(u))$ (this easily follows from condition (6)). By Theorem 6 we find that $\forall p \in C \cap P$ $\mu^{pp} = 0$, where C is the complement to $U = \bigcup_{I \in \mathcal{I}} I$. Then, using Lemma 1, we derive that for a.e. (t,x) the set $[a(t,x), b(t,x)] \cap C$ lays in the complement $\mathbb{R} \setminus P$ and, therefore, is at most countable. Hence, for a.e. (t,x) each function $\psi_{0k}(y, \lambda)$ takes at most countable set of values on $[a(t,x), b(t,x)]$ and, therefore, $\psi_{0k}(y, \lambda) = \text{const} = v_{0k}(y)$ on this segment. This, in turn, implies that the sequence $\psi_{0k}(y, u^\varepsilon(t,x;y))$ converges strongly, and according to Theorem 3(iii) we conclude that the strong trace property is satisfied.

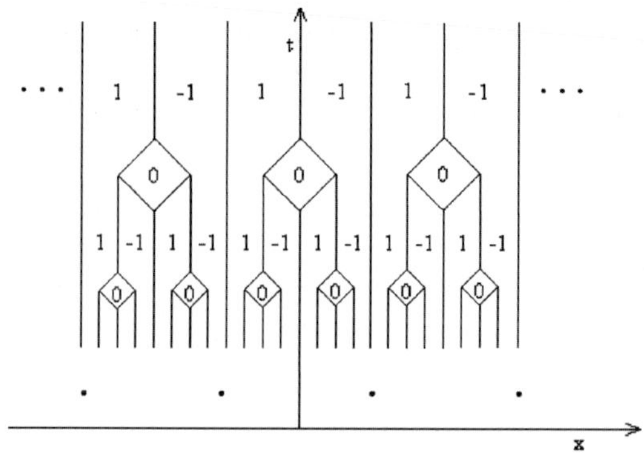

Fig. 1. A weak solution to the Burgers equation, with a weak-star but not strong trace at initial time

5 Example of a Weak Solution Without a Strong Trace

Consider the Burgers equation $u_t + (u^2)_x = 0$, $(t, x) \in \Pi = \mathbb{R}_+ \times \mathbb{R}$. To construct the desired weak solution $u(t, x)$ we introduce the function

$$w(t,x) = \begin{cases} 0, & |x| + |t-6| \le 1 \\ -\text{sign}\, x, & |x| + |t-6| > 1,\ t \in (6,8] \\ \text{sign}\,(1-x)\text{sign}\, x, & |x| + |t-6| > 1,\ t \in (4,6) \end{cases},$$

defined in the square $t \in (4, 8]$, $-2 \le x < 2$. We extend this function in the whole layer $t \in (4, 8]$, as a 4-periodic function v over the variable x. In the half-space Π we define the piecewise constant function $u(t, x) = v(2^k t, 2^k x)$ if $t \in (4 \cdot 2^{-k}, 8 \cdot 2^{-k}]$, $k \in \mathbb{Z}$, see Fig. 1.

As easily verified, on the discontinuity lines the Rankine–Hugoniot condition is satisfied and therefore $u(t, x)$ is a weak solution of the Burgers equation. As $t \to 0$ $u(t, \cdot) \to 0$ weakly-$*$ in $L^\infty(\mathbb{R})$ but there is no strong limit of $u(t_k, x)$ for any choice of a sequence $t_k \to 0$. Remark also that for the constructed solution the condition from Definition 1 is satisfied with the measures $\gamma_k \in M_{loc}(\Pi)$ $\forall k \in \mathbb{R}$, but certainly $\gamma_k \notin \bar{M}_{loc}(\Pi)$.

Acknowledgments

This work was carried out under financial support of Russian Foundation for Basic Research (grant No. 06-01-00289) and DFG project 436 RUS 113/895/0-1.

References

1. Bardos, C., LeRoux, A.-Y., Nédélec, J.C.: First order quasilinear equations with boundary conditions. Comm. in Partial Differential Equations, **6**, No. 9, 1017–1034 (1979)

2. Chen, G.-Q., Frid, H.: Divergence-Measure fields and hyperbolic conservation laws. Arch. Ration. Mech. Anal., **147**, 89–118 (1999)
3. Chen, G.-Q., Rascle, M.: Initial layers and uniqueness of weak entropy solutions to hyperbolic conservation laws. Arch. Ration. Mech. Anal., **153**, No. 3, 205–220 (2000)
4. DiPerna, R.J.: Measure-valued solutions to conservation laws. Arch. Rational Mech. Anal., **88**, 223–270 (1985)
5. Gérard, P.: Microlocal defect measures. Comm. Partial Diff. Equat., **16**, 1761–1794 (1991)
6. Panov, E. Yu.: On sequences of measure-valued solutions of first-order quasilinear equations. Matemat. Sbornik. **185**, No. 2, 87–106 (1994); English transl. in *Russian Acad. Sci. Sb. Math.,* **81**, No. 1, 211–227 (1995)
7. Panov, E. Yu.: On strong precompactness of bounded sets of measure valued solutions for a first order quasilinear equation. Matemat. Sbornik, **186**, No. 5, 103–114 (1995); Engl. transl. in *Sbornik: Mathematics,* **186** 729–740 (1995)
8. Panov, E. Yu.: Property of strong precompactness for bounded sets of measure valued solutions of a first-order quasilinear equation. Matemat. Sbornik, **190**, No. 3, 109–128 (1999); Engl. transl. in *Sbornik: Mathematics,* **190**, No. 3, 427–446 (1999)
9. Panov, E. Yu.: Existence of Strong Traces for Generalized Solutions of Multidimensional Scalar Conservation Laws. Journal of Huperbolic Differential Equations, **2**, No. 4, 885–908 (2005)
10. Tartar, L.: Compensated compactness and applications to partial differential equations. Research notes in mathematics, nonlinear analysis, and mechanics. Heriot-Watt Symposium, **4**, 136–212 (1979)
11. Tartar, L.: H-measures, a new approach for studying homogenisation, oscillations and concentration effects in partial differential equations. Proceedings of the Royal Society of Edinburgh, **115A**, No. 3–4, 193–230 (1990)
12. Vasseur, A.: Strong Traces for Solutions of Multidimensional Scalar Conservation Laws. Arch. Ration. Mech. Anal., **160**, 181–193 (2001)

Path-Conservative Numerical Schemes for Nonconservative Hyperbolic Systems

C. Parés

1 Introduction

In the present study, one-dimensional hyperbolic systems in nonconservative form:

$$\frac{\partial W}{\partial t} + \mathcal{A}(W)\frac{\partial W}{\partial x} = 0, \quad x \in \mathbf{R},\ t > 0, \tag{1}$$

are considered. The nonconservative product $\mathcal{A}(W)W_x$ makes difficult the definition of weak solutions for this kind of systems. After the theory developed by Dal Maso et al. [DLM95] a notion of weak solutions as Borel measures can be given, which is based on the selection of a family of paths in the phases space.

Systems of conservation laws with source terms (or balance laws)

$$\frac{\partial W}{\partial t} + \frac{\partial F}{\partial x}(W,\sigma) = S(W,\sigma)\frac{d\sigma}{dx}, \tag{2}$$

$\sigma(x)$ being a known function, can be viewed as particular cases of (1) if the trivial equation

$$\frac{\partial \sigma}{\partial t} = 0,$$

is added to the system. Besides, using the same idea, more general types of coupled systems of conservation laws with source terms can be written in the form (1). In particular, one-dimensional hyperbolic shallow water systems with source terms due to bed elevations or breadth variations, as well as two-layer shallow water systems of immiscible fluids can be formulated under the form (1).

In [Par06], a theoretical framework for the numerical approximation of weak solutions of strictly hyperbolic systems of the form (1) was introduced. The main concept is that of path-conservative numerical scheme, which is a natural generalization of the notion of conservative scheme for conservation laws. Numerical schemes based on approximate Riemann solvers, as

Godunov methods or the generalized Roe methods introduced in [Tou92], fit into this framework. This theory is being useful to design new numerical schemes for hyperbolic systems with nonconservative terms, and to analyze their well-balance properties, specially in the context of shallow water models (see [CGG04, PC04, CGP06, CGGP06, CGVP06]).

The organization of this contribution is as follows: in Sect. 2 a brief resume of the theory developed in [DLM95] is presented. Next, the concept of path-conservative numerical schemes is introduced. Section 4 is devoted to the generalization of the concept of approximate Reimann solvers for (1). Finally, in Sect. 5 an abstract definition of well-balanced numerical scheme is given and some general results concerning the well-balance properties of numerical schemes based on approximate Riemann solvers are mentioned.

2 Weak Solutions

Consider the problem

$$\frac{\partial W}{\partial t} + \mathcal{A}(W)\frac{\partial W}{\partial x} = 0, \quad x \in \mathbf{R},\ t > 0, \tag{3}$$

where $W(x,t)$ belongs to Ω, an open convex subset of $\mathbf{R}^N$, and $W \in \Omega \mapsto \mathcal{A}(W) \in \mathcal{M}_{N\times N}(\mathbf{R})$ is a smooth locally bounded map. We suppose that system (3) is strictly hyperbolic, that is, for each $W \in \Omega$, $\mathcal{A}(W)$ has N real distinct eigenvalues $\lambda_1(W) < \cdots < \lambda_N(W)$, with associated eigenvectors $R_1(W)$,...,$R_N(W)$. We also suppose that for each $i = 1,\ldots,N$, the characteristic field $R_i(W)$ is either genuinely nonlinear,

$$\nabla\lambda_i(W)\cdot R_i(W) \neq 0, \quad \forall\ W \in \Omega,$$

or linearly degenerate,

$$\nabla\lambda_i(W)\cdot R_i(W) = 0, \quad \forall\ W \in \Omega.$$

The theory developed by Dal Maso et al. (see [DLM95]) allows one to give a rigorous definition of nonconservative products associated with the choice of a family of paths in Ω.

Definition 1. *A* family of paths *in $\Omega \subset \mathbf{R}^N$ is a locally Lipschitz map*

$$\Phi\colon [0,1]\times\Omega\times\Omega \mapsto \Omega,$$

such that:

- $\Phi(0; W_L, W_R) = W_L$ *and* $\Phi(1; W_L, W_R) = W_R$, *for any* $W_L, W_R \in \Omega$
- *For every arbitrary bounded set* $\mathcal{O} \subset \Omega$, *there exists a constant* k *such that*

$$\left|\frac{\partial\Phi}{\partial s}(s; W_L, W_R)\right| \leq k|W_R - W_L|,$$

for any $W_L, W_R \in \mathcal{O}$ *and almost every* $s \in [0,1]$

- *For every bounded set $\mathcal{O} \subset \Omega$, there exists a constant K such that*

$$\left| \frac{\partial \Phi}{\partial s}(s; W_L^1, W_R^1) - \frac{\partial \Phi}{\partial s}(s; W_L^2, W_R^2) \right| \le K(|W_L^1 - W_L^2| + |W_R^1 - W_R^2|),$$

for any $W_L^1, W_R^1, W_L^2, W_R^2 \in \mathcal{O}$ and almost every $s \in [0,1]$

The choice of the family of paths is important as it determines the speed of propagation of discontinuities. For scalar balance laws, rigorous justifications of the choice of the family of paths can be given using different techniques based on weak limits. But, in general, this choice has to be based on the physical background (see, for instance, [LeF93], [Rav95], [Ber02] for instance).

Suppose that a family of paths Φ in Ω has been chosen. Then, for $W \in (L^\infty(\mathbf{R} \times \mathbf{R}^+) \cap BV(\mathbf{R} \times \mathbf{R}^+))^N$, the nonconservative product can be interpreted as a Borel measure denoted by $[\mathcal{A}(W)W_x]_\Phi$. W is then a weak solution if (3) is satisfied in the sense of measures. According to this definition, a weak solution must satisfy the generalized Rankine–Hugoniot condition across a discontinuity:

$$\int_0^1 (\xi \mathcal{I} - \mathcal{A}(\Phi(s; W^-, W^+))) \frac{\partial \Phi}{\partial s}(s; W^-, W^+)\, ds = 0, \tag{4}$$

where ξ is the speed of the discontinuity; W^-, W^+, the left and right limits of the solution; and $\mathcal{I}$, the identity matrix. In the particular case of a system of conservation laws (that is, if $\mathcal{A}(W)$ is the Jacobian matrix of some flux function $F(W)$), (4)) is independent of the family of paths and it reduces to the usual Rankine–Hugoniot condition.

As it occurs in the conservative case, not every discontinuity is admissible. Therefore, a definition of entropic solution has to be assumed, as the classical Lax's concept or a definition related to an entropy pair.

3 Path-Conservative Numerical Schemes

In [Par06] the following definition was introduced:

Definition 2. *Given a family of paths Φ, a numerical scheme is said to be Φ-conservative if it can be written under the form*

$$W_i^{n+1} = W_i^n - \frac{\Delta t}{\Delta x}(D_{i-1/2}^+ + D_{i+1/2}^-), \tag{5}$$

where

$$D_{i+1/2}^\pm = D^\pm(W_{i-q}^n, \ldots, W_{i+p}^n),$$

D^- and D^+ being two continuous functions from Ω^{p+q+1} to Ω satisfying:

$$D^\pm(W, \ldots, W) = 0, \quad \forall\, W \in \Omega, \tag{6}$$

and

$$
\begin{aligned}
&D^-(W_{-q}, \ldots, W_p) + D^+(W_{-q}, \ldots, W_p) \\
&\quad = \int_0^1 \mathcal{A}(\Phi(s; W_0, W_1)) \frac{\partial \Phi}{\partial s}(s; W_0, W_1)\, ds,
\end{aligned}
\tag{7}
$$

for every $W_i \in \Omega$, $i = -q, \ldots, p$.

As usual, Δt and Δx represent the time step and the space step, which are supposed to be constant for simplicity; W_i^n represents the approximation of the average of the exact solution at the ith cell $I_i = [x_{i-1/2}, x_{i+1/2}]$ at time $t^n = n\Delta t$.

The underlying idea is the following: at time t^n the numerical scheme produces a piecewise constant approximation W^n of $W(\cdot, t^n)$; according to the definition of the nonconservative products, $\mathcal{A}(W^n)W^n_x$ is a measure consisting of a sum of Dirac masses placed at the intercells whose weights are calculated by means of the family of paths Φ. In order to advance in time, these Dirac masses are decomposed into two terms $D^{\pm}_{i+1/2}$, one contributing to the cell I_i and the other to the cell I_{i+1}.

This definition generalizes the usual concept of a conservative numerical scheme for a system of conservation laws:

Proposition 1. *Let us suppose that* (3) *is a system of conservation laws, i.e.,* $\mathcal{A}$ *is the Jacobian of a flux function* F. *Then, every numerical scheme which is* Φ*-conservative for some family of paths* Φ *is consistent and conservative in the usual sense. Conversely, a consistent conservative numerical scheme is* Φ*-conservative for every family of paths* Φ.

4 Approximate Riemann Solvers

This section is devoted to generalize the notion of approximate Riemann solvers introduced in [HLV83] for conservative systems and extended in [Bou04] for balance laws.

Definition 3. *Given a family of paths* Φ, *a* Φ*-approximate Riemann solver for* (3) *is a function* $\widetilde{V} : \mathbf{R} \times \Omega \times \Omega \mapsto \Omega$ *satisfying the following:*

(i) *For every* $W \in \Omega$,

$$
\widetilde{V}(v; W, W) = W \quad \forall\, v \in \mathbf{R}
\tag{8}
$$

(ii) *For every* W_L, $W_R \in \Omega$ *there exist* $\lambda_{min}(W_L, W_R)$, $\lambda_{max}(W_L, W_R)$ *in* $\mathbf{R}$ *such that,*

$$
\begin{aligned}
\widetilde{V}(v; W_L, W_R) &= W_L, \quad \text{if } v < \lambda_{min}(W_L, W_R), \\
\widetilde{V}(v; W_L, W_R) &= W_R, \quad \text{if } v > \lambda_{max}(W_L, W_R)
\end{aligned}
$$

(iii) *For every* $W_L, W_R \in \Omega$,

$$\int_0^1 \mathcal{A}\left(\Phi(s; W_L, W_R)\right) \frac{\partial \Phi}{\partial s}(s; W_L, W_R)\, ds + \int_0^{\infty} \left(\widetilde{V}(v; W_L, W_R) - W_R\right) dv + \int_{-\infty}^{0} \left(\widetilde{V}(v; W_L, W_R) - W_L\right) dv = 0 \tag{9}$$

Given a Φ-approximate Riemann solver for (3) a numerical scheme can be constructed as follows:

$$W_i^{n+1} = \frac{1}{\Delta x}\left(\int_{x_{i-1/2}}^{x_i} \widetilde{V}\left(\frac{x - x_{i-1/2}}{\Delta t}; W_{i-1}^n, W_i^n\right) dx + \int_{x_i}^{x_{i+1/2}} \widetilde{V}\left(\frac{x - x_{i+1/2}}{\Delta t}; W_i^n, W_{i+1}^n\right) dx\right). \tag{10}$$

Under a CFL condition 1/2, the numerical scheme can also be written under the form (5) with

$$D_{i+1/2}^{-} = -\int_{-\infty}^{0} \left(\widetilde{V}(v; W_i^n, W_{i+1}^n) - W_i^n\right) dv, \tag{11}$$

$$D_{i+1/2}^{+} = -\int_{0}^{\infty} \left(\widetilde{V}(v; W_i^n, W_{i+1}^n) - W_{i+1}^n\right) dv. \tag{12}$$

It is straightforward to prove the following result:

Proposition 2. *A numerical scheme* (5) *based on a* Φ*-approximate Riemann solver is* Φ*-conservative.*

A particular case is obtained by considering exact Riemann solvers, i.e. Godunov's methods. As in the case of systems of conservation laws, approximate Riemann solvers can be constructed, for instance, by means or relaxation techniques or linear approximations of the Riemann problems. In this latter case, $\widetilde{V}(x/t; W_L, W_R)$ is the solution of a Riemann problem for a linear equation associated to a linearization $\mathcal{A}(W_L, W_R)$ of $\mathcal{A}(W)$. It can be easily shown that this is a Φ-approximate Riemann solver, if and only if, $\mathcal{A}(W_L, W_R)$ is a Roe linearization in the sense defined by Toumi in [Tou92].

Definition 4. *Given a family of paths* Φ*, a function* $\mathcal{A}_\Phi \colon \Omega \times \Omega \mapsto \mathcal{M}_{N \times N}(\mathbf{R})$ *is called a* Roe linearization *if it verifies the following properties:*

1. *For each* $W_L, W_R \in \Omega$*,* $\mathcal{A}_\Phi(W_L, W_R)$ *has* N *distinct real eigenvalues*
2. $\mathcal{A}_\Phi(W, W) = \mathcal{A}(W)$*, for every* $W \in \Omega$
3. *For any* $W_L, W_R \in \Omega$

$$\mathcal{A}_\Phi(W_L, W_R)(W_R - W_L) = \int_0^1 \mathcal{A}(\Phi(s; W_L, W_R)) \frac{\partial \Phi}{\partial s}(s; W_L, W_R)\, ds. \tag{13}$$

Once a Roe linearization $\mathcal{A}_\Phi$ has been chosen, some straightforward calculations allow one to show that, under a CFL condition 1/2, the numerical scheme can be written under the form (5) with

$$D^-_{i+1/2} = \mathcal{A}^-_{i+1/2}(W^n_{i+1} - W^n_i),$$
$$D^+_{i+1/2} = \mathcal{A}^+_{i+1/2}(W^n_{i+1} - W^n_i),$$

where

$$\mathcal{A}_{i+1/2} = \mathcal{A}_\Phi(W^n_i, W^n_{i+1}),$$

and, as usual,

$$\mathcal{L}^\pm_{i+1/2} = \begin{bmatrix} (\lambda_1^{i+1/2})^\pm & & 0 \\ & \ddots & \\ 0 & & (\lambda_N^{i+1/2})^\pm \end{bmatrix}, \quad \mathcal{A}^\pm_{i+1/2} = \mathcal{K}_{i+1/2}\mathcal{L}^\pm_{i+1/2}\mathcal{K}^{-1}_{i+1/2} \tag{14}$$

being $\mathcal{L}_{i+1/2}$ the diagonal matrix whose coefficients are the eigenvalues of $\mathcal{A}_{i+1/2}$

$$\lambda_1^{i+1/2} < \lambda_2^{i+1/2} < \cdots < \lambda_N^{i+1/2},$$

and $\mathcal{K}_{i+1/2}$ is a $N \times N$ matrix whose columns are associated eigenvectors.

As in the case of systems of conservation laws, an entropy-fix technique has to be added to the numerical scheme.

The construction of Roe methods for systems of the form (2) has been studied in [PC04].

5 Well-Balancing

Well-balancing is related to the numerical approximation of equilibria, i.e., steady-state solutions (see [Gos01], [Bou04]). Notice that system (3) can only have nontrivial steady-state solutions if it has some linearly degenerate fields: if $W(x)$ is a regular steady-state solution

$$\mathcal{A}(W(x)) \cdot W'(x) = 0 \quad \forall\, x \in \mathbf{R},$$

such that $W'(x) \neq 0$, then 0 is an eigenvalue of $\mathcal{A}(W(x))$ and $W'(x)$ is an associated eigenvector. Therefore, $x \mapsto W(x)$ can be interpreted as a parameterization of an integral curve of a linearly degenerate characteristic field whose corresponding eigenvalue takes the value 0 through the curve. In order to define the concept of well-balancing, let us introduce the set Γ of all the integral curves γ of a linearly degenerate field of $\mathcal{A}(W)$ such that the corresponding eigenvalue vanishes on Γ. According to [PC04] we introduce the following definitions.

Definition 5. *Given a curve* $\gamma \in \Gamma$, *a numerical scheme for solving* (3)

$$W_j^{n+1} = W_j^n + \frac{\Delta t}{\Delta x} H(W_{j-q}^n, \dots, W_{j+p}^n) \tag{15}$$

is said to be exactly well-balanced for γ *if, given any* $\mathcal{C}^1$ *function* $x \in (\alpha, \beta) \subset \mathbf{R} \mapsto W(x) \in \Omega$ *such that*

$$W(x) \in \gamma, \quad \forall\, x \in (\alpha, \beta), \tag{16}$$

and $p+q+1$ *points in* (α, β) $x_{-q}, \dots, x_p$ *such that*

$$x_{-q} < \cdots < x_p; \quad x_{i+1} - x_i = \Delta x, \; i = -q, \dots, p-1, \tag{17}$$

then

$$H(W(x_{-q}), \dots, W(x_p)) = 0. \tag{18}$$

The scheme is said to be well-balanced with order k *for* γ *if, given any* $\mathcal{C}^{k+1}$ *function* W *and any set of points* $\{x_{-q}, \dots, x_p\}$ *satisfying* (16), (17), *then*

$$|H(W(x_{-q}), \dots, W(x_p))| = O\left(\Delta x^{k+1}\right). \tag{19}$$

Finally, the scheme is said to be exactly well-balanced or well-balanced with order k *if these properties are satisfied for any curve of* Γ.

We have only considered 1-level schemes and uniform meshes in order to avoid an excess of notation, but the definition can be easily extended to more general schemes.

The well-balance property of a scheme is strongly related to its ability to approximate stationary contact discontinuities. We can state for instance the following proposition (see [Par06]):

Proposition 3. *Given a numerical scheme of the form* (5) *with* $q = 0$ *and* $p = 1$ *and a curve* γ *of* Γ, *the numerical scheme is exactly well-balanced for* γ *if and only if it solves exactly every stationary contact discontinuity linking two states belonging to* γ.

Godunov's methods are obviously exactly well-balanced. Concerning methods based on linear approximate Riemann solvers, in [PC04] it has been shown that a Roe scheme based on a family of paths Φ is exactly well-balanced for a curve $\gamma \in \Gamma$ if, given two states W_L and W_R in γ, the path $\Phi(s; W_L, W_R)$ is a parameterization of the arc of γ linking these states. The numerical scheme is well-balanced with order k if $\Phi(s; W_L, W_R)$ approximates with order $k+1$ a regular parameterization of the arc of γ linking the states. In particular, a Roe scheme based on the family of segments is always well-balanced with order 2. Moreover, it is exactly well-balanced for curves of Γ that are straight lines.

References

[Ber02] Berthon, C.: Schéma nonlinéaire pour l'aproximation numérique d'un système hyperbolique non conservatif. C.R. Acad. Sci. Paris Ser. I **335**, 1069–1072 (2002)

[Bou04] Bouchut F.: Nonlinear Stability of Finite Volume Methods for Hyperbolic Conservation Laws and Well-Balanced Schemes for Sources. Birkhäuser, Basel, Switzerland (2004)

[CGG04] Castro M.J., García J.A., González-Vida J.M., Macías J., Parés C., Vázquez-Cendón M.E.: Numerical simulation of two layer shallow water flows through channels with irregular geometry. J. Comput. Phys. **195**, 202–235 (2004)

[CGP06] Castro M.J., Gallardo J.M., Parés C.: High order finite volume schemes based on reconstruction of states for solving hyperbolic systems with nonconservative products. Applications to shallow water systems. Math. Comp., **75**, 1103–1134 (2006)

[CGGP06] Castro M.J., García J.A., González-Vida J.M., Parés C.:. A parallel 2d finite volume scheme for solving systems of balance laws with nonconservative products: application to shallow flows. Comp. Meth. Appl. Mech. Eng., **196**, 2788–2815 (2006)

[CGVP06] Castro M.J., González-Vida J.M., Parés C.: Numerical treatment of wet/dry fronts in shallow flows with a modified Roe scheme. Math. Mod. Meth. App. Sci. **16**, 897–931 (2006)

[DLM95] Dal Maso G., LeFloch P.G., Murat F.: Definition and weak stability of nonconservative products. J. Math. Pures Appl., **74**, 483–548 (1995)

[Gos01] Gosse L.: A well-balanced scheme using nonconservative products designed for hyperbolic systems of conservation laws with source terms. Math. Mod. Meth. App. Sci. **11**, 339–365 (2001)

[HLV83] Harten A., Lax P.D., Van Leer B.: On upstream differencing and Godunov-type schemes for hyperbolic conservation laws. SIAM Rev. **25**, 35–61 (1983)

[LeF93] LeFloch P.G.: Propagating phase boundaries. Formulation of the problem and existence via the Glimm method. Arch. Rational Mech. Anal. **123**, 153–197 (1993)

[PC04] Parés C., Castro M.J.: On the well-balance property of Roe's method for nonconservative hyperbolic systems. Applications to shallow-water systems. ESAIM-Math. Model. Num. **38**, 821–852 (2004)

[Par06] Parés C.: Numerical methods for nonconservative hyperbolic systems: a theoretical framework. SIAM J. Num. An., **44**, 300–321 (2006)

[Rav95] Raviart P.A., Sainsaulieu L.: A nonconservative hyperbolic system modeling spray dynamics. I. Solution of the Riemann problem. Math. Models Methods Appl. Sci. **5**, 297–333 (1995)

[Tou92] Toumi I.: A weak formulation of Roe's approximate Riemann solver. J. Comput. Phys., **102**, 360–373, (1992)

Numerical Modeling of Two-Phase Gravitational Granular Flows with Bottom Topography

M. Pelanti, F. Bouchut, A. Mangeney, and J.-P. Vilotte

Summary. We study a depth-averaged model of gravity-driven mixtures of solid grains and fluid moving over variable basal surface. The particular application we are interested in is the numerical description of geophysical flows such as avalanches and debris flows, which typically contain both solid material and interstitial fluid. The depth-averaged mass and momentum equations for the solid and fluid components form a nonconservative system, where nonconservative terms involving the derivatives of the unknowns couple together the sets of equations of the two phases. The system can be shown to be hyperbolic at least when the difference of velocities of the two constituents is sufficiently small.

We numerically solve the model equations in one dimension by a finite volume scheme based on a Roe-type Riemann solver. Well-balancing of topography source terms is obtained via a technique that includes these contributions into the wave structure of the Riemann solution.

1 Introduction

Geophysical flows such as avalanches and debris flows are gravity-driven granular masses, typically composed of a mixture of solid grains and interstitial fluid. Following the pioneering work of Savage and Hutter [SH89], in recent years great advances have been made in the mathematical and numerical modeling of these natural processes by means of depth-averaged (thin layer) models, which are based on the small aspect ratio of typical flows. Early studies [SH89, SH91], and part of recent literature as well [MVB+03, PH03], were limited to dry (one-phase) granular flows. Iverson [Ive97] and Iverson and Denlinger [ID01] first addressed the need of accounting for interstitial fluid effects in the flowing mass, and developed a solid–fluid mixture theory based on the simplifying assumptions of constant porosity and equality of fluid and solid velocities. Some numerical results are reported in [DI01]. See also recent numerical applications to debris flow modeling based on a similar mixture theory approach in [PWH05]. Making a step forward, Pitman and Le [PL05] have recently presented a depth-averaged two-phase model for debris flows and avalanches that contains mass and momentum equations for both the solid and fluid component, thus providing equations for porosity and velocity of both phases. However, in this work the authors propose a numerical method only for a reduced model that ignores fluid inertial terms.

With the objective of contributing to develop a numerical model describing accurately the coexistence and the interaction between solid and fluid constituents, here we study a two-phase granular flow model that follows the work of [PL05]. The mathematical equations are presented in Sects. 2, and in Sect. 3 we analyze eigenvalues and hyperbolicity. Then, in Sect. 4, we illustrate the numerical technique we use to solve the model system, and we report some numerical results in Sect. 5. Some prospected work is finally mentioned in Sect. 6.

2 Mathematical Model

Following Pitman and Le [PL05], we consider a thin layer of a mixture of solid granular material and fluid moving over a variable basal surface. Solid and fluid components are assumed incompressible, with constant specific densities ρ_s and $\rho_f < \rho_s$, respectively. Under the shallow flow assumption, and for slowly varying topography, depth-averaged mass and momentum equations for the two phases can be derived in the following form (here in one dimension):

$$\partial_t(\varphi\, h) + \partial_x(\varphi\, h\, v_s) = 0, \tag{1a}$$

$$\begin{aligned}&\partial_t(\varphi\, h\, v_s) + \partial_x\left(\varphi\, h\, v_s^2 + \tfrac{g}{2}(1-\gamma)\,\varphi\, h^2\right) + \gamma\,\tfrac{g}{2}\,\varphi\,\partial_x h^2 \\ &\quad = -g\,\varphi\, h\partial_x b + K_{xz}\, g(1-\gamma)\,\varphi\, h + \gamma D h\,(v_f - v_s),\end{aligned} \tag{1b}$$

$$\partial_t((1-\varphi)h) + \partial_x((1-\varphi)h\, v_f) = 0, \tag{1c}$$

$$\begin{aligned}&\partial_t((1-\varphi)h\, v_f) + \partial_x((1-\varphi)h\, v_f^2) + (1-\varphi)\tfrac{g}{2}\partial_x h^2 \\ &\quad = -g(1-\varphi)\, h\,\partial_x b - Dh\,(v_f - v_s).\end{aligned} \tag{1d}$$

Above h is the flow depth, φ the solid volume fraction, v_s and v_f are the solid and fluid velocities, respectively, and $b(x)$ represents the bottom topography. Moreover, D is a drag function, K_{xz} a friction coefficient, g the gravity constant, and $\gamma = \frac{\rho_f}{\rho_s} < 1$. The two-phase model (1) is a variant of the Pitman–Le model [PL05], and it differs from the original work of [PL05] in the description of the fluid and mixture momentum balance. See [BPM] for more details. In particular, contrarily to [PL05], our model has the property of recovering a conservative equation for the momentum of the mixture, which has the form:

$$\begin{aligned}&\partial_t(\varphi h v_s + \gamma(1-\varphi)h v_f) + \partial_x\left(\varphi h v_s^2 + \gamma(1-\varphi)h v_f^2 + \tfrac{g}{2}(\varphi + \gamma(1-\varphi))h^2\right) \\ &= -g(\varphi + \gamma(1-\varphi))h\partial_x b + K_{xz} g(1-\gamma)\varphi h.\end{aligned} \tag{2}$$

We now rewrite system (1) by expressing quantities containing the variables φ and h in terms of the conserved quantities $h_s \equiv \varphi h$ and $h_f \equiv (1-\varphi)h$. We neglect friction and drag, which will not be considered hereafter. Modeling of drag will be discussed in [BPM]. Manipulating suitably (1), and setting $q = (h_s, h_s v_s, h_f, h_f v_f)^{\mathrm{T}}$, we obtain the system

$$\partial_t q + \partial_x f(q) + s(q, \partial_x q) = \psi^{\mathrm{b}}(q), \quad \text{with} \tag{3a}$$

$$f(q) = (h_s\, v_s,\, h_s\, v_s^2 + \tfrac{g}{2}\, h_s^2 + \tfrac{g(1-\gamma)}{2}\, h_s\, h_f,\, h_f\, v_f,\, h_f\, v_f^2 + \tfrac{g}{2}\, h_f^2)^{\mathrm{T}}, \tag{3b}$$

$$s(q, \partial_x q) = (0,\, \gamma\, g\, h_s\, \partial_x h_f,\, 0,\, g\, h_f\, \partial_x h_s)^{\mathrm{T}}, \tag{3c}$$

$$\text{and} \quad \psi^{\mathrm{b}}(q) = -\,(0,\, g\, h_s\, \partial_x b,\, 0,\, g\, h_f\, \partial_x b)^{\mathrm{T}}. \tag{3d}$$

Above, we have put into evidence the conservative portion of the system $\frac{\partial f(q)}{\partial x}$, and the nonconservative term $s(q, \partial_x q)$. An interesting feature of this formulation in terms of h_s, h_f is its formal similarity to the classical two-layer shallow water model system, e.g., [CMP01]. The only difference is the additional conservative cross term $\frac{\partial}{\partial x}\left(\frac{g}{2}(1-\gamma)h_s h_f\right)$ in the solid momentum equation of our two-phase model. Let us finally write system (3) in quasilinear form. We have $\partial_t q + A(q)\partial_x q = \psi^{\mathrm{b}}(q)$, where

$$A(q) = \begin{pmatrix} 0 & 1 & 0 & 0 \\ -v_s^2 + g h_s + \frac{g(1-\gamma)}{2} h_f & 2v_s & \frac{g(1+\gamma)}{2} h_s & 0 \\ 0 & 0 & 0 & 1 \\ g h_f & 0 & -v_f^2 + g h_f & 2v_f \end{pmatrix}. \tag{4}$$

3 Eigenvalue Analysis and Hyperbolicity

In general, explicit expressions of the eigenvalues λ_k, $k = 1, \ldots, 4$, of the matrix A of the system cannot be derived. In the particular case of equality of solid and fluid velocities, $v_f = v_s \equiv v$, the eigenvalues are real and distinct ($\varphi \neq 1$) and given by $\lambda_{1,4} = v \mp a$, and $\lambda_{2,3} = v \mp a\beta$, where we have introduced the quantities

$$a = \sqrt{gh} \qquad \text{and} \qquad \beta = \sqrt{\tfrac{1}{2}(1-\varphi)(1-\gamma)} < 1. \tag{5}$$

Other particular cases are: (i) $\varphi = 0$, for which the eigenvalues are $v_f \mp a$, $v_s \mp a\beta$, with $\beta = \sqrt{\frac{1-\gamma}{2}}$; (ii) $\varphi = 1$, for which we find the two distinct eigenvalues $v_s \mp a$ and the double eigenvalue v_f.

In general, for $h > 0$, we can state the following result (see proof in [BPM]):

Proposition 1. *Matrix A has always at least two real eigenvalues $\lambda_{1,4}$, and moreover, the eigenvalues λ_k of A, $k = 1, \ldots, 4$, satisfy:*

$$\min(v_f, v_s) - a \le \lambda_1 \le \mathfrak{Re}(\lambda_2) \le \mathfrak{Re}(\lambda_3) \le \lambda_4 \le \max(v_f, v_s) + a, \tag{6}$$

where $\Re\mathfrak{e}(\cdot)$ *denotes the real part. Furthermore:*

(i) If $|v_s - v_f| \leq 2a\beta$ *or* $|v_s - v_f| \geq 2a$ *then all the eigenvalues are real. If these inequalities are strictly satisfied, and if* $\varphi \neq 1$*, then the eigenvalues are also distinct, and system (3) is strictly hyperbolic.*

(ii) If $2a\beta < |v_s - v_f| < 2a$ *then the internal eigenvalues* $\lambda_{2,3}$ *may be complex.*

3.1 Eigenvectors

The right and left eigenvectors of the matrix $A(q)$ can be easily written in terms of the eigenvalues λ_k. Let us assume here $h_s, h_f > 0$. Then the right eigenvectors r_k, $k = 1, \ldots, 4$, can be expressed as $r_k = (1,\, \lambda_k,\, d_k, d_k\lambda_k)^{\mathrm{T}}$, with $d_k = 2\frac{(\lambda_k - v_s)^2 - g(h_s + h_f(1-\gamma)/2)}{g(1+\gamma)h_s} = \frac{gh_f}{(\lambda_k - v_f)^2 - gh_f}$. The left eigenvectors l_k of A can be taken as $l_k = \frac{n_k}{P'(\lambda_k)}$, where $P(\lambda)$ is the characteristic polynomial of A and $n_k = (c_{s,k}\,(\lambda_k - 2v_s),\, c_{s,k},\, c_f\,(\lambda_k - 2v_f),\, c_f)$, with $c_{s,k} = (\lambda_k - v_f)^2 - gh_f$ and $c_f = g\frac{(1+\gamma)}{2}h_s$. Here we have normalized the eigenvectors l_k so that $L = R^{-1}$, where R is the matrix with columns r_k, and L the matrix with rows l_k.

4 Numerical Solution

We assume $h_s, h_f > 0$ during the flow evolution, and that the difference between solid and fluid velocities is small enough so that the model system is strictly hyperbolic (see Proposition 1). We develop a numerical solution method for (3) in the framework of finite volume schemes based on Riemann solvers.

4.1 A Roe-Type Scheme

Let us first consider system (3) without topography terms, $\partial_t q + \partial_t f(q) + s(q, \partial_x q) = 0$. We numerically solve these equations by employing a Roe-type method [Roe81]. Following the usual technique, at every time step and at each interface between left and right states q_ℓ, q_r, we solve a Riemann problem for a linearized system $\partial_t q + \hat{A}(q_\ell, q_r)\partial_x q = 0$. The constant coefficient matrix $\hat{A}(q_\ell, q_r)$ is defined so as to guarantee conservation for the mass of each phase and for the momentum of the mixture. That is, we need $\Delta f^{(p)} = \hat{A}^{(p,:)}\Delta q$, for row indexes $p = 1$ and $p = 3$, and $\Delta f^{(2)} + \gamma \Delta f^{(4)} = (\hat{A}^{(2,:)} + \gamma \hat{A}^{(4,:)})\Delta q$, where $\Delta q = q_r - q_\ell$, and $\Delta f = f(q_r) - f(q_\ell)$. This can be satisfied by taking $\hat{A}$ as the original matrix $A(q)$ evaluated in an average state $\hat{q} = \hat{q}(\hat{h}_s, \hat{h}_f, \hat{v}_s, \hat{v}_f)$, where

$$\hat{h}_\theta = \frac{h_{\theta,\ell} + h_{\theta,r}}{2} \quad \text{and} \quad \hat{v}_\theta = \frac{\sqrt{h_{\theta,\ell}}\, v_{\theta,\ell} + \sqrt{h_{\theta,r}}\, v_{\theta,r}}{\sqrt{h_{\theta,\ell}} + \sqrt{h_{\theta,r}}}, \qquad \theta = s, f. \tag{7}$$

4.2 F-Wave Formulation

The algorithm we employ is formulated in the framework of the wave-propagation methods of [LeV97] (see also [LeV02]). In particular, we adopt the so-called *f-wave formulation* [BLMR02] of these methods, since this approach will be useful to incorporate topography source terms (see Sect. 4.3).

The f-wave technique is designed for conservative systems, that is systems endowed with a flux function, and its main idea is to define the structure of the Riemann solution by decomposing into waves (*f-waves*) the flux difference between neighboring cells. Although our system (3) contains the nonconservative products $s(q, \partial_x q)$, we can still employ this approach by linearizing $s(q, \partial_x q)$ and by defining locally an approximate flux $\tilde{f}$ consistent with the Roe linearization. Here we take $\tilde{f}(q) = f(q) + (0,\, \gamma\, g\, \hat{h}_s\, h_f,\, 0,\, g\, \hat{h}_f\, h_s)^{\mathrm{T}}$, with $\hat{h}_s$, $\hat{h}_f$ as in (7). Note that $\Delta\tilde{f} = \hat{A}\Delta q$. Then, in our algorithm we project the difference $\Delta\tilde{f} = \tilde{f}(q_r) - \tilde{f}(q_\ell)$ onto the eigenvectors $\hat{r}_k$ of the Roe matrix, $\Delta\tilde{f} = \sum_{k=1}^{4} \zeta_k \hat{r}_k$, and we use the f-waves $\mathcal{Z}_k \equiv \zeta_k \hat{r}_k$ with corresponding speeds $\hat{\lambda}_k$ (eigenvalues of $\hat{A}$) to update cell averages. Second-order correction terms and limiters are applied to these f-waves.

4.3 Topography Source Terms

We now consider system (3) with bottom topography source terms included. A well-known difficulty in the approximation of hyperbolic systems with sources (e.g., [Bou04]) is the preservation of steady state conditions at the discrete level, and the efficient modeling of small perturbations from steady states. In particular, for the system under study we are concerned with the steady state conditions at rest $h + b =$ const., $\varphi =$ const., $v_s = v_f = 0$. To build a well-balanced scheme, we follow the approach of [BLMR02, LP01, LG04], which uses the f-wave formulation framework described above. The idea is to incorporate the effect of bottom topography terms into the Riemann solution, by taking interface values $\Psi^{\mathrm{b}}_{\ell,r}$ of the topography source term $\psi^{\mathrm{b}}(q)$ and by including contributions $\Psi^{\mathrm{b}}_{\ell,r}\, \Delta x$ into the wave splitting that we have defined for the solution of the homogeneous system. We now decompose $\Delta\tilde{f} - \Psi^{\mathrm{b}}_{\ell,r}\, \Delta x = \sum_{k=1}^{4} \zeta_k\, \hat{r}_k$. The interface source term $\Psi^{\mathrm{b}}_{\ell,r}$ must be defined so that the discrete steady state condition $\Delta\tilde{f}/\Delta x = \Psi^{\mathrm{b}}_{\ell,r}$ holds whenever initial Riemann data correspond to equilibrium, that is $(h + b)_\ell = (h + b)_r$, $\varphi_\ell = \varphi_r$, $v_{s,\ell} = v_{s,r} = v_{f,\ell} = v_{f,r} = 0$. To satisfy this requirement we take $\Psi^{\mathrm{b}}_{\ell,r}\Delta x = -(0,\, g\, \hat{h}_s\, \Delta b,\, 0,\, g\, \hat{h}_f\, \Delta b)^{\mathrm{T}}$, with $\Delta b = b_r - b_\ell$.

5 A Numerical Test: Perturbation of a Steady State

We have implemented our algorithm by using the basic Fortran 77 routines of the CLAWPACK software [LeV]. Since explicit formulas for the system's eigenvalues are not available, the external eigenvalues $\hat{\lambda}_{1,4}$ are computed numerically through Newton's iteration method. The internal eigenvalues $\hat{\lambda}_{2,3}$ can be then found analytically.

We present here the results of a numerical experiment that is an extension of LeVeque's classical test [LeV98] for (one-phase) shallow water equations with bottom topography. In this problem we observe the behavior of a small perturbation of steady state conditions at rest over a bottom topography defined by $b(x) = 0.25(\cos(\pi(x - 0.5)/0.1) + 1)$ if $x \in (0.4, 0.6)$, and $b(x) = 0$ otherwise. Initially, we take a small perturbation of the flow depth h and of the solid volume fraction φ: $h(x, 0) = h_0 + \tilde{h}$ and $\varphi(x, 0) = \varphi_0 - \tilde{\varphi}$ for $x \in (-0.6, -0.5)$, with $h_0 = 1$, $\varphi_0 = 0.6$, and $\tilde{h} = \tilde{\varphi} = 10^{-3}$. The computational domain is $[-0.9, 1.1]$, and free flow boundary conditions are used. Moreover, we take $\gamma = 1/2$ and $g = 1$. We compute the solution with 100 grid cells and compare it with a fine grid reference solution obtained with 1,000 grid cells. Second-order corrections with the MC limiter [LeV02] are applied. In Fig. 1 we display results at four different times for $h + b$ and φ (top and bottom subplot of each sub-figure, respectively). The bold line over the x-interval (0.4, 0.6) in the plots of $h + b$ indicates the region of the domain where $b(x) \neq 0$. As we can

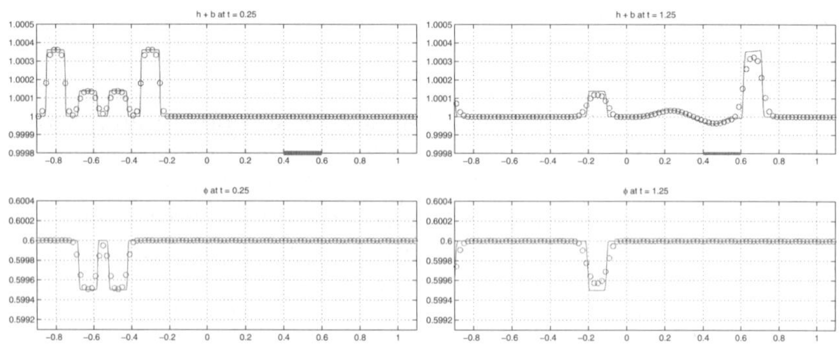

(a) $t = 0.25$ (b) $t = 1.25$

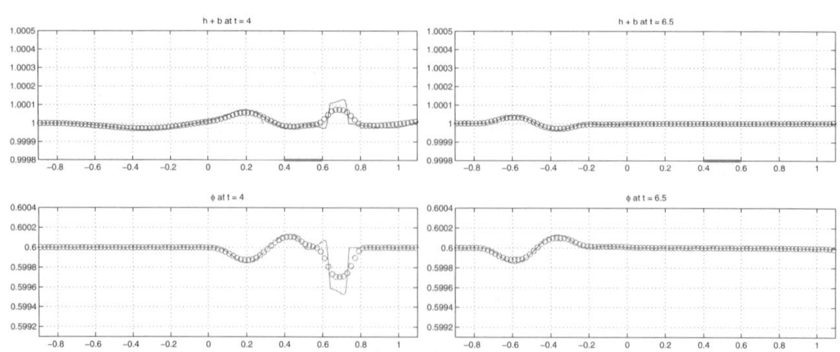

(c) $t = 4.0$ (d) $t = 6.5$

Fig. 1. Numerical test of Sect. 5. Solution for $h + b$ and φ (top and bottom subplot of each sub-figure). *Circles*: solution computed with 100 grid cells; *continuous line*: reference solution computed with 1,000 grid cells. CFL = 0.9

observe from the first couple of plots (Fig. 1a), the initial perturbation splits into four waves. These are approximately linear waves propagating at the characteristic speeds corresponding to the background state, that is $\pm\sqrt{gh_0}$ for the external waves, and $\pm\sqrt{\frac{g}{2}h_0(1-\varphi_0)(1-\gamma)}$ for the internal ones. Note that in this problem φ appears to vary only across the internal waves. This is related to the fact that if $v_s \approx v_f$, then φ is almost invariant across the first and fourth characteristic fields [BPM]. Figure 1b shows the time at which the right-going external wave has just passed over the obstacle at the bottom, and it has been partially reflected. Similarly, Fig. 1c shows when the right-going internal wave has now moved past the hump and has produced a reflected wave. At this time the reflected wave generated by the external wave has left the domain from the left boundary, after passing through the incoming internal wave. The last plots (Fig. 1d) show the situation in which all the waves have exited from the domain, except the disturbance produced by the internal wave, which will eventually leave from the left boundary. No spurious disturbances are observed in this test.

6 Prospected Work

We have presented a new numerical model for grain–fluid mixtures over variable topography. This is only a first stage toward the development of a model applicable to realistic geophysical flows. The primary issue on which we are currently focusing our research efforts is preservation of positivity of the flow depth ($h \geq 0$), to be able to handle interfaces between flow fronts and bed dry states. Planned work includes also the formulation of a two-dimensional model for arbitrary complex topography.

Acknowledgment

Part of this work was done by M. Pelanti during a postdoctoral stay at the Institut de Physique du Globe de Paris, financed by the City of Paris.

References

[BLMR02] D. Bale, R.J. LeVeque, S. Mitran, and J.A. Rossmanith. A wave-propagation method for conservation laws and balance laws with spatially varying flux functions. *SIAM J. Sci. Comput.*, 24:955–978, 2002.

[Bou04] F. Bouchut. *Nonlinear stability of finite volume methods for hyperbolic conservation laws, and well-balanced schemes for sources.* Birkhäuser-Verlag, 2004.

[BPM] F. Bouchut, M. Pelanti, and A. Mangeney. A Roe-type scheme for two-phase granular flows with bottom topography. *In Preparation.*

[CMP01] M.J. Castro, J. Macías, and C. Parés. A Q-scheme for a class of systems of coupled conservation laws with source term. Application to a two-layer 1-D shallow water system. *ESAIM-Math. Model. Num.*, 35:107–127, 2001.

[DI01] R.P. Denlinger and R.M. Iverson. Flow of variably fluidized granular masses across three-dimensional terrain: 2. Numerical predictions and experimental tests. *J. Geophys. Res.*, 106:553–566, 2001.

[ID01] R.M. Iverson and R.P. Denlinger. Flow of variably fluidized granular masses across three-dimensional terrain: 1. Coulomb mixture theory. *J. Geophys. Res.*, 106:537–552, 2001.

[Ive97] R.M. Iverson. The physics of debris flows. *Rev. Geophys.*, 35:245–296, 1997.

[LeV] R.J. LeVeque. CLAWPACK. `http://www.amath.washington.edu/~claw`.

[LeV97] R.J. LeVeque. Wave propagation algorithms for multi-dimensional hyperbolic systems. *J. Comput. Phys.*, 131:327–353, 1997.

[LeV98] R.J. LeVeque. Balancing source terms and flux gradients in high-resolution Godunov methods: The quasi-steady wave-propagation algorithm. *J. Comput. Phys.*, 146:346–365, 1998.

[LeV02] R.J. LeVeque. *Finite Volume Methods for Hyperbolic Problems*. Cambridge University Press, 2002.

[LG04] R.J. LeVeque and D.L. George. High-resolution finite volume methods for the shallow water equations with bathymetry and dry states. In P. Liu, editor, *Proceedings of Long-Wave Workshop, Catalina*, page to appear, 2004.

[LP01] R.J. LeVeque and M. Pelanti. A class of approximate Riemann solvers and their relation to relaxation schemes. *J. Comput. Phys.*, 172:572–591, 2001.

[MVB+03] A. Mangeney, J.-P. Vilotte, M.O. Bristeau, B. Perthame, F. Bouchut, C. Simeoni, and S. Yernini. Numerical modeling of avalanches based on Saint-Venant equations using a kinetic scheme. *J. Geophys. Res.*, 108(B11):2527, 2003.

[PH03] S.P. Pudasaini and K. Hutter. Rapid shear flows of dry granular masses down curved and twisted channels. *J. Fluid Mech.*, 495:193–208, 2003.

[PL05] E.B. Pitman and L. Le. A two-fluid model for avalanche and debris flows. *Phil. Trans. R. Soc. A*, 363:1573–1601, 2005.

[PWH05] S.P. Pudasaini, Y. Wang, and K. Hutter. Modelling debris flows down general channels. *Natural Hazards and Earth System Sciences*, 5:799–819, 2005.

[Roe81] P.L. Roe. Approximate Riemann solvers, parameter vectors, and difference schemes. *J. Comput. Phys.*, 43:357–372, 1981.

[SH89] S.B. Savage and K. Hutter. The motion of a finite mass of granular material down a rough incline. *J. Fluid. Mech.*, 199:177–215, 1989.

[SH91] S.B. Savage and K. Hutter. The dynamics of avalanches of granular materials from initiation to runout, part I. Analysis. *Acta Mech.*, 86:201–223, 1991.

Linear Lagrangian Systems of Conservation Laws

Y.-J. Peng

1 Introduction

Consider the Cauchy problem for a system of conservation laws:

$$\partial_t u + \partial_x f(u) = 0, \quad t > 0,\ x \in I\!R, \tag{1}$$

with an initial condition:

$$t = 0 \ : \quad u = u^0(x), \quad x \in I\!R. \tag{2}$$

Here $u = (u_1, u_2, \cdots, u_n)^t$, $u^0 = (u_1^0, u_2^0, \cdots, u_n^0)^t \in L^\infty(I\!R)$, and $f = (f_1, f_2, \cdots, f_n)^t$ is a smooth function from a domain $U \subset I\!R^n$ to $I\!R^n$.

Let $u \in L^\infty(I\!R^+ \times I\!R)$ be an entropy solution of the system (1) with $u_1(t,x) \geq \underline{u}_1 > 0$. Using the equation for u_1 in (1), we may take an Euler–Lagrange transformation $(t,x) \longmapsto (s,y)$:

$$(s,y) = (t, Y(t,x)) \quad \text{and} \quad dy = u_1 dx - f_1(u)dt. \tag{3}$$

Then in Lagrangian coordinates (s,y), the system (1) reads (see [14]):

$$\begin{cases} \partial_s\Big(\dfrac{1}{\tilde{u}_1}\Big) - \partial_y\Big(\dfrac{f_1(\tilde{u})}{\tilde{u}_1}\Big) = 0, \\ \partial_s\Big(\dfrac{\tilde{u}_i}{\tilde{u}_1}\Big) + \partial_y\Big(f_i(\tilde{u}) - \dfrac{f_1(\tilde{u})\tilde{u}_i}{\tilde{u}_1}\Big) = 0 \qquad (2 \leq i \leq n), \end{cases} \tag{4}$$

for $s > 0$ and $y \in I\!R$. Here and in what follows we denote by $\tilde{u}$ the variable u in Lagrangian coordinates, i.e., $\tilde{u}(s,y) = u(t,x)$.

Set

$$Y(0,x) = Y_0(x) \stackrel{\text{def}}{=} \int_0^x u_1^0(\xi)d\xi. \tag{5}$$

Then for each given $u \in L^\infty(I\!R^+ \times I\!R)$, (3) and (5) define a unique E–L transformation. Obviously, for $u_1^0(x) \geq \underline{u}_1$, Y_0 is strictly increasing, Lipschitzian and bijective from $\mathbb{R}$ to $\mathbb{R}$. So is its inverse function X_0. Similarly, for all $t > 0$, $x \longmapsto Y(t,x)$ is bijective from $\mathbb{R}$ to $\mathbb{R}$ and $(t,x) \longmapsto Y(t,x)$ is Lipschitzian.

From (3) we have $\tilde{u}(t, Y(t,x)) = u(t,x)$. Denote by $X(t,\cdot)$ the inverse function of $Y(t,\cdot)$. Then $\tilde{u}(t,y) = u(t, X(t,y))$ in which taking $t = 0$ gives $\tilde{u}(0,y) = u^0(X_0(y))$. Finally, we obtain the following equivalence result for the Cauchy problems whose proof can be found in [9].

Theorem 1. *Let $u \in L^\infty(\mathbb{R}^+ \times \mathbb{R})$ be an entropy solution of (1)–(2) with $u_1(t,x) \geq \underline{u}_1$ and $(t,x) \longmapsto (s,y)$ be the unique E–L transformation defined by (3) and (5). Then $\tilde{u}(t,y) = u(t, X(t,y))$ is an entropy solution of (4) satisfying the initial condition $\tilde{u}|_{s=0} = u^0(X_0)$.*

Furthermore, if the entropy solution of the Cauchy problem for the system (4) is unique, then the entropy solution $u \in L^\infty(\mathbb{R}^+ \times \mathbb{R})$ of (1)–(2) with $u_1(t,x) \geq \underline{u}_1$ is unique.

2 Structure of a Linear Lagrangian System

Definition 1. *A system of conservation laws (1) is called a linear Lagrangian system if it is linear in Lagrangian coordinates (s,y) defined by the E–L transformation (3).*

The conservative variables of the system (4) are $v = (v_1, v_2, \cdots, v_n)^t$ with

$$v_1 = \frac{1}{u_1}, \quad v_i = \frac{u_i}{u_1} \quad (2 \leq i \leq n). \tag{6}$$

Therefore, the system (4) is linear if and only if there are real constants a_{ij} $(1 \leq i,j \leq n)$ such that

$$-\frac{f_1(u)}{u_1} = \sum_{j=1}^{n} a_{1j} v_j, \quad f_i(u) - \frac{f_1(u) u_i}{u_1} = \sum_{j=1}^{n} a_{ij} v_j \quad (2 \leq i \leq n).$$

This gives the general form of a linear Lagrangian system:

$$\begin{cases} \partial_t u_1 - \partial_x \Big(\sum_{j=2}^{n} a_{1j} u_j \Big) = 0, \\ \partial_t u_i + \partial_x \Big[\dfrac{1}{u_1} \Big(a_{i1} + u_i f_1(u) + \sum_{j=2}^{n} a_{ij} u_j \Big) \Big] = 0 \quad (2 \leq i \leq n), \end{cases} \tag{7}$$

which can also be written as a linear system in Lagrangian coordinates:

$$\partial_s \tilde{v} + A \partial_y \tilde{v} = 0, \tag{8}$$

where $A = (a_{ij})_{1 \leq i,j \leq n}$ is a constant matrix.

Let λ_i $(1 \le i \le n)$ be the eigenvalues of A and l_i be an ith left eigenvector of A associated to λ_i. In what follows we suppose that the linear Lagrangian system (7) is hyperbolic (not necessarily strictly hyperbolic). By the equivalence of hyperbolicity [12], the linear system (8) is also hyperbolic. Then $(l_1, l_2, \cdots, l_n)$ is a basis of $\mathbb{R}^n$, $w_i = l_i v$ is an ith classical Riemann invariant of (7) and satisfies

$$\partial_s \tilde{w}_i + \lambda_i \partial_y \tilde{w}_i = 0 \quad (1 \le i \le n).$$

We deduce that each entropy–entropy flux pair $(\tilde{E}, \tilde{F})$ of the system (8) has the form:

$$(\tilde{E}(v), \tilde{F}(v)) = \sum_{j=1}^{n} (1, \lambda_j) g_j(w_j) = \sum_{j=1}^{n} (1, \lambda_j) g_j(l_j v), \tag{9}$$

where g_j $(1 \le j \le n)$ are arbitrary continuous functions. From the equivalence relation [14], we obtain an explicit expression of the entropy–entropy flux pair of the linear Lagrangian system as:

$$(E(u), F(u)) = \sum_{j=1}^{n} (u_1, \lambda_j + f_1(u)) g_j\Big(\frac{1}{u_1} l_j (1, u_2, \cdots, u_n)^t\Big). \tag{10}$$

Moreover, let l_{ij} be the ith component of l_j, we have

$$\frac{\partial^2 \tilde{E}(v)}{\partial v_i \partial v_k} = \sum_{j=1}^{n} g_j''(w_j) l_{ji} l_{jk} \stackrel{\text{def}}{=} \alpha_{ik}^g(w), \quad \text{with } w = (w_1, \cdots, w_n)^t.$$

Therefore, by the equivalence of the convexity of entropies [14], $E(u)$ defined by (10) is convex if and only if the symmetric matrix $\Delta_g(w) \stackrel{\text{def}}{=} (\alpha_{ik}^g(w))_{1 \le i,k \le n}$ is positively definite for all $w \in \mathbb{R}^n$. In particular, since $(l_1, l_2, \cdots, l_n)$ is linearly independent, the constant matrix $(< l_i, l_k >)_{1 \le i,k \le n}$ is positively definite, where $< \cdot, \cdot >$ denotes the inner product of $\mathbb{R}^n$. Thus a strictly convex entropy $E(u)$ is constructed by choosing $g_j(\xi) = \xi^2$ for all $j = 1, 2, \cdots, n$.

3 Explicit Solutions of the Cauchy Problem

Consider the Cauchy problem for the linear Lagrangian system (7) with the initial condition (2). Let $v^0 = (1, u_2^0, \cdots, u_n^0)^t / u_1^0$. By Theorem 1 we have

$$s = 0 \ : \quad \tilde{w}_i = w_i^0(X_0(y)) = l_i v^0(X_0(y)), \quad y \in \mathbb{R}. \tag{11}$$

Therefore, the unique solution of the Cauchy problem (8) and (11) is given by

$$\tilde{w}_i(s, y) = w_i^0(X_0(y - \lambda_i s)).$$

It follows from Theorem 1 that the Cauchy problems (7) and (2) admit a unique entropy solution u. Denote by r_i the right eigenvector of A associated to λ_i normalized as $l_i r_j = \delta_{ij}$, where δ_{ij} stands for the Kronecker's symbol. Then we may write

$$\tilde{v}(s,y) = \sum_{i=1}^{n} (l_i v^0)(X_0(y - \lambda_i s)) r_i. \tag{12}$$

On the other hand, from (3) we have

$$dx = \tilde{v}_1(t,y)dy - \sum_{j=1}^{n} a_{1j}\tilde{v}_j(t,y)dt,$$

which defines a unique function $x = X(t,y)$ together with $X(0,y) = X_0(y)$. A straightforward calculation using the first equation in (8) gives

$$X(t,y) = X_0(y) - \int_0^t \sum_{j=1}^{n} a_{1j}\tilde{v}_j(\tau,y)d\tau.$$

Let r_{ji} be the jth component of r_i. It follows from (12) that

$$X(t,y) = X_0(y) - \int_0^t \sum_{i=1}^{n} \lambda_i r_{1i}(l_i v^0)(X_0(y - \lambda_i \tau))d\tau. \tag{13}$$

Now suppose that

$$u^0 \in L^\infty(\mathbb{R}), \quad u_1^0(x) \geq \underline{u}_1 > 0, \quad \text{a.e. } x \in \mathbb{R} \tag{14}$$

and

$$\sum_{i=1}^{n} (l_i v^0)(X_0(y - \lambda_i t)) r_{1i} \geq \underline{v}_1 > 0, \quad \text{a.e. } (t,y) \in \mathbb{R}^+ \times \mathbb{R}. \tag{15}$$

Then the function $y \longmapsto X(t,y)$ is invertible and bi-Lipschitzian from $\mathbb{R}$ to $\mathbb{R}$ for all $t \geq 0$. Let $y = Y(t,x)$ be its inverse function satisfying $Y(0,x) = Y_0(x)$. Thus, (12) gives

$$v(t,x) = \sum_{i=1}^{n} (l_i v^0)(X_0(Y(t,x) - \lambda_i t)) r_i, \tag{16}$$

which implies, together with (15), that $v_1(t,x) \geq \underline{v}_1$, a.e. $(t,x) \in \mathbb{R}^+ \times \mathbb{R}$. In particular,

$$w_i(t,x) = w_i^0(X_0(Y(t,x) - \lambda_i t)) \tag{17}$$

which shows that the maximum principle holds for each w_i.

On the other hand, the expression $\tilde{w}_i(s,y) = \tilde{w}_i^0(y - \lambda_i s)$ shows that the entropy equality $\partial_s \tilde{E}(v) + \partial_y \tilde{F}(v) = 0$ is satisfied for all entropy–entropy flux pairs $(\tilde{E}(v), \tilde{F}(v))$ given in (9). This equality is equivalent to $\partial_t E(u) + \partial_x F(u) = 0$.

Theorem 2. *Let the assumptions (14)–(15) hold and the linear Lagrangian system (7) be hyperbolic. Then the Cauchy problem (7) and (2) admit a unique solution satisfying the entropy equality $\partial_t E(u) + \partial_x F(u) = 0$ for all entropy–entropy flux pairs given in (10). It is given by the explicit expression (16) and (6).*

Remark 1. From (13), we have

$$(\partial_{tt}X(t,y), \partial_{yy}X(t,y)) = \sum_{i=1}^{n}(\lambda_i^2, 1)r_{1i}l_i(v^0 \circ X_0)'(y - \lambda_i t).$$

Hence, when $|\lambda_i| = \lambda$ for all $i = 1, 2, \cdots, n$, X satisfies a linear wave equation

$$\partial_{tt}X - \lambda^2\partial_{yy}X = 0.$$

4 Weak Stability

Consider a hyperbolic linear Lagrangian system (7). Let $(u_\varepsilon^0)_{\varepsilon>0}$ be a bounded sequence in $L^\infty(I\!R)$ satisfying (14)–(15) uniformly with respect to ε. We denote by $(u_\varepsilon)_{\varepsilon>0}$ the corresponding sequence of entropy solutions of the system (7) with the initial data $(u_\varepsilon^0)_{\varepsilon>0}$. From (14) and(15) and the explicit formulas (16)and (17), we know that the sequence $(u_\varepsilon)_{\varepsilon>0}$ is bounded in $L^\infty(I\!R^+ \times I\!R)$ and there is a constant $\underline{u} > 0$, independent of $\varepsilon \to 0$, such that $u_{1\varepsilon}(t,x) \geq \underline{u}$, a.e. $(t,x) \in I\!R^+ \times I\!R$.

Therefore, up to subsequences (not relabeled), we have

$$u_\varepsilon \rightharpoonup u, \quad \text{in } L^\infty(I\!R^+ \times I\!R) \text{ weakly-}*$$

and

$$u_\varepsilon^0 \rightharpoonup u^0, \quad \text{in } L^\infty(I\!R) \text{ weakly-}*,$$

with $u \in L^\infty(I\!R^+ \times I\!R)$ and $u^0 \in L^\infty(I\!R)$ satisfying $u_1^0(x) \geq \underline{u}_1$ and $u_1(t,x) \geq \underline{u}$, a.e. $(t,x) \in I\!R^+ \times I\!R$. The weak stability means that the weak limit u is a unique entropy solution of (7) with the initial datum u^0.

Indeed, since f_1 is an affine function, we have

$$f_1(u_\varepsilon) \rightharpoonup f_1(u), \quad \text{in } L^\infty(I\!R^+ \times I\!R) \text{ weakly-}*.$$

Applying the div–curl lemma (see [13]) to the first and ith ($2 \leq i \leq n$) equations in (7), we obtain, in the sense of distributions,

$$\lim_{\varepsilon\to 0}(u_{1\varepsilon}f_i(u_\varepsilon) - u_{i\varepsilon}f_1(u_\varepsilon)) = u_1 \lim_{\varepsilon\to 0} f_i(u_\varepsilon) - u_i \lim_{\varepsilon\to 0} f_1(u_\varepsilon).$$

It follows from the definition of f_i that, for all $2 \leq i \leq n$,

$$\lim_{\varepsilon\to 0} f_i(u_\varepsilon) = \frac{1}{u_1}\Big(a_{i1} + u_i f_1(u) + \sum_{j=2}^{n} a_{ij}u_j\Big) = f_i(u).$$

This allows to pass to the limit in the weak formulation of the Cauchy problem for u_ε and deduce that u is a weak solution of (7) and the initial datum of u is just u^0.

Notice that the sequence $(Y_{0\varepsilon})_{\varepsilon>0}$ defined by

$$Y_{0\varepsilon}(x) = \int_0^x u^0_{1\varepsilon}(\xi)\, d\xi$$

is bounded in $W^{1,\infty}_{loc}(I\!R)$. Then up to a subsequence, still denoted by $(Y_{0\varepsilon})_{\varepsilon>0}$, we have $Y_{0\varepsilon} \longrightarrow Y_0$ in $L^p_{loc}(I\!R)$ strongly for any $p \geq 1$, where Y_0 is given in (5). Let $\phi \in C_0^\infty(I\!R)$ and $X_{0\varepsilon} = Y_{0\varepsilon}^{-1}$. Then $\phi(Y_{0\varepsilon}) \longrightarrow \phi(Y_0)$ in $L^p_{loc}(I\!R)$ strongly. It follows that

$$\int_{I\!R} (v^0_\varepsilon \circ X_{0\varepsilon})(y)\phi(y)dy \longrightarrow \int_{I\!R} (v^0 \circ X_0)(y)\phi(y)dy,$$

with $v^0 = (1, u^0_2, \cdots, u^0_n)^t/u^0_1$. Here we have used the changes of variables $y = Y_{0\varepsilon}(x)$ and $y = Y_0(x)$ in the computation. This shows that

$$v^0_\varepsilon \circ X_{0\varepsilon} \longrightarrow v^0 \circ X_0, \quad \text{in } L^\infty(I\!R) \text{ weakly-} *.$$

Since u^0_ε satisfies (15) uniformly with respect to ε, i.e.,

$$\sum_{i=1}^n (l_i v^0_\varepsilon)(X_{0\varepsilon}(y - \lambda_i t)) r_{1i} \geq \underline{v}_1 > 0, \quad \text{a.e. } (t, y) \in I\!R \times I\!R,$$

we deduce that u satisfies (15). Thus, u is a unique entropy solution of the Cauchy problem to the linear Lagrangian system (7).

Theorem 3. *The linear Lagrangian system is weakly stable in $L^\infty(I\!R^+ \times I\!R)$.*

5 Examples

Example 1. Brenier's augmented Born–Infeld system

In one space dimension, the augmented Born–Infeld system reads [2]:

$$\begin{cases} \partial_t h + \partial_x (h v_2) = 0, \\ \partial_t (h v_2) + \partial_x \left(h v_2^2 - a^2 h^{-1}\right) = 0, \\ \partial_t (h v_3) + \partial_x \left(h v_2 v_3 - D_1 v_7 + v_6\right) = 0, \\ \partial_t (h v_4) + \partial_x \left(h v_2 v_4 - D_1 v_8 - v_5\right) = 0, \\ \partial_t (h v_5) + \partial_x \left(h v_2 v_5 - B_1 v_7 - v_4\right) = 0, \\ \partial_t (h v_6) + \partial_x \left(h v_2 v_6 - B_1 v_8 + v_3\right) = 0, \\ \partial_t (h v_7) + \partial_x \left(h v_2 v_7 - D_1 v_3 - B_1 v_5\right) = 0, \\ \partial_t (h v_8) + \partial_x \left(h v_2 v_8 - D_1 v_4 - B_1 v_6\right) = 0, \end{cases} \tag{18}$$

where $h > 0$ is the energy density, B_1, D_1, and $a = \sqrt{1 + B_1^2 + D_1^2} > 0$ are constants. This is a linear Lagrangian system with $n = 8$ and $u_1 = h$. The function $y = Y(t, x)$ satisfies $dy = hdx - hv_2 dt$ and in Lagrangian coordinates (t, y), the matrix A of the corresponding linear system is given by

$$A = \begin{pmatrix} 0 & -1 & 0 & 0 & 0 & 0 & 0 & 0 \\ -a^2 & 0 & 0 & 0 & 0 & 0 & 0 & 0 \\ 0 & 0 & 0 & 0 & 0 & 1 & -D_1 & 0 \\ 0 & 0 & 0 & 0 & -1 & 0 & 0 & -D_1 \\ 0 & 0 & 0 & -1 & 0 & 0 & -B_1 & 0 \\ 0 & 0 & 1 & 0 & 0 & 0 & 0 & -B_1 \\ 0 & 0 & -D_1 & 0 & -B_1 & 0 & 0 & 0 \\ 0 & 0 & 0 & -D_1 & 0 & -B_1 & 0 & 0 \end{pmatrix}.$$

It is easy to see that A is diagonalizable. Then the system (18) is hyperbolic and Theorems 2 and 3 can be applied. Note that the condition (15) becomes:

$$\inf_{t>0,\, y\in I\!R} \Big((v_2^0 + a(h^0)^{-1})(X_0(y+t)) - (v_2^0 - a(h^0)^{-1})(X_0(y-t)) \Big) > 0,$$

where (h^0, v_2^0) is the initial datum of (h, v_2), $X_0 = Y_0^{-1}$ and

$$Y_0(x) = \int_0^x h^0(\xi)\, dx.$$

From these results, we deduce analogous results for the Born–Infeld system which is not a linear Lagrangian system [8, 9].

Remark that the E–L transformation depends only on the first two equations of (18), called the system of Chaplygin gas dynamics (see [2]) with the pressure being given by the Von Kármán–Tsien law [4]: $p(h) = -a^2 h^{-1}$. It is also a linear Lagrangian system on which results can be found in [6, 10, 7]. In particular, its eigenvalues in Lagrangian coordinates are -1 and 1. Therefore, from Remark 1 the function $X(t, \cdot) = Y^{-1}(t, \cdot)$ satisfies the linear wave equation:

$$\partial_{tt} X - \partial_{yy} X = 0.$$

Example 2. The system of pressureless gas dynamics

Consider the system of pressureless gas dynamics [1, 5, 3]:

$$\begin{cases} \partial_t \rho + \partial_x(\rho u) = 0, \\ \partial_t(\rho u) + \partial_x(\rho u^2) = 0, \quad t > 0,\ x \in I\!R. \end{cases}$$

It is a linear Lagrangian system with $u_1 = \rho > 0$ being the density. Although this system is not hyperbolic, its local solution can be obtained by the same technique introduced in Section 3. It is given by the expression [9]

$$\rho(t,x) = \frac{\rho^0(X_0(Y(t,x)))}{1 + t(\rho^0(u^0)')(X_0(Y(t,x)))}, \quad u(t,x) = u^0(X_0(Y(t,x))),$$

where (ρ^0, u^0) is the initial datum satisfying $\rho^0 \in L^\infty(\mathbb{R})$ with $\rho^0(x) \geq \underline{\rho} > 0$, $u^0 \in W^{1,\infty}(\mathbb{R})$, $X_0 = Y_0^{-1}$, $Y(t,\cdot) = X^{-1}(t,\cdot)$ and

$$Y_0(x) = \int_0^x \rho^0(\xi)d\xi, \quad X(t,y) = X_0(y) + tu^0(X_0(y)).$$

References

1. Bouchut, F.: On zero pressure gas dynamics. Advances in kinetic theory and computing, 171–190, Ser. Adv. Math. Appl. Sci., **22**, World Sci. Publ., River Edge, NJ (1994)
2. Brenier, Y.: Hydrodynamic structure of the augmented Born-Infeld equations. Arch. Rat. Mech. Anal., **172**, 65–91 (2004)
3. Brenier, Y., Grenier, E.: Sticky particles and scalar conservation laws. SIAM J. Numer. Anal., **35**, 2317–2328 (1998)
4. Courant, R., Friedrichs, K.O.: Supersonic Flow and Shock Waves. Interscience publishers, New-York (1948)
5. E.W., Rykov, Y.G., Sinai, Y.G.: Generalized variational principles, global weak solutions and behavior with random initial data for systems of conservation laws arising in adhesion particle dynamics. Comm. Math. Phys., **177**, 349–380 (1996)
6. Lochak, G.: Sur la théorie non linéaire des ondes. C.R. Acad. Sci. Paris, Série III, **250**, 1986–1988 (1960)
7. Peng, Y.J.: Explicit solutions for 2×2 linearly degenerate systems. Appl. Math. Letters, **11**, 75–78 (1998)
8. Peng, Y.J.: Entropy solutions of Born-Infeld systems in one space dimension. Rend. Circ. Mat. Palermo., Serie II, **78**, 259–271 (2006)
9. Peng, Y.J.: Euler-Lagrange change of variables in conservation laws. preprint (2005)
10. Serre, D.: Intégrabilité d'une classe de système de lois de conservation. Forum Math., **4**, 607–623 (1992)
11. Serre, D.: Systèmes de Lois de Conservation II. Diderot, Paris (1996)
12. Sévennec, B.: Géométrie des Systèmes Hyperboliques de Lois de Conservation. Mém. Soc. Math. France, No. **56** (1994)
13. Tartar, L.: Compensated compactness and applications to partial differential equations. Nonlinear analysis and mechanics: Heriot-Watt symposium, Vol. **IV**, 136–212, Research Notes in Math., **39**, Pitman (1979)
14. Wagner, D.H.: Equivalence of the Euler and Lagrangian equations of gas dynamics for weak solutions. J. Differential Equations, **68**, 118–136 (1987)

Normal Modes Analysis of Subsonic Phase Boundaries in Elastic Materials

H. Freistühler and R.G. Plaza

1 Introduction

1.1 Equations and Assumptions

Consider the equations of nonthermal elasticity with no external forces,

$$\begin{aligned} U_t - \nabla_x V &= 0, \\ V_t - \operatorname{div}_x \sigma(U) &= 0, \end{aligned} \tag{1}$$

where $(x,t) \in \mathbb{R}^d \times [0,+\infty)$, $d \geq 2$, $U \in \mathbb{R}_+^{d\times d}$ is the local deformation gradient, and $V \in \mathbb{R}^d$ is the local velocity. Equation (1) is subject to the constraints

$$\operatorname{curl}_x U = 0. \tag{2}$$

In (1), $\sigma = \sigma(U)$ denotes the first Piola–Kirchhoff stress and is supposed to derive from a stored-energy density function $W : \mathbb{R}_+^{d\times d} \to \mathbb{R}$ as $\sigma(U) = \partial W/\partial U$. System (1) is hyperbolic at U if W is rank-one convex at U [Ci88], i.e., if the acoustic tensor $\mathcal{N}(U,\xi) := D^2W(U)(\xi,\xi)$ is positive definite for all $\xi \in \mathbb{R}^d$. We are interested in the stability of subsonic phase boundaries, which are weak solutions to (1) of form

$$(U,V)(x,t) = \begin{cases} (U^-,V^-), & x\cdot N < st, \\ (U^+,V^+), & x\cdot N > st, \end{cases} \tag{3}$$

with direction of propagation $N \in S^{d-1}$, and speed s satisfying the subsonicity condition

$$0 \leq s^2 < \min\{\text{eigenvalues of}\,(\mathcal{N}(U^\pm,N))\}. \tag{4}$$

Note that $s = 0$ is included in definition (4). Configurations (3) are subject to the classical Rankine–Hugoniot jump conditions

$$\begin{aligned} -s[U] - [V]\otimes N &= 0, \\ -s[V] - [\sigma(U)]N &= 0, \end{aligned} \tag{5}$$

expressing conservation across the front. In addition, solutions (3) are required to satisfy an additional jump condition or *kinetic rule*

$$g((U^-, V^-), (U^+, V^+), s, N) = 0, \tag{6}$$

where g is real-valued and continuously differentiable. We can assemble both the Rankine–Hugoniot and the kinetic jump conditions into one vector relation $h((U^-, V^-), (U^+, V^+), s, N) = 0$, where h takes values in $\mathbb{R}^{d^2+d+1}$.

Due to its importance for applications, we pay special attention to the static configuration

$$(U, V)(x, t) = \begin{cases} (U^A, 0), & x \cdot N^* < 0, \\ (U^B, 0), & x \cdot N^* > 0, \end{cases} \tag{7}$$

again with $N^* \in S^{d-1}$, and where $U^A \neq U^B \in \mathbb{R}_+^{d\times d}$ corresponds to martensitic wells, local minima of W, and, by physical considerations [Mu99], rank one connected: rank $(U^A - U^B) = 1$.

For applicability of the existing stability theory for subsonic undercompressive shocks [Me90, Co03] we assume also the constant-multiplicity condition of Métivier [Me00], namely, that the eigenvalues of $\mathcal{N}(U, \xi)$ are all semi-simple and their multiplicities are independent of U and ξ, locally; and the nondegeneracy of the jump relations, introduced by Coulombel [Co03], requiring the matrix $(d_{(U^+,V^+)}h,\, d_{(U^-,V^-)}h)$ to have full rank.

1.2 The Lopatinski Determinant

Thanks to the fundamental work of Majda and Métivier [Ma83, Me90, Me01], the nonlinear stability of shock fronts like (3) is determined by the Lopatinski conditions of linear hyperbolic problems [K71]. The Majda–Métivier theory has been extended to general undercompressive shocks [F98, Co03], and subsonic phase boundaries like (3) fit into this setting. The starting point of these analyses is the Fourier decomposition in normal modes of the constant coefficients linearized problem at the end states, leading to the analysis of the so-called *Lopatinski determinant.* If $A_\xi^\pm = \sum A_j^\pm \xi_j$ denotes the linearization of system (1) in direction $\xi \in \mathbb{R}^d$ at $(U^\pm, V^\pm)$, one may verify (see [F98]) that for subsonic phase boundaries (3) with $s \neq 0$, the Lopatinski (or stability) function [Ma83] takes the form of the $(d^2+d+1) \times (d^2+d+1)$ determinant

$$\Delta(\lambda, \xi) = \det \begin{pmatrix} R_-^s & Q & R_+^u \\ -(d_{u^-}g)(A_N^- - sI)^{-1} R_-^s & q & (d_{u^+}g)(A_N^+ - sI)^{-1} R_+^u \end{pmatrix}, \tag{8}$$

where $(\lambda, \xi) \in \Gamma_N := \{\mathrm{Re}\,\lambda \geq 0, \xi \cdot N = 0, |\lambda|^2 + |\xi|^2 = 1\}$; Q and q are the "jump vector" fields associated to conditions (5) and (6), respectively (see [FP05]); $R_\pm^{s,u}(\lambda, \xi)$ denotes the right invariant stable/unstable subspaces of $\mathcal{A}_\pm(\lambda, \xi) = (\lambda I + i \sum \xi_j A_j^\pm)(A_N^\pm - s)^{-1}$. In [FP05] we establish the following,

Theorem 1. *For $((U^+, V^+), (U^-, V^-), s, N)$, g and W satisfying hyperbolicity, constant-multiplicity, and subsonicity assumptions, together with jump conditions (5) and (6), the stability behavior of (3) is determined by the stability function*

$$\hat{\Delta}(U^-, U^+) = \det \begin{pmatrix} \hat{R}^s_{s,N}(U^-) & \hat{Q} & \hat{R}^u_{s,N}(U^+) \\ \hat{p}^- & \hat{q} & \hat{p}^+ \end{pmatrix} : \Gamma_N \to \mathbb{C}, \tag{9}$$

in which

$$\begin{aligned} \hat{Q}(\lambda, \xi) &:= \begin{pmatrix} [U]N \\ -(\lambda s[U]N + i[\sigma(U)]\xi) \end{pmatrix}, \\ \hat{q}(\lambda, \xi) &:= -\lambda(d_s g) + i(\xi \cdot d_N) g, \\ \hat{p}^-(\lambda, \xi) &:= -(d_{(U^-,V^-)} g)\mathcal{K}_{s,N}(U^-)\hat{R}^s_{s,N}(U^-), \\ \hat{p}^+(\lambda, \xi) &:= (d_{(U^+,V^+)} g)\mathcal{K}_{s,N}(U^+)\hat{R}^u_{s,N}(U^+), \end{aligned}$$

in the sense that if $\hat{\Delta}$ has no zero on Γ_N, then (3) is nonlinearly stable; if $\hat{\Delta}$ vanishes for some $(\lambda, \xi) \in \Gamma_N$ with $\operatorname{Re}\lambda > 0$, then (3) is strongly unstable; and, if $\hat{\Delta}$ does not vanish on $\Gamma_N \cap \{\operatorname{Re}\lambda > 0\}$ then (3) is weakly stable. $\hat{\Delta}$ is a $(2d+1) \times (2d+1)$ determinant, in which $\hat{R}^s$ and $\hat{R}^u$ represent the right stable and unstable spaces of a matrix field $\mathbb{M}_{s,N}(U) : \Gamma_N \to \mathbb{C}^{2d\times 2d}$, and $\mathcal{K}_{s,N}$ denotes a continuous mapping $\mathcal{K}_{s,N} : \Gamma_N \to \mathbb{C}^{(d^2+d)\times 2d}$. $\mathbb{M}$ and $\mathcal{K}$ are given by explicit formulae in terms of the first and second derivatives of W. Moreover, $\mathbb{M}, \hat{R}^s, \hat{R}^u$ and $\mathcal{K}$ depend continuously on (U, s, N) in their domains of definition, which is given by

$$s^2 < \min \{\text{eigenvalues of } \mathcal{N}(U, N)\}$$

(including $s = 0$).

Corollary 1. *If W, g, U^A, and U^B satisfy the hypotheses of Theorem 1 for the equilibrium configuration with $s = 0$, then the dynamic stability of the phase boundary (7) is uniformly controlled by the static-case Lopatinski function*

$$\hat{\Delta}(U^A, U^B) : \Gamma_{N^*} \to \mathbb{C},$$

in the sense detailed in Theorem 1.

Remark 1. Theorem 1 states that it suffices to perform the normal modes analysis on a sub-bundle $\mathbb{G}$ of amplitudes. This simplifies the analysis greatly. In particular, $\mathbb{G}$ is also regular in the characteristic limit $s = 0$, including static configurations into the analysis. This last feature is highlighted in Corollary 1 because of its importance. The theorem and the corollary offer a contribution to the problem of modeling phase-boundary dynamics in real materials, in the sense that any kinetic rule which does not satisfy such a multidimensional stability condition can hardly be accepted as a mathematical description of stably moving boundaries.

2 The Normal Modes Analysis

Let us specify the notation used in [FP05]. U_j will denote the jth column of U, and, accordingly, the stress tensor has columns $\sigma(U)_j = W_{U_j}$. $\{e_j\}$ denotes the canonical basis of $\mathbb{R}^d$. We gather the second derivatives of W into the following $\mathbb{R}^{d\times d}$ matrices $B_i^j(U) := \partial\sigma_j/\partial U_i$, whose (l,k)-component is $\partial^2 W/\partial U_{lj}\partial U_{ki}$. Clearly, B_i^i is symmetric, and $(B_j^i)^\top = B_i^j$. Then, the fluxes and Jacobians of system (1) are expressed as,

$$f_j(U,V) := -\begin{pmatrix} 0 \\ \vdots \\ V \\ \vdots \\ 0 \\ \sigma(U)_j \end{pmatrix}, \qquad A_j(U) = Df_j(U) = -\begin{pmatrix} & & & 0 \\ & & & \vdots \\ & 0 & & I \\ & & & \vdots \\ & & & 0 \\ B_1^j(U) & \cdots & B_d^j(U) & 0 \end{pmatrix}.$$

Notice the high multiplicity characteristic zero mode. Under hyperbolicity and symmetrizability with constant multiplicity, for any $N \in \mathbb{R}^d \setminus \{0\}$, the characteristic speeds of $\sum A_j(U)N_j$ are: $a_0(N,U) = 0$ with constant algebraic multiplicity $\alpha_0 = d^2 - d$, and $a_j^\pm(N,U) = \pm\sqrt{\kappa_j(N,U)}$, $j = 1,\ldots,m$, where κ_j are the m distinct semi-simple eigenvalues of $\mathcal{N}$, $m \le d$, with constant multiplicities α_j, satisfying $\sum \alpha_j = d$. Hence, the planar front corresponds to an undercompressive shock with degree of undercompressivity $l = 1$ (see [F95]).

2.1 The Space $\mathbb{G}$

Without loss of generality assume $N = e_1$. Suppose W is hyperbolic at U, s subsonic with respect to (e_1, U). The squares of the characteristic speeds are therefore the positive eigenvalues of $B_1^1(U)$. The set of time–space frequencies is thus $\Gamma = \{(\lambda,\tilde{\xi}) \in \mathbb{C}\times\mathbb{R}^{d-1} : \operatorname{Re}\lambda \ge 0, |\lambda|^2 + |\tilde{\xi}|^2 = 1\}$. We are interested in the matrix field

$$\mathcal{A}(U,s,\lambda,\tilde{\xi}) = C(s)^{-1}(\lambda I + i\sum_{j\neq 1}\xi_j A_j(U))(A_1(U) - sI)^{-1}C(s),$$

where

$$C(s) := \begin{pmatrix} I_d & & \\ & sI_{d^2-d} & \\ & & I_d \end{pmatrix}.$$

Denote $\hat{B}(s) := (s^2 - B_1^1)^{-1}$ (its invertibility is guaranteed by subsonicity, and it is regular at $s = 0$). It is easy to verify that the matrix fields $C(s)^{-1}(A_1 - sI)$ and $(A_1 - sI)^{-1}C(s)$ are analytic for all subsonic s, including $s = 0$ [FP05]. Define on Γ the $2d$-dimensional bundle

$$\mathbb{G}(\lambda,\tilde{\xi}) := \left\{(\lambda Y, i\xi_2 Y, \ldots, i\xi_d Y, Z)^\top \;:\; Y, Z \in \mathbb{C}^d\right\}. \tag{10}$$

Lemma 1 ([FP05]). *For $(\lambda, \tilde{\xi}) \in \Gamma$ and s subsonic, $\mathbb{G}$ is an invariant subspace for $\mathcal{A}$, and the $2d \times 2d$ matrix field $\mathbb{M}(U, s, \lambda, \tilde{\xi}) = \begin{pmatrix} M_1^1 & M_1^2 \\ M_2^1 & M_2^2 \end{pmatrix}$ that expresses the action of $\mathcal{A}$ on $\mathbb{G}$ has the $d \times d$-block components*

$$M_1^1 := -\hat{B}(\lambda s I + i \sum_{j\neq 1} \xi_j B_j^1), \qquad M_1^2 := \hat{B},$$

$$M_2^1 := (\lambda s I + i \sum_{j\neq 1} \xi_j B_1^j)\hat{B}(\lambda s I + i \sum_{j\neq 1} \xi_j B_j^1) - \lambda^2 I - \sum_{i,j\neq 1} \xi_i \xi_j B_j^i,$$

$$M_2^2 := -(\lambda s I + i \sum_{j\neq 1} \xi_j B_1^j)\hat{B},$$

being $\mathbb{M}$ well defined and smooth for all subsonic s including 0. Moreover, the eigenvalues $-i\mu$ of $\mathbb{M}(U, s, \lambda, \tilde{\xi})$ satisfy $\det(\mathcal{N}(U, \mu, \tilde{\xi}) + (i\mu s - \lambda)^2 I) = 0$, with the property that for $\operatorname{Re}\lambda > 0$, d of these eigenvalues (counting multiplicities) have $\operatorname{Im}\mu > 0$, while the remaining d have $\operatorname{Im}\mu < 0$. Finally, $(Y, Z)^\top \in \mathbb{C}^{2d}$ is an eigenvector of $\mathbb{M}$ if and only if

$$Y \in \ker(\mathcal{N}(U, \mu, \tilde{\xi}) + (i\mu s - \lambda)^2 I), \ Y \neq 0 \ \text{and}$$
$$Z = \Big(s(\lambda - i\mu s)I + i\mu B_1^1(U) + i \sum_{j\neq 1} \xi_j B_j^1(U)\Big)Y.$$

Remark 2. The restriction $\mathbb{M} = \mathcal{A}_{|\mathbb{G}}$ has a unique analytic extension to $s = 0$, even though $\mathcal{A}$ is not defined there. As a result, we only investigate the modes of $\mathcal{A}(U, s, \cdot, \cdot)$ whose amplitudes lie in $\mathbb{G}$. Another feature of $\mathbb{M}$ is the following.

Lemma 2 ([FP05]). *The matrix $\mathbb{M}$ satisfies the block structure assumption of Majda [Ma83].*

Remark 3. The last lemma follows from adapting the results of Métivier [Me00] to our setting. In [Cor93], Corli showed that $\mathcal{A}$ satisfies the block structure condition for noncharacteristic speeds s, i. e. excluding $s = 0$. (This is also a direct consequence of a later general theorem in [Me00].) Note that $\mathcal{A}$ has a singular eigenvalue $\beta_* = -\lambda/s$, and that the corresponding mode is exactly what is avoided by our restriction to $\mathbb{G}$.

The significance of Lemma 2 is that the restriction $\mathcal{A}_{|\mathbb{G}}$ satisfies the block structure also in the characteristic limit $s = 0$, allowing the construction of Kreiss' symmetrizers [K71, Me90, Co03] for the static case.

Therefore, the stable/unstable bundles of $\mathbb{M}$, namely $\hat{R}^{s,u}(\lambda, \xi)$, have dimension $2d$ each, and have continuous representations in all Γ (including $\operatorname{Re}\lambda = 0$), and for all subsonic $s \geq 0$, as required in Theorem 1. The key point regarding the case $s = 0$ now is the fact that thanks to the curl-free constraint (2), the Fourier analysis can be performed on the $2d$-dimensional bundle $\mathbb{G}$.

Indeed, if $(U, V)(x, t) = (\hat{U}(x_1 - st), \hat{V}(x_1 - st)) \exp(i\tilde{\xi} \cdot \tilde{x} + \lambda t)$ are normal modes solutions to (1) and (2) (where $x = (x_1, \tilde{x})$, $\tilde{x} \in \mathbb{R}^{d-1}$ and $(\lambda, \tilde{\xi}) \in \Gamma$), then necessarily, $C(s)^{-1}(A_1 - sI)(\hat{U}(\cdot), \hat{V}(\cdot))^\top \in \mathbb{G}$ (see [FP05]). One checks that (9) represents (8) in the restriction to $\mathbb{G}$. Then Theorem 1 and Corollary 1 follow from existing nonlinear theory [Co03, Me90, Ma83]. See [FP05] for details.

3 An Example: Martensite Twins

As an illustration, consider a crystal in two dimensions which is described by an energy function of form

$$W(U) = \tfrac{1}{8}(\beta_1 - (1+\delta^2))^2 + (\beta_2 - 1)^2 + \gamma(\beta_3^2 - \delta^2)^2, \tag{11}$$

with $\beta_1 := |U_1|^2$, $\beta_2 := |U_2|^2$, $\beta_3 := U_1^\top U_2$, with $\gamma > 0$, $\delta \neq 0$. W is rank-one convex at the two rank-one connected ("martensitic") wells

$$U^A = I - \delta e_2 \otimes e_1, \quad U^B = I + \delta e_2 \otimes e_1.$$

(Family (11) is also used to model orthorhombic to monoclinic phase transformations [Lu96]. The interest here is in martensite–martensite phase boundaries, however.) Each such W satisfies all previous hypotheses, plus "material frame-indifference". In particular (11) has constant multiplicity near U^A and U^B in an open nonempty set of the parameters (γ, δ) (see [FP06]). The choice of the kinetic relation is crucial. Motivated by the classical Hugoniot rule of fluid dynamics (see for example [W49] and, for use in connection with phase boundaries [B98, B99]), we consider a *generalized Hugoniot rule*

$$g = [W(U)] - N^\top [U]^\top \langle \sigma(U) \rangle N.$$

This rule expresses conservation of energy across the front. Here $\langle \sigma \rangle = \frac{1}{2}(\sigma^+ + \sigma^-)$. Also, we are interested in perturbations of the form $g \to g + \tilde{g}$, where $\tilde{g} \in C^1$ satisfies, (a) $\tilde{g} = 0$ for $s = 0$; (b) $\tilde{g} > 0$ for $s < 0$, $\tilde{g} < 0$ for $s > 0$; and (c) $d_s \tilde{g} < 0$. The family of perturbations is thus compatible with energy considerations. As a paradigmatic (though artificial) example, we reckon $\tilde{g} = -\epsilon s$ with $\epsilon > 0$, as artificial energy dissipation at rate ϵ.

Using Lemma 1 to represent the stable/unstable bundles associated to the static phase boundary, one can numerically compute the Lopatinski determinant. As a first step in the stability analysis, we look at the mapping $\lambda \mapsto \hat{\Delta}(\lambda, \pm 1)$ for an appropriately normalized version of $\hat{\Delta}$ of Corollary 1, and λ along the imaginary axis. Figure 1 shows the computed values of $\hat{\Delta}$ for $(\gamma, \delta) = (1, 1)$ and $\lambda = i\tau$, $\tau \in [-2, 2]$. The graph on the right shows the values of $\hat{\Delta}(i\tau, +1)$, with dots for $\tau > 0$, and with circles for $\tau < 0$. It corresponds to conservation of energy as kinetic rule, depicting two zeroes along the imaginary axis (weak stability). The graph on the right corresponds to a perturbation of conservation of energy under $\tilde{g} = -\epsilon s$ with $\epsilon = 0.75$, and the zeroes are left out of the contour, suggesting strong stability. These observations (and the concurrent treatment in [FP06]) seem to indicate that in

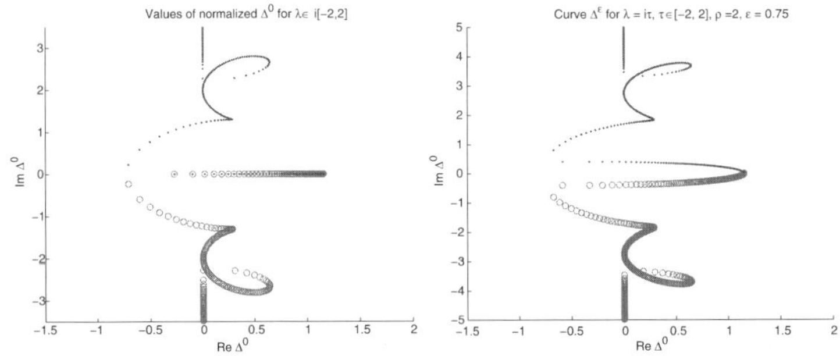

Fig. 1. Computed values of a normalized $\hat{\Delta}(i\tau, +1)$ for $\tau \in [-2, 2]$ under generalized Hugoniot kinetics or conservation of energy (*left*), and corresponding values of $\hat{\Delta}$ under small perturbations of the kinetic rule (*right*)

the case of the generalized Hugoniot rule, the static boundary is dynamically weakly stable, while it is strongly stable in the case of the above-mentioned perturbations of the generalized Hugoniot rule. This is similar to the picture for two-phase fluids [B98, B99]. Details will be provided in [FP06].

References

[B98] Benzoni-Gavage, S.: Stability of multi-dimensional phase transitions in a van der Waals fluid. Nonlinear Anal. **31**, no. 1-2, 243–263 (1998)

[B99] Benzoni-Gavage, S.: Stability of subsonic planar phase boundaries in a van der Waals fluid. Arch. Rational Mech. Anal. **150**, 23–55 (1999)

[Ci88] Ciarlet, P.G.: Mathematical elasticity. Vol. I: Three-dimensional elasticity. Vol. 20 of Studies in Mathematics and its Applications, North-Holland Publishing Co., Amsterdam (1988)

[Cor93] Corli, A.: Weak shock waves for second-order multi-dimensional systems. Boll. Un. Mat. Ital. B (7) **7**, no. 3, 493–510 (1993)

[Co03] Coulombel, J.-F.: Stability of multidimensional undercompressive shock waves. Interfaces Free Bound. **5**, no. 4, 360–390 (2003)

[F95] Freistühler, H.: A short note on the persistence of ideal shock waves. Arch. Math. (Basel) **64**, no. 4, 344–352 (1995)

[F98] Freistühler, H.: Some results on the stability of non-classical shock waves. J. Partial Diff. Eqs. **11**, no. 1, 25–38 (1998)

[FP05] Freistühler, H., Plaza, R.G.: Normal modes and nonlinear stability behaviour of dynamic phase boundaries in elastic materials. Arch. Rational Mech. Anal., to appear.

[FP06] Freistühler, H., Plaza, R.G.: In preparation.

[K71] Kreiss, H.O.: Initial boundary value problems for hyperbolic systems. Comm. Pure Appl. Math. **23**, 277–298 (1970)

[Lu96] Luskin, M.: On the computation of crystalline microstructure. In: Acta numerica, 1996, vol. 5 of Acta Numer., Cambridge Univ. Press, Cambridge (1996)

[Ma83] Majda, A.: The stability of multi-dimensional shock fronts. Mem. Amer. Math. Soc. **41**, no. 275 (1983)

[Me90] Métivier, G.: Stability of multidimensional weak shocks. Comm. Partial Differential Equations **15**, no. 7, 983–1028 (1990)

[Me00] Métivier, G.: The block structure condition for symmetric hyperbolic systems. Bull. London Math. Soc. **32**, no. 6, 689–702 (2000)

[Me01] Métivier, G.: Stability of multidimensional shocks. In Freistühler, H. and Szepessy, A. (eds.) Advances in the Theory of Shock Waves, vol. 47 of Progress in Nonlinear Differential Equations and Their Applications, Birkhäuser, Boston (2001)

[Mu99] Müller, S.: Variational models for microstructure and phase transitions. In Calculus of variations and geometric evolution problems (Cetraro, 1996), vol. 1713 of Lecture Notes in Math., Springer, Berlin (1999)

[W49] Weyl, H.: Shock waves in arbitrary fluids. Comm. Pure Appl. Math. **2**, 103–122 (1949)

Large Time Step Positivity-Preserving Method for Multiphase Flows

F. Coquel, Q.-L. Nguyen, M. Postel, and Q.-H. Tran

Summary. Using a relaxation strategy in a Lagrangian–Eulerian formulation, we propose a scheme in local conservation form for approximating weak solutions of complex compressible flows involving wave speeds of different orders of magnitude. Explicit time integration is performed on slow transport waves for the sake of accuracy while fast acoustic waves are dealt with implicitly to enable large time stepping. A CFL condition based on the slow waves is derived ensuring positivity properties on the density and the mass fraction. Numerical benchmarks validate the method.

1 Statement of the Problem

The present work treats the numerical approximation of the discontinuous solutions of the following PDE system

$$\left|\begin{array}{ll} \partial_t(\rho) \quad + \partial_x(\rho u) & = 0, \\ \partial_t(\rho Y) + \partial_x(\rho Y u) & = 0, \\ \partial_t(\rho u) \ + \partial_x(\rho u^2 + P) & = 0. \end{array}\right. \tag{1}$$

Here and with classical notations, ρ stands for the total density of a compressible material, u its velocity, and Y a gas mass fraction. The pressure P is a given nonlinear function of the unknown $\mathbf{u} = (\rho, \rho Y, \rho u)$, in the form $P(\tau, Y)$ with $\tau = 1/\rho$, which in the present setting (oil production) turns to be highly nonlinear. For the closure laws $P(\mathbf{u})$ to be dealt with, the system of conservation laws (1) is typically hyperbolic over

$$\Omega = \Big\{\mathbf{u} \in \mathbb{R}^3; \rho > 0, 0 \leq Y \leq 1, u \in \mathbb{R}\Big\}. \tag{2}$$

The PDE system (1) governs solutions made up of distinct waves propagating with the next three distinct eigenvalues: $u - c(\mathbf{u}) < u < u + c(\mathbf{u})$, where the definition of the sound speed $c(\mathbf{u})$ follows from the prescribed pressure law $P(\mathbf{u})$. The closure laws under concern yields sound speed $c(\mathbf{u})$ with

$c(\mathbf{u}) >> |u|$ in the setting of solutions $\mathbf{u}$ we address. Such a smallness of the Mach number renders particularly stiff the accurate time resolution of the slow wave phenomena we are actually primarily interested in. Here, stiffness in the time approximation means difficulties in preserving the hyperbolic domain Ω in (2) when invoking large time steps, *i.e.*, comparable with the slow wave characteristic times.

The severe nonlinearities involved in the pressure law $P(\mathbf{u})$ makes such an issue already challenging within the frame of purely time explicit methods, *i.e.*, here subject to (very) small time steps. To conveniently deal with these nonlinearities, we suggest, motivated by [2, 3, 1], to approximate the solutions of the original system (1) by those of the following relaxation system in the regime of a large relaxation parameter $\lambda > 0$

$$\left| \begin{array}{ll} \partial_t(\rho)^\lambda & + \partial_x(\rho u)^\lambda = 0, \\ \partial_t(\rho Y)^\lambda & + \partial_x(\rho Y u)^\lambda = 0, \\ \partial_t(\rho u)^\lambda & + \partial_x(\rho u^2 + \Pi)^\lambda = 0, \\ \partial_t(\rho \Pi)^\lambda & + \partial_x(\rho \Pi u + a^2 u)^\lambda = \lambda \rho^\lambda [P(\tau, Y)^\lambda - \Pi^\lambda]. \end{array} \right. \tag{3}$$

Indeed, recasting the last equation as

$$\Pi^\lambda = P(\tau^\lambda, Y^\lambda) - \frac{1}{\lambda \rho^\lambda} \left\{ \partial_t(\rho \Pi)^\lambda + \partial_x(\rho \Pi u + a^2 u)^\lambda \right\} \tag{4}$$

yields the following formal limit $\lim_{\lambda \to \infty} \Pi^\lambda = P(\rho, Y)$ so that the solutions of (3) may serve to approximate those of (1). But to prevent such an approximation procedure from instabilities in the limit $\lambda \to \infty$, the coefficient $a > 0$ entering (4) must be chosen so as to meet the following subcharacteristic condition:

$$a^2 > \rho^2 c^2(\mathbf{u}) = -\partial_\tau P(\tau, Y), \tag{5}$$

for all the $\mathbf{u}$ under consideration. The Jacobian matrix of (3) is diagonalizable with eigenvalues $u - a/\rho$, u (double) and $u + a/\rho$, all the associated fields being linearly degenerate. Due to (5), note that approximating the solutions $\mathbf{u}$ of the original system (1) with $c(\mathbf{u}) \gg |u|$ via those of the relaxation model (3) yields $a/\rho \gg |u|$.

The key issue stays in the derivation of a large time step conservative method for approximating the weak solutions of (1) with the mandatory positivity preserving properties (see (2)): $\rho > 0$ and $Y \in [0, 1]$. We show in this chapter that these difficulties can be jointly solved when invoking a Lagrangian–Eulerian formulation of the relaxation version (3) of the original system (1).

To that purpose, let us equivalently rewrite (3) in a new general referential configuration, in which the coordinates are denoted by χ. This configuration is neither the material (Lagrangian) configuration X nor the Eulerian configuration x but moves at an imposed speed s with respect to the latter. The velocity in this moving frame is thus $u - s$ so that if $x = x(\chi, t)$ denotes the correspondence between the moving frame and the Eulerian frame, we have $\partial_t x_{|\chi} = u - s$. Let $J = \partial_\chi x$ be the dilatation rate. Then, it is a classical matter to show that $\partial_t J = J\partial_x(u - s)$ so as to infer that the system (3) recasts in the new conservation form:

$$\left|\begin{array}{llll} \partial_t(J) & + \partial_\chi(s) & - \partial_\chi(u) & = 0, \\ \partial_t(\rho J) & + \partial_\chi(\rho s) & & = 0, \\ \partial_t(\rho Y J) & + \partial_\chi(\rho Y s) & & = 0, \\ \partial_t(\rho u J) & + \partial_\chi(\rho u s) & + \partial_\chi(\Pi) & = 0, \\ \partial_t(\rho \Pi J) & + \underbrace{\partial_\chi(\rho \Pi s)}_{\text{projection}} & + \underbrace{\partial_\chi(a^2 u)}_{\text{Lagrange}} & = \lambda\rho[P(\rho, Y) - \Pi]. \end{array}\right. \tag{6}$$

Here and from now on, the superscript λ is omitted for simplicity but let us keep in mind that the solutions of (6) are addressed in the limit $\lambda \to \infty$. The interest in the proposed reformulation stems from the property that the Lagrangian part of these equations essentially deals with the fast waves while the remaining part is uniquely concerned with the slow wave when s is chosen close to the velocity u. This suggests a natural conservative splitting strategy we now describe.

2 Numerical Scheme

The interval $[0, L]$ is discretized in N cells $\Omega_i = [x_{i-1/2}, x_{i+1/2}]$ of size Δx_i with $\sum_{i=0}^{N-1} \Delta x_i = L$. Approximate solutions of (1) are classically sought for under the form of piecewise constant functions $\mathbf{u}_h(x, t)$, $h = \max_i(\Delta x_i)$. Assume that an approximate solution $\mathbf{u}_h(x, t^n)$ is known at time t^n, then this one is advanced to the next time level $t^{n+1} = t^n + \Delta t$ in two steps. In the present work and for the sake of simplicity, we use Neumann boundary conditions.

2.1 Lagrange Step $t^n \to t^{n+1/2}$: Implicit Treatment of the Fast Waves

Let us build from $\mathbf{u}_h(x, t^n)$ the function $\mathbf{v}_h(x, t^n) = (\mathbf{u}_h, (\rho\Pi))(x, t^n)$ when defining the relaxation pressure at equilibrium: $\Pi(x, t^n) = P(\mathbf{u}_h(x, t^n))$. This piecewise constant function serves as an initial data to the following system

$$\left|\begin{array}{lll} m\partial_t \tau & - \partial_\chi(v) & = 0, \\ m\partial_t Y & & = 0, \\ m\partial_t v & + \partial_\chi(\Pi) & = 0, \\ m\partial_t \Pi & + \partial_\chi(a^2 v) & = \lambda\rho[P(\tau, Y) - \Pi], \end{array}\right. \tag{7}$$

deduced from the Lagrangian part of (6) with $m(x) = \rho J$ since $\partial_t \rho J = 0$ in this step. Since initially $J(x,t^n) = 1$ we get $m(x) = \rho_h(x,t^n)$. For $\lambda > 0$ fixed, the solution $\mathbf{v}^\lambda(x, t^{n+1/2})$ of the above Cauchy problem is approximated when first addressing the following time implicit discrete form of (7)

$$\left| \begin{aligned} &\rho_i^n \frac{\tau_i^{n+1/2} - \tau_i^n}{\Delta t} - \frac{\widetilde{u}_{i+1/2}^{n+1/2} - \widetilde{u}_{i-1/2}^{n+1/2}}{\Delta x_i} = 0, \\ &\rho_i^n \frac{Y_i^{n+1/2} - Y_i^n}{\Delta t} = 0, \\ &\rho_i^n \frac{u_i^{n+1/2} - u_i^n}{\Delta t} + \frac{\widetilde{\Pi}_{i+1/2}^{n+1/2} - \widetilde{\Pi}_{i-1/2}^{n+1/2}}{\Delta x_i} = 0, \\ &\rho_i^n \frac{\Pi_i^{n+1/2} - \Pi_i^n}{\Delta t} + a^2 \frac{\widetilde{u}_{i+1/2}^{n+1/2} - \widetilde{u}_{i-1/2}^{n+1/2}}{\Delta x_i} = \lambda \rho_i^n \left[P(\tau_i^{n+1/2}, Y_i^n) - \Pi_i^{n+1/2} \right], \end{aligned} \right. \tag{8}$$

where the numerical flux functions $\widetilde{u}_{i+1/2}^{n+1/2}$ and $\widetilde{\Pi}_{i+1/2}^{n+1/2}$ denote implicit version of the Godunov fluxes for (7)

$$\left| \begin{aligned} \widetilde{\Pi}_{i+1/2}^{n+1/2} &= \frac{\Pi_i^{n+1/2} + \Pi_{i+1}^{n+1/2}}{2} - a \frac{u_{i+1}^{n+1/2} - u_i^{n+1/2}}{2}, \\ \widetilde{u}_{i+1/2}^{n+1/2} &= \frac{u_i^{n+1/2} + u_{i+1}^{n+1/2}}{2} - \frac{\Pi_{i+1}^{n+1/2} - \Pi_i^{n+1/2}}{2a}. \end{aligned} \right. \tag{9}$$

To handle safely the family of solutions $\left\{ \mathbf{v}_h^\lambda(x, t^{n+1/2}) \right\}_{\lambda > 0}$ of (7) in the limit $\lambda \to \infty$, we choose a at each time t^n according to the stability condition (5)

$$a^2 = \max_i (-\partial_\tau P(\tau_i^n, Y_i^n)). \tag{10}$$

Then formally passing to the limit $\lambda \to \infty$ in the discrete analog of (5)

$$\Pi_i^{n+1/2} = P(\tau_i^{n+1/2}, Y_i^n) - \frac{1}{\lambda \rho_i^n} \left\{ \frac{\Pi_i^{n+1/2} - \Pi_i^n}{\Delta t} + a^2 \frac{\widetilde{u}_{i+1/2}^{n+1/2} - \widetilde{u}_{i-1/2}^{n+1/2}}{\Delta x_i} \right\} \tag{11}$$

yields the guess $\Pi_i^{n+1/2} = P(\tau_i^{n+1/2}, Y_i^n)$ which we Taylor expand in time to propose $\Pi_i^{n+1/2} = P(\tau_i^n, Y_i^n) + \partial_\tau P(\tau_i^n, Y_i^n)(\tau_i^{n+1/2} - \tau_i^n)$ in order to lower the overall computational cost. We eventually define at time $t^{n+1/2}$ the approximate relaxation pressure by

$$\Pi_i^{n+1/2} = P(\tau_i^n, Y_i^n) - a^2(\tau_i^{n+1/2} - \tau_i^n), \tag{12}$$

when invoking the upper-bound (10) so as to further simplify the computations. Indeed with the notation $\delta X_i = X_i^{n+1/2} - X_i^n$, the unknowns $\{\tau_i^{n+1/2}, u_i^{n+1/2}\}_i$ can be obtained thanks to (12) by solving the following linear problems in the unknowns $\delta C_i^+ = (\delta u_i + a\delta\tau_i)/2$ and $\delta C_i^- = (\delta u_i - a\delta\tau_i)/2$

$$\begin{cases} \begin{pmatrix} 0 & 1+\dfrac{a\Delta t}{\rho_i^n \Delta x_i} & \dfrac{-a\Delta t}{\rho_i^n \Delta x_i} \end{pmatrix} \begin{pmatrix} \delta C_{i-1}^+ \\ \delta C_i^+ \\ \delta C_{i+1}^+ \end{pmatrix} = \dfrac{\Delta t}{2\rho_i^n \Delta x_i} (R_i^n(u) + aR_i^n(\tau)) \\ \begin{pmatrix} \dfrac{-a\Delta t}{\rho_i^n \Delta x_i} & 1+\dfrac{a\Delta t}{\rho_i^n \Delta x_i} & 0 \end{pmatrix} \begin{pmatrix} \delta C_{i-1}^- \\ \delta C_i^- \\ \delta C_{i+1}^- \end{pmatrix} = \dfrac{\Delta t}{2\rho_i^n \Delta x_i} (R_i^n(u) - aR_i^n(\tau)) \end{cases} \tag{13}$$

with

$$R_i^n(\tau) = \widetilde{u}_{i+1/2}^n - \widetilde{u}_{i-1/2}^n \qquad \text{and} \qquad R_i^n(u) = -\widetilde{\Pi}_{i+1/2}^n + \widetilde{\Pi}_{i-1/2}^n, \tag{14}$$

where $\widetilde{u}_{i+1/2}^n$ and $\widetilde{\Pi}_{i+1/2}^n$ are computed from the analog of (9) at time t^n.

2.2 Projection Step $t^{n+1/2} \to t^{n+1}$: Explicit Treatment of the Slow Wave

Let us consider the following Cauchy problem with initial data $\mathbf{v}_h(x, t^{n+1/2})$

$$\left| \begin{array}{l} \partial_t(J) \;\; + \partial_\chi(s) \;\; = 0, \\ \partial_t(\mathbf{v}J) + \partial_\chi(\mathbf{v}s) = 0. \end{array} \right. \tag{15}$$

Note that, by combining equations of (15), we come up with a system of advection equations

$$\partial_t \mathbf{v} + \frac{u}{J}\partial_\chi \mathbf{v} = 0, \tag{16}$$

where to avoid moving the mesh, we have specified $s = u$. We then propose the following time explicit discrete form of (16)

$$\begin{aligned} \mathbf{v}_i^{n+1} &= \tfrac{\Delta t}{\Delta x_i}(\widetilde{u}_{i-1/2}^{n+1/2})^+ \mathbf{v}_{i-1}^{n+1/2} + [1 - \tfrac{\Delta t}{\Delta x_i}((\widetilde{u}_{i-1/2}^{n+1/2})^+ - (\widetilde{u}_{i+1/2}^{n+1/2})^-)]\mathbf{v}_i^{n+1/2} \\ &- \tfrac{\Delta t}{\Delta x_i}(\widetilde{u}_{i+1/2}^{n+1/2})^- \mathbf{v}_{i+1}^{n+1/2} \end{aligned} \tag{17}$$

which can be actually inferred from a conservative and consistent discretization of the system (15).

To conclude the method, we define the required update $\mathbf{u}_i^{n+1}$ from the first three components of $\mathbf{v}_i^{n+1} = (\mathbf{u}_i^{n+1}, (\rho\Pi)_i^{n+1})$. The last component $(\rho\Pi)_i^{n+1}$ is set to equilibrium using the pressure law $P(\mathbf{u}_i^{n+1})$. The proposed time implicit–explicit method actually falls within the frame of finite volume methods in local conservation form in view of

Lemma 1. *Let us define $\mathbf{u}_i^{n+1/2}$ from $\mathbf{v}_i^{n+1/2} = (\mathbf{u}_i^{n+1/2}, (\rho\Pi)_i^{n+1/2})$ in the first step so as to introduce the following numerical flux function*

$$F_{i+1/2}^{n+1/2} = (\widetilde{u}_{i+1/2}^{n+1/2})^+ \mathbf{u}_i^{n+1/2} + (\widetilde{u}_{i+1/2}^{n+1/2})^- \mathbf{u}_{i+1}^{n+1/2} + \left(0, 0, \widetilde{\Pi}_{i+1/2}^{n+1/2}\right)^T. \tag{18}$$

Then

$$\mathbf{u}_i^{n+1} = \mathbf{u}_i^n - \frac{\Delta t}{\Delta x_i}\left(F_{i+1/2}^{n+1/2} - F_{i-1/2}^{n+1/2}\right), \quad i \in \{1, \ldots, N\}, n \geq 0. \tag{19}$$

2.3 Definition of a CFL-Like Condition

The time step Δt is the same one for the two steps and must be selected beforehand in order to preserve the phase space Ω in (2) at the discrete level. A natural CFL condition arises when imposing that the time explicit formula (17) in the second step is a convex combination. The next statement shows that this condition is explicitly tractable thanks to the simple algebraic structure of the linear problems (13) in the first step

Proposition 1. *Let us define at time* t^n

$$m_i^{n,-} = \max_{j\leq i} \frac{\left|R_j^n(u) - aR_j^n(\tau)\right|}{2\rho_j^n \Delta x_j}, \qquad m_i^{n,+} = \max_{j\geq i} \frac{\left|R_j^n(u) + aR_j^n(\tau)\right|}{2\rho_j^n \Delta x_j}, \tag{20}$$

so as to introduce the following CFL-like condition

$$\Delta t < \min_i \frac{\min(\Delta x_{i-1}, \Delta x_i)}{\left\{\left|\widetilde{u}_{i-1/2}^n\right| + \sqrt{\left|\widetilde{u}_{i-1/2}^n\right|^2 + 2(m_i^{n,+} + m_{i-1}^{n,-}) \min(\Delta x_{i-1}, \Delta x_i)}\right\}}. \tag{21}$$

If $\rho_i^n > 0$ *and* $Y_i^n \in [0,1]$ *for all* $i \in 1, \ldots, N$ *then the time implicit–explicit method (19) with a time step* Δt^n *prescribed under the condition (21) obeys* $\rho_i^{n+1} > 0$ *and* $Y_i^{n+1} \in [0,1]$ *for all* $i \in \{1, \ldots, N\}$.

It is not obvious from the above expression that the time step limitation is indeed ruled by the slow wave velocity, since the relaxation coefficient a comes into play in (20). We will actually show in the numerical experiments that the time step is of the same order as the one dictated by the standard CFL condition on the slow transport wave.

3 Numerical Simulation

The benchmark compares the proposed scheme with both a semi-implicit time method and time explicit method based on the relaxation system in Eulerian coordinates (see [3] for the details). The simulation consists in a Riemann problem set in a 8 km long pipeline. At the initial time the density of the mixture is $400\,\mathrm{kg\,m^{-3}}$ until $x = 4\,\mathrm{km}$ and $500\,\mathrm{kg\,m^{-3}}$ beyond. The gas mass fraction is, respectively, 0.4 and 0.2 and the speed is, respectively, $-10\,\mathrm{m\,s^{-1}}$ and $+10\,\mathrm{m\,s^{-1}}$. The pressure law is $P(\rho, Y) = a_g^2 \frac{\rho_\ell \rho Y}{\rho_\ell - \rho(1-Y)}$, with $\rho_\ell = 10^3\,\mathrm{kg\,m^{-3}}$ and $a_g = 316\,\mathrm{m\,s^{-1}}$.

The transport wave moves slowly toward the right at a speed $20\,\mathrm{m\,s^{-1}}$ given by Rankine–Hugoniot while two acoustic waves are visible on the density and speed components moving in opposite directions at roughly $\pm 254\,\mathrm{m\,s^{-1}}$. The Fig. 1 displays the density and velocity fields at times $t = 0\,\mathrm{s}$ and $t = 6\,\mathrm{s}$, obtained with a uniform discretization of 1,000 cells using the relaxation schemes described in [3] and the Lagrange projection algorithm described

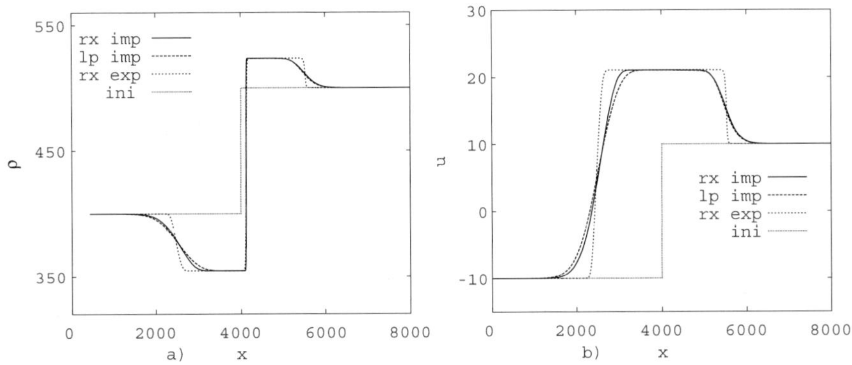

Fig. 1. Comparison of Lagrange projection and Euler relaxation schemes on a Riemann problem. Solution at time t = 6 s. (**a**) Density (**b**) velocity

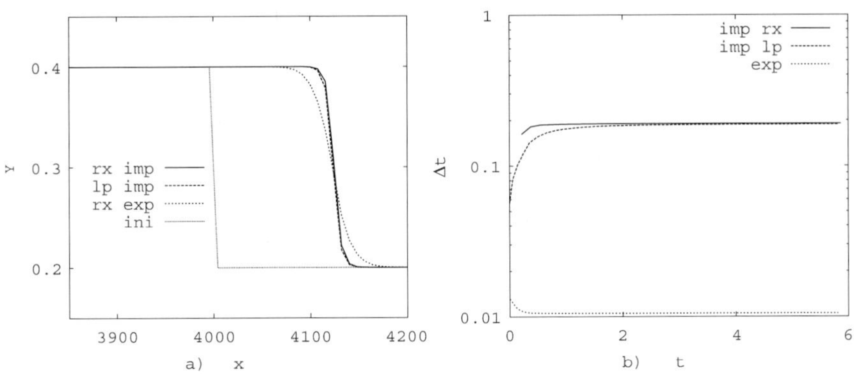

Fig. 2. (**a**) Gas mass fraction at t = 6 s. (**b**) Time step evolution using different schemes

above. The Fig. 2 displays the gas mass fraction at times $t = 0$ s and $t = 6$ s and the evolution of the time step during the simulation for the different schemes. All computations are performed with a second-order enhancement in space – using limited reconstructions on variables ρ, Y, and u in the Lagrange step, or Eulerian relaxation scheme, and limited reconstruction on the conservative variables during the Euler projection step. Second order in time is achieved using Heun quadrature in the explicit case.

Of course the dissipation on the acoustic part of the solution is much larger in the solutions obtained with implicit schemes. It is also slightly stronger using the Lagrange projection method rather than the standard relaxation one, because it uses a global relaxation coefficient a which amounts to strengthen the diffusive kernel. Beside this discrepancy the solutions obtained with the implicit schemes are not distinguishable on the transport profile (Fig. 2a), even though for this component Y the graph is zoomed in the region of interest. We notice however that the explicit solution has more dispersion on this component because the time step used for the explicit scheme is about 10 times smaller than for the implicit ones, as displayed on Fig. 2b). The time step evolution for the two implicit schemes are very similar. The time step for the semi-implicit relaxation scheme is bounded by a CFL condition on the slow wave speed, while the Lagrange projection scheme uses the stability condition (21). From the computational performance point of view, the Lagrange projection method is twice as fast as the Euler relaxation one, because in its case the implicit part of the scheme consists in solving two bidiagonal linear systems instead of a 4×4 block tridiagonal system in the Euler case.

In this short chapter we have presented numerical simulations on fixed grids although the method is already implemented within the framework of the multiresolution as developed in [3, 1]. The time step used in this approach is constant throughout the domain and dictated by the size of the smallest cell in the adaptive grid which enters into the CFL stability condition (21). The gain in computing time is roughly a factor of 10 compared with the uniform grid computation. An important advantage of the Lagrange projection scheme is that it is locally conservative and can therefore be used in the framework of a local time stepping multiresolution method such as the one proposed in [4]. This work is currently under progress.

Acknowledgments

This work was supported by the Ministère de la Recherche under grant ERT-20052274: *Simulation avancée du transport des hydrocarbures* and by the Institut Français du Pétrole. We are also grateful to L. Linise for the numerical simulations during her master degree internship at IFP.

References

1. N. Andrianov, F. Coquel, M. Postel, and Q.H. Tran. A relaxation multi-resolution scheme for accelerating realistic two-phase flows calculations in pipelines. *Internat. J. Numer. Methods Fluids* 54(2):207–236, 2007.
2. M. Baudin, C. Berthon, F. Coquel, R. Masson, and Q.H. Tran. A relaxation method for two-phase flow models with hydrodynamic closure law. *Numer. Math.*, 99:411–440, 2005.
3. F. Coquel, M. Postel, N. Poussineau, and Q.H. Tran. Multiresolution technique and explicit-implicit scheme for multicomponent flows. *Journal of Numerical Mathematics. Special Issue on Breaking Complexity: Multiscale Methods for Efficient PDE Solvers*, 2005.
4. S. Müller and S.M. Youssef Stiriba. Fully adaptive multiscale schemes for conservation laws employing locally varying time stepping. *Journal for Scientific Computing*, 2006. Report No. 238, IGPM, RWTH Aachen.

Velocity Discretization in Numerical Schemes for BGK Equations

A. Alaia, S. Pieraccini, and G. Puppo

1 Introduction

The need for accurate numerical solutions to kinetic equations has sharply increased in recent times due to the fact that the dynamics of gas in micro structures largely occurs in the kinetic regime, when the Knudsen number Kn, representing the ratio between the mean free path of molecules and the physical dimensions of the computational domain, cannot be neglected. In particular much interest has focused on models approximating the Boltzmann equations for small to moderate Knudsen numbers. One such model is the BGK model introduced in [2].

Several numerical schemes for the BGK model have been proposed in recent years, see for instance [1, 3, 6, 5] or [4]. All schemes mentioned rely on numerical quadrature in velocity space to evaluate the macroscopic variables that enter the construction of the local Maxwellian which appears in the collision term. Moreover, the macroscopic variables satisfy conservation laws which, in the hydrodynamic limit, approach the compressible Euler equations. Thus it is important that conservation be satisfied in order to capture the correct hydrodynamic limit. Clearly, the number of quadrature points used in velocity space determines the precision of the quadrature formula, but on the other hand at each node in velocity space the evolution equation for the distribution function f must be solved, and therefore the numerical complexity of the scheme is proportional to the number of quadrature points.

An interesting approach to velocity discretization appears in [4]. Here the Maxwellian is constructed in order to satisfy exact conservation. Thus the number of quadrature points in velocity space, in this case, is determined only through accuracy considerations, and in general it can be smaller than for the other schemes. However this particular construction must be carried out at each grid point in space and it is very expensive, since it involves the solution of nonlinear systems of equations.

In this chapter we study the performance of several approaches to velocity discretization, studying the enforcement of conservation and the efficiency of the resulting schemes. We consider the trapezoidal rule, Gauss–Hermite quadrature and the device proposed in [4], applied to the trapezoidal rule. All these techniques are used in conjunction with the IMEX-BGK schemes recently proposed in [5], which are based on an explicit–implicit time discretization and we compare the behavior of the different velocity discretizations. We find that the error in conservation does not depend on the accuracy of the space–time discretization, as was proven in [5], but it depends only on the velocity quadrature. Moreover our data suggest that the best trade-off between accuracy and efficiency can be obtained with Gauss–Hermite integration.

Let us consider the classical BGK model [2]

$$\frac{\partial f}{\partial t}(x,\,v,\,t)+v\cdot\nabla_x f(x,\,v,\,t)=\frac{1}{\tau}(f_M(x,\,v,\,t)-f(x,\,v,\,t))\qquad t\geq 0,\ (1)$$
$$f(x,v,0)=f_0(x,v),$$

where $x\in\mathbb{R}^d$, $v\in\mathbb{R}^N$, $f_0(x,v)\geq 0$ and f_M is the Maxwellian obtained from the moments of f:

$$f_M(x,\,v,\,t)=\frac{\rho(x,\,t)}{(2\pi RT(x,\,t))^{N/2}}\exp\left(-\frac{\|v-u(x,\,t)\|^2}{2RT(x,\,t)}\right).$$

The quantities ρ, u, and T are, respectively, the macroscopic density, velocity and temperature of the gas, and they are obtained from the moments of f, which are defined by

$$\begin{pmatrix}\rho\\ m\\ E\end{pmatrix}=\left\langle f\begin{pmatrix}1\\ v\\ \frac{1}{2}\|v\|^2\end{pmatrix}\right\rangle,\tag{2}$$

where $\langle g\rangle$ denotes (componentwise) the quantity $\int_{\mathbb{R}^N} g(v)\,dv$. Dependence from x and t is dropped for simplicity. Here m is momentum, so that the macroscopic velocity is $u=m/\rho$, while E is the total energy, which is linked to temperature T and internal energy e through the relations $\rho e=E-\frac{1}{2}\rho u^2$, $e=NRT/2$. In (1) the relaxation time τ is assumed to depend in general on the moments of f. Since f and f_M have the same moments, multiplying (1) by $\phi(v)=(1,v,1/2||v||^2)^T$ and integrating in velocity space, it is found that the macroscopic moments of f are conserved, in the sense that:

$$\partial_t\langle f\rangle+\nabla_x\cdot\langle fv\rangle=0,\tag{3a}$$
$$\partial_t\langle fv\rangle+\nabla_x\cdot\langle v\otimes vf\rangle=0,\tag{3b}$$
$$\partial_t\left\langle\frac{1}{2}\|v\|^2 f\right\rangle+\nabla_x\cdot\left\langle\frac{1}{2}\|v\|^2 vf\right\rangle=0.\tag{3c}$$

A numerical scheme for (1) should be able not only to yield an accurate solution to (1), but also to satisfy the conservation equations in some discretized form.

2 Velocity Discretization

The main difficulty arising from the discretization of velocity is that the moments of f are computed approximately through a quadrature formula. As a consequence, it is no longer true that f and f_M have the same moments. Thus the right-hand side of (3) is no longer zero, and a spurious source term arises for mass, momentum, and energy.

A second problem is due to the difficulty of selecting a suitable grid in velocity space. On one hand, a fine grid ensures a good accuracy of the velocity integrals which yield the macroscopic variables. On the other hand, f must be updated on each velocity grid point, which means that the scheme increases rapidly in computational cost as the velocity grid is refined. Moreover, the solution is quite sensitive to the choice of velocity grid points, which seems to be highly problem dependent, see for instance [1], where the velocity grid changes at almost each test problem.

In this work we consider the behavior of the BGK-IMEX schemes developed in [5] under several velocity discretizations, investigating the error in conservation, the error on macroscopic variables, and the scheme complexity in each case. As test problems, we consider a set up that yields a smooth solution, and a Riemann problem.

Choose a grid in velocity space, and let $\{v_k\}$, $k \in K$, be the set of the grid points, where k is a multi-index if $N > 1$. Given a function g, let:

$$\langle g \rangle_K = Q(g, \{v_k\}) = \sum_k g(v_k) w_k \simeq \int_{\mathbb{R}^N} g(v)\, dv,$$

i.e., $\langle g \rangle_K$ denotes the approximation of $\langle g \rangle$ obtained by means of a suitable quadrature rule $Q(\,\cdot\,, \{v_k\})$ built on the nodes $\{v_k\}$ with weights w_k; for a vector function g, $\langle \cdot \rangle_K$ is meant componentwise.

The macroscopic variables now depend on the quadrature rule and on the grid used in velocity space. Let:

$$\begin{pmatrix} \rho_K \\ m_K \\ E_K \end{pmatrix} = \langle f \phi \rangle_K$$

be the moments computed from f with the quadrature $Q(\,\cdot\,, \{v_k\})$. Now construct an approximate Maxwellian with the formula:

$$M_K(f)(x, v, t) = \frac{\rho_K(x, t)}{(2\pi R T_K(x, t))^{N/2}} \exp\left(-\frac{||v - u_K(x, t)||^2}{2 R T_K(x, t)}\right). \tag{4}$$

The problem is that there is no reason why f and $M_K(f)$ should have the same discrete moments, that is in general:

$$\langle (f - M_K(f))\, \phi \rangle_K \neq 0,$$

as pointed out in [4]. However, again in [4], it is proved that it is possible to find a discrete Maxwellian $\mathcal{M}$ such that:

$$\langle (f - \mathcal{M}(f))\, \phi \rangle_K = 0, \qquad \mathcal{M}(f)(x,v,t) = \exp\left(\alpha(x,t)\cdot\phi(v)\right), \tag{5}$$

where $\alpha(x,t)$ is an unknown vector that depends on the macroscopic quantities. Note that α is computed precisely solving the nonlinear system defined by (5).

In this work, we consider the following strategies to compute the moments of f.

2.1 Trapezoidal Rule

In this case, we pick a uniform grid in velocity space, with spacing Δv, and the grid points are distributed symmetrically around 0 up to a distance V_M from the origin.

2.2 Gauss–Hermite Integration

Here, the nodes are the zeroes of Gauss–Hermite polynomials. The grid is symmetric around 0, and the values of the nodes depend on the degree of the formula. Gauss–Hermite quadrature with N_{v} nodes is exact for the integration of functions of the form $p(v)\exp(-v^2)$ on $\mathbb{R}$, where $p(v)$ is a polynomial of degree $d \le 2N_{\mathrm{v}} - 1$. For a general Maxwellian f_M, Gauss–Hermite quadrature computes the moments exactly with only two nodes, under the change of variables $\xi = (v-u)/\sqrt{2RT}$. However, it is not convenient to use it in this form, because the change of variables depends on x through the macroscopic variables u and T, thus, at each space grid point, the velocity grid would be different. This in turn would require complex interpolation formulas to evaluate $v\partial_x f$.

For this reason, we use Gauss–Hermite quadrature centered at 0. Since f is not a Maxwellian, we compute the moments of f, using a previously computed Maxwellian f_M at the same grid point in space, but from a previous stage of the time integrator:

$$\langle f\phi\rangle_K \simeq \int f_M \phi\, \mathrm{d}v + \int \phi \operatorname{sign}(f - f_M)\exp\left(\log|f - f_M| + v^2\right)\exp(-v^2)\mathrm{d}v. \tag{6}$$

2.3 Conservative Rule

This rule implements the device proposed in [4]. Namely, we compute the discrete Maxwellian $\mathcal{M}$ defined in (5). The exponent α must be computed solving the nonlinear system appearing in (5), which can be carried out with Newton scheme. The Jacobian matrix of this system has the form:

$$J = \sum_k \phi(v_k)\phi(v_k)^T w_k \exp(\alpha\cdot\phi(v_k)),$$

which depends on the quadrature rule in velocity space. The Newton method converges fast with a uniform grid for the trapezoidal rule, but apparently it does not converge on a Gaussian grid, probably because the matrices $\phi(v_k)\phi(v_k)^T$ are not well scaled for nodes which are very close to zero. For this reason, we use Mieussens' method only on the trapezoidal rule. We find errors in conservation which are within machine precision, but the accuracy of macroscopic variables remains comparable to the accuracy obtained with the underlying trapezoidal rule.

3 Numerical Results

We run all tests with the second-order BGK2 scheme of [5], except the last 2D test which was carried out with the first-order BGK1. The results we find for the behavior of the velocity discretization do not seem to depend on the order of the scheme.

We first consider a test which results in a smooth solution. The initial condition is locally Maxwellian, with density and temperature uniformly equal to 1, while the macroscopic speed is given by $u_0(x) = \frac{1}{\sigma}(\exp(-(\sigma x - 1)^2) - 2\exp(-(\sigma x + 3)^2))$, with $\sigma = 10$, so that the perturbation in velocity is localized initially close to $x = 0$. The computational domain is $[-1, 1]$ and we use free flow boundary conditions up to $t_f = 0.04$. The collisional time is constant, with $\tau = \mathrm{Kn} = 10^{-5}$.

Table 1 shows the errors in mass conservation obtained with the three velocity discretization techniques compared in this work. These errors are computed evaluating total mass at the final time, and comparing with the total mass present at the beginning of the computation.

As expected, the error does not depend on N_{x}, since the equations are discretized in space with a conservative scheme. The trapezoidal rule yields the largest errors, which converge to zero as N_{v} grows, while both Gauss–Hermite and the conservative algorithm give errors which are within machine precision.

Table 2 contains the maximum error obtained for the density, measured against a reference solution. Here the error converges to zero as both N_{v} and $N_{\mathrm{x}} \to \infty$. When N_{v} grows, the error reaches a *plateau*: Now the dominant term is the error due to space discretization, and in fact the error decreases as N_{x} increases. We note that this *plateau* has already been reached on all velocity grids tested for Gauss–Hermite quadrature: In this case, the scheme is approximately conservative because the moments are accurately computed. On the other hand, the errors computed with the conservative technique on the trapezoidal rule are of the same order of magnitude of the errors obtained with the trapezoidal rule alone. In other words: The fact that conservation is exactly imposed does not translate in accurate values for the macroscopic variables. We obtain analogous results for momentum and energy.

Table 1. Error in mass conservation

N_v	$N_x = 100$	$N_x = 200$	$N_x = 300$
	Trapezoidal rule		
10	3.8421e-02	3.8844e-02	3.8989e-02
12	1.7298e-02	1.7684e-02	1.7817e-02
14	2.9204e-03	3.0269e-03	3.0643e-03
16	2.1491e-04	2.2400e-04	2.7221e-04
18	9.3833e-06	9.7880e-06	9.9312e-06
20	2.7277e-07	2.8457e-07	2.8874e-07
	Gauss–Hermite rule		
10	8.0047e-16	2.0517e-15	2.1990e-15
12	1.7197e-15	2.2815e-15	1.2057e-15
14	4.2188e-17	2.7978e-16	3.5823e-16
16	1.9329e-15	1.5199e-15	1.6853e-15
18	2.8411e-15	2.1061e-15	2.3352e-15
20	2.4236e-15	2.1294e-15	2.7630e-15
	Conservative rule		
10	1.0436e-16	2.7645e-16	7.3349e-16
12	8.6930e-16	5.4401e-16	1.4877e-16
14	8.2268e-16	1.2734e-15	1.7157e-15
16	3.2407e-15	2.6101e-15	2.2804e-15
18	3.5216e-15	3.6682e-15	4.2329e-15
20	1.3223e-15	1.6243e-15	2.1612e-15

Table 2. L_∞ error in density

N_v	$N_x = 100$	$N_x = 200$	$N_x = 300$
	Trapezoidal rule		
10	8.4646e-02	8.5345e-02	8.5498e-02
12	2.5441e-02	2.5911e-02	2.6016e-02
14	4.8155e-03	5.0442e-03	5.0777e-03
16	5.5482e-04	5.7510e-04	5.8144e-04
18	6.2025e-04	1.1741e-04	4.9184e-05
20	6.1935e-04	1.2182e-04	4.4556e-05
	Gauss–Hermite rule		
10	6.8219e-04	1.4597e-04	5.5290e-05
12	6.5932e-04	1.3895e-04	5.2461e-05
14	6.5082e-04	1.3405e-04	5.0416e-05
16	6.4383e-04	1.3155e-04	4.8885e-05
18	6.3801e-04	1.2950e-04	4.7976e-05
20	6.3312e-04	1.2779e-04	4.7222e-05
	Conservative rule		
10	5.6079e-02	5.6237e-02	5.6225e-02
12	1.1543e-02	1.1626e-02	1.1623e-02
14	1.8392e-03	1.9424e-03	1.9545e-03
16	5.8891e-04	2.6025e-04	2.6607e-04
18	6.2351e-04	1.1998e-04	4.2842e-05
20	6.1950e-04	1.2189e-04	4.4635e-05

Next, we consider a Riemann problem, between two Maxwellian states. The left and right Maxwellians are determined by $(\rho_L, u_L, T_L) = (2.25, 0, 1.125)$ and $(\rho_R, u_R, T_R) = (3/7, 0, 1/6)$, respectively. This test is derived from [3]. The collision time is again constant, $\tau = \mathrm{Kn}$. Figure 1 shows the results obtained with the trapezoidal rule, with and without the conservative strategy by Mieussens, $\mathrm{Kn} = 10^{-5}$. The trapezoidal rule alone fails with $N_\mathrm{v} < 20$, while the conservative scheme yields a solution even in this case. Moreover, on coarse grids the conservative technique yields better results.

Figure 2 contains the results obtained with the Gauss–Hermite rule, for $\mathrm{Kn} = 10^{-5}$ and $\mathrm{Kn} = 10^{-2}$. Starting from $N_\mathrm{v} = 8$, the solution has already reached convergence, in both cases. Thus, even without enforcing exact conservation, the solution is accurate with very few nodes.

Finally, to measure the efficiency of the techniques described, we run a 2D Riemann problem on a monoatomic gas, with three degrees of freedom in velocity space. We first solved the four 1D Riemann problems forming the

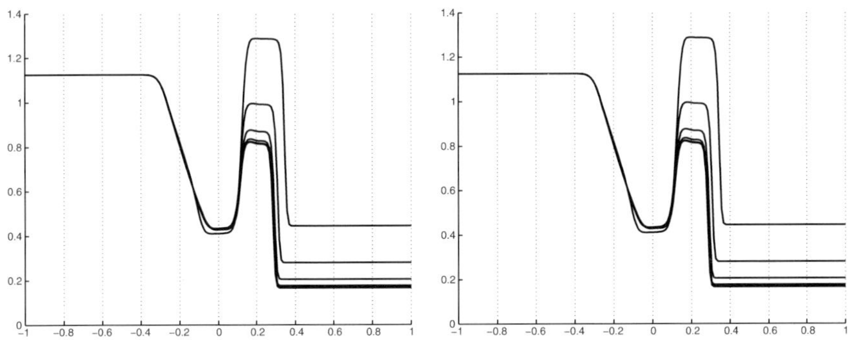

Fig. 1. Temperature. *Left*, trapezoidal rule, $N_\mathrm{v} = 20, 24, 28, \ldots 40, 50$ from the *top* to the *bottom* curve. *Right*, conservative scheme $N_\mathrm{v} = 16, 20, 24, \ldots 40, 50$ from the *top* to the *bottom* curve. $\mathrm{Kn} = 10^{-5}$

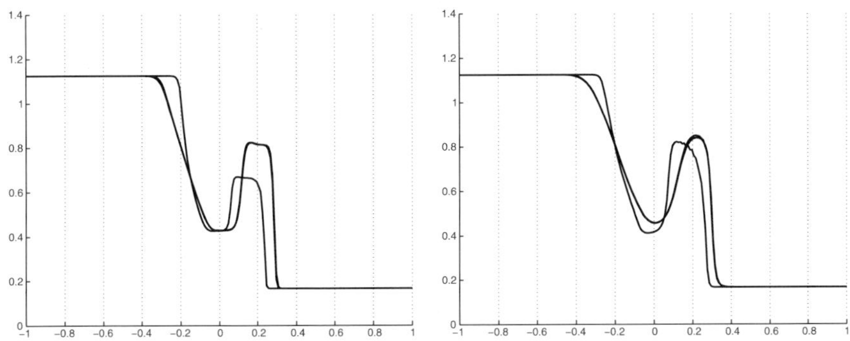

Fig. 2. Temperature, Gauss–Hermite rule, $N_\mathrm{v} = 4, 8, 12, \ldots 40$ from the *bottom* to the *top* curve. *Left*, $\mathrm{Kn} = 10^{-5}$, *right*, $\mathrm{Kn} = 10^{-2}$

Table 3. CPU time on a 2D test problem

Trapezoidal rule	Gauss–Hermite rule	Conservative rule
$N_v = (65, 65)$	$N_v = (28, 28)$	$N_v = (16, 16)$
20556.2 s	3290.1 s	5774.8 s

2D test, and we found for each technique the number of velocity grid points needed to reach an error in conservation smaller than the threshold 10^{-10}. With this number of grid points we ran the 2D test, measuring the CPU time, see Table 3. Note that here only conservation is enforced: We expect that the accuracy reached by the Gauss–Hermite quadrature will be higher, due to the considerations reached from Fig. 2, so that the CPU time scored by the Gaussian rule is overestimated. This notwithstanding, the Gaussian rule appears to be the fastest algorithm for this problem.

References

1. K. Aoki, Y. Sone, and T. Yamada, *Numerical analysis of gas flows condensing on its plane condensed phase on the basis of kinetic theory*, Physics of Fluids A, 2 (1990), pp. 1867–1878.
2. P.L. Bhatnagar, E.P. Gross, and M. Krook, *A model for collision processes in gases. Small amplitude processes in charged and neutral one-component systems*, Physical Reviews, 94 (1954), pp. 511–525.
3. F. Coron and B. Perthame, *Numerical passage from kinetic to fluid equations*, SIAM J. Numer. Anal., 28 (1991), pp. 26–42.
4. L. Mieussens, *Discrete velocity model and implicit scheme for the BGK equation of rarefied gas dynamics*, Mathematical Models and Methods in Applied Sciences, 10 (2000), pp. 1121–1149.
5. S. Pieraccini and G. Puppo, *Implicit-Explicit schemes for BGK kinetic equations.* To appear on *Journal of Scientific Computing*, 2007.
6. J.Y. Yang and J.C. Huang, *Rarefied flow computations using nonlinear model Boltzmann equations*, Journal of Computational Physics, 120 (1995), pp. 323–339.

A Space–Time Conservative Method for Hyperbolic Systems of Relaxation Type

S. Qamar and G. Warnecke

Summary. We propose a higher-order space–time conservative method for hyperbolic systems of relaxation type. In the present model the relaxation time may vary from order of one to a very small value. These small values make the relaxation term stronger and highly stiff. In such situations under-resolved numerical schemes may produce spurious numerical results. However, our present scheme has the capability to correctly capture the behavior of the physical phenomena with high order accuracy even if the initial layer and the small relaxation time are not numerically resolved. The scheme treats the space and time in a unified manner. The flow variables and their slopes are the basic unknowns in the scheme. The source term is treated by its volumetric integration over the space–time control volume and is a direct part of the overall space–time flux balance.

1 Introduction

There are variety of physical phenomena where hyperbolic systems with relaxation arise, namely nonlinear waves [WH74], gas flows with relaxation [CL78], multiphase and phase transitions [GL86], etc. The development of efficient numerical scheme for these hyperbolic systems is quite challenging, since in many applications the relaxation time of the source term varies from one to very small values if one compares it with the characteristic time scale of the hyperbolic system.

The study of relaxation problems was initiated by Whitham [WH74] for linear problems and later on extended to nonlinear hyperbolic systems of two equations by Liu [LI87] and Chen et al. [CH94]. More recently Liotta et al. [LR78] extended the central schemes [NT90] to hyperbolic systems with source terms. Later on, Pareschi [PA01] extended the schemes in order to treat the stiff sources in more efficient manner. Apart from central schemes, Pareschi and Russo [PR00] developed implicit–explicit (IMEX) Runge–Kutta methods up to order three. Furthermore, Yu and Chang [YC97] have extended the CE/SE methods of Chang [CH95] to hyperbolic systems with source terms.

In this article we present a second order accurate scheme for hyperbolic systems of relaxation type; see Qamar and Warnecke [QW06] for further details. The scheme uses the basic concept of finite volume schemes and conservation laws. The scheme utilizes the advantages of both the CE/SE method of Chang et al. [CH95, YC97] and central schemes of Nessyahu and Tadmor [NT90]. However, unlike the CE/SE method the present scheme is Jacobian-free and hence like the central schemes can also be applied to any hyperbolic system. The scheme uses space–time control volumes in order to compute the conservative flow variables and their slopes. The source term is included as volumetric integral over space–time control volume such that it becomes an integral part of the overall space–time flux balance. In particular, the treatment of the source term is done in two ways. The first way is suitable for moderately stiff source terms with relaxation time up to the order of 10^{-5}. However, the method will fail when it is used to solve very stiff relaxation systems. This failure is due to amplification effects by the source terms over the differences of the flow properties at adjacent nodes at the same time level. Hence it means that the source term effects become more important in the steady state situation. For highly stiff source terms the failure is removed in such a way that the source-term effects are only restricted to the mesh nodes at the updated time level. As a result we can write our scheme in the ϑ-scheme form with ϑ ranging from 1/2 to 1. Hence for nonstiff source terms one takes $\vartheta = 1/2$, while for highly stiff source terms $\vartheta = 1$.

The outline of the paper is the following. In Sect. 2, we give a brief overview of our scheme for the one-dimensional hyperbolic systems with relaxation; see [QW06] for more details. In Sect. 3, we present the numerical results of our scheme.

2 The One-Dimensional SP-Method

Let us consider the initial value problem

$$\partial_t \mathbf{U} + \partial_x \mathbf{F}(\mathbf{U}) = \frac{1}{\varepsilon}\mathbf{R}(\mathbf{U}), \quad x \in \mathbb{R}, \quad t \in \mathbb{R}^+, \tag{1}$$

$$\mathbf{U}(x,0) = \Phi(x)\,, \quad -\infty < x < \infty \tag{2}$$

where $\mathbf{U} \in \mathbb{R}^m$, $\mathbf{f}(\mathbf{U}) : \mathbb{R}^m \longrightarrow \mathbb{R}^m$, $\mathbf{R}(\mathbf{U}) : \mathbb{R}^m \longrightarrow \mathbb{R}^m$, for $m \geq 1$.

Here $\mathbf{U}(x,t)$ is a vector of conserved variables and $\mathbf{F}(\mathbf{U})$ is the corresponding vector of fluxes and $\mathbf{R}(\mathbf{U})$ is a vector of source terms. For flows in one spatial dimension using a fixed spatial domain for CV's, the integral form of the above equation is

$$\oint_{S(V)}^{c.c} (F_k dt - U_k dx) = \frac{1}{\varepsilon}\int_V R_k(\mathbf{U})\, dV\,, \quad \forall\, k = 1, 2, \cdots, m. \tag{3}$$

where *c.c.* indicates that the line integration is carried out in the counter-clockwise direction. Here, U_k, F_k, R_k are the components of vectors $\mathbf{U}$, $\mathbf{F}$, and $\mathbf{R}$, respectively. Equation (3) is enforced over a space–time region V, called control volume (CV), in which discontinuities of the flow variables are allowed.

In the present approach, like the original *a*-scheme of Chang [CH95], the number of CV's associated with each grid point is identical to the number of unknowns. In one dimension the unknowns are $\mathbf{U}$ and $\mathbf{U}_x$; therefore, we need two control volumes to determine them. To proceed we divide the entire computational spatial domain into nonoverlapping cells, which are line elements of mesh length Δx in this case, see Fig. 1. The dashed lines indicate the boundaries of the CVs. We introduce the grid points

$$x_i = i \cdot \Delta x \quad \text{for} \quad i \in \mathbb{Z} \quad \text{and} \quad x_{i+\frac{1}{2}} = x_i + \frac{\Delta x}{2}.$$

The grid points actually used are denoted by small dots, where the hollow dots represent the grid points at the previous time step, while the filled dot is the node at the updated time. The mesh is staggered in time with a time interval $\Delta t = t^{n+1} - t^n$, $\forall n \in \mathbb{N}_0$. At each grid point, we construct two CV's as

$$\text{CV-I} = [x_i, x_{i+\frac{1}{2}}] \times [t^n, t^{n+1}], \quad \text{CV-II} = [x_{i+\frac{1}{2}}, x_{i+1}] \times [t^n, t^{n+1}].$$

We approximate the flow variables in each cell I_i^n by

$$\mathbf{U}(x,t) = \sum [\mathbf{U}_i^n + (\mathbf{U}_x)_i^n (x - x_i) + (\mathbf{U}_t)_i^n (t - t^n)]\, \chi_i(x). \tag{4}$$

Here, $\chi_i(x)$ is the characteristic function of the cell $I_i^n := \{\xi \mid |\xi - x_i| \leq \frac{\Delta x}{2}\}$, centered around $x_i = i \cdot \Delta x$. By using (1) one can replace the $(\mathbf{U}_t)_i^n$ in (4) by space derivatives as

$$\mathbf{U}(x,t) = \sum \left[\mathbf{U}_i^n + (\mathbf{U}_x)_i^n (x - x_i) + \left(\frac{1}{\varepsilon}\mathbf{R}_i^n - (\mathbf{F}_x)_i^n\right)(t - t^n)\right] \chi_i(x). \tag{5}$$

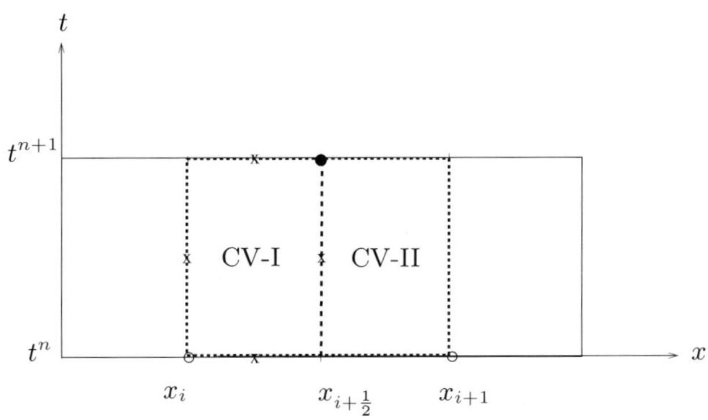

Fig. 1. Geometrical representation of the Control volumes

Note that the third terms in (4) and (5) are redundant with respect to later integration. But, we have chosen to retain them in the presentation. Applying (3) to the control volumes CV-I and CV-II along with the midpoint rule on the time integrals we get the following ϑ-scheme (for details see Qamar and Warnecke [QW06])

$$\begin{aligned}\mathbf{U}_{i+\frac{1}{2}}^{n+1} &= \frac{1}{2}(\mathbf{U}_i^n + \mathbf{U}_{i+1}^n) + \frac{\Delta x}{8}\left[(\mathbf{U}_x)_i^n - (\mathbf{U}_x)_{i+1}^n\right] + \lambda\left[\mathbf{F}(\mathbf{U}_i^{n+\frac{1}{2}}) - \mathbf{F}(\mathbf{U}_{i+1}^{n+\frac{1}{2}})\right] \\ &+ \frac{\Delta t}{2\varepsilon}\left[(1-\vartheta)\mathbf{R}(\mathbf{U}_i^n) + 2\vartheta\mathbf{R}(\mathbf{U}_{i+\frac{1}{2}}^{n+1}) + (1-\vartheta)\mathbf{R}(\mathbf{U}_{i+1}^n)\right], \qquad (6)\end{aligned}$$

and

$$\begin{aligned}\Delta x(\mathbf{U}_x)_{i+\frac{1}{2}}^{n+1} &= 2(\mathbf{U}_{i+1}^n - \mathbf{U}_i^n) - \frac{\Delta x}{2}\left[(\mathbf{U}_x)_i^n + (\mathbf{U}_x)_{i+1}^n\right] \\ &- 4\lambda\left[\mathbf{F}(\mathbf{U}_i^{n+\frac{1}{2}}) - 2\mathbf{F}(\tilde{\mathbf{U}}_{i+\frac{1}{2}}^{n+\frac{1}{2}}) + \mathbf{F}(\mathbf{U}_{i+1}^{n+\frac{1}{2}})\right] \qquad (7) \\ &+ 2(1-\vartheta)\frac{\Delta t}{\varepsilon}\left[\mathbf{R}(\mathbf{U}_{i+1}^n) - \mathbf{R}(\mathbf{U}_i^n)\right],\end{aligned}$$

where either $\theta = 1$ or $\theta = \frac{1}{2}$. We still need in (6) and (7) the predicted values $\mathbf{U}_i^{n+\frac{1}{2}}$ in each cell I_i as well as the cells interface values $\tilde{\mathbf{U}}_{i+\frac{1}{2}}^{n+\frac{1}{2}}$. First we define

$$\mathbf{U}_i^{n+\frac{1}{2}} = \mathbf{U}_i^n + \frac{\Delta t}{2}(\mathbf{U}_t)_i^n = \mathbf{U}_i^n + \frac{\Delta t}{2}\left[\frac{1}{\varepsilon}\mathbf{R}(\mathbf{U}_i^{n+\frac{1}{2}}) - \frac{\Delta\mathbf{F}(\mathbf{U}_i^n)}{\Delta x}\right]. \qquad (8)$$

Note that the use of $R(\mathbf{U}_i^{n+\frac{1}{2}})$ is important in the case of stiff source terms, such as hyperbolic systems near fluid dynamic regime. The flux differences $\Delta\mathbf{F}_i^n = \Delta\mathbf{F}(\mathbf{U}_i^n)$ can be computed as follows. For any grid function $\{\mathbf{F}_i^n\}$ and parameter $1 \le \alpha \le 2$ we set

$$\Delta\mathbf{F}_i^n = MM\left[\alpha\Delta_-\mathbf{F}_i^n, \frac{(\Delta_-\mathbf{F}_i^n + \Delta_+\mathbf{F}_i)}{2}, \alpha\Delta_+\mathbf{F}_i^n\right], \quad \Delta_\pm\mathbf{F}_i^n = \pm\left(\mathbf{F}_{i\pm1}^n - \mathbf{F}_i^n\right)$$

and MM denotes the minmod nonlinear limiter [NT90]. Next we set

$$\tilde{\mathbf{U}}_{i+\frac{1}{2}}^{n+\frac{1}{2}} = \mathbf{U}_{i+\frac{1}{2}}^{n+1} - \frac{\Delta t}{2}(\mathbf{U}_t)_{i+\frac{1}{2}}^{n+1} = \mathbf{U}_{i+\frac{1}{2}}^{n+1} - \frac{\Delta t}{2}\left[\frac{1}{\varepsilon}\mathbf{R}(\mathbf{U}_{i+\frac{1}{2}}^{n+1}) - \frac{\tilde{\Delta}\mathbf{F}(\mathbf{U}_{i+\frac{1}{2}}^{n+1})}{\Delta x}\right]. \qquad (9)$$

The updated values of the flow variables $\mathbf{U}_{i+\frac{1}{2}}^{n+1}$ have been calculated in (6) and our formula (8) is only needed in (7) afterwards. The approximation (9) is the predicted value at the backward half time step with respect to the data at the updated time step. Here

$$\tilde{\Delta}\mathbf{F}(\mathbf{U}_{i+\frac{1}{2}}^{n+1}) = MM\left(\alpha\tilde{\Delta}_-\mathbf{F}_{i+\frac{1}{2}}^{n+1}, \frac{(\tilde{\Delta}_-\mathbf{F}_{i+\frac{1}{2}}^{n+1} + \tilde{\Delta}_+\mathbf{F}_{i+\frac{1}{2}}^{n+1})}{2}, \alpha\tilde{\Delta}_+\mathbf{F}_{i+\frac{1}{2}}^{n+1}\right), \qquad (10)$$

where

$$\tilde{\Delta}_{-}\mathbf{F}^{n+1}_{i+\frac{1}{2}} = \mathbf{F}(\mathbf{U}^{n+1}_{i+\frac{1}{2}}) - \mathbf{F}((\mathbf{U}')^{n+1}_{i}), \quad \tilde{\Delta}_{+}\mathbf{F}^{n+1}_{i+\frac{1}{2}} = \mathbf{F}((\mathbf{U}')^{n+1}_{i+1}) - \mathbf{F}(\mathbf{U}^{n+1}_{i+\frac{1}{2}})$$

and

$$(\mathbf{U}')^{n+1}_{i} = \mathbf{U}^{n}_{i} - \Delta t \left[\frac{1}{\varepsilon}\mathbf{R}(\mathbf{U}^{n+\frac{1}{2}}_{i}) - \frac{\Delta \mathbf{F}(\mathbf{U}^{n}_{i})}{\Delta x} \right], \quad \forall\, i \in \mathbb{Z}. \tag{11}$$

The values $(\mathbf{U}')^{n+1}_{i}$ are predicted at the updated time step. The equations (6)–(11) form a complete scheme. Note that (6) and (8) are implicit, hence in case of nonlinear source terms one needs to use some root finding routine.

In the above scheme for $\vartheta = \frac{1}{2}$ we get a scheme for the nonstiff source terms and for $\vartheta = 1$ a scheme for stiff source term.

The above scheme is a nondissipative scheme. However, in case of problems with discontinuities we need some limiter to suppress the oscillations [QW06]. Here, we use two different procedures to limit our slopes. Let us define for each $k = 1, 2, \cdots, m$

$$(V_k)^{n+1}_{i+\frac{1}{2}} = \left((U'_k)^{n+1}_{i+1} - (U_k)^{n+1}_{i+\frac{1}{2}} \right), \quad (W_k)^{n+1}_{i+\frac{1}{2}} = \left((U_k)^{n+1}_{i+\frac{1}{2}} - (U'_k)^{n+1}_{i} \right), \tag{12}$$

where $(U'_k)^{n+1}$ and U^{n+1}_k are, respectively, the component wise predicted values as given by (11) and updated values at t^{n+1} as given by (6).

Limiter 1: Direct procedure for limiting derivatives in (7)

Using (7) and (12) we can get for the parameter $1 < \alpha < 2$ the following expression for the discrete slopes at updated time

$$\Delta x (U_{kx})^{n+1}_{i+\frac{1}{2}} = \Delta x (U_{kx})^{n+1}_{i+\frac{1}{2}} \cdot \varphi \left(\frac{\alpha\, (W_k)^{n+1}_{i+\frac{1}{2}}}{\Delta x (U_{kx})^{n+1}_{i+\frac{1}{2}}} \right) \cdot \varphi \left(\frac{\alpha\, (V_k)^{n+1}_{i+\frac{1}{2}}}{\Delta x (U_{kx})^{n+1}_{i+\frac{1}{2}}} \right), \tag{13}$$

where $(U_{kx})^{n+1}_{i+\frac{1}{2}}$ represent the components of vector $(\mathbf{U}_x)^{n+1}_{i+\frac{1}{2}}$ and $\varphi(r)$ is some limiting function, e.g., the minmod function [QW06] as given by

$$\varphi(r) = \begin{cases} 0, & \text{for } r < 0, \\ r, & \text{for } 0 < r < 1, \\ 1, & \text{for } r > 1. \end{cases} \tag{14}$$

The optimal α depends on the problem at hand. However, our experiments indicate that the value of α between 1.4 and 1.7 is a good choice.

Limiter 2: Central finite difference procedure for the slopes

Here, instead of (7), we use a finite difference procedure to calculate the slopes U_{kx}. Using (12) we get

$$\Delta x\, (U_{kx})^{n+1}_{i} = \frac{|V_k|^{\alpha}\, W_k \,+\, |W_k|^{\alpha}\, V_k}{|V_k|^{\alpha} \,+\, |W_k|^{\alpha}} \qquad \forall\, k = 1, 2, \cdots, m. \tag{15}$$

Here $1 \leq \alpha \leq 2$; however, in our calculation we take $\alpha = 1$.

3 Numerical Simulations

In this section we apply our scheme to Broadwell model of rarefied gas dynamics [BR64, PA01]. For this kinetic model we have

$$\mathbf{U} = (\rho, m, z), \ \mathbf{F}(\mathbf{U}) = (m, z, m), \ \mathbf{R}(\mathbf{U}) = \left(0, 0, \frac{1}{2}[\rho^2 + m^2 - 2\rho z]\right). \quad (16)$$

Here ε represents the mean-free path of particles. The density ρ and momentum m are the only conserved quantities. In the fluid dynamics limit $\varepsilon \to 0$ we have

$$z = z_E \equiv \frac{\rho^2 + m^2}{2\rho},$$

and the Broadwell system is approximated by the reduced system with

$$\mathbf{U} = (\rho, \rho u), \quad \mathbf{G}(\mathbf{U}) = \left(\rho u, \frac{1}{2}[\rho + \rho u^2]\right) \quad u = \frac{m}{\rho}. \quad (17)$$

We consider the following two Riemann problems [LR78, PA01, PR00] for this model

$$\begin{aligned} \text{RIEM1:} \quad & \rho_l = 2, \quad m_l = 1, \quad z_l = 1, \quad x < 0.2, \\ & \rho_r = 1, \quad m_r = 0.13962, \quad z_r = 1, \quad x > 0.2. \end{aligned} \quad (18)$$

$$\begin{aligned} \text{RIEM2:} \quad & \rho_l = 1, \quad m_l = 0, \quad z_l = 1, \quad x < 0, \\ & \rho_r = 0.2, \quad m_r = 0, \quad z_r = 1, \quad x > 0. \end{aligned} \quad (19)$$

The results for different values of relaxation time ε are shown in Figs. 2–4 with CFL condition $\Delta t = 0.3\Delta x$. All the numerical results with symbols $\circ, *, +$ are calculated on a mesh $N = 200$ points, while the reference solutions with thin continuous line are obtained by using the same scheme on a 2,000 grid points. Here the symbols $\circ, *, +$ refer to ρ, z, and m, respectively. In Fig. 2 the first two results for the initial data RIEM1 with $\varepsilon = 1$ and $\varepsilon = 0.02$ are obtained from the scheme (6)–(9) where we choose $\vartheta = 1/2$. The last result for $\varepsilon = 10^{-8}$ is obtained from the same scheme with $\vartheta = 1$. The last results in Fig. 2 demonstrate that the solution of our scheme is projected to the equilibrium solution for $\varepsilon = 10^{-8}$. The left hand plot in Fig. 3 shows that our scheme (6)–(9) with $\vartheta = 1/2$ works well up to $\varepsilon = 10^{-5}$ and the solution is an equilibrium solution for this value of ε. However, the right hand plot shows that the solution does not converge to the equilibrium solution when $\varepsilon = 10^{-8}$. This experiment supports our claim that it is better to use the scheme with $\vartheta = 1$ when the relaxation is smaller than 10^{-5}. The results in Fig. 4 for RIEM2 problem are obtained from the same scheme with $\vartheta = 1$ and $\varepsilon = 10^{-8}$. All the results presented here show that our scheme gives highly resolved solutions.

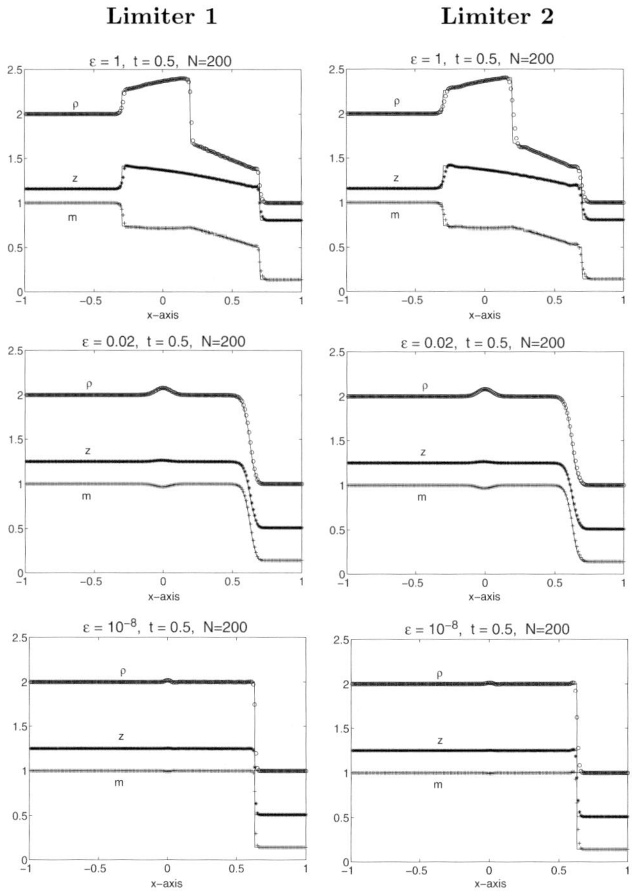

Fig. 2. Solution of Broadwell model for RIEM1

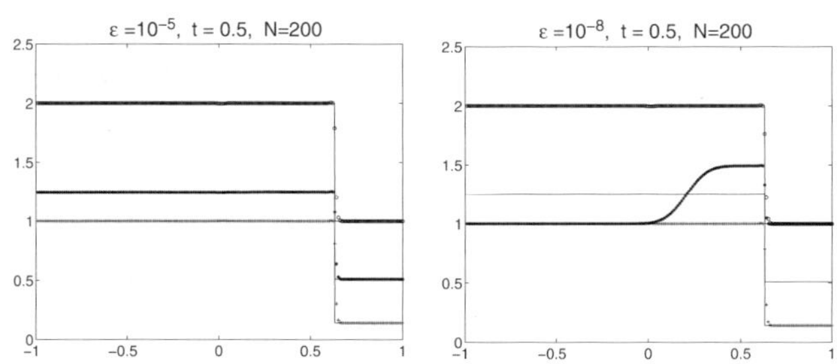

Fig. 3. Solution of RIEM1 with a scheme for nonstiff source terms

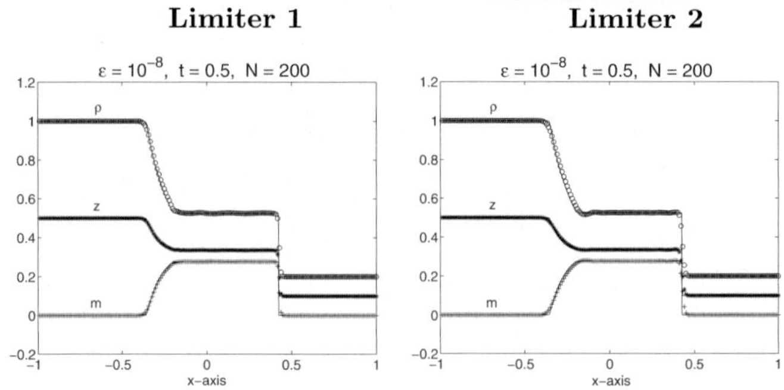

Fig. 4. Solution of RIEM2 with a scheme for stiff source terms

References

[BR64] Broadwell, J.E.: Shock structure in a simple discrete velocity gas. Phys. Fluid **7**, 1243–1247 (1964).

[CH95] Chang, S.C.: The method of space time conservation element and solution element -A new approach for solving the Navier Stokes and Euler equations. J. Comput. Phys. **119**, 295–324 (1995).

[CH94] Chen, G.-Q., Levermore, C.D., Liu, T.-P.: Hyperbolic conservation laws with stiff relaxation terms and entropy. Commun. Pure & Appl. Math. **47**, 787–830 (1994).

[CL78] Clarke, J.F.: Gas dynamics with relaxation effects. Rep. Prog. Phys. **41**, 807–864 (1978).

[GL86] Glimm, J.: The continuous structure of discontinuities. Lect. Notes in Phys. **344**, 177–1861 (1986).

[LR78] Liotta, S.F., Romano, V., Russo, G.: Central schemes for balance laws of relaxation type. SIAM J. Numer. Anal. **38**, 1337–1356 (1978).

[LI87] Liu, T.-P.: Hyperbolic conservation laws with relaxation. Comm. Math. Phys. **108**, 153–175 (1987).

[NT90] Nessyahu, H., Tadmor, E.: Non-oscillatory central differencing fo hyperbolic conservation laws. SIAM J. Comput. Phys. **87**, 408–448 (1990).

[PA01] Pareschi, L.: Central difference based numerical schemes for hyperbolic conservation laws with relaxation terms. SIAM J. Numer. Anal. **39**, 1395–1417 (2001).

[PR00] Pareschi, L., Russo, G.: Implicit-explicit Runge-Kutta schemes and applications to hyperbolic systems with relaxation. Recent Trends in Numerical Analysis, Edited by L. Brugnano and D. Trigiante, Vol. 3, 269–289 (2000).

[QW06] Qamar, S., Warnecke, G.: A space-time conservative method for hyperbolic systems with stiff and non stiff source terms. Commun. in Comput. Phys. (CiCP) **1**, 451–480 (2006).

[WH74] Whitham, G.B.: Linear and nonlinear waves. Wiley, New York, (1974).

[YC97] Yu, S.T., Chang, S.C.: Treatments of stiff source terms in conservation laws by the method of space-time conservation element and solution element. AIAA 97-0435, (1997).

A Numerical Scheme Based on Multipeakons for Conservative Solutions of the Camassa–Holm Equation

H. Holden and X. Raynaud

Summary. We present a convergent numerical scheme based on multipeakons to compute conservative solutions of the Camassa–Holm equation.

1 Introduction

The Camassa–Holm equation [3, 4] reads

$$u_t - u_{xxt} + 3uu_x - 2u_x u_{xx} - uu_{xxx} = 0. \tag{1}$$

It has a bi-Hamiltonian structure, is completely integrable, and has infinitely many conserved quantities. In this chapter we study the Cauchy problem $u|_{t=0} = \bar{u} \in H^1(\mathbb{R})$ for (1). A highly interesting property of the equation is that for a wide class of initial data the solution experiences wave breaking in finite time in the sense that the solution u remains bounded pointwise while the spatial derivative u_x becomes unbounded pointwise. However, the H^1 norm of u remains finite, see, e.g., [7, 8].

The extension of the solution beyond wave breaking is nontrivial and can be illustrated by studying multipeakons that are solutions of the form

$$u(t,x) = \sum_{i=1}^{n} p_i(t) \mathrm{e}^{-|x-q_i(t)|}, \tag{2}$$

where the $(p_i(t), q_i(t))$ satisfy the explicit system of ordinary differential equations

$$\dot{q}_i = \sum_{j=1}^{n} p_j \mathrm{e}^{-|q_i - q_j|}, \quad \dot{p}_i = \sum_{j=1}^{n} p_i p_j sgn(q_i - q_j) \mathrm{e}^{-|q_i - q_j|}. \tag{3}$$

Observe that the solution (2) is not smooth even with continuous functions $(p_i(t), q_i(t))$.

Peakons interact in a way similar to that of solitons of the Korteweg–de Vries equation, and wave breaking may appear when at least two of the q_i's coincide. If all the $p_i(0)$ have the same sign, the peakons move in the same direction and we have a global solution and no wave breaking. Higher peakons move faster than the smaller ones, and when a higher peakon overtakes a smaller, there is an exchange of mass, but no wave breaking takes place. Furthermore, the $q_i(t)$ remain distinct. However, if some of $p_i(0)$ have opposite sign, wave breaking may incur. For simplicity, consider the case with $n = 2$ and one peakon $p_1(0) > 0$ (moving to the right) and one antipeakon $p_2(0) < 0$ (moving to the left). In the symmetric case ($p_1(0) = -p_2(0)$ and $q_1(0) = -q_2(0) < 0$) the solution will vanish pointwise at the collision time t^* when $q_1(t^*) = q_2(t^*)$, that is, $u(t^*, x) = 0$ for all $x \in \mathbb{R}$. Clearly, at least two scenarios are possible; one is to let $u(t, x)$ vanish identically for $t > t^*$, and the other possibility is to let the peakon and antipeakon "pass through" each other in a way that is consistent with the Camassa–Holm equation. In the first case the energy $\int(u^2 + u_x^2)\,dx$ decreases to zero at t^*, while in the second case, the energy remains constant except at t^*. The first solution is denoted a dissipative solution, while the second one is called conservative. Other solutions are also possible. Global dissipative solutions of a more general class of equations were recently derived by Coclite et al. [6, 5], while global conservative solutions have been derived by Bressan and coworkers [1, 2] and Holden and Raynaud [10, 11].

Multipeakons are fundamental building blocks for general solutions. Indeed, if the initial data $\bar{u}$ is in H^1 and $\bar{m} := \bar{u} - \bar{u}''$ is a positive Radon measure, then it can be proved, see [13], that one can construct a sequence of multipeakons that converges in $L^\infty_{\text{loc}}(\mathbb{R}; H^1_{\text{loc}}(\mathbb{R}))$ to the unique global solution of the Camassa–Holm equation. Here we extend this analysis to avoid the sign constraint on $\bar{m}$ for given initial data $\bar{u}$. We are thereby able to include the case where one may have wave breaking, and we select the conservative solution. This is in part based on [13] and a reformulation of the formulas for multipeakons [9]. In Bressan–Fonte [2] the approximation of general initial data by multipeakons is proved in an abstract way, while we in this chapter provide a constructive approach amenable to numerical computations. The method is illustrated on concrete examples.

A completely different numerical approach based on a semi discrete finite difference scheme can be found in [12].

2 Description of the Method

It turns out to be useful to reformulate the Camassa–Holm equation as the following system

$$u_t + uu_x + P_x = 0, \qquad P - P_{xx} = u^2 + \frac{1}{2}u_x^2. \tag{4}$$

In [10] a continuous semigroup of conservative solutions for the Camassa–Holm equation is constructed. The semigroup is defined on the set $\mathcal{D}$ of pair (u,μ) where $u \in H^1(\mathbb{R})$ and μ is a Radon measure satisfying $\mu_{\rm ac} = (u^2+u_x)\,dx$. The conservative solutions are obtained by a change of variable to Lagrangian coordinates. The new set of coordinates (y,U,H) correspond to the characteristics, $y_t(t,\xi) = u(t,y(t,\xi))$, Lagrangian velocity, $U(t,\xi) = u(t,y(t,\xi))$, and the Lagrangian energy distribution, $H(t,\xi) = \int_{-\infty}^{y(t,\xi)}(u^2+u_x^2)\,dx$, respectively. They satisfy the following system of ordinary differential equations in some Banach space E:

$$y_t = U, \qquad U_t = -Q, \qquad H_t = U^3 - 2PU, \tag{5}$$

where

$$Q(t,\xi) = -\frac{1}{4}\int_{\mathbb{R}} sgn(\xi-\eta)\exp\big(-sgn(\xi-\eta)(y(\xi)-y(\eta))\big)\big(U^2y_\xi + H_\xi\big)(\eta)\,d\eta,$$
$$P(t,\xi) = \frac{1}{4}\int_{\mathbb{R}} \exp\big(-sgn(\xi-\eta)(y(\xi)-y(\eta))\big)\big(U^2y_\xi + H_\xi\big)\,d\eta.$$

The set $\mathcal{D}$ is in bijection with a closed subset of E, see [10], and the topology on $\mathcal{D}$ is the topology of E transported to $\mathcal{D}$ by this bijection. In [10], we prove that if $u_n \to u$ in $H^1(\mathbb{R})$ then $(u_n,(u_n^2+u_{n,x}^2)\,dx) \to (u,(u^2+u_x^2)\,dx)$ and if $(u_n,\mu_n) \to (u,\mu)$ in $\mathcal{D}$ then $u_n \to u$ in $L^\infty(\mathbb{R})$ and $\mu_n \overset{*}{\rightharpoonup} \mu$.

Multipeakons are particular solutions of the Camassa–Holm equation given by (2)–(3) up to collision time, that is, a time t_c such that $q_i(t_c) = q_{i+1}(t_c)$ for some i. At collision time, the system (3) blows up and we have $\lim_{t\to t_c} p_i(t) = -\lim_{t\to t_c} p_{i+1}(t) = \infty$.

An equivalent definition of a multipeakon with n peaks is given by the set of pairs (q_i,u_i) for $i=1,\dots,n$, where q_i represents the position of the ith peak and u_i denotes its height. Naturally, we must impose that if $q_i = q_j$ for some i and j, then $u_i = u_j$. The multipeakon function u corresponding to (2) for a given time is then given on the whole real axis as the solution of the boundary value problem: $u - u_{xx} = 0$, $u(x_i) = u_i$, $u(x_{i+1}) = u_{i+1}$. We denote by $\mathcal{M}_n$ the set of all such functions, i.e., functions that correspond to initial data of multipeakon solutions with n peaks. Furthermore, we set $\mathcal{M} = \cup_{n=1}^{\infty}\mathcal{M}_n$. The two representations are equivalent, and later we explain how one goes from one to the other. The advantage of the second representation is that it fits directly into the Lagrangian approach. In [9], we follow this direction and present a system of ordinary differential equations for the conservative multipeakon solutions of the Camassa–Holm equation. Conservative solutions require that we take into account the energy density $(u^2+u_x^2)\,dx$. We introduce the variables H_i, for $i=1,\dots,n$, which correspond to energy contained between $-\infty$ and q_i, i.e., $H_i = \int_{-\infty}^{q_i}(u^2+u_x^2)\,dx$, and the differences $\delta H_i = H_{i+1} - H_i$. When $q_i \neq q_{i+1}$, δH_i is given by

$$\delta H_i = H_{i+1} - H_i = \int_{q_i}^{q_{i+1}} (u^2 + u_x^2)\,dx. \tag{6}$$

Because of the special structure of multipeakons, one can compute δH_i in terms of q_i and u_i. We have

$$\delta H_i = 2\bar{u}_i^2 \tanh(\delta q_i) + 2\delta u_i^2 \coth(\delta q_i) \tag{7}$$

where we for convenience have introduced the variables

$$\bar{q}_i = \frac{1}{2}(q_i+q_{i+1}),\quad \delta q_i = \frac{1}{2}(q_{i+1}-q_i),\quad \bar{u}_i = \frac{1}{2}(u_i+u_{i+1}),\quad \delta u_i = \frac{1}{2}(u_{i+1}+u_i).$$

In [9], we prove that the structure of multipeakons is preserved by the continuous semigroup of conservative solutions of the Camassa–Holm equation and the equations for (q_i, u_i, H_i) are given by the following system

$$\frac{dq_i}{dt} = u_i,\qquad \frac{du_i}{dt} = -Q_i,\qquad \frac{dH_i}{dt} = u_i^3 - 2P_i u_i, \tag{8}$$

with $P_i = \sum_{j=0}^{n} P_{ij}$, $Q_i = -\sum_{j=0}^{n} \kappa_{ij} P_{ij}$, and

$$P_{ij} = \begin{cases} \mathrm{e}^{(q_1-q_i)}\frac{u_1^2}{4} & \text{for } j=0,\\ \frac{\mathrm{e}^{-\kappa_{ij}q_i}\mathrm{e}^{\kappa_{ij}\bar{q}_j}}{8\cosh(\delta q_j)}\big[2\delta H_j \cosh^2(\delta q_j) & \\ \quad + 8\kappa_{ij}\bar{u}_j \delta u_j \sinh^2(\delta q_j) + 4\bar{u}_j^2 \tanh(\delta q_j)\big] & \text{for } j = 1,\dots,n-1,\\ \mathrm{e}^{(q_i-q_n)}\frac{u_n^2}{4} & \text{for } j=n. \end{cases} \tag{9}$$

The system (8) is well-posed and has global solutions.

Given initial data $\bar{u} \in H^1(\mathbb{R})$, we denote by (u,μ) the conservative solution of the Camassa–Holm equation with initial data $(\bar{u}, (\bar{u}^2 + \bar{u}_x^2)\,dx)$. We will prove that there exists a sequence $\bar{u}_n$ in $\mathcal{M}$, which converges to $\bar{u}$ in $H^1(\mathbb{R})$. We know that $(\bar{u}_n, (\bar{u}_n^2 + \bar{u}_{n,x}^2)\,dx)$ then converges to $(\bar{u}, (\bar{u}^2 + \bar{u}_x^2)\,dx)$ in $\mathcal{D}$. We compute the multipeakon solution (u_n, μ_n) with initial data $(\bar{u}_n, \bar{u}_n^2 + \bar{u}_{n,x}^2)\,dx)$ by solving the system of ordinary differential equations (8). Since the semigroup is continuous, $(u_n, \mu_n) \to (u, \mu)$ in $C([0,T], \mathcal{D})$ for any $T > 0$ and therefore $u_n \to u$ in $L^\infty(\mathbb{R})$.

3 Approximation of the Initial Data

In [2], Bressan and Fonte show that $\mathcal{M}$ is dense in $H^1(\mathbb{R})$. The proof is short and elegant but it is not constructive and therefore not suited to numerical applications. Here we define constructive approximation procedure.

For any integer n and any points x_i, $i = 1, \dots, n$, such that $x_i \leq x_{i+1}$, we denote by $\mathcal{P}$ the partition corresponding to those points. Let $L_{\mathcal{P}}$ be the span of the partition, i.e., $L_{\mathcal{P}} = x_n - x_1$ and $\Delta_{\mathcal{P}}$ the maximum step size, i.e., $\Delta_{\mathcal{P}} = \max_{i=1,\dots,n-1}(x_{i+1} - x_i)$. Given a partition $\mathcal{P} = \{x_1, \dots, x_n\}$, we define by $I_{\mathcal{P}}$ the operator from $H^1(\mathbb{R})$ to $\mathcal{M}$ defined as follows: For any $f \in H^1(\mathbb{R})$, we set $f_i = f(x_i)$ and $u = I_{\mathcal{P}}(f)$ where u is solution of the boundary problem

$$u - u_{xx} = 0, \; u(x_i) = f_i, \; u(x_{i+1}) = f_{i+1} \tag{10}$$

in (x_i, x_{i+1}). The definition (10) extends naturally to the intervals $(-\infty, x_1)$ and (x_n, ∞) by setting $x_0 = -x_{n+1} = -\infty$ and $u_0 = u_{n+1} = 0$. We have the following theorem.

Theorem 1. *For any $f \in H^1(\mathbb{R})$, $I_{\mathcal{P}}(f)$ converges to f in $H^1(\mathbb{R})$ when $\Delta_{\mathcal{P}}$ tends to zero and $L_{\mathcal{P}}$ tends to infinity.*

Proof. Given a partition $\mathcal{P}$, the map $I_{\mathcal{P}} \colon \mathcal{M} \to H^1(\mathbb{R})$ is linear and continuous. Let us prove that $I_{\mathcal{P}}$ is uniformly continuous, that is, there exists a constant C such that

$$\|I_{\mathcal{P}}(f)\|_{H^1(\mathbb{R})} \leq C \|f\|_{H^1(\mathbb{R})} \tag{11}$$

for any $f \in H^1(\mathbb{R})$ and any partition $\mathcal{P}$ such that $\Delta_{\mathcal{P}} \leq 1$. The $H^1(\mathbb{R})$ norm of $I_{\mathcal{P}}(f)$ can be obtained from (6) and (7):

$$\|I_{\mathcal{P}}(f)\|^2_{H^1(\mathbb{R})} = f_1^2 + f_n^2 + \sum_{i=1}^{n-1} \Big(2\Big(\frac{f_i + f_{i+1}}{2}\Big)^2 \tanh\Big(\frac{x_{i+1} - x_i}{2}\Big) + 2\Big(\frac{f_{i+1} - f_i}{2}\Big)^2 \coth\Big(\frac{x_{i+1} - x_i}{2}\Big)\Big). \tag{12}$$

Let $\alpha = \inf_{x \in [x_i, x_{i+1}]} f(x)^2$ and $x_{\min}$ be such that $f(x_{\min})^2 = \alpha$. Since f is continuous, $x_{\min}$ and α are well-defined. It follows from the definition of α that

$$\alpha(x_{i+1} - x_i) \leq \|f\|^2_{L^2(x_i, x_{i+1})}. \tag{13}$$

We have, for any $x \in [x_i, x_{i+1}]$,

$$\begin{aligned} f(x)^2 &= \alpha + 2 \int_{x_{\min}}^{x} f(t) f'(t)\, dt \leq \alpha + 2 \|f\|^2_{H^1(x_i, x_{i+1})} \\ &\leq \frac{1}{x_{i+1} - x_i} \|f\|^2_{L^2(x_i, x_{i+1})} + 2 \|f\|^2_{H^1(x_i, x_{i+1})}, \end{aligned}$$

from (13), which implies $f(x)^2(x_{i+1}-x_i) \leq 3\,\|f\|^2_{H^1(x_i,x_{i+1})}$ since, by assumption, $x_{i+1}-x_i \leq 1$. Hence, $\|f\|^2_{L^\infty(x_i,x_{i+1})}(x_{i+1}-x_i) \leq 3\,\|f\|^2_{H^1(x_i,x_{i+1})}$. We can estimate the first term of the sum in (12):

$$2\Big(\frac{f_i+f_{i+1}}{2}\Big)^2 \tanh\Big(\frac{x_{i+1}-x_i}{2}\Big) \leq 2L_1(x_{i+1}-x_i)\,\|f\|^2_{L^\infty(x_i,x_{i+1})},$$

where $L_1 = \sup_{z\in[0,1]}\Big(\frac{\tanh(z/2)}{z}\Big)$ and therefore

$$2\Big(\frac{f_i+f_{i+1}}{2}\Big)^2 \tanh\Big(\frac{x_{i+1}-x_i}{2}\Big) \leq 6L_1\,\|f\|^2_{H^1(x_i,x_{i+1})}. \tag{14}$$

As far as the other term in the sum is concerned, we have, by the Cauchy–Schwarz inequality, $|f_{i+1}-f_i| \leq \int_{x_i}^{x_{i+1}}|f_x|\,dx \leq \sqrt{x_{i+1}-x_i}\,\|f\|_{H^1(x_i,x_{i+1})}$ and therefore

$$2\Big(\frac{f_{i+1}-f_i}{2}\Big)^2 \coth\Big(\frac{x_{i+1}-x_i}{2}\Big) \leq \frac{L_2}{2}\,\|f\|^2_{H^1(x_i,x_{i+1})}, \tag{15}$$

where $L_2 = \sup_{z\in[0,1]}\coth(z/2)z$. Because of the Sobolev embedding $H^1(\mathbb{R}) \subset L^\infty(\mathbb{R})$, there exists a constant L_0 such that $f_1^2 + f_n^2 \leq L_0\,\|f\|^2_{H^1(\mathbb{R})}$. Hence, gathering (12), (14), and (15), we obtain (11). Let us prove now that for any smooth function ϕ with compact support, $I_\mathcal{P}(\phi) \to \phi$ in $H^1(\mathbb{R})$ when $\Delta_\mathcal{P} \to 0$ and $L_\mathcal{P} \to \infty$. We denote $v_\mathcal{P} = \phi - I_\mathcal{P}(\phi)$. We have that $v_\mathcal{P}$ satisfies in each interval $[x_i, x_{i+1}]$, $v_\mathcal{P} - v_{\mathcal{P},xx} = \phi - \phi_{xx}$ and $v_\mathcal{P}(x_i) = v_\mathcal{P}(x_{i+1}) = 0$. By integration by part, we obtain

$$\begin{aligned}\|v_\mathcal{P}\|^2_{H^1(x_i,x_{i+1})} &= \int_{x_i}^{x_{i+1}}(v_\mathcal{P} - v_{\mathcal{P},xx})v_\mathcal{P}\,dx = \int_{x_i}^{x_{i+1}}(\phi-\phi_{xx})v_\mathcal{P}\,dx \\ &\leq \|\phi-\phi_{xx}\|_{L^2(x_i,x_{i+1})}\,\|v_\mathcal{P}\|_{L^2(x_i,x_{i+1})},\end{aligned} \tag{16}$$

after using Cauchy–Schwarz. Hence,

$$\|v_\mathcal{P}\|_{H^1(x_i,x_{i+1})} \leq \|\phi-\phi_{xx}\|_{L^2(x_i,x_{i+1})} \leq C\sqrt{\Delta_\mathcal{P}}, \tag{17}$$

where $C = \sup_{x\in\mathbb{R}}(\phi-\phi_{xx})$. Since $v_\mathcal{P}(x_i) = 0$, we have $v_\mathcal{P}(x)^2 = 2\int_{x_i}^x v_\mathcal{P}v_{\mathcal{P},x}\,dx \leq 2\,\|v_\mathcal{P}\|^2_{H^1(x_i,x_{i+1})}$ and therefore, after using (17), we obtain $\|v_\mathcal{P}\|_{L^\infty(x_i,x_{i+1})} \leq 2C\sqrt{\Delta_\mathcal{P}}$, which, together with (16), implies

$$\|v_\mathcal{P}\|^2_{H^1(x_i,x_{i+1})} \leq 2C\sqrt{\Delta_P}\int_{x_i}^{x_{i+1}}|\phi-\phi_{xx}|\,dx. \tag{18}$$

When $L_\mathcal{P}$ is large enough, ϕ and $I_\mathcal{P}(\phi)$ vanish in $(-\infty, x_1)$ and (x_n, ∞) so that $\|v_\mathcal{P}\|^2_{H^1(\mathbb{R})} = \sum_{i=1}^{n-1}\|v_\mathcal{P}\|^2_{H^1(x_i,x_{i+1})}$. From (18), it follows that $\|v_\mathcal{P}\|^2_{H^1(\mathbb{R})} \leq 2C\sqrt{\Delta_\mathcal{P}}\,\|\phi-\phi_{xx}\|_{L^1(\mathbb{R})}$. Hence, we have proved that $I_\mathcal{P}(\phi) \to \phi$ in $H^1(\mathbb{R})$ when $\Delta_\mathcal{P} \to 0$ and $L_\mathcal{P} \to \infty$. We conclude the proof of the theorem by a standard approximation argument. For any $\varepsilon > 0$, there exists a smooth function ϕ with compact support such that $\|f - \phi\|_{H^1(\mathbb{R})} \leq \varepsilon$. For any $\Delta_{\mathcal{P}} \leq 1$ small enough and $L_{\mathcal{P}}$ large enough, we have $\|\phi - I_{\mathcal{P}}(\phi)\|_{H^1(\mathbb{R})} \leq \varepsilon$. Hence, by the uniform continuity of $I_{\mathcal{P}}$, we have

$$\begin{aligned}\|f - I_{\mathcal{P}}(f)\|_{H^1(\mathbb{R})} &\leq \|f - \phi\|_{H^1(\mathbb{R})} + \|\phi - I_{\mathcal{P}}(\phi)\|_{H^1(\mathbb{R})} + \|I_{\mathcal{P}}(\phi) - I_{\mathcal{P}}(f)\|_{H^1(\mathbb{R})} \\ &\leq (2 + C)\varepsilon. \qquad \square\end{aligned}$$

We choose a sequence of partition $\mathcal{P}_n$ such that $\Delta_{\mathcal{P}_n} \to 0$ and $L_{\mathcal{P}_n} \to 0$. We denote by $x_{i,n}$ the points of this partition, for example, we can take $x_{i,n} = i/n - n$ for $i = 1, \ldots, n^2$. The numerical scheme described in the following theorem converges.

Theorem 2. *For any initial data $\bar{u} \in H^1(\mathbb{R})$, let us define $\bar{u}_n \in \mathcal{M}$ by $\bar{u}_n = I_{\mathcal{P}_n}(\bar{u})$. We denote by u_n the solution of (8) with initial condition given by $(\bar{q}_{i,n} = x_{i,n}, \bar{u}_{i,n} = \bar{u}_n(x_{i,n}), \bar{H}_{i,n})$, where the $\bar{H}_{i,n}$ are computed from $\bar{u}_{i,n}$ and $\bar{q}_{i,n}$ using (7). Then, u_n converges to the global conservative solution u of the Camassa–Holm equation in $C([0,T], L^\infty(\mathbb{R}))$, for any $T > 0$.*

Proof. By Theorem 1, we have that $\bar{u}_n \to \bar{u}$ in $H^1(\mathbb{R})$. From [10, Proposition 5.1] it follows that $(\bar{u}_n, (\bar{u}_n^2 + \bar{u}_{x,n}^2)\,dx)$ converges to $(\bar{u}_n, (\bar{u}^2 + \bar{u}_x^2)\,dx)$ in $\mathcal{D}$. The conservative solution of the Camassa–Holm equation constitues a continuous semigroup in $\mathcal{D}$. Hence, $(\bar{u}_n, (u_n^2 + u_{n,x}^2)\,dx)$ converges to $(u, (u^2 + u_x^2)\,dx)$ in $C([0,T], \mathcal{D})$ and therefore in $C([0,T], L^\infty(\mathbb{R}))$, by [10, Proposition 5.2]. $\square$

4 Numerical Example

We consider the initial data $\bar{u}$ given by $\bar{u}(x) = x(x^2 - 1)e^{-x^2/4}$. The computed solution is shown in Fig. 2 for different times t. Figure 1 shows that the scheme is remarkably stable as, after a number of collisions, the solution computed

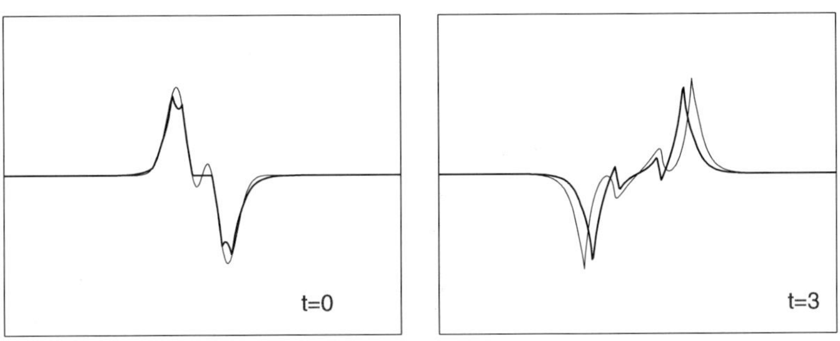

Fig. 1. *Left*: Approximation of the function $\bar{u}(x) = x(x^2 - 1)e^{-x^2/4}$ by $n = 10$ equidistributed peakons. *Right*: Comparison of the computed solutions for $n = 10$ peakons (in bold) and $n = 100$ peakons at time $t = 3$

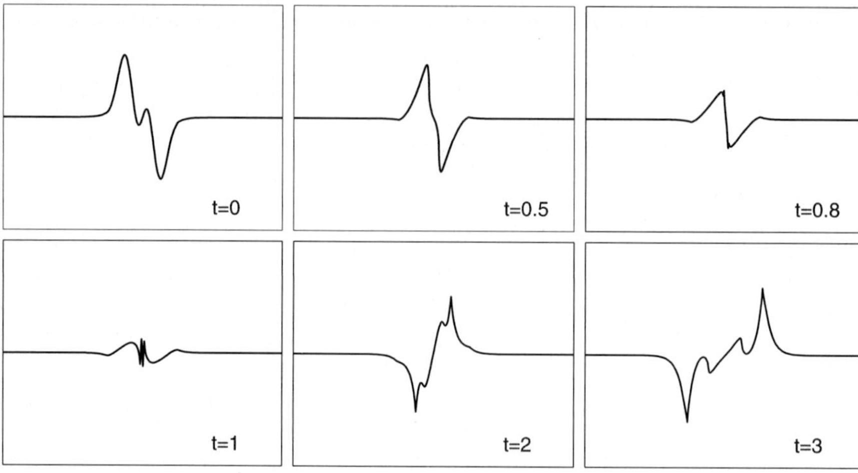

Fig. 2. The computed solution for $n = 100$ peakons shown at different times

with only $n = 10$ peakons remains close to the solution computed with much higher accuracy.

Acknowledgments

The research is supported by the Research Council of Norway.

References

1. Alberto Bressan and Adrian Constantin. Global conservative solutions of the Camassa–Holm equation. *Arch. Ration. Mech. Anal.*, 183(2):215–239, 2007.
2. Alberto Bressan and Massimo Fonte. An optimal transportation metric for solutions of the Camassa–Holm equation. *Methods Appl. Anal.*, 12:191–220, 2005.
3. Roberto Camassa and Darryl D. Holm. An integrable shallow water equation with peaked solitons. *Phys. Rev. Lett.*, 71(11):1661–1664, 1993.
4. Roberto Camassa, Darryl D. Holm, and James Hyman. A new integrable shallow water equation. *Adv. Appl. Mech.*, 31:1–33, 1994.
5. G.M. Coclite, H. Holden, and K.H. Karlsen. Global weak solutions to a generalized hyperelastic-rod wave equation. *SIAM J. Math. Anal.*, 37(4):1044–1069 (electronic), 2005.
6. Giuseppe Maria Coclite, Helge Holden, and Kenneth Hvistendahl Karlsen. Wellposedness for a parabolic-elliptic system. *Discrete Contin. Dyn. Syst.*, 13(3):659–682, 2005.
7. Adrian Constantin. Existence of permanent and breaking waves for a shallow water equation: a geometric approach. *Ann. Inst. Fourier (Grenoble)*, 50(2):321–362, 2000.

8. Adrian Constantin and Joachim Escher. Wave breaking for nonlinear nonlocal shallow water equations. *Acta Math.*, 181(2):229–243, 1998.
9. Helge Holden and Xavier Raynaud. Global conservative multipeakon solutions of the Camassa–Holm equation. *J. Hyperbolic Differ. Equ.*, 4(1):39–64, 2007.
10. Helge Holden and Xavier Raynaud. Global conservative solutions of the Camassa–Holm equation – a lagrangian point of view. *Comm. Partial Differential Equations, to appear.*
11. Helge Holden and Xavier Raynaud. Global conservative solutions of the generalized hyperelastic-rod wave equation. *J. Differential Equations*, 233(2):448–484, 2007.
12. Helge Holden and Xavier Raynaud. Convergence of a finite difference scheme for the camassa–holm equation. *SIAM Journal on Numerical Analysis*, 44(4):1655–1680, 2006.
13. Helge Holden and Xavier Raynaud. A convergent numerical scheme for the Camassa-Holm equation based on multipeakons. *Discrete Contin. Dyn. Syst.*, 14(3):505–523, 2006.

Consistency of the Explicit Roe Scheme for Low Mach Number Flows in Exterior Domains

F. Rieper and G. Bader

Summary. This chapter presents an analysis of the explicit Roe scheme for the Euler equations of a perfect gas in the low Mach number regime. The CFL condition in explicit schemes is essential in the discrete asymptotic analysis, since it introduces a Strouhal number $\mathrm{Str} = 1/\mathrm{M}$ into the equations, where M is the reference Mach number. As a result, we can show the consistency of the Roe scheme with the asymptotic Euler equations on the acoustic timescale. Numerical experiments confirm these theoretical results down to a Mach number of $\mathrm{M} = 10^{-7}$.

1 Introduction

1.1 The Euler Equations

The Euler equations in primitive variables consist of the equations for the density ρ, the velocity $\mathbf{u}$, and the pressure p. They model the flow of an ideal fluid, i.e., a fluid without viscosity nor thermal conductivity. In nondimensional form

$$\mathrm{Str}\,\rho_t + \nabla \cdot (\rho \mathbf{u}) = 0 \tag{1a}$$

$$\mathrm{Str}\,\mathbf{u}_t + \mathbf{u} \cdot \nabla \mathbf{u} + \frac{1}{\mathrm{M}^2}\frac{1}{\rho}\nabla p = 0 \tag{1b}$$

$$\mathrm{Str}\,p_t + \nabla \cdot (\mathbf{u}p) + (\gamma - 1)p\nabla \cdot \mathbf{u} = 0, \tag{1c}$$

they include the Mach number $\mathrm{M} = u_{\mathrm{ref}}/a_{\mathrm{ref}}$ and the Strouhal number $\mathrm{Str} = l_{\mathrm{ref}}/(t_{\mathrm{ref}}\, u_{\mathrm{ref}})$. The quantities ρ_{ref}, u_{ref}, l_{ref}, t_{ref}, and p_{ref} are the globally defined reference quantities for density, flow velocity, length scale, timescale, and pressure, respectively. For the reference speed of sound, we set $a_{\mathrm{ref}}^2 = p_{\mathrm{ref}}/\rho_{\mathrm{ref}}$, valid for perfect gases.

1.2 The Strouhal Number

If we focus on the *flow*, the timescale is linked to the length scale and the *flow* velocity by $t_{\mathrm{ref}} = l_{\mathrm{ref}}/u_{\mathrm{ref}}$, and as a consequence we obtain $\mathrm{Str} = 1$.

For *acoustic phenomena*, time- and length scale are linked by the speed of sound $l_{\text{ref}}/t_{\text{ref}} = a_{\text{ref}}$, which leads to a different value of the Strouhal number $\text{Str} = 1/\text{M}$. In *explicit numerical schemes*, the choice of the time step Δt for a given cell size δ is restricted by the Courant–Friedrichs–Lewy (CFL) condition: the "grid velocity" $\delta/\Delta t$ is limited by the fastest signal speed – approximately the speed of sound a. This carries over to the reference quantities $l_{\text{ref}}/t_{\text{ref}} = (O)(a_{\text{ref}})$, so that for the acoustic Strouhal number it follows: $\text{Str} = 1/\text{M}$.

2 Continuous Asymptotic Analysis

The analysis and interpretation of the asymptotic equations on the timescale of the flow were done by Klein [2, 3]. For the acoustic timescale, we substitute the Strouhal number $\text{Str} = 1/\text{M}$ into the nondimensional Euler equations (1a)–(1c), replace the physical quantities by their asymptotic expansion

$$\phi = \phi^{(0)} + \text{M}\phi^{(1)} + \text{M}^2\phi^{(2)} \dots$$

and obtain for the relevant terms:

$$\rho_t^{(0)} = 0 \tag{2a}$$

$$\frac{1}{\rho^{(0)}} \nabla p^{(0)} = 0 \tag{2b}$$

$$\mathbf{u}_t^{(0)} + \left[\frac{\nabla p}{\rho}\right]^{(1)} = 0 \tag{2c}$$

$$\mathbf{u}_t^{(1)} + \mathbf{u}^{(0)} \cdot \nabla \mathbf{u}^{(0)} + \left[\frac{\nabla p}{\rho}\right]^{(2)} = 0 \tag{2d}$$

$$p_t^{(0)} = 0 \tag{2e}$$

$$p_t^{(1)} + \mathbf{u}^{(0)} \cdot \nabla p^{(0)} + \gamma p^{(0)} \nabla \cdot \mathbf{u}^{(0)} = 0 \tag{2f}$$

$$p_t^{(2)} + [\mathbf{u} \cdot \nabla p]^{(1)} + \gamma [p \nabla \cdot \mathbf{u}]^{(1)} = 0 \tag{2g}$$

The expressions in [ ]-brackets are asymptotic terms as a whole and can be expanded. Equation (2a) implies that the leading order density is constant in time

$$\rho^{(0)} = \rho^{(0)}(x)$$

and the mass transport due to convection is frozen. From (2b) and (2e), it follows that the background pressure is constant in space *and* time

$$p^{(0)} = \text{const.}$$

Wave Equation for $p^{(1)}$ and $\mathbf{u}^{(0)}$

Simplifying (2c) and (2f) using the constancy of $p^{(0)}$

$$\mathbf{u}_t^{(0)} + \frac{\nabla p^{(1)}}{\rho^{(0)}} = 0, \tag{3}$$

$$p_t^{(1)} + \gamma p^{(0)} \nabla \cdot \mathbf{u}^{(0)} = 0, \tag{4}$$

illustrates that $p^{(1)}$ and $\mathbf{u}^{(0)}$ form an *acoustic pressure–velocity couple*: the gradient of $p^{(1)}$ is the accelerating agent for $\mathbf{u}^{(0)}$, and the compression $\nabla \cdot \mathbf{u}^{(0)}$ leads to a pressure increase in $p^{(1)}$.

Wave Equation for $p^{(2)}$ and $\mathbf{u}^{(1)}$

In the evolution equation (2g) for $p^{(2)}$, the terms $\nabla p^{(1)}$ and $\nabla \cdot \mathbf{u}^{(0)}$ can be dropped if we assume no acoustic waves in $p^{(1)}$, e.g., in the absence of global compression. The remaining part

$$p_t^{(2)} = \gamma p^{(0)} \nabla \cdot \mathbf{u}^{(1)} \tag{5}$$

couples pressure changes in $p^{(2)}$ with the first-order compression $\nabla \cdot \mathbf{u}^{(1)}$. Expanding (2d) and dropping $\nabla p^{(0)}$, we obtain

$$\mathbf{u}_t^{(1)} + \mathbf{u}^{(0)} \cdot \nabla \mathbf{u}^{(0)} + \frac{1}{\rho^{(0)}} \nabla p^{(2)} - \frac{\nabla p^{(1)} \rho^{(1)}}{(\rho^{(0)})^2} = 0, \tag{6}$$

that shows that the gradient of $p^{(2)}$ acts as an accelerating agent for the velocity of order $\mathcal{O}(\mathrm{M})$. Therefore $\mathbf{u}^{(1)}$ and $p^{(2)}$ form a further *acoustic pressure–velocity couple.* The nonlinear source term $\mathbf{u}^{(0)} \cdot \nabla \mathbf{u}^{(0)}$ reflects the coupling with the flow.

3 Discrete Asymptotic Analysis of the Roe Scheme

An asymptotic analysis of the semidiscrete first-order Roe scheme was done by Viozat and Guillard [1]. Our analysis is different, as we restrict the analysis to explicit schemes with a Strouhal number of $1/\mathrm{M}$.

3.1 Asymptotic Equations

To make reading easier to those familiar with [1], we use the authors' notation for the equidistant Cartesian grid of cell size δ. The index of the grid node with the coordinates $(x, y) = (i\delta, j\delta)$ is denoted by $\mathbf{i} = (i, j)$. For neighboring nodes, we use the symbol $\nu(\mathbf{i}) = \{(i-1, j), (i+1, j), (i, j-1), (i, j+1)\}$ or $\nu = \{N, S, W, E\}$. The volume of the grid cell corresponding to node $\mathbf{i}$ is

$C_{\mathbf{i}} = [(i-1/2)\delta, (i+1/2)\delta] \times [(j-1/2)\delta, (j+1/2)\delta]$. The unit normal vector of the edge between cells **i** and **l** is $\mathbf{n_{il}}$.

After introducing the asymptotic expansions of the quantities in the Roe scheme, we obtain:

Equation of Order M^{-1}

$$p_{\mathrm{N}}^{(0)} - p_{\mathrm{S}}^{(0)} = 0; \quad p_{\mathrm{E}}^{(0)} - p_{\mathrm{W}}^{(0)} = 0. \qquad (\mathrm{C}_{-1})$$

Equations of Order M^0

$$\frac{\mathrm{d}}{\mathrm{d}t}\rho_{\mathbf{i}}^{(0)} + \sum_{\mathbf{l}\in\nu(\mathbf{i})} \frac{\Delta_{\mathbf{il}}p^{(0)}}{2\delta \cdot a_{\mathbf{il}}^{(0)}} = 0, \qquad (\mathrm{C}_0)$$

$$\frac{\mathrm{d}}{\mathrm{d}t}\rho_{\mathbf{i}}^{(0)}u_{\mathbf{i}}^{(0)} + \sum_{\mathbf{l}\in\nu(\mathbf{i})} \frac{p_{\mathbf{l}}^{(1)}(n_x)_{\mathbf{il}}}{2\delta} + \sum_{\mathbf{l}\in\nu(\mathbf{i})} \left\{ \frac{(U^{(0)}n_x + u^{(0)})_{\mathbf{il}}}{a_{\mathbf{il}}^{(0)}} \frac{\Delta_{\mathbf{il}}p^{(0)}}{2\delta} + \frac{\rho_{\mathbf{il}}^{(0)}a_{\mathbf{il}}^{(0)}(n_x)_{\mathbf{il}}\Delta_{\mathbf{il}}U^{(0)}}{2\delta} \right\} = 0, \qquad (\mathrm{M}_{x0})$$

$$\frac{\mathrm{d}}{\mathrm{d}t}\rho_{\mathbf{i}}^{(0)}e_{\mathbf{i}}^{(0)} + \sum_{\mathbf{l}\in\nu(\mathbf{i})} \frac{h_{\mathbf{il}}^{(0)}}{a_{\mathbf{il}}^{(0)}} \frac{\Delta_{\mathbf{il}}p^{(0)}}{2\delta} = 0. \qquad (\mathrm{E}_0)$$

Equations of Order M^1

$$\frac{\mathrm{d}}{\mathrm{d}t}\rho^{(1)} + \sum_{\mathbf{l}\in\nu(\mathbf{i})} \left\{ \frac{\Delta_{\mathbf{il}}p^{(1)}}{a_{\mathbf{il}}^{(0)} \cdot 2\delta} + \frac{\rho_{\mathbf{l}}^{(0)}\mathbf{u}_{\mathbf{l}}^{(0)} \cdot \mathbf{n_{il}}}{2\delta} + \frac{1}{2\delta}|U_{\mathbf{il}}^{(0)}|(\Delta_{\mathbf{il}}\rho^{(0)} - \frac{\Delta_{\mathbf{il}}p^{(0)}}{(a_{\mathbf{il}}^{(0)})^2}) \right\} = 0, \qquad (\mathrm{C}_1)$$

$$\begin{aligned} \frac{\mathrm{d}}{\mathrm{d}t}(\rho_{\mathbf{i}}u_{\mathbf{i}})^{(1)} &+ \sum_{\mathbf{l}\in\nu(\mathbf{i})} \frac{p_{\mathbf{l}}^{(2)}(n_x)_{\mathbf{il}}}{2\delta} \\ &+ \sum_{\mathbf{l}\in\nu(\mathbf{i})} \left\{ \frac{[(Un_x + u)_{\mathbf{il}}\Delta_{\mathbf{il}}p]^{(1)}}{a_{\mathbf{il}}^{(0)} \cdot 2\delta} + \frac{[\rho_{\mathbf{il}}a_{\mathbf{il}}(n_x)_{\mathbf{il}}\Delta_{\mathbf{il}}U]^{(1)}}{2\delta} \right. \\ &+ \frac{\rho_{\mathbf{l}}^{(0)}u_{\mathbf{l}}^{(0)}\mathbf{u}_{\mathbf{l}}^{(0)} \cdot \mathbf{n_{il}}}{2\delta} + \frac{1}{2\delta} \cdot |U_{\mathbf{il}}^{(0)}|(\Delta_{\mathbf{il}}\rho^{(0)} \\ &\left. - \frac{\Delta_{\mathbf{il}}p^{(0)}}{(a_{\mathbf{il}}^{(0)})^2})u_{\mathbf{il}}^{(0)} - \frac{\rho_{\mathbf{il}}^{(0)}|U_{\mathbf{il}}^{(0)}|(n_y)_{\mathbf{il}}\Delta_{\mathbf{il}}V^{(0)}}{2\delta} \right\} = 0, \end{aligned} \qquad (\mathrm{M}_{x1})$$

$$\frac{\mathrm{d}}{\mathrm{d}t}(\rho_{\mathbf{i}}e_{\mathbf{i}})^{(1)} + \sum_{\mathbf{l}\in\nu(\mathbf{i})} \left\{ \frac{(h_{\mathbf{il}}\Delta_{\mathbf{il}}p)^{(1)}}{a_{\mathbf{il}}^{(0)} \cdot 2\delta} + \frac{(\rho_{\mathbf{l}}^{(0)}e_{\mathbf{l}}^{(0)} + p_{\mathbf{l}}^{(0)})\mathbf{u}_{\mathbf{l}}^{(0)} \cdot \mathbf{n_{il}}}{2\delta} \right\} = 0. \qquad (\mathrm{E}_1)$$

3.2 Consistency

From (C_{-1}), it follows that the pressure $p^{(0)}$ has a four-value structure and it can easily be shown that fluctuations in $p^{(0)}$ will be damped out with time, since $p^{(0)}$ satisfies a discrete form of a diffusion equation.

Discrete Wave Equation for $p^{(1)}$ and $\mathbf{u}^{(0)}$

In (E_1) we replace the energy density by the pressure

$$(\rho_{\mathbf{i}} e_{\mathbf{i}})^{(1)} = \varepsilon_{\mathbf{i}}^{(1)} = \frac{1}{\gamma - 1} p_{\mathbf{i}}^{(1)}, \tag{7}$$

using the thermodynamic law of perfect gas. The assumption $p^{(0)} = \text{const}$ implies $h^{(0)} = \text{const}$, which simplifies the Roe averages to $h_{\mathbf{il}}^{(0)} = h_{\mathbf{i}}^{(0)}$ and $a_{\mathbf{il}}^{(0)} = a_{\mathbf{i}}^{(0)}$. After some algebra we obtain:

$$\underbrace{\frac{\mathrm{d}}{\mathrm{d}t} p_{\mathbf{i}}^{(1)}}_{\frac{\partial}{\partial t} p^{(1)}} + \underbrace{\frac{\gamma p_{\mathbf{i}}^{(0)}}{a_{\mathbf{i}}^{(0)}} \frac{\delta}{2} \sum_{\mathbf{l} \in \nu(\mathbf{i})} \frac{\Delta_{\mathbf{il}} p^{(1)}}{\delta^2}}_{\sim -\delta(\nabla^2 p^{(1)})_{\mathbf{i}}} + \underbrace{\gamma p_{\mathbf{i}}^{(0)} \sum_{\mathbf{l} \in \nu(\mathbf{i})} \frac{\mathbf{u}_{\mathbf{l}}^{(0)} \cdot \mathbf{n}_{\mathbf{il}}}{2\delta}}_{\gamma p^{(0)} \nabla \cdot \mathbf{u}^{(0)}} = 0. \tag{8}$$

Identifying the differential operators and letting $\delta \to 0$, we see the consistency with (4).

Evolution Equation for $\mathbf{u}^{(0)}$

In (M_{x0}), we omit $\Delta_{\mathbf{il}} p^{(0)}$ and replace the Roe averages of constant quantities by their node values $\rho_{\mathbf{il}}^{(0)} = \rho_{\mathbf{i}}^{(0)}$ and $a_{\mathbf{il}}^{(0)} = a_{\mathbf{i}}^{(0)}$ to obtain

$$\underbrace{\frac{\mathrm{d}}{\mathrm{d}t} u_{\mathbf{i}}^{(0)}}_{\frac{\partial}{\partial t} u^{(0)}} + \underbrace{\frac{1}{\rho_{\mathbf{i}}^{(0)}} \sum_{\mathbf{l} \in \nu(\mathbf{i})} \frac{p_{\mathbf{l}}^{(1)} (n_x)_{\mathbf{il}}}{2\delta}}_{\frac{1}{\rho^{(0)}} (\nabla p^{(1)})_x} + \underbrace{\frac{a_{\mathbf{i}}^{(0)}}{\rho_{\mathbf{i}}^{(0)}} \sum_{\mathbf{l} \in \nu(\mathbf{i})} \frac{\rho_{\mathbf{il}}^{(0)} \Delta_{\mathbf{il}} U^{(0)}}{2\delta} (n_x)_{\mathbf{il}}}_{\sim -\delta \nabla^2 u^{(0)}} = 0. \tag{9}$$

An analogous equation can be obtained for the other components of $\mathbf{u}$. If we let $\delta \to 0$, the numerical viscosity of the correct order $\mathcal{O}(\delta)$ disappears and we retrieve the x-component of the continuous equation (3).

Discrete Wave Equation for $p^{(2)}$

To show the consistency of the explicit Roe scheme concerning the evolution of the $p^{(2)}$ and $\mathbf{u}^{(1)}$ field, we subtract the kinetic energy $\frac{1}{2} \rho_{\mathbf{i}}^{(0)} (u_{\mathbf{i}}^2 + v_{\mathbf{i}}^2)^{(0)}$

from the *Total Energy Equation of Order* M^2 and obtain after some algebra the discrete evolution equation for $p^{(2)}$:

$$\frac{1}{\gamma-1}\frac{\mathrm{d}}{\mathrm{d}t}p_{\mathbf{i}}^{(2)} - u_{\mathbf{i}}^{(0)} \sum_{\mathbf{l}\in\nu(\mathbf{i})} \frac{p_{\mathbf{l}}^{(1)}(n_x)_{\mathbf{il}}}{2\delta} - v_{\mathbf{i}}^{(0)} \sum_{\mathbf{l}\in\nu(\mathbf{i})} \frac{p_{\mathbf{l}}^{(1)}(n_y)_{\mathbf{il}}}{2\delta}$$
$$+ \frac{1}{2\delta}\frac{\gamma}{\gamma-1} \sum_{\mathbf{l}\in\nu(\mathbf{i})} \left\{ p_{\mathbf{l}}^{(0)}\mathbf{u}_{\mathbf{l}}^{(1)} \cdot \mathbf{n}_{\mathbf{il}} + p_{\mathbf{l}}^{(1)}\mathbf{u}_{\mathbf{l}}^{(0)} \cdot \mathbf{n}_{\mathbf{il}} \right\} + T_\eta = 0. \qquad (\mathrm{P}_2)$$

The abbreviated term T_η can be simplified using the assumption of constant $p^{(0)}$ and applying the definition $U = \mathbf{u}\cdot\mathbf{n}$ and $U_{\mathbf{il}}^{(0)} = \mathbf{u}_{\mathbf{il}}^{(0)} \cdot \mathbf{n}_{\mathbf{il}}$, giving

$$T_\eta = \sum_{\mathbf{l}\in\nu(\mathbf{i})} \frac{\rho_{\mathbf{il}}^{(0)} U_{\mathbf{il}}^{(0)} h_{\mathbf{il}}^{(0)}}{a_{\mathbf{il}}^{(0)}} \frac{\Delta_{\mathbf{il}} U^{(0)}}{2\delta}$$
$$+ \frac{1}{2\delta} \sum_{\mathbf{l}\in\nu(\mathbf{i})} \left\{ \frac{h_{\mathbf{il}}^{(0)}}{a_{\mathbf{il}}^{(0)}} \Delta_{\mathbf{il}} p^{(2)} + \frac{h_{\mathbf{il}}^{(1)}}{a_{\mathbf{il}}^{(0)}} \Delta_{\mathbf{il}} p^{(1)} - \frac{h_{\mathbf{il}}^{(0)}}{(a_{\mathbf{il}}^{(0)})^2} a_{\mathbf{il}}^{(1)} \Delta_{\mathbf{il}} p^{(1)} \right\}.$$

All terms of the form

$$\sum_{\mathbf{l}\in\nu(\mathbf{i})} \alpha_{\mathbf{l}} \frac{\Delta_{\mathbf{il}}\phi}{2\delta} \qquad (10)$$

are numerical viscosity terms of order $\mathcal{O}(\delta)$ that can be shown by expanding $\alpha(\mathbf{x})$ in a Taylor series about grid node $\mathbf{x_i}$. In the limit of vanishing grid size $\delta \to 0$, the numerical viscosity terms disappear and we obtain the continuous equation (5) showing consistency for the evolution of $p^{(2)}$.

Evolution Equation for $\mathbf{u}^{(1)}$

We apply to (M_{x1}) the product rule and the rules for expanding asymptotic expressions to obtain:

$$\rho_{\mathbf{i}}^{(0)} \frac{\mathrm{d}}{\mathrm{d}t} u_{\mathbf{i}}^{(1)} + u_{\mathbf{i}}^{(1)} \frac{\mathrm{d}}{\mathrm{d}t} \rho_{\mathbf{i}}^{(0)} + \rho_{\mathbf{i}}^{(1)} \frac{\mathrm{d}}{\mathrm{d}t} u_{\mathbf{i}}^{(0)} + u_{\mathbf{i}}^{(0)} \frac{\mathrm{d}}{\mathrm{d}t} \rho_{\mathbf{i}}^{(1)}$$
$$+ \sum_{\mathbf{l}\in\nu(\mathbf{i})} \frac{p_{\mathbf{l}}^{(2)}(n_x)_{\mathbf{il}}}{2\delta} + \sum_{\mathbf{l}\in\nu(\mathbf{i})} \frac{\rho_{\mathbf{l}}^{(0)} u_{\mathbf{l}}^{(0)} \mathbf{u}_{\mathbf{l}}^{(0)} \cdot \mathbf{n}_{\mathbf{il}}}{2\delta} + T_\eta = 0. \qquad (11)$$

The term T_η is a sum of terms containing $\Delta_{\mathbf{il}}(\cdot)$ and can be identified as numerical viscosities in analogy to the previous paragraph. Under the assumption of constant $p^{(0)}$, all gradients of $p^{(0)}$ vanish and the continuity equation (C_0)

$$\frac{\mathrm{d}}{\mathrm{d}t}\rho_{\mathbf{i}}^{(0)} = 0$$

makes the second term of (11) disappear. To eliminate the third term, we use the x-momentum equation (I_x0) and for the fourth term of (11), we use the first-order continuity equation (C_1) and obtain after some tedious algebra:

$$\frac{\mathrm{d}}{\mathrm{d}t}u_{\mathbf{i}}^{(1)} - \frac{u_{\mathbf{i}}^{(0)}}{\rho_{\mathbf{i}}^{(0)}} \sum_{\mathbf{l}\in\nu(\mathbf{i})} \frac{\rho_{\mathbf{l}}^{(0)}\mathbf{u}_{\mathbf{l}}^{(0)} \cdot \mathbf{n}_{\mathbf{il}}}{2\delta} + \frac{1}{\rho_{\mathbf{i}}^{(0)}} \sum_{\mathbf{l}\in\nu(\mathbf{i})} \frac{\rho_{\mathbf{l}}^{(0)} u_{\mathbf{l}}^{(0)}\mathbf{u}_{\mathbf{l}}^{(0)} \cdot \mathbf{n}_{\mathbf{il}}}{2\delta}$$
$$+ \frac{1}{\rho_{\mathbf{i}}^{(0)}} \sum_{\mathbf{l}\in\nu(\mathbf{i})} \frac{p_{\mathbf{l}}^{(2)}(n_x)_{\mathbf{il}}}{2\delta} - \frac{\rho_{\mathbf{i}}^{(1)}}{(\rho^{(0)})^2} \sum_{\mathbf{l}\in\nu(\mathbf{i})} \frac{p_{\mathbf{l}}^{(1)}(n_x)_{\mathbf{il}}}{2\delta} + \tilde{T}_\eta = 0.$$

Apart from $\mathbf{u}^{(0)} \cdot \nabla\mathbf{u}^{(0)}$ all terms in this discrete equation can easily be identified with the x-component of the continuous equation for $\mathbf{u}^{(1)}$:

$$\mathbf{u}_t^{(1)} + \mathbf{u}^{(0)} \cdot \nabla\mathbf{u}^{(0)} + \frac{1}{\rho^{(0)}}\nabla p^{(2)} - \frac{\nabla p^{(1)}\rho^{(1)}}{(\rho^{(0)})^2} = 0.$$

To identify $\mathbf{u}^{(0)} \cdot \nabla\mathbf{u}^{(0)}$ we write the term

$$T = \sum_{\mathbf{l}\in\nu(\mathbf{i})} \frac{\rho_{\mathbf{l}}^{(0)} u_{\mathbf{l}}^{(0)}\mathbf{u}_{\mathbf{l}}^{(0)} \cdot \mathbf{n}_{\mathbf{il}}}{2\delta} - u_{\mathbf{i}}^{(0)} \sum_{\mathbf{l}\in\nu(\mathbf{i})} \frac{\rho_{\mathbf{l}}^{(0)}\mathbf{u}_{\mathbf{l}}^{(0)} \cdot \mathbf{n}_{\mathbf{il}}}{2\delta}$$

in an explicit way:

$$T = \frac{\rho_{\mathrm{E}}u_{\mathrm{E}}^2 - \rho_{\mathrm{W}}u_{\mathrm{W}}^2 + \rho_{\mathrm{N}}u_{\mathrm{N}}v_{\mathrm{N}} - \rho_{\mathrm{S}}u_{\mathrm{S}}v_{\mathrm{S}} - \{\rho_{\mathrm{E}}u_{\mathrm{E}}u_{\mathbf{i}} - \rho_{\mathrm{W}}u_{\mathrm{W}}u_{\mathbf{i}} + \rho_{\mathrm{N}}v_{\mathrm{N}}u_{\mathbf{i}} - \rho_{\mathrm{S}}v_{\mathrm{S}}u_{\mathbf{i}}\}}{2\delta}.$$

Using Taylor series expansions, we obtain

$$T = \overline{(\rho u)}_{\mathbf{i}}\frac{\partial u}{\partial x} + \overline{(\rho v)}_{\mathbf{i}}\frac{\partial u}{\partial y} + \mathcal{O}(\delta),$$

and in the limit $\delta \to 0$, we can replace the averages by their node values:

$$\lim_{\delta\to 0} T = \rho_{\mathbf{i}}u_{\mathbf{i}}\frac{\partial u}{\partial x} + \rho_{\mathbf{i}}v_{\mathbf{i}}\frac{\partial u}{\partial y}.$$

This expression can be identified with the x-component of expression $\rho(\mathbf{u}\cdot\nabla\mathbf{u})$. The y-component can be derived in an analogous way.

4 Numerical Results

We compare ideal potential flow around a cylinder with results of a first-order Roe scheme on an unstructured mesh with 9,822 cells. Figure 1 shows lines of constant pressure for the analytical and the numerical solution. Further numerical experiments with different inflow Mach number show that the pressure fluctuations produced are of order $\mathcal{O}(\mathrm{M}^2)$.

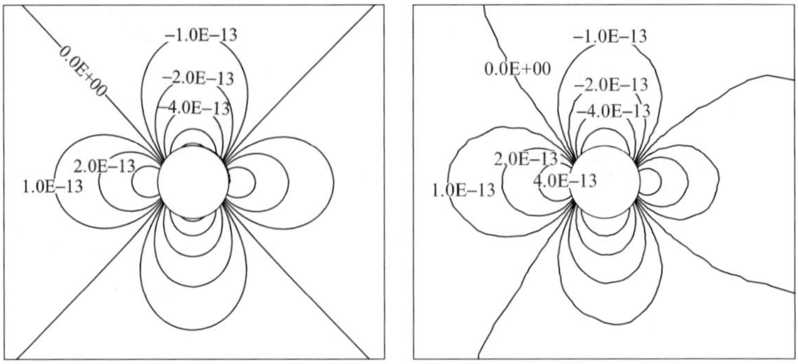

Fig. 1. Isolines of pressure for the flow around a cylinder; incompressible potential flow at $u_0 = \sqrt{\gamma}10^{-6}$ (*left*) and the first-order Roe scheme at $\mathrm{M} = 10^{-6}$ (*right*)

5 Conclusion and Outlook

We have shown that the evolution equations for the relevant asymptotic terms of the Roe scheme are consistent with the equations gained from the continuous analysis of the Euler equations. The explicit first-order Roe scheme – like the Euler equations – supports pressure waves of order $\mathcal{O}(\mathcal{M})$, but in steady flow only the pressure of order $\mathcal{O}(\mathrm{M}^2)$ remains. In a paper to appear in 2007, we show that there exist upwind schemes that produce an $\mathcal{O}(\mathrm{M})$ pressure field – albeit for a different reason, related to the numerical viscosity on the shear wave.

Acknowledgments

We thank the IAG Stuttgart, especially Dr. Roller, for making the HYDSOL code available to us.

References

1. Hervé Guillard and Cécile Viozat. On the behaviour of upwind schemes in the low Mach number limit. *Comput. Fluids*, 28(1):63–86, 1999.
2. Sergiu Klainerman and Andrew Majda. Compressible and incompressible fluids. *Commun. Pure Appl. Math.*, 35:629–651, 1982.
3. R. Klein. Semi-implicit extension of a Godunov-type scheme based on low Mach number asymptotics. I: One-dimensional flow. *J. Comput. Phys.*, 121(2):213–237, 1995.

Weak and Classical Solutions for a Model Problem in Radiation Hydrodynamics

C. Rohde, N. Tiemann, and W.-A. Yong

Summary. It has been observed for a long time that radiation effects can prevent the development of singularities of shock-wave type in solutions for mathematical models for compressible flows. We consider a multidimensional model problem in the form of a system of nonlinear hyperbolic balance laws and prove that the associated Cauchy problem can have smooth global solutions provided that the initial data are sufficiently close to an equilibrium state. Numerical experiments confirm this result but also show that shock waves can develop for large amplitude initial data.

1 Introduction

A widely used mathematical model for radiation-driven ideal compressible flows, in particular in astrophysics [4, 5], is given in $\mathbb{R}^d \times (0,T)$, $T>0$, $d \in \{1,2,3\}$, by the equations of gas dynamics coupled via an integral-type source term to a family of radiation transport equations:

$$\begin{aligned} \rho_t + \operatorname{div}(\rho v) &= 0, \\ (\rho v)_t + \operatorname{div}(\rho v \otimes v + p\, Id) &= 0, \\ (\rho e)_t + \operatorname{div}(\rho e v + p v) &= \rho \oint_{\mathcal{S}^{d-1}} \kappa(I(\cdot,\omega) - B(\theta))d\omega, \\ I(\cdot,\omega)_t + c\omega\cdot\nabla I(\cdot,\omega) &= c\rho(B(\theta) - I(\cdot,\omega)). \end{aligned} \tag{1}$$

Here the unknowns are the density of the fluid $\rho = \rho(x,t) > 0$, the velocity $v = v(x,t) \in \mathbb{R}^d$, the temperature $\theta = \theta(x,t) > 0$, and the radiation intensity $I = I(x,t,\omega) \geq 0$ for $(x,t) \in \mathbb{R}^d \times [0,\infty)$ and $\omega \in \mathcal{S}^{d-1}$. The vector ω denotes the direction of radiation. System (1) is closed by the relations

$$p = p(\rho,\theta), \quad B = B(\theta), \quad e = \epsilon(\rho,\theta) + \frac{1}{2}|v|^2,$$

where p denotes the given pressure function, $\kappa > 0$ a positive constant, B the Planck function, and ε the specific energy, respectively. $c > 0$ is the speed of light. Under standard thermodynamical assumptions, (1) turns out to be an infinite-dimensional hyperbolic balance law. We have to face the problem that solutions of (1) contain singularities of shock-wave type even if we consider the Cauchy problem for (1) with smooth initial data. However, it has been observed early (see, e.g., [1]) that the Cauchy problem for (1) might possess global smooth solutions due to radiative damping. In this contribution we will focus on the discrete-ordinate approximation for (1). The most basic approximation with two directions has been analyzed for the case of one-space dimension in the seminal work of Kawashima and coworkers ([2] and references therein, see also the recent work [3] on profiles). Rigorous results for the general discrete-ordinate approximation can be found in [9]. However, up to our knowledge no results on global smooth solutions in multiple-space dimensions are available.

With $\mathbf{f} := (f^1, \dots, f^d)^T \in C^\infty(\mathbb{R}_{>0}, \mathbb{R}^d)$, a multidimensional model problem for (1) in $\mathbb{R}^d \times (0, \infty)$ and for $\omega \in \mathcal{S}^{d-1}$ is given by

$$\begin{aligned} u_t + f^1(u)_{x_1} + \dots + f^d(u)_{x_d} &= \kappa \oint_{\mathcal{S}^{d-1}} (I(\cdot, \omega) - B(u))\, d\omega, \\ I(\cdot, \omega)_t + c\omega \cdot \nabla I(\cdot, \omega) &= c\rho(B(u) - I(\cdot, \omega)). \end{aligned} \tag{2}$$

This model relates to (1) as a scalar nonlinear conservation law to the gas dynamics equations. The unknown u takes the role of a lumped quantity of density, velocity, and temperature. Since (2) is still a (more difficult to handle) infinite-dimensional system, we introduce a finite-dimensional approximation. For this sake, let for each $L \in \mathbb{N}$ a partition of the unit sphere $\mathcal{S}^{d-1}$ into L subsets $\Omega_1, \Omega_2, \dots, \Omega_L$ of equal size be given such that it satisfies properties

$$\mathcal{S}^{d-1} = \Omega_1 \cup \dots \cup \Omega_L,\ \mathring{\Omega}_l \cap \mathring{\Omega}_k = \emptyset\ \forall\ k \neq l. \tag{3}$$

Choose for $l = 1, \dots, L$ the discrete ordinates $\omega^l \in \Omega_l$ as arbitrary vectors and define

$$\sigma_L := |\Omega_l| = L^{-1} |\mathcal{S}^{d-1}|. \tag{4}$$

Now, consider for the unknown $U = (u, I^1, \dots, I^L)^T : \mathbb{R}^d \times [0, \infty) \to \mathcal{U} := \mathbb{R}^{L+1}_{>0}$ the Cauchy problem

$$\begin{aligned} u_t + f^1(u)_{x_1} + \dots + f^d(u)_{x_d} &= \kappa \sigma_L \sum_{l=1}^{L} (I_l - B(u)) \\ I^1_t + c\omega^1 \cdot \nabla I^1 &= c(B(u) - I^1) \qquad \text{in } \mathbb{R}^d \times (0, \infty) \\ &\dots \\ I^L_t + c\omega^L \cdot \nabla I^L &= c(B(u) - I^L) \end{aligned} \tag{5}$$

with

$$u(\cdot, 0) = u_0,\ I^1(\cdot, 0) = I^1_0, \dots, I^L(\cdot, 0) = I^L_0 \text{ in } \mathbb{R}^d. \tag{6}$$

For the Planck function $B \in C^\infty(\mathbb{R}_{>0})$, we only assume

$$B(u) > 0,\ B'(u) > 0 \qquad (u \in \mathbb{R}_{>0}). \tag{7}$$

With obvious definitions for $F_1, \dots, F_d, Q : \mathcal{U} \to \mathbb{R}^{L+1}$, we write (5) in the compact form

$$U_t + F_1(U)_{x_1} + \cdots + F_d(U)_{x_d} = Q(U). \tag{8}$$

Let us give a short outline of this chapter. In Sect. 2, we review an existence result for global weak entropy solutions for (5) and (6) in the framework of BV-techniques taken from [8]. In Sect. 3, the main result of this chapter, namely Theorem 3, is presented and proven.

In the concluding section (Sect. 4), we display several numerical experiments which show more details about the mechanism of radiative smoothing and confirm our analytical findings.

2 The Existence of Global Weak Solutions

A weak solution of (5) and (6) is a function $(u, I^1, \dots, I^L) \in L^\infty_{loc}(\mathbb{R}^d \times [0,\infty))^{L+1}$ such that

$$\begin{aligned}
&\int_{\mathbb{R}^d}\int_0^\infty u\varphi_t + \mathbf{f}(u)\cdot\nabla\varphi\, dxdt \\
&\quad = -\int_{\mathbb{R}^d} u_0\varphi(\cdot,0)\, dx - \int_{\mathbb{R}^d}\int_0^\infty \varphi\kappa\sigma_L \sum_{l=1}^{L} \left(I^l - B(u)\right) dxdt, \\
&\qquad\qquad \int_{\mathbb{R}^d}\int_0^\infty I^l\varphi_t + cI^l\omega^l\cdot\nabla\varphi\, dxdt \\
&\quad = -\int_{\mathbb{R}^d} I_0^l\varphi(\cdot,0)\, dx - c\int_{\mathbb{R}^d}\int_0^\infty (B(u) - I^l)\varphi\, dxdt
\end{aligned}$$

holds for $l = 1, \dots, L$ and all $\varphi \in C_0^\infty(\mathbb{R}^d \times [0,\infty))$.

Theorem 1. *Let $u_0, I_0^1, \dots, I_0^L \in L^\infty(\mathbb{R}^d) \cap BV(\mathbb{R}^d)$ and $b > a > 0$ with*

$$(u_0(x), I_0^1(x), \dots, I_0^L(x)) \in [a,b] \times [B(a), B(b)]^L \quad \text{for almost all } x.$$

Then there is a weak solution $(u, I^1, \dots, I^L)$ of the Cauchy problem (5) and (6) that satisfies for almost all (x,t)

$$(u(x,t), I^1(x,t), \dots, I^L(x,t)) \in [a,b] \times [B(a), B(b)]^L,$$

$$|u(\cdot,t)|_{BV(\mathbb{R}^d)} + \frac{\kappa\sigma_L}{c}\sum_{l=1}^{L} |I^l(\cdot,t)|_{BV(\mathbb{R}^d)} \le |u_0|_{BV(\mathbb{R}^d)} + \frac{\kappa\sigma_L}{c}\sum_{l=1}^{L} |I_0^l|_{BV(\mathbb{R}^d)}. \tag{9}$$

The proof of this theorem via a numerical approximation can be found in [8]. We remark in passing that the estimates on u in (9) are uniformly with respect to $c \to \infty$. This allows to perform this nonrelativistic limit. In [8] it is shown that the u-component of the limit satisfies a hyperbolic integrodifferential equation.

3 The Existence of Global Classical Solutions

In this section, we analyze the regularity of the weak solutions for (5) and (6) which exist due to the result in Sect. 2. We show that the Cauchy problem (5) and (6) admits global classical solutions provided the initial data are close to an equilibrium state in an appropriate Sobolev norm.

To obtain this result, we shall apply a more general theory on smooth solutions for multidimensional hyperbolic balance laws by Yong [7]. More precisely, consider the Cauchy problem for a system of balance laws of the form

$$\begin{aligned} \hat{U}_t + \hat{F}_1(\hat{U})_{x_1} + \cdots + \hat{F}_d(\hat{U})_{x_d} &= \hat{Q}(\hat{U}) := \begin{pmatrix} 0 \\ \hat{q}(\hat{u}, \hat{v}) \end{pmatrix} \text{ in } \mathbb{R}^d \times (0, \infty), \\ \hat{U}(\cdot, 0) &= \hat{U}_0 \text{ in } \mathbb{R}^d. \end{aligned} \tag{10}$$

Here $\hat{U} = (\hat{u}^T, \hat{v}^T)^T : \mathbb{R}^d \times [0, \infty) \to \hat{\mathcal{U}} \subset \mathbb{R}^n$, $n \in \mathbb{N}$, is the unknown split up into the components $\hat{u} : \mathbb{R} \times [0, \infty) \to \mathbb{R}^{n-r}$ and $\hat{v} : \mathbb{R} \times [0, \infty) \to \mathbb{R}^r$. $\hat{F}_1, \ldots, F_d$, and $\hat{Q} \in C^\infty(\mathcal{U}, \mathbb{R}^n)$ are the flux and source functions where we assume that the first $n-r$ components of $\hat{Q}$ vanish and the last r-components are given by the function $\hat{q} : \hat{\mathcal{U}} \to \mathbb{R}^r$. $\hat{U}_0 : \mathbb{R} \to \hat{\mathcal{U}}$ is the initial function.

Theorem 2 (From [7]). *Let $\hat{U}_e \in \hat{\mathcal{U}}$ be such that $\hat{Q}(\hat{U}_e) = 0$. We assume that we have:*

(i) The Jacobian $D_{\hat{v}}\hat{q}(\hat{U}_e)$ is regular.
(ii) There is a strictly convex function $\hat{\eta} \in C^\infty(\hat{\mathcal{U}}, \mathbb{R})$ and functions $\Psi_1, \ldots, \Psi_d \in C^\infty(\hat{\mathcal{U}}, \mathbb{R})$ such that

$$\nabla_{\hat{U}}\hat{\eta}(\hat{U})^T D_{\hat{U}}\hat{F}_i(\hat{U}) = \nabla_{\hat{U}}\hat{\Psi}_i(\hat{U})^T \quad (\hat{U} \in G,\, i = 1, \ldots, n). \tag{11}$$

(iii) For all $G \subset\subset \hat{\mathcal{U}}$, there is a constant $c_G > 0$ such that

$$\Big(\nabla_{\hat{U}}\hat{\eta}(\hat{U}) - \nabla_{\hat{U}}\hat{\eta}(\hat{U}_e)\Big) \cdot \hat{Q}(\hat{U}) \le -c_G|\hat{Q}(\hat{U})|^2 \quad (\hat{U} \in G).$$

(iv) The kernel of the Jacobian $D_{\hat{U}}\hat{Q}(\hat{U}_e)$ contains no eigenvector of the matrix $n_1 D_{\hat{U}}\hat{F}_1(\hat{U}_e) + \cdots + n_d D_{\hat{U}}\hat{F}_d(\hat{U}_e)$ for any $n \in \mathcal{S}^{d-1}$.

Then, for $s \ge [d/2] + 2$, there exists a constant $C > 0$ such that for all initial functions $\hat{U}_0$ satisfying

$$\|\hat{U}_e - \hat{U}_0\|_{H^s(\mathbb{R}^d)} \le C$$

there is a unique classical solution $\hat{U} - \hat{U}_e \in C([0, \infty); H^s(\mathbb{R}^d))$ of (10).

To apply Theorem 2 to our radiation model problem (5), we reformulate (5) in the new variables

$$\hat{U} = (\hat{u}, \hat{v}_1, \dots, \hat{v}_d) := \Big(u + \frac{\kappa\sigma_L}{c} \sum_{l=1}^{L} I^l, I^1, \dots, I^L\Big)^T \in \hat{\mathcal{U}}. \qquad (12)$$

We obtain then a system in the form (10) which is equivalent to (5) (resp., (8)) by the choices

$$\hat{F}_1(\hat{U}) := AF_1(A^{-1}\hat{U}), \dots, \hat{F}_d(\hat{U}) := AF_d(A^{-1}\hat{U}),\ \hat{Q}(\hat{U}) := AQ(A^{-1}\hat{U}), \quad (13)$$

where $A \in \mathbb{R}^{(L+1)\times(L+1)}$ is the matrix with the first row $(1, (\kappa\sigma_L)/c, \cdots, (\kappa\sigma_L)/c)$ and the ith row vanishes except for the entry 1 at position i, $i = 2, \dots, L+1$. Note that we have $r = L$ and

$$\hat{q}_i(\hat{u}, \hat{v}) = c\left(\hat{v}_i - B\Big(\hat{u} - \frac{\kappa\sigma_L}{c} \sum_{l=1}^{L} \hat{v}_l\Big)\right) \qquad (i = 1, \dots, L). \qquad (14)$$

The main result of this section is the following.

Theorem 3. *Let $L > 1$ and $s \geq [d/2] + 2$. Consider an equilibrium point $U_e := (u_e, I_e, \dots, I_e)^T \in \mathcal{U}$ of Q, i.e., with $B(u_e) = I_e$.*

Then there exists a constant $C > 0$ such that for all initial data with

$$\|U_0 - U_e\|_{H^s(\mathbb{R}^d)} \leq C$$

there is a classical solution $U - U_e \in C([0,\infty); H^s(\mathbb{R}^d))$ of (5) and (6).

Proof. We have to verify the conditions (i)–(iv) from Theorem 2. So, consider system (10) with the choices (12) and (13). For (i) a straightforward computation for (14) leads to

$$\det\big(D_{\hat{v}}\hat{q}(\hat{u}, \hat{v})\big) = c^L\Big(1 + L\frac{\kappa\sigma_L}{c}B'\Big(\hat{u} - \frac{\kappa\sigma_L}{c}\sum_{l=1}^{L}\hat{v}_l\Big)\Big) \qquad ((\hat{u}, \hat{v}^T)^T \in \hat{\mathcal{U}}).$$

We conclude that $\det\big(D_{\hat{v}}\hat{q}(\hat{U}_e)\big) \neq 0$ with

$$\hat{U}_e := \Big(u_e + \frac{|S^{d-1}|}{c}B(u_e), B(u_e), \dots, B(u_e)\Big)^T$$

holds since we have $B' \geq 0$ due to (7).

Turning to (ii) we observe: if $(\eta, \psi_1, \dots, \psi_d)$ is an entropy tuple for (5), then $(\hat{\eta}, \hat{\psi}_1, \dots, \hat{\psi}_d)$ with

$$\hat{\eta}(\hat{U}) = \eta(A^{-1}\hat{U}), \quad \hat{\psi}_i(\hat{U}) = \psi_i(A^{-1}\hat{U}) \quad (\hat{U} \in \hat{\mathcal{U}}), \qquad (15)$$

is an entropy tuple for (10), i.e., $\hat{\eta}$ is convex and (11) holds. Therefore it suffices to construct an entropy tuple for (5) which we choose as

$$\begin{aligned} \eta(U) &= -\ln(u) + \frac{\kappa\sigma_L}{c}\sum_{l=1}^{L}\pi(I^l), \\ \psi_i(U) &= -\int_k^u \frac{f_i'(w)}{w}\,dw + \kappa\sigma_L\sum_{l=1}^{L}\omega_i^l\pi(I^l),\ k>0. \end{aligned} \tag{16}$$

The function $\pi \in C^\infty(\mathbb{R}_{>0},\mathbb{R})$ is given such that we have for $I > 0$

$$\pi'(I) = -\frac{1}{B^{-1}(I)}. \tag{17}$$

Obviously (11) is satisfied. The convexity of η follows due to (7) from

$$\pi''(I) = \frac{B^{-1\prime}(I)}{\left(B^{-1}(I)\right)^2} > 0 \quad (I > 0).$$

Thus we have verified (ii) in Theorem 2.

Since we can add any linear function to an entropy and still obtain an entropy, we can assume w.l.o.g. $\nabla\eta(U_e) = 0$. Furthermore we have $\nabla\eta(U)\cdot Q(U) = \nabla_{\hat{U}}\hat{\eta}(\hat{U})\cdot\hat{Q}(\hat{U})$ for all $U \in G$, G is any compact subset of $\mathcal{U}$, and $\hat{U} = A^{-1}U$. Therefore it is enough to check for (iii) that there is a constant $c_G > 0$ such that the estimate

$$\nabla_U\eta(U)\cdot Q(U) \leq -c_G|Q(U)|^2 \tag{18}$$

holds. We compute with (16) and (17)

$$\begin{aligned} \nabla_U\eta(U)\cdot Q(U) &= \kappa\sigma_L\sum_{l=1}^{L}(I^l - B(u))\frac{u - B^{-1}(I^l)}{uB^{-1}(I^l)} \\ &= -\kappa\sigma_L\sum_{l=1}^{L}\frac{(B(u)-I^l)^2}{uB^{-1}(I^l)}\frac{B^{-1}(B(u)) - B^{-1}(I^l)}{B(u)-I^l} \\ &\leq -c_G|Q(U)|^2. \end{aligned}$$

The last inequality follows for some $c_G > 0$ from (7), $u > 0$, the definition of Q, and the compactness of the set G.

Finally we have to consider condition (iv). Again it suffices to show that the kernel of $D_UQ(U_e)$ contains no eigenvector of $n_1D_UF_1(U_e) + \cdots + n_dD_U F_d(U_e)$ for no $n \in \mathcal{S}^{d-1}$. We compute

$$D_UQ(U) = \begin{pmatrix} -\kappa\sigma_L LB'(u_e) & \kappa\sigma_L & \cdots & \kappa\sigma_L \\ cB'(u_e) & -c & \cdots & 0 \\ \vdots & & \ddots & \\ cB'(u_e) & 0 & \cdots & -c \end{pmatrix}.$$

and thus the kernel of $D_U Q(U_e)$ is given by $\{\alpha(1, B'(u_e), \dots, B'(u_e))^T \mid \alpha \in \mathbb{R}\}$. A short calculation shows that an element of this set can only be an eigenvector provided

$$cn \cdot \omega^l = n \cdot (f_1'(u_e), \dots, f_d'(u_e))^T$$

holds for $l = 1, \dots, L$. Due to $L > 1$ and (3), this cannot be true and we have verified (iv). □

4 Numerical Experiments for the Model Problem

Theorem 3 gives sufficient conditions for the existence of smooth global solutions. We underline this finding by a series of numerical experiments. We consider the model (5) for $d = 2$ in the domain $(0,1)^2$ and choose the (discontinuous) initial data

$$u_0(x) = \begin{cases} 1.5 & : \ |x - (0.5, 0.5)| \leq 0.125 \\ 1 & : \ \text{elsewhere} \end{cases}, \qquad I_0^l(x) = 1.0 \quad (l = 1, \dots, L).$$

As boundary conditions we assume periodic ones. For c we choose $c = 10$, the fluxes and B are given by

$$f^1(u) = f^2(u) = \frac{u^2}{2}, \quad B(u) = u^4.$$

The initial/boundary value problem is solved by a first-order standard finite volume scheme on a uniform Cartesian mesh with mesh width 0.025. As numerical flux we use the Engquist–Osher flux in each component.

In Fig. 1, we display a contour plot of u for $\kappa = 0$ and $\kappa = 0.25$ computed with eight equidistributed ordinates.

One observes clearly the strong damping effect of the radiation. More results are displayed in Fig. 2 where we display u restricted to the diagonal line connecting the coordinate points $(0,1)$ with $(1,0)$. We observe that with increasing the value of κ, that means to increase the radiative damping effect,

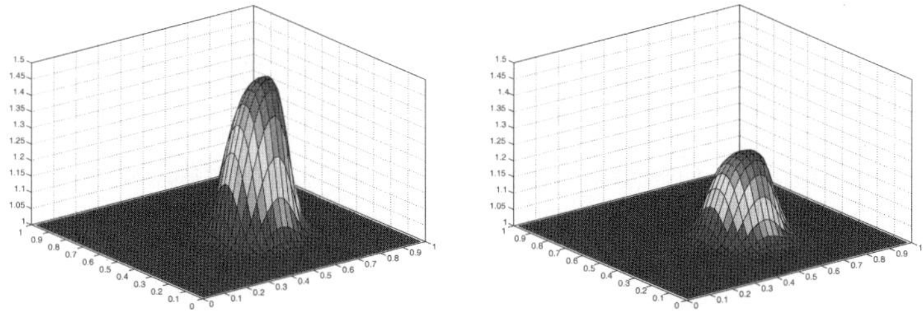

Fig. 1. Computational results for u with $\kappa = 0$ (*left*) and $\kappa = 8$, $L = 8$ (*right*) at time $t = 0.0625$

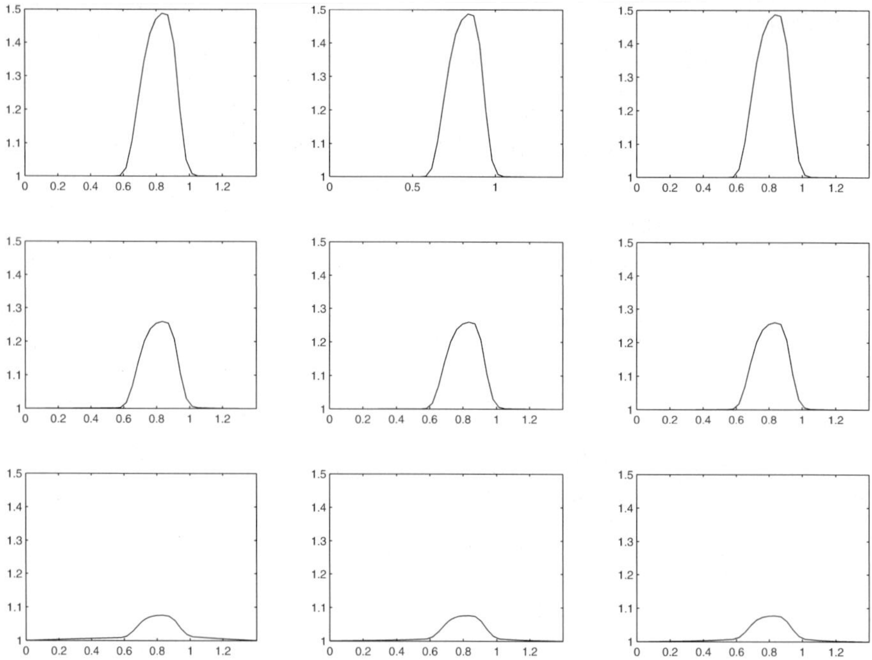

Fig. 2. Computational results for u restricted to the diagonal with $L = 4, 8, 16$ ordinates (*from left to right*) and $\kappa = 0.25, 0.5, 1.0$ (*from top to bottom*) at time $t = 0.0625$

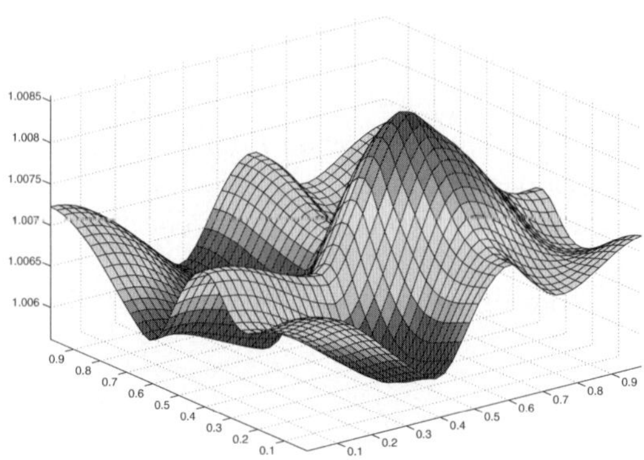

Fig. 3. Computational results for u with $\kappa = 1$ and $L = 16$ at time $t = 0.2$

the solution component u itself decays and the initial discontinuity seems to be smeared out. Almost not visible in Fig. 2 is the difference which results from the choice of different numbers of ordinates. Finally we present in Fig. 3 the results for 16 ordinates at a later time. We see that the lumped quantity u is already quite close to the equilibrium value 1 and moreover the solution appears to be very smooth (note the scaling in vertical direction in Fig. 3).

References

1. Heaslet, M., Baldwin, B.: Predictions of the structure of radiation-resisted shock waves. Phys. Fluids **6**, 781–791 (1963).
2. Kawashima, S., Nikkuni, Y., Nishibata, S.: The initial value problem for hyperbolic-elliptic coupled systems and applications to radiation hydrodynamics. In: Freistühler, H. (ed.) Analysis of Systems of Conservation Laws. Chapman & Hall, 87–127 (1998).
3. Lin, C., Coulombel, J.-F., Goudon, T.: Shock profiles for non-equilibrium radiating gases. Phys. D **218**, no. 1, 83–94 (2006).
4. Mihalas, D., Mihalas, B.: Foundations of Radiation Hydrodynamics. Oxford University Press, New York (1984).
5. Pomraning, G.C.: The Equations of Radiation Hydrodynamics. Pergamon Press, Oxford, New York (1973).
6. Rohde, C.: Entropy solutions for weakly coupled hyperbolic systems in several space dimensions. Z. Angew. Math. Phys. **49**, no. 3, 470–499 (1998).
7. Yong, W.-A.: Entropy and global existence for hyperbolic balance laws. Arch. Rat. Mech. Anal. **172** (2004), 247–266.
8. Rohde, C., Yong, W.-A.: The Nonrelativistic Limit in Radiation Hydrodynamics. I. Weak Entropy Solutions for a Model Problem. Preprint, 2005.
9. Rohde, C., Yong, W.-A.: Shock structure for radiation hydrodynamics. In preparation.

Spectral Analysis of Coupled Hyperbolic–Parabolic Systems on Finite and Infinite Intervals

J. Rottmann-Matthes

1 Introduction

In many applied problems in biology, physics, or chemistry, traveling waves arise as solutions of systems of partial differential equations of the form

$$U_t = f(U, U_x, U_{xx}) \quad \text{in } [0, \infty) \times \mathbb{R}. \tag{1}$$

A traveling wave has the special property that it is constant if one looks at it in a comoving frame. More precisely this means if U is a traveling wave solution of (1) with speed c, the function $\tilde{U}(t, x) := U(t, x + ct)$ is a steady state of the transformed PDE

$$0 = \frac{\mathrm{d}}{\mathrm{d}t} U(t, x + ct) = f(\tilde{U}, \tilde{U}_x, \tilde{U}_{xx}) + c\tilde{U}_x \quad \text{in } [0, \infty) \times \mathbb{R}. \tag{2}$$

For the stability analysis of traveling waves, it is important to know where the point spectrum of the linearized right-hand side of (2) lies. We will show how this can be approximated by computing the spectrum of boundary value problems.

2 General Assumptions on the Problem's Structure

We consider a linear coupled hyperbolic–parabolic PDE of the form

$$\begin{pmatrix} u \\ v \end{pmatrix}_t = P \begin{pmatrix} u \\ v \end{pmatrix}_t := \begin{pmatrix} A & 0 \\ 0 & 0 \end{pmatrix} \begin{pmatrix} u \\ v \end{pmatrix}_{xx} + \begin{pmatrix} B_{11} & B_{12} \\ 0 & B_{22} \end{pmatrix} \begin{pmatrix} u \\ v \end{pmatrix}_x + \begin{pmatrix} C_{11} & C_{12} \\ C_{21} & C_{22} \end{pmatrix} \begin{pmatrix} u \\ v \end{pmatrix}, \tag{3}$$

where $P : H^2(\mathbb{R}, \mathbb{C}^n) \times H^1(\mathbb{R}, \mathbb{C}^m) \to L^2(\mathbb{R}, \mathbb{C}^n) \times L^2(\mathbb{R}, \mathbb{C}^m)$. This structure, for example, may arise by linearizing the nonlinear PDE (2) at the wave's profile (see also Sect. 4).

Assumptions. We now state the basic assumptions on the operator P:

(0) The matrix-valued functions B_{ij} and C_{ij} are twice continuously differentiable and converge to constant limit matrices:
- $B_{11}(x) \to B_{11\pm} \in \mathbb{C}^{n,n}$, $C_{11}(x) \to C_{11\pm} \in \mathbb{C}^{n,n}$ as $x \to \pm\infty$.
- $B_{22}(x) \to B_{22\pm} \in \mathbb{C}^{m,m}$, $B_{22,x}(x) \to 0 \in \mathbb{C}^{m,m}$, and $C_{22}(x) \to C_{22\pm} \in \mathbb{C}^{m,m}$ as $x \to \pm\infty$, $\|B_{22,xx}\|_\infty < \infty$, $\|C_{22,x}\|_\infty < \infty$.
- $B_{12}(x) \to B_{12\pm}$, $C_{12}(x) \to C_{12\pm} \in \mathbb{C}^{n,m}$, and $C_{21}(x) \to C_{21\pm} \in \mathbb{C}^{m,n}$, as $x \to \pm\infty$, $\|B_{12,x}\|_\infty < \infty$.

(P) The matrix $A \in \mathbb{C}^{n,n}$ satisfies $A + A^* \geq \alpha I > 0$ as a quadratic form for some $\alpha \in \mathbb{R}$.

(H) The matrix function B_{22} is real diagonal valued and there exist $b_0, \gamma > 0$ such that for all $x \in \mathbb{R}$ the diagonal elements satisfy
$|b_{ii}(x) - b_{jj}(x)| \geq \gamma$, for $i \neq j$,
$b_{ii}(x) \geq b_0$, for $1 \leq i \leq r$ and $-b_{ii}(x) \geq b_0$, for $r+1 \leq i \leq m$.
Furthermore the real part of the diagonal entries of the limit matrices $C_{22\pm}$ is bounded from above by -2δ for some $\delta > 0$.

(D) There exists $\delta > 0$ such that for all $\omega \in \mathbb{R}$ and for all $s \in \mathbb{C}$ the equality

$$\det\left(-\omega^2\begin{pmatrix} A & 0 \\ 0 & 0 \end{pmatrix} + i\omega\begin{pmatrix} B_{11\pm} & B_{12\pm} \\ 0 & B_{22\pm} \end{pmatrix} + \omega\begin{pmatrix} C_{11\pm} & C_{12\pm} \\ C_{21\pm} & C_{22\pm} \end{pmatrix}\right) = 0 \quad (4)$$

implies $\Re s \leq -\delta$.

Note that assumption (0) is generically satisfied if the profile of the traveling wave is a smooth connecting orbit of rest states of the nonlinear PDE (1).

For our analysis of the spectral properties of P, we consider the resolvent equation

$$(sI - P)\begin{pmatrix} u \\ v \end{pmatrix} = \begin{pmatrix} f \\ g \end{pmatrix} \quad \text{in } L^2(\mathbb{R}, \mathbb{C}^n) \times L^2(\mathbb{R}, \mathbb{C}^m) \quad (5)$$

and transform it into a first-order system using the variables $(u, Au_x, v) = z$

$$L(s)z = z_x - M(x,s)z = (0, -f + B_{12}B_{22}^{-1}g, -B_{22}^{-1}g)^T. \quad (6)$$

Here we consider the operator $L(s)$ as a mapping $H^2 \times H^1 \times H^1 \to H^1 \times L^2 \times L^2$. The matrix M is of the form

$$M(\cdot, s) = \begin{pmatrix} 0 & A^{-1} & 0 \\ B_{12}B_{22}^{-1}C_{21} + (sI - C_{11}) & -B_{11}A^{-1} & -C_{12} - B_{12}B_{22}^{-1}(sI - C_{22}) \\ -B_{22}^{-1}C_{21} & 0 & B_{22}^{-1}(sI - C_{22}) \end{pmatrix}. \quad (7)$$

Recall that an operator $L : z \mapsto z_x - M(x)z$, $M(x) \in \mathbb{C}^{l,l}$, is said to have an exponential dichotomy on a closed interval J if there exist positive numbers K, β, and for every $x \in J$ there is a projection $\pi(x) \in \mathbb{C}^{l,l}$ such that

$$\begin{aligned} S(x,y)\pi(y) &= \pi(x)S(x,y),\ \forall x,y\in J,\\ |S(x,y)\pi(y)| &\le Ke^{-\beta(x-y)},\ \forall x\ge y\in J,\\ |S(x,y)(I-\pi(y))| &\le Ke^{-\beta(y-x)},\ \forall x< y\in J, \end{aligned} \tag{8}$$

where $S(\cdot,\cdot)$ is the solution operator to L. We call (K,β,π) the data of the exponential dichotomy. For general results about systems with exponential dichotomies, see [4].

By assumption (0), the limit matrices $M_\pm(s) := \lim_{x\to\pm\infty} M(x,s)$ exist and the dispersion relation (D) implies that these are hyperbolic matrices for all $s\in\mathbb{C}$ with $\Re s > -\delta$. More generally $M_\pm(s)$ are nonhyperbolic if and only if the number s lies on the algebraic curves defined by (4). But for the stability analysis, we are only interested in the rightmost part of the spectrum.

We denote by $V_\pm^I(s)$ and $V_\pm^{II}(s)$ bases of the stable and unstable subspace of $M_\pm(s)$, respectively. By the Roughness theorem [4, 2], the hyperbolicity of the limit matrices implies that for each $s\in\{\Re s>-\delta\}$ the operator $L(s)$ has exponential dichotomies on $\mathbb{R}_+$ and $\mathbb{R}_-$.

Using the transformation $\tilde z = \begin{pmatrix} I & 0 & 0\\ 0 & \frac{1}{\sqrt{s}}I & 0\\ 0 & 0 & I\end{pmatrix}\begin{pmatrix} I & 0 & 0\\ 0 & I & B_{12\pm}\\ 0 & 0 & I\end{pmatrix} z$ for $\Re s$ sufficiently large, it is shown in [9] that the dimensions of $V_+^I(s)$ and $V_-^I(s)$, and of $V_+^{II}(s)$ and $V_-^{II}(s)$ coincide for $\Re s > -\delta'$. Then a result from [8] shows that $L(s)$ is a Fredholm operator of index zero for $\Re s > -\delta$. Utilizing the relation of $L(s)$ and $sI-P$, we obtain that for all $s\in\mathbb{C}$ with $\Re s > -\delta$ the operator $sI-P$ is Fredholm of index zero.

It is well known (e.g., [6]) that the operator $sI-P$ is one to one for $s\in\mathbb{R}$ sufficiently large, such that the following lemma is implied.

Lemma 1. *The operator P only has isolated eigenvalues of finite algebraic multiplicity in the right half-plane $\{\Re s > -\delta\}$.*

Hence we can decompose the spectrum of P according to $\sigma(P) = \sigma_P \dot\cup \sigma_{\text{ess}}$, where σ_P is the point spectrum which consists of all isolated eigenvalues of finite multiplicity, and $\sigma_{\text{ess}} := \sigma(P)\backslash\sigma_P$ is the essential spectrum.

3 Statement of the Main Result

As explained in Sect. 2, there only is point spectrum to the right of the axis $\{\Re s = -\delta\}$. Since it is in general not possible to determine the location of the point spectrum analytically, one has to approximate it numerically. One possibility is to implement the "Evans function" which is an analytic function whose roots coincide with eigenvalues of the operator, for details see [5] and the references therein. Another possibility is to restrict the operator P to a finite but large interval and to impose suitable boundary conditions, see for example [2]. Then one uses the spectrum of these boundary value problems as an approximation.

We follow the latter approach and consider the restriction of the resolvent equation (5) to a finite interval $J = [x_-, x_+]$

$$(sI - P_{|J}) \begin{pmatrix} u_J \\ v_J \end{pmatrix} = \begin{pmatrix} f_J \\ g_J \end{pmatrix} \quad \text{in } L^2(J, \mathbb{C}^n) \times L^2(J, \mathbb{C}^m), \tag{9}$$

where $P_{|J}$ is defined the same way as P, but only on the interval J. To obtain a well-posed problem on the finite interval J, we impose boundary conditions

$$R \begin{pmatrix} u_J \\ v_J \end{pmatrix} = \eta \in \mathbb{C}^{2n+m}, \tag{10}$$

where R is a two-point boundary operator of the form

$$R \begin{pmatrix} u_J \\ v_J \end{pmatrix} := \begin{pmatrix} R_-^I & R_-^{II} & R_-^{III} \end{pmatrix} \begin{pmatrix} u_J(x_-) \\ u_{J,x}(x_-) \\ v_J(x_-) \end{pmatrix} + \begin{pmatrix} R_+^I & R_+^{II} & R_+^{III} \end{pmatrix} \begin{pmatrix} u_J(x_+) \\ u_{J,x}(x_+) \\ v_J(x_+) \end{pmatrix}. \tag{11}$$

The crucial assumption we must make on the boundary operator is the determinant condition

$$D(s) := \det \left[\begin{pmatrix} R_-^I & R_-^{II} A^{-1} & R_-^{III} \end{pmatrix} V_-^{II}(s), \begin{pmatrix} R_+^I & R_+^{II} A^{-1} & R_+^{III} \end{pmatrix} V_+^I(s) \right]. \tag{12}$$

This condition basically states that the stable and unstable subspaces of the solutions can be controlled at the left and right endpoint of the interval, respectively. As above $V_\pm^{I/II}(s)$ are bases of the stable and unstable subspaces of $M_\pm(s)$.

Using a general convergence result, Proposition 1, which gives quantitative resolvent estimates of the finite interval problem, is proven in [3]. It is also possible to prove this proposition by applying techniques from the theory of exponential dichotomies as is done in [2] for the purely parabolic case.

Proposition 1. *Let $\Omega \subset \{\Re s > -\delta'\}$ be a compact subset of the resolvent set $\rho(P)$ of P and assume $D(s) \neq 0$ for all $s \in \Omega$.*

Then there is a compact interval J_0 and $K_0 > 0$ such that for all $J = [x_-, x_+] \supset J_0$ and for all $s \in \Omega$ the finite interval problem (9) and (10) has for every right-hand side $f_J \in L_2(J, \mathbb{C}^n)$, $g_J \in L_2(J, \mathbb{C}^m)$, $\eta \in \mathbb{C}^{2n+m}$ a unique solution $(u_J, v_J) \in H^2(J, \mathbb{C}^n) \times H^1(J, \mathbb{C}^m)$. Moreover, the solution can be estimated by

$$\|u_J\|_{H^2} + \|v_J\|_{H^1} + |u_J|_\Gamma + |u_{J,x}|_\Gamma + |v_J|_\Gamma \leq K_0 \big(\|f\|_{L_2} + \|g\|_{L_2} + |\eta| \big),$$

where $|u|_\Gamma^2 := |u(x_-)|^2 + |u(x_+)|^2$.

Together with Lemma 1, Proposition 1 also gives a qualitative result that at most the point spectrum of the operator P is approximated in the right half-plane. But note that it does not say whether each eigenvalue is really approximated, and how close the approximate spectrum is to the spectrum of P. This will be the statement of our Theorem 1. Before we state this theorem, we introduce some notation. As usual we denote by $\mathcal{N}(A)$ the kernel of an operator A. Furthermore we will use the Banach spaces

$$\begin{aligned} E &:= H^2(\mathbb{R}, \mathbb{C}^n) \times H^1(\mathbb{R}, \mathbb{C}^m),\ F := L^2(\mathbb{R}, \mathbb{C}^n) \times L^2(\mathbb{R}, \mathbb{C}^m), \\ E_J &:= H^2(J, \mathbb{C}^n) \times H^1(J, \mathbb{C}^m), \quad F := L^2(J, \mathbb{C}^n) \times L^2(J, \mathbb{C}^m) \times \mathbb{C}^{2n+m}. \end{aligned}$$

Finally we denote by

$$\mathcal{A}(s) := sI - P \in L(E, F) \ \text{ and } \ \mathcal{A}_J(s) := \begin{pmatrix} sI - P_{|J} \\ R \end{pmatrix} \in L(E_J, F_J)$$

the operator polynomials, corresponding to the all line operator and to its finite interval approximation. Note that the boundary operator R is included in the definition of the finite interval approximation.

For an element s_0 in the point spectrum σ_P of P, choose $\varepsilon > 0$ such that s_0 is the only element of the spectrum of P in the closed ball $\overline{B_\varepsilon(s_0)}$. Then we call $\sigma_J := \{s \in \overline{B_\varepsilon(s_0)} : s \text{ is an eigenvalue } \mathcal{A}_J(\cdot)\}$ the s_0-group of eigenvalues of $\mathcal{A}_J$.[1]

Theorem 1. *Let the assumptions (0), (P), (H), and (D) hold and let Σ be an open neighborhood of the isolated eigenvalue s_0.*

Assume $D(s) \neq 0$ for all $s \in \Sigma$ and choose $\varepsilon > 0$ such that $\overline{B_\varepsilon(s_0)} \subset \Sigma$. Let $\beta_\pm$ denote the dichotomy exponents of $L(s_0)$ on $\mathbb{R}_\pm$.

Then there is a compact interval $J_0 \subset \mathbb{R}$ such that for every interval $J = [x_-, x_+] \supset J_0$ the following properties hold.

The s_0-group of eigenvalues σ_J converges to the eigenvalue s_0, in the sense that for every $0 < \beta' < \min(\beta_-, \beta_+)$ there is $c = c(\beta') > 0$ such that

$$\max_{s \in \sigma_J} |s - s_0| = \operatorname{dist}(\sigma_J, s_0) \le c e^{-\frac{\beta'}{\kappa} \min(x_+, -x_-)}, \tag{13}$$

where κ is the maximal multiplicity of an eigenvector of $\mathcal{A}(\cdot)$ to the eigenvalue s_0.

The eigenspace of $\mathcal{A}_J(\cdot)$ for σ_J converges to the eigenspace of $\mathcal{A}(\cdot)$ to the eigenvalue s_0. Moreover, the following estimate holds

$$\sup_{\left[\begin{array}{c} \left\| \binom{u_J}{v_J} \right\|_{E_J} = 1 \\ s_J \in \sigma_J, \\ \binom{u_J}{v_J} \in \mathcal{N}(\mathcal{A}_J(s_J)) \end{array}\right]} \inf_{\binom{u_0}{v_0} \in \mathcal{N}(\mathcal{A}(s_0))} \left\| \begin{pmatrix} u_J \\ v_J \end{pmatrix} - \begin{pmatrix} u_{0|J} \\ v_{0|J} \end{pmatrix} \right\|_{E_J} \le c e^{-\frac{\beta'}{\kappa} \min(x_+, -x_-)}. \tag{14}$$

Furthermore, the dimensions of the generalized eigenspace of $\mathcal{A}(\cdot)$ to s_0 and of $\mathcal{A}_J(\cdot)$ to the s_0-group σ_J coincide.

[1] For a definition of eigenvalue, eigenvector, and multiplicity of eigenvectors for operator polynomials, see for example [7].

The theorem is stated in a more quantitative version in [9], where also a closeness result for the generalized eigenspaces is stated. The proof is in [9] and it applies the abstract theory of discrete approximations (see [10]). The main problem in the proof is to show that for every $s \in \Sigma$ the operators $\mathcal{A}_J(s)$ regularly converge to the all line operator $\mathcal{A}(s)$. This is shown in [3].

4 Approximating the Point Spectrum of the FitzHugh–Nagumo System

As an example we consider the FitzHugh–Nagumo system. The system reads

$$u_t = u_{xx} + u - \frac{1}{3}u^3 - v, \quad v_t = \Phi(u + a - bv), \tag{15}$$

where we choose the parameter values $a = 0.7$, $b = 0.8$, and $\Phi = 0.08$. It is well known that for this choice of parameters the system (15) has a stable and an unstable traveling pulse. We denote by $(\overline{u}, \overline{v})^T(x)$ the profile of the pulse and by c its speed. Furthermore let $(\overline{u}_\infty, \overline{v}_\infty)^T$ denote the limit for $x \to \infty$.

Linearizing the system in the comoving frame at the profile leads to the linear PDE (cf. (3))

$$\begin{pmatrix} u \\ v \end{pmatrix}_t = \begin{pmatrix} 1 & 0 \\ 0 & 0 \end{pmatrix} \begin{pmatrix} u \\ v \end{pmatrix}_{xx} + \begin{pmatrix} c & 0 \\ 0 & c \end{pmatrix} \begin{pmatrix} u \\ v \end{pmatrix}_x + \begin{pmatrix} 1 - \overline{u}^2 & -1 \\ \Phi & -\Phi b \end{pmatrix} \begin{pmatrix} u \\ v \end{pmatrix}. \tag{16}$$

Obviously the assumptions (0), (P), and (H) are satisfied. It follows from the following observation that also assumption (D) is fulfilled.

If there are matrices $H_\pm = \begin{pmatrix} H_\pm^i & 0 \\ 0 & H_\pm^{ii} \end{pmatrix}$, $H_\pm = H_\pm^* > 0$ and $H_\pm^{ii}$ diagonal, with

$$\begin{gathered} H_\pm^i A + A^* H_\pm^i > 0, \quad H_\pm B_\pm - B_\pm^* H_\pm = 0, \\ H_\pm C_\pm + C_\pm^* H_\pm < -2\delta H \quad \text{for some } \delta > 0 \end{gathered}$$

then assumption (D) holds.

Choosing $H = \mathrm{diag}(1, \Phi^{-1})$ shows that the FitzHugh–Nagumo system also satisfies (D).

For our computations, we approximate the unstable traveling wave on a large interval $J = [0, 65]$ with projection boundary conditions, see [1]. For the computation of the spectrum, we then linearize at this approximation and discretize using central differences and suitable boundary conditions. For the figures shown in this chapter, we always used periodic boundary conditions since they obviously satisfy the determinant condition (12) for all s with $\Re s > -\delta$ if the profile is a pulse.

In Fig. 1, we plot the spectrum of (16) with periodic boundary conditions corresponding to the unstable wave. One can see the isolated eigenvalue at 0 which always appears for traveling waves and also that the unstable eigenvalue is present in the spectrum.

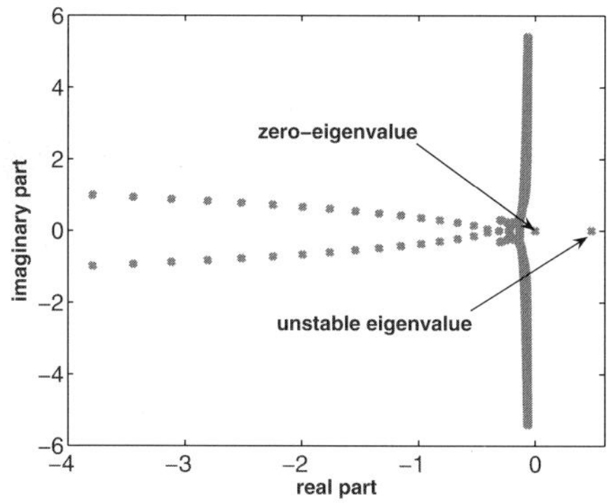

Fig. 1. The spectrum of the finite interval operator using central differences and periodic boundary conditions

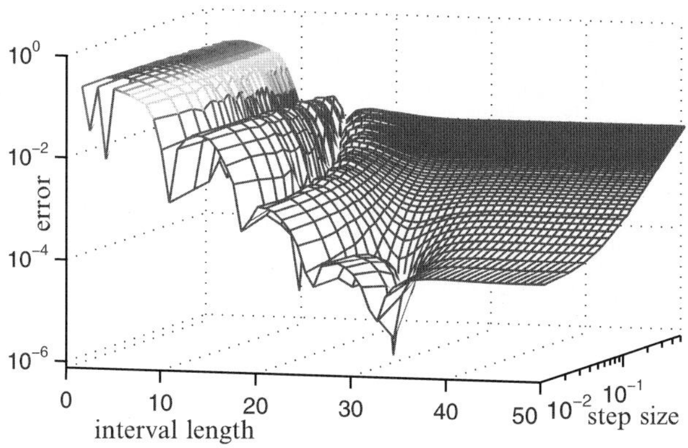

Fig. 2. Convergence of the zero eigenvalue for the unstable traveling wave for different step sizes and interval lengths

In Fig. 2, we compute the length of the eigenvalue closest to zero of the discretized operator for different step sizes and interval lengths. Similarly in Fig. 3, we compute the distance of the eigenvalue with maximal real part of the discretized operator to the unstable eigenvalue of the pulse. Since the exact value of the unstable eigenvalue is not known explicitly, we treated the eigenvalue with maximal real part of the operator on the interval $J = [0, 65]$ with step size 0.005 as the exact unstable eigenvalue.

One can observe the exponential rate of convergence of the eigenvalues depending on the interval length as predicted in Theorem 1. Furthermore one can see that the eigenvalues seem to converge exponentially fast, also in the step size of the discretization. Note that the convergence to the unstable eigenvalue seems to be much better. This is, because we only use an approximation of it which is not exact. So the picture is probably to optimistic.

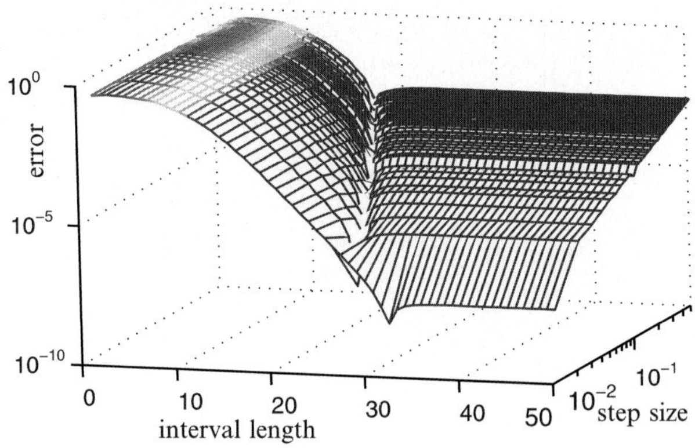

Fig. 3. Convergence of the unstable eigenvalue

Acknowledgments

This is part of the author's Diploma thesis [9], written under the supervision of Wolf-Jürgen Beyn at the University of Bielefeld. This research was supported by CRC 701 "Spectral Structures and Topological Methods in Mathematics."

References

1. W.-J. Beyn. The numerical computation of connecting orbits in dynamical systems. *IMA J. Numer. Anal.*, 10(3):379–405, 1990.
2. W.-J. Beyn and J. Lorenz. Stability of traveling waves: dichotomies and eigenvalue conditions on finite intervals. *Numer. Funct. Anal. Optim.*, 20(3–4): 201–244, 1999.
3. W.-J. Beyn and J. Rottmann-Matthes. Resolvent estimates for boundary value problems on large intervals via the theory of discrete approximations. Preprint http://www.math.uni-bielefeld.de/sfb701/preprints/sfb05016.ps.gz, submitted to Numer. Funct. Anal. Optim. 2006.
4. W.A. Coppel. *Dichotomies in stability theory.* Springer-Verlag, Berlin, 1978. Lecture Notes in Mathematics, Vol. 629.
5. J. Humpherys, B. Sandstede, and K. Zumbrun. Efficient computation of analytic bases in Evans function analysis of large systems. *Numer. Math.*, 103(4):631–642, 2006.
6. G. Kreiss, H.-O. Kreiss, and N.A. Petersson. On the convergence to steady state of solutions of nonlinear hyperbolic-parabolic systems. *SIAM J. Numer. Anal.*, 31(6):1577–1604, 1994.
7. A.S. Markus. *Introduction to the spectral theory of polynomial operator pencils*, volume 71 of *Translations of Mathematical Monographs.* American Mathematical Society, Providence, RI, 1988.
8. K.J. Palmer. Exponential dichotomies and transversal homoclinic points. *J. Differential Equations*, 55(2):225–256, 1984.

9. J. Rottmann. Spectral properties of mixed hyperbolic-parabolic systems. Diplomathesis, Universität Bielefeld, Bielefeld, 2005.
10. G. Vainikko. *Funktionalanalysis der Diskretisierungsmethoden.* B.G. Teubner Verlag, Leipzig, 1976.

Toward an Improved Capture of Stiff Detonation Waves

O. Rouch, M.-O. St-Hilaire, and P. Arminjon

1 Introduction

Detonations can be modeled by a compressible reactive flow involving a single exothermic reaction between two chemical states: the unburnt gas and the burnt gas. The detonation wave can be modeled as a powerful nonreactive shock followed by a deflagration wave, where the gas is burnt. The shock propagates in the unburnt gas, heating it; if the ignition temperature is reached through this purely mechanical shock, then a chemical reaction is triggered. One differentiates detonations from other types of combustion by the great quantity of energy they release, thus making negligible any potential contribution from viscosity, conduction, or heat radiation. Neglecting these local effects makes the reactive Euler equations the most adequate way to describe a detonation process at large scale.

The reactive Euler equations form a hyperbolic system of conservation laws with a source term representing a chemical reaction

$$U_t + \nabla \cdot F(U) = -\frac{1}{\tau} G(U) \tag{1}$$

with $U \in \mathbb{R}^n$, $\tau \in \mathbb{R}^+$, and $F, G : \mathbb{R}^n \to \mathbb{R}^n$. The source term G is governed by a rate function K depending on the gas temperature T

$$K(T) : \text{Unburnt gas} \to \text{Burnt Gas},$$

usually defined by an Arrhenius chemistry law or a Dirac function, called the *ignition temperature model.*

Despite the fact that such a source G can also be used in combustion problems, in real situations, detonation waves move with great speed – more than $1\,\mathrm{km\,s^{-1}}$ – and the reactive mixture is consumed 10^3–10^6 times faster than in a flame [4]. This translates into a very stiff source term ($\tau \ll 1$) in the governing equations.

This simultaneous presence of two timescales, macroscopic for the fluid flow and microscopic for the chemical reaction, makes numerical simulation of detonation waves very delicate.

Resolved simulations, where the small timescale of chemistry is fully taken into account by the space/time mesh, effectively capture the wave in details. However, this becomes way too expensive in computing time and random access memory when dealing with detonations, especially for multidimensional problems. On the other hand, an underresolved approach, where the discretization is proportional to the macroscopic scale, while being economic, is unfortunately unable to capture stiff detonation waves. Indeed, the smearing of the shock, due to numerical viscosity induced by about all conservative numerical methods, raises temperature in front of the compression shock so that the (violent) chemical reaction is prematurely triggered in stiff cases. Simulations may then produce waves that look reasonable at first glance, but propagate at nonphysical speed, purely related to numerical artifacts [3]. Of course, results can be improved by modifying the numerical method used, according to each particular problem, but this will certainly not produce reliable results for real life problems.

2 Underresolved Detonation Capture Without Numerical Viscosity

We propose a family of accurate operator-splitting methods, numerically stable, allowing underresolved calculations and requiring neither the resolution of the Riemann problem nor the knowledge of the characteristic structure of the flux Jacobian matrix, while converging to the physical solution. With a refinement of the grid, these methods moreover efficiently capture the unstable character of the detonation and provide the exact front structure of the wave. It is realistic to claim that such methods can also solve many other hyperbolic systems with source term. We thus elaborate "black box"-type methods, while the majority of the existing schemes for the detonation problem use properties of the analytical solution itself.

The leading idea of our work is to minimize numerical viscosity induced by homogeneous hyperbolic system solvers, as it is responsible for the smearing of the shock.

To achieve this goal, we use our own version of artificial compression methods (ACMs), based on the one introduced by Harten [6], which provides a shock capture in only one grid point. For optimal results, our ACM can be combined with a detector of discontinuities (DoDs), tuned with physical, not solution dependent, information about the particular problem that we are solving. ACM and DoD plug-ins can be added to about any homogeneous hyperbolic system solver, without compromising the accuracy and stability. We chose to use it on central finite volume methods from [8, 1], as they are Riemann-solver-free, second-order accurate and also developed for two- and three-space dimensions. The source term will be treated by operator splitting.

As ACM provides a numerical viscosity free shock capture, it is indeed the perfect tool to verify the hypothesis that holds that the smearing of detonation shocks is responsible for the unphysical behavior of numerical schemes. Through the elaboration of a robust method for the detonation capture, we will investigate this widely accepted assumption.

2.1 Artificial Compression Method

The main idea of our version of artificial compression is to focus ACM's application on regions of the domain, where discontinuities are potentially present. This is achieved through what we call a detector of discontinuities, which is designed independently of ACM itself, but provides a filter to prevent overcompression in smooth regions.

The version of ACM introduced by Harten [6] is based on flux augmentation, and relies on two main observations. The first one is that, in the neighborhood of a discontinuity between two states U_L and U_R, any solution of $U_t + \nabla \cdot F(U) = 0$ is also the solution of

$$U_t + \nabla \cdot \big(F(U) + K(U)\big) = 0$$

where $K(U)$ is nonzero only for U being a physically admissible intermediate state between U_L and U_R. At this point we note that, when coupling ACM with a DoD that locates discontinuities accurately enough, the determination of those intermediate states becomes superfluous, and one can build K with significantly more freedom.

The second observation concerns viscous profiles. Any physical shock occurring in the solution of $U_t + \nabla \cdot F(U) = 0$, captured with a numerical scheme (that will inevitably introduce diffusion), admits a steady viscous profile. Therefore, the number $W(U_-, U_+)$ of cells occupied by intermediate states between U_- (close to U_L) and U_+ (close to U_R) can be approximated by

$$W(U_-, U_+) = \int_{\text{Physical path from } U_- \text{ to } U_+} \frac{\beta(w)}{||F(U) + K(U) - C||} dw$$

where β represents the numerical viscosity implied by the method and C depends on the nature, strength, and speed of the shock. The key idea here is that we can amplify the denominator's value in W, through our design of K, thus reducing W, the width of discontinuity capture. Even contact discontinuities will artificially be transformed to shocks, and will therefore also be captured sharply, and not diffuse with time.

As soon as K is correctly chosen (see [6], [9]), the operator relative to the flux F and the one relative to K can be separated, giving a standard advection step H, performed with any desired numerical scheme, followed by a stationary ACM step:

$$U^n = H(U^{n-1}),\ \text{then}\ U_K^n = \text{ACM}(U^n)$$

As powerful as this tool is, it can still suffer from its very strength. When applied in smooth regions of the solution U, ACM introduces undesirable nonphysical steps. Our DoD prevents this to happen, in addition to save the computational cost associated with this supplementary useless compression.

A DoD is a boolean function which returns *false* when the considered region is known to be smooth and *true* in the neighborhood of discontinuities. In our case, as sharp capture of the detonation front is critical, we use the unburnt gas ratio. The DoD consists of testing where the unburnt gas ratio is (numerically) equal to 1, and extending the regions where this condition is satisfied, according to the main scheme H pattern. Of course, this DoD is especially written for detonations, but other DoDs based on entropy, successive derivatives comparison, fluid speed, or any other suitable value can be designed. The advantage of this approach is that DoDs are easy to design and independent. Some examples are presented in [9].

These two tools, namely ACM and DoD, can now be combined to give a, computationally speaking, not too costly step, that will guarantee a one-point capture of discontinuities:

$$U_K^n = \big(\text{DoD} \cdot \text{ACM}\big)(U^n).$$

Therefore, our approach optimally combines, through a modular structure, central schemes, ACM with DoD strategies, and ODE solvers, leading to a family of schemes accurately capturing detonation waves with the advantage to circumvent the resolution of the Riemann problem.

3 Numerical Experiments

The complete reactive Euler equations are

$$\begin{aligned} \rho_t + (\rho u)_x &= 0 \\ (\rho u)_t + (\rho u^2 + p)_x &= 0 \\ E_t + (u(E+p))_x &= 0 \\ (\rho Z)_t + (\rho u Z)_x &= -\rho Z K(T) \end{aligned} \tag{2}$$

where the variable Z is the following mass fraction

$$0 \le Z = \frac{\text{mass of the unburnt gas}}{\text{total mass of the gas}} \le 1,$$

and the other quantities are the velocity u, the pressure p, the density ρ, and the total energy E.

The gas temperature is given by $T = p/\rho$, where the pressure is defined by the equation of state

$$p = (\gamma - 1)\big(E - \frac{1}{2}\rho u^2 - q_0 \rho z\big) \qquad (3)$$

with γ the specific heat ratio and q_0 the chemical heat release.

We solve this system by separating the advection operator from the chemical one. The homogeneous part of the system,

$$U_t + \nabla . F(U) = 0,$$

is solved by a central type scheme, while the underlying ODE,

$$U_t = -UK(T), \qquad (4)$$

can be solved by one's favorite ODE solver or even "exactly." We observe that the last equation of (2) can be reduced to

$$Z_t + uZ_x = -ZK(T)$$

by using the first conservation relation. Since all physical quantities except Z are constant through the chemical reaction, the system (4) can be reduced to the scalar equation

$$Z_t = -ZK(T). \qquad (5)$$

Assuming that the temperature stays constant during the chemical reaction, (5) can be solved exactly. This assumption on temperature is not quite true, since Z is a nonconserved variable, but is harmless for the splitting strategy (see [7]). We now present some numerical examples.

Example 1. Let us consider the one-dimensional classical test of an Arrhenius chemistry, with normalized characteristic length $L_{1/2} = 1$. The reaction rate of (4) is $K(T) = K_0 \exp(-E^+/T)$ and $L_{1/2}$ is the length of the zone where half of the burning occurs. This test is also used by Bourlioux [2] for the study of instabilities. Initially, the unburnt gas U_U is separated from the burnt gas U_B by a discontinuity

$$(\rho, p, u, Z)_U = (1, 1, 0, 1), K_0 = 165.6762, E^+ = 50, q_0 = 50, \gamma = 1.2.$$

The overdrive [4] is $f = 1.74$, so that the detonation is stable [2]. The burnt gas state U_B is analytically calculated using standard ZND model. The analytical solution on a coarse grid consists of two constant states U_B and U_U separated by a discontinuity traveling at the speed $s = 8.9823$. Figure 1 shows that the scheme using ACM and DoD accurately captures the physical discontinuity and its speed.

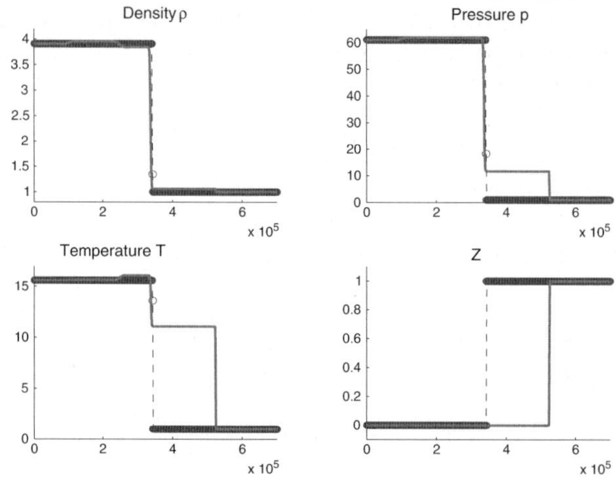

Fig. 1. Example 1, $\Delta x = 2{,}100 = 2.1 \times 10^3 L_{1/2}$, $t = 38{,}136$ (500 time steps). *Dashed line*: analytical underresolved solution, *continuous line*: without ACM and DoD, *circles*: with ACM and DoD

4 Limits of the Numerical Viscosity Hypothesis

The unphysical solution cannot be avoided without cleaning all of the numerical viscosity in the shock fitting. But, even if the precombustion shock is captured in only one (or even zero) mesh cell, there will still be a limit in the discretization's coarseness, beyond which the physical solution cannot be reached. It can easily be understood by considering the ignition temperature model. Let us call $z_c^* \in (0,1]$ the unique point of capture of the discontinuity in Z for a split method $\{u^{n+1} = ODE(u^*); u^* = ACM(u'); u' = Euler(u^n)\}$, after the compression step, just before resolution of the underlying ODE (5).

In the case where $T_c^* < T_{ign}$, no burning occurs, but if $T_c^* \geq T_{ign}$,

$$z_c^{n+1} = z_c^* \exp\left(\frac{-\Delta t}{\tau}\right). \tag{6}$$

As $\lim\limits_{\frac{\Delta t}{\tau} \to \infty} z_c^{n+1} = 0$, a limit discretization $\left(\frac{\Delta t}{\tau}\right)'$ such that z_c^{n+1} is numerically null always exists. For Δt large enough with respect to τ, all the remaining unburnt gas in the capture cell will burn, at each time step, leading to an unphysical combustion wave. The failure discretization $\left(\frac{\Delta t}{\tau}\right)'$ depends on the numerical method chosen for the advection step. Note that for a Glimm scheme [5], where z_c^* is always 1, the failure point would be larger than for our ACM approach, but still exists.

Our experiments show that cleaning the numerical diffusion introduced by the shock capture is necessary to obtain the physical detonation profile, but not sufficient for very stiff cases.

Unphysical behavior for schemes with minimal numerical viscosity is due to the inappropriate numerical treatment of the source term in the conservation law, for which the following approximation is performed

$$\int_{t^n}^{t^{n+1}} \Big\langle g(u(x,t)) \Big\rangle_{x_j}^{x_{j+1}} dt \cong \int_{t^n}^{t^{n+1}} g\Big(\langle u(x,t) \rangle_{x_j}^{x_{j+1}} \Big) dt, \qquad (7)$$

where $\langle \ \rangle_a^b := \frac{1}{|b-a|}\int_a^b$. By doing so, we use the mean value of temperature to compute the amount of unburnt gas after the chemical reaction. For very stiff cases where $\frac{\Delta t}{\tau} > \left(\frac{\Delta t}{\tau}\right)'$, this mean value can be above the ignition temperature, and then cause the burning of all the gas in the cell.

We propose to limit the amount of gas that can be burnt, $\delta z_{j+1/2} = z^*_{j+1/2} - z^{n+1}_{j+1/2}$, by a mean value of order 1, obtained from the quantity of gas that has been burnt at the edges of the cell. By taking

$$z^{n+1}_{j+1/2} = z^*_{j+1/2} - (\delta z)_{j+1/2},$$

with

$$(\delta z)_{j+1/2} = \min\left(\delta z_{j+1/2}, \frac{\delta z_j + \delta z_{j+1}}{2} \right),$$

the left-hand side of (7) is then well approximated:

$$\begin{aligned} \overline{(\delta z)}_{j+1/2} &= \int_{t^n}^{t^{n+1}} \frac{z(x_j,t)K(T(x_j,t)) + z(x_{j+1},t)K(T(x_{j+1},t))}{2} dt \\ &\cong \int_{t^n}^{t^{n+1}} \Big\langle g(z) \Big\rangle_{x_j}^{x_{j+1}} dt. \end{aligned} \qquad (8)$$

A proof of consistency is presented in [10].

Example 2. This example describes a strong detonation modeled by an ignition chemistry law

$$(\rho, p, u, Z)_U = (1,1,0,1), q_0 = 50, \gamma = 1.2, f = 1.6.$$

Table 1 shows the parameters' values beyond which our method fails, for different settings of our modular tools, with a discretization $\Delta x = 0.25$ fixed and $CFL = 0.4$, where CMV stands for our correction of mean value.

Table 1. Failure discretization for Example 2, at $T_{ign} = 1.01$, for different combinations of our tools

T_{ign}	$(\Delta t/\tau)'$	$\Delta x/L_{1/2}$	CMV	ACM
1.01	0.0016678	0.051651	Disabled	Disabled
	0.0079715	0.25808	Enabled	Disabled
	0.6589	19.8208	Disabled	Enabled
	2.66×10^{17}	2.88×10^{18}	Enabled	Enabled

References

1. P. Arminjon, M.-C. Viallon, A. Madrane, *A Finite Volume Extension of the Lax-Friedrichs and Nessyahu-Tadmor Schemes for Conservation Laws on Unstructured Grids*, Int. J. of Comp. Fluid Dyn., 9 (1997), 1–22.
2. A. Bourlioux, *Numerical Study of Unstable Detonations*, Ph.D. Thesis, Princeton University, Princeton, NJ, 1991.
3. P. Colella, A. Majda, V. Roytburd, *Theoretical and Numerical Structure for Reacting Shock Waves*, SIAM J. Stat. Comp., 7 (1986), 1059–1080.
4. W. Fickett, W.C. Davis, *Detonation, Theory and Experiment*, Dover Publications, Mineola, New York, 1979.
5. J. Glimm, *Solutions in the large for nonlinear hyperbolic systems of equations*, Comm. Pure Appl. Math., 18 (1965), 695–715.
6. A. Harten, *The Artificial Compression Method for Computation of Shocks and Contact Discontinuities. III. Self-Adjusting Hybrid Schemes*, Math. Comp., 32 (1978), 363–389.
7. R.J. LeVeque, *Nonlinear Conservation Laws and Finite Volume Methods*, Edited by O. Steiner and A. Gautschy, Comp. Meth. for Astrophysical Fluid Flow, Saas-Fee Advanced Course 27, Springer-Verlag, Berlin, 1998.
8. H. Nessyahu, E. Tadmor, *Non-oscillatory Central Differencing for Hyperbolic Conservation Laws*, J. of Comp. Phys., 87 (1990), 408–463.
9. O. Rouch, P. Arminjon, *Artificial Compression Method: A Modular Approach*, to appear.
10. M.-O. St-Hilaire, *Méthodes de volumes finis pour la simulation sous-résolue de détonations*, Thesis, Université de Montréal, Canada, 2006.

Generalized Momenta of Mass and Their Applications to the Flow of Compressible Fluid

O. Rozanova

1 Main Equations

The motion of compressible viscous, heat-conductive, polytropic fluid in $\mathbb{R} \times \mathbb{R}^n$, $n \geq 1$, is governed by the compressible Navier–Stokes (NS) equations

$$\partial_t \rho + \mathrm{div}_x(\rho v) = 0, \tag{1.1}$$

$$\partial_t(\rho v) + \mathrm{div}_x(\rho v \otimes v) + \nabla_x p = \mathrm{Div} T, \tag{1.2}$$

$$\partial_t \left(\frac{1}{2}\rho|v|^2 + \rho e \right) + \mathrm{div}_x \left((\frac{1}{2}\rho|v|^2 + \rho e + p)v \right) = \mathrm{div}(Tv) + k\Delta_x \theta, \tag{1.3}$$

where ρ, $u = (u_1, ..., u_n)$, p, e, θ denote the density, velocity, pressure, internal energy, and absolute temperature, respectively; T is the stress tensor given by the Newton law $T = T_{ij} = \mu(\partial_i v_j + \partial_j v_i) + \lambda \, \mathrm{div} u \, \delta_{ij}$, with constant coefficients of viscosity μ and λ ($\mu \geq 0$, $\lambda + \frac{2}{n}\mu \geq 0$), $k \geq 0$ is the coefficient of heat conduction. We denote Div and div the divergency of tensor and vector, respectively. The state equations are $p = (\gamma - 1)\rho e$, and $p = R\rho\theta$, where $\gamma > 1$ and $R > 0$ are the specific heat ratio and the universal gas constant, respectively. Thus, we can consider NS as a system for unknown ρ, u, p. Indeed, NS and the state equations give

$$\partial_t p + (v, \nabla_x p) + \gamma p \, \mathrm{div} v = (\gamma - 1) \sum_{i,j=1}^{n} T_{ij} \partial_j v_i + \frac{k}{R} \Delta \frac{p}{\rho}. \tag{1.4}$$

For classical solutions NS is equivalent to system (1.1), (1.2), and (1.4), denoted NS* for short.

NS* is supplemented with the initial data

$$(\rho, v, p)\Big|_{t=0} = (\rho_0(x), v_0(x), p_0(x)) \in C^2(\mathbb{R}^n).$$

If $\mu = \lambda = k = 0$, we get the gas dynamic (GD) equations.

2 Conservation Laws, Generalized Momenta of Mass and the Decay Rate

System NS is the differential form of conservation laws for the material volume $\mathcal{V}(t)$ (i.e., the volume, consisting from the same particles). NS expresses conservation of mass $m = \int\limits_{\mathcal{V}(t)} \rho\, dx$, momentum $P = \int\limits_{\mathcal{V}(t)} \rho v\, dx$, and total energy $\mathcal{E} = \int\limits_{\mathcal{V}(t)} \left(\frac{1}{2}\rho|v|^2 + \rho e\right) dx = E_k(t)+E_i(t)$. Here $E_k(t)$ and $E_i(t)$ are the kinetic and internal components of energy, respectively.

When $\mathcal{V}(t) = \mathbb{R}^n$, the conservation of mass, angular momentum and energy take place provided the components of solution decrease at infinity sufficiently quickly.

Definition 1. *We say that the classical solution to NS belongs to the class $K(M(t),\alpha)$, if there exist a positive vector-function $M(t) = (M_v(t), M_\rho(t), M_p(t))$ with components from $C([0,\infty))$, a constant vector $\alpha = (\alpha_v, \alpha_{Dv}, \alpha_\rho, \alpha_p, \alpha_\theta)$ and constants $R_0 > 0$, $T \geq 0$, such that $|v(t,x(t))| \leq M_v(t)|x(t)|^{\alpha_v}$, $|Dv(t,x(t))| \leq M_{Dv}(t)|x(t)|^{\alpha_{Dv}}$, $\rho(t,x(t)) \leq M_\rho(t)|x(t)|^{\alpha_\rho}$, $p(t,x(t)) \leq M_p(t)|x(t)|^{\alpha_p}$, $\theta(t,x(t)) \leq M_\theta(t)|x(t)|^{\alpha_\theta}$ for all trajectories $x(t)$ such that $|x(t)| > R_0$, $t > T$.*

Let us introduce a functional

$$G_\phi(t) = \int\limits_{\mathcal{V}(t)} \rho(t,x)\phi(|x|)\, dx.$$

If $\phi(|x|) = \frac{1}{2}|x|^2$, then $G_\phi(t)$ is the usual momentum of mass. By analogy for others ϕ we call $G_\phi(t)$ the generalized momentum of mass.

2.1 Decay Rate for NS Equation

Let us choose the appropriate class $K(M(t),\alpha)$ to guarantee conservation of mass, momentum, energy, and convergence of the mass momentum $G(t)$ for NS system. It is sufficient to set $\alpha = (\alpha_v, \alpha_{Dv}, \alpha_\rho, \alpha_p, \alpha_\theta) = (-n, -n-1, -n-2-\varepsilon, -n-\varepsilon, -n)$ with a constant $\varepsilon > 0$. We denote this class K_{NS}.

If the heat conductivity is zero, we do not need to require the decay of θ, i.e., $\alpha = (-n, -n-1, -n-2-\varepsilon, -n-\varepsilon, \alpha_\theta)$ with an arbitrary last component. We denote this class K_{NS_0}.

2.2 Decay Rate for GD Equations

In this case the behavior of velocity is less restrictive. Here it is sufficient to set $\alpha = (\alpha_v, \alpha_{Dv}, -n-2-\varepsilon, -n-\varepsilon, \alpha_\theta)$ with $\varepsilon > 0$, $\alpha_v \leq 1$ and arbitrary α_{Dv} and α_θ. The components of velocity may rise as $|\mathbf{x}| \to \infty$. We denote this class K_{GD}.

3 Behavior of Generalized Momenta of Mass on Solutions

We denote $\sigma = (\sigma_1, ..., \sigma_K)$ the vector with components $\sigma_k = v_i x_j - v_j x_i$, $i > j$, $i, j = 1, ..., n$, $k = 1, ..., K$, $K = \mathrm{C}_n^2$.

The following lemma describes the behavior of generalized momenta for GD system.

Lemma 1. *Let us suppose that $\phi(|x|)$ belongs to the class C^2 inside $\mathcal{V}(t)$. For classical solution to system GD the following equalities take place:*

$$G'_\phi(t) = \int\limits_{\mathcal{V}(t)} \frac{\phi'(|x|)}{|x|}(v, x)\rho\, dx,$$

$$G''_\phi(t) = I_{1,\phi}(t) + I_{2,\phi}(t) + I_{3,\phi}(t) + I_{4,\phi}(t),$$

where

$$I_{1,\phi}(t) = \int\limits_{\mathcal{V}(t)} \frac{\phi''(|x|)}{|x|^2}|(v, x)|^2\rho\, dx, \qquad I_{2,\phi}(t) = \int\limits_{\mathcal{V}(t)} \frac{\phi'(|x|)}{|x|^3}|\sigma|^2\rho\, dx,$$

$$I_{3,\phi}(t) = \int\limits_{\mathcal{V}(t)} (\phi''(|x|)+(n-1)\frac{\phi'(|x|)}{|x|})p\, dx, \quad I_{4,\phi}(t) = -\int\limits_{\partial\mathcal{V}(t)} \frac{\phi'(|x|)}{|x|}(x, \nu)p\, d\partial\mathcal{V},$$

where ν is a unit outer normal to $\mathcal{V}$.

The *proof* is a direct application of the Stokes formula.

Let us consider $\phi(|x|) = \frac{1}{2}|x|^2$ and $\mathcal{V}(t) = \mathbb{R}^n$. We denote the respective functional $G_\phi(t)$ as $G(t)$. In this case the following lemma is true.

Lemma 2. *For classical solutions to NS, $k = 0$, of the class K_{NS_0} and solutions to GD of the class K_{GD} the following equality holds:*

$$G''(t) = 2E_k(t) + n(\gamma - 1)E_i(t). \tag{3.1}$$

Proof. First of all, in NS case $G''(t) = I_1(t) + I_2(t) + I_3(t) + I_4(t) + I_5(t)$, where $I_1(t) = \int\limits_{\mathbb{R}^n} \frac{|(u,x)|^2}{|x|^2}\rho\, dx$, $I_2(t) = \int\limits_{\mathbb{R}^n} \frac{|\sigma|^2}{|x|^2}\rho\, dx$, $I_3(t) = n\int\limits_{\mathbb{R}^n} p\, dx$, $I_{4,\phi}(t) = -\lim\limits_{R\to\infty} \int\limits_{\partial B_R} (x, \nu)p\, d\partial B_R$, $I_5(t) = \int\limits_{\mathbb{R}^n} (\mathrm{Div}T, x)\, dx = \lim\limits_{R\to\infty} \int\limits_{\partial B_R} T_{ij}x_i\nu_j\, d\partial B_R - (2\mu + n\lambda)\int\limits_{B_R} \mathrm{div} v\, dx = \lim\limits_{R\to\infty} \int\limits_{\partial B_R} (T_{ij}x_i - (2\mu + n\lambda)v_i\,\delta_{ij})\nu_j\, d\partial B_R$, where $B_R : |x| \le R$, we sum the repeated indices. Owing to respective assumptions on the decay of solution as $|\mathbf{x}| \to \infty$ all improper integrals converge and $I_4(t) = I_5(t) = 0$. To end the proof we note that $I_1(t)+I_2(t) = 2E_k(t)$, $I_3(t) = n(\gamma - 1)E_i(t)$. In GD case the integral I_5 is missing. □

Remark 1. Equality (3.1) for GD was obtained firstly in [1].

Remark 2. It follows from (3.1) that $n(\gamma-1)\mathcal{E} \le G''(t) \le 2\mathcal{E}$. The inequality implies the same two-sided estimate of $G(t)$ that was obtained in three-dimensional for the sum of squared semi-axis of a nonrotating gas ellipsoid next to vacuum [2].

4 Necessary Conditions for Existence of Global Solution

If the initial density ρ_0 is compact, then in arbitrary space dimensions no solution to NS from $C^1([0,\infty), H^m(\mathbb{R}^n))$, $m \ge \left[\frac{n}{2}\right]+2$, exists [3]. This blow-up result depends crucially on the assumption about compactness of support of the initial density. Thus, the question remains: is it true that the global in t smooth solution exists for any smooth initial data in the case where the support of initial density coincides with the whole space? Below we find necessary conditions of existence of the global solution with prescribed decay rate as $|x| \to \infty$.

Theorem 1. *If the global in t solution to NS of the class K_{NS_0} exists, then the solution components grow as $t \to \infty$ at least such that*

$$\int_0^t M_v(\tau)\,d\tau = O(t^{1-\alpha_v}), \quad M_\rho(t) = O(t^{2+\epsilon}), \quad \epsilon > 0, \quad t \to \infty. \tag{4.1}$$

Proof. It follows from Lemma 2 that $G''(t) \ge (\gamma-1)\mathcal{E}$,

$$G(t) \ge \frac{(\gamma-1)n\mathcal{E}}{2}t^2 + G'(0)t + G(0). \tag{4.2}$$

Let us get the estimate of $G(t)$ from above.

We consider a material volume $\mathcal{V}(t)$ that initially coincides with the ball $|\mathbf{x}| \le R_0$ (see Definition 1). We denote $B_{R(t)} = (x : |x| \le R(t))$ a ball that contains $\mathcal{V}(t)$. From the definition of K_{NS} we have

$$\frac{dR(t)}{dt} \le R^{\alpha_v}(t)\,M_v(t), \quad R(t) \le \left((1-\alpha_v)\int_0^t M_v(\tau)\,d\tau + R_0^{1-\alpha_v}\right)^{\frac{1}{1-\alpha_v}}.$$

Further,

$$\begin{aligned} G(t) &= \frac{1}{2}\left(\int\limits_{B_{R(t)}} \rho|x|^2\,dx + \int\limits_{\mathbb{R}^n\setminus B_{R(t)}} \rho|x|^2\,dx\right) \\ &\le \frac{R^2(t)}{2}m + \frac{1}{2}M_\rho(t)\int\limits_{\mathbb{R}^n\setminus B_{R(t)}} |x|^{2+\alpha_\rho}\,dx \end{aligned}$$

$$\le O\left(\left(\int_0^t M_v(\tau)\,d\tau\right)^{\frac{2}{1-\alpha_v}}\right) + O\left(M_\rho(t)\left(\int_0^t M_v(\tau)\,d\tau\right)^{\frac{-\epsilon}{1-\alpha_v}}\right).$$

If the growth of $M_v(t)$ and $M_\rho(t)$ is less then prescribed in (4.1), then the latter inequality contradicts to (4.2). □

Remark 3. In particular, Theorem 1 implies that there exists no smooth solution to NS from the class K_{NS_0} with $M_v(t) = \text{const.}$ and/or $M_\rho(t) = \text{const.}$

The analog of Theorem 1 in the case of GD equations is the following.

Theorem 2. *If the global in t solution to GD from the class K_{GD} exists then the functions $M_\rho(t)$ are $M_v(t)$, restricting the density and velocity behavior as $t \to \infty$, grow at least such that $M_\rho(t) = O(t^{2+\epsilon})$, $\epsilon > 0$, and $\int_0^t M_v(\tau)\, d\tau = O(\ln t)$ (for $\alpha_v = 1$) or as prescribed in (4.6) (for $\alpha_v < 1$).*

The statement can be proved exactly as Theorem 1.

5 Motion with Uniform Deformation

We dwell in this section on GD equations, however due to the specific choice of the velocity field the results remain true for NS with $k = 0$.

The motion with uniform deformation (i.e., with linear profile of velocity $v(t, x) = A(t)\, x$, $A(t)$ is a matrix $(n \times n)$) was considered in many works. Let us mention [4], where this special solution to GD was firstly constructed in three-dimensional and [5], where many applications are given. Generally speaking, for this solution the velocity, density and pressure may blow up as $t \to T < \infty$ and $|\mathbf{x}| \to \infty$. Below we show how the solution with uniform deformation can be constructed by means of the generalized momenta of mass. Moreover, the requirement of finiteness of energy and mass momentum prohibits the blowup, therefore the respective solution is globally in time smooth.

Below we compare the procedure of constructing of the simplest solution with the linear velocity profile

$$v(t, x) = a(t)\, x \tag{5.1}$$

for a different choice of generalized momenta.

5.1 Basing on the Momentum of Mass

Let us consider once more the particular case $\phi(|x|) = \frac{1}{2}|x|^2$, the respective functional is the usual momentum of mass, $G(t)$.

Lemma 2 and (1.4) imply

$$G'(t) = 2a(t)G(t), \qquad a'(t) = -a^2(t) + KG^{-\frac{(\gamma-1)n+2}{2}}(t), \tag{5.2}$$

with a constant $K > 0$ depending on initial data. If the initial data satisfy the compatibility condition

$$p_0'(|x|) = -(\gamma - 1)G^{-1}(0)E_i(0)\rho_0(|x|), \tag{5.3}$$

then the density and pressure can be found from (1.1) and (1.4) as

$$\rho(t,|x|) = \exp(-n\int_0^t a(\tau)d\tau)\rho_0(|x|\exp(-\int_0^t a(\tau)d\tau)), \tag{5.4}$$

$$p(t,|x|) = \exp(-n\gamma\int_0^t a(\tau)d\tau)p_0(|x|\exp(-\int_0^t a(\tau)d\tau)). \tag{5.5}$$

It follows from (5.2) that $a(t)$ solves

$$a'(t) = -a^2(t) + K_1 \exp\left(-((\gamma-1)n+2)\int_0^t a(\tau)\,d\tau\right), \tag{5.6}$$

with $K_1 = K(G(0))^{-((\gamma-1)n+2)/2}$. Since $K_1 > 0$, the solution $a(t)$ remains bounded for all $t \ge 0$ and $a(t) = O(t^{-1})$, $t \to \infty$ [7].

Remark 4. In the case of globally smooth solution discussed above, the velocity $v(t,x) = a(t)\,x$, therefore $\alpha_v = 1$. Here $M_v(t) = a(t) = O(t^{-1})$ and $\int_0^t M_v(\tau)\,d\tau = O(\ln t)$, $M_\rho(t) = O(t^{2+\epsilon})$, $t \to \infty$. Thus, according Theorem 1, the solution presents a less possible rate of its component in t.

Remark 5. In [6] the solutions with linear profile of velocity for GD equations (including the presence of the Coriolis force and damping) were constructed for a general matrix $A(t)$ in two-dimensional and for particular cases in three-dimensional.

5.2 The "Excluding Pressure" Case

Let us choose as ϕ a function, proportional to the fundamental solution of the Laplace operator, namely

$$\phi(|x|) = \ln|x|, \quad n = 2, \qquad \phi(|x|) = |x|^{2-n}, \quad n \neq 2.$$

In this case Lemma 1 implies

$$G_\phi'(t) = \lambda_1(n)\int_{\mathbb{R}^n} \frac{(v,x)}{|x|^n}\rho\,dx,$$

$$G_\phi''(t) = \lambda_2(n)\int_{\mathbb{R}^n} \frac{|(v,x)|^2}{|x|^{n+2}}\rho\,dx + \lambda_3(n)\int_{\mathbb{R}^n} \frac{|\sigma|^2\rho}{|x|^{n+2}}\,dx + \frac{p(t,0)}{\omega_{n-1}(2-n)},$$

where

$$\lambda_1(n) = 1, \quad n = 1,2 \quad \text{and} \quad 2-n, \quad n \ge 3,$$

$$\lambda_2(n) = 0, \quad n = 1; \ -1, \quad n = 2, \quad \text{and} \quad (1-n)(2-n), \quad n \ge 3,$$

$$\lambda_3(n) = 0, \quad n = 1; \quad 1, \quad n = 2, \quad \text{and} \quad 2 - n, \quad n \geq 3.$$

Here we take as $\mathcal{V}(t)$ the space $\mathbb{R}^n$, all improper integrals are supposed convergent both at the origin and at infinity. Actually, this signifies that $v(t,0) = 0$.

We use the value of pressure only in the origin; in this sense the pressure in the remaining space is excluded.

Let us consider the case $n \geq 3$. Here $G'_\phi(t) = \lambda_1(n)\, a(t) G_\phi(t)$,

$$G''_\phi(t) = \lambda_2(n)\, a^2(t)\, G_\phi(t) + \frac{p(0,0)}{\omega_{n-1}(2-n)} \exp\left(-\gamma n \int_0^t a(\tau)\, d\tau\right).$$

The latter system implies

$$a'(t) = -a^2(t) + K_2 \exp\left(-((\gamma-1)n+2) \int_0^t a(\tau)\, d\tau\right), \tag{5.7}$$

with $K_2 = \frac{P(0,0) G_\phi^{n-2}(0)}{\omega_{n-1}(2-n)^2}$. We can see that (5.6) coincides with (5.7), the only difference is in constants K_1 and K_2. Provided initial data satisfy the compatibility condition

$$G_\phi(0) p'_0(|x|) = -\frac{p(0,0)}{\omega_{n-1}(2-n)^2} \rho_0(|x|)|x|, \tag{5.8}$$

the density and pressure can be found by formulas (5.4), (5.5).

The analysis of (5.6) (or (5.7)) shows that if the constant K_1 (or K_2) is positive, as in our case, then the solution exists for all $t \geq 0$ and $\alpha(t) = O(t^{-1})$ as $t \to \infty$.

We see that in the second case ($\phi = |x|^{2-n}$) we get a more wide class of solutions. Indeed, in Sect. 5.1 we need to require the significant decay of solutions as $|x| \to \infty$ to guarantee the finiteness both of energy and momentum of mass, i.e. we restrict ourselves by solutions from K_{GD} with $\alpha_v = 1$. When we construct solutions using the generalized momentum, we do not require the finiteness of energy and the behavior of smooth pressure and density, prescribed by convergence of $G_\phi(t)$ and condition (5.8) is the following:

$$p(|x|) = O(|x|^q), \quad q < 0, \quad \rho(|x|) = O(|x|^s), \quad s = q - 2 < 0, \quad |x| \to \infty.$$

6 Behavior of Boundary of a Liquid Volume

We mention here very briefly one more application of the generalized momenta of mass to GD equations (see [8] for details). Namely, the expansion of boundary of a material volume inside of a smooth flow of gas can be studied by this method. The last question is connected with a problem of air pollution: one can be interested if under usual meteorological conditions a polluted cloud will attain some geographical object.

We suppose that initially a point x_0 do not belong to $\mathcal{V}(t)$. Let us set the following question: what conditions we have to impose on initial data provided they are known only inside of $\mathcal{V}(0)$, to guarantee that the boundary of given material volume within a smooth flow will attain a given ε – neighborhood of point x_0? It is clear that for the answer to this question certain assumptions on the thermodynamic variables in the whole space have to be done. To formulate these assumptions we need the following definition.

Definition 2. *We say that the pressure in the moment t is* distributed *along the boundary $\partial\mathcal{V}(t)$ of domain $\mathcal{V}(t)$, $x_0 \notin \mathcal{V}(t)$,* regularly with constant $M \geq 0$, *if* $\left| \int\limits_{\partial\mathcal{V}(t)} \left(\frac{x-x_0}{|x-x_0|}, \nu\right) p\, d\partial\mathcal{V} \right| \leq M$, *where ν is a unit outer normal.*

If $p(t,x)$ is constant, then the integral in the left-hand side of the latter inequality is equal to zero. Setting M sufficiently small we assume that the material volume will not meet a zone of large gradient of pressure.

To prove the following theorem we use the generalized momentum of mass with $\phi(|x|) = |x - x_0|^q$, $q < 0$.

Theorem 3. [8] *Let a finite material volume $\mathcal{V}(t)$ of compressible liquid with C^1 – smooth boundary $\partial\mathcal{V}(t)$ do not contain a point x_0. Suppose that the flow is C^1 – smooth for all $t \in [0,T]$, $T \leq \infty$, and the pressure along the boundary $\partial\mathcal{V}(t)$ is distributed regularly with a constant M uniformly in t for $t \in [0,T]$.*

Let us choose some real numbers $q < -n - \frac{2}{\gamma-1}$ and ε, $0 < \varepsilon <$ dist$(\partial\mathcal{V}(0), x_0)$. *Then for all initial data there exists such constant $\delta \leq 0$, depending on initial data, ε, q, T, M, n, γ that if initially*

$$\int\limits_{\mathcal{V}(0)} |x - x_0|^{q-2} \left(v(0,x), x - x_0\right) \rho_0(x)\, dx < \delta,$$

then within a time t_1, later then T, the boundary of given material volume will attain the ε – neighborhood of point x_0.

Acknowledgment

This work was supported by DFG Project 436 RUS 113/823/0-1.

References

1. J.-Y.Chemin, *Dynamique des gaz à masse totale finie,* Asymptotic Analysis. **3**(1990), 215–220.
2. S.I. Anisimov, Yu. I. Lysikov, *Applied Math. Mech.* **34**(5)(1970), 926.
3. Z.P. Xin, *Blowup of smooth solutions to the compressible Navier-Stokes equation with compact density,* Comm. Pure Appl. Math. **51**(1998), 229–240.

4. L.V. Ovsyannikov, New solution to the hydrodynamics equations, *Academy of Science of URSS. Doklady,* **111**(1)(1956), 47–49.
5. O.I. Bogoyavlensky, Methods in the qualitative theory of dynamic systems in astrophysics and gas dynamics. *Springer Series in Soviet Mathematics*, Springer, IX, 1985.
6. O.S. Rozanova, *Classes of smooth solutions to multidimensional balace law of gas gynamic type on riemannian manifolds,* in: "Trends in Mathematical Physics Research" ed. C.V. Benton, New York, Nova Science Publishers, pp. 155–204.
7. O.S. Rozanova, *Solutions with linear profile of velocity to the Euler equations in several dimensions.* Hyperbolic problems: theory, numerics, applications, 861–870, Springer, Berlin, 2003.
8. O.S.Rozanova, *Behavior of a boundary of moving volume inside of a smooth flow of compressible liquid,* Differ. Equat., N8 (2006), 192–212, (e-Print archive: http://arxiv.org/math-ph/0511088)

ADER–Runge–Kutta Schemes for Conservation Laws in One Space Dimension

G. Russo, E.F. Toro, and V.A. Titarev

1 Introduction

ADER is a recent Godunov-type approach for constructing arbitrarily high-order finite-volume schemes for hyperbolic conservation laws. The idea was first proposed for the constant coefficient linear advection equation in multiple space dimensions [12]. The extension to nonlinear systems is based on the approximate solution procedure for the so-called derivative Riemann problem [13, 14] for nonlinear hyperbolic systems with reactive source terms. For the resulting schemes see [11, 9, 2] and references therein.

The ADER flux is obtained by integrating the DRP solution flux at the edge of the cell, by the use of a quadrature formula [11]. Construction of this DRP solution involves the Taylor time expansion along the cell edge. Here we propose an alternative approach, in which the Taylor expansion is replaced by a Runge–Kutta procedure, which can be used to compute the stage values at the edge of each cell, and, from these, the (conservative) numerical solution at the new time step. This procedure, called Central Runge–Kutta, has been used for the construction of high-order central schemes on staggered grids [6]. Central Runge–Kutta and ADER have been previously combined in the context of staggered central schemes [7]. There the pointwise value of the unknown field is reconstructed from cell averages on a staggered grid. Although that approach is promising, it still suffers from the usual large dissipation of central schemes.

The rest of the chapter is organized as follows. In Sect. 2 we review the original ADER approach. In Sect. 3 the construction of ADER–RK scheme will be described. In Sect. 4 we present numerical examples for the one-dimensional compressible Euler equations, and draw some conclusions.

2 Conventional ADER Schemes

In this section we review the construction of conventional ADER schemes as applied to the one-dimensional systems [11]. Extension to multiple space dimensions and source terms can be found in [9].

Consider a hyperbolic system in conservation form given by

$$\partial_t \mathbf{Q} + \partial_x \mathbf{F}(\mathbf{Q}) = \mathbf{0}, \tag{1}$$

along with initial and boundary conditions. Here $\mathbf{Q}$ is the vector of unknown conservative variables and $\mathbf{F}(\mathbf{Q})$ is the physical flux vector. Integrating (1) over a space–time control volume in $x - t$ space $[x_{i-1/2}, x_{i+1/2}] \times [t^n, t^{n+1}]$ of dimensions $\Delta x = x_{i+1/2} - x_{i-1/2}$, $\Delta t = t^{n+1} - t^n$, we obtain the following one-step finite-volume scheme:

$$\bar{\mathbf{Q}}_i^{n+1} = \bar{\mathbf{Q}}_i^n + \frac{\Delta t}{\Delta x}\left(\bar{\mathbf{F}}_{i-1/2}^{n+1/2} - \bar{\mathbf{F}}_{i+1/2}^{n+1/2}\right), \tag{2}$$

$$\bar{\mathbf{Q}}_i^n = \frac{1}{\Delta x}\int_{x_{i-1/2}}^{x_{i+1/2}} \mathbf{Q}(x, t^n)\, dx, \quad \bar{\mathbf{F}}_{i+1/2}^{n+1/2} = \frac{1}{\Delta t}\int_{t^n}^{t^{n+1}} \mathbf{F}(\mathbf{Q}(x_{i+1/2}, t))\, dt. \tag{3}$$

The flux average is obtained by a quadrature formula at the edge of the cell. Simpson rule, for example, can be used for third- and fourth-order schemes:

$$\mathbf{F}_{i+1/2} = \frac{1}{6}\left(\mathbf{F}(\mathbf{Q}_{i+1/2}(0)) + 4\mathbf{F}(\mathbf{Q}_{i+1/2}(\frac{1}{2}\Delta t)) + \mathbf{F}(\mathbf{Q}_{i+1/2}(\Delta t))\right) \tag{4}$$

where $\mathbf{Q}_{i+1/2}(\tau) \approx \mathbf{Q}(x_{i+1/2}, t_n + \tau)$, $\tau \in [0, \Delta t]$ denotes an approximation of the state values at cell edge. This approximate state $\mathbf{Q}_{i+1/2}(\tau)$ is obtained by solving semi-analytically the Derivative Riemann problem at cell interface position [13, 14]:

$$\begin{aligned} &\partial_t \mathbf{Q} + \partial_x \mathbf{F}(\mathbf{Q}) = \mathbf{0}, \\ &\mathbf{Q}(x, 0) = \begin{cases} \mathbf{Q}_L(x) = \mathbf{p}_i(x), & x < x_{i+1/2}, \\ \mathbf{Q}_R(x) = \mathbf{p}_{i+1}(x), & x > x_{i+1/2} \end{cases} \end{aligned} \tag{5}$$

where $\mathbf{p}_i(x)$ and $\mathbf{p}_{i+1}(x)$ are polynomial reconstructions in cells i and $i+1$, obtained by WENO (see [4, 1] and references therein). The solution of (5) is sought as a Taylor expansion in time:

$$\mathbf{Q}_{i+1/2}(\tau) = \mathbf{Q}_{i+1/2}(0+) + \sum_{k=1}^{r-1}\left[\frac{\partial^k}{\partial t^k}\mathbf{Q}(x, t)(x_{i+1/2}, t_n+)\right]\frac{\tau^k}{k!}. \tag{6}$$

The term of order zero in the expansion, $\mathbf{Q}_{i+1/2}(0+)$, is obtained by solving a standard Riemann problem with left and right states, respectively, $\mathbf{Q}_L = \mathbf{p}_i(x_{i+1/2})$, $\mathbf{Q}_R = \mathbf{p}_{i+1}(x_{i+1/2})$, while the higher order derivatives

$$\mathbf{Q}^{[k]} \equiv \frac{\partial^k}{\partial x^k}\mathbf{Q}, \quad 1 \le k \le r-1$$

at time $\tau = 0^+$ are obtained as solution of the following linear conventional Riemann problems:

$$\partial_t\left(\mathbf{Q}^{[k]}\right) + \mathbf{A}_{i+1/2}\,\partial_x\left(\mathbf{Q}^{[k]}\right) = \mathbf{0}, \quad \mathbf{A}_{i+1/2} = \mathbf{A}(\mathbf{Q}_{i+1/2}(0+)),$$

$$\mathbf{Q}^{[k]}(x,0) = \begin{cases} \dfrac{\partial^k}{\partial x^k}\mathbf{Q}_L(x_{i+1/2}),\ x < x_{i+1/2}, \\ \dfrac{\partial^k}{\partial x^k}\mathbf{Q}_R(x_{i+1/2}),\ x > x_{i+1/2}. \end{cases} \tag{7}$$

Finally, time derivatives are computed from spatial derivatives, by time differentiation of the original evolution equation (1). For example

$$\mathbf{Q}_t = -\mathbf{A}\mathbf{Q}_x, \quad \mathbf{Q}_{tx} = -\mathbf{B}\mathbf{Q}_x\mathbf{Q}_x - \mathbf{A}\mathbf{Q}_{xx}, \quad \mathbf{Q}_{tt} = -\mathbf{B}\mathbf{Q}_t\mathbf{Q}_x - \mathbf{A}\mathbf{Q}_{tx}, \tag{8}$$

where

$$\mathbf{B} = \left(\frac{\partial^2 \mathbf{F}}{\partial \mathbf{Q}^2}\right).$$

Further details about standard ADER schemes are illustrated in [11, 9].

3 ADER–RK Schemes

As can be seen from (8), the Cauchy–Kowalewski procedure is a rather complicated and cumbersome part of the conventional ADER approach. For example, for fourth order accuracy one needs to compute three third-order derivatives.

Therefore, it would be desirable to reduce the number of high order derivatives used in (8) and to reduce the number of matrix evaluations. To do so, we adopt, with proper modifications, an idea from Central Runge–Kutta WENO schemes [6]. The procedure is outlined below.

Starting from the evolution equation (1), an exact evolution equation for cell averages is derived, by integrating over cell i, and dividing by the cell size:

$$\frac{d\bar{\mathbf{Q}}_i}{dt} + \frac{1}{\Delta x}(\mathbf{F}(\bar{\mathbf{Q}}(x_{i+1/2},t) - \mathbf{F}(\bar{\mathbf{Q}}(x_{i-1/2},t)) = 0. \tag{9}$$

Such equation is discretized in time using a CRK approach [6, 7], i.e., the numerical solution for the cell average corresponding to a r-stage Runge–Kutta method is obtained as

$$\bar{\mathbf{Q}}_i^{n+1} = \bar{\mathbf{Q}}_i^n - \frac{\Delta t}{\Delta x}\sum_{l=0}^{r-1} b_l(\mathbf{F}(\mathbf{Q}_{i+1/2}^{(l)}) - \mathbf{F}(\mathbf{Q}_{i-1/2}^{(l)})), \tag{10}$$

and the stage values $\mathbf{Q}^{(l)}_{i+1/2}$ are obtained by applying the Runge–Kutta procedure to the evolution equation for $\mathbf{Q}$ written in the form

$$\frac{\partial \mathbf{Q}}{\partial t} = -\mathbf{A}(\mathbf{Q})\frac{\partial \mathbf{Q}}{\partial x}, \tag{11}$$

which, in turn is given by

$$\mathbf{Q}^{(j)}(x) = \mathbf{Q}^{(0)}(x) - \Delta t \sum_{l=0}^{j-1} a_{jl}\mathbf{A}^{(l)}\mathbf{Q}^{(l)}_x(x), \quad j = 1, \ldots r-1 \tag{12}$$

evaluated at $x = x_{j-1/2}$ or $x = x_{j+1/2}$. Here $\mathbf{A}^{(l)} \equiv \mathbf{A}(\mathbf{Q}^{(l)})$. The quantities $\mathbf{Q}^{(l)}_{i+1/2}$ are the so-called stages values of the conservative variable at position $x_{i+1/2}$; the coefficients a_{jl} and b_l are the usual Runge–Kutta coefficients. The ones we used for third- and fourth-order schemes are the same used in [6], and can be found in any textbook on numerical methods for ordinary differential equations. The number r of stages may be equal to the order of the scheme up to $r = 4$.

Once the stage values are computed, the scheme can be cast in the form (1), with

$$\bar{\mathbf{F}}^{n+1/2}_{i+1/2} = \sum_{l=0}^{r-1} b_l \mathbf{F}\left(\mathbf{Q}^{(l-1)}_{i+1/2}\right). \tag{13}$$

In the Central Runge–Kutta methods [6, 7], the initial value $\mathbf{Q}^{(0)}$ is obtained by performing pointwise reconstruction from cell averages on a staggered grid using Central WENO reconstruction [5]. Here we use the ADER approach, and compute $\mathbf{Q}^{(0)}$ as the solution of the classical Riemann problem for (1) at each cell edge.

For a moment we keep the spatial variable x continuous and write down the Runge Kutta method as follows. Assume that we are given the initial value $\mathbf{Q}^{(0)}(x)$ and all its spatial derivatives. Then we can write

$$\mathbf{Q}^{(j)}(x) = \mathbf{Q}^{(0)}(x) - \Delta t \sum_{l=0}^{j-1} a_{jl}\mathbf{A}^{(l)}\mathbf{Q}^{(l)}_x(x), \quad j = 1, \ldots r-1. \tag{14}$$

The computation of stage values for $1 \leq j \leq r-1$ requires the unknown spatial derivatives $\mathbf{Q}_x$ of the solution at stage levels $0 \leq j \leq r-2$. To find them we differentiate (14) with respect to x to obtain

$$\mathbf{Q}^{(j)}_x(x) = \mathbf{Q}^{(0)}_x(x) - \Delta t \sum_{l=0}^{j-1} a_{lj}\left(\mathbf{A}^{(l)}\mathbf{Q}^{(l)}_x(x)\right)_x, \quad j = 1, \ldots r-2. \tag{15}$$

It is again obvious, that the computation of stage values of derivatives for $1 \leq j \leq r-2$ requires the unknown second-order spatial derivatives $\mathbf{Q}_{xx}$ of the solution at stage levels $0 \leq j \leq r-3$. We again take the derivative of (15) and get

$$\mathbf{Q}_{xx}^{(j)}(x) = \mathbf{Q}_{xx}^{(0)}(x) - \Delta t \sum_{l=1}^{j-1} a_{lj} \left(\mathbf{A}^{(l-1)} \mathbf{Q}_x^{(l-1)}(x) \right)_{xx}, \quad j = 1, \ldots r-3. \quad (16)$$

Therefore, we obtain a triangular table of stage values of the conservative variable and its derivatives. We now return from the continuous case to the discretized one and rewrite (14)–(16) in terms of spatial derivatives $\mathbf{Q}^{(k,j)} := \partial_x^k \mathbf{Q}^{(j)}$, $k = 0, \ldots r-1$ at the interface position $x_{i+1/2}$:

$$\begin{aligned}
\mathbf{Q}^{(0,j)} &= \mathbf{Q}^{(0,0)} - \Delta t \sum_{l=0}^{j-1} a_{jl} (\mathbf{A}\mathbf{Q}^{[1]})^{(l)}, \\
\mathbf{Q}^{(1,j)} &= \mathbf{Q}^{(1,0)} - \Delta t \sum_{l=0}^{j-1} a_{jl} \left(\mathbf{B}\mathbf{Q}^{[1]}\mathbf{Q}^{[1]} + \mathbf{A}\mathbf{Q}^{([2]} \right)^{(l)}, \\
\mathbf{Q}^{(2,j)} &= \mathbf{Q}^{(2,0)} - \Delta t \sum_{l=0}^{j-1} a_{jl} \left(\mathbf{C}\mathbf{Q}^{[1]}\mathbf{Q}^{[2]}\mathbf{Q}^{[1]} + \mathbf{B}(\mathbf{Q}^{[2]}\mathbf{Q}^{[1]} \right. \\
&\qquad \left. + 2\mathbf{Q}^{[1]}\mathbf{Q}^{[2]}) + \mathbf{A}\mathbf{Q}^{[3]} \right)^{(l)},
\end{aligned} \quad (17)$$

where $\mathbf{Q}^{(0,j)} \equiv \mathbf{Q}_{i+1/2}^j$. Notice that the values at the zero stage $\mathbf{Q}^{(k,0)}$ are obtained exactly as in the original ADER approach by solving the Derivative Riemann problem (5), (7) at the cell interface position for the state $\mathbf{Q}_{i+1/2}(\tau)$ and for the derivatives, while $\mathbf{Q}(k,j)$, $j > 0$, are obtained by the RK procedure.

By comparing (8) and (17) we see that the new form of the temporal discretization uses a smaller part of the conventional Cauchy–Kowalewski procedure, which resulting a simplification. However, the right-hand side in (17) has to be calculated not only at initial time but also for other stages, which increases the cost. However, it is still cheaper than the original version.

Linear stability of the scheme, and the extension of the method to more spatial dimensions will be dealt with in a forthcoming chapter [3].

4 Numerical Results

Here we study numerically the convergence properties of the new ADER–RK schemes and compare them with the original state-expansion ADER schemes. We present the results of third- ($r = 3$) and fourth-order ($r = 4$) schemes. For both versions of ADER weighted essentially nonoscillatory reconstruction is used [4, 8]. We apply our methods to the time-dependent, one-dimensional compressible Euler equations for a polytropic gas of the form (1) with

$$\mathbf{Q} = \begin{pmatrix} \rho \\ \rho u \\ E \end{pmatrix}, \quad \mathbf{F} = \mathbf{Q}u + \begin{pmatrix} 0 \\ p \\ pu \end{pmatrix}, \quad (18)$$

$$p = (\gamma - 1)(E - \frac{1}{2}\rho(u^2)).$$

Here ρ, u, p, and E are density, components of velocity in the x coordinate direction, pressure and total energy, respectively; γ is the ratio of specific heats. We use $\gamma = 1.4$ throughout.

4.1 Smooth Problem

This test problem is chosen to study the convergence properties of the schemes. We solve (18) for the following initial condition, defined on $[-1, 1]$:

$$u = p = 1, \quad \rho = 2 + (\sin \pi x)^4, \tag{19}$$

so that the exact solution is $\rho(x,t) = 2 + [\sin(\pi(x-t))]^4$, $u = p = 1$. Periodic boundary conditions are used. The error of *cell averages* is measured at the output time $t = 4$ (two-time periods). Table 1 shows convergence rates and errors for density in different norms.

We observe that both versions of the schemes produce virtually identical results. The designed order of accuracy is reached in the integral norm. In the maximum norm the observed order of accuracy for fourth-order schemes is somewhat below the expected one which is due to the nature of the problem. As is well known [4], profiles of $(\sin x)^4$ type are difficult for schemes based on ENO/WENO type reconstructions.

Table 1. Convergence study for various schemes as applied to (18) with initial condition (19) at output time $t = 4$

Method	N	L_∞ error	L_∞ order	L_1 error	L_1 order
ADER3	25	1.29×10^{-1}		1.09×10^{-1}	
	50	1.04×10^{-2}	3.62	8.66×10^{-3}	3.65
	100	1.86×10^{-3}	2.49	9.51×10^{-4}	3.19
	200	2.23×10^{-4}	3.06	8.94×10^{-5}	3.41
ADER3–RK	25	1.29×10^{-1}		1.09×10^{-1}	
	50	1.04×10^{-2}	3.62	8.66×10^{-3}	3.65
	100	1.86×10^{-3}	2.49	9.51×10^{-4}	3.19
	200	2.23×10^{-4}	3.06	8.95×10^{-5}	3.41
ADER4	25	3.43×10^{-2}		3.07×10^{-2}	
	50	4.59×10^{-3}	2.90	2.48×10^{-3}	3.63
	100	5.75×10^{-4}	3.00	1.90×10^{-4}	3.70
	200	6.56×10^{-5}	3.13	1.26×10^{-5}	3.92
ADER4–RK	25	3.43×10^{-2}		3.07×10^{-2}	
	50	4.59×10^{-3}	2.90	2.48×10^{-3}	3.63
	100	5.75×10^{-4}	3.00	1.90×10^{-4}	3.71
	200	6.56×10^{-5}	3.13	1.23×10^{-5}	3.95

All schemes are used with CFL = 0.95

4.2 Modified Shock/Turbulence Interaction

Here we use a test problem from [10] that is a modification of the shock/turbulence problem proposed in [4]. The initial condition defined on $[-5, 5]$ is given by

$$(\rho, u, p) = \begin{cases} (1.515695, 0.523346, 1.80500), & x < -4.5, \\ (1 + 0.1 \sin 20\pi x, 0.0, 1), & x > -4.5. \end{cases} \tag{20}$$

The test problem consists of a right facing shock wave of Mach number 1.1 running into a high-frequency density perturbation. As time evolves, the shock moves into this perturbation, which spreads upstream. We compute the solution at the output time $t = 5$.

Results of the new ADER4–RK scheme are shown in Fig. 1. Symbols denote the numerical solution and the solid line denotes the reference solution, computed on a very fine mesh. The solution contains *physical oscillations* which have to be resolved by the numerical method. For the calculations shown here we choose a coarse mesh of 1, 500 cells and again use $CFL = 0.95$. It can be checked that the results are essentially the same as those obtained by the original ADER4 scheme, and presented in [10].

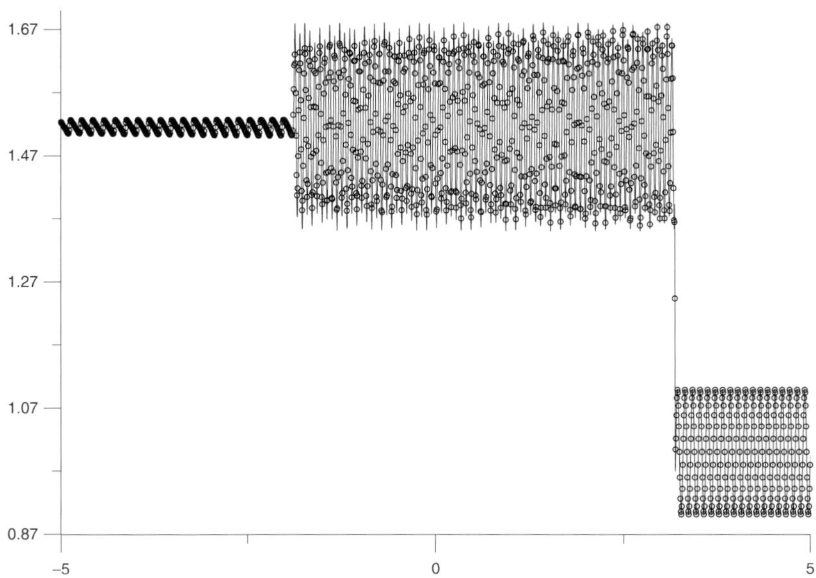

Fig. 1. New ADER4–RK scheme for the modified shock-turbulence interaction problem with 1,500 cells

Acknowledgment

The third author acknowledges the financial support provided by the PRIN programme (2004–2006) of the Italian Ministry of Education and Research (MIUR).

References

1. D.S. Balsara and C.-W. Shu. Monotonicity preserving weighted essentially non-oscillatory schemes with increasingly high order of accuracy. *J. Comput. Phys.*, 160:405–452, 2000.
2. M. Dumbser and C.D. Munz. Building blocks for arbitrary high order discontinuous Galerkin schemes. *Journal of Scientific Computing*, 27:215–230, 2006.
3. G. Russo, V.A. Titarev, and E.F. Toro. ADER-Runge-Kutta schemes for systems of conservation laws. *JCP.* in preparation.
4. G.S. Jiang and C.W. Shu. Efficient implementation of weighted ENO schemes. *J. Comput. Phys.*, 126:202–212, 1996.
5. D. Levy, G. Puppo, and G. Russo. Central WENO schemes for hyperbolic systems of conservation laws. *M2AN Math. Model. Numer. Anal.*, 33:547–571, 1999.
6. L. Pareschi, G. Puppo, and G. Russo. Central Runge-Kutta schemes for conservation laws. *SIAM J. Sci. Comput.*, 26(3):979–999, 2005.
7. R.M. Pidatella and G. Russo. Solving conservation laws by ADER central Runge-Kutta schemes. In *Applied and industrial mathematics in Italy*, number 69 in Ser. Adv. Math. Appl. Sci., pages 428–439. World Sci. Publ., Hackensack, NJ, 2005.
8. J. Shi, C. Hu, and C.-W. Shu. A technique for treating negative weights in WENO schemes. *J. Comput. Phys.*, 175:108–127, 2002.
9. V.A. Titarev and E.F. Toro. ADER schemes for three-dimensional nonlinear hyperbolic systems. *J. Comput. Phys.*, 204(2):715–736, 2005.
10. V.A. Titarev and E.F. Toro. WENO schemes based on upwind and centred TVD fluxes. *Computers and Fluids*, 34(6):705–720, 2005.
11. V.A. Titarev and E.F. Toro. ADER: arbitrary high order Godunov approach. *J. Sci. Comput.*, 17:609–618, 2002.
12. E.F. Toro, R.C. Millington, and L.A.M. Nejad. Towards very high order Godunov schemes. In E.F. Toro, editor, *Godunov Methods. Theory and Applications*, pages 907–940. Kluwer/Plenum Academic Publishers, 2001.
13. E.F. Toro and V.A. Titarev. Solution of the generalised Riemann problem for advection-reaction equations. *Proc. Roy. Soc. London*, 458 (2018):271–281, 2002.
14. E.F. Toro and V.A. Titarev. Derivative Riemann solvers for systems of conservation laws and ADER methods. *J. Comput. Phys.*, 212(1):150–165, 2006.

Strong Boundary Traces and Well-Posedness for Scalar Conservation Laws with Dissipative Boundary Conditions

B. Andreianov and K. Sbihi

1 Introduction

The aim of this chapter is to give sense to the following formal problem for a scalar conservation law with boundary condition (BC, in the sequel):

$$(H)\begin{cases} u_t + \operatorname{div}\, \varphi(u) = f & \text{in } \; Q := (0,T)\times\Omega \\ u(0,\cdot) = u_0 & \text{on } \; \Omega \\ \varphi_\nu(u) := \varphi(u)\cdot\nu \in \beta(u) & \text{on } \; \Sigma := (0,T)\times\partial\Omega. \end{cases}$$

Here $\Omega = \mathbb{R}^+ \times \mathbb{R}^{N-1}$ $(N \geq 1)$, $T > 0$, ν is the unit outward normal vector on $\partial\Omega$, $\varphi : \mathbb{R} \to \mathbb{R}^N$ is continuous, and β is a maximal monotone graph on $\mathbb{R}$. Assume $\varphi_\nu(0) - \beta(0) \ni 0$ and normalize φ, β by $\varphi(0) = 0$, $0 \in \beta(0)$. The classical Neumann (zero flux) and Dirichlet homogeneous BC correspond to the graphs $\beta = \mathbb{R} \times \{0\}$ and $\beta = \{0\} \times \mathbb{R}$, respectively.

To show existence, we restrict our attention to the case of Ω with flat boundary, to $L^1 \cap L^\infty$ data u_0,f (see (12) for the precise assumption on the data), and make the following simplifying assumptions:

$$\text{there exists a constant } C \text{ such that } |\beta(z)| \geq \operatorname{sign}(z)\,\varphi_\nu(z) \;\; \forall |z| > C; \tag{1}$$

$$\varphi \text{ is Lipschitz continuous, and } \varphi_\nu = \varphi\cdot\nu \text{ is piecewise monotone;} \tag{2}$$

$$\forall \xi \in \mathbb{R}^N \setminus \{0\}, \text{ the function } z \mapsto \xi\cdot\varphi(z) \text{ is nonconstant on any interval.} \tag{3}$$

In the conclusion, possible generalizations are indicated.

It is well known that existence for (H) generally fails if one interprets the Dirichlet BC literally (cf. [BLN]). This is also the case for general β, except for some particular situations (cf. e.g., [BFK]). If one approximates (H) by a sequence of problems (H^ϵ) with the BC understood literally (e.g., parabolic "viscous" approximations, or numerical schemes), a boundary layer can form in the corresponding solutions u^ϵ. The convergence of u^ϵ then takes place only locally inside the domain, and the limiting function u satisfies to the scalar conservation law with a different BC, which we call "effective BC."

It is the goal of this chapter to investigate the form of the effective BC for a general BC given by a maximal monotone graph β. Note that the monotonicity is necessary, if we hope the boundary condition to be dissipative in L^1 (in particular, if we hope for an L^1 contraction principle for solutions of (H), as in the classical situations of $\Omega = \mathbb{R}^N$, or of bounded Ω with Dirichlet BC).

We suggest that the effective BC is given by a monotone graph $\tilde{\beta}$ defined by

$$\tilde{\beta} := \left\{ (z, \varphi_\nu(z));\ \begin{array}{c} \exists b \in \overline{\mathrm{Range}(\beta)} \text{ such that } \varphi_\nu(z) = b \text{ and} \\ \text{if } z < m := \inf \beta^{-1}(b), \text{ then } \varphi_\nu(k) \geq b\ \forall k \in [z, m[\\ \text{if } z > M := \sup \beta^{-1}(b), \text{ then } \varphi_\nu(k) \leq b\ \forall k \in]M, z] \end{array} \right\}, \quad (4)$$

which can be visualized as the horizontal projection of β on the graph of φ_ν. The graph $\tilde{\beta}$ is monotone, and $\tilde{\tilde{\beta}} = \tilde{\beta}$ (thus operation $\tilde{\cdot}$ is indeed a projection).

Example 1. (i) If $\beta = \{0\} \times \mathbb{R}$, then $\tilde{\beta}$ is the Bardos–LeRoux–Nédélec graph:

$$\tilde{\beta} = \left\{ (z, \varphi_\nu(z)) \mid \mathrm{sign}(z)(\varphi_\nu(z) - \varphi_\nu(k)) \geq 0 \quad \forall k \in [0 \wedge z, 0 \vee z] \right\}. \quad (5)$$

(ii) If $\beta = \mathbb{R} \times \{0\}$, then $\tilde{\beta} = \{(z, \varphi_\nu(z)) \mid \varphi_\nu(z) = 0\}$. Assumption (1) is restrictive in this case; a similar assumption is made in [BFK]. For the general case, one has to complete β to a maximal monotone graph on $[-\infty, +\infty]$ before defining $\tilde{\beta}$ by (4); see Sect. 4 and the forthcoming chapter [AS].

2 Strong Boundary Traces for Entropy Solutions

The Bardos–LeRoux–Nédélec pointwise formulation of the Dirichlet BC, which can be expressed by means of the graph (5) was initially given for *BV* solutions (compare to the work of Otto in [MNRR], where a weak formulation of the Dirichlet boundary condition is given, valid for any L^∞ solution). It has recently been realized that any L^∞ entropy solution actually has strong L^1_{loc} initial and boundary traces in a fairly general situation (see Panov [P05, P06] and the previous works of Chen-Rascle and Vasseur). In particular, the non-degeneracy condition (3) is not needed. The concept of strong trace (in L^1_{loc}) is stated in Definition 1. Set $x = (x_1, x'), x_1 \in \mathbb{R}^+, x' \in \mathbb{R}^{N-1}$.

Definition 1. *A function $\tilde{v} \in L^1_{loc}(\mathbb{R}^{N-1})$ is a strong trace of function $v \in L^1_{loc}(\mathbb{R}^+ \times \mathbb{R}^{N-1})$ on $\{x_1 = 0\}$ if for all $\xi \in C_c(\mathbb{R}^{N-1})$, $\xi \geq 0$*

$$ess\text{-}\lim_{x_1 \to 0} \int_{\mathbb{R}^{N-1}} \xi(x') |v(x_1, x') - \tilde{v}(x')| dx' = 0.$$

In the sequel $\gamma : L^1_{loc}(\mathbb{R}^+ \times \mathbb{R}^{N-1}) \to L^1_{loc}(\mathbb{R}^{N-1})$ is the strong trace operator in the sense of Definition 1. Traces of $L^1_{loc}(Q)$ functions on $\Sigma = (0, T) \times \mathbb{R}^{N-1}$ and on $\{t = 0\}$ are defined similarly. Clearly, the strong trace operators are unbounded. The following proposition, inferred by the result of Panov [P05] on traces of entropy (quasi-)solutions at $\{t = 0\}$, is therefore remarkable.

Proposition 1. *Assume (2). Let u be a quasisolution for $u_t + \operatorname{div}\varphi(u)$.*[1] *Then there exists a strong trace $\gamma V_{\varphi_\nu}(u)$ of the function $V_{\varphi_\nu}(u)$ on Σ in the sense of Definition 1, where $V_{\varphi_\nu}(\cdot)$ is the variation function of φ_ν:*

$$V_{\varphi_\nu}(z) = \operatorname{sign}(z)\operatorname{Var}_{[0\wedge z, 0\vee z]}\varphi_\nu(\cdot).$$

In particular, $\operatorname{sign}^\pm(u-k)(\varphi(u)-\varphi(k))\cdot\nu$*, which can be rewritten as*

$$Q^\pm(V_{\varphi_\nu}(u), V_{\varphi_\nu}(k)) := \operatorname{sign}^\pm(V_{\varphi_\nu}(u) - V_{\varphi_\nu}(k))(\Psi_\nu(V_{\varphi_\nu}(u)) - \Psi_\nu(V_{\varphi_\nu}(k)))$$

with $\Psi_\nu := \varphi_\nu \circ V_{\varphi_\nu}^{-1}$, has the strong trace $Q^\pm(\gamma V_{\varphi_\nu}(u), V_{\varphi_\nu}(k))$ on Σ.

Notice that the above result does not depend on (1), (3). It still holds if Ω is a domain with piecewise C^1-smooth boundary and $\varphi \in BV$ (see Panov [P06]).

3 Entropy Solutions and Well-Posedness

Definition 2. *A function $u \in L^\infty(Q)$ is said an entropy solution for Problem (H) if $\forall k \in \mathbb{R}$, $\forall \xi \in C_c^\infty(Q)$, $\xi \geq 0$ the local entropy inequalities hold:*

$$\int_Q (u-k)^\pm \xi_t + \int_Q \operatorname{sign}^\pm(u-k)(\varphi(u)-\varphi(k))\cdot D\xi + \int_Q f\operatorname{sign}^\pm(u-k)\xi \geq 0, \quad (6)$$

u has the strong trace u_0 on $\{t=0\}$, and for $\mathcal{H}^N$-a.e. $(t,x) \in \Sigma$ the strong traces $\tilde{w} = \gamma\varphi_\nu(u), \tilde{v} = \gamma V_{\varphi_\nu}(u)$ on Σ of the functions $\varphi_\nu(u), V_{\varphi_\nu}(u)$ verify

$$(\tilde{v}(t,x), \tilde{w}(t,x)) \in \tilde{\beta} \circ V_{\varphi_\nu}^{-1}. \quad (7)$$

Notice that if V_{φ_ν} is invertible (which is the case under assumption (3)), then requirement (7) is equivalent to the requirement that $(\tilde{u}(t,x), \tilde{w}(t,x)) \in \tilde{\beta}$, where $\tilde{u}$ is the strong trace of u on Σ, $\tilde{u} = V_{\varphi_\nu}^{-1}(\tilde{v})$.

Definition 2 can be reformulated so that to extend the entropy inequalities up to the boundaries Σ and $\{t=0\}$. Indeed, we have

Proposition 2. *A function $u \in L^\infty(Q)$ such that strong traces $\tilde{v} := \gamma V_{\varphi_\nu}(u)$, $\tilde{w} := \gamma\varphi_\nu(u)$ on Σ exist and satisfy (7) is an entropy solution for (H) if and only if it satisfies $\forall k \in \mathbb{R}$, $\forall \xi \in C_c^\infty([0,T) \times \mathbb{R}^N)$, $\xi \geq 0$:*

$$\begin{aligned} 0 \leq &\int_Q (u-k)^\pm \xi_t + \int_\Omega (u_0 - k)^\pm \xi(0) + \int_Q \operatorname{sign}^\pm(u-k) f\xi \\ &+ \int_Q \operatorname{sign}^\pm(u-k)(\varphi(u)-\varphi(k))\cdot D\xi - \int_\Sigma Q^\pm(\tilde{v}, V_{\varphi_\nu}(k))\xi. \quad (8) \end{aligned}$$

[1] A function $u \in L^\infty(Q)$ is called a quasisolution if $\forall k \in \mathbb{R}$ $\eta_k^\pm(u)_t + \operatorname{div} q_k^\pm(u) = -\mu_k^\pm$ in $\mathcal{D}'(Q)$, where $(\eta_k^\pm, q_k^\pm)$ are the Kruzhkov entropy-flux pairs and $\mu_k^\pm$ are Borel measures on Q, locally finite up to the boundary; see [P05, P06].

For the proof, one truncates $\xi \in C_c^\infty([0,T) \times \mathbb{R}^N)$ in a neighborhood of the boundaries, and passes to the limit using Definition 1.

Notice that both formulations (6), (7) and (8), (7) make sense for L^∞ (and even more general) data u_0, f, for general domain Ω with Lipschitz deformable boundary, and without assumptions (1)–(3). Uniqueness and comparison results of Sect. 3.1 remain valid in this general framework. For the existence part, we use a chain of approximations of (H) enjoying convenient compactness properties (this is the aim of our simplifying assumptions). We show that they converge to an entropy solution inside Q, then deduce the existence of strong traces $\tilde{v}$, $\tilde{w}$ by Proposition 1, and finally, we identify the couple $(\tilde{v}, \tilde{w})$ as belonging to $\tilde{\beta} \circ V_{\varphi_\nu}^{-1}$.

3.1 Comparison and Uniqueness of Entropy Solutions

Theorem 1 (The Kato inequality). *For $i = 1, 2$, let u_i be an entropy solution for Problem (H) with data $(u_0^i, f_i) \in L^1(\Omega) \times L^1(Q)$. Then for all $\xi \in C_c^\infty([0,T) \times \mathbb{R}^N), \xi \geq 0$*

$$\int_Q (u_1 - u_2)^+ \xi_t + \int_Q \operatorname{sign}^+(u_1 - u_2)(\varphi(u_1) - \varphi(u_2)) \cdot D\xi$$
$$+ \int_Q \operatorname{sign}^+(u_1 - u_2)(f_1 - f_2)\xi \geq \int_\Sigma (\tilde{w}_1 - \tilde{w}_2)^+ \xi. \tag{9}$$

Proof. We use the Kruzhkov method of doubling of variables. As $u_1(t,x)$, resp., $u_2(s,y)$ is an entropy solution with data $u_0^1(x)$ and $f_1(t,x)$, resp., $u_0^2(y)$ and $f_2(s,y)$, then for all $\phi = \phi(t,x,s,y) \in C_c^\infty(Q \times Q)$ one has

$$\int_{Q\times Q} (u_1 - u_2)^+(\phi_t + \phi_s) + \operatorname{sign}^+(u_1 - u_2)(\varphi(u_1) - \varphi(u_2)) \cdot (D_x\phi + D_y\phi)$$
$$+ \int_{Q\times Q} \operatorname{sign}^+(u_1 - u_2)(f_1 - f_2) \geq 0. \tag{10}$$

Let $\xi \in C_c^\infty([0,T) \times \mathbb{R}^N), \xi \geq 0$ and ρ_n, resp., ρ_m, be a classical sequence of mollifiers in $\mathbb{R}^N$, resp., in $\mathbb{R}$. Define $\phi(t,x,s,y) = \mu_\delta(x)\mu_\eta(y)\rho_n(x-y)\rho_m(t-s)$. Using ϕ as a test function in (10) and passing to the limit with $\delta, \eta \to 0$ yields

$$0 \leq \int_{Q\times Q} \rho_m\rho_n(u_1 - u_2)^+\xi_t + \int_{Q\times Q} \rho_m\rho_n \operatorname{sign}^+(u_1 - u_2)(\varphi(u_1) - \varphi(u_2)) \cdot D_x\xi$$
$$+ \int_{Q\times Q} \operatorname{sign}^+(u_1 - u_2)(f_1 - f_2)\rho_m\rho_n\xi$$
$$- \int_{\Sigma\times Q} Q^+(\tilde{v}_1, V_{\varphi_\nu}(u_2))\rho_m\rho_n\xi - \int_{Q\times\Sigma} Q^+(V_{\varphi_\nu}(u_1), \tilde{v}_2)\rho_m\rho_n\xi. \tag{11}$$

By Proposition 1, each of the last two terms converges to $\dfrac{1}{2}\displaystyle\int_\Sigma Q^+(\tilde{v}_1, \tilde{v}_2)$ as $n, m \to \infty$. Relations (7) for u_1, u_2 and the definitions of Q^+ and $\tilde{\beta}$ yield (10). □

Corollary 1. *Assume (2). For data* (u_0^i, f_i), $i = 1, 2$, *satisfying*

$$\begin{cases} u_0 \in (L^1 \cap L^\infty)(\Omega), \ f \in L^1(Q) \ \text{with} \\ f(t,\cdot) \in L^\infty(\Omega) \ \text{a.e.} \ t \in (0,T), \ \int_0^T \|f(t,\cdot)\|_\infty dt < \infty, \end{cases} \tag{12}$$

let u_i *an entropy solution for Problem* (H). *Then for all* $t \in (0,T)$,

$$\int_0^t \int_{\partial\Omega} (\tilde{w}_1 - \tilde{w}_2)^+ + \int_\Omega (u_1 - u_2)^+(t) \leq \int_\Omega (u_0^1 - u_0^2)^+ + \int_0^T \int_\Omega (f_1 - f_2)^+.$$

Thus if $u_0^1 \leq u_0^2$ *a.e. on* Ω *and if* $f_1 \leq f_2$ *a.e. on* Q, *then* $u_1 \leq u_2$ *a.e. on* Q. *In particular, an entropy solution of* (H) *with data (12) is unique.*

Proof. Take in (9), $\xi(t,x) = \xi_\alpha(x)\kappa(t)$, where $\kappa \in C_c^\infty([0,T)), \kappa \geq 0$, $\xi_\alpha \to 1$ in Ω, $|D\xi_\alpha| \leq C$ and $\mathrm{Supp}(D\xi_\alpha) \subset \{x \,|\, \alpha < |x| < \alpha + 1\}$. As $\alpha \to 0$ the claim follows, because $u_i \in L^1(Q)$ under assumption (12), and φ is Lipschitz. □

Notice that the comparison and uniqueness result of Corollary 1 holds true also in the L^∞ framework (see the results of Kruzhkov–Panov, Bénilan–Kruzhkov).

3.2 Existence of Entropy Solutions

We infer the existence of an entropy solution to Problem (H) using the tools of the nonlinear semigroup theory (cf. e.g. [BCP]). We first study the boundary problem in Ω associated with (H) (known as the "stationary" problem)

$$(S)(f) \begin{cases} u + \mathrm{div}\ \varphi(u) = f & \text{on} \ \ \Omega, \\ \varphi(u) \cdot \nu \in \beta(u) & \text{on} \ \ \partial\Omega. \end{cases}$$

Definition 3. *A function* $u \in L^\infty(\Omega)$ *is an entropy solution of Problem* $(S)(f)$ *if* $\forall k \in \mathbb{R}, \forall \phi \in C_c^\infty(\Omega), \phi \geq 0$ *the local entropy inequalities*

$$\int_\Omega \mathrm{sign}^\pm(u-k)(\varphi(u) - \varphi(k)) \cdot D\phi + \int_\Omega \mathrm{sign}^\pm(u-k)(f-u)\phi \geq 0$$

hold, and traces $\tilde{v} = \gamma V_{\varphi_\nu}(u)$, $\tilde{w} = \gamma\varphi_\nu(u)$ *verify (7) for* $\mathcal{H}^{N-1}$ *– a.e.* $x \in \partial\Omega$.

Notice that Definition 3 can be reformulated in the way of Proposition 2. Define the operator $\mathcal{A}$ associated with the problem $(S)(f)$ by its graph:

$$(u, f) \in \mathcal{A} \Leftrightarrow u \text{ is an entropy solution of } (S)(f+u).$$

Theorem 2. *Let (1), (2), and (3) hold. Then the operator* $\mathcal{A}$ *is T-accretive with dense domain in* $L^1(\Omega)$, *and we have* $(L^1 \cap L^\infty)(\Omega) \subset Range(I + \mathcal{A})$. *Moreover, for all* $(u_i, f_i) \in \mathcal{A}$, $i = 1, 2$ *we have*

$$\int_{\partial\Omega} (\tilde{w}_1 - \tilde{w}_2)^\pm + \int_\Omega (u_1 - u_2)^\pm \leq \int_\Omega \mathrm{sign}^\pm(u_1 - u_2)(f_1 - f_2) + \int_{[u_1 = u_2]} (f_1 - f_2)^\pm. \tag{13}$$

A proof of this theorem is given in [S] (see also [AS]).

Proof. The T-accretivity and (13) are obtained in the same way as Corollary 1. Let us sketch the proof of existence with data $f \in (L^1 \cap L^\infty)(\Omega)$. We use the standard vanishing viscosity approximation of the equation in $(S)(f)$ together with a Lipschitz regularization β^ε of the graph β. Let $(u^\varepsilon)_\varepsilon$ denote the corresponding sequence of approximate solutions; $\beta^\varepsilon(u^\varepsilon)$ are the corresponding normal fluxes on the boundary. Assumption (1) yields a uniform L^∞ bound on u^ε. Assumption (3) is sufficient for the strong precompactness of $(u^\varepsilon)_\varepsilon$ in $L^1_{loc}(\Omega)$, by the result of Panov [P94] (see also the well-known result of Lions–Perthame–Tadmor). This is sufficient to get (6) and deduce (by the "stationary" analogue of Proposition 1) the existence of boundary traces $\tilde{v}, \tilde{w}$ of $V_{\varphi_\nu}(u), \varphi_\nu(u)$, respectively, where u is an accumulation point of $(u^\varepsilon)_\varepsilon$. It remains to show (7). Due to the flatness of $\partial\Omega$, problem $(S)(f)$ is invariant (taking into account translations of f) with respect to translations in directions $(0, h')$, $h' \in \mathbb{R}^{N-1}$. Using the analogue of (13) which holds true for $\varepsilon > 0$, by the Fréchet–Kolmogorov theorem we deduce that the sequence $(\beta^\varepsilon(u^\varepsilon))_\varepsilon$ is strongly compact in $L^1(\partial\Omega)$. We then deduce (7) by using the argument sketched in the proof of Lemma 3 below (with $\tilde{\tilde{\beta}}$ replaced by $\tilde{\beta}$, and $\tilde{\beta}$ replaced by β). Finally, $\overline{D(\mathcal{A})} = L^1(\Omega)$, since we show that the solution u_α to $u + \alpha\,\mathcal{A}(u) = f \in C_c^\infty(\Omega)$ converges to f in $L^1(\Omega)$ as $\alpha \to 0$. □

Theorem 2 means that the closure of $\mathcal{A}$ is an m-T-accretive densely defined operator in $L^1(\Omega)$ (see e.g. [BCP]).

By the Crandall–Liggett theorem, it generates a T-contractive semigroup on $L^1(\Omega)$; more exactly, we have

Theorem 3. *Let (1), (2), and (3) hold. Then for any $f \in L^1(Q)$, $u_0 \in L^1(\Omega)$ there exists a unique mild solution of the abstract Cauchy problem*

$$u_t + \mathcal{A}u \ni f, \ u(0) = u_0. \tag{14}$$

We deduce existence for Problem (H) by showing that any mild solution of problem (14) is also a solution of (H) in the sense of Definition 2.

Theorem 4. *Assume (1), (2), and (3). If data (u_0, f) satisfy (12), then the mild solution of the Cauchy problem (14) is an entropy solution of Problem (H).*

Proof. For $m \in \mathbb{N}^*$, set $\varepsilon = \frac{T}{m}$. For $i = 0, 1, ..., m$, set $t_i = \varepsilon i$ and $f_i^\varepsilon(x) = \frac{1}{\varepsilon}\int_{t_{i-1}}^{t_i} f(t,x)dt$ a.e. $x \in \Omega$. We have $\varepsilon \sum_{i=1}^m \|f_i^\varepsilon\|_{L^\infty(\Omega)} \leq \int_0^T \|f(t)\|_{L^\infty(\Omega)}dt$ and $\sum_{i=1}^m \int_{t_{i-1}}^{t_i} \|f(t) - f_i^\varepsilon\|_{L^1(\Omega)} \to 0$ as $\varepsilon \to 0$. Take $u_0^\varepsilon \in D(\mathcal{A})$ such that $\|u_0^\varepsilon - u_0\|_{L^1} \leq \varepsilon$. Let u_i^ε be the solution of $\varepsilon f_i^\varepsilon + u_{i-1}^\varepsilon \in (I + \varepsilon\mathcal{A})(u_i^\varepsilon)$, $i = 1, ..., m$. Set $u^\varepsilon(t) = u_i^\varepsilon, f^\varepsilon(t) = f_i^\varepsilon, \tilde{v}^\varepsilon(t) = \tilde{v}_i^\varepsilon, \tilde{w}^\varepsilon(t) = \tilde{w}_i^\varepsilon$ if $t_{i-1} \leq t \leq t_i$ $(1 \leq i \leq m)$. By the nonlinear semigroup theory (see e.g. [BCP]), the sequence $(u_\varepsilon)_\varepsilon$ is precompact in $L^\infty(0,T; L^1(\Omega))$. Hence there exists a subsequence, still denoted by u_ε, and a function $u \in C(0,T; L^1(\Omega))$ such that

$\|u^\varepsilon - u\|_{L^\infty(0,T;L^1(\Omega))} \to 0$ as $\varepsilon \to 0$. By construction, u admits u_0 for the strong trace on $\{t=0\}$. Thanks to (12) and Assumption (1), $(u^\varepsilon)_\varepsilon$ is bounded in $L^\infty(Q)$. The continuation of the proof is divided into three lemmas.

Lemma 1. *The function u verifies inequality (6).*

Proof. By Definition 3, for all $\phi \in C_c^\infty(Q),\ \phi \geq 0,\ k \in \mathbb{R}$, u_ε satisfies

$$0 \leq \int_Q \Big(\frac{1}{\varepsilon}(u^\varepsilon(t-\varepsilon) - u^\varepsilon(t)) + f^\varepsilon(t)\Big)\mathrm{sign}^\pm(u^\varepsilon(t) - k)\phi(t) + \int_Q \mathrm{sign}^\pm(u^\varepsilon(t)-k)(\varphi(u^\varepsilon(t)) - \varphi(k)) \cdot D\phi(t). \tag{15}$$

As $\varepsilon \to 0$ the result follows, because $\mathrm{sign}^\pm(\cdot - k)$ belongs to $\partial(\cdot - k)^\pm$:

$$\mathrm{sign}^\pm(u^\varepsilon(t) - k)(u^\varepsilon(t-\varepsilon) - u^\varepsilon(t)) \leq (u^\varepsilon(t-\varepsilon) - k)^\pm - (u^\varepsilon(t) - k)^\pm.$$

Lemma 2. *The sequence $(\tilde{w}^\varepsilon(t,x))_\varepsilon$ of strong traces of $\varphi_\nu(u^\varepsilon)$ on Σ converges (up to a subsequence) $\mathcal{H}^N$-a.e. on Σ and in $L^1_{loc}(\Sigma)$ to $b(t,x) \in \overline{Range(\tilde{\beta})}$.*

Proof. We prove that the sequence $(\tilde{w}^\varepsilon)_\varepsilon$ is bounded in $L^1(\Sigma)$, and

$$\int_0^{T-\Delta t} \|\tilde{w}^\varepsilon(t+\Delta t) - \tilde{w}^\varepsilon(t)\|_{L^1(\partial\Omega)} \leq \tilde{\psi}(\Delta t), \quad \lim_{h\to 0} \tilde{\psi}(h) = 0.$$

A similar estimate of the space translates of $\tilde{w}^\varepsilon(t,x')$ follows from (13) and the translation invariance of $\partial\Omega$. We then apply the Fréchet–Kolmogorov theorem.

Lemma 3. *Strong traces $\tilde{v} = \gamma V_{\varphi_\nu}(u)$, $\tilde{w} = \gamma\varphi_\nu(u)$ on Σ of functions $V_{\varphi_\nu}(u)$ and $\varphi_\nu(u)$ exist. Moreover, $(\tilde{v}(t,x), \tilde{w}(t,x)) \in \tilde{\beta} \circ V_{\varphi_\nu}^{-1}$ $\mathcal{H}^N$-a.e. $(t,x) \in \Sigma$.*

Proof. By Proposition 1, strong traces of functions $V_{\varphi_\nu}(u)$ and $\varphi_\nu(u)$ exist. Now let us prove that $(\tilde{v}, \tilde{w}) \in \tilde{\beta} \circ V_{\varphi_\nu}^{-1}$ a.e. on Σ. By choosing in inequality (3) the test function $\phi(1-\mu_\delta)$ with $\phi \in C_c^\infty([0,T) \times \mathbb{R}^N), \phi \geq 0, (\mu_\delta)_\delta \in C^2(\overline{\Omega}), \mu_\delta \to 1$ on Ω, $\mu_\delta = 0$ on $\partial\Omega$, and by letting $\varepsilon \to 0$ then $\delta \to 0$ we get

$$\int_\Sigma Q^\pm(\tilde{v}, V_{\varphi_\nu}(k))\,\phi \geq \int_\Sigma (b - \varphi \circ V_{\varphi_\nu}^{-1}(V_{\varphi_\nu}(k)))\,\theta_k^\pm\,\phi, \tag{16}$$

where $\theta_k^\pm, k \in \mathbb{Q}$ is the weak-star limit in $L^\infty(\Sigma)$ of $\mathrm{sign}^\pm(u_\varepsilon - k)$. But for a.e. $(t,x) \in \Sigma$, we have $b(t,x) = \lim_{\varepsilon\to 0} \tilde{w}^\varepsilon(t,x)$. By construction, $(\tilde{v}^\varepsilon, \tilde{w}^\varepsilon) \in \tilde{\beta} \circ V_{\varphi_\nu}^{-1}$; therefore for all $k \notin \tilde{\beta}^{-1}(b(t,x))$ we can identify $\theta_k^\pm(t,x)$ with $\mathrm{sign}^\pm(b(t,x) - \tilde{\beta}(k))$. For simplicity, let us use assumption (3) (notice that Lemma 3 holds without (3)). We can use $\tilde{u} = V_{\varphi_\nu}^{-1}(\tilde{v})$; then (16) implies $\mathrm{sign}^\pm(\tilde{u}-k)(\tilde{w}-\varphi_\nu(k)) \geq \mathrm{sign}^\pm(b-k)(b-\varphi_\nu(k))\ \forall k \notin \tilde{\beta}^{-1}(b)$, $\mathcal{H}^N$-a.e. on Σ. Considering different values of k, we get $(\tilde{u}, \tilde{w}) \in \tilde{\tilde{\beta}}$. Since $\tilde{\tilde{\beta}} = \tilde{\beta}$, (7) holds. $\square$

4 Conclusion and Generalizations

We justify that the entropy formulation (6), (7) or, equivalently, (8), (7) is an adequate interpretation of the formal boundary value problem (H). Uniqueness, comparison and L^1 contraction properties hold for this formulation.

Furthermore, the proof of Lemma 3 actually shows how the "effective BC" graph $\tilde{\beta}$ appears from the "formal BC" graph β (or β^ε, if the graphs are perturbed). In our proof, this passage requires strong L^1 compactness of the sequence $(\beta^\varepsilon(u^\varepsilon))_\varepsilon$ of the associated boundary fluxes.

Using the same techniques and the fact that $\tilde{\tilde{\beta}} = \tilde{\beta}$, one deduces existence of a solution verifying (6), (7) under weaker assumptions. For instance, if we approximate $u_0 \in L^\infty(\Omega)$ by $u_0^{m,n} = u_0^+ 1\!\!1_{\{\|x\|<n\}} + u_0^- 1\!\!1_{\{\|x\|<m\}} \in (L^1 \cap L^\infty)(\Omega)$, the sequence of the resulting solutions $u^{m,n}$ of (H) is monotone in each of the indices m, n, by (13). Thus we have convergence of (a subsequence of) $u^{m,n}$ in $L^1_{loc}(Q)$, of $\tilde{\beta}(u^{m,n})$ in $L^1_{loc}(\Sigma)$. Whence existence for (H) in the L^∞ framework follows (cf. [S]). Similarly, perturbation of φ by piecewise strictly monotone φ^ε permits to bypass (3), by using the techniques of L^1 continuous dependence of entropy solutions on the flux function (see e.g. the papers by Bouchut–Perthame and Chen–Karlsen). Further, approximation of β by quickly growing at infinity graphs β^ε such that $|\beta^\varepsilon| \downarrow |\beta|$ permits to get rid of hypothesis (1) and justify the remark on the extension of β in Example 1(ii).

Fine translation techniques permit to consider inhomogeneous and mixed BC. Finally, to extend the well-posedness results to the L^1 data, one has to apply the same techniques to renormalized solutions of (H) (see Bénilan–Carrillo–Wittbold) instead of the entropy solutions.

References

[AS] Andreianov, B., Sbihi, K.: On boundary conditions for scalar conservation laws. In preparation

[BLN] Bardos, C., Le Roux, A.Y., Nédélec, J.C.: First order quasilinear equations with boundary conditions. Comm. Partial Diff. Equ., **4**, 1017–1034 (1979)

[BCP] Bénilan, Ph., Crandall, M.G., Pazy, A.: Evolution equations governed by accretive operators. Preprint book

[BFK] Bürger, R., Frid, H., Karlsen, K.H.: On the well-posedness of entropy solutions to conservation laws with a zero-flux boundary condition. To appear in J. Math. Anal. Appl.

[MNRR] Malek, J., Nečas, J., Rokyta, M., Ružička, M.: Weak and Measure-valued Solutions to Evolutionary PDEs., Chapman & Hall (1996)

[P94] Panov, E. Yu.: On sequences of measure-valued solutions of a first-order quasilinear equation. Mat. Sb., **185**, no. 2, 87–106 (1994)

[P05] Panov, E. Yu.: Existence of strong traces for generalized solutions of multidimensional scalar conservation laws. J. Hyp. Diff. Eq., **4**, 885–908(2005)

[P06] Panov, E. Yu.: communication at HYP06 Lyon, paper in preparation

[S] Sbihi, K.: Etude de quelques E.D.P. non linéaires dans L^1 avec des conditions générales sur le bord. Thesis, Strasbourg University, France (2006)

A Relaxation Method for the Coupling of Systems of Conservation Laws

A. Ambroso, C. Chalons, F. Coquel, E. Godlewski, F. Lagoutière, P.-A. Raviart, and N. Seguin

Summary. This work deals with the coupling of barotropic systems through a fixed interface. An approach based on a relaxation approximation of the models is proposed. It allows us to construct a numerical method that is consistent with different models of coupling.

1 Introduction

The present work investigates the spatial coupling of different hyperbolic models from the theoretical and numerical points of view. The problem of coupling often occurs in an industrial context. Consider a complex system, such as a nuclear reactor. Different phenomena can appear according to the region and thus, different models (and numerical methods) are used. The difficulty lies in letting the different models communicate, since they are not necessarily compatible.

Hence we study the spatial coupling of two different hyperbolic systems of conservation laws through the interface $\{x = 0\}$:

$\partial_t \mathbf{u} + \partial_x \mathbf{f}_L(\mathbf{u}) = 0$ $x < 0$	$\partial_t \mathbf{u} + \partial_x \mathbf{f}_R(\mathbf{u}) = 0$ $x > 0$

$$x = 0$$

Through this interface of coupling, some conditions on the solution $\mathbf{u}(t, x)$ must be added, namely between its traces $\mathbf{u}(t, 0^-)$ and $\mathbf{u}(t, 0^+)$. Several models of coupling can be proposed, using the framework developed in [GLTR05]. Such a problematic can also appear in the frame of traffic flows on road network (see for instance [BNP06, GP06b, HR06]).

In this work, an alternative model of coupling is presented, using the approximation by relaxation developed in [CGP01] in order to overcome nonlinear effects such as resonance.

Here, the hyperbolic systems we couple are two barotropic models of gas dynamics

$$\begin{cases} \partial_t \rho + \partial_x \rho u = 0, \\ \partial_t \rho u + \partial_x(\rho u^2 + p) = 0, \end{cases} \qquad t > 0, x \in \mathbb{R} \setminus \{0\}, \tag{1}$$

where the pressure law is

$$p(\rho) = \begin{cases} P_L(\rho) \text{ if } x < 0, \\ P_R(\rho) \text{ if } x > 0, \end{cases} \tag{2}$$

and we take for the P_α's usual pressure laws. We define the set of admissible states by $\Omega_{\mathbf{u}} =]0, \infty[\times \mathbb{R} = \{\rho > 0, \rho u \in \mathbb{R}\}$. Though these models are simple, their coupling is not so easy. Actually, since the models are different, the solution will greatly depend on the coupling conditions we impose. Here, the main goal is to propose different mathematical methods of coupling according to physical requirements (such as conservativity or continuity) we want to impose on the solution, so that we can speak of coupling models.

2 Coupling Conditions at the Interface

Different models of coupling are proposed, some of them have already been introduced in [GR04] and [GLTR05].

2.1 Continuity of the Conservative Variable

This model of coupling has been defined in [GR04] and [GLTR05]. It consists in imposing the continuity of the conservative variable $\mathbf{u} = (\rho, \rho u)$ at the coupling interface:

$$\mathbf{u}(0^-, t) = \mathbf{u}(0^+, t), \quad \forall t > 0. \tag{3}$$

However, since we are dealing with hyperbolic models, this condition can lead to nonexistence. It must be weakened, and we follow the spirit of weak boundary conditions proposed by Dubois and LeFloch [DL88]:

$$\begin{cases} \mathbf{u}(0^-, t) \in \mathcal{O}_L(\mathbf{u}(0^+, t)), \ \forall t > 0, \\ \mathbf{u}(0^+, t) \in \mathcal{O}_R(\mathbf{u}(0^-, t)), \ \forall t > 0, \end{cases} \tag{4}$$

where

$$\mathcal{O}_L(\mathbf{u}_b) = \{\mathcal{W}_L(0^-; \mathbf{u}, \mathbf{u}_b), \mathbf{u} \in \Omega_{\mathbf{u}}\},$$
$$\mathcal{O}_R(\mathbf{u}_b) = \{\mathcal{W}_R(0^+; \mathbf{u}_b, \mathbf{u}), \mathbf{u} \in \Omega_{\mathbf{u}}\},$$

and for $\alpha = L, R$, $\mathcal{W}_\alpha(x/t; \mathbf{u}_\ell, \mathbf{u}_r)$ denotes the solution of the Riemann problem for (1), with the same pressure law P_α on the whole line $x \in \mathbb{R}$, whose data are $\mathbf{u}_\ell$ for $x < 0$ and $\mathbf{u}_r$ for $x > 0$. Note that other coupling conditions for such systems have been proposed by Banda et al. [BH06] for specific configurations of flow.

2.2 Global Conservative Model

The aim of this model of coupling is to recover the global conservation of $\mathbf{u}$. We then replace the definition (2) of the pressure law by

$$p(\rho, x) = (1 - H(x))P_L(\rho) + H(x)P_R(\rho), \quad \forall x \in \mathbb{R}, \tag{5}$$

where H is the Heaviside function, while the system (1) is now defined on the whole spatial domain $\mathbb{R}$. Noting $\mathbf{f}_\alpha(\mathbf{u}) = (\rho u, \rho u^2 + P_\alpha(\rho))$, $\alpha = L, R$, it leads to the following coupling condition:

$$\mathbf{f}_L(\mathbf{u}(0^-, t)) = \mathbf{f}_R(\mathbf{u}(0^+, t)). \tag{6}$$

Contrary to the previous case, this condition can be fulfilled for almost every positive t, no weak form of (6) has to be introduced.

Because of the *genuine nonlinearity* of the model (1)–(2) (assuming for instance that P_L and P_R are usual perfect gas pressure laws), the coupling conditions (4), (12), and (6) are very difficult to handle, even when restricting to the study of Riemann problems. In particular, when a wave of one of the two models interacts with the interface of coupling, that is to say in the case of *resonance*, the solution can be very difficult to compute. Hence, we are going to perform the coupling through some approximate models, which are wholly *linearly degenerate*. This approximation relies on a relaxation process [CGP01].

3 Coupling by Nonlinear Relaxation

3.1 Global Relaxation Approximation

The relaxation model we consider is a variant of Suliciu's relaxation system (we refer for instance to [B04])

$$\begin{cases} \partial_t \rho + \partial_x \rho u = 0, \\ \partial_t \rho u + \partial_x (\rho u^2 + \pi) = 0, & t > 0, \\ \partial_t \rho \mathcal{T} + \partial_x \rho \mathcal{T} u = \lambda(1 - \rho \mathcal{T}), \end{cases} \tag{7}$$

with the relaxed pressure law $\pi(\rho, \mathcal{T}) = P(1/\mathcal{T}) + a^2(\mathcal{T} - 1/\rho)$, where a is large enough to satisfy the Whitham condition which writes, setting $\overline{P}(\tau) = P(\rho), \tau = 1/\rho$,

$$a^2 > \max(-\overline{P}'(s)). \tag{8}$$

This condition is known to ensure the positivity of ρ and entropy inequalities [B04]. This system is hyperbolic and all the characteristic fields, corresponding to the eigenvalues u and $u \pm a/\rho$, are linearly degenerate.

We now focus on the coupling of two such systems, one in $x<0$ and one in $x>0$, with different pressure laws

$$\pi=\pi_\alpha(\rho,\mathcal{T})=\begin{cases}P_L(1/\mathcal{T})+a^2(\mathcal{T}-1/\rho) \text{ if } x<0,\\ P_R(1/\mathcal{T})+a^2(\mathcal{T}-1/\rho) \text{ if } x>0,\end{cases} \tag{9}$$

where again a is large enough to satisfy the Whitham condition. Then, using a Chapman–Enskog expansion, one may check the compatibility of both systems (7)–(9) and (1)–(2) when $\lambda\to+\infty$ in the sense of [CLL94]. We will use the shortened notation

$$\partial_t\tilde{\mathbf{u}}+\partial_x\tilde{\mathbf{f}}_\alpha(\tilde{\mathbf{u}})=\lambda\tilde{\mathbf{r}}(\tilde{\mathbf{u}}) \tag{10}$$

($\alpha=L,R$) for system (7)–(9) where $\tilde{\mathbf{u}}=(\rho,\rho u,\rho\mathcal{T})^T$, $\tilde{\mathbf{f}}_\alpha(\tilde{\mathbf{u}})=(\rho u,\rho u^2+\pi_\alpha,\rho\mathcal{T}u)^T$, $\tilde{\mathbf{r}}(\tilde{\mathbf{u}})=(0,0,1-\rho\mathcal{T})^T$, and define the set $\Omega_{\tilde{\mathbf{u}}}=]0,\infty[\times\mathbb{R}\times]0,\infty[=\{\rho>0,\rho u\in\mathbb{R},\rho\mathcal{T}>0\}$.

The study of the approximation by relaxation of the coupling problem is given by the following so called *global relaxation approximation.*

1. Let $\mathbf{u}_0(.)=(\rho_0,\rho_0u_0)(.)$ be an initial datum for system (1)–(2). Then, we define the extended initial datum $\tilde{\mathbf{u}}_0(.)=(\rho_0,\rho_0u_0,\rho_0\mathcal{T}_0)(.)$ for the relaxation system defining $\mathcal{T}_0\equiv 1/\rho_0$.
2. We solve exactly the Cauchy problem (7)–(9) with the initial data $\tilde{\mathbf{u}}_0$ and with an appropriate coupling condition (that we will detail in the sequel), without the relaxation term in (7), that is to say setting $\lambda=0$ in (7).
3. The source term is taken into account, simply setting $\mathcal{T}\equiv 1/\rho$, that is to say by projecting the solution on the equilibrium set of system (7)–(9).

Different models of coupling are used in step 2, in relation with the ones defined above and we present now the different counterparts of the previous models of coupling in the context of this relaxation model (7)–(9). The coupling conditions for the relaxed problem follow the spirit of (4). However, we can choose for the variables, which are transmitted a as set that is different from the conservative variables as presented in [ACC05]. Indeed, let ϕ_L and ϕ_R be two changes of variable from $\Omega_{\tilde{\mathbf{u}}}$ into some set $\Omega_{\tilde{\mathbf{v}}}$. If we aim at obtaining at the interface the condition of transmission

$$\phi_L(\tilde{\mathbf{u}}(t,0^-))=\phi_R(\tilde{\mathbf{u}}(t,0^+)), \tag{11}$$

the associated weak form of (11) is given by

$$\begin{cases}\tilde{\mathbf{u}}(0^-,t)\in\tilde{\mathcal{O}}_L(\phi_L^{-1}(\phi_R(\tilde{\mathbf{u}}(0^+,t)))),\ \forall t>0,\\ \tilde{\mathbf{u}}(0^+,t)\in\tilde{\mathcal{O}}_R(\phi_R^{-1}(\phi_L(\tilde{\mathbf{u}}(0^-,t)))),\ \forall t>0,\end{cases} \tag{12}$$

where

$$\tilde{\mathcal{O}}_L(\tilde{\mathbf{u}}_b)=\{\tilde{\mathcal{W}}_L(0^-;\tilde{\mathbf{u}},\tilde{\mathbf{u}}_b),\tilde{\mathbf{u}}\in\Omega_{\tilde{\mathbf{u}}}\},$$
$$\tilde{\mathcal{O}}_R(\tilde{\mathbf{u}}_b)=\{\tilde{\mathcal{W}}_R(0^+;\tilde{\mathbf{u}}_b,\tilde{\mathbf{u}}),\tilde{\mathbf{u}}\in\Omega_{\tilde{\mathbf{u}}}\},$$

and $\tilde{\mathcal{W}}_\alpha(x/t;\tilde{\mathbf{u}}_\ell,\tilde{\mathbf{u}}_r)$ is the solution of the Riemann problem for (7) with the relaxed pressure law π_α, whose data are $\tilde{\mathbf{u}}_\ell$ for $x<0$ and $\tilde{\mathbf{u}}_r$ for $x>0$, with $\alpha=L,R$.

3.2 Continuity of Variable $(\rho, \rho u, \rho \mathcal{T})$

We first investigate the coupling in conservative variables, following (3). The idea is to use the coupling conditions (4) with $\phi_L \equiv \phi_R \equiv \mathrm{Id}$.

The correspondence with the coupling in conservative variables $(\rho, \rho u)$ for system (1)–(2) is given by the following result:

Proposition 1. *Assume that the initial datum $\mathbf{u}_0 = (\rho_0, \rho_0 u_0)$ is constant and consider the associated initial datum $(\rho_0, \rho_0 u_0, \rho_0 \mathcal{T}_0)$ for the relaxation problem, where $\mathcal{T}_0 = 1/\rho_0$. Then, the solution given by the global relaxation approximation with the coupling in variable $(\rho, \rho u, \rho \mathcal{T})$ is stationary.*

The proof is easily obtained by showing that each step of the relaxation approximation does not modify the initial datum.

3.3 Continuity of Variable $(\rho, \rho u, \pi)$

Here, we define in (12) the two changes of variable by

$$\phi_\alpha(\widetilde{\mathbf{u}}) = (\rho, \rho u, \pi)^T, \tag{13}$$

where π depends on $\widetilde{\mathbf{u}}$ through (9): $\pi = \pi_\alpha(\rho, \mathcal{T})), \ \alpha = L, R$. The mapping ϕ_α is indeed an admissible change of variable since we can write $\mathcal{T} = \mathcal{T}_\alpha(\rho, \pi)$. This comes from the relation $P_\alpha(1/\mathcal{T}) + a^2\mathcal{T} = \pi + a^2/\rho$ and the fact that $\mathcal{T} \to P_\alpha(1/\mathcal{T}) + a^2\mathcal{T}$ can be inverted under Whitham's condition. This model of coupling allows to obtain a conservative coupling. Indeed, the continuity of $(\rho, \rho u, \pi)$ at the interface implies the continuity of the flux of the two first equations of (7), which leads to the conservation of ρ and ρu.

We now give an existence and uniqueness result concerning the coupled Riemann problem, i.e., the Cauchy problem for the coupled homogeneous parts of the systems (7)–(9), $\partial_t \widetilde{\mathbf{u}} + \partial_x \widetilde{\mathbf{f}}_\alpha(\widetilde{\mathbf{u}}) = 0$ ($\alpha = L$ in $x < 0$, $\alpha = R$ in $x > 0$) with initial Riemann data

$$\widetilde{\mathbf{u}}(x, 0) = \begin{cases} \widetilde{\mathbf{u}}_L \text{ if } x < 0, \\ \widetilde{\mathbf{u}}_R \text{ if } x > 0, \end{cases} \tag{14}$$

and coupling condition (12), (13), which means coupling in variable $(\rho, \rho u, \pi)$.

Theorem 1. *Assume $\lambda = 0$, and consider the coupling of systems (7)–(9) in variable $(\rho, \rho u, \pi)$. The associated coupled Riemann problem admits a unique solution.*

This very important result will be commented in the following. But before going on, let us state this consistency result concerning the global relaxation approximation described above:

Proposition 2. *The solution provided by the global relaxation approximation is conservative with respect to ρ and ρu.*

It is easy to show that since $(\rho, \rho u, \pi)$ are transmitted at the coupling interface, the solution of the second step of the relaxation approximation is conservative, that is to say $\widetilde{\mathbf{f}}_L(\widetilde{\mathbf{u}}(0^-,t)) = \widetilde{\mathbf{f}}_R(\widetilde{\mathbf{u}}(0^+,t))$ for almost every $t > 0$. Moreover, the third step keeps ρ and ρu unchanged, which leads to their conservation by the whole algorithm.

The result of uniqueness of Theorem 1 is very important. Indeed, it is well-known that the system (1)–(2) with the coupling condition (6) may admit an infinity of solutions if no entropy condition is prescribed at the interface [IT92]. It means that this procedure of approximation selects only one of these solutions, without ambiguity (it would then be interesting to characterize this solution).

4 Numerical Method of Coupling

Using the fact that the model (7)–(9) is formally equivalent to the model (1)–(2) if $\lambda = +\infty$, the relaxation approximation enables us to build a simple numerical scheme for the coupling problem (1)–(2), based on the splitting of the relaxed model (7) at each time step:

1. The homogeneous part of system (7)–(9) is computed, with one of the possible choices of coupling condition.
2. The source term is taken into account: $\mathcal{T} = 1/\rho$.

Let us be more precise. Let Δt and Δx be the time and space steps. The mesh is composed by cells $K_{i+1/2} = (i\Delta x, (i+1)\Delta x)$, $i \in \mathbb{Z}$, and $u^n_{i+1/2}$ represents the approximation of the solution in cell $K_{i+1/2}$ at time $n\Delta t$, $n \in \mathbb{N}$. The numerical scheme is given by

$$\begin{aligned} \mathbf{u}^{n+1}_{i+1/2} &= \mathbf{u}^n_{i+1/2} - \frac{\Delta t}{\Delta x}((\widetilde{\mathbf{h}}_L)^n_{i+1} - (\widetilde{\mathbf{h}}_L)^n_i),\ i < 0, \\ \mathbf{u}^{n+1}_{i+1/2} &= \mathbf{u}^n_{i+1/2} - \frac{\Delta t}{\Delta x}((\widetilde{\mathbf{h}}_R)^n_{i+1} - (\widetilde{\mathbf{h}}_R)^n_i),\ i \geq 0, \end{aligned} \tag{15}$$

where

$$\widetilde{\mathbf{h}}^n_i = \begin{cases} \widetilde{\mathbf{h}}_L(\mathbf{u}^n_{i-1/2}, \mathbf{u}^n_{i+1/2}) \text{ if } i < 0, \\ \widetilde{\mathbf{h}}_R(\mathbf{u}^n_{i-1/2}, \mathbf{u}^n_{i+1/2}) \text{ if } i > 0, \end{cases} \tag{16}$$

and $\widetilde{\mathbf{h}}_L$ and $\widetilde{\mathbf{h}}_R$ are usual numerical fluxes (Godunov, Roe...), respectively, consistent with $\mathbf{f}_L$ and $\mathbf{f}_R$. It remains to define the numerical fluxes at the interface of coupling $(\widetilde{\mathbf{h}}_L)^n_0$ and $(\widetilde{\mathbf{h}}_R)^n_0$.

These numerical fluxes are computed using the global relaxation approximation:

1. Define $\widetilde{\mathbf{u}}^n_{-1/2} = (\mathbf{u}^n_{-1/2}, 1)$ and $\widetilde{\mathbf{u}}^n_{1/2} = (\mathbf{u}^n_{1/2}, 1)$.
2. Solve the coupled Riemann problem for the relaxed model (7)–(9) with $\lambda = 0$ and initial data $\widetilde{\mathbf{u}}^n_{-1/2}$ and $\widetilde{\mathbf{u}}^n_{1/2}$, using the chosen coupling condition at the interface. Let us note $\widetilde{\mathcal{W}}_c(x/t; \widetilde{\mathbf{u}}^n_{1/2}, \widetilde{\mathbf{u}}^n_{-1/2})$ the corresponding self-similar solution.

3. Define the numerical flux $(\widetilde{\mathbf{h}}_L)_0^n$ (respectively $(\widetilde{\mathbf{h}}_R)_0^n$) by the two first components of $\widetilde{\mathbf{f}}_L(\mathcal{W}_c(0^-;\widetilde{\mathbf{u}}_{1/2}^n,\widetilde{\mathbf{u}}_{-1/2}^n))$ (resp. $\widetilde{\mathbf{f}}_R(\mathcal{W}_c(0^+;\widetilde{\mathbf{u}}_{1/2}^n,\widetilde{\mathbf{u}}_{-1/2}^n))$).

Using the previous results, this numerical method fulfills the following properties, according to the choice of transmitted variable in the coupling condition. The first one refers to the coupling in variable $(\rho, \rho u)$.

Proposition 3. *Consider system (1)–(2) and the coupling in variable $(\rho, \rho u)$. Let $\mathbf{u}_0(.) = \overline{\mathbf{u}} \in \Omega_{\mathbf{u}}$ be a constant initial datum. Then, the associated numerical method, with transmission of variable $(\rho, \rho u, \rho T)$ in the global relaxation approximation, satisfies for all $n \in \mathbb{N}$ and $i \in \mathbb{Z}$*

$$\mathbf{u}_{i+1/2}^n = \overline{\mathbf{u}}.$$

The second result concerns the conservative model of coupling.

Proposition 4. *Consider the conservative coupling (6) for system (1)–(2). Then, the associated numerical method, with transmission of variable $(\rho, \rho u, \pi)$ in the global relaxation approximation, is conservative, that is to say*

$$(\widetilde{\mathbf{h}}_L)_0^n = (\widetilde{\mathbf{h}}_R)_0^n, \quad \forall n \in \mathbb{N}.$$

It is worth noting that, since the Riemann problems we solve in the second step of the scheme correspond to the relaxed model, their resolution can be performed explicitly.

5 Conclusion

The relaxation approximation has allowed us to provide different models of coupling in a natural way. It also gives Riemann problems that are simpler to solve than in the nonlinear case and enables us to construct a simple numerical method. Several developments based on the relaxation approximation are in progress:

- Study of the coupling of more complex systems (polytropic Euler equations, bifluid models [ACC06]...).
- Use of more complex models of coupling (coupling according to the dynamics, optimized coupling...).
- Comparison with other models and numerical methods of coupling.

This work falls within the scope of an ingoing joint research program on multiphase flows between CEA and University Pierre et Marie Curie–Paris 6, in the framework of the Neptune project. The topics of this work will be further investigated in a forthcoming paper [ACC07].

References

[ACC05] Ambroso, A., Chalons, C., Coquel, F., Godlewski, E., Lagoutière, F., Raviart, P.-A., Seguin, N.: Extension of interface coupling to general Lagrangian systems, Enumath 2005 proceedings, 817–824 Springer-Verlag (2006)

[ACC06] Ambroso, A., Chalons, C., Coquel, F., Godlewski, E., Lagoutière, F., Raviart, P.-A., Seguin, N.: Numerical mathematics and advanced applications, 852–860, Springer, Berlin, 2006

[ACC07] Ambroso, A., Chalons, C., Coquel, F., Godlewski, E., Lagoutière, F., Raviart, P.-A., Seguin, N.: A relaxation procedure for the interface coupling of two Euler systems (in preparation)

[BH06] Banda, M.K., Herty, M., Klar, A.: Coupling conditions for gas networks governed by the isothermal Euler equations. Netw. Heterog. Media **1** 295–314 (2006)

[B04] Bouchut, F.: Nonlinear stability of finite volume methods for hyperbolic conservation laws and well-balanced schemes for sources. Frontiers in Mathematics. Birkhäuser Verlag, Basel (2004)

[BNP06] Bretti, G., Natalini, R., Piccoli, B.: Numerical approximations of a traffic flow model on networks. Netw. Heterog. Media **1**, 5784 (2006)

[CLL94] Chen, G.Q., Levermore, C.D., Liu, T.P.: Hyperbolic conservation laws with stiff relaxation terms and entropy. Comm. Pure Appl. Math., **47**, 787–830 (1994)

[CGP01] Coquel, F., Godlewski, E., Perthame, B., In, A., Rascle, P.: Some new Godunov and relaxation methods for two-phase flow problems. In: Godunov methods (Oxford, 1999), 179–188. Kluwer/Plenum, New York (2001)

[DL88] Dubois, F., LeFloch, P.G.: Boundary conditions for nonlinear hyperbolic systems of conservation laws. J. Differential Equations, **71**, 93–122 (1988)

[GP06b] Garavello, M., Piccoli, B.: Traffic flow on network. Applied Math Series no. 1, American Institute of Mathematical Sciences (2006)

[GLTR05] Godlewski, E., Le Thanh, K.-C., Raviart, P.-A.: The numerical interface coupling of nonlinear hyperbolic systems of conservation laws. II. The case of systems. M2AN Math. Model. Numer. Anal., **39**, 649–692 (2005)

[GR04] Godlewski, E., Raviart, P.-A.: The numerical interface coupling of nonlinear hyperbolic systems of conservation laws. I. The scalar case. Numer. Math., **97**, 81–130 (2004)

[HR06] Herty, M., Rascle, M.: Coupling conditions for a class of second-order models for traffic flow. SIAM J. Math. Anal. **38**, 595–616 (2006)

[IT92] Isaacson, E., Temple, B.: Nonlinear resonance in systems of conservation laws. SIAM J. Appl. Math., **52**, 1260–1278 (1992)

Increasing Efficiency Through Optimal RK Time Integration of Diffusion Equations

F. Cavalli, G. Naldi, G. Puppo, and M. Semplice

Summary. The application of Runge–Kutta schemes designed to enjoy a large region of absolute stability can significantly increase the efficiency of numerical methods for PDEs based on a method of lines approach. In this work we investigate the improvement in the efficiency of the time integration of relaxation schemes for degenerate diffusion problems, using SSP Runge–Kutta schemes and computing the maximal CFL coefficients. This technique can be extended to other PDEs, linear and nonlinear, provided the space operator has eigenvalues with a nonzero real part.

1 Introduction

The integration of evolution PDEs through a method of lines approach leads to the solution of large systems of ODEs. Often such ODEs are stiff or moderately stiff; therefore, the possibility of increasing the stability region of the time integrator can lead to a significant increase in efficiency for explicit or semi-implicit time integration.

Specifically, we consider the system of PDEs

$$u_t + f_x(u) = Dp_{xx}(u), \tag{1}$$

where $f(u)$ is hyperbolic, i.e., the Jacobian of f is provided with real eigenvalues and a basis of real eigenvectors for each u, while $p(u)$ is a nondecreasing Lipschitz continuous function, with Lipschitz constant μ. We assume that the system has been fully discretized in space on a grid of N points $x_j, j = 1, \dots N$, and we denote with $U(t) = [U_1(t), \dots, U_N(t)]^T$ the vector of the grid values of the numerical solution at time t. The space discretized system can be written in the form

$$\frac{dU}{dt} = L(U(t)), \tag{2}$$

leading to an autonomous system of N nonlinear first order ODEs. If we consider for instance a system of conservation laws, $D = 0$, the operator L obtained from a conservative space discretization will be written as

$$L(U) = -\frac{1}{h}\left(F_{j+1/2} - F_{j-1/2}\right),$$

where $F_{j+1/2}$ is the numerical flux consistent with the physical flux $f(u)$ in the usual sense of the Lax–Wendroff theorem and h is the grid spacing.

In recent years, much research has focused on the efficient integration of the semidiscrete system (2). In particular, in this work we will concentrate on the performance of optimal Runge–Kutta schemes, characterized by a large stability region, introduced in [SR02].

Optimal Runge–Kutta schemes are built choosing an accuracy order p and a number of stages s, with $s \geq p$. In principle, once s and p are fixed, the coefficients of the Butcher tableaux defining the Runge–Kutta schemes are computed maximizing in some sense the stability region and keeping as constraints the fulfilment of the accuracy requirements. To compute their optimal schemes, Spiteri and Ruuth in [SR02] start from a strong stability condition, which requires that the operator L be nonlinearly stable, with respect to a suitable norm, for a certain CFL with Forward Euler integration, namely

$$||U^n + \Delta t L(U^n)|| \leq ||U^n||, \quad \forall \Delta t \leq \Delta t_{FE}.$$

Once this assumption is satisfied, the optimal schemes proposed in [SR02] do yield considerable savings in CPU time for a given accuracy.

The idea is that an s stages Runge–Kutta scheme applied to (2) can be written as a convex combination of s Forward Euler steps as (see also [GST01])

$$U^{(1)} = U^n \tag{3}$$

$$U^{(i)} = \sum_{k=1}^{i-1} \alpha_{ik} \left[U^{(k)} + \Delta t \frac{\beta_{ik}}{\alpha_{ik}} L(U^{(k)}) \right] \tag{4}$$

$$U^{(n+1)} = U^{(s)}. \tag{5}$$

Thus if the space discretized operator L is strongly stable for $\Delta t \leq \Delta t_{FE}$ with Forward Euler time integration, then the scheme in (5) will be strongly stable for

$$\Delta t \leq \lambda\, \Delta t_{FE}, \qquad \lambda = \min \frac{\alpha_{ik}}{\beta_{ik}}. \tag{6}$$

Note that (6) is only a sufficient condition for stability: clearly it may be possible to violate (6) and still find a stable scheme.

The problem is that several high order space discretization operators L *are not* stable under the Forward Euler scheme, and therefore it is not possible to use the estimate (6) to guarantee that optimal SSP schemes will improve the efficiency of the time integration of (2).

In the case of pure convection, that is for $D = 0$ in (1), high order space discretizations of $f_x(u)$ have purely imaginary eigenvalues ν up to high order powers of the mesh width h, that is $\Re(\nu) = O(h)^p$ and therefore are not stable under Forward Euler time integration. This is the case for instance of the fifth order WENO space discretization, see also [Spi], but we conjecture that the same holds for other widely used high order space discretizations for convective operators, such as ENO.

On the other hand, when $D \neq 0$, the eigenvalues of the semidiscrete operator L do have a negative real part of order 1, and therefore can be made stable under Forward Euler for a nonzero Δt_{FE}. In this work we study the stability of a family of high order numerical fluxes coupled with optimal RK-SSP schemes for (1), in the case of pure diffusion, i.e., $f \equiv 0$. In this case we find that several widespread space discretization schemes are stable with Forward Euler time integration, and therefore optimal RK-SSP schemes do yield a significant increase in the allowable CFL. Since in this case the eigenvalues of the exact operator are real, we even find that the stability estimate in (6) may be quite pessimistic, because it underestimates the stability region of some SSP schemes. In these cases, it is easy to compute numerically the maximal CFL, obtaining a further improvement in the efficiency of the scheme.

2 Diffusive Relaxation

We consider high order approximations of the degenerate parabolic equation

$$u_t = Dp_{xx}(u), \tag{7}$$

using relaxation schemes, a technique initially proposed in [JX95] for conservation laws. Following [NP00] and [CNPS06], we rewrite the system as a hyperbolic system with a stiff source term depending on a parameter ε, which formally relaxes on the original parabolic equation as $\varepsilon \to 0$, namely

$$\begin{cases} \dfrac{\partial u}{\partial t} + \dfrac{\partial v}{\partial x} = 0 \\[2ex] \dfrac{\partial v}{\partial t} + \Phi^2 \dfrac{\partial w}{\partial x} = -\dfrac{1}{\varepsilon} v + \left(\Phi^2 - \dfrac{D}{\varepsilon} \right) \dfrac{\partial w}{\partial x} \\[2ex] \dfrac{\partial w}{\partial t} + \dfrac{\partial v}{\partial x} = -\dfrac{1}{\varepsilon}(w - p(u)), \end{cases} \tag{8}$$

where Φ^2 is a suitable positive parameter. Formally, as $\varepsilon \to 0$, $w \to p(u)$, and $v \to -\partial_x p(u)$ and the original equation (7) is recovered. We integrate the system above with an IMEX Runge–Kutta scheme [PR05]. In this fashion, the stiff source term is implicit and does not require restrictive stability conditions, while the linear convective term is explicit.

In this work we consider only the relaxed scheme, which is obtained setting $\varepsilon = 0$ in the discretized equations. Let $\tilde{a}_{i,k}$ and $\tilde{b}_i$ be the coefficients forming the Butcher tableaux of the explicit Runge–Kutta scheme in the IMEX pair. The computation of the first stage value of the Runge–Kutta scheme reduces to

$$\begin{aligned} u^{(1)} &= u^n \\ w^{(1)} &= p(u^n) \\ v^{(1)} &= -D\partial_x w^{(1)} \end{aligned} \quad . \tag{9}$$

For the following stages the first equation is

$$u^{(i)} = u^n - \Delta t \sum_{k=1}^{i-1} \tilde{a}_{i,k} \partial_x v^{(k)}. \tag{10}$$

In the other equations the convective terms are dominated by the source terms and thus $v^{(i)}$ and $w^{(i)}$ are given by

$$w^{(i)} = p(u^{(i)}), \qquad v^{(i)} = -D\partial_x w^{(i)}. \tag{11}$$

In this fashion, because of the particular structure of the relaxation scheme we are considering only the explicit tableaux of the Implicit–Explicit Runge–Kutta scheme enters the actual computation, while the coefficients of the implicit scheme drop out as the relaxation step is computed. For this reason we can apply any explicit Runge–Kutta scheme to advance in time the solution u. Finally the updated solution is given by

$$u^{n+1} = u^n - \Delta t \sum_{i=1}^{\nu} \tilde{b}_i \partial_x v^{(i)}. \tag{12}$$

The overall accuracy of the scheme depends on the accuracy of the numerical flux used to approximate $\partial_x v^{(i)}$ and on the accuracy of the Runge–Kutta explicit scheme.

A nonlinear stability analysis for the first order version of this scheme (upwind numerical flux on a piecewise constant space reconstruction to evaluate $\partial_x v$ and Forward Euler in time) yields a parabolic stability restriction of the form

$$\Delta t \leq \frac{2h^2}{\mu} \frac{1}{1 + 2h\phi}, \tag{13}$$

where μ is the Lipschitz continuity constant of $p(u)$, see [CNPS06]. The parabolic CFL implies that to obtain a scheme of order p one should use a pth order accurate numerical flux for $\partial_x v$ and a $p/2$ order accurate Runge–Kutta method in time.

A linear stability analysis of several higher order numerical fluxes yields again a parabolic CFL of the form $\Delta t \leq C_1 h^2 (1 - C_2 h\Phi)/\mu$, with constant C_1 given by Table 1. Note that the scheme is unstable for $\Delta t = C_1 h^2/\mu$, but for a sufficiently small h it is enough to pick $\Delta t = (C_1 - \delta)h^2/\mu$ for a

Table 1. CFL constant C_1 for a few space reconstruction algorithms (from linear analysis) for standard first-, second-, and third-order RK schemes

	RK1	RK2	RK3
P-wise constant	2	2	2.51
P-wise linear	0.94	0.94	1.18
WENO5	0.79	0.79	1

Table 2. Improved stability coefficients for several s stages order p Runge–Kutta schemes (from [SR02])

	$s = 1$	$s = 2$	$s = 3$	$s = 4$	$s = 5$
$p = 1$	1	2	3	4	5
$p = 2$		1	2	3	4
$p = 3$			1	2	2.65

suitable positive δ, i.e., the scheme is stable provided C_1 is slightly decreased, see Table 3. Note that the stability requirement becomes more strict as space accuracy increases, while it loosens as time accuracy increases.

3 Diffusive Relaxation and SSP Schemes

We start fixing a standard notation for s stages Runge–Kutta schemes. As in [SR02] we denote SSP(s,p) the optimal strongly stable Runge–Kutta scheme of order p with s stages. We point out that SSP(1,1) is the Forward Euler scheme, SSP(2,2) the Heun scheme, and SSP(3,3) the TVD third order Runge–Kutta method of [SO88], which is probably the high order Runge–Kutta scheme most frequently used in conjunction with high order space discretizations for (1).

From the first column of Table 1 we note that all the numerical fluxes considered are at least linearly stable with the Forward Euler scheme. Thus the theory of strongly stable Runge–Kutta schemes can be applied in this case. In particular, we consider the schemes introduced in [SR02], for which the improved stability coefficients λ of (6) can be found in Table 2. To indicate the fact that these coefficients are found through the theory of strongly stable Runge-Kutta schemes and to identify to which scheme they apply, we will label these coefficients as $\lambda_{\text{SSP}}(s,p)$.

The coefficients $\lambda_{\text{SSP}}(s,p)$ are not optimal in the case of pure diffusion. They can be further improved considering the actual stability region of SSP schemes and recalling that for the purely diffusive operator we are considering, only the intersection of the stability region with the real line is relevant.

In Fig. 1 we find the regions of absolute stability for some SSP schemes given in [SR02]. On the top left of the figure we find the stability regions of schemes of order 1 with 1, 2, and 3 stages. We point out that here the improvement in the CFL constant is exactly balanced by the increased computational

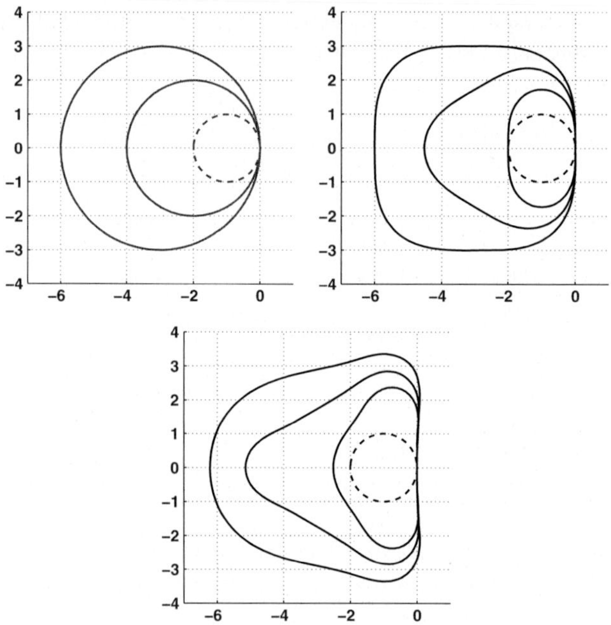

Fig. 1. Regions of absolute stability for optimal SSP schemes. On the top left: order 1, stages 1,2,3. On the top right: order 2, stages 2,3,4. On the bottom: order 3, stages 3,4,5. The dashed line is the Forward Euler scheme

effort due to the higher number of stages: there is no gain in efficiency with respect to the standard Forward Euler scheme. In the top right of Fig. 1 there are the stability regions of schemes of order 2 with 2, 3, and 4 stages and, for comparison, of the Forward Euler scheme (dashed line). The SSP theory underestimates the effective CFL and direct inspection of the stability plot suggests that the stability coefficient λ appearing in Table 3 can be increased finding the intersection of the stability curve with the real axis. Finally, on the bottom of Fig. 1 we show the stability regions of schemes of order 3 with 3, 4, and 5 stages. Again the CFL gain of Table 2 can be improved by direct inspection of the graph.

Let $\eta_{s,p}$ be the abscissa of the intersection of the stability curve with the negative real axis. Then the maximal gain in CFL with respect to Forward Euler, for a problem with real eigenvalues, is given by

$$\lambda_{\mathrm{OPT}}(s,p) = \frac{|\eta_{s,p}|}{|\eta_{1,1}|} = \frac{|\eta_{s,p}|}{2}.$$

For several schemes in the figure it is easy to see that $\lambda_{\mathrm{OPT}}(s,p) > \lambda_{\mathrm{SSP}}(s,p)$.

The optimal CFL number for a given scheme can be found multiplying the coefficient $C_1 - \delta$ determined by the space discretization by the proper stability coefficient λ, i.e., $\Delta t \leq (C_1 - \delta)\lambda_{\mathrm{OPT}}(s,p)h^2/\mu$.

Table 3. Order of convergence and number of numerical flux function evaluations N_f (with 80 grid points) for the SSP-CFL $\lambda_{\rm SSP}(s,p)$ (*left*) and the maximal CFL $\lambda_{\rm OPT}(s,p)$ (*right*) for several RK schemes

SSP(2,s) + WENO5

Stages	CFL	Order	N_f
2	0.78	4	810
3	2 × 0.78	4	606
4	3 × 0.78	4	540

SSP(2,s) + WENO5

Stages	CFL	Order	N_f
2	1 × 0.78	4	810
3	**2.259**×0.78	4	537
4	3 × 0.78	4	540

SSP(3,s) + WENO5

Stages	CFL	Order	N_f
3	0.78	4.5	1230
4	2 × 0.78	5.2	820
5	2.65 × 0.78	5.2	770

SSP(3,s) + WENO5

Stages	CFL	Order	N_f
3	**1.256**×0.78	4.8	978
4	**2.574**×0.78	5.6	636
5	**3.106**×0.78	5.4	660

To measure the computational complexity of a scheme we compute the number N_f of numerical flux evaluations needed to reach a fixed integration time. Thus, for a given numerical flux, the most efficient scheme has the lowest value of N_f. Table 3 shows the different values of N_f obtained with CFL chosen according to the values of $\lambda_{\rm SSP}(s,p)$ on the left and $\lambda_{\rm OPT}(s,p)$ on the right. The grid spacing h is the same for all values of N_f, namely, $h = 1/80$. The table contains also data on the accuracy of the space–time scheme. The accuracy was evaluated using four different grids, and modifying the time step according to h and the chosen value of λ. We do not show data for first order schemes since $\lambda_{\rm SSP}(s,1) = \lambda_{\rm OPT}(s,1)$. The table shows the increased efficiency of the SSP(s,p) schemes with $s > p$ (left column) and the gain obtained by a better estimate of λ (right column). In the table, the optimal values of λ are indicated in boldface, when they are sensitively larger than the corresponding $\lambda_{\rm SSP}(s,p)$. In particular, the improvement between the standard SSP(3,3) and the five stages SSP(5,3) with the optimal CFL is quite striking.

The data in Table 3 refer to a linear diffusion problem, although we find analogous results for a nonlinear degenerate diffusion equation. We performed tests on the self-similar Barenblatt solution of the porous media equation. In this case the order of accuracy is limited by the nonregularity of the solution, but we find that the errors with respect to the exact solution slightly decrease using the SSP(s,p) with $s > p$ and the optimal λ.

4 Final Remarks

We have shown that the theory of [SR02] can be applied to diffusion equations in the relaxation framework to improve the efficiency of the time integration. Moreover, the fact that the eigenvalues of the semidiscrete operator are real numbers permits to further improve the efficiency of the schemes. We expect that analogous results can be obtained for convection–diffusion operators, thanks to the nonzero real part of the eigenvalues of the discrete operators. This allows to achieve stability under Forward Euler integration.

We also note that the same framework can be applied to other space discretizations besides the numerical fluxes obtained via relaxation schemes: the key factor is the localization of the eigenvalues of the differential operator, which must have a nonzero real part. On the other hand, the plots of absolute stability regions show that Runge–Kutta schemes with number of stages $s > p$ can be built with improved stability regions, notwithstanding stability under Forward Euler. These schemes can be applied to semidiscrete operators even in the convective regime, but their stability conditions cannot be derived from the behavior of the operator under the Forward Euler scheme.

References

[CNPS06] F. Cavalli, G. Naldi, G. Puppo, and M. Semplice. High order relaxation schemes for non linear diffusion problems. Arxiv preprint math. NA/0604572, 2006.

[GST01] S. Gottlieb, C. Shu, and E. Tadmor. Strong stability-preserving high-order time discretization methods. *SIAM Rev.*, 43(1):89–112, 2001.

[JX95] S. Jin and Z. Xin. The relaxation schemes for systems of conservation laws in arbitrary space dimension. *Comm. Pure and Appl. Math.*, 48:235–276, 1995.

[NP00] G. Naldi and L. Pareschi. Numerical schemes for hyperbolic systems of conservation laws with stiff diffusive relaxation. *SIAM J. Numer. Anal.*, 37:1246–1270, 2000.

[PR05] L. Pareschi and G. Russo. Implicit-explicit Runge-Kutta schemes and applications to hyperbolic systems with relaxation. *J. Sci. Comp.*, 25:129–155, 2005.

[SO88] Chi-Wang Shu and Stanley Osher. Efficient implementation of essentially nonoscillatory shock-capturing schemes. *J. Comput. Phys.*, 77(2):439–471, 1988.

[Spi] Raymond J. Spiteri. Private communication.

[SR02] Raymond J. Spiteri and Steven J. Ruuth. A new class of optimal high-order strong-stability-preserving time discretization methods. *SIAM J. Numer. Anal.*, 40(2):469–491 (electronic), 2002.

Numerical Simulation of Relativistic Flows Described by a General Equation of State

S. Serna

1 Introduction

In this research work we focus on the numerical simulation of the dynamics and thermodynamics of relativistic flows described in terms of the Euler equations of special relativity for fluids whose intrinsic properties are described by means of an equation of state (EOS) for a relativistic perfect gas due to Synge, ([Syn57, FK96]).

The ideal gas EOS is standard in relativistic fluid dynamics ([AIMM99, MIM91, MMFIM97, MM03, NW86]. The use of the ideal gas EOS is computationally convenient but is not the most appropriate EOS to simulate the thermodynamics intrinsic to relativistic flows. The ideal gas EOS forces the gas to keep constant the ratio of specific heats γ along the evolution. In addition, the ideal gas EOS is not consistent with the relativistic formulation of the kinetic theory of gases and allows superluminal speed of sound for fixed adiabatic exponents $\gamma > \frac{5}{3}$.

Synge proposed in 1957 ([Syn57]) the exact form of a relativistic ideal gas EOS to overcome the limitations of the ideal gas EOS. The Synge EOS is given by means of an analytic expression of the relativistic enthalpy as a function of temperature in terms of the Bessel functions of second kind, K_2 and K_3. The use of the Synge EOS allows the gas to reach states thermodynamically relativistic as a direct result of wave dynamics making possible variations of the γ value in a natural way. At the same time, small variations of the γ value produce significant differences on the wave structure along the evolution compared with the case where ideal gas EOS is used.

The main drawback of the Synge EOS is that it is computationally expensive and, as a consequence, it has not been frequently used in numerical simulations of relativistic flows, [FK96].

In this research work we propose a simple and accurate rational approximation to the Synge enthalpy function that is computationally efficient and describes precisely the thermodynamics of relativistic perfect gas. Preliminary numerical tests in one space dimension are presented by using a high order accurate shock capturing scheme suitable for the simulation of relativistic flows.

In Sect. 2 we display the equations of special relativity. In Sect. 3 we introduce and analyze an approximation to the Synge EOS. In Sect. 4 we describe the numerical algorithm and we compute numerical tests in 1D.

2 Hydrodynamic Equations for Special Relativity in One Space Dimension

Let us consider the special relativistic Euler equations in one spatial dimension,

$$\mathbf{u}_t + \mathbf{f}(\mathbf{u})_x = 0, \tag{1}$$

where

$$\mathbf{u} = \begin{bmatrix} D \\ S \\ \tau \end{bmatrix}, \quad \mathbf{f}(\mathbf{u}) = \begin{bmatrix} Du \\ Su + P \\ S - Du \end{bmatrix}$$

where $D = \rho W$ is the rest mass density, ρ is the density, $S = \rho h W^2 u$ is the momentum, u is the velocity, P is the pressure, and $\tau = \rho h W^2 - P - D$ is the energy. We assume the speed of light, c, is normalized to $c = 1$.

The flow Lorentz factor is defined as

$$W = \frac{1}{\sqrt{1-u^2}}$$

and the specific relativistic enthalpy as

$$h = 1 + \epsilon + \frac{P}{\rho}, \tag{2}$$

where ϵ is the specific internal energy.

The system of (1) becomes hyperbolic if the speed of sound c_s defined as

$$c_s^2 = \frac{1}{h} \cdot \frac{\partial P}{\partial \rho}\Big|_S \tag{3}$$

is a real number and the Jacobian of the flux $\mathbf{f}$ is diagonalizable with real eigenvalues and a complete system of eigenvectors.

For a general EOS, the eigenvalues of $\frac{\partial \mathbf{f}}{\partial u}$ (the characteristic velocities) are

$$\begin{aligned} \lambda_{\pm} &= \frac{u \pm c_s}{1 \pm u c_s}, \\ \lambda_0 &= u. \qquad \text{(simple)} \end{aligned} \tag{4}$$

A complete system of the right and left eigenvectors is displayed in [FIMM94, MMFIM97, MM03].

The Euler equations for special relativity are strongly coupled through the Lorentz factor $W(\geq 1)$ and the relativistic specific enthalpy, $h(\geq 1)$.

The system of conservation laws is closed with a general equation of state of the form

$$P = P(\rho, \epsilon)$$

to compute the pressure in terms of ρ and the specific internal energy ϵ.

3 Relativistic Perfect Gases: An Approximate Equation of State for Relativistic Gases

Taub showed in [Tau48] that the choice of the EOS to describe a relativistic gas is not arbitrary. It is necessary to ensure that the EOS is *causal*, i.e., superluminal wave propagation speed is not allowed.

We can formulate a perfect gas EOS in terms of the temperature T as

$$P = \rho T \tag{5}$$

from a given function of the specific enthalpy, $h(T)$.

Given $h(T)$ we define the function

$$g(T) = \frac{(h(T) - 1)}{T} - 1. \tag{6}$$

From the expression of the relativistic specific enthalpy (2) we can deduce the expression for the specific internal energy as

$$\epsilon(T) = Tg(T). \tag{7}$$

If we define

$$\gamma(T) := 1 + \frac{1}{g(T)}, \tag{8}$$

we can rewrite (5) as

$$P = (\gamma(T) - 1)\, \rho\, \epsilon(T) \tag{9}$$

that resembles the analytical expression of the ideal gas EOS.

From the Rankine–Hugoniot equations for the system (1), Taub ([Tau48]) proved that an equation of state described by means of a specific enthalpy function $h(T)$ is *causal* if it satisfies the so-called Taub's inequality

$$(h(T) - T)(h(T) - 4T) \geq 1. \tag{10}$$

We can recover the ideal gas equation of state with a constant ratio of specific heat $\gamma > 1$ from the following specific enthalpy function:

$$h_{ID} = 1 + \frac{\gamma}{\gamma - 1} T, \tag{11}$$

from which the expression for the pressure is

$$P = (\gamma - 1)\,\rho\,\epsilon. \tag{12}$$

The ideal gas EOS is *causal* for all T if and only if $\frac{4}{3} \leq \gamma \leq \frac{5}{3}$ as follows from (10) and (11). Thus, for a fixed adiabatic exponent γ out of that interval, the flow described by (1) with an ideal gas EOS (12) might reach wave propagation speeds larger than the speed of light.

The perfect gas EOS by Synge ([Syn57]) can be formulated from the following expression of the specific relativistic enthalpy

$$h_s(T) := \frac{K_3(\frac{1}{T})}{K_2(\frac{1}{T})}; \quad T \geq 0 \tag{13}$$

in terms of the modified Bessel functions of second kind K_2 and K_3.

We denote by $g_s(T)$, $\epsilon_s(T)$, and $\gamma_s(T)$ the corresponding expressions to (6), (7), and (8) obtained from (13), respectively. The resulting Synge EOS is consistent with the thermodynamical limits ($\gamma_s \longrightarrow \frac{5}{3}$ if $T \longrightarrow 0$ and $\gamma_s \longrightarrow \frac{4}{3}$ if $T \longrightarrow \infty$) and it satisfies Taub's inequality ([Syn57]).

The calculation of the specific enthalpy for the Synge EOS and therefore, the calculation of the thermodynamic variables depending on it, is computationally expensive and inaccurate. Falle and Komissarov in [FK96] proposed to construct a "look-up table" for $\gamma_s(T)$ to reduce the computational cost.

For practical purposes, a global and simple analytical approximation for the Synge EOS is highly convenient and useful.

We propose a rational approximation to the Synge specific enthalpy $h_s(T)$ of the form

$$h_s(T) \approx \frac{\alpha_1\,T^2 + \alpha_2\,T + \alpha_3}{\alpha_4\,T + \alpha_5}. \tag{14}$$

Proposition 1. *The rational function*

$$h_a(T) = \frac{4\,T^2 + (1 + \frac{5}{2}\lambda)\,T + \lambda}{T + \lambda} \tag{15}$$

is consistent with the thermodynamic limits when $0 < \lambda \leq \frac{2}{3}$ *and satisfies Taub's inequality (10) when* $\frac{1}{3} \leq \lambda \leq \frac{2}{3}$.

Proof. The corresponding function $g_a(T)$, $g_a(T) := \frac{h_a(T)-1}{T} - 1$, is strictly monotone in $[0, +\infty[$. Since $\lim_{T \longrightarrow \infty} g_a(T) = 3$ and $g_a(0) = \frac{3}{2}$ we have that $\lim_{T \longrightarrow \infty} \gamma_a(T) = \frac{4}{3}$ and $\gamma(0) = \frac{5}{3}$ and then, there is consistency with the thermodynamical limits.

Taub's inequality is satisfied if $1 <= 3 * \lambda + (1 - (3/2) * \lambda) * T$, and this is true if $\frac{1}{3} \leq \lambda \leq \frac{2}{3}$. □

From (15) with $\lambda \in [\frac{1}{3}, \frac{2}{3}]$ and (6)–(8) we propose an EOS (9) that describes consistently the thermodynamics of a relativistic perfect gas. The proposed EOS approximates Synge EOS with a maximum error of about one percent.

4 Numerical Scheme

The numerical scheme to evolve in time the system of equations (1) is based on the so-called conservation form.

We introduce the computational grid: $x_j = j\Delta x$, (Δx the spatial step), $t_n = n\Delta t$, the time discretization, (Δt is the time step), $I_j = [x_{j-\frac{1}{2}}, x_{j+\frac{1}{2}}]$ is the spatial cells where $x_{j+\frac{1}{2}} = x_j + \frac{\Delta x}{2}$, and $C_j^n = I_j \times [t_n, t_{n+1}]$ is the computational cell.

Let $\mathbf{u}_j^n$ be an approximation of the mean value in I_j

$$\frac{1}{\Delta x} \int_{x_{j-\frac{1}{2}}}^{x_{j+\frac{1}{2}}} \mathbf{u}(x, t_n) dx \tag{16}$$

of the exact solution $\mathbf{u}(x, t_n)$ of the initial value problem (1), obtained from a finite difference scheme in conservation form

$$\mathbf{u}_{j,k}^{n+1} = \mathbf{u}_{j,k}^n - \frac{\Delta t}{\Delta x}(\tilde{\mathbf{f}}_{j+\frac{1}{2}} - \tilde{\mathbf{f}}_{j-\frac{1}{2}}), \tag{17}$$

where the numerical flux, $\tilde{\mathbf{f}}$, is a function of $k + l$ variables

$$\tilde{\mathbf{f}}_{j+\frac{1}{2}} = \tilde{\mathbf{f}}(\mathbf{u}_{j-l+1}^n, \cdots \mathbf{u}_{j+k}^n), \tag{18}$$

which is consistent with the flux of the equation (1), $\tilde{\mathbf{f}}(\mathbf{u}, \cdots, \mathbf{u}) = \mathbf{f}(\mathbf{u})$.

To obtain an explicit scheme of first-order accuracy in space and time we need to specify an explicit formula to calculate the numerical fluxes at the spatial interfaces $x_{j+\frac{1}{2}}$. In our case we use the Marquina Flux Formula (MFF) for relativistic equations ([MMFIM97]) that requires the use of two linearizations at both sides of each interface. The characteristic fluxes at the interfaces are computed from the solution of the Riemann problem. The flux is chosen either selecting the upwind one depending on the sign of the characteristic speed or by local flux decomposition using local Lax–Friedrichs, [MMFIM97, MM03].

Higher order of accuracy in space is obtained by applying a high-order reconstruction procedure to extrapolate the linearizations and the characteristic numerical fluxes at the interfaces. In our calculations we use the third order accurate Power PHM method, [Ser06].

High-order accuracy in time is obtained by using the third-order accurate Runge–Kutta method by Shu and Osher, [SO89].

In the case of the system of special relativistic hydrodynamics, we also need to recover the physical variables from the conserved ones in each stage of the Runge–Kutta iteration by obtaining the solution of a system of nonlinear algebraic equations where the EOS plays an important role. To this end we use an extension of the Newton method proposed in [AIMM99] to obtain the unique solution of the equation $\phi(P) = 0$, where the function ϕ is defined as

$$\phi(P) = (\gamma(T) - 1)\rho\,\epsilon(T) - P$$

for $P \in [P_{\min}, \infty[$ where $P_{min} = |S| - \tau - D$.

4.1 Norman and Winkler Blast Wave

This one-dimensional blast wave due to Norman and Winkler ([NW86]) is generated when evolving in time a Riemann problem with two constant states where the interface is located at the center of the interval $[0,1]$. The initial data is given by

$$(\rho_L, v_L, P_L) = (1, 0, 1000) \;\; x \leq 0.5,$$
$$(\rho_R, v_R, P_R) = (1, 0, 0.01) \;\; x > 0.5.$$

The evolution in time of this initial data generates a rarefaction wave that propagates to the left while a contact discontinuity separates the gases and a shock wave propagates to the right. The specific internal energy is transformed into kinetic energy in the thin layer of high density generated in the shocked gas. In this layer the gas loses temperature and the energy is transformed into kinetic energy where the velocity reaches a value close to $0.96c$.

The final simulation time is $t = 0.4$. We compute the solution for two cases. One case considering the ideal EOS with $\gamma = \frac{5}{3}$ and the other case considering an approximation to the Synge EOS computed from the approximation to the relativistic enthalpy (15) with $\lambda = \frac{2}{3}$. For this case, the initial values of γ deduced from the approximation to the Synge EOS are $\gamma_L = 1.33344$ and $\gamma_R = 1.65695$.

In Figs. 1 and 2 we plot the density, the velocity, the adiabatic exponent γ, and the temperature profiles for the ideal EOS case and for the general EOS

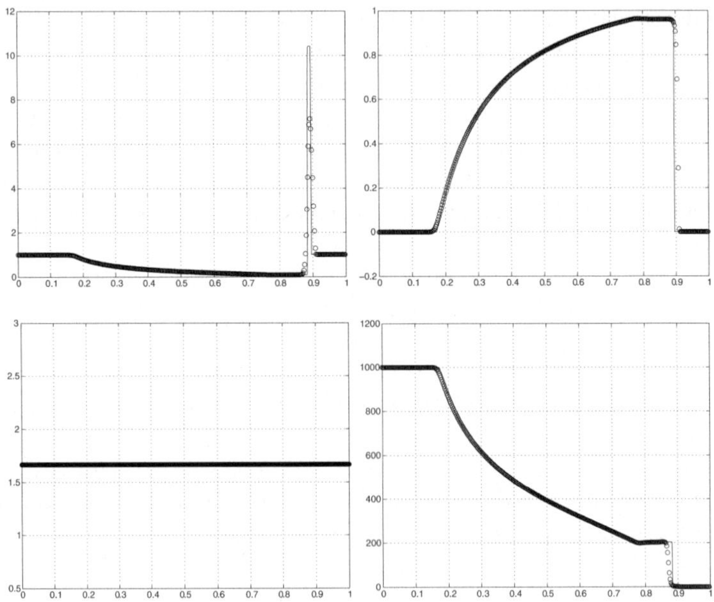

Fig. 1. Ideal EOS case, $\gamma = 5/3$. *Top left*, density; *top right*, velocity; *bottom left*, adiabatic exponent γ, and *bottom right*, temperature

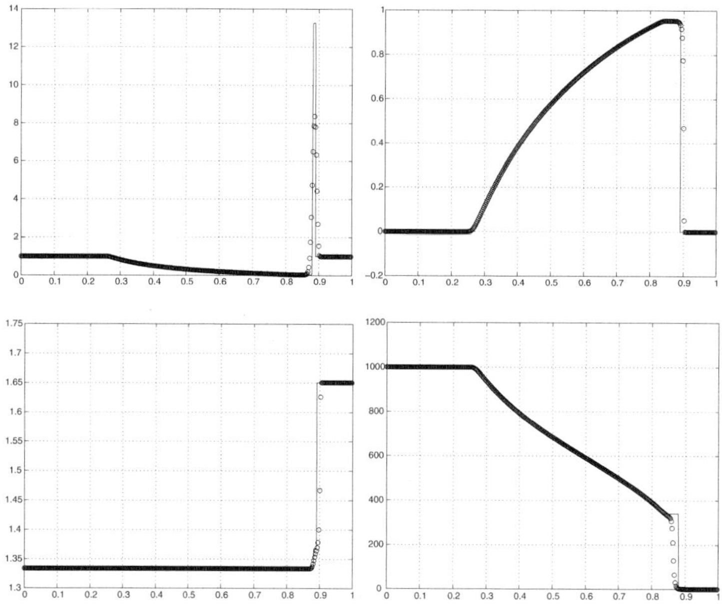

Fig. 2. General EOS case. *Top left*, density; *top right*, velocity; *bottom left*, adiabatic exponent γ and *bottom right*, temperature

case, respectively. In both figures we display the numerical approximation computed with 400 points vs. the exact solution.

We observe differences between both experiments. The density and temperature at the thin layer between the contact wave and the shock wave (*shocked gas*) are substantially larger for the approximate Synge EOS compared to the ideal gas case. The locations of the head and tail of the rarefaction wave are neatly different, showing that the wave speeds are strongly affected by small variations of the γ value. Long-time calculations can make those differences significant.

5 Conclusion

In this research work a simple and accurate approximation to the Synge equation of state to describe a relativistic perfect gas is presented. The formulation of the hydrodynamics in special relativity of compressible flows governed by this equation of state provides an analytic tool that is computationally efficient and accurate to describe precisely the thermodynamics of a relativistic perfect gas.

Acknowledgment

The author thank Professor Antonio Marquina for very useful suggestions and discussions. Also, the author thank IPAM at UCLA for supporting the stay at the Institute where this research work was initiated. Grant MTM2005-07708 is acknowledged.

References

[AIMM99] Aloy, M.A., Ibañez, J.M, Marti J.M. and Muller, E: Genesis: A high resolution code for three-dimensional relativistic hydrodynamics. ApJ Suppl. Series, 122, 151–166 (1999)

[FK96] Falle, S.A.E.G. and Komissarov, S.S.: An upwind numerical scheme for relativistic hydrodynamics with a general equation of state. MNRAS, vol 278, 586 (1996).

[FIMM94] Font, J.A., Ibañez, J.M., Marquina, A. and Marti, J.M.: Multidimensional relativistic hydrodynamics: characteristic fields and modern high-resolution shock-capturing schemes. Astron. Astrophys, vol282, pp 304–314 (1994).

[MIM91] Marti, J.M., Ibañez, J.M. and Miralles, J.A.: Numerical relativistic hydrodynamics: Local characteristic approach. Phys. Rev. D, 43, 3794–3801 (1991)

[MM03] Marti, J.M. and Muller, E. : Living Rev. Relativity, 7 (2003)

[MMFIM97] Marti, J.M., Muller, E, Font J.A., Ibañez, J.M and Marquina, A.: Morphology and Dynamics of Relativistic Jets. ApJ, 479, 151–163 (1997).

[NW86] Norman, M.L., and Winkler, K.-H.A.: Why Ultrarelativistic Hydrodynamics is Difficult, Astrophysical Radiation Hydrodynamics, 449–476, Reidel, Dordrecht, (1986).

[Ser06] Serna,S.: A class of extended limiters applied to Piecewise Hyperbolic methods. SIAM Journal of Scientific Computing, 28 (1), 123–140, (2006).

[SO89] Shu, C.W. and Osher S.J.: Efficient Implementation of Essentially Non-Oscillatory Shock Capturing Schemes II,. J. Comput. Phys. 83, 32–78 (1989).

[Syn57] Synge, S.L. The Relativistic Gas. North-Holland. Amsterdam (1957).

[Tau48] Taub A.H.: Relativistic Rankine-Hugoniot Equations. Phys. Rev, vol 74, 328–334 (1948).

On Delta-Shocks and Singular Shocks

V.M. Shelkovich

1 Introduction

It is well known that there are "nonclassical" situations where, in contrast to Lax's and Glimm's results, the Cauchy problem for a system of conservation laws does not possess a weak L^∞-solution except for some particular initial data. To solve the Cauchy problem in this "nonclassical" situation, it is necessary to introduce new singular solutions called *δ-shocks and singular shocks*. The components of these solutions contain delta functions [ASh05], [B94], [DSh03]– [LW02], [S02]– [Sh04], [TZZ94]. The exact structure of such type solutions is given below in (2), (7) and Definition 1. The theory of *δ-shocks and singular shocks* has been intensively developed in the last 10 years. In particular, in numerous papers δ-shock type solutions of "zero-pressure gas dynamics" have been studied. Moreover, in the recent papers [PSh06], [Sh06] the theory of *δ'-shocks* was established, and a concept of *$\delta^{(n)}$-shocks* was introduced, $n = 2, 3, \dots$. They are *new type singular solutions* such that their components contain delta functions and their derivatives.

To deal with δ- and δ'-shocks we must devise some way to define a *singular superposition* of distributions (for example, a *product of the Heaviside function and the delta function*); discover a proper notion of weak solution; and define *in which sense* a distributional solution satisfies a nonlinear system. In view of these facts, in the δ-shock (singular shock) and δ'-shock theories there are many open and complicated problems. Study of this area gives a new perspective in the theory of conservation law systems.

In this paper we consider one of the complicated problems mentioned above, which is related to the concept of *singular shocks*.

Let us remind that some problems related with singular shocks were studied in [K99]– [KK90], [S02], [S03]. A model system admitting a *singular shock* is the well-known Keyfitz–Kranzer system

$$u_t + (u^2 - v)_x = 0, \quad v_t + \left(\frac{1}{3}u^3 - u\right)_x = 0, \tag{1}$$

which was studied in [KK95], [KK90]. In the exellent paper [KK95], to construct *approximate solutions*, the Colombeau theory approach as well as the Dafermos–DiPerna regularization (under the assumption that Dafermos profiles exist) and the box approximations are used. However, the notion of a *singular solution* has *not been defined*. Later, in [Sc04], the existence of Dafermos profiles for singular shocks was proved. But it was not clear in which sense a singular shock satisfies the system (1).

In [KSS03], [KSZ04], [S02], [S03] for system of conservation laws $w_t + \big(q(w)\big)_x = 0$, $x \in \mathbf{R}$, $w(x,t) \in \mathbf{R}^n$, where $q : \mathbf{R}^n \to \mathbf{R}^n$ is a smooth function, a *singular shock solution* is a measure of the form

$$w(x,t) = \omega(x,t) + \sum_i M_i \chi_i(t) \delta(x - x_i(t)), \tag{2}$$

where ω is a classical weak solution away from the singularities, χ_i is the characteristic function of interval $[A_i, B_i)$; $M_i \in W^\infty$ and $x_i \in W^{1,\infty}$. The function w is the weak limit of a sequence w^ε with $w^\varepsilon(\cdot,t) \in L^1_{loc}$ uniformly with respect to ε, pointwise in t; satisfying $w^\varepsilon(\cdot,t) \to w(\cdot,t)w(x,t)$ and

$$\big(w^\varepsilon(\cdot,t)\big)_t + \big(q(w^\varepsilon(\cdot,t))\big)_x - \varepsilon(A(w^\varepsilon(\cdot,t))_x)_x \to 0, \quad \varepsilon \to 0, \tag{3}$$

weakly in the space of measures on $\mathbf{R}$, pointwise with respect to t, for some positive definite matrix A. In the above papers some modifications of this definition are also used. Note that since $w^\varepsilon \to w$ weakly, Definition (2), (3) can be used without the term $\varepsilon(A(w^\varepsilon(\cdot,t))_x)_x$ (this was done in [S02]). These authors ([K99]– [KSZ04], [S02], [S03]) *distinguish* between δ-shocks and singular shocks. In fact, the main distinction of a *singular shock* is that its *flux function is not defined.* As said in [K99, p.106], “unlike the delta-shocks..., the singular shocks which are needed to solve (1) are truly nonlinear objects that cannot be defined in the context of classical distribution theory.” According to [K99]– [KSZ04], [S02], [S03], some model problems for δ-shocks are described in [B94], [ERS96], [LW02], [TZZ94]. Here for “zero-pressure gas dynamics” the *measure-valued solution* approach is used, and flux-functions ρu, ρu^2 are well-defined measures.

We would like to stress that Definition (2), (3) of a *singular shock* and the other ones from [KSS03], [KSZ04], [S02], [S03] do not connect the *limiting function* (2) with the system $w_t + \big(q(w)\big)_x = 0$; they only connect the *regularizing function* w^ε with the *regularizing system* (3). Thus it is not defined in which sense a singular shock (2) satisfies to nonlinear system. In this way only approximating (viscosity) solutions and their structure can be studied. (Note that a more general and strict definition of the type (2), (3) was introduced in [DSh03].)

To deal with δ- and δ'-shocks, the *weak asymptotics method* was developed in [DSh03]– [DSh06], [Sh03-1], [Sh04]. In [ASh05], [DSh03]– [DSh06], [Sh03], [Sh03-1] the definition of δ-shock type solutions to systems (8) (see Definition 1) and (9) were introduced, and the corresponding *δ-shock Rankine–Hugoniot conditions* derived. These definitions give *natural* generalizations of the classical definition of the weak L^∞-solutions. According to them, δ-shocks are *Schwartz distributional solutions.* In these papers some Cauchy problems admitting (exact) δ-shocks were solved. In particular, the Cauchy problems for the Keyfitz–Kranzer system (1) and its generalization

$$L_{21}[u,v] = u_t + \big(f(u) - v\big)_x = 0, \quad L_{22}[u,v] = v_t + \big(g(u)\big)_x = 0 \qquad (4)$$

were first solved in [Sh03], [Sh03-1] (see also [ASh05]), where $f(u)$ and $g(u)$ are polynomials of degree n and $n+1$, respectively; n is even.

In this paper, by using our results [ASh05], [DSh05], [DSh06], [Sh03] – [Sh04] we show that both *singular shock* and *δ-shock are solutions of the same type* (in the sense of Definition 1). To prove our assertion we compare singular solutions that have δ-singularities for the systems (1) and (4) and the system

$$L_{31}[u] = u_t + \big(f(u)\big)_x = 0, \quad L_{32}[u,v] = v_t + \big(g(u)v\big)_x = 0. \qquad (5)$$

According to [K99]– [KSZ04], [S02], [S03], systems (1), (4), and (5) are model problems for *singular shocks* and δ-shocks, respectively. For these systems we consider the front-problem with the initial data of the form

$$\begin{aligned} u^0(x) &= u^0_+(x) + [u^0(x)]H(-x), \\ v^0(x) &= v^0_+(x) + [v^0(x)]H(-x) + e^0\delta(-x), \end{aligned} \qquad (6)$$

where $[u^0] = u^0_- - u^0_+$, $[v^0] = v^0_- - v^0_+$, and $u^0_\pm$, $v^0_\pm$ are given smooth functions, e^0 is a given constant, $H(x)$ is the Heaviside function, $\delta(x)$ is the delta-function.

Our arguments are the following. (*i*) According to Theorems 2, 3 (from the papers [ASh05], [DSh05], [DSh06], [Sh03]– [Sh04]), δ-shock wave type solutions of the Cauchy problems (1), (6); (4), (6); (5), (6) have the form

$$\begin{aligned} u(x,t) &= u_+(x,t) + [u(x,t)]H(-x+\phi(t)), \\ v(x,t) &= v_+(x,t) + [v(x,t)]H(-x+\phi(t)) + e(t)\delta(-x+\phi(t)), \end{aligned} \qquad (7)$$

(where $u_\pm(x,t)$, $v_\pm(x,t)$, $e(t)$, $\phi(t)$ are desired functions, $x = \phi(t)$ is the discontinuity curve) and satisfy corresponding systems of conservation laws in the sense of the same Definition 1 (see Sects. 2 and 3). (*ii*) According to Theorem 1, the Rankine–Hugoniot conditions for the above δ-shock wave type solutions are given by the identical formula (12) (see Sect. 2). (*iii*) For these problems the flux-functions of δ-shocks (15), (16) and (17), (18) are well-defined *Schwartz distributions* and have the identical structure (see Sect. 4).

Nevertheless, flux-functions of δ-shocks for the Keyfitz–Kranzer system (1) and its generalization have some specific and "strange" properties. The point is that δ-shocks constitute the universe with unusual and "strange" properties, and the Keyfitz–Kranzer system is an excellent model example that demonstrates this. Note that it is *impossible* to construct δ-shocks for systems (1) and (4) by using the *nonconservative product* [DLM95] as well as the *measure-valued solutions approach* [B94], [TZZ94] (for details, see [ASh05]).

2 δ-Shocks and the Rankine–Hugoniot Conditions

Consider two particular systems of conservation laws:

$$L_1[u,v] = u_t + \big(F(u,v)\big)_x = 0, \qquad L_2[u,v] = v_t + \big(G(u,v)\big)_x = 0, \tag{8}$$

$$L_1[u,v] = v_t + \big(G(u,v)\big)_x = 0, \qquad L_2[u,v] = (uv)_t + \big(H(u,v)\big)_x = 0, \tag{9}$$

where $F(u,v)$, $G(u,v)$, $H(u,v)$ are smooth functions, *linear* with respect to v; $u = u(x,t)$, $v = v(x,t) \in \mathbf{R}$; $x \in \mathbf{R}$. As far as we know, all one-dimensional systems of conservation laws admitting δ-shocks are particular cases of systems (8) and (9). Our model examples (1), (4), (5) are particular cases of (8); the "zero-pressure gas dynamics" is a particular case of (9).

Suppose that $\Gamma = \{\gamma_i : i \in I\}$ is a graph in the upper half-plane $\{(x,t) : x \in \mathbf{R},\ t \in [0,\infty)\} \in \mathbf{R}^2$ containing smooth arcs γ_i, $i \in I$, and I is a finite set. Arcs of Γ have orientation corresponding to increase of time t. By I_0 we denote a subset of I such that an arc γ_k for $k \in I_0$ starts from points of the x-axis. Let $\Gamma_0 = \{x_k^0 : k \in I_0\}$ be the set of initial points of arcs γ_k, $k \in I_0$.

Consider *δ-shock type initial data*

$$(u^0(x), v^0(x)), \quad v^0(x) = \widehat{v}^0(x) + e^0\delta(\Gamma_0), \qquad u^0, \widehat{v}^0 \in L^\infty(\mathbf{R};\mathbf{R}), \tag{10}$$

where $e^0\delta(\Gamma_0) \stackrel{def}{=} \sum_{k\in I_0} e_k^0 \delta(x - x_k^0)$, e_k^0 are constants, $k \in I_0$.

Definition 1. ([DSh05], [DSh06]) A pair of distributions $(u(x,t), v(x,t))$ and a graph Γ, where $v(x,t)$ has the form of the sum

$$v(x,t) = \widehat{v}(x,t) + e(x,t)\delta(\Gamma), \qquad u, \widehat{v} \in L^\infty\big(\mathbf{R}\times(0,\infty);\mathbf{R}\big),$$

$e(x,t)\delta(\Gamma) \stackrel{def}{=} \sum_{i\in I} e_i(x,t)\delta(\gamma_i)$, $e_i(x,t) \in C(\Gamma)$, $i \in I$, and is called a *δ-shock wave type solution* of the Cauchy problem (8), (10) if the integral identities

$$\int_0^\infty \int \Big(u\varphi_t + F(u,\widehat{v})\varphi_x\Big)\,dx\,dt + \int u^0(x)\varphi(x,0)\,dx = 0,$$

$$\begin{aligned} \int_0^\infty \int \Big(\widehat{v}\varphi_t + G(u,\widehat{v})\varphi_x\Big)\,dx\,dt + \sum_{i\in I}\int_{\gamma_i} e_i(x,t)\frac{\partial\varphi(x,t)}{\partial \mathbf{l}}\,dl \\ + \int \widehat{v}^0(x)\varphi(x,0)\,dx + \sum_{k\in I_0} e_k^0\varphi(x_k^0,0) = 0 \end{aligned} \tag{11}$$

hold for all test functions $\varphi(x,t) \in \mathcal{D}(\mathbf{R}\times[0,\infty))$, where $\dfrac{\partial\varphi(x,t)}{\partial \mathbf{l}}$ is the tangential derivative on the Γ, $\int_{\gamma_i} \cdot\,dl$ is the line integral over the arc γ_i.

Suppose that the arcs of the graph $\Gamma = \{\gamma_i : i \in I\}$ have the form $\gamma_i = \{(x,t) : x = \phi_i\}$, $\phi_i(t) \in C^1(0,+\infty)$, $i \in I$. In this case $\mathbf{n} = (\nu_1, \nu_2) =$

$\frac{(1,-\dot{\phi}_i(t))}{\sqrt{1+(\dot{\phi}_i(t))^2}}$ is the unit oriented normal to the curve γ_i, $\mathbf{l}=(-\nu_2,\nu_1)$. Here $\frac{\partial\varphi(x,t)}{\partial\mathbf{l}}\Big|_{\gamma_i}=\frac{\varphi_t(\phi_i(t),t)+\dot{\phi}_i(t)\varphi_x(\phi_i(t),t)}{\sqrt{1+(\dot{\phi}_i(t))^2}}=\frac{\frac{d\varphi(\phi_i(t),t)}{dt}}{\sqrt{1+(\dot{\phi}_i(t))^2}}$.

By using Definition 1 we derive the *δ-shock Rankine–Hugoniot conditions.*

Theorem 1. ([Sh03-1], [Sh04]) *Assume that $\Omega\subset\mathbf{R}\times(0,\ \infty)$ is some region cut by a curve $\Gamma=\{(x,t):x=\phi(t)\}$, $\phi(t)\in C^1(0,+\infty)$ into left- and right-hand parts $\Omega_{\mp}=\{(x,t):\pm(x-\phi(t))>0\}$; (u,v), and Γ is a δ-shock solution of the system* (8), *and (u,v) is smooth in $\Omega_{\pm}$ and have one-sided limits $u_{\pm}$, $\widehat{v}_{\pm}$, on Γ. Then the Rankine–Hugoniot conditions for the δ-shock*

$$\begin{aligned}\dot{\phi}(t)&=\frac{[F(u,v)]}{[u]}\Big|_{x=\phi(t)},\\ \dot{e}(t)&=\Big([G(u,v)]-[v]\frac{[F(u,v)]}{[u]}\Big)\Big|_{x=\phi(t)}\end{aligned}\qquad(12)$$

hold along Γ, where $\big[a(u,v)\big]=a(u_-,v_-)-a(u_+,v_+)$ is a jump of the function $a(u(x,t),v(x,t))$ across the discontinuity curve Γ, $e(t)\stackrel{def}{=}e(\phi(t),t)$.

The first equation in (12) is the *standard* Rankine–Hugoniot condition; the right-hand side of the second equation in (12) is the *Rankine–Hugoniot deficit* in v.

3 The Cauchy Problems

The eigenvalues of the characteristic matrix of system (4) are $\lambda_{\pm}(u)=\frac{1}{2}\big(f'(u)\pm\sqrt{\big(f'(u)\big)^2-4g'(u)}\big)$, $\big(f'(u)\big)^2\ge 4g'(u)$. For system (5) the eigenvalues of the characteristic matrix are $\lambda_-(u)=f'(u)$, $\lambda_+(u)=g(u)$. Let $f''(u)>0$, $g'(u)>0$, $f'(u)\le g(u)$. We assume that the "overcompression" conditions are satisfied.

Theorem 2. ([Sh03]–[Sh04], see also [ASh05]) *Suppose that $\lambda_+(u_+^0(0))\le\frac{[f(u^0)]-[v^0]}{[u^0]}\Big|_{x=0}\le\lambda_-(u_-^0(0))$. Then there exists $T>0$ such that for $t\in[0,\ T)$ the Cauchy problem* (4), (6) *has a unique solution* (7) *that satisfies the integral identities* (11), *where $\Gamma=\{(x,t):x=\phi(t),\ t\in[0,\ T)\}$, and functions $u_{\pm}(x,t)$, $v_{\pm}(x,t)$, $\phi(t)$, $e(t)$ are defined by the system*

$$\begin{aligned}L_{21}[u_{\pm},v_{\pm}]&=0,\quad \pm x>\pm\phi(t),\\ L_{22}[u_{\pm},v_{\pm}]&=0,\quad \pm x>\pm\phi(t),\\ \dot{\phi}(t)&=\frac{[f(u)]-[v]}{[u]}\Big|_{x=\phi(t)},\\ \dot{e}(t)&=\Big([g(u)]-[v]\frac{[f(u)]-[v]}{[u]}\Big)\Big|_{x=\phi(t)},\end{aligned}\qquad(13)$$

with the initial data defined from (6), $\phi(0)=0$.

Theorem 3. ([DSh05], [DSh06]) *Let* $[u^0(0)] > 0$. *Then there exists* $T > 0$ *such that for* $t \in [0, T)$ *the Cauchy problem* (5), (6) *has a unique solution* (7), *which satisfies the integral identities* (11), *where* $\Gamma = \{(x,t) : x = \phi(t),\, t \in [0,\ T)\}$, *and functions* $u_\pm(x,t)$, $v_\pm(x,t)$, $\phi(t)$, $e(t)$ *are defined by the system*

$$\begin{aligned} L_{31}[u_\pm] &= 0, \quad \pm x > \pm\phi(t),\\ L_{32}[u_\pm, v_\pm] &= 0, \quad \pm x > \pm\phi(t),\\ \dot\phi(t) &= \frac{[f(u)]}{[u]}\Big|_{x=\phi(t)},\\ \dot e(t) &= \Big([vg(u)] - [v]\frac{[f(u)]}{[u]}\Big)\Big|_{x=\phi(t)}, \end{aligned} \tag{14}$$

with the initial data defined from (6), $\phi(0) = 0$.

The last two equations in (13) and (14) give the corresponding Rankine–Hugoniot conditions. They are particular cases of (12).

Recall that $O_{\mathcal{D}'}(\varepsilon^\alpha)$, $\varepsilon \to +0$ ($\alpha \in \mathbf{R}$) is a collection of distributions (with respect to x) $f(x,t,\varepsilon) \in \mathcal{D}'(\mathbf{R}_x)$, $x \in \mathbf{R}$, $t \in [0,T]$, $\varepsilon > 0$ such that $\langle f(\cdot,t,\varepsilon), \psi(\cdot)\rangle = O(\varepsilon^\alpha)$, $\varepsilon \to +0$, for any test function $\psi(x) \in \mathcal{D}(\mathbf{R})$, $x \in \mathbf{R}$; $\langle f(\cdot,t,\varepsilon), \psi(\cdot)\rangle$ is a continuous function in t; the estimate $O(\varepsilon^\alpha)$ is understood in the standard sense, being uniform with respect to t. The notation $o_{\mathcal{D}'}(\varepsilon^\alpha)$ is understood in a corresponding way.

According to [DSh03]– [DSh06], a pair of functions $\big(u_\varepsilon(x,t), v_\varepsilon(x,t)\big)$ smooth as $\varepsilon > 0$, $t \in [0,T]$ is called a *weak asymptotic solution* of the Cauchy problem (8), (10) if $L_1[u_\varepsilon, v_\varepsilon] = o_{\mathcal{D}'}(1)$, $L_2[u_\varepsilon, v_\varepsilon] = o_{\mathcal{D}'}(1)$, $u_\varepsilon(x,0) = u^0(x) + o_{\mathcal{D}'}(1)$, $v_\varepsilon(x,0) = v^0(x) + o_{\mathcal{D}'}(1)$, $\varepsilon \to +0$, where the first two estimates are uniform in $t \in [0,T]$. Since within the *vanishing viscosity method* a *viscosity term* admits an estimate of the form $o_{\mathcal{D}'}(1)$, a *viscosity solution* can be considered as a *weak asymptotic solution.*

Within the framework of the *weak asymptotics method* [ASh05], [DSh03]–[DSh06], [Sh03] – [Sh04], in order to prove Theorems 2, 3, first of all we construct a *weak asymptotic solution* $(u_\varepsilon, v_\varepsilon)$ to the corresponding Cauchy problem. Next, we find a δ-shock wave type solution of the Cauchy problem (4), (6) or (5), (6) as the weak limit $u(x,t) = \lim_{\varepsilon\to+0} u_\varepsilon(x,t)$, $v(x,t) = \lim_{\varepsilon\to+0} v_\varepsilon(x,t)$ of the *weak asymptotic solution.* In the end, multiplying the relations $L_1[u_\varepsilon, v_\varepsilon] = o_{\mathcal{D}'}(1)$, $L_2[u_\varepsilon, v_\varepsilon] = o_{\mathcal{D}'}(1)$, by a test function $\varphi(x,t) \in \mathcal{D}(\mathbf{R} \times [0,\infty))$, integrating these relations by parts and then passing to the limit as $\varepsilon \to +0$, we will see that the pair of limit distributions (u,v) of the form (7) satisfies the integral identities (11). Thus we prove that the left-hand sides of the relations $\lim_{\varepsilon\to+0}\int_0^\infty\int L_1[u_\varepsilon, v_\varepsilon]\varphi\,dx\,dt = 0$, $\lim_{\varepsilon\to+0}\int_0^\infty\int L_2[u_\varepsilon, v_\varepsilon]\varphi\,dx\,dt = 0$ coincide with the left-hand sides of the integral identities (11) for all $\varphi(x,t) \in \mathcal{D}(\mathbf{R} \times [0,\infty))$. To prove that systems (13) and (14) are solvable, we use Majda's "free boundary" approach.

4 Flux-Functions of δ-Shocks

Using a *weak asymptotic solution* $(u_\varepsilon, v_\varepsilon)$ to the Cauchy problem (see Sect. 3) one can define flux-functions of δ-shocks, i.e., construct explicit unique formulas for the *"right" singular superpositions*: $F(u,v) \stackrel{def}{=} \lim_{\varepsilon\to+0} F(u_\varepsilon, v_\varepsilon)$, $G(u,v) \stackrel{def}{=} \lim_{\varepsilon\to+0} G(u_\varepsilon, v_\varepsilon)$ (see [ASh05], [DSh06], [Sh03-1], [Sh04]). For the solution (7) of the Cauchy problem (4), (6) we have

$$f\big(u(x,t)\big) - v(x,t) \stackrel{def}{=} \lim_{\varepsilon\to+0} \big(f(u_\varepsilon) - v_\varepsilon\big)$$

$$= f(u_+) - v_+ + \big[f(u) - v\big]H(-x+\phi(t)), \tag{15}$$

$$g\big(u(x,t)\big) \stackrel{def}{=} \lim_{\varepsilon\to+0} \big(g(u_\varepsilon)\big)$$

$$= g(u_+) + \big[g(u)\big]H(-x+\phi(t)) + e(t)\frac{[f(u)]}{[u]}\delta(-x+\phi(t)). \tag{16}$$

For the solution (7) of the Cauchy problem (5), (6) we have

$$f\big(u(x,t)\big) \stackrel{def}{=} \lim_{\varepsilon\to+0} f\big(u_\varepsilon\big) = f(u_+) + \big[f(u)\big]H(-x+\phi(t)), \tag{17}$$

$$v(x,t)g\big(u(x,t)\big) \stackrel{def}{=} \lim_{\varepsilon\to+0} v_\varepsilon g\big(u_\varepsilon)\big) = v_+ g(u_+)$$

$$+\big[vg(u)\big]H(-x+\phi(t)) + e(t)\frac{[f(u)]}{[u]}\delta(-x+\phi(t)). \tag{18}$$

In fact, by (18) we *define* the *unique "right" product* of the Heaviside function and the δ-function in the context of the Cauchy problem (5), (6). In contrast to system (5), formulas (15), (16) *do not define (!)* the *product of the Heaviside function and the δ-function.* Moreover, although (according to (7)) $u(x,t)$ *does not depend (!)* on the term $e(t)\delta(-x+\phi(t))$, the right-hand side of the *"right" singular superposition* (16) *does depend (!)* on this term. Thus one can say that the term $e(t)\delta(-x+\phi(t))$ "appears in (16) from nothing." Analogously, the left-hand side in (15) *depends* on $e(t)\delta(-x+\phi(t))$, while the right-hand side *does not depend* on this term. Nevertheless, in the context of solving the Cauchy problem, a flux-function is determined *uniquely.*

Acknowledgments

The author was supported in part by DFG Projects 436 RUS 113/823 and 436 RUS 113/895, and Grant 05-01-00912 of Russian Foundation for Basic Research.

References

[ASh05] Albeverio, S., Shelkovich, V.M.: On the delta-shock front problem. In: Rozanova, O.S. (ed) Analytical Approaches to Multidimensional Balance Laws, Ch. 2, , Nova Science Publishers, Inc., 45–88 (2005)

[B94] Bouchut, F.: On zero pressure gas dynamics. in: "Advances in Kinetic Theory and Computing", Series on Advances in Mathematics for Applied Sciences, Vol. 22, World Scientific, Singapore, 171–190 (1994)

[DLM95] Dal Maso, G., Le Floch, P.G., Murat, F.: Definition and weak stability of nonconservative products. J. Math. Pures Appl., **74**, 483–548 (1995)

[DSh03] Danilov, V.G., Shelkovich, V.M.: Propagation and interaction of delta-shock waves of a hyperbolic system of conservation laws. In: Hou, T.Y., Tadmor, E. (eds) Hyperbolic Problems: Theory, Numerics, Applications. Proc. 9th Int. Conf. on Hyperbolic Problems held in CalTech, Pasadena, March 25-29, 2002, Springer Verlag, 483–492 (2003)

[DSh05] Danilov, V.G., Shelkovich, V.M.: Delta-shock wave type solution of hyperbolic systems of conservation laws. Quart. Appl. Math., **63**, no. 3, 401–427 (2005)

[DSh06] Danilov, V.G., Shelkovich, V.M.: Dynamics of propagation and interaction of delta-shock waves in conservation law systems. J. Diff. Eqns., **211**, 333–381 (2005)

[ERS96] E., Weinan, Rykov, Yu., Sinai, Ya.G.: Generalized variational principles, global weak solutions and behavior with random initial data for systems of conservation laws arising in adhesion particle dynamics. Comm. Math. Phys., **177**, 349–380 (1996)

[K99] Keyfitz, B.L.: Conservation laws, delta-shocks ans singular shocks. In: Grosser, M., Horman, G., Kunzinger, M., Oberguggenberger, M. (eds) Nonlinear theory of generalized functions. Boca Raton: Chpman and Hall/CRC press, 99–111 (1999)

[KK95] Keyfitz, B.L., Kranzer, H.C.: Spaces of weighted measures for conservation laws with singular shock solutions. J. Diff. Eqns., **118**, 420–451 (1995)

[KSS03] Keyfitz, B.L., Sanders, R., Sever, M.: Lack of hyperbolicity in the two-fluid model for two-phase incompreddible flow. Discrete and continuous dynamical systems–Series B, **3**, no. 4, 541–563 (2003)

[KSZ04] Keyfitz, B.L., Sever, M., Zhang Fu: Viscous singular shock structure for nonohyperbolic two-fluid model. Nonlinearity, **17**, 1731–1747 (2004)

[KK90] Kranzer H.C., Keyfitz, B.L.: A strictly hyperbolic system of conservation laws admitting singular shocks. Nonlinear Evolution Equations That Change Type, IMA Vol. Math. Appl. **27**, Springer-Verlag, 107–125 (1990)

[LW02] Li, J., Warnece, G.: On measure solutions to the zero-pressure gas model and their uniquenness. Mathematica Bohemica, **127**, no. 2, 265–273 (2002)

[PSh06] Panov, E.Yu., Shelkovich, V.M.: δ'-Shock waves as a new type of solutions to systems of conservation laws. J. Diff. Eqns., **228** , 49–86 (2006)

[Sc04] Schecter S.: Existence of Dafermos profiles for singular shocks. J. Diff. Eqns., **205**, 185–210 (2004)

[S02] Sever, M.: Viscous structure of singular shocks. Nonlinearity, **15**, 705–725 (2002)

[S03] Sever, M.: Distribution solutions of nonolinear systems of conservation laws. Preprint Hebrew University, Jerusalem, (2003)

[Sh03] Shelkovich, V.M.: Delta-shock waves of a class of hyperbolic systems of conservation laws. In: Abramian, A., Vakulenko, S., Volpert V. (eds) Patterns and Waves, AkademPrint, St. Petersburg, 155–168 (2003)

[Sh03-1] Shelkovich, V.M.: A specific hyperbolic system of conservation laws admitting delta-shock wave type solutions. Preprint 2003-059 at the url: http://www.math.ntnu.no/conservation/2003/059.html

[Sh04] Shelkovich, V.M.: Delta-shocks, the Rankine–Hugoniot conditions, and singular superposition of distributions. Proc. Int. Seminar Days on Difraction'2004, June 29–July 2, 2004, St.Petersburg, 175–196 (2004)

[Sh06] Shelkovich, V.M.: The Riemann problem admitting δ-, δ'-shocks, and vacuum states (the vanishing viscosity approach). J. Diff. Eqns., **231**, 459–500 (2006)

[TZZ94] Tan, Dechun, Zhang, Tong, Zheng, Yuxi: Delta-shock waves as limits of vanishing viscosity for hyperbolic systems of conservation laws. J. Diff. Eqns., **112**, 1–32 (1994)

Finite Dimensional Representation of Solutions of Viscous Conservation Laws

W. Shen

1 Introduction

We consider the finite dimensional representation of solutions of a scalar conservation law with viscosity

$$u_t + f(u)_x = \varepsilon u_{xx}.$$

We assume that the flux f is smooth and genuinely nonlinear, so that $f''(u) \geq \kappa > 0$ for every u.

Generally speaking, one may ask the following question: Assume that a particular solution $u = u(t, x)$ has already been computed. If we are allowed only a finite number of parameters in order to describe its most relevant features, what is the best way to compress the information? For the problem we are considering, it is natural to focus the attention on the viscous traveling shock profiles. Our main interest is thus how to identify the emergence of viscous shocks in a solution, and how to optimally trace their locations and strengths.

In the literature, the problem of finite dimensional approximation of a dynamical system has been studied mainly by looking at ω-limit sets [9]. For example, several results, valid for evolution equations of parabolic type, provide estimates on the dimension of an attractor. Of course, this yields a bound on the number of parameters needed to describe the evolution of the system asymptotically as $t \to +\infty$.

In our case, however, the focus is different. Here we seek a finite dimensional description that is not only accurate in the asymptotic limit as $t \to +\infty$, but also in the "transient" regime. It is well-known that, for solutions to a scalar viscous conservation law, the ω-limit set is rather trivial if the flux function is convex. In fact, the asymptotic limit of any solution $t \mapsto u(t, \cdot)$ can be described in terms of the solution of a Riemann problem, i.e., either a single rarefaction or a viscous shock wave. Therefore, the transient behavior of the solutions is actually the most interesting feature that can be observed.

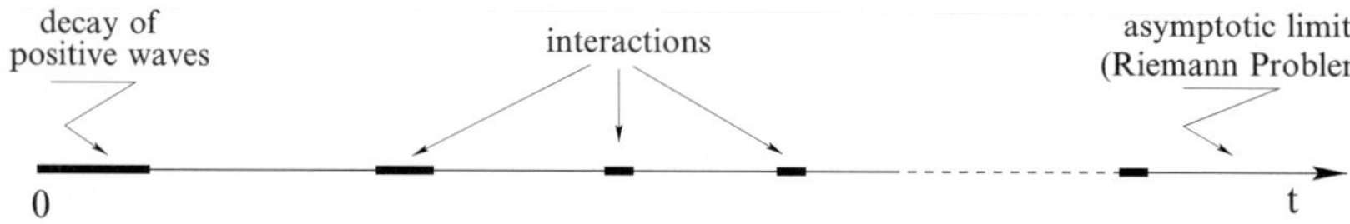

Fig. 1. The exceptional set of times where the finite dimensional representation in terms of viscous traveling shocks is not accurate

The problem of optimal location of viscous shock profiles was mentioned also in [10]. In our work [7], we introduce a scalar functional whose minimizers identify the strengths and locations of viscous shock profiles present in the solution. We also prove that, outside a set of times with finite measure, at all other times our functional has very small values. In other words, the description of the solution profile $u(t,\cdot)$ in terms of finitely many viscous shocks is accurate, for most times t. The exceptional set consists of an initial time interval and times at which shock interactions occur, see Fig. 1.

2 The Main Results

Without loss of generality (by a suitable stretching in the space variable), we consider here the single conservation law with unit viscosity

$$u_t + f(u)_x = u_{xx}. \tag{1}$$

We fix $M > 0$ and let $\mathcal{F}_M$ denote the set of all solutions to the Cauchy problem for (1) with initial data

$$u(0,x) = \bar{u}(x) \tag{2}$$

satisfying

$$\text{Tot.Var.}\{\bar{u}\} \leq M, \qquad \|\bar{u}\|_{\mathbf{L}^\infty} \leq M. \tag{3}$$

We shall assume that the flux f is $\mathcal{C}^2$ and strictly convex, so that $f''(u) > 0$ for all $u \in I\!R$. In particular, this implies that there exist constants κ, κ',

$$0 < \kappa \leq f''(u) \leq \kappa' \qquad \text{for all } u \in [-M, M]. \tag{4}$$

Our main claim is the following. Apart from a small set of times $J \subset [0, \infty[$, the profile $u(t,\cdot)$ of any solution of (1) can be accurately described in terms of the "superposition" of finitely many traveling viscous shocks. Indeed, the assumption (4) of genuine nonlinearity implies that all rarefaction waves (positive waves) will decay within an initial time interval. Meanwhile, in regions where the gradient u_x is large and negative, viscous shock profiles will form. These profiles can travel for a long time without much changing their shape, except when they interact with each other. In fact, in any solution the only features that can survive for arbitrarily long time are viscous shock profiles.

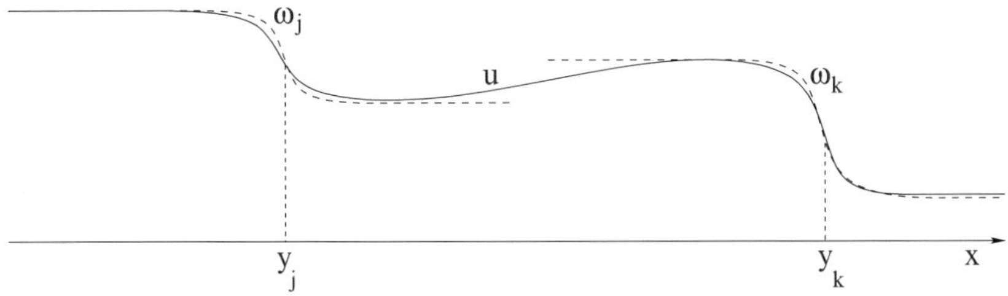

Fig. 2. Fitting two viscous shocks ω_j, ω_k in a solution

The set J of "exceptional times" where our description is not accurate will thus include an initial time interval, and also the intervals where wave interactions occur. Much of the following analysis aims at making rigorous the above claims.

For every $u^- > u^+$ and $y \in I\!R$, let $\omega^{(u^\pm, y)}$ be the unique viscous shock profile joining the states u^-, u^+, centered at y. This profile can be found as the unique solution to the O.D.E (Fig. 2).

$$\omega' = f(\omega) - \sigma\,\omega - \big[f(u^-) - \sigma u^-\big], \qquad \sigma = \frac{f(u^-) - f(u^+)}{u^- - u^+},$$

satisfying the additional conditions

$$\omega''(y) = 0, \qquad \omega(-\infty) = u^-, \qquad \omega(+\infty) = u^+.$$

Given any solution $u \in \mathcal{F}_M$ of the conservation law, for each $t > 0$ we introduce a description based on optimal location of shock profiles. Fix an integer $N \geq 1$ and let $\omega_i = \omega^{(u_i^\pm, y_i)}$ be the ith viscous shock profile we try to fit in. We consider the functional

$$\mathcal{J}\big(u(t),\,\omega_1,\ldots,\omega_N\big) \doteq \sum_{i=1}^{N} \int_{I\!R} \Big|u(t,x) - \omega_i(x)\Big| \cdot \big|\omega_{i,x}(x)\big|^2\,dx + \int_{I\!R} \Big|u_x(t,x) - \sum_{i=1}^{N} \omega_{i,x}(x)\Big|^2\,dx. \tag{5}$$

Notice that the first integral measures the distance between u and the traveling viscous shock ω_i, multiplied by a weight function $|\omega_{i,x}|^2$, which is vanishingly small away from the center of the ith shock. The second integral measures how well the derivative u_x is approximated by derivatives of traveling shock profiles.

If we fix a priori the complexity of our description, i.e., the integer N, our problem can be formulated as a minimization problem

$$\inf_{\omega_1,\ldots,\omega_N} \mathcal{J}\big(u(t),\,\omega_1,\ldots,\omega_N\big), \tag{6}$$

where the infimum is taken over all N-tuples of traveling shock profiles $\omega_i = \omega^{(u_i^{\pm}, y_i)}$, for some states $u_i^- > u_i^+$ and $y_i \in I\!R$. Notice that if we choose $\omega_i \equiv 0$ for $i = 1, \ldots, N$ (i.e., all traveling waves of zero amplitude), then the first integral in (5) vanishes because trivially $\omega_{i,x} \equiv 0$. However, in this case the second integral equals $\left\| u_x(t, \cdot) \right\|_{\mathbf{L}^2}^2$, which is of order $\{\text{Tot.Var.}(u)^3\}$ due to regularization and can be large.

To estimate the quantity in (6), an intuitive argument goes as follows. Set $\delta = M/N$, where M is given in (3). Since the total variation of $u(t)$ is bounded by M, there can be at most N shock profiles of strength $\geq \delta$. Each one of these can be traced accurately. In addition, there may be an arbitrarily large number of smaller shocks, say of strengths σ_j, $j \geq 1$, with

$$\sigma_j \leq \delta, \qquad\qquad \sum_j \sigma_j \leq M. \tag{7}$$

Each shock that is not traced produces an error in the second integral of (5) of the order

$$\int \left| \omega_{j,x}(x) \right|^2 dx = \mathcal{O}(1) \cdot \sigma_j^3 .$$

Because of (7) we thus expect that the minimum of $\mathcal{J}$ is approximately

$$\mathcal{J}_{\min} \approx \mathcal{O}(1) \cdot M\delta^2 = \mathcal{O}(1) \cdot \frac{M^3}{N^2}. \tag{8}$$

The estimate (8) should indeed hold outside an initial time interval, where positive waves will decay, and away from interaction times. Our main results are as follows.

Theorem 1. *Assume $f''(u) \geq \kappa > 0$ for every $u \in I\!R$. Let $u \in \mathcal{F}_M$ be a solution of the viscous conservation law (1), and fix $N \geq 1$. Then, for every $t > 0$, the minimization problem (6) has at least one solution.*

Theorem 2. *There exist constants α (uniformly valid for all $N \geq 1$ and $u \in \mathcal{F}_M$) and $\beta = \beta_{N,M}$ (depending only on N and M) such that*

$$\mathcal{J}_{\min}\big(u(t)\big) \leq \alpha \cdot \frac{1}{N^2} \tag{9}$$

for all $t \in [0, \infty[\setminus I^u$, for an exceptional set I^u of times, with meas$(I^u) \leq \beta$.

Theorem 1 states the existence of a minimizer for the scalar function $\mathcal{J} : R^{3N} \mapsto R$. Since $\mathcal{J}$ is continuous and positive, the result would be trivial if $\mathcal{J}(y) \to \infty$ as $|y| \to \infty$. However, it is easily seen that this coercivity condition fails. The heart of the proof consists in showing that, if $\{X^{(m)}\}_{m\geq 1}$ is a minimizing sequence with $|X^{(m)}| \to \infty$, then a second minimizing sequence $\tilde{X}^{(m)}$ can be defined (in terms of $X^{(m)}$) whose elements remain uniformly bounded. We refer to [7] for details.

The proof of Theorem 2 involves a deeper argument. In next section we will give the sketch for the proof of the qualitative part, i.e., the functional $\mathcal{J}_{\min}$ is small for most of the time outside an initial and interaction time. A complete proof can be found in [7].

3 Sketch of the Proof for Theorem 2

The proof of Theorem 2 relies on some Lyapunov functionals associated with shortening of curves, introduced in [1, 2, 3]. Introducing a new set of variables:

$$v = f(u) - u_x, \qquad \tau = t, \qquad \eta = u. \tag{10}$$

For each fixed time $t > 0$, the solution of (1)–(2) is smooth. The map

$$x \mapsto \gamma^t(x) \doteq \big(u(t,x),\ v(t,x)\big)$$

parameterizes a curve γ^t in the u-v plane. In particular, for a viscous shock profile, this curve is a segment of a straight line, with end-points on the graph of the flux function.

To see how this curve evolves in time in general cases, from (1) one obtains

$$v_t + f'(u)v_x = v_{xx}.$$

On regions where $u_x \neq 0$ we can now use (τ, η) as independent variables, instead of (t, x). After some computations, we get

$$v_\tau = (u_x)^2 v_{\eta\eta} = \big(v - f(\eta)\big)^2 v_{\eta\eta}. \tag{11}$$

In particular, the curve $\gamma = \gamma(\tau, \eta) = (\eta, v(\tau, \eta))$ evolves in the direction of the curvature and its total length is monotone decreasing in time (Fig. 3).

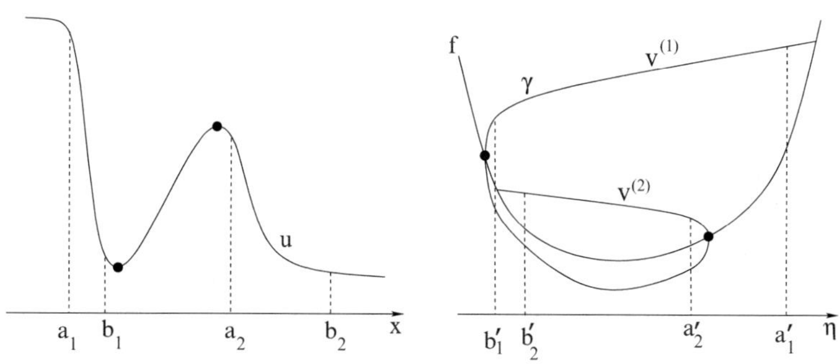

Fig. 3. A solution $u(t, x)$ and the corresponding curve γ in the new coordinate

Another functional that is monotonically decreasing in time is the area functional

$$Q(\gamma) \doteq \frac{1}{2} \iint_{\eta<\tilde{\eta}} \Big|\gamma_\eta(\eta) \wedge \gamma_\eta(\tilde{\eta})\Big| \, d\eta d\tilde{\eta},$$

defined as the double integral of a wedge product. By parabolic regularization estimate, at time $t = 1$ we now have

$$Q\big(u(1)\big) \leq C_1$$

for some constant C_1, uniformly valid for all solutions $u \in \mathcal{F}_M$. Hence

$$\int_1^\infty - \left\{\frac{d}{dt} Q(u(t))\right\} dt \;\leq\; C_1$$

and

$$\left|\frac{d}{dt} Q(\gamma((t))\right|$$

must be very small for most times $t \in [0, \infty[$.

As proved in [2], the decrease of the functional Q controls the area swept by the curve γ in its motion. Therefore,

$$\int |v_t| \;\leq\; \left|\frac{d}{dt} Q(\gamma(t))\right|$$

must also be very small for most times $t \in [0, \infty[$. Recalling (11), this implies that, for most times t, we must have either $v - f(\eta) = 0$ (the curve is on the graph of the flux function) or $v_{\eta\eta} = 0$ (the curve is a straight line). We conclude that the curve γ must be close to a polygonal, with edges near the graph of $v = f(u)$. Going back to the original variables, this in turn implies that the solution $u(t, \cdot)$ must be close to a finite sequence of traveling wave profiles, and therefore the cost functional $\mathcal{J}_{\min}(u(t))$ is small. The only times when our representation is not accurate are the interaction times and an initial time interval.

4 Concluding Remarks

Remark 1. We remark that, in (6), the integer N is fixed. Of course, one could let N vary and look at the minimization problem

$$\min_{N\geq 0} \inf_{\omega_1,\ldots,\omega_N} \Big\{\epsilon N + \mathcal{J}\big(u(t), \omega_1, \ldots, \omega_N\big)\Big\}. \tag{12}$$

Here the first term penalizes the complexity of the description, adding a cost for each new viscous profile. The small constant $\epsilon > 0$ acts as a threshold parameter. Small viscous shock waves, whose strength $\|\omega_x\|^2_{\mathbf{L}^2}$ is of order $< \epsilon$, will not be traced. From Theorem 1 it immediately follows that the problem

(12) also admits a global minimizer. This can be interpreted as an "optimal description" of the solution profile $u(t, \cdot)$ as "superposition" of traveling viscous shocks.

Remark 2. For solutions to the conservation law (1), the transient behavior is nontrivial and can last an arbitrary long time. This happens because we are considering solutions defined on the whole real line. On the other hand, if the equation is restricted to a bounded interval, say

$$u_t + f(u)_x = u_{xx} \qquad\qquad x \in]a, b[, \tag{13}$$

with boundary conditions

$$u(a) = \alpha, \qquad\qquad u(b) = \beta, \tag{14}$$

then all solutions would converge at an exponential rate to a unique steady state $w(\cdot)$.

Indeed, from basic theory of parabolic equations [5] it follows that there exists a unique function $w : [a, b] \mapsto I\!R$, which satisfies the two-point boundary value problem

$$f(w)_x = w_{xx}, \qquad w(a) = \alpha, \qquad w(b) = \beta. \tag{15}$$

Linearizing (13) around the steady state w, one obtains the existence of some $\delta > 0$ such that, for every initial data $\bar{u} \in \mathbf{L}^2$ the corresponding solution of (13)–(14) satisfies

$$\|u(t) - w\|_{\mathcal{C}^k([a,b])} \leq C\,e^{-\delta t}. \tag{16}$$

Here one can choose a constant C uniformly valid on bounded subsets of $\mathbf{L}^2$. After an initial time interval, the long term behavior of the solution is thus described by the exponential limit (16).

References

1. Bianchini, S., Bressan, A.: A case study in vanishing viscosity, *Discrete and Continuous Dynamic Systems*, **7:3** (2001), 449–476.
2. Bianchini, S., Bressan, A.: On a Lyapunov functional relating shortening curves and viscous conservation laws, *Nonlinear Analysis, T.M.A.* **51** (2002), 649–662.
3. Bianchini, S., Bressan, A.: Vanishing viscosity solutions of nonlinear hyperbolic systems, *Annals of Mathematics* **161** (2005), 223–342.
4. Bressan, A.: *Hyperbolic Systems of Conservation Laws. The One Dimensional Cauchy Problem*, Oxford University Press (2000).
5. Henry, D.: *Geometric Theory of Semilinear Parabolic Equations*, Lecture Notes in Mathematics **840**, Springer-Verlag, New York (1981).
6. Serre, D.: *Systems of Conservation Laws I*, Cambridge University Press (2000).

7. Shen, W., Park, M.: Optimal tracing of viscous shocks in solutions of viscous conservation laws, *SIAM J. Math. Anal.* **38:5** (2006/07), 1474–1488.
8. Smoller, J.: *Shock Waves and Reaction-Diffusion Equations*, Springer-Verlag, New York (1983).
9. Temam, R.: *Infinite Dimensional Dynamical Systems in Mathematics and Physics*, Springer-Verlag, New York (1988).
10. Whitham, G.: *Linear and Nonlinear waves*, Wiley-Interscience (1974).

A Moving-Boundary Tracking Algorithm for Inviscid Compressible Flow

K.-M. Shyue

1 Introduction

This chapter is concerned with the development of a Cartesian-grid approach for the numerical simulation of general (single or multicomponent) compressible flow problems with complex moving geometries. As a preliminary, in this work, we are interested in a class of moving objects that undergo solely rigid-body motion with the propagation speeds determined by either a given function of time or the Newton's second law of motion. We use the two-dimensional Euler equations of gas dynamics as a model system in which the principal motion of the single-component flow is governed by

$$\frac{\partial}{\partial t}\begin{pmatrix}\rho\\ \rho u\\ \rho v\\ \rho E\end{pmatrix}+\frac{\partial}{\partial x}\begin{pmatrix}\rho u\\ \rho u^2+p\\ \rho uv\\ \rho Eu+pu\end{pmatrix}+\frac{\partial}{\partial y}\begin{pmatrix}\rho v\\ \rho uv\\ \rho v^2+p\\ \rho Ev+pv\end{pmatrix}=0. \tag{1}$$

Here ρ denotes the density, u and v are the particle velocities in the x- and y-direction, respectively, p is the pressure, and E is the specific total energy. To close the system, a stiffened gas (or called Tammann) equation of state,

$$p(\rho, e) = (\gamma - 1)\,\rho e - \gamma\mathcal{B}, \tag{2}$$

is employed to describe the thermodynamic behavior of the fluid component of interests, where e represents the specific internal energy, γ is the usual ratio of specific heats ($\gamma > 1$), and $\mathcal{B}$ is a prescribed pressure-like constant (cf. [10] and references cited therein). Then, we have $E = e + (u^2 + v^2)/2$ as usual.

Our approach to model the presence of moving objects in the problem formulation takes essentially the same idea as reported in [2, 4, 5, 8, 11] in that an underlying uniform Cartesian grid is used with some rectangles subdivided by the tracked interfaces into two parts, where approximate locations of the moving objects are expected. In each time step, in accordance with the prescribed rule for the moving-object motion, appropriate boundary states are set at the tracked interfaces so that not only the fluid–rigid body interaction, but also the new location of the tracked moving object, can be computed accurately by the method at the end of the time step. See Sect. 3 for more discussion, and [8] for a concise survey of the other approaches.

Rather than using the standard conservative flux-difference or flux-splitting method as in the aforementioned work for moving-boundary problems which can be done but is typically limited to applications governed by conservation laws, we implement the algorithm based on a finite volume method in wave-propagation form, see Sect. 2. One of the advantages of the method is that reasonable time steps can be taken even if some of the subcells created by the tracked interfaces are orders of magnitude smaller than the underlying Cartesian cells (cf. [6, 7]). In addition, generalization of the method from single to multicomponent flows can be made in a straightforward manner when it is used in conjunction with a quasiconservative model system for numerical approximation (cf. [10]). Numerical results presented in Sect. 4 gives some indications of the feasibility of the proposed method for practical applications.

2 Wave-Propagation Methods

We begin by reviewing a finite volume method in wave-propagation form for the numerical approximation of two-dimensional hyperbolic conservation laws,

$$\frac{\partial q}{\partial t} + \frac{\partial f(q)}{\partial x} + \frac{\partial g(q)}{\partial y} = 0,$$

which is one of the fundamental steps in our algorithm devised in Sect. 3. Here $q \in \mathbb{R}^m$ denotes the vector of m conservative quantities, and $f(q)$, $g(q)$ denote the flux vectors, see (1) for an example. We will describe the method on a slightly nonuniform grid denoted by $\mathcal{G}$ that is composed of a set of regular cells with a fixed mesh spacing Δx and Δy in the x- and y-direction, respectively, and some irregular cells which are formed by inserting tracked boundaries into the grid, see Fig. 1a for an illustration.

We use a standard finite volume formulation of the method that the approximate value of the cell average of the solution over any given cell S at time t_n can be written as

$$Q_S^n \approx \frac{1}{\mathcal{M}(S)} \int_S q(x, y, t_n)\, dx\, dy,$$

where $\mathcal{M}(S)$ is the measure (area) of this cell. The methods we use are based on solving one-dimensional Riemann problems at each cell edge, and the resulting waves (discontinuities moving at constant speeds) are employed to update the cell averages in the cells neighboring each edge (cf. [6]).

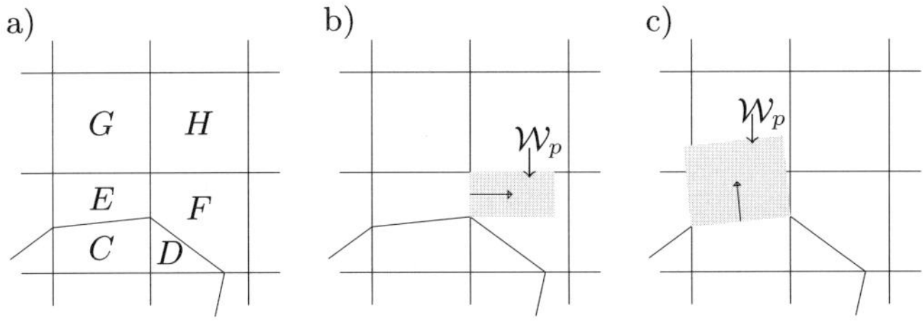

Fig. 1. (**a**) A typical grid created by our algorithm. A wave propagating normally from the cell edge: (**b**) between cells E and F updates the solution in cell F, and (**c**) between cells C and E updates four neighboring cell averages. Here $\mathcal{W}_p$ represents the region swept out by the p-wave over a time step Δt, for $p = 1, 2, \cdots, m$

Consider the cell edge between irregular cells E and F as illustrated in Fig. 1a, for example. We solve the one-dimensional Riemann problem normal to this face, which in this case will be

$$\frac{\partial q}{\partial t} + \frac{\partial f(q)}{\partial x} = 0,$$

with the initial data given by Q_E^n and Q_F^n. If Roe's approximate solver (cf. [9]) is chosen to solve the Riemann problem, the jump $Q_F^n - Q_E^n$ would be decomposed into eigenvectors of A which is a localization of the Jacobian matrix of $f(q)$, $Q_F^n - Q_E^n = \sum_{p=1}^m \alpha_p r_p$, where r_p is the pth eigenvector of A, $Ar_p = \lambda_p r_p$ with λ_p the corresponding eigenvalue, and α_p gives the wave strength.

Figure 1b shows a typical p-wave with $\lambda_p > 0$, where part of cell F is affected by this wave over a time step Δt. In the simplest first-order version of the wave-propagation method, this cell average is updated by

$$Q_S^{n+1} := Q_S^{n+1} - \frac{\mathcal{M}(\mathcal{W}_p \cap S)}{\mathcal{M}(S)} \alpha_p r_p \tag{3}$$

for $S = F$. Here $\mathcal{W}_p$ represents the region swept out by the wave and $\alpha_p r_p$ is the jump across the wave. Note that we have employed a standard initialization procedure, $Q_S^{n+1} := Q_S^n$, for all S.

Now, if we are concerned with the cell edge between cells C and E (this is one segment of the tracked moving object with its normal oriented toward the outside), a Riemann problem would be solved in the direction normal to this face with appropriate chosen data Q_C^n and Q_E^n that takes account of the moving-boundary conditions on this face. Here the p-wave indicated in Fig. 1c overlaps four cells, and using (3) in a similar manner gives modification of the cell average in each of these cells by the jump across this wave, weighted by the fraction of the cell area covered by the wave.

Note that because waves are allowed to propagate through more than one grid cell and are not confined to stay within neighboring cells, this first-order method is typically stable as long as the time step Δt satisfies the CFL (Courant–Friedrichs–Lewy) condition relative to the regular Δx and Δy,

$$\nu = \frac{\Delta t \, \max_{p=1}^{m} |\lambda_p|}{\min(\Delta x, \Delta y)} \le \frac{1}{2}, \tag{4}$$

even when there are very small irregular cells near the tracked interfaces. To improve the stability condition of the method to $\nu \le 1$ as well as the accuracy of the method to high resolution, it is straightforward to apply the techniques developed in [7] to the current instance.

3 Moving-Boundary Tracking Algorithm

In each time step, our moving-boundary tracking method for compressible fluid dynamics problems in two dimensions consists of the following steps:

(1) With a given moving-object velocity at the current time, advance the location of the tracked interfaces on the grid system $\mathcal{G}^n$ in a Lagrangian manner over a time step Δt, see Fig. 2a for an illustration.
(2) Find the new tracked-interface location using the result obtained in step 1, yielding the new grid system $\mathcal{G}^{n+1}$, see Fig. 2b. Some cells will be subdivided, and the values in each new irregular cell must be initialized.
(3) Employ the finite volume wave-propagation method on the grid $\mathcal{G}^{n+1}$ created in step 2 to update the cell averages of our physical model for single or multicomponent compressible flow over Δt.
(4) Determine the new moving-object velocity at the next time step.

Note that steps 1 and 2 constitute the basic surface-moving procedure (cf. [7]) for the propagation of rigid object, step 3 is the construction of approximate solutions to the physical problems of interests on a grid containing moving object, see Sect. 2, and step 4 is simply the calculation of the velocity for the moving object at the next time step.

Here the initialization of the irregular cell values is done by

$$Q_S^n := \sum_{\chi \in \mathcal{G}^n} \frac{\mathcal{M}(\chi \cap S)}{\mathcal{M}(S)} Q_\chi^n \tag{5}$$

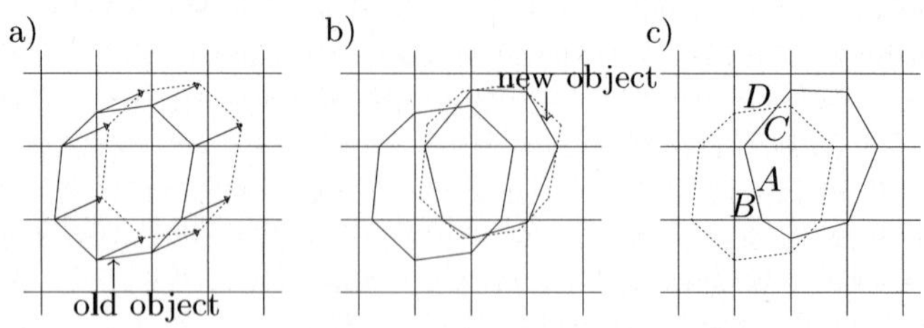

Fig. 2. (**a**) With a given moving-object velocity at the current time, the tracked interfaces are advanced in a Lagrangian manner over a time step Δt. (**b**) Final location of tracked interfaces are set by where the new segments obtained in (**a**) intersect the grid line. (**c**) Sample cases for the initialization of new irregular cells

for all $S \in \mathcal{G}^{n+1}$, when the value Q^n_χ is known a priori on $\mathcal{G}^n$. As in many existing moving-boundary methods (cf. [1, 5, 11]), an extrapolation procedure should be devised to assign state values that are in the interior region of the moving object, and incorporate that in (5) for the evaluation, see the cases for $S = A$, B, C, and D as illustrated in Fig. 2c. In the present work, we take an approach developed by Forrer and Berger [5].

It should be mentioned that in case the motion of the object is induced by the flow field, such as the cylinder liftoff problem studied in Sect. 4, the velocity of the object $(u, v)_b$ at the time step t_{n+1} can be computed based on the Newton's second law of motion by

$$(u, v)_b^{n+1} := (u, v)_b^n + \frac{\Delta t}{M_b}(F_x, F_y)_b^n,$$

where M_b is the mass of the object, and the vector $(F_x, F_y)_b^n$ is the force on the object which can be calculated by a numerical integration of the flow pressure times the normal vector along the surface of the object.

4 Numerical Examples

We now present sample numerical results obtained using our algorithm for compressible flow problems with moving geometries. Unless stated otherwise, we have done all the tests using a high-resolution wave-propagation method with the MINMOD limiter and the Courant number $\nu = 0.9$ as defined in (4). Moreover, the material-dependent parameters in the equation of state (2) are set by $(\gamma,\ \mathcal{B}) = (1.4,\ 0)$ and $(7,\ 3{,}000\,\text{bar})$ for the gas- and liquid-phase, respectively. The boundary conditions of the computational domain are solid walls on the top and bottom, and nonreflecting outflow on the left and right.

Example 4.1. We begin by considering a benchmark test that involves a one-dimensional piston moving at constant velocity in a two-dimensional domain. Initially, inside a shock tube with dimensions $1 \times 0.2\,\text{m}^2$, we have a piston of size: $(x, y) \in [0.4, 0.44] \times [0, 0.2]\,\text{m}^2$, that is about to move vertically from left to right with a constant velocity $(u, v)_b = (0, 300)\,\text{m}\,\text{s}^{-1}$ into a region of quiescent perfect gas with $\rho = 1\,\text{kg}\,\text{m}^{-3}$ and $p = 10^5\,\text{Pa}$. It is known that the exact solution of the problem would consist of a rightward-going shock wave in the front and a leftward-going rarefaction wave in the rear of the moving piston. In Fig. 3, we show the computed solution for the density, velocity in the x-direction, and pressure at time $t = 800\,\mu\text{m}$ along the line $y = 0.1\,\text{m}$, using a 100×20 uniform grid. From the displayed profiles, it is clear that our numerical result is in good agreement with the exact solution.

Example 4.2. Our next example is concerned with the so-called cylinder liftoff problem (cf. [1, 3, 4, 5]) in which the Newton's second law of motion is applied to model the movement of the moving object. In this problem, the initial condition consists of a planar Mach 3 shock wave in air located at $x = 0.08\,\text{m}$ traveling from left to right (the preshock state is at rest with

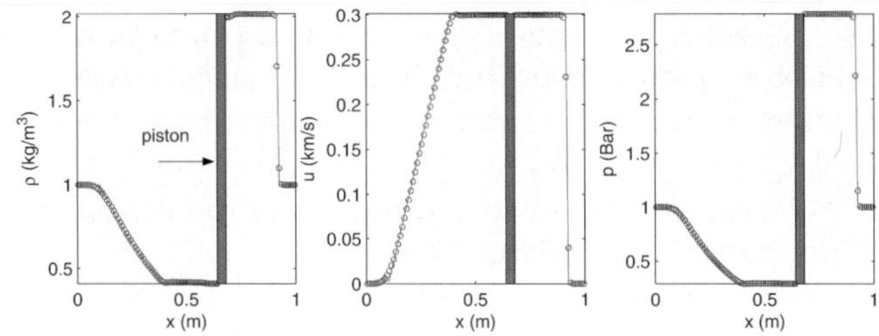

Fig. 3. Cross-sectional plots of the density, x-component velocity, and pressure for the one-dimensional moving piston problem at time $t = 800\,\mu s$ along the line $y = 0.1$ m. The *solid line* is the exact solution

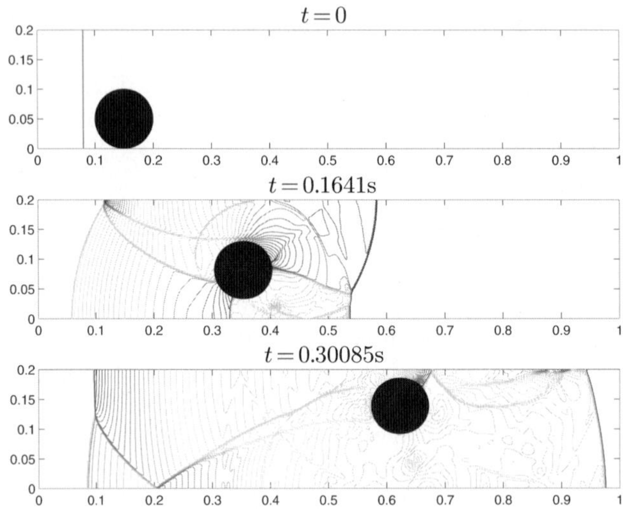

Fig. 4. Contour plots of the pressure for the cylinder liftoff problem at three different times $t = 0$, 0.1641, and 0.30085 s by using a 1,000 × 200 grid

density $1.4\,\mathrm{kg\,m^{-3}}$ and pressure 1 Pa), and a stationary rigid circular cylinder (which has density $10.77\,\mathrm{kg\,m^{-3}}$) with center $(x_0, y_0) = (0.15, 0.05)$ m and of radius $r_0 = 0.05$ m in front of the shock wave.

Figure 4 shows the contours of pressure at three different times, $t = 0$, 0.1641, and 0.33085 s obtained using our method with a 1,000 × 200 grid. From the figure, it is easy to notice the asymmetric reflection of the incident shock wave by the cylinder at the bottom side. When the time proceeds, the reflected pressure would be higher near the bottom, causing the liftoff effect to the cylinder. In view of the global structure of the solution, our results agree quite well with the results displayed in [1]. In Table 1, we present a convergence study of the center of the cylinder and the relative mass loss at the final stopping time $t = 0.30085$ s, observing a linear convergence on the latter quantity, when the mesh is refined. It should be noted that when comparing this results with the one appeared in [1], we see a slight difference

Table 1. A convergence study of the center of cylinder and the relative mass loss for the cylinder liftoff problem at the final stopping time $t = 0.30085$ s

Mesh size	Center of cylinder (m)	Relative mass loss
250×50	$(0.618181, 0.134456)$	-0.257528
500×100	$(0.620266, 0.136807)$	-0.131474
1000×200	$(0.623075, 0.138929)$	-0.066984

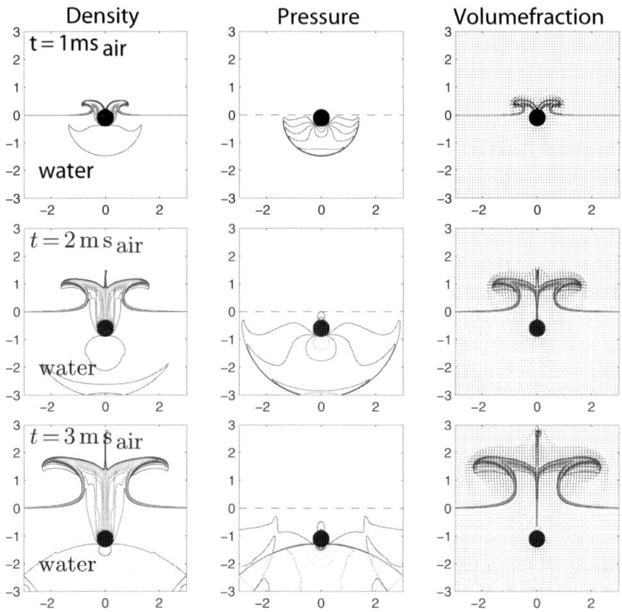

Fig. 5. Contours of the density, pressure, and volume fraction for a flying projectile into water problem at three different times $t = 1$, 2, and 3 ms. The *dashed line* is the initial location of the air–water interface, and the *arrows* are the computed velocities

on the center of the cylinder. More work should be done to clear up the discrepancy.

Example 4.3. Finally, as an example to show how our algorithm performs for multicomponent problems with moving geometries, we are interested in the flying of a supersonic projectile into a partially filled water container. Here we take a projectile with a circular shape of radius 0.3 m, and consider a constant velocity motion $(u, v)_b = (0, -500)\,\mathrm{m\,s^{-1}}$ of the projectile where its initial location is at $(x_0, y_0) = (0, 0.4)$ m in a square domain with dimensions $6 \times 6\,\mathrm{m}^2$. Above the horizontal air–water interface at $y = 0$, the fluid is a motionless air with the standard atmospheric condition, while below the interface the fluid is water at rest with density 1,000 $\mathrm{kg\,m^{-3}}$ and pressure 1 bar. The boundary conditions on the computational domain are solid walls on the left, right, and bottom, and nonreflecting outflow on the top.

To solve this two-fluid problem numerically, our moving-boundary tracking algorithm uses a mixture type of the model equations proposed in [10] as the basis for approximation. Results of a sample run obtained using the algorithm with an uniform 600×600 grid are shown in Fig. 5, where contours of the density, pressure, and volume fraction are shown at three different times $t = 1$, 2, and 3 ms. From the figure, it is interesting to see the time evolution of the air–water interface, upon the projectile entering the body of water. In the future, we plan to validate this result by making comparison with either the laboratory experiments or the ones obtained using other numerical methods.

Acknowledgments

This work was supported in part by National Science Council of Republic of China Grants NSC-93-2115-M-002-010 and 94-2115-M-002-016.

References

1. M. Arienti, P. Hung, E. Morano, and J.E. Shepherd. A level-set approach to Eulerian-Lagrangian coupling. *J. Comput. Phys.*, 185:213–251, 2003.
2. S.A. Bayyuk, K.G. Powell, and B. van Leer. An algorithm for simulation of flows with moving boundaries and fluid-structure interactions. *AIAA paper CP-97-1771*, 1997.
3. M.-H. Chung. A level set approach for computing solutions to inviscid compressible flow with moving solid boundary using fixed Cartesian grids. *Int. J. Numer. Meth. Fluids*, 36:373–389, 2001.
4. J. Falcovitz, G. Alfandary, and G. Hanoch. A two-dimensional conservation laws scheme for compressible flows with moving boundaries. *J. Comput. Phys.*, 138:83–102, 1997.
5. H. Forrer and M. Berger. Flow simulations on Cartesian grids involving complex moving geometries. In *Hyperbolic Problems: Theory, Numerics, Applications*, pages 315–324. Birkhäuser-Verlag, 1999. Intl. Ser. of Numer. Math., Vol. 130.
6. R.J. LeVeque. High resolution finite volume methods on arbitrary grids via wave propagation. *J. Comput. Phys.*, 78:36–63, 1988.
7. R.J. LeVeque and K.-M. Shyue. Two-dimensional front tracking based on high resolution wave propagation methods. *J. Comput. Phys.*, 123:354–368, 1996.
8. S.M. Murman, M.J. Aftosmis, and J. Berger. Implicit approaches for moving boundaries in a 3-D Cartesian method. *AIAA-2003-1119.*
9. P.L. Roe. Approximate Riemann solvers, parameter vector, and difference scheme. *J. Comput. Phys.*, 43:357–372, 1981.
10. K.-M. Shyue. An efficient shock-capturing algorithm for compressible multicomponent problems. *J. Comput. Phys.*, 142:208–242, 1998.
11. G. Yang, D.M. Causon, D.M. Ingram, R. Saunders, and P. Batten. A Cartesian cut cell method for compressible flows- Part B: moving body problems. *Aeronaut. J*, 101:57–65, 1997.

Transparent Boundary Conditions for the Elastic Waves in Anisotropic Media

I.L. Sofronov and N.A. Zaitsev

Summary. A numerical method of generating nonreflecting boundary conditions on open circular boundary for anisotropic media is proposed. The approach is based on the conception of transparent boundary conditions. A two-dimensional wave propagation problem in polar coordinates for orthotropic elastic media is considered to demonstrate transparency and efficiency of the proposed boundary conditions. It is shown that the method gives accurate and stable boundary conditions including cases when PML fails.

1 Introduction

We propose a numerical approach to construct nonreflecting boundary conditions (NRBC) on a circular boundary for two-dimensional orthotropic elastodynamics. The method is a result of elaborating the conception of *transparent boundary conditions* (TBC) toward generalization for anisotropy and inhomogeneities in governing equations.

Anisotropic media is a source of challenges in trying to generate NRBC. Indeed, considering TBC and PML – two mostly advanced approaches based on physically reasonable principles of *transparency* and *absorption* – one can see: On the one hand, operators of known TBC, e.g., [1, 2], use *closed analytical representations* of Green's functions, which are mainly available for *isotropic* governing equations only. On the other hand, the popular PML approach can fundamentally fail because of instability of solutions of auxiliary equations in correspondent "absorbing" layers, see [3, 4].

However, unlike PML, the TBC conception does not change original governing equations; therefore it has no physical restrictions to construct desired NRBC. The aforementioned drawback with Green's functions is overcome numerically, cf. idea of *discrete TBC* in [5].

Our approach consists of two main parts. First, after the Laplace transform of governing equations with respect to time, we *calculate Dirichlet-to-Neumann maps* of auxiliary well-posed external elliptic problems parameterized by the dual variable s. Using Fourier basis functions on the circle to represent functions and normal derivatives, we set down the maps by a matrix in the Fourier space with entries varying with s. In the second part of the algorithm, each entry of the matrix *is approximated by rational functions* with respect to s in order to represent time-dependent component of our TBC operator in terms of sums-of-exponentials after the inverse Laplace transform.

The method extends [1, 6] toward *numerical* evaluation of required Green's functions and *mutual influence* of azimuth harmonics. It has been announced in [7]. Version of *discrete* TBC for anisotropy is considered in [8].

2 Governing Equations

We consider two-dimensional equations of elastodynamics in polar system of coordinate for orthotropic anisotropy:

$$\frac{\partial^2 f}{\partial t^2} = A^{11}\frac{\partial^2 f}{\partial r^2} + A^{22}\frac{\partial^2 f}{\partial \theta^2} + A^{12}\frac{\partial^2 f}{\partial r \partial \theta} + A^1\frac{\partial f}{\partial r} + A^2\frac{\partial f}{\partial \theta} + A^0 f; \quad (1)$$

here $f = (v_r, v_\theta)$ is vector of the radial and azimuth velocities; the 2×2 matrices A^{ij}, A^i of variable coefficients are obtained after the Cartesian-to-polar coordinate transformation of the original system with constant elasticity coefficients,

$$\frac{\partial^2 v}{\partial t^2} = \begin{pmatrix} c^{11} & 0 \\ 0 & c^{33} \end{pmatrix}\frac{\partial^2 v}{\partial x^2} + \begin{pmatrix} 0 & c^{33}+c^{12} \\ c^{33}+c^{12} & 0 \end{pmatrix}\frac{\partial^2 v}{\partial x \partial y} + \begin{pmatrix} c^{33} & 0 \\ 0 & c^{22} \end{pmatrix}\frac{\partial^2 v}{\partial y^2}, \quad (2)$$

written for Cartesian velocities $v = (v_x, v_y)$.

3 Derivation of the TBC Operator

The procedure of generating TBC operator on the boundary Γ of a circle C is described by several main stages. Let us use (1) written in the operator form:

$$\frac{\partial^2 f}{\partial t^2} - Lf = 0. \quad (3)$$

Stage 1. Denote by $\{\varphi^m(\theta)\}_{m=0}^{\infty}$ the circular basis functions (constructed from $\{\sin m\theta, \cos m\theta\}$) and consider the set of auxiliary external initial boundary-valued problems outside C, $m = 0, 1, ...$:

$$\begin{cases} \mathcal{E}^m_{tt} - L\mathcal{E}^m = 0 & \text{in } R^2/C \\ \mathcal{E}^m|_{t=0} = 0 & \\ \mathcal{E}^m|_\Gamma = \delta(t)\varphi^m(\theta), & \end{cases}$$

where $\delta(t)$ is Dirac's delta function. (Function $\mathcal{E}^m$ is usually called *Green's function of the Dirichlet problem.*)

Making Laplace transform we pass to elliptic boundary-valued problems parameterized by s:

$$\begin{cases} s^2\hat{\mathcal{E}}^m - L\hat{\mathcal{E}}^m &= 0 \quad \text{in } R^2/C \\ \hat{\mathcal{E}}^m|_{t=0} &= 0 \\ \hat{\mathcal{E}}^m|_{\Gamma} &= \varphi^m(\theta). \end{cases}$$

Stage 2. We take a finite interval $[0, S_{\max}]$ and form set of points $\{s_j\}_{j=1}^J \in [0, S_{\max}]$, J is an integer, using distribution law of Chebyshev's grid nodes. Then we solve numerically $J \times (M+1)$ problems (4) – for each $s = s_j$ with a desired amount $M+1$ of azimuth harmonics ($m = 0, 1, ..., M$), and evaluate $\psi^m \equiv \frac{\partial}{\partial n}\hat{\mathcal{E}}^m$ on Γ. Thus the Neumann data ψ of basis functions are obtained as function of two indices (j, m):

$$\{\varphi^m(\theta), s_j\} \mapsto \{\psi^m(\theta, s_j)\}.$$

Taking now arbitrary Dirichlet data on Γ, $\hat{f}(s,\theta) = \sum_m \hat{c}^m(s)\varphi^m(\theta)$, we write the representation of correspondent Neumann data $N[\hat{f}]$ by using the expansion in terms of the basis $\{\varphi^m(\theta)\}$:

$$N[\hat{f}(s,\theta)] = \sum_m \hat{c}^m(s)\psi^m(\theta, s) = \sum_m \hat{c}^m(s) \sum_n P_n^m(s)\varphi^n(\theta).$$

Thus we obtain the Poincaré–Steklov operator on the Fourier coefficients space:

$$\hat{d}^n(s) = \sum_n \hat{P}_n^m(s)\hat{c}^m(s)$$

or, in matrix form:

$$\hat{\mathbf{d}}(s) = \hat{\mathbf{P}}(s)\hat{\mathbf{c}}(s), \tag{4}$$

$$\hat{\mathbf{c}} \equiv \{\hat{c}^0,\ \hat{c}^1,\ ...\}^{\mathrm{T}}, \quad \hat{\mathbf{d}} \equiv \left\{\hat{d}^0,\ \hat{d}^1,\ ...\right\}^{\mathrm{T}}, \quad \frac{\partial}{\partial n}\hat{f}(s,\theta) \equiv \sum_m \hat{d}^m(s)\varphi^m(\theta, s).$$

Stage 3. We make inverse Laplace transform for the Poincaré–Steklov operator (4) as follows. First, the matrix $\hat{\mathbf{P}}(s)$ is represented by sum of three matrices to take into account asymptotics as $s \to \infty$:

$$\hat{\mathbf{P}}(s) = \mathbf{P}_1 s + \mathbf{P}_0 + \hat{\mathbf{K}}(s); \quad \mathbf{P}_1, \mathbf{P}_0 \text{ are } \mathit{consts}, \quad \hat{\mathbf{K}}(s) = o(1).$$

These constant matrices $\mathbf{P}_1$, $\mathbf{P}_0$ are estimated from rational approximations

$$R_n^m(s_j) \approx \hat{P}_n^m(s_j)$$

that are evaluated by using Chebyshev–Padé algorithm on the interval $[0, S_{\max}]$.

Then we calculate rational approximations of each entry in $\hat{\mathbf{K}}(s)$ so that all poles have negative real parts, i.e.,

$$\hat{K}_n^m(s) \approx \hat{\tilde{K}}_n^m(s) \equiv \sum_{l=1}^{L_n^m} \frac{\alpha_{n\ l}^m}{s - \beta_{n\ l}^m}, \quad \mathrm{Re}(\beta_{n\ l}^m) < 0,$$

where L_n^m is a number of poles (it exerts significant influence on accuracy of the approximation). Again the Chebyshev–Padé algorithm on the interval $[0, S_{\max}]$ is the key tool for the approximations.

Finally, the inverse Laplace transform of (4) gives:

$$\mathbf{d}(t) \approx \mathbf{P}_1 \frac{\partial \mathbf{c}(t)}{\partial t} + \mathbf{P}_0 \mathbf{c}(t) + \tilde{\mathbf{K}}(t) * \mathbf{c}(t). \tag{5}$$

Important property of (5) is that kernels $\tilde{K}_n^m(t)$ having the explicit form

$$\tilde{K}_n^m(t) = \sum_{l=1}^{L_n^m} \alpha_{n\ l}^m \exp(\beta_{n\ l}^m t), \qquad \mathrm{Re}(\beta_{n\ l}^m) < 0$$

permit to treat convolutions by *stable recurrent* formulas.

Stage 4. Denote by $\mathbf{Q}$, $\mathbf{Q}^{-1}$ operators of Fourier expansion with respect to basis $\{\varphi^m(\theta)\}_{m=0}^M$, i.e.,

$$\mathbf{Q} : f(t,\theta) \to \{c^m(t)\}, \qquad \mathbf{Q}^{-1} : \{c^m(t)\} \to f.$$

Ignoring an approximate character of the relationship (5) we are in position to write out the desired operator of TBC in the physical space:

$$\mathbf{Q}^{-1}\mathbf{P}_1\mathbf{Q}\frac{\partial f}{\partial t} - \frac{\partial f}{\partial r} + \mathbf{Q}^{-1}\mathbf{P}_0\mathbf{Q}f + \mathbf{Q}^{-1}\left\{\tilde{\mathbf{K}}(t)*\right\}\mathbf{Q}f = 0. \tag{6}$$

Actually, the scheme of generating (6) looks very similar to [1], where analytical derivation of TBC operator for the wave equation is proposed; the discrimination is that all *analytical* stages are now replaced by *numerical* counterparts (Stages 2,3). It is worth to emphasize that the numerical stages require highly accurate calculations, in particular we use pseudospectral methods while solving (4).

Notice that for the isotropic case of (1) the matrices $\mathbf{P}_0$, $\mathbf{P}_1$, and $\tilde{\mathbf{K}}(t)$ are diagonal and we can obtain $\hat{\mathbf{P}}(s)$ both analytically – in terms of Bessel functions, and numerically – current approach.

4 Numerical Tests

We consider test tasks in an annulus $\Omega : R_0 \leq r \leq R_\Gamma$. At the internal boundary R_0 we prescribe Dirichlet boundary conditions to initiate elastic waves (initial data of solution in the annulus are zero); at the external boundary R_Γ

we put our TBC (6). Equation (1) are discretized by an explicit second-order central-difference scheme on the uniform polar grid with N_r and N_θ intervals. Difference counterpart of (6) is approximated on R_Γ also with the second order of accuracy, the convolution integrals are evaluated by recurrence formulas with respect to time due to the sum-of-exponentials convolution kernels $\tilde{\mathbf{K}}(t)$.

Parameters of grids: $R_0 = 2$, $R_\Gamma = 10$; Grid1: $N_r = 50$, $N_\theta = 100$; Grid2: $N_r = 150$, $N_\theta = 300$. Dirichlet data at R_0 generate time pulse on the interval $0 < t < 2$ for sum of first three functions $\varphi^m(\theta)$, $m = 0, 1, 2$, (notice that because of anisotropy all other azimuth harmonics are present in solutions as well).

Reference solutions U_{ref} are calculated on the extended domain with $R_{ext} = 8R_\Gamma$ using very fine grid $N_r = 4,950$, $N_\theta = 900$.

In order to check the accuracy of difference solution itself on the domain Ω, we introduce U_{ext}, solutions calculated on the extended domain $R_0 < r < R_{ext}$ with the same grid spacings as in Grid1 and Grid2. Solutions on Ω with our TBC are denoted U_{TBC}. In the described calculations, we use TBC matrices (6) with the basis $\{\varphi^m(\theta)\}_{m=0}^{36}$ for both Grid1 and Grid2.

The extended domain permits to observe unperturbed solutions in Ω until time $T = 25$, when reflections from the boundary $R_{ext} = 8R_\Gamma$ come back to R_Γ.

We test all four anisotropic cases introduced in [3]: case I ($c^{11} = 4,\ c^{12} = 3.8$); case II ($c^{11} = 20,\ c^{12} = 3.8$); case III ($c^{11} = 4,\ c^{12} = 4.9$); and case IV ($c^{11} = 4,\ c^{12} = 7.5$). Other coefficients $c^{22} = 20,\ c^{33} = 2,\ c^{13} = c^{23} = 0$.

Graphs with respect to time for the relative residuals in L_1-norm

$$\delta U_{appr} = \frac{\|U_{ext} - U_{ref}\|}{\max\limits_t \|U_{ref}\|}, \quad \delta U_{TBC} = \frac{\|U_{TBC} - U_{ref}\|}{\max\limits_t \|U_{ref}\|}, \quad \|U\| = \frac{1}{N_r N_\theta} \sum_{\Omega} |U|$$

are given in Figs. 1 and 2.

Because of the second-order central differences, the residual δU_{appr} caused by the difference scheme must decrease about nine times for Grid2 comparing Grid1; we observe this behavior for the plots marked by symbols.

The second residual δU_{TBC} contains two errors: The approximation one δU_{appr} and the error of reflections generated by the TBC. Evidently, NRBC are good enough if

$$\max_t \|\delta U_{TBC}\| \approx \max_t \|\delta U_{appr}\| \,. \tag{7}$$

Let us look at Fig. 1, left. We see that the approximation residual δU_{appr} (circle and cross symbols) has a maximum at $1 < t < 8$ for both grids. Then the error leaves the domain Ω and its amplitude drops. The residual of solutions with TBC (dashed and solid lines) follows the approximation residual until $t \approx 8$, and then remains almost at the same level (because of spurious reflections for high-frequency azimuth harmonics with $m > M = 36$ produced by TBC). The most important thing here is that for both Grid1 and Grid2 the

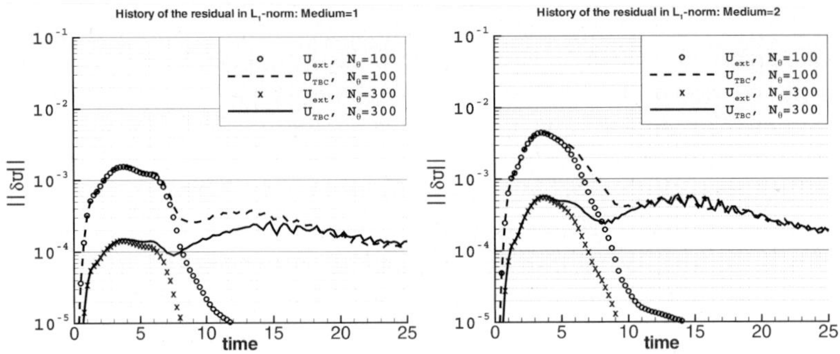

Fig. 1. Residuals δU_{appr} (*symbols*) and δU_{TBC} (*lines*) for case1 (*at the left*) and case2 (*at the right*)

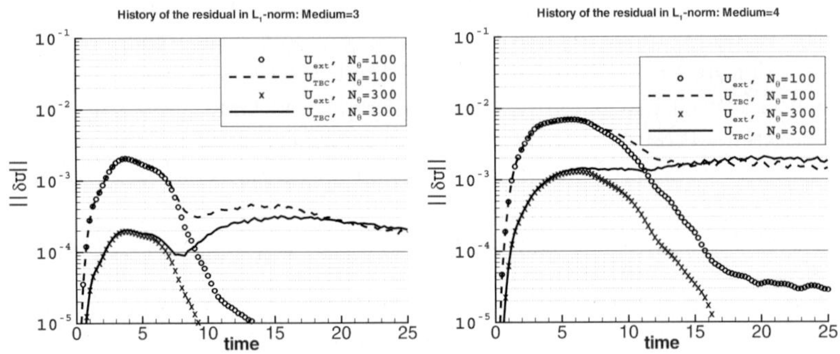

Fig. 2. Residuals δU_{appr} (*symbols*) and δU_{TBC} (*lines*) for case3 (*at the left*) and case4 (*at the right*)

residual δU_{TBC} is not greater then the maximum of the residual of solution U_{TBC} (except for interval $10 < t < 25$ for Grid2, where a little elevation is observed), cf. criterium (7). This means that our TBC have enough accuracy for the considered test calculations.

The similar behavior is observed for all other cases, Figs. 1 and 2.

In order to check long-time stability we run calculations until $T = 100$ on Grid2, see Fig. 3. The residual for difference $U_{TBC} - U_{ref}$ until $T = 25$ continued by U_{TBC} at $T > 25$ (where $||U_{ref}|| \ll ||U_{TBC}||$) is shown. All cases are stable.

Calculation of matrices in the operator (6) can require several hours on a single 3 MHz processor. First of all because of massive solutions of elliptic problems (4). However, exploration of already precalculated TBC is quite fast, e.g., while running aforementioned test problems, the difference counterpart of (6) with $M = 36$ consumes not greater than 30% of the total computations per time step.

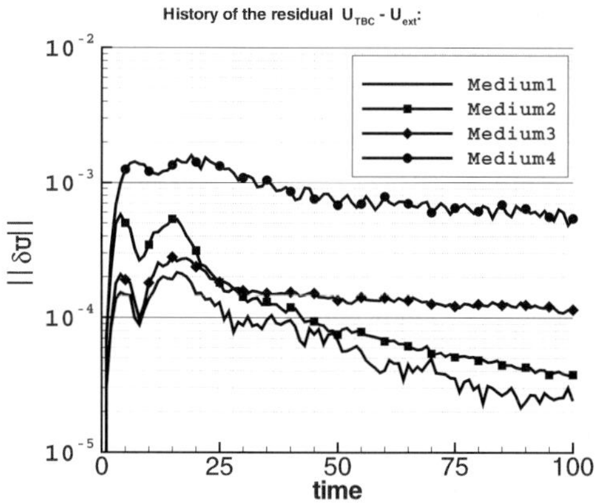

Fig. 3. Long-time stability of test calculations, Grid2

5 Conclusions

- Novel approach of generating NRBC for anisotropic elastodynamics is proposed and implemented
- It does work in all tested anisotropic media, even where PML approach fails
- Similarly to analytical TBC, our operator couples functions and derivatives on the boundary directly. I.e., the approach is independent on the mesh inside the computational domain; this is advantage comparing the discrete version [8]
- The algorithm is highly parallelizable due to massive calculations of independent elliptic problems (4) defined by $\{s_j, \varphi^m\}$
- One can generate *a library* of operators (6) in advance by parameterizing, e.g., with respect to media coefficients $\{c^{11} \div c^{33}\}$, radii R_Γ of the computational domain, and numbers M of treated azimuth harmonics

Acknowledgments

The work is supported by RFBR grant No. 07-01-00476 and CRDF grant No. RUM1-1605-MO-05. The authors are grateful to Alexander N. Daryin and Olga I. Voskoboinikova for the help in algorithmic implementation.

References

1. Sofronov, I.L.: Artificial boundary conditions of absolute transparency for two- and three-dimensional external time-dependent scattering problems. Euro. J. Appl. Math., **9**, No.6, 561–588 (1998)

2. Hagstrom, T.: Radiation boundary conditions for the numerical simulation of waves., Acta Numerica, Cambridge: Cambridge University Press, **8**, 47–106 (1999)
3. Becache, E., Fauqueux, S., Joly, P.: Stability of perfectly matched layers group velocities and anisotropic waves, JCP, **188**, 399–433 (2003)
4. Appelo, D., Kreiss, G.: A new absorbing layer for elastic waves, JCP, **215**, No. 2., 642–660 (2006)
5. Ryaben'kii, V.S.: Exact transfer of difference boundary conditions, Functional Anal. Appl., **24**, 251–253 (1990)
6. Sofronov, I.L.: Conditions for complete transparency on the sphere for the three-dimensional wave equation, Russian Acad. Sci. Dokl. Math. **46**, 397–401 (1993)
7. Sofronov, I.L., Zaitsev, N.A.: Non-reflecting boundary conditions for 2D anisotropic elastodynamics. In: PAMM, Special Issue: GAMM Annual Meeting 2006 – Berlin, **6**, Iss. 1, 611–612 (2007)
8. Sofronov, I.L., Podgornova, O.V.: A spectral approach for generating non-local boundary conditions for external wave problems in anisotropic media. J. Sci. Comp., **27**, No. 1–3, 419–430 (2006)

Counterflow Combustion in a Porous Medium

A.J. de Souza

Summary. We determine the Riemann solution for a model of counterflow combustion due to injection of air into a porous medium containing solid fuel. Assuming that the combustion front is modeled by a traveling wave, we obtain a family of Riemann solutions for the same initial and injection conditions.

1 Introduction

Air injection and in situ combustion have long been considered as potential techniques for displacement and recovery of heavy oil, [3, 8]. An operational advantage of these thermal recovery techniques is the abundance of air independently of location. They utilizes heavy and immobile components of the crude oil as fuel producing in place the heat necessary for the recovery of upgraded crude oil.

A large number of studies on the structure of the combustion front have been reported since the 1950s, see [1, 2, 4, 10], for instance. Despite the advantages and a long history, only a small fraction of the total thermal recovery utilizes this technique. Some reasons are technical, such as the possibility of front extinction and the necessity of (re-)ignition for sustained propagation within in situ combustion in the presence of external heat losses, [1]. Thus mathematical analysis of this problem is important to predict these events.

We consider a porous rock cylinder with a homogeneously distributed solid fuel, initially filled with air that is injected at constant rate on the left end of the cylinder. A bimolecular reaction is assumed to take place between the injected oxygen and the solid fuel, hence the region of reaction behaves as a source of heat as well as a sink for the oxygen and the fuel. We neglect air compressibility and heat losses. In a previous work for the same model, [5], we considered the coflow combustion problem, where ignition is started on the left end of the cylinder and the combustion front propagates in the same direction of the injected gas. In that work we discussed the possible wave sequences in the Riemann solution for various combustion zone conditions. This work is complementary to that one in the sense that here we consider the counterflow combustion case, in which ignition is started at time zero on the right end of the cylinder and the combustion front propagates toward the incoming oxygen with negative speed, [9]. For prescribed injection and initial conditions we obtain a one parameter family of Riemann solutions rather than a unique solution.

2 The Nondimensionalized Equations

Following the same approach as in [1, 5], we assume that air is injected at the left end of a porous rock cylinder containing solid fuel, so that all wave propagation is one dimensional. Balance equations are written for the total energy, the total gas mass, the oxygen mass, and the fuel mass. For the latter, the fuel density per total volume ρ_f is introduced and the extent of conversion depth, $\eta(\tilde{x},\tilde{t}) = 1-\rho_f(\tilde{x},\tilde{t})/\rho_f^o$, where ρ_f^o is the initial fuel concentration, such that $\eta = 0$ corresponds to complete availability and $\eta = 1$ to the complete lack of fuel.

In the formulation of the conservation equations the following assumptions are made: the pore space and the solid matrix are in thermal equilibrium so that a one-temperature model is used for the energy balance; heat transfer by radiation, energy source terms due to pressure increase, work from surface and body forces, and heat losses are all negligible; the ideal gas law is the equation of state for the gas phase; thermodynamic and transport properties, such as conductivity, diffusivity, heat capacity of the solid, heat of reaction, etc., all remain constant. It is assumed that gas heat capacity is negligible with respect to the rock heat capacity. It is assumed also that pressure changes within the system are negligible compared to the average pressure of the system, so that in the ideal gas law and in other physical properties the pressure appears as a constant. Under these assumptions, introducing dimensionless space and time variables x and t, respectively, and dimensionless temperature θ, gas density ρ, oxygen mass fraction in the gas phase Y, and velocity v, we arrive at the following nondimensionalized PDE system (1)–(4) plus (5) in the domain $\Omega = \{\theta > 0,\ 0 \le Y \le 1,\ 0 \le \eta \le 1,\ v > 0\}$:

$$\frac{\partial \theta}{\partial t} + \frac{\partial(a\rho v\theta)}{\partial x} = \frac{\partial^2 \theta}{\partial x^2} + q\Phi(\theta, Y, \eta), \tag{1}$$

$$\frac{\partial(\phi Y\rho)}{\partial t} + \frac{\partial(\rho v Y)}{\partial x} = \frac{1}{L_e}\frac{\partial}{\partial x}\left(\rho\frac{\partial Y}{\partial x}\right) - \mu\Phi(\theta, Y, \eta), \tag{2}$$

$$\phi\frac{\partial \rho}{\partial t} + \frac{\partial(\rho v)}{\partial x} = \mu_g\Phi(\theta, Y, \eta), \tag{3}$$

$$\frac{\partial \eta}{\partial t} = \Phi(\theta, Y, \eta), \tag{4}$$

$$\rho\theta = 1, \tag{5}$$

where

$$\Phi(\theta, Y, \eta) = \alpha e^{-\gamma/(\theta-1)} Y(1-\eta), \quad \text{for } \theta > 1 \text{ and } \Phi = 0 \text{ for } 0 < \theta \leq 1 \tag{6}$$

is obtained from a modified Arrhenius' law and the second-order law of mass action, while (5) is obtained from the ideal gas law. Typical values of the constants a, q, ϕ, L_e, μ, μ_g, α, and γ are given in Table 1 of Appendix.

Remark 1. In the standard Arrhenius' law the exponent denominator $(\theta - 1)$ in (6) is taken only as θ. The modification in the Arrhenius' law is introduced to avoid technical problems related to the fact that the exponential factor becomes extremely small at the ambient temperature $\theta = 1$, but the formalism we use requires it to be exactly zero to model complete extinction due to low temperature, see [6].

3 The Wave Structure

We proceed next with the study of combustion wave propagation. The combustion wave encounters *unburned states* upstream and leaves *burned states* downstream.

We look for counterflow combustion fronts as steady traveling waves of system (1)–(6) with propagation speed $V < 0$ by setting $\xi = x - Vt$. In these moving nondimensional coordinates, after using (5) to eliminate ρ, (1)–(4) read:

$$\frac{d}{d\xi}(av - V\theta) = \frac{d^2\theta}{d\xi^2} - q\frac{d(V\eta)}{d\xi}, \tag{7}$$

$$\frac{d}{d\xi}\left(\frac{1}{\theta}(v - \phi V)Y\right) = \frac{1}{L_e}\frac{d}{d\xi}\left(\frac{1}{\theta}\frac{dY}{d\xi}\right) + \mu\frac{d(V\eta)}{d\xi}, \tag{8}$$

$$\frac{d}{d\xi}\left(\frac{1}{\theta}(v - \phi V)\right) = -\mu_g\frac{d(V\eta)}{d\xi}, \tag{9}$$

$$\frac{d(V\eta)}{d\xi} = -\Phi(\theta, Y, \eta). \tag{10}$$

In the case studied here, upstream of the combustion zone there is abundant fuel and injected oxygen. The only possibility for the combustion zone to cease is a very low temperature. Thus upstream of the combustion zone we consider the *temperature-controlled case*, setting the injection conditions as:

$$\theta = \theta^u = 1, \qquad Y = Y^u = 1, \qquad \eta = \eta^u = 0, \qquad v = v^i > 0, \tag{11}$$

where the superscript u means *unburned* and v^i is the gas injection rate.

Downstream of the combustion zone reaction ceases either due to the complete lack of fuel, *the fuel-deficient case*, or to the complete lack of oxygen, *the oxygen-deficient case*. In both cases the temperature and the velocity downstream need to be calculated; in the fuel (oxygen)-deficient case the oxygen mass fraction (conversion depth) needs to be calculated as well. Thus downstream of the combustion zone we have generically the conditions:

$$\theta = \theta^b > 0, \qquad Y = Y^b > 0, \qquad \eta = \eta^b = 1, \qquad v = v^b > 0, \tag{12}$$

or

$$\theta = \theta^b > 0, \qquad Y = Y^b = 0, \qquad \eta = \eta^b > 0, \qquad v = v^b > 0, \tag{13}$$

where the superscript b means *burned.*

In order to model the combustion front as a traveling wave, we consider the condition (11) for $\xi \to -\infty$ and either the boundary conditions (12) or (13) for $\xi \to +\infty$. With this in mind, integrating (7)–(9) from $-\infty$ to ξ once and making some manipulations we get the algebraic equation

$$\frac{1}{\theta}(v - \phi V) - (v^i - \phi V) + \mu_g V \eta = 0, \tag{14}$$

obtained from (9), plus the ODE system

$$\frac{d\theta}{d\xi} = a\big(v^i\theta - ((\theta - 1)\phi + \mu_g\eta\theta)V - v^i\big) - V(\theta - 1 - q\eta), \tag{15}$$

$$\frac{dY}{d\xi} = L_e\theta\big((v^i - (\phi + \mu_g\eta)V)Y - \mu V\eta - v^i + \phi V\big), \tag{16}$$

$$\frac{d\eta}{d\xi} = -\frac{\alpha}{V}Y(1-\eta)e^{-\gamma/(\theta-1)}, \quad \text{for } \theta > 1 \text{ and } \frac{d\eta}{d\xi} = 0 \text{ for } 0 < \theta \leq 1, \tag{17}$$

obtained from (7), (8), and (10), respectively.

Remark 2. As it is shown in [5], neglecting second-derivative terms and source terms in (1)–(4) we find that in the hyperbolic framework the characteristic speeds of the corresponding system are all nonnegative and correspond to contact discontinuities. Thus in the counterflow combustion case studied here where the combustion wave speed is negative, there are no resonances between the combustion wave with the other waves moving from left to right.

3.1 The Rankine–Hugoniot Equations for the Combustion Wave

Downstream of the combustion zone we consider the two possibilities given in (12) or in (13) for $\xi \to +\infty$ and we take into account that $d\theta/d\xi$, $dY/d\xi$, and $d\eta/d\xi$ tend to zero as ξ tends to $\pm\infty$. Thus, considering $\xi \to +\infty$ in (14)–(17) we get the Rankine–Hugoniot conditions:

$$v^b = v^i\theta^b + V(\phi(1-\theta^b) - \mu_g\eta^b\theta^b), \tag{18}$$

$$a\big(v^i\theta^b - ((\theta^b - 1)\phi + \mu_g\eta^b\theta^b)\,V - v^i\big) - V\left(\theta^b - 1 - q\eta^b\right) = 0, \tag{19}$$

$$L_e\theta^b\big(\,(v^i - (\phi + \mu_g\eta^b)V)\,Y^b - v^i + (\phi - \mu\eta^b)V\big) = 0, \tag{20}$$

$$\alpha Y^b(1-\eta^b)e^{-\gamma/(\theta^b-1)} = 0, \quad \text{for } \theta^b > 1. \tag{21}$$

Rankine–Hugoniot for the Fuel-Deficient Case

For the condition (12) at $+\infty$, in which $\eta^b = 1$, (21) becomes a trivial identity and after some manipulations (19) and (20) give θ^b and Y^b in terms of V and v^i as follows:

$$\theta^b = \frac{(1 + a\phi + q)V - av^i}{(1 + a\phi + a\mu_g)V - av^i} \qquad \text{and} \qquad Y^b = \frac{v^i - (\phi - \mu)\,V}{v^i - (\phi + \mu_g)V}, \tag{22}$$

while (18) gives

$$v^b = \theta^b(v^i - \phi V - \mu_g V) + \phi V. \tag{23}$$

Equation (22a) represents θ^b in terms of V as a hyperbola shown schematically in Fig. 1a. Since $V < 0$ and $\theta^b > 0$, we are interested only in part of this hyperbola in the second quadrant with $\theta^b \geq 1$.

Equation (22b) represents Y^b in terms of V also as a hyperbola shown schematically in Fig. 1b. Since $V < 0$ and $0 \leq Y^b \leq 1$, we are interested only in part of this hyperbola in the second quadrant connecting the points $(0, 1)$ to $(V_M, 0)$, where according to the values in Table 1 of Appendix, V_M is the negative number:

$$V_M = \frac{v^i}{\phi - \mu}. \tag{24}$$

Corresponding to this value of V_M, we have

$$\theta^b_M \equiv \theta^b(V_M) = \frac{1 + q + a\mu}{1 + a\mu_g + a\mu}. \tag{25}$$

We conclude that in the downstream fuel-deficient case, the combustion front speed, the scaled temperature, and the oxygen mass fraction must vary, respectively, in the following ranges corresponding to the bold segments on the coordinate axes in Fig. 1:

$$V_M \leq V \leq 0, \quad 1 \leq \theta^b \leq \theta^b_M \quad 0 \leq Y^b \leq 1. \tag{26}$$

Remark 3. We notice that for $|V|$ decreasing from $|V_M|$ to 0, the scaled temperature θ^b decreases from $\theta^b = \theta^b_M$ to $\theta^b = 1$ and Y^b increases from $Y^b = 0$ to $Y^b = 1$.

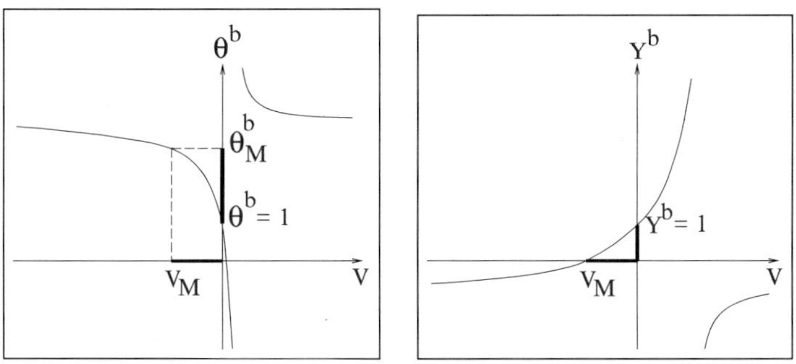

Fig. 1. (**a**) $\theta^b(V)$ and (**b**) $Y^b(V)$ in the fuel-deficient case

Rankine–Hugoniot for the Oxygen-Deficient Case

For the condition (13) at $+\infty$, in which $Y^b = 0$, with the same procedure as in the fuel-deficient case, after some manipulations we get from (18)–(20):

$$\theta^b = \frac{(\mu + a\mu\phi + q\phi)V - v^i(a\mu + q)}{(\mu + a\mu\phi + a\mu_g\phi)V - v^i(a\mu + a\mu_g)}, \qquad \eta^b = \frac{\phi V - v^i}{\mu V}, \tag{27}$$

$$v^b = \theta^b(v^i - \phi V - \mu_g \eta^b V) + \phi V. \tag{28}$$

Equation (27a) represents $\theta^b(V)$ as a hyperbola, which is shown schematically in Fig. 2a. As before we are interested only on part of this hyperbola in the second quadrant, with $V < 0$ and θ^b above the asymptotic value θ^b_a, where

$$\theta^b_a = \frac{\mu(1 + a\phi) + q\phi}{\mu(1 + a\phi) + a\mu_g\phi} > 1. \tag{29}$$

Equation (27b) represents $\eta^b(V)$ as a hyperbola, which is shown schematically in Fig. 2b. Since the values of η^b lie in the interval [0, 1], according to the graph of $\eta^b(V)$, we have $-\infty < V \leq V_M$, where V_M is given in (24). According to the graph of θ^b in Fig. 2a, for V decreasing from V_M to $-\infty$ we have θ^b decreasing from θ_M to θ^b_a, where θ_M is given in (25) and θ^b_a in (29). Thus the combustion front speed, the scaled temperature, and the fuel conversion depth must vary, respectively, in the following ranges corresponding to the bold segments on the coordinate axes in Fig. 2:

$$V \leq V_M, \quad \theta^b_a < \theta^b \leq \theta^b_M, \quad \phi/\mu < \eta^b \leq 1. \tag{30}$$

Remark 4. We notice that there is no counterflow combustion in the downstream oxygen-deficient case for $V_M < V \leq 0$, $\theta^b > \theta^b_M$ and $0 \leq \eta^b \leq \phi/\mu$. See Fig. 2.

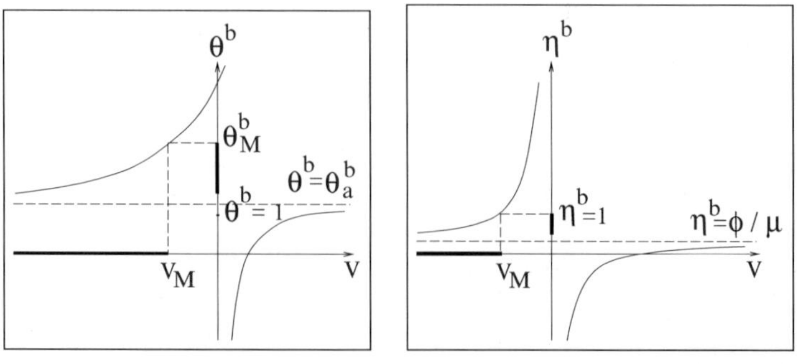

Fig. 2. **(a)** $\theta^b(V)$ and **(b)** $\eta^b(V)$ in the oxygen-deficient case

4 The Riemann Solutions

In Sect. 3.1 we have determined states along the Hugoniot locus that are admissible as burned states for fixed unburned states surrounding the combustion front. A detailed discussion on admissibility can be found in [6, 7, 9], for instance.

By direct inspection in (22), (23) for the fuel-deficient case and in (27), (28) for the oxygen-deficient case we can prove the following theorems, which provide formulae for all unknow constant states in the Riemann solution in each case.

Theorem 1. *(The fuel-deficient case). Given the injection conditions* $\theta = 1$, $Y = 1$, $\eta = 0$, *and* $v = v^i > 0$, *assume that there is a counterflow combustion wave with propagation speed* V *in the admissible range* $[V_M, 0]$ *defined in* (26) *and surrounding states given by (11) and (12). Then the values of* θ^b, Y^b *and* v^b *are uniquely determined.*

Theorem 2. *(The oxygen-deficient case). Given the injection conditions* $\theta = 1$, $Y = 1$, $\eta = 0$, *and* $v = v^i > 0$, *assume that there is a counterflow combustion wave with finite propagation speed* V *in the admissible range* $(-\infty, V_M]$ *defined in* (30) *and surrounding states given by (11) and (13). Then the values of* θ^b, η^b *and* v^b *are uniquely determined.*

5 Conclusions

As we have seen in Theorems 1 and 2, for the one-dimensional model of gas-solid counterflow combustion studied here the injection conditions together with the initial data are not sufficient to determine the Riemann solution uniquely, but they determine a family of Riemann solutions depending on the parameter V, which are analogous to the results of [5] for coflow combustion. This nonuniqueness could be eliminated by considering an entropy condition for the combustion front. According to [7] such entropic solution could be related to the uniqueness of a strong connection between the equilibria of the ODE system (15)–(17), defined by condition (11) at $-\infty$ and condition (12) or (13) at $+\infty$.

Acknowledgements

The author is grateful to professor D. Marchesin for his valuable contribution to this chapter. This work was supported in part by CNPq under grants 306609/2004-5 and 620017/2004-0, and Instituto do Milênio Avanço Global e Integrado da Matemática Brasileira, IM-AGIMB.

Appendix: Typical Values of Parameters

In Table 1 we list the constant values that appeared in the nondimensionalized PDE system (1)–(4).

Table 1. (Source: [1])

$a = 6.13E\text{–}4$	$\phi = 0.3$	$\mu = 205.8$	$\mu_g = 68.19$
$q = 1.0121$	$L_e = 0.214$	$\alpha = 0.027$	$\gamma = 23.69$

References

1. AKKUTLU, I.Y., YORTSOS, Y.C., *The Dynamics of In-situ Combustion Fronts in Porous Media*, J. of Combustion and Flame, 134, 229–247, 2003.
2. BENHAM, A.L., POETTMANN, F.H., The Thermal Recovery Process – An Analysis of Laboratory Combustion Data, Petroleum Trans. AIME 213: 406–408, 1958.
3. BOBERG, T.C., *Thermal Methods of Oil Recovery*, An Exxon Monograph Series, 1988.
4. BOUSAID, I.S., RAMEY, JR., H.J., *Oxidation of Crude Oil in Porous Media*, Soc. Pet. Eng. J., 8, 2:137–148, 1968.
5. DE SOUZA, A.J., AKKUTLU, I.Y., MARCHESIN, D., *Wave sequences for solid fuel adiabatic in-situ combustion in porous media*, Comput. Appl. Math. 25, 1:27–54, 2006.
6. MOTA, J.C., MARCHESIN, D., DANTAS, W.B., *Combustion Fronts in Porous Media*, SIAM Journal on Applied Mathematics, 62, 6:2175–2198, 2002.
7. MOTA, J.C., SCHECTER, S., *Combustion Fronts in a Porous Medium with Two Layers*, J. Dynam. Differential Equations, 18:615–665, 2006.
8. PRATS, M., *Thermal Recovery*, SPE Monograph Series SPE of AIME, 1982.
9. SCHULT, D.A., BAYLISS, A., MATKOWSKY, B.J., Traveling Waves in Natural Counterflow Filtration Combustion and Their Stability, SIAM J. Appl. Math, 58, 3:806–852, 1998.
10. WILSON, L.A., REED, R.A., REED, D.W., CLAY, R.R., *Some Effects of Pressure in Forward and Reverse Combustion*, Soc. Pet. Eng. J., 3, 2:127–135, 1963.

Global Attractor and its Dimension for a Klein–Gordon–Schrödinger System

M.N. Poulou and N.M. Stavrakakis

In this chapter we prove the existence of a global attractor for the following evolution system of Klein–Gordon–Schrödinger type

$$\begin{aligned} i\psi_t + \kappa\psi_{xx} + i\alpha\psi &= \phi\psi + f, \\ \phi_{tt} - \phi_{xx} + \phi + \lambda\phi_t &= -Re\psi_x + g, \\ \psi(x,0) = \psi_0(x),\ \phi(x,0) &= \phi_0(x),\ \phi_t(x,0) = \phi_1(x), \\ \psi(x,t) = \phi(x,t) = 0, &\quad x \in \partial\Omega,\ t > 0, \end{aligned}$$

where $x \in \Omega$, $t > 0$, $\kappa > 0$, $\alpha > 0$, $\lambda > 0$, f, and g are driving terms and Ω (bounded) $\subset R$. Finally, we make some comments concerning the finite dimensionality of the global attractor.

1 Introduction

The aim of this chapter is to prove the existence of a global attractor for the following Klein–Gordon–Schrödinger type system in a bounded domain, $\Omega \subset R$

$$i\psi_t + \kappa\psi_{xx} + i\alpha\psi = \phi\psi + f, \tag{1}$$

$$\phi_{tt} - \phi_{xx} + \phi + \lambda\phi_t = -Re\psi_x + g, \tag{2}$$

$$\begin{aligned} \psi(x,0) = \psi_0(x),\ \phi(x,0) &= \phi_0(x),\ \phi_t(x,0) = \phi_1(x), \\ \psi(x,t) = \phi(x,t) = 0, &\quad x \in \partial\Omega,\ t > 0, \end{aligned}$$

where $x \in \Omega$, $t > 0$, $\kappa > 0$, $\alpha > 0$, $\lambda > 0$, and f, g are driving terms. The variable ψ stands for the dimensionless low-frequency electron field, whereas the (real) variable ϕ denotes the dimensionless low frequency density. For more details on the physical interpretation of the equation the reader may refer to [9, 18]. The details of the proofs are presented in the paper [14].

In [2] Biler studied the long-time behavior of solutions for a bounded domain and proved the existence of a weak global attractor in $H_0^1(\Omega)\times H_0^1(\Omega)$.

Recent papers improving this result are [3, 6, 12, 17]. Also in [13] authors proved the existence of a global attractor in the space $H^k(R^n) \times H^k(R^n)$ for $k \geq 1, n \leq 3$. In addition we mention papers [4] and [5] that have studied similar questions for the Zakharov problem.

This chapter is divided in four sections. In Sect. 2 we derive some useful estimates on the solutions of the system (1) and (2) in $(H_0^1(\Omega) \cap H^2(\Omega))^2 \times H_0^1(\Omega)$. Section 3 is based on a method first introduced by Ball in [1]. We are going to use the energy equations of the problem to prove the continuity of solutions on the initial data for the space $(H_0^1(\Omega) \cap H^2(\Omega))^2 \times H_0^1(\Omega)$. In Sect. 4, we show the asymptotic compactness of the dynamical system and prove the existence of a global attractor. In Sect. 4, we sketch the basic steps of the finite dimensionality of the attractor.

Notation: For *simplicity reasons* sometimes we use H^s, L^s for $H^s(\Omega)$, $L^s(\Omega)$, and $\int dx$ to denote the integration over the domain Ω. By $||.||$ we denote the standard L^2 norm. Finally, C is a general symbol for any positive constant.

2 Global Existence

Let us introduce the transformation $\theta = \phi_t + \delta\phi$, where θ is real, with δ a small positive constant to be specified later. Then, systems (1) and (2) takes the form

$$i\psi_t + \kappa\psi_{xx} + i\alpha\psi = \phi\psi + f, \tag{3}$$

$$\phi_t + \delta\phi = \theta, \tag{4}$$

$$\theta_t + (\lambda - \delta)\theta - \phi_{xx} + (1 - \delta(\lambda - \delta))\phi = -Re\psi_x + g. \tag{5}$$

Also the initial and boundary conditions take the form

$$\psi(x,0) = \psi_0(x),\ \phi(x,0) = \phi_0(x),\ \theta(x,0) = \theta_0(x), \quad x \in \Omega, \tag{6}$$

$$\psi(x,t) = \phi(x,t) = 0, \quad x \in \partial\Omega,\ t > 0. \tag{7}$$

Lemma 1. *Let $||\psi_0(t)|| \leq R$ for some $R > 0$, and suppose that f belongs to $L^2(\Omega)$. Every solution of (3)–(7) satisfies*

$$||\psi(t)|| \leq R^*, \quad t \geq t_1,$$

where R^* is a constant depending on α, $||f||$; t_1 depending on α, $||f||$ and R.

Proof. The proof is analogue to the proof of Lemma 2.1 in [13]. □

Lemma 2. *Assume that $f,\ g \in L^2(\Omega)$, let $||(\psi_0, \phi_0, \theta_0)||_{H_0^1 \times H_0^1 \times L^2} \leq R$, where $R > 0$. Then, every solution (ψ, ϕ, θ) of problem (3)–(7) satisfies*

$$||\psi(t)||_{H_0^1} + ||\phi(t)||_{H_0^1} + ||\theta(t)|| \leq M_1, \quad t \geq t_2,$$

where M_1 depends on $\alpha, \kappa, \lambda, \delta, ||f||, ||g||$ and t_2 on α, κ, λ, δ, $||f||$, $||g||$, and R.

Proof. Multiplying (3) by $-\psi_t$, integrating and taking the real part gives

$$\frac{1}{2}\frac{d}{dt}\left(\kappa||\psi_x||^2 + \int \phi|\psi|^2 + 2Re\int f\bar{\psi}\right) + \kappa\alpha||\psi_x||^2 + \left(\alpha + \frac{\delta}{2}\right)\int \phi|\psi|^2$$
$$= \frac{1}{2}\int \theta|\psi|^2 + \alpha Re\int f\bar{\psi}. \tag{8}$$

Next, multiplying (5) by θ and substituting θ from (4) implies

$$\frac{1}{2}\frac{d}{dt}(||\theta||^2 + ||\phi_x||^2 + (1-\delta(\lambda-\delta))||\phi||^2) + (\lambda-\delta)||\theta||^2 + \delta||\phi_x||^2$$
$$+\delta(1-\delta(\lambda-\delta))||\phi||^2 = -Re\int \theta\psi_x + \int g\theta. \tag{9}$$

Adding relations (2) and (9) we get $F_1'(t)+\delta F_1(t) = G_1(t)$, where for simplicity reasons the following quantities are introduced

$$F_1 := \kappa||\psi_x||^2 + \int \phi|\psi|^2 dx + ||\theta||^2 + ||\phi_x||^2 + (1-\delta(\lambda-\delta))||\phi||^2 + 2Re\int f\bar{\psi},$$
$$G_1 := (\delta - 2\kappa\alpha)||\psi_x||^2 - 2\alpha\int \phi|\psi|^2 + (3\delta-2\lambda)||\theta||^2 - \delta(1-\delta(\lambda-\delta))||\phi||^2$$
$$-\delta||\phi_x||^2 + \int \theta|\psi|^2 + 2(\delta-\alpha)Re\int f\bar{\psi} - 2Re\int \theta\psi_x + 2\int g\theta.$$

Taking δ small enough such that $\delta - 2\kappa\alpha < 0,\ \ 3\delta - 2\lambda < 0,\ \ 1-\delta(\lambda-\delta) > 0$, one can render several terms of G_1 negative. Let us proceed with majorizing the integrals of G_1 as follows

$$\left|\int \theta|\psi|^2\right| \le ||\theta||||\psi||_4^2 \le ||\theta||||\psi||^{1/2}||\psi||^{3/2} \le \frac{\epsilon_1}{2}||\theta||^2 + \frac{\epsilon_2}{2}||\psi_x||^2 + C.$$

$$\left|2\alpha\int \phi|\psi|^2\right| \le \epsilon_3||\phi||^2 + \frac{\epsilon_2}{2}||\psi_x||^2 + C,\quad \left|\int \theta\psi_x\right| \le ||\psi_x||||\theta|| \le \frac{\epsilon}{2}||\psi_x||^2 + \frac{1}{2\epsilon}||\theta||^2.$$

$$\left|2(\delta-\alpha)\int f\bar{\psi}\right| \le C||f||\ ||\psi|| \le C,\quad \left|2\int g\theta\right| \le 2||g||||\theta|| \le \frac{\epsilon_1}{2}||\theta||^2 + C.$$

The next step is to estimate the arbitrary positive constants $\epsilon_1, \epsilon_2, \epsilon$, such that the following two inequalities hold simultaneously true $\epsilon_1 + \frac{1}{2\epsilon} \le -(3\delta - 2\lambda)$, $\epsilon_2 + \frac{\epsilon}{2} \le -(\delta - 2\kappa\alpha)$. Let $\nu > 0, \nu \ne \frac{1}{2}, \bar{\alpha} = -(3\delta-2\lambda)$ and $\bar{\beta} = -(\delta - 2\kappa\alpha)$. Setting $\epsilon_1 = \dfrac{\bar{\alpha}}{2\nu}$, $\epsilon_2 = \frac{\bar{\beta}}{2\nu}$ we have the following necessary condition: $\bar{\alpha}\bar{\beta} \ge \frac{\nu^2}{(2\nu-1)^2}$. Since $\bar{\alpha}, \bar{\beta} > 0$ the inequality is always true for sufficiently small ν. Finally, taking ϵ_3 small enough, so that $\epsilon_3 < -\delta(1-\delta(\lambda-\delta))$ implies $F_1'(t) + \delta F_1(t) \le C$. Applying Gronwall's Inequality completes the proof. □

Lemma 3. *Assume that* $f,\ g \in H_0^1(\Omega)$. *Let* $||(\psi_0, \phi_0, \theta_0)||_{(H_0^1 \cap H^2)^2 \times H_0^1} \leq R$, *where* $R > 0$. *Then, every solution* (ψ, ϕ, θ) *of the problem (3)–(7) satisfies*

$$||\psi(t)||_{H_0^1 \cap H^2} + ||\phi(t)||_{H_0^1 \cap H^2} + ||\theta(t)||_{H_0^1} \leq M_2 \quad t \geq t_3,$$

where M *depends on* $\alpha, \kappa, \lambda, \delta, ||f||_{H_0^1}, ||g||_{H_0^1}$ *and* t_3 *depends on* α, κ, λ, δ, $||f||_{H_0^1}$, $||g||_{H_0^1}$, *and* R.

Proof. The proof follows the same procedure with Lemma 2. □

Repeating a similar procedure we obtain results on a finite-time interval.

Lemma 4. *Assume that* $f,\ g \in L^2(\Omega)$ *and let* $||\psi_0, \phi_0, \theta_0||_{H_0^1 \times H_0^1 \times L^2} \leq R$, *where* $R > 0$. *Then, every solution* (ψ, ϕ, θ) *of the problem (3)–(7) with* L_1 *depending on* $\alpha, \kappa, \lambda, \delta, ||f||, ||g||$, *and* T) *satisfies*

$$||\psi(t)||_{H_0^1} + ||\phi(t)||_{H_0^1} + ||\theta|| \leq L_1 \quad 0 \leq t \leq T.$$

Lemma 5. *Assume that* $f, g \in H_0^1(\Omega)$, *and* $||\psi_0, \phi_0, \theta_0||_{(H_0^1 \cap H^2)^2 \times H_0^1} \leq R$, *where* $R > 0$. *Then, every solution* (ψ, ϕ, θ) *of the problem (3)–(7) with* L_2 *depending on* $(\alpha, \kappa, \lambda, \delta, ||f||_{H_0^1}, ||g||_{H_0^1}$ *and* T) *satisfies*

$$||\psi(t)||_{H_0^1 \cap H^2} + ||\phi(t)||_{H_0^1 \cap H^2} + ||\theta||_{H_0^1} \leq L_2 \quad 0 \leq t \leq T.$$

Finally we state the uniqueness result in the following way.

Theorem 1. *Let* $f,\ g \in L^\infty(R^+, H_0^1(\Omega))$ *and assume that*

$$(\psi_0, \phi_0, \theta_0) \in (H_0^1(\Omega) \cap H^2(\Omega))^2 \times H_0^1(\Omega).$$

Then, there exists a unique solution for the system (3), (7) such that

$$\begin{aligned}
&\psi \in L^\infty(0, \infty; H_0^1(\Omega) \cap H^2(\Omega)),\ \psi_t \in L^\infty(0, \infty; L^2(\Omega)),\\
&\phi \in L^\infty(0, \infty; H_0^1(\Omega) \cap H^2(\Omega)),\ \phi_t \in L^\infty(0, \infty; H_0^1(\Omega)),\\
&\phi_{tt} \in L^\infty(0, \infty; L^2(\Omega)),\\
&\psi(x, 0) = \psi_0(x),\ \phi(x, 0) = \phi_0(x),\ \phi_t(x, 0) = \phi_1(x),\ x \in \Omega.
\end{aligned}$$

3 The Solution Semigroup

Let B_1, B_2 denote the following balls of center zero and radius M_1, M_2, respectively

$$B_1 = \{(\psi, \phi, \theta) \in H_0^1 \times H_0^1 \times L^2 : ||\psi||_{H_0^1} + ||\phi||_{H_0^1} + ||\theta|| \leq M_1\}, \tag{10}$$

$$B_2 = \{(\psi, \phi, \theta) \in (H_0^1 \cap H^2)^2 \times H_0^1 : ||\psi||_{H_0^1 \cap H^2} + ||\phi||_{H_0^1 \cap H^2} + ||\theta||_{H_0^1} \leq M_2\}, \tag{11}$$

where M_1, M_2 are the constants introduced in Lemmas 2 and 3, respectively. Therefore B_1, B_2 are bounded absorbing sets for (3)–(7). Since B_1 is bounded, we see that there exists a constant $T(B_1)$ depending on B_1 such that $S(t)B_1 \subset B_1$, $\forall\, t \geq T(B_1)$.

Lemma 6. *If* $(\psi_n, \phi_n, \theta_n) \to (\psi, \phi, \theta)$ *weakly in* $H_0^1(\Omega) \times H_0^1(\Omega) \times L^2(\Omega)$, *then for every* $T > 0$, *we have*

$$S(\cdot)(\psi_n, \phi_n, \theta_n) \to S(\cdot)(\psi, \phi, \theta), \ \text{weakly in } L^2(0, T; H_0^1 \times H_0^1 \times L^2) \tag{12}$$

$$S(t)(\psi_n, \phi_n, \theta_n) \to S(t)(\psi, \phi, \theta), \ \text{weakly in } H_0^1 \times H_0^1 \times L^2, \ \ 0 \le t \le T. \tag{13}$$

Proof. The proof follows similar arguments with [13]. □

Theorem 2. *Assume that* f *and* g *belong to* $L^2(\Omega)$. *The solutions* $(\psi, \phi, \theta) \in C(R^+, H_0^1 \times H_0^1 \times L^2)$ *of the problem (3)–(7) depend continuously on the initial data in* $H_0^1 \times H_0^1 \times L^2$.

Proof. Assume that $(\psi_{0,n}, \phi_{0,n}, \theta_{0,n}) \to (\psi_0, \phi_0, \theta_0)$ in $H_0^1 \times H_0^1 \times L^2$, therefore we need to prove that $S(t)(\psi_n, \phi_n, \theta_n) \to S(t)(\psi, \phi, \theta), \ \forall\ t > 0$, as $n \to \infty$. Given $t > 0$, we choose $T > t$. From the statement above we know that $(\psi_{0,n}, \phi_{0,n}, \theta_{0,n})$ is bounded and therefore by Lemma 4 there exists a solution such that

$$||\psi_n(\tau)||_{H_0^1} + ||\phi_n(\tau)||_{H_0^1} + ||\theta_n(\tau)|| \le C, \quad 0 \le \tau \le T, \tag{14}$$

will hold where $(\psi_n(\tau), \phi_n(\tau), \theta_n(\tau)) = S(\tau)(\psi_{0,n}, \phi_{0,n}, \theta_{0,n})$. But from the system (3)–(7) we can see that $\left\|\frac{\partial}{\partial\tau}\psi_n\right\|_{L^2(0,T;H_0^{-1})} + \left\|\frac{\partial}{\partial\tau}\phi_n\right\|_{L^2(0,T;L^2)} \le C.$ Hence, there exists a $(\psi(\tau), \phi(\tau), \theta(\tau)) \in L^\infty(0, T; H_0^1 \times H_0^1 \times L^2)$, such that

$$(\psi_n(\tau), \phi_n(\tau), \theta_n(\tau)) \to (\psi(\tau), \phi(\tau), \theta(\tau)), \ \text{weakly in } \ L^2(0, T; H_0^1 \times H_0^1 \times L^2),$$

$$\frac{\partial}{\partial\tau}\psi_n \to \frac{\partial}{\partial\tau}\psi \ \text{weakly} \ \in L^2(0, T; H_0^{-1}), \ \frac{\partial}{\partial\tau}\phi_n \to \frac{\partial}{\partial\tau}\phi \ \text{weakly} \ \in L^2(0, T; L^2). \tag{15}$$

Thereby, using the relations above and standard compactness results we have

$$(\psi_n, \phi_n) \to (\psi, \phi), \ \text{strongly} \ \in L^2(0, T; L^2 \times L^2). \tag{16}$$

Using similar arguments as the ones above and (14) we can deduce with the help of Lemma 2 that for a fixed t there exists a $(\psi(t), \phi(t), \theta(t))$ such that

$$(\psi_n(t), \phi_n(t), \theta_n(t)) \to (\psi(t), \phi(t), \theta(t)), \ \text{weakly} \ \in H_0^1 \times H_0^1 \times L^2,$$

where $(\psi(t), \phi(t), \theta(t))$ is the solution of the problem (3)–(7) with initial conditions $(\psi_0, \phi_0, \theta_0)$. Taking into consideration Lemma 6, the arguments above imply that

$$S(t)(\psi_n, \phi_n, \theta_n) \to S(t)(\psi, \phi, \theta), \ \text{weakly in} \ \ H_0^1 \times H_0^1 \times L^2. \tag{17}$$

Our aim is to prove the strong convergence for the statement above. In order to do so we are going to use the energy equation derived by Lemma 3 and the fact that every solution of the system (3)–(7) verifies the energy equation, hence

$$F_1(S(t)(\psi_0, \phi_0, \theta_0)) = e^{-\delta t} F_1(\psi_0, \phi_0, \theta_0) + \int_0^t e^{-\delta(t-\tau)} G_1(S(t)(\psi_0, \phi_0, \theta_0)) d\tau, \tag{18}$$

where $(\psi(t), \phi(t), \theta(t)) = S(t)(\psi_0, \phi_0, \theta_0)$. The same will also hold for the solution $S(t)(\psi_{0,n}, \phi_{0,n}, \theta_{0,n})$, i.e.,

$$F_1(S(t)(\psi_{0,n}, \phi_{0,n}, \theta_{0,n})) = e^{-\delta t} F_1(\psi_{0,n}, \phi_{0,n}, \theta_{0,n}) + \int_0^t e^{-\delta(t-\tau)} G_1(S(t)(\psi_{0,n}, \phi_{0,n}, \theta_{0,n})) d\tau. \quad (19)$$

Proving that the right-hand side of (18), (19) are equal completes the proof of the theorem. □

Theorem 3. *Assume that f and g belong to $H_0^1(\Omega)$. The solutions $(\psi, \phi, \theta) \in C(R^+, (H_0^1 \cap H^2)^2 \times H_0^1)$ of the problem (3)–(7) depend continuously on the initial data in $(H_0^1 \cap H^2)^2 \times H_0^1$.*

Proof. The proof is analogue to the proof of Theorem 3.3 in [12]. □

4 Existence of a Global Attractor

The aim of this section is to prove the existence of a global attractor for the dynamical system $S(t)$ in the space $(H_0^1 \cap H^2)^2 \times H_0^1$.

Theorem 4. *Let f, g belong to $L^2(\Omega)$. Then the dynamical system $S(t)$ is asymptotically compact in $H_0^1 \times H_0^1 \times L^2$ that is, if $(\psi_n, \phi_n, \theta_n)$ is bounded in $H_0^1 \times H_0^1 \times L^2$ and $t_n \to \infty$, then $S(t)(\psi_n, \phi_n, \theta_n)$ is precompact in the same space.*

Proof. The proof follows the similar arguments with Theorem 2. □

Theorem 5. *Assume that f and g belong to $H_0^1(\Omega)$. Then, the dynamical system $S(t)$ is asymptotically compact in $(H_0^1 \cap H^2)^2 \times H_0^1$ that is, if $(\psi_n, \phi_n, \theta_n)$ is bounded in $(H_0^1 \cap H^2)^2 \times H_0^1$ and $t_n \to \infty$, then $S(t)(\psi_n, \phi_n, \theta_n)$ is precompact in the same space.*

Proof. The proof is omitted as it follows similar steps to Theorem 3. □

Proposition 1. *Assume that X is a metric space and $\{S(t)\}_{t \geq 0}$ is a semigroup of continuous operators in X. If $\{S(t)\}_{t \geq 0}$ has a bounded absorbing set and is asymptotically compact, then $\{S(t)\}_{t \geq 0}$ possesses a global attractor which is a compact invariant set and attracts every bounded set in X.*

Theorem 6. *Assume that f and g belong to $H_0^1(\Omega)$. Then the problems (3)–(7) possesses a strong compact global attractor X_1 in $(H_0^1 \cap H^2)^2 \times H_0^1$, which is a compact invariant subset and attracts every bounded set of $(H_0^1 \cap H^2)^2 \times H_0^1$, with respect to the norm topology.*

Proof. Taking into consideration the asymptotic compactness of $S(t)$ in $H_0^1 \times H_0^1 \times L^2$ (Theorem 4), the asymptotic compactness of $S(t)$ in $(H_0^1 \cap H^2)^2 \times H_0^1$, (Theorem 5) and Proposition 1 the proof is completed. □

Remark 1. It is possible to prove that the global attractor X_1 of the problems (3)–(7) has finite Hausdorff and fractal dimensions (See, work in preparation [15]).

Acknowledgments

This work was partially financially supported by a grant from the `Pythagoras` Basic Research Program No. 68/831 of the Ministry of Education of the Hellenic Republic.

References

1. J.M. Ball, *Global Attractors for Damped Semilinear Wave Equations*, Discrete and Continuous Dynamical Systems, 10 (1–2) (2004), 31–52.
2. P.Biler, *Attractors for the system of the Schrödinger and Klein-Gordon Equations with Yukawa coupling*, SIAM J. Math. Anal. 21 (1990), 1190–1212.
3. M.M. Cavalcanti and V.N. Domingos Cavalcanti, *Global Existence and Uniform Decay for the Coupled Klein-Gordon-Schrödinger Equations*, Nonlinear Differential Equations and Applications (NoDEA), 7 (2000), 285–307.
4. I. Flahaut, *Attractors for the Dissipative Zakharov System*, Nonlinear Analysis, TMA, 16, No. 7/8 (1991), 599–633.
5. O. Goubet and I. Moise, *Attractors for Dissipative Zakharov System*, Nonlinear Analysis, TMA, 31, No. 7 (1998), 823–847.
6. B. Guo and Y. Li, *Attractors for Klein-Gordon-Schrödinger Equations in R^3*, Journal of Differential Equations, 136 (1997), 356–377.
7. J.K. Hale, *Asymptotic Behaviour of Dissipative Systems*, American Mathematical Society, Providence, (1988).
8. A. Haraux and E. Zuazua, *Decay estimates for some semilinear dampted hyperbolic problems*, Archive Rational Mechanics Analysis 100, (1988), No. 2, 191.
9. N. Karachalios, N.M. Stavrakakis and P. Xanthopoulos, *Parametric Exponential Energy Decay for Dissipative Electron-Ion Plasma Waves*, ZAMP, 56 (2) (2005), 218–238.
10. N. Karachalios and Stavrakakis, *Existence of a Global Attractor for Semilinear Dissipative Wave Equations on R*, Journal of Differential Equations, 157 (1999), 183–205.
11. O. Ladyzhenskaya, *Attractors for Semigroups and Evolution Equations*, Cambridge University Press, Cambridge, UK, (1991).
12. H. Lange and B. Wang, *Attractors for the Klein-Gordon-Schrödinger Equation*, Journal Mathematical Physics, Vol 40, No. 5 (1999), 2445–2457.
13. K. Lu and B. Wang, *Global Attractors for the Klein-Gordon-Schrödinger Equation in Unbounded Domains*, Journal of Differential Equations, 170 (2001), 281–316.
14. M.N. Poulou and N.M. Stavrakakis, *Global Attractor for a Klein-Gordon-Schrödinger Type System*, submitted.
15. M.N. Poulou and N.M. Stavrakakis, *Finite Dimensionality of the Global Attractor for a Klein - Gordon - Schrödinger Type System*, in preparation.

16. R. Teman, *Infinite-Dimensional Dynamical Systems in Mechanics and Physics* (2nd Edition), Appl. Math. Sc., 68, Springer-Verlag, New York, (1997).
17. B. Wang, *Attractors for reaction-diffusion equations in unbounded domains*, Physica D, 128 (1999), 41–52.
18. P. Xanthopoulos, *Modeling and Asympotic behavior of Nonlinear Dispersive Systems in Magnetic Fussion* (in Greek), Ph.D. Thesis, Dept. of Mathematics, National Technical University of Athens, October 2003.

A Few Remarks About a Theorem by J. Rauch

F. Sueur

1 Vanishing Viscosity for Multidimensional Systems

We consider a symmetric hyperbolic linear operator:

$$\mathcal{H}(t,x,\partial) := A_0(t,x)\partial_t + \sum_{1\leq j\leq n} A_j(t,x)\partial_j + B(t,x).$$

The $N \times N$ matrices $(A_j)_{0\leq j\leq n}$, B are symmetric. Their coefficients are real valued, C^∞. The matrix A_0 is positive definite. We denote by Ω an open subset of $\mathbb{R}^n$ with smooth, compact boundary Γ such that Ω lies on one side of Γ. The real $T > 0$ is fixed in all the papers. We introduce the boundary matrix: $A_{\mathfrak{n}} := \sum_{1\leq j\leq n} \mathfrak{n}_j A_j$ where $\mathfrak{n} := (\mathfrak{n}_1, \ldots, \mathfrak{n}_n)$ is the interior unit normal to Γ.

Assumption 1. The dimension $d_0(t,x) := dim \ker A_{\mathfrak{n}}(t,x)$ does not depend of $(t,x) \in (0,T) \times \Gamma$.

Notice that in Assumption 1, we only consider $x \in \Gamma$ (and not $x \in \Omega$). Thus, the boundary is either characteristic of constant multiplicity (if $d_0 = 0$) or noncharacteristic (if $d_0 \neq 0$).

Let us first consider a symmetric viscosity tensor

$$\mathcal{E}(t,x,\partial) := \sum_{1\leq i,j\leq n} \partial_i E_{i,j}(t,x)\partial_j, \tag{1}$$

where the $N \times N$ matrices $(E_{i,j})_{1\leq i,j\leq n}$ are symmetric, their coefficients are real and C^∞, and verifies:

$$\exists c > 0 / \ \forall \zeta \in \mathbb{R}^d, \ \forall (t,x) \in (0,T) \times \Omega, \quad \sum_{1\leq i,j\leq n} \zeta_i \zeta_j E_{i,j}(t,x) \geqslant c|\zeta|^2 Id.$$

We look at the solutions $(u^\varepsilon)_\varepsilon$ of

$$\mathcal{P}^\varepsilon(t,x,\partial)u^\varepsilon = F(t,x,u^\varepsilon) + f(t,x) \quad \text{when } (t,x) \in (-1,T) \times \Omega, \tag{2}$$

$$u^\varepsilon = 0 \quad \text{when } (t,x) \in (-1,T) \times \Gamma, \tag{3}$$

$$u^\varepsilon = 0 \quad \text{when } (t,x) \in (-1,0) \times \Omega, \tag{4}$$

here we have denoted $(\mathcal{P}^\varepsilon)_\varepsilon$ the family of parabolic operators $\mathcal{P}^\varepsilon = \mathcal{H} - \varepsilon\mathcal{E}$, F is a C^∞ function with $F(t,x,0) = 0$ and f is a C^∞ source term with $f|_{t\leq 0} = 0$. For $\varepsilon \to 0^+$, Bardos et al. [1] find the limit hyperbolic initial boundary value problem:

$$\mathcal{H}(t,x,\partial)u^0 = F(t,x,u) + f(t,x) \quad \text{when } (t,x) \in (-1,T) \times \Omega, \tag{5}$$

$$Mu^0 = 0 \quad \text{when } (t,x) \in (-1,T) \times \Gamma, \tag{6}$$

$$u^0 = 0 \quad \text{when } (t,x) \in (-1,0) \times \Omega, \tag{7}$$

where M is a $N \times N$ matrix, C^∞, on the boundary Γ such that

$$\forall (t,x) \in (0,T) \times \Gamma, \quad \ker M(t,x) = E_{\leq 0}(E^{-1}(t,x)A_{\mathfrak{n}}(t,x)), \tag{8}$$

where $E_{\leq 0}(E^{-1}A_{\mathfrak{n}})$ is the sum of the eigenspaces of $E^{-1}A_{\mathfrak{n}}$, associated to nonpositive eigenvalues of $E^{-1}A_{\mathfrak{n}}$. Thus the boundary conditions M in (6) are strictly dissipative, which means that there is a real $c > 0$ such that, for all $(t,x) \in (0,T) \times \Omega$, for all $u \in \ker M(t,x)$,

$$\langle A_{\mathfrak{n}}(t,x)u, u\rangle \leq -c||(Id - \Pi_0(t,x))u||^2,$$

where $\Pi_0(t,x)$ is the orthogonal projector on $\ker A_{\mathfrak{n}}(t,x)$. According to [6, 2, 7] the problem (5)–(7) is locally well posed: there exists $T_0 > 0$ and a unique solution in $u^0 \in C^\infty((-1,T_0) \times \Omega)$.

The goal of chapter [8] we present here is to show that it is indeed possible to obtain any dissipative boundary condition including the conservative ones, which answers to a question explicitly raised by Rauch in paper [5]. To do this we will relax the conditions on the viscosity $\mathcal{E}$. It is well known (cf. [4]) that for every $\varepsilon > 0$, the semi-linear problem (2)–(4) is locally Well posed in H^s, $s > \frac{d}{2}$, if $\mathcal{E}$ is a viscosity tensor of the form (1) which verifies the following uniform strong ellipticity assumption. This means that for every $\varepsilon \in]0,1]$, there are a real $T_\varepsilon > 0$ and one and only one $u^\varepsilon \in H^\infty((-1,T_\varepsilon) \times \Omega)$ solution of (2)–(4).

Assumption 2. There is $c > 0$ such that for all $\zeta \in \mathbb{R}^n - \{0\}$, for all $(t,x) \in (0,T) \times \Omega$, the eigenvalues $\mu(t,x)$ of

$$\sum_{1\leq i,j\leq n} \zeta_i\zeta_j (A_0^{-1}E_{i,j})(t,x)$$

verify $Re(\mu(t,x)) \geqslant c|\zeta|^2$.

Notice that in Assumption 2, the matrices $E_{i,j}$ need not be symmetric. To insure some suitable estimates uniform in ε, we make the following original assumption.

Assumption 3 [*Weakly dissipation*]. There exists $c > 0$ such that for all $t \in (0,T)$, for all $u \in C_0^\infty(\Omega, \mathbb{R}^N)$,

$$-\int_{x\in\Omega} u(x)^t \cdot \mathcal{E}(t,x,\partial)u(x)dx \geqslant -c\int_{x\in\Omega} |u(x)|^2 dx. \tag{9}$$

Assumption 3 is a key point in the success of our method. To deal with characteristic components, we need the following assumption about the normal viscosity matrix E:

Assumption 4 [*Strong characteristic dissipation*]. There exists $c > 0$ such that for all $(t, x) \in (0, \infty) \times \Omega$, for all $u \in \mathbb{R}^N$,

$$\langle \Pi_0^t E \Pi_0 u, u \rangle \geqslant c ||\Pi_0 u||^2. \tag{10}$$

We can now state our first main theorem.

Theorem 1. *Assume that $\mathcal{E}$ is a C^∞ viscosity tensor verifying the assumptions (8), 2, 3, and 4. There exist $T_1 \in]0, T_0]$ and a real $\varepsilon_0 \in]0, 1]$ such that for all $\varepsilon \in]0, \varepsilon_0]$ the problem (2)–(4) admits one and only one solution $u^\varepsilon \in H^\infty((-1, T_1) \times \Omega)$. Moreover, the solutions $(u^\varepsilon)_{\varepsilon \in]0,1]}$ converge in $H^s((-1, T_1) \times \Omega)$ to u^0 when $\varepsilon \to 0^+$ for all $s \in [0, \frac{1}{2}[$.*

Indeed the proof even gives some optimal small viscosity uniform estimate of $u^\varepsilon - u^0$. We prove in [7] that there exists a C^∞ viscosity tensor $\mathcal{E}$ verifying the assumptions (8), 2, 3, and 4. Thus the previous theorem is not empty. To simplify, we will assume from now on that $\Omega := \{x_n > 0\}$. We consider the spaces $\mathcal{N}_\theta := H^\infty((-1, T) \times \Omega, \mathcal{S}(\mathbb{R}_\theta^+))$ and $\mathcal{N}_z := H^\infty((-1, T) \times \Omega, \mathcal{S}(\mathbb{R}_z^+))$, where $\mathcal{S}$ is the Schwartz space of C^∞ rapidly decreasing functions, and profile space

$$\mathcal{P}(T) := \{\mathcal{U}(t, x, z, \theta) = \mathcal{U}_a(t, x) + \mathcal{U}_b(t, x, \theta) + \mathcal{U}_c(t, x, z), \quad \text{where} \\ \mathcal{U}_a \in H^\infty((-1, T) \times \Omega),\ \mathcal{U}_b \in \mathcal{N}_\theta \text{ and } \mathcal{U}_c \in \mathcal{N}_z\}.$$

The function $\mathcal{U}_a$ is the regular part or the interior part of the profile, $\mathcal{U}_b$ is a characteristic boundary layer and $\mathcal{U}_c$ is a noncharacteristic boundary layer. We define the set

$$\beta(T) := \{(u^\varepsilon)_{\varepsilon \in]0,1]} \in (L^2((-1, T) \times \Omega))^{]0,1]} / \ \forall l \in \mathbb{N}\ \exists \varepsilon' \in]0, 1] / \\ \sup_{\varepsilon \in]0, \varepsilon']} (||(\varepsilon \partial_{t,x})^l u^\varepsilon||_{L^2((0,T) \times \Omega)} < \infty\}.$$

We denote Z for a tangential derivative chosen between $\partial_0 := \partial_t, ..., \partial_{n-1}$. Theorem 1 is a consequence of the next theorem which shows that we can describe the solutions u^ε of the perturbated problems as boundary layer expansions at all orders.

Theorem 2. *There are a real $T_1 \in]0, T_0]$ and a real $\varepsilon_0 \in]0, 1]$ such that for all $\varepsilon \in]0, \varepsilon_0]$, problem (2)–(4) admits one and only one solution $u^\varepsilon \in H^\infty((-1, T_1) \times \Omega)$. Moreover there are some profiles $(\mathcal{U}^j)_{j \geqslant 0}$ in $\mathcal{P}(T_1)$ such that for all $\varepsilon \in]0, 1]$, for all $k \in \mathbb{N}$, there is $R^\varepsilon \in \beta(T_1)$ such that*

$$u^\varepsilon(t, x) = \sum_{j=0}^{k} \sqrt{\varepsilon}^j\, \mathcal{U}^j(t, x, \frac{x_n}{\varepsilon}, \frac{x_n}{\sqrt{\varepsilon}}) + \sqrt{\varepsilon}^{k+1} R^\varepsilon(t, x).$$

First we prove that there is a real $T_1 \in]0, T_0]$ such that for all $k \in \mathbb{N}$, there are $(\mathcal{U}^j)_{0 \le j \le k}$ in $\mathcal{P}(T_1)$ such that the family $(a^\varepsilon)_{\varepsilon \in]0,1]}$ defined by

$$a^\varepsilon(t,x) := \sum_{j=0}^{k} \sqrt{\varepsilon}^j\, \mathcal{U}^j(t,x,\frac{x_n}{\varepsilon},\frac{x_n}{\sqrt{\varepsilon}}) \tag{11}$$

verifies $\mathcal{P}^\varepsilon a^\varepsilon = F(t,x,a^\varepsilon) + f^\varepsilon + \varepsilon^M g^\varepsilon$ with $a^\varepsilon = 0$ when $t = 0$ or when $x \in \Gamma$, with $(g^\varepsilon)_{\varepsilon \in]0,1]} \in \beta(T_1)$. The general process is the same as in [3, 7, 9]. The main difference with [7] lies in the lack of symmetry of these problems. However thanks to Assumption 4 the profile problems are well posed. Then we complete the proof of theorem 2 thanks to the following convergence theorem.

Theorem 3. *Let $T > 0$, $m > \frac{1+n}{2}$ and $M' \geqslant 2m$. Let $(a^\varepsilon)_{\varepsilon \in]0,1]}$ is a family of approximate regular solutions of the problems (2)–(4), i.e.,*

$$\begin{aligned}
\mathcal{P}^\varepsilon a^\varepsilon = F(t,x,a^\varepsilon) + f + \varepsilon^{M'} g^\varepsilon \quad &\text{when } (t,x) \in (-1,T) \times \Omega,\\
a^\varepsilon = 0 \quad &\text{when } (t,x) \in (-1,T) \times \Gamma,\\
a^\varepsilon = 0 \quad &\text{when } (t,x) \in (-1,T) \times \Omega,
\end{aligned}$$

with $(\varepsilon^m g^\varepsilon)_{\varepsilon \in]0,1]}$ bounded in $H^m((0,T)\times\Omega)$. Assume that the family $(a^\varepsilon)_{\varepsilon \in]0,1]}$ verifies the following estimates:

$$\forall \alpha \in \mathbb{N}^{n+1}, \forall k \in \mathbb{N} \quad \sup_{\varepsilon \in]0,1]} ||Z^\alpha (\varepsilon \partial_n)^k a^\varepsilon||_{L^\infty((-1,T)\times\Omega)} < \infty. \tag{12}$$

Then there is a real $\varepsilon_0 \in]0,1]$ such that for all $\varepsilon \in]0,\varepsilon_0]$, problem (2)–(4) has one exact solution and $(\varepsilon^{-m}(u^\varepsilon - a^\varepsilon))_{\varepsilon \in]0,1]}$ is bounded in $H^m((-1,T) \times \Omega)$.

2 Proof of Theorem 3

There is a smooth $N \times N$ matrix G such that for all (t,x,u,v), $F(t,x,u+v) = F(t,x,u) + G(t,x,u,v).v$. We define $w^\varepsilon := \varepsilon^{-M}(u^\varepsilon - a^\varepsilon)$, where $M > 0$ will be chosen later, and obtain the equivalent problem:

$$\begin{cases}
\mathcal{P}^\varepsilon w^\varepsilon = G(t,x,a^\varepsilon,\varepsilon^M w^\varepsilon).w^\varepsilon + \varepsilon^{M'-M} g^\varepsilon \quad \text{when } (t,x) \in (-1,T) \times \Omega,\\
w^\varepsilon = 0 \quad \text{when } (t,x) \in (-1,T) \times \Gamma,\\
w^\varepsilon = 0 \quad \text{when } (t,x) \in (-1,T) \times \Omega.
\end{cases} \tag{13}$$

The classical theory of parabolic IBVP gives the existence for all $\varepsilon > 0$ of a strictly nonnegative time $T_\varepsilon > 0$ and of a regular solution w^ε. We have to obtain ε-uniform estimates. We introduce a new function unknown. We denote by Π the constant matrix

$$\Pi := \begin{bmatrix} I_{N-d_0} & 0 \\ 0 & 0 \end{bmatrix}.$$

There are invertible C^∞ $N \times N$ matrices $\Gamma_1(t,x)$ and $\Gamma_2(t,x)$ such that for all $(t,y) \in (0,T) \times R^{n-1}$, $(\Gamma_2 A_n \Gamma_1)(t,y,0) = \Pi$. For each $\varepsilon \in]0,1]$, we denote by $\tilde{w}^\varepsilon := \Gamma_1^{-1} w^\varepsilon$ the new unknown. Then for each $\varepsilon \in]0,1]$, the function w^ε verifies problem (13) if and only if $\tilde{w}^\varepsilon$ verifies $\tilde{\mathcal{P}}^\varepsilon \tilde{w}^\varepsilon = \tilde{G}(t,x,a^\varepsilon,\varepsilon^M \tilde{w}^\varepsilon).\tilde{w}^\varepsilon + \varepsilon^{M'-M}\tilde{g}^\varepsilon$ with $\tilde{w}^\varepsilon = 0$ when $t \leqslant 0$ or $x \in \Gamma$, where $\tilde{\mathcal{P}}^\varepsilon := \Gamma_2 \mathcal{P}^\varepsilon \Gamma_1$, $\tilde{G}(t,x,a,\tilde{w}) := \Gamma_2 G(t,x,a,\Gamma_1 \tilde{w})$ and $\tilde{g}^\varepsilon := \Gamma_2 g^\varepsilon$. The operator $\tilde{\mathcal{P}}^\varepsilon$ is of the form $\tilde{\mathcal{P}}^\varepsilon = \tilde{\mathcal{H}} - \varepsilon\tilde{\mathcal{E}}$, where $\tilde{\mathcal{H}} := \tilde{A}_0\partial_t + \sum_{1\leq j\leq n} \tilde{A}_j Z_j + \Pi\partial_n + \tilde{B}$ and $\tilde{\mathcal{E}} := \sum_{1\leq i,j\leq n} \partial_i \tilde{E}_{i,j}\partial_j + \sum_{1\leq i\leq n} \tilde{E}_i\partial_i$. Thanks to this change of unknown function, the commutator of the normal matrix of $\tilde{\mathcal{H}}$ with conormal derivatives will be simplified. We consider the family of iterative schemes $(\tilde{w}^{\varepsilon,\nu})_{\nu\in\mathbb{N},\varepsilon\in]0,1]}$ defined by $\tilde{w}^{\varepsilon,0} = 0$ and $\tilde{\mathcal{P}}^\varepsilon \tilde{w}^{\varepsilon,\nu+1} = \tilde{B}^{\varepsilon,\nu}\tilde{w}^{\varepsilon,\nu+1} + \varepsilon^{M'-M}\tilde{g}^\varepsilon$ with $\tilde{w}^{\varepsilon,\nu+1} = 0$ when $t \leq 0$ or when $x \in \Gamma$, where $\tilde{B}^{\varepsilon,\nu}(t,x) := \tilde{G}(t,x,a^\varepsilon,\varepsilon^M \tilde{w}^{\varepsilon,\nu})$. To obtain estimates for the $(\tilde{w}^{\varepsilon,\nu})_{\nu\in\mathbb{N},\varepsilon\in]0,1]}$, we first look at the family of linear problems:

$$\tilde{\mathcal{P}}^\varepsilon w^\varepsilon = \mathfrak{B}^\varepsilon w^\varepsilon + \mathfrak{g}^\varepsilon \quad \text{when } (t,x) \in (-1,T) \times \Omega, \tag{14}$$

$$w^\varepsilon = 0 \quad \text{when } (t,x) \in (-1,T) \times \Gamma \text{ or when } (t,x) \in (-1,0) \times \Omega, \tag{15}$$

where $(\mathfrak{B}^\varepsilon)_{\varepsilon\in]0,1]}$ is a family of $N \times N$ matrices defined by $\mathfrak{B}^\varepsilon := \tilde{G}(t,x,a^\varepsilon,\varepsilon^M \mathfrak{w}^\varepsilon)$. The family of functions $\mathfrak{w}^\varepsilon$ is given. We denote $\langle .,. \rangle$ the scalar product of $L^2(\mathbb{R}^n_+)$ and $||.||_2$ the associated norm. We introduce, for $\mu > 0$, $\lambda \geqslant 1$ and $m \in \mathbb{N}$, the weighted norms $|w||_{0,\lambda} := ||e^{-\lambda t} w||_{L^2((0,T)\times\Omega)}$, $|w||_{m,\mu,\lambda} := \sum_{0\leq k\leq m} \mu^{m-k} |Z^k w||_{0,\lambda}$, $||w||_{m,\mu,\lambda} := \sum_{0\leq k\leq m} \mu^{m-k} |\partial^k w||_{0,\lambda}$. We denote $\langle .,. \rangle_{0,\lambda}$ the scalar product of $L^2((0,T) \times \Omega)$ endowed with the measure $e^{-2\lambda t} dtdx$. Next proposition sums up estimates for the family of linear problems (14)–(15).

Proposition 1. *Let $R > 0$. For all $m \in \mathbb{N}$, there are $\lambda_m, C_m \geqslant 1$, such that for all $\varepsilon \in]0,1]$, if $\varepsilon^M ||\mathfrak{w}^\varepsilon||_\infty \leq R$, then for all $\lambda \geqslant \lambda_m$,*

$$\varepsilon^m ||w^\varepsilon||_{m,\frac{\lambda}{\varepsilon},\lambda} \leq \frac{C_m}{\lambda}(||\mathfrak{g}^\varepsilon||_{m,\lambda,\lambda} + (\varepsilon^M ||w^\varepsilon||_\infty).\varepsilon^m ||\mathfrak{w}^\varepsilon||_{m,\frac{\lambda}{\varepsilon},\lambda}).$$

Notice that there is a lost of a factor ε^m from the source term $\mathfrak{g}^\varepsilon$ to the solution w^ε. More precisely, each derivative of w^ε needs a factor ε to be bounded by the correspondent derivative of $\mathfrak{g}^\varepsilon$. This explains that we use two different weights: $\frac{\lambda}{\varepsilon}$ and λ in the norms which appear.

Proof. We reason by induction. To initiate the induction ($m = 0$), we start with L^2 estimates. A key point is to use two forms of (14) and the two kinds of estimates associated to these forms. The first step mainly uses the symmetry of the operator $\mathcal{H}$ and Assumption 3. The second step uses Assumption 2. Thanks to Assumption 2, there exists some matrices $P_0(t,x,\zeta)$, smooth in its arguments, for $\zeta \neq 0$, homogeneous of degree 0 in ζ, positive definite (i.e., $P_0 \geqslant cI > 0$), such that $-((P_0 A_0^{-1}\mathrm{e})(t,x,\zeta) + ((A_0^{-1}\mathrm{e})^* P_0)(t,x,\zeta))$ is also positive definite, i.e.,

$$-((P_0 A_0^{-1}\mathrm{e})(t,x,\zeta) + ((A_0^{-1}\mathrm{e})^* P_0)(t,x,\zeta)) \geqslant c|\zeta|^2 I > 0. \tag{16}$$

To simplify the notations, we will write P_0 instead of $P_0(t,x)$. We choose $K > 0$ such that $Q := \frac{1}{2}(P_0 + P_0^*) + K\Lambda^{-1}$ is positive definite on $L^2((0,T)\times\Omega)$ endowed with the scalar product $\langle .,. \rangle_{0,\lambda,T}$. We have denoted by Λ the operator associated to the symbol $\langle\zeta\rangle := (1+\zeta^2)^{\frac{1}{2}}$. We denote by S the $N \times N$ matrix $S := \Gamma_1^t \Gamma_2^{-1}$.

Proposition 2. *There is $\lambda_0 \geqslant 1$, such that for all $\varepsilon \in]0,1]$, if $\varepsilon^M ||\mathfrak{w}^\varepsilon||_\infty \leq R$, then for all $\lambda \geqslant \lambda_0$,*

$$\begin{aligned}\lambda|w^\varepsilon||^2_{0,\lambda} + \varepsilon^2|\partial_x w^\varepsilon||^2_{0,\lambda} \leq \lambda_0(|\langle w^\varepsilon, S\mathfrak{g}^\varepsilon\rangle_{0,\lambda}| \\ + \varepsilon|\langle A_0^{-1}\Gamma_2^{-1}\mathfrak{g}^\varepsilon, Q\Gamma_1 w^\varepsilon\rangle_{0,\lambda}|).\end{aligned} \tag{17}$$

Proof. We proceed in two steps. We begin with a first L^2 estimate of w^ε.

Lemma 1. *There is $\lambda_0 \geqslant 1$, such that for all $\varepsilon \in]0,1]$, if $\varepsilon^M ||\mathfrak{w}^\varepsilon||_\infty \leq R$, then for all $\lambda \geqslant \lambda_0$,*

$$\lambda|w^\varepsilon||^2_{0,\lambda} \leq \lambda_0|\langle w^\varepsilon, S\mathfrak{g}^\varepsilon\rangle_{0,\lambda}|. \tag{18}$$

Proof. For all $t \in (0,T)$, we multiply (14) on the left by the matrix S. We introduce the following symmetric operators $\check{\mathcal{H}} := \Gamma_1^t \tilde{\mathcal{H}} \Gamma_1$, $\check{\mathcal{E}} := \Gamma_1^t \tilde{\mathcal{E}} \Gamma_1$, and $\check{\mathcal{P}}^\varepsilon := \check{\mathcal{H}}(t,x,\partial) - \varepsilon\check{\mathcal{E}}(t,x,\partial)$. The operators $\check{\mathcal{H}}$ and $\check{\mathcal{E}}$ are of the form

$$\check{\mathcal{H}}(t,x,\partial) = \check{A}_0(t,x)\partial_t + \sum_{1\leq j\leq n} \check{A}_j(t,x)\partial_j + \check{B}(t,x),$$

$$\check{\mathcal{E}}(t,x,\partial) = \sum_{1\leq i,j\leq n} \partial_i \check{E}_{i,j}(t,x)\partial_j + \sum_{1\leq i\leq n} \check{E}_i(t,x)\partial_i.$$

We introduce the matrices $\check{\mathfrak{B}}^\varepsilon := S\mathfrak{B}^\varepsilon$ and the functions $\check{\mathfrak{g}}^\varepsilon := S\mathfrak{g}^\varepsilon$. Thus we get the following equation: $\check{\mathcal{P}}^\varepsilon w^\varepsilon = \check{\mathfrak{B}}^\varepsilon w^\varepsilon + \check{\mathfrak{g}}^\varepsilon$. Then we do a scalar product with w^ε then a space integration. We get

$$\begin{aligned}&\int_{x\in\mathbb{R}^n_+} (w^\varepsilon)^t \cdot \check{\mathcal{H}} w^\varepsilon - \varepsilon \int_{x\in\mathbb{R}^n_+} (w^\varepsilon)^t \cdot \check{\mathcal{E}} w^\varepsilon \\ &\qquad = \int_{x\in\mathbb{R}^n_+} (w^\varepsilon)^t \cdot \check{\mathfrak{B}}^\varepsilon w^\varepsilon + \int_{x\in\mathbb{R}^n_+} (w^\varepsilon)^t \cdot \check{\mathfrak{g}}_\varepsilon.\end{aligned}$$

Thanks to the symmetry and the homogenous Dirichlet conditions (15), we get

$$\int_{x\in\mathbb{R}^n_+} (w^\varepsilon)^t \cdot \check{\mathcal{H}} w^\varepsilon = \frac{1}{2}\partial_t\Big(\int_{x\in\mathbb{R}^n_+} |\check{A}_0 w^\varepsilon|^2\Big) - \frac{1}{2}\int_{x\in\mathbb{R}^n_+} \sum_{0=1}^{n} (w^\varepsilon)^t \cdot (\partial_i \check{A}_i).w^\varepsilon.$$

Thus with Assumption 3 we have, for all $t \in (0,T)$, $\partial_t||w^\varepsilon||_2^2 \leq C(||w^\varepsilon||_2^2 + |\langle\breve{\mathfrak{g}}^\varepsilon, w^\varepsilon\rangle|^2)$. Then, multiply by $e^{-2\lambda t}$, integrate for $t \in (0,T)$, and use a Gronwall lemma which yield estimate (18).

We now use Assumption 2 and its consequence (16) to get an estimate of $\partial_x w^\varepsilon$.

Lemma 2. *There is $\lambda_0 \geqslant 1$, such that for all $\varepsilon \in]0,1]$, if $\varepsilon^M||\mathfrak{w}^\varepsilon||_\infty \leq R$, then for all $\lambda \geqslant \lambda_0$,*

$$\varepsilon^2|\partial_x w^\varepsilon||_{0,\lambda}^2 \leq \lambda_0(\varepsilon|\langle A_0^{-1}\Gamma_2^{-1}\mathfrak{g}^\varepsilon, Q\Gamma_1 w^\varepsilon\rangle_{0,\lambda}| + |\langle w^\varepsilon, S\mathfrak{g}^\varepsilon\rangle_{0,\lambda}|). \tag{19}$$

Proof. We introduce $\underline{Q} := \Gamma_1^t Q\Gamma_1$ and define the operator $\hat{\mathcal{P}}^\varepsilon := \tilde{A}_0^{-1}\tilde{\mathcal{P}}^\varepsilon = \hat{\mathcal{H}} - \varepsilon\hat{\mathcal{E}}$, where $\hat{\mathcal{H}} := \partial_t + \sum_{1\leq j\leq n}\hat{A}_j\partial_j + \hat{B}$ and $\hat{\mathcal{E}} := \sum_{1\leq i,j\leq n}\partial_i\tilde{E}_{i,j}\partial_j + \sum_{1\leq i\leq n}\tilde{E}_i\partial_i$. We introduce the matrices $\hat{\mathfrak{B}}^\varepsilon := \tilde{A}_0^{-1}\tilde{\mathfrak{B}}^\varepsilon$ and the functions $\hat{\mathfrak{g}}^\varepsilon := \tilde{A}_0^{-1}\tilde{\mathfrak{g}}^\varepsilon$ and multiply (14) by $\tilde{A}_0^{-1}$ to find

$$\begin{aligned}\langle\partial_t w^\varepsilon, \underline{Q}w^\varepsilon\rangle := -\sum_{1\leq j\leq n}\langle\hat{A}_j\partial_j w^\varepsilon, \underline{Q}w^\varepsilon\rangle - \langle(\hat{B}+\hat{\mathfrak{B}}^\varepsilon)w^\varepsilon, \underline{Q}w^\varepsilon\rangle \\ + \varepsilon\langle\hat{\mathcal{E}}(t,x,\partial)w^\varepsilon, \underline{Q}w^\varepsilon\rangle + \langle\hat{\mathfrak{g}}^\varepsilon, \underline{Q}w^\varepsilon\rangle.\end{aligned}$$

Thanks to property (16), there are $c > 0$ and $C \in \mathbb{R}$ such that

$$-\langle\hat{\mathcal{E}}(t,x,\partial)w^\varepsilon, \underline{Q}w^\varepsilon\rangle = -\langle A_0^{-1}\mathcal{E}\Gamma_1 w^\varepsilon, \underline{Q}\Gamma_1 w^\varepsilon\rangle \geqslant -c||\partial_x w^\varepsilon||_2^2 + C||w^\varepsilon||_2^2.$$

Because the lack of symmetry, we can deal with the hyperbolic part as we have done in Lemma 1 integrating by parts. Instead, for $1 \leq i \leq n$, we use Cauchy–Schwarz inequality to obtain the singular estimate:

$$|\langle\hat{A}_i\partial_i w^\varepsilon, \underline{Q}w^\varepsilon\rangle| \leq C||w^\varepsilon||_2||\partial_i w^\varepsilon||_2 \leq \frac{1}{\varepsilon}C||w^\varepsilon||_2^2 + \frac{c\varepsilon}{2}||\partial_i w^\varepsilon||_2^2.$$

Thus, we get

$$\frac{d}{dt}\langle w^\varepsilon, \underline{Q}w^\varepsilon\rangle + \frac{c\varepsilon}{2}||\partial_x w^\varepsilon||_2^2 \leq C(|\langle A_0^{-1}\Gamma_2^{-1}\mathfrak{g}^\varepsilon, Q\Gamma_1 w^\varepsilon\rangle| + \frac{1}{\varepsilon}||w^\varepsilon||_2^2).$$

Multiply by $e^{-2\lambda t}$, integrate for $t \in (0,T)$, use a Gronwall lemma, the definite positiveness of $\underline{Q}$ and Lemma 1 to find estimate (19).

Mix estimates (18) and (19) to find estimate (17) and prove proposition 2.

To get higher order estimates, we proceed in two steps. First, we get conormal estimates and then normal derivatives estimates. Classic methods yield the end of the proof of Theorem 3. We do not detail them. □

References

1. C. Bardos, D. Brézis, and H. Brezis. Perturbations singulières et prolongements maximaux d'opérateurs positifs. *Arch. Rational Mech. Anal.*, 53:69–100, 1973/74.
2. Olivier Guès. Problème mixte hyperbolique quasi-linéaire caractéristique. *Comm. Partial Differential Equations*, 15(5):595–645, 1990.
3. Olivier Guès. Perturbations visqueuses de problèmes mixtes hyperboliques et couches limites. *Ann. Inst. Fourier (Grenoble)*, 45(4):973–1006, 1995.
4. Heinz-Otto Kreiss and Jens Lorenz. *Initial-boundary value problems and the Navier-Stokes equations*, volume 136 of *Pure and Applied Mathematics*. Academic Press Inc., Boston, MA, 1989.
5. J. Rauch. Boundary value problems as limits of problems in all space. In *Séminaire Goulaouic-Schwartz (1978/1979)*, pages Exp. No. 3, 17. École Polytech., Palaiseau, 1979.
6. Jeffrey Rauch. Symmetric positive systems with boundary characteristic of constant multiplicity. *Trans. Amer. Math. Soc.*, 291(1):167–187, 1985.
7. Franck Sueur. Couches limites semilinéaires. *Annales de la faculté des sciences de Toulouse, Sér. 6, 15 no. 2 (2006), p. 323–380.*
8. Franck Sueur. A few remarks on a theorem by J. Rauch. *Indiana Univ. Math. J.*, 54(4):1107–1143, 2005.
9. Franck Sueur. Couches limites: un problème inverse. *Comm. Partial Differential Equations*, 31(1-3):123–194, 2006.

A Riemann Solver Approach for Conservation Laws with Discontinuous Flux

M. Garavello, R. Natalini, B. Piccoli, and A. Terracina

1 Introduction

We consider the hyperbolic conservation law with discontinuous flux

$$\begin{cases} u_t + (H(x)f(u) + (1 - H(x))g(u))_x = 0, & \text{in } \mathbf{R} \times (0, T), \\ u(x, 0) = u_0(x), & \text{in } \mathbf{R} \times \{0\}, \end{cases} \tag{1}$$

where H denotes the Heaviside function. There are different physical models in which this kind of problem can arise (e.g., continuous sedimentation, two phases flow in porous medium, traffic flow).

The Cauchy problem (1) was extensively studied in literature both from numerical and theoretical point of view (see [AJG04, AMG05, BKRT04, Dih95, GR92, KKR04, KRT02, KRT03, KT04, SV03, Tow01, Tow02]). In particular in [AMG05], the authors considered the case where the fluxes f and g have a unique maximum (or minimum) point and they proved that is possible to give different definitions (depending on applications) of entropy solutions, which produce contractive semigroups in L^1. Moreover, by using singular mapping methods, they also showed the convergence of the Godunov-type approximations to an entropy solution u in the domains $(-\infty, 0) \times (0, T)$ and $(0, \infty) \times (0, T)$. The solution u satisfies also an entropy formulation along the line $x = 0$, provided that the discontinuities of u lies in a finite number of Lipschitz curves.

In this chapter, we are motivated by a model of traffic flow for which there is no a priori preferable physical solution. Thus we are interested in giving a unifying point of view for all the possible entropy formulations. We assume that the fluxes $f : [0, 1] \to \mathbf{R}$, $g : [0, 1] \to \mathbf{R}$ satisfy the following properties:

1. f and g are strictly concave functions
2. $f(0) = g(0) = f(1) = g(1) = 0$
3. There exists $\sigma_g \in]0, 1[$ such that $g(\sigma_g) \geq g(u)$ for every $u \in [0, 1]$
4. There exists $\sigma_f \in]0, 1[$ such that $f(\sigma_f) \geq f(u)$ for every $u \in [0, 1]$

For simplicity, we assume that $g(\sigma_g) \leq f(\sigma_f)$.

The Cauchy problem (1) with concave flux functions f and g derives from considering a fluidodynamic description for traffic flow on a simple road network, composed by two roads connected together by a junction (see [CGP05, GP06] for more general networks). In this chapter, we use the scalar model, introduced by Lighthill, Whitham, and Richards (see [Ric56, Whi74, LW55]), with different flux functions in the two roads.

Using a different approach with respect to [AMG05], we deduce the same classes of entropy solutions. Our method is based in the study of all the possible Riemann solvers along the junction $x = 0$ (see Sect. 2). We prove, by using the wave-front tracking method, the existence of an entropy solution for the Cauchy problem (1) when the initial datum u_0 has bounded total variation. The key point is to obtain a uniform BV estimate for the sequence of approximating fluxes $f(u_n(\cdot, 0+))$, where u_n are wave-front tracking approximations. In general, it is not possible to obtain BV estimates directly for u_n, since interactions of waves with the junction can increase the total variation of the conserved quantity.

2 The Riemann Problem

Consider the following Riemann problem

$$\begin{cases} u_t + g(u)_x = 0, & \text{if } x < 0,\, t > 0, \\ u_t + f(u)_x = 0, & \text{if } x > 0,\, t > 0, \\ u(0,x) = u_l, & \text{if } x < 0, \\ u(0,x) = u_r, & \text{if } x > 0, \end{cases} \tag{2}$$

where $u_l \in [0,1]$ and $u_r \in [0,1]$. From standard theory for scalar conservation laws with boundary conditions, a solution to (2) is completely determined if we know the traces $u^- = u(t, 0-)$ and $u^+ = u(t, 0+)$. Then, solving the Riemann problem (2) is equivalent to find a function, which associates to every couple of initial data (u_l, u_r) a couple of traces (u^-, u^+) and which also satisfies some properties.

Definition 1. *A Riemann solver for the Riemann problem (2) is a function* $R : [0,1] \times [0,1] \to [0,1] \times [0,1]$, $R(u_l, u_r) = (R_1(u_l, u_r), R_2(u_l, u_r)) = (u^-, u^+)$, *such that:*

(H1) $g(u^-) = f(u^+)$

(H2) The wave (u_l, u^-) *has negative speed, while the wave* (u^+, u_r) *has positive speed*

(H3) The function $(u_l, u_r) \mapsto (g(u^-), f(u^+))$ *is continuous*

(H4) $R(R(u_l, u_r)) = R(u_l, u_r)$ *for every* $u_l \in [0,1]$ *and* $u_r \in [0,1]$

(H5) For every $(u_l, u_r) = R(u_l, u_r)$ *and* $\tilde{u}$ *such that the wave* $(\tilde{u}, R_1(u_l, u_r))$ *has positive speed the following holds*

$$g(R_1(\tilde{u}, u_r)) \in [\min\{g(u_l), g(\tilde{u})\}, \max\{g(u_l), g(\tilde{u})\}] \tag{3}$$

(H6) For every $(u_l, u_r) = R(u_l, u_r)$ *and* $\tilde{u}$ *such that the wave* $(R_2(u_l, u_r), \tilde{u})$ *has negative speed the following holds*

$$f(R_2(u_l, \tilde{u})) \in [\min\{f(u_r), f(\tilde{u})\}, \max\{f(u_r), f(\tilde{u})\}] \tag{4}$$

Definition 2. *A couple* (u_l, u_r) *is said an equilibrium if* $R(u_l, u_r) = (u_l, u_r)$.

Remark 1. Observe that conditions (H1) and (H2) are physically motivated by the conservation of mass at the junction and by the fact that waves originated at $x = 0$ in $(-\infty, 0)$ (resp. in $(0, \infty)$) must travel with negative (resp. positive) speed.

Condition (H3) is a regularity property for the Riemann solver, while (H4) is a stability condition, in the sense that the image of R is a fixed point of the same function.

Finally conditions (H5) and (H6) are the key assumptions for some important estimates for the existence of solutions to Cauchy problems, as we see in Sect. 3.

For every fixed Riemann solver we give the following definition of solution to the Cauchy problem (1).

Definition 3. *Fix a Riemann solver* R *and an initial condition* $u_0 \in BV(\mathbf{R})$. *We say that* u *is an entropy solution to the Cauchy problem (1), related to the Riemann solver* R, *if and only if:*

1. u, *restricted to* $(0, T) \times (-\infty, 0)$, *is an entropy solution to*

$$u_t + g(u)_x = 0$$

2. u, *restricted to* $(0, T) \times (0, +\infty)$, *is an entropy solution to*

$$u_t + f(u)_x = 0$$

3. $u(0, x) = u_0(x)$ *for almost every* $x \in \mathbf{R}$
4. $R(u(t, 0-), u(t, 0+)) = (u(t, 0-), u(t, 0+))$ *for almost every* $t \in (0, T)$

To characterize all the possible Riemann solvers, introduce the set

$$X := \{s \in [0, g(\sigma_g)] : (u_l, u_r) \in [\sigma_g, 1] \times [0, \sigma_f] \text{ equilibrium}, \ g(u_l) = f(u_r) = s\} \tag{5}$$

The following lemma is a simple consequence of the properties $(H_1) - (H_6)$ in the definition of Riemann solver.

Lemma 1. X *is a closed, nonempty and connected set. Thus* $X = [\bar{\gamma}_1, \bar{\gamma}_2]$, *with* $0 \leq \bar{\gamma}_1 \leq \bar{\gamma}_2 \leq g(\sigma_g)$.

It is easy to show that, if the set X is composed by more than one element, then the Cauchy problem (1) does not have a unique solution, in the sense of Definition 3. In fact let $\gamma_1 < \gamma_2 \in X$ and take $u_1^-, u_2^- \in [\sigma_g, 1]$, $u_1^+, u_2^+ \in [0, \sigma_f]$ such that

$$g(u_1^-) = f(u_1^+) = \gamma_1, \qquad g(u_2^-) = f(u_2^+) = \gamma_2.$$

Clearly both (u_1^-, u_1^+) and (u_2^-, u_2^+) are equilibria. Then for every initial data

$$u_0(x) = \begin{cases} \underline{u}, & \text{if } x \leq 0, \\ \overline{u}, & \text{if } x > 0, \end{cases}$$

where $\underline{u} \in [\sigma_g, 1]$ and $\overline{u} \in [0, \sigma_f]$, we can find two admissible solutions u_1 and u_2 to (1) in the sense of Definition 3, such that the traces at $x = 0$ are, respectively, (u_1^-, u_1^+) and (u_2^-, u_2^+).

For this reason, in the following we consider only the case in which X is a singleton. If $X = \{\gamma\}$, then we denote the associate Riemann solver with R_γ. We describe completely the Riemann solver R_γ (see [GNPT06]). We have the following possibilities:

1. $u_l \in [\sigma_g, 1]$ and $u_r \in [0, \sigma_f]$. In this case the solution to the Riemann problem satisfies $u^- \in [\sigma_g, 1]$, $u^+ \in [0, \sigma_f]$ and $g(u^-) = f(u^+) = \bar{\gamma}$.
2. $u_l \in [0, \sigma_g[$ and $u_r \in [0, \sigma_f]$. If $g(u_l) > \bar{\gamma}$, then the solution to the Riemann problem satisfies $u^- \in [\sigma_g, 1]$, $u^+ \in [0, \sigma_f]$ and $g(u^-) = f(u^+) = \bar{\gamma}$.
   If $g(u_l) \leq \bar{\gamma}$, then the solution to the Riemann problem satisfies: $u^- = u_l$, $u^+ \in [0, \sigma_f]$ and $g(u^-) = f(u^+)$.
3. $u_l \in [\sigma_g, 1]$ and $u_r \in]\sigma_f, 1]$. The situation is completely symmetric to the previous case.
4. $u_l \in [0, \sigma_g[$ and $u_r \in]\sigma_f, 1]$. If $\min\{g(u_l), f(u_r)\} > \bar{\gamma}$, then the solution to the Riemann problem satisfies $u^- \in [\sigma_g, 1]$, $u^+ \in [0, \sigma_f]$ and $g(u^-) = f(u^+) = \bar{\gamma}$.
   If $\min\{g(u_l), f(u_r)\} \leq \bar{\gamma}$ and $g(u_l) = f(u_r)$, then the solution to the Riemann problem is (u_l, u_r).
   If $\min\{g(u_l), f(u_r)\} \leq \bar{\gamma}$ and $g(u_l) < f(u_r)$, then the solution to the Riemann problem satisfies $u^- = u_l$, $u^+ \in [\sigma_f, 1]$ and $g(u_l) = f(u^+)$.
   If $\min\{g(u_l), f(u_r)\} \leq \bar{\gamma}$ and $g(u_l) > f(u_r)$, then the solution to the Riemann problem satisfies $u^- \in [\sigma_g, 1]$, $u^+ = u_r$ and $g(u^-) = f(u_r)$.

Remark 2. If $f(\sigma_f) = g(\sigma_g)$ and $X = \{g(\sigma_g)\}$, then the Riemann solver is completely identical to that used in [AJG04] and [CGP05].

Remark 3. If there exists a unique $u^* \in]0, 1[$ such that $f(u^*) = g(u^*)$ and if $X = \{f(u^*)\}$, then the Riemann solver is identical to that used in [Dih95, KT04].

Using the previous analysis we obtain the following conditions about admissible equilibria for the Riemann solver R_γ.

Proposition 1. *A function $u \in L^\infty((0,T) \times \mathbf{R})$ is an admissible solution to (1) in the sense of Definition 3 for the Riemann solver R_γ if and only if u verifies conditions* 1 *and* 2 *of Definition 3 and*

3′ *for almost every $t \geq 0$ the couple $(u(t,0-), u(t,0+))$ satisfies*
   (i) $g(u(t,0-)) = f(u(t,0+)) \leq \gamma$;
   (ii) if $(u(t,0-), u(t,0+)) \in [\sigma_g, 1) \times (0, \sigma_f]$, then

$$(u(t,0-), u(t,0+)) = (a_\gamma, b_\gamma)$$

where a_γ and b_γ are determined by $g(a_\gamma) = f(b_\gamma) = \gamma$.

The previous characterization permits to prove the following proposition.

Proposition 2. *Consider a Riemann solver R_γ. For every two equilibria (u_1^-, u_1^+) and (u_2^-, u_2^+) of R_γ it holds*

$$\operatorname{sgn}(u_1^- - u_2^-)\left[g(u_1^-) - g(u_2^-)\right] \geq \operatorname{sgn}(u_1^+ - u_2^+)\left[f(u_1^+) - f(u_2^+)\right]. \tag{6}$$

This can be easily proved by considering all the different possibilities.

By using Kruzkov techniques and previous proposition, it is possible to prove the following result about the L^1 stability of the solution with respect to the initial data.

Theorem 1. *Fix $\gamma \in (0, g(\sigma_g)]$ and u_0, $v_0 \in BV(\mathbf{R})$. Let u and v be admissible solutions to the Cauchy problem (1) in the sense of Definition 3 with respect to the Riemann solver R_γ, where the initial conditions are respectively u_0 and v_0. Then, for every $C > 0$ and for almost every $t \in (0,T)$,*

$$\int_{-C}^{C} |u(t,x) - v(t,x)|\, dx \leq \int_{-C-Mt}^{C+Mt} |u_0(x) - v_0(x)|\, dx, \tag{7}$$

where

$$M = \max\left\{ \max_{u \in [-a,a]} |f'(u)|, \max_{u \in [-a,a]} |g'(u)| \right\}$$

and $a = \max\{\|u\|_{L^\infty}, \|v\|_{L^\infty}\}$.

3 Existence of Entropy Solutions

In this section we consider the problem of existence of an entropy solution to (1) according to Definition 3. Fix the Riemann solver R_γ such that $X = \{\gamma\}$.

We use the wave-front tracking method. We fix a sequence of piecewise constant approximations $u_{0,\nu}$ of the initial datum u_0, such that $\text{Tot.Var.}u_{0,\nu} \leq \text{Tot.Var.}u_0$. We solve the Riemann problems at any point of discontinuity and at the junction $x = 0$. Therefore we have a solution until two waves interact together or a wave hits the junction. Then we repeat inductively this procedure (see [Bre00]). We have the following result.

Theorem 2. *Given $\gamma \in (0, \sigma_g]$ and $u_0 \in BV(\mathbf{R})$, there exists an admissible solution u to problem (1) in the sense of Definition 3.*

We give only a sketch of the proof (for details see [GNPT06]). First of all we need global BV estimates, that allow compactness. As mentioned in the introduction, it is not possible to obtain BV estimate of u_ν, while we can control the total variation of $h(x, u_\nu)$ where $h(x, u) = H(x)f(u) + (1 - H(x))g(u)$. The following lemma holds.

Lemma 2. *Fix an approximate wave front tracking solution $\bar{u}$. For every $t \geq 0$, it holds*

$$Tot.Var.(h(\cdot, \bar{u}(t, \cdot))) \leq Tot.Var.(h(\cdot, \bar{u}(0+, \cdot))). \tag{8}$$

Proof (Sketch). When two waves interact together, the total variation of u_ν decreases. If a wave interacts with the junction, then conditions (H5)–(H6) imply that the total variation of the fluxes does not change. □

Let us consider an equilibrium (u^-, u^+) for an approximate wave front tracking solution. We say that u^- (resp. u^+) is "good" if $u^- \in [a_\gamma, 1]$ (resp. $u^+ \in [0, b_\gamma]$). If u^- (resp. u^+) is not good, then we say u^- (resp. u^+) is "bad." Using these definitions, we introduce four classes of equilibrium:

1. u^- and u^+ are "good": we denote the equilibrium by $G|G$. In this case u^- and u^+ are equal to a_γ and b_γ
2. u^- is "good" and u^+ is "bad": we denote the equilibrium by $G|B$
3. u^- is "bad" and u^+ is "good": we denote the equilibrium by $B|G$
4. u^- and u^+ are "bad": we denote the equilibrium by $B|B$

By considering the characterizations of the admissible equilibria given in Proposition 1, the following remark can be easily proved (see [GNPT06]).

Remark 4. If the trace u^- (resp. u^+) is good, then it can become bad only when a wave interacts with the junction from the left (resp. from the right).

This remark is important for the following lemma.

Lemma 3. *For every initial datum $u_0 \in BV(\mathbf{R})$ there exists an entropy solution $u(t, x)$ that satisfies points 1 and 2 of Definition 3.*

Proof. By using the previous results, we can extract from $\{h(x, u_\nu)\}$ a subsequence converging almost everywhere to a function $\overline{h}$. The convergence of the subsequence of $\{u_\nu\}$ can be proved by using the procedure of [CGP05]. It consists in considering two regions for each roads. In the first region the solution depends both on the junction and on the initial condition, while in the second region the solution depends only on the initial condition. In the second region, the convergence of the subsequence of u_n can be easily deduces with standard methods. In the first region instead, it is necessary to do considerations similar to those of Remark 4. □

To prove Theorem 2, we need to show that the limit function u satisfies condition $3'$ of Proposition 1. The first step in this direction consists to prove a priori BV estimates for the functions $f(u_\nu(t,0+))$.

Lemma 4. *Let $\{u_\nu\}$ be an approximate wave-front tracking sequence. Then for every ν we have*

$$TV f(u_\nu^+(\cdot),(0,T)) \leq 2\,TV(h(\cdot,u_0(\cdot)),\mathbf{R}), \tag{9}$$

where $u_\nu^+(\cdot) := u_\nu(t,0+)$.

The proof can be found in [GNPT06]. Again Remark 4 is crucial for studying the cases of interactions between waves and the junction at $x=0$.

Sketch of the proof of Theorem 2. By using Lemma 3, we obtain a function u that verifies 1 and 2 of Definition 3. Since for every ν the couple $(u_\nu(t,0-),u_\nu(t,0+))$ is an admissible equilibrium of problem (1), we can prove, by using Lemma 9, Proposition 1 and by passing to the limit along a subsequence, that $(u(t,0-),u(t,0+))$ it is also an admissible equilibrium.

References

[AJG04] Adimurthi, Jaffré, J., Gowda, G.D.V.: Godunov-type methods for conservation laws with a flux function discontinuous in space, SIAM J. Numer. Anal. **42**, No. 1 179–208 (2004).

[AMG05] Adimurthi, Mishra, S., Gowda, G.D.V.: Optimal entropy solutions for conservation laws with discontinuous flux-functions, J. Hyperbolic Differ. Equ. **2**, No. 4, 783–837 (2005).

[Bre00] Bressan, A.: Hyperbolic Systems of Conservation Laws – The one-dimensional Cauchy Problem, Oxford Univ. Press, Oxford (2000).

[BKRT04] Burger, R., Karlsen, K.H., Risebro, N.H., Towers, J.D.: Well-posedness in BV_t and convergence of a difference scheme for continuous sedimentation in ideal clarifier-thickener units, Numer. Math. **97**, 25–65 (2004).

[CP05] Chitour, Y., Piccoli, B.: Traffic circles and timing of traffic lights for car flow, Discrete Contin. Dyn. Syst. Ser. B **5**, No. 3, 599–630 (2005).

[CGP05] Coclite, G.M., Garavello, M. , Piccoli, B.: Traffic flow on a road network, SIAM J. Math. Anal., **36**, No. 6 1862–1886 (2005).

[Dih95] Diehl, S.: On scalar conservation laws with point source and discontinuous flux function, SIAM J. Math. Anal., **26**, 1425–1451 (1995).

[GNPT06] Garavello, M., Natalini, R., Piccoli, B., Terracina, A.: Conservation Laws with Discontinuous Flux, *to appear in* Networks and Heterogenous Media.

[GR92] Gimse, T., Risebro, N.H.: Solution of the Cauchy problem for a conservation law with a discontinuous flux function, SIAM J. Math. Anal., **23**, 635–648 (1992).

[GP06] Garavello, M., Piccoli, B.: Traffic Flow on Networks, Applied Mathematics Series Vol. 1, American Institute of Mathematical Sciences, (2006).

[KKR04] Karlsen, K.H., Klingeberg, C., Risebro, N.H.: A relaxation scheme for conservation laws with a discontinuous coefficient, Math. Comp., **73**, 1235–1259 (2004).

[KRT02] Karlsen, K.H., Risebro, N.H., Towers, J.D.: Upwind difference approximations for degenerate parabolic convection-diffusion equations with a discontinuous coefficient, IMA J. Numer. Anal., **22**, 623–664 (2002).

[KRT03] Karlsen, K.H., Risebro, N.H.,Towers, J.D.: L^1 stability for entropy solutions of nonlinear degenerate parabolic convection-diffusion equations with discontinuous coefficients, K. Nor. Vidensk Selsk, **3**, 1–49 (2003).

[KT04] Karlsen, K.H., Towers, J.D.: *Convergence of the Lax-Friedrichs scheme and stability for conservation laws with a discontinuous space-time dependent flux*, Chinese Ann. Math. Ser. B, **25**, 287–318 (2004).

[LW55] Lighthill, M.J., Whitham, G.B.: On kinematic waves. II. A theory of traffic flow on long crowded roads, Proc. Roy. Soc. London. Ser. A., **229**, 317–345 (1955).

[Ric56] Richards, P.I.: Shock waves on the highway, Operations Res., **4**, 42–51 (1956).

[SV03] Seguin, N., Vovelle, J.: Analysis and approximation of a scalar conservation law with a flux function with discontinuous coefficients, Math. Models Methods Appl. Sci., **13**, 708–721 (2003).

[Tow01] Towers, J.D.: Convergence of a difference scheme for conservation laws with a discontinuous flux, SIAM J. Numer. Anal. **38**, 681–698 (2000).

[Tow02] Towers, J.D.: Difference scheme for conservation laws with a discontinuous flux: the nonconvex case, SIAM J. Numer. Anal. **39**, 1197–1218 (2001).

[Whi74] Whitham, G.B.: Linear and Nonlinear Waves, Pure and Applied Math., Wiley–Interscience, New York, 1974.

The Strong Shock Wave in the Problem on Flow Around Infinite Plane Wedge

D.L. Tkachev, A.M. Blokhin, and Y.Y. Pashinin

Summary. We consider the flow of an inviscid nonheat-conducting gas at the thermodynamical equilibrium around a plane infinite wedge and study the stationary solution to this problem associated with the so-called strong shock wave, when the flow behind the shock is subsonic.

We find a solution to the linearized problem and prove that its trace on the shock wave is a superposition of direct and reflected waves. Moreover, and that is most important, we prove the asymptotic Lyapunov's stability of the strong shock wave provided that the uniform Lopatinsky condition is fulfilled, the initial data are compactly supported, and some solvability conditions are satisfied.

1 Introduction

As is well known [1,2], in a supersonic uniform flow of an inviscid nonheat-conducting gas around an infinite plane wedge two regimes with a adjoint shock wave are theoretically possible: the flow with the weak shock wave, when the flow behind the shock wave is, generally speaking, supersonic, i.e., $u_0^2 + v_0^2 > c_0^2$, and the flow with the strong shock wave, when the flow behind the shock wave is subsonic, i.e., $u_0^2 + v_0^2 < c_0^2$. Here u_0 *and* v_0 are components of the gas velocity in the shock wave, c_0 is the sound speed. Besides, $U_\infty > c_\infty$ in the incoming flow, where c_∞ is the sound speed.

Practically, only the solution with the weak shock wave is realized unless experiments or numerical calculations [3] organized in a special way. The reason of that is still unknown. Consequently, it is unclear in advance which of two solutions is realized for a concrete flow.

A possible approach to tackling the problem is suggested by Courant and Friedrichs [4] and consists in studying the stability of steady regimes of gas flows with respect to small perturbations, in other words, studying the asymptotics of solutions to linearized initial-boundary-value problems (see (2.1)–(2.5) in Sect. 2) as $t \to \infty$.

Moreover, Courant and Friedrichs actually have proposed a hypothesis that the regime with the weak shock wave is stable with respect to small perturbations whereas the regime with the strong shock is unstable.

This hypothesis is strictly justified in [5,6] for small perturbations depending on a single "spatial" variable.

It has been shown in [7] for the general two-dimensional case that the main solution corresponding to the supersonic flow with the weak shock wave is stable with respect to small perturbations if the gas flow behind the shock is supersonic and $M_1(\Theta) > 1$ for $\sigma \le \Theta \le \Theta_s$.

Here

$$M_1(\Theta) = \frac{u_0 \cos\Theta + v_0 \sin\Theta}{c_0}.$$

At the same time, it has been stated in [8] that the linearized initial-boundary-value problem is also well posed if

$$u_0^2 + v_0^2 < c_0^2$$

(at least for small wedge angles σ, see Fig. 1). However, stability of such flow regimes has not been proven in [8].

In a series of articles (see, i.e., [9,10]), the absence of steady regimes with strong shock waves for pointed bodies of a finite thickness was established with a qualitative reasoning. Reliable arguments are also given in [11,12].

In this work we continue the study of stability of the solution with the strong shock started in [13,14] and formulate some stability conditions on the initial data for the linearized initial-boundary-value problem.

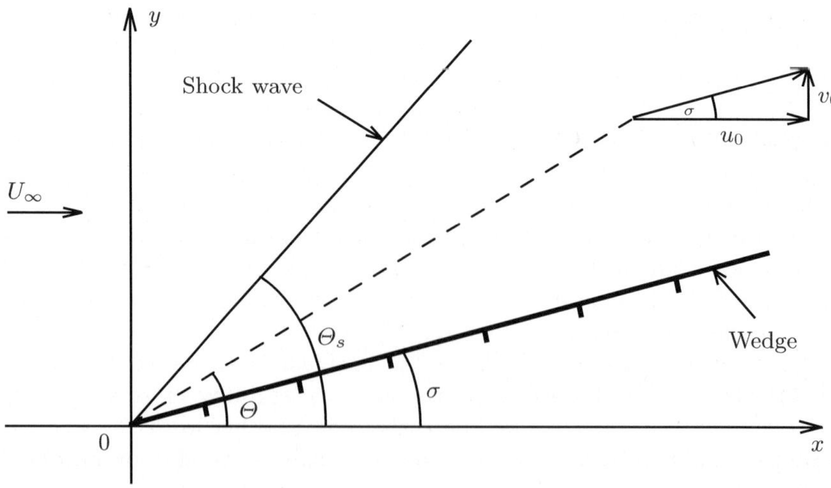

Fig. 1. A uniform steady flow past a wedge

In conclusion, we note that justification of stability for a chosen stationary solution is very important from the numerical point of view because it gives a basis for the application of the popular stabilization method [15].

2 Preliminaries, Formulation of Problem, and Primary Results

The mathematical formulation of the problem on supersonic flow around a wedge is given in [8]: for $t, x > 0$ and $y > x \tan\sigma$, we seek a solution to the acoustics equations

$$AU_t + BU_x + C_\sigma U_y = 0, \tag{2.1}$$

satisfying the following boundary conditions at the shock wave ($x = 0$) and at the wedge surface ($y = x \tan\sigma$):

$$u_1 + du_3 = 0,\ u_3 + u_4 = 0,\ u_2 = \frac{\lambda}{\mu}F_y,\ F_t + F_y\ \tan\sigma = \mu u_3,\ x = 0; \tag{2.2}$$

$$u_2 = u_1\ \tan\sigma,\ y = x\ \tan\sigma, \tag{2.3}$$

and the initial condition at $t = 0$

$$U(0,x,y) = U_0(x,y),\ F(0,y) = F_0(y). \tag{2.4}$$

Here $U(t,x,y) = (u_1,u_2,u_3,u_4)^T$ ($u_1 = \delta u$ and $u_2 = \delta v$ are small perturbations of the velocity components; $u_3 = \delta p$ and $u_4 = \delta s$ are small perturbations of the pressure and the entropy); $x = F(t,y)$ is a small perturbation of the shock front. Moreover,

$$F(t,0) = F_0(0) = 0. \tag{1}$$

The matrices A, B, and C_σ are the following:

$$A = diag(M^2, M^2, 1, 1), B = \begin{pmatrix} M^2 & 0 & 1 & 0 \\ 0 & M^2 & 0 & 0 \\ 1 & 0 & 1 & 0 \\ 0 & 0 & 0 & 1 \end{pmatrix},\ C = \begin{pmatrix} 0 & 0 & 0 & 0 \\ 0 & 0 & 1 & 0 \\ 0 & 1 & 0 & 0 \\ 0 & 0 & 0 & 0 \end{pmatrix},$$

$$C_\sigma = C + \tan\sigma \cdot A;$$

$M^2 = \frac{u_0^2}{c_0^2} < 1$, the constants d, λ and μ are given in [8].

Remark 2.1. *The initial-boundary-value problem (2.1)–(2.5) is formulated for the case when the shock wave is directed toward the axes* (Oy) *and the flow around the wedge is the main solution.*

Provided that $U(t,x,y)$ and $F(t,y)$ are sufficiently smooth, we get the initial-boundary-value problem for u_3: for $t, x > 0$, and $y > x\tan\sigma$, we seek a solution to the wave equation

$$\{M^2L_1^2 - L_2^2 - (\partial_y)^2\}u_3 = 0, \tag{2.6}$$

satisfying the boundary conditions on the shock wave ($x = 0$) and on the wedge surface ($y = x\ \tan\sigma$):

$$\{m_1L_1^2 + nL_2^2 - \frac{\beta_1}{M^2}L_1L_2\}u_3 = 0,\ x = 0; \tag{2.7}$$

$$\{\cos\sigma\cdot\eta - \sin\sigma\cdot\xi\}u_3 = 0,\ y = x\ \tan\sigma \tag{2.8}$$

and the initial data at $t = 0$:

$$u_3|_{t=0} = u_0(x,y), (u_3)_t|_{t=0} = u_1(x,y). \tag{2.9}$$

Here

$$L_1 = \frac{l_1}{\beta_1},\quad l_1 = \partial_t + \tan\sigma\cdot\partial_y,\quad L_2 = \beta_1\partial_x - \frac{M^2}{\beta_1}l_1,$$

$$\beta_1^2 = 1 - M^2\ (M^2 < 1),\ n = -\frac{\lambda}{\beta_1},\quad m = \beta_1 d + \frac{\lambda M^2}{\beta_1}.$$

We rewrite (2.6)–(2.9) in terms of new independent variables

$$x' = x,\ y' = y - x\ \tan\sigma.$$

Then, omitting primes and taking $u = u_3$, we obtain

$$\left\{M^2\left(\frac{\partial}{\partial t} + \frac{\partial}{\partial x}\right)^2 - \left(\frac{\partial}{\partial x} - \tan\sigma\frac{\partial}{\partial y}\right)^2 - \left(\frac{\partial}{\partial y}\right)^2\right\}u = 0, \tag{2.10}$$

$$t, x, y > 0;$$

$$\begin{aligned}\Bigg\{\left(\frac{\partial}{\partial t} + \tan\sigma\frac{\partial}{\partial y}\right)\Bigg(\frac{\partial}{\partial t} + \frac{\partial}{\partial x} + d\left(\frac{\partial}{\partial t} + \tan\sigma\frac{\partial}{\partial y}\right) -\\ -\frac{1}{M^2}\left(\frac{\partial}{\partial x} - \tan\sigma\frac{\partial}{\partial y}\right)\Bigg) + \lambda\left(\frac{\partial}{\partial y}\right)^2\Bigg\}u = 0,\end{aligned} \tag{2.11}$$

$$x = 0,\ t, y > 0;$$

$$\left\{\frac{\partial}{\partial y} + \sin\sigma\cos\sigma\frac{\partial}{\partial x}\right\}u = 0,\ x = 0,\ t, y > 0; \tag{2.12}$$

$$u(x,y,0) = u_0(x,y),\quad u_t(x,y,0) = u_1(x,y). \tag{2.13}$$

Using the notations

$$\alpha = M^2 \ (\alpha < 1), \quad \beta = \tan\sigma > 0, \quad D_1 = 1 + d,$$
$$D_2 = 1 - \frac{1}{\alpha}, \quad D_3 = d\beta + \frac{\beta}{\alpha} + \beta, \quad D_4 = \beta - \frac{\beta}{\alpha},$$
$$D_5 = \beta^2 d + \frac{\beta^2}{\alpha} + \lambda, \quad D_6 = -\sin\sigma\cos\sigma,$$

we finally rewrite problems (2.10)–(2.13) as follows:

$$Lu = \left\{\alpha\frac{\partial^2}{\partial t^2} + 2\alpha\frac{\partial^2}{\partial t\partial x} - \beta_1^2\frac{\partial^2}{\partial x^2} + 2\beta\frac{\partial^2}{\partial x\partial y} - (1+\beta^2)\frac{\partial^2}{\partial y^2}\right\}u = 0, \quad \text{if } t, x, y > 0; \tag{2.14}$$

$$\left\{D_1\frac{\partial^2}{\partial t^2} + D_2\frac{\partial^2}{\partial t\partial x} + D_3\frac{\partial^2}{\partial t\partial y} + D_4\frac{\partial^2}{\partial y\partial x} + D_5\frac{\partial^2}{\partial y^2}\right\}u = 0, \quad \text{if } x = 0, t, y > 0; \tag{2.15}$$

$$\left(\frac{\partial}{\partial y} + D_6\frac{\partial}{\partial x}\right)u = 0, \quad y = 0, \quad t, x > 0; \tag{2.16}$$

$$u(x, y, 0) = u_0(x, y), \quad u_t(x, y, 0) = u_1(x, y). \tag{2.17}$$

We suppose that the initial data have a compact support, i.e., $u_0(x, y)$, $u_1(x, y) \in C_0^\infty(R_+^2)$, where $R_+^2 = \{(x, y) \in R^2 | x, y > 0\}$.

By $W_{2,\sigma_0}^2(R_+^3)$ we denote a Hilbert space such that $u \in W_{2,\sigma_0}^2(R_+^3)$ if the following norm is finite

$$\|u\|_{W_{2,\sigma_0}^2(R_+^3)} = \sqrt{\sum_{\substack{k=k_1+k_2+k_3,\\ 0\le k\le 2}} \int_{R_+^3} e^{-2\sigma_0 t}\left|\frac{\partial^k u}{\partial t^{k_1}\partial x^{k_2}\partial y^{k_3}}\right|^2 dt\, dx\, dy}.$$

Here σ_0 is a positive number, $R_+^3 = \{(t, x, y) | t, x, y > 0\}$.

Using the polynomial

$g(\xi, \eta, s) = -s^2\big(\beta_1^2 D_1 + 2\alpha D_2\big) - is\xi D_2\beta_1^2 - \xi\eta D_4\beta_1^2 + \eta^2(2\beta D_4 + \beta_1^2 D_5) + is\eta\big(2\beta D_2 + D_3\beta_1^2 + 2\alpha D_4\big)$,

we can formulate the uniform Lopatinsky condition for problems (2.14)–(2.17) as follows:

$$g(\xi|_{\xi=\xi_*(\eta)}, \eta, s) \neq 0, \tag{2.18}$$

where $\xi_*(\eta) = \frac{(\beta\eta + i\alpha s) + i\sqrt{\alpha(\beta\eta + is)^2 - \beta_1^2\eta^2}}{\beta_1^2}$, $Im\ \xi_*(\eta) > 0$; $\eta \in R$, $s \in C$, $Re\ s \geq 0$; $\eta^2 + |s|^2 \neq 0$.

Here ξ, η, s are dual variables for x, y, t, respectively, for the Fourier and Laplace transforms $F_{\substack{\xi\to x,\\ \eta\to y}}$, $L_{s\to t}$.

Let the following solvability conditions for problem (2.14)–(2.17) are satisfied:

$$\int_0^{+\infty} \frac{L(\eta^2, s)\eta d\eta}{(\eta^2+s^2)^{-k} X^+(s,\eta)} d\eta = 0, \;\; k = 0, 1. \tag{2.19}$$

Here

$$L(\lambda^2, s) = -\frac{iD_4\eta(\lambda) - D_2 s}{g(\xi_*(\eta(-\lambda)), \eta(-\lambda), s)} (\bar{F}(\xi(\eta(\lambda)), \eta(\lambda), s) + \bar{F}(\xi_*(\eta(-\lambda)), \eta(-\lambda), s)),$$
$$\eta(\lambda) = \frac{i\beta\alpha s}{1-\alpha-\alpha\beta^2} - \frac{\beta_1}{\sqrt{(1+\beta^2)(1-\alpha-\alpha\beta)}}\lambda, \;\; \bar{F} = (\alpha s - 2i\alpha\xi)\widehat{u_0} + \alpha\widehat{u_1},$$

$X^+(s,\eta)$ is the trace of the canonical function for the Riemann problem.

The main results of the article are the following.

Theorem 1. *Let the uniform Lopatinsky condition (2.18) on the boundary $x = 0$ and the solvability conditions (2.19) are satisfied. Then a solution $u \in W^2_{2,\sigma_0}(R^3_+)$ to problem (2.14)–(2.17) exists, it has a compact support for every $t \geq 0$, i.e., supp $u \subset \bar{R}^2_+$, and is given in the form*

$$u(x,y,t) = L^{-1}_{s\to t} F^{-1}_{\substack{\xi\to x,\\ \eta\to y}} \left[\frac{1}{\alpha s^2 - 2i\alpha s\xi + \beta_1^2\xi^2 - 2\beta\xi\eta + (1+\beta^2)\eta^2} \right.$$
$$\times \left[\widehat{u(0,y,t)} \left\{ i\beta_1^2\xi + 2\alpha s - 2i\beta\eta + \frac{\beta_1^2(D_5\eta^2 + iD_3 s\eta - D_1 s^2)}{iD_4\eta - D_2 s} \right\} \right.$$
$$\left.\left. + \widehat{u(x,0,t)} \left\{ i(1+\beta^2)\eta - i\beta\xi \right\} + (\alpha s - 2i\alpha\xi)\widehat{u_0(x,y)} + \alpha\widehat{u_1(x,y)} \right]\right], \tag{2.20}$$

where the symbols $L^{-1}_{s\to t}, F^{-1}_{\substack{\xi\to x,\\ \eta\to y}}$ stand for the inverse Laplace and Fourier transforms, respectively, the functions $\widehat{u(x,0,t)}$, $\widehat{u(0,y,t)}$ are Fourier–Laplace transforms of the traces; $\widehat{u_0(x,y)}, \widehat{u_1(x,y)}$ are Fourier transforms of the initial data.

Theorem 2. *Let the conditions of the Theorem 1 be fulfilled. Then*

$$u(0,y,t') = -\frac{1}{\pi\gamma}\frac{\partial}{\partial t'} \int_{R^2_+} Re\left\{ \left.\frac{(\alpha - 2\alpha\xi_*(k))(iD_4 k - D_2)}{g(\xi_*(k),k,1)}\right|_{k=\widetilde{k^{(1)}}} \frac{d\widetilde{k^{(1)}}}{dt'} \right\}$$
$$\times\, \Theta(t' - t_*^{(1)}) u_0(z_1,z_2) dz_1 dz_2 - \frac{\alpha}{\pi\gamma} \int_{R^2_+} Re\left\{ \left.\frac{iD_4 k - D_2}{g(\xi_*(k),k,1)}\right|_{k=\widetilde{k^{(1)}}} \frac{d\widetilde{k^{(1)}}}{dt'} \right\}$$
$$\times\, \Theta(t' - t_*^{(1)}) u_1(z_1,z_2) dz_1 dz_2$$

$$-\frac{1}{\pi\gamma}\frac{\partial}{\partial t'}\int\limits_{R^2_+} Re\left\{\frac{(\alpha-2\alpha\xi_*(k))(iD_4k-D_2)}{g(\xi_*(k),k,1)}\Bigg|_{k=\widetilde{k^{(2)}}}\frac{d\widetilde{k^2}}{dt'}\right\}\Theta(t'-t_*^{(2)})$$

$$\times\, u_0(z_1,z_2)dz_1dz_2-\frac{\alpha}{\pi\gamma}\int\limits_{R^2_+} Re\left\{\frac{iD_4k-D_2}{g(\xi_*(k),k,1)}\Bigg|_{k=\widetilde{k^{(2)}}}\frac{d\widetilde{k^{(2)}}}{dt'}\right\}$$

$$\times\,\Theta(t'-t_*^{(2)})u_1(z_1,z_2)dz_1dz_2, \tag{2.21}$$

where $t'=t-\frac{\beta\alpha}{1-\alpha-\alpha\beta^2}y\geq 0$, $\gamma=\frac{\beta_1}{\sqrt{(1+\beta^2)(1-\alpha-\alpha\beta^2)}}$, $\widetilde{k^{(1)}},\widetilde{k^{(2)}},t_*^{(1)},t_*^{(2)}$ are some functions of y,t', and Θ is the Heaviside function.

By this, the solution trace on the shock wave $u(0,y,t-\frac{\beta\alpha}{1-\alpha-\alpha\beta^2}y)$ is a superposition of the direct and reflected from $y=0$ waves.

From (2.21) we can obtain an information on the asymptotic behavior of the function $u(0,y,t)$ as $t\to+\infty$.

More precisely, we have the following theorem.

Theorem 3. *Let conditions of Theorem 2 be fulfilled. Then we have the asymptotics*

$$u\left(0,y,t-\frac{y\beta\alpha}{1-\alpha-\alpha\beta^2}\right)=c_0\left(u_0,u_1,y\right)\frac{1}{t}+o\left(\frac{1}{t}\right)\ \text{at}\ t\to\infty, \tag{2.22}$$

where $c_0(u_0,u_1,y)$ is a functional depending on the functions u_0, u_1, and the variable y.

Acknowledgments

This work was supported by the RFBR grant No. 04-01-00900. The authors are also indebted to RFBR (the grant no. 06-01-10611-z), the Organizing Committee of the "Eleventh International Conference on Hyperbolic Problems. Theory, Numerics, Applications," and the Mathematical Department of the Novosibirsk State University which financial support made possible the participation of one of the authors (D.L.T.) in the above conference.

References

1. L.V. Ovsyannikov, *Lectures on fundamentals of gas dynamics*, Moscow-Izhevsk: Institute of computer investigations, 2003, 336 Pp.
2. G.G. Chernyj, *Gas dynamics*, Moscow, Nauka, 1988, 424 Pp.
3. M.D. Salas, B.D. Morgan, *Stability of shock waves attached to wedges and cones*, AIAA Jornal, Vol. 21, N 12, 1983, Pp. 1611–1617.
4. R. Courant, K.D. Friedrichs, *Supersonic flow and shock waves*, New York, Interscience Publishers, 1948.

5. A.M. Blokhin, E.N. Romensky, *Stability of limit stationary solution in problem on flowing around a circular cone*, Izv. SB AS USSR. Ser. tech. scie., N 13, iss. 3, 1978, Pp. 87–97.
6. V.V. Rusanov, A.A. Sharakshane, *Stability of flows around an infinite wedge or cone situated in supersonic gas flow. In: Current problems in mathematical physics and numerical mathematics*, Moscow, Nauka, N 338, 1982, Pp. 268–272.
7. A.M. Blokhin, *Well-posedness of linear mixed problem on supersonic flowing around a wedge*, Sib. Math. J., Vol. 29, N 5, 1988, Pp. 48–58.
8. A.M. Blokhin, *Energy integrals and their applications to problem of gas dynamics*, Novosibirsk, Nauka, 1986, 240 p.
9. A.I. Rylov, *On regimes of flowing around peaked bodies of finite thickness for arbitrary supersonic sounds of incoming flow*, Prikl. Mat. i Mech., Vol. 55, iss. 1, 1991, Pp. 95–99.
10. A.A. Nikolsky, *On plane turbulent gas flows*, Theoretical study in mechanics of gas and liquid: Pr. CAGI., iss. 2122, 1981, Pp. 74–85.
11. B.M. Bulach, *Nonlinear conic gas flows*, Moscow, Nauka, 1970, 344 p.
12. B.L. Rozhdestvensky, *Revision of the theory on flowing around a wedge by a inviscid supersonic gas flow*, Math. modelling, Vol. 1, N 8, 1989, Pp. 99–102.
13. Yu.Yu. Pashinin, D.L. Tkachev, *Uniqueness of solution to the problem on flowing around a wedge. Strong shock wave*, Vestnik NGU, Vol. III, iss. 4, 2003, Pp. 33–50.
14. A.M. Blokhin, D.L. Tkachev, L.O. Baldan, *Study of the stability in the problem on flowing around a wedge. The case of strong wave*, J. Math. Anal. and Appl., Vol. 319, 2006, Pp. 248–277.
15. A.N. Lubimov, V.V. Rusanov, *Gas flow around pointed bodies*, Moscow, Nauka, 1970.

The Derivative Riemann Problem for the Baer–Nunziato Equations

E.F. Toro and C.E. Castro

1 Introduction

We solve the derivative Riemann problem (DRP) for the Baer–Nunziato (BN) equations for compressible two phase flows [1]. The DRP is the Cauchy problem in which the initial condition consists of two smooth vectors, typically high-degree polynomials, with a discontinuity at the origin. In the classical Riemann problem these polynomials are two constant vectors. The technique to solve the DRP for the BN equations is an extension of that reported in [4] and [3]. The solution $\mathbf{Q}_{LR}(\tau)$ is sought at the interface as a function of time. It is assumed that $\mathbf{Q}_{LR}(\tau)$ may be expressed as a time series expansion in which the leading term $\mathbf{Q}(0, 0_+)$ is the solution of the classical Riemann problem, evaluated at the interface, for $t = 0_+$. The coefficients of the higher order terms are time derivatives of the vector of unknowns, all to be evaluated at $x = 0$ and $t = 0_+$. Use of the Cauchy–Kowalewski method allows us to express all time derivatives as functions of space derivatives. These spatial derivatives at $x = 0$ and $t = 0_+$ are found by first defining new evolution equations for spatial derivatives and then solving classical Riemann problems. The scheme reduces the solution of the derivative Riemann problem with polynomial data of two polynomials of degree at most K to the problem of solving one classical nonlinear Riemann problem for the leading term and K classical linear Riemann problems for spatial derivatives.

The computation of the leading term $\mathbf{Q}(0, 0_+)$ requires a classical Riemann solver for the Baer–Nunziato equations. Here we make use of the recently proposed (classical) Riemann solver proposed by Schwendeman et al. [2]. To compute the higher order terms we need a method to solve linear Riemann problems in terms of conserved variables. These solvers are developed in this paper. We assess the technique on a number of test problems, using as a reference solution the one computed numerically by a second-order Godunov method on a very fine mesh.

The rest of the paper is as follows: Sect. 2 gives the governing equations and their eigenstructure. In Sect. 3 we solve the derivative Rieman problem. Test problems are presented in Sect. 4 and conclusions are drawn in Sect. 5.

2 Governing Equations and Their Eigenstructure

2.1 The Equations

Consider two compressible fluids denoted by suffixes $k = 1,\ 2$. The homogeneous Baer–Nunziato equations [1] in one space dimension are

$$\partial_t \mathbf{Q} + \partial_x \mathbf{F}(\mathbf{Q}) + \mathbf{A}(\mathbf{Q})\, \partial_x \mathbf{W}(\mathbf{Q}) = \mathbf{0}, \tag{1}$$

where vector $\mathbf{Q}$ is the unknown conservative vector, $\mathbf{F}(\mathbf{Q})$ is a flux vector, $\mathbf{A}(\mathbf{Q})$ is a coefficient matrix. Equation (1) is a first-order system with nonconservative terms.

$$\mathbf{Q} = \begin{bmatrix} \alpha_1\rho_1 \\ \alpha_1\rho_1 u_1 \\ \alpha_1 E_1 \\ \alpha_2\rho_2 \\ \alpha_2\rho_2 u_2 \\ \alpha_2 E_2 \\ \alpha_1 \end{bmatrix} \qquad \mathbf{F}(\mathbf{Q}) = \begin{bmatrix} \alpha_1\rho_1 u_1 \\ \alpha_1[\rho_1 u_1^2 + p_1] \\ \alpha_1 u_1[E_1 + p_1] \\ \alpha_2\rho_2 u_2 \\ \alpha_2[\rho_2 u_2^2 + p_2] \\ \alpha_2 u_2[E_2 + p_2] \\ 0 \end{bmatrix} \tag{2}$$

$$\mathbf{A}(\mathbf{Q}) = \begin{bmatrix} 0\,0\,0\,0\,0\,0 & 0 \\ 0\,0\,0\,0\,0\,0 & -p_2 \\ 0\,0\,0\,0\,0\,0 & -p_2 u_1 \\ 0\,0\,0\,0\,0\,0 & 0 \\ 0\,0\,0\,0\,0\,0 & p_2 \\ 0\,0\,0\,0\,0\,0 & p_2 u_1 \\ 0\,0\,0\,0\,0\,0 & u_1 \end{bmatrix} \qquad \mathbf{W}(\mathbf{Q}) = \begin{bmatrix} \rho_1 \\ u_1 \\ p_1 \\ \rho_2 \\ u_2 \\ p_2 \\ \alpha_1 \end{bmatrix}. \tag{3}$$

Here ρ_k is density, u_k is particle velocity, p_k is pressure, $E_k = \rho_k(\frac{1}{2}u_k^2 + e_k)$ is total energy, e_k is specific energy, and α_k is void fraction, with $\sum \alpha_k = 1$. The stiffened equation of state (EOS) is assumed for each fluid by providing appropriate values for γ_k and p_k^o. The specific energy e_k and sound speed a_k are

$$e_k = \frac{p_k + \gamma_k p_k^o}{\rho_k(\gamma_k - 1)} \qquad a_k^2 = (\gamma_k - 1)(\frac{p_k}{\rho_k} + e_k). \tag{4}$$

2.2 Eigenstructure for the Conservative Variables

The equations in terms of conservative variables may be written thus

$$\partial_t \mathbf{Q} + \mathbf{B}(\mathbf{Q})\partial_x \mathbf{Q} = \mathbf{0}, \tag{5}$$

where $\mathbf{Q}$ is the vector of conservative variables and $\mathbf{B}(\mathbf{Q})$ is a coefficient matrix. The eigenvalues of the system are

$$\lambda_1 = u_1 - a_1,\ \lambda_2 = u_2 - a_2,\ \lambda_3 = u_2,\ \lambda_4 = \lambda_5 = u_1,\ \lambda_6 = u_2 + a_2,\ \lambda_7 = u_1 + a_1. \tag{6}$$

For the conservative variables the right eigenvectors corresponding to the seven real eigenvalues (6) are found to be

$$\mathbf{R}^{(1)} = \begin{bmatrix} 1 \\ u_1 - a_1 \\ \frac{u_1^2}{2} + \frac{a_1^2}{\gamma_1 - 1} - u_1 a_1 \\ 0 \\ 0 \\ 0 \\ 0 \end{bmatrix}, \mathbf{R}^{(2)} = \begin{bmatrix} 0 \\ 0 \\ 0 \\ 1 \\ u_2 - a_2 \\ \frac{u_2^2}{2} + \frac{a_2^2}{\gamma_2 - 1} - a_2 u_2 \\ 0 \end{bmatrix}, \mathbf{R}^{(3)} = \begin{bmatrix} 1 \\ u_1 \\ \frac{u_1^2}{2} \\ 0 \\ 0 \\ 0 \\ 0 \end{bmatrix},$$

$$\mathbf{R}^{(4)} = \begin{bmatrix} 0 \\ 0 \\ \frac{\left(e_1\rho_1 - \frac{p_1 - p_2}{\gamma_1 - 1}\right)\left((u_1 - u_2)^2 - a_2^2\right)}{\rho_2 (u_1 - u_2)^2} \\ 1 \\ \frac{u_2(u_1 - u_2) + a_2^2}{u_1 - u_2} \\ \frac{\left(\left(\frac{a_2^2}{\gamma_2 - 1} + \frac{u_2^2}{2}\right)(u_1 - u_2) + a_2 u_2\right)}{(u_1 - u_2)} \\ \frac{(u_1 - u_2)^2 - a_2^2}{\rho_2 (u_1 - u_2)^2} \end{bmatrix},$$

$$\mathbf{R}^{(5)} = \begin{bmatrix} 0 \\ 0 \\ 0 \\ 1 \\ u_2 \\ \frac{u_2^2}{2} \\ 0 \end{bmatrix}, \mathbf{R}^{(6)} = \begin{bmatrix} 0 \\ 0 \\ 0 \\ 1 \\ u_2 + a_2 \\ \frac{u_2^2}{2} + \frac{a_2^2}{\gamma_2 - 1} + a_2 u_2 \\ 0 \end{bmatrix}, \mathbf{R}^{(7)} = \begin{bmatrix} 1 \\ u_1 + a_1 \\ \frac{u_1^2}{2} + \frac{a_1^2}{\gamma_1 - 1} + u_1 a_1 \\ 0 \\ 0 \\ 0 \\ 0 \end{bmatrix}.$$

This eigenstructure will be used to solve linearized Riemann problems for spatial derivatives in Sect. 3.

3 The Derivative Riemann Problem

The Derivative Riemann Problem is the initial-value problem

$$\left.\begin{array}{l} \text{PDEs: } \partial_t \mathbf{Q} + \partial_x \mathbf{F}(\mathbf{Q}) = \mathbf{S}(\mathbf{Q}),\ x \in (-\infty, \infty),\ \ t > 0, \\ \text{ICs:} \quad \mathbf{Q}(x, 0) = \begin{cases} \mathbf{Q}_L(x) \text{ if } x < 0, \\ \mathbf{Q}_R(x) \text{ if } x > 0. \end{cases} \end{array}\right\} \tag{7}$$

The PDEs with source terms are a general system of hyperbolic balance laws. The initial conditions $\mathbf{Q}_L(x)$ and $\mathbf{Q}_R(x)$ are two vectors, the components of which are assumed to be smooth functions of x, with K continuous nontrivial spatial derivatives away from zero. We denote by DRP_K the Cauchy problem (7). DRP_0 corresponds to the *classical* piece-wise constant data Riemann problem.

3.1 A Method of Solution

Here we apply the method proposed by Toro and Titarev to solve the DRP [4], [3]. Their method first expresses the sought solution $\mathbf{Q}_{LR}(\tau)$ at the interface $x = 0$ as the power series expansion in time

$$\mathbf{Q}_{LR}(\tau) = \mathbf{Q}(0, 0_+) + \sum_{k=1}^{K} \left[\partial_t^{(k)} \mathbf{Q}(0, 0_+)\right] \frac{\tau^k}{k!}, \tag{8}$$

where

$$\mathbf{Q}(0, 0_+) = \lim_{t \to 0_+} \mathbf{Q}(0, t).$$

The determination of the leading term $\mathbf{Q}(0, 0_+)$ and of the coefficients of the high-order terms determined by the time derivatives $\partial_t^{(k)} \mathbf{Q}(0, 0_+)$ involves the following steps:

Leading Term

To compute the leading term one solves exactly or approximately the conventional Riemann problem

$$\left.\begin{array}{ll} \text{PDEs:} & \partial_t \mathbf{Q} + \partial_x \mathbf{F}(\mathbf{Q}) = \mathbf{0}, \\ \text{ICs:} & \mathbf{Q}(x, 0) = \left\{\begin{array}{ll} \mathbf{Q}_L(0_-) & \text{if } x < 0, \\ \mathbf{Q}_R(0_+) & \text{if } x > 0, \end{array}\right. \end{array}\right\} \tag{9}$$

with

$$\mathbf{Q}_L(0_-) = \lim_{x \to 0_-} \mathbf{Q}_L(x), \quad \mathbf{Q}_R(0_+) = \lim_{x \to 0_+} \mathbf{Q}_R(x). \tag{10}$$

The similarity solution of (9) is denoted by $\mathbf{D}^{(0)}(x/t)$ and the leading term in (8) is

$$\mathbf{Q}(0, 0_+) = \mathbf{D}^{(0)}(0). \tag{11}$$

To compute the leading term we use the recently proposed scheme of [2].

Higher Order Terms

There are three sub-steps here.

1. *Time derivatives in terms of spatial derivatives:* Use the Cauchy–Kowalewski method to express time derivatives in (8) in terms of functions of space derivatives

$$\partial_t^{(k)}\mathbf{Q}(x,t) = \mathbf{G}^{(k)}(\partial_x^{(0)}\mathbf{Q}, \partial_x^{(1)}\mathbf{Q}, \ldots, \partial_x^{(k)}\mathbf{Q}). \quad (12)$$

   The source terms $\mathbf{S}(\mathbf{Q})$ in (7) are all included in the arguments of the functions $\mathbf{G}^{(k)}$. The problem now is that of determining the arguments of $\mathbf{G}^{(k)}$, namely the spatial derivatives at the interface.
2. *Evolution equations for derivatives:* Construct evolution equations for spatial derivatives

$$\partial_t(\partial_x^{(k)}\mathbf{Q}(x,t)) + \mathbf{A}(\mathbf{Q})\partial_x(\partial_x^{(k)}\mathbf{Q}(x,t)) = \mathbf{H}^{(k)}, \quad (13)$$

   where $\mathbf{A}(\mathbf{Q})$ is the Jacobian matrix of the PDEs in (7).
3. *Riemann problems for spatial derivatives:* Simplify (13) by neglecting the right-hand side terms $\mathbf{H}^{(k)}$ and linearizing the evolution equations. Then pose classical, homogeneous linearized Riemann problems for spatial derivatives

$$\left.\begin{array}{ll} \text{PDEs:} & \partial_t(\partial_x^{(k)}\mathbf{Q}(x,t)) + \mathbf{A}_{LR}^{(0)}\partial_x(\partial_x^{(k)}\mathbf{Q}(x,t)) = \mathbf{0}, \\ \text{ICs:} & \partial_x^{(k)}\mathbf{Q}(x,0) = \begin{cases} \partial_x^{(k)}\mathbf{Q}_L(0_-) \text{ if } x < 0, \\ \partial_x^{(k)}\mathbf{Q}_R(0_+) \text{ if } x > 0, \end{cases} \end{array}\right\} \quad (14)$$

   with $\mathbf{A}_{LR}^{(0)} = \mathbf{A}(\mathbf{Q}(0,0_+))$. Solve these Riemann problems to obtain similarity solutions $\mathbf{D}^{(k)}(x/t)$ and set

$$\partial_x^{(k)}\mathbf{Q}(0,0_+) = \mathbf{D}^{(k)}(0). \quad (15)$$

The Solution

Form the solution as the power series expansion:

$$\mathbf{Q}_{LR}(\tau) = \mathbf{C_0} + \mathbf{C_1}\tau + \mathbf{C_2}\tau^2 + \ldots + \mathbf{C_K}\tau^K, \quad (16)$$

with $\mathbf{C}_0$ as in (11) and

$$\mathbf{C}_k \equiv \frac{\partial_t^{(k)}\mathbf{Q}(0,0_+)}{k!} = \frac{\mathbf{G}^{(k)}(\mathbf{D}^{(0)}(0), \mathbf{D}^{(1)}(0), \ldots, \mathbf{D}^{(k)}(0))}{k!}, \quad (17)$$

for $k = 1, \ldots, K$.

This solution technique for the Derivative Riemann Problem DRP_K reduces the problem to that of solving $K+1$ classical homogeneous Riemann problems, one (generally nonlinear) Riemann problem to compute the leading term and K linearized Riemann problems to determine the higher order terms.

The leading term requires the availability of a *Riemann solver*, exact or approximate. For the Baer–Nunziato equations we use the exact Riemann solver of Schwendeman et al. [2]. The K linearized Riemann problems (14) can be solved analytically, as detailed further.

3.2 Higher-Order Terms: Linear Riemann Problems

A linear Riemann solver based on conserved variables is required for solving linear Riemann problems for spatial derivatives. The solution is required at the interface $x = 0$. Assuming the initial condition (piece-wise constant) is given by the two vectors $\mathbf{Q}_L$, $\mathbf{Q}_R$ we first decompose the initial jump $\mathbf{\Delta} = \mathbf{Q}_R - \mathbf{Q}_L$ as

$$\sum_{i=1}^{7} \alpha^{(i)}\mathbf{R}^{(i)} = \Delta, \tag{18}$$

where $\mathbf{R}^{(i)}$ are constant eigenvectors and the wave strengths $\alpha^{(i)}$ are to be found. Algebraic manipulations give explicit expressions for $\alpha^{(i)}$, for $i = 1, \ldots, 7$, not reproduced here.

4 Test Problems

We have applied the method to a range of test problems and have compared results against a reference numerical solution obtained by applying a second-order Godunov method on a very fine mesh. Here we show results for one test problem. Densities are set equal to 1 and 0.9 for fluid 1 and 2, respectively. Velocities for both phases are set equal to 0. The pressures and void fractions are given by polynomials to the left (L) of $x = 0$ and to the right (R) of $x = 0$:

$$\left.\begin{array}{ll} p_{1L} = 1.091 + 0.012x - 0.0106x^2 & p_{1R} = 1.027 + 0.012x + 0.0106x^2 \\ p_{2L} = 8.725 + 1.9x + 0.9x^2 & p_{2R} = 7.275 + 1.9x - 0.9x^2 \\ \alpha_{1L} = 0.5 + 0.1x^2 + 0.2x^3 & \alpha_{1R} = 0.5 + 0.1x^2 + 0.2x^3 \end{array}\right\} . \tag{19}$$

Figure 1 shows a plot of the initial pressure distribution for fluid 1 and of the reference (numerical) solution at time $t = 0.1$ s. Our method gives the solution at the interface $x = 0$, as a function of time.

Figure 2 shows a comparison of the solution of the derivative Riemann problem at $x = 0$, as a function of time. Results are compared with the reference solution (broken line). Figure 3 shows plots of the corresponding errors. It is seen that for short times our solution method provides accurate results, and these get better as the order of accuracy of the DRP increases.

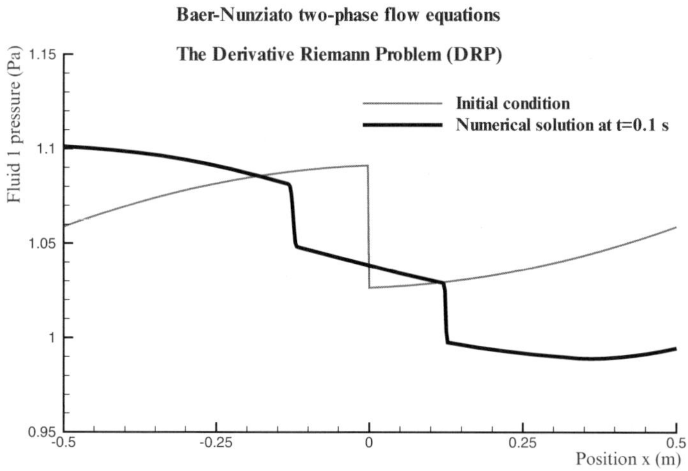

Fig. 1. Plot of initial pressure distribution and reference solution at time $t = 0.1\,\mathrm{s}$

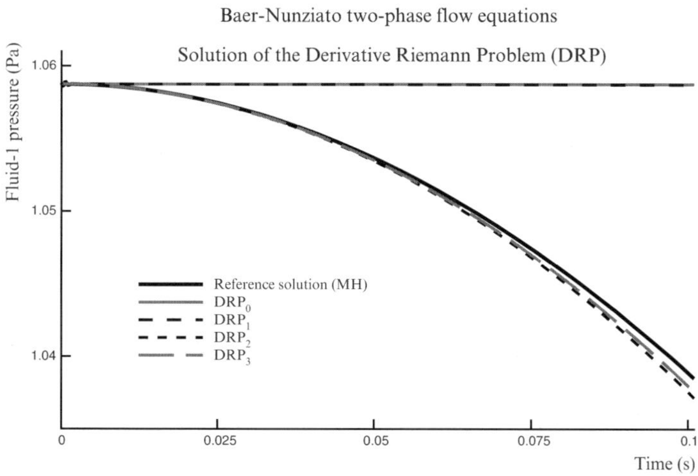

Fig. 2. Derivative Riemann Problem solution for the Baer–Nunziato two-phase flow equations, at $x = 0$ as a function of time

5 Conclusions and Further Work

We have solved the derivative Riemann problem for the Baer–Nunziato equations for two phase compressible flows. To provide the leading term of our DRP solution we have used the solver of Schwendemann et al. [2], thus extending their recent work. The same technique employed here could be applied to solve the DRP with source terms; then the complete solution can be used to construct very high-order numerical methods to solve the general initial-boundary

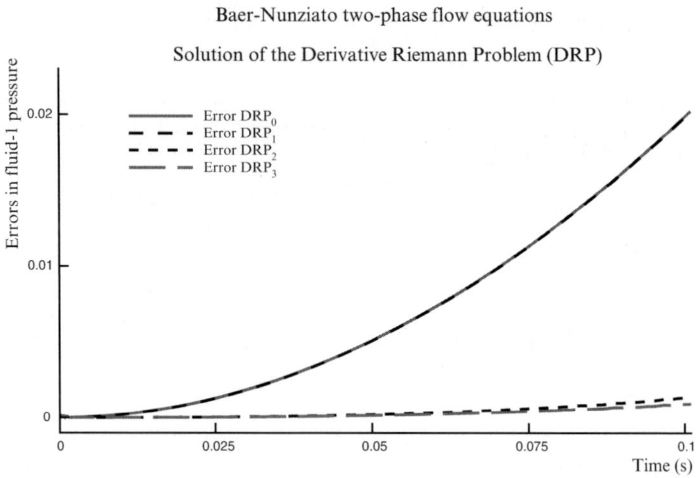

Fig. 3. Errors in DRP solution for Baer–Nunziato equations

value problem for the Baer–Nunziato equations with source terms. These topics are the subject of current investigations by the authors.

References

1. M. Baer and J. Nunziato. A two-phase mixture theory for the deflagration to detonation transition (DDT) in reactive granular materials. *International Journal Multiphase Flows*, 12:861–889, 1986.
2. D.W. Schwendeman, C.W. Wahle, and A.K. Kapila. The Riemann problem and high-resolution Godunov method for a model of compressible two-phase flow. *J. Comp. Phys.*, 212:490–526, 2006.
3. E.F Toro and Titarev V.A. Derivative Riemann Solvers for Systems of Conservation Laws and ADER Methods. *J. Comput Phys.*, 212(1):150–165, 2006.
4. E.F. Toro and V.A. Titarev. Solution of the Generalised Riemann Problem for Advection-Reaction Equations. *Proc. Roy. Soc. London A*, 458:271–281, 2002.

Stability of Contact Discontinuities for the Nonisentropic Euler Equations in Two-Space Dimensions

A. Morando and P. Trebeschi

1 Introduction

We study the nonisentropic Euler equations of a perfect polytropic ideal gas in the plane $\mathbb{R}^2$

$$\begin{aligned} &\partial_t p + \mathbf{u} \cdot \nabla p + \gamma p \nabla \cdot \mathbf{u} = 0, \\ &\rho(\partial_t \mathbf{u} + (\mathbf{u} \cdot \nabla)\mathbf{u}) + \nabla p = 0, \\ &\partial_t S + \mathbf{u} \cdot \nabla S = 0, \end{aligned} \tag{1}$$

where $x = (x_1, x_2) \in \mathbb{R}^2$ are the space variables, $p = p(t,x) \in \mathbb{R}$ is the pressure, $\mathbf{u} = \mathbf{u}(t,x) \in \mathbb{R}^2$ is the velocity of the flow, $S = S(t,x) \in \mathbb{R}$ is the entropy, ρ is the density, obeying the constitutive law $\rho(p,S) = Ap^{\frac{1}{\gamma}} e^{-\frac{S}{\gamma}}$ with $A > 0$ given, and $\gamma > 1$ is the adiabatic number.

We study the linear stability of "contact discontinuities" and, under a "supersonic" condition precluding violent instability, we give an energy estimate for the linearized boundary value problem. This is a crucial step toward proving the existence of nonconstant compressible vortex sheets (solutions which allow discontinuities in the velocity or in the entropy).

Since the interface of discontinuities is a part of the unknowns, the existence of compressible vortex sheets is a nonlinear free boundary value problem; moreover, for a contact discontinuity such an interface is a characteristic boundary with respect to both the smooth states. A general existence theorem is not available for such a problem. In the noncharacteristic case, a complete analysis of existence and stability of a *single shock wave* was made by Majda [M83] and Métivier [M01]. The existence of *rarefaction waves* was obtained by Alinhac [A89], the existence of *sound waves* was obtained by Métivier [M91], and the uniform existence of *shock waves* with small strength was proved by Francheteau and Métivier [FM00] under a uniform stability condition. In the characteristic case, the existence and stability of *vortex sheets* for the *isentropic* Euler equations were studied recently by Coulombel–Secchi [CS04] and [CS05]: in this chapter we summarize the result we obtained in [MT06] where the result proved by Coulombel–Secchi is generalized to the *nonisentropic* case.

2 Discontinuities

Let $\Gamma := \{x_2 = \varphi(t, x_1)\}$ be a smooth hypersurface and

$$(p, \mathbf{u}, S) := \begin{cases} (p^+, \mathbf{u}^+, S^+) & \text{if } x_2 > \varphi(t, x_1) \\ (p^-, \mathbf{u}^-, S^-) & \text{if } x_2 < \varphi(t, x_1), \end{cases}$$

a smooth function on either side of Γ. We say that $(p, \mathbf{u}, S)$ is a *contact discontinuity solution of (1)* if it is a classical solution of (1) on both sides of Γ and satisfies the Rankine–Hugoniot jump conditions at Γ

$$\partial_t \varphi = \mathbf{u}^+ \cdot \nu = \mathbf{u}^- \cdot \nu, \qquad p^+ = p^-, \tag{2}$$

where $\nu := (-\partial_{x_1}\varphi, 1)$ is a (space) normal vector to Γ. These conditions yield that the normal velocity and pressure are continuous across the interface Γ, while the tangential velocity and the entropy of the fluid may undergo any jump. Thus, a contact discontinuity is a vortex sheet. Particular solutions $(p, \mathbf{u}, S)$ of (1) are the so-called *planar contact discontinuities*. They consist of functions

$$(p, \mathbf{u}, S) = \begin{cases} (p_r, \mathbf{u}_r, S_r), & \text{if } x_2 > \sigma t + n x_1, \\ (p_l, \mathbf{u}_l, S_l), & \text{if } x_2 < \sigma t + n x_1, \end{cases}$$

where $\mathbf{u}_{r,l} = (v_{r,l}, u_{r,l})^T$ are fixed vectors in $\mathbb{R}^2$ and $p_{r,l} > 0, S_{r,l}, \sigma, n$ are fixed real numbers. The previous quantities are related by the Rankine–Hugoniot jump conditions

$$\sigma + v_{r,l} n - u_{r,l} = 0, \qquad p_r = p_l =: p.$$

Without loss of generality, we may assume $\sigma = u_r = u_l = 0$, $n = 0$, and $v_r + v_l = 0$ with $v_r \neq 0$. This corresponds to the following function $U_{r,l} = (p, v_{r,l}, 0, S_{r,l})^T$, with $v_r + v_l = 0$.

The linear stability of planar contact discontinuities in dimensions 2 and 3 has been analyzed a long time ago by Fejer–Miles [FM63] and more recently by Coulombel–Morando [CM04]. The result can be summarized by the following.

Theorem 1. *Consider a d-dimensional planar contact discontinuity*

$$U_{r,l} = (p, v_{r,l}, 0, S_{r,l})^T$$

with $v_r + v_l = 0$ *(*$v_{r,l}$ *are the tangential velocities) and linearize the Euler equations and the jump conditions around this solution:*

1. *If* $d = 3$*, the linearized equations do not satisfy the Lopatinskii condition, then we have violent instability.*

2. *If* $d = 2$ *and* $\frac{|v_r - v_l|}{2} < \frac{1}{2}\left(c_r^{2/3} + c_l^{2/3}\right)^{3/2}$*, the linearized equations do not satisfy the Lopatinskii condition, then we have violent instability.*
3. *If* $d = 2$ *and* $\frac{|v_r - v_l|}{2} > \frac{1}{2}\left(c_r^{2/3} + c_l^{2/3}\right)^{3/2}$*, the linearized equations satisfy the weak Lopatinskii condition, then we have weak stability.*

$c(p, S)$*, defined by* $\frac{1}{c^2} = \frac{\partial \rho}{\partial p}(p, S) = \frac{\rho}{\gamma p}$*, is the sound speed and* c_r, c_l *are the constant values of* $c(p, S)$ *in both sides of* Γ*.*

3 Energy Estimates in Two-Space Dimensions

Let $U_{r,l} = (p, v_{r,l}, 0, S)^T$ be a two-dimensional supersonic planar contact discontinuity according to point *3* of Theorem 1. We derive an energy estimate for the linearized Euler equations around such discontinuity. Here below we list the main steps to obtain such an estimate.

3.1 Reformulation of the Problem in a Fixed Domain

To work in a fixed domain, it is convenient introducing in (1) and (2) the change of variables $(t, x_1) = (\tau, y_1)$, $x_2 = \Phi(\tau, y_1, y_2)$, where $\Phi : \mathbb{R}^3 \to \mathbb{R}$, $\Phi(\tau, y_1, 0) = \varphi(\tau, y_1)$, and $\partial_{y_2}\Phi(\tau, y_1, y_2) \geq \kappa > 0$. We define the new unknowns

$$(p_\sharp^\pm, \mathbf{u}_\sharp^\pm, S_\sharp^\pm)(\tau, y_1, y_2) := (p, \mathbf{u}, S)(\tau, y_1, \Phi(\tau, y_1, \pm y_2)).$$

The functions $p_\sharp^\pm$, $\mathbf{u}_\sharp^\pm$, and $S_\sharp^\pm$ are smooth on the fixed domain $\{y_2 > 0\}$. For convenience, we drop the $\sharp$ index and only keep the $\pm$ exponents; finally, we write (t, x_1, x_2) instead of (τ, y_1, y_2). Let us set $\mathbf{u} = (v, u)$. The existence of compressible vortex sheets amounts to prove the existence of smooth solutions to the first-order system

$$\begin{aligned}
&\partial_t p^+ + v^+\partial_{x_1} p^+ + (u^+ - \partial_t\Phi^+ - v^+\partial_{x_1}\Phi^+)\,\frac{\partial_{x_2} p^+}{\partial_{x_2}\Phi^+}\\
&+\gamma p^+\partial_{x_1} v^+ + \gamma p^+ \frac{\partial_{x_2} u^+}{\partial_{x_2}\Phi^+} - \gamma p^+ \frac{\partial_{x_1}\Phi^+}{\partial_{x_2}\Phi^+}\partial_{x_2} v^+ = 0,\\
&\partial_t v^+ + v^+\partial_{x_1} v^+ + (u^+ - \partial_t\Phi^+ - v^+\partial_{x_1}\Phi^+)\,\frac{\partial_{x_2} v^+}{\partial_{x_2}\Phi^+}\\
&+\frac{1}{\rho^+}\partial_{x_1} p^+ - \frac{1}{\rho^+}\frac{\partial_{x_1}\Phi^+}{\partial_{x_2}\Phi^+}\partial_{x_2} p^+ = 0,\\
&\partial_t u^+ + v^+\partial_{x_1} u^+ + (u^+ - \partial_t\Phi^+ - v^+\partial_{x_1}\Phi^+)\,\frac{\partial_{x_2} u^+}{\partial_{x_2}\Phi^+}\\
&+\frac{1}{\rho^+}\frac{\partial_{x_2} p^+}{\partial_{x_2}\Phi^+} = 0,\\
&\partial_t S^+ + v^+\partial_{x_1} S^+ + (u^+ - \partial_t\Phi^+ - v^+\partial_{x_1}\Phi^+)\,\frac{\partial_{x_2} S^+}{\partial_{x_2}\Phi^+} = 0,
\end{aligned} \tag{3}$$

in the fixed domain $\{x_2 > 0\}$, $((p^-, v^-, u^-, S^-, \Phi^-)$ must solve a similar system) together with the boundary conditions

$$\begin{aligned}&\Phi^+_{|x_2=0} = \Phi^-_{|x_2=0} = \varphi,\\ &\quad \partial_t\varphi = -v^+_{|x_2=0}\partial_{x_1}\varphi + u^+_{|x_2=0} = -v^-_{|x_2=0}\partial_{x_1}\varphi + u^-_{|x_2=0},\\ &p^+_{|x_2=0} = p^-_{|x_2=0}.\end{aligned} \tag{4}$$

Here $\Phi^\pm(t,x_1,x_2) := \Phi(t,x_1,\pm x_2)$. An additional equation relating Φ, U, and φ is needed to close the system. We may prescribe that $\Phi^\pm$ solve in the domain $\{x_2 > 0\}$ the *eikonal* equations

$$\partial_t\Phi^\pm + v^\pm\partial_{x_1}\Phi^\pm - u^\pm = 0. \tag{5}$$

With this choice, the boundary matrix of the system for $U^\pm$ has constant rank in the whole domain $\{x_2 \geq 0\}$ and not only on the boundary $\{x_2 = 0\}$.

3.2 The Linearized Equations

Let us consider a planar contact discontinuity $U_{r,l} = (p.v_{r,l}, 0, S_{r,l})^T$ and $\Phi_{r,l} = \pm x_2$. We assume that $v_r + v_l = 0$, $v_r > 0$. Let us consider solution to (3)–(5) of the form $U_{r,l} + \varepsilon\dot{U}_\pm$, $\pm x_2 + \varepsilon\dot{\psi}_\pm$, where $\dot{U}_\pm = (\dot{p}_\pm, \dot{\mathbf{u}}_\pm, \dot{S}_\pm)$, $\dot{\psi}_\pm$ represent some small perturbations of the exact solution $U_{r,l}, \Phi_{r,l}$. Up to second order, $\dot{U}_\pm = (\dot{p}_\pm, \dot{\mathbf{u}}_\pm, \dot{S}_\pm)^T$ must solve the linear system

$$\begin{aligned}&L'\dot{U} = 0, \text{ in } \{x_2 > 0\},\\ &B(\dot{U},\psi) = 0, \text{ on } \{x_2 = 0\},\end{aligned}$$

where we have set for shortness $\dot{U} := (\dot{U}_+, \dot{U}_-)^T$,

$$L'\dot{U} := \partial_t\dot{U} + \begin{pmatrix} A_1(U_r) & 0\\ 0 & A_1(U_l)\end{pmatrix}\partial_{x_1}\dot{U} + \begin{pmatrix} A_2(U_r) & 0\\ 0 & -A_2(U_l)\end{pmatrix}\partial_{x_2}\dot{U},$$

$A_i(U)$, $i = 1,2$, are 4×4 real matrices, $\psi = \dot{\psi}_+ = \dot{\psi}_-$ on $\{x_2 = 0\}$ and

$$B(\dot{U},\psi) := \begin{pmatrix}(v_r - v_l)\partial_{x_1}\psi - (\dot{u}_+ - \dot{u}_-)\\ \partial_t\psi + v_r\partial_{x_1}\psi - \dot{u}_+\\ p_+ - p_-\end{pmatrix}.$$

3.3 Symmetric Form of the System

We study $L'\dot{U} = f$ on $\{x_2 > 0\}$ with the boundary condition $B(\dot{U},\psi) = g$ on $\{x_2 = 0\}$. The linear change of unknowns

$$W_1 := \dot{v}_+,\ W_2 := \frac{1}{2}\left(-\frac{\dot{p}_+}{\gamma p} + \frac{\dot{u}_+}{c_r}\right)\ W_3 := \frac{1}{2}\left(\frac{\dot{p}_+}{\gamma p} + \frac{\dot{u}_+}{c_r}\right),\ W_4 := \dot{S}_+,$$

$$W_5 := \dot{v}_-,\ W_6 := \frac{1}{2}\left(-\frac{\dot{p}_-}{\gamma p} + \frac{\dot{u}_-}{c_l}\right)\ W_7 := \frac{1}{2}\left(\frac{\dot{p}_-}{\gamma p} + \frac{\dot{u}_-}{c_l}\right),\ W_8 := \dot{S}_-.$$

puts the system $L'\dot{U} = f$ in the symmetric hyperbolic form below

$$\begin{aligned} \mathcal{L}W &= f, \text{ in } \{x_2 > 0\} \\ \mathcal{B}(W^{\mathrm{nc}}, \psi) &= g, \text{ on } \{x_2 = 0\}, \end{aligned} \tag{6}$$

where f and g are new data, and $\mathcal{L}W := \mathcal{A}_0\partial_t W + \mathcal{A}_1\partial_{x_1} W + \mathcal{A}_2\partial_{x_2} W$, $\mathcal{B}(W^{\mathrm{nc}}, \psi) := \underline{M}W^{\mathrm{nc}} + \underline{b}\left(\partial_t\psi \partial_{x_1}\psi\right)$, with suitable real symmetric matrices $\mathcal{A}_i$, $i = 0, 1, 2$, and $\mathcal{A}_0$ positive definite. We also set $W := (W_1, \cdots, W_8)^T$, $W^{\mathrm{c}} := (W_1, W_4, W_5, W_8)^T$, and $W^{\mathrm{nc}} := (W_2, W_3, W_6, W_7)^T =$ (linear combination of p, u)

$$\underline{M} := \begin{pmatrix} -c_r & -c_r & c_l & c_l \\ -c_r & -c_r & 0 & 0 \\ -1 & 1 & 1 & -1 \end{pmatrix}, \qquad \underline{b} := \begin{pmatrix} 0 & v_r - v_l \\ 1 & v_r \\ 0 & 0 \end{pmatrix}.$$

We want to find an L^2 a priori estimate of the solution to the linearized problem in $\Omega := \{(t, x_1, x_2) \in \mathbb{R}^3 \text{ s.t. } x_2 > 0\} = \mathbb{R}^2 \times \mathbb{R}^+$; the boundary $\partial\Omega = \{x_2 = 0\}$ is identified to $\mathbb{R}^2$. Define $H^s_\gamma(\mathbb{R}^2) := \{u \in \mathcal{D}'(\mathbb{R}^2) \text{ s.t. } \exp(-\gamma t)u \in H^s(\mathbb{R}^2)\}$, equipped with the norm $\|u\|_{H^s_\gamma(\mathbb{R}^2)} := \|\exp(-\gamma t)u\|_{H^s(\mathbb{R}^2)}$. Define similarly the space $H^s_\gamma(\Omega)$. The space $L^2(\mathbb{R}^+; H^s_\gamma(\mathbb{R}^2))$ is the space of all functions $v = v(t, x_1, x_2)$ in Ω such that the following norm $|||v|||^2_{L^2(H^s_\gamma)} := \int_0^{+\infty} \|v(\cdot, x_2)\|^2_{H^s_\gamma(\mathbb{R}^2)} dx_2$ is finite.

3.4 The Main Result

Theorem 2. *Let $(U_{r,l}, \Phi_{r,l})$ be a planar contact discontinuity (unperturbed solution):*

(i) If $v_r - v_l > \left(c_r^{\frac{2}{3}} + c_l^{\frac{2}{3}}\right)^{\frac{3}{2}}$ and $v_r - v_l \neq \sqrt{2}\,(c_r + c_l)$, then there exists $C > 0$ such that for all $\gamma \geq 1$ and all $(W, \psi) \in H^2_\gamma(\Omega) \times H^2_\gamma(\mathbb{R}^2)$

$$\begin{aligned} &\gamma|||W|||^2_{L^2_\gamma(\Omega)} + \|W^{\mathrm{nc}}_{|x_2=0}\|^2_{L^2_\gamma(\mathbb{R}^2)} + \|\psi\|^2_{H^1_\gamma(\mathbb{R}^2)} \\ &\leq C\left(\tfrac{1}{\gamma^3}|||\mathcal{L}W|||^2_{L^2(H^1_\gamma)} + \tfrac{1}{\gamma^2}\|\mathcal{B}(W^{\mathrm{nc}}, \psi)\|^2_{H^1_\gamma(\mathbb{R}^2)}\right). \end{aligned}$$

(ii) If $v_r - v_l = \sqrt{2}\,(c_r + c_l)$, then there exists $C > 0$ such that for all $\gamma \geq 1$ and all $(W, \psi) \in H^3_\gamma(\Omega) \times H^3_\gamma(\mathbb{R}^2)$

$$\begin{aligned} &\gamma|||W|||^2_{L^2_\gamma(\Omega)} + \|W^{\mathrm{nc}}_{|x_2=0}\|^2_{L^2_\gamma(\mathbb{R}^2)} + \|\psi\|^2_{H^1_\gamma(\mathbb{R}^2)} \\ &\leq C\left(\tfrac{1}{\gamma^5}|||\mathcal{L}W|||^2_{L^2(H^2_\gamma)} + \tfrac{1}{\gamma^4}\|\mathcal{B}(W^{\mathrm{nc}}, \psi)\|^2_{H^2_\gamma(\mathbb{R}^2)}\right). \end{aligned}$$

In (i) the energy estimate entails the loss of *one* derivative for W, no loss of derivatives for the front function ψ; in the case (ii) we have a loss of *two* derivatives for W, loss of *one* derivative for the front function ψ; note that in both estimates only the trace of the noncharacteristic part of the solution can be evaluated since the problem is characteristic; thus the loss of control regards the tangential velocity and the entropy.

If we are in the "isentropic case" $c_r = c_l := c$, both values $(c_r^{\frac{2}{3}} + c_l^{\frac{2}{3}})^{\frac{3}{2}}$ and $\sqrt{2}(c_r + c_l)$ appearing in (i) and (ii) actually reduce to the same value $2\sqrt{2}c$. In this case, the assumption $v_r - v_l > 2\sqrt{2}c$ in (i) prevents the occurrence of (ii). If we are in the "full nonisentropic" situation $c_r \neq c_l$, the value $\sqrt{2}\,(c_r + c_l)$ is strictly greater than $(c_r^{\frac{2}{3}} + c_l^{\frac{2}{3}})^{\frac{3}{2}}$ so that (ii) has to be accounted.

4 Main Steps of the Proof

Here below the main steps of the proof of the energy estimates in Theorem 2 are summarized.

4.1 Reformulating the Problem

It is convenient rewriting the system (6) in terms of the new unknown $\widetilde{W} := \exp(-\gamma t)W$ and $\widetilde{\psi} := \exp(-\gamma t)\psi$. Then we consider the following system

$$\mathcal{L}^\gamma \widetilde{W} := \gamma \mathcal{A}_0 \widetilde{W} + \mathcal{L}\widetilde{W} = \exp(-\gamma t)f,$$
$$\mathcal{B}^\gamma(\widetilde{W}^{\rm nc}, \widetilde{\psi}) := \underline{M}\widetilde{W}^{\rm nc}_{|x_2=0} + \underline{b}\begin{pmatrix} \gamma\widetilde{\psi} + \partial_t\widetilde{\psi} \\ \partial_{x_1}\widetilde{\psi} \end{pmatrix} = \exp(-\gamma t)g.$$

It is enough to find an energy estimate for the problem (we drop "tilde")

$$\begin{aligned} &\mathcal{L}^\gamma W := 0, \\ &\mathcal{B}^\gamma(W^{\rm nc}, \psi) := G \end{aligned} \tag{7}$$

where $G = G(W^{nc}, \psi, f)$. Since $\mathcal{A}_j$ are symmetric and $\mathcal{A}_0$ is positive definite, taking the scalar product of $(7)_1$ with W, we get $\gamma|||W|||_0^2 \leq C||W^{nc}_{|x_2=0}||_0^2$. Hence it remains to find an estimate for $W^{nc}_{|x_2=0}$ and the front ψ.

4.2 Eliminating the Front ψ

Let W and ψ satisfy problem (7). Perform a Fourier transform in (t, x_1). Let us denote the dual variables by (δ, η) and write also $\tau := \delta + i\gamma$. The Fourier transform $(\widehat{W}, \widehat{\psi})$ of (W, ψ) must solve the following system of algebraic–differential equations

$$(\tau\mathcal{A}_0 + i\eta\mathcal{A}_1)\widehat{W} + \mathcal{A}_2\frac{d\widehat{W}}{dx_2} = 0, \quad \text{if } x_2 > 0 \tag{8}$$

$$b(\tau, \eta)\widehat{\psi} + \underline{M}\widehat{W}^{\rm nc}(0) = \widehat{G}, \tag{9}$$

where $b(\tau, \eta)$ is defined by

$$b(\tau, \eta) := \underline{b} \begin{pmatrix} \tau \\ i\eta \end{pmatrix} = \begin{pmatrix} 2i\eta v_r \\ \tau + i\eta v_r \\ 0 \end{pmatrix}.$$

It can be shown that $b(\tau, \eta)$ satisfies an *elliptic bound*, from which we derive that $\|\psi\|_{1,\gamma}^2 \leq C\left(\|W^{\mathrm{nc}}_{|x_2=0}\|_0^2 + \|G\|_0^2\right)$. As a consequence of the ellipticity of b, the boundary condition (9) can be written as $\beta(\tau, \eta)\widehat{W}^{\mathrm{nc}}(0) = \widehat{h}$ where the front ψ disappears. We focus on the following problem

$$\begin{aligned} &(\tau \mathcal{A}_0 + i\eta \mathcal{A}_1)\widehat{W} + \mathcal{A}_2 \frac{d\widehat{W}}{dx_2} = 0, \quad x_2 > 0, \\ &\beta(\tau, \eta)\widehat{W}^{\mathrm{nc}}(0) = \widehat{h}, \end{aligned} \tag{10}$$

where the new boundary condition involves only the noncharacteristic part of $\widehat{W}$. Because of the characteristic boundary, some equations in (10) do not entail differentiation with respect to x_2. Expliciting these equations we obtain the ODE system

$$\begin{aligned} &\frac{d\widehat{W}^{\mathrm{nc}}}{dx_2} = \mathcal{A}(\tau, \eta)\widehat{W}^{\mathrm{nc}}, \quad x_2 > 0, \\ &\beta(\tau, \eta)\widehat{W}^{\mathrm{nc}} = \widehat{h}, \quad x_2 = 0, \end{aligned} \tag{11}$$

where only the noncharacteristic part of the unknown $\widehat{W}$ appears. We have to prove an energy estimate for the problem (11) in which the boundary is characteristic with constant multiplicity. Under the supersonic assumption $v_r - v_l > (c_r^{\frac{2}{3}} + c_l^{\frac{2}{3}})^{\frac{3}{2}}$, the problem satisfies the Kreiss–Lopatinskii condition in a weak sense: in other words, the Lopatinskii determinant associated to (11) vanishes at some boundary frequencies $(\tau, \eta) \neq (0, 0)$ with $\Re\tau = 0$. The failure of the uniform Kreiss–Lopatinskii condition yields a loss of derivatives with respect to the source terms in the energy estimate: the loss is strictly related to the order of vanishing of the Lopatinskii determinant. The technique of the proof of the energy estimate is the construction of a degenerate Kreiss symmetrizer.

4.3 Construction of the Kreiss Symmetrizer

The construction of the symmetrizer is microlocal and is performed in a neighborhood of different classes of frequency points (τ, η):

(a) Interior points $(\tau, \eta) : \Re\tau > 0$. Here standard Kreiss symmetrizer techniques can be employed and an L^2 estimate without loss of derivatives is obtained.

(b) Boundary points $(\tau, \eta) \neq (0, 0)$ with $\Re\tau = 0$. We need to distinguish the following classes:

- (b1) Points where $\mathcal{A}(\tau, \eta)$ is diagonalizable and the Kreiss–Lopatinskii condition is satisfied; classical Kreiss symmetrizer techniques still apply and give an L^2 estimate with no loss of derivatives.

(b2) Points where $\mathcal{A}(\tau,\eta)$ is diagonalizable and the Lopatinskii condition breaks down (i.e., $\Delta(\tau,\eta)=0$); here a degenerate Kreiss symmetrizer can be constructed: this yields an L^2 estimate with loss of derivatives. The loss of derivatives depends on the multiplicity of the roots of the Lopatinskii determinant Δ according to the following scheme:
- Roots of multiplicity 1 give a loss of one derivative.
- Roots of multiplicity 2 give a loss of two derivatives.

(b3) Points where $\mathcal{A}(\tau,\eta)$ is not diagonalizable. In those points, the Lopatinskii condition is satisfied (i.e., $\Delta(\tau,\eta)\neq 0$). The construction of the symmetrizer follows as in case (a), and an L^2 energy estimate without loss of derivatives is obtained.

(b4) Poles of $\mathcal{A}(\tau,\eta)$. At those points, the Lopatinskii condition is satisfied. We construct a symmetrizer by working on the full system (10) and the energy estimate resulting is without loss of derivatives.

References

[A89] Alinhac, S.: Existence d'ondes de raréfaction pour des systémes quasi-linéaires hyperboliques multidimensionnels. Comm. Partial Differential Equations, **14**, 173–230, (1989)

[FM63] Fejer, J.A., Miles, J.W.: On the stability of a plane vortex sheet with respect to three-dimensional disturbances. J. Fluid Mech., **15**, 335–336, (1963)

[CM04] Coulombel, J.F., Morando, A.: Stability of contact discontinuities for the nonisentropic Euler equations. Ann. Univ. Ferrara Sez. VII (N.S.), **50**, 79–90, (2004)

[CS04] Coulombel J.F., Secchi P.: The Stability of Compressible Vortex Sheets in Two Space Dimensions. Indiana Univ. Math. J. **53, no. 4**, 941-1012, (2004)

[CS05] Coulombel J.F., Secchi P.: Nonlinear Compressible Vortex Sheets in Two Space Dimensions. Preprint Quad2005n28 Seminario Matematico, Brescia

[FM00] Francheteau J., Métivier G.: Existence de chocs faibles pour des systémes quasi-linéaires hyperboliques multidimensionnels. Astérique, 2000. MR 2002c:35183 (French)

[M83] Majda, A.: The existence of multidimensional shock fronts. Mem. Amer. Math. Soc. **43**, (1983)

[M91] Métivier G.: Ondes soniques. (French) [Sound waves]. J. Math. Pures Appl. (9) **70 no. 2**, 197–268, (1991)

[M01] Métivier G.: Stability of multidimensional shocks. Advances in the Theory of Shock Waves, Progr. Nonlinear Differential Equations Appl., vol. 47, Birkhäuser, Boston (2001)

[MT06] Morando A., Trebeschi P.: Two-dimensional vortex sheets for the nonisentropic Euler equations: linear stability. In progress

Three-Dimensional Numerical MHD Simulations of Solar Convection

S.D. Ustyugov

Summary. Three-dimensional magnetohydrodynamical large eddy simulations of solar surface convection using realistic model physics are conducted. The effects of magnetic fields on thermal structure of convective motions into radiative layers, the range of convection cell sizes, and penetration depths of convection are investigated. We simulate some portion of the solar photosphere and the upper layers of the convection zone, a region extending 30×30 Mm horizontally from 0 Mm down to 18 Mm below the visible surface. We solve equations of the fully compressible radiation magnetohydrodynamics with dynamical viscosity and gravity. For numerical simulation we use (1) realistic initial model of Sun and equation of state and opacities of stellar matter, (2) high-order conservative TVD scheme for solution magnetohydrodynamics, (3) diffusion approximation for solution radiative transfer, and (4) calculation dynamical viscosity from subgrid scale modeling. Simulations are conducted on horizontal uniform grid of 320×320 and with 144 nonuniformly spaced vertical grid points on the 128 processors of supercomputer MBC-1500 with distributed memory multiprocessors in Russian Academy of Sciences.

1 Introduction

Convection near solar surface has strongly nonlocal and dynamical character. Hence, numerical simulation provides useful information on the structure spatial scales by convection and helps to construct consistent models of the physical processes underlying the observed solar phenomena. We investigate effects of compressibility and of magnetic field on formation nonlocal structure of convection using realistic physics and conservative TVD numerical scheme of Godunov type. The previous simulations were confined by small computational domain and studied processes on scales order size of granulation [1]. To investigate collective interaction of convective modes at different scales and process of formation of supergranulation, we conducted calculation in three-dimensional computational box by size 30 Mm in horizontal direction and by size 18 Mm in vertical direction.

2 Numerical Method

We take distribution of the main thermodynamic variables by radius due to Standard Solar Model [2] with parameters $(X, Z, \alpha) = (0.7385, 0.0181, 2.02)$, where X and Y are hydrogen and helium abundance by mass, and α is the ratio of mixing length to pressure scale height in convection region. We use OPAL opacities and equations of state for solar matter [3].

We solve fully compressible nonideal magnetohydrodynamics equations

$$\frac{\partial \rho}{\partial t} + \nabla \cdot \rho \mathbf{v} = 0$$

$$\frac{\partial \rho \mathbf{v}}{\partial t} + \nabla \cdot \left[\rho \mathbf{v}\mathbf{v} + \left(P + \frac{B^2}{8\pi} \right) I - \frac{\mathbf{B} \cdot \mathbf{B}}{4\pi} \right] = \rho \mathbf{g} + \nabla \cdot \tau$$

$$\frac{\partial E}{\partial t} + \nabla \cdot \left[\mathbf{v} \left(E + P + \frac{B^2}{8\pi} \right) - \frac{\mathbf{B}\,(\mathbf{v} \cdot \mathbf{B})}{4\pi} \right]$$

$$= \frac{1}{4\pi} \nabla \cdot (\mathbf{B} \times \eta \nabla \times \mathbf{B}) + \nabla \cdot (\mathbf{v} \cdot \tau) + \rho\,(\mathbf{g} \cdot \mathbf{v}) + Q_{rad}$$

$$\frac{\partial \mathbf{B}}{\partial t} + \nabla \cdot (\mathbf{v}\mathbf{B} - \mathbf{B}\mathbf{v}) = -\nabla \times (\eta \nabla \times \mathbf{B})$$

where $E = e + \rho v^2/2 + B^2/8\pi$ is the total energy, Q_{rad} is the energy transferred by radiation, and τ is the viscous stress tensor.

We assume that small scales are independent of resolved scales (large eddy simulations) and rate dissipation is defined from buoyancy and shear production terms [4]. The numerical method that we used was an explicit Godunov-type conservative TVD difference scheme [5]

$$U^{n+1}_{i,j,k} = U^n_{i,j,k} - \Delta t L(U^n_{i,j,k}),$$

where $\Delta t = t^{n+1} - t^n$ and operator L is

$$L(U_{i,j,k}) = \frac{\tilde{F}_{i+1/2,j,k} - \tilde{F}_{i-1/2,j,k}}{\Delta x_i} + \frac{\tilde{G}_{i,j+1/2,k} - \tilde{G}_{i,j-1/2,k}}{\Delta y_j} + \frac{\tilde{H}_{i,j,k+1/2} - \tilde{H}_{i,j,k-1/2}}{\Delta z_k} + S_{i,j,k}$$

Term $S_{i,j,k}$ is the accounted effect of gravitation forces and radiation.

Flux along each direction, for example x, was defined by local-characteristic method as follows

$$\tilde{F}_{i+1/2,j,k} = \frac{1}{2} \left[F_{i,j,k} + F_{i+1,j,k} + R_{i+1/2} W_{i+1/2} \right]$$

where $R_{i+1/2}$ is the matrix whose columns are right eigenvectors of $\partial F / \partial U$ evaluated at generalized Roe average for real gases of $U_{i,j,k}$ and $U_{i+1,j,k}$. The $W_{i+1/2}$ is the matrix of numerical dissipation.

The element l of vectors $W_{i+1/2}$ is

$$(\phi^l_{i+1/2})^U = \frac{1}{2}\psi(a^l_{i+1/2})(g^l_{i+1} + g^l_i) - \psi(a^l_{i+1/2} + \gamma^l_{i+1/2})\alpha^l_{i+1/2}$$

where

$$\gamma^l_{i+1/2} = \frac{1}{2}\psi(a^l_{i+1/2})\begin{cases} (g^l_{i+1} - g^l_i)/\alpha^l_{i+1/2}, & \alpha^l_{i+1/2} \neq 0 \\ 0, & \alpha^l_{i+1/2} = 0 \end{cases}$$

The amplitude of waves $\alpha^l_{i+1/2}$ is elements of vectors

$$\alpha_{i+1/2} = R^{-1}_{i+1/2}(U_{i+1,j,k} - U_{i,j,k}).$$

The function ψ is

$$\psi(z) = \begin{cases} |z|, & |z| \geq \delta_1 \\ (z^2 + \delta_1^2)/2\delta_1, & |z| < \delta_1 \end{cases}$$

and define entropy correction by $|z|$, where δ_1 is the small positive parameter. The function limiter g^l_i was defined as follows:

$$g^l_i = minmod(\alpha^l_{i-1/2}, \alpha^l_{i+1/2})$$

$$g^l_i = (\alpha^l_{i+1/2}\alpha^l_{i-1/2} + |\alpha^l_{i+1/2}\alpha^l_{i-1/2}|)/(\alpha^l_{i+1/2} + \alpha^l_{i-1/2})$$

The one step of time integration is defined by Runge–Kutta method [6] as

$$U^{(1)} = U^n + \Delta t L(U^n)$$

$$U^{(2)} = \frac{3}{4}U^n + \frac{1}{4}U^{(1)} + \frac{1}{4}\Delta t L(U^{(1)})$$

$$U^{n+1} = \frac{1}{3}U^n + \frac{2}{3}U^{(2)} + \frac{2}{3}\Delta t L(U^{(2)})$$

The scheme is second order by space and time. For approximation viscous terms, we used central differences. For evaluation radiative term in energy equation, we used the diffusion approximation

$$Q_{rad} = \nabla \cdot \left[\frac{4acT^3}{3k\rho}\nabla T\right]$$

We use uniform grid in $x - y$ directions and nonuniform grid in vertical z one. We apply periodic boundary condition in horizontal planes and choose on the top and bottom as follows:

$$v_{z,k} = -v_{z,k\pm1}, v_{x,k} = v_{x,k\pm1}, v_{y,k} = v_{y,k\pm1}$$

$$dp/dz = \rho g_z, p = p(\rho), e = const$$

$$B_x = B_y = 0, dB_z/dz = 0$$

In initial moment magnetic field equals 50 G and has just one vertical component. For the magnetic diffusivity, we take constant value $\eta = 1.1 \times 10^{11}\,\mathrm{cm^2\,s^{-1}}$.

3 Results

In Figs. 1–4, results of development convection after 12 solar hour MHD numerical simulation are shown. We found that magnetic field concentrates on the boundary of convective cells in forms of magnetic flux and sheets.

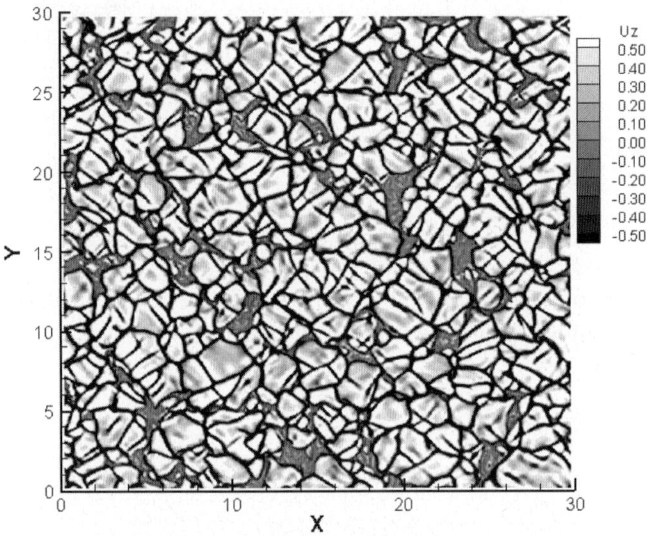

Fig. 1. Image of contours of the vertical velocity in horizontal plane near solar surface

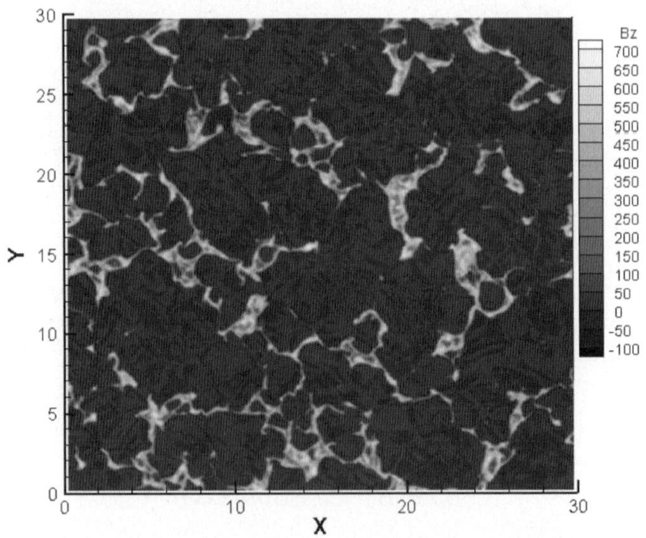

Fig. 2. Image of contours of the vertical component of magnetic field at the same plane as that for Fig. 1

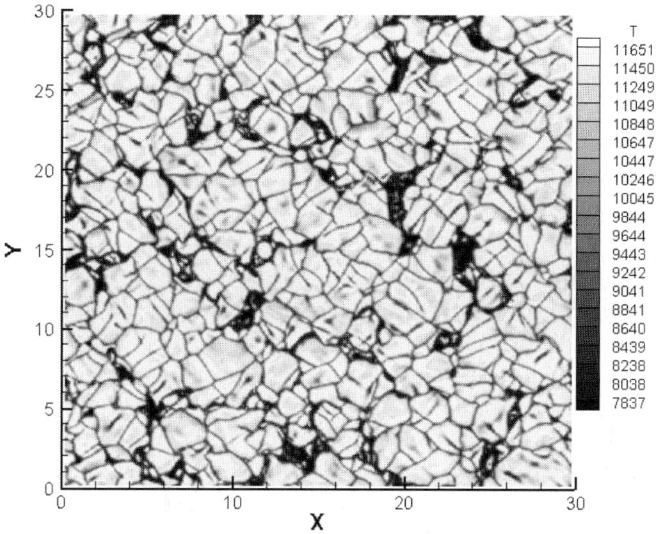

Fig. 3. Image of contours of the temperature in the horizontal plane near solar surface

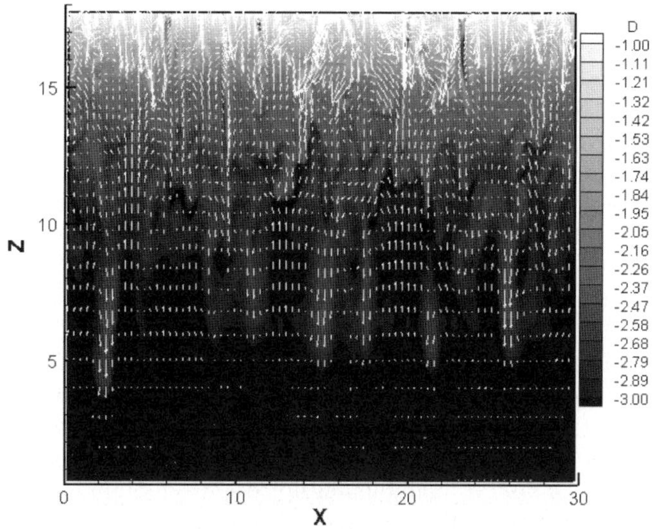

Fig. 4. Image of contours of the fluctuation density and field of velocity in the vertical plane

Diverging of convective flows from center supergranular expel weak magnetic field on the edges of convective cell. Average size of supergranular cells is 10–20 Mm with lifetime about 8–10 h. We apply procedure averaging by interval time 2 h and find value of velocity from center supergranular equal to 1–1.5 km s^{-1}. In places of action of strong magnetic field with strength

700–900 G we observe effect of suppression of convection and decreasing of fluctuation of temperature. Magnetic pressure in regions of magnetic flux prevents inflow of matter. Transfer of radiation energy in these places is suppressed. We found from simulation that the maximum value of magnetic field in computational domain is equal to 1,300 G.

Inside of supergranular, we have usual picture of evolution of convection on scale of granulation with average sizes of cells about 1–2 Mm and lifetimes about 10 min. Here we see wider upflows of warm, low density, and entropy neutral matter; and downflows of cold, converging into filamentary structures, dense material. We observe continuous picture formation and destruction of granules. Granules with highest pressure grow and push matter against neighboring granules that then shrink and disappear. Ascending flow increases pressure in center of granule and upflowing fluids decelerate motion. This process reduces heat transport to surface and allows material above the granule to cool, become denser, and move down by the action of gravity. We observe the formation of new cold intergranule lane splitting the original granule.

From Fig. 4, we see distinctly the development of three different regions of convection. In near of solar surface to the depth 4 Mm, we found zone of turbulent convection. In this region, cold blobs of matter move down with velocity in maximum about 4 km s^{-1} with maximum Mach number equal to 1. The downdrafts have different and very complicated vertical structure. Some ones travel small distance from surface and become weak enough to be broken up by the surrounding fluid motion. Other ones conserve motion with high velocity and move on distance about 6 Mm. We observe that different nature of such behavior is due to initial condition of downdrafts formation. In place of confluence of convective cells to one point, more energy is released.

In region from 5 to 8 Mm of depth, we reveal more quiet character of convective flow than in turbulent zone. Below 8 Mm we see clearly separated large-scale density fluctuations and streaming flow of matter similar jets with average velocity about 1 km s^{-1}. The magnetic field in these places has values about 300 G. Distance between different narrow jets gives size of supergranular cells. Places of intersections of path jets with horizontal plane are vertices of huge convective cells. From distribution isosurface of magnetic field with strength 500 G, we have discovered that pumping of magnetic field occupies more part of all computational domain up to bottom boundary.

Magnetic field near solar surface in our numerical simulation has very complicated structure. Inside supergranular cells, we see formation, growth, and evolution with time, many arising to surface loops of magnetic field. There are places in vertices of huge cells with vorticity motion that provide quick rate magnetic helicity transport across solar photosphere (Fig. 5). On the boundary supergranular cell, magnetic field has just vertical component. In these parts, we observe quick changing sign and big values of current helicity (Fig. 6).

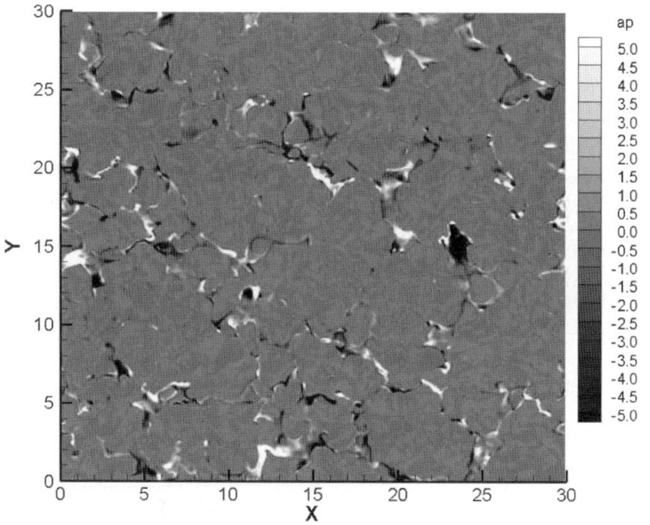

Fig. 5. Image of contours of the magnetic helicity transport rate in the horizontal plane near solar surface as that for Fig. 1

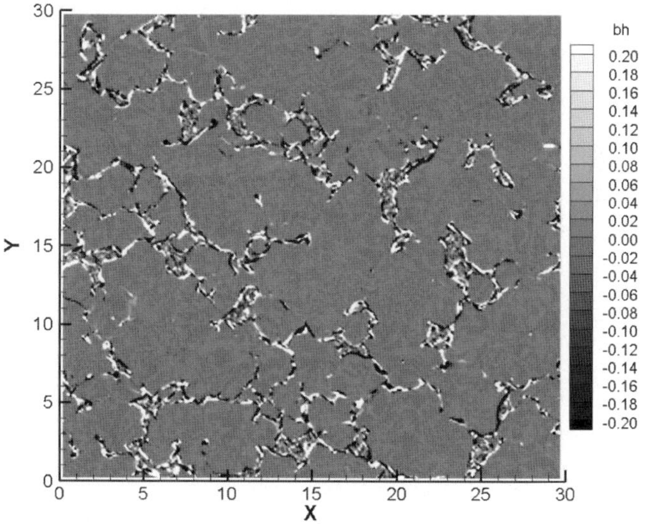

Fig. 6. Image of contours of the current helicity at the same plane as that for Fig. 1

Acknowledgments

I would like to thank Prof. Denis Serre and Prof. Sylvie Benzoni from École Normale Supérieure de Lyon for financial support for participating me in conference HYP2006.

References

1. Stein, R.F., & Nordlund, A., Astrophys. J., **642**, 1246, 2006.
2. Christensen-Dalsgaard, J., Rev. Mod. Phys., **74**, 1073, 2003.
3. Iglesias, C.A., & Rogers, F.J., Astrophys. J., **464**, 943, 1996.
4. Canuto, V.M., Minotti, F.O., & Schilling, J.L., Astrophys. J., **425**, 303, 1994.
5. Yee, H.C., Kloppfer, G.H., & Montagné, J.-L., J. Comput. Phys., **88**, 31, 1990.
6. Shu, C.W., & Osher, S., J. Comput. Phys., **77**, 439, 1988.

Large-Time Behavior of Entropy Solutions to Scalar Conservation Laws on Bounded Domain

J. Vovelle and S. Martin

Summary. We study the large-time behavior of entropy solutions to scalar conservation laws on a one-dimensional bounded domain. We show that, if the boundary data are stabilizing (in the sense that they allow for the existence of a stationary solution), then the entropy solution converges to a stationary solution to the problem.

1 Introduction

Let $A \in \mathrm{Lip}_{\mathrm{loc}}(\mathbf{R})$. Let $u(t)$ be the entropy solution to the following Cauchy–Dirichlet problem

$$u_t + (A(u))_x = 0, \, t > 0, 0 < x < 1, \tag{1}$$

$$u(x,0) = u_0(x), \qquad 0 < x < 1 \tag{2}$$

$$u(0,t) = \bar{u}_1(t), \quad u(1,t) = \bar{u}_2(t), \qquad t > 0. \tag{3}$$

We investigate the asymptotic behavior $[t \to +\infty]$ of $u(t)$. The data u_0 and u_α are supposed to be measurable and bounded a.e.

1.1 Large-Time Behavior of the Entropy Solution to the Cauchy Problem

The behavior as $[t \to +\infty]$ of $u(t)$, the entropy solution to the Cauchy problem

$$u_t + (A(u))_x = 0, \quad x \in \mathbf{R}, t > 0 \tag{4}$$

$$u(x,0) = u_0(x), \quad x \in \mathbf{R}, \tag{5}$$

has been extensively analyzed since the first works of Lax ([Lax57], and we also refer to Ilin and Oleinik [IO60], DiPerna [DiP75], Liu and Pierre [LP84], DaFermos [Daf85], and Kim [Kim03]). In particular, if the flux A is strictly convex and if $u_0 \in L^1 \cap L^\infty(\mathbf{R})$, the entropy solution converges to a so-called N-wave with specific rates in $L^1(\mathbf{R})$ and $L^\infty(\mathbf{R})$.

A related problem is the question of the stability of the solution traveling waves of (4) and (5) or, more generally, the question of the stability of solution as profiles to conservation laws. Let us refer to the works of Kawashima and Matsumura [KM85], Matsumura and Nishihara [MN94], Liu and Nishihira [LN97], Freistühler and Serre [FS98], and Serre [Ser04]. We state in particular the following results of Freistühler and Serre.

Theorem 1 (Freistühler and Serre). *Assume that A is strictly convex. Let w be a (nontrivial) shock profile of (4). Then, for all $u_0 \in w + L^1(\mathbf{R})$,*

$$\|u(t) - w(\cdot + \delta)\|_{L^1(\mathbf{R})} \to 0, \quad \delta = \frac{1}{w_+ - w_-} \int_{\mathbf{R}} u_0 - w.$$

Remark 1. It is a result of orbital stability; the value of δ is determined by the conservation of the mass. The result is false as soon as A is not convex.

1.2 Large-Time Behavior of the Entropy Solution to the Cauchy–Dirichlet Problem

The Cauchy–Dirichlet problem is of course different from the Cauchy problem since the boundary data can possibly act on the values of the solution at any time and prevent any kind of convergence. There are much less references in this topic: see Mascia and Terracina [MT99] (asymptotic behavior driven by a source term) and also the work of Freistühler and Serre [FS01] related to stability of boundary layers in the viscous approximation of a first-order scalar conservation law. We consider a.e. bounded measurable data and, without loss of generality, we actually consider data (hence solutions) with values in $[0,1]$. Our result is the following one: let

$$X_0 := L^\infty(0,1;[0,1]), \quad \overline{X} := L^\infty(\mathbf{R}_+;[0,1])^2$$

be endowed with the L^1 norm. These are closed subspaces of $L^1(0,1)$ and $L^1(\mathbf{R}_+)^2$, respectively (because the limit of a converging sequence in L^p is also the limit of a a.e. converging subsequence), hence Banach spaces. On the space $X_0 \times \overline{X}$, we define the flow $S(t)$ associated to the resolution of the Cauchy–Dirichlet problem (1)–(2)–(3) (see Sect. 2). We then introduce the notion of stationary solution and stabilizing boundary data.

Definition 1. *A function $w \in X_0$ is said to be a stationary solution to (1)–(2)–(3) if there exists $(\overline{u}_1, \overline{u}_2) \in \overline{X}$ such that $w = S(t)(w,(\overline{u}_1,\overline{u}_2))$: we say that w is a stationary solution associated to the boundary data $(\overline{u}_1,\overline{u}_2)$. A set of boundary data $(\overline{u}_1,\overline{u}_2) \in \overline{X}$ is said to be stabilizing if there exists a stationary solution associated to it.*

Theorem 2. *Suppose that A is a strictly convex function on $[0,1]$. Let*

$$(\overline{u}_1, \overline{u}_2) \in \overline{X}$$

be a stabilizing set of boundary data. Then for all $u_0 \in X_0$, $S(t)(u_0, (\overline{u}_1, \overline{u}_2))$ converges to a stationary solution associated to $(\overline{u}_1, \overline{u}_2)$ when $t \to +\infty$.

We refer to [MV06] for a similar result in case of an nonautonomous equation (A depending on the variable x).

The proof of Theorem 2 uses the tools developed by Freistühler and Serre for the study of the stability of profiles to conservation laws, it also exploits the principle of comparison for sub- and supersolutions for the Cauchy–Dirichlet problem (1)–(2)–(3).

2 Sub-, Supersolutions, and Flow

Definition 2. *Let ω be an open subset of $\mathbf{R}$. Let $u_0 \in X_0$, $(\overline{u}_1, \overline{u}_2) \in \overline{X}$. A function $u \in L^\infty((0,1) \times \mathbf{R}_+)$ is an entropy subsolution to the problem (1)–(2)–(3) on ω if: for all $k \in \mathbf{R}$, for all $\varphi \in \mathcal{C}_c^\infty(\omega \times \mathbf{R}_+)$, $\varphi \geq 0$,*

$$\begin{aligned}\int_0^1 \int_0^{+\infty} (u-k)^+ \varphi_t + \mathrm{sgn}_+(u-k)(A(u) - A(k))\varphi_x dx dt \\ + \int_0^1 (u_0 - k)^+ \varphi(x,0) dx \\ + \int_0^{+\infty} \mathrm{sgn}_+(\bar{u}_1 - k)(\hat{A}_1(\bar{u}_1) - \hat{A}_1(k))\varphi(0,t) dtt \\ + \int_0^{+\infty} \mathrm{sgn}_+(\bar{u}_2 - k)(\hat{A}_2(\bar{u}_2) - \hat{A}_2(k))\varphi(1,t) dt \geq 0,\end{aligned} \tag{6}$$

with $\hat{A}_1(u) := \int_0^u (A'(\xi))^+ d\xi$ and $\hat{A}_2(u) := \int_0^u (A'(\xi))^- d\xi$. A supersolution on ω is defined similarly, using the entropy $(u-k)^-$. In case u is both a sub- and supersolution on $\omega \supset [0,1]$, we say that u is an entropy solution.

Theorem 3. *The entropy solution u of the Cauchy–Dirichlet problem (1)–(2)–(3) is continuous in time with values in $L^1(0,1)$ and thus defines a flow $S(t)$ which has the following properties: it is monotone:*

$$(u_0, (\overline{u}_1, \overline{u}_2)) \leq (v_0, (\overline{v}_1, \overline{v}_2))$$

implies

$$S(t)(u_0, (\overline{u}_1, \overline{u}_2)) \leq S(t)(v_0, (\overline{v}_1, \overline{v}_2))$$

for all $t > 0$; it is L^1 nonexpansive:

$$\begin{aligned}\|S(t)(u_0, (\overline{u}_1, \overline{u}_2)) - S(t)(v_0, (\overline{v}_\alpha))\|_{L^1(0,1)} \leq \|u_0 - v_0\|_{L^1(0,1)} \\ + \mathcal{L}\|\bar{u}_1 - \bar{v}_1\|_{L^1(0,t)} + \mathcal{L}\|\bar{u}_2 - \bar{v}_2\|_{L^1(0,t)},\end{aligned} \tag{7}$$

where $\mathcal{L} := \|A'\|_{L^\infty}$; *it preserves the space* BV: *if*

$$\|u_0\|_{BV(0,1)}, \|\bar{u}_1\|_{BV(\mathbf{R}_+)}, \|\bar{u}_2\|_{BV(\mathbf{R}_+)} \leq C,$$

then

$$\|S(t)(u_0, (\overline{u}_1, \overline{u}_2))\|_{BV(0,1)} \leq C'. \tag{8}$$

The property of monotony and (7) can be deduced from the following result of comparison (and more generally we refer to [BLN79, Ott93, Ott96] for the proof of Theorem 3).

Proposition 1. *Let* ω *be an open subset of* $\mathbf{R}$. *Let* $u, v \in L^\infty((0,1)\times\mathbf{R}_+)$ *be, respectively, an entropy sub- and supersolution to the problem (1)–(2)–(3) on* ω. *Then, for all* $\varphi \in C_c^\infty(\omega \times \mathbf{R}_+)$, $\varphi \geq 0$,

$$\begin{aligned}\int_0^1\int_0^{+\infty} (u-v)^+\varphi_t + \mathrm{sgn}_+(u-v)(A(u)-A(v))\varphi_x dxdt \\ + \int_0^1 (u_0-v_0)^+\varphi(x,0)dx \\ + \int_0^{+\infty} \mathrm{sgn}_+(\bar{u}_1-\bar{v}_1)(\hat{A}_1(\overline{u}_1)-\hat{A}_1(\bar{v}_1))\varphi(0,t)dt \\ + \int_0^{+\infty} \mathrm{sgn}_+(\bar{u}_2-\bar{v}_2)(\hat{A}_2(\overline{u}_2)-\hat{A}_2(\bar{v}_2))\varphi(1,t)dt \geq 0,\end{aligned} \tag{9}$$

We will use Proposition 1 to derive Lyapunov functionals for the evolution along the flow S, but first we have to analyze the stationary solutions associated to given stabilizing boundary data.

3 Stationary Solution

Let $(\overline{u}_1, \overline{u}_2) \in \overline{X}$ be stabilizing boundary data. Let $\Gamma(\overline{u}_1, \overline{u}_2)$ be the set of stationary solutions associated to $(\overline{u}_1, \overline{u}_2)$. The set $\Gamma(\overline{u}_1, \overline{u}_2)$ is nonempty by hypothesis: let $w \in \Gamma(\overline{u}_1, \overline{u}_2)$. The weak form of the equation shows that the flux $q := A(w)$ is constant. Since A is strictly convex on $[0,1]$, the equation $q = A(w)$ admits at most two *constant* solutions, say $\lambda \leq \mu$. By use of the entropy conditions in their weak form, we show that

$$\Gamma(\overline{u}_1, \overline{u}_2) \subset \left\{w_z := \lambda\mathbf{1}_{(z,1]} + \mu\mathbf{1}_{[0,z)},\ z \in [0,1]\right\}$$

(to sum up, the proof of this result uses Proposition 1 on $\omega = (0,1)$ and amounts to compare an element of $\Gamma(\overline{u}_1, \overline{u}_2)$ with specific stationary solutions with flux $\tilde{q} \neq q$). Then, by analysis of the boundary conditions, we prove the following result.

In the following, and to give self-consistent proofs in this short chapter (see Sect. 4), we will assume:

there is a nontrivial stationary shock in $\Gamma(\overline{u}_1, \overline{u}_2)$. (10)

This amounts to suppose $\lambda < \mu$ and that there is $z_* \in (0,1)$ such that $w_{z_*} \in \Gamma(\overline{u}_1, \overline{u}_2)$. Notice that, although we do this hypothesis, Theorem 2 is true in its generality. Under hypothesis (10), the result of Theorem 2 is to be related to the result of Theorem 1. Indeed, in that case, we have

$$\Gamma(\overline{u}_1, \overline{u}_2) = \left\{ w_z := \lambda \mathbf{1}_{(z,1]} + \mu \mathbf{1}_{[0,z)},\ z \in [0,1] \right\},$$

i.e., every stationary solution is a shift of w_{z_*}.

4 Convergence in Large Time

We give the proof of Theorem 2 under hypothesis (10). Let $(\overline{u}_1, \overline{u}_2) \in \overline{X}$ be fixed stabilizing boundary data. We use the notations of Sect. 3. For $u_0 \in X_0$, we also denote $(S(t)u_0)$ the trajectory $(S(t)(u_0, (\overline{u}_1, \overline{u}_2))$ since the boundary data are fixed.

Lemma 1. *Let $u_0 \in X_0$. The trajectory $(S(t)u_0)$ is relatively compact in $L^1(0,1)$.*

The relative compactness of the bounded sets of $BV(0,1)$ in $L^1(0,1)$ and the property (8) show that the result in case u_0 has, additionally, a bounded variation on $(0,1)$. We then deduce the lemma from the fact that $S(t)$ is L^1 nonexpansive. Indeed, if $\varepsilon > 0$, then there exists $u_0^\varepsilon \in X_0 \cap BV(0,1)$ such that $\|u_0 - u_0^\varepsilon\|_{L^1(0,1)} < \varepsilon/2$. We then have $\|S(t)u_0 - S(t)u_0^\varepsilon\|_{L^1(0,1)} < \varepsilon/2$ for all $t > 0$. Since $\{S(t)u_0^\varepsilon\}$ is relatively compact in X_0, there is a finite number of balls of radii $\varepsilon/2$ covering it. We deduce that $\{S(t)u_0\}$ is covered by a finite number of balls of radii ε. Since ε is arbitrary and X_0 is a Banach space, $\{S(t)u_0\}$ is relatively compact in X_0.

A corollary of Lemma 1 is the fact that each ω limit set

$$\omega(u_0) := \bigcap_{t>0} \overline{\{S(\tau)(u_0); \tau \geq t\}}$$

is not empty. In a second step, we show that $\omega(u_0) \subset \Gamma(\overline{u}_1, \overline{u}_2)$. To this purpose, we first prove Lemma 2.

Lemma 2. *Set*

$$\omega(X_0) := \bigcup_{u_0 \in X_0} \omega(u_0) \text{ and } X_1 := \{w \in X_0, \lambda \leq w \leq \mu \text{ a.e.}\}.$$

Then $\omega(X_0)$ is a subset of X_1 and X_1 is invariant by the flow.

That X_1 is invariant by the flow follows from the result of comparison of Proposition 1 between $S(t)u_0$ and the constant solutions λ and μ. To show the inclusion $\omega(X_0) \subset X_1$, we also use comparison between $S(t)u_0$ and the constant solution μ (or λ): for every $\varphi \in \mathcal{C}_c^\infty(\mathbf{R} \times (0,+\infty))$, $\varphi \geq 0$,

$$\int_0^1 \int_0^{+\infty} (S(t)u_0 - \mu)^+ \varphi_t + \mathrm{sgn}_+(S(t)u_0 - \mu)(A(S(t)u_0) - A(\mu))\varphi_x dxdt \geq 0. \tag{11}$$

Since $\lambda < \mu$ and λ, μ, are both constant solutions of the equation $A(w) = q$ and A is convex, we have $\gamma := A'(\mu) > 0$ and

$$\mathrm{sgn}_+(S(t)u_0 - \mu)(A(S(t)u_0) - A(\mu)) \geq \gamma(S(t)u_0 - \mu)^+.$$

Setting $\varphi(x,t) := e^{-x}\alpha(t)$, $\alpha \in \mathcal{C}_c^\infty((0,+\infty)$, $\alpha \geq 0$ in (11), we derive the inequality

$$\dot{\eta} \leq -\gamma\eta, \quad \eta(t) := \int_0^1 (S(t)u_0 - \mu)^+ e^{-x} dx$$

in $\mathcal{D}'(0,+\infty)$, from which follows the exponential decrease of η. Similarly, we show that

$$t \mapsto \int_0^1 (S(t)u_0 - \lambda)^- e^x dx$$

decreases exponentially fast to 0. These two results show that $\omega(X_0) \subset X_1$.

The next step of the proof is to use the set of solutions $\{w_z, z \in [0,1]\}$ to derive Lyapunov functionals. Indeed, the result of comparison of Proposition 1 (used with $\omega = \mathbf{R}$ and φ independent on x) shows that, for any $z \in [0,1]$, the map

$$u \mapsto \|(w_z - u)^+\|_{L^1(\mathbf{R})}$$

is nonincreasing along the trajectories. As a consequence of the LaSalle' principle, we have, for any $a \in \omega(u_0)$,

$$\|(w_z - S(t)a)^+\|_{L^1(\mathbf{R})} = \mathrm{cst} = \|(w_z - a)^+\|_{L^1(\mathbf{R})}.$$

But by Lemma 2, we have

$$\|(w_z - S(t)a)^+\|_{L^1(\mathbf{R})} = \int_0^z (\mu - S(t)a)dx$$

and we deduce

$$\int_0^z (\mu - S(t)a)dx = \int_0^z (\mu - a)dx, \forall z \in (0,1),$$

i.e., $\int_0^z (S(t)a - a)dx = 0$ for all $z \in (0,1)$. This shows that $S(t)a = a$ and that a is a stationary solution. We then conclude to the convergence of the trajectory. First, there is at most one element in $\omega(u_0)$ for if $a \in \omega(u_0)$, the trajectory can only get closer and closer to a because a is a stationary solution and $S(t)$ is nonexpansive. Second, $\omega(u_0)$ is nonempty by compactness, it is therefore reduced to one element and $(S(t)u_0)$ converges to this element $a \in \Gamma(\overline{u}_1, \overline{u}_2)$.

References

[BLN79] C. Bardos, A.Y. LeRoux, and J.-C. Nédélec, *First order quasilinear equations with boundary conditions*, Comm. Partial Differential Equations **4** (1979), no. 9, 1017–1034.

[Bre83] Y. Brenier, *Résolution d'équations d'évolution quasilinéaires en dimension N d'espace à l'aide d'équations linéaires en dimension $N+1$*, J. Differential Equations **50** (1983), no. 3, 375–390.

[Daf85] C. M. Dafermos, *Regularity and large time behaviour of solutions of a conservation law without convexity*, Proc. Roy. Soc. Edinburgh Sect. A **99** (1985), no. 3-4, 201–239.

[DiP75] R.J. DiPerna, *Decay and asymptotic behavior of solutions to nonlinear hyperbolic systems of conservation laws*, Indiana Univ. Math. J. **24** (1974/75), no. 11, 1047–1071.

[FS98] H. Freistühler and D. Serre, *L^1 stability of shock waves in scalar viscous conservation laws*, Comm. Pure Appl. Math. **51** (1998), no. 3, 291–301.

[FS01] H. Freistühler and D. Serre, *The L^1-stability of boundary layers for scalar viscous conservation laws*, J. Dynam. Differential Equations **13** (2001), no. 4, 745–755.

[GM83] Y. Giga and T. Miyakawa, *A kinetic construction of global solutions of first order quasilinear equations*, Duke Math. J. **50** (1983), no. 2, 505–515.

[IO60] A.M. Il′in and O.A. Oleĭnik, *Asymptotic behavior of solutions of the Cauchy problem for some quasi-linear equations for large values of the time*, Mat. Sb. (N.S.) **51 (93)** (1960), 191–216.

[Kim03] Y.J. Kim, *Asymptotic behavior of solutions to scalar conservation laws and optimal convergence orders to N-waves*, J. Differential Equations **192** (2003), no. 1, 202–224.

[KM85] S. Kawashima and A. Matsumura, *Asymptotic stability of traveling wave solutions of systems for one-dimensional gas motion*, Comm. Math. Phys. **101** (1985), no. 1, 97–127.

[Lax57] P. D. Lax, *Hyperbolic systems of conservation laws. II*, Comm. Pure Appl. Math. **10** (1957), 537–566.

[LN97] T.-P. Liu and K. Nishihara, *Asymptotic behavior for scalar viscous conservation laws with boundary effect*, J. Differential Equations **133** (1997), no. 2, 296–320.

[LP84] T.-P. Liu and M. Pierre, *Source-solutions and asymptotic behavior in conservation laws*, J. Differential Equations **51** (1984), no. 3, 419–441.

[LPT94] P.-L. Lions, B. Perthame, and E. Tadmor, *A kinetic formulation of multidimensional scalar conservation laws and related equations*, J. Amer. Math. Soc. **7** (1994), no. 1, 169–191.

[MV06] S. Martin, J. Vovelle, *Large-time behaviour of the entropy solution of a scalar conservation law with boundary conditions*, accepted for publication in Quart. J. Mech. Appl. Math.

[MN94] A. Matsumura and K. Nishihara, *Asymptotic stability of traveling waves for scalar viscous conservation laws with non-convex nonlinearity*, Comm. Math. Phys. **165** (1994), no. 1, 83–96.

[MT99] C. Mascia and A. Terracina, *Large-time behavior for conservation laws with source in a bounded domain*, J. Differential Equations **159** (1999), no. 2, 485–514.

[Ott93] F. Otto, *Initial-boundary value problem for a scalar conservation law*, Ph. D. Thesis (1993).

[Ott96] F. Otto, *Initial-boundary value problem for a scalar conservation law*, C. R. Acad. Sci. Paris Sér. I Math. **322** (1996), no. 8, 729–734.

[Per02] B. Perthame, *Kinetic formulation of conservation laws*, Oxford Lecture Series in Mathematics and its Applications, vol. 21, Oxford University Press, Oxford, 2002.

[Ser04] D. Serre, *L^1-stability of nonlinear waves in scalar conservation laws*, Evolutionary equations. Vol. I (Amsterdam), Handb. Differ. Equ., North-Holland, Amsterdam, 2004, pp. 473–553.

A Second-Order Improved Front Tracking Method for the Numerical Treatment of the Hyperbolic Euler Equations

J.A.S. Witteveen

Summary. Front tracking methods can be used to accurately resolve discontinuities in numerical simulations of Euler flows. They usually result in first-order error convergence due to their piecewise constant approximation of the flow conditions. In this chapter, a piecewise linear reconstruction of the solution is proposed based on wave types which track the physical phenomena that the fronts represent. It is demonstrated numerically that this approach results in second-order error convergence. A verification and validation study is performed by comparing the results with those of the Godunov method and experimental data.

1 Introduction

Front tracking can be used as a numerical method for resolving discontinuities in the simulation of hyperbolic conservation laws. In this chapter, front tracking is considered as a tool for the numerical simulation of the Euler equations of inviscid gas dynamics. The Euler equations allow for discontinuous solutions such as shock waves, slip lines, and two-fluid interfaces. Front tracking is often used for resolving these discontinuities in addition to a background grid for resolving the continuous flow phenomena. Richtmyer and Morton [RM67] have initiated this branch of front tracking methods. Important contributions and applications have been presented by Moretti [Mor87], Swartz and Wendroff [SW86], and Glimm, Grove, Chern, Holmes, and coworkers [CGM86, HGS95].

Front tracking can also be used for resolving both the discontinuous and continuous flow phenomena. In that case the continuous phenomena are approximated by a piecewise constant function. This type of front tracking methods has been developed in the context of gas dynamics by Holden, Lie, Risebro, Tveito, and coworkers [HLR99, RT92]. The latter class of front tracking methods without background grid is considered in the current chapter.

Front tracking methods in one dimension are based on the piecewise constant approximation of the solution. In the simulation of an initial–boundary value problem, the initial conditions are approximated by a piecewise constant function. At the discontinuities in this approximation, the flow conditions resemble the initial conditions of a local Riemann problem. After solving the local Riemann problems, their solutions are also approximated by a piecewise constant function. The position of the moving discontinuities in this approximation is tracked by the fronts. At an intersection of two fronts, the solution again resembles the initial condition of a local Riemann problem. The piecewise constant approximation of the solution of the local Riemann problem results in the creation of new fronts and so on until the solution in the space–time domain has been found.

One-dimensional front tracking is of interest for, for example, pipe flows and shock tube problems. A similar approach can also be employed to simulate two-dimensional supersonic flows. In addition, front tracking in one dimension can be used for a basic comparison of different front tracking algorithms and as a test for the implementation in a multidimensional code as a tool in dimensional splitting. Extensions of front tracking methods to multiple spatial dimensions can be found in for example [HLR99].

Due to the piecewise constant approximation of the solution, front tracking methods usually result in first-order error convergence [RT92]. An example of a second-order moving mesh method is the method by Lucier [Luc86] for scalar conservation laws.

In this chapter, a second-order front tracking method for the system of Euler equations is proposed. In [Wit06], it has been demonstrated numerically that the position of the fronts is approximated with second-order accuracy even for a first-order front tracking method. The first-order convergence is caused by the piecewise constant approximation of the solution. Therefore, a piecewise linear reconstruction of the solution of a first-order front tracking method can result in second-order error convergence. For a correct piecewise linear reconstruction, wave types are employed which track the physical phenomena that the fronts represent. In [WKB06], these wave types are used for an improved front interaction modeling, which results in a physically more correct simulation.

This chapter is organized as follows. In Sect. 2, the second-order improved front tracking method is introduced. In Sect. 3, numerical results are compared with those of the Godunov method and experimental data. The conclusions are summarized in Sect. 4.

2 A Second-Order Front Tracking Method

In this section, a second-order improved front tracking method for the Euler equations is proposed based on a piecewise linear reconstruction of the solution of a first-order front tracking method. For the reconstruction the wave types of the improved front tracking method [WKB06] are employed. It is demonstrated in numerical experiments in Sect. 3 that the piecewise linear solution results in second-order error convergence.

2.1 Wave Types

In standard front tracking methods without background grid, the front interactions are resolved by solving standard local Riemann problems at the intersection points. This approach has some limitations; for example, it cannot resolve isentropic compressions as truly isentropic phenomena. In [WKB06], an improved front interaction modeling is proposed, which employs wave front types to track the physical phenomena that the fronts represent for a physically more correct modeling of the Euler equations. The wave types are also employed for neglecting insignificant waves to avoid the build up of an infinite number of fronts. The same wave type labels are employed here for a piecewise linear reconstruction of the solution.

The basic waves in an Euler flow are shock waves, rarefaction or isentropic compression waves, and contact waves. For a detailed review of wave phenomena and classifications, the reader is referred to standard works as [CM79, Smo94].

In the front tracking method, the discontinuous phenomena such as a shock wave and a contact discontinuity are discretized by a single discontinuity. The wave front types that are assigned to the fronts representing these discontinuities are the shock wave (sw) and contact discontinuity (cd) wave types, respectively.

The continuous phenomena such as a rarefaction or isentropic compression wave are discretized by a series of small discontinuities. The fronts discretizing these continuous phenomena resemble characteristics. To represent both the solution inside the rarefaction fan and its spatial dimension accurately, three characteristic wave types are used: the left most characteristic (lch), the right most characteristic (rch), and an interior characteristic (ich) of a fan of characteristics. The same wave types are used in discretizing isentropic compression waves.

In contrast with a discontinuous change of entropy at a contact discontinuity, a continuous change of entropy is discretized by a series of contact waves. They are represented by similar wave types as rarefaction fans: the left most contact wave (lcw), the right most contact wave (rcw), and an interior contact wave (icw) of a continuous change in entropy.

Finally, a wave family type is assigned to the fronts to track the velocity of the wave relative to the flow. Left running (−1), right running (+1), and waves with no relative velocity with respect to the flow (0) are used.

2.2 Piecewise Linear Reconstruction

A piecewise linear reconstruction of the solution of a front tracking method is not straightforward due to the large variations in cell size from a cell to another. Based on the wave types, it can be identified whether the flow conditions in a cell are truly uniform or not. In the latter case, a piecewise linear reconstruction of the flow conditions is used based on the wave types.

Consider the piecewise constant solution of the improved front tracking method with n_f fronts and n_Ω cells in space–time. Let f_j denote the fronts and Ω_i denote the cells with flow conditions U_i in the space–time solution. Then the cell numbers of the cell to the left and the right of front f_j are $i_{\mathrm{f}_j,\mathrm{l}}$ and $i_{\mathrm{f}_j,\mathrm{r}}$, respectively. The numbers of the fronts to the left and the right of cell Ω_i are $j_{\Omega_i,\mathrm{l}}(t)$ and $j_{\Omega_i,\mathrm{r}}(t)$, respectively.

Consider the simple example of a single discontinuity in the initial condition at which only a centered rarefaction fan is created (see Fig. 1). The piecewise constant approximation results in n_Ω uniform domains separated by $n_\mathrm{f} = n_\Omega - 1$ discontinuities. At the local Riemann problem, the created cells are labeled as domains with truly uniform or nonuniform flow conditions based on the position of the cell in the solution of the Riemann problem. In this case, the flow conditions in cells Ω_1 and Ω_{n_Ω} are labeled as uniform, where the cells are numbered from the left to the right. These cells are not affected by the piecewise linear reconstruction. In the other cells Ω_i with $i = 2, \ldots, n_\Omega - 1$, which discretize the rarefaction wave, the flow conditions are labeled as nonuniform.

Front f_1 in Fig. 1 has wave type lch, front f_{n_f} has wave type rch, and fronts f_j with $j = 2, \ldots, n_\mathrm{f} - 1$ have wave type ich, where the fronts are numbered from the left to the right. These wave types are used in the piecewise linear reconstruction of the solution in the cells Ω_i with $i = 2, \ldots, n_\Omega - 1$. At all characteristic wave types, the flow conditions vary continuously. At front f_1 with wave type lch, the flow conditions are equal to those in the cell to its left $U_{i_{\mathrm{f}_1,\mathrm{l}}}$. The flow conditions at front f_{n_f} with wave type rch are equal to those in the cell to its right $U_{i_{\mathrm{f}_{n_\mathrm{f}},\mathrm{r}}}$. At the fronts with wave type ich, f_j with $j = 2, \ldots, n_\mathrm{f} - 1$, the flow conditions are the average of those in the cell to its left and its right $(U_{i_{\mathrm{f}_j,\mathrm{l}}} + U_{i_{\mathrm{f}_j,\mathrm{r}}})/2$.

Since the gradient of the flow conditions changes in time and in general, the cells Ω_i can be bounded by different fronts $f_{j_{\Omega_i,\mathrm{l}}(t)}$ and $f_{j_{\Omega_i,\mathrm{r}}(t)}$ in time, the piecewise linear flow conditions in the cells Ω_i with $i = 2, \ldots, n_\Omega - 1$ are represented as follows. The flow velocity in a centered rarefaction wave, which originates in a single point, varies exactly linear in space. Therefore, linear flow conditions in the cells Ω_i are reconstructed as if the cell is a centered

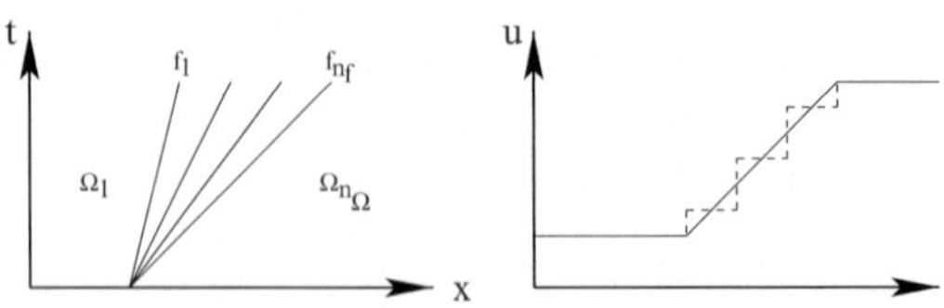

Fig. 1. The piecewise linear reconstruction of the solution of a rarefaction fan

rarefaction wave with origin $\mathcal{O}_i$ at $(x_{\mathcal{O}_i}, t_{\mathcal{O}_i})$. The origin of the centered rarefaction wave might be the actual origin of the cell or a virtual origin. The flow conditions at the fronts bounding the cell initially at the left and the right are linearly projected onto flow conditions at lines with velocity $u^- = -1$ and $u^+ = 1$ through the cell origin $\mathcal{O}_i$, U_i^- and U_i^+, respectively. The piecewise linear flow conditions $U_i(x,t)$ in cell Ω_i are then based on the linear relation through U_i^- and U_i^+:

$$U_i(x,t) = \frac{1}{2}(U_i^+ + U_i^-) + \frac{1}{2}(U_i^+ - U_i^-)\frac{x - x_{\mathcal{O}_i}}{t - t_{\mathcal{O}_i}}, \tag{1}$$

with $(x,t) \in \Omega_i$. The result of this linearization is for the centered rarefaction fan shown in Fig. 1. This piecewise linear reconstruction can mostly be employed as a postprocessing step. Special attention has to be paid at the cells at the final time level and the piecewise linear reconstruction of the flow conditions in nonsimple waves.

3 Numerical Results

In this section, the second-order front tracking method is verified by comparison of numerical results with those of Godunov method for a one-dimensional unsteady Riemann problem. A validation study is performed by comparing results for a supersonic wing section flow with experimental data.

3.1 Verification

The second-order front tracking method is applied to Sod's Riemann problem [Sod78] to verify the convergence in comparison with the first-order Godunov finite volume method [Tor97]. The initial left and right states of Sod's Riemann problem are defined as $p_{\text{left}} = \rho_{\text{left}} = 1$, $p_{\text{right}} = 0.1$, $\rho_{\text{right}} = 0.125$, and zero velocity.

In Fig. 2, the solution for the density at $t = 1$ is given for the front tracking (FT) method with $n_{\text{f}} = 8$ and for the finite volume (FV) method with $n_{\text{x}} = 128$ spatial cells, and cfl $= 0.5$. The front tracking method results in an approximation of the shock wave and the contact discontinuity as true discontinuities, since the uniform domains in the solution are unaffected by the piecewise linear reconstruction. The flow conditions in the rarefaction fan are approximated accurately by the piecewise linear representation using eight characteristics. The most profound difference with the finite volume results is the numerical smearing of especially the contact discontinuity in the finite volume results.

The L_1 error convergence with respect to n_{f} and n_{x} is shown in Fig. 3 for the piecewise constant and piecewise linear front tracking method, and the finite volume method. The piecewise linear reconstruction of the solution

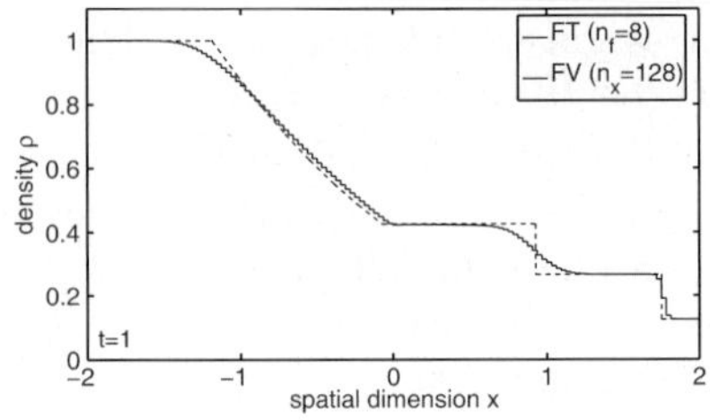

Fig. 2. Front tracking (FT) and finite volume (FV) solution for the density of Sod's Riemann problem at $t = 1$

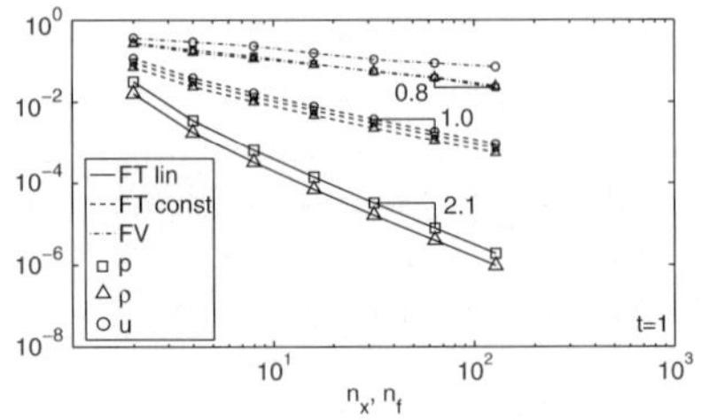

Fig. 3. Error convergence of the front tracking (FT) method and the finite volume (FV) method for Sod's Riemann problem at $t = 1$

results in second-order error convergence for the front tracking method. This results in significantly lower error than for the first-order piecewise constant front tracking method. The finite volume method converges approximately with first-order accuracy toward the front tracking solution. The absolute error is, however, larger than for the first-order front tracking method due to the numerical smearing of the contact discontinuity.

3.2 Validation

A similar front tracking algorithm as for one-dimensional unsteady Euler flow can also be applied to two-dimensional supersonic flow problems. Supersonic flow over a circular arc wing section with a maximum thickness of 12% in a free stream flow with Mach number 2 and 2.5 is considered. The numerical results for the drag of the wing section are validated by comparison with experimental data.

In Figs. 4 and 5, the front tracking solution for the Mach number field is shown for free stream Mach numbers of $M_\infty = 2$ and $M_\infty = 2.5$, respectively. The surface of the wing section is approximated by $n_f = 41$ points. The leading and trailing edge shock wave are slightly curved due to their interaction with the expansion created at the wing section surface. The front tracking method also resolves a slight gradient of the Mach number downstream of the trailing edge shock wave due to the entropy gradient caused by the curved shocks.

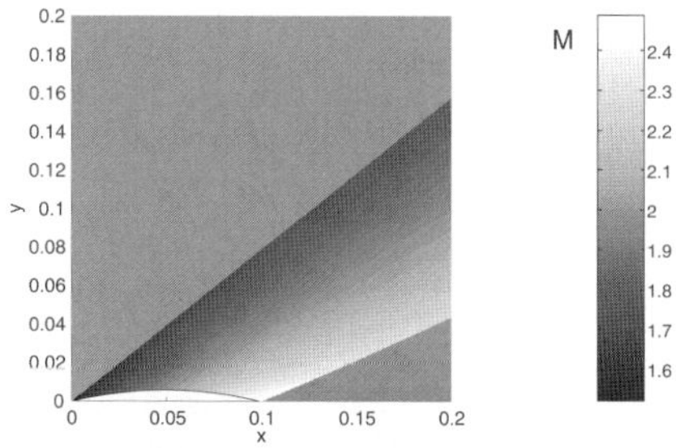

Fig. 4. Front tracking solution for the Mach number field of the supersonic wing section flow with $M_\infty = 2$

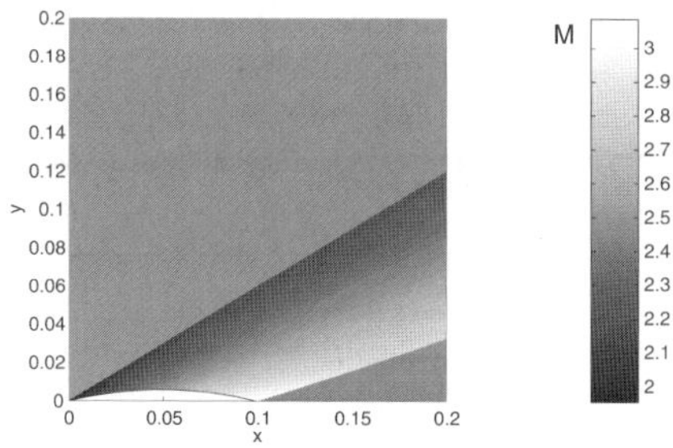

Fig. 5. Front tracking solution for the Mach number field of the supersonic wing section flow with $M_\infty = 2.5$

Table 1. Drag of the supersonic wing section obtained from front tracking simulations and experiments

Mach	Cells	Computed	Measured
2	2,028	0.0450	0.0462
2.5	8,142	0.0339	0.0330

The case of $M_\infty = 2.5$ in Fig. 5 results in a slightly higher range of Mach numbers and steeper shock wave angles compared to $M_\infty = 2$ in Fig. 4.

In Table 1, the drag computed by the front tracking method is compared with experimental data. The experimental data are obtained by pressure hole measurements performed by Souverein, Van Oudheusden, and Scarano [Sou77, SOS07]. The numerical results for the drag show good agreement with the measured values for both $M_\infty = 2$ and $M_\infty = 2.5$.

4 Conclusions

A piecewise linear reconstruction of the solution of the improved front tracking method is proposed. A comparison with the Godunov method demonstrated second-order error convergence for Sod's Riemann problem. The results of the application to two-dimensional supersonic wing section flow showed good agreement with experimental data. More challenging test problems including multidimensional unsteady problems will be considered in future work.

References

[CGM86] Chern, I.-L., Glimm, J., McBryan, O., Plohr, B., Yaniv, S., Front tracking for gas dynamics, J. Comput. Phys., **62**, 83 (1986)

[CM79] Chorin, A.J., Marsden, J.E.: A mathematical introduction to fluid mechanics. Springer-Verlag, New York (1979)

[HLR99] Holden, H., Lie, K.-A., Risebro, N.H.: An unconditionally stable method for the Euler equations, J. Comput. Phys., **150**, 76 (1999)

[HGS95] Holmes, R.L., Grove, J.W., Sharp, D.H.: Numerical investigation of Richtmyer-Meshkov instability using front tracking, J. Fluid Mech., **301**, 51 (1995)

[Luc86] Lucier, B.J.: A moving mesh numerical method for hyperbolic conservation laws, Math. Comp., **173**, 59 (1986)

[Mor87] Moretti, G.: Computations of flows with shocks. Ann. Rev. Fluid Mech., **19**, 313 (1987)

[RM67] Richtmyer, R., K. Morton: Difference methods for intial value problems. Interscience, New York (1967)

[RT92] Risebro, N.H., Tveito, A.: A front tracking method for conservation laws in one dimension, J. Comput. Phys., **101**, 130 (1992)

[Smo94] Smoller, J.: Shock waves and reaction-diffusion equations (2nd ed) Springer-Verlag, New York (1994)

[Sod78] Sod, G.A.: A survey of several finite difference methods for systems of nonlinear hyperbolic conservation laws, J. Comput. Phys., **27**, 1 (1978)

[Sou77] Souverein, L.J.: PIV based aerodynamic loads determination in supersonic flows. MSc Thesis, Delft University of Technology, Delft (2006)

[SOS07] Souverein, L.J., van Oudheusden, B.W., Scarano, F.: Particle image velocimetry based loads determination in supersonic flows. AIAA-2007-50, 45th AIAA Aerospace Sciences Meeting and Exhibit, Reno, NV (2007)

[SW86] Swartz, B.K., Wendroff, B.,: AZTEC: A front tracking code based on Godunov's method. Appl. Numer. Math., **2**, 385 (1986)

[Tor97] Toro, E.F.: Riemann solvers and numerical methods for fluid dynamics, Springer Verlag, Berlin (1997)

[WKB06] Witteveen, J.A.S., Koren, B., Bakker, P.G.: An improved front tracking method for the Euler equations. J. Comput. Phys., in press

[Wit06] Witteveen, J.A.S.: A second-order front tracking method applied to the Euler equations. AIAA-2006-1277, 44th AIAA Aerospace Sciences Meeting and Exhibit, Reno, NV (2006)

Simulation of Field-Aligned Ideal MHD Flows Around Perfectly Conducting Cylinders Using an Artificial Compressibility Approach

M.S. Yalim, D.V. Abeele, and A. Lani

Summary. Ideal magnetohydrodynamics (MHD) simulations are known to have problems in satisfying the solenoidal constraint ($\nabla \cdot \mathbf{B} = 0$) and diverge unless appropriate numerical techniques are used to enforce this constraint. In this chapter, a technique inspired by the artificial compressibility concept is developed. This technique is thought to be a cure for the drawbacks of the families of solenoidal constraint-satisfying techniques presented in the literature so far: incorrect shock capturing and poor performance of the convective stabilization mechanism in regions of stagnant flow for Powell's source term method, exceedingly complex implementation for staggered-grid approaches, computationally expensive nature due to the necessity of a Poisson solver combined with hyperbolic and elliptic numerical methods for classical projection schemes. The new technique is in principle equivalent to the hyperbolic divergence cleaning technique advocated by other authors. However, the underlying derivation and justification are quite different. We apply our approach to a test case relevant to space weather predictions, a field-aligned (i.e., free-stream flow velocity and magnetic field are aligned) superfast-mode plasma around a quarter of a perfectly conducting cylinder, in which a complex dimpled bow shock topology occurs. The performance of our approach is demonstrated in terms of accuracy, robustness, and convergence speed.

1 Introduction

The equations of ideal magnetohydrodynamics (MHD) describe the behavior of a certain class of plasmas, primarily encountered in astrophysics and fusion devices. They consist of a hyperbolic system of eight partial differential equations (PDEs), for density, energy, and three components of the velocity and the magnetic field. At the PDE level, it can easily be shown that the solenoidal constraint ($\nabla \cdot \mathbf{B} = 0$) is exactly satisfied at all times once satisfied initially by means of taking the divergence of the magnetic induction equation. However, this is not the case in numerical implementations, for which this constraint is in general satisfied up to a discretization error only, since the divergence of a curl ($\nabla \cdot \nabla \times$) is in general not exactly zero numerically. It is well known that, unless care is taken, these errors accumulate and cause numerical instabilities.

A variety of techniques have been proposed to suppress this instability. The most popular approach, proposed by Powell [9], relies on a nonconservative form of the equations, for which errors in the divergence of the magnetic field are swept away by the flow. This approach suffers from two drawbacks (1) the convective stabilization mechanism does not work well in regions of stagnant flow and (2) nonconservative discretizations suffer from incorrect shock capturing [3].

Less common are projection techniques [8], in which the gradient of a scalar potential field is added to the magnetic field at each iteration, to ensure that the solenoidal constraint be satisfied up to machine accuracy at each discrete time step. The disadvantage of this approach is that a costly Poisson problem needs to be solved for the potential at each time step; moreover, the fact of combining elliptic and hyperbolic solution methods appears to pose numerical problems.

In a recently proposed alternative technique [5], the hydrodynamic variables, magnetic and electric fields are computed on different staggered meshes. This arrangement can be shown to satisfy the solenoidal constraint up to machine accuracy. The main disadvantage of this technique is its exceedingly complex implementation.

We instead advocate the use of an artificial compressibility-type approach, in which the aforementioned Poisson problem is solved by means of a pseudotime stepping procedure for an additional potential variable, similar to the potential variable used in projection techniques. The modified system of nine equations thus obtained is fully hyperbolic and conservative. Therefore, conventional upwind finite volume (FV) technology can be readily applied. Moreover, it can be shown that errors in the divergence of the magnetic field are now radiated away with a constant wave speed, thereby solving the problems of Powell's scheme in regions of low-speed flow. Last but not least, the technique is very easy to implement.

2 Governing Equations

The ideal MHD equations can be written in the conservative differential form, as follows

$$\frac{\partial}{\partial t}\begin{bmatrix}\rho\\ \rho\mathbf{u}\\ \mathbf{B}\\ e\end{bmatrix}+\nabla\cdot\begin{bmatrix}\rho\mathbf{u}\\ \rho\mathbf{u}\mathbf{u}+\hat{I}(p+\frac{\mathbf{B}\cdot\mathbf{B}}{2})-\mathbf{B}\mathbf{B}\\ \mathbf{u}\mathbf{B}-\mathbf{B}\mathbf{u}\\ (e+p+\frac{\mathbf{B}\cdot\mathbf{B}}{2})\mathbf{u}-\mathbf{B}(\mathbf{u}\cdot\mathbf{B})\end{bmatrix}=0 \tag{1}$$

where $\hat{I}$ is a 3×3 identity matrix, ρ is the density, $\mathbf{u}$ is the velocity vector, $\mathbf{B}$ is the magnetic field vector, and p is the static pressure. The specific total energy, e, is defined as

$$e=\frac{p}{\gamma-1}+\frac{1}{2}\rho|\,\mathbf{u}\,|^2+\frac{1}{2}|\,\mathbf{B}\,|^2 \tag{2}$$

where γ is the ratio of specific heats. The units in (1) have been chosen such that the magnetic permeability, μ, is unity. Moreover, the plasma is assumed to obey the ideal gas law, which is a good approximation for most space and solar plasmas [6]. The ideal gas law together with (2) are necessary constitutive relations to close the set of ideal MHD equations. Finally, there is the solenoidal constraint, which is written as follows:

$$\nabla \cdot \mathbf{B} = 0 \tag{3}$$

3 Numerical Method

3.1 Artificial Compressibility Approach

Inspired by Chorin's artificial compressibility method (ACM) [4] for incompressible flows, we write the ideal MHD equations under the following modified form (underlined terms are the modifications made to the original ideal MHD system):

$$\frac{\partial}{\partial t}\begin{bmatrix} \rho \\ \rho\mathbf{u} \\ \mathbf{B} \\ e \\ \underline{\phi} \end{bmatrix} + \nabla \cdot \begin{bmatrix} \rho\mathbf{u} \\ \rho\mathbf{u}\mathbf{u} + \hat{I}(p + \frac{\mathbf{B}\cdot\mathbf{B}}{2}) - \mathbf{B}\mathbf{B} \\ \mathbf{u}\mathbf{B} - \mathbf{B}\mathbf{u} + \underline{\hat{I}\phi} \\ (e + p + \frac{\mathbf{B}\cdot\mathbf{B}}{2})\mathbf{u} - \mathbf{B}(\mathbf{u}\cdot\mathbf{B}) \\ \underline{V_{ref}^2\mathbf{B}} \end{bmatrix} = 0 \tag{4}$$

where ϕ is an artificially introduced scalar potential and V_{ref} is a reference velocity.

The eigenvalues of the Jacobian matrix of (4) are $u \pm c_f$, $u \pm c_s$, $u \pm c_A$, u, and $\pm V_{ref}$, where c_f, c_s, and c_A are the speeds of sound corresponding to fast and slow magnetoacoustic waves and Alfvén wave, respectively [9]. In terms of right and left eigenvectors, if we examine (4), the differences from the original system occur only in the equations corresponding to the evolution of the magnetic field component normal to the cell interface and in the potential field. It is possible to treat these equations in an isolated manner as if we had a system of just two PDEs which are written as follows [1]:

$$\frac{\partial}{\partial t}\begin{bmatrix} \mathbf{B_x} \\ \phi \end{bmatrix} + \begin{pmatrix} 0 & 1 \\ V_{ref}^2 & 0 \end{pmatrix} \frac{\partial}{\partial x}\begin{bmatrix} \mathbf{B_x} \\ \phi \end{bmatrix} = 0 \tag{5}$$

After applying the method of characteristics to (5), the right and left eigenvectors corresponding to $\pm V_{ref}$ are

$$\begin{bmatrix} 1 \pm V_{ref} \end{bmatrix}^T \quad \begin{bmatrix} 1 \pm \frac{1}{V_{ref}} \end{bmatrix} \tag{6}$$

For the right and left eigenvectors of (4), if we now combine (5) with (1), for all eigenvalues except $\pm V_{ref}$, the entries in the eigenvectors corresponding to the equations in (5) are zero and the rest of the entries is the same as in [9] whereas for $\pm V_{ref}$, the entries except the ones corresponding to the equations in (5) are zero and the rest of the entries is given in (6).

The $\nabla\cdot\mathbf{B}$ error for each cell is computed by integrating the flux function for the evolution of the potential field in (4) over all cell interfaces using Gauss' theorem. For instance, for the simplest scalar dissipation scheme, the equation is as follows

$$(\nabla\cdot\mathbf{B})_L = \frac{1}{V_{ref}^2 \Omega_L} \sum_{faces} \frac{1}{2}\left[V_{ref}^2 (B_{nL} + B_{nR}) - |\lambda_{maxLR}| (\phi_R - \phi_L)\right] \triangle s \tag{7}$$

where Ω is the volume of a cell, λ_{max} is the maximum eigenvalue, and $\triangle s$ is the area of a cell interface. Moreover, L, R, and LR stand for the cell center, the neighboring cell center, and the corresponding cell interface, respectively.

If we look at the wave equation for $\nabla\cdot\mathbf{B}$, we see that the $\nabla\cdot\mathbf{B}$ errors are swept away by speeds independent from the flow speed

$$\frac{\partial^2(\nabla\cdot\mathbf{B})}{\partial t^2} - \nabla\cdot\left[V_{ref}^2 \nabla(\nabla\cdot\mathbf{B})\right] = 0. \tag{8}$$

This ensures that ACM works even in the regions of stagnant flow. However, it should be noted that the higher the V_{ref} imposed, the bigger will be the numerical dissipation introduced. Similarly, a too low value for V_{ref} would not sweep away $\nabla\cdot\mathbf{B}$ errors sufficiently fast and can therefore not suppress the instabilities due to the accumulation of $\nabla\cdot\mathbf{B}$ errors. Based upon practical experience, we recommend to assign the value of the free-stream speed to V_{ref}.

Consistency Proof

The consistency proof consists of answering the following questions (1) do the discretized equations converge to a correct weak solution of the modified PDEs? and (2) do the modified PDEs converge toward a correct solution of the original ideal MHD system?

For simplicity, we consider only steady flows in our consistency proof, without loss of generality.

At the *discrete* level, we have a usual *conservative*, cell-centered FV discretization which converges to a correct weak solution of the underlying PDEs on regular meshes [10].

For *smooth* flows, taking the divergence of the modified induction equation, we obtain

$$\nabla^2\phi = \nabla\cdot[\nabla\times(\mathbf{u}\times\mathbf{B})] = 0 \tag{9}$$

with $\phi = 0$ at the outlets and $\partial\phi/\partial n = 0$ at all other boundaries as the boundary conditions. By using separation of variables, (9) yields $\phi = 0$ everywhere. Hence, the artificial $\nabla\phi$ term in the induction equation vanishes and we get the correct solution.

In the presence of *shocks*, the jump relations will be as follows:

$$F_L + \phi_L = F_R + \phi_R \tag{10}$$

Clearly, an entropy-satisfying solution to (1), complemented by $\phi_L = \phi_R$, is still an entropy-satisfying weak solution. There is the possibility that other weak solutions with $\phi_L \neq \phi_R$ would also be allowed by (10). However, any such solution is entropy violating. To see this, let us consider the resistive form of the induction equation

$$\frac{\partial \mathbf{B}}{\partial t} = \nabla\times(\mathbf{u}\times\mathbf{B}) - \nabla\phi - \nabla\times[\alpha(\nabla\times\mathbf{B})] \tag{11}$$

For *smooth* flows, taking the divergence of both sides of (11), for each value of the resistivity α, with the previously defined boundary conditions, $\nabla\phi$ disappears and ACM converges to the correct entropy-satisfying solution of the original ideal MHD system. In the limit of $\alpha \to 0$, shocks appear and clearly only the weak solution with $\phi_L = \phi_R$ is approached asymptotically.

4 Results

We performed a field-aligned superfast-mode MHD simulation around quarter of a perfectly conducting cylinder [7]. The MHD parameters are selected such that a complex dimpled bow shock topology occurs, including fast switch-on shocks, various intermediate shocks, a hydrodynamic shock, and a tangential discontinuity. This shock topology is thought to be observed in fast solar Coronal Mass Ejections and other phenomena in space plasmas.

The mesh consists of 83,008 triangular elements. The inflow Mach numbers based on the speed of sound and the Alfvén wave speed of sound are $M = 1.5\sqrt{3}$ and $M_A = 1.5$, respectively. The corresponding inflow magnetic beta is $\beta = 0.4$. Finally, γ is set to 5/3.

We performed FV calculations using both Powell's source term approach and ACM with a total variation diminishing (TVD) scalar dissipation (SD) scheme (7) combined with Barth–Jespersen's limiter. The simulations are performed in parallel and implicitly using a backward Euler time integration.

In Fig. 1, the shock profiles, especially in terms of the length of the hydrodynamic shock at the bottom, obtained with Powell's approach and ACM seem to be different. However, both are still in reasonable agreement with the corresponding reference solutions obtained with Powell's source term method [6] and with the projection scheme [7], respectively. On the other hand, we can observe that ACM is slightly more dissipative from the resolution of the intermediate shocks just behind the dimple. In Fig. 2a, the density variation in the hydrodynamic shock followed by a slow rarefaction wave is predicted better by ACM than Powell's approach. However, afterwards, the uniform flow, intermediate shock, and continuous diverging compression regions appear to be better

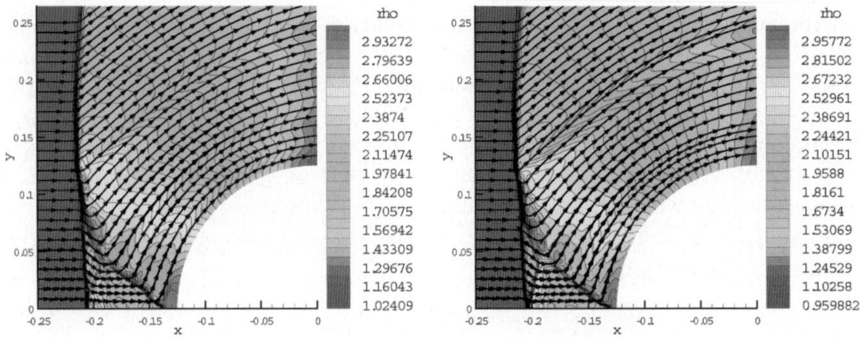

(a) *ACM* (b) *Powell*

Fig. 1. Density contours with streamlines of magnetic field (*zoomed view*)

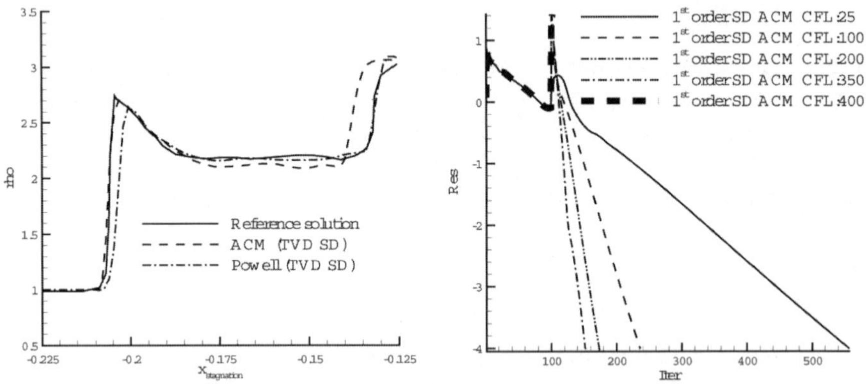

(a) (b)

Fig. 2. (**a**) Density variation along the stagnation line for the simulations in Fig. 1 and the reference solution in [7]. (**b**) Convergence speed and robustness test for ACM using backward Euler with CFL = 10 for the first 100 iterations and with a mesh consisting of 17,052 triangular elements

predicted by Powell's approach. Especially, at the jump on the right-hand side of the figure, we can once more observe the dissipative nature of ACM.

In Fig. 2b, results of the convergence speed and robustness test for ACM are presented. Using the first-order SD scheme together with backward Euler time integration, ACM could converge up to high CFL numbers for this challenging test case in which magnetic effects dominate over hydrodynamic effects (low-β flow).

5 Conclusions

In this chapter, a new approach in satisfying the solenoidal constraint based upon the artificial compressibility concept is introduced. We hope that it can be a cure to the drawbacks of the families of solenoidal constraint-satisfying techniques presented in the literature so far. It is purely hyperbolic, stable, consistent, and works even in the vicinity of stagnant flow regions, is easy to implement, robust, has good shock-capturing properties, and a remarkable performance with implicit schemes. These properties were demonstrated in a quite challenging test case of the space weather applications. The implementation of the solver utilized was made in the object-oriented software platform COOLFluiD [2] that is under development at the VKI.

The mathematical principles underlying our approach are the same as those used for the hyperbolic divergence cleaning technique [1]. However, we believe that the artificial compressibility perspective is more intuitive and has a cleaner mathematical formulation (i.e., consistency proof). A detailed study of these differences will be the topic of our next publication in this area.

Acknowledgments

The work in this chapter is made as part of the ESA PRODEX-8 project entitled "Solar Drivers of Space Weather II." The authors acknowledge the valuable discussions and suggestions of Prof. Herman Deconinck of the VKI and Prof. Stefaan Poedts of the Center for Plasma Astrophysics (CPA) at KU Leuven, as well as the support of the COOLFluiD team at the VKI.

References

1. Dedner A., Kemm F., Kröner D., Munz C.-D., Schnitzer T., and Wesenberg M. Hyperbolic divergence cleaning for the MHD equations. *Jour. of Comp. Phys.*, 175:645–673, 2002.
2. Lani A., Quintino T., Kimpe D., Deconinck H., Vandewalle S., and Poedts S. The COOLFluiD framework: Design solutions for high-performance object oriented scientific computing software. In *Computational Science - ICCS 2005, LNCS 3514*, volume 1, pages 281–286. Springer-Verlag, May 2005.
3. Csìk A.G. *Upwind residual distribution schemes for general hyperbolic conservation laws and application to ideal magnetohydrodynamics.* PhD thesis, VKI & KU Leuven, 2002.
4. Chorin A.J. A numerical method for solving incompressible viscous flow problems. *Jour. of Comp. Phys.*, 2:12–26, 1967.
5. Balsara D.S. Second-order-accurate schemes for magnetohydrodynamics with divergence-free reconstruction. *The Astrophys. Jour. Supp. Series*, 151:149–184, 2004.
6. De Sterck H. *Numerical simulation and analysis of magnetically dominated MHD bow shock flows with applications in space physics.* PhD thesis, KU Leuven, 1999.
7. De Sterck H., Low B.C., and Poedts S. Complex magnetohydrodynamic bow shock topology in field-aligned low-β flow around a perfectly conducting cylinder. *Phys. of Plasmas*, 5(11):4015–4027, 1998.
8. Brackbill J.U. and Barnes D.C. The effect of nonzero $\nabla \cdot \mathbf{B}$ on the numerical solution of the magnetohydrodynamic equations. *Jour. of Comp. Phys.*, 35: 426–430, 1980.

9. Powell K.G., Roe P.L., Myong R.S., Gombosi T., and De Zeeuw D. An upwind scheme for magnetohydrodynamics. In *12th AIAA CFD Conf.*, pages 661–674. AIAA, June 1995. AIAA paper 95-1704.
10. LeVeque R.J. *Numerical methods for conservation laws.* Lectures in Mathematics ETH Zuerich. Birkhäuser Verlag, 1992.

Vanishing at Most Seventh-Order Terms of Scalar Conservation Laws

N. Fujino and M. Yamazaki

1 Introduction

Let us consider the sequence $\{u^\varepsilon\}$ of smooth solutions to the following scalar conservation laws having second- and higher-order terms

$$\partial_t u + \partial_x f(u) = \varepsilon \partial_x^2 u + \sum_{n=1}^{N} (-1)^n \delta_n \partial_x^{2n+1} u, \quad (x,t) \in \mathbf{R} \times (0,\infty), \tag{1}$$

$$u(x,0) = u_0^\varepsilon(x), \quad x \in \mathbf{R}, \tag{2}$$

where $\varepsilon > 0$, $\delta_n = \delta_n(\varepsilon) > 0$, and $N \in \mathbf{N}$. In this note, we assume that the flux function is a given smooth function $f : \mathbf{R} \to \mathbf{R}$ which satisfies the following assumption:

$$\text{(I)} \exists C > 0,\ m > 1 \text{ s.t. } |f'(u)| \le C(1 + |u|^{m-1}) \text{ for any } u \in \mathbf{R},$$

which is introduced in [18]. We also assume that the initial data u_0^ε are smooth functions with compact support and there exits a function $u_0 \in L^1(\mathbf{R}) \cap L^q(\mathbf{R})$ for $q > 1$ such that

$$\lim_{\varepsilon \to 0} u_0^\varepsilon = u_0 \qquad \text{in} \quad L^1(\mathbf{R}) \cap L^q(\mathbf{R}). \tag{3}$$

The objective is to show that, if $\varepsilon, \delta_1, \cdots, \delta_N$ tend to zero, the sequence $\{u^\varepsilon\}$ of solutions to (1) and (2) converges to the unique entropy solution to the hyperbolic scalar conservation law

$$u_t + f(u)_x = 0, \quad (x,t) \in \mathbf{R} \times (0,\infty), \tag{4}$$

$$u(x,0) = u_0(x), \quad x \in \mathbf{R}. \tag{5}$$

When $N = 1$, $\varepsilon = 0$, and the flux function $f(u) = u^2/2$, (1) is reduced to the Korteweg–de Vries (KdV) equation. On the other hand, when $\delta_n = 0$ $(n = 1, \cdots, N)$, (1) is reduced to a parabolic equation.

The limit of (1) as $\varepsilon, \delta_1, \cdots, \delta_N \to 0$ is a singular perturbation limit and it involves many serious mathematical problems. The dispersion limit of the KdV equation ($N = 1$ and $\varepsilon = 0$, $\delta_1 \to 0$) is studied by many authors. Amazingly, the sequence of the solutions to (1) does not converge to the solution to (4) in general (cf. Lax [15] and Lax–Levermore [16, 17]). It is related to the existence of traveling wave solutions.

In the case of the diffusion limit ($\varepsilon \to 0$ and $\delta_n = 0\,(n = 1, \cdots, N)$), it is rather trivial that the sequence $\{u^\varepsilon\}$ of solutions to (1) converges to the solution to the hyperbolic conservation law (4) by virtue of the classical vanishing viscosity method.

For such reasons, the good balance in relation between ε and $\delta_1, \cdots, \delta_N$ is demanded so that the sequence of solutions to the conservation law with diffusion and higher dispersion terms converges to the solution to the hyperbolic conservation law. This balance of ε and $\delta_1, \cdots, \delta_N$ is considered by several authors and there are many previous results for this issue.

Schonbek [20] gives a convergence result for a scalar conservation law with linear second- and third-order terms ($N = 1$) under the assumption that either $\delta_1 = O(\varepsilon^2)$ for $f(u) = u^2/2$ or $\delta_1 = O(\varepsilon^3)$ for arbitrary subquadratic flux functions f. This result is generalized by Kondo–LeFloch [14] for the flux with linear growth at infinity: $|f'(u)| \leq M$ for $\forall u \in \mathbf{R}$ with some $M > 0$ and the convergence of the sequence $\{u^\varepsilon\}$ in $L^k(\mathbf{R}_+; L^p(\mathbf{R}))$ ($1 < k < \infty$ and $1 < p < 2$) to a weak solution under the assumption $\delta_1 = O(\varepsilon^2)$ is obtained. It is also shown that the above limit is the unique entropy solution in the sense of Kružkov under the stronger condition $\delta_1 = o(\varepsilon^2)$. They also obtain a convergence result for multispace-dimensional conservation laws. In [9, 11], we give convergence results on (1) for $N = 2$ with flux $f(u) = u^2/2$ (Burgers' type) if the condition $\delta_1 = O(\varepsilon^{\frac{3}{3-q}})$ and $\delta_2 = O(\varepsilon^4 \delta_1^{\frac{8q-7}{9}})$ holds for $q \in (2, 3)$ or $\delta_1 = O(\varepsilon^{\frac{12}{19-4q}} \delta_2^{\frac{3}{19-4q}})$ and $\delta_2 = O(\varepsilon^4 \delta_1^{\frac{8q-7}{9}})$ holds for $q \in [3, \frac{16}{5})$. Moreover, if the sequence $\{u^\varepsilon\}$ is bounded in $L^\infty(0, T^*; L^q(\mathbf{R}))$, then the same conclusion is obtained [9, 11] for any $q > 2$ provided that $\delta_1 = o(\varepsilon^3)$ and $\delta_2 = o(\varepsilon^4 \delta_1)$.See also a study for systems in Hayes–LeFloch [12].

There are also investigations of the convergence for a scalar conservation law with nonlinear diffusion and nonlinear dispersion terms

$$u_t + f(u)_x = \varepsilon b(u_x)_x - \delta \left((u_x)^{2\ell-1} \right)_{xx}, \quad \ell \geq 1 \tag{6}$$

under the conditions that the flux satisfies Condition (I) and a nondecreasing smooth function b satisfies $b(0) = 0$, $b(\lambda)\lambda \geq 0$ (for any $\lambda \in \mathbf{R}$), and

$$C_1 |\lambda|^{(2\ell+1)r} \leq b(\lambda)\lambda \leq C_2 |\lambda|^{(2\ell+1)r} \text{ for any } |\lambda| \geq R$$

for C_1, C_2, $R > 0$, $r \geq 1$. The case of $\ell = 1$ (linear dispersion term) is studied by LeFloch–Natalini [18]. They show that, if the sequence $\{u^\varepsilon\}$ of solutions to (6) is bounded in $L^\infty(0, T^*; L^q(\mathbf{R}))$ for $q > m$ and $\delta = o(\varepsilon^{\frac{1}{r}})$, then the sequence $\{u^\varepsilon\}$ converges to the unique entropy solution $u \in L^\infty(0, T^*; L^q(\mathbf{R}))$

in $L^k(0,T^*;L^p(\mathbf{R}))$ $(k<\infty,\ p<q)$. In the case of $m<5-\frac{1}{r}(=:q')$, they show that the sequence $\{u^\varepsilon\}$ is bounded in $L^\infty(0,T^*;L^{q'}(\mathbf{R}))$ and hence they obtain the result that the sequence $\{u^\varepsilon\}$ converges to the unique entropy solution $u\in L^\infty(0,T^*;L^{q'}(\mathbf{R}))$ in $L^k(0,T^*;L^{p'}(\mathbf{R}))$ $(k<\infty,\ p'<q')$ for $\delta=O(\varepsilon^{\frac{5-m}{r(5-m)-1}})$ $(r\geq 1)$.

In the case of general $\ell\geq 1$ for (6), Fujino–Yamazaki [8, 10] prove that, if $m<q'$, then the sequence $\{u^\varepsilon\}$ of solutions to (6) converges for $\delta=O(\varepsilon^{\frac{6\ell-m-1}{r(6\ell-m-1)-1}})$ $(\forall\ell\geq 1)$ to the unique entropy solution $u\in L^\infty(0,T^*;L^{q'}(\mathbf{R}))$ in $L^k(0,T^*;L^{p'}(\mathbf{R}))$ $(k<\infty,\ p'<q')$.

In this note, we consider the scalar conservation laws (1) for $N=3$ having a general flux. We carry out successfully new a priori estimates for general $N(\geq 1)$. The main result is that, in the case $N=3$ (i.e., at most seventh order), if the sequence $\{u^\varepsilon\}$ is bounded in $L^\infty(0,T^*;L^q(\mathbf{R}))$, then the sequence $\{u^\varepsilon\}$ of solutions to (1) and (2) for any $q(>m>1)$ converges to the unique entropy solution $u\in L^\infty(0,T^*;L^q(\mathbf{R}))$ in $L^k(0,T^*;L^p(\mathbf{R}))$ $(k<\infty,\ p<q)$ provided that $\delta_1=o(\varepsilon^{\frac{5-m}{3-m}})$, $\delta_2=o(\varepsilon^4\delta_1^{\frac{3m+1}{5-m}})$, and $\delta_3=o(\varepsilon^2\delta_1^{\frac{3(m+3)}{2(5-m)}}\delta_2)$. In particular, our theorem extends the result on (1) for $N=2$ [9, 11].

We recall a generalization of the Young measures and entropy measure-valued (m.-v.) solutions.

Lemma 1. *Let $\{u_j\}$ be a uniformly bounded sequence in $L^\infty((0,\infty);L^q(\mathbf{R}))$. Then there exists a subsequence $\{u_{j'}\}$ and a weakly $\star$ measurable mapping $\nu:\mathbf{R}\times(0,\infty)\to\mathrm{Prob}(\mathbf{R})$ such that, for any functions $g\in C(\mathbf{R})$ satisfying*

$$g(u)\leq C(1+|u|^{q''})\ \text{for some}\ q''\in(0,q), \tag{7}$$

the following limit holds

$$g(u_{j'})\to\langle\nu_{(x,t)},g(\cdot)\rangle:=\int_{\mathbf{R}}g(\lambda)d\nu_{(x,t)}(\lambda)\ \text{as}\ j\to\infty$$

in $L^s((0,\infty))$ for $s\in(1,\frac{q}{q''})$.

The above measure-valued mapping $\nu_{(x,t)}$ is called a *Young measure* associated with the sequence $\{u_{j'}\}$.

A fundamental notion in our chapter is the measure-valued solutions to the Cauchy problem (4) and (5) which are investigated by DiPerna [7], LeFloch–Natalini [18], and Szepessy [21].

Definition 1. *Let $f\in C(\mathbf{R})$ satisfy the growth condition (7) and $u_0\in L^1(\mathbf{R})\cap L^q(\mathbf{R})$. A Young measure ν associated with the sequence $\{u_j\}$, which is assumed to be uniformly bounded in $L^\infty((0,\infty);L^q(\mathbf{R}))$, is then called an entropy measure-valued solution of Cauchy problem (4) and (5) if*

$$\partial_t\langle\nu_{(\cdot)},\ |\lambda-k|\rangle+\partial_x\langle\nu_{(\cdot)},\ \mathrm{sgn}(\lambda-k)(f(\lambda)-f(k))\rangle\leq 0, \tag{8}$$

in $\mathcal{D}'(\mathbf{R} \times (0, \infty))$ *for any* $k \in \mathbf{R}$ *and*

$$\lim_{T \to 0+} \frac{1}{T} \int_0^T \int_K \langle \nu_{(x,t)}, |\lambda - u_0(x)| \rangle dx dt = 0 \tag{9}$$

for any compact sets $K \subset \mathbf{R}$.

Our main convergence tool which is given in [18] is the following.

Theorem 1 ([18]). *Let f satisfy (7) and $u_0 \in L^1(\mathbf{R}) \cap L^q(\mathbf{R})$. Suppose that ν is a Young measure associated with a sequence $\{u_j\}$ and that the sequence $\{u_j\}$ is uniformly bounded in $L^\infty((0, \infty); L^q(\mathbf{R}))$ for $q \geq 1$. If ν is an entropy m.-v. solution of (4) and (5), then*

$$\lim_{j \to \infty} u_j = u \text{ in } L^\infty((0, \infty); L^{q''}_{loc}(\mathbf{R})) \text{ for any } q'' \in [1, q),$$

where $u \in L^\infty((0, \infty); L^q(\mathbf{R}))$ is the unique entropy solution of (4) and (5).

2 A Priori Estimates

Suppose that there exists a sequence of the smooth solutions $\{u^\varepsilon\}$ to (1) and (2) defined on $\mathbf{R} \times (0, T^*)$, vanishing at infinity and associated with initial data $\{u_0^\varepsilon\}$ for some $T^* \in (0, \infty]$. We derive a priori estimates for smooth approximate solutions u^ε to (1) and (2). We assume the following uniform estimate concerning the initial data u_0^ε with some constant $C_0 > 0$ independent of $\varepsilon, \delta_1, \cdots, \delta_N$:

$$||u_0||_{L^2(\mathbf{R})} + \sum_{n=1}^N \delta_n^{\frac{1}{2}} ||\partial_x^n u_0||_{L^2(\mathbf{R})} \leq C_0. \tag{10}$$

To simplify the notation of u^ε and u_0^ε, we omit "ε" throughout this section. Then multiplying (1) by u or $f(u) + \sum_{n'=1}^N (-1)^{n'+1} \delta_{n'} \partial_x^{2n'} u$, we obtain:

Lemma 2. *For every $T \in (0, T^*)$, we have*

$$\int_{\mathbf{R}} u^2(x, T) dx + 2\varepsilon \int_0^T \int_{\mathbf{R}} \left(\partial_x u(x, t)\right)^2 dx dt = \int_{\mathbf{R}} u_0^2(x) dx \tag{11}$$

and

$$\begin{aligned} &\sum_{n=1}^N \delta_n \left(\int_{\mathbf{R}} (\partial_x^n u(x, T))^2 dx + 2\varepsilon \int_0^T \int_{\mathbf{R}} \left(\partial_x^{n+1} u(x, t)\right)^2 dx dt \right) \\ &= \sum_{n=1}^N \delta_n \int_{\mathbf{R}} (\partial_x^n u_0(x))^2 dx - 2 \int_{\mathbf{R}} F(u_0(x)) dx + 2 \int_{\mathbf{R}} F(u(x, T)) dx \\ &\quad + 2\varepsilon \int_0^T \int_{\mathbf{R}} f'(u(x, t)) \left(\partial_x u(x, t)\right)^2 dx dt \end{aligned} \tag{12}$$

where F is defined by $F'(u) = f(u)$.

Combining (11), (12), and Condition (I), we get the following estimate for every $T \in (0, T^*)$

$$\sum_{n=1}^{N} \delta_n \left(\int_{\mathbf{R}} (\partial_x^n u(x,T))^2 dx + 2\varepsilon \int_0^T \int_{\mathbf{R}} \left(\partial_x^{n+1} u(x,t)\right)^2 dx dt \right)$$
$$\leq C \left(1 + \sup_{t \in [0,T]} ||u(\cdot,T)||_{L^\infty(\mathbf{R})}^{m-1} \right) \tag{13}$$

with some $C > 0$. Using (13), we get an estimate of u in the L^∞ norm.

Lemma 3. *Assume $1 < m < 5$ in Condition (I), then for the solution to (1), there exists $C > 0$ such that*

$$\sup_{t \in (0,T^*)} ||u(\cdot,t)||_{L^\infty} \leq C \delta_1^{-\frac{1}{5-m}}. \tag{14}$$

Proof. From (13), we are led to the inequality of $\partial_x u$ in the L^2 norm:

$$||\partial_x u(\cdot,T)||_{L^2(\mathbf{R})} \leq C \delta_1^{-\frac{1}{2}} (1 + ||u(\cdot,T)||_{L^\infty(\mathbf{R})}^{m-1})^{\frac{1}{2}}.$$

Using the Cauchy–Schwarz inequality, we obtain

$$\begin{aligned} |u(x,T)|^2 &\leq 2 \int_{-\infty}^{x} |u(y,T) \partial_x u(y,T)| dy \\ &\leq 2 ||u(\cdot,T)||_{L^2(\mathbf{R})} ||\partial_x u(\cdot,T)||_{L^2(\mathbf{R})} \\ &\leq C \delta_1^{-\frac{1}{2}} \left(1 + ||u(\cdot,T)||_{L^\infty(\mathbf{R})}^{m-1} \right)^{\frac{1}{2}}. \end{aligned}$$

Hence we obtain for $\forall t \in (0, T_*)$

$$||u(\cdot,t)||_{L^\infty(\mathbf{R})}^4 \leq C \delta_1^{-1} \left(1 + ||u(\cdot,t)||_{L^\infty(\mathbf{R})}^{m-1} \right).$$

When setting $y = ||u(\cdot,t)||_{L^\infty(\mathbf{R})}$ and considering the algebraic inequality $y^4 \leq C \delta_1^{-1} (1 + y^{m-1})$, we obtain the estimate (14). □

Substituting the uniform boundedness (14) of u into the inequality (13), we can easily obtain:

Lemma 4. *For any $T \in (0, T_*)$, there exists a constant $C > 0$ such that*

$$\sum_{n=1}^{N} \delta_n \left(\int_{\mathbf{R}} (\partial_x^n u(x,T))^2 dx + 2\varepsilon \int_0^T \int_{\mathbf{R}} \left(\partial_x^{n+1} u(x,t)\right)^2 dx dt \right) \leq C \delta_1^{\frac{1-m}{5-m}}. \tag{15}$$

From inequality (15), we can write the following form:

$$||\partial_x^n u(\cdot,T)||_{L^2(\mathbf{R})} \leq C \delta_1^{\frac{1-m}{2(5-m)}} \delta_n^{-\frac{1}{2}} \ (n = 1, 2, \cdots, N). \tag{16}$$

Using inequality (15) (or estimate (16)), we have the estimate of higher order in the L^∞ norm by the same argument in Lemma 3.

Lemma 5. *Assume $1 < m < 5$ in Condition (I), then for the solution to (1), there exists $C > 0$ such that*

$$\sup_{t\in(0,T^*)} ||\partial_x^n u(\cdot,t)||_{L^\infty} \le C\delta_1^{\frac{1-m}{2(5-m)}}\delta_n^{-\frac{1}{4}}\delta_{n+1}^{-\frac{1}{4}}\,(n=1,2,\cdots,N-1). \tag{17}$$

3 Convergence Result

By making use of a priori estimates obtained in Sect. 2, we consider (1) in the case $N = 3$, i.e., scalar conservation law having at most seventh-order terms

$$\partial_t u + \partial_x f(u) = \varepsilon\partial_x^2 u + \sum_{n=1}^{3}(-1)^n\delta_n\partial_x^{2n+1}u, \quad (x,t)\in \mathbf{R}\times(0,\infty) \tag*{(1)'}$$

and show that the sequence $\{u^\varepsilon\}$ of solutions to $(1)'$ and (2) converges to the unique entropy solution to (4) and (5) as $\varepsilon, \delta_1, \cdots, \delta_3 \to 0$.

Our main result of this note is the following.

Theorem 2. *Assume Condition (I). Suppose that there exists a sequence of the smooth solutions $\{u^\varepsilon\}$ to $(1)'$ and (2) defined on $\mathbf{R}\times(0,T^*)$ $(T^*>0)$, vanishing at infinity and associated with initial data $\{u_0^\varepsilon\}$ satisfying (3) and*

$$||u_0||_{L^2(\mathbf{R})} + \sum_{n=1}^{3}\delta_n^{\frac{1}{2}}||\partial_x^n u_0||_{L^2(\mathbf{R})} \le C_0. \tag*{(10)'}$$

If the sequence $\{u^\varepsilon\}$ of solutions to $(1)'$ and (2) is bounded in $L^\infty(0,T^*;L^q(\mathbf{R}))$ for $q > m$, then the sequence $\{u^\varepsilon\}$ of solutions converges to the unique entropy solution $u \in L^\infty(0,T^*;L^q(\mathbf{R}))$ to (4) and (5) in $L^k(0,T^*;L^p(\mathbf{R}))$ $(k < \infty$ and $p < q)$ provided that $\delta_1 = o(\varepsilon^{\frac{5-m}{3-m}})$, $\delta_2 = o(\varepsilon^4\delta_1^{\frac{3m+1}{5-m}})$, and $\delta_3 = o(\varepsilon^2\delta_1^{\frac{3(m+3)}{2(5-m)}}\delta_2)$.

Outline of proof. To apply Theorem 1, we show that the Young measure ν associated with the sequence $\{u^\varepsilon\}$ is the entropy m.-v. solution to (4) and (5). Let U be any convex function $U(u)$ such that $U', U'', U^{(3)}, U^{(4)}$ are uniformly bounded on $\mathbf{R}$. Consider the distribution

$$\Lambda^\varepsilon := \partial_t U(u^\varepsilon) + \partial_x\tilde{F}(u^\varepsilon), \tag{18}$$

where $\tilde{F}' = f'U'$. We show that Λ^ε converges to a nonpositive measure in $\mathcal{D}'(\mathbf{R}\times\mathbf{R}_+)$. We have

$$
\begin{aligned}
\Lambda^{\varepsilon} &= \varepsilon U'(u)u_{xx} + \sum_{n=1}^{3}(-1)^n \delta_n U'(u)\partial_x^{2n+1}u \\
&= \varepsilon(U'(u)u_x)_x - \varepsilon U''(u)u_x^2 \\
&\quad + \sum_{n=1}^{3}(-1)^{2n-k}\delta_n \sum_{k=0}^{n} \frac{n!}{k!(n-k)!}\partial_x^k \left(\partial_x^{n-k}U'(u)\partial_x^{n+1}u\right) \\
&=: \Lambda_1 + \Lambda_2 + \Lambda_3.
\end{aligned}
$$

Using (11) and the Cauchy–Schwarz inequality, we estimate Λ_1. For any smooth function $\theta \in C_0^\infty(\mathbf{R} \times (0, T^*))$ $(\theta \geq 0)$, we get

$$
\begin{aligned}
|\langle \Lambda_1, \theta\rangle| &= \left| \int_0^{T^*} \int_{\mathbf{R}} \varepsilon U'(u)u_x\theta_x dxdt \right| \\
&\leq C\varepsilon \left(\int_0^{T^*} \int_{\mathbf{R}} |u_x|^2 dxdt \right)^{\frac{1}{2}} \left(\int_0^{T^*} \int_{\mathbf{R}} |\theta_x|^2 dxdt \right)^{\frac{1}{2}} \\
&\leq C\varepsilon \cdot \varepsilon^{-\frac{1}{2}} \cdot ||\theta_x||_{L^2(\mathbf{R}\times(0,T^*))} \\
&\leq C\varepsilon^{\frac{1}{2}} \to 0 \quad (\varepsilon \to 0)
\end{aligned}
\tag{19}
$$

with some $C > 0$.

Secondly the term Λ_2 is nonpositive:

$$
\langle \Lambda_2, \theta\rangle = -\int_0^{T^*} \int_{\mathbf{R}} \varepsilon U''(u)u_x^2 \theta dxdt \leq 0. \tag{20}
$$

One can similarly handle Λ_3. Hence it follows that Λ^ε converges to a nonpositive measure in $\mathcal{D}'(\mathbf{R} \times \mathbf{R}_+)$. That is, from the definition of the Young measure, it follows that $U(u^\varepsilon) \to \langle \nu, U\rangle$, and $\tilde{F}(u^\varepsilon) \to \langle \nu, \tilde{F}\rangle$ in the sense of distribution as $\varepsilon \to 0$. Therefore for any convex entropy pairs, we obtain

$$
\partial_t\langle \nu_{(\cdot)}, U(\lambda)\rangle + \partial_x\langle \nu_{(\cdot)}, \tilde{F}(\lambda)\rangle \leq 0. \tag{21}
$$

By virtue of (21), the regularization of $|u - k|$ (for any $k \in \mathbf{R}$) gives (8). We omit the proof for the initial condition (9). Then we just apply Theorem 1 to show that the Young measure ν associated with the sequence $\{u^\varepsilon\}$ is entropy m.-v. solution to (4) and (5). □

References

1. N. Bedjaoui, P.G. LeFloch: Diffusive-dispersive traveling waves and kinetic relations. I. Nonconvex hyperbolic conservation laws. J. Differential Equations **178**, 574–607 (2002)
2. J.L. Bona, M.E. Schonbek: Travelling-wave solutions to the Korteweg-de Vries-Burgers equation. Proc. Roy. Soc. Edinburgh Sect. **A 101**, 207–226 (1985)

3. J.L. Bona, R. Smith: The initial-value problem for the Korteweg-de Vries equation. Philos. Trans. Roy. Soc. London Ser. **A 278**, 555–601 (1975)
4. G.Q. Chen, Y.G. Lu: A study of approaches to applying the theory of compensated compactness. Kexue Tongbao **33**, 641–644 (1988)
5. G.Q. Chen, Y.G. Lu: Convergence of the approximate solutions to isentropic gas dynamics. Acta Math. Sci. **10**, 39–45 (1990)
6. G.M. Coclite, K.H. Karlsen: A Singular Limit Problem for Conservation Laws Related to the Camassa Holm Shallow Water Equation. Comm. Partial Differential Equations **31**, 1253–1272 (2006)
7. R.J. DiPerna: Measure-valued solutions to conservation laws. Arch. Rational Mech. Anal. **88**, 223–270 (1985)
8. N. Fujino, M. Yamazaki: Hyperbolic conservation laws with nonlinear diffusion and nonlinear dispersion. J. Differential Equations **228**, 171–190 (2006)
9. N. Fujino, M. Yamazaki: A Convergence Result on the Burgers Equation of Conservation Laws. Hyperbolic Problems: Theory, Numerics and Applications I, Tenth International Conference in Osaka, Yokohama Publishers, 407–414 (2006)
10. N. Fujino, M. Yamazaki: A result on the equation of conservation laws having second and third order terms. Nonlinear Dispersive Equations, T. Ozawa & Y. Tsutsumi eds., Gakuto International Series, Mathematical Sciences and Applications **26**, 19–34, Gakkotosho (2006)
11. N. Fujino, M. Yamazaki: Burgers' type equation with vanishing higher order, to appear in Commun. Pur. Appl. Anal.
12. B.T. Hayes, P.G. LeFloch: Nonclassical shocks and kinetic relations: strictly hyperbolic systems. SIAM J. Math. Anal. **31**, 941–991 (2000)
13. D. Jacobs, W.R. McKinney, M. Shearer: Traveling wave solutions of the modified Korteweg-de Vries-Burgers equation. J. Differential Equations **116**, 448–467 (1995)
14. C.I. Kondo, P.G. LeFloch: Zero diffusion-dispersion limits for scalar conservation laws. SIAM J. Math. Anal. **33**, 1320–1329 (2002)
15. P.D. Lax: The zero dispersion limit, a deterministic analogue of turbulence. Comm. Pure Appl. Math. **44**, 1047–1056 (1991)
16. P.D. Lax, C.D. Levermore: The small dispersion limit of the Korteweg-de Vries equation. I III. Comm. Pure Appl. Math. **36**, no. 3, 253–290, no. 5, 571–593, no. 6, 809–829 (1983)
17. P.D. Lax, C.D. Levermore: The zero dispersion limit for the Korteweg-de Vries KdV equation. Proc. Nat. Acad. Sci. U.S.A. **76**, 3602–3606 (1979)
18. P.G. LeFloch, R. Natalini: Conservation laws with vanishing nonlinear diffusion and dispersion. Nonlinear Anal. **36**, 213–230 (1999)
19. Y. Lu: Hyperbolic conservation laws and the compensated compactness method. Chapman & Hall/CRC Monographs and Surveys in Pure and Applied Mathematics, 128, Chapman & Hall/CRC, Boca Raton, FL, (2003)
20. M.E. Schonbek: Convergence of solutions to nonlinear dispersive equations. Comm. Partial Differencial Equations **7**, 959–1000 (1982)
21. A. Szepessy: An existence results for scalar conservation laws using measure valued solutions. Comm. Partial Differencial Equations **14**, 1329–1350 (1989)

Large-Time Behavior for a Compressible Energy Transport Model

L. Hsiao and Y. Li

1 Introduction

In this chapter, we consider the following energy transport model arising from the semiconductor science and plasma physics,

$$\begin{cases} \dfrac{\partial}{\partial t}n(\mu,T)+\mathrm{div}J_1=0, \\ \dfrac{\partial}{\partial t}U(\mu,T)+\mathrm{div}J_2=\nabla V\cdot J_1+W(\mu,T), \\ \lambda^2\triangle V=n-\mathcal{C}(x), \end{cases} \tag{1}$$

with

$$\begin{cases} J_1=-L_{11}\Big(\nabla\Big(\dfrac{\mu}{T}\Big)-\dfrac{\nabla V}{T}\Big)-L_{12}\nabla\Big(-\dfrac{1}{T}\Big), \\ J_2=-L_{21}\Big(\nabla\Big(\dfrac{\mu}{T}\Big)-\dfrac{\nabla V}{T}\Big)-L_{22}\nabla\Big(-\dfrac{1}{T}\Big), \end{cases} \tag{2}$$

where the parameters μ and T are the chemical potential of the electrons and the electron temperature, respectively, V is the electrostatic potential, $n(\mu,T)$ is the electron density, $U(\mu,T)$ is the density of the internal energy, $W(\mu,T)$ is the energy relaxation term satisfying $W(\mu,T)(T-T^{(0)})\le 0$, where the positive constant $T^{(0)}$ is the lattice temperature, J_1 is the carrier flux density, J_2 is the energy flux density, or heat flux, L is the diffusion matrix which is nonnegative definite, λ is the scaled Debye length, and $\mathcal{C}(x)$ is the doping profile which represents the background of the device. For more details on the energy transport model, we refer to the references [AD96, Ju01, MRS90].

The expressions for n, U, L, and W are constitutive relations. In a parabolic band structure, the relations for $n(\mu,T)$ and $U(\mu,T)$ derived from the Boltzmann statistics are

$$n(\mu,T)=T^{\frac{1}{\gamma-1}}\exp\Big(\frac{\mu}{T}\Big),\quad U(\mu,T)=\frac{1}{\gamma-1}T^{\frac{\gamma}{\gamma-1}}\exp\Big(\frac{\mu}{T}\Big). \tag{3}$$

The coefficient matrix L and energy relaxation term W in consideration are

$$L=(L_{ij})=\mu_0 T^{\frac{\gamma}{\gamma-1}}\exp\left(\frac{\mu}{T}\right)\begin{pmatrix}1 & \frac{\gamma T}{\gamma-1}\\ \frac{\gamma T}{\gamma-1} & \frac{\gamma^2T^2}{(\gamma-1)^2}\end{pmatrix}, \tag{4}$$

$$W(\mu,T)=\frac{1}{\gamma-1}T^{\frac{1}{\gamma-1}}\exp\left(\frac{\mu}{T}\right)\frac{T^{(0)}-T}{\tau}, \tag{5}$$

where μ_0 is the mobility constant, τ is the energy relaxation time, and $\gamma>1$ is the adiabatic exponent.

Since the structure of system (1)–(5) is completely different from the weakly coupled case so that the usual method including the maximum principle and the regularity theory for parabolic equations cannot be used. There are some global existence results for different coefficient matrices, where the coefficient matrix L is uniformly positive definite in [DGJ97, Ju00] and is only positive definite in [CH02, CHL04, CHL05].

Degond [De00] has derived the energy transport model from the spherical harmonics expansion (SHE) model in which the electron–electron collision operator dominates over the phonon collision operator, with the following coefficient matrix

$$L=(L_{ij})=\mu_0 nT\begin{pmatrix}1 & \frac{\gamma T}{\gamma-1}\\ \frac{\gamma T}{\gamma-1} & \frac{\gamma^2T^2}{(\gamma-1)^2}+\frac{\kappa_0T^2}{\gamma-1}\end{pmatrix}, \tag{6}$$

where κ_0 is the heat conductivity. When $\kappa_0>0$, the matrix (6) is positive definite, Chen et al. [CL06] have studied this case. In our case, we consider the coefficient matrix in (4) (namely for $\kappa_0=0$ in (6)), which is positive semidefinite. Moreover, it is easy to know that the determinant of the coefficient matrix is always zero. This is totally different from all the previous cases studied before and this contains much more difficulties to be dealt with.

The model (1)–(5) can be rewritten as follows, where the unknowns are n, T, and V,

$$\begin{cases} n_t-\operatorname{div}[\nabla(nT)-n\nabla V]=0,\\ \left(\frac{nT}{\gamma-1}\right)_t-\operatorname{div}\left(\frac{\gamma T}{\gamma-1}[\nabla(nT)-n\nabla V]\right)\\ \qquad\qquad=[n\nabla V-\nabla(nT)]\cdot\nabla V-\frac{n(T-T^{(0)})}{\tau(\gamma-1)},\\ \lambda^2\Delta V=n-\mathcal{C}(x),\end{cases} \tag{7}$$

which has a strong crossdiffusion and is a degenerate parabolic system coupling a Poisson equation for the electric field. The model (7) can be derived from the nonisentropic Euler–Poisson equations, which is used to describe the hydrodynamic model for semiconductors [An95, BK70], and is given by

$$\begin{cases} n_t + \nabla \cdot (nu) = 0, \\ (nu)_t + \nabla \cdot (nu \otimes u) + \nabla(nT) = n\nabla V - \dfrac{nu}{\tau_p}, \\ \Big(\dfrac{n}{2}|u|^2 + \dfrac{nT}{\gamma - 1}\Big)_t + \nabla \cdot \Big(\dfrac{nu}{2}|u|^2 + \dfrac{\gamma nT}{\gamma - 1}u\Big) \\ \qquad\qquad = nu \cdot \nabla V - \dfrac{1}{\tau_w}\Big(\dfrac{n}{2}|u|^2 + \dfrac{n(T - T^{(0)})}{\gamma - 1}\Big), \\ \lambda^2 \Delta V = n - \mathcal{C}(x), \end{cases} \tag{8}$$

where τ_p is the momentum relaxation time and τ_w is the energy relaxation time. As far as the relaxation time limits are concerned, the convergence from (8) to energy transport model and drift-diffusion model have been obtained in [GN99] and [Li07], respectively.

Without loss of generality, we assume that $\lambda = \tau = T^{(0)} = 1$. Then for one-dimensional case, the system (7) can be rewritten as

$$\begin{cases} n_t - [(nT)_x - nE]_x = 0, \\ (nT)_t - (\gamma T[(nT)_x - nE])_x = (\gamma - 1)[nE - (nT)_x]E - n(T - 1), \\ E_x = n - \mathcal{C}(x), \qquad E = V_x, \end{cases} \tag{9}$$

in a domain $Q_t = \Omega \times (0, t] = (0, 1) \times (0, t]$.

Denote $J = nE - (nT)_x$. We consider the insulated boundary conditions

$$J(0, t) = J(1, t) = 0, \qquad T_x(0, t) = T_x(1, t) = 0, \qquad E(0, t) = 0, \tag{10}$$

and the following initial values

$$n(x, 0) = n_I(x), \qquad T(x, 0) = T_I(x), \qquad x \in \Omega, \tag{11}$$

where the initial density $n_I(x)$ is chosen to satisfy

$$\int_\Omega (n_I - \mathcal{C})(x)dx = 0, \tag{12}$$

which implies $E(1, t) = 0$ as well.

Thus from (10) and (12), we have

$$n_x(0, t) = n_x(1, t) = 0, \quad T_x(0, t) = T_x(1, t) = 0, \quad E(0, t) = E(1, t) = 0.$$

2 Main Results

Our aim is to study the global existence and the large-time behavior of solutions to (7) when the initial data are around a stationary solution to the corresponding linear drift-diffusion model. We consider the smooth solution of (7) around a typical stationary solution $(\mathcal{N}, 1, \mathcal{V})$. The corresponding stationary problem is

$$\begin{cases} \nabla\mathcal{N} - \mathcal{N}\nabla\mathcal{V} = 0, & x \in \Omega, \\ \Delta\mathcal{V} = \mathcal{N} - \mathcal{C}(x), & x \in \Omega, \end{cases} \tag{13}$$

with boundary condition

$$\nabla\mathcal{V} \cdot \gamma|_{\partial\Omega} = 0, \tag{14}$$

where γ is the unit outer normal vector to $\partial\Omega$.

For the above *isothermal* stationary problem (13) and (14), we have the following theorem.

Theorem 1 ([CHL05]). *Assume that* $0 < \underline{C} \le \mathcal{C}(x) \le \overline{C}$, *then the problem (13) and (14) has a solution* $(\mathcal{N}, \mathcal{V})$, *for which the following estimates hold*

$$0 < \underline{C} \le \mathcal{N}(x) \le \overline{C}, \qquad x \in \Omega,$$
$$\underline{c} \le \mathcal{V}(x) \le \overline{c}, \qquad x \in \Omega,$$
$$|\Delta\mathcal{V}(x)|,\ |\nabla\mathcal{V}(x)|,\ |\nabla\mathcal{N}(x)|,\ |\Delta\mathcal{N}(x)| \le C, \qquad x \in \Omega,$$

where C is a positive constant, and $\underline{c}$ and $\overline{c}$ are constants.

Our main theorems on the global existence and exponential decay for the smooth solution are as follows. For one-dimensional case, we have

Theorem 2 ([HL06]). *Suppose* $0 < \underline{C} \le \mathcal{C}(x) \le \overline{C}$, $\mathcal{C}(x) \in C^2(\Omega)$, *and* $(n_I(x), T_I(x)) \in H^3(\Omega)$. *There exists a positive constant δ_1 such that if*

$$\|n(x,0) - \mathcal{N}(x)\|_{H^3} + \|T(x,0) - 1\|_{H^3} \le \delta_1,$$

then the problem (9)–(12) has a unique solution (n, T, E) in $\Omega \times [0, \infty)$ satisfying

$$\|E(\cdot,t) - \mathcal{V}_x(\cdot)\|_{H^3} + \|n(\cdot,t) - \mathcal{N}(\cdot)\|_{H^3} + \|T(\cdot,t) - 1\|_{H^3}$$
$$\le c_0(\|n(\cdot,0) - \mathcal{N}(\cdot)\|_{H^3} + \|T(\cdot,0) - 1\|_{H^3}) \exp(-\eta t)$$

for some positive constants c_0 and η.

For multidimensional case, we concern with the Neumann boundary conditions, and we obtain Theorem 3.

Theorem 3 ([JL06]). *Suppose* $0 < \underline{C} \le \mathcal{C}(x) \le \overline{C}$, $\mathcal{C}(x) \in C^3(\Omega)$, *and* $(n_I(x), T_I(x)) \in H^4(\Omega)$. *There exists a positive constant δ_1, such that if*

$$\|n_I(x) - \mathcal{N}(x)\|_{H^4} + \|T_I(x) - 1\|_{H^4} \le \delta_1,$$

then the problem has a unique solution (n, T, E) in $\Omega \times [0, \infty)$ satisfying

$$\|E(\cdot,t) - \nabla\mathcal{V}(\cdot)\|_{H^4} + \|n(\cdot,t) - \mathcal{N}(\cdot)\|_{H^4} + \|T(\cdot,t) - 1\|_{H^4}$$
$$\le c_0(\|n(\cdot,0) - \mathcal{N}(\cdot)\|_{H^4} + \|T(\cdot,0) - 1\|_{H^4}) \exp(-\eta t)$$

for some positive constants c_0 and η.

Remark 1. In our theorems, we do not need the smallness for the doping profile $\mathcal{C}(x)$, which means that we could obtain these theorems for large-doping profile. This is different from the previous results in [CHL04, CHL05].

3 The A Priori Estimates

The crucial step in getting the large-time behavior for the smooth solution is to establish the a priori estimates. Throughout this section, we consider the system (7) in one-space dimension.

Since the coefficient matrix in (4) is just a positive semidefinite matrix, and one eigenvalue is always zero, much more difficulties occur. This makes it very hard to deal with the system (7) directly. Fortunately, we are able to use some ideas from the hyperbolic theory, since the system (7) can be derived from the nonisentropic Euler–Poisson equations. To avoid treating the degenerate system directly, we introduce two new functions – the "flux" and the entropy function to deal with the system – and then we could change the system (7) from the degenerate quasilinear crossdiffusion parabolic coupling system into a new system which is a hyperbolic–parabolic–elliptic coupling system. In this way, the new system can be understood as the compressible model for the "fluid" basically, coupling the Poisson equation. Thus, we are able to modify some energy functionals and employ it in establishing of the a priori estimates by virtue of some suitable weighted energy estimates.

To understand the system (7) more plainly in the viewpoint of hyperbolic theory, we firstly introduce the "flux" variable nu as

$$nu = nE - (nT)_x = nV_x - (nT)_x.$$

Then the system (9) is changed to

$$\begin{cases} n_t + (nu)_x = 0, \\ (nT)_t + \gamma(nTu)_x = (\gamma-1)nuE - n(T-1), \\ E_x = n - \mathcal{C}(x). \end{cases} \tag{15}$$

This system (15) can be understood as the "full" compressible model (namely with a temperature equation) coupling a Poisson equation,

$$\begin{cases} n_t + (nu)_x = 0, \\ (nu)_t + (\gamma nu^2)_x + (nT)_x = \gamma(nTu_x)_x - [2+\mathcal{C}(x)]nu + 2nE - n_x, \\ T_t + uT_x + (\gamma-1)Tu_x = (\gamma-1)u^2 - (T-1), \\ E_x = n - \mathcal{C}(x). \end{cases}$$

Secondly, we introduce the entropy function s as

$$T = n^{\gamma-1}s.$$

Then (15) can be changed to the following system

$$\begin{cases} n_t + (nu)_x = 0, \\ s_t + us_x = (\gamma - 1)n^{1-\gamma}u^2 + n^{1-\gamma} - s, \\ E_x = n - \mathcal{C}(x), \end{cases} \tag{16}$$

with

$$u = E - n^{\gamma-2}(ns_x + \gamma s n_x).$$

As these mentioned above, we make the following variable transformations

$$n = \mathcal{N}(1+\rho), \qquad T = 1 + \theta, \qquad E = \mathcal{E} + \varphi = \mathcal{V}_x + \phi_x,$$

and introduce the entropy function s as follows:

$$\theta = (1+\rho)^{\gamma-1}(1+s) - 1.$$

Then the system (9) can be rewritten as

$$\begin{cases} \rho_t + [(1+\rho)u]_x = -\mathcal{E}(1+\rho)u, \\ s_t + us_x = \dfrac{\gamma-1}{(1+\rho)^{\gamma-1}}|u|^2 + (\gamma-1)\mathcal{E}(1+s)u - \dfrac{(1+\rho)^{\gamma-1}(1+s)-1}{(1+\rho)^{\gamma-1}}, \\ \varphi_x = \mathcal{N}\rho, \end{cases} \tag{17}$$

where

$$u = \varphi - \frac{1}{1+\rho}[(1+\rho)^{\gamma}(1+s)]_x - \mathcal{E}[(1+\rho)^{\gamma-1}(1+s) - 1].$$

The corresponding initial data and the boundary conditions are

$$\rho(x,0) = \frac{n_I(x)}{\mathcal{N}(x)} - 1, \qquad s(x,0) = \mathcal{N}(x)^{\gamma-1}T_I(x)n_I(x)^{-(\gamma-1)} - 1,$$

$$\rho_x(0,t) = \rho_x(1,t) = 0, \quad s_x(0,t) = s_x(1,t) = 0, \quad \varphi(0,t) = \varphi(1,t) = 0.$$

For the smooth solutions of the above problem, we have the following a priori estimates.

Lemma 1 ([HL06]). *There exists a positive constant δ_1 such that for any $T_1 > 0$, if the solution (ρ, s, φ) exists on $\Omega \times [0, T_1]$ satisfying*

$$A(t) = \sup_{0 \le t \le T_1} \|(\rho, s)(\cdot, t)\|^2_{H^3} \le \delta_1^2,$$

then

$$\begin{aligned} &\frac{d}{dt}\int \sum_{i=0}^{3}(|\partial_x^i \rho|^2 + |\partial_x^i s|^2 + |\partial_x^i \varphi|^2) + \frac{d}{dt}\int \sum_{i=0}^{1}(|\partial_x^i \rho_t|^2 + |\partial_x^i s_t|^2 + |\partial_x^i \varphi_t|^2) \\ &\qquad + C_1 \int \sum_{i=0}^{3}(|\partial_x^i \rho|^2 + |\partial_x^i s|^2 + |\partial_x^i \varphi|^2) \\ &\qquad + C_2 \int \sum_{i=0}^{1}(|\partial_x^i u_t|^2 + |\partial_x^i \rho_t|^2 + |\partial_x^i s_t|^2 + |\partial_x^i \varphi_t|^2) \le 0 \end{aligned}$$

for any $t \in [0, T_1]$, where C_1 and C_2 are positive constants.

To establish the above a priori estimates, we introduce a number of energy functionals. Here we just give the zero-order estimates, for more detail see [HL06].

Lemma 2 ([HL06]). *Denote functionals*

$$H_0(t) = \int \Big\{ \frac{\mathcal{N}}{\gamma - 1}[(1+\rho)^\gamma - 1 - \gamma\rho](1+s) + \mathcal{N}\rho s + \frac{\mathcal{N}}{2(\gamma-1)}(1+\rho)s^2 + \frac{1}{2}(1+\frac{\lambda_0}{\mathcal{N}})\varphi^2 \Big\},$$

$$G_0(t) = \int \Big\{ \mathcal{N}[u^2 + (\gamma - 1 + \gamma\lambda_0)\rho^2 + (2+\lambda_0)\rho s + \frac{1}{\gamma-1}s^2] + \lambda_0\varphi^2 - \lambda_0\mathcal{E}\varphi[(\gamma-1)\rho + s] \Big\},$$

where λ_0 is a positive constant to be determined later. We obtain the following estimates by choosing δ_1 small enough,

$$\frac{d}{dt}H_0(t) + G_0(t) \le 0.$$

Remark 2. For one-dimensional case, with the help of system (17), we could directly obtain the following equation for the electric field:

$$E_t + \mathcal{N}(1+\rho)u = 0.$$

But for multidimensional case, we just have

$$\nabla \cdot E_t + \nabla \cdot [\mathcal{N}(1+\rho)u)] = 0.$$

So we need to overcome more difficulties than one-space dimension and to treat more careful with the energy density functionals.

It can be shown that there exists λ_0 satisfying

$$0 < \lambda_0 < \frac{4}{\gamma - 1}\inf\Big(\frac{\mathcal{N}}{\mathcal{N} + \mathcal{E}^2}\Big),$$

such that

$$G_0(t) \ge c_1 \int (|u|^2 + |\rho|^2 + |s|^2 + |\varphi|^2).$$

Also it is easy to show that

$$c_2\|(\rho, s, \varphi)\|_{L^2}^2 \le H_0(t) \le c_3\|(\rho, s, \varphi)\|_{L^2}^2,$$

where c_i $(i = 1, 2, 3)$ are positive constants. Then we obtain

$$\frac{d}{dt}\int (|\rho|^2 + |s|^2 + |\varphi|^2) + C\int (|u|^2 + |\rho|^2 + |s|^2 + |\varphi|^2) \le 0.$$

Acknowledgments

L. Hsiao is supported by the Funds of NSAF of China (No. 10276036) and the Funds of NSFC (No. 10431060); Y. Li is supported by the Tianyuan Mathematics Fund of NSFC (No. 10626030) and the HYP2006 organization.

References

[AD96] Abdallah, N. Ben and Degond, P.: On a hierarchy of macroscopic models for semiconductor. J. Math. Phys., **37**, 3333–3383 (1996)

[ADG96] Abdallah, N. Ben and Degond, P. and Génieys, S.: An energy-transport model for semiconductors derived from the Boltzmann equation. J. Stat. Phys., **84**, 205–231 (1996)

[An95] Anile, A.M.: An extended thermodynamic framework for the hydrodynamic modeling of semiconductors. in: P. Marcati *et al.* (Ed.), Mathematical Problem in semiconductors physics, Pitman Research Notes In Mathematics Series Vol. **340**, Longman, 3–41 (1995).

[BK70] Blölekjaer, K.: Transport equations for electrons in two-valley semiconductors. IEEE Trans. Electron Device, **17**, 38–47 (1970)

[CH02] Chen, L. and Hsiao, L.: The solution of *Lyumiks* energy transport model in semiconductor science. Math. Meth. Appl. Sci., **26**, 1421–1433 (2003)

[CHL04] Chen, L., Hsiao, L. and Li, Y: Global existence and asymptotic behavior to the solutions of 1-D *Lyumkis* energy transport model for semiconductors. Quart. Appl. Math., **LXII** No.2, 337–358 (2004)

[CHL05] L. Chen, L. Hsiao and Y. Li, Large time behavior and energy relaxation time limit of the solutions to an energy transport model in semiconductors. J. Math. Anal. Appl., **312**, 596–619 (2005)

[CL06] Chen, L. and Li, Y: Global existence and exponential decay for a multidimensional compressible energy transport model with heat conductivity. Preprint

[De00] Degond, P.: Mathematical modelling of microelectronics semiconductor devices, Proceedings of the Morningside Mathematical Center. AMS/IP Studies in Advanced mathematics, AMS Society and Internatinal Press, Beijng, 77–109 (2000)

[DGJ97] Degond, P., Génieys, S. and Jüngel, A.: A system of parabolic equations in nonequilibrium thermodynamics including thermal and electrical effects. J. Math. Pures Appl., **76**, 991–1015 (1997)

[GN99] Gasser, I. and Natalini, R.: The energy transport and the drift diffusion equations as relaxation limits of the hydrodynamic model for Semiconductors. Quart. Appl. Math., **LVII** No. 2, 269–282 (1999)

[HL06] Hsiao, L. and Li, Y: Global existence and asymptotic behavior for a compressible energy transport model. Preprint.

[JL06] Jian, H.Y. and Li, Y: Large time behavior of smooth solutions for a multidimensional compressible energy transport model. Preprint.

[Ju00] Jüngel, A.: Regularity and uniqueness of solutions to a parabolic system in nonequilibrium thermodynamics. Nonlin. Anal., **41**, 669–688 (2000)

[Ju01] Jüngel, A.: Quasi-hydrodynamic Semiconductor Equations. Basel, Boston, Berlin, Birkhäuser (2001)

[Li07] Li, Y.: Relaxation time limits for smooth solutions to a multidimensional nonisentropic Euler-Poisson equations. To appear in Acta Math. Sci., **27B** No. 2, (2007)

[MRS90] Markowich, P.A., Ringhofer, C. and Schmeiser, C.: Semiconductors Equations. Springer-Verlag, Vienna, New York (1990)

Novel Entropy Stable Schemes for 1D and 2D Fluid Equations

E. Tadmor and W. Zhong

Summary. We present a systematic study of the novel entropy stable approximations of a variety of nonlinear conservation laws, from the scalar Burger's equation to 1D Navier–Stokes and 2D shallow water equations. This new family of second-order difference schemes avoids using artificial numerical viscosity, in the sense that their entropy dissipation is dictated solely by physical dissipation terms. The numerical results of 1D compressible Navier–Stokes equations provide us a remarkable evidence for different roles of viscosity and heat conduction in forming sharp monotone profiles in the immediate neighborhoods of shocks and contacts. Further implementation in 2D shallow water equations is realized dimension by dimension.

1 The Inviscid Burger's Equation

1.1 Entropy Conservative Schemes

We begin with the inviscid Burger's equation

$$\frac{\partial u}{\partial t} + \frac{\partial}{\partial x} f(u) = 0, \quad f(u) = \frac{1}{2}u^2. \tag{1}$$

It is equipped with a family of entropy functions, $U_p(u) = u^{2p}$, $p = 1, 2, \cdots$, such that solutions of (1) satisfy, at the *formal* level,

$$\frac{\partial}{\partial t} U_p(u) + \frac{\partial}{\partial x} F_p(u) = 0. \tag{2}$$

These are additional conservation laws balanced by the corresponding entropy flux functions $F_p(u) = 2pu^{2p+1}/(2p+1)$. Spatial integration then yields (ignoring boundary contributions)

$$\int_x u^{2p}(x,t)\, dx = \int_x u^{2p}(x,0)\, dx. \tag{3}$$

We turn to the discrete framework. Discretization in space yields the semidiscrete scheme

$$\frac{d}{dt}u_\nu(t) + \frac{1}{\Delta x}\left(f_{\nu+\frac{1}{2}} - f_{\nu-\frac{1}{2}}\right) = 0. \tag{4}$$

Here, $u_\nu(t)$ denotes the discrete solution along the grid line $(x_\nu,\, t)$ with $x_\nu := \nu\Delta x$, Δx being the uniform mesh size, and $f_{\nu+\frac{1}{2}} := f(u_{\nu-r+1}, \cdots, u_{\nu+r})$ is a consistent numerical flux based on a stencil of $2r+1$ neighboring grid values that makes (4) conservative in the sense that $\sum_\nu u_\nu(t)\Delta x = \sum_\nu u_\nu(0)\Delta x$. Fix p. We seek a semidiscrete scheme that conserves the entropy $U_p(u) = u^{2p}$ in the sense of satisfying the discrete analogue of $(2)_p$

$$\frac{d}{dt}U_p(u_\nu(t)) + \frac{1}{\Delta x}(F_{\nu+\frac{1}{2}} - F_{\nu-\frac{1}{2}}) = 0.$$

Here, $F_{\nu+\frac{1}{2}}$ is a consistent numerical entropy flux. According to [Ta1987, Theorem 5.2], such three-point scalar *entropy conservative* schemes are uniquely determined by the entropy conservative numerical flux $f_{\nu+\frac{1}{2}} = f^*_{\nu+\frac{1}{2}}$ given by

$$f_{\nu+\frac{1}{2}} = f^*_{\nu+\frac{1}{2}} = \frac{2p-1}{2(2p+1)} \cdot u_\nu^2 \cdot \frac{(u_{\nu+1}/u_\nu)^{2p+1} - 1}{(u_{\nu+1}/u_\nu)^{2p-1} - 1}. \tag{5}$$

The resulting scheme (4) and (5) is entropy conservative, in the sense that the discrete analogue of total entropy conservation (3) is satisfied

$$\sum_\nu u_\nu^{2p}(t)\,\Delta x = \sum_\nu u_\nu^{2p}(0)\,\Delta x$$

Of course, all the above manipulations are at the formal level. To recover the physical relevant *entropy inequality*, $\partial_t U_p(u) + \partial_x F_p(u) \leq 0$, one can add numerical dissipation

$$\frac{d}{dt}u_\nu(t) + \frac{1}{\Delta x}\left(f^*_{\nu+\frac{1}{2}} - f^*_{\nu-\frac{1}{2}}\right) = \frac{\epsilon}{(\Delta x)^2}\Big(d(u_{\nu+1}) - 2d(u_\nu) + d(u_{\nu-1})\Big), \quad \epsilon > 0.$$

This serves as an approximation to the vanishing viscosity regularization $u_t + f(u)_x = \epsilon d(u)_{xx}$, $d'(u) > 0$. Sum this scheme against $v_\nu := U_p'(u_\nu) = 2pu^{2p-1}$: the resulting entropy balance that follows reads

$$\frac{d}{dt}\sum_\nu U_p(u_\nu(t))\Delta x = -\frac{\epsilon}{\Delta x}\sum_\nu \frac{\Delta d_{\nu+\frac{1}{2}}}{\Delta v_{\nu+\frac{1}{2}}}(\Delta v_{\nu+\frac{1}{2}})^2 \leq 0, \tag{6}$$

since $\frac{\Delta d_{\nu+\frac{1}{2}}}{\Delta v_{\nu+\frac{1}{2}}} := \frac{d(u_{\nu+1}) - d(u_\nu)}{v_{\nu+1} - v_\nu} > 0$ for $d'(u) > 0$. Observe that the amount of entropy dissipation on the right is completely determined by the dissipation term $\epsilon d(u)$. No artificial viscosity is introduced by the convective term. If we exclude any dissipative mechanism ($\epsilon = 0$), the entropy conservative solutions admit dispersive oscillations interesting for their own sake, consult [La1986, LL1996].

1.2 Numerical Experiments

The semidiscrete entropy conservative scheme (4) and (5) is integrated with the following third-order Runge–Kutta (RK3), consult [GST2001]

$$\begin{aligned} u^{(1)} &= u^n + \Delta t L(u^n), \\ u^{(2)} &= \frac{3}{4}u^n + \frac{1}{4}u^{(1)} + \frac{1}{4}\Delta t L(u^{(1)}), \\ u^{n+1} &= \frac{1}{3}u^n + \frac{2}{3}u^{(2)} + \frac{2}{3}\Delta t L(u^{(2)}), \end{aligned} \tag{7}$$

where $[L(u)]_\nu = -\frac{1}{\Delta x}(f^*_{\nu+\frac{1}{2}} - f^*_{\nu-\frac{1}{2}})$. We note that this explicit RK3 time discretization produces a negligible amount of entropy dissipation. For a general framework of entropy conservative fully discrete schemes, consult [LMR2002].

We solve the inviscid Burger's equation with the sine initial condition, $u(0,x) = \sin(2\pi x)$, and periodic boundary. In Fig. 1, we display the numerical solutions for (7) with the numerical flux (5) for different choices of p. Observe that the amplitude of the spurious dispersive oscillations decreases

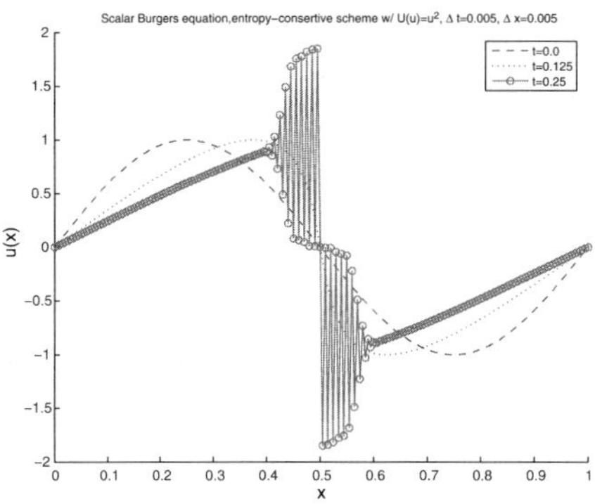

(a) p = 1

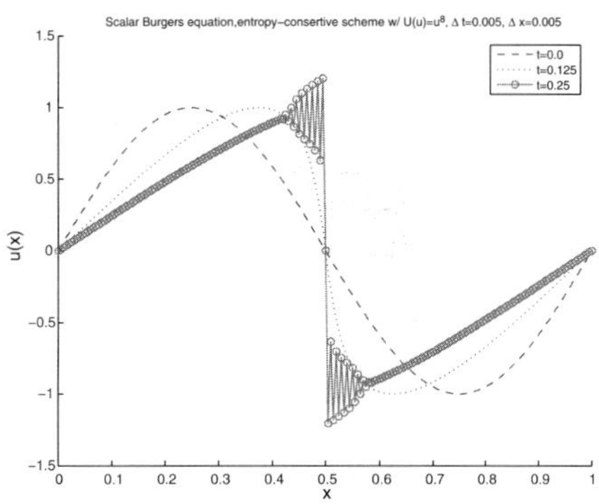

(b) p = 4

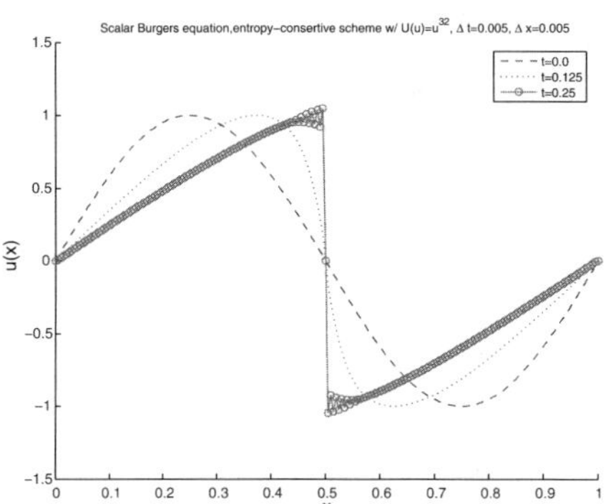

(c) p = 16

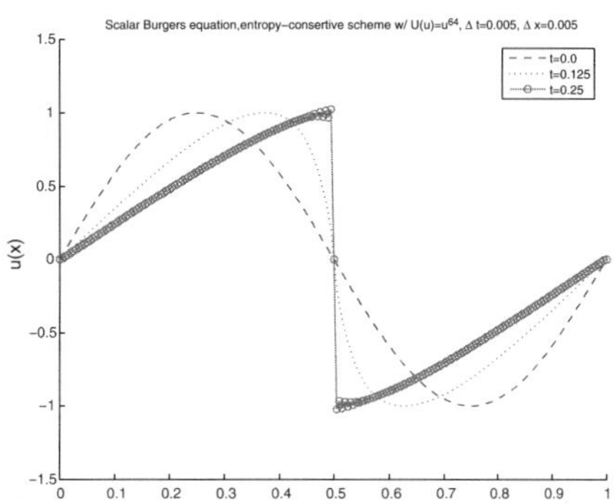

(d) p = 32

Fig. 1. Scalar Burger's equation, sine initial condition, entropy conservative schemes, 200 spatial grids, $U(u) = u^{2p}$, $\Delta t = 5 \times 10^{-3}$, and $\Delta x = 5 \times 10^{-3}$

as p increases. Indeed, as we increase p, the control of a constant U_p entropy $\left[\sum_\nu u_\nu^{2p}(t)\,\Delta x\right]^{\frac{1}{2p}}$ approaches the control of L^∞-norm.

2 The 1D Navier–Stokes Equations

2.1 Entropy Dissipation

We consider the Navier–Stokes equations (NSE) governing the density $\rho = \rho(x,t)$, momentum $m = m(x,t)$, and energy $E = E(x,t)$

$$\frac{\partial}{\partial t}\mathbf{u} + \frac{\partial}{\partial x}\mathbf{f}(\mathbf{u}) = \epsilon\frac{\partial^2}{\partial x^2}\mathbf{d}(\mathbf{u}), \quad \mathbf{u} = [\rho, m, E]^\top. \tag{8}$$

They are driven by the convective flux $\mathbf{f}(\mathbf{u}) = \left[m, qm + p, q(E+p)\right]^\top$, together with the dissipative flux $\epsilon\mathbf{d}(\mathbf{u}) = (\lambda+2\mu)\left[0, q, q^2/2\right]^\top + \kappa\left[0,0,\theta\right]^\top$ which stands for the combined viscous and heat fluxes. Here, ϵ represents the amplitudes of the viscosity and conductivity. These fluxes involve the velocity $q := m/\rho$, the pressure $p = p(x,t) = (\gamma-1)e$ where $e := E - \frac{m^2}{2\rho}$, and the absolute temperature $\theta = \theta(x,t) > 0$, such that $C_v\rho\theta = e$. Here $\gamma > 1, \lambda, \mu$ are fixed and $\kappa \mapsto \kappa/C_v$ with $C_v = 1$.

The viscous and heat fluxes are dissipative, in the sense that they are responsible for the dissipation of total *entropy*, $U(\mathbf{u}) = -\rho S$ with $S = \ln(p\rho^{-\gamma})$ being the specific entropy,

$$\frac{\partial}{\partial t}(-\rho S) + \frac{\partial}{\partial x}(-mS + \kappa(\ln\theta)_x) = -(\lambda+2\mu)\frac{q_x^2}{\theta} - \kappa\left(\frac{\theta_x}{\theta}\right)^2 \leq 0. \tag{9}$$

Spatial integration of (9) then yields the second law of thermodynamics

$$\frac{d}{dt}\int_x(-\rho S)\,dx = -(\lambda+2\mu)\int_x\frac{q_x^2}{\theta}dx - \kappa\int_x\left(\frac{\theta_x}{\theta}\right)^2 dx \leq 0. \tag{10}$$

In fact, (10) specifies the precise entropy decay rate. In the case of the Euler equations, $\lambda = \mu = \kappa = 0$, total entropy is precisely conserved

$$\int_x -\rho S(x,t)\,dx = \int_x -\rho S(x,0)\,dx.$$

This corresponds to the scalar entropy conservation (3).

2.2 Entropy Stable Schemes for the Navier–Stokes Equations

We now turn our attention to the construction of entropy stable schemes for NSE (8). We use the conservative differences for convective flux and standard centered differences for the dissipative terms on the RHS

$$\frac{d}{dt}\mathbf{u}_\nu(t) + \frac{1}{\Delta x}\left(\mathbf{f}^*_{\nu+\frac12} - \mathbf{f}^*_{\nu-\frac12}\right) = \frac{\epsilon}{(\Delta x)^2}\Big(\mathbf{d}_{\nu+1} - 2\mathbf{d}_\nu + \mathbf{d}_{\nu-1}\Big), \tag{11}$$

Here, $\mathbf{d}_\nu := \mathbf{d}(\mathbf{u}_\nu)$. As in the scalar case, we seek entropy conservative flux $\mathbf{f}^*_{\nu+\frac{1}{2}}$, so that the entropy decay will be dictated *solely* by the dissipation terms on the right of (11). The construction of the entropy conservative flux $\mathbf{f}^*_{\nu+\frac{1}{2}}$ follows [Ta2004] and [TZ2006], and is summarized in the following.

Algorithm 1. *If* $\mathbf{u}_\nu = \mathbf{u}_{\nu+1}$, then $\mathbf{f}^*_{\nu+\frac{1}{2}} = \mathbf{f}(\mathbf{v}_\nu)$*; else*

- *Set* $\Delta\mathbf{u}_{\nu+\frac{1}{2}} := \mathbf{u}_{\nu+1} - \mathbf{u}_\nu$. *Starting with* $\mathbf{u}^1_{\nu+\frac{1}{2}} := \mathbf{u}_\nu$, *compute recursively the intermediate states*

$$\mathbf{u}^{j+1}_{\nu+\frac{1}{2}} = \mathbf{u}^{j}_{\nu+\frac{1}{2}} + \left\langle \widehat{\boldsymbol{\ell}}^j_{\nu+\frac{1}{2}}, \Delta\mathbf{u}_{\nu+\frac{1}{2}} \right\rangle \widehat{\mathbf{r}}^j_{\nu+\frac{1}{2}}, \quad j = 1,2,3. \tag{12}$$

Here, $\{\widehat{\boldsymbol{\ell}}^j_{\nu+\frac{1}{2}}\}$ *and* $\{\widehat{\mathbf{r}}^j_{\nu+\frac{1}{2}}\}$ *are, respectively, the left and right eigensystems of the Roe matrix* $A(\mathbf{u}_\nu, \mathbf{u}_{\nu+1})$ *(see [Roe1981]).*

- *Set* $\mathbf{r}^j_{\nu+\frac{1}{2}} := \mathbf{v}(\mathbf{u}^{j+1}_{\nu+\frac{1}{2}}) - \mathbf{v}(\mathbf{u}^{j}_{\nu+\frac{1}{2}})$ *and let* $\{\boldsymbol{\ell}^j\}^3_{j=1}$ *be the corresponding orthogonal system. Compute the entropy conservative numerical flux*

$$\mathbf{f}^*_{\nu+\frac{1}{2}} = (\gamma-1)\sum_{j=1}^{3} \frac{m^{j+1}_{\nu+\frac{1}{2}} - m^{j}_{\nu+\frac{1}{2}}}{\left\langle \boldsymbol{\ell}^j_{\nu+\frac{1}{2}}, \Delta\mathbf{v}_{\nu+\frac{1}{2}} \right\rangle} \boldsymbol{\ell}^j_{\nu+\frac{1}{2}}, \quad \left\langle \boldsymbol{\ell}^j_{\nu+\frac{1}{2}}, \mathbf{v}^{j+1}_{\nu+\frac{1}{2}} - \mathbf{v}^{j}_{\nu+\frac{1}{2}} \right\rangle = \delta_{jk} \tag{13}$$

Here, $\mathbf{v}(\mathbf{u}) := U_{\mathbf{u}}(\mathbf{u}) = [-E/e - S + \gamma + 1, q/\theta, -1/\theta]^\top$ are the entropy variables and $\{m^j\}$ are the intermediate momentum values along the path. We now arrive at our main result of NSE corresponding to the Burger's statement (6).

Theorem 2 ([TZ2006, Theorem 3.6]). *The semidiscrete scheme* (11) *with the entropy conservative numerical flux* $\mathbf{f}^*_{\nu+\frac{1}{2}}$ *in* (12) *and* (13) *and* $\mathbf{d}(\mathbf{u}_\nu)$ *being the dissipative NS flux is entropy stable in the sense that*[1]

$$\begin{aligned}\frac{d}{dt}\sum_\nu [-\rho_\nu(t)S_\nu(t)]\,\Delta x &= -\sum_\nu \frac{\epsilon}{\Delta x}\left\langle \Delta\mathbf{v}_{\nu+\frac{1}{2}}, \frac{\Delta\mathbf{d}_{\nu+\frac{1}{2}}}{\Delta\mathbf{v}_{\nu+\frac{1}{2}}}\Delta\mathbf{v}_{\nu+\frac{1}{2}} \right\rangle \\ &= -(\lambda+2\mu)\sum_\nu \left(\frac{\Delta q_{\nu+\frac{1}{2}}}{\Delta x}\right)^2 \left(\widehat{1/\theta}\right)_{\nu+\frac{1}{2}} \Delta x \\ &\quad -\kappa\sum_\nu \left(\frac{\Delta\theta_{\nu+\frac{1}{2}}}{\Delta x}\right)^2 \left(\widetilde{1/\theta}\right)^2_{\nu+\frac{1}{2}} \Delta x \le 0.\end{aligned} \tag{14}$$

This statement is a discrete analogue of the entropy balance statement (10).

Here is our main point: we introduce no excessive entropy dissipation due to spurious, *artificial* numerical viscosity. According to (14), the semidiscrete scheme contains the precise amount of numerical viscosity to enforce the correct entropy dissipation dictated by NSE. More can be found in [Ta2004, TZ2006].

[1] We let $\widehat{z}_{\nu+\frac{1}{2}} = (z_\nu + z_{\nu+1})/2$ and $\widetilde{z}_{\nu+\frac{1}{2}} = \sqrt{z_\nu z_{\nu+1}}$.

2.3 Numerical Experiments

We consider ideal polytropic gas equations as an approximation of air with

$$\gamma = 1.4, \quad C_v = 716, \quad \kappa = 0.03, \quad \lambda + 2\mu = 2.28 \times 10^{-5}.$$

We simulate the Sod's shock tube problem, where the Euler and NSE are solved over the interval [0, 1] subject to Riemann initial conditions

$$(\rho, m, E)_{t=0} = \begin{cases} (1.0,\ 0.0,\ 2.5) & 0 < x \leq 0.5 \\ (0.125,\ 0.0,\ 0.25) & 0.5 < x < 1. \end{cases}$$

In Fig. 2, we display the numerical solutions for the fully discrete scheme (11) with RK3 method (7) and the numerical flux (13). The density fields of four different cases are recorded.

Density field of the Euler equations (Fig. 2a) demonstrates the purely dispersive character of the entropy conservative schemes. Dispersive oscillations on the mesh scale are observed in shocks and contact regions due

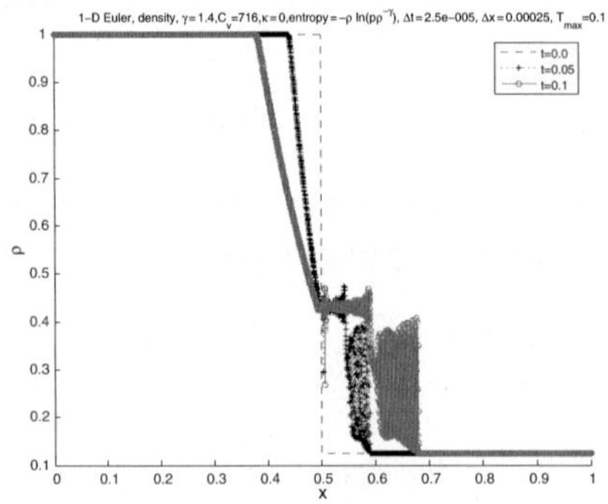

(a) Euler

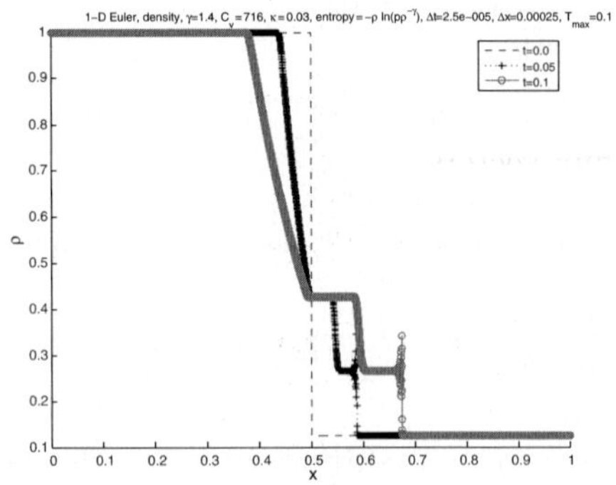

(b) Navier-Stokes with heat only

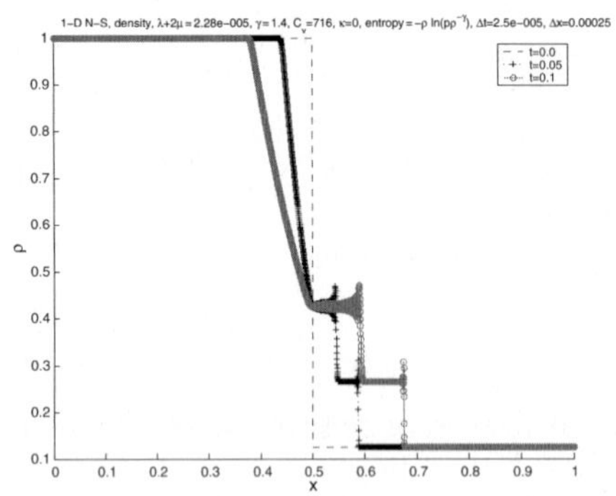

(c) Navier-Stokes with viscosity only

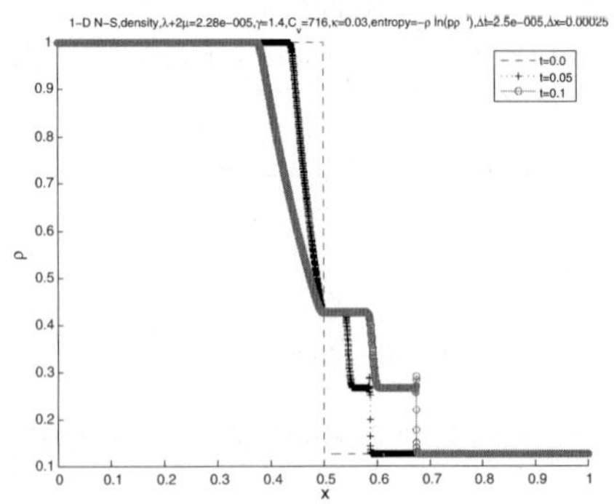

(d) Navier-Stokes with viscosity and heat

Fig. 2. Density field of Sod's problem with 4,000 spatial grid points, $U(\mathbf{u}) = -\rho \ln\left(p\rho^{-\gamma}\right)$, $\Delta t = 2.5 \times 10^{-5}$, and $\Delta x = 2.5 \times 10^{-4}$

to the absence of any dissipative mechanism. These oscillations approach a modulated wave envelop, consult [La1986, LL1996] for more discussions on dispersive oscillations.

For the results of NSE in Fig. 2b–d, the presence of heat flux causes the oscillations to be dramatically reduced around the contact discontinuity and the shock in Fig. 2b. The viscous flux in NSE, on the other hand, is doing a better job than heat flux in damping oscillations around the shock in Fig. 2c, but we still can observe an oscillatory behavior around the contact discontinuity. In Fig. 2d, not only the oscillations around the shock are damped out by viscosity, but also the oscillations around the contact discontinuity are significantly reduced due to the heat flux.

3 The 2D Shallow Water Equations

We turn to 2D shallow water equations

$$\frac{\partial}{\partial t}\mathbf{u} + \frac{\partial}{\partial x}\mathbf{f}(\mathbf{u}) + \frac{\partial}{\partial y}\mathbf{g}(\mathbf{u}) = \eta\frac{\partial}{\partial x}\left(h\frac{\partial}{\partial x}\mathbf{d}(\mathbf{u})\right) + \eta\frac{\partial}{\partial y}\left(h\frac{\partial}{\partial y}\mathbf{d}(\mathbf{u})\right), \tag{15}$$

with $\mathbf{u} = [h,\, uh,\, vh]^\top$ being the vector of conserved variables balanced by the flux vectors

$$\mathbf{f} = [uh,\, u^2h + gh^2/2,\, uvh]^\top, \quad \mathbf{g} = [vh,\, uvh,\, v^2h + gh^2/2]^\top,$$

and the viscous flux vector $\mathbf{d} = [0,\, u,\, v]^\top$. Here, $h = h(x,t)$ is the total water height and $(u(x,t), v(x,t))$ are the depth-averaged velocities along x and y direction. Finally, g is the constant acceleration due to gravity and $\eta > 0$ is the constant eddy viscosity which models the turbulence stress in the flow.

The total energy $U(\mathbf{u}) = (gh^2 + u^2h + v^2h)/2$ serves as an entropy function

$$\frac{\partial}{\partial t}U(\mathbf{u}) + \frac{\partial}{\partial x}F(\mathbf{u}) + \frac{\partial}{\partial y}G(\mathbf{u}) = -\eta h(u_x^2 + u_y^2 + v_x^2 + v_y^2), \tag{16}$$

where $F(\mathbf{u}) = guh^2 + \frac{u^3h + uv^2h}{2} - huu_x$ and $G(\mathbf{u}) = gvh^2 + \frac{u^2vh + v^3h}{2} - hvv_y$ are the entropy fluxes. Spatial integration of (16) yields

$$\frac{d}{dt}\int_y\int_x U(\mathbf{u})\, dxdy = -\eta\int_y\int_x h(u_x^2 + u_y^2 + v_x^2 + v_y^2)\, dxdy. \tag{17}$$

For the inviscid case ($\eta = 0$), the global entropy conservation is satisfied

$$\int_y\int_x U(\mathbf{u}(x,t))\, dxdy = \int_y\int_x U(\mathbf{u}(x,0))\, dxdy$$

Arguing along the same line as the above NSE dimension by dimension, we obtain the entropy stable semidiscrete schemes (recall $\widehat{z_{\nu+\frac{1}{2}}} := (z_{\nu+1} + z_\nu)/2$)

$$\frac{d}{dt}\mathbf{u}_{\nu,\,\mu}(t)+\frac{1}{\Delta x}(\mathbf{f}^*_{\nu+\frac{1}{2},\,\mu}-\mathbf{f}^*_{\nu-\frac{1}{2},\,\mu})+\frac{1}{\Delta y}(\mathbf{g}^*_{\nu,\,\mu+\frac{1}{2}}-\mathbf{g}^*_{\nu,\,\mu-\frac{1}{2}})$$
$$=\frac{\eta}{\Delta x}(\widehat{h_{\nu+\frac{1}{2},\,\mu}}\frac{\mathbf{d}_{\nu+1,\,\mu}-\mathbf{d}_{\nu,\,\mu}}{\Delta x}-\widehat{h_{\nu-\frac{1}{2},\,\mu}}\frac{\mathbf{d}_{\nu,\,\mu}-\mathbf{d}_{\nu-1,\,\mu}}{\Delta x})$$
$$+\frac{\eta}{\Delta y}(\widehat{h_{\nu,\,\mu+\frac{1}{2}}}\frac{\mathbf{d}_{\nu,\,\mu+1}-\mathbf{d}_{\nu,\,\mu}}{\Delta x}-\widehat{h_{\nu,\,\mu-\frac{1}{2}}}\frac{\mathbf{d}_{\nu,\,\mu}-\mathbf{d}_{\nu,\,\mu-1}}{\Delta x}),\quad (18a)$$

with the entropy conservative fluxes $\mathbf{f}^*_{\nu+\frac{1}{2},\,\mu}$ and $\mathbf{g}^*_{\nu,\,\mu+\frac{1}{2}}$ outlined in Algorithm 1 along x and y direction, respectively,

$$\mathbf{f}^*_{\nu+\frac{1}{2},\,\mu}=\frac{g}{2}\sum_{j=1}^{3}\frac{(h^{j+1}_{\nu+\frac{1}{2},\,\mu})^2u^{j+1}_{\nu+\frac{1}{2},\,\mu}-(h^{j}_{\nu+\frac{1}{2},\,\mu})^2u^{j}_{\nu+\frac{1}{2},\,\mu}}{\left\langle \boldsymbol{\ell}^{x^j}_{\nu+\frac{1}{2},\,\mu},\,\Delta\mathbf{v}_{\nu+\frac{1}{2},\,\mu}\right\rangle}\boldsymbol{\ell}^{x^j}_{\nu+\frac{1}{2},\,\mu},\quad (18b)$$

$$\mathbf{g}^*_{\nu,\,\mu+\frac{1}{2}}=\frac{g}{2}\sum_{j=1}^{3}\frac{(h^{j+1}_{\nu,\,\mu+\frac{1}{2}})^2u^{j+1}_{\nu,\,\mu+\frac{1}{2}}-(h^{j}_{\nu,\,\mu+\frac{1}{2}})^2u^{j}_{\nu,\,\mu+\frac{1}{2}}}{\left\langle \boldsymbol{\ell}^{y^j}_{\nu,\,\mu+\frac{1}{2}},\,\Delta\mathbf{v}_{\nu,\,\mu+\frac{1}{2}}\right\rangle}\boldsymbol{\ell}^{y^j}_{\nu,\,\mu+\frac{1}{2}},\quad (18c)$$

Here, $u_{\nu,\,\mu}(t)$ denotes the discrete solution at the grid point (x_ν, y_ν, t), $\mathbf{d}_{\nu,\,\mu} := \mathbf{d}(\mathbf{u}_{\nu,\,\mu})$, and $\mathbf{v} := U_{\mathbf{u}} = [gh - \frac{1}{2}(u^2+v^2),\, u,\, v]^\top$ is the entropy variable. Numerical fluxes $\mathbf{f}^*$ and $\mathbf{g}^*$ are constructed separately along two different phase pathes dictated by two sets of vectors $\{\boldsymbol{\ell}^{x^j}\}$ and $\{\boldsymbol{\ell}^{y^j}\}$. $\{h^j\}$ and $\{u^j\}$ are, respectively, the intermediate values of height and velocity along the path. The above difference scheme (18a)–(18c) is an entropy stable scheme with no artificial viscosity, in the sense that the following discrete entropy balance is satisfied

$$\frac{d}{dt}\sum_{\nu,\,\mu}U(\mathbf{u}_{\nu,\,\mu}(t))\Delta x\Delta y=-\eta\sum_{\nu,\,\mu}\Bigg\{\widehat{h_{\nu+\frac{1}{2},\,\mu}}\left[\left(\frac{\Delta\mathbf{u}_{\nu+\frac{1}{2},\,\mu}}{\Delta x}\right)^2+\left(\frac{\Delta\mathbf{v}_{\nu+\frac{1}{2},\,\mu}}{\Delta x}\right)^2\right]$$
$$+\widehat{h_{\nu,\,\mu+\frac{1}{2}}}\left[\left(\frac{\Delta\mathbf{u}_{\nu,\,\mu+\frac{1}{2}}}{\Delta y}\right)^2+\left(\frac{\Delta\mathbf{v}_{\nu,\,\mu+\frac{1}{2}}}{\Delta y}\right)^2\right]\Bigg\}\Delta x\Delta y.\quad (19)$$

Equation (19) is a discrete analogue of the entropy balance statement (17).

Equipped by RK3, we test the entropy stable scheme (18a)–(18c) by the 2D partial dam-break problem with free-slip boundary described in [FC1990]. Both the inviscid and viscous case are tested. The water surface profiles at time $t = 7.2\,$s are recorded in Fig. 2.

Comparing Fig. 3b to Fig. 3a, we observe the improvements in smoothness of the numerical solutions. There is no analytical reference solution for this test case, but other numerical results are available in [FC1990].

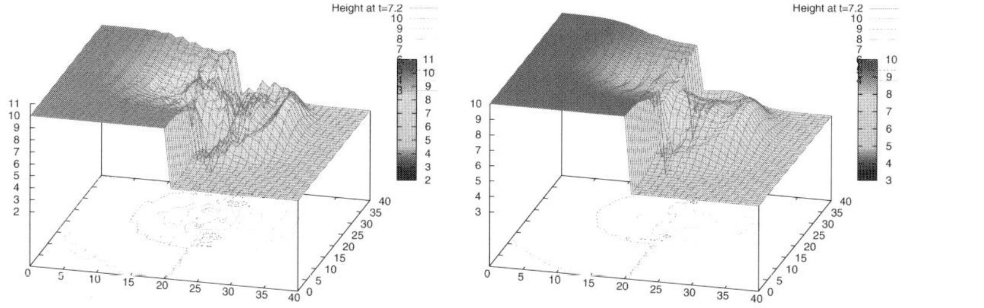

(a) $\eta = 0$ (b) $\eta = 10\,\mathrm{m}^2\,\mathrm{s}^{-1}$

Fig. 3. Shallow water equations, dam-break, $200 \times 200\,\mathrm{m}^2$ basin, free-slip boundary, $\Delta x = \Delta y = 5\,\mathrm{m}$, and $\Delta t = 5 \times 10^{-3}\,\mathrm{s}$

References

[FC1990] Fennema, R.J., Chaudhry, M.H.: Explicit methods for 2D transient free-surface flows. J. Hydraul. Eng. ASCE, **116**, 1013–1034 (1990)

[GST2001] Gottlieb, S., Shu, C.-W., and Tadmor, E.: Strong stability-preserving high-order time discretization methods. SIAM Review, **43**, 1, 89–112 (2001)

[La1986] Lax, P.D.: On dispersive difference schemes. Phys. D, **18**, 250–254 (1986)

[LMR2002] LeFloch, P.G., Mercier, J.M., and Rohde, C.: Full discrete, entropy conservative schemes of arbitrary order. SIAM J. Numer. Anal., **40**, 5, 1968–1992 (2002)

[LL1996] Levermore, C.D., Liu, J.-G.: Oscillations arising in numerical experiments. Phys. D, **99**, 191–216 (1996)

[Roe1981] Roe, P.L.: Approximate Riemann solvers, parameter vectors and difference schemes. J. Comp. Phys., **43**, 357–372 (1981)

[Ta1987] Tadmor, E.: The numerical viscosity of entropy stable schemes for systems of conservation laws, I. Math. Comp., **49**, 91–103 (1987)

[Ta2004] Tadmor, E.: Entropy stability theory for difference approximations of nonlinear conservation laws and related time-dependent problems. Acta Numerica, 451–512 (2004)

[TZ2006] Tadmor, E. and Zhong, W.: Entropy stable approximations of Navier-Stokes equations with no artificial numerical viscosity. J. of Hype. Diff. Equa., **3**, 3, 529–559 (2006)

Index

Printing: Krips bv, Meppel, The Netherlands
Binding: Stürtz, Würzburg, Germany